Nonlinear Evolution Equations and Related Topics

Dedicated to Philippe Bénilan

Wolfgang Arendt
Haïm Brézis
Michel Pierre
Editors

Birkhäuser Verlag
Basel · Boston · Berlin

Contains reprints from *Journal of Evolution Equations* Volume 3 (2003), No. 1–4, and Volume 4 (2004), No. 2, 273–295.

Editors

Wolfgang Arendt
Department of Applied Analysis
University of Ulm
89069 Ulm
Germany
e-mail: arendt@mathematik.uni-ulm.de

Haïm Brézis
Analyse Numérique
Université Pierre et Marie Curie, B.C. 187
4, place Jussieu
75252 Paris Cedex 05
France
e-mail: brezis@ccr.jussieu.fr

and

Michel Pierre
Antenne de Bretagne de l'ENS Cachan
Campus de Ker Lann
35170 Bruz
France
e-mail: pierre@bretagne.ens-cachan.fr

Rutgers University
Department of Mathematics
Hill Center, Busch Campus
110 Frelinghuysen RD
Piscataway, NJ 08854
USA
e-mail: brezis@math.rutgers.edu

2000 Mathematical Subject Classification 35xx, 46xx, 31xx

A CIP catalogue record for this book is available from the
Library of Congress, Washington D.C., USA

Bibliographic information published by Die Deutsche Bibliothek
Die Deutsche Bibliothek lists this publication in the Deutsche Nationalbibliografie; detailed bibliographic data is
available in the Internet at <http://dnb.ddb.de>.

ISBN 3-7643-7107-2 Birkhäuser Verlag, Basel – Boston – Berlin

© 2004 Birkhäuser Verlag, P.O. Box 133, CH-4010 Basel, Switzerland
Part of Springer Science+Business Media
Printed on acid-free paper produced of chlorine-free pulp. TCF ∞
Printed in Germany
ISBN 3-7643-7107-2

9 8 7 6 5 4 3 2 1

www.birkhauser.ch

Preface

The present volume is dedicated to *Philippe Bénilan,* a most original and charismatic mathematician who had a deep and decisive impact on the theory of *nonlinear evolution equations.* It contains research papers related to Bénilan's work and covers a wide range of nonlinear and linear equations.

Special topics are Hamilton–Jacobi equations, the porous medium equation, reaction diffusion systems, integro-differential equations and visco-elasticity, maximal regularity for elliptic and parabolic equations, and the Ornstein–Uhlenbeck operator.

The book gives insight into new developments of nonlinear analysis and its applications in physics, mechanics, chemistry, biology and others.

Most contributions appeared in the regular issues of the *Journal of Evolution Equations,* Volume 3, in 2003.

Wolfgang Arendt
Haïm Brézis
Michel Pierre

Contents

Ph.D.-Students of Philippe Bénilan

1. Picard, Colette: *Opérateurs ϕ-accrétifs et génération de semi-groupes non linéaires*, Orsay, 1972.

2. de Gromard, Thierry: *Valeurs-frontière des fonctions à variation bornée sur un ouvert de $\mathbb{R}^N$*, Orsay, 1973.

3. Pierre, Michel: *Génération et perturbations de semi-groupes de contraction non linéaires*, Paris VI, 1976.

4. Ha, Ki Sik: *Sur des semi-groupes non linéaires dans les espaces $L^\infty(\Omega)$*, Paris VI, 1976.

5. Án, Lê Châu-Ho: *Etude de la classe des opérateurs m-accrétifs dans $L^\infty(\Omega)$*, Paris VI, 1977.

6. Menou, Segla: *Mesurabilité des familles d'opérateurs maximaux monotones*, Besançon, 1979.

7. Bathélemy, Louise: *Application de l'accrétivité dans L^∞ à l'étude d'inéquations quasi-variationnelles en contrôle impulsionnel*, Besançon, 1980.

8. Gagneux, Gérard: *Sur des problèmes unilatéraux dégénérés de la théorie des écoulements diphasiques en milieu poreux*, Besançon, 1982.

9. Touré, Hamidou: *Etude des équations $u_t + f(u)_x = \varphi(u)_{xx}$ par la théorie des semi-groupes non linéaires*, Besançon, 1982.

10. Perriot, André: *Résolution d'équations modélisant le mouvement d'un fluide dans un conluiner*, Besançon, 1982.

11. Idrissi, Mekki: *Sur une résolution directe de problèmes paraboliques dans des ouverts non cyclindriques — Existence de solutions dans $W^{2,p}(\Omega)$ d'équations quasi-linéaires elliptiques*, Besançon, 1983.

12. Ismail, Samir: *Formule de Lie-Trotter et théorie des semi-groupes non linéaires*, Besançon, 1986.

13. Simondon, Frédérique: *Solutions fortes pour $u_t = \varphi(u)_{xx} - f(t)\psi(u)_x$*, Besançon, 1986.

14. Abourjaily, Chaouki: *Comparaison de deux problèmes paraboliques*, Besançon, 1987.

15. Labani, Halima: *Contribution à l'étude des systèmes de réaction diffusion*, Besançon, 1989.

16. Amraoui-Benyaich, Saida: *Contribution à l'étude des bifurcations dans le Brusselator*, Besançon, 1990.

17. Laurençot, Philippe: *Etude de quelques problèmes aux dérivées partielles non linéaires*, Besançon, 1993.

18. Wittbold, Petra: *Absorptions non linéaires*, Besançon, 1994.

19. Alaarabiou, El Hachem: *Etude de quelques relations entre l'analyse fonctionnelle linéaire et non linéaire*, Besançon, 1994.

20. Maliki, Mohamed: *Solution généralisée d'équations paraboliques dégénérées*, Besançon, 1995.

21. Boulaamayel, Bennasser: *Sous-potentiels d'opérateurs différentiels*, Besançon, 1995.

22. Igbida, Nourreddin: *Limite singulière de problèmes d'évaluation non linéaires*, Besançon, 1997.

23. Andreianov, Boris: *Quelques problèmes de la théorie des systèmes paraboliques dégénérés non-linéaires et des lois de conservation*, Besançon, 2000.

24. Bouhsiss, Fouzia: *Quelques resultats d'unicité pour des problèmes elliptiques et paraboliques*, Besançon, 2001.

Philippe Bénilan
October 6, 1940 – February 17, 2001

J.evol.equ. 3 (2003) 01 – 09
1424–3199/03/010001 – 09
© Birkhäuser Verlag, Basel, 2003

Journal of Evolution Equations

Introduction

The present Volume 3 (2003) of JEE is devoted to Philippe Bénilan.

It consists entirely of invited contributions. Several of them were presented at the conference *"Journées d'Analyse Non-Linéaire"* which was organized in the French Jura in October 2000 on the occasion of his 60th birthday.

Philippe Bénilan died on February 17, 2001. He was a most original and charismatic mathematician who had a deep and decisive impact on the theory of Evolution Equations. He was one of the founders of this journal.

Here we describe some of his essential contributions. They reflect an important part of the development of Nonlinear Analysis and the Theory of Evolution Equations.

At the beginning of his œuvre stands a work which became a bestseller in the domain: His Thèse d'Etat at the University of Paris Orsay in 1972 under the supervision of Haïm Brézis and Jacques Deny with the title: "Equations d'évolution dans un espace de Banach quelconque et applications". Here he proved his famous theorem on the uniqueness of mild solutions (*"bonnes solutions"* in French) for the evolution equation governed by a (non-linear multivalued) accretive operator in an arbitrary Banach space. In the early seventies, it was a great challenge to understand how the well-known theory of strongly continuous semigroups of linear operators in Banach spaces (e.g. the Hille-Yosida theorem) could be extended to nonlinear operators. In the case of Hilbert spaces there was a complete theory (of maximal monotone operators) which could even be extended to uniformly convex spaces. However, nearly everything went wrong in general Banach spaces, even in spaces such as L^1 and L^∞, which are natural for many models. In particular, no notion of solution was known which was suitable for the Cauchy problem $u'(t) + Au(t) = f(t)$ governed by a "good" (i.e. *accretive*) operator on such spaces. The notion of *mild solution* is "simply" defined as the uniform limit of the implicit Euler approximation scheme of the differential equation. The uniqueness result of Philippe Bénilan shows that this limit is independent of the process of approximation. Existence of such solutions was provided by the (equally famous) Crandall-Liggett Theorem under an additional surjectivity hypothesis (*m-accretivity*) which asserts that the exponential formula $(I + t/nA)^{-n}$ converges strongly on the closure of the domain of the operator A.

His Thèse d'Etat contains further very interesting abstract results, for example another deep extension of the Hille-Yosida Theorem to the nonlinear framework. Whereas in Hilbert space there is a perfect bijection between nonlinear contraction semigroups and m-accretive operators, a whole bunch of counterexamples were produced in that period to show that such results are not true in general Banach spaces. Now Philippe Bénilan showed that there is a bijective correspondance between the set of all m-accretive operators and the solutions of the *complete* evolution equation $u'(t) + Au(t) = y$ where y is an arbitrary element of the Banach space.

Concrete applications were also presented in his thesis. He shows in particular how the abstract framework can be applied to two popular nonlinear equations, namely the conservation law

$$u + div\phi(u) = \psi(u),$$

and the porous medium equation

$$u_t - \Delta\Phi(u) = v \text{ on }]0, T[\times\Omega, \ \partial\Phi(u)/\partial v = \beta(u) \text{ on }]0, T[\times\partial\Omega.$$

The right *m*-accretive operators governing these two equations are described in detail. It is absolutely remarkable that their definition is based on the notion of entropic solution in the sense of Kruzhkov.

Already his thesis gives the flavour of his favorite way of doing mathematics: to go back and forth from an abstract theory to concrete applications and models. Behind this is his deep need of understanding the basic ideas and the generality of an approach leading to the solution of a particular problem. As a leitmotiv he would typically ask: *"What is the right abstract framework?"*

Philippe Bénilan devoted much of his scientific activity to the problem of finding a good definition of solutions for nonlinear partial differential equations, which turned out to be surprisingly delicate. He introduced the notion of *entropic solution* for elliptic equations of the type

$$diva(\cdot, u, Du) = f.$$

One knows, that even in the linear case, if the coefficients are merely bounded, the distributional solutions are not unique. For the p-Laplacian uniqueness of the solutions with initial value in L^1 is still an open problem. The framework of entropic solutions, though, leads to a well-posed problem (it is related to the notion of renormalized solution introduced independently by P. L. Lions and F. Murat in the spirit of the type of solutions considered by R. Di Perna and P. L. Lions for the Boltzmann equation). Meanwhile Bénilan's results have been extended to diverse versions of the very nonlinear parabolic equation

$$b(u)_t - diva(\cdot, u, Du) = f.$$

It was most surprising and illuminating that the entropic solutions coincide with the abstract mild solutions for accretive operators in Banach spaces.

Going back in time, we have to mention the famous article of Aronson-Bénilan [13] from 1979 where it is shown that the positive solutions of the porous medium equation $u_t = \Delta u^m$ in $\mathbb{R}^N$ satisfy the pointwise inequality

$$u_t \geq -cu/t, \tag{1}$$

for a certain constant c. This very precise estimate was a breakthrough at the time. In fact, it signifies a regularization effect. For example, it implies that

$$\|u_t(t)\|_{L^1} \leq \frac{\hat{c}}{t}\|u(0)\|_{L^1}, \tag{2}$$

an estimate which in the linear case is equivalent to the analyticity of the semigroup. The Aronson-Bénilan paper was the starting point for numerous works on such regularizing effects for nonlinear equations, and also for the investigation of the regularity of solutions. A direct consequence of the pointwise estimate (1) is that u_t is a measure, this leads to many other nice regularity properties. Surprisingly, these pointwise inequalities remain valid for equations governed by an arbitrary homogeneous m-accretive operator A (i.e. A satisfies $A(\lambda u) = \lambda^m Au$ with $m \neq 1$). This was shown later by Bénilan and Crandall [17]. The result yields very strong regularization effects if the underlying space is L^∞, for example.

We have seen above, that L^1 is the right space for many examples. Philippe Bénilan particularly liked and mastered this space, which shows at the same time good and bad behaviour in the class of general Banach spaces. And indeed, he showed us the important role the space L^1 plays in the theory of non-linear partial differential equations. Most popular is his article [8] jointly with H. Brézis and M. G. Crandall on this subject. In the same spirit he solved a large number of open questions for solutions of nonlinear equations with a measure as initial value. One has also to mention his recent papers joint with S. N. Kruzhkov [73], [79], and Andreianov and Kruzhkov [92] on the conservation law with merely continuous non-linearity. Finally, he gave essential contributions to understanding the role of the space L^∞ exploiting the maximum principle for nonlinear equations.

It is indeed difficult to give an account on an œuvre going deeply into so many different areas of Nonlinear Analysis. Let us add a (non-exhaustive) list of key words and equations corresponding to contributions of Philippe Bénilan not mentioned so far: the Lie-Trotter product formula [22], [31], the equation $u'' + Au = f$ [72], completely m-accretive operators [53], T-accretivity [44], Kato's inequality [55], the equation $u_t + A\beta u = 0$ [10], the equation $u_t + \beta Au = 0$ [18], the equation $u_t + \max A_i u = f$ [19], Thomas-Fermi equations [58], reaction diffusion systems [57], [61], [82], asymptotic behaviour of the equation $u_t = \Delta_p u^m$ as m tends to ∞ [51], [68], [74], [91], the equation $u_t = a(\cdot, u, \varphi(\cdot, u)_x)_x$ [70], [71], the equation $u_t = \alpha(u)\Delta u + |Du|^2 + F(u)$ [76], Schrödinger semigroups [54], and many more.

Philippe Bénilan wrote important contributions to the fundamental treatise edited by Dautray-Lions [26], [27], [28], [39], [40], [41], [50]. In particular, his contribution [25],

[50]: *"The Laplacian"*, is a very successful modern treatment of Potential Analysis via the systematic use of distributions.

Let us come back to the initial subject, *nonlinear semigroups on Banach spaces*. One can say that under the essential influence of Philippe Bénilan it became a fascinating and very elegant theory which is best documented in his treatise *"Nonlinear Evolution Equations Governed by Accretive Operators"* jointly with M. G. Crandall and A. Pazy [93]. Bénilan's book is famous and widely quoted in the community *"Evolution Equations"*, an apparent contradiction since it is not published and not even finished. We greatly hope that this will be done very soon.

The high esteem Philippe Bénilan obtained all over the mathematical community is not only due to his outstanding mathematical oeuvre, but also to his extraordinary personality. He gave the highest value to human relations and he immersed himself completely and without any reserve in all his activities. Characteristic for him is the abundance of international contacts and the popularity which he acquired all over the world.

The community of Evolution Equations and Nonlinear Analysis lost a great mathematician and a wonderful colleague.

The scientific committee of this special volume,

Haïm Brézis (President), Wolfgang Arendt, Louise Barthélemy, Michael Crandall,
Jose Carillo, J. Ildefonso Diaz, Jerome A. Goldstein, Michel Pierre,
Juan Luis Vazquez.

PUBLICATIONS OF PHILIPPE BÉNILAN

[1] BÉNILAN, PH., *Opérateurs localement lipschitziens*. C. R. Acad. Sci. Paris Série A-B *269* (1969), A1125–A1127.

[2] BÉNILAN, PH., *Une remarque sur la convergence des semi-groupes non linéaires*. C. R. Acad. Sci. Paris Série A-B *272* (1971), A1182–A1184.

[3] BÉNILAN, PH., *Solutions intégrales d'équations d'évolution dans un espace de Banach*. C. R. Acad. Sci. Paris Série A-B *274* (1972), A47–A50.

[4] BÉNILAN, PH., and BRÉZIS, H., *Solutions faibles d'équations d'évolution dans les espaces de Hilbert*. Ann. Inst. Fourier *22* (1972), 311–329.

[5] ATTOUCH, H., BÉNILAN, PH., DAMLAMIAN, A., and PICARD, C., *Equations d'évolution avec conditions unilatérale*. C. R. Acad. Sci. Paris Série A *279* (1974), 607–609.

[6] BÉNILAN, PH., *Semi-groupes invariants par un cône de fonctions convexes*. Lecture Notes in Econom. and Math. Systems *102* (1974), 49–65.

[7] BÉNILAN, PH., *Opérateurs m-accrétifs hémicontinus dans un espace de Banach quelconque*. C. R. Acad. Sci. Paris Série A *278* (1974), 1029–1032.

[8] BÉNILAN, PH., BRÉZIS, H., and CRANDALL, M. G., *A semilinear equation in $L^1(\mathbb{R}^N)$*. Ann. Scuola Norm. Sup. Pisa *2* (1975), 523–555.

[9] BÉNILAN, PH., DENY, J., and HIRSCH, F., *Séminaire sur les semi-groupes et les équations d'évolution*. Années 1972–1973/1974–1975, Publication Mathématique d'Orsay, Univ. Paris XI, n° 166–7547, 1975.

[10] BÉNILAN, PH., and HA, K., *Equations d'évolution du type $(du/dt) + \beta\upsilon\phi(u) \ni 0$ dans $L^\infty(U)$*. C. R. Acad. Sci. Paris Série A–B Math. n°22, *281* (1975), A947–A950.

[11] BÉNILAN, PH., Existence de solutions fortes pour l'équation des milieux poreux. C. R. Acad. Sci. Paris Série A-B Math. n°16, *285* (1977), A1029–A1031.

[12] BÉNILAN, PH. (ed.), ROBERT, J., (ed.), *Journées d'analyse non linéaire*. Proceedings d'une conférence tenue à Besançon, Juin 1977, Lecture Notes in Mathematics *665* (1978), 256.

[13] ARONSON, D.G., and BÉNILAN, PH., *Régularité des solutions de l'équation des milieux poreux dans $\mathbb{R}^N$*. C. R. Acad. Sci. Paris Série I Math. n°2, *288* (1979), A103–A105.

[14] BÉNILAN, PH., and CATTÉ, F., *Sur le traitement d'une équation de la théorie des plasmas par des techniques d'accrétivité dans L^∞*. Publications Mathématiques de Besançon, Analyse non linéaire, 1979, 1–8.

[15] BÉNILAN, PH., and PICARD, C., *Quelques aspects non-linéaires du principe du maximum*. Séminaire de Théorie du Potentiel, N°4 (Paris, 1977/1978), *Lecture Notes in Mathematics 713* (1979), 1–37.

[16] BÉNILAN, PH., and PIERRE, M., *Inéquations différentielles ordinaires avec obstacles irréguliers*. Ann. Fac. Sci. Toulouse Math. n°1, *1*(5) (1979), 1–8.

[17] BÉNILAN, PH., and CRANDALL, M. G., *Regularizing effects of homogeneous evolution equations*. Contributions to analysis and geometry (Baltimore, Md., 1980), Johns Hopkins Univ. Press 1981, 23–39.

[18] BÉNILAN, PH., and CRANDALL, M. G., *The continuous dependence on φ of solutions of $u^t - \Delta\varphi(u) = 0$*. Indiana Univ. Math. J. *30* (1981), 161–177.

[19] BÉNILAN, PH., and CATTÉ, F., *Equation d'évolution du type $(du/dt) + \max_i A_i u = f$ par la théorie des semi-groupes non linéaires dans L^∞*. C. R. Acad. Sci. Paris Série I Math. *295* (1982), 447–450.

[20] BÉNILAN, PH., and DÍAZ, J. I., *Comparison of solutions of nonlinear evolution problems with different nonlinear terms*. Israel J. Math. *42* (1982), 241–257.

[21] BÉNILAN, PH., *A strong regularity L^p for solution of the porous media equation*. Contributions to nonlinear partial differential equations (Madrid, 1981), Res. Notes in Math. *89* (1983), 39–58.

[22] BÉNILAN, PH., and ISMAIL, S., *Une généralisation d'un résultat de Kato-Masuda sur la formule de Trotter*. Publications Mathématiques de Besançon, Analyse non linéaire n° 7, 1983, 17.

[23] BÉNILAN, PH., CRANDALL, M.G., and PIERRE, M., *Solutions of the porous medium equation in $\mathbb{R}^N$ under optimal conditions on initial values*. Indiana Univ. Math. J. *33* (1984), 51–87.

[24] BÉNILAN, PH., and TOURÉ, H., *Sur l'équation générale $u_t = \varphi(u)_{xx} - \psi(u)_x + v$*. C. R. Acad. Sci. Paris Série I Math. *299* (1984), 919–922.

[25]   BÉNILAN, PH., *L'opérateur de Laplace*. Chapitre II, p. 255–769, in: Dautray, R., Lions, J. -L., *Analyse mathématique et calcul numérique pour les sciences et les techniques*, Tome I, Masson, Paris, 1984.

[26]   BÉNILAN, PH., *Opérateurs differentiels linéaires*. Chapitre V, p. 947–1086, in: Dautray, R., Lions, J.-L., Analyse mathématique et calcul numérique pour les sciences et les techniques, Tome I, Masson, Paris, 1984.

[27]   AUTHIER, M., ARTOLA, M., BÉNILAN, PH., CESSENAT, H. LANCHON, M., and MERCIER, B., *Problèmes variationnelles linéaires. Régularité*. Chapitre VII, p 1215–1306, in: Dautray, R., Lions, J.-L., Analyse mathématique et calcul numérique pour les sciences et les techniques, Tome I, Masson, Paris, 1984.

[28]   BÉNILAN, PH., *Intégrales singulières*. Annexe du Chapitre XIII, p. 969–1040, in: Dautray, R., Lions, J. -L., Analyse mathématique et calcul numérique pour les sciences et les techniques, *Tome II*, Masson, Paris, 1985.

[29]   BÉNILAN, PH., and PIERRE, M., *Quelques remarques sur la "localité" dans L^1 d'opérateurs differentiels*. Semesterberichte Funktionalanalysis Tübingen, Wintersemester 1987/88, p. 23–29.

[30]   BÉNILAN, PH., and BERGER, J., *Estimation uniforme de la solution de $u_t = \Delta\phi(u)$ et caractérisation de l'effet régularisant*. C. R. Acad. Sci. Paris Série I Math. *300* (1985) 573–576.

[31]   BÉNILAN, PH., and ISMAIL, S., *Générateur des semi-groupes non linéaires et la formule de Lie-Trotter*. Ann. Fac. Sci. Toulouse Math. n°2, *7*(5) (1985), 151–160.

[32]   BÉNILAN, PH., *An introduction to partial differential equations*. Semigroups, theory and applications, Vol. II (Trieste, 1984), Pitman Research Notes in Mathematics Series *152* (1986), 33–92.

[33]   BÉNILAN, PH., and SPITÉRI, P., *Discrétisation par schéma implicite d'un problème abstrait $d^2u/dt^2 \in Au$ sur* $[0, 1], u(0) = u_0, u(1) = u_1$. Publications Mathématiques de Besançon, Analyse non linéaire *10* (1987), 93–100.

[34]   BÉNILAN, PH., and VÁZQUEZ, J. L., *Concavity of solutions of the porous medium equation*. Trans. Am. Math. Soc. *299* (1987), 81–93.

[35]   BARTHÉLEMY, L., and BÉNILAN, PH., *Phénomène de couche limite pour la pénalisation d'équations quasi-variationnelles elliptiques avec conditions de Dirichlet au bord*. Nonlinear partial differential equations and their applications. Collège de France Seminar, Vol. VIII (Paris, 1984–1985), Pitman Research Notes in Mathematics Series *166* (1988), 47–68.

[36]   BARTHÉLEMY, L., and BÉNILAN, PH., *Sous-potentiels d'un opérateur non-linéaire*. Israel J. Math. *61* (1988), 85–112.

[37]   BÉNILAN, PH., CRANDALL, M.G., and PAZY, A., *"Bonnes solutions" d'un problème d'évolution semi-linéaire*. C. R. Acad. Sci. Paris Série I Math. *306* (1988), 527–530.

[38]   BÉNILAN, PH., CRANDALL, M. G., and SACKS, P., *Some L^1 existence and dependence results for semi-linear elliptic equations under nonlinear boundary conditions*. Appl. Math. Optimization *17* (1988), 203–224.

[39]   BÉNILAN, PH., *Linear differential operators*. Chapter V, p. 148–268, in: Dautray, R., Lions, J.-L., Mathematical analysis and numerical methods for science and technology, Vol. 2, Functional and Variational Methods, Springer-Verlag, Berlin, 1988.

[40]   AUTHIER, M., ARTOLA, M., BÉNILAN, PH., CESSENAT, LANCHON, H., and MERCIER, B., *Linear variational problems. Regularity*. Chapitre VII, p. 375–456, in: Dautray, R., Lions, J. -L., Mathematical analysis and numerical methods for science and technology, vol. 2, Functional and Variational Methods, Springer-Verlag, Berlin, 1988.

[41]   BÉNILAN, PH. (ed.), CHIPOT, M. (ed.), EVANS, L.C. (ed.), and PIERRE, M. (ed.), *Recent advances in nonlinear elliptic and parabolic problems*. Proceedings of an International Conference held in Nancy, March 1988, Pitman Research Notes in Mathematics Series *208* (1989), 340.

[42]   BÉNILAN, PH., and LABANI, H., *Existence of attractors for the Brusselator*. Recent advances in nonlinear elliptic and parabolic problems (Nancy, 1988), Pitman Research Notes in Mathematics Series *208* (1989), 139–151.

[43]   BÉNILAN, PH., and RIEDER, G. R., *Introduction to multivalued evolution equations*. Semesterbericht Funktionalanalysis, eds. G. Greiner, R. Nagel, F. Räbiger, U. Schlotterbeck, Tübingen, 1989, 3–14.

[44] BÉNILAN, PH., and DE LEÓN, S. S., *Accretivity, T-accretivity: the uniqueness-comparison theorem for strong solutions*. Semesterbericht Funktionalanalysis, eds. G. Greiner, R. Nagel, F. Räbiger, U. Schlotterbeck, Tübingen, 1989, 15–24.

[45] BÉNILAN, PH., and EGBERTS, P., *Mild solutions*. Semesterbericht Funktionalanalysis, eds. G. Greiner, R. Nagel, F. Räbiger, U. Schlotterbeck, Tübingen, 1989, 25–36.

[46] BÉNILAN, PH., *An introduction to concrete accretive operators in L^p*. Semesterbericht Funktionalanalysis, eds. G. Greiner, R. Nagel, F. Räbiger, U. Schlotterbeck, Tübingen, 1989, 37–48.

[47] BÉNILAN, PH., and KÉYANTUO, V., *The main theorems for mild solutions*. Semesterbericht Funktionalanalysis, eds. G. Greiner, R. Nagel, F. Räbiger, U. Schlotterbeck, Tübingen, 1989, 49–58.

[48] BÉNILAN, PH., BLANCHARD, D., and GHIDOUCHE, H., *On a nonlinear system for shape memory alloys*. Contin. Mech. Thermodyn. 2 (1990), 65–76.

[49] BÉNILAN, PH., *Singular integrals*. Appendix of Chapter XIII, p. 371–426, in: Dautray, R., Lions, J. -L.: Mathematical analysis and numerical methods for science and technology. vol 2, Volume 4, Spectral Theory and Applications, Springer-Verlag, Berlin, 1990.

[50] BÉNILAN, PH., *The Laplace operator*. Chapter II, p. 220–690, in: Dautray, R., Lions, J.-L.: Mathematical analysis and numerical methods for science and technology. Vol 1, Physical Origins and Classical Methods, Springer-Verlag, Berlin, 1990.

[51] BÉNILAN, PH., BOCCARDO, L., and HERRERO, M. A., *On the limit of solutions of $u_t = \Delta u^m$ as $m \to \infty$. Some topics in nonlinear PDEs*. Rend. Sem. Mat. Univ. Politec. Torino *1989* (1991), 1–13.

[52] BÉNILAN, PH., CORTÁZAR, C., and ELGUETA M., M., *Uniqueness and non uniqueness of the solutions of a mixed boundary value problem for the porous medium equation*. Rev. Union Mat. Argentina *37* (1991), 10–15.

[53] BÉNILAN, PH., and CRANDALL, M. G., *Completely accretive operators*. Semigroup theory and evolution equations (Delft, 1989), Lecture Notes in Pure and Appl. Math. *135* (1991), 41–75.

[54] ARENDT, W., BATTY, C. J. K., and BÉNILAN, PH., *Asymptotic stability of Schrödinger semigroups on $L^1(\mathbb{R}^N)$*. Math. Z. *209* (1992), 511–518.

[55] ARENDT, W., and BÉNILAN, PH., *Inégalités de Kato et semigroupes sous-markoviens*. Rev. Mat. Univ. Complutense Madrid 5 (1992), 279–308.

[56] BARTHÉLEMY, L., and BÉNILAN, PH., *Subsolutions for abstract evolution equations*. Potential Anal. *1* (1992), 93–113.

[57] BÉNILAN, PH., GOLDSTEIN, J. A., and RIEDER, G. R., *A nonlinear elliptic system arising in electron density theory*. Commmun. Partial Differential Equations *17* (1992), 2079–2092.

[58] BÉNILAN, PH., GOLDSTEIN, J. A., and RIEDER, G. R., *The Fermi-Amaldi correction in spinpolarized Thomas-Fermi theory*. Differential equations and mathematical physics (Birmingham, AL, 1990), Math. Sci. Eng. *186* (1992), 25–37.

[59] BÉNILAN, PH., and WITTBOLD, P., *Absorptions non linéaires*. J. Funct. Anal. *114* (1993), 59–96.

[60] BÉNILAN, PH., *Solutions d'équations d'évolution non linéaires: quelques problèmes ouverts*. Publications Mathématiques de Besançon, Analyse non linéaire n° 14, 1994, 10.

[61] BÉNILAN, PH., and LABANI, H., *Systèmes de réaction-diffusion abstraits*. Publications Mathématiques de Besançon, Analyse non linéaire n° 14, 1994, 18.

[62] BÉNILAN, PH., and LACROIX, M.-TH., *Estimation L^∞ du gradient d'un problème de Dirichlet dans un coin arrondi*. Publications Mathématiques de Besançon, Analyse non linéaire n° 14, 1994, 10.

[63] BÉNILAN, PH., Touré, H., *Solution entropique pour une équation parabolique-hyperbolique non linéaire*. Ann. Fac. Sci. Toulouse Math. 3 (1994), 63–80.

[64] BÉNILAN, PH., and WITTBOLD, P., *Nonlinear evolution equations in Banach spaces: basic results and open problems*. Funktionalanalysis (Essen, 1991), Lecture Notes in Pure and Appl. Math. *150* (1994), 1–32.

[65] ALAARABIOU, E. H., and BÉNILAN, PH., *Une version abstraite du héorème de Crandall-Liggett*. C. R. Acad. Sci. Paris Série I Math. *320* (1995), 923–928.

[66] BÉNILAN, PH., BOCCARDO, L., GALLOUËT, TH., GARIEPY, R., PIERRE, M., and VÁZQUEZ, J.L., *An L^1-theory of existence and uniqueness of solutions of nonlinear elliptic equations*. Ann. Scuola Norm. Sup. Pisa *22* (1995), 241–273.

[67] BÉNILAN, PH., and GARIEPY, R., *Strong solutions in L^1 of degenerate parabolic equations*. J. Differential Equations *119* (1995), 473–502.

[68] BÉNILAN, PH., and IGBIDA, N., *La limite de la solution de $u_t = \Delta_p u^m$ lorsque $m \to \infty$*. C. R. Acad. Sci. Paris Série I Math. *321* (1995), 1323–1328.

[69] BÉNILAN, PH., and KRUZHKOV, S. N., *First-order quasilinear equations with continuous nonlinearities*. Russian Acad. Sci. Dokl. Math. *50* (1995), 391–396.

[70] BÉNILAN, PH., and TOURÉ, H., *Sur l'équation générale $u_t = a(., u, \varphi(., u)_x)_x + v$ dans L^1. I. Etude du problème stationnaire*. Evolution equations (Baton Rouge, LA, 1992) Lecture Notes in Pure and Appl. Math. *168* (1995), 35–62.

[71] BÉNILAN, PH., and TOURÉ, H., *Sur l'équation générale $u_t = a(., u, \varphi(., u)_x)_x + v$ dans L^1 : II. Le problème d'évolution*. Ann. I. H. P. Nonlinear An. *12* (1995), 727–761.

[72] ALAARABIOU, E. H., and BÉNILAN, PH., *Sur le carré d'un opérateur non linéaire*. Archiv Math. *66* (1996), 335–343.

[73] BÉNILAN, PH., and BOUILLET, J. E., *On a parabolic equation with slow and fast diffusions*. J. Nonlinear Anal. *26* (1996), 813–822.

[74] BÉNILAN, PH., and IGBIDA, N., *Singular limit of perturbed nonlinear semigroups*. Communications in Appl. Nonlinear Anal. *3* (1996), 23–42.

[75] BÉNILAN, PH., and KRUZHKOV, S.N., *Conservation laws with continuous flux functions*. Nonlinear Differential Equations and Applications *3* (1996), 395–419.

[76] BÉNILAN, PH., and MALIKI, M., *Approximation de la solution de viscosité d'un problème d'Hamilton-Jacobi*. Rev. Mat. Univ. Complutense Madrid *9* (1996), 369–383.

[77] BÉNILAN, PH., and WITTBOLD, P., *On mild and weak solutions of elliptic-parabolic problems*. Advances in Differential Equations *1* (1996), 1053–1073.

[78] BÉNILAN, PH., *Quelques remarques sur la convergence singulière des semi-groupes linéaires*. Publications Mathématiques de Besançon, Analyse non linéaire n° 15, 1997, 1–11.

[79] BÉNILAN, PH., and BOUHSISS, F., *Un contre-exemple dans le cadre des équations hyperboliques paraboliques dégénérées*. Publications Mathématiques de Besançon, Analyse non linéaire n° 15, 1997, 123–126.

[80] BÉNILAN, PH., and BOUHSISS, F., *Une remarque sur l'unicité des solutions pour l'opérateur de Serrin*. C. R. Acad. Sci. Paris Série I Math. *325* (1997), 611–616.

[81] BÉNILAN, PH., MOUGEOT, M., and NAEGELEN, F., *Enchères asymétriques: contribution à la détermination numérique des stratégies d'équilibre Bayésien*. Ann. d'Economie et de Statistiques, n°46 1997, 225–251.

[82] BÉNILAN, PH., and WRZOSEK, D., *On an infinite system of reaction-diffusion equations*. Advances in Mathematical Sciences and Applications *7* (1997), 351–366.

[83] ABOURJAILY, C., and BÉNILAN, PH., *Symmetrization of quasilinear parabolic problems*. Dedicated to the memory of Julio E. Bouillet, Revista Union Mat. Argentina *41* (1998), 1–13.

[84] BÉNILAN, PH., and LAMI DOZO, E., *In memoriam : Julio E. Bouillet (1942–1994)*. Revista Union Mat. Argentina *41* (1998), i–iii.

[85] ARENDT, W., and BÉNILAN, PH., *Wiener regularity and heat semigroups on spaces of continuous functions*. Topics in nonlinear analysis, Progress in Nonlinear Differential Equations Appl. *35*, Birkhäuser, Basel, 1999, 29–49.

[86] BÉNILAN, PH., and BOULAAMAYEL, B., *Sous-solutions d'équations elliptiques dans L^1*. Potential Anal. *10* (1999), 215–241.

[87] BÉNILAN, PH., and WITTBOLD, P., *Sur un problème parabolique-elliptique*. M2AN. Mathematical Modelling Numerical Analysis *33* (1999), 121–127.

[88] ANDREIANOV, B., BÉNILAN, PH., and KRUZHKOV, S. N., *L^1-theory of scalar conservation law with continuous flux function*. J. Funct. Anal. *171* (2000), 15–33.

[89] AUSCHER, P., BARTHÉLEMY, L., BÉNILAN, PH., and OUHABAZ, E.M., *Absence de la L^∞-contractivité pour les semi-groupes associés aux opérateurs elliptiques complexes sous forme divergence*. Potential Anal. *12* (2000), 169–189.

[90] BÉNILAN, PH., CARRILLO, J., and WITTBOLD, P., *Renormalized entropy solutions of scalar conservation laws*. Ann. Scuola Norm. Sup. Pisa *29* (2000), 313–327.

[91] BÉNILAN, PH., and IGBIDA, N., *Limite de la solution de $u_t - \Delta u^m + \operatorname{div} F(u) = 0$ lorsque $m \to \infty$*. Rev. Mat. Complut. *13* (2000), 195–205.

[92] BÉNILAN, PH., and KAMYNIN, V.L., *A priori estimates for and the asymptotic behavior of solutions of semilinear integrodifferential parabolic equations on a plane*. Mathematical physics, mathematical modeling and approximate methods, (Obninsk, 2000), Mat. Model. *13* (2001), 10–22.

[93] BÉNILAN, PH., CRANDALL, M., and PAZY, A., *Nonlinear Evolution Equations Governed by Accretive Operators*, Book Manuscript, Besançon, 2001.

[94] BÉNILAN, PH., EVANS, L. C., and GARIEPY, R. F., *On singular limits of homogeneous semigroups*, J. evol. equ. *3* (2003).

[95] BÉNILAN, PH., and IGBIDA, N., *Singular limits of changing sign solutions of the porous medium equation*, J. evol. equ. *3* (2003).

[96] BÉNILAN, PH., and DIAZ, J. I., *Pointwise gradient estimates to one-dimensional nonlinear parabolic equations*, J. evol. equ. *3* (2003).

[97] BÉNILAN, PH., and BRÉZIS, H., *Nonlinear problems to the Thomas-Fermi equation*, J. evol. equ. *3* (2003).

J.evol.equ. 3 (2003) 11 – 25
1424–3199/03/010011 – 15
© Birkhäuser Verlag, Basel, 2003

**Journal of Evolution
Equations**

Intrinsic metrics and Lipschitz functions

FRANCIS HIRSCH

Abstract. We study the notions of measurable metric and Lipschitz function which were introduced by N. Weaver ([12]), in the framework of Dirichlet spaces. To this respect, we bring some precisions and complements to [15], notably concerning links with the notion of intrinsic metric ([2]). In the particular case of an abstract Wiener space, we establish the relationship between these notions and that of H-metric ([5]) and μ-a.e. H-Lipschitz continuous function ([4]).

1. Introduction

In the paper [12], N. Weaver introduced the notion of *measurable metric* and that of *Lipschitz function* with respect to such a metric. The study of these notions was pursued by the same author in subsequent papers ([13, 15, . . .]) and in the book [14]. Moreover, in [15] an exterior differential calculus was constructed in this general setting.

On the other hand, M. Biroli and U. Mosco defined in [2, 3] an *intrinsic metric* associated with a local Dirichlet form. Various topics related to this metric gave rise to many papers in the last decade.

In [15], N. Weaver studied, in the framework of Dirichlet structures satisfying the hypotheses of locality and of existence of a carré du champ operator (in the sense of [1]), the relationship between intrinsic metric and measurable metric. One of the aims of the present paper (see Section 3) is to come back to this study with essentially the same basic ideas, but in bringing some modifications and complements which seem useful.

In the particular case of abstract Wiener spaces (see e.g. [8, 11, 6, 1, . . .]) another notion of metric, the *H-metric*, was introduced and studied by S. Fang in [5] (see also [10] and [7]). Moreover, in [9, 10], S. Kusuoka introduced a class of functions that O. Enchev and D. W. Stroock called μ-a.e. *H-Lipschitz continuous* in [4]. Various characterizations were given ([9, 10, 4]). In this paper (Section 4), we state the links between these notions and the previous ones defined for general Dirichlet forms.

2. Measurable metrics and Lipschitz functions

This section essentially comes from [14, 15, 16]. The notion of measurable metric was first formulated in [12].

In the following, we consider a σ-finite measure space (X, μ). All spaces L^p will be related to this measure space. All functions that we consider in this paper are real functions. If S is a norm bounded subset of L^∞, S has a least upper bound or *join* denoted by $\bigvee S$ and a greatest lower bound or *meet* denoted by $\bigwedge S$. We denote by Ω the collection of all positive measure subsets of X. If A and B belong to Ω, we denote by $A \sim B$ the fact that A and B only differ from a null set ($\mu(A \setminus B) = \mu(B \setminus A) = 0$).

2.1. *Measurable pseudometrics*

DEFINITION 2.1. A *measurable pseudometric* is a map $\rho : \Omega^2 \longrightarrow [0, \infty]$ such that

1. $A' \sim A \Longrightarrow \rho(A', B) = \rho(A, B)$
2. $\rho(A, A) = 0$
3. $\rho(A, B) = \rho(B, A)$
4. $\rho(\cup_{n=1}^\infty A_n, B) = \inf_{n \geq 1} \rho(A_n, B)$
5. $\rho(A, C) \leq \sup_{B' \subset B}(\rho(A, B') + \rho(B', C))$

where all subsets which appear above belong to Ω.

We now give two classes of examples which will be of special interest.

2.1.1

Let ρ_0 be a map from X^2 into $[0, \infty]$ satisfying, for any $x, y \in X$,

$$\rho_0(x, x) = 0 \quad \text{and} \quad \rho_0(x, y) = \rho_0(y, x).$$

For $A, B \in \Omega$, we set

$$\rho_0(A, B) = \inf\{\rho_0(x, y) : x \in A, y \in B\},$$

and

$$\rho(A, B) = \sup\{\rho_0(A', B') : A' \sim A, B' \sim B\}.$$

For $A \in \Omega$, we also denote by ρ_A the function

$$\rho_A(x) = \rho_0(A, \{x\}) = \inf\{\rho_0(x, y) : y \in A\}.$$

REMARKS. 1. In the definition of $\rho(A, B)$, the sup is actually a max: There exist $A' \sim A$ and $B' \sim B$ with $A' \subset A$, $B' \subset B$ and $\rho(A, B) = \rho_0(A', B')$.

2. Clearly,

$$A_1 \subset A_2 \Longrightarrow \rho(A_1, B) \geq \rho(A_2, B).$$

3. We also have

$$\rho(A, B) = \max\{C \geq 0 : C \leq \rho_0(x, y) \text{ for } \mu\text{-a.e. } x \in A \text{ and for } \mu\text{-a.e. } y \in B\}.$$

PROPOSITION 2.2. *The map ρ satisfies properties 1 to 4 of Definition 2.1*

Proof. Straightforward.   □

The following proposition extends [14, Example 6.1.5].

PROPOSITION 2.3. *Suppose that moreover ρ_0 satisfies both following assumptions:*

1. $\forall x, y, z \in X \quad \rho_0(x, z) \leq \rho_0(x, y) + \rho_0(y, z)$.
2. *For all $A \in \Omega$, there exists $A' \in \Omega$ with $A' \subset A$, $A' \sim A$ and $\rho_{A'}$ measurable.*

Then ρ is a measurable pseudometric.

Proof ([16]). By Proposition 2.2, we only have to prove property 5 of Definition 2.1.

Let $A, B, C \in \Omega$ and $\varepsilon > 0$. We may assume that $\rho(A, B)$ is finite. Moreover, by Remark 1 above, there exist A_1, B_1, C_1 in Ω, respectively contained in A, B, C, such that $A \sim A_1, B \sim B_1, C \sim C_1$ and $\rho(A, B) = \rho_0(A_1, B_1)$, $\rho(A, C) = \rho_0(A_1, C_1)$. By assumption 2, we may also suppose that ρ_{A_1} is measurable. Set

$$B' = \{x \in B_1 : \rho_{A_1}(x) \leq \rho(A, B) + \varepsilon\}.$$

We then have $\rho_0(A_1, B_1 \setminus B') \geq \rho(A, B) + \varepsilon$, and hence $B' \in \Omega$.

There exist $y \in B'$ and $z \in C_1$ such that $\rho_0(y, z) \leq \rho_0(B', C_1) + \varepsilon$ and there exists $x \in A_1$ such that $\rho_0(x, y) \leq \rho(A, B) + 2\varepsilon$. Then

$$\rho_0(x, z) \leq \rho(A, B') + \rho(B', C) + 3\varepsilon$$

and therefore

$$\rho(A, C) \leq \rho(A, B') + \rho(B', C) + 3\varepsilon,$$

which yields the result.   □

REMARK. Property 2 in the previous Proposition is in particular satisfied under the following assumptions: X is a topological space, ρ_0 is a lower semicontinuous function on $X \times X$, and μ is a Borel measure which is inner regular, which means that, for any Borel set A,

$$\mu(A) = \sup\{\mu(K) : K \text{ compact and } K \subset A\}.$$

2.1.2

Let u be a μ-class of measurable real functions. If $A \in \Omega$, we denote by $F_A(u)$ the support of the image measure $(u_A)_*(\mu_A)$, where the index A indicates the restriction to A.

This support $F_A(u)$ also is the essential image of the restriction u_A of u to A. We set, for $A, B \in \Omega$,

$$\rho_u(A, B) = d(F_A(u), F_B(u))$$

where we denote by d the usual distance in $\mathbb{R}$:

$$\rho_u(A, B) = \inf\{|x - y| : x \in F_A(u), y \in F_B(u)\}.$$

REMARKS. 1. Define, for a particular representative of u,

$$\rho_0(x, y) = |u(x) - u(y)|.$$

Then ρ_u is nothing but the map ρ associated with ρ_0 by the method of the previous example 2.1.1.

2. Let φ be a Lipschitz continuous function from $\mathbb{R}$ into $\mathbb{R}$ with Lipschitz constant C. Then

$$\rho_{\varphi \circ u} \leq C \, \rho_u.$$

PROPOSITION 2.4. *The map ρ_u is a measurable pseudometric.*

The proof is easy.

Consider now a collection Φ of classes and set

$$\rho_\Phi(A, B) = \sup\{\rho_u(A, B) : u \in \Phi\}.$$

PROPOSITION 2.5. *The map ρ_Φ satisfies properties 1, 2, 3 and 5 of Definition 2.1.*

The following proposition is essentially [14, Example 6.2.5].

PROPOSITION 2.6. *Let F be a subspace of L^∞ containing the constant function 1, and let $\| \quad \|_0$ be a seminorm on F such that $\|1\|_0 = 0$. We assume that, for any $M, N \geq 0$, for any subset S of the set*

$$F_{M,N} = \{f \in F : \|f\|_\infty \leq M \text{ and } \|f\|_0 \leq N\},$$

the join $\bigvee S$ belongs to $F_{M,N}$. Then, denoting by Φ the set $\{f \in F : \|f\|_0 \leq 1\}$, ρ_Φ is a measurable pseudometric.

By Proposition 2.5, it is enough to prove that ρ_Φ satisfies property 4. This is done (see [14, Example 6.2.5]) by using the following lemma, which is a reformulation of [14, Lemma 6.2.4]):

LEMMA 2.7. *Let F be as above. Then, for any $f \in F$, for any $A, B \in \Omega$, there exists $g \in F$ such that*

$$\|g\|_0 \le \|f\|_0, \quad 0 \le g \le \rho_f(A, B), \quad \rho_g(A, B) = \rho_f(A, B),$$
$$g = 0 \text{ on } A, \quad g = \rho_f(A, B) \text{ on } B.$$

2.2. *Lipschitz functions*

We now consider a triplet $M = (X, \mu, \rho)$ where ρ is a measurable pseudometric on the σ-finite measure space (X, μ).

DEFINITION 2.8. Let u be a μ-class of measurable real functions. Then u is said to be a *Lipschitz function on M* if there exists a constant $C \ge 0$ such that

$$\forall A, B \in \Omega \quad \rho_u(A, B) \le C\rho(A, B).$$

In this case,

$$L(u) = \sup\{\rho_u(A, B)/\rho(A, B) : A, B \in \Omega \text{ and } \rho(A, B) > 0\}$$

is finite and called the *Lipschitz constant* of u.

We shall denote by $Lip(M)$ the set of all Lipschitz functions on M, and by $\mathrm{Lip}^p(M)$ the set $Lip(M) \cap L^p$.

DEFINITION 2.9. The measurable pseudometric ρ is called a *measurable metric* if $\mathrm{Lip}^\infty(M)$ is weak*-dense in L^∞. In this case, M is called a *measurable metric space*.

For an equivalent definition, we refer to [15, Proposition 14].

We then have the following result.

PROPOSITION 2.10. *Suppose that ρ is the measurable pseudometric ρ_Φ defined in Proposition 2.6. Then $\mathrm{Lip}^\infty(M) = F$ and, for every $f \in F$, $L(f) = \|f\|_0$.*

Proof. By the definitions, it is clear that $F \subset \mathrm{Lip}^\infty(M)$ and, for every $f \in F$, $L(f) \le \|f\|_0$.

The proof of the converse is very similar to that of [14, Theorem 6.3.6]. Let $f \in \mathrm{Lip}^\infty(M)$ satisfying $f \ge 0$ and $L(f) \le 1$. For any $A \in \Omega$, we set $a_A = \mathrm{ess\,inf}\,(f_A)$ (where, as previously, f_A denotes the restriction of f to A). Let $A, B \in \Omega$. For any n, there exists $f_n \in F$ with $\|f_n\|_0 \le 1$ so that

$$\rho(A, B) = \sup_n(\rho_{f_n}(A, B)).$$

By Lemma 2.7, there exists $g_n^{A,B} \in F$ such that

$$\|g_n^{A,B}\|_0 \le 1, \quad 0 \le g_n^{A,B} \le \rho_{f_n}(A,B), \quad \rho_{g_n^{A,B}}(A,B) = \rho_{f_n}(A,B),$$

$$g_n^{A,B} = 0 \text{ on } A, \quad g_n^{A,B} = \rho_{f_n}(A,B) \text{ on } B.$$

We then set

$$g^A = \bigvee \{(g_n^{A,B} \wedge a_A) : n \in \mathbb{N}, \ B \in \Omega\}.$$

We have

$$g^A \in F, \quad \|g^A\|_0 \le 1, \quad 0 \le g^A \le a_A, \quad g^A = 0 \text{ on } A, \quad g^A \ge \rho(A,B) \wedge a_A \text{ on } B.$$

Finally, we set

$$h = \bigvee \{a_A - g^A : A \in \Omega\}.$$

Of course, $h \in F$ and $\|h\|_0 \le 1$. As $L(f) \le 1$, $\rho(A,B) \ge a_A - \text{ess sup}(f_B)$ and hence $h \le \text{ess sup}(f_B)$ on B. This shows that, for any $B \in \Omega$, ess sup $(h_B) \le$ ess sup (f_B), which implies $h \le f$.

Now, $a_A - g^A = a_A$ on A and therefore ess inf $(h_A) \ge$ ess inf (f_A). Since this holds for any $A \in \Omega$, we obtain $h \ge f$. Consequently, $h = f$ and hence $f \in F$ and $\|f\|_0 \le 1$.

This result extends to any $f \in \text{Lip}^\infty(M)$ with $L(f) \le 1$ by applying the previous result to $f - \text{ess inf}(f)$. Finally we obtain that $\text{Lip}^\infty(M) \subset F$ and, for any $f \in \text{Lip}^\infty(M)$, $\|f\|_0 \le L(f)$. $\qquad\square$

3. Dirichlet forms

We consider in this section a Dirichlet structure $(X, \mathcal{B}, \mu, \mathbb{D}, \mathcal{E})$ in the sense of [1] to which we refer for the main definitions and properties. In particular, μ is as before a σ-finite measure on X equipped with the σ-algebra $\mathcal{B}$, and $\mathcal{E}$ is a Dirichlet form on L^2 whose domain is $\mathbb{D}$. We assume that the Dirichlet form $\mathcal{E}$ is local (which means e.g.

$$\forall f, g \in \mathbb{D} \ \forall a \in \mathbb{R} \quad (f+a)g = 0 \Longrightarrow \mathcal{E}(f,g) = 0)$$

and there exists a carré du champ operator Γ (which means that there exists a continuous map $\Gamma : \mathbb{D} \times \mathbb{D} \longrightarrow L^1$ such that

$$\forall f, g, h \in \mathbb{D} \cap L^\infty \quad \mathcal{E}(fh,g) + \mathcal{E}(gh,f) - \mathcal{E}(fg,h) = \int h\, \Gamma(f,g)\, \mathrm{d}\mu).$$

In what follows, we write $\Gamma(f)$ instead of $\Gamma(f,f)$. We set

$$\mathbb{D}^\infty = \{f \in \mathbb{D} \cap L^\infty : \Gamma(f) \in L^\infty\}.$$

In this section, we bring some precisions to [15, Section 6], using a somewhat different approach that we believe more direct.

3.1. *Case* $1 \in \mathbb{D}$

We first assume that the constant function 1 belongs to $\mathbb{D}$. This implies of course that the measure μ is finite.

PROPOSITION 3.1. *The space* $F = \mathbb{D}^{\infty}$, *equipped with the seminorm* $\|f\|_0 = \|\Gamma(f)^{1/2}\|_{\infty}$ *satisfies the assumptions of Proposition 2.6.*

Proof. By [1], $1 \in \mathbb{D}^{\infty}$ ($\Gamma(1) = 0$) and $\|f\|_0 = \|\Gamma(f)^{1/2}\|_{\infty}$ is a seminorm on $F = \mathbb{D}^{\infty}$. Moreover, if $f_1, \ldots, f_n$ are elements of $\mathbb{D}^{\infty}$, then $f = \sup\{f_j : 1 \le j \le n\}$ belongs to $\mathbb{D}^{\infty}$ and

$$\Gamma(f) \le \sup\{\Gamma(f_j) : 1 \le j \le n\}$$

(see, for instance, [1, Chapter I, Exercise 7.2]). Consider now a subset S of $F_{M,N}$ (notation of Proposition 2.6). We denote by $\widetilde{S}$ the set of joins of finite subsets of S. Of course, $\bigvee S = \bigvee \widetilde{S}$ and $\widetilde{S} \subset F_{M,N}$. Let $f = \bigvee \widetilde{S}$. Then f is the w^*-limit, in L^{∞}, of the directed net $\widetilde{S}$. As μ is finite, we also have that f is the weak limit, in L^2, of $\widetilde{S}$. Consequently, there exists a sequence $(F_n)_{n \ge 1}$ of convex combinations of elements of $\widetilde{S}$ such that $\lim_{n \to \infty} F_n = f$ strongly in L^2. The sequence $(F_n)_{n \ge 1}$ is contained in $F_{M,N}$ and hence is bounded in $\mathbb{D}$. By weak compactness, $f \in \mathbb{D}$ and $\lim_{n \to \infty} F_n = f$ weakly in $\mathbb{D}$. Taking again convex combinations, we may suppose $\lim_{n \to \infty} F_n = f$ strongly in $\mathbb{D}$. This implies that $\lim_{n \to \infty} \Gamma^{1/2}(F_n) = \Gamma^{1/2}(f)$ in L^2 and therefore $\|\Gamma(f)^{1/2}\|_{\infty} \le N$. Hence, $f \in F_{M,N}$. $\qquad\square$

We now define the *intrinsic metric* ρ by

$$\forall A, B \in \Omega \quad \rho(A, B) = \sup\{\rho_f(A, B) : f \in \mathbb{D}^{\infty} \text{ and } \Gamma(f) \le 1\}.$$

Then, by Proposition 3.1 and Proposition 2.10, we have:

COROLLARY 3.2. $M = (X, \mu, \rho)$ *is a measurable pseudometric space*, $\mathrm{Lip}^{\infty}(M) = \mathbb{D}^{\infty}$ *and, for any* $f \in \mathbb{D}^{\infty}$,

$$L(f) = \|\Gamma(f)^{1/2}\|_{\infty}.$$

We also have:

COROLLARY 3.3.

$$\mathrm{Lip}^2(M) = \{f \in \mathbb{D} : \Gamma(f) \in L^{\infty}\}$$

and, for any $f \in \mathrm{Lip}^2(M)$, $L(f) = \|\Gamma(f)^{1/2}\|_{\infty}$.

Proof. Let $f \in \mathbb{D}$ such that $\Gamma(f) \in L^\infty$. Set $f_n = (f \wedge n) \vee (-n)$. Of course $f_n \in \mathbb{D}^\infty$ and $\Gamma(f_n) \leq \Gamma(f)$. Therefore by Corollary 3.2, $f_n \in Lip(M)$ and $L(f_n) \leq \|\Gamma(f)^{1/2}\|_\infty$. Let $C_n = f^{-1}(]-n, n[)$ and $A, B \in \Omega$. For any n, m and $p > \max(n, m)$, $f_p = f$ on $C_n \cup C_m$. Hence,

$$\rho_f(A, B) \leq \rho_f(A \cap C_n, B \cap C_m) = \rho_{f_p}(A \cap C_n, B \cap C_m)$$
$$\leq \|\Gamma(f)^{1/2}\|_\infty \, \rho(A \cap C_n, B \cap C_m).$$

Taking the infimum, first with respect to n, then with respect to m, and using property 4 of Definition 2.1, we obtain

$$\rho_f(A, B) \leq \|\Gamma(f)^{1/2}\|_\infty \, \rho(A, B).$$

Then $f \in \mathrm{Lip}^2(M)$ and $L(f) \leq \|\Gamma(f)^{1/2}\|_\infty$.

Suppose conversely that $f \in \mathrm{Lip}^2(M)$ and set as before, $f_n = (f \wedge n) \vee (-n)$. By Remark 2 in Paragraph 2.1.2, $f_n \in \mathrm{Lip}^\infty(M)$. Hence $f_n \in \mathbb{D}^\infty$ and $\|\Gamma(f_n)^{1/2}\|_\infty \leq L(f)$. Going to the limit, clearly $f \in \mathbb{D}$ and $\|\Gamma(f)^{1/2}\|_\infty \leq L(f)$. $\qquad\square$

REMARK. Using the same argument as in the first part of the previous proof, we see easily that the intrinsic metric ρ may also be defined by

$$\rho(A, B) = \sup\{\rho_f(A, B) : f \in \mathbb{D} \text{ and } \Gamma(f) \leq 1\}.$$

3.2. *General case*

We no longer assume $1 \in \mathbb{D}$. However, we may use the following fact: There exists $\phi \in \mathbb{D}$ such that $\phi(x) > 0$ μ-a.e. This is easy to see, using the σ-finiteness of μ and the density of $\mathbb{D}$ in L^2. Let now $\varphi_p = (p\,\phi) \wedge 1$ and $A_p = \{\varphi_p = 1\}$. Then $\varphi_p \in \mathbb{D}$, $0 \leq \varphi_p \leq 1$ and $\cup_p A_p = X$ a.e. We set

$$\widetilde{\mathbb{D}} = \{f \in L^\infty : \forall \varphi \in \mathbb{D} \;\; \varphi f \in \mathbb{D}\}.$$

Then, $\widetilde{\mathbb{D}} \subset \mathbb{D}_{\mathrm{loc}}$, where the space $\mathbb{D}_{\mathrm{loc}}$ of functions locally in $\mathbb{D}$ is defined in [1]: Actually, if $f \in \widetilde{\mathbb{D}}$, then $f = \varphi_p f$ on A_p, $\varphi_p f \in \mathbb{D}$ and $\cup_p A_p = X$ a.e. By the hypothesis of locality, the carré du champ operator can be extended to $\mathbb{D}_{\mathrm{loc}} \times \mathbb{D}_{\mathrm{loc}}$ and therefore to $\widetilde{\mathbb{D}} \times \widetilde{\mathbb{D}}$ by setting

$$\forall f, g \in \widetilde{\mathbb{D}} \quad \Gamma(f, g) = \Gamma(\varphi f, \psi g) \text{ on } \{\varphi = 1\} \cap \{\psi = 1\}$$

for any $\varphi, \psi \in \mathbb{D}$. We then set

$$\widetilde{\mathbb{D}}^\infty = \{f \in \widetilde{\mathbb{D}} : \Gamma(f) \in L^\infty\}$$

where, as previously, $\Gamma(f)$ is put for $\Gamma(f, f)$.

PROPOSITION 3.4. $1 \in \widetilde{\mathbb{D}}^\infty$, $\mathbb{D}^\infty \subset \widetilde{\mathbb{D}}^\infty$ *and, if* $1 \in \mathbb{D}$, $\mathbb{D}^\infty = \widetilde{\mathbb{D}}^\infty$.

Proof. Suppose that $f \in \mathbb{D}^\infty$. Then, for any $\varphi \in \mathbb{D} \cap L^\infty$,

$$\varphi f \in \mathbb{D} \cap L^\infty \quad \text{and} \quad \Gamma(\varphi f) \le 2 \, (f^2 \, \Gamma(\varphi) + \varphi^2 \, \Gamma(f)).$$

By approximation, if $\varphi \in \mathbb{D}$ then $\varphi f \in \mathbb{D}$. Therefore $f \in \widetilde{\mathbb{D}}^\infty$.

Obviously, if $1 \in \mathbb{D}$, then $\widetilde{\mathbb{D}} \subset \mathbb{D}$ and hence $\widetilde{\mathbb{D}}^\infty \subset \mathbb{D}^\infty$. $\qquad\square$

PROPOSITION 3.5. *The space* $F = \widetilde{\mathbb{D}}^\infty$, *equipped with the seminorm* $\|f\|_0 = \|\Gamma(f)^{1/2}\|_\infty$ *satisfies the assumptions of Proposition 2.6.*

Proof. The proof is similar to that of Proposition 3.1. We see as before that $\|f\|_0 = \|\Gamma(f)^{1/2}\|_\infty$ is a seminorm on $F = \widetilde{\mathbb{D}}^\infty$ and, if $f_1, \ldots, f_n$ are elements of $\widetilde{\mathbb{D}}^\infty$, then $f = \sup\{f_j : 1 \le j \le n\}$ belongs to $\widetilde{\mathbb{D}}^\infty$ and

$$\Gamma(f) \le \sup\{\Gamma(f_j) : 1 \le j \le n\}.$$

Moreover, if $f \in \widetilde{\mathbb{D}}^\infty$ and $\varphi \in \mathbb{D}$,

$$\Gamma(\varphi f) = \varphi^2 \, \Gamma(f) + f^2 \, \Gamma(\varphi) + 2 f \, \varphi \, \Gamma(f, \varphi) \le (|f| \, \Gamma(\varphi)^{1/2} + |\varphi| \, \Gamma(f)^{1/2})^2.$$

Consider now a subset S of $F_{M,N}$ (notation of Proposition 2.6). We denote by $\widetilde{S}$ the set of joins of finite subsets of S. Of course, $\bigvee S = \bigvee \widetilde{S}$ and $\widetilde{S} \subset F_{M,N}$. Let $f = \bigvee \widetilde{S}$. Then, f is the w^*-limit, in L^∞, of the directed net $\widetilde{S}$. As a consequence, for any $\varphi \in \mathbb{D}$, φf is the weak limit, in L^2, of $\varphi \widetilde{S}$. Therefore, there exists a sequence $(F_n)_{n \ge 1}$ of convex combinations of elements of $\widetilde{S}$ such that $\lim_{n \to \infty} \varphi F_n = \varphi f$ strongly in L^2. For any n

$$|\varphi F_n| \le M \, |\varphi| \quad \text{and} \quad \Gamma(\varphi F_n) \le (M \, \Gamma(\varphi)^{1/2} + N \, |\varphi|)^2$$

and hence the sequence $(\varphi F_n)_{n \ge 1}$ is bounded in $\mathbb{D}$. By weak compactness, $\varphi f \in \mathbb{D}$ and $\lim_{n \to \infty} \varphi F_n = \varphi f$ weakly in $\mathbb{D}$. Taking again convex combinations, we may suppose $\lim_{n \to \infty} \varphi F_n = \varphi f$ strongly in $\mathbb{D}$. This implies that $\lim_{n \to \infty} \Gamma^{1/2}(\varphi F_n) = \Gamma^{1/2}(\varphi f)$ in L^2 and therefore $\Gamma(\varphi f)^{1/2} \le M \, \Gamma(\varphi)^{1/2} + N \, |\varphi|$. In particular, $\Gamma(\varphi f)^{1/2} \le N$ on $\{\varphi = 1\}$. Hence, $f \in F_{M,N}$. $\qquad\square$

We now define the *intrinsic metric* ρ by

$$\forall A, B \in \Omega \quad \rho(A, B) = \sup\{\rho_f(A, B) : f \in \widetilde{\mathbb{D}}^\infty \text{ and } \Gamma(f) \le 1\}.$$

Using again Proposition 2.10 and the previous proposition, we obtain:

COROLLARY 3.6. $M = (X, \mu, \rho)$ *is a measurable pseudometric space,* $\mathrm{Lip}^\infty(M) = \widetilde{\mathbb{D}}^\infty$ *and, for any* $f \in \widetilde{\mathbb{D}}^\infty$,

$$L(f) = \|\Gamma(f)^{1/2}\|_\infty.$$

REMARK. Suppose that $\widetilde{\mathbb{D}}^{\infty}$ is w^*-dense in L^{∞}. Then it is possible to show that Γ with domain $\widetilde{\mathbb{D}}^{\infty} \times \widetilde{\mathbb{D}}^{\infty}$ is an L^{∞}-diffusion form in the sense of [15, Definition 52]. Using [15, Theorem 54], this gives another proof, in this case, of corollary 3.6.

4. Abstract Wiener spaces

We consider in this section the particular setting of an abstract Wiener space (X, H, μ) in the sense of L. Gross ([8]). We recall that, in particular, X is a separable Banach space, μ is a centered Gaussian probability measure on X with support X, and $(H, |\ \ |_H)$ is a Hilbert space compactly embedded in X, which is called the Cameron-Martin space. We refer to [1] for a precise definition and for the notation we adopt here. Actually, we could take the more general framework considered in [6].

As shown in [1], there is a canonical Dirichlet structure $(X, \mathcal{B}, \mu, \mathbb{D}, \mathcal{E})$, where $\mathcal{B}$ denotes the Borel σ-algebra of X, which is local and admits a carré du champ Γ. Moreover, $1 \in \mathbb{D}$ and if we set

$$\mathbb{D}_0 = \{ f(l_0, \ldots, l_n) : n \in \mathbb{N}, \ l_0, \ldots, l_n \in X' \text{ and } f \in C_b^{\infty}(\mathbb{R}^{n+1}) \},$$

where $C_b^{\infty}(\mathbb{R}^{n+1})$ denotes the set of C^{∞}-functions on $\mathbb{R}^{n+1}$ which are bounded as well as all their derivatives, then $\mathbb{D}_0 \subset \mathbb{D}^{\infty}$ and $\mathbb{D}_0$ is dense in L^2. As a consequence, $\mathbb{D}^{\infty}$ is weak*-dense in L^{∞}.

The results of Subsection 3.1 are available and we can define the intrinsic metric:

$$\begin{aligned}
\rho(A, B) &= \sup\{\rho_f(A, B) : f \in \mathbb{D}^{\infty} \text{ and } \Gamma(f) \leq 1\} \\
&= \sup\{\rho_f(A, B) : f \in \mathbb{D} \text{ and } \Gamma(f) \leq 1\}.
\end{aligned}$$

We still denote by M the measurable metric space (X, μ, ρ) and, according to Corollary 3.2 and Corollary 3.3,

$$\mathrm{Lip}^{\infty}(M) = \mathbb{D}^{\infty}, \quad \mathrm{Lip}^2(M) = \{ f \in \mathbb{D} : \Gamma(f) \in L^{\infty} \}.$$

4.1. μ-a.e. H-Lipschitz continuous functions

In [4], O. Enchev and D.W. Stroock introduced another notion of Lipschitz function (see also [9, 10]):

DEFINITION 4.1. Let f be a μ-class of real Borel functions. Then f is said μ-a.e. H-Lipschitz continuous if there is a representative $\tilde{f}$ of f and a real constant C such that

$$\forall x \in X \ \forall h \in H \quad |\tilde{f}(x + h) - \tilde{f}(x)| \leq C \, |h|_H.$$

In this case, the *H-Lipschitz constant* of f is defined by

$$L_H(f) = \operatorname{ess\,sup}\left\{\sup_{h \in H \setminus \{0\}} \frac{1}{|h|_H} |\tilde{f}(x+h) - \tilde{f}(x)| : x \in X\right\}.$$

The main result of [4] is the following:

THEOREM 4.2. *Let $f \in L^2$. Then f is μ-a.e. H-Lipschitz continuous if and only if $f \in \mathbb{D}$ and $\Gamma(f) \in L^\infty$.*

In this case, $L_H(f) = \|\Gamma(f)^{1/2}\|_\infty$.

A direct consequence of the above theorem and Corollary 3.3 is the following:

COROLLARY 4.3.

$$\operatorname{Lip}^2(M) = \{f \in L^2 : f \text{ is } \mu\text{-a.e. } H\text{-Lipschitz continuous}\}$$

and, if $f \in \operatorname{Lip}^2(M)$, then $L(f) = L_H(f)$.

4.2. *H-metric*

We shall now give an equivalent definition of the intrinsic metric in terms of the Cameron-Martin space. We set, for $x, y \in X$,

$$\begin{aligned} \rho_0^H(x, y) &= |x - y|_H \quad \text{if } x - y \in H \\ &= +\infty \quad \text{otherwise} \end{aligned}$$

and, for $B \subset X$, $\rho_B^H(x) = \inf\{\rho_0^H(x, y) : y \in B\}$.

This metric was first considered by S. Kusuoka ([10]). In [5], S. Fang proved (see also [7]) the following result.

PROPOSITION 4.4. *Let B be a K_σ (i.e. a countable union of compact sets) of X, with positive measure. Then $\rho_B^H \in L^2$.*

Moreover, we have the following easy result:

PROPOSITION 4.5. *Let B be as in the previous proposition. Then ρ_B^H is μ-a.e. H-Lipschitz continuous and $L_H(\rho_B^H) \le 1$.*

Proof. Clearly $\rho_B^H(x) < +\infty$ if and only if $x \in B + H$. Therefore, by the previous Proposition 4.4, $B + H$ is a K_σ whose complement is μ-negligible. Set now $\varphi(x) = \rho_B^H(x)$ if $x \in B + H$ and $\varphi(x) = 0$ if $x \notin B + H$. We have

$$\forall x \in X \ \forall h \in H \quad |\varphi(x+h) - \varphi(x)| \le |h|_H,$$

which yields the result. $\qquad\qquad\square$

We now define, like in Paragraph 2.1.1, for $A, B \in \Omega$,

$$\rho_0^H(A, B) = \inf\{\rho_0^H(x, y) : x \in A, y \in B\},$$

and

$$\rho^H(A, B) = \sup\{\rho_0^H(A', B') : A' \sim A, B' \sim B\}.$$

By the fact that, for any B a K_σ of positive measure, ρ_B^H is a Borel function, and by Proposition 2.3, ρ^H is a measurable pseudometric.

THEOREM 4.6. *For any $A, B \in \Omega$,*

$$\rho^H(A, B) = \rho(A, B).$$

Proof. Let first $f \in \mathbb{D}^\infty$ with $\Gamma(f) \leq 1$. By Theorem 4.2, there exist a representative $\tilde{f}$ of f and a Borel set R whose complement is μ-negligible, such that

$$\forall x \in R \; \forall h \in H \quad |\tilde{f}(x + h) - \tilde{f}(x)| \leq |h|_H.$$

Consider now $A, B \in \Omega$ and set

$$A' = A \cap \tilde{f}^{-1}(F_A(f)) \cap R, \quad B' = B \cap \tilde{f}^{-1}(F_B(f)).$$

We have $A' \subset A, B' \subset B, A' \sim A, B' \sim B$ and

$$\forall x \in A' \; \forall y \in X \quad |\tilde{f}(x) - \tilde{f}(y)| \leq |x - y|_H.$$

Then

$$\rho_f(A, B) \leq \inf\{|\tilde{f}(x) - \tilde{f}(y)| : x \in A', \; y \in B'\} \leq \rho_0^H(A', B') \leq \rho^H(A, B).$$

Therefore, for any $A, B \in \Omega$, $\rho(A, B) \leq \rho^H(A, B)$.

Suppose now $A, B \in \Omega$. There exist B', a K_σ in X with $B' \subset B$ and $B' \sim B$, and $A' \subset A$ with $A' \sim A$, such that $\rho^H(A, B) = \rho_0^H(A', B')$ (see Remark 1 in 2.1.1). Let $n \geq \rho^H(A, B)$ and $f = \rho_{B'}^H \wedge n$. By Theorem 4.2 and Proposition 4.5, $f \in \mathbb{D}^\infty$ and $\Gamma(f) \leq 1$. We set $A'' = A' \cap f^{-1}(F_A(f))$. Of course $A'' \sim A$ and $\overline{f(A'')} = F_A(f)$. We have

$$\rho^H(A, B) = \rho_0^H(A'', B') = \inf_{x \in A''} \rho_{B'}^H(x).$$

Therefore

$$\rho^H(A, B) = \inf_{x \in A''} f(x).$$

On the other hand, $F_B(f) = \{0\}$. Hence,

$$\rho_f(A, B) = \inf F_A(f) = \inf f(A'') = \inf_{x \in A''} f(x).$$

Finally,

$$\rho^H(A, B) = \rho_f(A, B) \le \rho(A, B).$$

$\square$

REMARKS.   1. By the second part of the above proof, if $A, B \in \Omega$ there exists $f \in \mathbb{D}^\infty$ such that $\Gamma(f) \le 1$ and $\rho(A, B) = \rho_f(A, B)$. Thus, the sup in the definition of the intrinsic metric actually is a max.

2. Using the previous theorem and Propositions 2.2 and 2.5 in Section 2, we have another proof that ρ and ρ^H are measurable pseudometrics. This proof uses neither Proposition 2.3 nor Proposition 2.6.

We deduce directly from Theorems 4.2 and 4.6 the following Corollary.

COROLLARY 4.7. *Let $f \in L^2$. Then f is μ-a.e. H-Lipschitz continuous if and only if there exists a constant $C \ge 0$ such that*

$$\forall A, B \in \Omega \quad \rho_f(A, B) \le C \, \rho^H(A, B).$$

Then,

$$L_H(f) = \sup\{\rho_f(A, B)/\rho^H(A, B) : A, B \in \Omega \text{ and } \rho^H(A, B) > 0\}.$$

Finally, the next proposition states that the H-metric ρ_0^H is also the *pointwise intrinsic metric*, in the usual sense, associated with the Dirichlet space $\mathbb{D}$.

PROPOSITION 4.8. *For every $x, y \in X$,*

$$\rho_0^H(x, y) = \sup\{|u(x) - u(y)| : u \in \mathbb{D} \cap \mathcal{C} \text{ and } \Gamma(u) \le 1\},$$

where $\mathcal{C}$ denotes the space of continuous functions on X.

Proof. Suppose first that $u \in \mathbb{D} \cap \mathcal{C}$ and satisfies $\Gamma(u) \le 1$. By Theorem 4.2, there exist a representative $\tilde{u}$ of u and a constant C such that

$$\forall x \in X \; \forall h \in H \quad |\tilde{u}(x + h) - \tilde{u}(x)| \le C \, |h|_H.$$

By the continuity of u and the quasi-invariance of μ, there exists a set R whose complement is negligible such that $\tilde{u} = u$ on $R + H$. We also may suppose that

$$\forall x \in R \; \forall h \in H \quad |\tilde{u}(x + h) - \tilde{u}(x)| \le |h|_H$$

and therefore the same holds for u. Finally, by continuity,

$$\forall x \in X \; \forall h \in H \quad |u(x+h) - u(x)| \le |h|_H.$$

This proves that, for $x, y \in X$,

$$\rho_0^H(x, y) \ge \sup\{|u(x) - u(y)| : u \in \mathbb{D} \cap \mathcal{C} \text{ and } \Gamma(u) \le 1\}.$$

On the other hand, if $l \in X'$ (the topological dual of X), then $l \in \mathbb{D} \cap \mathcal{C}$ and $\Gamma(l) = \|l\|_{H'}^2$. Now, it is easy to see that, for any $x \in X$,

$$\sup\{|l(x)| : l \in X' \text{ and } \|l\|_{H'} \le 1\}$$

is equal to $|x|_H$ if $x \in H$ and is equal to $+\infty$ if $x \notin H$. Hence, for $x, y \in X$,

$$\rho_0^H(x, y) \le \sup\{|u(x) - u(y)| : u \in \mathbb{D} \cap \mathcal{C} \text{ and } \Gamma(u) \le 1\}.$$

$\square$

REMARK. As far as we know, it is an open problem whether, in the general situation of Section 3, what we called the intrinsic metric coincides, as in the case of Wiener spaces, with the measurable pseudometric associated with the usual pointwise intrinsic metric.

Acknowledgement

Many thanks to Nik Weaver who kindly answered a lot of questions. His valuable help is gratefully acknowledged.

REFERENCES

[1] BOULEAU, N. and HIRSCH, F., *Dirichlet forms and analysis on Wiener space*, de Gruyter, Berlin, 1991.

[2] BIROLI, M. and Mosco, U., *Formes de Dirichlet et estimations structurelles dans les milieux discontinus*, C.R. Acad. Sci. Paris, t. *313*, Série I, 1991, p. 593–598.

[3] BIROLI, M. and MOSCO, U., *A Saint-Venant type principle for Dirichlet forms on discontinuous media*, Ann. Mat. Pura Appl., *169* (1995), 125–181.

[4] ENCHEV, O. and STROOCK, D.W., *Rademacher's theorem for Wiener functionals*, Ann. Probab. *21* (1993), 25–33.

[5] FANG, S., *On the Ornstein-Uhlenbeck process, Stoch. and Stoch. Rep.*, *46* (1994), 3–4, 141–159.

[6] FEYEL, D. and DE LA PRADELLE, A., *Espaces de Sobolev gaussiens*, Ann. Inst. Fourier, 39–4 (1989), 875–908.

[7] FEYEL, D. and DE LA PRADELLE. A. *Démonstration géométrique d'une loi de tout ou rien*, C.R. Acad. Sci. Paris, t. 316, Série I, 1993, p. 229–232.

[8] GROSS, L., Abstract Wiener Spaces, in: *Proc. Fifth Berkeley Sympos. Math. Statist. Proba.*, p. 31–42, Univ. of California Press, 1965.

[9] KUSUOKA, S., *Dirichlet forms and diffusion processes on Banach spaces*, J. Fac. Sc. Univ. Tokyo, Sect. IA, *29* (1982), 79–95.

[10] KUSUOKA, S., *The nonlinear transformation of Gaussian measure on Banach space and its absolute continuity* (I), J. Fac. Sc. Univ. Tokyo, Sect. IA, *29* (1982), 567–597.

[11] WATANABE, S., *On stochastic differential equations and Malliavin calculus*, Tata Intitute of Fund. Research, vol. 73, Springer, 1984.

[12] WEAVER, N., *Nonatomic Lipschitz Spaces*, Studia Math., *115* (1995), 277–289.

[13] WEAVER, N., *Lipschitz Algebras and Derivations of von Neumann Algebras*, J. Funct. Anal., *139* (1996), 261–300.

[14] WEAVER, N., *Lipschitz Algebras*, World Scientific, Singapore, 1999.

[15] WEAVER, N., *Lipschitz Algebras and Derivations II. Exterior Differentiation*, J. Funct. Anal., *178* (2000), 64–112.

[16] WEAVER, N., *Personal communication.*

Francis Hirsch
Equipe d'Analyse et Probabilités
Université d'Évry – Val d'Essonne
Boulevard F. Mitterrand
F-91025 Evry Cedex
France
e-mail: hirsch@maths.univ-evry.fr

To access this journal online:
http://www.birkhauser.ch

J.evol.equ. 3 (2003) 27 – 37
1424–3199/03/010027 – 11
© Birkhäuser Verlag, Basel, 2003

Decay estimates for "anisotropic" viscous Hamilton-Jacobi equations in $\mathbb{R}^N$

SAÏD BENACHOUR AND PHILIPPE LAURENÇOT

Dédié à la mémoire de Philippe Bénilan

Abstract. The large time behaviour of the L^q-norm of nonnegative solutions to the "anisotropic" viscous Hamilton-Jacobi equation

$$u_t - \Delta u + \sum_{i=1}^{m} |u_{x_i}|^{p_i} = 0 \quad \text{in } \mathbb{R}_+ \times \mathbb{R}^N,$$

is studied for $q = 1$ and $q = \infty$, where $m \in \{1, \ldots, N\}$ and $p_i \in [1, +\infty)$ for $i \in \{1, \ldots, m\}$. The limit of the L^1-norm is identified, and temporal decay estimates for the L^∞-norm are obtained, according to the values of the p_i's. The main tool in our approach is the derivation of L^∞-decay estimates for $\nabla \left(u^\alpha\right)$, $\alpha \in (0, 1]$, by a Bernstein technique inspired by the ones developed by Bénilan for the porous medium equation.

1. Introduction

In this note, we study the large time behaviour of nonnegative and integrable solutions to the Cauchy problem for the following "anisotropic" viscous Hamilton-Jacobi equation

$$u_t - \Delta u + \sum_{i=1}^{N} \varepsilon_i \, |u_{x_i}|^{p_i} = 0 \quad \text{in } \mathbb{R}_+ \times \mathbb{R}^N, \tag{1}$$

$$u(0) = u_0 \quad \text{in } \mathbb{R}^N, \tag{2}$$

where $\mathbb{R}_+ := (0, +\infty)$, u_0 is a nonnegative function in $L^1(\mathbb{R}^N) \cap W^{1,\infty}(\mathbb{R}^N)$, $u_0 \not\equiv 0$, and

$$\varepsilon_i \in [0, +\infty), \quad p_i \in [1, +\infty), \quad i \in \{1, \ldots, N\}. \tag{3}$$

When $\varepsilon_i = 0$ for $i \in \{1, \ldots, N\}$, (1) reduces to the classical linear heat equation and the main purpose of this work is to figure out the influence of the nonlinear reaction term on the large time dynamics. Observe that the reaction term is nonnegative so that it acts as

Mathematics Subject Classification (2000): 35B40, 35B45, 35K55.
Key words: Viscous Hamilton-Jacobi equation, temporal decay estimates, gradient estimates.

an absorption term for nonnegative solutions. Actually, it readily follows from (1) that, if u is a nonnegative and integrable solution to (1)–(2), the L^1-norm of $u(t)$ decreases to a nonnegative limit I_∞ as $t \to +\infty$. Without the reaction term, it is well-known that $I_\infty = \|u_0\|_{L^1}$ and a natural question is to determine whether the reaction term implies $I_\infty = 0$ or $I_\infty > 0$, according to the values of the p_i's. Another straightforward remark is that the nonnegativity of the reaction term entails that u is a subsolution to the linear heat equation. Consequently, the L^∞-norm of $u(t)$ is bounded from above by $Ct^{-N/2}$ and the next question to be considered herein is whether the reaction term induces a faster decay rate for large times. A description of the large time behaviour of the L^q-norms of nonnegative and integrable solutions to (1)–(2) for $q = 1$ and $q = \infty$ is thus the main goal of this note and is to be found in Theorems 2.3 and 2.4 below.

Similar questions have already been answered for an "isotropic" counterpart of (1), namely

$$v_t - \Delta v + |\nabla v|^p = 0 \quad \text{in } (0, +\infty) \times \mathbb{R}^N, \tag{4}$$

where $|\nabla v|$ denotes the euclidean norm in $\mathbb{R}^N$ of ∇v and $p \in [1, +\infty)$, and we summarize the known results hereafter. Introducing $p_c := (N + 2)/(N + 1)$ and letting v be a nonnegative and integrable solution to (4), the L^1-norm of v converges to zero as $t \to +\infty$ if $1 \le p \le p_c$ and to a positive limit otherwise [1, 2, 4]. As for the L^∞-norm, it was shown in [5] that, for $t \in (0, +\infty)$, there holds

$$\|v(t)\|_{L^\infty} \le \kappa\, t^{-N/(p(N+1)-N)} \quad \text{if } p \in [1, p_c],$$

$$\|v(t)\|_{L^\infty} \le \kappa\, t^{-N/2} \quad \text{if } p > p_c,$$

where κ is a positive constant depending solely on N, p and $\|u_0\|_{L^1}$. Observe that $N/(p(N + 1) - N) \ge N/2$ for $p \in [1, p_c]$, so that the large time dynamics is diffusion-dominated if $p > p_c$ and the effect of the nonlinear absorption term only becomes effective for $p \in [1, p_c]$. Not surprisingly, the critical exponent $p_c = (N + 2)/(N + 1)$ also plays an important rôle in the analysis of (1).

2. Main results

Before stating precisely our results, we outline the well-posedness of the Cauchy problem (1)–(2) in the positive cone of $L^1(\mathbb{R}^N) \cap W^{1,\infty}(\mathbb{R}^N)$. The regularity assumptions on the initial data could certainly be weakened but a general theory of well-posedness is beyond the scope of this note. Let us however mentioned that well-posedness has been thoroughly investigated for (4), see [3, 4, 9] and the references therein.

PROPOSITION 2.1. *Assume that* (3) *holds true and consider a nonnegative function* u_0 *in* $L^1(\mathbb{R}^N) \cap W^{1,\infty}(\mathbb{R}^N)$. *There is a unique nonnegative function*

$$u \in \mathcal{C}([0, +\infty); L^1(\mathbb{R}^N)) \cap L^\infty(0, +\infty; W^{1,\infty}(\mathbb{R}^N))$$

satisfying, for each $t \geq 0$,

$$u(t) = G(t)\, u_0 - \int_0^t G(t-s)\left(\sum_{i=1}^N \varepsilon_i\, |u_{x_i}(s)|^{p_i} \right) ds. \tag{5}$$

In addition, $u(t) \leq G(t)u_0$ for $t \geq 0$ and

$$t \mapsto \|u(t)\|_{L^q},\ q \in [1, \infty],\ \text{and } t \mapsto \|\nabla u(t)\|_{L^\infty}\ \text{are nonincreasing functions.} \tag{6}$$

Here and below $G(t)$ denotes the linear heat semigroup in $\mathbb{R}^N$, that is,

$$(G(t)u)(x) = (4\pi t)^{-N/2} \int \exp\left(-\frac{|x-y|^2}{4t} \right) u(y)\, dy.$$

The proof of Proposition 2.1 is classical and follows the same lines as the existence proofs for (4) to which we refer [1, 3, 9]. It relies on a fixed point method, together with comparison principle arguments, observing that u is actually a classical solution to (1) for positive times.

We now fix a nonnegative function u_0 in $L^1(\mathbb{R}^N) \cap W^{1,\infty}(\mathbb{R}^N)$, $u_0 \not\equiv 0$, and denote by u the corresponding solution to (1)–(2) given by Proposition 2.1. Without loss of generality (after a possible renumbering of the coordinates in $\mathbb{R}^N$), we may assume that there is $m \in \{1, \ldots, N\}$ such that

$$\varepsilon_i = 0 \ \text{ for } \ i > m \ \text{ and } \ p_i \leq p_{i+1} \ \text{ for } \ i \in \{1, \ldots, m-1\}. \tag{7}$$

The cornerstone of our analysis is the following gradient estimates.

THEOREM 2.2. (i) *For $\alpha \in (0, 1]$, there is a constant $C(\alpha)$ depending only on α such that, for $i \in \{1, \ldots, N\}$ and $t \in \mathbb{R}_+$, there holds*

$$\|(u^\alpha)_{x_i}(t)\|_{L^\infty} \leq C(\alpha)\, \|u_0\|_{L^\infty}^\alpha\, t^{-1/2}. \tag{8}$$

(ii) *For $i \in \{1, \ldots, m\}$ such that $p_i > 1$ and $t \in \mathbb{R}_+$, there holds*

$$\|(u^{(p_i-1)/p_i})_{x_i}(t)\|_{L^\infty} \leq \frac{(p_i-1)^{(p_i-1)/p_i}}{p_i}\, (\varepsilon_i\, t)^{-1/p_i}. \tag{9}$$

The proof of Theorem 2.2 will be performed in Section 3. It relies on a Bernstein technique and is very much inspired by the ones developed by Bénilan to establish pointwise estimates for the derivatives of nonlinear functions of the solutions to the porous medium equation [6]. It is worth pointing out here that (8) results from the diffusive part of (1), while (9) stems from the Hamilton-Jacobi term (and is of course not available when $\varepsilon_i = 0$). Similar L^∞-estimates involving $\nabla v^{(p-1)/p}$ and ∇v have been obtained for (4) in [4] and [9], respectively.

As a first consequence of Theorem 2.2, we may determine when the L^1-norm of $u(t)$ has a non-vanishing limit as $t \to +\infty$, the case where the limit is zero being handled by a different method relying partly on arguments developed in [2]. We have actually the following result which will be proved in Section 4.

THEOREM 2.3. *We have*

$$I_\infty := \lim_{t \to +\infty} \|u(t)\|_{L^1} = 0 \iff p_1 = \min_{i \in \{1,\dots,m\}} p_i \leq p_c = \frac{N+2}{N+1}.$$

Theorem 2.2 also allows us to establish temporal decay estimates for the L^∞-norm of u. More precisely, we prove the following result in Section 5.

THEOREM 2.4. (i) *There is a constant C depending only on N and (p_i) such that*

$$\|u(t)\|_{L^\infty} \leq C \|u_0\|_{L^1} t^{-N/2}, \quad t \in \mathbb{R}_+. \tag{10}$$

(ii) *If $p_1 > p_c$, there holds*

$$\lim_{t \to +\infty} t^{N(1-1/q)/2} \|u(t) - I_\infty G(t) \delta_0\|_{L^q} = 0$$

for every $q \in [1, \infty]$, where $I_\infty > 0$ is defined in Theorem 2.3 and δ_0 denotes the Dirac mass at $x = 0$.

(iii) *If $1 < p_1 \leq p_c$, there is a constant C depending only on N and (p_i) such that, for $t \in \mathbb{R}_+$,*

$$\|u(t)\|_{L^\infty} \leq C \|u_0\|_{L^1}^{(N-k+1)/(N+1-\sigma)} t^{-(2\sigma+(N+1)(N-k))/(2(N+1-\sigma))}, \tag{11}$$

where

$$k := \max\{i \in \{1, \dots, m\}, \ p_i \leq p_c\} \quad and \quad \sigma := \sum_{i=1}^{k} \frac{1}{p_i} < N+1.$$

Observe that, under the assumptions of Theorem 2.4. (iii), $\sigma \geq (k\,(N+1))/(N+2)$, which guarantees that

$$\frac{2\sigma + (N+1)(N-k)}{2(N+1-\sigma)} \geq \frac{N}{2},$$

whence a faster decay rate for large times as expected.

Throughout this paper, we denote by C any positive constant depending only on N, p_i and ε_i, $i \in \{1, \dots, m\}$. The dependence of C upon additional parameters will be indicated explicitly.

3. Anisotropic gradient estimates in L^∞

This section is devoted to the proof of Theorem 2.2. As already mentioned, it relies on a Bernstein technique in the spirit of that developed in [6]. Let us first mention that the computations performed below are mainly formal but can be justified by classical approximation-regularisation procedures (see, e.g., the proof of [4, Theorem 1] or [9]).

Proof of Theorem 2.2 We consider $\alpha \in (0, 1)$, $i \in \{1, \ldots, N\}$ and put $f(r) := r^{1/\alpha}$, $r \in [0, +\infty)$. Introducing $w := u^\alpha$ and $z_i := w_{x_i}^2$, we realize that w solves

$$w_t - \Delta w - \left(\frac{f''}{f'}\right)(w)\,|\nabla w|^2 + \sum_{j=1}^{m} \varepsilon_j\,|f'(w)|^{p_j-2}\,f'(w)\,|w_{x_j}|^{p_j} = 0,$$

where $|\nabla w|$ denotes the euclidean norm of ∇w in $\mathbb{R}^N$. Using the previous equation, we evaluate $z_{i,t}$ and obtain

$$\begin{aligned}
z_{i,t} \;=\; & 2\,w_{x_i}\,\Delta w_{x_i} + 2\left(\frac{f''}{f'}\right)'(w)\,|\nabla w|^2\,z_i + 2\left(\frac{f''}{f'}\right)(w)\,\nabla w.\nabla z_i \\[2mm]
& - 2\sum_{j=1}^{m}\varepsilon_j\,(p_j-1)\,|f'(w)|^{p_j-2}\,f''(w)\,|w_{x_j}|^{p_j}\,z_i \\[2mm]
& - \sum_{j=1}^{m}\varepsilon_j\,|f'(w)|^{p_j-2}\,f'(w)\,p_j\,|w_{x_j}|^{p_j-2}\,w_{x_j}\,z_{i,x_j}\,,
\end{aligned}$$

whence, since $2\,w_{x_i}\,\Delta w_{x_i} \le \Delta z_i$,

$$\begin{aligned}
\mathcal{L}z_i - 2\left(\frac{f''}{f'}\right)'(w)\,|\nabla w|^2\,z_i & \\[2mm]
+\, 2\sum_{j=1}^{m}\varepsilon_j\,(p_j-1)\,|f'(w)|^{p_j-2}\,f''(w)\,|w_{x_j}|^{p_j}\,z_i & \le 0,
\end{aligned} \qquad (12)$$

with

$$\begin{aligned}
\mathcal{L}z_i := \; & z_{i,t} - \Delta z_i - 2\left(\frac{f''}{f'}\right)(w)\,\nabla w.\nabla z_i \\[2mm]
& + \sum_{j=1}^{m}\varepsilon_j\,|f'(w)|^{p_j-2}\,f'(w)\,p_j\,|w_{x_j}|^{p_j-2}\,w_{x_j}\,z_{i,x_j}.
\end{aligned}$$

Since f is convex and satisfies

$$-2\left(\frac{f''}{f'}\right)'(w) = \frac{2(1-\alpha)}{\alpha\,w^2} \ge \frac{1-\alpha}{\alpha}\,\|u_0\|_{L^\infty}^{-2\alpha} \ge 0\,,$$

$$|f'(w)|^{p_j-2}\,f''(w) \ge \frac{1-\alpha}{\alpha^{p_j}}\,w^{((1-\alpha)(p_j-1)/\alpha)-1} \ge 0$$

by (6), we deduce from (12) that

$$\mathcal{L}z_i + \frac{1-\alpha}{\alpha}\,\|u_0\|_{L^\infty}^{-2\alpha}\,z_i^2 \le 0 \tag{13}$$

for $i \in \{1, \ldots, N\}$ and

$$\mathcal{L}z_i + 2\,\varepsilon_i\,(p_i - 1)\,\frac{1-\alpha}{\alpha^{p_i}}\,w^{((1-\alpha)(p_i-1)/\alpha)-1}\,z_i^{(2+p_i)/2} \le 0 \tag{14}$$

for $i \in \{1, \ldots, m\}$. The assertion (8) for $\alpha \in (0, 1)$ then readily follows from (13) by the comparison principle (see, e.g., [8, Ch. V, § 5.2, Prop. 1, p. 1073]). For $i \in \{1, \ldots, m\}$ such that $p_i > 1$, we take $\alpha = (p_i - 1)/p_i$ in (14) and obtain

$$\mathcal{L}z_i + 2\,\varepsilon_i\,\left(\frac{p_i}{p_i - 1}\right)^{p_i - 1}\,z_i^{(2+p_i)/2} \le 0,$$

whence (9) by the comparison principle.

To prove (8) for $\alpha = 1$, we put $f(r) = (\|u_0\|_{L^\infty} - r)^2$ for $r \in [0, \|u_0\|_{L^\infty}]$ as in [9]. A similar computation with $w := f^{-1}(u)$ and $z_i := w_{x_i}^2$ leads again to (12) which now yields

$$\mathcal{L}z_i + \|u_0\|_{L^\infty}^{-2}\,z_i^2 \le 0,$$

thanks to the convexity of f and (6). The assertion (8) for $\alpha = 1$ is then again a straightforward consequence of the comparison principle and the proof of Theorem 2.2 is complete.

A useful consequence of Theorem 2.2 is the following result:

COROLLARY 3.1. *Consider $i \in \{1, \ldots, m\}$ such that $p_i > 1$. There is a constant C depending only on N and p_i such that, for $t \in \mathbb{R}_+$,*

$$\|(u^{(p_i-1)/p_i})_{x_i}(t)\|_{L^\infty} \le C\,\|u_0\|_{L^1}^{(p_i-1)/p_i}\,t^{-(p_i(N+1)-N)/(2p_i)}. \tag{15}$$

Proof. Since (1) is an autonomous equation, we deduce from (8) with $\alpha = (p_i - 1)/p_i$ that

$$\|(u^{(p_i-1)/p_i})_{x_i}(t)\|_{L^\infty} \le C(p_i)\,\|u(t/2)\|_{L^\infty}^{(p_i-1)/p_i}\,\left(\frac{t}{2}\right)^{-1/2}.$$

But, u being a subsolution of the heat equation, we also have

$$\|u(t/2)\|_{L^\infty} \le \|G(t/2)u_0\|_{L^\infty} \le C\,\|u_0\|_{L^1}\,t^{-N/2}.$$

Combining the above two inequalities yields (15). $\square$

4. Large time behaviour of the L^1-norm

We begin this section by stating two preliminary estimates which are needed in the proof of Theorem 2.3. We first recall a Morrey-type inequality established in [2, Eq. (2.1)].

LEMMA 4.1. [2] *If* $\varphi \in W^{1,1}(\mathbb{R}^N)$, $R > 0$ *and* $i \in \{1, \dots, N\}$, *there holds*

$$\|\varphi\|_{L^1} \le 2R \int_{\{|x| \le 3R\}} |\varphi_{x_i}(x)| \, dx + 2 \int_{\{|x| > R\}} |\varphi(x)| \, dx . \tag{16}$$

Next, since u is a subsolution to the linear heat equation, a control of $u(t, x)$ for large values of x and t is available and is a consequence of [2, Lemma 2.1].

LEMMA 4.2. [2] *If* $r \in \mathcal{C}([0, +\infty))$ *is a nonnegative function such that*

$$\lim_{t \to +\infty} r(t) \, t^{-1/2} = +\infty , \tag{17}$$

then

$$\lim_{t \to +\infty} \int_{\{|x| \ge r(t)\}} u(t, x) \, dx = 0 . \tag{18}$$

Proof of Theorem 2.3. We first notice that (7) ensures that

$$\min_{i \in \{1, \dots, m\}} p_i = p_1 .$$

Let us thus assume that $p_1 \le p_c$. Observe that (1), (3) and the nonnegativity of u imply that

$$\int_0^\infty \int |u_{x_1}(t, x)|^{p_1} \, dxdt \le \|u_0\|_{L^1} \tag{19}$$

after integration of (1) over $(0, +\infty) \times \mathbb{R}^N$. We then put

$$\omega(t) = \left(\int_{t/2}^\infty \int |u_{x_1}(s, x)|^{p_1} \, dxds \right)^{1/p_1}$$

for $t \ge 0$ and notice that $\omega \in \mathcal{C}([0, +\infty))$ is a nonincreasing function which satisfies

$$\lim_{t \to +\infty} \omega(t) = 0, \tag{20}$$

thanks to (19). Consider $t \ge 1$, $s \in (t/2, t)$ and $R \in \mathbb{R}_+$. We infer from Lemma 4.1 with $i = 1$ and the Hölder inequality that

$$\|u(s)\|_{L^1} \le CR^{(p_1(N+1)-N)/p_1} \|u_{x_1}(s)\|_{L^{p_1}} + 2 \int_{\{|x| > R\}} |u(s, x)| \, dx.$$

Since $s \mapsto \|u(s)\|_{L^1}$ is a nonincreasing function on $(t/2, t)$, it follows from the previous estimate and the Hölder inequality that

$$
\begin{aligned}
\|u(t)\|_{L^1} &\leq \frac{2}{t} \int_{t/2}^t \|u(s)\|_{L^1} \, ds \\
&\leq CR^{(p_1(N+1)-N)/p_1} \, t^{-1/p_1} \, \omega(t) \\
&\quad + \frac{4}{t} \int_{t/2}^t \int_{\{|x|>R\}} |u(s,x)| \, dxds.
\end{aligned}
$$

Since $p_1 \geq 1$, we choose $\delta \in (0, p_1/(p_1(N+1)-N))$ and take

$$
R = R(t) := t^{1/2} \, \omega(t)^{-\delta}
$$

in the above estimate. Noticing that $R(t) \geq R(s)$ for $s \in (t/2, t)$, we finally obtain

$$
\begin{aligned}
\|u(t)\|_{L^1} &\leq Ct^{(p_1(N+1)-(N+2))/(2p_1)} \, \omega(t)^{(p_1-\delta(p_1(N+1)-N))/p_1} \\
&\quad + \frac{4}{t} \int_{t/2}^t \int_{\{|x|>R(s)\}} |u(s,x)| \, dxds. \tag{21}
\end{aligned}
$$

On the one hand, since $R(s)s^{-1/2} \to +\infty$ as $s \to +\infty$ by (20), Lemma 4.2 ensures that the second term of the right-hand side of (21) converges to zero as $t \to +\infty$. On the other hand, the fact that $p_1 \leq p_c$ together with (20) imply that the first term of the right-hand side of (21) also converges to zero as $t \to +\infty$. Consequently, $I_\infty = 0$.

We next consider the case where $p_1 > p_c$ and proceed along the lines of [4, Theorem 6] to show that $I_\infty > 0$. Consider $s \in \mathbb{R}_+$ and $t \in (s, +\infty)$. Integrating (1) over $(s,t) \times \mathbb{R}^N$ yields

$$
\|u(s)\|_{L^1} = \|u(t)\|_{L^1} + \sum_{i=1}^m \varepsilon_i \int_s^t \int |u_{x_i}|^{p_i} \, dxd\tau. \tag{22}
$$

Let $i \in \{1, \ldots, m\}$. Owing to Corollary 3.1 and the time monotonicity of $\|u(s)\|_{L^1}$, we have

$$
\begin{aligned}
\int_s^t \int |u_{x_i}|^{p_i} \, dxd\tau &\leq C \, \|u(s)\|_{L^1} \int_s^t \|(u^{(p_i-1)/p_i})_{x_i}\|_{L^\infty}^{p_i} \, dxd\tau \\
&\leq C(u_0) \, \|u(s)\|_{L^1} \int_s^t \tau^{-(p_i(N+1)-N)/2} \, d\tau \\
&\leq C(u_0) \, s^{-(p_i(N+1)-(N+2))/2} \, \|u(s)\|_{L^1}, \tag{23}
\end{aligned}
$$

since $p_i \geq p_1 > p_c$. We may then let $t \to +\infty$ in (22) and use (23) to deduce that

$$
\|u(s)\|_{L^1} \left(1 - \sum_{i=1}^m \varepsilon_i \, C(u_0) \, s^{-(p_i(N+1)-(N+2))/2}\right) \leq I_\infty,
$$

whence

$$I_\infty \geq \frac{\|u(s)\|_{L^1}}{2} \quad \text{for } s \text{ large enough} \tag{24}$$

Now, if $I_\infty = 0$, we have $u(s) \equiv 0$ for s large enough. Since u is a classical solution to (1) for positive times, this implies that $u \equiv 0$ which contradicts the fact that $u_0 \neq 0$. Consequently, $I_\infty > 0$ and the proof of Theorem 2.3 is complete.

5. Temporal decay estimates for the L^∞-norm

We first recall the following interpolation inequality.

LEMMA 5.1. *There is a constant C depending only on N such that*

$$\|\varphi\|_{L^\infty} \leq C \, \|\varphi\|_{L^1}^{1/(N+1)} \prod_{i=1}^{N} \|\varphi_{x_i}\|_{L^\infty}^{1/(N+1)} \tag{25}$$

for every $\varphi \in L^1(\mathbb{R}^N) \cap W^{1,\infty}(\mathbb{R}^N)$.

Proof. Lemma 5.1 follows at once from the identity

$$\varphi(x)^{N+1} = (N+1)! \int_{-\infty}^{x_1} \cdots \int_{-\infty}^{x_N} \varphi(y) \prod_{i=1}^{N} \varphi_{x_i}(z_i) \, dy_N \ldots dy_1,$$

where $x = (x_1, \ldots, x_N)$, $y = (y_1, \ldots, y_N)$ and $z_i = (x_1, \ldots, x_{i-1}, y_i, \ldots, y_N)$ for $i \in \{1, \ldots, N\}$. □

Proof of Theorem 2.4 (i) & (ii). Let $t \in \mathbb{R}_+$. Since $0 \leq u(t) \leq G(t)u_0$, the assertion (10) follows at once from the properties of $G(t)$. In addition, we deduce from (8) with $\alpha = 1$ that, for $i \in \{1, \ldots, N\}$ and $t \in \mathbb{R}_+$, there holds

$$\|u_{x_i}(t)\|_{L^\infty} \leq C \, \|G(t/2)u_0\|_{L^\infty} \left(\frac{t}{2}\right)^{-1/2} \leq C \, \|u_0\|_{L^1} \, t^{-(N+1)/2}. \tag{26}$$

We may then proceed as in [7, Section 6] or [10, Proposition 2.2] to prove Theorem 2.4 (ii) for $q = 1$. Indeed, it readily follows from (6) and Corollary 3.1 that, for $i \in \{1, \ldots, m\}$,

$$\|u_{x_i}(t)\|_{L^{p_i}}^{p_i} \leq C(u_0) \, t^{-(p_i(N+1)-N)/2}, \quad t \in \mathbb{R}_+,$$

whence, since $p_i \geq p_1 > p_c$ for $i \in \{1, \ldots, m\}$,

$$F := \sum_{i=1}^{m} \varepsilon_i \, |u_{x_i}|^{p_i} \in L^1((1, +\infty) \times \mathbb{R}^N).$$

We now fix $t_0 > 0$. By (5), we have, for $t \geq t_0$,

$$
\begin{aligned}
\|u(t) - I_\infty G(t)\delta_0\|_{L^1} &\leq \|u(t) - G(t - t_0)u(t_0)\|_{L^1} \\
&\quad + \|G(t - t_0)u(t_0) - \|u(t_0)\|_{L^1} \, G(t)\delta_0\|_{L^1} \\
&\quad + |\|u(t_0)\|_{L^1} - I_\infty| \\
&\leq \int_{t_0}^{\infty} \int F \, dx \, ds + |\|u(t_0)\|_{L^1} - I_\infty| \\
&\quad + \|G(t - t_0)u(t_0) - \|u(t_0)\|_{L^1} \, G(t)\delta_0\|_{L^1} .
\end{aligned}
$$

Letting $t \to \infty$ and using a classical property of the heat semigroup in $L^1(\mathbb{R}^N)$, we obtain

$$
\limsup_{t \to +\infty} \|u(t) - I_\infty G(t)\delta_0\|_{L^1} \leq \int_{t_0}^{\infty} \int F \, dx \, ds + |\|u(t_0)\|_{L^1} - I_\infty|.
$$

Recalling the definition of I_∞, we may now pass to the limit as $t_0 \to +\infty$ to conclude that

$$
\lim_{t \to +\infty} \|u(t) - I_\infty G(t)\delta_0\|_{L^1} = 0.
$$

This last fact, together with (26) and classical properties of the heat semigroup, finally yield the second assertion of Theorem 2.4 by interpolation.

Proof of Theorem 2.4 (iii). Let $t \in \mathbb{R}_+$. On the one hand, we have

$$
\|u_{x_i}(t)\|_{L^\infty} \leq C \, \|u_0\|_{L^1} \, t^{-(N+1)/2}
$$

by (26) for $i \in \{k+1, \ldots, N\}$. On the other hand, for $i \in \{1, \ldots, k\}$, we have $p_i \in (1, p_c]$ and (9) entails that

$$
\|u_{x_i}(t)\|_{L^\infty} \leq C \, \|u(t)\|_{L^\infty}^{1/p_i} \, t^{-1/p_i}.
$$

Inserting these estimates in (25) and using (6) yield

$$
\|u(t)\|_{L^\infty} \leq C \, \|u_0\|_{L^1}^{(N-k+1)/(N+1)} \, \|u(t)\|_{L^\infty}^{\sigma/(N+1)} \, t^{-(2\sigma + (N+1)(N-k))/(2N+2)}
$$

$$
\|u(t)\|_{L^\infty}^{1-\sigma/(N+1)} \leq C \, \|u_0\|_{L^1}^{(N-k+1)/(N+1)} \, t^{-(2\sigma + (N+1)(N-k))/(2N+2)},
$$

from which (11) follows since $\sigma < N + 1$.

REMARK 5.2. If $m = N$, $p_i = p$ for some $p \in (1, p_c]$ and $\varepsilon_i = 1$ for $i \in \{1, \ldots, N\}$, then $\sigma = N/p$ and (11) reads

$$
\|u(t)\|_{L^\infty} \leq C \, \|u_0\|_{L^1}^{p/(p(N+1)-N)} \, t^{-N/(p(N+1)-N)}, \quad t \in \mathbb{R}_+.
$$

The L^∞-decay rate thus obtained is the same as the one for the solutions to (4).

REFERENCES

[1]   AMOUR, L. and BEN-ARTZI, M., *Global existence and decay for viscous Hamilton-Jacobi equations*, Nonlinear Anal. *31* (1998), 621–628.

[2]   BEN-ARTZI, M. and KOCH, H., *Decay of mass for a semilinear parabolic equation*, Comm. Partial Differential Equations *24* (1999), 869–881.

[3]   BEN-ARTZI, M., SOUPLET, PH. and WEISSLER, F.B., *The local theory for viscous Hamilton-Jacobi equations in Lebesgue spaces*, J. Math. Pures Appl. *81* (2002), 343–378.

[4]   BENACHOUR, S. and LAURENÇOT, PH., *Global solutions to viscous Hamilton-Jacobi equations with irregular initial data*, Comm. Partial Differential Equations *24* (1999), 1999–2021.

[5]   BENACHOUR, S., LAURENÇOT, PH. and SCHMITT, D., *Extinction and decay estimates for viscous Hamilton-Jacobi equations in $\mathbb{R}^N$*, Proc. Amer. Math. Soc. *130* (2002), 1103–1111.

[6]   BÉNILAN, PH., *Evolution equations and accretive operators*, Lecture notes taken by S. Lenhardt, University of Kentucky, Spring 1981.

[7]   BILER, P., GUEDDA, M. and KARCH, G., *Asymptotic properties of solutions of the viscous Hamilton-Jacobi equation*, prépublication LAMIFA , Université de Picardie, 2000.

[8]   DAUTRAY, R. and LIONS, J. L., *Analyse mathématique et calcul numérique pour les sciences et les techniques*, vol. 3, with the collaboration of Philippe Bénilan, Michel Cessenat, Bertrand Mercier and Claude Zuily, Masson, Paris, 1987.

[9]   GILDING, B. H., GUEDDA, M. and KERSNER, R., *The Cauchy problem for $u_t = \Delta u + |\nabla u|^q$*, prépublication LAMFA *28*, Université de Picardie, 1998.

[10] LAURENÇOT, PH. and SOUPLET, PH., *On the growth of mass for a viscous Hamilton-Jacobi equation*, J. Anal. Math., to appear.

Saïd Benachour
Institut Elie Cartan – Nancy
Université de Nancy 1
BP 239
F–54506 Vandœuvre-lès-Nancy cedex
France
e-mail: benachou@iecn.u-nancy.fr

Philippe Laurençot
Mathématiques pour l'Industrie et la Physique
CNRS UMR 5640
Université Paul Sabatier – Toulouse 3
118 route de Narbonne
F–31062 Toulouse cedex 4
France
e-mail: laurenco@mip.ups-tlse.fr

To access this journal online:
http://www.birkhauser.ch

J.evol.equ. 3 (2003) 39 – 65
1424–3199/03/010039 – 27
© Birkhäuser Verlag, Basel, 2003

**Journal of Evolution
Equations**

The Cauchy problem for linear growth functionals

F. ANDREU, V. CASELLES AND J. M. MAZÓN

Dedicated to Ph. Bénilan

1. Introduction and preliminaries

In this paper we are interested in the Cauchy problem

$$
\begin{cases}
\dfrac{\partial u}{\partial t} = \operatorname{div} \mathbf{a}(x, Du) & \text{in} \quad Q = (0, \infty) \times \mathbb{R}^N \\[2mm]
u(0, x) = u_0(x) & \text{in} \quad x \in \mathbb{R}^N,
\end{cases}
\tag{1.1}
$$

where $u_0 \in L^1_{loc}(\mathbb{R}^N)$ and $\mathbf{a}(x, \xi) = \nabla_\xi f(x, \xi)$, $f : \mathbb{R}^N \times \mathbb{R}^N \to \mathbb{R}$ being a function with linear growth as $\|\xi\| \to \infty$ satisfying some additional assumptions we shall precise below. An example of function $f(x, \xi)$ covered by our results is the nonparametric area integrand $f(x, \xi) = \sqrt{1 + \|\xi\|^2}$; in this case the right-hand side of the equation in (1.1) is the well-known mean-curvature operator. The case of the total variation, i.e., when $f(\xi) = \|\xi\|$ is not covered by our results. This case has been recently studied by G. Bellettini, V. Caselles and M. Novaga in [8]. The case of a bounded domain for general equations of the form (1.1) has been studied in [3] and [4] (see also [18], [11] and [15]). Our aim here is to introduce a concept of solution of (1.1), for which existence and uniqueness for initial data in $L^1_{loc}(\mathbb{R}^N)$ is proved.

Due to the linear growth condition on the Lagrangian, the natural energy space to study (1.1) is the space of functions of bounded variation. Let Ω be an open subset of $\mathbb{R}^N$. A function $u \in L^1(\Omega)$ whose gradient Du in the sense of distributions is a vector valued Radon measure with finite total variation in Ω is called a function of bounded variation. The class of such functions will be denoted by $BV(\Omega)$. Thus, if $u \in BV(\Omega)$, then Du is a Radon measure that decomposes into its absolutely continuous and singular parts $Du = D^a u + D^s u$. Then $D^a u = \nabla u \, \mathcal{L}^N$ where ∇u is the Radon-Nikodym derivative of the measure Du with respect to the Lebesgue measure $\mathcal{L}^N$. Moreover, we have the polar decomposition $D^s u = \overrightarrow{D^s u}|D^s u|$ where $|D^s u|$ is the total variation measure of $D^s u$. Finally, we denote by $BV_{loc}(\Omega)$ the sspace of functions $u \in L^1_{loc}(\Omega)$ such that $u\varphi \in BV(\Omega)$ for all $\varphi \in C_0^\infty(\Omega)$. For information concerning functions of bounded variation we refer to [1], [13] and [20].

By $L^1_w(0, T; BV(\mathbb{R}^N))$ we denote the space of functions $w : [0, T] \to BV(\mathbb{R}^N)$ such that $w \in L^1(]0, T[\times\mathbb{R}^N)$, the maps

$$t \in [0, T] \mapsto \int_{\mathbb{R}^N} \phi \, d Dw(t)$$

are measurable for every $\phi \in C_0^1(\mathbb{R}^N, \mathbb{R}^N)$ and $\int_0^T |Dw(t)|(\mathbb{R}^N) \, dt < \infty$. By $L^1_w(0, T; BV_{loc}(\mathbb{R}^N))$ we denote the space of functions $w : [0, T] \to BV_{loc}(\mathbb{R}^N)$ such that $w\varphi \in L^1_w(0, T, BV(\mathbb{R}^N))$ for all $\varphi \in C_0^\infty(\mathbb{R}^N)$.

Following [5], let

$$X_p(\mathbb{R}^N) = \{z \in L^\infty(\mathbb{R}^N, \mathbb{R}^N) \; : \; \mathrm{div}(z) \in L^p(\mathbb{R}^N)\}. \tag{1.2}$$

If $z \in X_p(\mathbb{R}^N)$ and $w \in BV(\mathbb{R}^N) \cap L^{p'}(\mathbb{R}^N)$ we define the functional $(z, Dw) : C_0^\infty(\mathbb{R}^N) \to \mathbb{R}$ by the formula

$$\langle (z, Dw), \varphi \rangle = -\int_{\mathbb{R}^N} w \, \varphi \, \mathrm{div}(z) \, dx - \int_{\mathbb{R}^N} w \, z \cdot \nabla\varphi \, dx. \tag{1.3}$$

Then (z, Dw) is a Radon measure in $\mathbb{R}^N$ and

$$\left| \int_B (z, Dw) \right| \le \int_B |(z, Dw)| \le \|z\|_\infty \int_B |Dw| \tag{1.4}$$

for any Borel set $B \subseteq \mathbb{R}^N$. Moreover, we have the following *Green's formula* ([5]), for $z \in X_p(\mathbb{R}^N)$ and $w \in BV(\mathbb{R}^N) \cap L^{p'}(\mathbb{R}^N)$:

$$\int_{\mathbb{R}^N} (z, Dw) + \int_{\mathbb{R}^N} w \, \mathrm{div}(z) \, dx = 0. \tag{1.5}$$

We define

$$z \cdot D^s w := (z, Dw) - (z \cdot \nabla w) \, d\mathcal{L}^N.$$

Then $z \cdot D^s w$ is a bounded measure which is absolutely continuous with respect to $|D^s w|$ [16], hence, it is singular, and we have $|z \cdot D^s w| \le \|z\|_\infty |D^s w|$.

2. The existence and uniqueness result

Our purpose in this section will be to define the notion of solution for the Cauchy problem (1.1) and to state an existence and uniqueness result for initial data in $L^1_{loc}(\mathbb{R}^N)$.

We shall assume that the Lagrangian $f : \mathbb{R}^N \times \mathbb{R}^N \to \mathbb{R}$ satisfies the following assumptions, which we shall refer collectively as (H):

(H$_1$) f is continuous on $\mathbb{R}^N \times \mathbb{R}^N$ and is a convex diffentiable function of ξ with continuous gradient for each fixed $x \in \mathbb{R}^N$. Furthermore we require f to satisfy the linear growth condition

$$C_0 \|\xi\| - C_1 \leq f(x, \xi) \leq M(\|\xi\| + C_2) \tag{2.1}$$

for some positive constants C_0, C_1, C_2. Moreover, f possesses an asymptotic function, i.e. for almost all $x \in \mathbb{R}^N$ there exists the finite limit

$$\lim_{t \to 0^+} tf\left(x, \frac{\xi}{t}\right) = f^0(x, \xi), \tag{2.2}$$

and $f^0(x, -\xi) = f^0(x, \xi)$ for all $\xi \in \mathbb{R}^N$ and all $x \in \mathbb{R}^N$.

(H$_2$) Let us consider the function $\tilde{f} : \mathbb{R}^N \times \mathbb{R}^N \times [0, +\infty[\to \mathbb{R}$ defined as

$$\tilde{f}(x, \xi, t) := \begin{cases} f(x, \frac{\xi}{t})t & \text{if} \quad t > 0 \\ f^0(x, \xi) & \text{if} \quad t = 0 \end{cases} . \tag{2.3}$$

We assume that $\tilde{f}(x, \xi, t)$ is continuous on $\mathbb{R}^N \times \mathbb{R}^N \times [0, +\infty[$ and convex in (ξ, t) for each fixed $x \in \mathbb{R}^N$.

We consider the function $\mathbf{a}(x, \xi) = \nabla_\xi f(x, \xi)$ associated to the Lagrangian f. By the convexity of f

$$\mathbf{a}(x, \xi) \cdot (\eta - \xi) \leq f(x, \eta) - f(x, \xi), \tag{2.4}$$

and the following monotonicity condition is satisfied

$$(\mathbf{a}(x, \eta) - \mathbf{a}(x, \xi)) \cdot (\eta - \xi) \geq 0. \tag{2.5}$$

Moreover, it is easy to see that

$$|\mathbf{a}(x, \xi)| \leq M \quad \forall\, (x, \xi) \in \mathbb{R}^N \times \mathbb{R}^N. \tag{2.6}$$

We consider the function $h : \mathbb{R}^N \times \mathbb{R}^N \to \mathbb{R}$ defined by

$$h(x, \xi) := \mathbf{a}(x, \xi) \cdot \xi.$$

From (2.4) and (2.1), it follows that

$$C_0 \|\xi\| - D_1 \leq h(x, \xi) \leq M \|\xi\| \tag{2.7}$$

for some positive constant D_1.

We assume that

(H$_3$) $h(x, \xi) \geq 0$ for $x, \xi \in \mathbb{R}^N$, h^0 exists and the function $\tilde{h}$ is continuous on $\mathbb{R}^N \times \mathbb{R}^N \times [0, +\infty[$.

We need to consider the mapping $\mathbf{a}^\infty$ defined by

$$\mathbf{a}^\infty(x, \xi) := \lim_{t \to +\infty} \mathbf{a}(x, t\xi).$$

Observe that

$$h^0(x, \xi) = \mathbf{a}^\infty(x, \xi) \cdot \xi \quad \text{and} \quad C_0 \|\xi\| \leq h^0(x, \xi) \leq M \|\xi\|.$$

(H$_4$) $\mathbf{a}^\infty(x, \xi) = \nabla_\xi f^0(x, \xi)$ for all $\xi \neq 0$ and all $x \in \mathbb{R}^N$.

In particular, as a consequence of Euler's Theorem, we have

$$f^0(x, \xi) = \mathbf{a}^\infty(x, \xi) \cdot \xi = h^0(x, \xi),$$

for all $\xi \in \mathbb{R}^N$ and all $x \in \mathbb{R}^N$, and consequently,

$$C_0 \|\xi\| \leq f^0(x, \xi) \leq M \|\xi\| \quad \forall \xi \in \mathbb{R}^N, \ \forall x \in \mathbb{R}^N. \tag{2.8}$$

(H$_5$) $\mathbf{a}(x, \xi) \cdot \eta \leq h^0(x, \eta)$ for all $x, \xi, \eta \in \mathbb{R}^N$.

Either from (H$_4$) or (H$_5$) it follows that $\mathbf{a}^\infty(x, \xi) \cdot \eta \leq h^0(x, \eta)$ for all $\xi, \eta \in \mathbb{R}^N, \xi \neq 0$, and all $x \in \mathbb{R}^N$. Indeed, it suffices to replace ξ by $t\xi$ in (H$_5$) and let $t \to +\infty$.

(H$_6$) $\mathbf{a}(x, 0) = 0$.

(H$_7$) We assume that

$$|a(x, \xi) - a(y, \xi)| \leq \omega(\|x - y\|) \tag{2.9}$$

for all $x, y \in \mathbb{R}^N$, and all $\xi \in \mathbb{R}^N$, where $\omega(r)$ is a modulus of continuity.

REMARK 2.1. Assumption (H$_7$) is only needed to prove uniqueness. The Lipschitz continuity in x of the flux is a common assumption to prove uniqueness of Kruzkov's solutions of scalar conservation laws ([17]).

We need to consider the space $BV(\mathbb{R}^N)_2$, defined as $BV(\mathbb{R}^N) \cap L^2(\mathbb{R}^N)$ endowed with the norm

$$\|w\|_{BV(\mathbb{R}^N)_2} := \|w\|_{L^2(\mathbb{R}^N)} + |Dw|(\mathbb{R}^N).$$

As usual, we denote by $BV(\mathbb{R}^N)_2^*$ the topological dual of $BV(\mathbb{R}^N)_2$. It is easy to see that $L^2(\mathbb{R}^N) \subset BV(\mathbb{R}^N)_2^*$ and

$$\|w\|_{BV(\mathbb{R}^N)_2^*} \leq \|w\|_{L^2(\mathbb{R}^N)} \quad \forall w \in L^2(\mathbb{R}^N). \tag{2.10}$$

We define the space

$$Z(\mathbb{R}^N) := \{(z, \xi) \in L^\infty(\mathbb{R}^N, \mathbb{R}^N) \times BV(\mathbb{R}^N)^* : \operatorname{div}(z) = \xi \ in \ \mathcal{D}'(\mathbb{R}^N)\}.$$

To make precise our notion of solution we need the following definitions.

DEFINITION 2.2. Let $\Psi \in L^1(0, T; BV(\mathbb{R}^N)_2)$ and $Q_T = (0, T) \times \mathbb{R}^N$ for $T > 0$. We say Ψ admits a *weak derivative* in the space $L^1_w(0, T; BV(\mathbb{R}^N)) \cap L^\infty(Q_T)$ if there is a function $\Theta \in L^1_w(0, T; BV(\mathbb{R}^N)) \cap L^\infty(Q_T)$ such that $\Psi(t) = \int_0^t \Theta(s) ds$, the integral being taken as a Pettis integral.

DEFINITION 2.3. Let $\xi \in (L^1(0, T; BV(\mathbb{R}^N)_2))^*$. We say that ξ is *the time derivative* in the space $(L^1(0, T; BV(\mathbb{R}^N)_2))^*$ of a function $u \in L^1(0, T; L^1_{loc}(\mathbb{R}^N))$ if

$$\int_0^T \langle \xi(t), \Psi(t) \rangle dt = -\int_0^T \int_{\mathbb{R}^N} u(t, x) \Theta(t, x) \, dx dt$$

for all test functions $\Psi \in L^1(0, T; BV(\mathbb{R}^N)_2)$ with compact support in time, which admit a weak derivative $\Theta \in L^1_w(0, T; BV(\mathbb{R}^N)) \cap L^\infty(Q_T)$ which is a function of compact support.

Observe that if $w \in L^1(0, T, BV(\mathbb{R}^N)_2) \cap L^\infty(Q_T)$ and $z \in L^\infty(Q_T, \mathbb{R}^N)$ is such that there exists $\xi \in (L^1(0, T; BV(\mathbb{R}^N)_2))^*$ with $\operatorname{div}(z) = \xi$ in $\mathcal{D}'(Q_T)$, associated to the pair (z, ξ), we define the distribution (z, Dw) in Q_T by

$$\langle (z, Dw), \phi \rangle := -\int_0^T \langle \xi(t), w(t)\phi(t) \rangle \, dt$$
$$- \int_0^T \int_{\mathbb{R}^N} z(t, x)w(t, x)\nabla_x \phi(t, x) \, dx dt \tag{2.11}$$

for all $\phi \in \mathcal{D}(Q_T)$.

DEFINITION 2.4. Let $\xi \in (L^1(0, T; BV(\mathbb{R}^N)_2))^*$, $z \in L^\infty(Q_T, \mathbb{R}^N)$. We say that $\xi = \operatorname{div}(z)$ in $(L^1(0, T; BV(\mathbb{R}^N)_2))^*$ if (z, Dw) is a Radon measure in Q_T such that

$$\int_{Q_T} (z, Dw) + \int_0^T \langle \xi(t), w(t) \rangle \, dt = 0,$$

for all $w \in L^1(0, T; BV(\mathbb{R}^N)_2) \cap L^\infty(Q_T)$.

We also need the following set of truncatures

$$\mathcal{P} := \{p \in W^{1,\infty}(\mathbb{R}) : p' \geq 0, \ \operatorname{supp}(p') \text{ compact}\}.$$

Our concept of solution for the Dirichlet problem (1.1) is the following.

DEFINITION 2.5. A measurable function $u : (0, T) \times \mathbb{R}^N \to \mathbb{R}$ is an *entropy solution* of (1.1) in Q_T if $u \in C([0, T]; L^1_{loc}(\mathbb{R}^N))$, $u(t)$ converges to u_0 in $L^1_{loc}(\mathbb{R}^N)$ as $t \to 0^+$, $p(u(\cdot)) \in L^1_w(0, T; BV_{loc}(\mathbb{R}^N))$ for all $p \in \mathcal{P}$, and there exists $\xi \in (L^1(0, T; BV(\mathbb{R}^N)_2))^*$ such that:

(i) $(\mathbf{a}(x, \nabla u(t)), \xi(t)) \in Z(\mathbb{R}^N)$ a.e. $t \in [0, T]$,

(ii) ξ is the time derivative of u in $(L^1(0, T; BV(\mathbb{R}^N)_2))^*$ in the sense of Definition 2.3,

(iii) $\xi = \mathrm{div}(\mathbf{a}(x, \nabla u(t)))$ in $(L^1(0, T; BV(\mathbb{R}^N)_2))^*$ in the sense of Definition 2.4,

(iv) the following inequality is satisfied

$$-\int_0^T \int_{\mathbb{R}^N} j(u(t) - l)\eta_t \, dxdt + \int_0^T \int_{\mathbb{R}^N} \eta(t)h(x, Dp(u(t) - l)) \, dt$$

$$+ \int_0^T \int_{\mathbb{R}^N} \mathbf{a}(x, \nabla u(t)) \cdot \nabla \eta(t)p(u(t) - l) \, dxdt \le 0$$

for all $l \in \mathbb{R}$, all $\eta \in C^\infty(]0, T[\times\mathbb{R}^N)$, with $\eta \ge 0$, $\eta(t, x) = \phi(t)\psi(x)$, being $\phi \in C_0^\infty(]0, T[)$, $\psi \in C_0^\infty(\mathbb{R}^N)$, and all $p \in \mathcal{P}$, where $j(r) := \int_0^r p(s) \, ds$.

Our main result is the following existence and uniqueness theorem.

THEOREM 2.6. *Assume we are under assumptions (H). Let $u_0 \in L^1_{loc}(\mathbb{R}^N)$. Then there exists a unique entropy solution of (1.1) in $[0, T] \times \mathbb{R}^N$ for all $T > 0$.*

3. The approximation problem with finite energy

To prove the existence part of Theorem 2.6. we approximate (1.1) by problems of the form

$$\begin{cases} \dfrac{\partial u}{\partial t} = \mathrm{div}\,(\varphi\mathbf{a}(x, Du)) & \text{in} \quad Q = (0, \infty) \times \mathbb{R}^N \\ u(0, x) = u_0(x) & \text{in} \quad x \in \mathbb{R}^N, \end{cases} \tag{3.1}$$

where $u_0 \in L^2(\mathbb{R}^N)$, $\varphi \in \mathcal{S}(\mathbb{R}^N)$ and $\mathcal{S}(\mathbb{R}^N)$ denotes the space of rapidly decreasing C^∞ functions in $\mathbb{R}^N$.

Let $\varphi \in \mathcal{S}(\mathbb{R}^N)$, $\varphi(x) > 0$ for every $x \in \mathbb{R}^N$. We define the space $BV(\mathbb{R}^N, \varphi \, dx)$ as the space of functions in $L^1_{loc}(\mathbb{R}^N)$ such that the distributional derivative Du is locally a Radon measure such that

$$\int_{\mathbb{R}^N} \varphi \, d|Du| < \infty.$$

By $W^{1,1}(\mathbb{R}^N, \varphi \, dx)$ we denote the space of functions in $BV(\mathbb{R}^N, \varphi \, dx)$ such that $Du \in L^1_{loc}(\mathbb{R}^N)$.

For simplicity, in what follows we shall assume that $\varphi \in \mathcal{S}(\mathbb{R}^N)$, $\varphi(x) > 0$ for all $x \in \mathbb{R}^N$, and satisfies the following property

$$|\varphi(x) - \varphi(y)| \leq C\varphi(y)\|x - y\| \tag{3.2}$$

for all $x, y \in \mathbb{R}^N$ such that $\|x - y\| \leq 1$ for some constant $C > 0$. It is easy to construct a function $\varphi(x) = \tilde{\varphi}(\|x\|)$ satisfying (3.2) if we take $\tilde{\varphi}$ a decreasing function such that $\tilde{\varphi}(r) = e^{-r}$ for $r \geq 1$ and such that φ is C^∞ in $\mathbb{R}^N$. Condition (3.2) enables us to prove the following Lemma.

LEMMA 3.1. *Assume that φ satisfies (3.2). Let $v \in BV(\mathbb{R}^N, \varphi\,dx) \cap L^p(\mathbb{R}^N)$, $1 \leq p < \infty$. Let $\eta \in C_0^\infty(\mathbb{R}^N)$, $\eta \geq 0$, with $supp(\eta) \subseteq B(0, 1)$, $\int_{\mathbb{R}^N} \eta(x)\,dx = 1$ and let $\tau_j \downarrow 0+$, $\eta_j = \frac{1}{\tau_j^N}\eta(\frac{x}{\tau_j})$. Then $v_j = \eta_j * v \in W^{1,p}(\mathbb{R}^N)$ satisfy*

$$v_j \to v \quad \text{in } L^p(\mathbb{R}^N) \text{ and} \quad \int_{\mathbb{R}^N} |Dv_j|\varphi\,dx \to \int_{\mathbb{R}^N} \varphi\,d|Dv| \quad \text{as } j \to \infty.$$

Proof. We only have to check that

$$\limsup_j \int_{\mathbb{R}^N} |Dv_j|\varphi\,dx \leq \int_{\mathbb{R}^N} \varphi\,d|Dv|. \tag{3.3}$$

For that, we write

$$\int_{\mathbb{R}^N} |Dv_j|\varphi\,dx \leq \int_{\mathbb{R}^N}\int_{\mathbb{R}^N} \eta_j(x - y)d|Dv|(y)\varphi(x)\,dx$$

$$= \int_{\mathbb{R}^N}\int_{\mathbb{R}^N} \eta_j(x - y)(\varphi(x) - \varphi(y))d|Dv|(y)\,dx$$

$$+ \int_{\mathbb{R}^N}\int_{\mathbb{R}^N} \eta_j(x - y)\varphi(y)d|Dv|(y)\,dx = \text{(I)} + \text{(II)}.$$

Observe that the second of the integrals above is convergent since η_j have compact support. By interchanging the order of integration in (II) we have

$$\text{(II)} = \int_{\mathbb{R}^N} \varphi\,d|Dv|.$$

Using (3.2) we have

$$\text{(I)} = \int_{\mathbb{R}^N}\int_{\mathbb{R}^N} \eta_j(x - y)(\varphi(x) - \varphi(y))d|Dv|(y)\,dx$$

$$\leq C \int_{\mathbb{R}^N}\int_{\mathbb{R}^N} \eta_j(x - y)\|x - y\|\varphi(y)d|Dv|(y)\,dx$$

$$= C \int_{\mathbb{R}^N} \eta_j(z)\|z\|\,dz \int_{\mathbb{R}^N} \varphi\,d|Dv|.$$

Letting $j \to \infty$, we obtain (3.3). $\qquad\square$

We denote by

$$X_\varphi(\mathbb{R}^N) := \{z \in X_2(\mathbb{R}^N) \ : \ z = \varphi z_1, \ \text{with } z_1 \in L^\infty(\mathbb{R}^N)\}.$$

If $z \in X_\varphi(\mathbb{R}^N)$ and $w \in BV(\mathbb{R}^N, \varphi\,dx) \cap L^2(\mathbb{R}^N)$ we define the functional $(z, Dw) :$ $C_0^\infty(\mathbb{R}^N) \to \mathbb{R}$ by

$$\langle(z, Dw), \phi\rangle = -\int_{\mathbb{R}^N} w\,\phi\,\mathrm{div}(z)\,dx - \int_{\mathbb{R}^N} w\,z \cdot \nabla\phi\,dx. \tag{3.4}$$

Using Lemma 3.1, instead of the one used by Anzellotti in [5], and some small modifications of the proofs given in [5], we can show that (z, Dw) is a Radon measure in $\mathbb{R}^N$ and

$$\left|\int_B (z, Dw)\right| \leq \|z_1\|_\infty \int_B \varphi\,d|Dw| \tag{3.5}$$

for all Borel set $B \subset \mathbb{R}^N$. Moreover, we also have the following *Green's formula* for $z \in X_\varphi(\mathbb{R}^N)$ and $w \in BV(\mathbb{R}^N, \varphi\,dx) \cap L^2(\mathbb{R}^N)$

$$\int_{\mathbb{R}^N} (z, Dw) + \int_{\mathbb{R}^N} w\,\mathrm{div}(z)\,dx = 0. \tag{3.6}$$

Our notion of solution for problem (3.1) when $u_0 \in L^2(\mathbb{R}^N)$ is the following:

DEFINITION 3.2. Let $u_0 \in L^2(\mathbb{R}^N)$. A measurable function $u : (0, T) \times \mathbb{R}^N \to \mathbb{R}$ is a *solution* of (3.1) in Q_T if $u \in C([0, T], L^2(\mathbb{R}^N))$, $u(0) = u_0$, $u'(t) \in L^2(\mathbb{R}^N)$, $u(t) \in BV(\mathbb{R}^N, \varphi\,dx) \cap L^2(\mathbb{R}^N)$, $\varphi\mathbf{a}(x, \nabla u(t)) \in X_2(\mathbb{R}^N)$ a.e. $t \in [0, T]$, and for almost all $t \in [0, T]$ $u(t)$ satisfies:

$$u'(t) = \mathrm{div}(\varphi\mathbf{a}(x, \nabla u(t)) \quad \text{in } \mathcal{D}'(\mathbb{R}^N) \tag{3.7}$$

$$\varphi(x)\mathbf{a}(x, \nabla u(t)) \cdot D^s u(t) = \varphi(x) f^0(x, D^s u(t)) = \varphi(x) f^0(x, \overrightarrow{D^s u})|D^s u|. \tag{3.8}$$

THEOREM 3.3. *Assume we are under assumptions (H) and φ satisfies (3.2). Given $u_0 \in L^2(\mathbb{R}^N)$, there exists a unique solution u of (3.1) in Q_T for every $T > 0$ such that $u(0) = u_0$.*

To prove Theorem 3.3 we shall use the nonlinear semigroup theory ([9]). For that we need to study the energy functional associated with the problem (1.1). In order to consider the relaxed energy we recall the definition of function of a measure ([6], [11]). Let $g : \mathbb{R}^N \times \mathbb{R}^N \to \mathbb{R}$ be a Carathéodory function such that

$$|g(x, \xi)| \leq M(1 + \|\xi\|) \quad \forall\, (x, \xi) \in \mathbb{R}^N \times \mathbb{R}^N, \tag{3.9}$$

for some constant $M \geq 0$. Furthermore, we assume that g possesses an asymptotic function g^0. It is clear that the function $g^0(x, \xi)$ is positively homogeneous of degree one in ξ.

We denote by $\mathcal{M}(\mathbb{R}^N, \mathbb{R}^N)$ the set of all $\mathbb{R}^N$-valued bounded Radon measures on $\mathbb{R}^N$. Given $\mu \in \mathcal{M}(\mathbb{R}^N, \mathbb{R}^N)$, we consider its Lebesgue decomposition $\mu = \mu^a + \mu^s$, where μ^a is the absolutely continuous part of μ with respect to the Lebesgue measure $\mathcal{L}^N$ of $\mathbb{R}^N$ and μ^s is singular with respect to $\mathcal{L}^N$. We denote by $\mu^a(x)$ the density of the measure μ^a with respect to $\mathcal{L}^N$ and by $(d\mu^s/d|\mu|^s)(x)$ the density of μ^s with respect to $|\mu|^s$.

For $\mu \in \mathcal{M}(\mathbb{R}^N, \mathbb{R}^N)$ and g satisfying the above conditions, we define the measure $g(x, \mu)$ on $\mathbb{R}^N$ as

$$\int_B g(x, \mu) := \int_B g(x, \mu^a(x)) \, dx + \int_B g^0\left(x, \frac{d\mu^s}{d|\mu|^s}(x)\right) d|\mu|^s \tag{3.10}$$

for all Borel set $B \subset \mathbb{R}^N$. In formula (3.10) we may write $(d\mu/d|\mu|)(x)$ instead of $(d\mu^s/d|\mu|^s)(x)$, because the two functions are equal $|\mu|^s$-a.e.

As it is proved in [6], if g is a Carathéodory function satisfying (3.9), then another way of writing the measure $g(x, \mu)$ is the following:

$$\int_B g(x, \mu) = \int_B \tilde{g}\left(x, \frac{d\mu}{d\alpha}(x), \frac{d\mathcal{L}^N}{d\alpha}(x)\right) d\alpha, \tag{3.11}$$

where α is any positive Borel measure such that $|\mu| + \mathcal{L}^N \ll \alpha$.

Let g be a function satisfying (3.9). Then for every $u \in BV(\mathbb{R}^N, \varphi \, dx)$ we have the measure $g(x, Du)\varphi$ defined by

$$\int_B g(x, Du)\varphi := \int_B g(x, \nabla u(x)) \, \varphi \, dx + \int_B g^0(x, \overrightarrow{D^s u}(x)) \, \varphi \, d|D^s u|$$

for all Borel set $B \subset \mathbb{R}^N$. Observe that if $\lambda = (\varphi Du, \varphi \mathcal{L}^N)$ and $\alpha = (|Du|, \mathcal{L}^N)$, then by Lemma 2.2 of [6] we have

$$g(x, Du)\varphi = \tilde{g}\left(x, \frac{d\lambda}{d\alpha}\right) = \tilde{g}(x, \lambda). \tag{3.12}$$

We define the energy functional

$$G_\varphi(u) := \int_{\mathbb{R}^N} g(x, Du) \, \varphi. \tag{3.13}$$

In [6], G. Anzellotti proves the lower semicontinuity of the functional G_φ in case of a bounded domain and $\varphi = 1$. Adapting the results in [6] or [4] we can also prove the lower semicontinuity of G_φ in our case. Indeed we have the following Lemma.

LEMMA 3.4. *Assume that $\tilde{g}(x, \xi, t)$ is lower-semicontinuous on $\mathbb{R}^N \times \mathbb{R}^N \times [0, +\infty[$, convex in (ξ, t) for each fixed $x \in \mathbb{R}^N$, and $g(x, \xi) \geq a\|\xi\| - b$ for all x and ξ. Then, for any sequence $u_n \in BV(\mathbb{R}^N, \varphi \, dx)$ such that $u_n \to u$ in $L^1_{loc}(\mathbb{R}^N)$ one has*

$$\liminf_{n \to \infty} G_\varphi(u_n) \geq G_\varphi(u).$$

We consider the energy functional associated with the problem (1.1) $\Phi_\varphi : L^2(\mathbb{R}^N) \to [0, +\infty]$ defined by

$$\Phi_\varphi(u) := \begin{cases} \displaystyle\iint_{\mathbb{R}^N} f(x, Du)\, \varphi, & \text{if } u \in BV(\mathbb{R}^N, \varphi\, dx) \cap L^2(\mathbb{R}^N) \\[2ex] +\infty & \text{if } u \in L^2(\mathbb{R}^N) \setminus BV(\mathbb{R}^N, \varphi\, dx). \end{cases}$$

Functional Φ_φ is clearly convex and has the form given in (3.13). Then, as a consequence of Lemma 3.4, we have that Φ_φ is lower-semicontinuous. Therefore, the subdifferential $\partial\Phi_\varphi$ of Φ_φ, is a maximal monotone operator in $L^2(\mathbb{R}^N)$ (see [9]). Consequently, the existence and uniqueness of a solution of the abstract Cauchy problem

$$\begin{cases} u'(t) + \partial\Phi_\varphi(u(t)) \ni 0 & t \in]0, \infty[\\[2ex] u(0) = u_0 & u_0 \in L^2(\mathbb{R}^N) \end{cases} \tag{3.14}$$

follows immediately from the nonlinear semigroup theory (see [9]). Now, to get the full strength of the abstract result derived from semigroup theory we need to characterize $\partial\Phi_\varphi$. To get this characterization, we introduce the following operator $\mathcal{B}_\varphi$ in $L^2(\mathbb{R}^N)$.

$$(u, v) \in \mathcal{B}_\varphi \iff u \in BV(\mathbb{R}^N, \varphi\, dx) \cap L^2(\mathbb{R}^N), v \in L^2(\mathbb{R}^N)$$

$\quad$ and $\varphi(x)\mathbf{a}(x, \nabla u) \in X_2(\mathbb{R}^N)$ satisfies:

$$-v = \mathrm{div}\,(\varphi\mathbf{a}(x, \nabla u)) \quad \text{in } \mathcal{D}'(\mathbb{R}^N) \tag{3.15}$$

$$\varphi\mathbf{a}(x, \nabla u) \cdot D^s u = \varphi f^0(x, D^s u) = \varphi f^0(x, \overrightarrow{D^s u})|D^s u|. \tag{3.16}$$

THEOREM 3.5. *Assume we are under assumptions (H) and φ satisfies (3.2), then the operator $\partial\Phi_\varphi$ has dense domain in $L^2(\mathbb{R}^N)$ and*

$$\partial\Phi_\varphi = \mathcal{B}_\varphi.$$

The proof of Theorem 3.5 follows the same approach used in [4] and we shall not include it here. Let us mention that one of the main tools needed is an approximation lemma similar to the one given by Anzellotti in [7]. The proof uses Lemma 3.1 and is similar to the proof in [7] (see also [4]).

LEMMA 3.6. *Assume that φ satisfies (3.2). If $v, u \in BV(\mathbb{R}^N, \varphi\, dx) \cap L^p(\mathbb{R}^N)$, $1 \le p < \infty$, then there exists a sequence $v_j \in C^1(\mathbb{R}^N) \cap W^{1,1}(\mathbb{R}^N, \varphi\, dx) \cap W^{1,p}(\mathbb{R}^N)$ such that*

$$v_j \to v \quad \text{in } L^p(\mathbb{R}^N), \tag{3.17}$$

$$\int_{\mathbb{R}^N} \sqrt{1 + |\nabla v_j(x)|^2}\, \varphi\, dx \to \int_{\mathbb{R}^N} \sqrt{1 + |Dv(x)|^2}\, \varphi\, dx, \tag{3.18}$$

$$\nabla v_j(x) \to \nabla v(x) \quad \mathcal{L}^N\text{-a.e. in } \mathbb{R}^N, \tag{3.19}$$

$$|\nabla v_j(x)| \to \infty \text{ and } \frac{\nabla v_j(x)}{|\nabla v_j(x)|} \to \frac{Dv(x)}{|Dv(x)|} \quad |Dv|^s\text{-a.e. in } \mathbb{R}^N, \tag{3.20}$$

$$|\nabla v_j(x)| \to \infty \text{ and } \frac{\nabla v_j(x)}{|\nabla v_j(x)|} \to \frac{Du(x)}{|Du(x)|} \quad |Du|^{ss}\text{-a.e. in } \mathbb{R}^N, \tag{3.21}$$

where $|Du|^{ss}$ denotes the part of the singular measure $|Du|^s$ which is singular with respect to $|Dv|^s$.

Standard semigroup theory (see [9]) and the characterization of Φ_φ given in Theorem 3.5 permits us to proof Theorem 3.3.

4. Proof of Theorem 1: Existence

We divide the proof into several steps.

STEP 1. Let $u_0 \in L^1_{\text{loc}}(\mathbb{R}^N)$. Let $u_{0n} \in L^2(\mathbb{R}^N) \cap L^\infty(\mathbb{R}^N)$ be such that $u_{0n} \to u_0$ in $L^1_{\text{loc}}(\mathbb{R}^N)$. Let $\varphi_n \in \mathcal{S}(\mathbb{R}^N)$ satisfying (3.2), $0 < \varphi_n \le 1$ and $\varphi_n(x) = 1$ for all $x \in B(0, n)$. By Theorem 3.3, for every $n \in \mathbb{N}$ there exists a solution u_n of (3.1) for $\varphi = \varphi_n$, corresponding to the initial conditions u_{0n}. Therefore, $u_n(t), u'_n(t) \in L^2(\mathbb{R}^N)$, $u_n(t) \in BV(\mathbb{R}^N, \varphi_n \, dx)$, $z_n(t) := \varphi_n \mathbf{a}(x, \nabla u_n(t)) \in X_2(\mathbb{R}^N)$ a.e. $t \in [0, T]$, and for almost all $t \in [0, T] \, u_n(t)$ satisfies:

$$u'_n(t) = \text{div}(z_n(t)) \quad \text{in } \mathcal{D}'(\mathbb{R}^N), \tag{4.1}$$

$$\begin{cases} z_n(t) \cdot D^s u_n(t) = \varphi_n f^0(x, D^s u_n(t)), \\ z_n(t) \cdot D^s p(u_n(t)) = \varphi_n f^0(x, D^s p(u_n(t))) \quad \forall \, p \in \mathcal{P}. \end{cases} \tag{4.2}$$

From (4.1) and (4.2), it follows that

$$-\int_{\mathbb{R}^N} (w - u_n(t)) u'_n(t) \, dx = \int_{\mathbb{R}^N} (z_n(t), Dw) - \int_{\mathbb{R}^N} \varphi_n h(x, Du_n(t)) \tag{4.3}$$

for every $w \in BV(\mathbb{R}^N, \varphi_n \, dx) \cap L^2(\mathbb{R}^N)$.

Let us prove that $\{u_n\}$ is a Cauchy sequence in $C([0, T]; L^1_{\text{loc}}(\mathbb{R}^N))$. Let $\alpha > N$, $T_k(r) := \max(\min(r, k), -k)$ $(k \ge 0)$ and let j_k^+ be the primitive of $p_k^+(r) = \alpha T_k^+(r)^{\alpha-1}$ vanishing at $r = 0$. We define j_k^- as the primitive of p_k^- which vanishes at $r = 0$, where $p_k^-(r) = -p_k^+(-r)$. If $N = 1$, we take $\alpha \ge 2$, so that $(j_k^\pm)' \in W^{1,\infty}(\mathbb{R})$.

Let $\phi \in C_0^\infty(\mathbb{R}^N)$, $\phi \geq 0$, and suppose that $m \geq n$. Then, by (4.1) and Green's formula, we have

$$\int_{\mathbb{R}^N} p_k^+(u_n(t) - u_m(t))\phi(u_n'(t) - u_m'(t))$$

$$= -\int_{\mathbb{R}^N} (z_n(t) - z_m(t), D(p_k^+(u_n(t) - u_m(t))\phi))$$

$$= -\int_{\mathbb{R}^N} \phi(z_n(t) - z_m(t), D(p_k^+(u_n(t) - u_m(t))))$$

$$- \int_{\mathbb{R}^N} \nabla\phi \cdot (z_n(t) - z_m(t))p_k^+(u_n(t) - u_m(t)).$$

Now, since $\phi\varphi_n = \phi$ for all $n \geq n(\phi)$, having in mind (4.2) and (2.5), it is easy to see that

$$\int_{\mathbb{R}^N} \phi(z_n(t) - z_m(t), D(p_k^+(u_n(t) - u_m(t)))) \geq 0.$$

Consequently, for every $m \geq n \geq n(\phi)$, we get

$$\frac{d}{dt} \int_{\mathbb{R}^N} j_k^+(u_n(t) - u_m(t))\phi \leq -\int_{\mathbb{R}^N} \nabla\phi \cdot (z_n(t) - z_m(t))p_k^+(u_n(t) - u_m(t))$$

$$\leq 2M \int_{\mathbb{R}^N} |\nabla\phi| \, |p_k^+(u_n(t) - u_m(t))|.$$

Then, choosing $\phi = \varphi^\alpha$, with $0 \leq \varphi \in C_0^\infty(\mathbb{R}^N)$,

$$\frac{d}{dt} \int_{\mathbb{R}^N} j_k^+(u_n(t) - u_m(t))\varphi^\alpha \leq 2\alpha M \int_{\mathbb{R}^N} |p_k^+(u_n(t) - u_m(t))| \, \varphi^{\alpha-1}|\nabla\varphi|$$

$$\leq 2\alpha M \left(\int_{\mathbb{R}^N} (|p_k^+(u_n(t) - u_m(t))| \, \varphi^{\alpha-1})^{\frac{\alpha}{\alpha-1}} \right)^{\frac{\alpha-1}{\alpha}} \left(\int_{\mathbb{R}^N} |\nabla\varphi|^\alpha \right)^{\frac{1}{\alpha}}$$

$$\leq 2\alpha^2 M \left(\int_{\mathbb{R}^N} |T_k^+(u_n(t) - u_m(t))|^\alpha \, \varphi^\alpha \right)^{\frac{\alpha-1}{\alpha}} \left(\int_{\mathbb{R}^N} |\nabla\varphi|^\alpha \right)^{\frac{1}{\alpha}}.$$

Now, we observe that $T_k^+(r)^\alpha \leq j_k^+(r)$ for all $r \in \mathbb{R}$. Hence

$$\frac{d}{dt} \int_{\mathbb{R}^N} j_k^+(u_n(t) - u_m(t))\varphi^\alpha$$

$$\leq 2\alpha^2 M \left(\int_{\mathbb{R}^N} j_k^+(u_n(t) - u_m(t)) \, \varphi^\alpha \right)^{\frac{\alpha-1}{\alpha}} \left(\int_{\mathbb{R}^N} |\nabla\varphi|^\alpha \right)^{\frac{1}{\alpha}}$$

and therefore,

$$\frac{d}{dt} \left(\int_{\mathbb{R}^N} j_k^+(u_n(t) - u_m(t))\varphi^\alpha \right)^{\frac{1}{\alpha}} \leq 2\alpha M \left(\int_{\mathbb{R}^N} |\nabla\varphi|^\alpha \right)^{\frac{1}{\alpha}}.$$

Setting $\psi_r(x) = \varphi(\frac{x}{r})$ instead of $\varphi(x)$ we get there exists $n_r \in \mathbb{N}$ such that for all $m \geq n \geq n_r$, we have

$$\frac{d}{dt}\left(\int_{\mathbb{R}^N} j_k^+(u_n(t) - u_m(t))\psi_r^\alpha\right)^{\frac{1}{\alpha}} \leq 2\alpha M \left(\int_{\mathbb{R}^N} |\nabla \psi_r|^\alpha\right)^{\frac{1}{\alpha}}$$

$$= 2M\alpha r^{\frac{N-\alpha}{\alpha}} \left(\int_{\mathbb{R}^N} |\nabla\varphi|^\alpha\right)^{\frac{1}{\alpha}}.$$

Integrating from 0 to s, with $0 \leq s \leq T$, we obtain

$$\left(\int_{\mathbb{R}^N} j_k^+(u_n(s) - u_m(s))\psi_r^\alpha\right)^{\frac{1}{\alpha}}$$

$$\leq \left(\int_{\mathbb{R}^N} j_k^+(u_{0n} - u_{0m})\psi_r^\alpha\right)^{\frac{1}{\alpha}} + 2TM\alpha r^{\frac{N-\alpha}{\alpha}} \left(\int_{\mathbb{R}^N} |\nabla\varphi|^\alpha\right)^{\frac{1}{\alpha}},$$

for all $m \geq n \geq n_r$ and $s \in [0, T]$. Given $\epsilon > 0$, since $\alpha > N$, we can find $r_\epsilon \in \mathbb{N}$ such that

$$2TM\alpha r^{\frac{N-\alpha}{\alpha}} \left(\int_{\mathbb{R}^N} |\nabla\varphi|^\alpha\right)^{\frac{1}{\alpha}} \leq \frac{\epsilon}{2} \quad \forall r \geq r_\epsilon.$$

Now, let $n_\epsilon \in \mathbb{N}$ be such that $n_\epsilon \geq n_{r_\epsilon}$ and

$$\left(\int_{\mathbb{R}^N} j_k^+(u_{0n} - u_{0m})\psi_{r_\epsilon}^\alpha\right)^{\frac{1}{\alpha}} < \frac{\epsilon}{2} \quad \forall m \geq n \geq n_\epsilon.$$

Then, we obtain that

$$\left(\int_{\mathbb{R}^N} j_k^+(u_n(s) - u_m(s))\psi_{r_\epsilon}^\alpha\right)^{\frac{1}{\alpha}} \leq \epsilon \quad \forall m \geq n \geq n_\epsilon \text{ and } s \in [0, T],$$

from where it follows that $\{u_n\}$ is a Cauchy sequence in $C([0, T]; L^1_{\text{loc}}(\mathbb{R}^N))$. Thus we may assume that $u_n \to u$ in $C([0, T]; L^1_{\text{loc}}(\mathbb{R}^N))$ for some function $u \in C([0, T]; L^1_{\text{loc}}(\mathbb{R}^N))$. In particular, we have that $u(t) \to u_0$ in $L^1_{\text{loc}}(\mathbb{R}^N)$ as $t \to 0+$.

STEP 2. Convergence of the derivatives and identification of the limit. Since the map $t \mapsto u'_n(t)$ is strongly measurable from $[0, T]$ into $L^2(\mathbb{R}^N)$ and, by (2.10),

$$\|u'_n(t)\|_{BV(\mathbb{R}^N)_2^*} \leq \|u'_n(t)\|_{L^2(\mathbb{R}^N)},$$

it follows that this map is strongly measurable from $[0, T]$ into $BV(\mathbb{R}^N)_2^*$. Moreover, for every $w \in BV(\mathbb{R}^N)_2$, if we take $u_n(t) - w$ as test function in (4.3), since

$$\int_{\mathbb{R}^N} \varphi_n h(x, Du_n(t)) = \int_{\mathbb{R}^N} (z_n(t), Du_n(t)),$$

we get

$$\int_{\mathbb{R}^N} u_n'(t)w\,dx = -\int_{\mathbb{R}^N} (z_n(t), Dw).$$

Hence

$$\left|\int_{\mathbb{R}^N} u_n'(t)w\,dx\right| \le M \int_{\mathbb{R}^N} |Dw| \le M_1 \|w\|_{BV(\mathbb{R}^N)_2} \quad \forall\, n \in \mathbb{N}.$$

Thus,

$$\|u_n'(t)\|_{BV(\mathbb{R}^N)_2^*} \le M_1 \quad \forall\, n \in \mathbb{N} \text{ and } t \in [0, T].$$

Consequently, $\{u_n'\}_{n\in\mathbb{N}}$ is a bounded sequence in $L^\infty(0, T; BV(\mathbb{R}^N)_2^*)$. Now, since the space $L^\infty(0, T; BV(\mathbb{R}^N)_2^*)$ is a vector subspace of the dual space $(L^1(0, T; BV(\mathbb{R}^N)_2))^*$, we can find a subnet $\{u_\alpha'\}$ such that

$$u_\alpha' \to \xi \in (L^1(0, T; BV(\mathbb{R}^N)_2))^* \quad \text{weakly}^*. \tag{4.4}$$

Since $\|z_n(t)\|_\infty \le M$ for all $n \in \mathbb{N}$ and a.e. $t \in [0, T]$, we may also assume that

$$z_n \to z \in L^\infty(Q_T, \mathbb{R}^N) \quad \text{weakly}^*. \tag{4.5}$$

Obviously, we have

$$\xi = \mathrm{div}_x(z) \quad \text{in } \mathcal{D}'(Q_T) \tag{4.6}$$

and

$$\xi(t) = \mathrm{div}_x(z(t)) \quad \text{in } \mathcal{D}'(\mathbb{R}^N) \text{ a.e. } t \in [0, T]. \tag{4.7}$$

Consequently, $(z(t), \xi(t)) \in Z(\mathbb{R}^N)$ for almost all $t \in [0, T]$.

With a similar proof to the one given for Lemma 4.1 of [3], we get the following result.

LEMMA 4.1. *ξ is the time derivative of u in the sense of the Definition 2.3.*

STEP 3. Next, we prove that $\xi = \mathrm{div}(z)$ in $(L^1(0, T, BV(\mathbb{R}^N)_2))^*$ in the sense of the Definition 2.4. To do that, let us first observe that (z, Dw), defined by (2.11), is a Radon measure in Q_T for all $w \in L^1_w(0, T, BV(\mathbb{R}^N)_2) \cap L^\infty(Q_T)$. Let $\phi \in \mathcal{D}(Q_T)$, then

$$\langle (z, Dw), \phi \rangle = -\int_0^T \langle \xi(t) - u_\alpha'(t), w(t)\phi(t) \rangle\, dt$$

$$-\int_{Q_T} w(z - z_\alpha) \cdot \nabla_x\phi\,dxdt + \int_0^T \langle (z_\alpha(t), Dw(t)), \phi(t) \rangle\, dt.$$

Taking limits in α, and using (4.4), we get

$$\langle (z, Dw), \phi \rangle = \lim_\alpha \int_0^T \langle (z_\alpha(t), Dw(t)), \phi(t) \rangle\, dt. \tag{4.8}$$

Therefore

$$\left| \langle (z, Dw), \phi \rangle \right| \le M \|\phi\|_\infty \int_0^T \int_{\mathbb{R}^N} |Dw(t)| \, dt.$$

Hence, (z, Dw) is a Radon measure in Q_T. Moreover, from (4.8), applying Green's formula we obtain that

$$\int_{Q_T} (z, Dw) = \lim_\alpha \int_0^T (z_\alpha(t), Dw(t)) \, dt$$

$$= -\lim_\alpha \int_0^T \int_{\mathbb{R}^N} \mathrm{div}(z_\alpha(t)) w(t) \, dx dt = -\int_0^T \langle \xi(t), w(t) \rangle \, dt,$$

that is,

$$\int_{Q_T} (z, Dw) + \int_0^T \langle \xi(t), w(t) \rangle \, dt = 0. \tag{4.9}$$

As a consequence of the boundedness of $\{u'_n\}$, (4.4) and the above statement, we have

$$u'_n \to \xi \in (L^1(0, T; BV(\mathbb{R}^N)_2))^* \quad \text{weakly}^*. \tag{4.10}$$

STEP 4. Convergence of the energy.

Let $0 \le \varphi \in C_0^\infty(\mathbb{R}^N)$. If n is large enough, we have $\varphi \varphi_n = \varphi$. From now on, we assume that this is the case. Multiplying (4.1) by $(w - p(u_n(t)))\varphi$, integrating in $\mathbb{R}^N$ and using (4.2), we have that

$$-\int_{\mathbb{R}^N} (w - p(u_n(t)))\varphi u'_n(t) \, dx$$

$$= \int_{\mathbb{R}^N} (z_n(t), Dw)\varphi - \int_{\mathbb{R}^N} h(x, Dp(u_n(t)))\varphi + \int_{\mathbb{R}^N} z_n(w - p(u_n))D\varphi \tag{4.11}$$

for every $w \in BV_{loc}(\mathbb{R}^N) \cap L^2_{loc}(\mathbb{R}^N)$ and all $p \in \mathcal{P}$.

First, we observe that setting $w = 0$ in (4.11) and integrating in $(0, T)$, we obtain

$$\int_{\mathbb{R}^N} J_p(u_n(T))\varphi dx + \int_0^T \int_{\mathbb{R}^N} h(x, Dp(u_n(t)))\varphi \, dx dt$$

$$= \int_{\mathbb{R}^N} J_p(u_{0,n})\varphi - \int_0^T \int_{\mathbb{R}^N} z_n p(u_n(t)) D\varphi \, dx dt,$$

where $J_p'(r) = p(r)$. In particular, we have

$$\int_0^T \int_{\mathbb{R}^N} J_p(u_n(t))\varphi\,dxdt \le C, \tag{4.12}$$

$$\int_0^T \int_{\mathbb{R}^N} h(x, Dp(u_n(t)))\varphi\,dxdt \le C. \tag{4.13}$$

Hence, by (2.7),

$$\int_0^T \int_{\mathbb{R}^N} \varphi|Dp(u_n(t))|\,dt \le C, \tag{4.14}$$

where C is a constant depending on u_0, φ, $\|p\|_\infty$ and the constants in (2.1). Since the functional $\Phi_\varphi : L^1_{loc}(\mathbb{R}^N) \to]-\infty, +\infty]$, defined by

$$\Phi_\varphi(w) = \begin{cases} \displaystyle\iint_{\mathbb{R}^N} \varphi\,d|Dw| & \text{if } w \in BV_{loc}(\mathbb{R}^N) \\ +\infty & \text{if } w \in L^1_{loc}(\mathbb{R}^N) \setminus BV_{loc}(\mathbb{R}^N), \end{cases} \tag{4.15}$$

is lower semicontinuous in $L^1_{loc}(\mathbb{R}^N)$, we have that

$$\Phi_\varphi(p(u(t))) \le \liminf_{n\to\infty} \Phi_\varphi(p(u_n(t)) = \liminf_{n\to\infty} \int_{\mathbb{R}^N} \varphi|Dp(u_n(t))|.$$

On the other hand, by Lemma 5 in [3], the map $t \mapsto \int_{\mathbb{R}^N} \varphi|Dp(u_n(t))|$ is measurable, then by the Fatou's Lemma and (4.14), it follows that

$$\int_0^T \Phi_\varphi(p(u(t))) \le \int_0^T \liminf_{n\to\infty} \left(\int_{\mathbb{R}^N} \varphi|Dp(u_n(t))|\right) dt$$

$$\le \liminf_{n\to\infty} \int_0^T \left(\int_{\mathbb{R}^N} \varphi|Dp(u_n(t))|\right) dt \le C. \tag{4.16}$$

As a consequence of (4.16), we obtain that $p(u(t)) \in BV_{loc}(\mathbb{R}^N)$ for almost all $t \in [0, T]$.

From Lemma 4.2 in [3], if $0 \le \eta = \psi(t)\varphi(x)$, $\psi(t) \in \mathcal{D}(]0, T[)$, $\varphi(x) \in C_0^\infty(\mathbb{R}^N)$, the map $t \mapsto p(u(t))\eta(t)$, from $[0, T]$ into $BV(\mathbb{R}^N)$, is weakly measurable.

Using the same technique than in the proofs of Lemmas 4.3 and Lemma 4.4 of [3], we obtain the following two results.

LEMMA 4.2. *For any $\tau > 0$, we define the function ψ^τ, as the Dunford integral (see [12])*

$$\psi^\tau(t) := \frac{1}{\tau} \int_{t-\tau}^t \eta(s)p(u(s))\,ds \in BV(\mathbb{R}^N)^{**},$$

that is,

$$\langle \psi^\tau(t), w \rangle = \frac{1}{\tau} \int_{t-\tau}^{t} \langle \eta(s) p(u(s)), w \rangle \, ds$$

for any $w \in BV(\mathbb{R}^N)^*$. Then $\psi^\tau \in C([0, T]; BV(\mathbb{R}^N))$. Moreover, $\psi^\tau(t) \in L^2(\mathbb{R}^N)$, thus, $\psi^\tau(t) \in BV(\mathbb{R}^N)_2$ and ψ^τ admits a weak derivative in $L_w^1(0, T, BV(\mathbb{R}^N)) \cap L^\infty(Q_T)$.

LEMMA 4.3. *For $\tau > 0$ small enough, we have*

$$\int_0^T \langle \psi^\tau(t), \xi(t) \rangle \, dt \le - \int_0^T \int_{\mathbb{R}^N} \frac{\eta(t-\tau) - \eta(t)}{-\tau} J_p(u(t)) \, dxdt. \tag{4.17}$$

We need the following result.

LEMMA 4.4. *Let*

$$A_n := \int_0^T \int_{\mathbb{R}^N} h(x, Dp(u_n)) \varphi \, dt,$$

$\eta = \psi(t)\phi(x)$, $\psi \in \mathcal{D}(0, T)$, $\phi \in C_0^\infty(\mathbb{R}^N)$, *and*

$$(\eta p(u))^\tau(t) = \frac{1}{\tau} \int_{t-\tau}^{t} \eta(s) p(u(s)) ds.$$

Then

$$\limsup_{n \to \infty} A_n \le \int_0^T \int_{\mathbb{R}^N} z(t) \cdot \nabla p(u(t)) \, \varphi \, dxdt$$

$$+ \liminf_{\eta \uparrow 1} \liminf_{\tau \to 0} \int_0^T (z(t), D^s(\eta p(u))^\tau(t)) \varphi \, dt.$$

Proof. Let $w \in W^{1,1}((0, T) \times \mathbb{R}^N)$. We use as test function $(\eta p(w(t)))^\tau$ in (4.11) and integrate in $(0, T)$ to obtain

$$- \int_0^T \int_{\mathbb{R}^N} (\eta p(w(t)))^\tau \varphi u_n'(t) \, dxdt + \int_0^T \int_{\mathbb{R}^N} p(u_n(t)) \varphi u_n'(t) \, dxdt$$

$$+ \int_0^T \int_{\mathbb{R}^N} h(x, Dp(u_n(t))) \varphi \, dt$$

$$= \int_0^T \int_{\mathbb{R}^N} (z_n(t), D(\eta p(w(t)))^\tau) \, \varphi dt + \int_0^T \int_{\mathbb{R}^N} z_n(t) D\varphi(\eta p(w(t)))^\tau dt$$

$$- \int_0^T \int_{\mathbb{R}^N} z_n(t) D\varphi p(u_n(t)),$$

where $\eta^\tau(t) = \frac{1}{\tau}\int_{t-\tau}^{t}\eta(s)ds$. Our purpose is to take limits in the above expression as $n \to \infty$, $w \to u$ in $L^1_{loc}((0,T)\times\mathbb{R}^N)$, $\tau \to 0$ and $\eta \uparrow 1$. We take $\tau > 0$ small enough. Let us analyze the first term

$$-\int_0^T\int_{\mathbb{R}^N}(\eta p(w(t)))^\tau \varphi u'_n(t)\,dxdt = \int_0^T\int_{\mathbb{R}^N}(\eta p(w(t)))_t^\tau \varphi u_n(t)\,dxdt$$

$$\to \int_0^T\int_{\mathbb{R}^N}(\eta p(w(t)))_t^\tau \varphi u(t)\,dxdt \quad \text{as } n \to \infty.$$

Now, using Lemma 4.7 and Lemma 4.9,

$$\int_0^T\int_{\mathbb{R}^N}(\eta p(w(t)))_t^\tau \varphi u = \int_0^T\int_{\mathbb{R}^N}\frac{\eta(t)p(w(t)) - \eta(t-\tau)p(w(t-\tau))}{\tau}\varphi u(t)\,dxdt$$

$$\to \int_0^T\int_{\mathbb{R}^N}\frac{\eta(t)p(u(t)) - \eta(t-\tau)p(u(t-\tau))}{\tau}\varphi u(t)\,dxdt, \quad \text{as } w \to u \text{ in } L^1_{loc},$$

$$= -\int_0^T\langle \xi(t), (\varphi\eta p(u(t)))^\tau\rangle\,dt \geq \int_0^T\int_{\mathbb{R}^N}\frac{\eta(t-\tau)-\eta(t)}{-\tau}\varphi J_p(u(t))\,dxdt$$

$$\to \int_0^T\int_{\mathbb{R}^N}\eta_t\varphi J_p(u(t))\,dxdt, \quad \text{as } \tau \to 0$$

$$\to \int_{\mathbb{R}^N}(J_p(u(0)) - J_p(u(T)))\varphi\,dx \quad \text{as } \eta \uparrow 1.$$

The analysis of the second term is easy. Letting $n \to \infty$ we have

$$\int_0^T\int_{\mathbb{R}^N}p(u_n(t))\varphi u'_n(t)\,dxdt = \int_0^T\frac{d}{dt}\int_{\mathbb{R}^N}J_p(u_n(t))\varphi\,dx$$

$$= \int_{\mathbb{R}^N}(J_p(u_n)(T) - J_p(u_n(0)))\varphi\,dx \to \int_{\mathbb{R}^N}(J_p(u(T)) - J_p(u(0)))\varphi\,dx.$$

Let us deal together the first two terms of the right hand side of the equality we are analyzing. Having in mind Steps 3, 4 and (4.10), taking limits as $n \to \infty$, $w \to u$ in L^1_{loc} and $\tau \to 0$, we get:

$$\int_0^T\int_{\mathbb{R}^N}(z_n(t), D[(\eta p(w))^\tau\varphi])\,dt \to$$

$$-\int_0^T\langle \xi(t), (\eta p(w))^\tau\varphi\rangle\,dt = \int_0^T\int_{\mathbb{R}^N}u(t)(\eta p(w))_t^\tau \varphi\,dxdt$$

$$= \int_0^T\int_{\mathbb{R}^N}u(t)\frac{\eta(t)p(w)(t) - \eta(t-\tau)p(w)(t-\tau)}{\tau}\varphi\,dxdt$$

$$\to \int_0^T\int_{\mathbb{R}^N}u(t)\frac{\eta(t)p(u)(t) - \eta(t-\tau)p(u)(t-\tau)}{\tau}\varphi\,dxdt$$

$$= \int_0^T \int_{\mathbb{R}^N} u(t)(\eta p(u))_t^\tau \varphi \, dx dt = - \int_0^T \langle \xi(t), (\eta p(u))^\tau \varphi \rangle \, dt$$

$$= \int_0^T \int_{\mathbb{R}^N} z(t) \cdot \nabla[(\eta p(u))^\tau \varphi] \, dx dt + \int_0^T \int_{\mathbb{R}^N} z(t) \cdot D^s[(\eta p(u))^\tau]\varphi \, dt$$

$$\rightarrow \int_0^T \int_{\mathbb{R}^N} z(t) \cdot \nabla(\eta p(u)\varphi) \, dx dt + \liminf_{\tau \to 0} \int_0^T \int_{\mathbb{R}^N} z(t) \cdot D^s(\eta p(u))^\tau \varphi \, dt$$

$$\rightarrow \int_0^T \int_{\mathbb{R}^N} z(t) \cdot \nabla[p(u(t))\varphi] \, dx dt$$

$$+ \liminf_{\eta \uparrow 1} \liminf_{\tau \to 0} \int_0^T \int_{\mathbb{R}^N} z(t) \cdot D^s(\eta p(u))^\tau \varphi \, dt$$

$$\text{as } \eta \uparrow 1.$$

The last term $\int_0^T \int_{\mathbb{R}^N} z_n(t) D\varphi p(u_n(t))$ easily converges to $\int_0^T \int_{\mathbb{R}^N} z(t) D\varphi p(u(t))$. $\qquad \square$

The lemma follows by collecting all these facts.

The proof of next Lemma is similar to the proof of Lemma 9 in [3].

LEMMA 4.5. *Let $\varphi \in C_0^\infty(\mathbb{R}^N)$. Let*

$$\Psi_{p,\varphi}(p(u(t))) = \int_{\mathbb{R}^N} f(x, Dp(u(t)))\varphi.$$

Then

$$\int_0^T \Psi_{p,\varphi}(p(u(t))) \, dt = \lim_{n \to \infty} \int_0^T \Psi_{p,\varphi}(p(u_n(t))) \, dt. \tag{4.18}$$

As a consequence, we also have that

$$\Psi_{p,\varphi}(p(u(t))) = \lim_{n \to \infty} \Psi_{p,\varphi}(p(u_n(t))) \quad \text{a.e. in } t. \tag{4.19}$$

From Lemma 4.5 it follows that

$$\int_{\mathbb{R}^N} h(x, Dp(u(t)))\varphi = \lim_{n \to \infty} \int_{\mathbb{R}^N} h(x, Dp(u_n(t)))\varphi, \tag{4.20}$$

a.e. in $t \in (0, T)$. Indeed, if we consider the $\mathbb{R}^N$-valued measures μ_n, μ on $\mathbb{R}^N$ which are defined by

$$\mu_n(B) := \int_B \varphi Dp(u_n), \quad \mu(B) := \int_B \varphi Dp(u)$$

for all Borel sets $B \subseteq \mathbb{R}^N$, we have

$$\mu_n \rightharpoonup \mu \quad \text{weakly as measures in } \mathbb{R}^N.$$

Moreover,

$$\Psi_{p,\varphi}(p(u)) = \int_{\mathbb{R}^N} \tilde{f}(x,\lambda) \quad \text{and} \quad \Psi_{p,\varphi}(p(u_n)) = \int_{\mathbb{R}^N} \tilde{f}(x,\lambda_n),$$

with $\lambda = (\mu, \varphi\mathcal{L}^N)$ and $\lambda_n = (\mu_n, \varphi\mathcal{L}^N)$. Hence, (4.18) yields

$$\lim_{n\to\infty} \int_{\mathbb{R}^N} \tilde{f}(x,\lambda_n) = \int_{\mathbb{R}^N} \tilde{f}(x,\lambda).$$

Then, applying Theorem 3 of [19] (see also [14], Theorem 1, page 90), it follows that

$$\int_{\mathbb{R}^N} \tilde{h}(x,\lambda) = \lim_{n\to\infty} \int_{\mathbb{R}^N} \tilde{h}(x,\lambda_n) = \lim_{n\to\infty} \int_{\mathbb{R}^N} h(x, Dp(u_n))\varphi.$$

Now, it is easy to see that

$$\int_{\mathbb{R}^N} \tilde{h}(x,\lambda) = \int_{\mathbb{R}^N} h(x, Dp(u))\varphi,$$

consequently, we obtain (4.20).

Let us now prove that

$$\int_0^T \int_{\mathbb{R}^N} z(t) \cdot \nabla p(u(t))\varphi \, dxdt \le \int_0^T \int_{\mathbb{R}^N} \mathbf{a}(x, \nabla u(t)) \cdot \nabla p(u(t))\varphi \, dxdt. \tag{4.21}$$

In fact, from the convexity of f in ξ, we have

$$\int_0^T \int_{\mathbb{R}^N} \mathbf{a}(x, \nabla p(u_n)) \cdot \nabla p(u)\varphi \, dxdt \le \int_0^T \int_{\mathbb{R}^N} \mathbf{a}(x, \nabla p(u_n)) \cdot \nabla p(u_n)\varphi \, dxdt$$

$$+ \int_0^T \int_{\mathbb{R}^N} f(x, \nabla p(u))\varphi \, dxdt - \int_0^T \int_{\mathbb{R}^N} f(x, \nabla p(u_n))\varphi \, dxdt$$

$$= \int_0^T \int_{\mathbb{R}^N} h(x, \nabla p(u_n))\varphi \, dxdt + \int_0^T \int_{\mathbb{R}^N} f^0(x, D^s p(u_n))\varphi \, dt$$

$$- \int_0^T \int_{\mathbb{R}^N} f^0(x, D^s p(u_n))\varphi \, dt + \int_0^T \int_{\mathbb{R}^N} f(x, \nabla p(u))\varphi \, dxdt$$

$$- \int_0^T \int_{\mathbb{R}^N} f(x, \nabla p(u_n))\varphi \, dxdt = \int_0^T \int_{\mathbb{R}^N} h(x, Dp(u_n))\varphi \, dt$$

$$- \int_0^T \int_{\mathbb{R}^N} f(x, Dp(u_n))\varphi \, dt + \int_0^T \int_{\mathbb{R}^N} f(x, \nabla p(u))\varphi \, dxdt.$$

Now, since

$$\lim_{n\to\infty} \int_0^T \int_{\mathbb{R}^N} [\mathbf{a}(x, \nabla u_n(t)) - \mathbf{a}(x, \nabla p(u_n(t)))] \cdot \nabla p(u(t))\varphi \, dxdt = 0,$$

we have

$$\lim_{n \to \infty} \int_0^T \int_{\mathbb{R}^N} \mathbf{a}(x, \nabla p(u_n(t))) \cdot \nabla p(u(t))\varphi \, dxdt$$

$$= \int_0^T \int_{\mathbb{R}^N} z(t) \cdot \nabla p(u(t))\varphi \, dxdt.$$

Then, letting $n \to \infty$, using Lemma 4.11 and (4.20), we deduce that

$$\int_0^T \int_{\mathbb{R}^N} z(t) \cdot \nabla p(u(t))\varphi \, dxdt \le \int_0^T \int_{\mathbb{R}^N} h(x, Dp(u))\varphi \, dt$$

$$- \int_0^T \int_{\mathbb{R}^N} f(x, Dp(u))\varphi \, dt + \int_0^T \int_{\mathbb{R}^N} f(x, \nabla p(u))\varphi \, dt$$

$$= \int_0^T \int_{\mathbb{R}^N} \mathbf{a}(x, \nabla u(t)) \cdot \nabla p(u(t)) \, dxdt,$$

and (4.21) holds.

STEP 5. Identification of the limit. Let us now prove that

$$z(t, x) = \mathbf{a}(x, \nabla u(t, x)) \qquad \text{a.e.} \ \ (t, x) \in (0, T) \times \mathbb{R}^N. \tag{4.22}$$

Let $0 \le \phi \in C_0^1((0, T) \times \mathbb{R}^N)$ and $g \in C^1([0, T] \times \mathbb{R}^N)$. We observe that

$$\int_0^T \int_{\mathbb{R}^N} \phi[(\mathbf{a}(x, \nabla u_n), Dp(u_n - g)) - \mathbf{a}(x, \nabla g)Dp(u_n - g)]$$

$$= \int_0^T \int_{\mathbb{R}^N} \phi[\mathbf{a}(x, \nabla u_n) - \mathbf{a}(x, \nabla g)] \cdot \nabla p(u_n - g) \, dxdt$$

$$+ \int_{\mathbb{R}^N} \phi[\mathbf{a}(x, \nabla u_n) - \mathbf{a}(x, \nabla g)] \cdot D^s p(u_n - g).$$

Since both terms at the right hand side of the above expression are positive, we have

$$\int_0^T \int_{\mathbb{R}^N} \phi[(\mathbf{a}(x, \nabla u_n), Dp(u_n - g)) - \mathbf{a}(x, \nabla g)Dp(u_n - g)] \ge 0. \tag{4.23}$$

Our purpose is to take limits as $n \to \infty$ in the above inequality. We assume that $\phi(t, x) = \eta(t)\psi(x)$, where $\eta \in \mathcal{D}(0, T)$, $\psi \in \mathcal{D}(\mathbb{R}^N)$, $\eta \ge 0$, $\psi \ge 0$. First, integrating by parts in the first term, we have

$$\int_0^T \int_{\mathbb{R}^N} \phi(\mathbf{a}(x, \nabla u_n), Dp(u_n - g)) \, dt = - \int_0^T \int_{\mathbb{R}^N} p(u_n - g)\nabla_x \phi \cdot \mathbf{a}(x, \nabla u_n) \, dxdt$$

$$- \int_0^T \int_{\mathbb{R}^N} \phi \mathrm{div}(\mathbf{a}(x, \nabla u_n))p(u_n - g) \, dxdt$$

$$= -\int_0^T \int_{\mathbb{R}^N} p(u_n - g)\nabla_x\phi \cdot \mathbf{a}(x, \nabla u_n)\,dxdt - \int_0^T \int_{\mathbb{R}^N} \phi u_n'(t)p(u_n - g)\,dxdt$$

$$= -\int_0^T \int_{\mathbb{R}^N} p(u_n - g)\nabla_x\phi \cdot \mathbf{a}(x, \nabla u_n)\,dxdt - \int_0^T \int_{\mathbb{R}^N} \phi\frac{d}{dt}J_p(u_n - g)\,dxdt$$

$$- \int_0^T \int_{\mathbb{R}^N} \phi g_t\, p(u_n - g)\,dxdt = -\int_0^T \int_{\mathbb{R}^N} p(u_n - g)\nabla_x\phi \cdot \mathbf{a}(x, \nabla u_n)\,dxdt$$

$$+ \int_0^T \int_{\mathbb{R}^N} \phi_t J_p(u_n - g)\,dxdt - \int_0^T \int_{\mathbb{R}^N} \phi g_t\, p(u_n - g)\,dxdt.$$

Letting $n \to \infty$ in (4.23), taking into account the above equalities, we obtain

$$- \int_0^T \int_{\mathbb{R}^N} p(u - g)\nabla_x\phi \cdot z\,dxdt + \int_0^T \int_{\mathbb{R}^N} \phi_t J_p(u - g)\,dxdt$$

$$- \int_0^T \int_{\mathbb{R}^N} \phi g_t\, p(u - g)\,dxdt - \int_0^T \int_{\mathbb{R}^N} \phi(a(x, \nabla g), Dp(u - g))\,dt \geq 0. \quad (4.24)$$

Now,

$$\int_0^T \int_{\mathbb{R}^N} \phi_t J_p(u - g)\,dxdt$$

$$= \lim_{\tau \to 0} \int_0^T \int_{\mathbb{R}^N} \frac{\phi(t - \tau) - \phi(t)}{-\tau} J_p(u - g)\,dxdt$$

$$= \lim_{\tau \to 0} \int_0^T \int_{\mathbb{R}^N} \psi(x)\frac{\eta(t - \tau) - \eta(t)}{-\tau} J_p(u - g)\,dxdt. \quad (4.25)$$

For simplicity, let us write $v = u - g$. Since

$$J_p(v(t)) - J_p(v(t + \tau)) \leq (v(t) - v(t + \tau))p(v(t)),$$

for τ small enough, we have

$$\int_0^T \int_{\mathbb{R}^N} \frac{v(t + \tau) - v(t)}{\tau}\eta(t)\psi(x)p(v(t))\,dxdt$$

$$\leq \int_0^T \int_{\mathbb{R}^N} \frac{J_p(v(t + \tau)) - J_p(v(t))}{\tau}\eta(t)\psi(x)\,dxdt$$

$$= \int_0^T \int_{\mathbb{R}^N} \frac{\eta(t - \tau) - \eta(t)}{\tau}\psi(x)J_p(v)\,dxdt. \quad (4.26)$$

By Lemma 4.1, we have

$$\int_0^T \int_{\mathbb{R}^N} \frac{v(t + \tau) - v(t)}{\tau}\eta(t)\psi(x)p(v(t))\,dxdt$$

$$= -\int_0^T \int_{\mathbb{R}^N} v(t)\frac{d}{dt}(\eta p(v))^\tau(t)\psi(x)\,dxdt$$

$$= \int_0^T \int_{\mathbb{R}^N} (\xi - g_t)(t)(\eta p(v))^\tau(t)\psi(x)\,dxdt. \quad (4.27)$$

Collecting these inequalities, we obtain

$$\int_0^T \int_{\mathbb{R}^N} \frac{\eta(t-\tau) - \eta(t)}{-\tau} \psi \, J_p(v) \, dxdt$$

$$\leq -\int_0^T \int_{\mathbb{R}^N} (\xi - g_t)(t)(\eta p(v))^{\tau}(t)\psi(x) \, dxdt$$

$$= -\lim_n \int_0^T \left\langle u_n'(t) - g_t, (\eta p(v))^{\tau}(t)\psi \right\rangle dt$$

$$= -\lim_n \int_0^T \frac{1}{\tau} \int_{t-\tau}^t \eta(s) \left\langle \operatorname{div}(z_n(t)) - g_t(t), p(v(s))\psi \right\rangle ds dt$$

$$= \lim_n \int_0^T \frac{1}{\tau} \int_{t-\tau}^t \eta(s) \left[\int_{\mathbb{R}^N} (z_n(t), D(p(v(s))\psi)) + \langle g_t, p(v(s))\psi \rangle \right] ds dt$$

$$= \lim_n \int_0^T \frac{1}{\tau} \int_{t-\tau}^t \eta(s) \int_{\mathbb{R}^N} (z_n(t), Dp(v(s)))\psi \, ds dt$$

$$+ \lim_n \int_0^T \frac{1}{\tau} \int_{t-\tau}^t \eta(s) \int_{\mathbb{R}^N} p(v(s))z_n(t) \cdot \nabla \psi \, ds dt$$

$$+ \int_0^T \frac{1}{\tau} \int_{t-\tau}^t \eta(s) \langle g_t, p(v(s))\psi \rangle \, ds dt.$$

Since

$$Dp(v(s)) = \nabla p(u(s) - g(s)) + D^s p(u(s) - g(s))$$

and

$$z_n(t) \cdot D^s p(u(s) - g(s)) = \mathbf{a}(x, \nabla u_n(t, x)) \cdot D^s p(u(s) - g(s))$$
$$\leq h^0(x, D^s p(u(s) - g(s))),$$

from the above inequality, it follows that

$$\int_0^T \int_{\mathbb{R}^N} \frac{\eta(t-\tau) - \eta(t)}{-\tau} \psi \, J_p(u - g) \, dxdt$$

$$\leq \int_0^T \frac{1}{\tau} \int_{t-\tau}^t \eta(s) \int_{\mathbb{R}^N} z(t) \cdot \nabla p(u(s) - g(s))\psi \, dxds dt$$

$$+ \int_0^T \frac{1}{\tau} \int_{t-\tau}^t \eta(s) \int_{\mathbb{R}^N} \psi h^0(x, D^s p(u(s) - g(s))) \, ds dt$$

$$+ \int_0^T \frac{1}{\tau} \int_{t-\tau}^t \eta(s) \int_{\mathbb{R}^N} p(u(s) - g(s))z(t) \cdot \nabla \psi \, dxds dt$$

$$+ \int_0^T \frac{1}{\tau} \int_{t-\tau}^t \eta(s) \int_{\mathbb{R}^N} g_t(t)p(u(s) - g(s))\psi(x) \, dxds dt. \tag{4.28}$$

Hence, letting $\tau \to 0$ in (4.28), we obtain

$$\int_0^T \int_{\mathbb{R}^N} \phi_t J_p(u-g) \le \int_0^T \int_{\mathbb{R}^N} \eta(t) z(t) \cdot \nabla p(u(t) - g(t)) \psi \, dx dt$$

$$+ \int_0^T \eta(t) \int_{\mathbb{R}^N} \psi h^0(x, D^s p(u(t) - g(t))) \, dt$$

$$+ \int_0^T \int_{\mathbb{R}^N} \eta(t) p(u(t) - g(t)) z(t) \cdot \nabla \psi \, dx dt$$

$$+ \int_0^T \int_{\mathbb{R}^N} \eta(t) g_t(t) p(u(t) - g(t)) \psi(x) \, dx ds dt. \tag{4.29}$$

Taking into account (4.24) and (4.29), we get

$$\int_0^T \int_{\mathbb{R}^N} \phi([z - \mathbf{a}(x, \nabla g)] \cdot \nabla p(u - g)$$

$$+ h^0(x, D^s p(u - g)) - \mathbf{a}(x, \nabla g) \cdot D^s p(u - g)) \ge 0$$

for all $\phi(t, x) = \eta(t) \psi(x)$, $\eta \in \mathcal{D}(0, T)$, $\psi \in \mathcal{D}(\mathbb{R}^N)$, $\eta, \psi \ge 0$. Thus, the measure

$$([z - \mathbf{a}(x, \nabla g)] \cdot \nabla p(u - g) + h^0(x, D^s p(u - g)) - \mathbf{a}(x, \nabla g) \cdot D^s p(u - g)) \ge 0.$$

Then its absolutely continuous part

$$[z - \mathbf{a}(x, \nabla g)] \cdot \nabla p(u - g) \ge 0 \quad \text{a.e. in } \mathbb{R}^N.$$

Hence

$$[z - \mathbf{a}(x, \nabla g)] \cdot \nabla (u - g) \ge 0 \quad \text{a.e. in } \mathbb{R}^N.$$

Since we may take a countable set dense in $C^1_{loc}([0, T] \times \mathbb{R}^N)$, we have that the above inequality holds for all $(t, x) \in S$, where $S \subseteq (0, T) \times \mathbb{R}^N$ is such that $\mathcal{L}^N(((0, T) \times \mathbb{R}^N) \setminus S) = 0$, and all $g \in C^1_{loc}([0, T] \times \mathbb{R}^N)$. Now, fixe $(t, x) \in S$, and given $y \in \mathbb{R}^N$ there is $g \in C^1_{loc}([0, T] \times \mathbb{R}^N)$ such that $\nabla g(t, x) = y$. Then

$$(z(t, x) - \mathbf{a}(x, y)) \cdot (\nabla u(t, x) - y) \ge 0 \quad \forall y \in \mathbb{R}^N,$$

and we get that

$$z(t, x) = \mathbf{a}(x, \nabla u(t, x)) \quad \text{a.e.} \quad (t, x) \in Q_T. \tag{4.30}$$

Then, we have

$$\text{div}(z(t)) = \text{div}(\mathbf{a}(x, \nabla u(t))) \quad \text{in } \mathcal{D}'(\mathbb{R}^N), \quad \text{a.e. } t \in [0, T].$$

STEP 6. Conclusion. Finally, we are going to prove that u verifies:

$$-\int_0^T \int_{\mathbb{R}^N} j(u(t) - l)\eta_t \, dxdt + \int_0^T \int_{\mathbb{R}^N} \eta(t)h(x, Dp(u(t) - l)) \, dt$$

$$+ \int_0^T \int_{\mathbb{R}^N} z(t) \cdot \nabla\eta(t)p(u(t) - l) \, dxdt \leq 0, \tag{4.31}$$

for all $\eta \in C^\infty(\overline{Q_T})$, with $\eta \geq 0$, $\eta(t, x) = \phi(t)\psi(x)$, being $\phi \in \mathcal{D}(]0, T[)$, $\psi \in C_0^\infty(\mathbb{R}^N)$, and $p \in \mathcal{T}$, where $j(r) = \int_0^r p(s) \, ds$.

Let $\eta \in C_0^\infty(Q_T)$, with $\eta \geq 0$, $\eta(t, x) = \phi(t)\psi(x)$, $\phi \in \mathcal{D}(]0, T[)$, $\psi \in C_0^\infty(\mathbb{R}^N)$, $p \in \mathcal{P}$ and $a \in \mathbb{R}$. Let $G_p(r) = \int_a^r p(s) \, ds$. Since $u'_n(t) = \mathrm{div}(z_n(t))$, multiplying by $p(u_n(t))\eta(t)$ and integrating, we obtain that

$$\int_0^T \int_{\mathbb{R}^N} \frac{d}{dt}G_p(u_n(t))\eta(t) \, dxdt = \int_0^T \int_{\mathbb{R}^N} p(u_n(t))u'_n(t)\eta(t) \, dxdt$$

$$= \int_0^T \int_{\mathbb{R}^N} \mathrm{div}(z_n(t))p(u_n(t))\eta(t) \, dxdt$$

$$= -\int_0^T \int_{\mathbb{R}^N} (z_n(t), D(p(u_n(t))\eta(t))) \, dt$$

$$= -\int_0^T \int_{\mathbb{R}^N} \eta(t)h(x, Dp(u_n(t))) \, dt - \int_0^T \int_{\mathbb{R}^N} z_n(t) \cdot \nabla\eta(t) \, p(u_n(t)) \, dxdt.$$

Hence, having in mind that $\eta(0) = \eta(T) = 0$, we get

$$\int_0^T \int_{\mathbb{R}^N} \eta(t)h(x, Dp(u_n(t))) \, dt = -\int_0^T \int_{\mathbb{R}^N} z_n(t) \cdot \nabla\eta(t) \, p(u_n(t)) \, dxdt$$

$$-\int_0^T \int_{\mathbb{R}^N} \frac{d}{dt}G_p(u_n(t))\eta(t) \, dxdt = -\int_0^T \int_{\mathbb{R}^N} z_n(t) \cdot \nabla\eta(t) \, p(u_n(t)) \, dxdt$$

$$+ \int_0^T \int_{\mathbb{R}^N} G_p(u_n(t)) \, \eta_t \, dxdt.$$

Now, observe that, by the Lemma 4.5, we have that

$$\int_{\mathbb{R}^N} \eta(t, x)f(x, Dp(u_n)) \rightarrow \int_{\mathbb{R}^N} \eta(t, x)f(x, Dp(u))$$

a.e. in $t \in (0, T)$, and, therefore,

$$\int_{\mathbb{R}^N} \eta(t, x)h(x, Dp(u_n)) \rightarrow \int_{\mathbb{R}^N} \eta(t, x)h(x, Dp(u)),$$

a.e. in $t \in (0, T)$. Hence, integrating in $(0, T)$ and using Fatou's Lemma, it follows that

$$\int_0^T \int_{\mathbb{R}^N} \eta(t) h(x, Dp(u(t))) \, dt \leq \lim_{n \to \infty} \int_0^T \int_{\mathbb{R}^N} \eta(t) h(x, Dp(u_n(t))) \, dt$$

$$= \lim_{n \to \infty} \left[-\int_0^T \int_{\mathbb{R}^N} z_n(t) \cdot \nabla \eta(t) \, p(u_n(t)) \, dxdt + \int_0^T \int_{\mathbb{R}^N} G_p(u_n(t)) \eta_t \, dxdt \right]$$

$$= -\int_0^T \int_{\mathbb{R}^N} z(t) \cdot \nabla \eta(t) \, p(u(t)) \, dxdt + \int_0^T \int_{\mathbb{R}^N} G_p(u(t)) \, \eta_t \, dxdt.$$

We have

$$-\int_0^T \int_{\mathbb{R}^N} G_p(u(t)) \eta_t \, dxdt + \int_0^T \int_{\mathbb{R}^N} \eta(t) h(x, Dp(u(t))) \, dt$$

$$+ \int_0^T \int_{\mathbb{R}^N} z(t) \cdot \nabla \eta(t) \, p(u(t)) \, dxdt \leq 0. \tag{4.32}$$

Finally, given $l \in \mathbb{R}$ and $p \in \mathcal{T}$, since $q(r) := p(r - l)$ is an element of $\mathcal{P}$, and taking $a = l$, we obtain (4.31) as a consequence of (4.32). The proof of the existence is finished.

5. Proof of Theorem 1: Uniqueness

Uniqueness of entropy solutions can be proved using the same technique used to prove uniqueness in the case of a bounded domain (see [2], [3]), a technique inspired by the doubling variables method introduced by Kruzhkov [17] (see also [10]) to prove the L^1-contraction estimate for entropy solutions for scalar conservation laws. This technique has been also applied in [8] to prove uniqueness of entropy solutions of the Total Variation flow in $\mathbb{R}^N$. Since the methods are similar to the above mentioned works, we shall not give the details here.

Acknowledgements

The first and third authors have been partially supported by the Spanish DGICYT, Project PB98-1442. The second author acknowledges partial support by the PNPGC project, reference BFM2000-0962-C02-01 and the TMR European Project 'Viscosity Solutions and their Applications', reference FMRX-CT98-0234.

REFERENCES

[1] AMBROSIO, L., FUSCO, N. and PALLARA, D., *Functions of Bounded Variation and Free Discontinuity Problems*, Oxford Mathematical Monographs, 2000.

[2] ANDREU, F., BALLESTER, C., CASELLES, V. and MAZÓN, J. M., *The Dirichlet problem for the total variational flow*, J. Funct. Anal. *180* (2001), 347–403.

[3] ANDREU-VAILLO, F., CASELLES, V. and MAZÓN, J. M., *Existence and uniqueness of solution for a parabolic quasilinear problem for linear growth functionals with L^1 data*. Math. Ann. *322* (2002), 139–206.

[4] ANDREU, F., CASELLES, V. and MAZÓN, J. M., *A parabolic quasilinear problem for linear growth functionals*. Rev. Mat. Iberoamericana *18* (2002), 135–185.

[5] ANZELLOTTI, G., *Pairings Between Measures and Bounded Functions and Compensated Compactness*, Ann. di Matematica Pura ed Appl. IV *135* (1983), 293–318.

[6] ANZELLOTTI, G., *The Euler equation for functionals with linear growth*, Trans. Amer. Math. Soc. *290* (1985), 483–500.

[7] ANZELLOTTI, G., *BV solutions of quasilinear PDEs in divergent form*, Commun. in Partial Differential Equations *12* (1987), 77–122.

[8] BELLETTINI, G., CASELLES, V. and NOVAGA, M., *The Total Variation Flow in $\mathbb{R}^N$*, To appear in J. Diff. Equat.

[9] BREZIS, H., *Operateurs Maximaux Monotones*, North Holland, Amsterdam, 1973.

[10] CARRILLO, J., *On the uniquenes of solution of the evolution dam problem*, Nonlinear Anal. *22* (1994), 573–607.

[11] DEMENGEL, F. and TEMAN, R., *Convex Functions of a Measure and Applications*, Indiana Univ. Math. J. *33* (1984), 673–709.

[12] DIESTEL, J. and UHL, JR., J. J., *Vector Measures*, Math. Surveys *15*, Amer. Math. Soc., Providence, 1977.

[13] EVANS, L. C. and GARIEPY, R. F., *Measure Theory and Fine Properties of Functions*, Studies in Advanced Math., CRC Press, 1992.

[14] GIAQUINTA, M., MODICA, G. and SOUCEK, J., *Cartesian Currents in the Calculus of Variations II, Variational Integrals*, Springer Verlag, 1997.

[15] HARDT, R. and ZHOU, X., *An evolution problem for linear growth functionals*, Commun. Partial Differential Equations *19* (1994), 1879–1907.

[16] KOHN, R. and TEMAM, R., *Dual space of stress and strains with application to Hencky plasticity*, Appl. Math. Optim. *10* (1983), 1–35.

[17] KRUZHKOV, S. N., *First order quasilinear equations in several independent variables*, Math. USSR-Sb. *10* (1970), 217–243.

[18] LICHNEWSKI, A. and TEMAM, R., *Pseudosolutions of the Time Dependent Minimal Surface Problem*, Journal of Differential Equations *30* (1978), 340–364.

[19] RESHETNYAK, YU. G., *Weak convergence of completely additive vector functions on a set*, Sibirsk. Mat. Z. *9* (1968), 1386–1394. (Translated)

[20] ZIEMER, W. P., *Weakly Differentiable Functions*, GTM 120, Springer Verlag, 1989.

F. Andreu
Departamento de Análisis Matemático
Universitat de Valencia
e-mail: Fuensanta.Andreu@uv.es

V. Caselles
Departament de Tecnologia
Universitat Pompeu-Fabra
e-mail: vicent.caselles@tecn.upf.es

J. M. Mazón
Deptartamento de Análisis Matemático,
Universitat de Valencia
e-mail: mazon@uv.es

J.evol.equ. 3 (2003) 67 – 118
1424–3199/03/010067 – 52
© Birkhäuser Verlag, Basel, 2003

**Journal of Evolution
Equations**

Asymptotic behaviour for the porous medium equation posed in the whole space

JUAN LUIS VÁZQUEZ

Dédié à la mémoire de Philippe Bénilan, ami et maître

This paper is devoted to present a detailed account of the asymptotic behaviour as $t \to \infty$ of the solutions $u(x, t)$ of the equation

$$u_t = \Delta(u^m) \tag{0.1}$$

with exponent $m > 1$, a range in which it is known as the *porous medium equation*, written here PME for short. The study extends the well-known theory of the *classical heat equation* (HE, the case $m = 1$) into a nonlinear situation, which needs a whole set of new tools. The space dimension can be any integer $n \geq 1$. We will also present the extension of the results to exponents $m < 1$ (*fast-diffusion equation*, FDE). For definiteness we consider the Cauchy Problem posed in $Q = \mathbb{R}^n \times \mathbb{R}^+$ with initial data

$$u(x, 0) = u_0(x), \qquad x \in \mathbb{R}^n \tag{0.2}$$

chosen in a suitable class of functions. In most of the paper we concentrate on the class $\mathcal{X}_0$ of integrable and nonnegative data,

$$u_0 \in L^1(\mathbb{R}^n), \quad u_0 \geq 0, \tag{0.3}$$

which is natural on physical grounds as the density or concentration of a diffusion process, the height of a ground-water mound, or the temperature of a hot medium (see a comment on the applications at the end). Consequently, we will deal mostly with nonnegative solutions $u(x, t) \geq 0$ defined in Q. An existence and uniqueness theory exists for this problem so that for every data u_0 we can produce an orbit $\{u(\cdot, t) : t > 0\}$ which lives in $L^1(\mathbb{R}^n) \cap L^\infty(\mathbb{R}^n)$ and describes the evolution of the process. The solution is not classical for $m > 1$, but it is proved that there exists a unique weak solution for all $m > 0$. It is also unique in the sense of *mild solution*, obtained as limit of the Implicit Time-Discretization Scheme, a method that was carefully studied by Bénilan and Crandall [BC1, BC2]. They also proved that the mass, $M = \int u(x, t)\, dx$, is an invariant of the motion for all $t > 0$ as long as $m \geq (n - 2)/n$, a number that will appear again below. The solution of the Cauchy Problem, (CP) for short, has a number of useful additional properties, in particular it is continuous in Q and bounded for $t \geq \tau > 0$.

The investigation of the existence of generalized solutions of the initial and boundary-value problems for the equations of nonlinear diffusion of this type and more general forms, like

$$u_t = \Delta \Phi(u) + f \tag{0.4}$$

(with Φ a monotone nondecreasing function and $f \in L^1(Q)$), was a main chapter of Ph. Bénilan's Thesis in 1972, [Be], and he continued during his life to enlarge the scope of the investigation to the whole range of second-order nonlinear equations of parabolic or hyperbolic-parabolic type. This is reflected in a large number of papers and in the book [Bk], unfortunately still not available to the wide community of researchers and students. In the best tradition of Applied Functional Analysis, Bénilan's way of attacking these equations was to provide an abstract framework, where the typical problems are well-posed and generate semigroups or flows in Banach spaces, mainly $L^1(\mathbb{R}^n)$. Concepts like integral solutions, mild solutions, renormalized solutions, entropy solutions appear in the work of Ph. Bénilan and his collaborators, many of the ideas were inspired by him and have become standard today in the study of nonlinear parabolic and hyperbolic phenomena. His work and teaching has illuminated the life of many friends and followers that contribute to this wide subject at the intersection of PDE's, Mechanics and Functional Analysis. Applications to problems in the oil industry and biology played a role in formulating new problems and questions in Bénilan's research and that of his collaborators, but I will not pursue this interesting direction here. Summing up this paragraph, I contend that the close combination of very concrete and important equations and very abstract and powerful methods is a most appealing way of doing PDEs, and we hope it will last in the work of younger generations.

The main part of the following material has been taught to graduate students at the Univ. Autónoma de Madrid in 1997 [V5]. Further progress is reported here, and a number of unpublished results are added at convenient places to close some arguments and solve some open problems. Let us mention that the general picture for (CP) is now quite well known, but some interesting problems still remain open!

1. Long-time asymptotics

Our main concern in this paper will be the study of the asymptotic behaviour of the solutions of equation (0.1). This is still a quite large subject, hence we will concentrate in the sequel on the initial-value problem (CP) posed in the whole space with integrable data. The behaviour is different for other classes of data or for boundary-value problems. A survey of results for the Dirichlet problem posed in a bounded domain is [V5], for the Neumann problem see [AR], for the exterior Dirichlet problem [QV]. On a general level, it has been pointed out in many papers and corroborated by numerical experiments that similarity solutions furnish the asymptotic representation for solutions of a wide range of problems in mathematical physics. The reader is referred to the book of G.I. Barenblatt

[B2] for a detailed discussion of this subject. Self-similar solutions and the forthcoming scaling techniques will play a prominent role in our asymptotic study.

The model. The asymptotic behaviour is a classical result for the linear heat equation, $m = 1$, and is now well known for the PME, $m > 1$: this is the theory that we want to present in a systematic way. In the first case, the classical result says that there is convergence of any solution of the Cauchy problem under conditions (0.3) towards the Gaussian kernel

$$u(x, t) \sim \frac{M}{(4\pi t)^{n/2}} \exp\left(-x^2/4t\right), \tag{1.1}$$

where $M = \int u_0(x)\, dx$ is the mass of the solution (space integration is performed by default in $\mathbb{R}^n$). In the case $m > 1$ the behaviour of our class of solutions can be described for large t by a one-parameter family of special solutions of (0.1)

$$\mathcal{U}(x, t; C) = t^{-\alpha} F(xt^{-\beta}; C), \tag{1.2}$$

with parameter $C > 0$. The functions $\mathcal{U}(x, t; C)$ are variously called *source-type solutions*, *fundamental solutions, Barenblatt solutions*, or *BZKP solutions*, cf. the original papers [ZK], [B1], [Pa]. They are given by the explicit formula

$$F(\eta) = (C - k\,\eta^2)_+^{\frac{1}{m-1}}, \quad \alpha = \frac{n}{n(m-1)+2}, \quad \beta = \frac{1}{n(m-1)+2}. \tag{1.3}$$

F is called the profile, α and β are the similarity exponents (that we call Barenblatt exponents). $C > 0$ is a free constant and k is fixed, $k = (m-1)\beta/2m$. We have $\mathcal{U}^{m-1} = (Ct^{2\beta} - k\,x^2)_+/t$. The fact that the fundamental solutions are self-similar is important in what follows, the fact that they are explicit is not.

Main result. We will prove that $u(x, t) \sim \mathcal{U}(x, t; C)$ for large t. For a given u there is a correct choice of the constant $C = C(u_0)$ in this asymptotic result which agrees with the rule of mass equality:

$$\int_{\mathbb{R}^n} u(x, t)\, dx = \int_{\mathbb{R}^n} \mathcal{U}(x, t; C)\, dx. \tag{1.4}$$

It follows that $C = c(m, n)\, M^{2(m-1)\beta}$. We also write $\mathcal{U}_M$ for the solution with mass M and F_M for its profile. This is the precise statement of the asymptotic convergence result:

THEOREM 1.1. *Let $u(x, t)$ be the unique weak solution of problem* (CP) *with initial data $u_0 \in L^1(\mathbb{R}^n)$, $u_0 \geq 0$. Let $\mathcal{U}_M$ be the Barenblatt solution with the same mass as u_0. Then as $t \to \infty$ we have*

$$\lim_{t \to \infty} \|u(t) - \mathcal{U}_M(t)\|_1 = 0. \tag{1.5}$$

Convergence holds also in L^∞-norm in the proper scale:

$$\lim_{t \to \infty} t^\alpha \|u(t) - \mathcal{U}_M(t)\|_\infty = 0 \tag{1.6}$$

with $\alpha = n/(n(m-1)+2)$. Moreover, for every $p \in (1, \infty)$ we have

$$\lim_{t \to \infty} t^{\alpha(p)} \|u(t) - \mathcal{U}_M(t)\|_{L^p(\mathbb{R}^n)} = 0, \tag{1.7}$$

with $\alpha(p) = \alpha(p-1)/p$.

The last result follows from (1.5) and (1.6) by simple interpolation, but (1.6) and (1.5) are (to an extent) independent. The main body of the paper is devoted to giving a detailed proof of this theorem, and to report on its many extensions. We will follow the "four-step method", a general plan to prove asymptotic convergence devised by Kamin and Vazquez in 1988, [KV1], who settled in this way the asymptotic behaviour both for the p-Laplacian equation, $u_t = \nabla \cdot (|\nabla u|^{p-2} \nabla u)$, and for the PME. But the first proof of convergence for the PME in several dimensions appeared in a celebrated paper by Friedman and Kamin in 1980 [FK]: it uses the method of lower barriers that will be presented in Section 14 and gets uniform convergence on compact expanding sets of the form $\{|x| \leq C t^\beta\}$, a weaker form of (1.6). The initial data belong to $L^1 \cap L^2$, but this is not a real restriction after the regularizing effect of Bénilan [Be2] and Véron [Ve]. A previous proof in one space dimension is due to S. Kamin, 1973 [K1].

The proof of the full result with uniform convergence in the whole space, formula (1.6), has first appeared in the notes [V5]. There, six different proofs of the main result are presented, all of them sharing a common 4-step basis, but concluding through quite different ideas, thus summing up the state of asymptotic convergence methods for the PME at the time. The interest in displaying all of these approaches lies in the fact that they call on many of the asymptotic methods available in the study of nonlinear parabolic problems. In the study of similar problems for other equations and/or classes of data some of these techniques will be applicable and have in fact been applied, but the best technique has to be chosen depending on the case.

Discussion. Immediate issues after Theorem 1.1. are discussing whether the class of data may be extended with the same type of convergence, or the rate of convergence can be improved. Both questions have been explored, but let us begin with two negative results.

For the first question, the restriction $u_0 \in L^1(\mathbb{R}^n)$ cannot be ignored if we want to keep the asymptotic size, since non-integrable data have a different behaviour.

COROLLARY 1.2. *If u is a nonnegative global solution of the PME with $u_0 \in L^1_{loc}(\mathbb{R}^n)$ and $\int u_0 \, dx = \infty$ then*

$$\lim_{t \to \infty} t^\alpha u(y \, t^\beta, t) = \infty \tag{1.8}$$

uniformly on compact sets $\{y \in K\} \subset \mathbb{R}^n$.

The proof of this result is very easy after approximating u_0 by an increasing sequence of integrable data u_{0n}, applying Theorem 1.1 and passing to the limit. Consequently, non-integrable solutions have other asymptotic size and shape, which has been partially investigated in [AR2] and other papers. We recall that a theory of global solutions of the PME with non-integrable data was constructed by Bénilan, Crandall and Pierre in [BCP] in 1984. Let us also refer to asymptotic results with a completely different flavor from Theorem 1.1: a general study of the asymptotic behaviour of the PME and other evolution equations in the framework of $\mathcal{X} = L^\infty(\mathbb{R}^n)$ has been performed recently in a paper with E. Zuazua [VZ]; the situation is quite different, there is no simple model of asymptotic attractor, the appropriate word is rather *complexity*, i.e., the possibility of chaos.

The second direction is improving the rates, and there we again find a limitation.

THEOREM 1.3. *The rates of convergence of Theorem* 1.1 *are optimal in the class of initial data* $\mathcal{X}_0 = \{u_0 \in L^1(\mathbb{R}^n),\ u_0 \geq 0\}$.

The main idea in constructing a counterexample consists of placing small bits of mass far enough at $t = 0$. The detailed proof is presented in Section 11. Better rates can be found if we restrict the class of data. This subject, which has been the object of an active interest in recent years, will discussed in Sections 16 to 19. A first natural idea is to consider initial data with compact support and to control the expansion of the support for large times. Then it is proved that the support of the solution $u(\cdot, t)$ approaches the size of the support of the source-type solution, a ball of radius $\mathcal{R}(t) = C_0\, t^\beta$ with $C_0 = c(m, N)\, M^{(m-1)\beta}$, cf. Section 9, Theorem 9.1. A larger class of solutions is considered in Section 19 using so-called entropy methods.

Another idea is to extend the convergence to the derivatives, i.e., convergence in C^k. The presence of a free boundary for compactly supported solutions makes the question tricky because the variable domain of positivity: though it converges to the ball of the source-type solution, it does not coincide with it. The question has been recently studied by Lee and Vazquez [LV], see details in Section 18.

Extensions. We have also made progress in extending the asymptotic study to cover the following three important directions:

(i) Signed solutions for the signed PME: $u_t = \Delta(|u|^{m-1}u)$, cf. Theorem 20.1 taken from the paper [KV2].

(ii) Equations with forcing, where we present a new result, Theorem 20.2 which gives the asymptotic behaviour for

$$u_t = \Delta(|u|^{m-1}u) + f, \qquad f \in L^1(Q). \tag{1.9}$$

This answers a question of Bénilan and is maybe the most natural extension of Theorem 1.1 in the spirit of his ideas.

(iii) Exponents m less than one (Fast Diffusion Equations). We introduce a new idea, *uniform convergence in relative error*, Theorem 21.1.

Though the theory has obvious counterparts for related equations, like the p-Laplacian equation, $u_t = \nabla \cdot (|\nabla u|^{p-2}\nabla u)$, the general Filtration Equation $u_t = \Delta\Phi(u)$ with Φ monotone nondecreasing, and reaction-diffusion equations like $u_t = \Delta u^m - u^p$, we will not pursue these directions here. In the first case the similarity of methods and results is impressive. The basic convergence result is proved in [KV1]. Rates of convergence have been recently obtained, see [DP2]. The survey paper [Kl], though a bit old is very informative about the theory of nonlinear diffusion equations.

Finally, the porous media equation looks like a variation of the heat equation, and perturbative techniques could be expected to be the way to the analysis. It must be stressed that it is not so, and the study performed below relies on the ideas and machinery of nonlinear analysis. Indeed, the PME is a good benchmark for a number of important nonlinear techniques.

Notation. In the whole paper $n \geq 1$ is the space dimension, and the exponents α and β will be fixed to the values (1.3): $\alpha = n/(2+n(m-1)) > 0$, $\beta = \alpha/n$. Given a solution $u(x,t)$, we will often write $u(t)$ to denote the function $t \mapsto u(\cdot, t)$, as in $\|u(t)\|_p = \|u(\cdot, t)\|_{L^p(\mathbb{R}^n)}$.

2. Definitions and preliminary results

Problem (0.1)–(0.2) does not possess classical solutions for general data in the class $\mathcal{X}_0 = \{u_0 \in L^1(\mathbb{R}^n), u_0 \geq 0\}$ (or even in a smaller class, like the set of smooth nonnegative and rapidly decaying initial data). This is due to the fact that the equation is parabolic only where $u > 0$, but degenerates at the level $u = 0$. Therefore, we need to introduce a concept of generalized solution and make sure that the problem is well-posed in that class.

Concept of solution. (I) By a solution of equation (0.1) we will mean a nonnegative function $u(x,t)$, defined for $(x,t) \in Q$ such that: (i) viewed as a map $t \to u(\cdot, t) = u(t)$ we have $u \in C((0, \infty) : L^1(\mathbb{R}^n))$; (ii) the functions u^m, u_t and Δu^m belong to $L^1(t_1, t_2 : L^1(\mathbb{R}^n))$ for all $0 < t_1 < t_2$; and (iii) equation (0.1) is satisfied in the sense of distributions in Q.

(II) By a solution of problem (CP) we mean a solution of (0.1) such that the initial data are taken in the following sense:

$$u(t) \to u_0 \quad \text{in } L^1(\mathbb{R}^n) \quad \text{as } t \to 0. \tag{2.1}$$

In other words, $u \in C([0, \infty) : L^1(\mathbb{R}^n))$ and $u(0) = u_0$.

This definition is usually called in the literature a **strong solution**. It is suitable for our purposes since problem (CP) is well-posed in this setting, but it is not the unique choice; we could have used the concept of **weak solution**, where we merely ask u^m and $\nabla_x u^m$ to be locally integrable functions in $\mathbb{R}^n \times [0, \infty)$ and the equation is satisfied in the sense that

$$\int\int \{u\varphi_t - \nabla_x u^m \cdot \nabla_x \varphi\}\, dxdt + \int u_0(x)\varphi(x, 0)\, dx = 0$$

holds for every smooth test function $\varphi \geq 0$ which vanishes for all large enough $|x|$ and t. See [V3] for a discussion of those equivalent alternatives. Viscosity solutions have been discussed in [CV].

THEOREM 2.1. *Problem (CP) is well-posed in the framework of strong solutions. Moreover, the maps* $S_t : u_0 \mapsto u(t)$ *are order-preserving contractions on* $X_0 = L^1_+(\mathbb{R}^n)$. *More precisely,*

$$\|(u_1(t) - u_2(t))_+\|_{L^1(\mathbb{R}^n)} \leq \|(u_1(0) - u_2(0))_+\|_{L^1(\mathbb{R}^n)}, \tag{2.2}$$

where $(\cdot)_+$ *denotes positive part,* $\max\{\cdot, 0\}$. *In particular, plain* L^1-*contraction holds*

$$\|u_1(t) - u_2(t)\|_{L^1(\mathbb{R}^n)} \leq \|u_1(0) - u_2(0)\|_{L^1(\mathbb{R}^n)}. \tag{2.3}$$

The Maximum Principle also follows from property (2.2).

Property (2.2) is called T-contraction in Bénilan's papers. Actually, the existence of a semigroup of T-contractions is true in the more general framework of signed solutions for the generalized PM equation $u_t = \Delta(|u|^{m-1}u)$, discussed in Section 20. Here are some additional important properties of the solutions.

PROPERTY 1. The solutions of problem (CP) satisfy the law of mass conservation

$$\int_{\mathbb{R}^n} u(x, t)\, dx = \int_{\mathbb{R}^n} u_0(x)\, dx \tag{2.4}$$

When $u \geq 0$ this means $\|u(t)\|_1 = \|u_0\|_1$ for all $t > 0$.

The property will also be true for solutions of any sign, but then it does not imply conservation of L^1-norm. It is also true for $0 < m < 1$ if $m \geq (n-2)/n$, but not below that value.

PROPERTY 2. The solutions are bounded for $t \geq \tau > 0$. Moreover, there exists a constant $C = C(m, n) > 0$ such that

$$0 \leq u(x, t) \leq C \|u_0\|_1^{2\beta}\, t^{-\alpha}. \tag{2.5}$$

PROPERTY 3. Energy estimates.
Another aspect of the regularization of the solutions in time is obtained by multiplying the equation by u^m and formally integrating by parts. We arrive at

$$\int_{\mathbb{R}^n} u^{m+1}(x, t)\, dx + (m+1) \int_\tau^t \int_{\mathbb{R}^n} |\nabla(u^m)|^2\, dxdt \leq \int_{\mathbb{R}^n} u^{m+1}(x, \tau)\, dx \tag{2.6}$$

for all $0 < \tau < t$. Since we know by the previous properties that $u(\tau) \in L^p(\mathbb{R}^n)$ for all $p > 1$, in particular $p = m + 1$, we conclude that ∇u^m is uniformly bounded in $L^2(\mathbb{R}^n \times (\tau, t))$ in terms of the mass of the initial data. The justification of the calculation can be found in [V4]. In the same spirit, multiplication by $(u^m)_t$ and integration by parts gives

$$\frac{8m}{(m+1)^2} \int_\tau^t \int_{\mathbb{R}^n} \left(\frac{\partial}{\partial t} u^{(m+1)/2}(x,t)\right)^2 dxdt$$
$$+ \int_{\mathbb{R}^n} |\nabla u^m(x,t)|^2 \, dx \leq \int_{\mathbb{R}^n} |\nabla u^m(x,\tau)|^2 \, dx. \tag{2.7}$$

Combining with the previous one, it gives a bound for $\partial(u^{(m+1)/2})/\partial t$ in $L^2(\mathbb{R}^n \times (\tau, t))$ in terms of the mass of the initial data, and a better bound for ∇u^m in $L^\infty(\tau, \infty : L^2(\mathbb{R}^n))$. These and other gradient estimates were developed by Bénilan, cf. [Be3].

The next estimate is due to Aronson-Bénilan [AB] and plays a big role in the study of the Cauchy problem for the PME.

PROPERTY 4. Fundamental regularity estimate and consequences.
Any nonnegative solution of the Cauchy problem (CP) satisfies the estimate

$$\Delta(u^{m-1}) \geq -\frac{C}{t}, \qquad C = \frac{\alpha(m-1)}{m}. \tag{2.8}$$

This implies another interesting estimate: $u_t \geq -\alpha u/t$. Moreover, conservation of mass is equivalent to $\int u_t \, dx = 0$, so that the last estimate leads to

$$\int |u_t(x,t)| \, dx \leq \frac{2\alpha}{t} \int u(x,t) \, dx. \tag{2.9}$$

The one-sided estimate (2.8) is exact precesily for the source-type solutions that play a key role in our theory. For a proof of all the above facts we refer to the text [V4] and its references. In the proof of better convergence we will use a further regularity result that can be found in [DB].

PROPERTY 5. Bounded solutions are uniformly Hölder continuous for $t \geq \tau > 0$.

Ph. Bénilan devoted much time to finding the best Hölder exponent and his Kentucky Notes are a proof of his efforts. In one dimension the answer is $\alpha = \min\{1, 1/(m-1)\}$, but the question is not completely settled for $n > 1$.

PROPERTY 6. Finite propagation property. If the initial function u_0 is compactly supported so are the functions $u(\cdot, t)$ for every $t > 0$. Under these conditions there exists a free boundary or interface which separates the regions $\{(x,t) \in Q : u(x,t) > 0\}$ and $\{(x,t) \in Q : u(x,t) = 0\}$. This interface is usually an n-dimensional hypersurface in $\mathbb{R}^{n+1}$.

Let us conclude this section by pointing out that the source-type solutions $\mathcal{U}(x, t; C)$ are strong solutions of (0.1), but, strictly speaking, they are **not** solutions of problem (CP) because they do not take L^1 initial data. Indeed, it is easy to check that $\mathcal{U}$ converges to a Dirac mass as $t \to 0$ (this is the reason for the name "source-type solutions"). We will use this fact strongly in the first proof of asymptotic convergence, cf. Section 5. The reader should note that the behaviour of a whole class of solutions of equation (0.1) is described in terms of a simple family of functions which *are not* solutions in the same class, but in a larger class. More precisely, the special solutions which represent the whole dynamics at the asymptotic level exhibit a *singularity* (at $x = 0$, $t = 0$). This is a curious and quite general feature in the theory of asymptotic analysis.

3. Outline of the next sections. Four-step method

Before proceeding with the actual proofs we explain the "four-step method" as a general plan to prove asymptotic convergence in the form written down by Kamin and Vazquez in 1988. The first three steps are the common basis of the different proofs of convergence using similarity techniques and are worked out in detail in Section 4.. They allow to produce out of the original orbit $u(t)$ a family of rescaled orbits $\widetilde{u}_\lambda(t)$, which upon passage to the limit $\lambda \to \infty$ produces one or several **limit orbits** or limit solutions, $U(t)$. These orbits represent the asymptotic dynamics that we want to study. The proof of the convergence results is then reduced to the identification of (a unique) limit orbit $U(t)$ and the description of the mode and rates of convergence.

The four-step method, which is a general procedure for the study of many asymptotic problems, can be described as follows.

STEP 1. Rescaling.

It produces out of an orbit $u(t)$ a family of *zoomed orbits*, $\{\widetilde{u}_\lambda(t)\}$. The orbit is viewed as a map $u(t) : t \mapsto u(\cdot, t)$ from $[0, \infty)$ into a suitable functional space.

STEP 2. Estimates and compactness.

Appropriate estimates allow to show that if we choose the correct rescaling in Step 1, we obtain a family of uniformly compact orbits (in suitable functional spaces). Failure of guessing the correct scaling size produces rescaled orbits that grow to infinity or go to zero. In either case the method stops.

STEP 3. Passing to the limit.

We replace the limit $t \to \infty$ by the limit in the scaling parameter λ for fixed t. We obtain one or several limit orbits $U(t)$.

STEP 4. Identifying the limit.

The function U obtained as a limit of the orbits in Step 3 has to be identified as a solution of an equation, usually (but not necessarily) the same equation we started with. Some additional characteristics allow to determine the limit function in a unique way.

There is a final mini-step, 5. Rephrasing the result.

That consists in undoing the λ-transformation and stating the result in the original variables.

We will perform all these steps for the Cauchy problem for the PME. The first three steps are common to all proofs below and, actually, these steps are applicable to a number of problems sharing the two properties of scale-invariance and regularity (smoothness) of the orbits. It is very important to make clear at this moment that the limit obtained after Step 3 must be nontrivial: neither zero nor infinity. Otherwise (and this happens quite often in preliminary analyses), we are missing the correct size of the asymptotic process and (probably) no relevant information is obtained from the analysis.

It is at the level of Step 4 that we can find very different ways of identifying the limit obtained in Step 3. Each of the different techniques will be discussed separately.

In Sections 5, 6 below we give the first proof of Theorem 1.1 by using the characterization of the limit solutions $U(t)$ through their initial data, which is shown to be a Dirac mass. This idea was explained in the paper [KV1].

In Section 7 the basic L^1 asymptotic result is improved into uniform convergence, which is the second part of Theorem 1.1; the result was announced in [V5] and appears here with a detailed proof using the *local regularizing effect*. We then estimate the asymptotics of supports and interfaces for compactly supported solutions, Section 9.

In Section 10 we introduce the *continuous rescaling* as an alternative scaling method that is used sometimes with advantage over the standard *fixed-rate rescaling* (i.e., the λ-scaling). This part of the notes concludes with the proof of optimality of the convergence rates, a question that has been a debated issue among experts: for restricted classes of data the rates can be improved, but not for the whole class $\mathcal{X}_0$ (even if we take smooth data the answer is still negative, it depends on the behaviour at infinity).

Subsequent sections are devoted to present alternative proofs of the main result: first, we have two methods based on the existence of a Lyapunov functional and the use of the Invariance Principle. Lyapunov methods are probably the most effective and popular tools in the study of asymptotic problems. Section 14 presents the method of lower barriers used by Friedman and Kamin [FK], a kind of nonstandard Lyapunov functional.

Afterwards, we turn to the ideas of asymptotic symmetry which allow to reduce the study of asymptotic dynamics to conditions of radial symmetry in the spatial variable, Section 15. There are then methods which use the special properties of one-dimensional problems. We present two of them, the method of Concentration Comparison, Section 16, and the method of Intersection Comparison, Section 17. For radial solutions good rates of convergence are obtained, Theorems 16.4 and 17.3.

The rest of the notes contains the analysis of further progress as mentioned at the end of Section 1. New results are presented in the three outlined directions.

4. The first steps

1. Rescaling.

In order to observe the asymptotic behaviour of the orbit of problem (CP) we rescale it according to the Barenblatt exponents. Let us see the whole story of scaling transformations in some detail. Let $u = u(x, t)$ be a solution of (0.1). We apply the group of dilations in all the variables

$$u' = Ku, \qquad x' = Lx, \qquad t' = Tt, \tag{4.1}$$

and impose the condition that u' so expressed as a function of x' and t', i.e.,

$$u'(x', t') = Ku\left(\frac{x'}{L}, \frac{t'}{T}\right), \tag{4.2}$$

has to be again a solution of (0.1). Then:

$$\frac{\partial u'}{\partial t'} = \frac{K}{T}\frac{\partial u}{\partial t}\left(\frac{x'}{L}, \frac{t'}{T}\right), \qquad \Delta_{x'}(u')^m = K^m L^{-2}\Delta_x(u^m)\left(\frac{x'}{L}, \frac{t'}{T}\right).$$

Hence, (4.2) will be a solution if and only if $KT^{-1} = K^m L^{-2}$, i.e.,

$$K^{m-1} = L^2 T^{-1}. \tag{4.3}$$

We thus obtain a two-parametric transformation group acting on the set of solutions of (0.1). Choosing as free parameters L and T it can be written as

$$u'(x', t') = L^{\frac{2}{m-1}} T^{\frac{-1}{m-1}} u(x, t) = \left(\frac{L^2}{T}\right)^{\frac{1}{m-1}} u\left(\frac{x'}{L}, \frac{t'}{T}\right).$$

Using standard letters for the independent variables and putting $u' = \mathcal{T}u$, we get:

$$(\mathcal{T}u)(x, t) = L^{\frac{2}{m-1}} T^{-\frac{1}{m-1}} u\left(\frac{x}{L}, \frac{t}{T}\right). \tag{4.4}$$

Moreover, we can use one of the parameters to force $\mathcal{T}$ to preserve some important behaviour of the orbit. Here we recall that $\mathcal{U}_M(x, t)$ has a constant mass; actually, this characterizes uniquely the solution (which is the ideal orbit we want to approach). Imposing thus the condition of mass conservation at $t = 0$ we get

$$\int_{\mathbb{R}^n} (\mathcal{T}u_0)(x)\, dx = \int_{\mathbb{R}^n} u_0(x)\, dx, \tag{4.5}$$

namely,

$$\int_{\mathbb{R}^n} Ku_0\left(\frac{x}{L}\right) dx = \int_{\mathbb{R}^n} u_0(x)\, dx.$$

It easily follows that $KL^n = 1$. This and (4.2) give the expressions

$$K = T^{-\alpha}, \qquad L = T^{\beta},\tag{4.6}$$

with the exponents given by (1.3). The transformation we are going to use is finally

$$(\mathcal{T}u)(x,t) = T^{-\alpha}u(x/T^{\beta}, t/T).$$

It is convenient to write the scaling factor in terms of $\lambda = 1/T$. Then, the solution is

$$\widetilde{u}_{\lambda}(x,t) = (\mathcal{T}_{\lambda}u)(x,t) = \lambda^{\alpha}u(\lambda^{\beta}x, \lambda t)\tag{4.7}$$

with initial data $\widetilde{u}_{0,\lambda}(x) = (\mathcal{T}_{\lambda}u_0)(x) = \lambda^{\alpha}u_0(\lambda^{\beta}x)$.

Visualization.

The scaled orbits at $t = 1$ are just a **zoomed** version of the original orbit at $t = \lambda$.

Property. The source-type solutions are invariant under the λ-rescaling, i.e. $\mathcal{U}_M(t) = \mathcal{T}_{\lambda}(\mathcal{U}_M(t))$.

2a. Uniform estimates. The family $\widetilde{u}_{\lambda}(t)$, $\lambda > 0$, is uniformly bounded in $L^1(\mathbb{R}^n)$ for t positive:

$$\int_{\mathbb{R}^n} \widetilde{u}_{\lambda}(x,t)dx = \int_{\mathbf{R}^n} \lambda^{\alpha}u(\lambda^{\beta}x, \lambda t)\,dx = \int_{\mathbb{R}^n} u(y, \lambda t)\,dy = M < \infty.\tag{4.8}$$

Using now (2.5) we get

$$\|\widetilde{u}_{\lambda}(\cdot, 1)\|_{\infty} = \lambda^{\alpha}\|u(\cdot, \lambda)\|_{\infty} \leq \lambda^{\alpha}\frac{M^{2\alpha/n}}{\lambda^{\alpha}}C = CM^{2\alpha/n}\tag{4.9}$$

independently of λ, and in the same way

$$\|\widetilde{u}_{\lambda}(\cdot, t_0)\|_{\infty} = \lambda^{\alpha}\|u(\cdot, t_0\lambda)\|_{\infty} \leq \lambda^{\alpha}\frac{M^{2\alpha/n}}{\lambda^{\alpha}t_0^{\alpha}}C = CM^{2\alpha/n}\,t_0^{-\alpha}.\tag{4.10}$$

Control of the norms $\|\cdot\|_1$ and $\|\cdot\|_{\infty}$ means control of all norms $\|\cdot\|_p$ for all $p \in [1, \infty]$. Thus, $\|\widetilde{u}_{\lambda}(\cdot, t)\|_p$ are equi-bounded for all $p \in [1, \infty]$.

Next, we take $t_0 > 0$ so that, by the regularizing effect, $u(t_0) \in L^{m+1}(\mathbb{R}^n)$. The energy estimates of Section 2 give

$$\int_{\mathbb{R}^n} |\nabla \widetilde{u}_{\lambda}^m(x,t)|^2 dx \leq C(t_0, \|\widetilde{u}_{\lambda}(x, t_0)\|_{L^{m+1}})\tag{4.11}$$

for $t \geq t_0 > 0$. Now, the $\|\widetilde{u}_{\lambda}(t)\|_{L^{m+1}}$ are equi-bounded, hence $\|\nabla \widetilde{u}_{\lambda}^m(x,t)\|_{L^2}$ are equi-bounded (for $t \geq t_0 > 0$).

Moreover, by (2.9) of Property 5, Section 2,

$$\left\|\frac{\partial \widetilde{u}_{\lambda}}{\partial t}(t)\right\|_{L^1} \leq C\frac{\|\widetilde{u}_{\lambda}(t)\|_{L^1}}{t}\tag{4.12}$$

Using the same argument we conclude that the norms $\|(\widetilde{u}_\lambda)_t(t)\|_{L^1}$ are equi-bounded if $t \geq t_0 > 0$.

2b. Compactness.

Let us recall the Rellich-Kondrachov Theorem. *Let Ω be a bounded domain with C^1 boundary. Then*

$$p < N \Rightarrow W^{1,p}(\Omega) \subset L^q(\Omega) \text{ for all } q \in [1, p^*), \quad \frac{1}{p^*} = \frac{1}{p} - \frac{1}{N},$$

$$p = N \Rightarrow W^{1,p}(\Omega) \subset L^q(\Omega) \text{ for all } q \in [1, +\infty),$$

$$p > N \Rightarrow W^{1,p}(\Omega) \subset C(\overline{\Omega}).$$

All these injections are compact. In particular, $W^{1,p}(\Omega) \subset L^p(\Omega)$ with compact injection for all $p \geq 1$. In $\Omega = \mathbb{R}^n$, the above injections are compact in local topology (convergence on compact subsets).

Let us now recall our situation for the family $\{\widetilde{u}_\lambda\}_{\lambda \geq 1}$ for $t \geq t_0 > 0$:

$$\widetilde{u}_\lambda(x,t) \in L^\infty_{x,t} \subset L^1_{loc}, \quad \frac{\partial \widetilde{u}_\lambda}{\partial t}(x,t) \in L^\infty_t(L^1_x) \subset L^1_{x,t} \ (t \in (t_0, t_1)),$$

and

$$\nabla_x \widetilde{u}^m_\lambda \in L^2_{x,t} \subset L^1_{x,t}.$$

All spaces in time are local in the sense that they exclude $t = 0$.

PROPOSITION 4.1. *The family $\{\widetilde{u}_\lambda\}_{\lambda > 1}$ is relatively compact locally in $L^1_{x,t}$. Also the family $\{\widetilde{u}^m_\lambda\}_{\lambda > 1}$.*

3. Passage to the limit.

We can now take a sequence $\lambda_k \to \infty$ and assert that $\widetilde{u}_{\lambda_k}$ converges in $L^1_{loc}(Q)$ to some function U:

$$\lim_{\lambda \to \infty} \widetilde{u}_\lambda(x,t) = U(x,t). \tag{4.13}$$

We need to study the properties of such **limit functions** $U(x,t)$.

LEMMA 4.2. *Any limit U is a nonnegative weak and strong solution of (0.1) satisfying uniform bounds in $L^1(\mathbb{R}^n)$ and $L^\infty(\mathbb{R}^n)$ for all $t \geq \tau > 0$.*

Proof. It is clear that, as a consequence of the passage to the limit U is nonnegative. Also, $U(t)$ is uniformly bounded in L^1 and L^∞ for $t \geq t_0 > 0$, according to formulas (2.4), (2.5). In order to check that it is a weak solution we review the sense in which $\widetilde{u}_\lambda$ is a weak solution:

$$\int\int \{\widetilde{u}_\lambda \varphi_t - \nabla(\widetilde{u}^m_\lambda) \cdot \nabla\varphi\} \, dxdt + \int \widetilde{u}_{0\lambda}(x)\varphi(x,0) \, dx = 0$$

for all φ test. We have already remarked that our uniform estimates are not good near $t = 0$. In view of this, we restrict the test functions to the class

$$\varphi \in C_0^\infty(\mathbb{R}^n \times (0, \infty)),$$

so that φ vanishes in a neighborhood of $t = 0$. Then

$$\iint \{\widetilde{u}_\lambda \varphi_t - \nabla \widetilde{u}_\lambda^m \nabla \varphi\} dxdt = 0. \tag{4.14}$$

With our estimates

$$\begin{cases} \widetilde{u}_\lambda \to U & \text{locally in } L^1_{x,t} \\ \widetilde{u}_\lambda \to U & \text{weak}^* \text{ in } L^\infty_{loc} \\ \nabla \widetilde{u}_\lambda^m \to \nabla U^m & \text{in } L^2_{x;t,loc} \text{weak,} \end{cases}$$

we may pass to the limit in this expression (along a subsequence $\lambda_n \to \infty$) to get

$$\iint \{U\varphi_t - \nabla_x U^m \cdot \nabla_x \varphi\}\, dxdt = 0. \tag{4.15}$$

This means that U is a weak solution of equation (0.1). In fact, if $\tau > 0$

$$\int_\tau^\infty \int_{\mathbf{R^n}} \{U\varphi_t - \nabla_x U^m \cdot \nabla_x \varphi\}\, dxdt + \int U(x, \tau)\varphi(x, \tau)\, dx = 0.$$

$\square$

5. Identification of the limit. Compact support

Thus far, we have posed the dynamics in the form of an initial value problem and we have introduced a method of rescaling which has allowed to obtain, after passage to the limit, one or several new solutions of the original problem. These solutions, that we call the **asymptotic dynamics**, form a special subset of the set of all orbits of our dynamical system and represent the (scaled) asymptotic behaviour of the original orbits. Their complete description becomes our main problem. The asymptotic dynamics turns out to be quite simple in the present case. The general framework in which this kind of ideas are studied is that of Dynamical Systems invariant under groups of transformations.

We proceed next with Step 4, i.e., the identification of the limit, in the form devised in [KV1]. Several other options will be examined later. We want to prove that the limit U along any sequence $\lambda_n \to \infty$ is necessarily $\mathcal{U}_M$. Both U and $\mathcal{U}_M$ are solutions of the PME for $t > 0$, enjoying a number of similar bounds. In order to identify them we only need to check their initial data and use a suitable uniqueness theorem for the Cauchy problem (CP). The necessary uniqueness theorem is available thanks to M. Pierre's work [Pi].

THEOREM 5.1. *Weak solutions of the PME in the class $u \in C((0,\infty) : L^1(\mathbb{R}^n))$, $u \geq 0$, which take a bounded and nonnegative measure $\mu(x)$ as initial data in the sense that*

$$\lim_{t\to 0} \int_{\mathbb{R}^n} u(x,t)\varphi(x)\,dx = \int_{\mathbb{R}^n} \varphi(x)\,d\mu(x), \tag{5.1}$$

for all $\varphi \in C_b(\mathbb{R}^n)$, $\varphi \geq 0$, are uniquely determined by the initial measure.

Let us then worry about the initial data. At first sight it looks easy:

LEMMA 5.2. *If $\lambda \to \infty \Rightarrow \lim \widetilde{u}_{0,\lambda}(x) \to M\delta(x)$ in the sense of bounded measures.*

Proof. As $\lambda \to \infty$, since $\alpha = n\beta > 0$,

$$\int_{\mathbb{R}^n} \widetilde{u}_{0\lambda}(x)\,\varphi(x)\,dx = \int_{\mathbb{R}^n} \lambda^\alpha u_0(\lambda^\beta x)\varphi(x)\,dx = \int_{\mathbb{R}^n} u_0(y)\varphi(y/\lambda^\beta)\,dy,$$

which converges to $\int_{\mathbb{R}^n} u_0(y)\,\varphi(0)\,dy$ for all $\varphi \in C_c^\infty(\mathbb{R}^n)$, $\varphi \geq 0$. We have used the mass value: $\int_{\mathbb{R}^n} u_0(y)\,dy = M$. $\qquad\square$

The problem of the double limit. Unfortunately, the fact that the initial data for $\widetilde{u}_\lambda$ converge to $M\delta(x)$ does not justify by itself that $U(t)$ takes initial data $M\delta(x)$, because we do not control the evolution of the $\widetilde{u}_\lambda$ near $t=0$ in a uniform way and a discontinuity might be taking place near $t=0$ in the limit $\lambda \to \infty$. This is a typical case of double limits,

$$\lim_{t\to 0}\lim_{\lambda\to\infty} \widetilde{u}_\lambda(x,t) = \lim_{\lambda\to\infty}\lim_{t\to 0} \widetilde{u}_\lambda(x,t)\,?$$

Preparing for a correct analysis, the first thing to do is to check that U and $\mathcal{U}_M$ have the same mass, i.e., that U has mass M. Since

$$\int_{\mathbb{R}^n} \widetilde{u}_\lambda(x,t)\,dx = M,$$

and $\widetilde{u}_{\lambda_k}$ converges to U in $L^1_{x,t}$-strong locally, we have $\widetilde{u}_{\lambda_k}(t) \to U(t)$ for a.e. t in $L^1_x(\mathbb{R}^n)$ locally and a.e. in $(x,t) \in Q$. By Fatou's Lemma

$$\int_{\mathbb{R}^n} U(t)\,dx \leq \lim_{k\to\infty} \int_{\mathbb{R}^n} \widetilde{u}_{\lambda_k}(x,t)\,dx = M,$$

hence the mass is equal or less. We have met again a difficulty. This difficulty is in principle essential. There are examples for rather simple equations in the nonlinear parabolic area where the initial data are not trivial but the whole solution disappears in the limit! Should such 'disaster' happen, we talk about an *initial layer of discontinuity*, an interesting object of study.

Compactly supported solutions. Here the only way the discontinuity can happen is by mass escaping to infinity, since there is only a mechanism at play, diffusion. In view of this difficulty we change tactics and try to establish the result under additional hypothesis:

• **Extra assumption.** *We take u_0 a bounded, $0 \leq u_0 \leq C$, and compactly supported function,* $\mathrm{supp}(u_0) \subset B_R(0)$.

Then, $\mathrm{supp}(\widetilde{u}_{0\lambda}) \subset B_{R/\lambda^\beta}(0)$. Moreover, there exists a source-type solution of the form $V(x,t) = \mathcal{U}_{M'}(x, t+1)$ with $M' >> M$ such that $V(x,0) = \mathcal{U}_{M'}(x,1) \geq u_0(x)$. Then,

$$\widetilde{u}_\lambda(x,0) = \lambda^\alpha u_0(x\lambda^\beta, 0) \leq \lambda^\alpha \mathcal{U}_{M'}(x\lambda^\beta, 1) = \mathcal{U}_{M'}\left(x, \frac{1}{\lambda}\right),$$

where in the last equality we have used the invariance of $\mathcal{U}$ under $\mathcal{T}_\lambda$. We conclude from the Maximum Principle that

$$\widetilde{u}_\lambda(x,t) \leq \mathcal{U}_{M'}\left(x, t+\frac{1}{\lambda}\right), \tag{5.2}$$

and in the limit $U(x,t) \leq \mathcal{U}_{M'}(x,t)$. The bound solves all our problems since it implies that the support of the family $\{\widetilde{u}_\lambda(t)\}$ is uniformly small for all λ large and t close to zero. Indeed, we observe the relation between the radii of the supports of a solution and its rescaling:

$$R_\lambda(t) = \frac{1}{\lambda^\beta} R(\lambda t), \tag{5.3}$$

It follows that the support of $\widetilde{u}_\lambda(t)$ is contained in a ball of radius

$$R = C \, (M')^{(m-1)\beta} \left(t + \frac{1}{\lambda}\right)^\beta \tag{5.4}$$

with $C = C(m, n)$. Now we can proceed.

LEMMA 5.3. *The limit U has mass M for all $t > 0$.*

This is a consequence of the Dominated Convergence Theorem since U is bounded above by a big source-type.

LEMMA 5.4. *Under the present assumptions on u_0 we have $U(x,t) \to M\delta(x)$ as $t \to 0$, i.e.,*

$$\lim_{t \to 0} \int_{\mathbb{R}^n} U(x,t)\varphi(x)\,dx = M\varphi(0) \tag{5.5}$$

for all test functions $\varphi \in C_c^\infty(\mathbb{R}^n)$.

Proof. Since $M = \int U(x, t)\, dx$ we have for $t > 0$

$$|\int U(x, t)\varphi(x) - M\varphi(0)\, dx| \leq \int |U(x, t)||\varphi(x) - \varphi(0)|dx$$

$$\leq \int_{|x|\leq\delta} |U(x, t)||\varphi(x) - \varphi(0)|dx + \int_{|x|>\delta} |U(x, t)||\varphi(x) - \varphi(0)|dx \leq (*)$$

By continuity there exists $\delta > 0$ such that $|\varphi(x) - \varphi(0)| \leq \varepsilon/2M$ if $|x| \leq \delta$. Besides, φ is bounded so that

$$|\varphi(x) - \varphi(0)| \leq 2C \quad (\varphi \in C_c^\infty).$$

Since U vanishes for $|x| \geq \delta$ if t is small enough we get

$$(*) \leq M\frac{\varepsilon}{2M} + 2C \int_{|x|>\delta} |U(x, t)| \leq K\varepsilon.$$

$\square$

Conclusion. Using the uniqueness result, Theorem 5.1, we can identify U. Hence, for $t = 1$ we have $\widetilde{u}_{\lambda_n}(x, 1) \to \mathcal{U}_M(x, 1)$ in $L^1_{loc}(\mathbb{R}^n)$. Now, the $\widetilde{u}_\lambda$ have compact support which is uniformly bounded in λ. It follows that

$$\widetilde{u}_{\lambda_n}(x, 1) \to \mathcal{U}_M(x, 1) \qquad \text{in } L^1\text{-strong.}$$

(We pass from local to global convergence). The limit is thus independent of the sequence λ_n. It follows that the whole family $\{\widetilde{u}_\lambda\}$ converges to $\mathcal{U}_M$ as $\lambda \to \infty$.

5. Rephrasing the result.

The argument has concluded, but we still have to write the conclusion in the original variables and scales. Let $F_M(x) = \mathcal{U}_M(x, 1)$. We have just proved that

$$\lim_{\lambda\to\infty} \|\lambda^\alpha u(\lambda^\beta x, \lambda) - F_M(x)\|_{L^1} = 0,$$

which means with $y = \lambda^\beta x$ that

$$\lim_{\lambda\to\infty} \int \lambda^\alpha |u(y, \lambda) - \lambda^{-\alpha} F_M(y/\lambda^\beta)|\lambda^{-\beta n}\, dy = 0.$$

Noting that $\mathcal{U}_M(y, \lambda) = \lambda^{-\alpha} F_M(y/\lambda^\beta)$ and that $\alpha = \beta n$, we arrive at

$$\lim_{\lambda\to\infty} \int |u(y, \lambda) - \mathcal{U}_M(y, \lambda)|\, dy = 0,$$

i.e., replacing λ by t

$$\lim_{t\to\infty} \|u(y, t) - \mathcal{U}_M(y, t)\|_{L^1_y} = 0.$$

This is the asymptotic formula (1.5). It has been proved for the class of bounded and compactly supported initial data.

6. General initial data

We now extend the result from compactly supported initial data to the whole class of data u_0 satisfying (0.3) by a general density argument. Given $\varepsilon > 0$ we construct an approximation $\tilde{u}_0$ which is bounded and compactly supported and such that

$$\|u_0 - \tilde{u}_0\|_1 \le \varepsilon, \qquad \int_{\mathbb{R}^n} \tilde{u}_0(x)\,dx = \tilde{M}.$$

To prove formula (1.5) for $u(x, t)$ we only have to use the triangle formula plus the contraction property (2.3):

$$\|u(t) - \mathcal{U}_M(t)\|_1 \le \|u(t) - \tilde{u}(t)\|_1 + \|\tilde{u}(t) - \mathcal{U}_{\tilde{M}}(t)\|_1 + \|\mathcal{U}_{\tilde{M}}(t) - \mathcal{U}_M(t)\|_1.$$

Now, $|M - \tilde{M}| \le \varepsilon$, hence $\|\mathcal{U}_{\tilde{M}}(t) - \mathcal{U}_M(t)\|_1 \le \varepsilon$. By the contraction principle,

$$\|u(t) - \tilde{u}(t)\|_1 \le \|u(0) - \tilde{u}(0)\|_1 \le \varepsilon.$$

Thus, we get $\|u(t) - \mathcal{U}_M(t)\|_1 \le \varepsilon + \delta(t) + \varepsilon = 2\varepsilon + \delta(t)$, where $\delta(t) \to 0$ as $t \to \infty$ according to the result proved for the special solutions. As $t \to \infty$ we get

$$\lim_{t \to \infty} \|u(t) - \mathcal{U}_M(t)\|_1 \le 2\varepsilon.$$

Since $\varepsilon > 0$ is arbitrary this completes the proof of the L^1 estimate in Theorem 1.1.

Comment on the method. The method we have used so far in the proof of Theorem 1.1 can be applied to different equations and systems as long as they possess good scaling properties that are relevant for the asymptotics, and as long as the identification step has some nice characteristic which enables us to determine the solution obtained as limit. We note that *no essential use is made of the Maximum Principle*, which is replaced as a main argument by compactness. This makes the method in principle well suited for systems and higher-order equations.

Alternative. In fact, we have used the maximum principle and the property of finite propagation in Step 4 in the case of compact support. This can be avoided with no much effort. We give next a direct proof of Step 4 suggested by the referee: the main point of the argument to show that the initial data are taken is to make sure that there is no mass escaping to infinity. Such a property is ensured by the following result.

LEMMA 6.1. *Let u be a solution with data $u_0 \in \mathcal{X}_0$. For $T, R > 0$ we have*

$$\limsup_{R \to \infty} \sup_{\substack{\lambda \ge 1 \\ t \in (0,T)}} \int_{\{|x| \ge R\}} \tilde{u}_\lambda(x, t)\,dx = \lim_{t \to 0, \lambda \to \infty} \int_{\{|x| \ge R\}} \tilde{u}_\lambda(x, t)\,dx = 0. \qquad (6.1)$$

Proof. Take a C^∞ cutoff function $\varrho = \varrho(|x|)$ such that $0 \le \varrho \le 1$, $\varrho(x) = 0$ for $|x| \le 1$, $\varrho(x) = 1$ for $|x| \ge 2$. Let $\varrho_R(x) = \varrho(x/R)$. Multiplying the equation by ϱ_R and integrating by parts we get

$$\frac{d}{dt} \int \varrho_R \widetilde{u}_\lambda(t)\, dx \le \frac{1}{R^2} \|\Delta \varrho\|_\infty \|\widetilde{u}_\lambda\|_\infty^{m-1} M \le \frac{C(m, n, \varrho, M)}{R^2}\, t^{-\alpha(m-1)},$$

by Property 2 of Section 2. We have $\alpha(m-1) < 1$, hence integration in time from 0 to $t \le T$ gives

$$\int \varrho_R \widetilde{u}_\lambda(t)\, dx \le \int \varrho_R \widetilde{u}_\lambda(0)\, dx + \frac{C(m, n, \varrho, M)}{R^2}\, T^{2\beta}.$$

Moreover, for $\lambda \ge 1$, $\int \varrho_R(x)\widetilde{u}_\lambda(x, 0)\, dx = \int \varrho(x/R\lambda^\beta)u_0(x)\, dx \le \int_{|x| \ge R\lambda^\beta} u_0(x)\, dx$, which goes uniformly to zero as $R\lambda^\beta \to \infty$. The two results follow. $\qquad\square$

With this result it is easy to check that the limit U has mass M for all $t > 0$, and with it the identification of Step 4 follows.

7. Uniform convergence. Local regularizing effect

Once the basic L^1- convergence result is proved, we turn next to the question of improved convergence. We will show that the asymptotic convergence takes actually place in L^∞, i.e., uniformly. The improvement of convergence is a consequence of the smoothness of the flow, in more precise form, of the local regularity properties which apply to our problem.

We know that the family $\{\widetilde{u}_\lambda\}$ is bounded for all $t \ge t_0 > 0$. The local regularity theory [DB] says that

LEMMA 7.1. *Equi-bounded families of solutions of the PME are equi-continuous with a uniform Hölder modulus of continuity, i.e., they are bounded in some space $C^\varepsilon(Q)$, $\varepsilon > 0$.*

This is a local result. When we apply it to the family $\{\widetilde{u}_\lambda\}$ it says that it is a bounded family in $C^\varepsilon(\Omega)$ for every compact subdomain of Q with

$$\|\widetilde{u}_\lambda\|_{C^\varepsilon} \le C(\|\widetilde{u}_\lambda\|_\infty, \Omega) \tag{7.1}$$

(where $\|\cdot\|_{C^\varepsilon}$ is the Hölder norm of the space $C^\varepsilon(\Omega)$). The same result applies in a bounded space-time domain and then C depends on the distance from Ω to the boundary of the total domain. We recall the usual definition of the Hölder semi-norm in parabolic domains

$$[u]_\varepsilon = \sup_{(x,t),(x',t')\in\Omega} \frac{|u(x, t) - u(x', t')|}{\|x - x'\|^\varepsilon + |t - t'|^{\varepsilon/2}}$$

The space is also written as $C_{x,t}^{\varepsilon,\varepsilon/2}$. Returning to our family, boundedness in the Hölder space implies equi-continuity. By the Ascoli-Arzelà theorem this implies compactness in the uniform norm, but on compact domains. We thus have

$$\widetilde{u}_\lambda(x, 1) \to \mathcal{U}_M(x, 1) \tag{7.2}$$

uniformly on compact subsets of Q. We again divide the end of the proof in two cases, according to their difficulty.

I. If we consider as before the subclass of solutions with bounded and compactly supported data we know that for $t = 1$ the $\widetilde{u}_\lambda$ have uniformly bounded supports, hence the convergence (7.2) is uniform in $x \in \mathbb{R}^n$. Rephrasing the result as before, we get

$$t^\alpha \|u(t, y) - t^{-\alpha} F_M(t^{-\beta} y)\|_\infty \to 0, \ t \to \infty,$$

which is the asymptotic formula (1.6). The uniform convergence of Theorem 1.1 is proved in this case.

II. For general u_0 things are not so simple. Arguing as before we know that

(i)   at $t = 1$ the rescaled family $\widetilde{u}_\lambda(x, 1)$ converges in $L^1(\mathbb{R}^n)$ to $\mathcal{U}_M$,
(ii) this convergence takes also place in the uniform norm on any ball $B_R(0) \subset \mathbb{R}^n$,
(iii) the same happens for every $t \in (1/2, 2)$ uniformly in time.

As a consequence, we get the following picture. Take a very large radius $R_1 \gg 1$, in particular larger than the radius of the support of $\mathcal{U}_M(x, 1)$. In the time interval $1/2 < t < 1$ we have uniform convergence of $\widetilde{u}_\lambda$ towards $\mathcal{U}_M$ in the ball of radius R_1. Now we have to examine the outer region, $\mathcal{O}_1 = \{|x| \geq R_1\}$. We know that $u_\lambda \geq \mathcal{U}_M$ because $\mathcal{U}_M$ vanishes identically there. Moreover, the mass $\int_{\mathcal{O}_1} \widetilde{u}_\lambda(x, t) \, dx$ is less than ε in that outer domain: the reason is that $\widetilde{u}_\lambda$ and $\mathcal{U}_M$ have the same total mass and we have shown that they are almost identical for $|x| \leq R_1$. Clearly, $\varepsilon \to 0$ as $\lambda \to \infty$. Under these circumstances we want to prove that there is a function $C(\varepsilon)$ with $C(\varepsilon) \to 0$ as $\varepsilon \to 0$ such that

$$\widetilde{u}_\lambda(x, 1) \leq C(\varepsilon) \qquad \text{for } |x| \geq R_1, \tag{7.3}$$

and the proof of Theorem 1.1 will be complete. In other words, we want to translate small L^1-norms into small L^∞-norms. The technical tool to do that is the following result that has an interest in itself.

LEMMA 7.2. (Local regularizing effect) *Let f be any nonnegative, smooth, bounded and integrable function in $B_1 = B_R(a) \subset \mathbb{R}^n$ such that*

$$\Delta(f^p) \geq -K$$

for some $p > 0$ and $K > 0$. Let $B_2 = B_{R/2}(a)$. Then $f \in L^\infty(B_2)$ and $\|f\|_{L^\infty(B_2)}$ depends only on p, K, n, R and $\|f\|_{L^1(B_1)}$. If $\|f\|_1$ is very small compared with R it takes the form

$$\|f\|_{L^\infty(B_2)} \le C(p,n) \, \|f\|^\rho_{L^1(B_1)} \, K^\sigma, \tag{7.4}$$

with $\rho = 2/(2+pn)$ and $\sigma = n/(2+pn)$.

The exact condition for (7.4) to hold is $\|f\|_1^{\rho \, p/2} \ll R K^{\rho/2}$. In the application B_1 is any ball contained in $\{|x| \ge R_1\}$, and we take

$$f = u(t), \quad p = m - 1, \quad K = \frac{\alpha \, (m - 1)}{mt}$$

in the case where the solution u is positive everywhere, hence smooth. The general case is done by approximation. This completes the proof of Theorem 1.1.

Observe that the LRE implies the regularizing effect (2.5) when in the application we take $R = \infty$.

8. A proof of the local regularizing effect

This is a kind of technical appendix that the reader may wish to jump in a first reading.

(i) To begin with, we may use scaling to reduce the number of parameters. Thus, if f satisfies the assumptions in the ball B_1 of radius R and center a, with constant K and integral $\|f\|_{L^1(B_1)} \le M$, then

$$f_1(x) = A \, f(Rx + a), \quad A^p = \frac{1}{KR^2},$$

satisfies the same assumptions with $R = K = 1$ and $a = 0$. Therefore, only this case must be considered, bearing in mind the transformation and the fact that

$$\|f_1\|_{L^1(B_1(0))} \le M_1 = MK^{-1/p} R^{-(np+2)/p}.$$

(ii) Under those assumptions, let $g(x) = f^p(x)$. Then $\Delta g \ge -1$. Therefore, for every $x_0 \in \mathbb{R}^n$ the function

$$G(x) = g(x) + \frac{1}{2n}|x - x_0|^2$$

is subharmonic in $B_1 = B_1(0)$. It follows that

$$G(x_0) \le \fint_B G(x) \, dx,$$

where B is any ball $B_r(x_0)$, $r > 0$ contained in B_1, and $\fint_B$ denotes average on B. The argument will continue in a different way for $p > 1$ and for $0 < p \le 1$.

(iii) If $p \leq 1$, we can use the last formula to estimate g at an arbitrary point $x_0 \in B_{1/2}(0)$ as follows. If $B = B_r(x_0)$ and $0 < r \leq 1/2$ then,

$$g(x_0) \leq \fint_B g(x)dx + \frac{1}{2n} \fint_B |x - x_0|^2 dx \leq \left(\fint_B g^{1/p} \, dx\right)^p + \frac{r^2}{2(n+2)},$$

$$\leq \|f\|_{L^1(B_1)}^p \left(\frac{1}{\omega_n r^n}\right)^p + \frac{r^2}{2(n+2)}.$$

where ω_n denotes the volume of the unit ball. Minimization of the last expression with respect to $0 < r < 1/2$ gives the bound. In particular, when $\|f\|_1$ is small enough the minimum happens at an interior point of the r-interval and then

$$g(x_0) \leq C \, \|f\|_1^{\frac{2p}{pn+2}},$$

Recall that this holds under the simplifying hypothesis, so that for $K, R \neq 1$ we have to undo the transformation. We ask the reader to check that it is really equivalent to (7.4).

iv) For $p > 1$ we proceed in two steps. In the first one we use the following result which can be easily established by repeated (but finite) iteration of the Moser integration technique.

LEMMA 8.1. *Let $G \geq 0$ be a locally integrable function in the ball $B = B_R(0)$ such that $\Delta G \geq 0$ and $G^{1/p} \in L^1(B)$ for some $p > 1$. Then*

$$\fint_{B_{R/2}} G \, dx \leq C \left(\fint_B G^{1/p} \, dx\right)^p, \tag{8.1}$$

where $C > 0$ depends on p and n.

Isf we prove it for $R = 1$ then it is easily translated to a ball of radius $R \neq 1$ by scaling. From this it follows that $g = f^p \in L^1(B_r)$ for all $r < 1/2$, and in fact

$$\fint_{B_r} f^p \, dx \leq C \left(\fint_{B_{2r}} f \, dx\right)^p + Cr^2 \tag{8.2}$$

with another $C = C(p, n)$. Then, for every $x_0 \in B_{1/2}(0)$ and $B = B_r(x_0)$ and $0 < r \leq 1/4$ we have

$$g(x_0) \leq \fint_B f^p(x) \, dx + \frac{1}{2n} \fint_B |x - x_0|^2 dx \leq Cr^{-np}\|f\|_1^p + Cr^2.$$

The end is as before.

A related version of the local regularizing effect is due to [BCP]. The idea of the present proof was announced in [V4], Chapter 3.

9. Convergence of supports and interfaces

We assume in this section that u_0 is compactly supported and describe the asymptotic shape and size of the support as $t \to \infty$. We may assume without loss of generality that u_0 is continuous and nontrivial and that 0 belongs to the positivity set of u_0. We introduce the minimal and maximal radius,

$$\begin{cases} r(t) = \sup\{r > 0 : \ u(x, t) > 0 \text{ in } B_r(0)\}, \\ R(t) = \inf\{r > 0 : \ \text{supp}(u(x, t)) \subset B_r(0)\}. \end{cases} \tag{9.1}$$

Since the source-type solution $\mathcal{U}_M(x, t)$ is given by formula (1.2)–(1.3), its support is the ball of radius

$$\mathcal{R}(t) = \xi_0(m, n)(M^{m-1} t)^\beta = C_0 t^\beta. \tag{9.2}$$

THEOREM 9.1. *As $t \to \infty$ we have*

$$\lim_{t \to \infty} \frac{r(t)}{\mathcal{R}(t)} = \lim_{t \to \infty} \frac{R(t)}{\mathcal{R}(t)} = 1. \tag{9.3}$$

Proof. The fact that the limits in (9.3) are equal or larger than 1 is a direct consequence of the uniform convergence of Theorem 1.1. On the contrary, the fact that for large t

$$R(t) \le (1 + \varepsilon)\,\mathcal{R}(t)$$

needs a proof. Of course, we know that a large Barenblatt solution with some delay is a super-solution, hence there is a constant $C > 1$ such that for all large t

$$R(t) \le C\mathcal{R}(t).$$

On the other hand, we know that the mass contained in exterior sets of the form

$$\Omega_\varepsilon = \{|x| > (1 + \varepsilon)\mathcal{R}(t)\}$$

is less than ε for all large t. By Lemma 7.2 there is uniform convergence to 0 in this region as $t \to \infty$, hence if the support is larger than the support of $\mathcal{U}$ the excess region takes the form of a *thin tail*.

We will show that the possible tail must disappear as time grows by means of a comparison with slow traveling waves. This is done as follows: if we define the ratio $s(t) = R(t)/t^\beta$, we must prove that

$$\lim \sup_{t \to \infty} s(t) = C_0.$$

Assume by contradiction that this limit is $C > C_0$ and take a very large time t_1 for which the ratio $s(t_1) \ge C - \varepsilon$, with ε very small. By scaling (4.7) we can reduce that time to

$t_1 = 1$. Since the ratio has limsup C we have $R(1/2) \leq (C + \varepsilon)/2^{\beta} = d < C$. On the other hand, by the uniform convergence $u \to \mathcal{U}$, we may also assume that $u \leq \varepsilon$ for $|x| \geq \mathcal{R}(1) + \varepsilon = C_0 + \varepsilon$, and $t \in [1/2, 1]$. Let $d_1 = \max\{d, C_0\}$, which we may take such that $d_1 < C - 4\varepsilon$ for ε small. Now, we compare u with the explicit traveling wave solution $\widehat{u}$ with small speed ε defined as

$$\widehat{u}^{m-1} = \frac{m-1}{m} \left(\varepsilon(t - 1/2) + \varepsilon + d_1 - x_1\right)_+$$

where x_1 is the first coordinate of x. Comparison takes place in the region: $\{t \in [1/2, 1]$, $x_1 \geq d_1\}$. By inspecting the parabolic boundary, we easily show that $u \leq \widehat{u}$ there. Since $\widehat{u}$ vanishes for $x_1 \geq d_1 + \varepsilon + \varepsilon(t - 1/2)$ we conclude that u vanishes at $t = 1$ for $x_1 \geq d + 2\varepsilon$. We may rotate the axes in the previous argument, hence we conclude that $u(x, 1) = 0$ for $|x| \geq d_1 + 2\varepsilon$ and this is a contradiction with $R(1) \geq C - \varepsilon$. The tail is eliminated. $\qquad\square$

Theorem 9.1 is a manifestation of the property of **asymptotic symmetrization**, which will be discussed in greater detail in Sections 15 and following.

10. Continuous rescaling and stationary solutions

A different way of implementing the scaling of the orbits of problem (CP) and proving the previous facts consists of using the *continuous rescaling*

$$\theta(\eta, \tau) = t^{\alpha} u(x, t), \quad \eta = x\, t^{-\beta}, \quad \tau = \log(t), \tag{10.1}$$

with α and β the standard similarity exponents given by (1.3). The new orbit $\theta(\tau)$ satisfies the equation

$$\boxed{\theta_{\tau} = \Delta(\theta^m) + \beta\eta \cdot \nabla\theta + \alpha\theta} \tag{10.2}$$

It is bounded uniformly in $L^1(\mathbb{R}^n) \cap L^{\infty}(\mathbb{R}^n)$. The source-type solutions transform into the stationary profiles F_M in this transformation, i.e., $F(\eta)$ solves the nonlinear elliptic problem

$$\Delta f^m + \beta\eta \cdot \nabla f + \alpha f = 0. \tag{10.3}$$

The boundedness and compactness arguments of Section 4.3 apply and we may pass to the limit and form the ω-limit, which is the set

$$\omega(\theta) = \{f \in L^1(\Omega) : \exists \tau_j \to \infty \text{ such that } \theta(\tau_j) \to f\}. \tag{10.4}$$

The convergence takes place in the topology of the functional space in question, here any $L^p(\Omega)$, $1 \leq p < \infty$ (strong). We have written a quite detailed account of this process in the Notes on the Dirichlet problem, [V5].

The rest of the proof consists in showing that the ω-limit is just the Barenblatt profile F_M. The argument can be translated in the following way. Corresponding to the sequence of scaling factors λ_n of Section 4 we take a sequence of **delays** s_n and define

$$\theta_n(\eta, \tau) = \theta(\eta, \tau + s_n). \tag{10.5}$$

The family $\{\theta_n\}$ is precompact in $L^\infty_{loc}(0, \infty : L^1(\mathbb{R}^n))$ hence, passing to a subsequence if necessary we have

$$\theta_n(\eta, \tau) \to \widehat{\theta}(\eta, \tau). \tag{10.6}$$

Again it is easy to see that $\widehat{\theta}$ is a weak solution of (10.2) satisfying the same estimates. The end of the proof is identifying it as a stationary solution, which was done in the previous proof by the other scaling method (*discrete rescaling*).

Theorem 1.1 can now be used to characterize the stationary solutions.

THEOREM 10.1. *The profiles F_M can be characterized as the unique solutions of equation (10.3) such that $f \in L^1(\mathbb{R}^n)$, $f^m \in L^1_{loc}(\mathbb{R}^n)$ and $f \geq 0$. The conditions $f^m \in H^1(\mathbb{R}^n)$, $f \in C(\mathbb{R}^n)$ are true, but not needed in the proof.*

Proof. Any other solution f can be taken as initial data for the evolution equation (10.2) and then Theorem 1.1 proves that the corresponding solution of (10.2) converges to the source-type solution with the same mass, F_M. Now, the solution $u(x, t) = t^{-\alpha} f(x t^{-\beta})$ is an admissible solution of the PME which converges in the rescaling to f. Therefore, $f = F_M$. $\qquad\qquad\square$

11. Optimality of the convergence rates in $\mathcal{X}_0$

We devote this section to a first exploration of the sharpness of the convergence rates, proving Theorem 1.3. A more precise formulation is as follows:

Counterexample. *Given any decreasing function $\rho(t) \to 0$, there exists a solution of the Cauchy Problem with integrable and nonnegative initial data of mass $M > 0$ such that*

$$\limsup_{t\to\infty} \frac{(u(0, t) - \mathcal{U}(0, t; M)) t^\alpha}{\rho(t)} = \infty. \tag{11.1}$$

Moreover, we can also get

$$\limsup_{t\to\infty} \frac{\|u(t) - \mathcal{U}(t; M)\|_1}{\rho(t)} = \infty. \tag{11.2}$$

We can also ask the solution to be radially symmetric with respect to the space variable.
Construction. (i) We recall that the proof need only be done for $M = 1$ since the scaling transformation

$$\widehat{u}_c(x, t) = c^{m-1} u(x, c t) \tag{11.3}$$

reduces a solution of mass $M > 0$ to a solution of mass 1 if $c = M^{-1/(m-1)}$. We take an initial function of the form

$$u_0(x) = \sum_{k=1}^{\infty} c_k \, \chi_k(x - a_k),$$

where $\chi_k(x)$ is the characteristic function of the ball of radius r_k centered at 0. The sequences a_k, c_k and r_k have to be determined in a suitable way. In the first place, we impose the conditions $c_k, r_k \geq 0$ and $c_k r_k^n = 2^{-k}/\omega$ (where ω is the volume of the ball of radius 1). Then, $M = \omega \sum_{1}^{\infty} c_k r_k^n = 1$.

(ii) We construct solutions u_k with initial data of the form

$$u_k(x, 0) = \sum_{1}^{k} c_i \, \chi_i(x - a_i), \tag{11.4}$$

and we proceed to choose c_k and a_k in an iterative way. In any case the mass of u_k is $M_k = 1 - 2^{-k}$, and we observe that (by the main convergence result) for every $\varepsilon > 0$ there must be a time $t_k(\varepsilon)$ (which depends also on the precise choice of the initial data) such that

$$t^{\alpha} |u_k(0, t) - \mathcal{U}(0, t; M_k)| \leq \varepsilon$$

for all $t \geq t_k(\varepsilon)$. We now recall that $\mathcal{U}(0, t; M) = c M^{2\beta} t^{-\alpha}$, so that the difference between $t^{\alpha}\mathcal{U}(0, t; M)$ and $t^{\alpha}\mathcal{U}(0, t; M')$ is constant in time, and in fact it can be estimated as larger than

$$t^{\alpha} (\mathcal{U}(0, t; M) - \mathcal{U}(0, t; M')) \geq k_1 (M - M')$$

with the same constant $k_1 > 0$ for all $1 \geq M > M' \geq 1/2$.

(iii) The iterative construction of the u_k starts as follows. We may take c_1 as we like, e.g., $c_1 = 1$, then $r_1 = (2\omega)^{-1/n}$, and find the solution $u_1(x, t)$ with data $u_1(x, 0) = c_1 \chi_1(x)$. Its mass is $M_1 = 1/2$ for all times. As said above, for sufficiently large times we have

$$t^{\alpha} |u_1(0, t) - \mathcal{U}(0, t; M_1)| \leq \varepsilon,$$

We can also find t_1 such that $\rho(t_1) < (1/2)k_1 (M - M_1) = k_1/4$. Using the estimate for the difference of source-type solutions and the triangular inequality, and taking ε small enough($\varepsilon \leq k_1/4$), we get for all $t \geq t_1$

$$t^{\alpha} |u_1(0, t) - \mathcal{U}(0, t; 1)| \geq t^{\alpha} |\mathcal{U}(0, t; 1)$$
$$-\mathcal{U}(0, t; M_1)| - t^{\alpha} |u_1(0, t) - \mathcal{U}(0, t; M_1)| \tag{11.5}$$
$$\geq k_1(1 - M_1) - \varepsilon \geq k_1/4 \geq \rho(t_1) \geq \rho(t). \tag{11.6}$$

(iv) Iteration step. Assuming that we have constructed $u_2, \ldots, u_{k-1}$ by solving the equation with data (11.4), we proceed to choose c_k, and a_k and construct u_k as follows. We can take any $c_k > 0$, and then find a_k large enough so that the support of the solution v_k with initial data $v_k(x, 0) = \chi_k(x - a_k)$ does not intersect the support of u_{k-1} until a time $t_k > 2 t_{k-1}$ (and we can even estimate how far a_k must be located for large t_k because we have a precise control of the support of u_{k-1} for large times, thanks to Theorem 9.1). Then, it is immediate to see that

$$u_k(x, t) = u_{k-1}(x, t) + v_k(x, t)$$

for all $x \in \mathbb{R}^n$ and $0 \le t \le t_k$ (i.e., superposition holds as long as the supports are disjoint). Indeed, this means that for all $0 \le t \le t_{k-1}$ we also have $u_k(0, t) = u_{k-2}(0, t)$, and by iteration we conclude that

$$u_k(0, t) = u_j(0, t) \quad \text{for all } 1 \le j < k \text{ and } 0 \le t \le t_{j+1}.$$

We now remark that t_k can be delayed as much as we like (on the condition of taking a_k far away). If we choose t_k large enough, the main asymptotic theorem implies the behaviour

$$t_k^\alpha u_k(0, t_k) = t_k^\alpha u_{k-1}(0, t_k) \sim t_k^\alpha \mathcal{U}(0, t_k; M_{k-1})$$

We want the error to be less than $k_1(1 - M_k)/2 = 2^{-(k+1)}k_1$. We also suggest to wait until $\rho(t_k) \le 2^{-(2k+1)} k_1$. Using again the triangle inequality: $|u_k - \mathcal{U}(M)| \ge |\mathcal{U}(M) - \mathcal{U}(M_{k-1})| - |u_k - \mathcal{U}(M_{k-1})|$ with $M = 1$, we get

$$t_k^\alpha |u_k(0, t_k) - \mathcal{U}(0, t_k; 1)| \ge 2^{-(k+1)}k_1 \ge 2^k \rho(t_k).$$

(v) In the final step we take the limit

$$u(x, t) = \lim_{k \to \infty} u_k(x, t).$$

By what was said before we may conclude that for $t \le t_k$ we have $u(0, t) = u_k(0, t)$, so that

$$\lim_{n \to \infty} t_k^\alpha \frac{u(0, t_k) - \mathcal{U}(0, t_k; 1)}{\rho(t_k)} = \infty.$$

This concludes the proof of the L^∞-estimate.

(vi) The construction can be easily modified so that the data u_0 are radially symmetric by defining χ_k to be the characteristic function of the annulus $A_k = \{x : a_k \le |x| \le a_k + r_k\}$ and imposing that c_k times the volume of A_k to equal 2^{-k}. The construction is repeated with the same attention to be given to a_k, i.e., to the far location of the A_k.

(vii) For the L^1 part we just observe that, taking t_k large enough we have at time $t = t_k$ and in a very large ball B_k (as large as we please by the iteration construction) the equality $u = u_k$ and the approximation

$$\|u_k - \mathcal{U}(x, t_k; 1)\| \ge \|u_k - \mathcal{U}(x, t_k; 1)\|_{L^1(\{x \notin B_k\})} \ge 2^{-k},$$

since the mass of u contained outside this ball is known (2^{-k}), and that of $\mathcal{U}$ is zero there. The result follows if the t_k have been chosen as before, $\rho(t_k)2^k \to 0$.

12. Proof of Theorem 1.1 by the Lyapunov method

We devote the next sections to derive alternative proofs of the main convergence result, Theorem 1.1. Two of them are based on standard implementations of the idea of *Lyapunov functional*. The third one introduces a non-standard version of this idea.

12.1. Lyapunov functional.

Given an orbit $\{u(t)\}$ with mass $M > 0$ we introduce the functional

$$J_u(t) = \int_{\mathbb{R}^n} |u(x, t) - \mathcal{U}_M(x, t)| \, dx. \tag{12.1}$$

It is clear from the Contraction Property that J is nonincreasing in t. We get the following result.

LEMMA 12.1. *There exists the limit* $J_\infty = \lim_{t \to \infty} J(t) \geq 0$.

Note that $J(t)$ becomes zero only if $u(t)$ coincides with the source-type solution for some $t_1 > 0$ and then the equality holds for all $t \geq t_1$ and the asymptotic result is trivial. Otherwise $J(t) > 0$ for all $t > 0$. We have to examine this case.

12.2. Limit solutions.

We perform Steps 1, 2 and 3 of the preceding proof to obtain a sequence $\lambda_k \to \infty$ such that

$$\widetilde{u}_{\lambda_k}(x, t) \to U(x, t) \tag{12.2}$$

in $L^1(\mathbb{R}^n \times (t_1, t_2))$. The limit U is again a solution of the PME. It is nontrivial and has mass M (this is easy for compactly supported solutions and then true for the rest by approximation, as we saw).

12.3. Invariance Principle.

One of the key features of the use of Lyapunov functionals is the following Asymptotic Invariance Property.

LEMMA 12.2. *The Lyapunov functional is constant on limit orbits, i.e.,* J_U *does not depend on* t.

Proof. The Lyapunov functional is translated to the rescaled family $\widetilde{u}_\lambda$ by the formula

$$J_{\widetilde{u}_\lambda}(t) = \int_{\mathbb{R}^n} |\widetilde{u}_\lambda(x, t) - \mathcal{U}_M(x, t)| \, dx = J_u(\lambda t). \tag{12.3}$$

It follows that for fixed $t > 0$ we have

$$\lim_{\lambda \to \infty} J_{\widetilde{u}_\lambda}(t) = \lim_{\lambda \to \infty} J_u(\lambda t) = J_\infty.$$

On the other hand, we see that J depends in a lower-semicontinuous form on u. Moreover, it is continuous under the passage to the limit that we have performed. That means that for every $t > 0$ we have $J_U(t) = J_\infty$. $\qquad\square$

12.4. A limit solution is a source-type.

In order to identify U, the next result we need is the following.

LEMMA 12.3. *Let be an orbit $u(t)$ with mass $M > 0$ and with connected support for $t \geq t_0$. Then the function $J(t)$ is strictly decreasing in any time interval (t_1, t_2), $t_0 < t_1 < t_2$, unless $u = \mathcal{U}_M$ or both solutions have disjoint supports in that interval.*

Proof. We consider for $t \geq t_1 > 0$ the solution w of the PME with initial data at $t = t_1$

$$w(x, t_1) = \max\{u(t_1), v(t_1)\}, \tag{12.4}$$

where we put $v = \mathcal{U}_M$ for easier notation. Clearly, $w \geq u$ and $w \geq v$, hence

$$w(t) \geq \max\{u(t), v(t)\}, \quad t > t_1.$$

Moreover, we have $w(x, t_1) - u(x, t_1) = (v(x, t_1) - u(x, t_1))_+$ and $w(x, t_1) - v(x, t_1) = (u(x, t_1) - v(x, t_1))_+$ so that

$$J_u(t_1) = \int_{\mathbb{R}^n} (w(t_1) - u(t_1))\, dx + \int_{\mathbb{R}^n} (w(t_1) - v(t_1))\, dx,$$

while for general $t > t_1$

$$J_u(t) + 2 \int_{\mathbb{R}^n} (w(t) - \max\{u(t), v(t)\})\, dx = \int_{\mathbb{R}^n} (w(t) - u(t))\, dx$$
$$+ \int_{\mathbb{R}^n} (w(t) - v(t))\, dx.$$

Both integrals on the right are nonincreasing in time by the contraction principle, hence constancy of J_u in an interval $[t_1, t_2]$ implies that

$$w(t_2) = \max\{u(t_2), v(t_2)\}. \tag{12.5}$$

In order to examine the consequences of this equality we use the Strong Maximum Principle. $\qquad\square$

LEMMA 12.4. *Two ordered solutions of the PME cannot touch for $t > 0$ wherever they are positive.*

This is a standard result for classical solutions of quasilinear parabolic equations, cf. [LSU]. It follows that (12.5) is then possible on any connected open set Ω where $w(\cdot, t_2) > 0$ under three circumstances:

(i) $w(t_2) = u(t_2) > v(t_2)$, or

(ii) $w(t_2) = v(t_2) > u(t_2)$, or

(iii) $w(t_2) = u(t_2) = v(t_2)$.

Since the support of the source-type solution is a ball and the support of u is also connected, we conclude the result of Lemma 12.3.

Note. If M is not the mass of u there is still another possibility for constant J, namely that the solutions are different but ordered: either $u(t) \geq \mathcal{U}_M(t)$ or $u(t) \leq \mathcal{U}_M(t)$.

We may now conclude the proof of Theorem 3.1 by this method in the case where u_0 has compact support, so that by standard properties of the propagation of support, it is connected after a certain time t_0. Since the source-type solution penetrates into the whole space eventually in time and U has a non-contracting support, it follows that for large t the supports of U and $\mathcal{U}_M$ do intersect. Since both solutions cannot be ordered because they have the same mass, $J_U(t)$ must be zero since it is not strictly decreasing by Lemma 12.2. We have thus proved that $J_\infty = 0$ and

$$U = \mathcal{U}_M, \tag{12.6}$$

which identifies all possible limits of rescalings as the unique source-type solution with the same mass. This ends the proof (see Section 5). The extension to general data is done by density as before.

12.5. Continuous rescaling.

One way of proving the previous facts is by using the continuous rescaling, formula (10.1). As explained in Section 10, taking a sequence of delays s_n we define

$$\theta_n(\eta, \tau) = \theta(\eta, \tau + s_n),$$

and passing to the limit

$$\theta_n(\eta, \tau) \to \widehat{\theta}(\eta, \tau). \tag{12.7}$$

Again it is easy to see that $\widehat{\theta}$ is a weak solution of (10.2) satisfying the same estimates. For θ the Lyapunov functional is translated into

$$J_\theta(t) = \int_{\mathbb{R}^n} |\theta(\eta, \tau) - F_M(\eta)| \, d\eta, \tag{12.8}$$

and we see that it is continuous under the passage to the limit we have performed. Let us examine now the situation when $J_\infty > 0$. Then $\widehat{\theta} \neq F_M$ and the orbit of $\widehat{\theta}$ has a strictly decreasing functional, so that for $\tau_2 = \tau_1 + h$ we have

$$J_{\widehat{\theta}}(\tau_1) - J_{\widehat{\theta}}(\tau_2) = c > 0.$$

Since $\widehat{\theta}$ is the limit of the θ_n we get for all large enough n

$$J_{\theta_n}(\tau_1) - J_{\theta_n}(\tau_1 + h) \geq c/2.$$

But this means that for all n large enough

$$J_\theta(\tau_1 + s_n) - J_\theta(\tau_1 + s_n + h) \geq c/2.$$

This contradicts the fact that J_θ has a limit. The proof is complete.

Comment. As we had announced, the proof of this section uses several steps of the former with a completely different end. It contains some fine regularity results that can make it difficult to apply in more general settings. However, some of these difficulties can be overcome by other means. Lasalle's Invariance Principle is a powerful tool in Dynamical Systems [Ls], worth knowing also in this context.

13. Another Lyapunov approach

A different Lyapunov approach was proposed in 1984 by Newman and developed in [N, R]. We can write the functional in continuously rescaled variables (cf. Section 10) as

$$\mathcal{J}_\theta(\tau) = \int_{\mathbb{R}^n} \{\theta(\eta, \tau)^m + \kappa \, |\eta|^2 \theta(x, \tau)\} \, d\eta, \qquad \kappa = \frac{1}{2}\beta(m - 1), \tag{13.1}$$

where β is the similarity exponent. The proof of convergence in this instance will be based on the possibility of calculating the value of $d\mathcal{J}/d\tau$ along an orbit.

LEMMA 13.1. *Let $\mathcal{J}$ be the Newman functional given by* (13.1). *Then for every rescaled orbit of problem (CP) we have*

$$\frac{d\mathcal{J}}{d\tau} = -\frac{m^2}{m-1} \int \theta \, \{\nabla(\theta^{m-1} + k\,\eta^2)\}^2 \, d\eta. \tag{13.2}$$

Proof. In order to analyze the evolution of $\mathcal{J}$ let us put for a moment

$$J(\theta) = \int_{\mathbb{R}^n} \{\theta(\eta, \tau)^m + \lambda \, |\eta|^2 \theta(\eta, \tau)\} \, d\eta,$$

with $\lambda > 0$. Let us perform the following formal computations:

$$\begin{aligned}
dJ/d\tau &= \int (m\theta^{m-1} + \lambda\,\eta^2)\,\theta_\tau \, d\eta \\
&= \int (m\theta^{m-1} + \lambda\,\eta^2)(\Delta\theta^m + \beta\nabla \cdot (\eta\,\theta)) \, d\eta \\
&= -\int \nabla(m\theta^{m-1} + \lambda\,\eta^2)\,(\nabla\theta^m + \beta\eta\theta) \, d\eta \\
&= -\int \theta \, \nabla(m\theta^{m-1} + \lambda\,\eta^2)\,\nabla\left(\frac{m}{m-1}\theta^{m-1} + \frac{\beta}{2}\eta^2\right) \, d\eta.
\end{aligned}$$

In case $\lambda = \beta(m-1)/2$ we can write this quantity as (13.2), which proves that J is a Lyapunov functional, i.e., it is monotone along orbits.

These computations are easily justified for classical solutions which decay quickly at infinity. The result for general solutions is then justified by a density argument using the regularity of the solutions of the PME. cf. [V4] (but we can also restrict the Lyapunov analysis to the above mentioned class of solutions since the proof of convergence for general solutions is then completed by a density argument as in Section 6). $\square$

Limit orbits and invariance. As in the previous section we pass to the limit along sequences $\theta_n(\tau) = \theta(\tau + s_n)$ to obtain limit orbits $\widehat{\theta}(\tau)$, on which the Lyapunov functional is constant, hence $d\mathcal{J}_{\widehat{\theta}}/d\tau = 0$.

Identification step. The proof of asymptotic convergence concludes in the present instance in a new way, by analyzing when $dJ/d\tau$ is zero. Here is the crucial observation that ends the proof: *the second member of* (13.2) *vanishes if and only if θ is a Barenblatt profile.*

The rate of convergence can be calculated by computing $d^2 J/dt^2$, which is not easy. We will continue with the subject in Section 19.

14. Proof of Theorem 3.1 using optimal barriers

14.1. We present still another kind of proof, based on the ideas of the paper of Friedman and Kamin [FK]. Given a solution $u(t)$ with mass $M > 0$ we consider for fixed $t > 0$ the set of source-type solutions which lie below $u(\cdot, t)$ and define a functional

$$\mathcal{M}(t; u) = \sup\{M' \geq 0 : \mathcal{U}_{M'}(x, t) \leq u(x, t)\} \tag{14.1}$$

Thus, $\mathcal{M}(t; u)$ is the mass of a certain *optimal barrier from below*, a source-type solution with mass $\mathcal{M}(t; u)$ which lies below u at time t. Indeed, the above definition is insufficient and we have to introduce a modification of the class of admissible barriers. For $\tau \in \mathbb{R}$ we define

$$\mathcal{M}(t; u, \tau) = \sup\{M' \geq 0 : \mathcal{U}_{M'}(x, t + \tau) \leq u(x, t)\} \tag{14.2}$$

(if $\tau < 0$ this definition applies for $t > -\tau$). The use of a delay τ is a tricky technicality involved in the argument of strict monotonicity which is essential in the proof.

It is clear from the Maximum Principle that

LEMMA 14.1. *For fixed u and τ the function $\mathcal{M}(t) = \mathcal{M}(t; u, \tau)$ is positive for every $t > 0$ and $\mathcal{M}(t) \leq M$. It is also nondecreasing in t so that there exists the limit*

$$\mathcal{M}_\infty(u) = \lim_{t \to \infty} \mathcal{M}(t; u, \tau), \quad 0 < \mathcal{M}_\infty(u) \leq M. \tag{14.3}$$

The monotonicity in time makes $\mathcal{M}(t)$ a kind of Lyapunov functional. Obviously, since $u(t)$ has mass M, the equality $\mathcal{M}(t) = M$ only happens if u is just the source-type solution $\mathcal{U}_M$ (with a delay). Otherwise, $0 < \mathcal{M}(t) < M$. It is also easy to prove that though $\mathcal{M}(t)$ depends on τ the limit $\mathcal{M}_\infty$ does not (hint: source-type solutions with same mass but some delay are very similar for large t).

14.2. Repeating the process of rescaling, compactness and passage to the limit of Section 4. on the orbit $u(t)$, solution of problem (CP), we have that along a sequence $\lambda_n \to \infty$

$$\widetilde{u}_{\lambda_n} \to U, \tag{14.4}$$

and U is a solution of equation (0.1). Note that the convergence is uniform on compact sets of space-time, which will imply at the end of the proof (Step 5 of the general plan) uniform convergence in sets of the form $|x| \le Ct^\beta$.

LEMMA 14.2. $\mathcal{M}(t; U, 0)$ *is constant and equals* $\mathcal{M}_\infty(u)$.

Proof. Let us put $\widehat{\mathcal{M}}(t) = \mathcal{M}(t; U, 0)$, the mass of the optimal lower barrier for U. Using the identity $\mathcal{M}(t; u, \tau) = \mathcal{M}(t/\lambda; \widetilde{u}_\lambda, \tau/\lambda)$, it is immediate from the passage to the limit in the rescalings and that $\widehat{\mathcal{M}}(t)$ is equal or larger than $\mathcal{M}_\infty(u)$ for all $t > 0$. Note that the delays scale like τ/λ, so that in passing to the limit the delay tends to 0.

Consider now for some $t > 0$ the barrier $\mathcal{U}_{\widehat{\mathcal{M}}(t)}(x, t)$ for $U(t)$. Since the convergence (14.4) is uniform on compact sets we conclude by approximation that for large λ_n there exists a lower barrier with mass $\widehat{\mathcal{M}}(t) - \varepsilon$. But the barrier for $\widetilde{u}_{\lambda_n}$ is approximately $\mathcal{M}_\infty(u)$. Hence, $\mathcal{M}_\infty(u) \ge \widehat{\mathcal{M}}(t)$. $\qquad\square$

14.3. We have to introduce the argument of strict monotonicity after which Lemma 14.2 cannot be true unless U is a source-type. We need first a lemma about the way source-type solutions evolve in time.

LEMMA 14.3. *Let* $R(t) = R(M)t^\beta$ *be the radius of the support of the source-type solution* $u = \mathcal{U}_M(x, t)$. *There exists a radius* R_* *smaller than* $R(M)$ *such that*

$$u_t < 0 \quad \text{if } |x| < R_* t^\beta, \qquad u_t > 0 \quad \text{if } R_* t^\beta < |x| < R t^\beta. \tag{14.5}$$

The proof of this result is a simple calculation with the explicit formula (1.2)–(1.3). We can now prove the technical result which was a main point in the [FK] paper.

LEMMA 14.4. *Let* $\mathcal{U}_{M_1}(x, t + \tau_1)$ *be a lower barrier for a solution* $u(x, t)$ *at a time* $t_1 > 0$. *Then either both functions coincide at* $t = t_1$ *or we can improve the barrier in the following way: given a sufficiently close delay* $\tau \le \tau_1$ *and a time* $t_2 > t_1$ *we can improve the mass to some* $M > M_1$ *and* $\mathcal{U}_M(x, t + \tau)$ *is a lower barrier for* u *at all times* $t \ge t_2$.

Proof. The inequality

$$u(x, t) \geq U_1(x, t) = \mathcal{U}_{\mathcal{M}_1}(x, t + \tau_1)$$

holds for all $t \geq t_1$. Let $t_2 > t_1$. Let B_2 be the support of the solution U_1 at $t = t_2$ and let R_2 be its radius. Since $u(x, t_2) \geq U_1(x, t_2) > 0$ in B_2 the Strong Maximum Principle which holds for classical solutions of nonlinear parabolic equations implies that u must be strictly larger than U_1 at t_2 inside B_2.

We construct a barrier with a larger mass at t_2 as follows. If the support of $u(\cdot, t_2)$ contains a larger ball than B_2 then it is immediate that there exists a lower barrier of the form

$$\mathcal{U}_{\mathcal{M}}(x, t_2 + \tau_1)$$

for some $\mathcal{M} > \mathcal{M}_1$. In case the ball cannot be expanded we have to perform the trick of changing the delay. Using Lemma 14.3 we see that a small decrease in the delay (to a value $\tau < \tau_1$) produces a source-type (Barenblatt) profile that is larger in the middle and smaller near the boundary with smaller support. This means that if τ is close enough to τ_1 then $\mathcal{U}_{\mathcal{M}}(x, t_2 + \tau)$ will be less than $u(x, t_2)$ inside the support the former. Hence, we can safely increase slightly the mass to some $\mathcal{M} > \mathcal{M}_1$ and the resulting function will still be a lower barrier

$$u(x, t_2) \geq \mathcal{U}_{\mathcal{M}}(x, t_2 + \tau). \tag{14.6}$$

$\square$

COROLLARY 14.5. *We have* $\mathcal{M}_\infty = M$ *and* $U = \mathcal{U}_M$.

Proof. Suppose that this is not true and $\mathcal{M}_\infty < M$. Then we are in the situation where the lower barrier for U can be increased from $\mathcal{M}_\infty$ to $\mathcal{M} = \mathcal{M}_\infty + \varepsilon$ for $t \geq t_2 > 0$ with some delay $\tau \leq 0$. The uniform convergence of $\tilde{u}_\lambda \to U$ on compact subsets means then that for large enough λ the function $\tilde{u}_\lambda$ admits a barrier of mass $\mathcal{M}' \geq \mathcal{M}_\infty + (\varepsilon/2)$ with the same delay. In terms of u it means a barrier of mass $\mathcal{M}'$ with a much larger delay τ. But this would mean that

$$\lim_{t \to \infty} \mathcal{M}(t; u, \tau) > \mathcal{M}_\infty,$$

a contradiction since $\mathcal{M}_\infty$ does not depend on τ.

$\square$

After some rephrasing, Theorem 1.1 is proved.

15. Asymptotic symmetry

We now turn to a different trend of ideas based on the exploitation of the special properties of the asymptotic dynamics, resuming the discussion started at the end of Section 4. The most general idea about asymptotic properties can be phrased as the obtention of **symmetries** possessed by the equation but absent from the initial data. We recall the wide sense given by physicists to the word 'symmetry'. Thus, our equation is invariant under plane symmetries and rotations in the space variable. This 'symmetry' is not true for general orbits because of the influence of the initial data. However, we have asymptotic symmetry which derives from the following **monotonicity lemma**, cf. [CVW].

LEMMA 15.1. *Let u a solution of the Cauchy problem* (CP) *with initial data supported in the ball $B_R(0)$, $R > 0$. Then for every x such that $|x| > 2R$ and every $r < |x| - 2R$, $r > 0$, we have*

$$u(x, t) \leq \inf_{|y|=r} u(y, t) \tag{15.1}$$

Proof. We use Alexandrov's Reflection Principle. We draw the hyperplane H which is mediatrix between the points x and y in the above situation. It is easy to see that H divides the space $\mathbb{R}^n$ into two half-spaces, one Ω_1 which contains y and the support of u_0 and another one, Ω_2, which contains x and where $u_0 = 0$. We consider now the initial and boundary-value problem in $\widehat{Q} = \Omega_1 \times (0, \infty)$. Two particular solutions of this problem are compared: one of them is u_1, the restriction of u to $\widehat{Q}$, another one is

$$u_2(z, t) = u(\pi(z), t), \quad z \in \Omega_1.$$

where π is the specular symmetry with respect to the hyperplane H. Thus, if we orient the coordinate axes so that $H = \{x_1 = 0\}$ then

$$\pi(x_1, \ldots, x_n) = (-x_1, \ldots, x_n).$$

Clearly, u_1 and u_2 are solutions of (0.1) in $\widehat{Q}$. Besides, $u_1 = u_2$ on the lateral boundary, $\Sigma = H \times (0, \infty)$. Finally, $u_1 \geq u_2$ for $t = 0$ since $u_2 = 0$ in Ω_1. By comparison for the mixed problem we have

$$u_1(z, t) \geq u_2(z, t) \qquad z \in \Omega_1, \ t > 0.$$

Putting $z = y$ we have $\pi(z) = x$ so that $u(y, t) \geq u(x, t)$ as desired. $\square$

The conclusion for the asymptotic orbits is immediate.

THEOREM 15.2. *Let U be a solution of problem* (CP) *obtained as limit of the rescaling discussed in Section 4. Then U must be radially symmetric in the space variable, $U = U(r, t)$, $r = |x|$. It is also a nonincreasing function of r for fixed t.*

Proof. By density we may argue with an original orbit $u(x, t)$ with initial data u_0 supported in the ball of radius $R > 0$. But the use of Lemma 15.1 implies that for every $x \in \mathbb{R}^n$, $x \neq 0$ we have

$$u_\lambda(x, t) \leq \inf_{|y|=r} u_\lambda(y, t)$$

as long as $|x| \geq 2R/\lambda^\beta$ and $0 < r < |x| - (2R/\lambda^\beta)$. In the limit $\lambda_n \to \infty$ we get

$$U(x, t) \leq \inf_{|y|=r} U(y, t), \qquad 0 < r < |x|. \tag{15.2}$$

The conclusion follows. The L^1 continuity allows to extend the result to general data. $\square$

Therefore, the asymptotic dynamics takes place under the conditions of radial symmetry and monotonicity in the radial variable. Of course, the reader will object that in our problem we already know, after Section 5, that U must be a source-type. There are two points we want to make in this connection: (i) the proof of Section 5 was based on a strong uniqueness result for the PME; (ii) the asymptotic property of Theorem 15.2 is a general fact that can be established under quite general assumptions and can be the base for alternative convergence proofs, applicable when uniqueness is not available in the form used in Section 5.

Let us also warn the reader that the influence of the initial data, though indirect, still exists in the form of class of data for which this process is true, which is given by the restriction (0.3). It is easy to see that other initial classes are incompatible with the process of asymptotic symmetry in this sense. For instance, data which are monotone in one direction preserve this property in time, and this is incompatible with radial symmetry.

Let us also mention another immediate consequence of Alexandrov's Reflection Principle.

LEMMA 15.3. *With the assumptions and notations of Section 9, the difference of maximal and minimal support radius, $R(t) - r(t)$, remains bounded in time.*

16. Convergence rates for radially symmetric solutions. Concentration comparison

16.1. The study of radially symmetric solutions $u = u(r, t)$ has an interest because we can use special techniques which produce error estimates with relative size $O(1/t)$. Combined with the idea of asymptotic symmetry of Section 15, it provides an alternative proof of the convergence results of Section 3. The ideas have been first explained in the paper [KV1]. It works as follows:

End of proof of Theorem 1.1 by this method. The proof starts as in Section 4 and proceeds through Steps 1, 2 and 3 so that we pass to the limit $\lambda_n \to \infty$ and obtain an asymptotic solution, U, which is radially symmetric by the results of Section 15. Assume that

Theorem 1.1 is proved for such solutions. Then we have convergence of U towards the source-type solution $\mathcal{U}_M(x,t)$. The triangle inequality gives then convergence along a subsequence

$$u(x,t_n) - \mathcal{U}_M(x,t_n) \to 0 \qquad \text{in } L^1(\mathbb{R}^n).$$

But the L^1-contraction implies that the whole family $\{u(t)\}_{t>0}$ converges. The proof is complete.

16.2. Therefore, we consider in the sequel the asymptotic behaviour of solutions of the Cauchy problem (CP) under the assumptions (0.3) plus

u_0 is bounded, radially symmetric and compactly supported. $\qquad\qquad$ (16.1)

A density argument allows to dispense with the later condition as in Section 6. The main technical tool that we shall use is the comparison principle introduced in [V3] (see also [V1]) and called **Concentration Comparison**. In the restricted form that we need here it reads as follows

THEOREM 16.1. *Let u_i, $i = 1, 2$ be a pair of radially symmetric solutions of problem (CP) with initial data u_{0i} in the class (0.3). Assume that for every $r > 0$*

$$\int_{B_r(0)} u_{01}(x)\,dx \le \int_{B_r(0)} u_{02}(x)\,dx. \qquad\qquad (16.2)$$

Then for every time $t > 0$ and radius $r > 0$

$$\int_{B_r(0)} u_1(x,t)\,dx \le \int_{B_r(0)} u_2(x,t)\,dx. \qquad\qquad (16.3)$$

In [V1], pp. 522, the notation $u_{02} \succ u_{01}$ is introduced to represent the situation of (16.2), and it is read as u_{02} *is more concentrated than* u_{01}. The result can then be phrased as saying that *the concentration relation $\succ$ is conserved in the evolution* (Note: the general formulation, [V3], allows to compare a non-symmetric solution $u_1(x,t)$ with a symmetric solution $u_2(x,t)$, after symmetrizing the first one, but such generality is not needed here).

As an immediate consequence of this result we obtain comparison of supports.

LEMMA 16.2. *Let u_1 and u_2 two solutions as before and let us assume that they are compactly supported and have the same mass. If $R_1(t)$ and $R_2(t)$ are the radii of their respective supports we have*

$$R_1(t) \ge R_2(t), \qquad \text{for every } t \ge 0. \qquad\qquad (16.4)$$

We only need to observe that if the common mass is M, then

$$R_i(t) = \sup\left\{r > 0 : \int_{B_r(0)} u_i(x,t)\,dx < M\right\}.$$

16.3. Let us now proceed with the analysis of the long-time behaviour of solutions and supports for compactly supported, radially symmetric solutions. To begin with, we have the following improved asymptotics for the free boundary.

THEOREM 16.3. *Let $R(t)$ and $\mathcal{R}(t)$ the radii of the supports of $u(t)$ and $\mathcal{U}_M(t)$. Then there exists $t_2 > 0$ such that $\mathcal{R}(t) \leq R(t) \leq \mathcal{R}(t + t_2)$. Therefore,*

$$R(t) = \mathcal{R}(t)\left(1 + O\left(\frac{1}{t}\right)\right). \tag{16.5}$$

Proof. It is easy to see that given an initial datum u_0 as assumed in (16.1) with mass $M > 0$ and such that u_0 is positive near $r = 0$, there exist times $0 < t_1 < t_2$ such that

$$\mathcal{U}_M(x, t_1) \succ u_0(x) \succ \mathcal{U}_M(x, t_2). \tag{16.6}$$

The result $\mathcal{U}_M(x, t + t_1) \succ u(x, t) \succ \mathcal{U}_M(x, t + t_2)$ follows. Moreover, t_1 can be taken as small as we please. In case u_0 is not positive at $r = 0$ we have to wait for a certain time until the positivity set spreads to cover that point. According to Section 9 this happens after some time t_3. Then relation (16.6) holds with u_0 replaced by $u(x, t_3)$. $\qquad\square$

We derive next an L^1-rate of convergence for the profiles.

THEOREM 16.4. *As $t \to \infty$ we have*

$$\|u(t) - \mathcal{U}_M(t)\|_1 = O\left(\frac{1}{t}\right). \tag{16.7}$$

Proof. We take for fixed $t \gg 1$ the source-type function $\widetilde{U}(x, t)$ with interface at $R(t)$. According to Theorem 16.3 and the formula for the free boundary of the source-type solution, it has a mass

$$\widetilde{M}(t) = M(1 + O(1/t))^{1/(\beta(m-1))} = M(1 + O(1/t)).$$

We have $\widetilde{U} \geq \mathcal{U}_M$. It is clear that $u(t)$ and $\widetilde{U}(t)$ have the same support at time t, the ball $\widetilde{B}$ of radius $R(t)$. Let $\phi(x) = u^{m-1}(x, t) - \widetilde{U}^{m-1}(x, t)$. According to the fundamental estimate we have $\Delta\phi \geq 0$ in $\widetilde{B}$, i.e., ϕ is subharmonic. Since ϕ is zero on the boundary of the ball, we conclude that $u \leq \widetilde{U}$. Therefore,

$$\|(u(t) - \mathcal{U}_M(t))_+\|_1 \leq \|(\widetilde{U}(t) - \mathcal{U}_M(t))_+\|_1 = \widetilde{M} - M = O\left(\frac{1}{t}\right).$$

Since u and $\mathcal{U}_M$ have the same mass, estimate (16.7) follows. $\qquad\square$

17. The technique of intersection comparison

Another technique that can be used in the study of radially symmetric problems is the extended version of the Maximum Principle which is called **Intersection Comparison** and consists in counting the number of sign changes of the difference of two solutions.

DEFINITION. Given two radially symmetric solutions, u_1 and u_2, of problem (CP) and for every time $t \geq 0$ we consider finite sequences of radii $0 < r_0 < r_1 < \cdots < r_n$ (of length n) such that

$$(u_1(r_i) - u_2(r_i))\,(u_1(r_{i+1}) - u_2(r_{i+1})) < 0 \tag{17.1}$$

for every $i = 0, \ldots, n - 1$. Then we define

$$N(t; u_1, u_2) = \max\{n : n \text{ is the length of a sequence satisfying (17.1)}\} \tag{17.2}$$

It is also important to record the list of signs of $u_1 - u_2$. The main result of this theory is

THEOREM 17.1. *For the solutions of the Cauchy problem for quasilinear parabolic equations of the type of equation PME posed in the whole space the counter N is non-increasing in time.*

This result stems from Sturm [St] and has been extended in recent years to wide classes of nonlinear parabolic equations by a number of authors, see the book [S4] and its references. The principle is also known as **lap number** theory, cf. [Ma]. More precisely, the decrease in N happens through loss of sign changes either at the boundary of the domain (if there is one) or inside (in this case two sign changes are lost at least at every time a change happens). The list of signs is conserved as long as N does not decrease.

By a careful geometrical inspection of our situation for the PME and using the regularity of the solutions, we get the following result.

LEMMA 17.2. *Given a radial solution u of problem* (CP) *with bounded, continuous and compactly supported data $u_0 \geq 0$, there exist time delays $0 \leq t_1 < t_2$ such that if $u_i(x, t) = \mathcal{U}_M(x, t + t_i)$, $i = 1, 2$, we have*

$$N(t, u, u_1) = N(t, u, u_2) = 1. \tag{17.3}$$

Moreover, the list of signs of $u - u_1$ is $-+$, while the list for $u - u_2$ is $+-$. Therefore, for all $r \approx 0$ we have for all large t

$$u_2(r, t) < u(r, t) < u_1(r, t). \tag{17.4}$$

Note that the counter $N(t, u, u_1)$ cannot decrease to 0 since that would imply the ordering $u \le u_1$, and by equality of mass $u = u_1$. Even if this happens a small change in delay would restore the value $N = 1$. The same applies to $u_2 - u$.

As an immediate consequence of these estimates we obtain the support radius estimates of Theorem 16.3. The proof of Theorem 16.4 remains unchanged. Moreover, the technique of Intersection Comparison allows for a strong improvement of Theorem 16.4.

THEOREM 17.3. *As $t \to \infty$ we have*

$$u^{m-1}(x, t) = \mathcal{U}_M^{m-1}(x, t) + O(t^{-\gamma}), \qquad \gamma = \alpha(m - 1) + 1. \tag{17.5}$$

Proof. Remark that $O(t^{-\gamma}) = \|\mathcal{U}_M(\cdot, t)\|_\infty^{m-1} O(t^{-1})$, therefore this is another estimate with relative error of size $O(1/t)$. Note that the delayed source-type solutions show this error, hence the stated convergence rate is optimal.

As for the proof, the estimate from above has been already established in the comparison $u \le \tilde{\mathcal{U}}$ of Theorem 16.4. In the estimate from below it is useful to write the error as $e(x, t) = u^{m-1}(x, t) - \mathcal{U}_M^{m-1}(x, t)$. For $r \approx 0$ and large t, the estimate $e \le O(t^{-\gamma})$ follows from (17.4). To extend it to all r we use the Aronson-Bénilan estimate that implies that $\Delta e \ge 0$ as long as $\mathcal{U}_M > 0$. Since the function is radially symmetric, $e = e(r, t)$, we have for every fixed $t > 0$

$$e(0, t) = 0, \quad e_r(0, t) = 0 \quad \text{and} \quad (r^{n-1} e_r)_r \ge 0 \text{ for } 0 < r < \mathcal{R}(t).$$

This implies that $e(r, t) \ge e(0, t)$ for $0 < r < \mathcal{R}(t)$. This and the estimate for $R(t)$ end the result. $\qquad \square$

No uniform rate can be given for general solutions without a decay estimate on u_0 at infinity, as we have already shown in Section 11, which applies even with radial symmetry. On the other hand, the convergence rates for radially symmetric solutions with compact support obtained in these two sections are optimal.

One dimension without radial symmetry. The problem without radial symmetry is rather well known in dimension $n = 1$, where we can still use the techniques of concentration (mass) comparison and intersection comparison used in these two last sections. As a first example of one-dimensional result, under the assumptions that $u_0 \in \mathcal{X}_0$ is compactly supported, not radially symmetric, it is proved in [V1] that there is an improved rate of convergence if we first calculate the *center of mass*, which is an invariant of the motion,

$$x_c = \frac{1}{M} \int_{\mathbb{R}} x\, u_0(x)\, dx = \frac{1}{M} \int_{\mathbb{R}} x\, u(x, t)\, dx,$$

where as usual $M = \int u_0(x)\, dx > 0$. Then, we have

THEOREM 17.4. *The left and right interfaces of u, $r_1(t) = \inf\{x : u(x, t) > 0\}$, $r_2(t) = \sup\{x : u(x, t) > 0\}$, satisfy*

$$r_i(t) = (-1)^i \mathcal{R}(t) + x_c + o(1) \tag{17.6}$$

as $t \to \infty$. This implies an asymptotic error in the pressures of the form

$$u^{m-1}(x, t) - \mathcal{U}_M^{m-1}(x - x_c, t) = O(t^{-m/(m+1)}). \tag{17.7}$$

18. Concavity and smooth convergence

The question of large time behaviour can be combined with the question of asymptotic geometry to obtain better asymptotic results. The first paper in that direction for the PME seems to be due to Bénilan and the author [BV], who prove that in dimension one and for compactly supported solutions of the PME concavity of the pressure is preserved in time (if $v = u^{m-1}$ and $v_{0,xx} \leq 0$ where $v_0 > 0$, then $v_{xx} \leq 0$ where $v > 0$). Aronson and the author then proved that *all* compactly supported solutions become eventually pressure-concave, which allows for better convergence estimates in 1D, [AV], Section 4:

$$\|u^{m-1}(x, t) - U_M^{m-1}(x - x_c, t)\|_{L^\infty(\mathbb{R}^n)} = O(t^{-2m/(m+1)}). \tag{18.1}$$

The result is inspired by the fact that in terms of the variable $v = u^{m-1}$ the source-type solution is an inverted parabola. A more precise asymptotics says that

$$v_{xx}(x, t) + \frac{m-1}{m(m+1)t} = O\left(\frac{1}{t^2}\right). \tag{18.2}$$

A simple geometrical argument shows that if such a function has the same mass as the parabola of the source-type solution pressure the error in size must be of order $O(1/t)$ times the height of the parabola, and this gives $O(t^{-2m/(m+1)})$ error for u^{m-1}. The error in the interface size is also of the same order, hence

$$r_i(t) = (-1)^i \mathcal{R}(t) + x_c + O(t^{-m/(m+1)}). \tag{18.3}$$

This reproduces in 1D the results for radially symmetric solutions of previous sections. A study of asymptotic geometry in 1D for more general diffusion equations is done in [GV] using Intersection Comparison.

The recent work of Lee and Vazquez [LV] extends these ideas to all dimensions $n \geq 1$: after a finite time the pressure of any solution with compact support becomes a concave function in the space variable, and it converges to all orders of differentiability to a truncated parabolic shape, Barenblatt's source-type profile. The assumptions on the initial data are u_0 nonnegative, compactly supported and what is called non-degenerate initial data, a technical condition. In particular, for large times the support of the solution is a convex subset of $\mathbb{R}^n$ which converges to a ball, and the convergence is smooth. Estimates are optimal. Here is a typical result from [LV].

THEOREM 18.1. *There is* $t_o > 0$ *such that* $v(x, t)$ *is concave in* $\Omega(t) = \{x : v(x, t) > 0\}$ *for* $t \geq t_o$. *More precisely,*

$$\lim_{t \to \infty} t \frac{\partial^2 v}{\partial x_i^2} = -\frac{(m-1)\beta}{m} \tag{18.4}$$

for any coordinate direction x_i, *uniformly in* $x \in supp(v)$. *The results also hold for the heat equation (where the pressure is defined as* $v = \log(u)$ *and we get asymptotic log-concavity) and fast diffusion (with* $v = 1/u^{1-m}$, $(n-2)/n < m < 1$, *and we get asymptotic pressure convexity). In these cases* $supp(v(t)) = \mathbb{R}^n$.

19. Improved rates by the entropy method

Obtaining better convergence rates than the plain ones given by the main result needs assumptions on the data, and this is the purpose of an extensive literature. We have seen the rates of convergence obtained under the assumptions of radial symmetry and compact support in Sections 16, 17 and 18. Without those assumptions, the Lyapunov approach of Newman and Ralston [N, R] has been improved by means of ideas of entropy and entropy-dissipation into a tool to obtain rates of convergence by a number of authors in [CT, C5, DP, Ot] for data in the class $\mathcal{X}_2 \subset \mathcal{X}_0$ of initial data having also finite second moment

$$\int |x|^2 u_0(x)\, dx < \infty. \tag{19.1}$$

THEOREM 19.1. *For every* $u_0 \in \mathcal{X}_2$ *we have*

$$\|u(x, t) - U_M(x, t)\|_{L^1(\mathbb{R}^n)} = O(t^{-\kappa}), \tag{19.2}$$

where κ *equals the Barenblatt exponent* β *for* $1 < m \leq 2$, *while* $\kappa = 2\beta/m$ *is obtained for* $m > 2$.

This theorem has been proved in the above-mentioned references using different techniques: entropy dissipation technique, mass transportation arguments, Riemannian calculus, variational calculus. By checking the error committed by a source-type solution displaced in space we can see that $\kappa = \beta$ is the optimal rate to be expected (for all $m \geq 1$). The argument holds also for $m < 1$ (fast diffusion, to be treated below) if $m > \max\{(n-1)/n, n/n+2\}$. The extension to the lower part of the range, $(n-1)/n > m > (n-2)/n, n \geq 2$, has been done recently in [CVa], with rate $\kappa = 1/2$, thus completing the range of exponents m where Barenblatt's source-type solutions play a role. We refer to these works for further references and to Section 21 for a general treatment of fast diffusion.

20. Extensions. Sign change and forcing

The generalization of the porous medium equation and other simple models into a general mathematical model was one of the leitmotives of Bénilan's mathematical activity, since in his opinion an analyst must always strive for the general ideas valid for wide classes of differential operators acting in general spaces, in an effort to discover what is the powerful mathematical idea and result. In that respect, he was always inclined to see the porous medium equation in the form $u_t = \Delta\Phi(u)$, and Φ must be allowed to be an arbitrary monotone function, or even a maximal monotone graph, like in the famous book [Br]. On the other hand, this and more general forms, like $u_t = \operatorname{div} a(x, t, u, Du)$, turned out to be of interest for the applications.

Signed solutions. Going back to the simple porous medium, this trend of ideas favors the elimination of the sign restriction on the solutions, so that the whole L^1 may enter the picture. In order to tackle negative values of u one must adapt the power in a convenient way so that the equation continues to be degenerate parabolic. The preferred version has a symmetrical nonlinearity:

$$u_t = \Delta(|u|^{m-1}u)\,. \tag{20.1}$$

(There is no a priori reason to do that, apart from mathematical simplicity, which is a strong reason in itself). This equation is treated by Bénilan and Crandall [BC2] in the framework of m-accretive operators in L^1 to obtain a mild solution, and the integral conditions derived by Bénilan in his thesis. The asymptotic behaviour of integrable solutions has been studied by Kamin and Vazquez [KV2] and the result is

THEOREM 20.1. *Let $u_0 \in L^1(\mathbb{R}^n)$ and let $M = \int u_0(x)\,dx$. If $M > 0$ then*

$$\lim_{t\to\infty} \|u(t) - \mathcal{U}(t; M)\|_{L^1(\mathbb{R}^n)} = 0, \tag{20.2}$$

If $M = 0$ the limit applies with $U = 0$. If $M = -M' < 0$ we put $\mathcal{U}(x, t; M) = -\mathcal{U}(x, t; M')$ and the result holds.

However, when the integral is zero the asymptotics is trivial in our scale, and the actual scaling where the asymptotics is nontrivial depends on the subclass of initial data; this is partially studied in [KV2] and [BHV], where a beautiful problem of *anomalous exponents* appears.

Equations with forcing. Another interesting result concerns the influence of a forcing term in the form of an integrable right-hand side in the equation.

THEOREM 20.2. *Let u be the mild solution of the Cauchy problem for*

$$u_t = \Delta(|u|^{m-1}u) + f, \tag{20.3}$$

with initial data $u_0(x) \in L^1(\mathbb{R}^n)$ and $f \in L^1(Q)$ (no sign restriction is imposed). Then,

$$\lim_{t \to \infty} \|u(t) - \mathcal{U}_{M'}(t)\|_1 = 0 \,, \tag{20.4}$$

where

$$M' = \int u_0(x)\,dx + \int\int_Q f\,dxdt. \tag{20.5}$$

Proof. We consider the solution $u_n(x, t)$ of the problem with same initial data and forcing term f_n such that

$$f_n(x, t) = f(x, t) \quad \text{if } t < n, \qquad f_n(x, t) = 0 \quad \text{if } t \geq n.$$

The mild solution becomes a standard weak continuous solution for $t \geq n$. We can think of this solution as having initial data at $t = n$ of the form $u_n(x, n) = \phi_n(x)$. According to Bénilan's analysis [Be]

$$M_n = \int \phi_n(x)\,dx = \int u_0(x)\,dx + \int_0^n \int f(x, t)\,dxdt.$$

Hence, for $t \gg n$ large enough, u_n approaches the source-type profile with mass

$$M_n = \int u_n(x, t)\,dx$$

(which is constant for $t \geq t_n$) with the modification explained above if $M_n \leq 0$. This means that

$$\lim_{t \to \infty} \|u_n(t) - \mathcal{U}_{M_n}(t)\|_1 = 0.$$

On the other hand, $\lim_{t \to \infty} M_n = M'$. Finally, the contraction property for mild solutions (or *bonnes solutions* in the sense of [Be]) implies that

$$\|u - u_n\|_1 \leq \int_0^t \int |f - f_n|\,dxdt = \int_n^\infty \int_{\mathbb{R}^n} |f|\,dxdt \to 0$$

as $n \to \infty$. The proof is complete. $\square$

21. Improved convergence for fast diffusion

The extension of the asymptotic results proved above to exponent $m = 1$ gives as a consequence results that are well-known for the classical heat equation. It is interesting to remark that the proof given here applies (with inessential minor changes), and is very different from the usual proofs based on the representation formula.

We can even go below $m = 1$ and prove similar results for some so-called **fast-diffusion equations** which are just equation (0.1) with $0 < m < 1$. To start with we need two basic ingredients.

(a) A theory of well-posedness for the Cauchy problem. The results of Section 1 apply also in this case with minor easy changes. The main novelty is that solutions are positive everywhere and C^∞-smooth, which is rather good news in this context.

(b) The second ingredient is the model of asymptotic behaviour. The source-type solution exists just for $m > m_c = (n-2)/n$ and it can be conveniently written in the form

$$\mathcal{U}_M(x,t) = \left(\frac{Ct}{|x|^2 + At^{2\beta}}\right)^{1/(1-m)} = \frac{K\,t^{-\alpha}}{(A + (x\,t^{-\beta})^2)^{1/(1-m)}} \tag{21.1}$$

where $\beta = (2 - n(1-m))^{-1}$ is positive precisely in that range, $\alpha = n\beta$, $C = 2m/\beta(1-m)$ is a fixed constant, $K = C^{1/(1-m)}$, and $A > 0$ is an arbitrary constant that can be determined as a decreasing function of the mass $M = \int \mathcal{U}(x,t)\,dx$, $A = k(m,n)\,M^{-\gamma}$ with $\gamma = 2(1-m)\beta$.

In dimensions $n = 1, 2$ the whole range $0 < m < 1$ is covered. However, the *critical exponent*, $m_c = 1 - (2/n)$, is larger than zero for $n \geq 3$. It is then proved that for $0 < m < m_c$ no solution of the Barenblatt type exists (i.e., self-similar with constant positive mass). The value $m_c = (n-2)/n$ is related to the Sobolev embedding exponents as the reader will easily realize.

The authors of [FK] claim that their result of uniform convergence uniformly on sets of the form $|x| \leq c\,t^\beta$, is also true for $m < 1$ in the range $m_c < m < 1$, where the Barenblatt solution exists. Indeed, *the convergence results of Theorem* 1.1 *hold true for* $m > m_c$ and the proofs given above are true but for minor details.

Relative error convergence.   However, the fact that the solutions of the fast diffusion equation do not have the property of conserving compact supports, but develop tails at infinity of a certain form gives rise to a very interesting estimate formulated in terms of *relative error*, or in other words, as *weighted convergence*, that we present next. It requires a suitable behaviour of the initial data as $|x| \to \infty$ (similar in decay to the Barenblatt solution).

THEOREM 21.1.   *Under the assumption that u_0 is bounded and $u_0(x) = O(|x|^{-2/(1-m)})$ as $|x| \to \infty$, we have the asymptotic estimate*

$$\lim_{t\to\infty} \frac{|u(x,t) - \mathcal{U}(x,t;M)|}{\mathcal{U}(x,t;M)} \to 0 \tag{21.2}$$

uniformly in $x \in \mathbb{R}^n$. The condition on the initial data can be weakened into the integral estimate

$$\int_{|y-x|\leq|x|/2} |u_0(y)|\,dy = O(|x|^{n - \frac{2}{1-m}}) \quad as \ |x| \to \infty. \tag{21.3}$$

In particular, we have for $\|u(t) - \mathcal{U}(t; M)\|_1 \to 0$ as $t \to \infty$ (like case $p = 1$ of Theorem 1.1), and $t^\alpha |u(x, t) - \mathcal{U}(x, t; M)| \to 0$, as $t \to \infty$ uniformly in x (case $p = \infty$), but estimate (21.2) is much more precise because the convergence is uniform with weight

$$\rho = (|y|^2 + c)^{1/(1-m)}, \qquad y = x\, t^{-\beta}.$$

Proof. (i) The standard theorem implies that we have uniform convergence

$$t^\alpha |u(x, t) - \mathcal{U}(x, t; M)| \to 0\,,$$

uniformly on sets of the form $|x| \le C\, t^\beta$. The problem is then to extend this estimate to the outer region $\{|x| \ge C\, t^\beta\}$ by means of a so-called *tail analysis*. This analysis uses two properties of the Barenblatt solutions. The first is the effect at infinity of a delay in time. Thus, it is easy to see that $\mathcal{U}(x, t + \tau; M)$ grows with τ for large values of $y = x\, t^{-\beta}$, and precisely

$$\mathcal{U}(x, t + \tau; M) - \mathcal{U}(x, t; M) = \frac{\tau}{1 - m}\, \mathcal{U}(x, t)\, t^{-1}(1 + O(t^{-1})),$$

uniformly for $|y| \gg 1$ (on the contrary, for small values of y the variation is opposite, $\partial U/\partial \tau < 0$). Next, we recall the fact that the asymptotic behaviour as $|x| \to \infty$ for fixed t of the source-type solutions is independent of the mass of the solution

$$\mathcal{U}(x, t; M) = (C\, t\, |x|^{-2})^{1/(1-m)}(1 + O(M^{-1/\gamma} y^{-2}))\,.$$

Next, we note that this kind of universal behaviour of some solutions of fast diffusion for large $|x|$ can be generalized to a general class of solutions $u(x, t)$.

(ii) The tail analysis goes as follows. We examine first the existence of an upper bound using the assumption in the strong form, $u_0(x) = O(|x|^{-2/(1-m)})$.

We have to use the trick of comparing our initial data with a source-type solution with negative value of A. Then the denominator of (21.1) vanishes for some value of $|y| = y_0(A)$, and the solution becomes infinite there (blow-up). It is finite for $|x| \ge y_0\, t^\beta$, and indeed, a smooth classical solution in that region. If we take any $\tau > 0$ then there exists a large value of A (hence a large value of y_0), such that we have the comparison

$$u_0(x) \le \mathcal{U}(x, \tau; -A) \qquad |x| \ge y_0\, \tau^\beta.$$

By comparison we have for all $t > 0$, $|x| \ge y_0\, (t + \tau)^\beta$

$$u(x, t) \le \mathcal{U}(x, t + \tau; -A).$$

In view of the behaviour of $\mathcal{U}$ as $|y| \to \infty$ independently of A and letting $\tau \to 0$ we get the estimate

$$\limsup_{t \to \infty} u^{1-m} \frac{x^2}{t} \le C + \varepsilon\,,$$

uniformly for $|x|\, t^{-\beta} = |y| \geq y_\varepsilon \gg 1$. In view of the form of $\mathcal{U}$, the reader will be able to check that this is the upper bound needed in the tail part of (21.2).

(iii) Thanks to the L^1-L^∞ regularizing effect the general assumption (21.3) is converted into the previous assumption for $t_1 > 0$. Indeed, it is known that

$$u(x, t_1) \leq C\, t_1^{-\alpha} \int_{B_{|x|/2}} u_0(x)\, dx$$

Therefore, under assumption (21.3) we have $u(x, t_1) = O(|x|^{-2/(1-m)})$, and we can take this function as initial function after displacing the axis of times.

(iv) We now turn to the lower bound at the tail. We will take a small $\tau > 0$ and large $a > 0$ and make a comparison of $u(x, t)$ and $\mathcal{U}(x, t - \tau; M')$ in the region

$$\mathcal{R} = \{(x, t) : \; \tau < t < 2\tau, \; |x| \geq a\}.$$

It is clear that $u(x, \tau) \geq \mathcal{U}(x, 0; M')$ for $|x| \geq a$ since the last quantity is zero. On the other hand, by the continuity of u we can get $u \geq \mathcal{U}$ on the parabolic boundary $|x| = a$ if the mass M' is chosen small enough. By the Maximum Principle we conclude that $u(x, t) \geq \mathcal{U}(x, t - \tau; M')$ in $\mathcal{R}$. In particular,

$$u(x, 2\tau) \geq \mathcal{U}(x, \tau; M') \sim (C\tau/x^2)^{1/(1-m)} \qquad x \to \infty.$$

But the positivity of u at $\tau > 0$ implies that the inequality is true for all $x \in \mathbb{R}^n$ if M' is small enough. Hence, the same comparison holds for all later times

$$u(x, t) \geq \mathcal{U}(x, t - \tau; M'), \quad x \in \mathbb{R}^n\,.$$

The uniform behaviour of $\mathcal{U}$ as $y \to \infty$ independently of M' and the fact that τ is arbitrary allow us to conclude much as before that

$$\liminf_{t \to \infty} u^{1-m} \frac{x^2}{t} \geq C - \varepsilon\,,$$

uniformly for $|x|\, t^{-\beta} = |y| \geq y_\varepsilon \gg 1$. $\qquad\qquad\square$

We remark that our result implies a uniform behaviour at the space far-field for all large t. More precisely,

COROLLARY 21.2. *Under the above assumptions we have the asymptotic limit as* $|y| \to \infty$, $t \to \infty$:

$$\lim (|x|^2/t)^{1/(1-m)}\, u(x, t) = K = (2m/\beta(1 - m))^{1/(1-m)}.$$

Further improvements. When u_0 has an exact decay at infinity, $u_0(x) \sim a\, |x|^{-2/(1-m)}$, we have a precise x-asymptotics.

LEMMA 21.3. *If* $\lim_{|x|\to\infty} |x|^{2/(1-m)} u_0(x) = a \geq 0$, *then* u *satisfies*

$$\lim_{|x|\to\infty} x^2 u^{1-m}(x, t) = C\,(t + T), \tag{21.4}$$

locally uniformly in time $t > 0$, *with* $T = a^{1-m}/C$ *and* C *as in* (21.1). *Moreover, there exists* $t_o > 0$ *such that* $v(x, t) = u^{m-1}$ *is convex as a function of* x *for* $t \geq t_o$.

This is proved in [LV]. On the other hand, for radial data in the class of the Theorem, Carrillo and Vazquez [CVa] show a relative rate of decay $O(t^{-1})$ in the results of Theorem 21.1, extending the results of Sections 16, 17 to fast diffusion.

Subcritical range.

Let us briefly point out that the breakdown of the asymptotic model implies in this case a complete change of asymptotic behaviour. Thus, it is proved that for the critical exponent, $m = m_c$, mass is conserved but the asymptotic behaviour is quite involved, cf. [Ki, GPV]. For $0 < m < m_c$ we get an even more drastic phenomenon: bounded integrable solutions disappear (vanish identically) in finite time and the relevant asymptotic profiles are self-similar solutions of the form

$$\mathcal{U}(x, t) = (T - t)^{\alpha} F(x\,(T - t)^{\beta}), \tag{21.5}$$

cf. [GP] and [PS]. We realize at once the crucial role played in the dynamics of these nonlinear heat equations by the existence of certain types of particular solutions.

On the other hand, in one space dimension, the standard theory of fast diffusion can be extended to the extra range $-1 < m \leq 0$ on the condition of writing the equation in the modified form $u_t = (u^{m-1}u_x)_x$ (the original formula is not parabolic!). Source-type solutions still exist and Theorems 1.1 and 21.3 hold on the condition of working with the special class of *maximal solutions,* for which the problem is well posed. We remind the reader that in this range of exponents the theory of the Cauchy Problem is quite special because of non-uniqueness: there are other classes of solutions determined by nontrivial Neumann conditions at infinity, cf. [ERV, RV]. For those other solutions different asymptotic descriptions have to be found.

22. Final comments on the literature and results

The PME was derived by Boussinesq in 1903/4 in connection with flows in porous media, [Bo]. Around 1950 Zel'dovich and his group studied this equation as a model for heat propagation in plasma [ZR]; the source-type solutions were constructed in a particular case by Zel'dovich and Kompaneets [ZK] and in all generality by Barenblatt [B1], 1952. They were re-discovered in the West by Pattle [Pa] in 1959. The first proof of existence and uniqueness of a generalized solution was done by Oleinik and collaborators [OKC] in 1958 (in one space dimension). There are many known applications of the PME: Muskat considered it for the flow of gases in porous media in 1937 [Mu], it was studied by Gurtin and McCamy [GMC] in a model in population dynamics, and it appears in thin viscous films moving under gravity studied by Buckmaster [Bu]. We refer to [Ar, Pe, V4, W4] for details about the mathematical theory.

The basis of this paper is taken from the Course Notes [V5]. The scheme of the proof of the main result follows [KV1], where convergence is proved in L^1, i.e., as in (1.5), and uniformly for solutions with compact support. This paper contains still another proof, based on asymptotic symmetry, detailed in Section 16. Newman's Lyapunov approach is studied in [N, R]. The simpler Lyapunov approach of Section 12 using the L^1-norm is more or less folklore, but I have not found it written. The convergence of the supports for compactly supported solutions is stated in [KV2]. This paper contains the proof of convergence for solutions with changing sign.

Barrier methods, like the one used by [FK] have proved to be very suitable in the study of nonlinear diffusion problems without a standard Lyapunov functional: a very interesting application occurs in the study of the equation of elastoplastic filtration, in [KPV], where no other method seems to be work by lack of known invariants.

Better rates of convergence for radial solutions are discussed in [V1] and [AV], but the main results of Sections 16, 17 are new. In one space dimension finer asymptotic results are known, cf. [ZB, V1, AV, An].

New developments include the optimality of the general rates in $\mathcal{X}_0$ and the weighted convergence for fast diffusion. Full proofs are given for both. We mention the new geometrical and smoothness developments in Section 18, the work by the energy method of Section 19. The study of complexity in L^∞ is only mentioned in passing. A number of less related results have been left out by lack of space.

Acknowledgment

The author is grateful to a number of colleagues for comments and suggestions, and very especially to the referee whose positive criticism has resulted in a clear improvement of the text. He also wants to apologize for unintended omissions of results.

REFERENCES

[AR]   ALIKAKOS, N. and ROSTAMIAN, R., *Large time behavior of solutions of Neumann boundary value problem for the porous medium equation*, Indiana Univ. Math. J. *30* (1981), 749–785.

[AR2]   ALIKAKOS, N. and ROSTAMIAN, R., *On the uniformization of the solutions of the porous medium equation in $\mathbb{R}^n$*, Israel J. Math. *47* (1984), 270–290.

[An]   ANGENENT, S., Large-time asymptotics of the porous media equation. *Nonlinear Diffusion Equations and their Equilibrium States I*, W. -M. Ni, L. A. Peletier and J. Serrin eds., (Berkeley, CA, 1986), Math. Sci. Res. Inst. Publ., Springer, New York, *12* (1988), 21–34.

[Ar] ARONSON, D. G., "The Porous Medium Equation", *Lecture Notes in Mathematics 1224*. Springer-Verlag, Berlin/New York, 1985.

[AB] ARONSON, D. G. and BÉNILAN, PH., *Régularité des solutions de l'équation des milieux poreux dans* $\mathbb{R}^n$, Comptes Rendus Ac. Sci. Paris, A *288* (1979), 103–105.

[AP] ARONSON, D. G. and PELETIER, L. A., *Large time behaviour of solutions of the porous medium equation in bounded domains*, J. Differ. Equat. *39* (1981), 378–412.

[AV] ARONSON, D. G. and VAZQUEZ, J. L. *Eventual C^∞-regularity and concavity of flows in one-dimensional porous media*, Archive Rat. Mech. Anal. *99* (1987), 329–348.

[B1] BARENBLATT, G. I., *On some unsteady motions of a liquid or a gas in a porous medium*, Prikl. Mat. Mekh. *16* (1952), 67–78 (in Russian).

[B2] BARENBLATT, G. I., *Scaling, self-similarity and intermediate asymptotics*, Cambridge Univ. Press, Cambridge, 1996.

[Be] BÉNILAN, PH., *Equations d'évolution dans un espace de Banach quelconque et applications*, Ph. D. Thesis, Univ. Orsay, 1972 (in French).

[Be2] BÉNILAN, PH., *Opérateurs accrétifs et semi-groupes dans les espaces L^p $(1 \leq p \leq \infty)$*, France-Japan Seminar, Tokyo, 1976.

[Be3] BENILAN, P., "A Strong Regularity L^p for Solutions of the Porous Media Equation," Research Notes in Math., vol. *89*, 39–58, Pitman, London, 1983.

[BC1] BÉNILAN, PH. and CRANDALL, M. G., The continuous dependence on φ of solutions of $u_t - \Delta\varphi(u) = 0$, Indiana Univ. Math. J. *30* (1981), 161–177.

[BC2] BÉNILAN, PH. and CRANDALL, M. G., Regularizing effects of homogeneous evolution equations, in *Contributions to Analysis and Geometry*, suppl. to Amer. Jour. Math., Baltimore, 1981, 23–39.

[Bk] BÉNILAN, PH., CRANDALL, M. G., and PAZY, A., *Evolution Equations Governed by Accretive Operators*, Book to appear.

[BCP] BÉNILAN, PH., CRANDALL, M. G., and PIERRE, M., *Solutions of the porous medium in $\mathbb{R}^N$ under optimal conditions on the initial values*, Indiana Univ. Math. Jour. *33* (1984), 51–87.

[BV] BÉNILAN, PH. and VAZQUEZ, J. L., *Concavity of solutions of the porous medium equation*, Trans. Amer. Math. Soc. *199* (1987), 81–93.

[BHV] BERNIS, F., HULSHOF, J., VAZQUEZ, J. L., *A very singular solution for the equation* $z_t = |z_{xx}|^{m-1}z_{xx}$ *and the asymptotic behaviour of general solutions*, J. Reine Ang. Math. *435* (1993), 1–31.

[Bo] BOUSSINESQ, J., *Recherches théoriques sur l'écoulement des nappes d'eau infiltrés dans le sol et sur le débit de sources*, Comptes Rendus Acad. Sci. / J. Math. Pures Appl. *10* (1903/04), 5–78.

[Br] BREZIS, H., "*Opérateurs maximaux monotones et semi-groupes de contractions dans les espaces de Hilbert*". North-Holland, 1973.

[Bu] BUCKMASTER, J., *Viscous sheets advancing over dry beds*, J. Fluid Mechanics, *81* (1977), 735–756.

[CF] CAFFARELLI, L. A. and FRIEDMAN, A., *Continuity of the density of a gas flow in a porous medium*, Trans. Amer. Math. Soc. *252* (1979), 99–113.

[CV] CAFFARELLI, L. A., J. L. V., *Viscosity solutions for the porous medium equation*. Proc. Symposia in Pure Mathematics, *65*, in honor of Profs. P. Lax and L. Nirenberg; M. Giaquinta et al. eds, 1999. pp. 13–26.

[CVW] CAFFARELLI, L. A., VAZQUEZ, J. L. and WOLANSKI, N. I., Lipschitz continuity of solutions and interfaces of the N-dimensional porous medium equation, Indiana Univ. Math. Jour. *36* (1987), 373–401.

[CT] CARRILLO, J. A. and TOSCANI, G., *Asymptotic L^1-decay of solutions of the porous medium equation to self-similarity*, Indiana Univ. Math. J. *49* (2000), 113–141.

[C5] CARRILLO, J. A. JÜNGEL, A. MARKOWICH, P. A. and TOSCANI, G., UNTERREITER, A. Entropy dissipation methods for degenerate parabolic systems and generalized Sobolev inequalities, Monatsh. Math. *133* (2001), 1–82.

[CVa] CARRILLO, L. A., VAZQUEZ, J. L., *Fine Asymptotics for Fast Diffusion Equations*, Preprint, 2002.

[CV]   CHASSEIGNE, E. and VÁZQUEZ, J. L., *Extended theory of fast diffusion equations in optimal classes of data. Radiation from singularities*, Arch. Rat. Mech. Anal., to appear.

[DB]   DIBENEDETTO, E., *Degenerate Parabolic Equations*, Springer Verlag, New York, 1993.

[DP]   DOLBEAULT, J. and DEL PINO, M., *Best constants for Gagliardo-Nirenberg inequalities and application to nonlinear diffusions*, Preprint Ceremade no. 0119; to appear in Journal Math. Pures Appl.

[DP2]   DOLBEAULT, J. and DEL PINO, M. *Nonlinear diffusions and optimal constants in Sobolev type inequalities: asymptotic behaviour of equations involving the p-Laplacian*, C. R. Math. Acad. Sci. Paris *334* (2002), no. 5, 365–370.

[ERV]   ESTEBAN, J. R., RODRÍGUEZ, A. and VAZQUEZ, J. L., *A nonlinear heat equation with singular diffusivity*, Commun. Partial Differ. Equat. *13* (1988), 985–1039.

[FK]   FRIEDMAN, A. KAMIN, S., *The asymptotic behavior of gas in an N-dimensional porous medium*, Trans. Amer. Math. Soc. *262* (1980), 551–563.

[GP]   GALAKTIONOV, V. A. and PELETIER, L. A., *Asymptotic behaviour near finite time extinction for the fast diffusion equation*, Arch. Rational Mech. Anal. *139* (1997), no. 1, 83–98.

[GPV]   GALAKTIONOV, V. A., PELETIER, L. A. and VAZQUEZ, J. L., *Asymptotics of the fast-diffusion equation with critical exponent*, SIAM J. Math. Anal. *31* (2000), no. 5, 1157–1174.

[GV]   GALAKTIONOV, V. A. and VAZQUEZ, J. L., *Geometrical Properties of the Solutions of One-dimensional Nonlinear Parabolic Equations*, Mathematische Annalen *303* (1995), 741–769.

[GMC]   GURTIN, M. E. and MCCAMY, R. C., *On the diffusion of biological populations*, Math. Biosc. *33* (1977), 35–49.

[Kl]   KALASHNIKOV, A. S., Some problems of the qualitative theory of non–linear degenerate second–order parabolic equations, Uspekhi Mat. Nauk *42*,(1987), No. 2, 135–176. English translation: Russian Math. Surveys *42* (1987), 169–222.

[K1]   KAMENOMOSTSKAYA (KAMIN), S., *The asymptotic behaviour of the solution of the filtration equation*, Israel J. Math. *14* (1973), 1, 76–87.

[K2]   KAMIN (KAMENOMOSTSKAYA), S., Similar solutions and the asymptotics of filtration equations, Arch. Rat. Mech. Anal. *60* (1976), 171–183.

[KPV]   KAMIN, S. PELETIER, L. A. and VAZQUEZ, J. L., *On the Barenblatt equation of elastoplastic filtration*, Indiana Univ. Math. Journal *40* (1991), no. 2, 1333–1362.

[KV1]   KAMIN, S. and VAZQUEZ, J. L., *Fundamental solutions and asymptotic behaviour for the p-Laplacian equation*, Rev. Mat. Iberoamericana *4* (1988), 339–354.

[KV2]   S. KAMIN, J. L. VAZQUEZ, *Asymptotic behaviour of the solutions of the porous medium equation with changing sign*, SIAM Jour. Math. Anal. *22* (1991), 34–45.

[Ki]   KING, J., *Self-similar behaviour for the equation of fast nonlinear diffusion*, Phil. Trans. Roy. Soc. London A *343* (1993), 337–375.

[LSU]   LADYZHENSKAYA, O. A., SOLONNIKOV, V. A. and URAL'TSEVA, N. N., *Linear and Quasilinear Equations of Parabolic Type,* Transl. Math. Monographs *23*, Amer. Math. Soc, Providence, 1968.

[Ls]   LASALLE, J. P., *The Stability of Dynamical Systems,* SIAM, Philadelphia, PA, 1976.

[LV]   LEE, K. A. and VAZQUEZ, J. L., *Geometrical properties of solutions of the Porous Medium Equation for large times,* Indiana Univ. Math. J., to appear.

[Ma]   MATANO, H., *Nonincrease of the lap number of a solution of a one-dimensional semilinear parabolic equation,* J. Fac. Sci. Univ. Tokyo, Sect. IA *29* (1982), 401–441.

[Mu]   MUSKAT, M., *The Flow of Homogeneous Fluids Through Porous Media,* McGraw-Hill, New York, 1937.

[N]   NEWMAN, W., *A Lyapunov functional for the evolution of solutions to the porous medium equation to self-similarity. I*, J. Math. Phys. *25* (1984), 3120–3123.

[OKC]   O. OLEĬNIK, KALASKNIKOV, S. A. and CZHOU, Y. L., *The Cauchy problem and boundary-value problems for equations of the type of unsteady filtration*, Izv. Akad. Nauk SSSR, Ser. Mat. 22 (1958), 667–704.

[Ot]   OTTO, F. The geometry of dissipative evolution equations: the porous medium equation, Comm. Partial Differential Equations *26* (2001), 101–174.

[Pa]   PATTLE, R. E., *Diffusion from an instantaneous point source with concentration dependent coefficient*, Quart. Jour. Mech. Appl. Math. *12* (1959), 407–409.

[Pe]   PELETIER, L. A., The Porous Medium Equation in *"Applications of Nonlinear Analysis in the Physical Sciences"*, H. Amann et al. eds., Pitman, London, 1981, 229–241.

[Pi]   PIERRE, M., Uniqueness of the solutions of $u_t - \Delta\phi(u) = 0$ with initial datum a measure, Nonlinear Anal. T. M. A. *6* (1982), 175–187.

[PS]   DEL PINO, M. and SAEZ, M., *On the Extinction Profile for Solutions of $u_t = \Delta u^{(N-2)/(N+2)}$*, Indiana Univ. Math. Journal *50* (2001), no. 2, 612–628.

[QV]   QUIRÓS, F. and VAZQUEZ, J. L., *Asymptotic behaviour of the porous media equation in an exterior domain*, Ann. Scuola Normale Sup. Pisa *28* (1999), 4, 183–227.

[R]   RALSTON, J., *A Lyapunov functional for the evolution of solutions to the porous medium equation to self-similarity. II*, J. Math. Phys. *25* (1984), 3124–3127.

[RV]   RODRÍGUEZ, A. and VAZQUEZ, J. L., *A well-posed problem in singular Fickian diffusion*, Archive Rat. Mech. Anal. *110* (1990), 2, 141–163.

[S]   SACKS, P., *Continuity of the solutions of a singular parabolic equation*, Nonlin. Anal. T.M.A. *7* (1983), 387–409.

[St]   STURM, C., *Mémoire sur une classe d'équations à différences partielles*, J. Math. Pure Appl. *1* (1836), 373–444.

[S4]   SAMARSKII, A. A., GALAKTIONOV, KURDYUMOV, S. P. and MIKHAILOV, A. P. *Blow- up in Quasilinear Parabolic Equations*, Nauka, Moscow, 1987 (in Russian); English translation: Walter de Gruyter, *19*, Berlin/New York, 1995.

[V1]   VAZQUEZ, J. L., *Asymptotic behaviour and propagation properties of the one-dimensional flow of gas in a porous medium*, Trans. Amer. Math. Soc. *277* (1983), 507–527.

[V2]   VAZQUEZ, J. L., *Behaviour of the velocity of one-dimensional flows in porous media*, Trans. Amer. Math. Soc. *286* (1984), 787–802.

[V3]   VAZQUEZ, J. L., *Symétrisation pour $u_t = \Delta\phi(u)$ et applications*, Comptes Rendus Acad. Sci. Paris *I*, *295* (1982), 71–74.

[V4]   VAZQUEZ, J. L., *An Introduction to the Mathematical Theory of the Porous Medium Equation*. In *Shape Optimization and Free Boundaries*, M. C. Delfour ed., Mathematical and Physical Sciences, Series C, vol. *380*, Kluwer Ac. Publ.; Dordrecht, Boston and Leiden; 1992. Pages 347–389.

[V5]   VAZQUEZ, J.L., "Asymptotic behaviour for the PME in a bounded domain. The Dirichlet problem", and "Asymptotic behaviour for the PME in the whole space", *Notes of the Ph. D.Course "Métodos Asintóticos en Ecuaciones de Evolución"*, UAM, spring 1997, *http://www.uam.es/~juanluis.vazquez*.

[VZ]   VAZQUEZ, J. L. and ZUAZUA, E., *Complexity of large time behaviour of evolution equations with bounded data*, Chinese Annals of Mathematics, 23, 2, ser. B (2002), 293-310. Volume in honor of J.L. Lions.

[Ve]   VÉRON, L., *Coercivité et propriétés régularisantes des semi-groupes non linéaires dans les espaces de Banach*, Ann. Fac. Sci. Toulouse *1* (1979), 171–200.

[W4]   ZHUOQUN WU, JINGXUE YIN, HUILAI LI, JUNNING ZHAO, *Nonlinear diffusion equations* World Scientific, Singapore, 2001.

[ZB]   ZEL'DOVICH, YA. B. and BARENBLATT, G. I., *The asymptotic properties of self-modelling solutions of the nonstationary gas filtration equations*, Soviet Phys. Doklady *3* (1958), 44–47 [Russian, Akad. Nauk SSSR, Doklady *118* (1958), 671–674].

[ZK]   ZEL'DOVICH, YA. B. and KOMPANEETS, A. S., *Towards a theory of heat conduction with thermal conductivity depending on the temperature. In "Collection of papers dedicated to 70th Anniversary of A. F. Ioffe"*, Izd. Akad. Nauk SSSR, Moscow, 1950, 61–72.

[ZR]   ZEL'DOVICH, YA. B. and RAIZER, YU. P., *Physics of Shock Waves and High-Temperature Hydrodynamic Phenomena* II, Academic Press, New York, 1966.

Juan Luis Vázquez
Dpto de Mathemáticas
Univ. Antónoma de Madrid
E-29049 Madrid
Spain
e-mail:juanluis.vazquuez@uam.es

J.evol.equ. 3 (2003) 119 – 135
1424–3199/03/010119 – 17
© Birkhäuser Verlag, Basel, 2003

**Journal of Evolution
Equations**

Dirichlet and Neumann boundary conditions:
What is in between?

Wolfgang Arendt and Mahamadi Warma*

Dédié à Philippe Bénilan

Abstract. Given an admissible measure μ on $\partial\Omega$ where $\Omega \subset \mathbb{R}^n$ is an open set, we define a realization Δ_μ of the Laplacian in $L^2(\Omega)$ with general Robin boundary conditions and we show that Δ_μ generates a holomorphic C_0-semigroup on $L^2(\Omega)$ which is sandwiched by the Dirichlet Laplacian and the Neumann Laplacian semigroups. Moreover, under a locality and a regularity assumption, the generator of each sandwiched semigroup is of the form Δ_μ. We also show that if $D(\Delta_\mu)$ contains smooth functions, then μ is of the form $d\mu = \beta d\sigma$ (where σ is the $(n-1)$-dimensional Hausdorff measure and β a positive measurable bounded function on $\partial\Omega$); i.e. we have the classical Robin boundary conditions.

0. Introduction

Let $\Omega \subset \mathbb{R}^n$ be an open set. Then it is standard to define selfadjoint realizations Δ^D and Δ^N of the Laplacian on $L^2(\Omega)$ with **Dirichlet boundary conditions**

$$u_{|\partial\Omega} = 0 \qquad \text{on} \quad \partial\Omega \tag{1}$$

or **Neumann boundary conditions**

$$\frac{\partial u}{\partial \nu}_{|\partial\Omega} = 0 \qquad \text{on} \quad \partial\Omega. \tag{2}$$

The exterior normal derivative $\frac{\partial u}{\partial \nu}$ may only exist in a weak form and actually the boundary of Ω may be so bad that no exterior normal can be defined.

Here we consider boundary conditions of the third kind

$$u d\mu + \frac{\partial u}{\partial \nu} d\sigma = 0 \qquad \text{on} \quad \partial\Omega \tag{3}$$

where μ is an (admissible) Borel measure on $\partial\Omega$ and σ is the surface measure if Ω is Lipschitz, or more generally the $(n-1)$-dimensional Hausdorff measure if Ω is arbitrary.

Mathematics Subject Classification (2000): 31C15, 31C25, 34D05, 35A15, 35J10, 47D07.

Key words: Dirichlet forms, Dirichlet, Neumann and Robin boundary conditions.

* This work is part of the DGF-Project: "Regularität und Asymptotik für elliptische und parabolische Probleme"

If $d\mu = \beta d\sigma$ for some $0 \leq \beta \in L^\infty(\partial\Omega, \sigma)$, then (3) reduces to the usual **Robin boundary conditions**

$$\beta u + \frac{\partial u}{\partial \nu} = 0 \qquad \text{on} \quad \partial\Omega. \tag{4}$$

We show by the method of quadratic forms that a selfadjoint realization Δ_μ of the Laplacian on $L^2(\Omega)$ can be associated to these kind of boundary conditions.

The semigroup $(e^{t\Delta_\mu})_{t\geq 0}$ generated by Δ_μ satisfies the following sandwich property

$$e^{t\Delta^D} \leq e^{t\Delta_\mu} \leq e^{t\Delta^N} \qquad (t \geq 0). \tag{5}$$

We show in this paper that conversely each symmetric semigroup T on $L^2(\Omega)$ satisfying $e^{t\Delta^D} \leq T(t) \leq e^{t\Delta^N}$ is of the form $T(t) = e^{t\Delta_\mu}$ provided a locality and a regularity assumption are satisfied.

This paper is based on some properties of admissible measures and relative capacity introduced in [AW]. But here we define the closed form directly with precise form domain. This is most convenient in order to establish the desired domination properties. On the way we prove a regularity result for such forms (Theorem 2.4) which is of independent interest. We also show that the measure μ is of the form $\beta d\sigma$ as soon as the domain of Δ_μ contains smooth functions.

Finally, we establish some asymptotic properties of the semigroup $e^{t\Delta_\mu}$ as $t \to \infty$ which are similar to those studied for Schrödinger semigroups in [ABB] and [Bat].

1. Preliminaries.

Let H be a Hilbert space over $\mathbb{R}$. A **positive form** on H is a bilinear mapping $a : D(a) \times D(a) \to \mathbb{R}$ such that $a(u, v) = a(v, u)$ and $a(u) := a(u, u) \geq 0$. Here $D(a)$ is a dense subspace of H, the **domain** of the form. The form a is called **closed** if $D(a)$ is complete for the norm $\|u\|_a = (a(u) + \|u\|_H^2)^{1/2}$. In that case we define the **operator A on H associated with** a by

$$\begin{cases} D(A) & = \{u \in D(a) : \exists v \in H, \ a(u, \varphi) = (v, \varphi)_H \ \forall \varphi \in D(a)\} \\ Au & = v. \end{cases}$$

Then A is selfadjoint and $-A$ generates a contraction C_0-semigroup $S = (S(t))_{t\geq 0}$ of symmetric operators on H. We also write $e^{-tA} = S(t)$ and call S **the semigroup associated with** a.

Now assume that $H = L^2(\Omega)$ where $(\Omega, \Sigma, \lambda)$ is a σ-finite measure space. We let $L^2(\Omega)_+ = \{f \in L^2(\Omega) : \ f \geq 0 \ \text{a.e.}\}$ and $F_+ = L^2(\Omega)_+ \cap F$ if F is a subspace of $L^2(\Omega)$.

Let S be the semigroup associated with a closed, positive form a on $L^2(\Omega)$. The **first Beurling-Deny criterion** [Dav, Theorem 1.3.2] asserts that S is positive (i.e., $S(t)L^2(\Omega)_+ \subset L^2(\Omega)_+$ for all $t \geq 0$) if and only if

$$u \in D(a) \quad \text{implies} \quad |u| \in D(a) \quad \text{and} \quad a(|u|) \leq a(u). \tag{6}$$

In that case, $u, v \in D(a)$ implies that $u \wedge v = \inf\{v, u\}$, $u \vee v = \sup\{v, u\} \in D(a)$ and the lattice operations are continuous in $(D(a), \|\cdot\|_a)$.

Assume that S is positive. Then the **second Beurling-Deny criterion** [Dav, Theorem 1.3.3] asserts that S is L^∞-contractive (i.e., if $f \in L^2(\Omega)$ satisfy $0 \leq f \leq 1$ then $0 \leq S(t)f \leq 1$ for all $t \geq 0$) if and only if

$$0 \leq u \in D(a) \quad \text{implies} \quad u \wedge 1 \in D(a) \quad \text{and} \quad a(u \wedge 1) \leq a(u). \tag{7}$$

In that case, the mapping $u \mapsto u \wedge 1$ is continuous from $D(a)_+$ into $D(a)_+$.

We say that a C_0-semigroup T on $L^2(\Omega)$ is **submarkovian** if it is positive and L^∞-contractive. For further information on this property we refer to [ArBe] and to [BC] in the nonlinear case. Notice that several authors ([Dav], [BH], [FOT]) call a symmetric Markov semigroup what we call a symmetric submarkovian semigroup on $L^2(\Omega)$. A **Dirichlet form** is a closed positive form satisfying the two Beurling-Deny criteria.

Now let b be a second closed, positive form on $L^2(\Omega)$ such that the associated semigroup T is positive. We say that $D(a)$ is an **ideal** of $D(b)$ if

a) $u \in D(a)$ implies $|u| \in D(a)$ and,
b) $0 \leq u \leq v$, $v \in D(a)$, $u \in D(b)$ implies $u \in D(a)$.

Ouhabaz's domination criterion [Ouh] says that

$$0 \leq S(t) \leq T(t) \qquad (t \geq 0) \tag{8}$$

if and only if $D(a)$ is an ideal of $D(b)$ and

$$a(u, v) \geq b(u, v) \quad \text{for all} \quad u, v \in D(a)_+. \tag{9}$$

2. Relative capacity and Robin boundary conditions

Let $\Omega \subset \mathbb{R}^n$ be an open set with boundary Γ. Let $H^1(\Omega) := \{u \in L^2(\Omega) : D_j u \in L^2(\Omega), \ j = 1, \ldots n\}$ be the first order Sobolev space and let $\widetilde{H}^1(\Omega)$ be the closure of $H^1(\Omega) \cap C_c(\bar{\Omega})$ in $H^1(\Omega)$. Here

$$C_c(\bar{\Omega}) = \{f : \bar{\Omega} \to \mathbb{R} \text{ continuous with compact support}\}.$$

If Ω has Lipschitz boundary, then $H^1(\Omega) = \widetilde{H}^1(\Omega)$. But e.g., $\widetilde{H}^1((0, 1) \cup (1, 2)) = H^1(0, 2) \neq H^1((0, 1) \cup (1, 2))$.

Using [GT, Lemma 7.6 p.152] one easily sees that $u \in \widetilde{H}^1(\Omega)$ implies that $|u| \in \widetilde{H}^1(\Omega)$ and $D_j|u| = (\operatorname{sign} u)D_j u$ $(j = 1, \ldots n)$. Hence $\| \, |u| \, \|_{\widetilde{H}^1(\Omega)} = \|u\|_{\widetilde{H}^1(\Omega)}$. This implies in particular that the mapping $u \mapsto |u|$ is continuous on $\widetilde{H}^1(\Omega)$. Also for $v \in \widetilde{H}^1(\Omega)$, the mappings $u \mapsto v \wedge u$ and $u \mapsto v \vee u$ are continuous.

We define the **relative capacity** $\operatorname{Cap}_{\bar{\Omega}}(A)$ of a subset A of $\bar{\Omega}$ by

$$\operatorname{Cap}_{\bar{\Omega}}(A) := \inf \{\|u\|^2_{H^1(\Omega)} : u \in \widetilde{H}^1(\Omega), \exists \, O \subset \mathbb{R}^n \text{ open such that}$$
$$A \subset O \text{ and } u(x) \geq 1 \text{ a.e. on } \Omega \cap O\}. \tag{10}$$

Here and elsewhere the word relative stands for relative with respect to $\bar{\Omega}$.

Then $\operatorname{Cap}_{\bar{\Omega}}$ is an outer measure on $\bar{\Omega}$. This notion of capacity is induced by the regular Dirichlet form $\mathcal{E}$ on $L^2(\Omega)$ given by

$$\mathcal{E}(u, v) := \int_\Omega \nabla u \nabla v \, dx$$

with domain $D(\mathcal{E}) = \widetilde{H}^1(\Omega)$, in the sense of [BH, I 8.1.1 p.52]. Here the underlying locally compact space is $X = \bar{\Omega}$, with Borel σ-algebra $\mathcal{B}(\bar{\Omega})$ and the measure $m(A) = \lambda(A \cap \Omega)$ $(A \in \mathcal{B}(\bar{\Omega}))$ (to make sure that $L^2(X, \mathcal{B}(\bar{\Omega}), m) = L^2(\Omega)$, the usual space with Lebesgue measure).

But now we may consider functions in $\widetilde{H}^1(\Omega)$ as defined on $\bar{\Omega}$.

We say that $A \subset \bar{\Omega}$ is **relatively polar** if $\operatorname{Cap}_{\bar{\Omega}}(A) = 0$. A property is said to hold **relatively quasi-everywhere (r.q.e.)** if it holds on $\bar{\Omega} \setminus N$ where $N \subset \bar{\Omega}$ is relatively polar.

A function $u : \bar{\Omega} \to \mathbb{R}$ is called **relatively quasi-continuous**, if for each $\varepsilon > 0$ there exists a relatively open set $G \subset \bar{\Omega}$ such that $\operatorname{Cap}_{\bar{\Omega}}(G) < \varepsilon$ and u is continuous on $\bar{\Omega} \setminus G$. Then by [BH, I, Proposition 8.2.1] for each $u \in \widetilde{H}^1(\Omega)$ there exists a relatively quasi-continuous function $\tilde{u} : \bar{\Omega} \to \mathbb{R}$ such that $u = \tilde{u}$ a.e. The function $\tilde{u}$ is relatively quasi-everywhere unique and we call it the **relatively quasi-continuous representative** of u. Moreover, $\tilde{u}$ may be chosen Borel measurable.

Finally we recall the following result which we shall use frequently (see [FOT, Theorem 2.1.4 p.69] or [BH, I, Proposition 8.2.5]).

PROPOSITION 2.1. *Let* $\lim_{m \to \infty} u_m = u$ *in* $\widetilde{H}^1(\Omega)$. *Then there exists a subsequence* $(\tilde{u}_{m_k})$ *such that* $\lim_{k \to \infty} \tilde{u}_{m_k}(x) = \tilde{u}(x)$ *r.q.e.*

REMARK 2.2. The notion of relative capacity was introduced in [AW]. It is clear that polar subsets of $\bar{\Omega}$ are relatively polar. The converse is true for subsets of Ω, and it is also true for subsets of $\bar{\Omega}$ if the boundary is Lipschitz. But if Ω is not regular, then there may exist relatively polar sets in the boundary which are not polar.

With the help of the relatively quasi-continuous representative the space $H_0^1(\Omega) := \overline{\mathcal{D}(\Omega)}^{H^1(\Omega)}$ may now be described as follows [AW, Theorem 2.3]

$$H_0^1(\Omega) = \{u \in \widetilde{H}^1(\Omega) : \tilde{u} = 0 \text{ r.q.e. on } \Gamma\}. \tag{11}$$

Now we define the class of measures by which we define general Robin boundary conditions.

DEFINITION 2.3. An **admissible measure** is a measure $\mu : \mathcal{B}(\Gamma_\mu) \to [0, \infty)$ where $\Gamma_\mu \subset \Gamma$ is relatively open, such that

   a) $\mu(K) < \infty$ for each compact set $K \subset \Gamma_\mu$; and
   b) $\mathrm{Cap}_{\bar\Omega}(A) = 0$ implies $\mu(A) = 0$ for each Borel set $A \subset \Gamma_\mu$.

The set Γ_μ is called **the domain of** μ. Here $\mathcal{B}(\Gamma_\mu)$ denotes the Borel σ-algebra of Γ_μ.

Let μ be an admissible measure with domain Γ_μ. Let $L^2(\Gamma_\mu) = L^2(\Gamma_\mu, \mathcal{B}(\Gamma_\mu), \mu)$. We define a form a_μ on $L^2(\Omega)$ by

$$D(a_\mu) = \left\{u \in \widetilde{H}^1(\Omega) : \tilde{u} = 0 \text{ r.q.e. on } \Gamma \setminus \Gamma_\mu, \int_{\Gamma_\mu} |\tilde{u}|^2 \, d\mu < \infty\right\},$$

$$a_\mu(u, v) = \int_\Omega \nabla u \nabla v \, dx + \int_{\Gamma_\mu} \tilde{u}\tilde{v} \, d\mu.$$

Here $\tilde{u}$ is the relatively quasi-continuous representative of u which we always choose Borel measurable. Since μ is admissible, a_μ is well-defined. In fact, let $u, u_1, v, v_1 \in D(a_\mu)$ such that $u = u_1, v = v_1$ a.e. on Ω. Then it follows from [BH, I, Proposition 8.1.6] that $\tilde{u} = \tilde{u}_1$ and $\tilde{v} = \tilde{v}_1$ r.q.e. on $\bar\Omega$. Since μ is admissible, it follows that $\tilde{u} = \tilde{u}_1$ and $\tilde{v} = \tilde{v}_1$ μ-a.e. on Γ_μ and thus $a_\mu(u, v) = a_\mu(u_1, v_1)$.

THEOREM 2.4. *Let μ be an admissible measure. Then a_μ is a Dirichlet form on $L^2(\Omega)$. The space $H^1(\Omega) \cap C_c(\Omega \cup \Gamma_\mu)$ is a form core of a_μ.*

   Proof. a) The mapping $D(a_\mu) \to \widetilde{H}^1(\Omega) \oplus L^2(\Gamma_\mu), u \mapsto (u, \tilde{u}_{|\Gamma_\mu})$ is isometric. In order to show that a_μ is closed, it suffices to show that the image of the mapping is closed. Let $u_m \in D(a_\mu)$ such that $u_m \to u$ in $\widetilde{H}^1(\Omega)$ and $\tilde{u}_m \to f$ in $L^2(\Gamma_\mu)$. Taking a subsequence, we can assume that $\tilde{u}_m \to \tilde{u}$ r.q.e. and $\tilde{u}_m \to f$ μ-a.e. on Γ_μ. Since μ is admissible it follows that $f = \tilde{u}$ in $L^2(\Gamma_\mu)$.
   b) Since $|u|\tilde{} = |\tilde{u}|$ and $(u \wedge 1)\tilde{} = \tilde{u} \wedge 1$ it follows that the two criteria of Beurling-Deny are satisfied.
   c) In order to show that $H^1(\Omega) \cap C_c(\Omega \cup \Gamma_\mu)$ is a form core, let $u \in D(a_\mu)$. We can assume that $u \geq 0$ a.e.

FIRST CASE. We assume that u is bounded; $u \leq c$, say. There exists a sequence $(u_m)_{m \in \mathbb{N}}$ in $H^1(\Omega) \cap C_c(\Omega \cup \Gamma_\mu)$ such that $u_m \to u$ in $\widetilde{H}^1(\Omega)$. We may assume that $u_m \to \tilde{u}$ r.q.e. Let $v_m = 0 \vee (u_m \wedge \tilde{u})$. Then $v_m \to u$ in $\widetilde{H}^1(\Omega)$ and, by the Dominated Convergence Theorem, $v_{m|_{\Gamma_\mu}} \to \tilde{u}_{|_{\Gamma_\mu}}$ in $L^2(\Gamma_\mu, \mu)$. Fix $m \in \mathbb{N}$. Let $O \subset \Gamma_\mu$ be relatively open such that $\bar{O}$ is compact, $\bar{O} \subset \Gamma_\mu$ and $\operatorname{supp}[v_m] \subset O \cup \Omega$. Since v_m is bounded and $\mu(O) < \infty$, it follows that $v_m \in D(a_\mu)$. By [FOT, Corollary 2.3.1] there exists a sequence $(w_k)_{k \in \mathbb{N}}$ in $H^1(\Omega) \cap C_c(\Omega \cup \Gamma_\mu)$ such that $w_k \to v_m$ $(k \to \infty)$ in $\widetilde{H}^1(\Omega)$ and r.q.e. Let $f_k = (0 \vee w_k) \wedge c$. Then $f_k \in H^1(\Omega) \cap C_c(\Omega \cup \Gamma_\mu)$ and $f_k \to v_m$ in $\widetilde{H}^1(\Omega)$ and $f_{k|_{\Gamma_\mu}} \to v_{m|_{\Gamma_\mu}}$ in $L^2(\Gamma_\mu)$ as $k \to \infty$. Hence u is in the closure of $H^1(\Omega) \cap C_c(\Omega \cup \Gamma_\mu)$ in $D(a_\mu)$.

SECOND CASE. The function u is not bounded. By the first case $u \wedge k$ can be approximated by functions in $H^1(\Omega) \cap C_c(\Omega \cup \Gamma_\mu)$. But $u_k \to u$ in $\widetilde{H}^1(\Omega)$ and $u_{k|_{\Gamma_\mu}} \to u_{|_{\Gamma_\mu}}$ in $L^2(\Gamma_\mu)$ as $k \to \infty$. This proves the claim of the theorem. $\qquad \square$

3. Monotonicity properties

Let μ be an admissible measure. We denote by A_μ the operator associated with a_μ. Then

$$A_\mu u = -\Delta u \qquad \text{in } \mathcal{D}(\Omega)' \tag{12}$$

for all $u \in D(A_\mu)$. In fact, let $A_\mu u = v$. Then it follows from the definition of the associated operator that

$$\int_\Omega \nabla u \nabla \varphi \, dx = a_\mu(u, \varphi) = \int_\Omega v\varphi \, dx$$

for all $\varphi \in \mathcal{D}(\Omega)$. This implies (12). We let $\Delta_\mu := -A_\mu$. Thus Δ_μ is a symmetric realization of the Laplacian in $L^2(\Omega)$.

If $\Gamma_\mu = \emptyset$, then $D(a_\mu) = H_0^1(\Omega)$ and $-A_\mu$ is just the **Dirichlet Laplacian** Δ^D given by

$$\begin{cases} D(\Delta^D) = \{u \in H_0^1(\Omega) : \ \Delta u \in L^2(\Omega)\} \\ \Delta^D u = \Delta u \qquad (\text{in } \mathcal{D}(\Omega)'). \end{cases}$$

If $\Gamma_\mu = \Gamma$ and $\mu = 0$, then $-A_\mu$ is the **Neumann Laplacian** Δ^N whose domain consists of all $u \in \widetilde{H}^1(\Omega)$ such that $\Delta u \in L^2(\Omega)$ and

$$\int_\Omega \Delta u\varphi \, dx = -\int_\Omega \nabla u \nabla \varphi \, dx \tag{13}$$

for all $\varphi \in \widetilde{H}^1(\Omega)$. In view of Green's formula, (13) may be seen as a weak formulation of Neumann boundary conditions

$$\frac{\partial u}{\partial \nu}\Big|_\Gamma = 0 \qquad \text{on } \ \Gamma.$$

Next, we show the following domination property.

THEOREM 3.1. *For each admissible measure μ, the semigroup $(e^{t\Delta_\mu})_{t\geq 0}$ satisfies*

$$e^{t\Delta^D} \leq e^{t\Delta_\mu} \leq e^{t\Delta^N} \tag{14}$$

for all $t \geq 0$ in the sense of positive operators.

Proof. 1) We show that $e^{t\Delta^D} \leq e^{t\Delta_\mu}$. By Ouhabaz's domination criterion, it suffices to prove that $H_0^1(\Omega)$ is an ideal of $D(a_\mu)$ and $a_\mu(u, v) \leq \int_\Omega \nabla u \nabla v \, dx$ for all $u, v \in H_0^1(\Omega)_+$. We may assume that functions in $\widetilde{H}^1(\Omega)$ are r.q.c.

a) We claim that $H_0^1(\Omega)$ is an ideal of $D(a_\mu)$. In fact, let $u \in H_0^1(\Omega)$ and $v \in D(a_\mu)$ such that $0 \leq v \leq u$. Since $\bar{\Omega}$ is relatively open, it follows from [FOT, Lemma 2.1.4] that $0 \leq v \leq u$ r.q.e. on $\bar{\Omega}$. Using the characterization of $H_0^1(\Omega)$ given by (11), we have that $u = 0$ r.q.e. on Γ and thus $v = 0$ r.q.e. on Γ. Therefore $v \in H_0^1(\Omega)$ which proves the claim.

b) Let $u, v \in H_0^1(\Omega)_+$. By the characterization of $H_0^1(\Omega)$, we have that $u = v = 0$ r.q.e. on Γ. Since μ is admissible, it follows that $u = v = 0$ μ a.e. on Γ_μ. We finally obtain that

$$\begin{aligned}
a_\mu(u, v) &:= \int_\Omega \nabla u \nabla v \, dx + \int_{\Gamma_\mu} uv \, d\mu \\
&= \int_\Omega \nabla u \nabla v \, dx
\end{aligned}$$

and the proof of this part is complete.

2) The proof of the inequality $e^{t\Delta_\mu} \leq e^{t\Delta^N}$ is a simple modification of the first part. $\square$

We will see in the next section that (14) characterizes the semigroups $(e^{t\Delta_\mu})_{t\geq 0}$. Before that we prove a monotonicity and uniqueness result.

THEOREM 3.2. *Let μ, ν be two admissible measures. The following assertions are equivalent.*

(i) $e^{t\Delta_\mu} \leq e^{t\Delta_\nu}$ $(t \geq 0)$.

(ii) (a) $\text{Cap}_{\bar{\Omega}}(\Gamma_\mu \setminus \Gamma_\nu) = 0$ *and*

    (b) $\mu(A) \geq \nu(A)$ *for each Borel set* $A \subset \Gamma_\mu \cap \Gamma_\nu$.

Proof. (i) $\Rightarrow$ (ii). (a) Let $K_m \subset \Gamma_\mu$ be compact sets such that $K_m \subset K_{m+1}$ and $\bigcup_{m\in\mathbb{N}} K_m = \Gamma_\mu$. Let $O \subset \mathbb{R}^n$ be open such that $\Gamma_\mu = O \cap \Gamma$. Let $u \in \mathcal{D}(\mathbb{R}^n)$ such that $0 \leq u \leq 1$, $u = 1$ on K_m and $\text{supp}[u] \subset O$. Then $u_{|\bar{\Omega}} \in D(a_\mu)$. Since $D(a_\mu) \subset D(a_\nu)$ by the domination criterion, it follows that $u(z) = 0$ r.q.e. on $\Gamma \setminus \Gamma_\nu$. In particular, $\text{Cap}_{\bar{\Omega}}(K_m \setminus \Gamma_\nu) = 0$. Thus $\text{Cap}_{\bar{\Omega}}(\Gamma_\mu \setminus \Gamma_\nu) = \lim_{n\to\infty} \text{Cap}_{\bar{\Omega}}(K_m \setminus \Gamma_\nu) = 0$.

(b) Let $A \subset \Gamma_\mu \cap \Gamma_\nu$ be a Borel set. Since by [Rud, 2.18 p.48] μ and ν are regular, it suffices to show that $\nu(K) \leq \mu(O)$ where $K \subset A$ is a compact set and O is a relatively open subset in $\Gamma_\mu \cap \Gamma_\nu$ containing A. Let $V \subset \mathbb{R}^n$ be open such that $V \cap \Gamma = \emptyset$. Let $u \in \mathcal{D}(\mathbb{R}^n)$ such that $\mathrm{supp}[u] \subset V, 0 \leq u \leq 1$ and $u = 1$ on K. Then $u_{|\bar{\Omega}} \in D(a_\mu)$. Hence

$$\int_\Omega |\nabla u|^2 \, dx + \int_{\Gamma_\mu} |u|^2 \, d\mu$$

$$= a_\mu(u) \geq a_\nu(u)$$

$$= \int_\Omega |\nabla u|^2 \, dx + \int_{\Gamma_\nu} |u|^2 \, d\nu$$

by the domination criterion. Consequently,

$$\nu(K) \leq \int_{\Gamma_\nu} |u|^2 \, d\nu \leq \int_{\Gamma_\mu} |u|^2 \, d\mu \leq \mu(O).$$

(ii) $\Rightarrow$ (i). Let $u \in D(a_\mu)$. Then (a) implies that $\tilde{u} = 0$ r.q.e. on $\Gamma \setminus \Gamma_\nu$. Since μ is admissible, $\mu(\Gamma_\mu \setminus \Gamma_\nu) = 0$. Hence

$$\int_{\Gamma_\nu} |\tilde{u}|^2 \, d\nu = \int_{\Gamma_\mu \cap \Gamma_\nu} |\tilde{u}|^2 \, d\nu$$

$$= \int_0^\infty \nu(\{z \in \Gamma_\mu \cap \Gamma_\nu : |\tilde{u}(z)|^2 > t\}) \, dt$$

$$\leq \int_0^\infty \mu(\{z \in \Gamma_\mu \cap \Gamma_\nu : |\tilde{u}(z)|^2 > t\}) \, dt$$

$$= \int_{\Gamma_\mu} |\tilde{u}(z)|^2 \, d\mu(z) < \infty.$$

Thus $u \in D(a_\nu)$. We have shown that $D(a_\mu) \subset D(a_\nu)$. Since $D(a_\mu)$ is an ideal of $\widetilde{H}^1(\Omega)$, it is also an ideal of $D(a_\nu)$. Let $u, v \in D(a_\nu)_+$. One proves similarly as in Theorem 3.1. that $a_\mu(u, v) \geq a_\nu(u, v)$. Now the domination criterion implies (i). $\square$

As corollary we note the following uniqueness theorem.

COROLLARY 3.3. *Let μ and ν be two admissible measures. The following assertions are equivalent.*

(i) $e^{t\Delta_\mu} = e^{t\Delta_\nu} \qquad (t \geq 0).$

(ii) $\mathrm{Cap}_{\bar{\Omega}}(\Gamma_\mu \Delta \Gamma_\nu) = 0$ *and* $\mu(A) = \nu(A)$ *for each Borel set* $A \subset \Gamma_\mu \cap \Gamma_\nu$.

4. Sandwiched semigroups.

In this section we show that the sandwich property (14) characterizes the semigroups $(e^{t\Delta_\mu})_{t\geq 0}$ under suitable conditions. Let Ω be an open subset of $\mathbb{R}^n$ with boundary Γ.

THEOREM 4.1. *Let T be a symmetric C_0-semigroup on $L^2(\Omega)$ associated with a positive closed form $(a, D(a))$. Then the following assertions are equivalent.*

(i) *There exists an admissible measure μ such that $a = a_\mu$.*

(ii) (a) *One has $e^{t\Delta^D} \le T(t) \le e^{t\Delta^N}$ $(t \ge 0)$;*

    (b) $\operatorname{supp}[u] \cap \operatorname{supp}[v] = \emptyset$ *implies $a(u, v) = 0$ for all $u, v \in D(a) \cap C_c(\bar{\Omega})$.*

    (c) $D(a) \cap C_c(\bar{\Omega})$ *is dense in $(D(a), \|\cdot\|_a)$.*

Proof. We know that the conditions in (ii) are necessary. In order to prove the converse assume that (ii) is satisfied. Then by the domination criterion, $D(a)$ is an ideal of $\tilde{H}^1(\Omega)$ containing $H_0^1(\Omega)$, and

$$b(u, v) := a(u, v) - \int_\Omega \nabla u \nabla v \, dx \tag{15}$$

is positive whenever $0 \le u, v \in D(a)$. Let $\Gamma_0 := \{z \in \Gamma : \exists u \in D(a) \cap C_c(\bar{\Omega}), u(z) \ne 0\}$ and let $Y = \Omega \cup \Gamma_0$. Notice that Y is a locally compact space. Since $D(a)$ is an ideal of $\tilde{H}^1(\Omega)$ one has

$$\tilde{H}^1(\Omega) \cap C_c(Y) = D(a) \cap C_c(Y) =: E_c. \tag{16}$$

The space E_c is a subalgebra of $C_c(Y)$ by [BH, I, Corollary 3.3.2] (or [FOT, Theorem 1.4.2 (ii)]). It follows from the Stone-Weierstrass Theorem that E_c is uniformly dense in $C_c(Y)$. From this follows that E_c is also dense in $C_c(Y)$ for the inductive topology. In fact, we observe first that a is a Dirichlet form since $T(t) \le e^{t\Delta^N}$ and $(e^{t\Delta^N})_{t\ge 0}$ is submarkovian.

Let $0 \le u \in C_c(Y)$ and $\varepsilon > 0$. There exists $0 \le v \in E_c$ such that $\|u - v\|_\infty \le \varepsilon$. Then $(v - \varepsilon)^+ \in E_c$, $\operatorname{supp}[(v - \varepsilon)^+] \subset \operatorname{supp}[u]$ and

$$\|u - (v - \varepsilon)^+\|_\infty \le \|u - v\|_\infty + \|v - (v - \varepsilon)^+\|_\infty \le 2\varepsilon.$$

This shows that u can be approximated in the inductive topology by functions in E_c.

Now b is a positve bilinear form on E_c (i.e., $b(u, v) \ge 0$ whenever $0 \le u, v \in E_c$). Thus b is continuous for the inductive topology. Hence there exists a unique positive bilinear form $\tilde{b}$ on $C_c(Y)$ extending b. Consequently, there exists a unique positive functional Φ on $C_c(Y \times Y)$ such that $\Phi(u \otimes v) = \tilde{b}(u, v)$ for all $u, v \in C_c(Y)$ (cf. [Bou, Chap. III., Section 4] or [Sch, p.297] and the proof of [FOT, Lemma 1.4.1]). Hence there exists a unique regular Borel measure ν on $Y \times Y$ such that

$$b(u, v) = \int_{Y \times Y} u(x) v(y) \, d\nu$$

for all $u, v \in E_c$. Observe that $\int_{Y \times Y} u(x) v(y) \, d\nu = 0$ for $u, v \in C_c(Y)$ such that $\operatorname{supp}[u] \cap \operatorname{supp}[v] = \emptyset$. In fact, if $u, v \in E_c$ this follows from the assumption. But in general, by [FOT, Lemma 1.4.2 (ii)] there exist $u_n, v_n \in E_c$ with $\operatorname{supp}[u_n] \subset \{y \in Y : u(y) \ne 0\}$

and $\mathrm{supp}[v_n] \subset \{y \in Y : \ v(y) \neq 0\}$ such that u_n, v_n converge uniformly to u and v, respectively. Hence $\int_{Y \times Y} u(x)v(y) \ dv = \lim_{n \to \infty} \int_{Y \times Y} u_n(x)v_n(y) \ dv = 0$. Thus $\mathrm{supp}[v] \subset \{(y, y) : \ y \in Y\} \subset Y \times Y$. Hence there exists a regular Borel measure μ on Y such that

$$b(u, v) = \int_Y u(x)v(x) \, d\mu$$

for all $u, v \in E_c$.

By the domination property (9), one has $b = 0$ on $H_0^1(\Omega) \times H_0^1(\Omega)$. Thus it follows that $\mathrm{supp}[\mu] \subset \Gamma_0$. We have shown that

$$a(u, v) = \int_\Omega \nabla u \nabla v \, dx + \int_{\Gamma_0} uv \, d\mu \tag{17}$$

for all $u \in E_c$. Next we show that

$$D(a) \cap C_c(\bar{\Omega}) = \left\{ u \in H^1(\Omega) \cap C_c(\bar{\Omega}) : \ u_{|\Gamma \setminus \Gamma_0} = 0, \ \int_{\Gamma_0} |u|^2 \, d\mu < \infty \right\} =: F_\mu \tag{18}$$

and that (17) remains true for all $u, v \in F_\mu$.

In order to prove (18) it suffices to consider positive functions. Let $0 \leq u \in F_\mu$. Then $(u - \varepsilon)^+ \in H^1(\Omega) \cap C_c(\Omega \cup \Gamma_0) = E_c$ (by (16)) for all $\varepsilon > 0$. Moreover, $(u - \varepsilon)^+ \to u$ in $H^1(\Omega)$ and $(u - \varepsilon)^+_{|\Gamma_0} \to u_{|\Gamma_0}$ in $L^2(\Gamma_0)$ as $\varepsilon \downarrow 0$. Hence $(u - \varepsilon)^+$ is a Cauchy net in $D(a)$. Thus $u \in D(a)$ and

$$a(u) = \lim_{\varepsilon \downarrow 0} a((u - \varepsilon)^+) = \lim_{\varepsilon \downarrow 0} \left(\int_\Omega |\nabla(u - \varepsilon)^+|^2 \, dx + \int_{\Gamma_0} ((u - \varepsilon)^+)^2 \, d\mu \right)$$

$$= \int_\Omega |\nabla u|^2 \, dx + \int_{\Gamma_0} |u|^2 \, d\mu.$$

Conversely, let $0 \leq u \in D(a) \cap C_c(\bar{\Omega})$. Since a is a Dirichlet form $(u - \varepsilon)^+$ converges to u in $D(a)$ as $\varepsilon \downarrow 0$. Moreover, $(u - \varepsilon)^+ \in F_\mu$. Hence

$$a(u) = \lim_{\varepsilon \downarrow 0} a((u - \varepsilon)^+) = \int_\Omega |\nabla u|^2 \, dx + \int_{\Gamma_0} |u|^2 \, d\mu.$$

We have proved (18) and (17) for $u = v$. The polarization identity shows that (17) holds for all $u, v \in F_\mu$. Since a is closed it follows from [AW, Theorem 2.3] that μ is admissible. Let $\Gamma_\mu = \Gamma_0$. Now Theorem 2.4 implies that $a = a_\mu$. $\qquad \square$

Next we characterize those sandwiched semigroups which come from a bounded measure.

COROLLARY 4.2. *Let Ω be bounded. Let T be a symmetric C_0-semigroup on $L^2(\Omega)$ associated with a positive closed form $(a, D(a))$. Then the following assertions are equivalent.*

(i) *There exists a bounded admissible measure μ on Γ such that $a = a_\mu$.*

(ii) (a) *One has $e^{t\Delta^D} \leq T(t) \leq e^{t\Delta^N}$ $(t \geq 0)$;*

    (b) $\operatorname{supp}[u] \cap \operatorname{supp}[v] = \emptyset$ *implies $a(u, v) = 0$ for all $u, v \in D(a) \cap C(\bar{\Omega})$.*

    (c) $1 \in D(a)$.

Proof. Assume that (ii) holds. We keep the notations of the proof of Theorem 4.1. Since $1 \in D(a)$, it follows from (18) that $\Gamma_\mu = \Gamma_0 = \Gamma$, that μ is a bounded admissible measure and that $D(a_\mu) \subset D(a)$ and $a(u, v) = a_\mu(u, v)$ for all $u, v \in D(a_\mu)$. Let $0 \leq u \in D(a)$. Then for $k \in \mathbb{N}$, $u \wedge k \in D(a_\mu)$ and by (7),

$$a_\mu(u \wedge k) = a(u \wedge k) = k^2 a\left(\frac{u}{k} \wedge 1\right) \leq k^2 a\left(\frac{u}{k}\right) = a(u).$$

Thus $(u \wedge k)$ is bounded in $(D(a_\mu), \|\cdot\|_{a_\mu})$ and converges to u in $L^2(\Omega)$. It follows that $(u \wedge k)$ converges weakly to u in $D(a)$. Thus $u \in D(a_\mu)$. We have shown that $D(a) = D(a_\mu)$. This proves (i). The other implication is clear. $\qquad\square$

We give several comments concerning Theorem 4.1 and Corollary 4.2. First of all, it is remarkable that in the situation of Corollary 4.2; i.e. assuming that $D(a)$ contains a stricty positive continuous function, the form a is automatically regular (i.e., $D(a) \cap C(\bar{\Omega})$ is dense in $D(a)$). In general, the situation is more complicated. Choosing Γ_μ open in Definition 2.3 we could prove in Theorem 2.4 that the form a_μ is regular. This shows in particular that condition (c) in Theorem 4.1 is satisfied for $a - a_\mu$. But we might consider the more general case where Γ_μ is merely a Borel set. In the following we do this for the special case where the measure μ is 0.

Let $\Omega \subset \mathbb{R}^n$ be an open set with boundary Γ.

EXAMPLE 4.3. (Dirichlet-Neumann boundary conditions) Let $\Gamma_0 \subset \Gamma$ be a Borel set. We define

$$J(\Gamma_0) := \{u \in \tilde{H}^1(\Omega) : \tilde{u} = 0 \text{ r.q.e. on } \Gamma \setminus \Gamma_0\}.$$

Then $J(\Gamma_0)$ is a closed ideal of $\tilde{H}^1(\Omega)$. Let $D(a) = J(\Gamma_0)$, $a(u, v) = \int_\Omega \nabla u \nabla v \, dx$. Then a is a Dirichlet form on $L^2(\Omega)$ and the associated semigroup T satisfies

$$e^{t\Delta^D} \leq T(t) \leq e^{t\Delta^N} \qquad (t \geq 0). \tag{19}$$

This follows from the domination criterion (9).

Now we describe under which conditions Γ_0 may be chosen relatively open in Γ. If $\Gamma_0 \subset \Gamma$ is relatively open, then it follows from [FOT, Corollary 2.3.1] that the space

$H^1(\Omega) \cap C_c(\Omega \cup \Gamma_0)$ is dense in $J(\Gamma_0)$. Conversely, assume that J is a closed ideal of $\widetilde{H}^1(\Omega)$ containing $H_0^1(\Omega)$. Assume that $J \cap C_c(\bar{\Omega})$ is a dense subspace of J. Let $\Gamma_0 = \{z \in \Gamma : \exists u \in J \cap C_c(\bar{\Omega}) \text{ such that } u(z) \neq 0\}$. Then $J = J(\Gamma_0)$. In fact, since J is an ideal, and $H_0^1(\Omega) \subset J$ it follows that $H^1(\Omega) \cap C_c(\Omega \cup \Gamma_0) \subset J \subset J(\Gamma_0)$. Now the claim follows from the preceding.

REMARK 4.4. By a result of Stollmann [Sto] each closed ideal J of $\widetilde{H}^1(\Omega)$ containing $H_0^1(\Omega)$ is of the form $J = J(\Gamma_0)$ for some Borel set $\Gamma_0 \subset \Gamma$.

Next we comment on the locality condition. It cannot be omitted as the following simple example shows.

EXAMPLE 4.5. (Non-local boundary conditions) Let $\Omega = (0, 1)$. Define the form a by $D(a) = H^1(0, 1)$,

$$a(u, v) = \int_0^1 u'v' \, dx + u(0)v(0) + u(1)v(0) + u(0)v(1) + u(1)v(1).$$

Then a is a closed positive form which is not local. Let T be the associated semigroup on $L^2(0, 1)$. Then condition (a) of Corollary 4.2 is satisfied by the domination criterion. However condition (b) is not satisfied.

For further properties of local forms we refer to [BH], [FOT] and [MR]. For locality properties of the Laplacian we refer to Bénilan-Pierre [BP].

5. The surface measure

Let $\Omega \subset \mathbb{R}^n$ be a bounded open set with Lipschitz boundary Γ. By $\sigma = \mathcal{H}^{n-1}$ we denote the surface measure on Γ. Then σ is admissible [AW, Proposition 4.1]. Recall that $H^1(\Omega) \cap C(\bar{\Omega})$ is dense in $H^1(\Omega)$ (i.e. $\widetilde{H}^1(\Omega) = H^1(\Omega)$) and the trace $u \mapsto u_{|_\Gamma}$ defined for $u \in H^1(\Omega) \cap C(\bar{\Omega})$ has a continuous extension from $H^1(\Omega)$ into $L^2(\Gamma)$. In order words, one has $\tilde{u} \in L^2(\Gamma)$ for all $u \in H^1(\Omega)$.

Let $u \in C^2(\bar{\Omega})$. Then

$$\int_\Omega \Delta u \varphi \, dx = - \int_\Omega \nabla u \nabla \varphi \, dx + \int_\Gamma \frac{\partial u}{\partial v} \varphi \, d\sigma \tag{20}$$

for all $\varphi \in H^1(\Omega)$, where $\frac{\partial u}{\partial v} = \langle \nabla u, v \rangle \in L^\infty(\Omega)$, $v(z)$ being the exterior normal at $z \in \Gamma$. We want to define the weak normal derivative $\frac{\partial u}{\partial v}$ of u. Let

$$D(\Delta_{\max}) := \{u \in H^1(\Omega) : \Delta u \in L^2(\Omega)\}.$$

For $u \in D(\Delta_{\max})$ we say that $\frac{\partial u}{\partial \nu}$ **exists weakly**, if there exists a function $b \in L^2(\Gamma)$ such that

$$\int_{\Omega} \Delta u \varphi \, dx = - \int_{\Omega} \nabla u \nabla \varphi \, dx + \int_{\Gamma} b\varphi \, d\sigma \tag{21}$$

for all $\varphi \in H^1(\Omega)$. In that case $b \in L^2(\Gamma)$ is unique and we write $\frac{\partial u}{\partial \nu} := b$.

Now let $0 \leq \beta \in L^{\infty}(\Gamma) := L^{\infty}(\Gamma, \sigma)$. Then the measure μ given by $d\mu = \beta d\sigma$ is admissible [AW]. The form $a_{\beta} := a_{\mu}$ is given by $D(a_{\beta}) = H^1(\Omega)$,

$$a_{\beta}(u, v) = \int_{\Omega} \nabla u \nabla v \, dx + \int_{\Gamma} \tilde{u}\tilde{v}\beta \, d\sigma.$$

Denote by $-\Delta_{\beta}$ the operator associated with a_{β}. We now can describe Δ_{β} as follows.

PROPOSITION 5.1. *One has*

$$\begin{cases} D(\Delta_{\beta}) = \{u \in D(\Delta_{\max}) : \ \frac{\partial u}{\partial \nu} \ exists \ weakly \ in \ L^2(\Gamma) \ and \ \frac{\partial u}{\partial \nu} + \beta u_{|\Gamma} = 0\}, \\ \Delta_{\beta} u = \Delta u \ in \ \mathcal{D}(\Omega)'. \end{cases} \tag{22}$$

Proof. Denote by A the operator associated with a_{β}. Let $u \in D(A)$ and $Au = v$. Then

$$\int_{\Omega} v\varphi \, dx = a_{\beta}(u, \varphi) = \int_{\Omega} \nabla u \nabla \varphi \, dx + \int_{\Gamma} u\varphi\beta \, d\sigma$$

for all $\varphi \in H^1(\Omega)$. Choosing $\varphi \in \mathcal{D}(\Omega)$ this implies that $v = -\Delta u$. This shows that $D(\Delta_{\beta})$ is included in the right-hand-side of (22).

Conversely, let $u \in D(\Delta_{\max})$ such that $\frac{\partial u}{\partial \nu}$ exists weakly and $\frac{\partial u}{\partial \nu} + \beta u_{|\Gamma} = 0$. Then one has for all $\varphi \in H^1(\Omega)$,

$$-\int_{\Omega} \Delta u\varphi \, dx = \int_{\Omega} \nabla u \nabla \varphi \, dx + \int_{\Gamma} u\varphi\beta \, d\sigma$$
$$= a_{\beta}(u, \varphi).$$

Hence $u \in D(A)$ and $Au = -\Delta u$. $\qquad\qquad\square$

In particular, in this case of *classical Robin boundary conditions* one has $\{u \in C^2(\bar{\Omega}) : \frac{\partial u}{\partial \nu} + \beta u_{|\Gamma} = 0\} \subset D(\Delta_{\beta})$.

Next we show that an admissible measure μ is necessarily of the form $\beta d\sigma$ whenever $D(\Delta_{\mu})$ contains smooth functions. More generally, we have the following.

PROPOSITION 5.2. *Let $\Omega \subset \mathbb{R}^n$ be a bounded open set of class C^1 with boundary Γ. Let T be a symmetric C_0-semigroup associated with a closed form a. Denote by A the generator of T. Assume that*

a) $e^{t\Delta^D} \leq T(t) \leq e^{t\Delta^N}$ $(t \geq 0)$, *that*

b) *a is **local**; i.e. $a(u, v) = 0$ whenever $u, v \in D(a) \cap C(\bar{\Omega})$ have disjoint support, and that*

c) *there exists $u \in D(A) \cap C^2(\bar{\Omega})$ such that $u(z) > 0$ for all $z \in \Gamma$.*

Then there exists a function $\beta \in C(\Gamma)_+$ such that $A = -\Delta_\beta$.

Proof. It follows from Theorem 4.1 that there exists an admissible measure μ on Γ such that $a = a_\mu$ and $A = -\Delta_\mu$. One considers the function u in c). Then for all $\varphi \in C^1(\bar{\Omega})$ one has

$$\int_\Omega \nabla u \nabla \varphi \, dx + \int_\Gamma u\varphi \, d\mu = a(u, \varphi)$$

$$= -\int_\Omega \Delta u \varphi \, dx$$

$$= \int_\Omega \nabla u \nabla \varphi \, dx - \int_\Gamma \frac{\partial u}{\partial \nu} \varphi \, d\sigma.$$

It follows from the Stone-Weierstrass Theorem that

$$\int_\Gamma u\varphi \, d\mu + \int_\Gamma \frac{\partial u}{\partial \nu} \varphi \, d\sigma = 0$$

for all $\varphi \in C(\Gamma)$. This implies that $d\mu = -\frac{1}{u}\frac{\partial u}{\partial \nu} \, d\sigma$. Thus the claim is proved with $\beta = -\frac{1}{u}\frac{\partial u}{\partial \nu}$. $\qquad\square$

Also a converse version of Proposition 5.2 holds.

PROPOSITION 5.3. *Assume that Ω is a bounded open set of class $C^{2,\alpha}$ where $0 < \alpha < 1$. Let $\beta \in C^{1,\alpha}(\Gamma)$ with $0 < \beta(z)$ $(z \in \Gamma)$. Then there exists $u \in D(\Delta_\beta) \cap C^{2,\alpha}(\bar{\Omega})$ such that $\inf_{x \in \bar{\Omega}} u(x) > 0$.*

Proof. By [GT, Theorem 6.31] there exists $u \in C^{2,\alpha}(\bar{\Omega})$ such that $-\Delta u = 1$ on $\bar{\Omega}$ and $\beta u + \frac{\partial u}{\partial \nu} = 0$ on Γ. Then $u \in D(\Delta_\beta)$ and $\Delta_\beta u = 1$. By Proposition 6.3 below, one has $0 \in \rho(\Delta_\beta)$. Thus, $u = R(0, \Delta_\beta)1$. It follows from the domination property (Theorem 3.2) that $u = R(0, \Delta_\beta)1 \geq R(0, \Delta^D)1$. Now it follows from the maximum principle (see e.g. [Are, Theorem 1.5]) that $u(x) > 0$ for all $x \in \Omega$. Assume that there exists $z_0 \in \Gamma$ such that $u(z_0) = 0$. Then by [RR, Lemma 4.7], it follows that $\frac{\partial u}{\partial \nu}(z_0) < 0$ which is impossible since u satisfies the boundary condition. Thus $u(x) > 0$ for all $x \in \bar{\Omega}$. $\qquad\square$

6. Asymptotics

Let $\Omega \subset \mathbb{R}^n$ be open and let μ be an admissible measure on Γ with domain Γ_μ. The semigroup $(e^{t\Delta_\mu})_{t\geq 0}$ on $L^2(\Omega)$ is submarkovian. Thus there exist consistent C_0-semigroups $(e^{t\Delta_{\mu,p}})_{t\geq 0}$ on $L^p(\Omega)$, $1 \leq p < \infty$, such that $\Delta_{\mu,2} = \Delta_\mu$ (cf. [Dav, Theorem 1.4.1]).

PROPOSITION 6.1. *Assume that Ω is connected. Assume that $\Gamma_\mu \neq \emptyset$ and $\mu \neq 0$. Then*

$$\lim_{t\to\infty} \|e^{t\Delta_{\mu,p}} f\|_{L^p(\Omega)} = 0 \tag{23}$$

for all $f \in L^p(\Omega)$ and $1 < p < \infty$.

Proof. a) We show that $a_\mu(u) = 0$ implies that $u = 0$ for all $u \in D(a_\mu)$. In fact, if $a_\mu(u) = 0$, then $\nabla u = 0$, hence u is a constant c since Ω is connected. It follows that $0 = a_\mu(u) = \int_{\Gamma_\mu} |u|^2 \, d\mu = \mu(\Gamma_\mu)c^2$. Thus $c = 0$.

b) Property (23) is true for $p = 2$. This follows from the spectral theorem. In fact the semigroup $(e^{t\Delta_\mu})_{t\geq 0}$ is unitarily equivalent to a semigroup T on $H = L^2(Y, v)$ given by $T(t)f = e^{tm}f$ where $m : Y \to [0, \infty)$ is measurable and (Y, v) is a σ-finite measure space. Via the unitary equivalence the form a_μ becomes the form a on H given by $a(u) = \int_Y |u|^2 m \, dv$ with $D(a) = \{u \in H : \int_Y |u|^2 m \, dv < \infty\}$, see e.g. [ABHN, Section 7.1]. By a) we have $a(u) = 0$ only if $u = 0$. Thus $m(y) > 0$ v-a.e. Now it follows from the Dominated Convergence Theorem that $\lim_{t\to\infty} T(t)f = 0$ in $H = L^2(Y, v)$.

c) Now the claim (23) follows from the interpolation inequality for arbitrary $1 < p < \infty$ as in [ABB, Proposition 3.1].

$\square$

COROLLARY 6.2. *Let $\Omega \subset \mathbb{R}^n$ be open, and connected of finite Lebesgue measure. Assume that $\Gamma_\mu \neq \emptyset$ and $\mu \neq 0$. Then*

$$\lim_{t\to\infty} \|e^{t\Delta_{\mu,1}} f\|_{L^1(\Omega)} = 0 \tag{24}$$

for all $f \in L^1(\Omega)$.

Proof. Since $L^2(\Omega) \hookrightarrow L^1(\Omega)$, (24) follows from (23) if $f \in L^2(\Omega)$. Since the semigroup $(e^{t\Delta_{\mu,1}})_{t\geq 0}$ is contractive on $L^1(\Omega)$ the claim follows from a density argument. $\square$

If Ω is a bounded, regular open set, then we obtain even exponential stability.

PROPOSITION 6.3. *Let Ω be a bounded open set in $\mathbb{R}^n$ with Lipschitz boundary. Let μ be an admissible measure on Γ. Then $\Delta_{\mu,p}$ has compact resolvent for $1 \leq p < \infty$ and* *the spectrum $\sigma(\Delta_{\mu,p})$ is independent of $p \in [1, \infty)$. Moreover, there exist $c > 0, \omega > 0$ such that*

$$\|e^{t\Delta_{\mu,p}}\|_{\mathcal{L}(L^p(\Omega))} \leq ce^{-\omega t} \qquad (t \geq 0)$$

for all $1 \leq p < \infty$.

Proof. Since Ω has Lipschitz boundary, one has $H^1(\Omega) \hookrightarrow L^{2n/(n-2)}(\Omega)$ if $n > 2$ and $H^1(\Omega) \hookrightarrow L^p(\Omega)$ for all $1 \leq p < \infty$ if $n = 1, 2$. It follows from [Dav, Section 2.4] that $e^{t\Delta_{\mu,1}}L^1(\Omega) \subset L^\infty(\Omega)$ and

$$\|e^{t\Delta_{\mu,1}}f\|_\infty \leq ct^{-n/2}\|f\|_1$$

for all $0 < t \leq 1$, $f \in L^1(\Omega)$. In particular, $e^{t\Delta_{\mu,2}}$ is a Hilbert-Schmidt operator and hence compact. Writing $e^{t\Delta_{\mu,1}} = e^{t/2\Delta_\mu}e^{t/2\Delta_\mu}$ one sees that $e^{t\Delta_{\mu,1}}$ is a compact operator on $L^1(\Omega)$ for $t > 0$. Now spectral p-independence follows from [Dav, Theorem 1.6.4]. It follows from [ArBa, Theorem 1.3] that $0 \notin \sigma(\Delta_{\mu,p})$. Thus $\Delta_{\mu,p}$ has negative spectral bound, which coincides with the growth bound of the semigroup. $\qquad\qquad\square$

REFERENCES

[Are]   ARENDT, W., Different domains induce different heat semigroups on $C_0(\Omega)$. In: *Evolution equations and Their Applications in Physics and Life Sciences*, G. Lumer, L. Weis eds. Marcel Dekker, (2001), 1–14.

[ArBa]   ARENDT, W. and BATTY, C. J. K., *Domination and ergodicity for positive semigroups*. Proc. Amer. Math. Soc. *114* (1992), 743–747.

[ABB]   ARENDT, W., BATTY, C. J. K., and BÉNILAN, PH., *Asymptotic stability of Schrödinger semigroups on $L^1(\mathbb{R}^N)$*. Math. Z. *209* (1992), 511–518.

[ABHN]   ARENDT, W., BATTY, C. J. K., HIEBER, M. and NEUBRANDER, F., *Vector-valued Laplace Transforms and Cauchy Problems*. Birkhäuser, Basel, 2001.

[ArBe]   ARENDT, W. and BÉNILAN, PH., *Inégalités de Kato et semi-groupes sous-markoviens*. Rev. Mat. Univ. Complutense Madrid *5* (1992), 279–308.

[AW]   ARENDT, W. and WARMA, M., *The Laplacian with Robin boundary conditions on arbitrary domains*. To appear in Potential Analysis, 2003.

[Bat]   BATTY, C. J. K., *Asymptotic stability of Schrödinger semigroups: path integral methods*. Math. Ann. *292* (1992), 457–492.

[BC]   BÉNILAN, PH. and GRANDALL, M. G., *Completely accretive operators*. Lect. Notes Pure Appl. Math., Ph. Clément, Ben de Pagter, E. Mitidieri eds. Marcel Dekker, *135* (1991), 41–75.

[BP]   BÉNILAN, PH. and PIERRE, M., *Quelques remarques sur la localité dans L^1 d'opérateurs différentiels*. Semesterbericht Funktionalanalysis, Tübingen *13* (1988), 23–29.

[BH]   BOULEAU, N. and HIRSCH, F., *Dirichlet Forms and Analysis on Wiener Space*. W. de Gruyter, Berlin, 1991.

[Bou]   BOURBAKI, N., *Eléments de Mathématique. Intégration. Vol. VI*. Hermann, Paris, 1965.

[Dan]   DANERS, D., *Robin boundary value problems on arbitrary domains*. Trans. Amer. Math. Soc. *352* (2000), 4207–4236.

[Dav]   DAVIES, E. B., *Heat kernels and Spectral Theory*. Cambridge University Press, Cambridge, 1989.

[EG]   EVANS, L. C. and GARIEPY, R. F., *Measure Theory and Fine Properties of Functions*. CRC. Press, Boca Raton, Florida, 1992.

[FOT]   FUKUSHIMA, M. OSHIMA, Y. and TAKEDA, M., *Dirichlet Forms and Symmetric Markov Processes*. Amsterdam: North-Holland, 1994.

[GT]   GILBARG, D. and TRUDINGER, N. S., *Elliptic Partial Differential Equations of Second Order*. Springer-Verlag, Berlin, 1986.

[MR]   MA, Z. M. and RÖCKNER, M., *Introduction to the Theory of Non-Symmetric Dirichlet Forms*. Springer-Verlag, Berlin, 1992.

[Maz]   MAZ'YA, V. G., *Sobolev Spaces*. Springer-Verlag, Berlin, 1985.

[Ouh]   OUHABAZ, E. M., *Invariance of closed convex sets and domination criteria for semigroups*. Potential Anal. 5 (1996), 611–625.

[RR]   RENARDY, M. and ROGERS, R. C., *An Introduction to Partial Differential Equations*. Springer-Verlag, Berlin, 1993.

[Rud]   RUDIN, W., *Real and Complex Analysis*. McGraw-Hill, Inc., 1966.

[Sch]   SCHAEFER, H. H., *Banach Lattices and Positive Operators*. Springer-Verlag, Berlin, 1974.

[Sto]   STOLLMANN, P., *Closed ideals in Dirichlet spaces*. Potential Anal. 2 (1993), 263–268.

[SV]   STOLLMANN, P. and VOIGT, J., *Perturbation of Dirichlet forms by measures*. Potential Anal. 5 (1996), 109–138.

Wolfgang Arendt and Mahamadi Warma
Abteilung Angewandte Analysis
Universität Ulm
D-89069 Ulm
Germany
e-mail: arendt@mathematik.uni-ulm.de
warma@mathematik.uni-ulm.de

To access this journal online:
http://www.birkhauser.ch

J.evol.equ. 3 (2003) 137 – 151
1424–3199/03/010137 – 15
© Birkhäuser Verlag, Basel, 2003

**Journal of Evolution
Equations**

The focusing problem for the Eikonal equation

S. B. ANGENENT AND D. G. ARONSON

Dedicated to the Memory of Philippe Bénilan

Abstract. We study the focusing problem for the eikonal equation

$$\partial_t u = |\nabla u|^2,$$

i.e., the initial value problem in which the support of the initial datum is outside some compact set in $\mathbf{R}^d$. The hole in the support will be filled in finite time and we are interested in the asymptotics of the hole as it closes. We show that in the radially symmetric case there are self-similar asymptotics, while in the absence of radial symmetry essentially any convex final shape is possible. However in $\mathbf{R}^2$, for generic initial data the asymptotic shape will be either a vanishing triangle or the region between two parabolas moving in opposite directions (a closing eye). We compare these results with the known results for the porous medium pressure equation which approaches the eikonal equation in the limit as $m \to 1$.

1. Introduction

In this paper we compare the focusing problem for the eikonal equation

$$\partial_t u = |\nabla u|^2 \tag{EE}$$

with the corresponding problem for the porous medium (pressure) equation

$$\partial_t u = (m-1)u\Delta u + |\nabla u|^2, \tag{PME}$$

in $\mathbf{R}^d \times \mathbf{R}^+$ for $d \geq 1$. Here ∇ denotes the gradient with respect to the spatial variables $(x_1, x_2, \ldots, x_d)$. The quantity $m > 1$ in (PME) is a constant.

Consider the initial value problem for either (EE) or (PME) with the initial datum $u(x, 0) = u_0(x)$. We assume that u_0 is compactly supported and has a "regular free boundary", i.e.,

$$u_0(x) = \max(\widetilde{u}_0(x), 0)$$

for some $\widetilde{u}_0 \in C^\infty(\mathbf{R}^d)$, where 0 is a regular value of $\widetilde{u}_0$ and $\{x \mid \widetilde{u}_0(x) \geq 0\}$ is compact. The viscosity solution to the initial value problem for (EE) with initial datum $u(x, 0) = u_0(x)$ is a continuous function of (x, t) (cf. [17]). The same is true for the generalized solution to the corresponding initial value problem for (PME) (cf. [8]). The zero set

$$Z_t = \{x \in \mathbf{R}^d : u(x, t) = 0\}$$

of either the viscosity or generalized solution has one unbounded component. Let K_t denote the union of all bounded components of Z_t. For the *focusing problem* we assume that K_0 is a compact simply connected set, i.e., that the support of u_0 "has a hole." The sets K_t form a nonincreasing and eventually strictly decreasing family of subsets of $\mathbf{R}^d$ which will become empty at some finite time

$$T = \inf \{t : K_t = \varnothing\}$$

which is called the *focusing time*. We say that the solution has focused (or filled the hole K_0) at time $t = T$.

The eikonal equation (EE) arises as a formal limit of (PME) as $m \searrow 1$, and indeed, it has been shown in [17] that weak solutions to the initial value problem for (PME) converge uniformly as $m \searrow 1$ to viscosity solutions of the corresponding problem for (EE). The absence of diffusion makes the eikonal equation considerably simpler than (PME), and in fact there is even an explicit formula for the viscosity solution in terms of its initial data,

$$u(x, t) = \sup_{y \in \mathbb{R}^d} \left\{ u_0(y) - \frac{|x - y|^2}{4t} \right\}. \tag{1}$$

This representation of the solution is called the *Lax-Hopf formula*. See [17] or [15] for a discussion. From this formula for the viscosity solution one immediately deduces the following representation of the free boundary. The free boundary is the graph in space-time of the *filling time*

$$T_*(x) = \inf_{u(x,t)>0} t = \inf_{u_0(y)>0} \frac{|x - y|^2}{4u_0(y)}. \tag{2}$$

The Lax-Hopf formula and the formula (2) for the filling time allow us to study focusing of solutions to (EE) in much more detail than can be done at present for (PME). In this note we compare the results on focusing for the eikonal equation with what is known for (PME).

In the non-generic case of radial symmetry, K_t is, of course, a d-dimensional ball. We show that in this case the viscosity solution u for (EE) behaves like a self-similar solution near focusing. More precisely, we show that

$$\lim_{\epsilon \searrow 0} \frac{1}{\epsilon} u(\epsilon \xi, T + \epsilon \tau) = c|\xi| + c^2 \tau \tag{3}$$

uniformly for bounded $\xi \in \mathbb{R}^d$ and bounded $\tau \in \mathbb{R}$. This extends our results [1] for (PME) to the case $m = 1$. In [1] it was shown for any $1 < m < \infty$ that for a radially symmetric weak solution of (PME) which focusses at $t = T$ one has

$$\lim_{\epsilon \searrow 0} \epsilon^{\alpha - 2} u(\epsilon \xi, T - \epsilon^{\alpha} \tau) = c^2 V_m \left(\frac{|\xi|}{c \tau^{1/\alpha}} \right),$$

in which V_m is the Aronson-Graveleau profile, $\alpha \in (1, 2)$ is the corresponding exponent, and $0 < c < \infty$ is a positive constant depending on the initial data.

In the nonradial case we show that the eikonal equation admits many more self-similar focusing solutions than (PME). In particular we show that *for any closed convex set $C \subset \mathbb{R}^d$ containing the origin there is a self-similar viscosity solution $S(x, t)$* (i.e. one which satisfies $S(\lambda x, \lambda t) = \lambda S(x, t)$ for all $\lambda > 0$) such that

$$\{x \in \mathbb{R}^d \mid S(x, t) = 0\} = (-t)C$$

for all $t < 0$. In other words, for the eikonal equation "a vanishing hole can have any convex shape."

This contrasts sharply with the porous medium equation. All self-similar focusing solutions for (PME) which have been constructed either analytically [2] or numerically [3] have some kind of discrete symmetry. Moreover, in [2] it is shown that self-similar focusing solutions satisfy a nonlinear elliptic free boundary problem which can be transformed to a nonlinear Fredholm equation. One therefore expects self-similar focusing solutions to (PME) to occur in discrete families, and not in infinite dimensional continua as is the case for the eikonal equation.

We next consider small perturbations of radially symmetric initial data. We show that, to leading order, ∂K_t propagates according to *Huygen's principle,* i.e., with constant normal velocity. More precisely, if $u^0(x, t) = U(|x|, t)$ is a radially symmetric solution which focusses at time $t = T$, then we observe that the viscosity solution u^ϵ with initial data $u^\epsilon(x, 0) = U(|x|) + \epsilon g(x)$ satisfies

$$\lim_{\epsilon \to 0} \frac{1}{\epsilon} u^\epsilon(\epsilon \xi, T + \epsilon \tau) = V(\xi, \tau) \tag{4}$$

where V is a viscosity solution of the Huygens-Hamilton-Jacobi equation

$$\partial_\tau V = c|\nabla V|. \tag{5}$$

The level sets of viscosity solutions to this equation are fronts which evolve by Huygens' principle, i.e. they propagate with constant normal velocity c. Moreover, we observe that *any solution of* (5) *can occur as the limit in* (4), provided $|\nabla V| \equiv c$.

This implies that radial focusing is unstable: a small perturbation will turn a radial focusing solution into one of the two generic focusing solutions described below. This instability also appears to be present in (PME). For (PME) radial hole filling is described by the Aronson-Graveleau profile, but if one perturbs a radial solution slightly, then numerical computations indicate that the perturbations will grow, and that the solution will generally not return to radial symmetry. However, for (PME) one expects a sufficiently *symmetric* peturbation to die out and disappear, at least if m is large enough. The instability of radial focusing for (EE) combined with Lions-Souganidis-Vázquez' convergence result [17] gives a heuristic reason for the occurrence of the infinite sequence of bifurcations found in [2]: for large values of m the diffusion term in (PME) is dominant, and the radial self-similar solution will only have a small number of unstable modes, but for $m = 1 + o(1)$ the solution to (PME) tries to follow the viscosity solution to (EE), where radial focussing has infinitely many unstable modes.

Finally, we consider generic solutions. For generic initial data, i.e. for an open and dense set of $\tilde{u}_0 \in C^\infty(\mathbb{R}^d)$, K_t shrinks to a point and, when $d = 2$, we can characterize the generic possibilities for the final form of ∂K_t. These are either

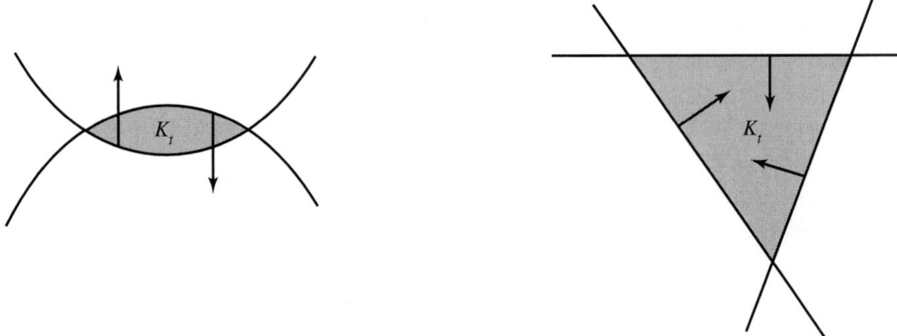

Figure 1   The closing eye and vanishing triangle solutions

A Closing Eye. K_t approximates the region between two parabolae

$$y = \pm\{c_\pm(T - t) - x^2\}.$$

The width of K_t is $O(\sqrt{T - t})$ while its height is $O(T - t)$. The aspect ratio of K_t tends to infinity as $t \nearrow T$ (cf. Figure 1).

A Vanishing Triangle. K_t is approximately the region enclosed by three straight lines which propagate with constant normal velocity, and which pass through a common point at $t = T$ (cf. Figure 1)

Numerics and asymptotic analysis strongly suggest that there are solutions to (PME) close to the closing eye ([3, 11]) for any $m > 1$, even though there is as yet no rigorous theory.

The situation with the vanishing triangle is different: numerics again suggest that there exists a self-similar solution whose free boundary approximates a triangle when $m - 1$ is small enough, but this triangle *must be an equilateral triangle*. Thus it seems that the second kind of generic hole filling for (EE) is highly nongeneric for (PME). Again, rigorous proofs for the statements about (PME) are lacking.

2. Radial solutions

If the initial function $u_0(x)$ is radially symmetric, i.e. $u_0(x) = U_0(r)$, $r = |x|$, then the viscosity solution $u(t, x)$ will be radially symmetric for all $t > 0$. Indeed, the Lax-Hopf formula reduces to

$$U(t, r) = \sup_{\rho > 0} \left\{ U_0(\rho) - \frac{(r - \rho)^2}{4t} \right\}. \tag{6}$$

This equation is obtained from (1) by maximizing over all y with $|y| = \rho$. So we see that the radial solution is exactly the same as the viscosity solution in one space dimension. This makes it easy to understand generic radial hole filling. Consider an initial function $U_0(r)$ as in Figure 2.

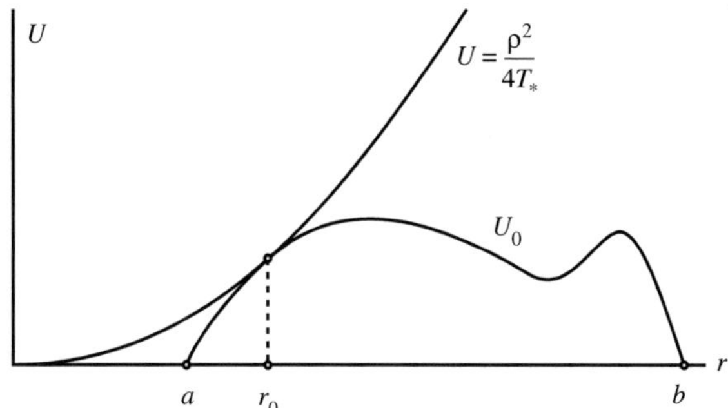

Figure 2   Computing the Radial Hole Filling Time

The unique hole in the center will fill up in time

$$T_* = \sup_{\rho > 0, U_0(\rho) > 0} \frac{\rho^2}{4U_0(\rho)}.$$

Let us assume that the function $\rho \mapsto \rho^2/4U_0(\rho)$ has a unique maximum at $\rho = r_0$ as will be the case for a generic smooth initial U_0. Then, at the hole filling time T_* the solution near $r = 0$ will be a smooth function of r, and if we allow $r < 0$ in (6), then it extends to a smooth function in a full neighborhood in $\mathbb{R}^2$ of $(r = 0, t = T_*)$. At $r = 0, t = T_*$ the method of characteristics tells us that $U_r = U'(r_0)$; the eikonal equation enforces $U_t(0, T_*) = (U_r)^2 = U'(r_0)^2$. Thus, locally we have

$$U(r, t) = U'(r_0)r + U'(r_0)^2(t - T_*) + \mathcal{O}(r^2 + (t - T_*)^2).$$

Hence

$$\lim_{\epsilon \searrow 0} \frac{1}{\epsilon} U(\epsilon R, T_* + \epsilon \tau) = U'(r_0)R + U'(r_0)^2 \tau.$$

As we noted in the Introduction, this is analogous to the radial hole-filling asymptotics for the porous medium equation. Here we see that that radial hole filling is described by a self-similar solution

$$U_c(r, t) = c(r + ct)_+,$$

and that these self-similar solutions occur in a one parameter family. For the porous medium equation the self-similar solutions are the Aronson-Graveleau solutions, for the eikonal equation they are simply plane waves converging upon the origin.

3. Nonradial self-similar solutions

For any vector $\xi \in \mathbb{R}^d$ the plane wave solution

$$W_\xi(x, t) = (\xi \cdot x + |\xi|^2 t)_+$$

is a self-similar solution which reaches the origin at $t = 0$.

LEMMA 3.1. *For any closed subset $F \subset \mathbb{R}^d$ the function*

$$S_F(x, t) = \sup_{\xi \in F} W_\xi(x, t) \tag{7}$$

is a self-similar solution which reaches the origin at time $t = 0$.

The "hole" in support of the self-similar solutions constructed here is the set

$$
\begin{aligned}
K_F(t) &= \{x \in \mathbb{R}^d \mid S_F(x, t) = 0\} \\
&= \{x \in \mathbb{R}^d \mid \forall_{\xi \in F} \xi \cdot x + |\xi|^2 t \leq 0\} \\
&= \left\{x \in \mathbb{R}^d \mid \forall_{\xi \in F} \frac{\xi}{|\xi|} \cdot x + |\xi| t \leq 0\right\} \\
&= \bigcap_{\xi \in F} \left\{x \mid \frac{\xi}{|\xi|} \cdot x \leq -|\xi| t\right\}. \tag{8}
\end{aligned}
$$

Thus we see that for $t < 0$ the hole $K_F(t)$ is a nonempty convex set whose support function $p : S^{d-1} \to \mathbb{R}$ is given by

$$p(\omega) = \inf \{p > 0 \mid -tp\omega \in F\}$$

Furthermore, *any compact convex set $C \subset \mathbb{R}^d$ containing the origin can be represented as $C = K_F(-1)$, with $K_F(-1)$ defined as in (8) for some suitably chosen $F \subset \mathbb{R}^d$*. Therefore there are *uncountably many* self similar solutions, and that self-similarity of a solution to the eikonal equation does not guarantee any kind of symmetry. This is in contrast with the porous medium equation. For (PME) not all self-similar solutions are known either rigrously or computationally. However all self-similar solutions whose existence has been proven analytically or by numerical computation have some kind of symmetry. This poses the following

Open problem: *Which of the self-similar solutions S_F of the eikonal equation arise as limits of self-similar solutions of (PME) as $m \searrow 1$?*

Proof of Lemma 3.1 We apply the Lax-Hopf formula to find the viscosity solution v at time $s > 0$ with initial datum $S_F(\cdot, t)$:

$$
\begin{aligned}
v(s, x) &= \sup_{y} \left\{ S_F(t, y) - \frac{|x - y|^2}{4s} \right\} \\
&= \sup_{y} \left\{ \sup_{\xi \in F} W_\xi(y, t) - \frac{|x - y|^2}{4s} \right\} \\
&= \sup_{\xi \in F} \left\{ \sup_{y} W_\xi(y, t) - \frac{|x - y|^2}{4s} \right\} \\
&= \sup_{\xi \in F} W_\xi(x, t + s) \\
&= S_F(x, t + s),
\end{aligned}
$$

where we have used the fact that W_ξ itself is a smooth solution to (EE) and hence also a viscosity solution. $\qquad\square$

4. Perturbation of radial solutions

In this section we observe that radial focusing is unstable.

Let $U_1 : \mathbb{R}^d \to \mathbb{R}$ be any given function, and consider the initial data

$$u^\epsilon(x) = U_0(r) + \epsilon U_1(r\theta),$$

where $x = r\theta$, $r > 0$ and $|\theta| = 1$, i.e. $\theta \in S^{d-1}$ is a unit vector.

The viscosity solution to the initial value problem for (EE) with initial datum u^ε is

$$u^\varepsilon(x, t) = \sup_{y \in \mathbb{R}^d} F(x, t; y, \varepsilon),$$

where

$$F(x, t; y, \varepsilon) = U_0(|y|) + \varepsilon U_1(y) - \frac{|x - y|^2}{4t}.$$

Set $x = \varepsilon\xi$, $t = T_* + \varepsilon\tau$, and $y = \rho\omega$ for $\omega \in S^{d-1}$, and maximize first over $\rho > 0$ keeping ω fixed. For $\varepsilon = 0$ the function to be maximized $\rho \mapsto F(0, T_*; \rho\omega, 0)$ has a unique nondegenerate maximum at $\rho = r_0$. Since this maximum describes focusing we have

$$F(0, T_*; r_0\omega, 0) = u^\varepsilon(0, T_*) = 0. \tag{9}$$

and

$$\omega \cdot F_y(0, T_*; r_0\omega, 0) = \frac{\partial}{\partial \rho} F(0, T_*; \rho\omega, 0)\Big|_{\rho=r_0} = 0. \tag{10}$$

By the Implicit Function Theorem $\rho \mapsto F(x, t; \rho\omega, \epsilon)$ will have a nondegenerate maximum at $\rho = r_\epsilon(x, t)$ where $r_\epsilon(x, t)$ depends smoothly on ϵ, x and t. We now compute

$$\frac{1}{\epsilon} u^\epsilon(x, t) = \max_{|\omega|=1} \epsilon^{-1} F(x, t; r_\epsilon(x, t)\omega, \epsilon)$$

by expanding F in a Taylor series. At the maximizing ω we have $F_y = 0$, so we get

$$F(x, t; r_\epsilon(x, t)\omega, \epsilon)$$
$$= F(\epsilon\xi, T_* + \epsilon\tau; (r_0 + \mathcal{O}(\epsilon))\omega, \epsilon)$$
$$= \epsilon\{\xi \cdot F_x(0, T_*; r_0\omega, 0) + \tau F_t(0, T_*; r_0\omega, 0) + F_\epsilon(0, T_*; r_0\omega, 0)\} + \mathcal{O}(\epsilon^2).$$

Computing the relevant derivatives of F and using (9) we get finally,

$$\frac{1}{\epsilon} u^\epsilon(x, t) = \max_{|\omega|=1} \left\{ U_1(r_0\omega) + \frac{r_0}{2T_*} \xi \cdot \omega \right\} + \frac{r_0^2}{4T_*^2} \tau + \mathcal{O}(\epsilon).$$

so that

$$\lim_{\epsilon \to 0} \frac{1}{\epsilon} u^\epsilon(x, t) = cP(\xi) + c^2\tau, \tag{11}$$

in which

$$P(\xi) = \max_{|\omega|=1} \frac{1}{c} U_1(r_0\omega) + \xi \cdot \omega$$

and

$$c = \frac{r_0}{2T_*}.$$

At time $\tau < 0$ the zeroset of the limiting solution in (11) is given by

$$K_\tau = \bigcap_{|\omega|=1} \left\{ \xi \mid \xi \cdot \omega \leq -c\tau - \frac{1}{c} U_1(r_0\omega) \right\}.$$

For suffcently large $\tau < 0$ we see that K_τ is a convex set with smooth boundary, whose support function is $p(\tau, \omega) = -c\tau - \frac{1}{c} U_1(r_0\omega)$. This support function decreases with constant rate $\partial_\tau p = -c$, so ∂K_τ shrinks with constant speed c. At some moment ∂K_τ will develop a singularity, and after that K_τ still shrinks with constant velocity in the sense of viscosity solutions.

Note that the support function $-c - c^{-1} U_1(r_0 \omega)$ of ∂K_{-1} can be arbitrary, so that *the limiting shape of the shrinking hole can be **any** convex front moving by Huygens' principle.*

5. Generic solutions and the method of characteristics

We briefly recall the method of characteristics. In this approach to the solution of (EE) one allows the solution $u(x, t)$ to become multiply valued and considers the graph of this solution together with its space-time gradient, i.e.

$$\Lambda_u = \{(x, t, u(x, t), \nabla u(x, t), u_t(x, t)) \mid (x, t) \in \Omega_t \times [0, \infty)\}$$

where $\Omega_t \subset \mathbb{R}^d$ is the domain of the possibly multiply valued function $x \mapsto u(x, t)$. In the language of geometric optics, one calls Λ_u a *front*.

This graph Λ_u is a subset of the 1-jet space

$$J^1(\mathbb{R}^{d+1}) = \{(x, t, u, p, \dot{u}) \mid x \in \mathbb{R}^d, t \in \mathbb{R}, u \in \mathbb{R}, p \in \mathbb{R}^d, \dot{u} \in \mathbb{R}\}$$

of $\mathbb{R}^{d+1}$. Points in $J^1(\mathbb{R}^{d+1})$ are called 1-*jets*, or also *contact elements*.

On the space of contact elements one has the contact form

$$\theta = du - p_1 dx_1 - \cdots - p_d dx_d - \dot{u} dt.$$

Λ_u is a $d + 1$ dimensional submanifold of $J^1(\mathbb{R}^{d+1})$, and it is even an integral manifold for θ.

The initial datum we prescribe is a smooth function $u_0 : \Omega_0 \to \mathbb{R}$, on some bounded domain $\Omega_0 \subset \mathbb{R}^d$ with smooth boundary $\partial \Omega_0$, such that $\nabla u_0 \neq 0$ on $\partial \Omega_0$, while $u_0(x) > 0$ on Ω_0. The front Λ_u is determined by the initial data through the characteristic flow: namely,

(i) Λ_u contains the initial surface

$$\Lambda_{u_0} = \{(x, 0, u_0(x), \nabla u_0(x), \dot{u}_0(x)) \mid x \in \mathbb{R}^d\},$$

where $\dot{u}_0(x) = |\nabla u_0(x)|^2$ is the initial velocity, and

(ii) Λ_u is invariant under the characteristic flow. For (EE) this flow is given by

$$\begin{cases} \dfrac{dx_k}{d\tau} = 2p_k, \quad \dfrac{dt}{d\tau} = 1, \quad \dfrac{dp_k}{d\tau} = 0, \\[2mm] \dfrac{du}{d\tau} = -\{p_1^2 + \cdots + p_d^2\}, \quad \dfrac{d\dot{u}}{d\tau} = 0 \end{cases} \tag{12}$$

(τ is the time variable for the flow.)

These equations are easily integrated, and they lead to the following parametrization of Λ_u

$$\begin{cases} X(t, y) = y - 2t\nabla u_0(y), \\ U(t, y) = u_0(y) - t|\nabla u_0(y)|^2, \\ P(t, y) = \nabla u_0(y) \end{cases} \tag{13}$$

LEMMA 5.1. *The map from $\Omega_0 \times [0, \infty)$ to $J^1(\mathbb{R}^{d+1})$ given by*

$$(y, t) \mapsto (X(t, y), t, U(t, y), P(t, y), \dot{U}(t, y))$$

with $\dot{U}(y) = |\nabla u_0(y)|^2$ is a Legendre embedding.

Proof. The map $(X, t, U, P) \mapsto (t, y) = (t, X + 2tP)$ is a differentiable inverse for our given map, which shows that it is an embedding.

To verify that it is a Legendre embedding one must show that $dU - P \cdot dX - \dot{U}dt = 0$. This can be done by a short direct computation, but it is also guaranteed by the method of characteristics. $\qquad\square$

Thus Λ_u is, in particular, a smooth submanifold of $J^1(\mathbb{R}^{d+1})$.

If one only looks at the graph of the solution u, rather than of $(u, \nabla u, u_t)$, then one finds the following set

$$\Gamma = \{(X(t, y), t, U(t, y)) \mid y \in \Omega_0, t \geq 0\}$$

This is a submanifold of $\mathbb{R}^d \times [0, \infty) \times \mathbb{R}$ which generally has singularities. The map

$$(t, y) \mapsto (X(t, y), t, U(t, y))$$

is called a *Legendre mapping* (see [7].)

If one extends the initial data u_0 by setting $u_0(x) = 0$ for $x \notin \Omega_0$, then the Lax-Hopf formula (1) provides us with a viscosity solution $u^*(x, t)$. One can recover u^* from Γ by considering the upper envelope of Γ,

$$u^*(x, t) = \max_{(x,t,u)\in\Gamma} u = \begin{cases} \max\limits_{\substack{X(t,y)=x}} U(t, y) & \text{if } \exists_{y\in\Omega_0} X(t, y) = x \\ 0 & \text{otherwise} \end{cases} \tag{14}$$

This follows from the Lax-Hopf formula (1): if $y \in \mathbb{R}^d$ maximizes $u_0(y) - |x - y|^2/4t$, then, since the viscosity solution is nonnegative one has $u_0(y) \geq |x - y|^2/4t$. Thus either $u_0(y) > 0$ and hence $y \in \Omega_0$, or $u_0(y) = 0$, $x = y$ and $u^*(x, t) = 0$. If $y \in \Omega_0$, then the fact that $u_0(y) - |x - y|^2/4t$ has a maximum at y implies $x = y - 2t\nabla u_0(y) = X(t, y)$ and $u(x, t) = U(t, y)$.

The singularities of Legendre mappings in general have been studied by the Arnol'd school [7]. The singularities which arise in generic solutions to (EE) and more general Hamilton-Jacobi equations, have been studied by I. A. Bogaevski [12, 13, 14]. He presents a classification the possible singularities by their topological type, at least in dimensions $d = 2$ and 3 (the case $d = 1$ is classical.) In higher dimensions the classification seems to be very complicated, and not completely known.

Before going on to the free boundary of a solution, we briefly recall some of the singularities found by Bogaevski.

For $d = 1$ a generic choice of initial data u_0 (i.e. for u_0 in an open and dense subset in the class of C^∞ functions) the viscosity solution u^* will only have the following singularities:

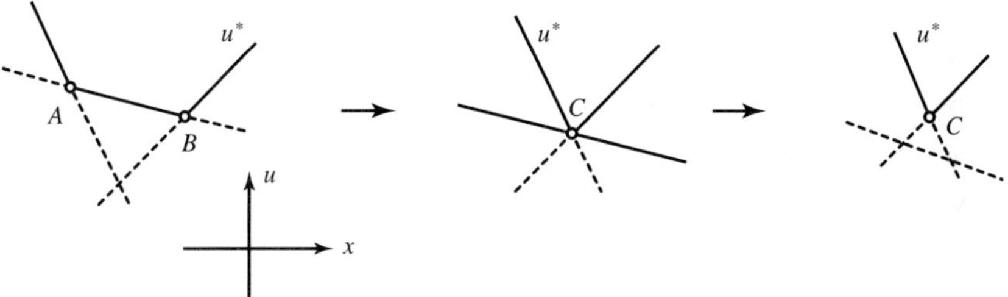

Figure 3   Two corners A and B merge into one new corner C $(d = 1)$

Except at a finite number of times the viscosity solution u^* has only simple corners, which come from transverse self-intersections of the Legendre map defined in (14). At isolated moments in time one such corner can be created in a so-called swallow tail singularity (see Figure 4), or two corners can meet and combine to form one new corner (Figure 3.)

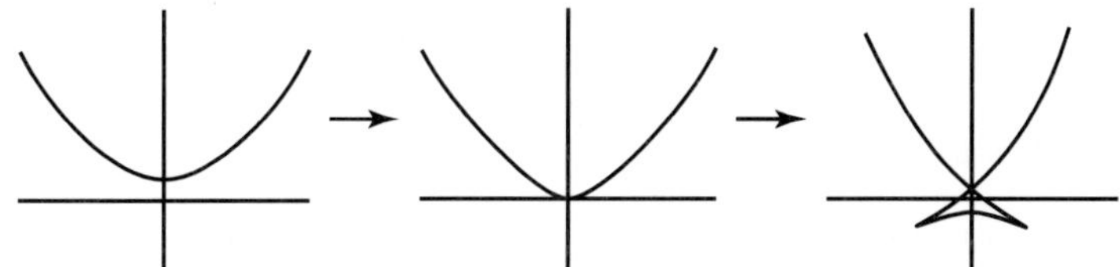

Figure 4   Birth of a corner through a swallowtail perestroika $(d = 1)$

The more complicated singularities which can only occur at isolated instances in time (swallowtail, and merging corners) have been called *perestroikas* by the Russian school since they describe how one constellation of simpler singularities can transform into another.

By combining the curves in Figure 4 into one surface one obtains the set Γ (Figure 5.) The graph of the viscosity solution is obtained by removing the curved tetrahedron $ABCD$ which is "at the bottom" of the surface Γ.

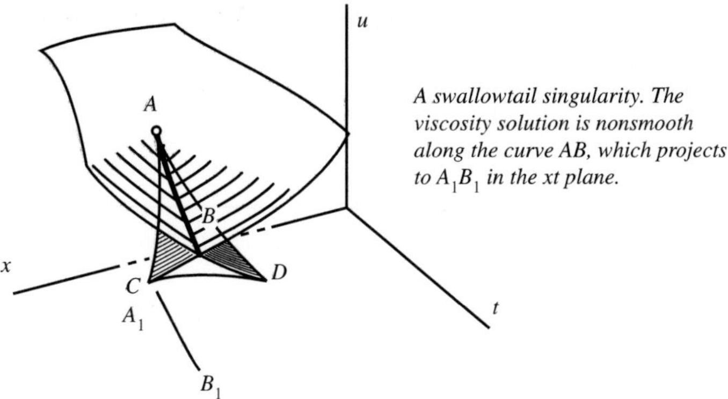

A swallowtail singularity. The viscosity solution is nonsmooth along the curve AB, which projects to $A_1 B_1$ in the xt plane.

Figure 5   A Typical Swallowtail Singularity

6. Generic hole filling

We now consider the free boundary of a solution. The free boundary of the viscosity solution is the graph of the filling time $T_*(x)$ defined in the introduction (See (2).) The filling time is, in general, not a smooth function. To analyze its singularities we use the method of characteristics which shows us that the free boundary is a subset of a Legendre map, and we find that (for generic initial data $u_0 \in C^\infty(\bar{\Omega}_0)$) its singularities are those of a generic Legendre map. The generic singularities of the filling time are then those obtained by Bogaevski for generic minimum functions [12].

The free boundary is given by setting $U(t, y) = 0$. This leads to an equation $(u_0(y) - t|\nabla u_0(y)|^2 = 0)$ for the time t it takes the contact element $(y, 0, u_0(y), \nabla u_0(y), \dot{u}_0(y))$ to reach the level $\{u = 0\}$ under the characteristic flow. Let

$$\Theta(y) = u_0(y)/|\nabla u_0(y)|^2$$

be this time. When the contact element reaches $\{u = 0\}$, it finds itself at $x = X(y)$, where

$$X(y) = y - \frac{2u_0(y)}{|\nabla u_0(y)|^2} \nabla u_0(y).$$

The map $y \mapsto (X(y), \Theta(y))$ then parametrizes the extended free boundary (extended, because it will also include those points where Γ hits the set $u = 0$.) In general Θ is not single valued, and the map $y \mapsto (X(y), \Theta(y))$ will have singularities. One can again analyze these by considering the graph of Θ and its gradient. This graph is contained in the 1-jet space $J^1(\mathbb{R}^d)$, it is parametrized by

$$\Theta(y) = \frac{u_0(y)}{|\nabla u_0(y)|^2}, \qquad N(y) = -\frac{\nabla u_0(y)}{|\nabla u_0(y)|^2}.$$

LEMMA 6.1. *Let $\Omega_0^* = \{y \in \Omega_0 \mid \nabla u_0(y) \neq 0\}$. The map Φ from Ω_0^* into $J^1(\mathbb{R}^d)$ given by*

$$\Phi : y \mapsto (X(y), \Theta(y), N(y))$$

is a smooth, proper Legendre embedding.

Proof. A smooth inverse is obtained by direct computation,

$$\nabla u_0 = -\frac{N}{|N|^2} \Rightarrow u_0 = \Theta|\nabla u_0|^2 = \Theta|N|^{-2} \Rightarrow y = X + \frac{2u_0}{|\nabla u_0|^2}\nabla u_0 = X + \frac{2\Theta}{|N|^2}N.$$

So the map is a smooth embedding.

As y approaches the zeroset of $\nabla u_0(y)$ the quantity $\Theta(y)$ becomes unbounded, so the map Φ is proper.

To verify that Φ is a Legendre embedding one must show $d\Theta - N \cdot dX = 0$. This can be done by an straightforward computation in which one verifies that $\partial\Theta/\partial y_j = N \cdot (\partial X/\partial y_j)$ for $j = 1, \ldots, d$. Alternatively one can deduce this from the fact that Φ parametrizes the Legendre submanifold of $J^1(\mathbb{R}^d)$ obtained by slicing Λ_u with $\{u = 0\}$. $\qquad\square$

The following (easy) observation is the key to the analysis of generic singularities of the free boundary.

LEMMA 6.2. *Let $u_0 \in C^\infty(\bar{\Omega}_0)$ be given, and let Φ_0 be its corresponding generalized free boundary. Then any sufficiently small perturbation of Φ_0 within the class of C^∞ Legendre immersions can be achieved by an appropriately chosen small perturbation of the initial data $u_0 \in C^\infty(\bar{\Omega}_0)$.*

We use the characteristic flow to solve the inverse problem in which one determines the initial data from a given free boundary. This is of course only possible for the generalized solution in the sense of Legendre submanifolds: For viscosity solutions the free boundary does not determine the initial data uniquely.

Proof. Let $\tilde{\Lambda} \subset J^1(\mathbb{R}^d)$ be any Legendre submanifold (such as the extended free boundary, i.e. the image of Φ). We embed this Legendre submanifold in $J^1(\mathbb{R}^{d+1})$ by mapping the contact element $(X, \Theta, N) \in \tilde{\Lambda} \subset J^1(\mathbb{R}^d)$ to the contact element

$$(x, t, u, p, \dot{u}) = \left(X, \Theta, 0, -\frac{N}{|N|^2}, \frac{1}{|N|^2}\right) \in J^1(\mathbb{R}^{d+1}).$$

The union of all characteristics passing throught the points in $J^1(\mathbb{R}^{d+1})$ thus obtained gives us a Legendre submanifold of $J^1(\mathbb{R}^{d+1})$ which satisfies $\dot{u} = |p|^2$. Slicing this large Legendre submanifold with the section $\{t = 0\}$ then gives a Legendre submanifold

$\Sigma \subset J^1(\mathbb{R}^d)$. If this last Legendre submanifold Σ is the graph of some smooth function u_1 and its gradient ∇u_1, then u_1 is the initial function whose free boundary is $\tilde{\Lambda}$.

In our situation $\tilde{\Lambda}$ is a small perturbation of the free boundary of the solution with initial function u_0. In this case the initial Legendre submanifold $\Sigma \subset J^1(\mathbb{R}^d)$ corresponding to the perturbed free boundary $\tilde{\Lambda}$ will be a small perturbation of the graph of $(u_0, \nabla u_0)$. Hence Σ will also be a graph of some function $x \mapsto (u_1(x), p(x))$. The Legendre condition forces $p(x) = \nabla u_1(x)$. $\qquad\qquad\qquad\qquad\qquad\qquad\qquad\qquad\qquad\qquad\qquad\qquad\qquad\square$

From this lemma we see that the filling time is just a generic minimum function, i.e. generic choices of u_0 lead to functions with the same kind of singularities as functions of the type

$$T(x) = \min_{y \in \Omega_0} F(x, y)$$

in which $F : \Omega_0 \times \Omega_0 \to \mathbb{R}$ is a generically chosen smooth function of two variables.

Close scrutiny of Bogaevski's list of possible singularities of minimum functions reveals that there are only two possible ways in which a hole can be filled when $d = 2$, namely, the vanishing triangle and the closing eye.

The "vanishing triangle" is locally described to leading order by

$$u(x, t) = \max\{0, w_1(x, t), w_2(x, t), w_3(x, t)\}$$

where the w_j are three plane waves converging upon the focusing point at time $t = T$. It is thus an asymptotically self-similar solution. In the language of singularity theory this singularity should be labelled "A_1^3" (there are three values of $y \in \mathbb{R}^2$ which minimize the quantity $Q^x(y) = |x - y|^2/4u_0(y)$ in the definition of the filling time, and these three minima each are nondegenerate critical points of Q^x.)

The "closing eye" is locally described by

$$u(x, t) = \max\{0, q_+(x, t), q_-(x, t)\}.$$

where near $(x, y, t) = (0, 0, 0)$ one has

$$q_\pm(x, t) = \pm\{y - c_\pm x^2 + (1 + 4c_\pm^2 x^2)tx^2 + \cdots\}.$$

In the language of singularity theory this singularity should be labelled "A_1^2" (there are two values of $y \in \mathbb{R}^2$ which minimize the quantity $Q^x(y) = |x - y|^2/4u_0(y)$, and these two minima each are nondegenerate critical points of Q^x.)

Acknowledgements

The authors would like to thank Ilya Bogaevsky for providing the (p)reprints [12], [13], [14] and his comments on an earlier version of this paper. The authors also would like to thank Craig Evans for some helpful elementary remarks on viscosity solutions.

REFERENCES

[1]   ANGENENT, S. B. and ARONSON, D. G., *The focussing problem for the readially symmetric porous medium equation,* Comm. P.D.E. *20* (1995), 1217–1240.

[2]   ANGENENT, S. B. and ARONSON, D. G., *Non-axial self-similar hole filling for the porous medium equation,* J.A.M.S. *14* (2001), 737–782.

[3]   ANGENENT, S. B. and ARONSON, D. G., BETELÚ, S., LOWENGRUB, J. S., *Focusing of an elongated hole in porous medium flow,* Physica D *151* (2001), no. 2–4, 228–252.

[4]   ARNOL'D, V. I., *Geometrical methods in the Theory of Ordinary Differential Equations,* Springer Grundlehren, *250* 1983.

[5]   ARNOL'D, V. I., *Catastrophe Theory,* Springer Verlag, 2nd edition, 1984.

[6]   ARNOL'D, V. I., *Mathematical Methods of Classical Mechanics,* Graduate Texts in Mathematics, Springer-Verlag, New York, *60* 1991.

[7]   ARNOL'D, V. I., SINGULARITIES OF CAUSTICS AND WAVE FRONTS, Mathematics and its Applications – Soviet series *62*, Kluwer 1990.

[8]   ARONSON, D. G., *The porous medium equation,* in: Some Problems in Nonlinear Diffusion (A. Fasano and M. Primicerio editors), Lecture Notes on Math. *1224*, Springer, Berlin 1986.

[9]   ARONSON, D. G. and GRAVELEAU, J., *A self-similar solution to the focusing problem for the porous medium equation,* Euro. J. Appl. Math. *4* (1993), 65–81.

[10]   BARENBLATT, G. I., *Similarity, Self-similarity and Intermediate Asymptotics,* Consultants Bureau, New York, 1978.

[11]   BETELÚ, S. I., ARONSON, D. G. and ANGENENT, S. B., *Renormalization study of two-dimensional convergent solutions of the porous medium equation,* Phys. D *138* (2000), no. 3–4, 344–359.

[12]   BOGAEVSKI, I. A., *Reconstructions of singularities of minimum functions, and bifurcations of shock waves of the Burgers equation with vanishing viscosity,* (Russian) Algebra i Analiz *4* (1989), 1, 1–16; translation in Leningrad Math. J. *4* (1990), 1, 807–823.

[13]   BOGAEVSKIĬ, I. A., *Perestroikas of fronts in evolutionary families,* (Russian) Trudy Mat. Inst. Steklov. Osob. Gladkikh Otobrazh. Dop. Strukt., *209* (1995), 65–83.

[14]   BOGAEVSKI, I. A., *Singularities of viscosity solutions of Hamilton-Jacobi equations,* Singularity theory and differential equations (Kyoto, 1999). Sūrikaisekikenkyūsho Kōkyūroku *1111* (1999), 138–143.

[15]   EVANS, L. C., *Partial Differential Equations,* Graduate Studies in Mathematics *19*, A.M.S. 1998.

[16]   JOHN, F., *Partial Differential Equations,* Applied Mathematical Sciences *1* 4th edition, Springer Verlag, 1982.

[17]   LIONS, P.-L., SOUGANIDIS, P. E. and VÁZQUEZ, J. L., *The relation between the porous medium and the eikonal equations in several space dimensions,* Rev. Mat. Iberoamericana *3* (1987), no. 3–4, 275–310.

S. B. Angenent
Mathematics Department
UW Madison
Wisconsin
e-mail: angenent@math.wisc.edu

D. G. Aronson
School of Mathematics
University of Minnesota
Minneapolis
e-mail: don@math.umn.edu

J.evol.equ. 3 (2003) 153 – 168
1424–3199/03/010153 – 16
© Birkhäuser Verlag, Basel, 2003

Journal of Evolution
Equations

Weak solutions and supersolutions in L^1 for reaction-diffusion systems

MICHEL PIERRE

A Philippe, mon maître et ami

Abstract. We prove here that limits of nonnegative solutions to reaction-diffusion systems whose nonlinearities are bounded in L^1 always converge to supersolutions of the system. The motivation comes from the question of global existence in time of solutions for the wide class of systems preserving positivity and for which the total mass of the solution is uniformly bounded. We prove that, for a large subclass of these systems, weak solutions exist globally.

1. Introduction

This paper is motivated by the general question of global existence in time of solutions to reaction-diffusion systems of the form:

$$\left.\begin{aligned}
u_t - d_1 \Delta u &= f(u, v) \quad \text{on} \quad Q, \\
v_t - d_2 \Delta v &= g(u, v) \quad \text{on} \quad Q, \\
u(0, \cdot) = u_0(\cdot) \geq 0, \quad v(0, \cdot) &= v_0(\cdot) \geq 0, \\
u, v \text{ satisfy some good boundary conditions on} \quad &\partial\Omega,
\end{aligned}\right\} \tag{1}$$

where $Q = (0, +\infty) \times \Omega$, Ω is a regular bounded open subset of $\mathbb{R}^N$, $d_1, d_2 > 0$ and f, g are regular functions whose nonlinear structure is such that two main properties occur:

- the nonnegativity of solutions of (1) is preserved in time,
- the total mass of the solutions is uniformly bounded in time.

The functions f, g may also depend on time and space variable ($f = f(t, x, u, v)$).

With good boundary conditions on $\partial\Omega$ like for instance $u = v = 0$ or $\partial_n u = \partial_n v = 0$ (where ∂_n is the normal derivative at the boundary), nonnegativity will be preserved in time, like for systems of ordinary differential equations, as soon as

$$\forall u, v \geq 0, \quad f(0, v) \geq 0, \quad g(u, 0) \geq 0. \tag{2}$$

Mathematics Subject Classification (2000): 35K10, 35K45, 35K57.
Key words: parabolic system, reaction-diffusion, blowup, global existence, semilinear system.

The second property will occur for instance when

$$f + g \leq 0. \tag{3}$$

Indeed, by just integrating the sum of the two equations, we obtain

$$\int_\Omega u(t) + v(t) \leq \int_\Omega u_0 + v_0.$$

Together with nonnegativity, this yields an L^1-bound of the solution uniformly in time. A general question is to understand how these two properties help to provide global existence in time of solutions.

Note that, if we had uniform L^∞-bounds rather than L^1-bounds, we would deduce global existence in time of "classical" solutions, by standard results for reaction-diffusion systems. By "classical" solution, we mean "bounded" solution, so that, by well-known regularity results, a "classical" solution also has classical derivatives at least $a.e.$ and the equations are understood pointwise.

The point here is that bounds are a priori only in L^1 and one cannot apply the L^∞-approach even if the initial data are regular.

This situation frequently comes out in applications where positivity of the unknowns u, v is implicit from their definition (they are densities, concentrations, normalized temperatures,...) and where the total mass is preserved or, at least, controlled in time. This explains why these systems have been studied in several places in the literature. Let us refer here to [11, 17] for a survey and references.

To help understand the situation, let us mention two particular examples of the nonlinearities we are considering:

$$f(u, v) = u^3 v^2 - u^2 v^3, \quad g(u, v) = -u^3 v^2 + \gamma u^2 v^3, \quad \text{where } 0 \leq \gamma \leq 1. \tag{4}$$

$$f(u, v) = c_1(x, t) u^\alpha v^\beta, \quad g(u, v) = c_2(x, t) u^\alpha v^\beta, \tag{5}$$

where $\alpha, \beta > 1$, and c_1, c_2 are regular functions such that

$$a.e. \, (t, x) \in Q, c_1(x, t) + c_2(x, t) \leq 0. \tag{6}$$

Obviously, for bounded initial data, we will have local existence of classical solutions. With some extra assumptions, like for instance $\gamma = 0$ in (4) or $c_2 \leq 0$ in (5), global existence of classical solutions may be proved. It is not straightforward: several approaches may be found in [13, 7, 8, 3, 14].

But for our purpose here, the main fact to remember is that, although one has an uniform L^1-bound in time, "classical" solutions may not globally exist when the diffusion coefficients d_1, d_2 are not equal (global existence obviously holds if they are equal). As surprisingly proved in [16, 17], it may indeed happen that, under assumptions (2), (3), solutions blow up in finite time in L^∞! In particular, classical bounded solutions do not exist globally in time.

We emphasize the fact that, in the examples of blow up provided in [16, 17], not only (3) holds, but even

$$f + \lambda_0 g \le 0 \text{ for some } \lambda_0 \in [0, 1). \tag{7}$$

Note that (3) together with (7) imply that

$$f + \lambda g \le 0 \quad \text{for } \lambda \in [\lambda_0, 1].$$

As we will see in more details later, under this more restrictive assumption, not only u, v are bounded in $L^1(\Omega)$, but the nonlinear terms $f(u, v)$, $g(u, v)$ are also bounded in $L^1(Q_T)$ for all $0 < T < \infty$ and $Q_T = (0, T) \times \Omega$.

A first main purpose of this paper is to prove that, under the latter stronger assumption, *global existence on* $[0, \infty)$ *of weak solutions holds* for the above considered systems (although these solutions may blow up in L^∞ at some time). By "weak" solution, we essentially mean solution in the sense of distributions or, equivalently here, solution in the sense of the variation of constant formula with the corresponding semigroups (see Appendix). In particular, classical derivatives may not exist. Such weak L^1-solutions had already been considered in [15, 10, 3] to handle initial data in L^1. However, an extra condition of "triangular" structure of the nonlinearities was required (which would, for instance, imply $\gamma = 0$ in example (4)).

Concerning the above examples, our result here means that weak solutions exist globally for the nonlinearities (4) if $\gamma \in [0, 1)$ and for the nonlinearities (5) if, moreover, $c_1 + \lambda_0 c_2 \le 0$ for some $\lambda_0 \ge 0, \lambda_0 \ne 1$. But, according to [16, 17], weak solutions in example (5) may blow up in finite time in $L^p(\Omega)$ for p large. We do not know specifically what happens for example (4), but we know that similar polynomial nonlinearities do lead to blow up in finite time [16, 17].

One of the main steps in the proof turns out to be interesting by itself for reaction-diffusion systems. One knows that maximum principle is valid for equations, but generally not for systems. It turns out that systems do nevertheless share some order properties with equations, no matter their structure: this is also a purpose of this paper to point it out.

To explain this point, let us first consider an equation and a sequence of *nonnegative* regular solutions of

$$\partial u_n / \partial t - \Delta u^n = F_n(u_n) \text{ on } Q_T$$

where $F_n : [0, +\infty) \to \mathbb{R}$ converges uniformly on bounded sets to the continuous function $F : [0, +\infty) \to \mathbb{R}$. Assume that

$$F_n(u_n) \text{ is bounded in } L^1(Q_T) \text{ independently of } n.$$

Assume also that u_n satisfies, for instance, the boundary condition $u_n(t, \cdot) = 0$ on $\partial\Omega$ and that $u_n(0, .)$ is bounded in L^1.

Then, up to a subsequence, u_n converges in $L^1(Q_T)$ to a *supersolution u* of the equation, namely

$$\partial u/\partial t - \Delta u \geq F(u) \text{ on } Q_T, \tag{8}$$

in the sense of distributions.

The proof of this fact goes essentially as follows. Thanks to the L^1 bound on the nonlinear term $F_n(u_n)$ and to the parabolic boundary conditions, u_n is relatively compact in $L^1(Q_T)$. Up to a subsequence, one can assume that u_n converges in $L^1(Q_T)$ and almost everywhere to a function u. Then, $F_n(u_n)$ converges pointwise to the integrable function $F(u)$. Unfortunately, this is not enough to pass to the limit in the equation.

Then, let us introduce a truncation procedure: for $k \geq 1$ and $r \geq 0$, set $\tau_k(r) = \min\{r, k\}$. By a simple computation, we obtain for all k, n:

$$\partial \tau_k(u_n)/\partial t - \Delta \tau_k(u^n) \geq \tau_k'(u_n) F_n(u_n) \text{ on } Q_T.$$

But, $\tau_k'(u_n) = 0$ where $u_n > k$. For k fixed, since $F_n(u_n)$ is bounded independently of n on the set where $u_n \leq k$, then, $\tau_k'(u_n) F_n(u_n)$ converges, not only pointwise, but also in $L^1(Q_T)$ to $\tau_k'(u) F(u)$, so that

$$\partial \tau_k(u)/\partial t - \Delta \tau_k(u) \geq \tau_k'(u) F(u) \text{ on } Q_T.$$

We now let k go to ∞ to obtain (8).

Obviously, this approach does not extend as such to a sequence u_n, v_n of solutions of a 2×2 system since, multiplying the first equation by $T_k'(u_n)$ does not take care of unbounded values of v_n. However, we are able to prove that the same result holds and this is another main goal of this paper: when the nonlinearities remain bounded in $L^1(Q_T)$, *the limit is a supersolution of the system.*

2. The main results

Let Ω be a bounded open subset of $\mathbb{R}^N$ with regular boundary. We denote $Q := (0, +\infty) \times \Omega$, and for $T \in (0, +\infty)$, $Q_T := (0, T) \times \Omega$.

Let $f, g : Q \times [0, +\infty)^2 \to \mathbb{R}$ satisfy the usual local Lipschitz conditions:

$$\left. \begin{array}{c} f, g \text{ are measurable, } \forall T > 0, \ f(\cdot, \cdot, 0, 0), \ g(\cdot, \cdot, 0, 0) \in L^1(Q_T), \\ \exists K : [0, +\infty) \to [0, +\infty) \text{ nondecreasing such that} \\ a.e. \ (t, x) \in [0, +\infty) \times \Omega, \ \forall M > 0, \forall r, s, \hat{r}, \hat{s} \in (0, M), \\ |f(t, x, r, s) - f(t, x, \hat{r}, \hat{s})| + |g(t, x, r, s) - g(t, x, \hat{r}, \hat{s})| \leq \dots \\ \dots K(r)(|r - \hat{r}| + |s - \hat{s}|). \end{array} \right\} \tag{9}$$

The condition (2) will take the form

$$a.e.\ (t, x) \in Q,\ \forall u, v \geq 0,\ f(t, x, 0, v) \geq 0,\ g(t, x, u, 0) \geq 0. \tag{10}$$

We will also assume that conditions (3)+(7) are satisfied in a weaker sense:

$$\left.\begin{array}{c} \exists \lambda_0 \in [0, 1),\ \text{such that}\ \forall \lambda \in [\lambda_0, 1], \\ a.e.(t, x) \in Q, \forall r, s \geq 0, \\ f(t, x, r, s) + \lambda\, g(t, x, r, s) \leq \sigma\, (r + s) + h(t, x), \\ \text{where}\ \sigma \geq 0,\ \text{and}\ h \in L^1(Q_T), \forall T > 0,\ h \geq 0. \end{array}\right\} \tag{11}$$

THEOREM 2.1. *Let f, g be given as in (9) and let $d_1, d_2 > 0$. Assume that f, g satisfy the positivity property (10) and the structure condition (11). Then, for all $u_0, v_0 \in L^1(\Omega), u_0, v_0 \geq 0$, there exists a global nonnegative solution (u, v) on $[0, +\infty)$ of*

$$\left.\begin{array}{c} u, v \in C([0, +\infty); L^1(\Omega)) \cap L^1_{loc}([0, +\infty); W_0^{1,1}(\Omega)), \\ u(0, \cdot) = u_0,\quad v(0, \cdot) = v_0, \\ \forall T > 0, f(\cdot, \cdot, u(\cdot, \cdot), v(\cdot, \cdot)),\ g(\cdot, \cdot, u(\cdot, \cdot), v(\cdot, \cdot)) \in L^1(Q_T), \\ u_t - d_1 \Delta u = f(t, x, u, v)\ \text{in}\ \mathcal{D}'(Q), \\ v_t - d_2 \Delta v = g(t, x, u, v)\ \text{in}\ \mathcal{D}'(Q). \end{array}\right\} \tag{12}$$

Here and hereafter, equations are understood in the sense of distributions $\mathcal{D}'(Q)$, that is, for all test-function φ in the space $C_0^\infty(Q)$ of infinitely differentiable functions with compact support in Q, we have:

$$- \int_Q u\, (\varphi_t + d_1 \Delta \varphi) = \int_Q \varphi\, f,$$

and similarly for v. It is well-known that (12) is equivalent to the variation of constant formula, that is to say (see Appendix)

$$u(t) = S_{d_1}(t)u_0 + \int_0^t S_{d_1}(t - s) f(s, \cdot, u(s, \cdot), v(s, \cdot))\, ds,$$

where $S_{d_1}(\cdot)$ is the semigroup generated in $L^1(\Omega)$ by the Laplacian operator with homogeneous boundary conditions (and the similar formula for v).

REMARK. The boundary condition $u = v = 0$ on $\partial\Omega$ is understood here in the sense that $a.e.t, u(t), v(t) \in W_0^{1,1}(\Omega)$. As usual, for all $1 \leq p < +\infty$, $W_0^{1,p}(\Omega)$ is the closure of the space $C_0^\infty(\Omega)$ equipped with the norm

$$\|w\|_{W_0^{1,p}} := \{\|w\|_{L^p(\Omega)}^p + \|\nabla w\|_{L^p(\Omega)^N}^p\}^{1/p}.$$

As it will be clear from the proof, a similar result could be stated for Neumann boundary conditions or for more general boundary conditions. One must however be careful when choosing two different boundary conditions for u and v (see [4, 12]).

As announced in the introduction, the second main result of this paper deals with limits of approximate solutions of systems when the nonlinearities are bounded in L^1.

THEOREM 2.2. *Let (u_n, v_n) be a sequence of (regular) nonegative solutions of*

$$\left.\begin{array}{c} u_n, v_n \in C([0, T]; L^2(\Omega)) \cap L^2((0, T); W_0^{1,2}(\Omega)), \\ u_{nt}, v_{nt}, \Delta u_n, \Delta v_n \in L^2(Q_T), \\ F_n(\cdot, \cdot, u_n(\cdot), v_n(\cdot)),\ \ G_n(\cdot, \cdot, u_n(\cdot), v_n(\cdot)) \in L^2(Q_T), \\ \partial u_n/\partial t - d_1 \Delta u_n = F_n(t, x, u_n, v_n) \text{ on } Q_T, \\ \partial v_n/\partial t - d_2 \Delta v_n = G_n(t, x, u_n, v_n) \text{ on } Q_T, \end{array}\right\} \tag{13}$$

where $F_n, G_n : Q \times [0, +\infty)^2 \to \mathbb{R}$ converge in the following sense to $F, G : Q \times [0, +\infty)^2 \to \mathbb{R}$ satisfying (9): for all $M > 0$, ϵ_M^n tends to zero in $L^1(Q_T)$ as $n \to +\infty$ where

$$\epsilon_M^n = \sup_{0 \le r, s \le M} \{|F_n(\cdot, \cdot, r, s) - F(\cdot, \cdot, r, s)| + |G_n(\cdot, \cdot, r, s) - G(\cdot, \cdot, r, s)|\}. \tag{14}$$

Assume that $F_n(\cdot, \cdot, u_n(\cdot), v_n(\cdot))$, $G_n(\cdot, \cdot, u_n(\cdot), v_n(\cdot))$ are bounded in $L^1(Q_T)$ independently of n. Assume also that $u_n(0), v_n(0)$ are bounded in L^1.

Then, up to a subsequence, u_n, v_n converge to u, v in $L^1(Q_T)$ satisfying

$$\left.\begin{array}{c} u, v \in L^\infty((0, T); L^1(\Omega)) \cap L^1((0, T); W_0^{1,1}(\Omega)), \\ F(\cdot, \cdot, u(\cdot, \cdot), v(\cdot, \cdot)),\ \ G(\cdot, \cdot, u(\cdot, \cdot), v(\cdot, \cdot)) \in L^1(Q_T), \\ \partial u/\partial t - d_1 \Delta u \ge F(t, x, u, v) \text{ in } \mathcal{D}'(Q_T), \\ \partial v/\partial t - d_2 \Delta v \ge G(t, x, u, v) \text{ on } \mathcal{D}'(Q_T). \end{array}\right\} \tag{15}$$

Moreover, if $u_n(0), v_n(0)$ converge to u_0, v_0 in $L^1(\Omega)$, then, for all nonnegative $\varphi \in C_0^\infty(\Omega)$, we have

$$\liminf_{t \to 0} \int_\Omega u(t)\varphi \ge \int_\Omega u_0\varphi, \quad \liminf_{t \to 0} \int_\Omega v(t)\varphi \ge \int_\Omega v_0\varphi. \tag{16}$$

REMARK. Although it is not essential, we assume here that the solutions are "regular", in the sense that they have derivatives $u_{nt}, \Delta u_n$ in L^2. This allows to make direct computations. Without L^2-regularity, we could also do it by using one more approximation process. Theorem 2 (which we prove first) will be sufficient for the approximate solutions considered in Theorem 1 which are regular by construction.

The above inequations are understood in the sense of distributions in Q: this means that for all nonnegative test-functions φ of $C_0^\infty(Q)$,

$$-\int_Q u\,(\varphi_t + d_1 \Delta\varphi) \ge \int_Q \varphi\,F,$$

and the same for v.

Obviously, estimate (16) says that the initial data of u, v are above the limit of the initial data of u_n, v_n. More precisely, since $u(t)$ is bounded in $L^1(\Omega)$, one may find a nonnegative Radon measure u_{0+} and a sequence t_k tending to zero as k tends to $+\infty$ such that $u(t_k)$ converges to u_{0+} in the sense of measures. Then, for all such limit u_{0+}, by (16), we have $u_{0+} \geq u_0$, and the same for v.

3. The proofs

We will use the following more or less classical compactness lemma for the heat operator (a proof may be found e.g. in [2]; comments are also given in Appendix)

LEMMA 3.1. *Let* $d > 0, w_0 \in L^1(\Omega), H \in L^1(Q_T)$. *Then there exists a unique solution of*

$$\left.\begin{aligned} & w \in C([0, T]; L^1(\Omega)) \cap L^1((0, T); W_0^{1,1}(\Omega)), \\ & \partial w/\partial t - d\Delta w = H \text{ in } \mathcal{D}'(Q_T), w(0, \cdot) = w_0. \end{aligned}\right\} \tag{17}$$

Moreover, for all $s, q \geq 1$ *with* $2s^{-1} + Nq^{-1} > N + 1$, *there exists* $C = C(q, s, \Omega, d)$ *such that*

$$\|w\|_{L^\infty(0,T;L^1(\Omega))} + \|w\|_{L^s(0,T;W_0^{1,q}(\Omega))} \leq C\,[\|H\|_{L^1(Q_T)} + \|w_0\|_{L^1(\Omega)}]. \tag{18}$$

Finally, the mapping $(H, w_0) \to w$ *is compact from* $L^1(Q_T) \times L^1(\Omega)$ *into* $L^1(Q_T)$.

Proof of Theorem 2.2. By Lemma 3.1, if (u_n, v_n) is the sequence considered in Theorem 2.2, up to a subsequence, we may assume that u_n, v_n converge in $L^1(Q_T)$ and $a.e.$ to $u, v \in L^\infty(0, T; L^1(\Omega)) \cap L^1(0, T; W_0^{1,1}(\Omega))$. According to the type of convergence of F_n, G_n to F, G, $F_n(t, x, u_n, v_n)$ converge a.e. to $F = F(t, x, u, v)$ and the same for G_n, G, but this pointwise convergence is not sufficient by itself to pass to the limit in the equations and this is where the more difficult step starts.

We introduce truncation functions. For technical reasons, we need them to be a little more regular than in the introduction. For all $k > 0$, we define a C^2- function T_k, such that

$$\left.\begin{aligned} & \forall r \in [0, k], T_k(r) = r \ ; \quad \forall r \geq k, \ T_k(r) \leq k + 1, \\ & \forall r \geq 0, \ 0 \leq T_k'(r) \leq 1 \ ; \quad \forall r \geq k + 1, T_k'(r) = 0, \\ & \qquad\qquad\qquad 0 \leq -T_k''(r) \leq C(k). \end{aligned}\right\}$$

For instance, we may choose T_k as $T_k(r) = r$ on $[0, k]$ and

$$\forall r \in [k, k+1], T_k(r) = (r - k)^4/2 - (r - k)^3 + r; \forall r > k+1, T_k(r) = k + 1/2.$$

Now, we fix $\eta > 0$ and we introduce $z_n := T_k(u_n + \eta\, v_n)$. Using the equations satisfied by u_n, v_n, we obtain

$$
\begin{aligned}
\partial z_n/\partial t - d_1 \Delta z_n &= T_k'(u_n + \eta\, v_n)(F_n + \eta G_n) \\
&\quad + \eta(d_2 - d_1)S_n^1 + d_1 S_n^2,
\end{aligned}
\tag{19}
$$

where

$$
S_n^1 = T_k'(u_n + \eta\, v_n)\Delta v_n, \quad S_n^2 = -T_k''(u_n + \eta\, v_n)|\nabla(u_n + \eta\, v_n)|^2 \geq 0.
\tag{20}
$$

The main point is to pass to the limit as n tends to ∞ in the equation (19), η and k being fixed. Let us look successively at the four terms involved.

Note first that the last term is nonnegative so that we may just forget it.

Since z_n tends to $z = T_k(u + \eta\, v)$ in $L^1(Q_T)$, $\partial z_n/\partial t - d_1 \Delta z_n$ converges in the sense of distributions to $\partial z/\partial t - d_1 \Delta z$.

Next, by the type of convergence of F_n, G_n to F, G (see (14)) and by the continuity property of F, G, $T_k'(u_n + \eta\, v_n)(F_n + \eta\, G_n)$ converges pointwise and in $L^1(Q_T)$ to $T_k'(u + \eta\, v)(F + \eta\, G)$: indeed, on one hand, $T_k'(u_n + \eta\, v_n) = 0$ on the set $[u_n + \eta\, v_n \geq k + 1]$. On the other hand, on the set $[u_n + \eta\, v_n \leq k + 1]$, we have:

$$
|F_n(t, x, u_n, v_n) - F(t, x, u_n, v_n)| \leq \epsilon_{(k+1)(1+n^{-1})}^n(t, x),
$$

and the right hand side tends to zero in $L^1(Q_T)$ and a.e. as n tends to $+\infty$. Moreover,

$$
F(t, x, u_n(t, x), v_n(t, x)) \to F(t, x, u(t, x), v(t, x)), \quad a.e.(t, x),
$$

and remains bounded on the set $[u_n + \eta\, v_n \leq k + 1]$ by

$$
\zeta(t, x) := \sup_{r \leq k+1, s \leq (k+1)\eta^{-1}} |F(t, x, r, s)|,
$$

which is in $L^1(Q_T)$ by the conditions (9) on F (and similarly for G_n, G).

Note that, by Fatou's Lemma and the L^1-bounds on F_n, G_n, F, G are in $L^1(Q_T)$.

Now we are left with the main step: estimating S_n^1. For this, we need the following lemma.

LEMMA 3.2. *There exists C depending only on the bounds on $\|F_n\|_{L^1(Q_T)}$, $\|G_n\|_{L^1(Q_T)}$, $\|u_0\|_{L^1(\Omega)}$, $\|v_0\|_{L^1(\Omega)}$ such that*

$$
\forall k \geq 1, \int_{[u_n \leq k]} |\nabla u_n|^2, \int_{[v_n \leq k]} |\nabla v_n|^2 \leq Ck.
\tag{21}
$$

Let us postpone the proof of this lemma and continue the proof of Theorem 2.2. We will denote by $C(k)$ all positive constants depending only on k. Let $\varphi \in C_0^\infty([0, T] \times \Omega)$. Then, for all $t \in [0, T]$

$$\int_{Q_t} \varphi S_n^1 = -\int_{Q_t} \nabla v_n[T_k'(u_n + \eta \, v_n)\nabla \varphi + \varphi \, T_k''(u_n + \eta \, v_n)\nabla(u_n + \eta \, v_n)],$$

so that, using the properties of T_k

$$|\textstyle\int_{Q_t} \varphi S_n^1| \leq \{\textstyle\int_{[u_n+\eta \, v_n \leq k+1]} |\nabla v_n|^2\}^{1/2}[\{\textstyle\int_{Q_T} |\nabla \varphi|^2\}^{1/2}$$
$$+ C(k)\|\varphi\|_{L^\infty(Q_T)}\{\textstyle\int_{[u_n+\eta \, v_n \leq k+1]} |\nabla u_n + \eta \, v_n|^2\}^{1/2}].$$

Note that $[u_n + \eta \, v_n \leq k + 1]$ is included in $[u_n \leq k + 1]$ and in $[v_n \leq (k + 1)\eta^{-1}]$. We bound the last term from above as follows:

$$\{\textstyle\int_{[u_n+\eta \, v_n \leq k+1]} |\nabla u_n + \eta \, v_n|^2\}^{1/2} \leq \{\textstyle\int_{[u_n \leq k+1]} |\nabla u_n|^2\}^{1/2}$$
$$+ \eta\{\textstyle\int_{[v_n \leq (k+1)\eta^{-1}]} |\nabla v_n|^2\}^{1/2}.$$

Setting $D(\varphi) := \{\textstyle\int_{Q_T} |\nabla \varphi|^2\}^{1/2} + \|\varphi\|_{L^\infty(Q_T)}$ and using Lemma 3.2, we deduce

$$\left|\int_{Q_t} \varphi S_n^1\right| \leq D(\varphi)C(k)\eta^{-1/2}[1 + \eta^{1/2}]. \tag{22}$$

We can now let n tend to $+\infty$ in (19). We will denote by $\langle Z, \varphi\rangle$ the result of a distribution Z of $\mathcal{D}'(Q_T)$ applied to the C_0^∞-test-function φ. We obtain for any *nonnegative* test-function φ

$$\langle z_t - d_1 \Delta z - T_k'(u + \eta \, v)(F + \eta \, G), \varphi\rangle \geq -C(k)\eta^{1/2}D(\varphi).$$

Now, we let η tend to zero in the above inequality. Since, $z = T_k(u + \eta \, v)$ tends to $T_k(u)$ in $L^1(Q_T)$ and since $T_k'(u + \eta \, v)$ remains uniformly bounded by 1 and tends a.e. to $T_k'(u)$, we can pass to the limit in the sense of distributions to find

$$\partial T_k(u)/\partial t - d_1 \Delta T_k(u) \geq T_k'(u) \, F \quad \text{in } \mathcal{D}'(Q_T).$$

Finally, we let k tend to $+\infty$: by monotonicity, $T_k(u)$ tends to u in $L^1(Q_T)$ and $T_k'(u)$ tends a.e. to 1; since $F \in L^1(Q_T)$, we can pass to the limit and obtain

$$\partial u/\partial t - d_1 \Delta u \geq F \quad \text{in } \mathcal{D}'(Q_T).$$

Let us now look at the **initial data:** we assume that $u_n(0)$, $v_n(0)$ tend in $L^1(\Omega)$ to u_0, v_0. We go back to equation (19) and multiply by $\varphi \in C_0^\infty(\Omega)$, nonnegative, to obtain

$$\int_\Omega [z_n(t) - z_n(0)]\varphi \geq \int_{Q_t} d_1 z_n \Delta \varphi + \varphi[\eta(d_2 - d_1)S_n^1 + T_k'(u_n + \eta \, v_n)(F_n + \eta \, G_n)].$$

We recall (22) which gives a bound from below for S_n^1. Letting n tend to ∞ leads to

$$\int_\Omega [z(t) - T_k(u_0 + \eta\, v_0)]\varphi \geq \int_{Q_t} d_1 z \Delta\varphi + \varphi T_k'(u + \eta\, v)(F + \eta\, G) - C(k, \varphi)\eta^{1/2}.$$

We let η tend to zero, then k tend to ∞: as before, using that $F \in L^1(Q_T)$, we may pass to the limit to obtain

$$\int_\Omega [u(t) - u_0]\varphi \geq \int_{Q_t} d_1 u \Delta\varphi + \varphi F.$$

Letting now t tend to zero gives

$$\liminf_{t \to 0} \int_\Omega u(t)\varphi \geq \int_\Omega u_0\varphi.$$

This is the statement of (16).

Proof of Lemma 3.2. We choose T_k as above. Multiplying the equation in u_n by $T_k(u_n)$ gives

$$\frac{\partial}{\partial t} \int_\Omega j_k(u_n) + d_1 \int_\Omega T_k'(u_n)|\nabla u_n|^2 = \int_\Omega T_k(u_n)\, F_n, \tag{23}$$

where $j_k(r) = \int_0^r T_k(s)\, ds$. Note that $j_k(r) \leq (k+1)\, r$. After integrating (23) in time, we obtain

$$d_1 \int_{[u_n \leq k]} |\nabla u_n|^2 \leq (k+1)\left\{ \int_{Q_T} |F_n| + \int_\Omega u_n(0) \right\},$$

whence the estimate (21) for u_n. The proof is the same for v_n.

Proof of Theorem 2.1. The first step is to truncate the data in order to solve an approximate problem. We set $u_{0n} := \inf\{u_0, n\}$, $v_{0n} := \inf\{v_0, n\}$. We truncate the nonlinearities f, g in such a way that they be bounded and that they keep satisfy the same conditions (9, 10, 11). For this, we introduce a C_0^∞ function $\psi_1 : [0, +\infty)^2 \to [0, 1]$ satisfying

$$\forall\, 0 \leq r, s \leq 1, \psi_1(r, s) = 1; \forall\, r, s \geq 2, \psi_1(r, s) = 0.$$

Next, we set $\psi_n(r, s) = \psi_1(r/n, s/n)$. With this choice, for all n, $0 \leq \psi_n \leq 1$, and ψ_n tends pointwise to 1 as n tends to ∞.

In order to take care of the fact that $f(\cdot, \cdot, 0, 0)$ is only in $L^1(Q_T)$, we also truncate it and, for technical reasons, we introduce

$$\mathcal{F}_n(t, x) := \tau_{K(n)2n}(f(t, x, 0, 0)), \quad \mathcal{G}_n(t, x) := \tau_{K(n)2n}(g(t, x, 0, 0)),$$

where $\tau_k(r) = r$ if $|r| \leq k$, $\tau_k(r) = k$ if $r > k$, $\tau_k(r) = -k$ if $r < -k$.

Finally, we define

$$f_n(t, x, r, s) := \psi_n(r, s)[f(t, x, r, s) - f(t, x, 0, 0)] + \mathcal{F}_n(t, x),$$
$$g_n(t, x, r, s) := \psi_n(r, s)[g(t, x, r, s) - g(t, x, 0, 0)] + \mathcal{G}_n(t, x). \tag{24}$$

One easily verifies that f_n, g_n converge to f, g in the sense (14) and that f_n, g_n satisfies the same conditions (9,10,11) as f, g. Note that f_n, g_n are even globally Lipschitz continuous with respect to r, s with a constant depending on n. Note also that f_n, g_n satisfy (11) with the same λ_0, σ, h as for f, g. For the nonegativity condition (10), we remark that, if $f(t, x, 0, 0) > K(n)\, 2n$, then

$$f_n(t, x, 0, s) \geq -K(n)\, 2n + \mathcal{F}_n(t, x) = 0,$$

and otherwise $\mathcal{F}_n(t, x) \geq f(t, x, 0, 0)$ so that

$$f_n(t, x, 0, s) \geq \psi_n(0, s) f(t, x, 0, s) + [1 - \psi_n(0, s)] f(t, x, 0, 0) \geq 0,$$

whence (10) and similarly for g_n.

Note finally that f_n, g_n are uniformy bounded since

$$\forall r, s, \ |f_n(t, x, r, s)| + |g_n(t, x, r, s)| \leq 3\, K(2n)\, 2n. \tag{25}$$

By a classical fixed point theorem (see e.g. Appendix), there exists a unique "classical" and nonnegative solution u_n, v_n of

$$\left.\begin{aligned}
&u_n, v_n \in C([0, +\infty); L^2(\Omega)) \cap L^2_{loc}([0, +\infty); W_0^{1,2}(\Omega)),\\
&\qquad u_n(0, \cdot) = u_{0n}, \quad v_n(0, \cdot) = v_{0n},\\
&\forall T \in (0, +\infty),\ u_{nt}, v_{nt}, \Delta u_n, \Delta v_n \in L^2(Q_T),\\
&f_n(\cdot, \cdot, u_n(\cdot, \cdot), v_n(\cdot, \cdot)),\ g_n(\cdot, \cdot, u_n(\cdot, \cdot), v_n(\cdot, \cdot)) \in L^\infty(Q_T),\\
&\qquad \partial u_n/\partial t - d_1 \Delta u_n = f(t, x, u_n, v_n)\ \text{in } Q,\\
&\qquad \partial v_n/\partial t - d_2 \Delta v_n = g(t, x, u_n, v_n)\ \text{on } Q.
\end{aligned}\right\} \tag{26}$$

Adding the two equations in u_n, λv_n, we obtain for all $\lambda \in [\lambda_0, 1]$,

$$\int_\Omega u_n(t) + \lambda v_n(t) - \int_{Q_t} f_n + \lambda g_n \leq \int_\Omega u_{0n} + \lambda v_{0n}. \tag{27}$$

(Here, we used, $\int_\Omega \Delta u_n \leq 0$ and the same for v_n which is true because u_n, v_n are nonnegative on Ω and equal to zero on $\partial\Omega$, see e.g. [6]). Using now (11), we deduce in particular that

$$\int_\Omega u_n(t) + \lambda v_n(t) - \int_{Q_t} \sigma\, (u_n(s) + v_n(s)) + h(s) \leq \int_\Omega u_{0n} + \lambda v_{0n}. \tag{28}$$

From this linear Gronwall type inequality, we deduce that

$$\sup_{t \in (0,T)} \{\|u_n(t)\|_{L^1(\Omega)} + \|v_n(t)\|_{L^1(\Omega)}\} \leq C(T, \|h\|_{L^1}, \sigma, \|u_{0n}\|_{L^1}, \|v_{0n}\|_{L^1}). \tag{29}$$

Now, the hypothesis (11) implies that

$$\|f_n + \lambda g_n\|_{L^1(Q_T)} \leq - \int_{Q_T} f_n + \lambda g_n + \|\sigma\,(u_n + v_n) + h\|_{L^1(Q_T)}.$$

Together with (27) and (29), this implies that $\|f_n + \lambda g_n\|_{L^1(Q_T)}$ is bounded for all $\lambda \in [\lambda_0, 1]$. Since, $\lambda_0 \neq 1$, this implies that f_n and g_n are separately bounded in $L^1(Q_T)$.

As a consequence, we are in the conditions of application of Theorem 2.2. It follows that, up to a subsequence, u_n, v_n converge a.e. on Q and in $L^1(Q_T)$ for all $T > 0$ to a supersolution (u, v) of our problem in the sense of (15) with F, G replaced by f, g.

To go from a supersolution to a solution, we argue as follows. Let φ be a C_0^∞, *nonnegative* test-function. We already know that

$$-\int_{Q_T} u\,(\varphi_t + d_1\Delta\varphi) \geq \int_{Q_T} \varphi\,f, \quad -\int_{Q_T} v\,(\varphi_t + d_2\Delta\varphi) \geq \int_{Q_T} \varphi\,g.$$

To obtain the reverse inequality for each, it is sufficient to prove that

$$-\int_{Q_T} (u + v)\varphi_t + (d_1 u + d_2 v)\Delta\varphi \leq \int_{Q_T} \varphi(f + g), \tag{30}$$

starting from

$$-\int_{Q_T} (u_n + v_n)\varphi_t + (d_1 u_n + d_2 v_n)\Delta\varphi = \int_{Q_T} \varphi(f_n + g_n). \tag{31}$$

By L^1-convergence of u_n, v_n, the left hand side of (31) does converge to the left hand side of (30). Since $f_n + g_n$ converges a.e. to $f + g$, and since, by (11), we have the pointwise estimate

$$\sigma\,(u_n + v_n) + h - (f_n + g_n) \geq 0,$$

by Fatou' Lemma, we deduce that

$$\liminf_{n\to\infty} \int_{Q_T} [\sigma\,(u_n + v_n) + h - (f_n + g_n)]\varphi \geq \int_{Q_T} [\sigma\,(u + v) + h - (f + g)]\varphi.$$

This gives the expected reverse inequality (30).

We now have to verify that u, v have the right initial data u_0, v_0. We already have one inequality by Theorem 2.2. (see 16). The bound in $L^\infty(0, T; L^1(\Omega))$ implies that $\{u(t), v(t), t \in (0, T)\}$ are compact for the weak convergence of measures on Ω (i.e. against continuous test-functions φ with compact support in Ω). If u_{0+}, v_{0+} is a weak-limit for a subsequence $u(t_k), v(t_k)$ where $\lim_{k\to+\infty} t_k = 0$, we already now (see Theorem 2.2) that

$$u_{0+} \geq u_0, \quad v_{0+} \geq v_0.$$

We will prove that

$$\limsup_{t \to 0} \int_\Omega u(t) + v(t) \le \int_\Omega u_0 + v_0. \tag{32}$$

It will then follow that $u_{0+} = u_0$, $v_{0+} = v_0$: this uniqueness of the possible weak limits and the fact that there is no loss of mass imply that $u(t)$, $v(t)$ converge as $t \to 0$, for the narrow convergence of measures, to u_0, v_0, namely (see Appendix)

$$\forall \varphi \in C_b(\Omega), \ \lim_{t \to 0} \int_\Omega \varphi u(t) = \int_\Omega \varphi u_0,$$

and the same for v. But, by the uniqueness Lemma 5.1 of the Appendix, we may then deduce that $u, v \in C([0, T]; L^1(\Omega))$ for all $T > 0$, which finishes the proof of Theorem 2.1.

To prove (32), we start from (28) with $\lambda = 1$ and we pass to the limit in n:

$$a.e.t, \ \int_\Omega u(t) + v(t) \le \int_{Q_t} \sigma\,(u(s) + v(s)) + h(s) + \int_\Omega u_0 + v_0.$$

Then, (32) follows directly from this inequality.

4. Some comments

The question of global existence for systems (1) when only (2) and (3) hold remains open (it is the case of example (4) when $\gamma = 1$ and of example (5)). A main new difficulty is that no more L^1 bound on the nonlinear terms f, g is available. It is likely that some kind of weak solutions exist globally in time, but to prove it would first require to introduce a quite weaker notion of solution. It could be possible that some notion of "renormalized" solution may work where the nonlinearities are truncated in the definition.

The analysis made here is not particular to 2×2 systems. Theorem 2 may be generalized to $N \times N$ systems where all the N nonlinearities are bounded in $L^1(Q_T)$. The idea is to replace in the proof $T_k(u + \eta\, v)$ by $T_k(u_1 + \eta[u_2 + \dots u_N])$ if we denote by $(u_1, ..., u_N)$ the unknown of the system. Theorem 1 also extends: we then have to assume that N linearly independent relations of the form $\sum_{i=1}^N \lambda_i f_i \le$ with $\lambda_i \ge 0$ hold. This will provide the L^1 bound on all the f_i.

One could also replace the Laplacian operators $d_1 \Delta$, $d_2 \Delta$ by more general elliptic operators.

Elliptic versions of the same results may be proved for systems of the form

$$\left.\begin{aligned} u + A_1 u &= f(u, v) + \mathcal{F} \ \text{ on } \ \Omega, \\ v + A_2 v &= g(u, v) + \mathcal{G} \ \text{ on } \ \Omega, \\ u, v \ \text{ satisfy some good boundary conditions on } \ \partial\Omega, \end{aligned}\right\}$$

where $\mathcal{F}, \mathcal{G}$ are given nonnegative functions on Ω, A_1, A_2 are good elliptic operators and f, g satisfy (2,3,7). As a nontrivial case, we might think for instance to the simple choice $A_1 = -\Delta$, $A_2 = -\Delta - du_{x_1 x_1}$ with d not too small. Here the difficulty is the proof of existence of solutions and the question is quite similar to proving global existence for the parabolic system (1). Then by the same technique, we can prove existence of weak solutions under similar hypotheses on the nonlinearities. One may also state an elliptic version of Theorem 2.2. The limit case (only (3)) is open as well in this elliptic situation.

5. Appendix

About Lemma 3.1 *let us first comment on the proof of Lemma* 3.1. *A starting point may be the L^2-theory: for $u_0 \in L^2(\Omega)$, $H \in L^2(Q_T)$, there exists a unique solution of*

$$\left. \begin{array}{c} u \in C([0, T]; L^2(\Omega)) \cap L^2((0, T); W_0^{1,2}(\Omega)) \\ u_t, \Delta u \in L^2(Q_T), \\ \partial u / \partial t - d\Delta u = H \text{ in } L^2(Q_T), \quad u(0) = u_0. \end{array} \right\} \tag{33}$$

Moreover,

$$u(t) = S_d(t) u_0 + \int_0^t S_d(t - s) H(s) \, ds, \tag{34}$$

where $S_d(\cdot)$ is the semigroup in $L^2(\Omega)$ whose infinitesimal generator is the Laplacian operator $-\Delta$, with domain $D(-\Delta) = H^2(\Omega) \cap H_0^1(\Omega)$.

This may be found in several places in the literature, as well as the contraction property

$$\forall p \in [1, +\infty], \quad \|S_d(t)u_0\|_{L^p(\Omega)} \le \|u_0\|_{L^p(\Omega)}, \tag{35}$$

(see e.g. [9, 1, 5]).

Thanks to the contraction property in $L^1(\Omega)$, the solution of (33) may be extended to $u_0 \in L^1(\Omega)$, $H \in L^1(Q_T)$ with $u \in C([0, T]; L^1(\Omega))$ at least. It is also given by the formula (34) where, now, $S_d(\cdot)$ is the realization of the heat semigroup in $L^1(\Omega)$ (see e.g. [6]) -we also denote it by $S_d(\cdot)$-. To obtain that $u \in L^1(0, T; W_0^{1,1}(\Omega))$, we need the estimates (18). They may be obtained by duality from the L^∞-estimates for the heat operator (see e.g. [9], Th. III.7.1), namely

$$\|u\|_{L^\infty(Q_T)} \le C \sum_{i=1}^{N} \|h_i\|_{L^{s'}(O, T; L^{q'}(\Omega))},$$

for the solution of (33) with $H = \sum_{i=1}^{N} \partial h_i / \partial x_i$.

For the uniqueness part, one has to be careful when working in an L^1-setting. Remember, for instance, that there is not uniqueness for the problem

$$u \in W_0^{1,1}(\mathcal{O}), \quad \Delta u = 0 \text{ in } D'(\mathcal{O}),$$

without any regularity on the open subset $\mathcal{O}$ of $\mathbb{R}^N$ (a counterexample is given by $u(x) = 1 - |x|^{2-N}$ for $N \geq 3$ and $\mathcal{O} = \{x \in \mathbb{R}^N \setminus \{0\}; |x| < 1\}$).

But we do have uniqueness if $\mathcal{O} = \Omega$ is regular -which we assumed throughout this paper- (see e.g. [6]). Again, this uniqueness result relies, by duality, on regularizing properties of the Laplacian in good domains.

We also have uniqueness for the parabolic problem. We will state it in a general way that we actually need in this paper. We denote by $C_b(\Omega)$ the continuous and bounded functions from Ω into $\mathbb{R}$.

LEMMA 5.1. *Let $w \in L^\infty(0, T; L^1(\Omega)) \cap L^1_{loc}((0, T], W_0^{1,1}(\Omega))$ be a solution of*

$$\left. \begin{array}{c} \partial w/\partial t - d\Delta w = H \quad in \ \ D'(Q_T), \\ \forall \varphi \in C_b(\Omega), \ \varphi \geq 0, \ \lim_{t \to 0} \int_\Omega \varphi \, w(t) = \int_\Omega \varphi \, w_0. \end{array} \right\} \tag{36}$$

Then,

$$w(t) = S_d(t)w_0 + \int_0^t S_d(t - s) \, H(s) \, ds.$$

In particular, $w \in C([0, T]; L^1(\Omega)) \cap L^1(0, T; W_0^{1,1}(\Omega))$.

We refer e.g. to [2] for details of the proof. As explained above, it is based on regularity property of the dual problem.

Note that the initial data are understood here in the sense of the "narrow" convergence for measures, that is to say that the test-functions φ in (36) have to be taken in $C_b(\Omega)$. The uniqueness *would not be true* if they were to be taken only among continuous functions with compact support in Ω.

About existence of regular solutions for systems: we consider the functions f_n, g_n defined in (24). Since they are defined only for $r, s \in [0, +\infty)^2$, we extend them in the variable r, s to $\mathbb{R}^2$ by $\Pi \circ f_n$, $\Pi \circ g_n$ where Π is the projection in $\mathbb{R}^2$ onto $[0, +\infty)^2$. We still denote by f_n, g_n this extension. The main point is that f_n, g_n are globally Lipschitz continuous on $\mathbb{R}^2$.

To prove existence of a classical solution to the system (26), for all $T \in (0, +\infty)$, we consider the mapping $\mathcal{S}$ from $X = C([0, T]; L^1(\Omega))^2$ into itself which to $(\hat{U}, \hat{V})$ associates (U, V) defined by

$$U(t) = S_{d_1}(t)u_{0n} + \int_0^t S_{d_1}(t - s) f_n(s, \cdot, \hat{U}(s), \hat{V}(s)) \, ds,$$

$$V(t) = S_{d_2}(t)v_{0n} + \int_0^t S_{d_2}(t - s) g_n(s, \cdot, \hat{U}(s), \hat{V}(s)) \, ds.$$

Since, f_n, g_n are globally Lipschitz in r, s, one easily proves that there exists p such that $\mathcal{S}^p$ is a strict contraction from X into itself, whence the existence of a unique weak solution to the system. By the above uniqueness results and the L^2-theory, since all the data u_{0n}, v_{0n}, f_n, g_n are uniformly bounded and therefore in L^2, the solution has the regularity announced in (26).

For the positivity, classically we multiply the equation in u_n by $-u_n^- =: inf\{u_n, 0\}$ to obtain

$$\partial/\partial t \int_\Omega \left(u_n^-\right)^2 \leq -2 \int_{[u_n<0]} u_n f_n(t, x, u_n, v_n).$$

But, by construction, on $[u_n < 0]$, $f_n(t, x, u_n, v_n) \geq 0$, since it is equal either to f_n $(t, x, 0, v_n)$ if $v_n \geq 0$ or to $f_n(t, x, 0, 0)$ if $v_n \leq 0$. We deduce that $u_n^- = 0$ and similarly $v_n^- = 0$.

REFERENCES

[1]   AMANN, H., *Global existence for semilinear parabolic problems*, J. Reine Angew. Math. *360* (1985), 47–83.

[2]   BARAS, P. and PIERRE, M., *Problèmes paraboliques semi-linéaires avec données mesures*, Applicable Analysis. *18* (1984), 111–149.

[3]   BOUDIBA, N., *Existence globale pour des systèmes de réaction-diffusion avec contrôle de masse*, Ph.D thesis, université de Rennes 1, France, 1999.

[4]   BEBERNES, J. and LACEY, A., *Finite time blow-up for semi-linear reactive-diffusive systems*, J. Diff. Equ. *95* (1992), 105–129.

[5]   BREZIS, H., *Analyse Fonctionnelle, Théorie et applications*, Masson, 1983.

[6]   BREZIS, H., and STRAUSS, W., *Semilinear elliptic equations in L^1*, J. Math. Soc. Japan *25* (1973), 565–590.

[7]   HARAUX, A. and YOUKANA, A., *On a result of K. Masuda concerning reaction-diffusion equations*, Tôhoku Math. J. *40* (1988), 159–163.

[8]   HOLLIS, S. L., MARTIN, R. H. and PIERRE, M., *Global existence and boundedness in reaction-diffusion systems*, SIAM J. Math. Ana. *18* (1987), 744–761.

[9]   LADYZENSKAJA, O. A., SOLONNIKOV, V. A. and URAL'CEVA, N. N., *Linear and Quasilinear Equations of Parabolic Type*, Transl. Math. Monogr. *23*, AMS, Providence RI, 1968.

[10]   LAAMRI, E., *Existence globale pour des systèmes de réaction-diffusion dans L^1*, Ph.D thesis, université de Nancy 1, France, 1988.

[11]   MARTIN, R. H. and PIERRE, M., *Nonlinear reaction-diffusion systems, in Nonlinear Equations in the Applied Sciences*, W. F. Ames and C. Rogers ed., Math. Sci. Eng. *185*, Ac. Press, New York, 1991.

[12]   MARTIN, R. H. and PIERRE, M., *Influence of mixed boundary conditions in some reaction-diffusion systems*, Proc. Roy. Soc. Edinburgh, section A *127* (1997), 1053–1066.

[13]   MASUDA, K., *On the global existence and asymptotic behavior of reaction-diffusion equations*, Hokkaido Math. J. *12* (1983), 360–370.

[14]   MORGAN, J., *Global existence for semilinear parabolic systems*, SIAM J. Math. Ana. *20* (1989), 1128–1144.

[15]   PIERRE, M., *An L^1-method to prove global existence in some reaction-diffusion systems*, in "Contributions to Nonlinear Partial Differential Equations", Vol.II, Pitman Research notes, J.I. Diaz and P.L. Lions ed., *155* (1987), 220–231.

[16]   PIERRE, M. and SCHMITT, D., *Blow up in Reaction-Diffusion Systems with Dissipation of Mass*, SIAM J. Math. Ana. *28* no 2. (1997), 259–269.

[17]   PIERRE, M. and SCHMITT, D., *Blow up in Reaction-Diffusion Systems with Dissipation of Mass*, SIAM Review *42* no 1. (2000), 93–106.

Michel Pierre
Antenne de Bretagne de l'ENS Cachan
et Institut de Recherche Mathématique de Rennes
Campus de Ker Lann
35170 - Bruz
France
e-mail: pierre@bretagne.ens-cachan.fr

J.evol.equ. 3 (2003) 169 – 201
1424–3199/03/020169 – 33
DOI 10.1007/s00028-003-0084-0
© Birkhäuser Verlag, Basel, 2003

**Journal of Evolution
Equations**

Global well-posedness and stability of a partial integro-differential equation with applications to viscoelasticity

S.-O. LONDEN H. PETZELTOVÁ* and J. PRÜSS

1. Introduction

In this paper we consider the equation

$$u_{tt}(t, x) = \int_0^t a(t - s)u_{txx}(s.x)ds + \frac{\mathrm{d}}{\mathrm{d}t} \int_0^t b(t - s)(g(u_x(s, x)))_x \, \mathrm{d}s + f(t, x), \quad (1.1)$$

$t > 0$, $x \in (0, 1)$, with boundary conditions

$$u(t, 0) = u(t, 1) = 0, \ t > 0, \tag{1.2}$$

and initial values

$$u(0, x) = u^0(x), \ u_t(0, x) = u^1(x). \tag{1.3}$$

Here $a(t)$ and $b(t)$ are creep kernels which behave like

$$a(t) \sim c_a t^{\alpha-1}, \quad b(t) \sim c_b t^{\beta-1}, \quad t \to 0,$$

where $\alpha, \beta \in (0, 1]$, $c_a, c_b > 0$ are constants and $\alpha < \beta$. Thus the order of the time-derivative of the linear term is strictly higher than that of the nonlinear term. The nonlinearity g belongs to the class $C^2(\mathbb{R})$ with normalization $g(0) = 0$ and roughly speaking its primitive $G(s) = \int_0^s g(r)dr$ satisfies

$$G(r) \sim c_1 r g(r) \sim c_2 |r|^{m+1}, \quad |r| \to \infty,$$

with positive constants $c_1, c_2 > 0$ and $m \geq 1$; note that g' may change sign. The data $f(t, x)$, $u^0(x)$ and $u^1(x)$ are given. The formulation of the precise assumptions is contained in the next section.

The motivation for this problem comes from the theory of viscoelastic material behaviour. As is well-known the governing equation in this field is *balance of momentum*, which in one dimension reads

$$u_{tt} + \sigma_x = h,$$

---

*Work supported by the grant A1019002 of GA AV ČR.

where u denotes displacement, σ means stress, and h accounts for body forces. This equation is supplemented by a constitutive law which reflects the properties of the material under consideration, the *stress-strain relation*. Here we consider the law

$$\sigma(t, x) = \int_0^\infty a(s)u_{xt}(t - s, x)ds + \frac{\partial}{\partial t} \int_0^\infty b(s)g(u_x(t - s, x))ds, \quad t \in \mathbb{R}.$$

Inserting this relation into balance of momentum we obtain (1.1), where f contains h as well as the stress history up to time $t = 0$.

Let us discuss a classical limiting case of our problem, i.e. the case $\alpha = 0, \beta = 1, f = 0$. This yields the problem

$$u_{tt} = a_0 u_{xxt} + g(u_x)_x, \quad t > 0, \ x \in (0, 1),$$
$$u(t, 0) = u(t, 1) = 0, \quad t > 0,$$
$$u(0, x) = u^0(x), \quad u_t(0, x) = u^1(x), \quad x \in (0, 1). \tag{1.4}$$

In case $a_0 = 0$ this is a quasilinear wave equation which in general does not admit global classical solutions because of breaking of waves, even if $g'(s) > 0$ for $s \in \mathbb{R}$. On the other hand, for $a_0 > 0$ it is an old result of Greenberg [6] that (1.4) does admit global classical solutions, without any restriction on the growth-order m of g. In other words, the damping induced by the parabolic term u_{xxt} in (1.4) is strong enough to control the "destructive" effects of the quasilinear term $g(u_x)_x$.

For problem (1.1) the situation is similar. The linear term $a * u_{xxt}$ is trying to stabilize while the nonlinear term $(b*g(u_x)_x)_t$ wants to destroy the regularity of the solution. Which part is going to win? In general the answer is not known. Here we can add the following result. If the order of the time-derivative in the linear term $(= 1 - \alpha)$ is greater than the order of the time-derivative in the nonlinear term $(= 1 - \beta)$, then the linear term $a * u_{txx}$ wins. Thus our key condition is $\beta > \alpha$. In addition, we require the problem to be *subcritical* in the sense that

$$\frac{m - 1}{m + 1} < 2\frac{\beta - \alpha}{1 + \alpha}.$$

We show below that if these two assumptions hold, then the problem (1.1) is globally well-posed; not only in a weak but also in a stronger L_p-setting, and that the L_∞-norm of u_x stays globally bounded in time.

Furthermore we show global asymptotic stability of the trivial solution, provided the steady state problem

$$g(u_x)_x = 0, \quad x \in (0, 1), \ u(0) = u(1) = 0,$$

admits only the trivial solution – which is equivalent to the sign-condition $sg(s) > 0$ for $s \neq 0$ – and provided the linearization at the trivial solution is asymptotically stable; the latter means

$$g'(0) > 0, \quad b_\infty := \lim_{t\to\infty} b(t) > 0.$$

Concerning the literature related to (1.1), we observe that the special case (1.4) was considered by Greenberg [6], Pego [10], Ball et al. [1]. Global existence of weak solutions has been obtained in Gripenberg, Londen and Prüss [7] under essentially the same assumptions as here, however, no uniqueness was shown and no global bounds were obtained. The special case $a(t) = t^{\alpha-1}$, $b(t) = 1$ was studied in the recent paper Petzeltová and Prüss [11], where the same results as in this paper are proved for this special case, assuming a stronger condition on α and m.

The plan for this paper is as follows. In Section 2 we introduce the precise assumptions and state the main results, Theorem 1 on global well-posedness, and Theorem 2 on asymptotic stability. The remaining sections are devoted to the proofs of Theorems 1 and 2.

2. Assumptions and main results

Recall that a function $a : (0, \infty) \to \mathbb{R}$ is called k-**monotone** if a is $k - 1$-times differentiable,

$$(-1)^j a^{(j)}(t) \geq 0 \quad \text{for all } t > 0, \ j = 0, \dots, k - 1,$$

and $(-1)^{k-1} a^{(k-1)}(t)$ is nonincreasing on $(0, \infty)$. A function $c : (0, \infty) \to \mathbb{R}$ is called **completely positive**, if there is $l_0 \geq 0$ and a 1-monotone function $l : (0, \infty) \to \mathbb{R}$ such that

$$l_0 c(t) + c * l(t) = l_0 c(t) + \int_0^t c(t - s)l(s)\,ds = 1, \quad t > 0.$$

For properties of k-monotone resp. completely positive functions we refer to Prüss [13].

The kernels a and b in equation (1.1) are assumed to satisfy the following assumptions.

$$\text{(a)} \quad \begin{cases} a \in L^1_{loc}(\mathbb{R}^+), \ a \text{ is } 3 - \text{monotone}, \\ |\arg \widehat{a}(\lambda)| \leq \theta_a < \frac{\pi}{2}, \ \operatorname{Re} \lambda > 0, \\ |\widehat{a}(\lambda)| \geq \frac{C_a}{1 + |\lambda|^\alpha}, \ \operatorname{Re} \lambda > 0, \text{ for some } \alpha \in [0, 1). \end{cases} \quad (2.1)$$

$$\text{(b)} \quad \begin{cases} b(t) = b_1 + b_\infty, \ b_\infty > 0, \\ b_1 \in L^1(\mathbb{R}^+), \ b_1 \text{ is } 3 - \text{monotone and completely positive}, \\ |\widehat{b}_1(\lambda)| \leq \frac{C_b}{1 + |\lambda|^\beta}, \ \operatorname{Re} \lambda > 0, \text{ for some } \beta \in (\alpha, 1]. \end{cases} \quad (2.2)$$

$$\text{(c)} \quad \begin{cases} \text{There exists } c \in L^1_{loc}(\mathbb{R}^+), \ 1\text{-monotone} \\ \text{and completely positive, such that} \quad b = a * c. \end{cases} \quad (2.3)$$

Note that, due to $\beta > \alpha$, (2.1) and (2.2) already imply the existence of an absolutely continuous function c such that $a * c = b$; in (2.3) we assume that a and b are connected through a kernel having the indicated structural properties. Typical examples of kernels satisfying these conditions are

$$a(t) = t^{\alpha-1}e^{-\omega_a t}, \quad b_1(t) = t^{\beta-1}e^{-\omega_b t}, \ t > 0,$$

where $0 \le \alpha < \beta \le 1, \omega_a \ge 0, \omega_b > 0.$

The basic assumptions on the nonlinearity that allow for global existence are

$$\text{(g)} \begin{cases} g \in C^2(\mathbb{R}), \ g(0) = 0, \ G(s) := \int_0^s g(r)dr. \\ \text{There are constants} \eta > 0, \ C_g > 0, \ p > 1 \text{ such that} \\ g'(s) \ge -\eta, \ s \in \mathbb{R}, \quad \frac{1}{p} < 2\frac{\beta-\alpha}{1+\alpha}, \\ |g(s)| + |g(s)|^p/|s|^p \le C_g(1 + G(s)). \end{cases} \tag{2.4}$$

To obtain bounded solutions we need in addition

$$\text{there are } \varepsilon > 0, \ R \ge 0, \text{ such that } g(s)s \ge \varepsilon|s|^2 \text{ for } |s| \ge R. \tag{2.5}$$

$$\lim_{|s|\to\infty} \frac{|g(s)|^p}{|s|^p G(s)} = 0. \tag{2.6}$$

Condition (2.5) with $R = 0$ allows us to prove the convergence result.

Note that in case $G(s) \sim c_1 sg(s) \sim c_2|s|^{m+1}$ as $|s| \to \infty$ the growth conditions on g are satisfied if $m \ge 1$ with $p < \frac{m+1}{m-1}$. This is the subcriticality assumption mentioned in Section 1.

For the function f we assume

$$\text{(f)} \quad f \in L^1_{loc}(\mathbb{R}^+; L^q(0, 1)); \quad q \ge \max\left\{2, \frac{1+\alpha}{1-\alpha}\right\}. \tag{2.7}$$

We will require $f \in L^1(\mathbb{R}^+; L^q(0, 1))$ for boundedness and convergence of solutions.

As to the initial data, we suppose

$$\text{(IC)} \begin{cases} u^0 \in X = W_0^{1,\infty}(0, 1) \cap W^{s,q}(0, 1); \text{ with } s > 2 - \frac{2\beta}{1+\alpha} \text{ if } \beta \le \frac{1+\alpha}{2}, \\ u^1 \in L^q(0, 1), \end{cases} \tag{2.8}$$

The assumptions on the kernels have the following consequences that will be used in the sequel.

- There exist $k_0 \ge 0$, and $k_1 \in L^1(\mathbb{R}^+)$, 1-monotone and completely positive, such that

$$k_0 b(t) + b * k_1(t) = 1, \quad t > 0. \tag{2.9}$$

- Let $l = k_0 a + a * k_1$. Then l is 1-monotone, completely positive, and

$$l \in L^1_{loc}(\mathbb{R}^+), \ \text{Re } \widehat{l}(\lambda) > 0 \text{ for all } \lambda \text{ with Re } \lambda \ge 0. \tag{2.10}$$

- Given $\theta \in (0, 1)$, for any $\eta > 0$ there exists $C_\eta > 0$ such that

$$\eta \operatorname{Re} \widehat{b}_1(\lambda) \leq \theta \operatorname{Re} \widehat{a}(\lambda) + C_\eta \operatorname{Re} \widehat{l}(\lambda), \quad \operatorname{Re} \lambda > 0. \tag{2.11}$$

- Assumption (2.1) implies that a is 2-regular, i.e., there is $c > 0$ such that

$$|\lambda^n \widehat{a}^{(n)}(\lambda)| \leq c|\widehat{a}(\lambda)| \text{ for } \operatorname{Re} \lambda > 0, \ n = 1, 2, \tag{2.12}$$

- and of strictly positive type, i.e., there exist $a_1 \in L^1(\mathbb{R}^+)$, $a_1 \not\equiv 0$, $C > 0$ such that

$$\int_0^t (a * \phi)(s)\phi(s) \, ds \geq C \int_0^t |a_1 * \phi(s)|^2 \, ds \text{ for all } t > 0, \ \phi \in C(\mathbb{R}^+). \tag{2.13}$$

Characterizations and some useful properties of kernels satisfying (2.1)–(2.2) are given in Appendix.

We define operators A_p, $1 \leq p \leq \infty$, by means of

$$\mathcal{D}(A_p) = W^{2,p}(0, 1) \cap W_0^{1,p}(0, 1), \quad A_p\phi = -\frac{d^2}{dx^2}\phi = -D^2\phi. \tag{2.14}$$

Note that $\mathcal{D}(A_p^\sigma) = H_0^{2\sigma,p}(0, 1)$ for $1 + 1/p > 2\sigma > 1/p$, in particular, $\mathcal{D}(A_p^{1/2}) = H_0^{1,p}(0, 1)$. Here $H_0^{2\sigma,p}(0, 1)$ denote the Bessel-potential spaces on $(0, 1)$ with zero trace. The following theorems contain the main results of this paper.

THEOREM 2.1. (Global existence). *Let the assumptions* (2.1)–(2.4), (2.7), (2.8) *be satisfied. Then there exists a unique weak solution u to the problem* (1.1)–(1.3) *such that*

$$u \in C(\mathbb{R}^+; W_0^{1,\infty}(0, 1)) \cap C^1(\mathbb{R}^+; L^q(0, 1)), \tag{2.15}$$

$$u_x \in C^{\frac{1-\alpha}{2}}(\mathbb{R}^+, L^q(0, 1)). \tag{2.16}$$

If, in addition, $u^0 \in \mathcal{D}(A_q)$, $u^1 \in \mathcal{D}(A_q^{\frac{1}{2}})$, $f \in L_{loc}^1(\mathbb{R}^+; \mathcal{D}(A_q^{\frac{1}{2}}))$ then

$$u \in C(\mathbb{R}^+; \mathcal{D}(A_q)) \cap C^1(\mathbb{R}^+; \mathcal{D}(A_q^{\frac{1}{2}})).$$

We want to mention that for global existence we may allow for $b_\infty = 0$ and $b_1 \notin L^1(\mathbb{R}^+)$, however, then we will not be able to get global bounds on the solution.

THEOREM 2.2. (Asymptotic stability). *Let the assumptions of Theorem* 2.1 *be satisfied. Moreover, let* (2.5) *and* (2.6) *hold and $f \in L^1(\mathbb{R}^+; L^q(0, 1))$. Then the unigue global solution of problem* (1.1)–(1.3) *is bounded and satisfies*

$$u \in BUC(\mathbb{R}^+; W_0^{1,\infty}(0, 1)) \cap BUC^1(\mathbb{R}^+; L^2(0, 1)).$$

If (2.5) *holds with* $R = 0$, *then* $u(t) \to 0$ *in* $W^{1,\infty}(0, 1)$, *and* $u_t(t) \to 0$ *in* $L^2(0, 1)$, *as* $t \to \infty$.

If the initial values satisfy in addition, for some $r \in [1, \infty)$, $u^0 \in W^{2,r}(0, 1)$, $u^1 \in W_0^{1,r}(0, 1)$, $f \in L^1(\mathbb{R}^+; \mathcal{D}(A_r^{\frac{1}{2}}))$, *then*

$$u \in BUC(\mathbb{R}^+; W^{2,r}(0, 1)) \cap BUC^1(\mathbb{R}^+; W_0^{1,r}(0, 1)),$$

and $u(t) \to 0$ *in* $W^{2,r}(0, 1)$, $u_t(t) \to 0$ *in* $W^{1,r}(0, 1)$ *as* $t \to \infty$.

The remaining part of the paper is devoted to the proofs of these theorems.

3. Preliminaries

Let us briefly outline the proof of local existence. First, the equation is rewritten in the form (3.2) and solved by the variation of parameters formula. This gives the equation (3.4). To apply the contraction method to this equation, we require some rather technical estimates on the solution operators. In particular, observe that in order to obtain the key estimates (3.16), (3.18) we need to work with the fractional powers A^δ (Lemma 3.2).

Let $\| \cdot \|_p$ denote the norm in $L^p(0, 1)$, $\langle \cdot, \cdot \rangle$ is the scalar product in $L^2(0, 1)$, $D = d/dx$, $\dot{v} = v_t$, $k * v = \int_0^t k(t - s)v(s)\, ds$; C stands for a generic positive constant. The Laplace transform of a function f we denote by $\widehat{f}$ or $\mathcal{L}(f)$.

We introduce equation (1.1) in a functional analytic setting.

It is well known that A_p defined by (2.14) is an invertible, sectorial operator on $L^p(0, 1)$, A_2 is selfadjoint, positive definite in $L^2(0, 1)$, $\sigma(A_p) \subset [\pi^2, +\infty)$ consists only of simple eigenvalues and there is $M(\theta) > 0$ such that

$$\|A_p^\delta(z + A_p)^{-1}\|_{B(L^p)} \leq \frac{M}{|\pi^2 + z|^{1-\delta}},$$

$$0 \leq \delta \leq 1, \ |\arg(z + \pi^2)| \leq \pi - \theta, \ \theta > 0. \tag{3.1}$$

For simplicity we drop the subscript p where there is no danger of confusion. Also, we will frequently write L^p instead of $L^p(0, 1)$.

We reformulate (1.1) as

$$\frac{d^2}{dt^2}u + a * \frac{d}{dt}Au = \frac{d}{dt}b * Dg(Du) + f \tag{3.2}$$

and proceed as in [11].

The Laplace transform of the equation (3.2) yields

$$\lambda(\lambda + \widehat{a}(\lambda)A)\widehat{u}(\lambda) = (\lambda + \widehat{a}(\lambda)A)u^0 + u^1 + \lambda\widehat{b}(\lambda)\mathcal{L}(Dg(u_x)) + \widehat{f}(\lambda).$$

Let S be the solution of the operator equation

$$\dot{S} + a * AS = 0, \quad S(0) = I.$$

By (a), $\widehat{a}(\lambda)$ can be extended to Re $\lambda \geq 0$, $\lambda \neq 0$, and for λ in this set we have $\lambda[\widehat{a}(\lambda)]^{-1} \notin (-\infty, 0]$. Therefore,

$$\widehat{S}(\lambda) = (\lambda + \widehat{a}(\lambda)A)^{-1}, \quad \text{Re } \lambda \geq 0, \ \lambda \neq 0$$

is well defined. Denote

$$S_1 = 1 * S, \quad S_b = b * S. \tag{3.3}$$

Then the solution u can be expressed by means of the variation of parameters formula

$$u(t) = u^0 + S_1(t)u^1 + S_b * Dg(Du)(t) + S_1 * f(t). \tag{3.4}$$

To prove existence of weak solutions we need to commute the operators S_b and D. To do so, we realize that

$$A = DA^N Q,$$

where $Q : L^p(0, 1) \to W^{1,p}(0, 1)$ is defined by

$$Q\phi(x) = \int_0^x y\phi(y) \, \mathrm{d}y - \int_x^1 (1 - y)\phi(y),$$

and A^N denotes the Laplace operator with Neumann boundary conditions defined on the space of zero mean functions, i.e.

$$\mathcal{D}(A_p^N) = \left\{ \phi \in W^{2,p} \, \middle| \, \int_0^1 \phi(x) \, \mathrm{d}x = 0, \phi_x(0) - \phi_x(1) = 0 \right\},$$

$$A_p^N \phi = -\frac{\mathrm{d}^2}{\mathrm{d}x^2}\phi = -D^2\phi.$$

We have

$$DQ\phi = \phi, \quad QD\phi = \phi - \int_0^1 \phi(x) \, \mathrm{d}x = P\phi,$$

where P denotes the projector of L^p to the subspace of zero mean functions. This implies

$$S(t) = DS^N(t)Q, \quad S_b(t) = DS_b^N(t)Q = Db * S^N(t)Q,$$

S^N being the solution operator corresponding to A^N, $\widehat{S}^N(\lambda) = (\lambda + \widehat{b}(\lambda)A^N)^{-1}$. So we can express u as follows:

$$u = u^0 + S_1 u^1 + DS_b^N * Pg(Du) + S_1 * f. \tag{3.5}$$

As all formulae, estimates and proofs for A, S and A^N, S^N are exactly the same, we omit the script ^{N} throughout the further text.

Now, we derive estimates for the solution operators S, S_1, S_b that we will need in the existence proofs.

The condition $(2.1)_2$ implies existence of a positive number $\theta < \frac{\pi}{2}$ such that

$$|\arg(\lambda/\widehat{a}(\lambda))| \leq \pi - \theta \text{ when } |\arg(\lambda)| \leq \frac{\pi}{2}.$$

Moreover, according to (3.1) and $(2.1)_3$, we have

$$\left\| \frac{1}{\widehat{a}(\lambda)} \left(\frac{\lambda}{\widehat{a}(\lambda)} + A \right)^{-1} \right\|_{\mathcal{B}(L^p)} \leq \frac{C}{1 + |\lambda|}.$$

We use the L^2- theory of the Laplace transform and express $S(t)$ by integrating over the imaginary axis.

$$S(t) = \frac{1}{2\pi i} \int_{-i\infty}^{i\infty} e^{\lambda t} (\lambda + \widehat{a}(\lambda)A)^{-1} \, d\lambda. \tag{3.6}$$

Set

$$H_\delta(\lambda) = A^\delta \widehat{S}(\lambda) = \frac{1}{\widehat{a}(\lambda)} A^\delta \left(\frac{\lambda}{\widehat{a}(\lambda)} + A \right)^{-1}, \quad H = H_0 = \widehat{S}(\cdot),$$

and estimate $\|A^\delta S(t)\|_{\mathcal{B}(L^p)}$. To this end, we shall take advantage of the following result on vector-valued analytic functions in $\mathbb{C}^+ = \{\lambda \in \mathbb{C}; \operatorname{Re}\lambda > 0\}$. (See [12, Theorem 1]).

LEMMA 3.1. *Suppose $h : \mathbb{C}^+ \to X$ is holomorphic and satisfies*

$$\|h(\lambda)\| + \|\lambda h'(\lambda)\| \leq c|\lambda|^{-\gamma}, \ \operatorname{Re}\lambda > 0,$$

for some $\gamma \in (0, 1)$. Then there is a continuous function $v : (0, \infty) \to X$ such that $\widehat{v}(\lambda) = h(\lambda)$ for $\operatorname{Re}\lambda > 0$. In addition, the following estimates hold.

$$\|v(t)\| \leq Mt^{\gamma-1}, \ t > 0,$$
$$\|tv(t) - sv(s)\| \leq M|t - s|^\gamma, \ 0 < s < t < \infty.$$

Here M denotes a constant depending only on γ and c.

We have for $\delta < 1/(1 + \alpha)$

$$\|H_\delta(\lambda)\phi\|_p = \left\| \frac{1}{\widehat{a}(\lambda)} A^\delta \left(\frac{\lambda}{\widehat{a}(\lambda)} + A \right)^{-1} \phi \right\|_p \leq C \frac{1}{|\widehat{a}(\lambda)|} \frac{1}{|1 + \lambda/\widehat{a}(\lambda)|^{1-\delta}} \|\phi\|_p$$

$$\leq C|\lambda|^{-1+\delta(1+\alpha)} \|\phi\|_p \text{ for } \operatorname{Re}\lambda > 0. \tag{3.7}$$

We get the same estimate for $\|A^\delta \lambda \frac{\mathrm{d}}{\mathrm{d}\lambda} \widehat{S}(\lambda)\phi\|_p$, because of (2.12). Applying Lemma 3.1 gives

$$\|A^\delta S(t)\phi\|_p \le Ct^{-\delta(1+\alpha)}\|\phi\|_p \quad \text{if } \delta < \frac{1}{1+\alpha}. \tag{3.8}$$

Next, we shall improve the estimate at infinity, specifically, we show that

$$A^\delta S \in L^1(\mathbb{R}^+; \mathcal{B}(L^p)) \quad \text{for } \delta < \frac{1}{1+\alpha}.$$

To do so, we decompose $S(t)$ into two parts

$$S(t) = \frac{1}{2\pi} \int_{-\infty}^{\infty} \psi(\rho)H(i\rho)e^{i\rho t}\,\mathrm{d}\rho + \frac{1}{2\pi} \int_{-\infty}^{\infty} [1 - \psi(\rho)]H(i\rho)e^{i\rho t}\,\mathrm{d}\rho,$$

where $\psi \in C^\infty(\mathbb{R})$, $\psi(\rho) = 1$ for $|\rho| \le N$, $\psi(\rho) = 0$ for $|\rho| \ge N + 1$, $0 \le \psi(\rho) \le 1$ elsewhere, $N > 0$ arbitrary fixed. After applying A^δ and performing two integrations by parts in the second integral we get (note that by (3.7) the substitution terms vanish)

$$A^\delta S(t) = \frac{1}{2\pi} \int_{-\infty}^{\infty} \psi(\rho)H_\delta(i\rho)e^{i\rho t}\,\mathrm{d}\rho - \frac{1}{2\pi t^2} \int_{-\infty}^{\infty} [(1 - \psi(\rho))H_\delta(i\rho)]''e^{i\rho t}\,\mathrm{d}\rho$$

$$= \mathcal{S}_1(t) + \mathcal{S}_2(t), \quad t \in \mathbb{R}^+.$$

Since $A^\delta S(t)$ satisfies (3.8) and $\mathcal{S}_1$ is obviously bounded on $\mathbb{R}^+$, $\mathcal{S}_2$ satisfies (3.8) for $t \in (0, 1)$ too. Also, (2.12), (2.1)$_3$ imply

$$\|H_\delta'(\lambda)\|_{\mathcal{B}(L^p)} \le |\lambda|^{-2+\delta(1+\alpha)}, \quad \|H_\delta''(\lambda)\|_{\mathcal{B}(L^p)} \le |\lambda|^{-3+\delta(1+\alpha)}.$$

which gives $\mathcal{S}_2(t) \le C/t^2$. These estimates on $\mathcal{S}_2$ imply $\mathcal{S}_2 \in L^1(\mathbb{R}^+; \mathcal{B}(L^p))$.

To show that $\mathcal{S}_1 \in L^1(\mathbb{R}^+; \mathcal{B}(L^p))$ we use local analyticity of $1/\widehat{a}(\lambda)$ on $i\mathbb{R}$, the fact that the product of a function that is locally analytic on $i\mathbb{R}$ with a compactly supported C^∞ function is the Fourier transform of a function in $L^1(\mathbb{R})$, and the Paley-Wiener theorem. We can proceed literally as in the proof of [13, Theorem 10.1].

To estimate Du we will need bounds for AS_b. We decompose AS_b by means of its Laplace transform as follows.

$$\widehat{AS_b}(\lambda) = \frac{\widehat{b}(\lambda)}{\widehat{a}(\lambda)}\left(I - \frac{\lambda}{\widehat{a}(\lambda)}\left(\frac{\lambda}{\widehat{a}(\lambda)} + A\right)^{-1}\right) = \widehat{c}(\lambda) - \widehat{T}(\lambda)$$

so

$$AS_b(t) = c(t) - T(t) \quad \text{where } T(t) = \frac{\mathrm{d}}{\mathrm{d}t}c * S(t), \ c \text{ given in (2.3)}, \tag{3.9}$$

and

$$T = T_1 + T_\infty, \quad \widehat{T}_1(\lambda) = \frac{\lambda \widehat{b_1}(\lambda)}{\widehat{a}(\lambda)}\widehat{S}(\lambda), \quad \widehat{T}_\infty(\lambda) = \frac{b_\infty}{\widehat{a}(\lambda)}\widehat{S}(\lambda). \tag{3.10}$$

In the following Lemma, we collect some estimates that will be used in the sequel.

LEMMA 3.2. *Let* S, S_1, *and* $T = T_1 + T_\infty$ *be given by the formulae* (3.6), (3.3), (3.9), (3.10), $S_{b_1} = b_1 * S$, $1 < p < \infty$, $0 \leq \delta < \frac{1}{1+\alpha}$. *Then there is a constant* $C > 0$ *such that*

$$\|A^\delta S(t)\phi\|_p \leq Ct^{-\delta(1+\alpha)}\|\phi\|_p, \tag{3.11}$$

$$A^\delta S(t) \in L^1(\mathbb{R}^+; \mathcal{B}(L^p)), \tag{3.12}$$

$$\|A^\delta S_1(t)\phi\|_p \leq C\min\{t^{1-\delta(1+\alpha)}, 1\}\|\phi\|_p, \tag{3.13}$$

$$\|A^\delta S_{b_1}(t)\phi\|_p \leq Ct^{\beta-\delta(1+\alpha)}\|\phi\|_p, \tag{3.14}$$

$$A^\delta S_{b_1} \in L^1(\mathbb{R}^+; \mathcal{B}(L^p)), \tag{3.15}$$

$$\|AS_b(t)\phi\|_p \leq \begin{cases} C(1 + t^{-1+\beta-\alpha})\|\phi\|_p, \text{ for } \phi \in L^p, \ \alpha < \beta < 1, \\ C(1 + t^{-\frac{1+\alpha}{2}})\|A^\gamma\phi\|_p, \text{ for } \phi \in \mathcal{D}(A^\gamma), \ \alpha < \beta \leq \frac{\alpha+1}{2}, \\ \gamma \in \left(\frac{1}{2} - \frac{\beta}{1+\alpha}, \frac{1}{2}\right), \end{cases} \tag{3.16}$$

$$\|D\dot{S}_{b_1}(t)\phi\|_p \leq C\min\{t^{-1+\beta-\frac{1+\alpha}{2}}, t^{-2+\beta-\alpha}\}\|\phi\|_p, \tag{3.17}$$

$$D\dot{S}_b \in L^1(\mathbb{R}^+, \mathcal{B}(L^p)) \text{ if } \beta > \frac{1+\alpha}{2}, \tag{3.18}$$

$$T \in L^1(\mathbb{R}^+, \mathcal{B}(L^p, L^\infty)) \text{ when } \frac{1}{2p} < \frac{\beta-\alpha}{1+\alpha}. \tag{3.19}$$

Proof. We have already proved (3.11) and (3.12). To obtain (3.13), we use (3.11), (3.12) and the definition $S_1 = 1 * S$. The estimates (3.14), (3.15) follow by $S_{b_1} = b_1 * S$, (3.11), (3.12), and by the size estimates that we can prove for b_1 (see Lemma 6.3).

The Laplace transform of $AS_b(t)\phi = AS_{b_1}(t)\phi + b_\infty AS_1\phi$ can be written as

$$\widehat{AS_b\phi}(\lambda) = \frac{\widehat{b_1}(\lambda)}{\widehat{a}(\lambda)}A^{1-\gamma}\left(\frac{\lambda}{\widehat{a}(\lambda)} + A\right)^{-1}A^\gamma\phi + \frac{b_\infty}{\lambda\widehat{a}(\lambda)}A\left(\frac{\lambda}{\widehat{a}(\lambda)} + A\right)^{-1}\phi.$$

Taking into account (2.1)$_3$, (2.2)$_3$ and (3.1) we conclude that the first part is bounded by $C|\lambda|^{\alpha-\beta-\gamma(\alpha+1)}$. Now take first $\gamma = 0$, then, in case $\beta \leq \frac{1+\alpha}{2}$, $\gamma = \frac{1}{2} - \frac{\beta}{1+\alpha}$. In both cases, apply Lemma 3.2 (in which the γ-exponents are, respectively, $\beta - \alpha > 0$ and $\frac{1-\alpha}{2} > 0$). This yields

$$\|AS_{b_1}(t)\phi\|_p \leq \begin{cases} Ct^{-1+\beta-\alpha}\|\phi\|_p, \\ Ct^{-\frac{1+\alpha}{2}}\|A^\gamma\phi\|_p \text{ if } \beta \leq \frac{1+\alpha}{2}, \ \gamma = \frac{1}{2} - \frac{\beta}{1+\alpha}. \end{cases}$$

The inverse Laplace transform of $\frac{1}{\lambda\widehat{a}(\lambda)}$ can be estimated by $C(1 + t^{-\alpha})$ combining proofs of Theorem 1 in [12] and Proposition 0.2 in [13]. Hence the inverse of the second term is bounded by $C(1 + t^{-\frac{1+\alpha}{2}})\|\phi\|_p$ and (3.16) follows.

To prove (3.17) distinguish two cases. If $\beta > \frac{1+\alpha}{2}$, the Laplace transform of $D\dot{S}_{b_1}$ satisfies

$$|\mathcal{L}(D\dot{S}_{b_1})(\lambda)| \leq C|\lambda|^{-\beta + \frac{1+\alpha}{2}}$$

and we can apply Lemma 3.1 directly to get the first part of (3.17). In the case $\beta \leq \frac{1+\alpha}{2}$ we take advantage of the fact that

$$\mathcal{L}(tf(t))(\lambda) = -\frac{\mathrm{d}}{\mathrm{d}\lambda}\widehat{f}(\lambda). \tag{3.20}$$

Hence

$$\mathcal{L}(tD\dot{S}_{b_1}(t))(\lambda) = -\lambda\frac{\mathrm{d}}{\mathrm{d}\lambda}\widehat{D\dot{S}_{b_1}}(\lambda) - \widehat{D\dot{S}_{b_1}}(\lambda)$$

and the estimates $\|\widehat{D\dot{S}_{b_1}}(\lambda)\|_{\mathcal{B}(L^p)} \leq C|\lambda|^{\alpha/2-\beta-1/2}$, $\|\widehat{D\dot{S}_{b_1}}(\lambda)\|_{\mathcal{B}(L^p)} \leq |\lambda|^{\alpha-\beta}$ respectively, together with the corresponding expressions for their derivatives and Lemma 3.1 yield the remaining assertion in (3.17) and, consequently also (3.18).

It remains to prove (3.19). It is sufficient to show that

$$A^\delta T \in L^1(\mathbb{R}^+, \mathcal{B}(L^p(0,1))) \quad \text{when} \quad \frac{1}{2p} < \delta < \frac{\beta-\alpha}{1+\alpha}.$$

Then

$$T \in L^1(\mathbb{R}^+, \mathcal{B}(L^p, C^\mu)) \quad \text{for} \quad \mu < 2\delta - \frac{1}{p}$$

due to the embedding $\mathcal{D}(A_p^\delta) \hookrightarrow C^\mu(0,1)$. We decompose T as in (3.10) and employ Lemma 3.1 again to show

$$\|A^\delta T_1(t)\phi\|_p \leq Ct^{-1+\beta-\alpha-\delta(1+\alpha)}\|\phi\|_p, \quad \text{for } \delta < \frac{\beta-\alpha}{1+\alpha},$$

$$\|A^\delta T_\infty(t)\phi\|_p \leq Ct^{-\alpha(1+\delta)-\delta}\|\phi\|_p \quad \text{for } \delta < \frac{1-\alpha}{1+\alpha}. \tag{3.21}$$

In the same way as we proved (3.12) we show

$$A^\delta T_\infty \in L^1(\mathbb{R}^+, \mathcal{B}(L^p(0,1))) \quad \text{when } \delta < \frac{1-\alpha}{1+\alpha}.$$

For T_1 we use (3.21), (3.20) and the estimate

$$\left\|\frac{\mathrm{d}}{\mathrm{d}\lambda}[A^\delta\widehat{T_1}(\lambda)\phi]\right\|_p \leq C|\lambda|^{\alpha-\beta}\|\phi\|_p.$$

The same holds also for $\lambda\frac{\mathrm{d}^2}{\mathrm{d}\lambda^2}[A^\delta\widehat{T_1}(\lambda)]$ and Lemma 3.1 applies to give

$$\|A^\delta T_1(t)\phi\|_p \leq Ct^{-2+\beta-\alpha}\|\phi\|_p.$$

The proof is complete. $\qquad\qquad\qquad\square$

We employ the last lemma to prove the local existence result.

PROPOSITION 3.3. *Let* $(2.1) - (2.2)$ *hold,* $g \in C^1(\mathbb{R})$, $g(0) = 0$, $q \geq \max\{2, \frac{1+\alpha}{1-\alpha}\}$, $\gamma = \max\{0, \frac{1}{2} - \frac{\beta}{1+\alpha}\}$. *Given any data* $u^0 \in W_0^{1,\infty} \cap H_0^{1+\gamma,q}$, $u^1 \in L^q$, $f \in L_{loc}^1(\mathbb{R}^+; L^q)$ *there exists a unique weak solution of* (1.1)

$$u \in C([0, t_0); W_0^{1,\infty}(0, 1)) \cap C^1([0, t_0); L^q(0, 1)), \tag{3.22}$$

defined on a maximal time interval $[0, t_0)$. *If* $t_0 < \infty$ *then* $\limsup_{t \to t_0} \|Du\|_\infty = \infty$.
In addition,

$$Du \in C^{\frac{1-\alpha}{2}}([0, t_0); L^q(0, 1)) \tag{3.23}$$

and

$$\|D\dot{u}(t)\|_q \leq C(T)t^{-\frac{1+\alpha}{2}} \text{ for } 0 < t \leq T < t_0. \tag{3.24}$$

Moreover, if $u^0 \in \mathcal{D}(A_q) = W^{2,q} \cap W_0^{1,q}$, $u^1 \in \mathcal{D}(A_q^{\frac{1}{2}}) = W_0^{1,q}$, *and*

$$f \in L_{loc}^1(\mathbb{R}^+; \mathcal{D}(A_q^{1/2})), then$$

$$u \in C([0, t_0); \mathcal{D}(A_q)) \cap C^1([0, t_0); \mathcal{D}(A_q^{\frac{1}{2}})).$$

Proof. According to (3.5) we have

$$u = u^0 + S_1 u^1 + DS_b * Pg(Du) + S_1 * f.$$

We apply the standard contraction method to the operator given by the right hand side of this equation in the space $C([0, t_1]; W_0^{1,\infty}(0, 1))$ with suitable $t_1 > 0$.
    We use estimates from Lemma 3.2, the relations

$$Du = Du^0 + DS_1 u^1 - AS_b * Pg(Du) + DS_1 * f, \tag{3.25}$$
$$\dot{u} = Su^1 + D\dot{S}_b * Pg(Du) + S * f, \tag{3.26}$$

obtained from (3.5) by differentiating, and the fact that

$$A^{\frac{1}{2}}Q, \ DA^{-\frac{1}{2}} \text{ are bounded operators in } L^q.$$

The inclusion

$$\mathcal{D}(A_q^\delta) \subset C^\mu(0, 1) \text{ if } 0 < \mu < 2\delta - \frac{1}{q} \tag{3.27}$$

with $\delta > \frac{1}{2q}$, allows for the estimate

$$\|DS_1(t)u^1\|_{L^\infty} \leq C\|A^\delta A^{\frac{1}{2}} S_1(t)u^1\|_q \leq Ct^{\frac{1-\alpha}{2} - \delta(1+\alpha)}\|u^1\|_q. \tag{3.28}$$

Observe that the exponent on the right hand side of this equation can be chosen positive provided $\frac{1}{q} < \frac{1-\alpha}{1+\alpha}$. In the same way we estimate $\|DS_1 * f\|_{C([0,t_1],L^\infty)}$. From (3.9), (3.19) we get $AS_b \in L^1_{loc}(\mathbb{R}^+; \mathcal{B}(L^\infty(0,1)))$, so $AS_b * Pg(Du) \in BUC([0,t_1]; L^\infty(0,1))$, $AS_b * Pg(Du)(0) = 0$.

To prove (3.24) we have to solve the following equation for

$$v = D\dot{u} \text{ in } L^r((0,t_0); L^q(0,1)),$$

$$v(t) = DS(t)u^1 - \int_0^t AS_b(t-s)P(g'(Du(s))v(s))\,\mathrm{d}s$$

$$-AS_b(t)Pg(Du^0) + DS * f(t). \tag{3.29}$$

We use (3.11) with $\delta = \frac{1}{2}$ and (3.16) to get

$$\|v(t)\|_q \le Ct^{-\frac{1+\alpha}{2}} + C\int_0^t (t-s)^{-1+\beta-\alpha}\|v(s)\|_q\,\mathrm{d}s\ .$$

The generalized Gronwall inequality then yields (3.24) and, consequently, (3.23).

To obtain the estimates for $\dot{u}$, use (3.26), (3.10) and (3.17) provided that $\beta > \frac{1+\alpha}{2}$. To prove that $u \in C^1([0,t_0); L^q)$ also for $\beta \le \frac{1+\alpha}{2}$, we make use of (3.23). We write $\dot{u}$ in the form

$$\dot{u}(t) = S(t)u^1 + S * f(t) + DS_b(t)Pg(Du^0)$$

$$+ \int_0^t DS_b(s)P(g'(Du(t-s))D\dot{u}(t-s))\,\mathrm{d}s$$

and use (3.13), (3.14), (3.24) and the fact that $g(Du^0) \in \mathcal{D}(A^\gamma)$.

We can continue the solution as long as $\|u(t)\|_{W^{1,\infty}}$ remains bounded, this implies boundedness of $\|\dot{u}(t)\|_q$ on the same interval.

The contraction principle applied in the space $C([0,t_1]; \mathcal{D}(A)) \cap C^1([0,t_1]; \mathcal{D}(A^{\frac{1}{2}}))$ yields the existence of a strong solution to (1.1). $\qquad\square$

To derive energy inequalities we will need the following lemma, which contains some well known facts on Volterra equations.

LEMMA 3.4. *Assume* $g \in C^2(\mathbb{R})$, $g(0) = 0$, $g'(s) \ge -\eta$ *for all* $s \in \mathbb{R}$, *let* a, b *be 2-monotone,* k *1-monotone and* $v \in W^{1,2}_{loc}(\mathbb{R}^+, L^2(0,1))$. *Then*

$$\int_0^T \left\langle \frac{\mathrm{d}}{\mathrm{d}t} k * v(t), v(t) \right\rangle \mathrm{d}t \ge \frac{1}{2} \int_0^T k(t)(\|v(T-t)\|_2^2 + \|v(t)\|_2^2)\,\mathrm{d}t \tag{3.30}$$

$$\int_0^T \left\langle \frac{\mathrm{d}}{\mathrm{d}t} b * g(v)(t), \dot{v}(t) \right\rangle \mathrm{d}t \ge -\eta \int_0^T \langle b * \dot{v}(t), \dot{v}(t) \rangle\,\mathrm{d}t$$

$$+ \int_0^T b(t)\langle g(v(0)), \dot{v}(t) \rangle\,\mathrm{d}t. \tag{3.31}$$

$$2 \int_0^T \langle a * \dot{v}(t), \dot{v}(t) \rangle \, \mathrm{d}t = \int_0^T \int_0^t \| v(t-s) - v(t) \|_2^2 \, \mathrm{d}\dot{a}(s) \, \mathrm{d}t$$

$$+ \int_0^T -\dot{a}(s)[\| v(s) - v(0) \|_2^2 + \| v(T-s) - v(T) \|_2^2] \, \mathrm{d}s$$

$$+ a(T) \| v(T) - v(0) \|_2^2. \tag{3.32}$$

Proof. First, we approximate the kernels by an increasing sequence of regular functions k_ϵ, b_ϵ of the same type and apply results from [8], Chapter 18. The Hilbert space version of Lemma 4.1, [8] gives

$$\int_0^T \left\langle \frac{\mathrm{d}}{\mathrm{d}t} k_\epsilon * v(t), v(t) \right\rangle \, \mathrm{d}t \geq \frac{1}{2} \int_0^T k_\epsilon(t) (\| v(T-t) \|_2^2 + \| v(t) \|_2^2) \, \mathrm{d}t.$$

To estimate $\int_0^T \langle \frac{\mathrm{d}}{\mathrm{d}t} b * g(v)(t), \dot{v}(t) \rangle \, \mathrm{d}t$, we denote $h(s) = g(s) + \eta s$ and make use of Lemma 5.4, [8]. The function h is increasing and we get

$$\int_0^T \left\langle \frac{\mathrm{d}}{\mathrm{d}t} b_\epsilon * h(v)(t), \dot{v}(t) \right\rangle \, \mathrm{d}t \geq \int_0^T \langle b_\epsilon(t) h(v(0)), \dot{v}(t) \rangle \, \mathrm{d}t.$$

Going back to the function $g(s) = h(s) - \eta s$, and passing to the limit with b_ϵ, k_ϵ we obtain (3.30), (3.31).

The third formula is proved, replacing $a(t)$ by $a(t+\varepsilon)$, by a direct computation and then by passing to the limit $\varepsilon \to 0$. $\qquad \square$

We close this section by results that enable us to prove global existence.

THEOREM 3.5. *Let* $g \in C^1(\mathbb{R})$, $g(s)s \geq \epsilon|s|^2$ *for* $|s| \geq R$ *and* c *be completely positive. Assume that* h_1, h_2 *are bounded continuous functions on* $J = [0, T]$ *and* v *is a solution of the equation*

$$v + c * g(v) = h_1 + c * h_2. \tag{3.33}$$

Then

$$\| v \|_{L^\infty(0,T)} \leq R + 2\| h_1 \|_{L^\infty(0,T)} + \frac{1}{\epsilon} \| h_2 \|_{L^\infty(0,T)}. \tag{3.34}$$

Proof. First, consider the equation

$$v_\sigma + c * g(v_\sigma) + \sigma g(v_\sigma) = h_1 + c * h_2 + \sigma h_2 \tag{3.33$_\sigma$}$$

with some $\sigma > 0$ fixed. This equation is equivalent to

$$\frac{\mathrm{d}}{\mathrm{d}t} r_\sigma * (v_\sigma - h_1) = -g(v_\sigma) + h_2,$$

with r_σ satisfying

$$\sigma r_\sigma + c * r_\sigma = 1, \quad \widehat{r_\sigma}(\lambda) = \frac{1}{\lambda(\widehat{c}(\lambda) + \sigma)}.$$

By Prüss [13, Proposition 4.5], r_σ is nonnegative and nonincreasing,

$$r_\sigma(0+) = 1/\sigma < \infty. \tag{3.35}$$

Set $\phi = v_\sigma - h_1$ and $\phi_\infty = R + \|h_1\|_{L^\infty(J)} + \frac{1}{\epsilon}\|h_2\|_{L^\infty(J)}$ and show $|\phi(t)| \leq \phi_\infty$ on J.

By the assumptions, ϕ is continuous and $|\phi(0)| < \phi_\infty$ (since, without loss of generality, we can take R large enough). Let $t_0 > 0$ be the first t such that $|\phi(t_0)| = \phi_\infty$ and suppose $\phi(t_0) = \phi_\infty$, the case $\phi(t_0) = -\phi_\infty$ being completely analogous. Then $\phi(t) < \phi_\infty$ for $t < t_0$.

Further

$$v_\sigma(t_0) = \phi(t_0) + h_1(t_0) \geq \phi_\infty - \|h_1\|_{L^\infty(J)} = R + \frac{\|h_2\|_{L^\infty(J)}}{\epsilon} \geq R,$$

so $g(v_\sigma(t_0)) \geq \epsilon v_\sigma(t_0) \geq \epsilon R + \|h_2\|_{L^\infty(J)}$ and

$$\frac{d}{dt} r_\sigma * \phi(t_0) = -g(v_\sigma(t_0)) + h_2(t_0) \leq -\epsilon R$$
$$-\|h_2\|_{L^\infty(J)} + h_2(t_0) \leq -\epsilon R < 0. \tag{3.36}$$

On the other hand,

$$\frac{1}{h}[r_\sigma * \phi(t_0) - r_\sigma * \phi(t_0 - h)]$$

$$= \frac{1}{h}[r_\sigma * (\phi - \phi_\infty)(t_0) - r_\sigma * (\phi - \phi_\infty)(t_0 - h)] + \frac{1}{h}\phi_\infty \int_{t_0-h}^{t_0} r_\sigma(s)\, ds$$

$$= \frac{1}{h}\int_0^{t_0-h} [r_\sigma(t_0 - s) - r_\sigma(t_0 - h - s)][\phi(s) - \phi_\infty]\, d$$

$$+ \frac{1}{h}\int_{t_0-h}^{t_0} r_\sigma(t_0 - s)(\phi(s) - \phi_\infty)\, ds]$$

$$+ \frac{1}{h}\phi_\infty \int_{t_0-h}^{t_0} r_\sigma(s)\, ds \geq \frac{1}{h}\int_{t_0-h}^{t_0} r_\sigma(t_0 - s)(\phi(s) - \phi_\infty)\, ds$$

$$\geq r_\sigma(0+)\frac{1}{h}\int_0^h (\phi(t_0 - s) - \phi_\infty)\, ds.$$

The last term tends to zero as $h \to 0$ due to the continuity of ϕ. This gives $\frac{d}{dt} r_\sigma * \phi(t_0) \geq 0$, a contradiction to (3.36).

Now, the assertion of the theorem follows by approximation of the solution v of (3.33) by solutions v_σ of (3.33)$_\sigma$. The function g is locally Lipschitz continuous which implies that solutions of (3.33), (3.33)$_\sigma$ are uniquely determined and $v_\sigma \to v$ as $\sigma \to 0$ uniformly on compact sets. Hence (3.34) holds also for v. $\qquad\square$

We will also use a modified version of the preceding theorem, specifically, we replace the assumption $g(s)s \geq \epsilon|s|^2$ by $g'(s) \geq -\eta$.

PROPOSITION 3.6. *Let $g \in C^1(\mathbb{R})$, $g(0) = 0$, $g'(s) \geq -\eta$, $\eta > 0$, $s \in \mathbb{R}$ and c be completely positive. Assume that h_1, h_2 are bounded continuous functions on $J = [0, T]$ and v is a solution of the equation*

$$v + c * g(v) = h_1 + c * h_2.$$

Then

$$\|v\|_{L^\infty(0,t)} \leq 2\|h_1\|_{L^\infty(0,t)} + \frac{1}{\epsilon}\|h_2\|_{L^\infty(0,T)}$$

$$+ 2(\eta + \epsilon) \int_0^t c(t - s)\|v\|_{L^\infty(0,s)} \, ds \tag{3.37}$$

holds for any $\epsilon > 0$, $t \in [0, T]$.

Proof. We write (3.33) in the form

$$v + c * g_1(v) = h_1 + c * h_2 + c * g_2(v) \tag{3.38}$$

with

$$g_1(s) = \frac{1}{s} \max\{sg(s), \epsilon s^2\}, \quad g_2(s) = g_1(s) - g(s).$$

The function g_1 satisfies the assumptions of the preceding Theorem with $R = 0$ and

$$|g_2(s)| \leq (\eta + \epsilon)|s|, \quad g_2(s)s \geq 0. \tag{3.39}$$

The formula (3.34) applied to (3.38) now gives that

$$\|v\|_{L^\infty(0,\tau)} \leq 2\|h_1\|_{L^\infty(0,\tau)} + \frac{1}{\epsilon}\|h_2\|_{L^\infty(0,\tau)} + 2 \sup_{0 \leq t \leq \tau} \int_0^t c(t - s)|g_2(v(s))| \, ds$$

holds for any $\tau \leq T$ which together with (3.39) leads to (3.37).

$\square$

4. A priori estimates and global existence

Now, we are going to deduce *a priori* estimates that allow us to prove global existence and boundedness of solutions. Observe that an important feature of (1.1) is the fact that energy inequalities do hold.

One common way to obtain energy estimates is to multiply the equation by $\dot{u}$ and integrate over the time-space interval. Here, because of the convolution of the nonlinearity with the kernel b, we would not get an estimate of u_x directly. To overcome this difficulty, we derive two types of energy inequalities.

First, we apply the inverse operator of $\frac{\mathrm{d}}{\mathrm{d}t} b*$ to (1.1), i.e., we convolve the equation with $k_0 \delta + k_1$ such that $k_0 b + k_1 * b = 1$, equivalently $k_0 + \hat{k}_1(\lambda) = \frac{1}{\lambda \hat{b}(\lambda)}$. The existence of such k_0, k_1 is stated in (2.9).

Denoting $l = k_0 a + k_1 * a$ we have

$$k_0 \ddot{u} + k_1 * \ddot{u} = l * D^2 u + Dg(Du) + k_0 f + k_1 * f.$$

Then we multiply formally the resulting equation by $\dot{u}$, integrate over $[0, \tau] \times [0, 1]$, and integrate by parts to get

$$\int_0^\tau \int_0^1 [k_0 \ddot{u} + k_1 * \ddot{u}] \cdot \dot{u} \, \mathrm{d}x \, \mathrm{d}t + \int_0^\tau \int_0^1 l * D\dot{u} \cdot D\dot{u} \, \mathrm{d}x \, \mathrm{d}t$$
$$+ \int_0^\tau \frac{\mathrm{d}}{\mathrm{d}t} \int_0^1 G(Du) \, \mathrm{d}x \, \mathrm{d}t = \int_0^\tau \int_0^1 [k_0 f + k_1 * f] \cdot \dot{u} \, \mathrm{d}x \, \mathrm{d}t,$$

with $G(s) = \int_0^s g(r) \, \mathrm{d}r$. We employ Lemma 3.3 to obtain

$$\int_0^\tau \langle k_1 * \ddot{u}(t), \dot{u}(t) \rangle \, \mathrm{d}t = \int_0^\tau \left\langle \frac{\mathrm{d}}{\mathrm{d}t} k_1 * \dot{u}(t), \dot{u}(t) \right\rangle \, \mathrm{d}t - \int_0^\tau \langle k_1(t) \dot{u}(0), \dot{u}(t) \rangle \, \mathrm{d}t$$
$$\geq \frac{1}{2} \int_0^\tau k_1(t) (\|\dot{u}(\tau - t)\|_2^2 + \|\dot{u}(t)\|_2^2) \, \mathrm{d}t - \int_0^\tau k_1(t) \langle \dot{u}(0), \dot{u}(t) \rangle \, \mathrm{d}t$$
$$\geq \frac{1}{4} \int_0^\tau k_1(t) \|\dot{u}(t)\|_2^2 \, \mathrm{d}t - \|k_1\|_{L^1(R^+)} \|u^1\|_2^2,$$

where we made use of the estimate

$$\int_0^\tau k_1(t) \langle \dot{u}(0), \dot{u}(t) \rangle \, \mathrm{d}t \leq \int_0^\tau k_1(t) \|u^1\|_2 \|\dot{u}(t)\|_2 \, \mathrm{d}t$$
$$\leq \int_0^\tau k_1(t) \|u^1\|_2^2 \, \mathrm{d}t + \frac{1}{4} \int_0^\tau k_1(t) \|\dot{u}(t)\|_2^2 \, \mathrm{d}t.$$

Thus we have

$$\frac{k_0}{2} \|\dot{u}(\tau)\|_2^2 + \frac{1}{4} \int_0^\tau k_1(t) \|\dot{u}(t)\|_2^2 \, \mathrm{d}t + \int_0^\tau \langle l * D\dot{u}(t), D\dot{u}(t) \rangle \, \mathrm{d}t$$
$$+ \int_0^1 G(Du(\tau, x)) \, \mathrm{d}x \leq \int_0^\tau \langle [k_0 f(t) + k_1 * f(t)], \dot{u}(t) \rangle \, \mathrm{d}t$$
$$+ \int_0^1 G(Du(0)) \, \mathrm{d}x + \left[\frac{k_0}{2} + \|k\|_{L^1(R^+)} \right] \|u^1\|_2^2. \tag{4.1}$$

The second energy inequality is obtained as follows. Multiplying (1.1) by $\dot{u}$, integrating over $[0, \tau] \times [0, 1]$ and again integrating by parts we obtain

$$\frac{1}{2} \int_0^\tau \frac{\mathrm{d}}{\mathrm{d}t} \|\dot{u}(t)\|_2^2 \, \mathrm{d}t + \int_0^\tau \langle a * D\dot{u}(t), D\dot{u}(t) \rangle \, \mathrm{d}t$$

$$+ \int_0^\tau \left\langle \frac{\mathrm{d}}{\mathrm{d}t} b * g(Du)(t), D\dot{u}(t) \right\rangle \mathrm{d}t = \int_0^\tau \langle f(t), \dot{u}(t) \rangle \, \mathrm{d}t.$$

With $b(t) = b_1(t) + b_\infty$ and using (3.31) we get the inequality

$$\|\dot{u}(\tau)\|_2^2 + 2 \int_0^\tau \langle a * D\dot{u}(t), D\dot{u}(t) \rangle \, \mathrm{d}t + 2b_\infty \int_0^1 G(Du(\tau, x)) \, \mathrm{d}x$$

$$\leq 2 \int_0^\tau \langle f(t), \dot{u}(t) \rangle \, \mathrm{d}t + \|u_1\|_2^2 + 2b_\infty \int_0^1 G(Du^0(x)) \, \mathrm{d}x$$

$$+ 2\eta \int_0^\tau \langle b_1 * D\dot{u}(t), D\dot{u}(t) \rangle \, \mathrm{d}t - 2 \int_0^\tau b_1(t) \langle g(Du^0), D\dot{u}(t) \rangle \, \mathrm{d}t. \tag{4.2}$$

Now, we multiply (4.1) by $2C_\eta$ where C_η is given by (2.11) and add to (4.2).

$$(1 + k_0 C_\eta) \|\dot{u}(\tau)\|_2^2 + 2 \int_0^\tau \langle (C_\eta l + \theta a - \eta b_1) * D\dot{u}(t), D\dot{u}(t) \rangle \, \mathrm{d}t$$

$$+ 2 \int_0^\tau (1 - \theta) \langle a * D\dot{u}(t), D\dot{u}(t) \rangle \, \mathrm{d}t$$

$$+ \frac{C_\eta}{2} \int_0^\tau k_1(t) \|\dot{u}(t)\|_2^2 \, \mathrm{d}t + 2(C_\eta + b_\infty) \int_0^1 G(Du(\tau, x)) \, \mathrm{d}x$$

$$\leq 2C_\eta \int_0^\tau \langle k_1 * f(t), \dot{u}(t) \rangle \, \mathrm{d}t + 2(1 + k_0 C_\eta) \int_0^\tau \langle f(t), \dot{u}(t) \rangle \, \mathrm{d}t$$

$$- 2 \int_0^\tau b_1(t) \langle g(Du^0), D\dot{u}(t) \rangle \, \mathrm{d}t + (1 + k_0 C_\eta + 2C_\eta \|k\|_1) \|u^1\|_2^2$$

$$+ 2(b_\infty + C_\eta) \int_0^1 G(Du^0(x)) \, \mathrm{d}x.$$

The second term on the left hand side of the last inequality is nonnegative due to (2.11).

For $f \in L^1((0, T); L^2(0, 1))$, we can estimate further

$$2C_\eta \int_0^\tau \langle k_1 * f(t), \dot{u}(t) \rangle \, \mathrm{d}t + 2(1 + k_0 C_\eta) \int_0^\tau \langle f(t), \dot{u}(t) \rangle \, \mathrm{d}t$$

$$\leq 2(1 + C_\eta k_0 + C_\eta \|k\|_{L^1(R^+)}) \|f\|_{L^1(0,T;L^2)} \sup_{t \in (0,\tau)} \|\dot{u}(t)\|_2.$$

The remaining term on the right hand side can be controlled in a simple way:

$$\int_0^\tau b_1(t) \langle g(Du^0), D\dot{u}(t) \rangle \, \mathrm{d}t = - \int_0^\tau b_1(t) \langle Dg(Du^0), \dot{u}(t) \rangle \, \mathrm{d}t,$$

however, this requires the additional condition $u^0 \in W^{2,2}$.

If we do not want to make this assumption we have to work harder, but we may proceed as follows. Integration by parts yields

$$\int_0^\tau b_1 \langle g(Du^0), D\dot{u} \rangle \, dt = b_1(\tau)\langle g(Du^0), Du(\tau) - Du^0 \rangle$$

$$+ \int_0^\tau (-\dot{b}_1(t)\langle g(Du^0), Du(t) - Du^0 \rangle \, dt, \tag{4.3}$$

hence

$$\left| \int_0^\tau b_1 \langle g(Du^0), D\dot{u} \rangle \, dt \right| \leq [b_1(\tau)\tau^\delta$$

$$+ \int_0^\tau |\dot{b}_1(t)| t^\delta \, dt] \, \|g(Du^0)\|_2 \|Du\|_{\dot{C}^\delta([0,\tau];L^2)},$$

where $\delta = (1-\alpha)/2$ and $\|\cdot\|_{\dot{C}^\delta([0,\tau]}$ denotes the homogeneous Hölder norm on $[0, \tau]$. For $\beta > (1+\alpha)/2$ this estimate is good enough to give control on this term for finite τ, say $\tau \leq 1$. In fact, we have by Lemma 6.3, after some simple estimates,

$$b_1(\tau)\tau^{(1-\alpha)/2} + \int_0^\tau |\dot{b}_1|(t)t^{(1-\alpha)/2} \, dt \leq C \int_0^1 b_1(t)t^{-(1+\alpha)/2} \, dt.$$

Condition **(a)** gives

$$\operatorname{Re} \widehat{a}(\lambda) \geq C\operatorname{Re} \frac{1}{(1+\lambda)^\alpha}, \quad \operatorname{Re} \lambda \geq 0,$$

hence by (3.32)

$$t^{\alpha-1}e^{-t}\|Du(t) - Du^0\|_2^2 \leq 2 \int_0^t \langle s^{\alpha-1}e^{-s} * D\dot{u}(s), D\dot{u}(s) \rangle \, ds$$

$$\leq C \int_0^t \langle a * D\dot{u}(s), D\dot{u}(s) \rangle \, ds, \tag{4.4}$$

which implies

$$\|Du\|_{\dot{C}^{(1-\alpha)/2}([0,\tau];L^2)}^2 \leq C \sup_{0 \leq t \leq 1} \int_0^t \langle a * D\dot{u}(s), D\dot{u}(s) \rangle \, ds.$$

In case $\beta \leq (1+\alpha)/2$, $\tau \leq 1$, choose $1/2 > \gamma > \frac{1}{2} - \frac{\beta}{1+\alpha}$, $\delta = \frac{1-\alpha}{2} + \gamma(1+\alpha)$, and apply the same reasoning as above to get the result

$$\left| \int_0^\tau b_1 \langle g(Du^0), D\dot{u} \rangle \, dt \right| \leq [b_1(\tau)\tau^\delta$$

$$+ \int_0^\tau |\dot{b}_1(t)| t^\delta \, dt]\|A^\gamma g(Du^0)\|_2 \|A^{-\gamma}Du\|_{\dot{C}^{1-\theta(1+\alpha)/2}([0,1];L^2)},$$

where $\theta = 1 - 2\gamma < 2\beta/(1 + \alpha)$, $\delta = 1 - \theta(1 + \alpha)/2 > 1 - \beta$. The norm $\|A^{-\gamma} Du\|_{\dot{C}^{1-\theta(1+\alpha)/2}([0,1];L^2)}$ can be controlled as follows. By the moment inequality we obtain

$$
\begin{aligned}
\|A^{-\gamma}(Du(t) - Du^0)\|_2 &\leq C\|A^{1/2-\gamma}(u(t) - u^0)\|_2 \\
&\leq C\|A^{1/2}(u(t) - u^0)\|_2^{1-2\gamma}\|u(t) - u^0\|_2^{2\gamma} \\
&\leq C\|D(u(t) - u^0)\|_2^{1-2\gamma}\left[\sup_{0\leq t\leq\tau}\|\dot{u}\|_2^{2\gamma}\right]t^{2\gamma} \\
&\leq Ct^{\delta}\left[\int_0^{\tau}\langle a * D\dot{u}, D\dot{u}\rangle\,dt\right]^{1/2-\gamma}\sup_{0\leq t\leq\tau}\|\dot{u}(t)\|_2^{2\gamma}.
\end{aligned}
$$

where the last inequality is a consequence of (4.4). Young's inequality shows that we have control also in this case.

For $\tau \geq 1$ we proceed differently and do not need to distinguish the cases $\beta > \frac{1+\alpha}{2}$ and $\beta \leq \frac{1+\alpha}{2}$. We have

$$
\operatorname{Re}\widehat{b}_1(\lambda) \leq \frac{C_b}{1 + |\lambda|^\beta} \leq \frac{C}{1 + |\lambda|^\alpha} \leq C\operatorname{Re}\widehat{a}(\lambda), \quad \operatorname{Re}\lambda > 0.
$$

This implies by (3.32) that the first term on the right hand side of (4.3) can be estimated as follows:

$$
\begin{aligned}
|b_1(t)\langle g(Du^0), D(u(t) - u^0)\rangle| &\leq \frac{b_1(t)}{2\varepsilon}\|g(Du^0)\|_2^2 + \frac{\varepsilon}{2}b_1(t)\|D(u(t) - u^0)\|_2^2 \\
&\leq \frac{b_1(t)}{2\varepsilon}\|g(Du^0)\|_2^2 + \varepsilon\int_0^t\langle b_1 * D\dot{u}, D\dot{u}\rangle\,ds \\
&\leq \frac{b_1(t)}{2\varepsilon}\|g(Du^0)\|_2^2 + \varepsilon C\int_0^t\langle a * D\dot{u}, D\dot{u}\rangle\,ds.
\end{aligned}
$$

In a similar way we get for the second term,

$$
\begin{aligned}
&\int_1^t|\dot{b}_1(s)|\langle g(Du^0), D(u(s) - u^0)\rangle\,ds \\
&\quad\leq \frac{b_1(1)}{2\varepsilon}\|g(Du^0)\|_2^2 + \varepsilon C\int_0^t\langle a * D\dot{u}, D\dot{u}\rangle\,ds.
\end{aligned}
$$

These arguments show how the bad term $\int_0^\tau b_1(t)\langle g(Du^0), D\dot{u}(t)\rangle\,dt$ can be dominated.

Summing up, we have

PROPOSITION 4.1. *Let* **(a)**, **(b)** *be satisfied, and assume* $g \in C^2(\mathbb{R})$, $g(0) = 0$, *is such that* $g'(s)$ *and* $G(s)$ *are bounded from below. Let* $u^1 \in L^2$, $u^0 \in X = W^{s,2} \cap$

$W_0^{1,\infty}$, $f \in L^1(0, T; L^2)$, *where* $s > 2 - 2\beta/(1 + \alpha)$. *Then there is a constant* $C = C(\|u^1\|_2, \|u^0\|_X, \|f\|_{L^1(0,T;L^2)})$ *such that any weak solution of* (1.1) *satisfies the inequality*

$$\left.\begin{aligned}
\|\ddot{u}(t)\|_2 + \int_0^t \langle a * D\dot{u}(s), D\dot{u}(s)\rangle \, ds + \int_0^1 G(Du(t, x)) \, dx \leq C, \quad 0 \leq t \leq T, \\
\int_0^t \|a_1 * D\dot{u}(s)\|_2^2 \, ds \leq C, \quad 0 \leq t \leq T.
\end{aligned}\right\} \tag{4.5}$$

Here a_1 *stems from strong positivity of* a.

Observe that once the first part of (4.5) holds, the second part follows by (2.13).

This is the starting point of our discussion of global existence and long time behaviour of the solutions of (1.1).

The growth assumption $(2.4)_3$ together with (4.5) yield

$$\|G(Du(t))\|_1 + \|g(Du(t))\|_1 \leq C, \ 0 \leq t < t_0, \tag{4.6}$$

and $\|g(Du(t))\|_p < \infty$ for each $t < t_0$. Another important consequence of (4.5) is

$$a_1 * D\dot{u} \in L^2(\mathbb{R}^+; L^2). \tag{4.7}$$

for every a_1 satisfying (2.13).

Now, we use (4.6) and Proposition 3.3 to obtain global existence.

Proof of Theorem 1.1. Let t_0 be the maximal existence time from the Proposition 3.3. We show that $\|Du(t)\|_\infty$ remains bounded on any interval $(0, T)$ with $T \leq t_0$. The decomposition (3.9) allows us to rewrite the formula (3.25) in the form

$$v + c * g(v) = h_1 + c * h_2 \tag{4.8}$$

with $v - Du$, $h_1 = Du^0 + DS_1 u^1 + DS_1 * f + T * Pg(Du)$, $h_2 = (I - P)g(Du)$. Observe that $\|h_2\|_\infty$ is bounded on $(0, t_0)$. Moreover, $\|Du^0 + DS_1 u^1 + DS_1 * f\|_\infty$ is bounded on $(0, t_0)$ due to (3.13), (3.27), (3.28). The convolution $T * Pg(Du)$ is estimated by

$$\|T * Pg(Du)(t)\|_\infty \leq C \int_0^t \|T(t - s)\|_{B(L^p; L^\infty)} \left\|\frac{g(Du)}{|Du|}(s)\right\|_p \|Du(s)\|_\infty \, ds.$$

Then we infer from $(2.4)_3$ and (4.6), that $\frac{g(Du)}{|Du|} \in L^\infty((0, T); L^p(0, 1))$. Theorem 3.5 and Proposition 3.3 apply and we obtain the estimate

$$\|Du\|_{L^\infty(0,t;L^\infty)} \leq C + C \int_0^t d(t - s)\|Du\|_{L^\infty(0,s;L^\infty)} \, ds, \tag{4.9}$$

where $d(t) = c(t) + \|T(t)\|_{B(L^p; L^\infty)}$. Since $d \in L^1_{loc}(R^+)$ due to (2.3) and (3.19), (4.9) gives boundedness of $\|Du(t)\|_\infty$ on $[0, T]$, and, consequently, global existence. As to strong solutions, having

$$u^0 \in \mathcal{D}(A_q), \ u^1 \in \mathcal{D}(A_q^{\frac{1}{2}}), \ f \in L^1_{loc}(\mathbb{R}^+, \mathcal{D}(A_q^{\frac{1}{2}})),$$

we get boundedness of $A_q u$, $A_q^{\frac{1}{2}} \dot{u}$, using expressions

$$Au(t) = Au^0 + AS_1(t)u^1 + AS_b * g'(Du)Au(t) + AS_1 * f(t),$$

(3.29), and Lemma 3.2. For both $\|Au(t)\|_q$, $\|A^{\frac{1}{2}} \dot{u}(t)\|_q$, we have estimates of the type

$$\|v(t)\|_q \leq C_1 + C_2 \int_0^t (t-s)^{-1+\beta-\alpha} \|v(t)\|_q \, ds$$

and may use the generalized Gronwall lemma.

Now, we are going to show that the global solution is bounded on $\mathbb{R}^+$, i.e. the first assertion of Theorem 2.2.

PROPOSITION 4.2. *Let the assumptions of Theorem 2.1 be satisfied. Moreover, let* $f \in L^1(\mathbb{R}^+, L^q)$ *and let g satisfy* (2.5), (2.6). *Then there exists a constant* $K = K(\|u^1\|_2,$ $\|u^0\|_{W^{s,2}}, \|f\|_{L^1(R^+,L^q)})$ *such that*

$$\|u(t)\|_{W^{1,\infty}} + \|\dot{u}(t)\|_2 \leq K, \quad t \geq 0. \tag{4.10}$$

Proof. We proceed in the same way as above, using the formula

$$Du + c * g(Du) = Du^0 + DS_1 u^1 + T * Pg(Du) + DS_1 * f + c * (I - P)g(Du),$$

rewritten in the form (3.33), Theorem 3.1 and the Gronwall inequality, this time with a kernel of L^1-norm less than 1. In fact, we denote $m(t) = \|T(t)P\|_{B(L^p;L^\infty)}$ and use (2.5) and the energy inequality (4.5), to estimate

$$\left\| \int_0^t T(t-s)Pg(Du(s)) \, ds \right\|_\infty .$$

For this purpose note that (2.6) implies that for each $\rho > 0$ there is constant $C_\rho \geq 0$ such that

$$|g(s)|^p \leq C_\rho + \rho|s|^p G(s), \quad s \in \mathbb{R}.$$

Fix any $\rho > 0$ small enough and estimate in the following way.

$$\left\| \int_0^t T(t-s)Pg(Du(s)) \, ds \right\|_\infty$$

$$\leq \int_0^t m(t-s)\|g(Du(s))\|_p \, ds$$

$$\leq \int_0^t m(t-s)(C_\rho + \rho\|Du(s)\|_\infty^p \|G(Du(s))\|_1)^{\frac{1}{p}} \, ds$$

$$\leq C_\rho^{1/p}|m|_1 + \rho^{1/p}\sup_s \|G(Du(s))\|_1^{1/p}\int_0^t m(t-s)\|Du(s)\|_\infty \, ds$$

$$\leq C_1 + \rho^{1/p}C_2\int_0^t m(t-s)\|Du(s)\|_\infty \, ds.$$

Now choose ρ so small that $\rho^{1/p}C_2\|m\|_{L^1(R+)} < 1$.

Applying Theorem 3.5 to Du (recall (4.8)), with $h_2 = (I - P)g(Du)$, we obtain for $z(t) = \|Du\|_{L^\infty((0,t);L^\infty)}$ the inequality

$$z(t) \leq C + \rho^{1/p}C_2 m * z(t)$$

which yields boundedness of z and hence of Du in the L^∞-norm on $\mathbb{R}^+$.

The time derivative of the solution is estimated in the following Lemma.

We set

$$a_1(t) = c_1 t^{\alpha/2-1}e^{-t} \tag{4.11}$$

where the constant c_1 is chosen such that

$$|\widehat{a}_1(i\rho)|^2 \leq \mathrm{Re}\,\widehat{a}(i\rho), \quad \rho \in \mathbb{R} \tag{4.12}$$

Then a is a_1-positive and we have

LEMMA 4.3. *Let a_1 be given by (4.11), (4.12), $f \in L^1(\mathbb{R}^+; L^2)$ and u be a solution of (1.1) satisfying the energy inequality (4.5). Then*

$$Du \in BUC(\mathbb{R}^+; L^2), \tag{4.13}$$

$$a_1 * \frac{d}{dt}g(Du) \in L^2(\mathbb{R}^+; L^2), \tag{4.14}$$

$$\dot{u}(t) \to 0 \text{ in } L^2(0, 1) \text{ as } t \to \infty. \tag{4.15}$$

Proof. The first assertion follows from the formula

$$Du = Du^0 + DS_1u^1 + DS_1 * f + T * Pg(Du) - c * Pg(Du).$$

In fact, it follows from assumptions **(IC)**, **(f)** and by uniform boundedness of Du that the first four terms are in $BUC(\mathbb{R}^+; L^2)$, by Lemma 3.2. It is easily checked that the last term belongs to this class since $c(t)$ is 1-monotone.

For the second statement we use the abbreviation

$$\langle\langle v \rangle\rangle(t) = \int_0^t \langle a * v(s), v(s)\rangle \, ds.$$

With this notation we have by (3.32) and (4.5)

$$\int_0^\tau \left\| a_1 * \frac{\mathrm{d}}{\mathrm{d}t} g(Du) \right\|_2^2 \, \mathrm{d}t \;\le\; C \left\langle\!\left\langle \frac{\mathrm{d}}{\mathrm{d}t} g(Du) \right\rangle\!\right\rangle (\tau)$$

$$\le\; C \sup_{|s|\le L} |g'(s)|^2 \langle\langle D\dot{u}\rangle\rangle(\tau) < \infty,$$

where $L = \sup_{t>0} \|Du(t)\|_\infty$.

Finally, the last statement follows as in [11] from $\dot{u} \in BUC(\mathbb{R}^+; L^2)$. To prove this we write

$$\dot{u} = Su^1 + S * f + b_\infty DS * Pg(Du) + T_2 * \frac{\mathrm{d}}{\mathrm{d}t} a_1 * Pg(Du),$$

where T_2 is defined by $\widehat{T}_2(\lambda) = D\widehat{S}_{b_1}(\lambda)/\widehat{a}_1(\lambda)$. Since $u_1 \in L^q$, $f \in L^1(\mathbb{R}^+; L^q)$, $q \ge 2$, by assumption, and $Pg(Du) \in L^\infty(\mathbb{R}^+; L^\infty)$ by Proposition 4.2, Lemma 3.2 implies that the first three terms belong even to $BUC(\mathbb{R}^+; L^q)$. By the second statement of this lemma we know that $\frac{\mathrm{d}}{\mathrm{d}t} a_1 * Pg(Du) \in L^2(\mathbb{R}^+; L^2)$, hence it remains to show that $T_2 \in L^2(\mathbb{R}^+; \mathcal{B}(L^2))$. To see this, observe that the estimate

$$|\widehat{T}_2(\lambda)|_{\mathcal{B}(L^2)} + |\lambda \widehat{T}_2'(\lambda)|_{\mathcal{B}(L^2)} \le C/|\lambda|^{\beta-\alpha+1/2}, \quad \mathrm{Re}\ \lambda > 0,$$

holds. Hence Lemma 3.3 yields

$$|T_2(t)|_{\mathcal{B}(L^2)} \le C t^{\beta-\alpha-1/2}, \quad t > 0,$$

which implies $T_2 \in L^2((0,1); \mathcal{B}(L^2))$, as well as $T_2 \in L^\infty((1,\infty); \mathcal{B}(L^2))$. A local analyticity argument as in the proof of Lemma 3.2 also yields $T_2 \in L^1((0,\infty); \mathcal{B}(L^2))$, hence $T_2 \in L^2((0,\infty); \mathcal{B}(L^2))$ follows. $\qquad\square$

5. Global Asymptotic Stability

In this section, we prove the remaining parts of Theorem 2.2. First, we show $Du \in L^r(1,\infty; L^2)$, for some $r \in [2,\infty)$, which together with (4.13) implies $Du \to 0$ in $L^2(0,1)$, i.e. $u(t) \to 0$ as $t \to \infty$ in $W^{1,2}(0,1)$. To this end, we require that (2.5) holds with $R = 0$. This assumption assures that the problem has a unique stationary state, the zero solution, and that $g'(0) > 0$.

We write Du as a sum corresponding to the decomposition $AS_b = c - T$,

$$Du(t) = Du^0 + DS_1(t)u^1 - AS_b * Pg(Du)(t) + DS_1 * f(t)$$

$$= w(t) - c * Pg(Du)(t), \tag{5.1}$$

where

$$w(t) = Du^0 + DS_1(t)u^1 + T * Pg(Du)(t) + DS_1 * f(t). \tag{5.2}$$

Let B denote the inverse operator to $c*$, i.e.,

$$\widehat{Bv}(\lambda) = \frac{1}{\widehat{c}(\lambda)}\widehat{v}(\lambda).$$

From (5.1) we have

$$B(Du - w) + Pg(Du) = 0.$$

The scalar product with $Du\|Du\|_2^{r-2}$ gives

$$\langle B(Du - w)(t), Du(t)\rangle\|Du(t)\|_2^{r-2} + \langle g(Du(t)), Du(t)\rangle\|Du(t)\|_2^{r-2} = 0.$$

We assume (2.5) with $R = 0$, so we have

$$\langle g(Du)(t), Du(t)\rangle\|Du(t)\|_2^{r-2} \geq \epsilon\|Du(t)\|_2^r.$$

It is well-known that the operator B is accretive in $L^p([0, T]; L^2(0, 1))$ for each $p \in [1, \infty]$, since $c(t)$ is completely positive; see Clément and Prüss [3]. This implies

$$\int_0^T \langle BDu(t), Du(t)\rangle\|Du(t)\|_2^{r-2}\, dt \geq 0.$$

Therefore we obtain

$$\epsilon\int_0^T \|Du(t)\|_2^r\, dt \leq \int_0^T \langle Bw(t), Du(t)\rangle\|Du(t)\|_2^{r-2}\, dt, \qquad \text{for all } T > 0. \tag{5.3}$$

We now show $Bw = w_1 + w_2$ with $w_1 \in L^1(\mathbb{R}^+; L^2)$ and $w_2 \in L^r(\mathbb{R}^+; L^2)$, which by the uniform bound on $\|Du(t)\|_\infty$ yields $Du \in L^r(\mathbb{R}^+; L^2)$.

LEMMA 5.1. *Let u be a weak solution of (1.1) and w be given by (5.2). Then*

$$Bw = w_1 + w_2, \; w_1 \in L^1(\mathbb{R}^+; L^2), \; w_2 \in L^r(\mathbb{R}^+; L^2), \tag{5.4}$$

for each $r > \frac{1}{\mu}$ with μ such that (6.6) holds.

Proof. By the definitions of $c(t)$, $l(t)$, and $T(t)$ we have

$$\begin{aligned}
Bw &= \frac{d}{dt}l * w = lDu^0 + l * DSu^1 + l * DS * f + \frac{d}{dt}l * \frac{d}{dt}c * S * Pg(Du) \\
&= a * (k_1 Du^0 + k_0 DSu^1 + k_1 * DSu^1 + k_0 DS * f + k_1 * DS * f) \\
&\quad + k_0 a Du^0 + SPg(Du^0) + S * \frac{d}{dt}Pg(Du).
\end{aligned}$$

The kernel a belongs to $L^r(1, \infty)$ with $r > \frac{1}{\mu}$, μ such that (6.6) holds, and it belongs to $L^1(0, 1)$. Hence a decomposes as $a = a_2 + a_3$ where $a_2 \in L^1(\mathbb{R}^+)$ and $a_3 \in L^r(\mathbb{R}^+)$. This gives the decomposition $Bw = w_1 + w_2$ where

$$
\begin{aligned}
w_1 &= a_2 * (k_1 Du^0 + k_0 DS * f + k_1 * DSu^1 + k_1 * DS * f) \\
&\quad + a_2 k_0 Du^0 + SPg(Du^0) \\
w_2 &= a_3 * (k_1 Du^0 + k_0 DS * f + k_1 * DSu^1 + k_1 * DS * f) \\
&\quad + a_3 k_0 Du^0 + S * \frac{\mathrm{d}}{\mathrm{d}t} Pg(Du).
\end{aligned}
$$

In view of $S \in L^1(\mathbb{R}^+; \mathcal{B}(L^2))$, $DS \in L^1(\mathbb{R}^+; \mathcal{B}(L^2))$, $f \in L^1(\mathbb{R}^+, L^2)$, $k_0 \geq 0$, and $k_1 \in L^1(\mathbb{R}^+)$, we have $w_1 \in L^1(\mathbb{R}^+; L^2)$ and, as soon as we show that $S * \frac{\mathrm{d}}{\mathrm{d}t} Pg(Du) \in L^r(\mathbb{R}^+; L^2)$, we get $w_2 \in L^r(\mathbb{R}^+; L^2)$.

To prove the latter, as in [11] we use the generalization of the Paley-Wiener theorem to the Laplace transform of Hilbert space valued L^2 functions. We have

$$
\mathcal{L}\left(S * \frac{\mathrm{d}}{\mathrm{d}t} g(u_x) \right) = \mathcal{L}\left(T_2 * a_1 * \frac{\mathrm{d}}{\mathrm{d}t} g(u_x) \right),
$$

where

$$
\widehat{T_2}(\lambda) = \frac{1}{\widehat{a_1}(\lambda)} \widehat{S}(\lambda), \quad a_1 \text{ given by (4.11).}
$$

The function $(1 + \lambda^{1 - \frac{\alpha}{2}}) \widehat{T_2}(\lambda)$ is bounded in $\mathrm{Re}\,\lambda > 0$ and $a_1 * \frac{\mathrm{d}}{\mathrm{d}t} g(u_x) \in L^2(\mathbb{R}^+, L^2)$ according to (4.14). Hence, by the vector-valued Paley-Wiener theorem,

$$
S * \frac{\mathrm{d}}{\mathrm{d}t} g(u_x) \in L^2(\mathbb{R}^+; L^2).
$$

We also know that $T_2 \in L^2(\mathbb{R}^+; \mathcal{B}(L^2))$ and so

$$
S * \frac{\mathrm{d}}{\mathrm{d}t} g(u_x) \in L^\infty(\mathbb{R}^+; L^2).
$$

Combining these results we get $w_2 \in L^r(\mathbb{R}^+; L^2)$. $\qquad\square$

From the relation (5.3) we have

$$
\begin{aligned}
\epsilon \int_0^T \|Du(t)\|_2^r \, \mathrm{d}t &\leq \int_0^T \langle Bw(t), Du(t) \rangle \|Du(t)\|_2^{r-2} \, \mathrm{d}t \\
&\leq \|w_1(t)\|_{L^1(R^+; L^2)} \sup_{t \in R^+} \|u_x(t)\|_{L^\infty}^{r-1} + \|w_2\|_{L^r(R^+; L^2)} \|Du\|_{L^r((0,T), L^2)}^{r-1} \\
&\leq C_1 + C_2 \|Du\|_{L^r((0,T), L^2)}^{r-1}
\end{aligned}
$$

for all $T > 0$. This implies $Du \in L^r(\mathbb{R}^+, L^2)$. By $Du \in BUC(\mathbb{R}^+; L^2)$ we obtain $Du(t) \to 0$ in L^2 as $t \to \infty$, and then also $Du(t) \to 0$ in every $L^q(0,1)$ because of uniform boundedness of Du in $L^\infty(0,1)$.

To prove the remaining assertions of Theorem 2.2 we write the problem (1.1)–(1.3) in the following form

$$\ddot{u} + a * A\dot{u} + \frac{d}{dt}b * vAu = f + \frac{d}{dt}b * (vAu + Dg(Du)); \ t > 0$$
$$u(0) = u^0, \ \dot{u}(0) = u^1. \tag{5.5}$$

This is rewritten as

$$\ddot{u} + (a + vb_1) * A\dot{u} + vb_\infty Au = f + \frac{d}{dt}b * (vAu + Dg(Du)) - vb_1 Au^0. \tag{5.6}$$

The solution of the linear problem

$$\ddot{u} + d * A\dot{u} + vb_\infty Au = f, \ u(0) = u^0, \ \dot{u}(0) = u^1, \ d = a + vb_1,$$

is given by the variation of parameters formula

$$u(t) = C(t)u^0 + R(t)u^1 + \int_0^t R(t-s)f(s)\,ds, \ t \geq 0, \tag{5.7}$$

where the operator families $C(t)$ and $R(t)$ correspond to the operator-cosine and -sine families in the case $d \equiv 0$. Their Laplace transforms are given by

$$\widehat{C}(\lambda) = (\lambda + \widehat{d}(\lambda)A)\widehat{R}(\lambda),$$

and

$$\widehat{R}(\lambda) = (\lambda^2 + \lambda\widehat{d}(\lambda)A + b_\infty A)^{-1},$$

for Re $\lambda > 0$. Some properties of operator families of this type are collected in [11], Lemma 4.4. They are based on results from Prüss [13], Fašangová and Prüss [4], [5]). Making use of these results and Lemma 3.1, we get the estimates

$$\|C(t)\|_{B(L^\infty)} \to 0, \quad \|AR(t)\|_{B(L^q)} \leq C\min\{t^{-\alpha}, t^{\alpha-2}\},$$
$$AR \in L^1(\mathbb{R}^+, \mathcal{B}(L^\infty)) \tag{5.8}$$

and, analogously to the decomposition of the operator AS_b, we write

$$AR(t) = k_v(t) - T_v(t)$$

where now

$$\widehat{k_v}(\lambda) = \frac{1}{\lambda\widehat{d}(\lambda) + vb_\infty}, \ \text{and} \ \widehat{T_v}(\lambda) = \lambda^2 \widehat{k_v}(\lambda)\widehat{R}(\lambda).$$

We may now proceed as we did above, to obtain the following equation for $Du(t)$:

$$Du(t) = C(t)Du^0 + DR(t)u^1 - AR * P$$
$$\left[\frac{d}{dt}b * (g(Du) - vDu) - vb_1Du^0\right](t) + DR * f(t).$$

Hence, denoting $k_v^b = \frac{d}{dt}b * k_v$, $\quad T_v^b = \frac{d}{dt}b * T_v$, we have

$$Du + k_v^b * [g(Du) - vDu] = w,$$

where

$$w(t) = C(t)Du^0 + DR(t)u^1 + T_v^b * P(g(Du) - vDu)(t)$$

$$+ k_v^b(I - P)g(Du)(t) + AR * b_1(t)Du^0 + DR * f(t).$$

By the asssumptions on g there is a constant $\epsilon > 0$ such that $sg(s) - \epsilon s^2 \geq 0$. The conditions satisfied by the kernels a, b, k_v with $v = \epsilon$ imply that $k_\epsilon^b \in L^1(\mathbb{R}^+)$. Therefore we may apply the second part of Theorem 3.5 of Chapter 20 in [8] to obtain $\|Du(t)\|_\infty \to 0$ as $t \to \infty$, once we have $\|w(t)\|_\infty \to 0$. This will follow from the fact that $\|Du(t)\|_r \to 0$ as $t \to \infty$, for any $r \in [1, \infty)$, and properties of the operator families $C(t)$, $R(t)$. In fact, we have $C(t)Du^0 \to 0$ by (5.8), $\|(I - P)g(Du)(t)\|_\infty \to 0$ implies $k_\epsilon^b * (I - P)g(Du)(t) \to 0$, and the estimates on $R(t)$ show

$$\|DR(t)u^1\|_\infty \leq \|DA^{-1}\|_{\mathcal{B}(L^q, L^\infty)}\|AR(t)u^1\|_q \leq Ct^{\alpha - 2}\|u^1\|_q \to 0.$$

In the same way we estimate $DR * f$. To conclude, observe that in the same way as for T we obtain

$$T_\epsilon^b \in L^1(\mathbb{R}^+; \mathcal{B}(L^r(0, 1), C^\mu[0, 1])), \quad \text{for } 0 < \mu < 2\frac{1 - \alpha}{1 + \alpha} - \frac{1}{r}.$$

This then implies convergence to 0 in $L^\infty(0, 1)$ of the third term in the definition of $w(t)$, hence $\|Du(t)\|_\infty \to 0$ as $t \to \infty$.

As to the convergence of strong solutions, with $v = g'(0) > 0$ we may rewrite problem (1.1)~(1.3) as

$$u(t) = C(t)u^0 + R(t)u^1 + R * f(t) - vR * b_1(t)Au^0$$

$$+ \int_0^t \frac{d}{dt}R * b(t - s)[g'(0) - g'(Du(s))]Au(s) \, ds, \quad t \geq 0,$$

which yields the following inequality for $\phi(t) = \|Au(t)\|_X$:

$$\phi(t) \leq \phi_0(t) + \int_0^t \gamma(t - s)\|g'(0) - g'(Du(s))\|_\infty \phi(s) \, ds, \quad t \geq 0,$$

where

$$\phi_0(t) = \|C(t)Au^0 + A^{1/2}R(t)A^{1/2}u^1 + A^{1/2}R(t) * A^{1/2}f(t) - \nu AR * b_1 Au^0\|_X$$

$$\text{and } \gamma(t) = \left\|\frac{\mathrm{d}}{\mathrm{d}t}b * AR(t)\right\|_{\mathcal{B}(X)} \in L^1(\mathbb{R}^+).$$

Because of $\|Du(t)\|_\infty \to 0$ as $t \to \infty$, and $\phi_0(t) \to 0$ this implies $\phi(t) = \|Au(t)\|_X \to 0$. Here X may be any of the spaces $X = L^p(0,1)$, $1 \le p < \infty$ or $X = C_0[0,1]$. In a similar way one also gets $\|A^{1/2}\dot{u}(t)\|_X \to 0$ as $t \to \infty$. This proves the last statement of Theorem 2.2. $\square$

6. Appendix

In this section, we collect some properties of the kernels satisfying hypotheses (2.1)–(2.3), in particular we give some necessary and sufficient conditions in the time domain. We begin with the argument condition for $a(t)$.

LEMMA 6.1. *Suppose a is 3-monotone, $a \not\equiv$ constant. Then the following assertions are equivalent:*

(i) $|\arg \widehat{a}(\lambda)| \le \theta_a < \frac{\pi}{2}$ *for* $\mathrm{Re}\,\lambda > 0$;
(ii) *There is a constant $c_a < 1$ such that* $ta(t) \le c_a \int_0^t a(s)\,\mathrm{d}s$
(iii) $\limsup_{t \to 0,\infty} \frac{ta(t)}{\int_0^t a(s)\,\mathrm{d}s} < 1.$

Proof. By [13, Proposition 3.10] we have

$$\frac{3}{5}\int_0^{\frac{1}{\rho}} -t\dot{a}(t)\,\mathrm{d}t \le \mathrm{Re}\,\widehat{a}(\mathrm{i}\rho) \le 4\int_0^{\frac{1}{\rho}} -t\dot{a}(t)\,\mathrm{d}t \tag{6.1}$$

as well as

$$\frac{1}{2\sqrt{2}}\int_0^{\frac{1}{\rho}} a(t)\,\mathrm{d}t \le |\widehat{a}(\mathrm{i}\rho)| \le 2\int_0^{\frac{1}{\rho}} a(t)\,\mathrm{d}t. \tag{6.2}$$

The equivalence between (i) and (ii) now follows easily:

$$|\arg \widehat{a}(\lambda)| \le \theta_a \iff \frac{|\mathrm{Im}\,\widehat{a}(\lambda)|}{\mathrm{Re}\,\widehat{a}(\lambda)} \le C \iff |\widehat{a}(\lambda)| \le C\mathrm{Re}\,\widehat{a}(\lambda)$$

$$\iff \int_0^t a(s)\,\mathrm{d}s \le C\int_0^t -s\dot{a}(s)\,\mathrm{d}s = C\left(-ta(t) + \int_0^t a(s)\,\mathrm{d}s\right)$$

$$\iff (C-1)\int_0^t a(s)\,\mathrm{d}s \ge Cta(t) \iff ta(t) \le \left(1 - \frac{1}{C}\right)\int_0^t a(s)\,\mathrm{d}s.$$

The equivalence between (ii) and (iii) follows since $ta(t) \leq \int_0^t a(s)\, ds$ for all $t > 0$ and equality holds if and only if $a(s) \equiv a(t)$ on $(0, t]$. Since a is convex, this yields a constant. $\qquad\qquad\square$

In the following corollary some sufficient conditions for Lemma 6.1 (iii) are given.

COROLLARY 6.2. *Suppose a is 3-monotone. Then*

(i) *If* $\lim_{t\to 0} \frac{a(t)}{t^{\alpha_0-1}} = a_{\alpha_0} \in (0, \infty)$ *exists for some* $\alpha_0 \in (0, 1)$
     then $\limsup_{t\to 0} \frac{ta(t)}{\int_0^t a(s)\, ds} = \alpha_0 < 1.$

(ii) *If* $\lim_{t\to\infty} \frac{a(t)}{t^{\alpha_1-1}} = a_{\alpha_1} \in (0, \infty)$ *exists for some* $\alpha_1 \in (0, 1)$
     then $\limsup_{t\to\infty} \frac{ta(t)}{\int_0^t a(s)\, ds} = \alpha_1 < 1.$

(iii) *If* $a \in L^1(\mathbb{R}^+)$ *then* $\limsup_{t\to\infty} \frac{ta(t)}{\int_0^t a(s)\, ds} = 0.$

The lower bound on $\widehat{a}$ can be characterized as follows.

LEMMA 6.3. *Suppose a is 2-monotone, $\alpha \in (0, 1]$, $a \not\equiv 0$. Then*

$$|\widehat{a}(\lambda)| \geq \frac{C}{1 + |\lambda|^\alpha}, \quad \mathrm{Re}\,\lambda > 0$$

iff

$$\frac{1}{t}\int_0^t a(s)\, ds \geq ct^{\alpha-1}, \quad t \in (0, 1),$$

for some constant $c > 0$.

Proof. By Prüss, [13, Proposition 3.3] a is 1-regular and [13, Lemma 8.1] shows that

$$C|\widehat{a}(|\lambda|)| \leq |\widehat{a}(\lambda)| \leq C^{-1}|\widehat{a}(|\lambda|)|, \quad \mathrm{Re}\,\lambda > 0. \tag{6.3}$$

Since a is nonnegative, $a \not\equiv 0$, we have $\widehat{a}(\lambda) > 0$ for $\lambda > 0$. Hence it is sufficient to show the equivalence

$$\widehat{a}(\lambda) \geq \frac{C}{\lambda^\alpha}, \ \lambda \geq 1 \iff \int_0^t a(s)\, ds \geq ct^\alpha, \ t \in (0, 1). \tag{6.4}$$

For $\lambda > 0$ we have

$$\frac{1}{e}\int_0^{\frac{1}{\lambda}} a(s)\, ds \leq \int_0^{\frac{1}{\lambda}} e^{-\lambda s} a(s)\, ds \leq \widehat{a}(\lambda),$$

$$\widehat{a}(\lambda) \leq \int_0^{\frac{1}{\lambda}} a(s)\, ds + a\left(\frac{1}{\lambda}\right)\int_{\frac{1}{\lambda}}^{\infty} e^{-\lambda s}\, ds$$

$$\leq \int_0^{\frac{1}{\lambda}} a(s)\, ds + a\left(\frac{1}{\lambda}\right)\frac{1}{\lambda}\frac{1}{e} \leq \left(1 + \frac{1}{e}\right)\int_0^{\frac{1}{\lambda}} a(s)\, ds.$$

This yields (6.4).   □

In a similar way the upper bound on $\widehat{b}$ can be characterized in the time domain.

LEMMA 6.4. *Suppose b is 2-monotone, $\beta \in (0, 1]$. Then*

$$|\widehat{b}(\lambda)| \leq \frac{C}{1 + |\lambda|^{\beta}}, \quad \mathrm{Re}\,\lambda > 0$$

iff

$$b \in L^1(\mathbb{R}^+) \quad and \quad \frac{1}{t} \int_0^t b(s)\,ds \leq Ct^{\beta-1}, \quad t \in (0, 1).$$

Proof. As in the proof of the preceding Lemma we have

$$\widehat{b}(\lambda) \leq \frac{C}{\lambda^{\beta}}, \ \lambda \geq 1 \iff \frac{1}{t} \int_0^t b(s)\,ds \leq Ct^{\beta-1}, \ t \in (0, 1). \tag{6.5}$$

On the other hand, since b is nonnegative, we have

$$\widehat{b}(\lambda) \leq \widehat{b}(0) = \int_0^\infty b(s)\,ds, \ \lambda > 0.$$

Hence $b \in L^1(\mathbb{R}^+) \iff \widehat{b}(\lambda) \leq C$ for $\lambda \in (0, 1]$.   □

REMARK.   In Lemma 6.4 we may replace $\frac{1}{t}\int_0^t b(s)\,ds$ by $b(t)$ while in Lemma 6.4 it is not clear.

The following upper bound on $\widehat{a}$ is used in the proof of Theorem 2.

LEMMA 6.5. *Suppose $a \in L^1_{loc}(\mathbb{R}^+)$ is 3-monotone and such that $|\arg \widehat{a}(\lambda)| \leq \theta_a < \frac{\pi}{2}$ for $\mathrm{Re}\,\lambda > 0$. Then there is $\mu > 0$ such that*

$$\widehat{a}(\lambda) \leq \frac{C}{\lambda^{1-\mu}} \ for \ 0 < \lambda \leq 1 \ and \ a(t) \leq Ct^{-\mu} \ for \ t \geq 1. \tag{6.6}$$

Proof. The argument condition implies

$$\mathrm{Re}\,\widehat{a}(i\rho) \leq |\widehat{a}(i\rho)| \leq \mathrm{Re}\,\widehat{a}(i\rho)/\cos\theta_a, \ \rho \in \mathbb{R}\setminus\{0\}.$$

Moreover, recall (6.1) and (6.2). Now for $0 < \lambda < 1$

$$\log\widehat{a}(\lambda) = \log\widehat{a}(1) + \int_\lambda^1 -\xi \frac{\widehat{a}'(\xi)}{\widehat{a}(\xi)} \frac{d\xi}{\xi},$$

and

$$-\xi \frac{\widehat{a}'(\xi)}{\widehat{a}(\xi)} = \xi \frac{\widehat{ta}(\xi)}{\widehat{a}(\xi)} = \frac{\widehat{\frac{d}{dt}(ta)}(\xi)}{\widehat{a}(\xi)} = 1 - \frac{-\widehat{ta}(\xi)}{\widehat{a}(\xi)}.$$

Further

$$\frac{-\widehat{t\dot{a}}(\xi)}{\widehat{a}(\xi)} \geq \frac{\frac{1}{e}\int_0^{\frac{1}{\xi}} -t\dot{a}(t)\,\mathrm{d}t}{2\int_0^{\frac{1}{\xi}} a(t)\,\mathrm{d}t} \geq \mu \text{ for all } \xi > 0,$$

with some constant $\mu > 0$; in the last step we used the argument condition once more. Therefore

$$\log\widehat{a}(\lambda) \leq \log\widehat{a}(1) + \int_\lambda^1 (1-\mu)\frac{\mathrm{d}\xi}{\xi} = \log\widehat{a}(1) + (\mu-1)\log\lambda,$$

which yields

$$\widehat{a}(\lambda) \leq \widehat{a}(1)\lambda^{\mu-1}, \quad 0 < \lambda < 1.$$

Finally,

$$\frac{1}{\lambda}a(1/\lambda) \leq \int_0^{\frac{1}{\lambda}} a(t)\,\mathrm{d}t \leq C\widehat{a}(\lambda) \leq C\lambda^{\mu-1}$$

gives

$$a(t) \leq Ct^{-\mu} \text{ for } t \geq 1,$$

taking $t = 1/\lambda$. $\square$

Acknowledgement

This work was carried out while H. P. and J. P. were visiting at Helsinki University of Technology. Thanks go to the Department of Mathematics and to the Academy of Science, Finland, for financial support. We appreciate the warm hospitality during this visit.

REFERENCES

[1] BALL, J. M., HOLMES, P. J., JAMES, R. D., PEGO, R. L. and STEWART, P. J., *On the dynamics of fine structure.* J. Nonlinear Science, *1* (1991), 17–70.

[2] CHILL, R., *Tauberian theorems for vector-valued Fourier and Laplace transforms.* Studia Math. *128* (1998), 55–69.

[3] CLÉMENT, PH. and PRÜSS, J., *Completely positive measures and Feller semigroups.* Math. Ann., *287* (1990), 73–105.

[4] FAŠANGOVÁ, E. and PRÜSS, J., *Asymptotic behaviour of a semilinear viscoelastic beam model.* Archiv Math. to appear 2001.

[5] FAŠANGOVÁ, E. and PRÜSS, J., *Evolution equations with dissipation of memory type.* Topics in Nonlinear Analysis, Birkhäuser (1998), 213–250.

[6] GREENBERG, J. M., *On the existence, uniqueness and stability of the equation* $\rho_0 x_{tt} = e(x_x)x_{xx} + \lambda x_{xxt}$. J. Math. Anal. Appl. *25* (1969), 575–591.

[7] GRIPENBERG, G., LONDEN, S.-O. and PRÜSS J., *On a fractional partial differential equation with dominating linear part*. Math. Meth. Appl. Sci., *20* (1997), 1427–1448.

[8] GRIPENBERG, G., LONDEN, S.-O. and STAFFANS, O., *Volterra integral and functional equations*. Cambridge University Press, Cambridge, 1990.

[9] HENRY, D., *Geometric theory of semilinear parabolic equations*. Lect. Notes in Math. Springer-Verlag, New York, *840* (1981).

[10] PEGO, R. L., *Phase transitions in one-dimensional viscoelasticity: admissibility and stability*. Arch. Rat. Mech. Anal., *97* (1987), 353–394.

[11] PETZELTOVÁ, H. and PRÜSS, J., *Global stability of a fractional partial differential equations*. J. Integral Equations Appl., *12* (2000), 323–347.

[12] PRÜSS J., Laplace transforms and regularity of solutions of evolutionary integral equations. Preprint (1996).

[13] PRÜSS J., *Evolutionary integral equations and applications*. Birkhäuser, Basel, Boston, Berlin, (1993).

[14] TRIEBEL, H., *Theory of function spaces*. Akademische Verlagsgesellschaft, Leipzig, (1983).

S.-O. Londen
Institute of Mathematics
Helsinki University of Technology
FIN-02150 Espoo
Finland

H. Petzeltová
Institute of Mathematics
Czech Academy of Sciences
Žitná 25
15 67 Praha 1
Czech Republic

J. Prüss
FB Mathematik und Informatik
Martin-Luther-Universität
Halle-Wittenberg
Theodor-Lieser-Str. 5
D-60120 Halle
Germany

To access this journal online:
http://www.birkhauser.ch

J.evol.equ. 3 (2003) 203 – 214
1424–3199/03/020203 – 12
DOI 10.1007/s00028-003-0087-x
© Birkhäuser Verlag, Basel, 2003

**Journal of Evolution
Equations**

On some singular limits of homogeneous semigroups

P. BÉNILAN, L. C. EVANS[1] AND R. F. GARIEPY

In memory of our friend Philippe Bénilan

"Mais là où les uns voyaient l'abstraction, d'autres voyaient la vérité"

Camus, *La Peste*

1. Introduction

This paper is based upon some handwritten notes Philippe sent us in late 1996, following
his visit to UC Berkeley. He was interested in a scaling argument from our paper with
Feldman [E-F-G], and in his notes extended this trick to cover general nonlinear evolutions
governed by homogeneous accretive operators. We reproduce his proof in §2 below, and
add some commentary and a few PDE examples.

The basic issue is this. Consider a sequence of first-order evolution equations in a Banach
space having the form

$$\begin{cases} \dot{u}_n + A_n(u_n) = 0 & \text{for } t > 0 \\ u_n(0) = x_n, \end{cases} \tag{1.1}$$

where the dot means $\frac{d}{dt}$ and A_n denotes some nonlinear operator satisfying the homogeneity
condition

$$A_n(\lambda u) = \lambda^{m_n} A_n(u) \quad \text{for all } \lambda > 0. \tag{1.2}$$

We ask what at first may seem an odd question: What happens when the degrees of
homogeneity m_n go to infinity? To gain some initial insight, we first note that since u_n
solves (1.1), the rescaled functions $v_n(t) := \lambda^{\frac{1}{m_n - 1}} u_n(\lambda t)$ solve

$$\begin{cases} \dot{v}_n + A_n(v_n) = 0 & \text{for } t > 0 \\ v_n(0) = \lambda^{\frac{1}{m_n - 1}} x_n. \end{cases} \tag{1.3}$$

We formally write

$$w_n := \frac{\partial}{\partial \lambda} v_n, \text{ evaluated at } \lambda = 1. \tag{1.4}$$

Mathematics Subject Classification 2000: 35K55 (35B40).
Key words: Singular limits, homogeneous semigroups.
[1] Supported in part by NSF Grant DMS-0070480 and by the Miller Institute for Basic Research in Science.

Let us differentiate (1.3) in λ, set $\lambda = 1$, and thereby obtain the linearized evolution

$$\begin{cases} \dot{w}_n + A'_n(v_n)w_n = 0 & \text{for } t > 0 \\ w_n(0) = \frac{1}{m_n - 1}x_n. \end{cases} \tag{1.5}$$

If $x_n \to x$, it presumably follows that

$$w_n \to 0 \quad \text{as } n \to \infty. \tag{1.6}$$

But (1.4) also implies

$$w_n = \frac{1}{m_n - 1}u_n + t\dot{u}_n;$$

and therefore (1.5) suggests that

$$\dot{u}_n \to 0 \quad \text{as } n \to \infty \tag{1.7}$$

for times $t > 0$. These purely formal computations lead us to guess that in the singular limit $m_n \to \infty$ the solutions u_n of (1.1) will converge to a constant value.

But which value? In many cases the limiting value will simply be the initial datum x, but in other situations the highly singular and nonlinear dynamics (1.1) will create an initial short-time layer during which the solution rapidly changes before settling down. The theorem proved in §2 below answers this question in many interesting cases. The point is to construct a new nonlinear evolution equation, having the general form

$$\dot{v} + A(v) \ni \frac{v}{t}, \tag{1.8}$$

such that the limiting value of $u_n(t)$ for times $t > 0$ is $v(1)$. Otherwise said, the dynamics (1.8) characterize a suitably rescaled limit of the fast changes undergone by solutions of (1.1) within a short initial time layer, as $n \to \infty$.

The paper Bénilan-Crandall [B-C] on regularizing properties of nonlinear semigroups generated by homogeneous accretive operators involves similar heuristics.

2. An abstract singular limit

Let X denote a real Banach space, with norm $\| \ \|$. If A is a nonlinear, possibly multivalued, m-accretive operator on X, we will sometimes write $(x, y) \in A$ to mean that $x \in D(A)$, the domain of A, and $y \in Ax$. We also let e^{-tA} denote the nonlinear semigroup generated by A.

Suppose that for each positive integer n, A_n is an m-accretive operator defined on X. Our key hypothesis is that each is also positively homogeneous of degree m_n, meaning

$$A_n(\lambda x) = \lambda^{m_n} A_n x \quad \text{for all } x \in D(A_n), \ \lambda > 0.$$

Define

$$C := \{x \in X \mid \text{there exist } (x_n, y_n) \in A_n \text{ with } x_n \to x, \ y_n \to 0\}$$

and let

$$X_0 := \overline{\bigcup_{\lambda > 0} \lambda C}.$$

We will assume that the degrees of homogeneity tend to infinity:

$$\lim_{n \to \infty} m_n = \infty, \tag{2.1}$$

and furthermore that the limit

$$Px := \lim_{n \to \infty} (I + A_n)^{-1} x \tag{2.2}$$

exists in X, for all $x \in X_0$.

THEOREM. *There exists a nonlinear operator* $Q : X_0 \to C$ *such that if* $x_n \in \overline{D(A_n)}$
for $n = 1, 2, \dots$ *and if* $x_n \to x \in X_0$, *then*

$$e^{-tA_n} x_n \to Qx,$$

uniformly for t *in compact subsets of* $(0, \infty)$.
More precisely, we assert

  (i) *C is a closed set, $\lambda C \subseteq C$ for each $\lambda \in [0, 1]$. The mapping P is a contraction of
X_0 onto C, with $Px = x$ for each $x \in C$.*

 (ii) *The operator*

$$A := P^{-1} - I$$

*is an accretive operator on X with $D(A) = C$, $R(I + \lambda A) \supseteq X_0$ and $(I + \lambda A)^{-1} = P$
on X_0 for each $\lambda > 0$.*

(iii) *Q is a contraction of X_0 onto C, and $Qx = x$ for $x \in C$. If $x \in \lambda C$ for some $\lambda > 1$,
we have*

$$Qx = v(1), \tag{2.3}$$

where v is the unique mild solution of the evolution equation

$$\begin{cases} \dot{v} + Av \ni \frac{v}{t} & \text{on } (\delta, \infty) \\ v(\delta) = \delta x \end{cases} \tag{2.4}$$

for $\delta := \lambda^{-1} \in (0, 1)$.

Proof. 1. Let $x \in X_0$ and set $x_n := (I + A_n)^{-1}x$. Define $\lambda_n := e^{-\frac{1}{\sqrt{m_n}}}$ and $x'_n := \lambda_n x_n$. Then $\lim x'_n = \lim x_n = Px$, $A_n x'_n \ni \lambda_n^{m_n}(x - x_n) \to 0$. It follows that $Px \in C$, and since each operator $(I + A_n)^{-1}$ is a contraction, so is P.

Suppose next that $x \in C$, $\lambda \in (0, 1]$, $(x_n, y_n) \in A_n$, $x_n \to x$, and $y_n \to 0$. We then have $\lambda x_n = (I + A_n)^{-1}(\lambda x_n + \lambda^{m_n} y_n)$, and therefore $\lambda x = P(\lambda x)$. This proves (i).

2. Next, fix $\lambda > 0$. For large n, we have $|\lambda^{-\frac{1}{m_n}} - 1| < 1$. Then

$$(I + \lambda A_n)^{-1} = (I + A_n(\lambda^{\frac{1}{m_n}} \cdot))^{-1}$$
$$= \lambda^{-\frac{1}{m_n}}(\lambda^{-\frac{1}{m_n}} I + A_n)^{-1}$$
$$= \lambda^{-\frac{1}{m_n}}(I + A_n)^{-1}(I + (\lambda^{-\frac{1}{m_n}} - 1)(I + A_n)^{-1})^{-1}.$$

Consequently

$$(I + \lambda A_n)^{-1}x \to Px \quad \text{for each } x \in X_0 \text{ and } \lambda > 0. \tag{2.5}$$

Statement (ii) follows, and one has $A \subset \liminf_{n \to \infty} A_n$.

3. Let $x_n \in \overline{D(A_n)}$, $x_n \to x$. If $x \in C = \overline{D(A)}$, then usual limit theory shows that $e^{-tA_n}x_n \to e^{-tA}x = x$, uniformly for t in bounded subsets of $[0, \infty)$.

Assume now $x \in \lambda C$ with $\lambda > 1$ and set $\delta := \lambda^{-1}$. We first demonstrate that

$$e^{-t_n A_n}x_n \to x \quad \text{for } t_n := \frac{\delta^{m_n}}{m_n}. \tag{2.6}$$

By definition of C, we can find $(x'_n, y_n) \in A_n$ such that $x'_n \to \delta x$, $y_n \to 0$. Since $\lambda^{m_n} y_n \in A_n(\lambda x'_n)$, we have $\|e^{-t_n A_n}(\lambda x'_n) - \lambda x'_n\| \le t_n \lambda^{m_n}\|y_n\| = \frac{\|y_n\|}{m_n}$. On the other hand,

$$\|e^{-t_n A_n}x_n - e^{-t_n A_n}(\lambda x'_n)\| \le \|x_n - \lambda x'_n\|;$$

and then (2.6) follows.

4. Let now $v_n(t) := te^{-\frac{t^{m_n}}{m_n}A_n}x_n$. Using the homogeneity of A_n and the definition of mild solution, it is clear that v_n is a mild solution of

$$\dot{v}_n + A_n v_n \ni \frac{v_n}{t} \quad \text{on } (0, \infty).$$

Using (2.6), we see that $v_n(\delta) \to \delta x \in C = \overline{D(A)}$. According to (ii), there exists a unique mild solution v of (2.4). Regular limit theory implies then that $v_n \to v$ in $\mathcal{C}([\delta, \infty); X)$.

Finally, $e^{-tA_n}x_n = \frac{v_n(\tau_m)}{\tau_m}$ with $\tau_m := (m_n t)^{\frac{1}{m_n}}$. Then $\tau_m \to 1$, and so $e^{-tA_n}x_n \to v(1)$, uniformly for t in compact subsets of $(0, \infty)$. This proves the existence of Q on $\bigcup_{\lambda > 0} \lambda C$. Since each mapping e^{-tA_n} is a contraction, we can extend the definition of Q to X_0. $\square$

REMARK. We note here explicitly that the foregoing construction of $Qx = v(1)$ does not depend upon the choice of $\lambda > 1$ for which $x \in \lambda C$.

3. Example 1: homogeneous Hamilton-Jacobi PDE

As a first, heuristic example, consider the initial-value problem

$$\begin{cases} u_t + \frac{1}{p}|Du|^p = 0 & \text{in } \mathbb{R}^n \times (0, \infty) \\ u = g & \text{on } \mathbb{R}^n \times \{t = 0\}, \end{cases} \tag{3.1}$$

where g is bounded and Lipschitz continuous.

We will work out for this problem the formal implications of the Theorem in §2, without providing complete justification. Careful proofs would entail some rather subtle issues concerning viscosity solutions for multivalued Hamiltonians, which we do not wish to address here. We will however later be able to check that the formalism in fact predicts the correct answer.

Let us take $X = BUC(\mathbb{R}^N)$, the space of bounded, uniformly continuous functions with the sup-norm, and define A_p to be the operator $\frac{1}{p}|Du|^p$ in the sense of viscosity solutions. Therefore

$$C = \{u \in W^{1,\infty}(\mathbb{R}^N) \mid |Du| \le 1 \; a.e.\}$$

and so $X_0 = X$. Now $u_p = (I + A_p)^{-1} f$ means

$$u_p + \frac{1}{p}|Du_p|^p = f \quad \text{in } \mathbb{R}^n$$

in the viscosity sense, and so also almost everywhere. We guess that as $p \to \infty$, we have $u_p \to u$ uniformly, where $|Du| \le 1$ a. e. and u is a solution of

$$u + \gamma(|Du|) \ni f \quad \text{in } \mathbb{R}^n, \tag{3.2}$$

for the multivalued graph

$$\gamma(z) := \begin{cases} \{0\} & \text{if } z < 1 \\ [0, \infty) & \text{if } z = 1 \\ \emptyset & \text{if } z > 1. \end{cases}$$

Then $Pf = u$.

The dynamics (2.4) therefore read

$$\begin{cases} v_t + \gamma(|Dv|) \ni \frac{v}{t} & \text{in } \mathbb{R}^n \times (\delta, \infty) \\ v = \delta g & \text{on } \mathbb{R}^n \times \{t = \delta\}. \end{cases} \tag{3.3}$$

This PDE is sufficiently simple that we can guess the solution

$$v(x, t) = \min_{y \in \mathbb{R}^n} \{|x - y| + t g(y)\}; \tag{3.4}$$

so that

$$Qg = v(\cdot, 1) = \min_{y \in \mathbb{R}^n} \{|x - y| + g(y)\}. \tag{3.5}$$

REMARK. We can quickly check this assertion, since the Hopf-Lax formula (cf. [E, Chapter 3]) provides us with a formula for the solution of (3.1):

$$u(x, t) = \min_{y \in \mathbb{R}^n} \left\{ \frac{t}{q} \frac{|x - y|^q}{t^q} + g(y) \right\},$$

with $\frac{1}{p} + \frac{1}{q} = 1$. Hence for each time $t > 0$, the solutions of (3.1) do in fact converge as $p \to \infty$ to $\min_{y \in \mathbb{R}^n} \{|x - y| + g(y)\}$, as predicted.

4. Example 2: the parabolic mesa problem[2]

We next investigate the limiting behavior as $m \to \infty$ of solutions to the porous medium equation

$$\begin{cases} u_{m,t} - \Delta(|u_m|^{m-1} u_m) = 0 & \text{in } \mathbb{R}^n \times (0, \infty) \\ u_m = g & \text{on } \mathbb{R}^n \times \{t = 0\}. \end{cases} \tag{4.1}$$

We take $X = L^1(\mathbb{R}^n)$ and define the operator A_m by $A_m u := -\Delta(|u|^{m-1} u)$ for u belonging to $D(A_m) = \{u \in L^1(\mathbb{R}^n) \mid \Delta(|u|^{m-1} u) \in L^1(\mathbb{R}^n)\}$. The operator A_m is m-accretive and is homogeneous of degree m. Furthermore,

$$C = \{u \in L^\infty(\mathbb{R}^n) \mid |u| \leq 1 \ a.e.\},$$

and therefore $X_0 = X$.

We compute P and Q. Firstly, $u_m = (I + A_m)^{-1} f$ means

$$u_m - \Delta(|u_m|^{m-1} u_m) = f \quad \text{in } \mathbb{R}^n. \tag{4.2}$$

We will need some compactness:

LEMMA. *The set of functions $\{u_m\}_{m=1}^\infty$ is precompact in $L^1(\mathbb{R}^n)$.*

Proof. 1. Assume first that f is bounded, nonnegative and has compact support, in which case $u \geq 0$. We simplify notation by removing the subscript m, and so write

$$u - \Delta(u^m) = f. \tag{4.3}$$

Standard L^1-contraction estimates imply the bounds

$$\|u\|_{L^1} \leq \|f\|_{L^1}, \quad \|u(\cdot) - u(\cdot + h)\|_{L^1} \leq \|f(\cdot) - f(\cdot + h)\|_{L^1}$$

for each $h \in \mathbb{R}^n$, and so $\{u_m\}_{m=1}^\infty$ is precompact in $L^1(K)$ for each compact set $K \subset \mathbb{R}^n$.

---

[2] After this paper was written we became aware that a paper by Noureddine Igbida and Philippe Bénilan which was also inspired by Philippe's notes and which addresses the parabolic mesa problem in a general domain, in a manner more direct than ours, had also been submitted to this memorial volume. That paper appears in this same issue, p. 215.

2. We must show that $\{u_m\}_{m=1}^{\infty}$ is tight, which is to say, that no mass moves to infinity as $m \to \infty$. For this, note first that our multiplying by u^m gives

$$\int_{\mathbb{R}^n} u^{m+1} + Du^m \cdot Du^m \, dx = \int_{\mathbb{R}^n} fu^m \, dx.$$

According to the Sobolev inequality, the second term on the left is greater than or equal to

$$C \left(\int_{\mathbb{R}^n} u^{\frac{2mn}{n-2}} \, dx \right)^{\frac{n-2}{n}}$$

for some positive constant C. Making some elementary estimates, we derive the inequality

$$\int_{\mathbb{R}^n} u^{m+1} \, dx \leq C \left(\int_{\mathbb{R}^n} f^{\frac{2n}{n+4}} \, dx \right)^{\frac{n+4}{n}},$$

where C does not depend on m.

Now fix $S > R > 0$, where the radius R is selected so large that supt $f \subseteq B(0, R)$. Take a smooth function ζ such that $0 \leq \zeta \leq 1$, $\zeta \equiv 0$ on $B(0, R)$, $\zeta \equiv 1$ on $\mathbb{R}^n - B(0, S)$, $|D\zeta| \leq \frac{C}{S}$, $|D^2\zeta| \leq \frac{C}{S^2}$. Our multiplying (4.3) by ζ gives

$$\int_{\mathbb{R}^n} u\zeta \, dx = \int_{\mathbb{R}^n} u^m \Delta\zeta \, dx \leq \frac{C}{S^2} \left(\int_{\mathbb{R}^n} u^{m+1} \, dx \right)^{\frac{m}{m+1}} |B(0, S)|^{\frac{1}{m+1}};$$

and this implies

$$\int_{\{|x| \geq S\}} u \, dx = O(S^{-2+\frac{n}{m+1}}) = o(1), \tag{4.4}$$

uniformly for $m \geq n$.

This proves tightness if f is bounded, nonnegative and has compact support. If f is bounded and has compact support, but can change sign, we compare u with solutions $u^{\pm}$ of (4.2) with $f^{\pm}$ replacing f. Note finally that bounded functions with compact support are dense in L^1 and the mapping $f \mapsto u$ is a contraction. From these observations it follows that for each $f \in L^1$, the sequence $\{u_m\}_{m=1}^{\infty}$ is precompact. $\qquad\square$

Using the Lemma, we can find a subsequence $m_j \to \infty$ for which $u_{m_j} \to u$ in L^1. We check that u is the unique solution of

$$u - \Delta\phi(u) \ni f \quad \text{in } \mathbb{R}^n,$$

for the multivalued graph

$$\phi(z) := \begin{cases} \emptyset & \text{if } z < -1 \\ (-\infty, 0] & \text{if } z = -1 \\ \{0\} & \text{if } -1 < z < 1 \\ [0, \infty) & \text{if } z = 1 \\ \emptyset & \text{if } z > 1. \end{cases}$$

By uniqueness, in fact $u_m \to u$ in L^1. We can restate our conclusion by noting that

$$Pf = u = f\chi_{\{w=0\}} + \chi_{\{w>0\}} - \chi_{\{w<0\}},$$

where $w \in \phi(u)$ is the solution of the mesa problem

$$\mathrm{sgn}(w) - \Delta w \ni f \quad \text{in } \mathbb{R}^n$$

and we write $\mathrm{sgn} = \phi^{-1}$. (See Bénilan-Igbida [B-I] for more details.)

In our case the evolution (2.4) becomes

$$\begin{cases} v_t - \Delta\phi(v) \ni \frac{v}{t} & \text{in } \mathbb{R}^n \times (\delta, \infty) \\ v = \delta g & \text{on } \mathbb{R}^n \times \{t = \delta\}, \end{cases} \tag{4.5}$$

and according to the theory from §2, we have $Qg = v(\cdot, 1)$.

Several authors, including Bénilan, Boccardo, and Herrero [B-B-H], have studied the situation that $g \geq 0$, and have shown that in this case the sequence of solutions of (4.1) converges for $t > 0$ to Pg. We now check that this conclusion accords with our theory.

PROPOSITION.   1. *We have*

$$Q = P \text{ on } L^1(\mathbb{R}^n)^+.$$

2. *In general, however*

$$Q \neq P \text{ on } L^1(\mathbb{R}^n).$$

Proof. 1. We modify some ideas from [B-B-H]. Assume that g is bounded and

$$g \geq 0. \tag{4.6}$$

We assert that

$$t \mapsto v(\cdot, t) \quad \text{is nondecreasing.} \tag{4.7}$$

This follows since (4.6) implies firstly that $v \geq 0$, and then that pointwise either $v = 1$ or else $v_t = \frac{v}{t}$.

2. Take $w \in \phi(v)$ so that

$$v_t - \Delta w = \frac{v}{t}; \tag{4.8}$$

and observe that since $v \geq 0$, we have $w \geq 0$. Define next

$$\hat{w} := \int_\delta^1 \frac{w}{t}\, dt.$$

Then (4.8) implies

$$\Delta \hat{w} = \int_{\delta}^{1} \frac{\Delta w}{t} \, dt = \int_{\delta}^{1} \frac{d}{dt}\left(\frac{v}{t}\right) dt = v(\cdot, 1) - g. \tag{4.9}$$

We claim that

$$\hat{w} \in \phi(v(\cdot, 1)). \tag{4.10}$$

Indeed, if $0 \le v(x, 1) < 1$, then (4.7) implies $0 \le v(x, t) < 1$ for all $\delta \le t \le 1$. Since $w \in \phi(v)$, we deduce that $w(x, t) = 0$ for $\delta \le t \le 1$. Consequently $\hat{w}(x) = 0 \in \phi(v(x, 1))$. On the other hand, if $v(x, 1) = 1$, then $0 \le \hat{w}(x) \in \phi(v(x, 1))$. This proves (4.10).

3. According to (4.9) and (4.10), we have

$$v(\cdot, 1) - \Delta \phi(v(\cdot, 1)) \ni g,$$

and so

$$Qg = v(\cdot, 1) = Pg.$$

4. In the next section we will discuss an example showing for $n = 1$ that in general $P \ne Q$. $\qquad \square$

5. Example 3: collapsing sandpiles

This example is based upon our paper with Feldman [E-F-G], where we suggest that the following provides a (very crude) model for sand dynamics.

We ask about the behavior as $p \to \infty$ of solutions to the parabolic p-Laplacian equation

$$\begin{cases} u_{p,t} - \operatorname{div}(|Du_p|^{p-2} Du_p) = 0 & \text{in } \mathbb{R}^n \times (0, \infty) \\ u_p = g & \text{on } \mathbb{R}^n \times \{t = 0\}. \end{cases} \tag{5.1}$$

We take $X = L^2(\mathbb{R}^n)$ and define the m-accretive operator A_p by $A_p u := -\operatorname{div}(|Du|^{p-1} Du)$ for u belonging to

$$D(A_p) = \{u \in L^2(\mathbb{R}^n) \mid \operatorname{div}(|Du|^{p-2} Du) \in L^2(\mathbb{R}^n)\}.$$

Then

$$C = \{u \in L^2(\mathbb{R}^n) \mid Du \in L^\infty, \ |Du| \le 1 \ a.e.\},$$

and hence $X_0 = X$.

Now $u_p = (I + A_p)^{-1} f$ means

$$u_p - \operatorname{div}(|Du_p|^{p-2} Du_p) = f \quad \text{in } \mathbb{R}^n.$$

The following assertion provides compactness:

LEMMA. *The set of functions $\{u_p\}_{p=1}^{\infty}$ is precompact in $L^2(\mathbb{R}^n)$.*

Proof. Assume firstly that f is bounded and the support of f lies in the ball $B(0, R)$. We drop the subscript p and so write

$$u - \operatorname{div}(|Du|^{p-2}Du) = f. \tag{5.2}$$

The estimates $\|u\|_{L^\infty} \leq \|f\|_{L^\infty}$ and $\int_{\mathbb{R}^n} |Du|^p \, dx \leq \|f\|_{L^2}$ follow. Also, L^2-contraction estimates imply the bounds

$$\|u\|_{L^2} \leq \|f\|_{L^2}, \quad \|u(\cdot) - u(\cdot + h)\|_{L^2} \leq \|f(\cdot) - f(\cdot + h)\|_{L^2}$$

for each $h \in \mathbb{R}^n$, and consequently $\{u_p\}_{p=1}^{\infty}$ is precompact in $L^2(K)$ for each compact set $K \subset \mathbb{R}^n$.

We must show tightness. For this, select a function ζ as in the proof of the Lemma in §4. Multiplying the PDE (5.2) by ζu and integrating by parts gives

$$\int_{\mathbb{R}^n} \zeta u^2 + \zeta |Du|^p \, dx = \int_{\mathbb{R}^n} u|Du|^{p-2}Du \cdot D\zeta \, dx \leq \frac{C}{S} \int_{B(0,S)} |Du|^{p-1} \, dx$$

$$\leq \frac{C}{S} \left(\int_{\mathbb{R}^n} |Du|^p \, dx \right)^{\frac{p-1}{p}} |B(0, S)|^{\frac{1}{p}}$$

$$= O(S^{-1+\frac{n}{p}}) = o(1),$$

uniformly in p.

This estimate establishes compactness in L^2 provided f is bounded and has compact support, and the general case follows, as such functions are dense. $\square$

Utilizing the Lemma, we deduce as in [E-F-G] that as $p \to \infty$, $u_p \to u$ in L^2, where u is the unique solution of

$$u + \partial I_C u \ni f.$$

Here ∂I_C denotes the subdifferential of the convex function

$$I_C[v] = \begin{cases} 0 & \text{if } v \in C \\ \infty & \text{otherwise} \end{cases}$$

In other words, $u = Pf$ is the L^2-projection onto the closed, convex set C. In this situation the evolution (2.4) becomes

$$\begin{cases} v_t + \partial I_C v \ni \frac{v}{t} & \text{in } \mathbb{R}^n \times (\delta, \infty) \\ v = \delta g & \text{on } \mathbb{R}^n \times \{t = \delta\}, \end{cases} \tag{5.3}$$

and $Qg = v(\cdot, 1)$.

An example where P $\neq$ Q. In the appendix of [E-F-G] we constructed a initial function g with compact support, for which $Qg \neq Pg$. We can as follows convert this into an example of $P \neq Q$ for the parabolic mesa problem, discussed earlier in §4.

We take $n = 1$ and consider the two problems

$$\begin{cases} u_t - (|u|^{m-1}u)_{xx} = 0 & \text{in } \mathbb{R} \times (0, \infty) \\ u = h & \text{on } \mathbb{R} \times \{t = 0\}. \end{cases} \tag{5.4}$$

and

$$\begin{cases} \hat{u}_t - (|\hat{u}_x|^{p-2}\hat{u}_x)_x = 0 & \text{in } \mathbb{R} \times (0, \infty) \\ \hat{u} = g & \text{on } \mathbb{R} \times \{t = 0\}. \end{cases} \tag{5.5}$$

For notational simplicity, we do not index u or $\hat{u}$ with a subscript m or p. Observe next that if $m = p - 1$ we have the transformation

$$\hat{u}_x = u, \tag{5.6}$$

provided

$$g_x = h. \tag{5.7}$$

The limit dynamics read, respectively,

$$\begin{cases} v_t - (\phi(v))_{xx} \ni \frac{v}{t} & \text{in } \mathbb{R} \times (\delta, \infty) \\ v = \delta h & \text{on } \mathbb{R} \times \{t = \delta\}. \end{cases} \tag{5.8}$$

and

$$\begin{cases} \hat{v}_t + \partial I_C \hat{v} \ni \frac{\hat{v}}{t} & \text{in } \mathbb{R} \times (\delta, \infty) \\ \hat{v} = \delta g & \text{on } \mathbb{R} \times \{t = \delta\}. \end{cases} \tag{5.9}$$

But according to the theory in §2, $\hat{v}_x = v$ and so

$$(\hat{Q}g)_x = Qh.$$

Likewise

$$(\hat{P}g)_x = Ph.$$

Since we have constructed in [E-F-G] initial data g for which

$$\hat{Q}g \neq \hat{P}g + C$$

for any constant C, it follows that $Qh \neq Ph$.

REMARK. Notice that in Examples 1 and 3 the sets C are basically the same, comprising functions with Lipschitz constant less than or equal to one. However the operators P and consequently the dynamics governed by A differ. In Example 1, the mapping P is not the L^2–projection onto C.

REFERENCES

[B-C]   BÉNILAN, P. and CRANDALL, M. G., *Regularizing effects of homogeneous evolution equations*, in Contributions to Analysis and Geometry 1981, 23–39.

[B-B-H]   BÉNILAN, P., BOCCARDO, L. and HERRERO, M., *On the limit of $u_t = \Delta u^m$ as $m \to \infty$*, in *Some topics in nonlinear PDEs (Turin, 1989)*, Rend. Sem. Mat. Univ. Politec. Torino 1989, Special Issue, 1991, 1–13.

[B-I]   BÉNILAN, P. and IGBIDA, N., *La limite de la solution de $u_t = \Delta_p u^m$ lorsque $m \to \infty$*, C. R. Acad. Sci. Paris Sr. I Math *321* (1995), 1323–1328.

[B-K-M]   BOUILLET, J. E., KORTEN, M. K. and MARQUEZ, V., *Singular limits and the mesa problem*, Rev. Un. Mat. Argentina. *41* (1998) 27–39.

[E]   EVANS, L. C., *Partial Differential Equations*, American Math Society, 1998.

[E-F-G]   EVANS, L. C. FELDMAN, M. and GARIEPY, R., *Fast/slow diffusion and collapsing sandpiles*, Journal of Differential Equations *137* (1997), 166–209.

[E-H-K-O]   ELLIOTT, C. M, HERRERO, M. A., KING, J. R. and OCKENDON, J. R., *The mesa problem: diffusion patterns*, IMA J. Appl. Math, *37* (1986), 147–154.

[F-K]   FRIEDMAN, A. and HÖLLIG, K., *On the mesa problem*, J. Math. Anal. Appl. *123* (1987) 564–571.

L. C. Evans
Department of Mathematics
University of California
Berkeley, CA 94720
USA

R. F. Gariepy
Department of Mathematics
University of Kentucky
Lexington, KY 40506
USA

To access this journal online:
http://www.birkhauser.ch

J.evol.equ. 3 (2003) 215 – 224
1424–3199/03/020215 – 10
DOI 10.1007/s00028-003-0088-9
© Birkhäuser Verlag, Basel, 2003

**Journal of Evolution
Equations**

Singular limit of changing sign solutions of the porous medium equation

PHILIPPE BENILAN AND NOUREDDINE IGBIDA

Abstract. In this paper, we study the limit as $m \to \infty$ of changing sign solutions of the porous medium equation: $u_t = \Delta : |u|^{m-1}u$ in a domain Ω of $\mathbb{R}^N$, with Dirichlet boundary condition.

1. Introduction and main results

We consider the initial boundary value problem:

$$
\begin{cases}
u_t = \Delta \, |u|^{m-1}u & \text{in } Q := (0, \infty) \times \Omega \\
u = 0 & \text{on } \Sigma := (0, \infty) \times \Gamma \\
u(0) = u_0
\end{cases}
\tag{$\mathrm{P_m}$}
$$

where Ω is an open domain of $\mathbb{R}^N$ not necessarily bounded, $m \geq 1$ and $u_0 \in L^1(\Omega)$. Throughout the paper we will use the notation r^m for $|r|^{m-1}r$, for any $r \in \mathbb{R}$. It is well known by now that $(\mathrm{P_m})$ has a unique strong solution u, that is $u \in \mathcal{C}([0, \infty); L^1(\Omega)) \cap L^\infty((\delta, \infty) \times \Omega) \cap W^{1,1}(\delta, \infty; L^1(\Omega))$, $u^m \in L^2(\delta, \infty; H_0^1(\Omega))$, for any $\delta > 0$, $\partial_t u = \Delta u^m$ in $\mathcal{D}'(Q)$ and $u(0) = u_0$ a.e. in Ω. Let us denote this solution by u_m. We are interested in the asymptotic behaviour of u_m, as $m \to \infty$. If $\|u_0\|_{L^\infty(\Omega)} \leq 1$, it is known (cf. [5] and [6]) that

$$
u_m \to u_0 \quad \text{in } \mathcal{C}([0, \infty), L^1(\Omega)).
$$

But, if $\|u_0\|_{L^\infty(\Omega)} > 1$, then one can prove that u_m is relatively compact in $\mathcal{C}((0, \infty), L^1(\Omega))$, but not in $\mathcal{C}([0, \infty), L^1(\Omega))$, an initial boundary layer appears at $t = 0$ when passing to the limit: the limit is singular. Indeed, since the nonlinearity $\varphi_m(r) = r^m$ in the equation $(\mathrm{P_m})$ converges in the graph sense to the maximal monotone graph φ_∞ given by

$$
\varphi_\infty(r) =
\begin{cases}
0 & \text{if } |r| < 1 \\
\pm[0, \infty) & \text{if } r = \pm 1,
\end{cases}
$$

Key words: Singular limit, porous medium equation, changing sign solution, Hele-Shaw problem, Mesa problem, Nonlinear semigroup of contractions.

then, the limiting equation is

$$u_t = \Delta w, \quad w \in \varphi_\infty(u) \quad \text{in} Q \tag{1.1}$$

for which a solution u satisfies $|u| \le 1$ and compatible initial data should live in $[-1, 1]$.

Our aim in this paper is to describe the limit of u_m, when the initial data $\|u_0\|_{L^\infty(\Omega)} > 1$. This kind of question attracts much attention, by its physical interest, since for large m, (P_m) appears in a variety of physical problems, for instance $m = 3$ for the spreading of a liquid film under gravity [23] and semiconductor fabrication [22] and $m \in (5.5, 6.5)$ in a radiation in ionized gazes [25], and also by its mathematical interest in the study of singular limits of linear and nonlinear semigroups (cf. [2] and [9]). Indeed, in a Banach space X, let us consider a family of m-accretive operator A_n, such that $\overline{D(A_n)} = X$ and, as $n \to \infty$, $A_n \to A$ in the sense of resolvent with $\overline{D(A)} \ne X$. For any $u_0 \in X$, the Cauchy problem

$$u_t + A_n u \ni 0 \quad \text{in } (0, \infty), \quad u(0) = u_0$$

has a unique mild solution u_n given by the Crandall-Ligget exponential formula

$$u_n(t) = L^1 - \lim_{k \to \infty} \left(I + \frac{t}{k} A_n \right)^{-k} u_0 \quad =: e^{-tA_n} u_0.$$

Letting $n \to \infty$, it is known that (cf. [11]) if $u_0 \in \overline{D(A)}$, then $u_n \to u$ in $\mathcal{C}([0, \infty), X)$ and u is the mild solution of

$$u_t + Au \ni 0 \quad \text{in } (0, \infty), \quad u(0) = u_0. \tag{1.2}$$

But, if $u_0 \in X \setminus \overline{D(A)}$, then (1.2) is not well posed and in general the limit of u_n may not exist. However for a large class of concrete problems the limit exists and it would be interesting to characterize it. In this case, a conjecture is that there exists $\underline{u}_0 \in \overline{D(A)}$, such that the limit u is the solution of

$$u_t + Au \ni 0 \quad \text{in}(0, \infty), \quad u(0) = \underline{u}_0,$$

but the characterization of $\underline{u}_0$ is not clear yet in general.

Coming back to the problem (P_m), $X = L^1(\Omega)$, and the family of operators A_m is given by

$$A_m u = -\Delta u^m \quad \text{in } \mathcal{D}'(\Omega) \tag{1.3}$$

with

$$D(A_m) = \{ z \in L^1(\Omega) \cap L^\infty(\Omega) ; \ z^m \in H_0^1(\Omega) \text{and } \Delta z^m \in L^1(\Omega)\}.$$

As $m \to \infty$, we prove (cf. Proposition 2.3) that A_m converges to A_∞ the multivalued operator given by

$$z \in A_\infty v \Leftrightarrow \begin{cases} v, z \in L^1(\Omega), \ \exists w \in H_0^1(\Omega), \ v \in \text{Sign}(w) \text{a.e. on} \Omega \\ \text{and} \int_\Omega \nabla w . \nabla \xi = \int_\Omega z \, \xi, \quad \forall \xi \in H_0^1(\Omega) \cap L^\infty(\Omega) \end{cases} \tag{1.4}$$

with $\overline{\mathcal{D}(A_\infty)} = \{u \in L^1(\Omega) \cap L^\infty(\Omega) \,; \, \|u\|_{L^\infty(\Omega)} \leq 1\}$. It is known that the mild solution u_m is the strong solution of the pde $(\mathrm{P_m})$. It is true that u_m is convergent in $L^1(\Omega)$, but as far as we know, the characterization of $\underline{u}_0$ is completely solved only in the case where $u_0 \geq 0$; it is known that

$$\underline{u}_0 = (I + A_\infty)^{-1} u_0. \tag{1.5}$$

More precisely, it was proved in [3] (see also[14] and [12]) that the limit of u_m is independent of t: it is equal to the mesa of height 1, that is $u_0 \, \chi_{[\underline{w}=0]} + \chi_{[\underline{w}>0]}$, where $\underline{w}$ is the unique solution of the so-called "mesa problem"

$$\underline{w} \in H^2(\Omega) \cap H_0^1(\Omega), \, \underline{w} \geq 0, \, 0 \leq \Delta\underline{w} + u_0 \leq 1, \, \underline{w}(\Delta\underline{w} + u_0 - 1) = 0 \text{ a.e } \Omega,$$

and one verifies easily that $u_0 \, \chi_{[\underline{w}=0]} + \chi_{[\underline{w}>0]}$ coincides with $(I + A_\infty)^{-1} u_0$. By the way, let us remark that, in this case, we can say

$$\lim_{m\to\infty} \lim_{k\to\infty} \left(I + \frac{t}{k}A_m\right)^{-k} u_0 = \lim_{k\to\infty} \lim_{m\to\infty} \left(I + \frac{t}{k}A_m\right)^{-k} u_0. \tag{1.6}$$

Recall that the proofs of these results use strongly the fact that $u_m \geq 0$, since they are based on the regularizing effect (cf. [4]) of the type

$$-u_t \leq \frac{u}{(m-1)t} \quad \text{a.e. in} Q,$$

which is fulfilled only if $u_0 \geq 0$. In this paper, our approach is completely different. Inspired by the paper [15] where the limit, as $p \to \infty$, of a solution of $u_t = \Delta_p u$ is studied, we characterize $\underline{u}_0$ in the case where u_0 may change sign. Our technics are general, we are using only, and strongly, the homogeneity of the equation in $(\mathrm{P_m})$. Let us notice, that Ph. Bénilan was able to develop most of the arguments we are using in a general setting of abstract nonlinear homogeneous semigroup (see the article [7]).

Before stating our main results, let us mention that the independence of time of the limit of u_m is due to the absence of a reaction in the equation and also to the homogeneity of the boundary conditions. In connection with the Hele Shaw problem, the equation (1.1) is used in the weak formulation of this problem (cf. [13]) and obviously $\underline{u}_0$ corresponds to a stationary solution; this is due to the absence of an injection and/or a suction. In the case of non null reaction (cf. [9, 8, 10]) and/or non null boundary conditions (cf. [20], [24] and [19, 18]), it is proved that the limit may depends on t and is a solution of the equation (1.1) with the corresponding nonhomogeneous terms and the (compatible) initial data $\underline{u}_0$.

Throughout the paper, we assume that Ω is an open domain, not necessarily bounded, $u_0 \in L^1(\Omega) \cap L^\infty(\Omega)$ and u_m is the mild solution of the Cauchy problem

$$u_t + A_m u = 0 \quad \text{in } (0, \infty), \quad u(0) = u_0, \tag{1.7}$$

with A_m given as in (1.3). To simplify the notation, we set

$$a = \begin{cases} 1 & \text{if } \|u_0\|_\infty \leq 1 \\ 1/\|u_0\|_\infty & \text{if } \|u_0\|_\infty > 1 \end{cases} \quad \text{and} \quad v_0 = a\, u_0, \quad \text{a.e. in } \Omega.$$

The main idea of the paper is to consider the change of variables

$$z_m(t) = t u_m(t^m/m)$$

and study the limit of z_m, as $m \to \infty$. So, our main result is

THEOREM 1.1. *As $m \to \infty$, we have*

$$z_m \to z \quad in \quad \mathcal{C}([0, \infty); L^1(\Omega))$$

where z is given as follows:

i - $z(t) = t\, u_0$ *for any $t \in [0, a]$.*
ii - *z is the unique mild solution of the evolution problem*

$$\begin{cases} z_t + A_\infty z \ni z/t & in (a, \infty), \\ z(a) = v_0. \end{cases} \tag{1.8}$$

As to the limit of $u_m(t)$, we can deduce now the following result.

COROLLARY 1.2. *As $m \to \infty$, we have*

$$u_m(t) \to z(1) \quad in\ L^1(\Omega), \quad uniformly\ for\ t\ in\ a\ compact\ set\ of\ (0, \infty)$$

where z is given as in Theorem 1.1.

It is clear that Corollary 1.2 implies that the limit of u_m, as $m \to \infty$, is independent of time t. Moreover, if $\|u_0\|_\infty > 1$, the convergence cannot be extended to 0; similarly it does not to ∞. This means that the question arises as to the asymptotic behavior of $u_m(x, t)$ when $t \to 0$ (or $t \to \infty$) as $m \to \infty$. Theorem 1.1, implies that with the new scale $\tau = t^m/m$, the limit of the solution u_m, depends on time and one can describe the asymptotic of the limit as $t \to 0$ or $t \to \infty$. This kind of results was first studied by A. Friedman and Sh. Huang [17] (see also [16] and [14]) in the case where $u_0 \geq 0$. In those papers, the new scale was $\tau = t^m$ suggested by the asymptotic behavior of the Barenblatt solutions, and the authors prove that the limit with the new scale depends on τ and coincides with the mesa of height $1/\tau$. In the case where u_0 is changing sign, the limit is not a mesa in general (see Remark 1.3 below).

REMARK 1.3. 1. Using the results of the appendix of [15], one can prove that, in contrast to the nonnegative case, the limit of u_m in general is not a projection on the closure of the domain of A_∞. In other words, (1.5) and (1.6) are not true in general.
2. By using the results of [9], [8] and [20], one can treat (P_m) with a reaction term and/or nonhomogeneous boundary condition of Dirichlet or Neumann type, exactly in the same way as in this paper.

2. Proofs

The main ingredient we use for the proof of the first part of the theorem is the following lemma.

LEMMA 2.1. *Let* $f \in L^1(\Omega) \cap L^\infty(\Omega)$ *such that* $\|f\|_\infty \leq 1$ *and, for* $\lambda > 0$, *let us consider* $f_m = (I + \lambda A_m)^{-1} f$. *As* $m \to \infty$, *we have*

$$f_m \to f \quad and \quad A_m f_m \to 0 \quad in \ L^1(\Omega).$$

Proof. Since $A_m f_m = f - f_m$, then it is enough to prove that $f_m \to f$ and the conclusion of the lemma follows. We begin by assuming that $\|f\|_{L^\infty(\Omega)} \leq c < 1$. By definition of A_m, f_m satisfies

$$f_m^m \in H_0^1(\Omega) \text{ and } -\lambda \Delta f_m^m = f - f_m \text{ in } \mathcal{D}'(\Omega); \tag{2.1}$$

and, moreover, we have

$$\|f_m\|_{L^p(\Omega)} \leq \|f\|_{L^p(\Omega)} \text{ for any } 1 \leq p \leq \infty. \tag{2.2}$$

Taking f_m^m as a test function and letting $m \to \infty$, we get

$$\lambda \int_\Omega |\nabla f_m^m|^2 = \int_\Omega (f - f_m)\, f_m^m$$
$$\leq 2\, c^m\, \|f\|_1 \to 0 \quad \text{as } m \to 0,$$

so that (2.1) implies that, $f_m \to f$ in $\mathcal{D}'(\Omega)$ and by using (2.2) we deduce that the convergence holds true in $L^1(\Omega)$. Indeed, thanks to (2.2), $f_m \to f$, in $L^p(\Omega)$-weak and $\|f\|_{L^p(\Omega)} = \lim_{m\to\infty} \|f_m\|_{L^p(\Omega)}$, for any $1 < p < \infty$, so that the convergence holds true in $L^p(\Omega)$, for any $1 < p < \infty$. Then, by using again (2.2), with $p = \infty$, and Lebesgue's dominated convergence theorem, we deduce, by choosing a subsequence that we denote again by m, that $f_m \to f$, in $L^1(\Omega)$. At last, if $\|f\|_\infty = 1$, then we consider a sequence $\{f_\varepsilon\}_{\varepsilon>0}$ in $L^1(\Omega)$ such that $\|f_\varepsilon\|_\infty < 1$ for any $\varepsilon > 0$ and, as $\varepsilon \to 0$, $f_\varepsilon \to f$ in $L^1(\Omega)$. Then, thanks to the L^1 contraction property of the operator $(I + \lambda A_m)^{-1}$, we deduce, by using the previous part of the proof, that $f_m \to f$ in $L^1(\Omega)$ and the proof is complete. $\square$

As a consequence of this lemma, we have

LEMMA 2.2. *As $m \to \infty$, we have*

$$u_m(t^m/m) \to u_0 \quad in \ L^1(\Omega) \ uniformly \ for \ t \in [0, a].$$

Proof. Set $f_m = (I + A_m)^{-1} v_0$ and $\tilde{f}_m = \frac{1}{a} f_m$. It is clear that $f_m \in \mathcal{D}(A_m)$, $\tilde{f}_m \in \mathcal{D}(A_m)$, and, thanks to Lemma 2.2, $\tilde{f}_m \to u_0$ and $A_m f_m \to 0$ in $L^1(\Omega)$, as $m \to \infty$. Using the fact that $u_m(t^m/m) = e^{-\frac{t^m}{m} A_m} u_0$, we can write

$$|u_m(t^m/m) - u_0| \le |e^{-\frac{t^m}{m} A_m} u_0 - e^{-\frac{t^m}{m} A_m} \tilde{f}_m|$$
$$+ |e^{-\frac{t^m}{m} A_m} \tilde{f}_m - \tilde{f}_m| + |\tilde{f}_m - u_0|,$$

so that, by using the L^1 contraction property of the semigroup generated by A_m and the homogeneity of A_m, we deduce that

$$\|u_m(t^m/m) - u_0\|_1 \le 2 \|u_0 - \tilde{f}_m\|_1 + \frac{t^m}{m} \|A_m \tilde{f}_m\|_1$$
$$\le 2 \|u_0 - \tilde{f}_m\|_1 + \frac{1}{m} (t/a)^m \|A_m f_m\|_1.$$

Letting $m \to \infty$, the second term of the last inequality tends to 0, uniformly for $t \in [0, a]$, and the result of the lemma follows. $\qquad\square$

At this stage, one sees that Lemma 2.2 gives the proof of the first part of Theorem 1.1. In other words, it characterizes the limit of $z_m(t)$, as $m \to \infty$, for $t \in [0, a]$. For the remaining part, i.e. for $t \in [a, \infty)$, the main ingredient we use, in this paper, is the convergence of A_m to A_∞ in the sense of resolvent. Recall that the result is well known by now in the case where $\Omega = \mathbb{R}^N$ (cf. [5]) and also in the case where Ω is a bounded domain of $\mathbb{R}^N$ (cf. [6]). As to the case of an open domain Ω, not equal to $\mathbb{R}^N$ and not necessarily bounded, this is done in [21]. For completeness, we give hereafter the proof.

PROPOSITION 2.3. *For any $f \in L^1(\Omega) \cap L^\infty(\Omega)$, as $m \to \infty$, we have*

$$(I + A_m)^{-1} f \to (I + A_\infty)^{-1} f \quad in \ L^1(\Omega).$$

To simplify the notation, set $u_m = (I + A_m)^{-1} f$, then u_m is the unique solution of

$$u_m^m \in H_0^1(\Omega) \ and \ -\Delta u_m^m = f - u_m \ in \ \mathcal{D}'(\Omega)$$

and, recall that (cf. [1]),

$$\|u_m\|_{L^p(\Omega)} \le \|f\|_{L^p(\Omega)} \ for \ any \ 1 \le p \le \infty.$$

To prove the proposition, we need to prove that $(u_m)_{m \ge 1}$ is relatively compact in $L^1(\Omega)$ and that $(u_m^m)_{m \ge 1}$ is weakly relatively compact in $H_0^1(\Omega)$. For this, we need the following technical lemma.

LEMMA 2.4. (cf. [21]) *For any $w \in H_0^1(\Omega)$ and $k > 0$, we have*

$$\|(|w| - k)^+\|_2 \le C \, |[|w| > k]|^{1/N} \left(\int_{[|w|>k]} |\nabla w|^2 \right)^{1/2}$$

where C is a constant depending only on Ω.

LEMMA 2.5. *The sequence $(u_m^m)_{m \ge 1}$ is bounded in $H^1(\Omega)$.*

Proof. First, thanks to Lemma 2.4, we see that it is enough to prove that

$$\int_{[|u_m^m| \ge 1]} |\nabla u_m^m|^2 \text{ is bounded.} \tag{2.3}$$

Indeed, it is clear that

$$|[|u_m^m| \ge 1]| = |[|u_m| \ge 1]| \le \|f\|_1,$$

and

$$\int_\Omega |u_m^m|^2 = \int_{[|u_m^m| \le 1]} |u_m^m|^2 + \int_{[|u_m^m| \ge 1]} |u_m^m|^2$$
$$\le \|f\|_1 + (\|(|u_m^m| - 1)^+\|_2 + |[|u_m^m| \ge 1]|^{1/2})^2,$$

so that, by using Lemma 2.4, and (2.3), we deduce that u_m^m is bounded in $L^2(\Omega)$. As to the boundness of $\|\nabla u_m^m\|_2$, this follows from the fact that

$$\int_\Omega |\nabla u_m^m|^2 = \int_\Omega (f - u_m) \, u_m^m$$
$$\le 2\|f\|_2 \, \|u_m^m\|_2.$$

Now, by taking $(|u_m^m| - 1)^+$ as a test function and using Lemma 2.4, we see that

$$\int_{[|u_m^m| \ge 1]} |\nabla u_m^m|^2 = \int (f - u_m) \, (|u_m^m| - 1)^+$$
$$\le 2 \, \|f\|_2 \, \|(|u_m^m| - 1)^+\|_2$$
$$\le 2 \, C \, \|f\|_2 \|f\|_1^{1/N} \left(\int_{[|u_m^m| \ge 1]} |\nabla u_m^m|^2 \right)^{1/2},$$

and (2.3) follows. $\square$

LEMMA 2.6. *The sequence $(u_m)_{m \ge 1}$ is relatively compact in $L^1(\Omega)$.*

Proof. First, we see that

$$\limsup_{\substack{|y|\to 0 \\ m\geq 1}} \int_{\Omega'} |u_m(.+y) - u_m(.)| = 0, \tag{2.4}$$

for any $\Omega' \subset\subset \Omega$. Indeed, for any $\xi \in H_0^1(\Omega')$ and $y \in \mathbb{R}^N$ such that $|y| \leq \text{dist}(\Omega', \Gamma)$, we have

$$\int_\Omega (u_m(.+y) - u_m(.))\, \xi$$
$$+ \int_\Omega \nabla(u_m^m(.+y) - u_m^m(.)) \cdot \nabla\xi = \int_\Omega (f(.+y) - f(.))\, \xi,$$

so that, by using standard arguments (see for instance [6]), we deduce that

$$\int_\Omega |u_m(.+y) - u_m(.)|\, \xi \leq \int_\Omega |u_m^m(.+y) - u_m^m(.)|\, \Delta\xi + \int_\Omega |f(.+y) - f(.)|\, \xi$$

and, by using Lemma 2.5, we get (2.4). This implies that $(u_m)_{m\geq 1}$ is relatively compact in $L_{\text{loc}}^1(\Omega)$. To end up the proof of the lemma, we show that

$$u_{m-} \leq u_m \leq u_{m+}\quad \text{a.e. in } \Omega \tag{2.5}$$

where u_{m-} (resp. u_{m+}) is the solution of

$$u_{m-} - \Delta u_{m-}^m = f^-\quad (\text{resp } u_{m-} - \Delta u_{m+}^m = f^+)\quad \text{in } \mathbb{R}^N,$$

in the sense that $u_{m-} \in L^\infty(\mathbb{R}^N) \cap L^1(\mathbb{R}^N)$ (resp. $u_{m+} \in L^\infty(\mathbb{R}^N) \cap L^1(\mathbb{R}^N)$), $u_{m-}^m \in H^1(\mathbb{R}^N)$ (resp. $u_{m+}^m \in H^1(\mathbb{R}^N)$) and the equation is satisfied in $\mathcal{D}'(\mathbb{R}^N)$. Indeed, since u_{m-} and u_{m+} are relatively compact in $L^1(\mathbb{R}^N)$ (cf. [5]), then (2.5) and the relative compactness of $(u_m)_{m\geq 1}$ in $L_{\text{loc}}^1(\Omega)$, ends up the proof of the lemma. So, let us prove (2.5). It is clear that

$$\int_\Omega (u_m - u_{m+})\, \xi + \int_\Omega \nabla(u_m^m - u_{m+}^m) \cdot \nabla\xi = \int_\Omega (f - f^+)\, \xi$$

for any $\xi \in H_0^1(\Omega)$. On the other hand, it is not difficult to see that, $(u_m^m - u_{m+}^m)^+ \in H_0^1(\Omega)$, so that

$$\int_\Omega (u_m - u_{m+})\, (u_m^m - u_{m+}^m)^+ + \int_\Omega |\nabla(u_m^m - u_{m+}^m)|^2$$
$$= \int_\Omega (f - f^+)\, (u_m^m - u_{m+}^m)^+$$
$$\leq 0$$

and we deduce that $u_m \leq u_{m+}$ a.e. in Ω. In the same way one can prove that $u_{m-} \leq u_m$ a.e. in Ω and (2.5) follows. $\square$

Proof of Proposition 2.3. Using Lemma 2.5 and Lemma 2.6, the proposition follows by standard arguments (cf. [5] and [6]). We omit the details of the proof.

Proof of Theorem 1.1. Now, it is clear that the first part of the theorem follows from Lemma 2.1. Let us prove part 2. It is clear that z_m is the mild solution of $\frac{du}{dt} + A_m u \ni u/t$ in (a, ∞), so that since $z_m(a) \to a u_0$ in $L^1(\Omega)$, as $m \to \infty$, and $a u_0 \in \mathcal{D}(A_\infty)$, then, by using Proposition 2.3 with classical theorem for regular perturbation of nonlinear semigroup (cf. [11]), we deduce that $z_m \to z$ in $C([a; \infty), L^1(\Omega))$ where z is the unique mild solution of (1.8), and the proof of the theorem is complete.

Proof of Corollary 1.2. It is clear that, for any $t > 0$, $u_m(t) = \dfrac{1}{(mt)^{1/m}} z_m((mt)^{1/m})$ and, as $m \to \infty$, $(mt)^{1/m} \to 1$, then the corollary is a simple conseqence of the L^1 convergence of $z_m(t)$ to $z(t)$, uniformly for t in a compact subset of $(0, \infty)$.

REFERENCES

[1] BÉNILAN, PH., *A Strong Regularity L^p For Solutions of the Porous Media Equation*, In C. Bardos et al., editor, *Contribution to Nonlinear PDE*. Pitman Research Notes, 1983.

[2] BÉNILAN, PH., *Quelques remarques sur la convergence singulière des semi-groupes linéaires*, Publ. Math. UFR Sci. Tech., Univ. Franche-Comté Besançon, *15* (1995/97), 1–11.

[3] BÉNILAN, PH., BOCCARDO, L. and HERRERO, M., *On the Limit of Solution of $u_t = \Delta u^m$ as $m \to \infty$*, In M. Bertch et al., editor, in Some Topics in Nonlinear PDE's, Torino, 1989. Proceedings Int.Conf.

[4] BÉNILAN, PH. and CRANDALL, M. G., *Regularizing effects of homogeneous evolution equation*, In Baltimore, editor, Contributions to Analysis and Geometry. (1981), 23–30.

[5] BÉNILAN, PH. and CRANDALL, M. G., *The Continuous Dependence on φ of Solutions of $u_t - \Delta\varphi(u) = 0$*, Ind. Uni. Math. J. (30) 2 (1981), 162–177.

[6] BÉNILAN, PH., CRANDALL, M. G. and SACKS, P., *Some L^1 Existence and Dependence Result for Semilinear Elliptic Equation Under Nonlinear Boundary Conditions*, Appl. Math. Optim. *17* (1988), 203–224.

[7] BÉNILAN, PH. EVANS, L. C. and GARIEPY, R. F., *On Some Singular Limits of Homogeneous Semigroups*, J. Evol. Equ. (the same volume).

[8] BÉNILAN, PH. and IGBIDA, N., *The mesa Problem for Neuman Boundary Value Problem*, (accepted in J. Differential Equations).

[9] BÉNILAN, PH. and IGBIDA, N., *Singular Limit for Perturbed Nonlinear Semigroup*, Comm. Applied Nonlinear Anal. (4) 3 (1996), 23–42.

[10] BÉNILAN, PH. and IGBIDA, N., *Limite de la solution de $u_t - \Delta u^m + \mathrm{div}\, F(u) = 0$ lorsque $m \to \infty$*, Rev. Mat. Complut. (13) 1 (2000), 195–205.

[11] BREZIS, H. and PAZY, A., *Convergence And Approximation of Semigroups of Nonlinear Operators in Banach Spaces*, J. Func. Anal. 9 (1972), 63–74.

[12] CAFFARELLI, L. A. and FRIEDMAN, A., *Asymptotic Behavior of Solution of $u_t = \Delta u^m$ as $m \to \infty$*, Indiana Univ. Math. J. (1987), 711–728.

[13] DIBENEDETTO, E. and FRIEDMAN, A., *The ill-posed Hele-Shaw model and the Stefan problem for super-cooled water*, Trans. Amer. Math. Soc. *282* (1984), 183–204.

[14] ELLIOT, C. M., HERRERO, M. A., KING, J. R. and OCKENDON, J. R., *The Mesa Patterns for $u_t = \nabla(u^m \nabla u)$ as $m \to \infty$*. IMA J. Appl. Math. *37* (1986), 147–154.

[15] EVANS, L. C., FELDMAN, M. and GARIEPY, R. F., *Fast/slow diffusion and collapsing sandpiles*, J. Differential Equations. *137* (1997), 166–209.

[16] FRIEDMAN, A. and HOLLIG, K. *On the Mesa Problem*, J. Math. Anal. Appl. *123* (1987), 564–571.

[17] FRIEDMAN, A. and HUANG, S., *Asymptotic Behavior of Solutions of $u_t = \Delta\varphi_m(u)$ as $m \to \infty$ with Inconsistent Initial Values*, In Analyse Mathématique et applications. Paris 1988.

[18] GIL, O. and QUIROS, F., *Boundary layer formation in the transition from porous medium to a Hele-Shaw flow*, Preprint UAM.

[19] GIL, O. and QUIROS, F., *Convergence of the porous media equation to Hele-Shaw*, Nonlinear Anal. TMA. (8) *44* (2001), 1111–1131.

[20] IGBIDA, N., *The mesa-limit of the porous medium equation and the Hele-Shaw problem*, Diff. and Integral Equations. (2) *15* (2002), 129–146.

[21] IGBIDA, N., *Limite Singulière de Problèmes d'Évolution Non Linéaires*, Thèse de doctorat, Université de Franche-Comté, Juin 1997.

[22] KING, J. R., D. phil. thesis, Oxford University, 1986.

[23] LACEY, A. A. OCKENDON, J. R. and TAYLER, J. B., *"Waiting time" solutions of a nonlinear diffusion equation*, SIAM J. Appl. Math. 6 (1982), 1252–1264.

[24] QUIROS, F. and VAZQUEZ, J. L., *Asymptotic convergence of the Stefan problem to Hele-Shaw*, Trans. Amer. Math. Soc. (2) *353* (2001), 609–634.

[25] ZELDOVICH, Y. N. and RAIZER, Y. P., *Physics of shock waves and high temperature hydrodynamics phenomena*, Academic Press 1986.

Noureddine Igbida
LAMFA, CNRS-UMR 6140
Université de Picardie Jules Verne
33 rue Saint Leu
80038 Amiens
France
e-mail: noureddine.igbida@u-picardie.fr

Philippe Bénilan
Equipe de Mathématiques
UMR-CNRS 6623
Université de Franche-Comté
Route de Gray
25030 Besançon
France

To access this journal online:
http://www.birkhauser.ch

J.evol.equ. 3 (2003) 225 – 236
1424–3199/03/020225 – 12
DOI 10.1007/s00028-003-0089-8
© Birkhäuser Verlag, Basel, 2003

Journal of Evolution Equations

On the regularizing effect of strongly increasing lower order terms

LUCIO BOCCARDO

Chiamato fui di là Ugo Ciappetta;
di me son nati i **Filippi** *e i Luigi*
per cui novellamente è Francia retta.
(Pur. XX)

Abstract. We show an existence result of bounded weak solutions for some semilinear Dirichlet problems, even if the right hand side belongs only to $L^1(\Omega)$. The model example is

$$\begin{cases} -\Delta u + h(u) = f(x) & \text{in } \Omega, \\ u = 0 & \text{on } \partial\Omega, \end{cases}$$

where Ω is a bounded open set in $\mathbb{R}^N$, $h(s)$ is a continuous and increasing function such that $\lim_{s \to \sigma} h(s) = +\infty$, for some $\sigma > 0$.

We also show a nonexistence result for some measures as data as in the model example

$$\begin{cases} -\Delta u + h(u) = \delta_{x_0} & \text{in } \Omega, \\ u = 0 & \text{on } \partial\Omega, \end{cases}$$

where δ_{x_0} is the Dirac mass in x_0 ($x_0 \in \Omega$).

1. Introduction

Let Ω be an open bounded set in $\mathbb{R}^N$ ($N > 2$), $p > 1$, and let A be a differential operator ($A : W_0^{1,p}(\Omega) \to W^{-1,p'}(\Omega)$, continuous, coercive and pseudomonotone). Let g be a Caratheodory function such that

$$\begin{cases} g(x,s) : \Omega \times \mathbb{R} \to \mathbb{R}, \ \rho(x) \in L^1(\Omega), \\ g(x,s)s \geq 0, \ |g(x,s)| \leq \rho(x)\gamma(s), \\ \gamma(s) \text{ continuous and increasing real function,} \end{cases}$$

It is known (see [1], [12], [13]) that the Dirichlet problem

$$u \in W_0^{1,p}(\Omega) : \ A(u) + g(x,u) = T \in W^{-1,p'}(\Omega), \tag{1}$$

Key words: Regularizing effect, singular data.

has a weak solution such that

$$g(x, u) \in L^1(\Omega), \quad ug(x, u) \in L^1(\Omega).$$

We recall that the main difficulty of the problem is due to the lower order term g. Indeed, if v belongs to $W_0^{1,p}(\Omega)$, $g(x, v)$, in general, does not belong to $L^1(\Omega)$.

However, in the simple case of lower order term of power type the existence theorem gives more information on the summability of u:

$$g(x, s) = |s|^{r-2}s, \ r > 1, \quad \text{implies that } u \in L^r(\Omega).$$

Thus, for r very large u is "almost bounded". Furthermore, we recall the following result.

THEOREM 1.1. ([11]) *Consider the boundary value problems*

$$u_r \in W_0^{1,p}(\Omega) \cap \in L^r(\Omega) : \ A(u_r) + |u_r|^{r-2}u_r = T \in W^{-1,p'}(\Omega). \tag{2}$$

As $r \to +\infty$, the sequence $\{u_r\}$ converges (up to a subsequence) in $W_0^{1,p}(\Omega)$ to a bounded function u^*, solution of the bilateral problem with operator A and datum T on the convex set

$$K = \{v \in W_0^{1,p}(\Omega) : |v| \le 1 \ in \ \Omega\}.$$

It is possible to study the behavior of the solutions of Dirichlet problems with the same differential operators, but with L^1 data (for any $s > p - 1$, the existence has been proved in [10]).

$$\begin{cases} A(u_s) + |u_s|^{s-1}u_s = f \in L^1(\Omega), \\ u_s \in W_0^{1,q}(\Omega) \cap L^s(\Omega), \quad q < \frac{ps}{s+1}. \end{cases} \tag{3}$$

Remark that $\frac{ps}{s+1} < p$ and $\frac{ps}{s+1} \to p$, as $s \to +\infty$.

THEOREM 1.2. ([15]) *As $s \to +\infty$, $\{u_s\}$ converges (up to a subsequence) in $W_0^{1,1}(\Omega)$ to a bounded function $u^{\#}$, solution of the bilateral problem with operator A and datum f on the convex set K.*

Theorems 1.1 and 1.2 say that the summability of the solutions increases as the power of the lower order term increases. Our goal is to prove that if the lower order term is defined through the composition with a continuous, but unbounded function on some real interval $[0, \sigma)$, the solutions are bounded.

Conversely, in [2], [4], is studied the limit, as $m \to \infty$, of the solutions of $u_t = \Delta u^m$ and $u_t - \Delta u^m + \text{div } F(u) = 0$, whose elliptic versions are the study of the behavior of the solutions of (1.2) and (1.3), as r, s tend to zero.

2. An existence result

Our setting is the following. The partial differential operator A is defined by

$$A(v) = -\operatorname{div}(a(x, v, \nabla v)),$$

where $a : \Omega \times \mathbb{R} \times \mathbb{R}^N \to \mathbb{R}^N$ is a Caratheodory function. We assume that there exist two real strictly positive continuous functions α, β, with $\alpha(s)$ decreasing, such that for every $s \in \mathbb{R}, \xi \in \mathbb{R}^N, \eta \in \mathbb{R}^N$ ($\xi \neq \eta$), and for almost every $x \in \Omega$,

$$\begin{cases} a(x, s, \xi)\xi \geq \alpha(s)|\xi|^p, \ p > 1, \\ |a(x, s, \xi)| \leq \beta(s)|\xi|^{p-1}, \\ [a(x, s, \xi) - a(x, s, \eta)][\xi - \eta] > 0. \end{cases} \tag{4}$$

On the lower order term g we assume that g is a Caratheodory function from $\Omega \times [0, \sigma) \to \mathbb{R}$ such that

$$\begin{cases} h(s) \leq g(x, s) \leq \rho(x)\gamma(s), \quad s \in [0, \sigma), \\ 0 \leq \rho \in L^1(\Omega), \\ h(0) = 0, \\ \lim_{s \to \sigma^-} h(s) = +\infty, \end{cases} \tag{5}$$

where $\gamma(s), h(s) : [0, \sigma) \mapsto \mathbb{R}^+$ are continuous, increasing real functions. Remark that the behavior of $h(s)$ implies that $\lim_{s \to \sigma^-} \gamma(s) = +\infty$.

Concerning the right hand side, we assume that

$$f \in L^1(\Omega), \quad f(x) \geq 0. \tag{6}$$

THEOREM 2.1. *Assume* (2.4), (2.5), (2.6). *Then there exists a bounded, weak solution of of the Dirichlet problem*

$$\begin{cases} A(u) + g(x, u) = f(x) \\ u \in W_0^{1,p}(\Omega) \cap L^\infty(\Omega), \\ 0 \leq u \leq \sigma, \ \operatorname{meas}\{x : u(x) = \sigma\} = 0. \end{cases} \tag{7}$$

Proof. Define $a_n(x, s, \xi) = a(x, T_n(s), \xi)$, $g_n(x, s) = T_n(g(x, s))$, and

$$A_n(v) = -\operatorname{div}(a_n(x, v, \nabla v)),$$

where, as usual,

$$T_k(t) = \begin{cases} t, & |t| \leq k, \\ k\frac{t}{|t|}. & |t| > k. \end{cases}$$

Thus, by (2.4), $A_n : W_0^{1,p}(\Omega) \to W^{-1,p'}(\Omega)$ is a continuous, coercive and pseudomonotone operator.

First step: $f \in L^\infty(\Omega)$.

Consider the boundary value problems

$$u_n \in W_0^{1,p}(\Omega) : A_n(u_n) + g_n(x, u_n) = f(x). \tag{8}$$

Thus, our differential problem is of the type (1.1) and we can apply the existence result of [12].

The use of $(u_n - h^{-1}(\|f\|_{L^\infty(\Omega)}))^+$ as test function in (2.8) implies that

$$\int_\Omega (g_n(x, u_n) - f(x))(u_n - h^{-1}(\|f\|_{L^\infty(\Omega)}))^+ \leq 0.$$

From the previous integral inequality we have

$$0 \geq \int_\Omega (T_n(h(u_n)) - f(x))(u_n - h^{-1}(\|f\|_{L^\infty(\Omega)}))^+$$

$$= \int_{\{h(u_n) \geq \|f\|_{L^\infty(\Omega)}\}} (T_n(h(u_n)) - f(x))(u_n - h^{-1}(\|f\|_{L^\infty(\Omega)}))$$

$$= \int_{\{n > h(u_n) \geq \|f\|_{L^\infty(\Omega)}\}} (h(u_n) - f(x))(u_n - h^{-1}(\|f\|_{L^\infty(\Omega)}))$$

$$+ \int_{\{h(u_n) \geq n \geq \|f\|_{L^\infty(\Omega)}\}} (n - f(x))(u_n - h^{-1}(\|f\|_{L^\infty(\Omega)})) \geq 0.$$

We thus obtain

$$0 \leq u_n \leq h^{-1}(\|f\|_{L^\infty(\Omega)}),$$

which implies, for some $\epsilon > 0$,

$$0 \leq u_n \leq \sigma - \epsilon, \tag{9}$$

$A_n(u_n) = A(u_n)$ (for $n > \sigma$) and (using u_n as test function in (8))

$$\alpha(\sigma) \|u_n\|_{W_0^{1,p}(\Omega)}^p \leq \sigma \|f\|_{L^1(\Omega)}, \tag{10}$$

and

$$\int_\Omega g_n(x, u_n) u_n \leq \sigma \|f\|_{L^1(\Omega)}. \tag{11}$$

Then there exists a subsequence (still denoted by u_{n_j}) and a function $u \in W_0^{1,p}(\Omega) \cap L^\infty(\Omega)$ such that u_{n_j} converges weakly in $W_0^{1,p}(\Omega)$ and a.e. to u.

In order to pass to the limit in the approximate equation, we are going to prove that

$$g_n(x, u_n) \to g(x, u) \qquad \text{strongly in } L^1(\Omega). \tag{12}$$

Since $g_n(x, u_n)$ converges almost everywhere to $g(x, u)$, it remains to prove the equi-integrability of the sequence $\{g_n(x, u_n)\}$.

For any measurable subset E of Ω, we have

$$\int_E g_n(x, u_n(x)) = \int_{E \cap \{x: s \le u_n(x) < \sigma - \epsilon\}} g_n(x, u_n(x))$$

$$\le \int_{E \cap \{x: s \le u_n(x) < \sigma - \epsilon\}} g(x, u_n(x)) \le \gamma(\sigma - \epsilon) \int_E \rho(x).$$

And so

$$\lim_{\text{meas}(E) \to 0} \int_E g_n(x, u_n(x)) = 0.$$

The above equi-integrability of $g_n(x, u_n(x))$ and the a.e. convergence to $g(x, u(x))$ imply that

$$g_n(x, u_n(x)) \to g(x, u(x)) \quad \text{in } L^1(\Omega).$$

Thus, we have proved that $\{g_n(x, u_n)\}$ is equi-integrable.

Moreover $\nabla u_{n_j}(x)$ converges a.e. to $\nabla u(x)$, since $A(u_n)$ is bounded in $L^1(\Omega)$ ($A(u_n) = f_n - g_n(x, u_n)$) (see [11]). The last two convergences imply that u_{n_j} converges strongly to u in $W_0^{1,q}(\Omega)$, $\forall q < p$. Then it is easy to pass to limit in (2.8), thus proving the existence of a solution u of (2.7).

Second step: $f \in L^1(\Omega)$.

We recall that the existence (and uniqueness) of solutions for Dirichlet problems with irregular data is studied in [3], [5], [6], [7], [8], [9], [10].

Consider the boundary value problems

$$u_n \in W_0^{1,p}(\Omega) : A(u_n) + g(x, u_n) = f_n(x), \tag{13}$$

where $f_n(x) = T_n(f(x))$. Step 1 gives the existence of u_n such that

$$0 \le u_n \le \sigma - \epsilon_n,$$

where ϵ_n may converge to zero.

In [5] the following inequality can be found:

$$\int_{\{x \in \Omega: s \le u_n(x) < \sigma\}} g(x, u_n) \le \int_{\{x \in \Omega: s \le u_n(x)\}} f_n, \quad \forall s < \sigma. \tag{14}$$

We use u_n as test function in (2.13) and we get estimates like (2.10), (2.11):

$$\alpha(\sigma)\|u_n\|^p_{W_0^{1,p}(\Omega)} \le \sigma \|f_n\|_{L^1(\Omega)} \le \sigma \|f\|_{L^1(\Omega)}, \tag{15}$$

$$\int_\Omega g(x, u_n) u_n \le \sigma \|f_n\|_{L^1(\Omega)} \le \sigma \|f\|_{L^1(\Omega)}. \tag{16}$$

Since

$$\int_{\{s \le u_n(x)\}} f_n = \int_{\{s \le u_n(x) < \sigma\}} f_n \le \int_{\{s \le u_n(x) < \sigma\}} f,$$

we deduce that

$$\int_{\{s \le u_n(x) < \sigma\}} g(x, u_n) \le \int_{\{s \le u_n(x) < \sigma\}} f.$$

For any measurable subset E of Ω, we have

$$\int_E g(x, u_n(x)) = \int_{E \cap \{x : s \le u_n(x) < \sigma\}} g(x, u_n(x)) + \int_{E \cap \{x : 0 \le u_n(x) < s\}} g(x, u_n(x))$$

$$\le \int_{\{s \le u_n(x) < \sigma\}} f + \gamma(s) \int_E \rho(x).$$

Now we use the first inequality of (2.5) in (2.16), which implies that

$$\int_{\{x : s \le u_n(x) < \sigma\}} h(s)\, s \le \int_{\{x : s \le u_n(x) < \sigma\}} g(x, u_n(x)) u_n(x)$$

$$\le \int_\Omega g(x, u_n) u_n \le \sigma \|f\|_{L^1(\Omega)}.$$

Thus

$$\lim_{s \to \sigma} \operatorname{meas}\{x : s \le u_n(x) < \sigma\} \le \lim_{s \to \sigma} \frac{\sigma}{h(s)\, s} \|f\|_{L^1(\Omega)} = 0,$$

thanks to the assumptions (2.5) on $h(s)$. Then, for any fixed $\epsilon > 0$, there exists $s_0 \in \mathbb{R}^+$ such that

$$\int_{\{s_0 \le u_n(x) < \sigma\}} f < \epsilon$$

and so

$$\lim_{\mathrm{meas}(E)\to 0} \int_E g(x, u_n(x)) \le \epsilon.$$

Thus, we have proved that $\{g(x, u_n)\}$ is equi-integrable. The above equi-integrability of $g(x, u_n(x))$ and the a.e. convergence to $g(x, u(x))$ imply that

$$g(x, u_n(x)) \to g(x, u(x)) \quad \text{in } L^1(\Omega).$$

Then there exists (again) a subsequence (still denoted by u_n) and a function $u \in W_0^{1,p}(\Omega)$ such that u converges weakly in $W_0^{1,p}(\Omega)$ and a.e. to u.

The setting is the same of (2.10), (2.11), (2.14), (2.12), with $g(x, u_n)$ instead of $g_n(x, u_n)$. Then the existence proof follows as in Step 1. $\qquad\square$

3. Measures data

In this section we shall study what happens if we try (as in the proof of Theorem 1.1) to approximate a measure λ with a sequence $\{f_n\}$ of $L^\infty(\Omega)$ functions, and to solve the corresponding problems with data f_n.

If λ is absolutely continuous with respect to the p–capacity, in [8] it is proved that $\lambda = f(x) + T$, with $f \in L^1(\Omega)$ and $T \in W^{-1,p'}(\Omega)$.

REMARKS 3.1. We point out that the existence result of of Theorem 2.1. still holds with the same proof if the right hand side of (2.7) is $f(x) + T$, with $f \in L^1(\Omega)$ and $T \in W^{-1,p'}(\Omega)$.

Thus the setting is new if λ is concentrated on a set E of zero p-capacity.

If we take the sequence $\{f_n\}$ bounded in $L^1(\Omega)$, then the same technique used in the proof of Theorem 2.1 yields that the corresponding sequence of solutions $\{u_n\}$ is bounded in $W_0^{1,p}(\Omega)$; hence, it converges weakly in $W_0^{1,p}(\Omega)$ towards some function. In the following theorem, we will see that $\{u_n\}$ converges to zero and zero cannot be a solution of the problem with datum λ. Hence a solution obtained by approximation does not exist.

We shall use the following result, proved in [14] (see also [9]), which gives a suitable sequence of cut-off functions.

In the following we will denote by ϵ_δ and $\epsilon_{n,\delta}$ respectively any sequence of real numbers such that

$$\lim_{\delta\to 0^+} \epsilon_\delta = 0, \qquad \lim_{\delta\to 0^+} \lim_{n\to+\infty} \epsilon_{n,\delta} = 0.$$

LEMMA 3.2. ([14]) *Let λ be a nonnegative measure in $\mathcal{M}_b(\Omega)$ which is concentrated on a set E of zero p-capacity. Then there exists a sequence $\{\psi_\delta\}$ of $C_0^\infty(\Omega)$ functions such that*

$$\int_\Omega |\nabla \psi_\delta|^p = \epsilon_\delta, \qquad 0 \le \psi_\delta \le 1, \qquad \int_\Omega (1 - \psi_\delta) d\lambda = \epsilon_\delta, \tag{17}$$

As a consequence of (3.17), we have that ψ_δ converges to zero both strongly in $W_0^{1,p}(\Omega)$, almost everywhere in Ω, and in the weak∗ topology of $L^\infty(\Omega)$.

The following nonexistence theorem strictly follows some results of [18] and [9].

THEOREM 3.3. *Let λ be a positive measure in $\mathcal{M}_b(\Omega)$, concentrated on a set E of zero p–capacity (that is such that $cap_p(E, \Omega) = 0$), and let $\{f_n\}$ be a sequence of positive $L^\infty(\Omega)$ functions such that*

$$\lim_{n \to +\infty} \int_\Omega f_n \varphi \, dx = \int_\Omega \varphi d\lambda \qquad \forall \varphi \in C^0(\bar{\Omega}). \tag{18}$$

Let u_n be a solution of the differential problem (2.13) with right hand side f_n. Then

$$T_k(u_n) \to 0 \qquad strongly \ in \ W_0^{1,p}(\Omega), \quad \forall k < \sigma.$$

Moreover,

$$\lim_{n \to +\infty} \int_\Omega g_n(x, u_n) \, \varphi \, dx = \int_\Omega \varphi d\lambda \qquad \forall \varphi \in C_0^1(\Omega) .$$

Proof. Since f_n is positive, u_n is positive. Thanks to (3.18), the sequence $\{f_n\}$ is bounded in $L^1(\Omega)$: $\|f_n\|_{L^1(\Omega)} \le c_0$. As in the previous theorem, we have that

$$0 \le u_n \le \sigma, \quad \text{meas}\{x \in \Omega : u_n(x) = \sigma\} = 0$$

and

$$\alpha(\sigma) \|u_n\|^p_{W_0^{1,p}(\Omega)} \le \sigma \|f\|_{L^1(\Omega)}.$$

On the other hand now the inequality (2.14) does not imply that $g(x, u_n)$ is L^1 compact. Anyway, there exists a subsequence, still denoted by u_n, and a function $u \in W_0^{1,p}(\Omega)$ such that u_n converges weakly in $W_0^{1,p}(\Omega)$ and a.e. to u and

$$a(x, u_n, \nabla u_n) \to H \qquad weakly \ in \ (L^{p'}(\Omega))^N;$$

the latter convergence is due to the fact that, by (2.4), $a(x, u_n, \nabla u_n)$ is bounded in $(L^{p'}(\Omega))^N$.

As first step, we choose as test function in the weak formulation of (2.13)

$$v = (k - T_k(u_n))\, \psi_\delta, \quad k < \sigma.$$

We obtain for $n > \sigma$:

$$-\int_\Omega a(x, T_k(u_n), \nabla T_k(u_n))\nabla T_k(u_n)\, \psi_\delta + \int_\Omega [a(x, u_n, \nabla u_n)\nabla\psi_\delta]\,(k - T_k(u_n))$$

$$\int_\Omega g_n(x, u_n)\,(k - T_k(u_n)))\, \psi_\delta = \int_\Omega f_n\,(k - T_k(u_n))\, \psi_\delta \geq 0.$$

Since $k - T_k(u_n)$ converges to $k - T_k(u)$ both in the weak$*$ topology of $L^\infty(\Omega)$ and almost everywhere in Ω, we have that $\nabla\psi_\delta\,(k - T_k(u_n))$ converges to $\nabla\psi_\delta\,(k - T_k(u))$ strongly in $(L^p(\Omega))^N$, and so

$$\int_\Omega [a(x, T_k(u_n), \nabla T_k(u_n))\nabla\psi_\delta]\,(k - T_k(u_n))$$

$$= \int_\Omega [H\nabla\psi_\delta]\,(k - T_k(u_n)) + \epsilon_n = \epsilon_{n,\delta}.$$

Moreover

$$\int_\Omega g_n(x, T_k(u_n))\,(k - T_k(u_n))\, \psi_\delta = \int_{\{0 \leq u_n < k\}} g_n(x, u_n)\,(k - u_n)\, \psi_\delta$$

and

$$\lim_{n\to\infty} \int_{\{0 \leq u_n < k\}} g_n(x, u_n)\,(k - u_n)\, \psi_\delta = \int_\Omega g(x, T_k(u))\,(k \quad T_k(u))\, \psi_\delta = c_\delta.$$

The conclusion of the first step is then

$$\int_\Omega a(x, T_k(u_n), \nabla T_k(u_n))\nabla T_k(u_n)\, \psi_\delta \leq \epsilon_{n,\delta}. \tag{19}$$

The second step consists in choosing as test function in (2.13) the function

$$v = T_k(u_n)(1 - \psi_\delta).$$

We get

$$\int_\Omega a(x, T_k(u_n), \nabla T_k(u_n))\nabla T_k(u_n)\,(1 - \psi_\delta) - \int_\Omega [a(x, u_n, \nabla u_n)\nabla\psi_\delta]\,T_k(u_n)$$

$$+ \int_\Omega g(x, u_n)\,T_k(u_n)\,(1 - \psi_\delta) = \int_\Omega f_n\,T_k(u_n)\,(1 - \psi_\delta).$$

Remark that

$$\int_\Omega g(x, u_n)\, T_k(u_n)\, (1 - \psi_\delta) \geq 0.$$

Working as before, we have

$$-\int_\Omega [a(x, u_n, \nabla u_n)\nabla \psi_\delta]\, T_k(u_n) = \epsilon_{n,\delta}.$$

On the other hand, we have

$$\int_\Omega f_n\, T_k(u_n)\, (1 - \psi_\delta) \leq k \int_\Omega f_n\, (1 - \psi_\delta)$$

and

$$\lim_{n\to\infty} k \int_\Omega f_n\, (1 - \psi_\delta) = k \int_\Omega (1 - \psi_\delta)d\lambda = k\,\epsilon_\delta$$

The conclusion of the second step is then

$$\int_\Omega a(x, T_k(u_n), \nabla T_k(u_n))\nabla T_k(u_n)\, (1 - \psi_\delta) \leq \epsilon_{n,\delta}. \tag{20}$$

Putting together (3.19) and (3.20) we easily obtain

$$\alpha(k) \int_\Omega |\nabla T_k(u_n)|^p \leq \int_\Omega a(x, T_k(u_n), \nabla T_k(u_n))\nabla T_k(u_n) \leq \epsilon_{n,\delta},$$

that is to say, $T_k(u_n)$, for every $k < \sigma$, converges strongly to zero in $W_0^{1,p}(\Omega)$. Since the limit is independent of the choice of the subsequence, the whole sequence $\{u_n\}$ is such that $T_k(u_n)$ converges to zero strongly in $W_0^{1,p}(\Omega)$. Thus, $u = 0$, and so u_n converges weakly to zero in $W_0^{1,p}(\Omega)$.

In order to prove the second part of the theorem, we observe that from the strong convergence to zero of $T_k(u_n)$ it follows the almost everywhere convergence to zero of ∇u_n, and this implies that $H \equiv 0$. Now we choose a test function φ in $C_0^1(\Omega)$ in (2.13); we obtain

$$\int_\Omega a(x, u_n, \nabla u_n)\nabla \varphi + \int_\Omega g(x, u_n)\, \varphi = \int_\Omega f_n\, \varphi. \tag{21}$$

Since $H = 0$, we have

$$\int_\Omega a(x, u_n, \nabla u_n)\nabla \varphi = \epsilon_n,$$

while

$$\int\limits_{\Omega} f_n\, \varphi = \int\limits_{\Omega} \varphi d\lambda + \epsilon_n.$$

Thus, from (3.21) we obtain by difference,

$$\lim_{n\to+\infty} \int\limits_{\Omega} g(x, u_n)\, \varphi = \int\limits_{\Omega} \varphi d\lambda,$$

for every φ in $C_0^1(\Omega)$, and this concludes the proof of the theorem. $\square$

Which contains the unpublished part of my lecture at the Fes Conference.

Acknowledgments

The author would like to thank Luigi Orsina (for the scientific help), Alessio Porretta, Chiara Leone (for several useful discussions on the subject of this paper) and the referee. Some results of this paper were presented in my lecture at the Fes Conference 1999.

Other problems with strongly increasing terms are studied in [16], [17].

REFERENCES

[1]   BÉNILAN, P., BREZIS, H., Haim, and Crandall, M.C., *A semilinear equation in* $L^1(R^N)$, Ann. Scuola Norm. Sup. Pisa Cl. Sci. (4) *2* (1975) no. 4, 523–555.

[2]   BÉNILAN, L., BOCCARDO, P., HERRERO, M.A., *On the limit of solutions of* $u_t = \Delta u^m$ *as* $m \to \infty$, Rend. Sem. Mat. Univ. Pol. Torino, fascicolo speciale (1989), 1–13.

[3]   BÉNILAN, P., BOCCARDO, L., GALLOUET, T., GARIEPY, R., PIERRE, M. and VAZQUEZ, J. L., *An* L^1 *theory of existence and uniqueness of solutions of nonlinear elliptic equations*, Annali Sc. Norm. Sup. Pisa. *22* (1995), 241–273.

[4]   BÉNILAN, P. and IGBIDA, N., *Limite de la solution de* $u_t - \Delta u^m + \operatorname{div} F(u) = 0$ *lorsque* $m \to \infty$, Rev. Mat. Complut. *13* (2000), 195–205.

[5]   BOCCARDO, L. and Gallouet, T., *Nonlinear elliptic and parabolic equations involving measure data*, J. Funct. Anal. *87* (1989), 149–169.

[6]   BOCCARDO, L. and Gallouet, T., *Nonlinear elliptic equations with right hand side measures*, Comm. P.D.E. *17* (1992), 641–655.

[7]   BOCCARDO, L. and Gallouet, T., *Strongly nonlinear elliptic equations having natural growth terms and* L^1 *data*, Nonlinear Anal. TMA. *19* (1992), 573–579.

[8]   BOCCARDO, L., GALLOUET, T. and ORSINA, L., *Existence and uniqueness of entropy solutions for nonlinear elliptic equations with measure data*, Ann. Inst. H. Poincaré Anal. Non Linéaire. *13* (1996), 539–551.

[9]   BOCCARDO, L., GALLOUET, T. and ORSINA, L., *Existence and nonexistence of solutions for some nonlinear elliptic equations*, J. Anal. Math. *73* (1997), 203–223.

[10] BOCCARDO, L., GALLOUET, T. and VAZQUEZ, J.L., *Nonlinear elliptic equations in* $\mathbb{R}^N$ *without growth restrictions on the data*, J. Diff. Eq. *105* (1993), 334–363.

[11] BOCCARDO, L. and MURAT, F., *Increasing powers leads to bilateral problems*, in Composite media and homogenization theory. G. Dal Maso and G. Dell'Antonio ed., World Scientific. 1995.

[12] BREZIS, H. and BROWDER, F.E., *Some properties of higher order Sobolev spaces*, J. Math. Pures Appl. *61* (1982), 245–259.

[13] BREZIS, H. and STRAUSS, W.A., *Semi-linear second-order elliptic equations in L^1*, J. Math. Soc. Japan 25 (1973), 565–590.

[14] DAL MASO, G., MURAT, F., ORSINA, L. and PRIGNET, A., *Renormalized solutions for elliptic equations with general measure data*, Ann. Sc. Norm. Sup. Pisa. 28 (1999) no. 4, 741–808.

[15] DALL'AGLIO, L. and ORSINA, L., *On the limit of some nonlinear elliptic equations involving increasing powers*, Asymptot. Anal. 14 (1997), 49–71.

[16] LEONE, C., Existence and non-existence of solutions to a class of Dirichlet problems involving measures, preprint.

[17] ORSINA, L., Existence results for some elliptic equations with unbounded coefficients, preprint.

[18] ORSINA, L. and PRIGNET, A., *Nonexistence of solutions for some nonlinear elliptic equations involving measures*, Proc. Roy. Soc. Edinburgh Sect. A 130 (2000) no. 1, 167–187.

Lucio Boccardo
Dipartimento di Matematica
Università di Roma I
Piazza A. Moro 2
00185 Roma
Italia
e-mail: boccardo@mat.uniroma1.it

To access this journal online:
http://www.birkhauser.ch

J.evol.equ. 3 (2003) 237 – 246
1424–3199/03/020237 – 10
DOI 10.1007/s00028-003-0092-0
© Birkhäuser Verlag, Basel, 2003

Journal of Evolution Equations

Global smooth solutions for a quasilinear fractional evolution equation

EMILIA BAZHLEKOVA AND PHILIPPE CLÉMENT

Dedicated to the memory of Philippe Bénilan

Abstract. The global existence of smooth solutions to a class of quasilinear fractional evolution equations is proved. The proofs are based on $L^p(L^q)$ maximal regularity results for the corresponding linear equations.

1. Introduction and main result

The purpose of this note is to establish the global existence of a smooth solution of the quasilinear partial integrodifferential equation:

$$u_t(t, x) = \int_0^t g_\beta(t - s)[\sigma(u_x(s, x))_x + f(s, x)]\, ds, \tag{1}$$

where $0 < t \leq T$, $-\infty < a < x < b < +\infty$, $\sigma : \mathbf{R} \to \mathbf{R}$ is a sufficiently smooth strictly increasing function and $f \in L^p(0, T; L^q(a, b))$. We consider kernels g_β of the form

$$g_\beta(t) := \frac{t^{\beta-1}}{\Gamma(\beta)}, \quad t > 0, \beta \in (0, 1),$$

so that (1) corresponds to a fractional differential equation of order $\beta + 1$. We prescribe zero boundary and initial conditions

$$u(t, a) = u(t, b) = 0, \ t \in (0, T]; \quad u(0, x) = 0, \ x \in [a, b]. \tag{2}$$

The reason for studying this equation is that it and some variants of it appear in mathematical models of viscoelasticity. It is also of interest in itself when one explores the borderline between the quasilinear diffusion ($\beta = 0$, then $g_0(t)$ is the "Dirac delta function") and quasilinear wave propagation ($\beta = 1$, $g_1(t) := 1$). The case $\beta = 1$ (the completely elastic case in mechanics) corresponds to the quasilinear wave equation

$$u_{tt}(t, x) = \sigma(u_x(t, x))_x + f(t, x) \tag{3}$$

Mathematics Subject Classification 2000: Primary 45K05; Secondary 35M99.
Key words: fractional derivative, integrodifferential equation, $L^p(L^q)$ maximal regularity.

together with the side conditions (2) and the additional initial condition $u_t(0, x) = 0$, $x \in [a, b]$. It is known that in this case smooth data may develop singularities in the derivatives in finite time. On the other hand, as is well known, the quasilinear diffusion equation

$$u_t(t, x) = \sigma(u_x(t, x))_x + f(t, x) \tag{4}$$

together with (2) (that corresponds to the completely viscous case) has always (under reasonable assumptions on σ) a global smooth solution. The question is whether the convolution with g_β, $\beta \in (0, 1)$, in the right hand-side of (4), which leads to (1), also will prevent the formation of singularities in the solution. Observe that when $\beta > 0$ the maximum principle does not hold and apriori estimates are more difficult to be established.

While the behaviour of the solution of (1) in the linear case $\sigma(y) = c_0 y$, $c_0 > 0$, is reasonably well understood, only partial results exist for the quasilinear equations of this type. The existence of global smooth solutions of (1) in a very general $L^p(L^q)$ setting is proved in [3] provided $2/p + 1/q < 1$ and $\beta < 1/3$ (although the second assumption is probably technical, it is not clear how to weaken it). In [5] the existence of global smooth solutions is proved for the Hilbert space case $L^2(L^2)$ under an assumption on the deviation of σ from a linear function. We make a similar assumption and apply $L^p(L^q)$ estimates to establish the existence of a global smooth solution of (1) in $L^p(0, T; L^q(a, b))$ for all $\beta \in (0, 1)$ and p and q satisfying $2/(\beta p + p) + 1/q < 1$. The proofs are based on $L^p(L^q)$ maximal regularity results for the corresponding linear problem.

To formulate precisely the main result for the quasilinear problem (1), (2), we need some auxiliary propositions concerning the linear case $\sigma(y) = c_0 y$, $c_0 > 0$.

First some notation has to be introduced and some basic assumptions have to be fixed. Denote for shortness $I := [0, T]$, $J := [a, b]$. Let X be a Banach space. By $\mathcal{B}(X)$ we denote the space of all bounded linear operators in X. Let $1 < p, q < \infty$. The usual notation $W^{m,q}$ for Sobolev spaces of vector-valued functions will be employed.

If $\alpha > 0$, the Bessel potential spaces $H^{\alpha,p}$ are defined as follows ($\widetilde{f}$ denotes the Fourier transform of f in the space of tempered distributions $\mathcal{S}'(\mathbf{R}; X)$):

$$H^{\alpha,p}(\mathbf{R}; X) := \{f \in L^p(\mathbf{R}; X) | \, \exists f_\alpha$$
$$\in L^p(\mathbf{R}; X) : \widetilde{f_\alpha} = (1 + | \cdot |^2)^{\alpha/2} \widetilde{f} \text{ in } \mathcal{S}'(\mathbf{R}; X)\}$$

with

$$\|f\|_{H^{\alpha,p}(\mathbf{R};X)} := \|f_\alpha\|_{L^p(\mathbf{R};X)}.$$

The Bessel potential spaces on I are defined by

$$H^{\alpha,p}(I; X) := \{f = g|_I, \, g \in H^{\alpha,p}(\mathbf{R}; X)\}$$

with

$$\|f\|_{H^{\alpha,p}(I;X)} := \inf_{g \in H^{\alpha,p}(\mathbf{R};X), g|_I = f} \|g\|_{H^{\alpha,p}(\mathbf{R};X)}.$$

If $\alpha - 1/p \neq 0, 1, 2, \ldots$, denote

$$H_0^{\alpha,p}(I; X) := \{f \in H^{\alpha,p}(I; X),\ f^{(k)}(0) = 0,\ k = 0, 1, \ldots, [\alpha - 1/p]\}.$$

For m integer we have $H^{m,p}(I; X) = W^{m,p}(I; X)$.

Recall for the sake of completeness the notion of nonnegative operator on X ([6]). A linear operator $A : D(A) \subset X \to X$ is nonnegative if $(-\infty, 0) \subset \rho(A)$ (the resolvent set of A) and

$$\sup_{\lambda > 0} \|\lambda(\lambda I + A)^{-1}\|_{\mathcal{B}(X)} < \infty.$$

Denote

$$\phi_A := \sup \left\{ \phi \in [0, \pi] \Big|\ \sup_{|\arg(\lambda)| \leq \phi, \lambda \neq 0} \|\lambda(\lambda I + A)^{-1}\|_{\mathcal{B}(X)} < \infty \right\}.$$

The number $\omega_A := \pi - \phi_A$ is usually called spectral angle of A.

If A is an operator on X with bounded imaginary powers (see e.g. [7]) and $\theta_A = \inf\{\theta \geq 0 : \exists C \geq 1,\ \|A^{is}\|_{\mathcal{B}(X)} \leq Ce^{\theta|s|},\ s \in \mathbf{R}\}$, we write $A \in \mathcal{BIP}(X; \theta_A)$.

First we investigate the linear equation corresponding to (1), where $c_0 > 0$:

$$u_t(t, x) = \int_0^t g_\beta(t - s)[c_0 u_{xx}(s, x) + f(s, x)]\,ds, \quad t \in I,\ x \in J, \tag{5}$$

with boundary and initial data (2). Denote

$$W_\#^{2,q}(J) := \{u \in W^{2,q}(J)|\ u(a) = u(b) = 0\}$$

and

$$E_{p,q,\beta} := H_0^{\beta+1,p}(I; L^q(J)) \cap L^p(I; W_\#^{2,q}(J)), \quad \beta \neq 1/p.$$

PROPOSITION 1.1. *Let* $1 < p, q < \infty$, $\beta \in (0, 1)$ *and* $\beta \neq 1/p$. *Then for any* $f \in L^p(I; L^q(J))$ *there is a unique solution* $u \in E_{p,q,\beta}$ *of* (5), (2). *Moreover, there exists a constant* $C_{p,q,\beta} > 1$ *independent of* T *and* c_0, *such that*

$$\|u\|_{H^{\beta+1,p}(I;L^q(J))} + \|c_0 u_{xx}\|_{L^p(I;L^q(J))} \leq C_{p,q,\beta}\|f\|_{L^p(I;L^q(J))}. \tag{6}$$

PROPOSITION 1.2. *Assume* $1 < p, q < \infty$, $\beta \in (0, 1)$, $\beta \neq 1/p$ *and*

$$\frac{2}{\beta p + p} + \frac{1}{q} < 1. \tag{7}$$

Then

$$E_{p,q,\beta} \hookrightarrow C(I; C^1(J)), \tag{8}$$

where the embedding is compact.

Next we formulate the main result of this paper concerning the quasilinear equation (1). Suppose that

$$\sigma \in C^1(\mathbf{R}) \text{ and } 0 < \sigma_0 \leq \sigma'(y) \leq \sigma_1 < \infty, \quad y \in \mathbf{R}, \tag{9}$$

for some constants σ_0 and σ_1, satisfying

$$\frac{\sigma_1}{\sigma_0} < \frac{C_{p,q,\beta} + 1}{C_{p,q,\beta} - 1}, \tag{10}$$

where $C_{p,q,\beta}$ is the constant from (6).

THEOREM 1.3. *Assume $\beta \in (0, 1)$ and let $1 < p, q < \infty$ be such that $\beta \neq 1/p$ and (7) holds. Let σ obey (9), where σ_0 and σ_1 satisfy (10). Then for any $f \in L^p(I; L^q(J))$ there exists a unique function $u \in E_{p,q,\beta}$ which is a solution to the problem (1), (2).*

2. Proofs

For the proofs it is convenient to rewrite problem (1) as a differential equation of fractional order $\alpha = \beta+1 \in (1, 2)$. To this aim we give first some definitions. Let X be an UMD space. The operators of fractional differentiation of order $\alpha \in (1, 2)$ are defined on $L^p(I; X)$, $1 < p < \infty$, by:

$$Ł_\alpha u(t) := \frac{d^2}{dt^2} \int_0^t g_{2-\alpha}(t - s)u(s)\, ds, \quad D(\mathcal{L}_\alpha) := H_0^{\alpha,p}(I; X).$$

The operators $\mathcal{L}_\alpha$ are nonnegative with spectral angles

$$\omega_{\mathcal{L}_\alpha} = \alpha \frac{\pi}{2}$$

and $0 \in \rho(\mathcal{L}_\alpha)$ (see e.g. [1], Lemma 1.8, or [7]).

Denote by $\mathcal{L}_{\alpha,\infty}$ the fractional derivative in $L^p(\mathbf{R}; X)$, $1 < p < \infty$, defined as follows. Let $u \in C_c^2(\mathbf{R}; X)$, the space of twice continuously differentiable functions with compact support, which is dense in $L^p(\mathbf{R}; X)$. Define

$$\mathcal{L}_{\alpha,\infty} u(t) := \frac{d^2}{dt^2} \int_{-\infty}^t g_{2-\alpha}(t - s)u(s)\, ds, \quad u \in C_c^2(\mathbf{R}).$$

This operator is closable, and we take its closure in $L^p(\mathbf{R}; X)$ as definition of fractional derivative on $L^p(\mathbf{R}; X)$. We use the same notation $\mathcal{L}_{\alpha,\infty}$ for the closure. We have (see e.g. [1] or [7]) $D(\mathcal{L}_{\alpha,\infty}) = H^{\alpha,p}(\mathbf{R}; X)$, $\mathcal{L}_{\alpha,\infty}$ is nonnegative, $\omega_{\mathcal{L}_{\alpha,\infty}} = \alpha\frac{\pi}{2}$, $(0 \notin \rho(\mathcal{L}_{\alpha,\infty}))$, $\mathcal{L}_{\alpha,\infty} \in \mathcal{BIP}(L^p(\mathbf{R}; X), \alpha\frac{\pi}{2})$.

If $A : D(A) \subset X \to X$ is closed then its extensions $\mathcal{A}$ to $L^p(I; X)$ and $\mathcal{A}_\infty$ to $L^p(\mathbf{R}; X)$ are defined as usual by

$$D(\mathcal{A}) := L^p(I; D(A)), \quad (\mathcal{A}u)(t) := Au(t), \ u \in D(\mathcal{A}), t \in I \text{ a.e. } ;$$

$$D(\mathcal{A}_\infty) := L^p(\mathbf{R}; D(A)), \quad (\mathcal{A}_\infty u)(t) := Au(t), \ u \in D(\mathcal{A}_\infty), t \in \mathbf{R} \text{ a.e. } .$$

If E is a Banach space, $\mathcal{A} : D(\mathcal{A}) \subset E \to E$ and $\mathcal{B} : D(\mathcal{B}) \subset E \to E$, the pair of closed operators $(\mathcal{A}, \mathcal{B})$ is called coercively positive in E if for any $f \in E$ and $\lambda > 0$ the problem

$$\lambda \mathcal{A}u + \mathcal{B}u = f$$

has a unique solution $u \in D(\mathcal{A}) \cap D(\mathcal{B})$ and there exists $M > 1$ such that for all $\lambda > 0$

$$\|\lambda \mathcal{A}u\| + \|\mathcal{B}u\| \leq M \|\lambda \mathcal{A}u + \mathcal{B}u\|.$$

The following result is well known (for a proof see e.g. [1], Cor. 4.12 or [7]).

LEMMA 2.1. *Let $\alpha \in (1, 2)$, X be an UMD space and $A \in \mathcal{BIP}(X; \theta_A)$ with $0 \in \rho(A)$ and*

$$\theta_A < \pi(1 - \alpha/2). \tag{11}$$

Then $(\mathcal{L}_\alpha, \mathcal{A})$ and $(\mathcal{L}_{\alpha,\infty}, \mathcal{A}_\infty)$ are coercively positive pairs in $L^p(I; X)$ and in $L^p(\mathbf{R}; X)$, respectively.

Proof of Proposition 1.1. Denote

$$Au := -c_0 u_{xx}, \quad D(A) = W^{2,q}_{\#}(J). \tag{12}$$

Consider first the problem:

$$\mathcal{L}_{\beta+1,\infty} u + \mathcal{A}_\infty u = f_0, \tag{13}$$

where f_0 is the extension of f with 0 outside $[0, T]$. Since

$$A \in \mathcal{BIP}(L^q(J), 0)$$

(see e.g. [8]), then according to Lemma 2.1, $(\mathcal{L}_{\beta+1,\infty}, \mathcal{A}_\infty)$ is a coercively positive pair of operators in $L^p(\mathbf{R}; L^q(J))$. Therefore, for any $f_0 \in L^p(\mathbf{R}; L^q(J))$ there is a unique solution $u \in H^{\beta+1,p}(\mathbf{R}; L^q(J)) \cap L^p(\mathbf{R}; W^{2,q}_{\#}(J))$ of (13), satisfying

$$\|\mathcal{L}_{\beta+1,\infty} u\|_{L^p(\mathbf{R};L^q(J))} + \|\mathcal{A}_\infty u\|_{L^p(\mathbf{R};L^q(J))} \leq C_{p,q,\beta} \|f_0\|_{L^p(\mathbf{R};L^q(J))}, \tag{14}$$

where $C_{p,q,\beta} > 1$ does not depend on c_0. Then the restriction of the solution of (13) to I (denote it again by u) is a solution of

$$\mathcal{L}_{\beta+1}u + \mathcal{A}u = f \tag{15}$$

and it satisfies

$$\|\mathcal{L}_{\beta+1}u\|_{L^p(I;L^q(J))} + \|\mathcal{A}u\|_{L^p(I;L^q(J))} \le C_{p,q,\beta}\|f\|_{L^p(I;L^q(J))} \tag{16}$$

with the same constant $C_{p,q,\beta}$. Since $\|\mathcal{L}_{\beta+1}u\|_{L^p(I;L^q(J))}$ is an equivalent norm in $H^{\beta+1,p}(I; L^q(J))$ (see e.g. [1], Section 1.4), estimate (6) holds for the solution of (15). We will show that this function satisfies also (5) and (2). Convolving both sides of (15) by $g_{\beta+1}$ and taking into account that $g_{\beta+1} * \mathcal{L}_{\beta+1}u = u$ when $u \in H_0^{\beta+1,p}(I; L^q(J))$ we obtain

$$u = g_{\beta+1} * (-\mathcal{A}u + f). \tag{17}$$

This implies $u(0, x) = 0$. Moreover, after differentiation of both sides w.r.t. t we obtain (5). It remains to prove uniqueness of solutions to (17). If $u \in L^p(I; W_{\#}^{2,q}(J))$ satisfies (17) with $f = 0$, then $u \in H_0^{\beta+1,p}(I; L^q(J))$ and satisfies (15) with $f = 0$. From Lemma 2.1 we obtain $u = 0$.

In the proof of Proposition 2 we apply the following result of Sobolevskii [9], a useful consequence of the property 'coercively positive pair':

LEMMA 2.2. *Let $\mathcal{A}$ and $\mathcal{B}$ be two resolvent commuting nonnegative operators densely defined in X with spectral angles $\omega_{\mathcal{A}}$ and $\omega_{\mathcal{B}}$. Assume the following conditions are satisfied:*

(a) $0 \in \rho(\mathcal{A}) \cap \rho(\mathcal{B})$;
(b) $\omega_{\mathcal{A}} + \omega_{\mathcal{B}} < \pi$;
(c) $(\mathcal{A}, \mathcal{B})$ is a coercively positive pair.

Then for any $\theta \in (0, 1)$, $D(\mathcal{A}) \cap D(\mathcal{B}) \subset D(\mathcal{A}^\theta \mathcal{B}^{1-\theta}) = D(\mathcal{B}^{1-\theta}\mathcal{A}^\theta)$ and there is $M_\theta > 0$ such that for all $u \in D(\mathcal{A}) \cap D(\mathcal{B})$ the following inequality holds:

$$\|\mathcal{A}^\theta \mathcal{B}^{1-\theta}u\| \le M_\theta \|\mathcal{A}u + \mathcal{B}u\|.$$

Proof of Proposition 1.2. Let as above $\mathcal{A}$ be the extension of the operator A, defined by (12), to $L^p(I; L^q(J))$. Then $\mathcal{A}$ is nonnegative, $0 \in \rho(\mathcal{A})$, the spectral angle $\omega_{\mathcal{A}} = 0$ and $\mathcal{A} \in \mathcal{BIP}(L^p(I; L^q(J)), 0)$. Therefore the pair $(\mathcal{L}_{\beta+1}, \mathcal{A})$ satisfies the conditions of Lemma 2.2. Hence, for any $\theta \in (0, 1)$ the following embedding holds:

$$E_{p,q,\beta} = D(\mathcal{L}_{\beta+1}) \cap D(\mathcal{A}) \hookrightarrow D(\mathcal{L}_{\beta+1}^\theta \mathcal{A}^{1-\theta}). \tag{18}$$

Recall that for any operator $\mathcal{B} : D(\mathcal{B}) \subset E \to E$ with bounded imaginary powers,

$$D(\mathcal{B}^\theta) = [E, D(\mathcal{B})]_\theta,$$

the complex interpolation space between E and $D(\mathcal{B})$ of order θ ([10], Section 1.15.3). Therefore

$$D(\mathcal{A}^{1-\theta}) = [L^p(I; L^q(J)), L^p(I; W_\#^{2,q}(J))]_{1-\theta}$$
$$= L^p(I; [L^q(J), W_\#^{2,q}(J)]_{1-\theta}) \hookrightarrow L^p(I; H^{2(1-\theta),q}(J))$$

and

$$D(\mathcal{L}_{\beta+1}^\theta) = [L^p(I; L^q(J)), H_0^{\beta+1,p}(I; L^q(J))]_\theta \hookrightarrow H^{(\beta+1)\theta,p}(I; L^q(J)).$$

Then (18) implies

$$E_{p,q,\beta} \hookrightarrow H^{(\beta+1)\theta,p}(I; H^{2(1-\theta),q}(J)).$$

Inequality (7) gives $0 < \frac{1}{(\beta+1)p} < \frac{q-1}{2q} < 1$. Take $\theta \in (\frac{1}{(\beta+1)p}, \frac{q-1}{2q})$. Then

$$\nu := (\beta+1)\theta - 1/p > 0, \quad \mu := 2(1-\theta) - 1/q > 1.$$

Therefore

$$H^{(\beta+1)\theta,p}(I; H^{2(1-\theta),q}(J)) \hookrightarrow C^\nu(I; C^\mu(J)) \hookrightarrow C(I; C^1(J)).$$

The first embedding follows from the fact that if X is an UMD space then (see e.g. [1], Section 1.4):

$$H^{\gamma,p}(I; X) \hookrightarrow C^{\gamma-1/p}(I; X), \quad \gamma > 1/p, \ \gamma - 1/p \notin \mathbf{N}.$$

The second embedding is compact and it is implied by the Arzelà-Ascoli theorem. Therefore the embedding (8) holds and it is compact.

Proof of Theorem 1.3. Take c_0 to be the mean of σ_0 and σ_1:

$$c_0 := \frac{\sigma_0 + \sigma_1}{2}. \tag{19}$$

For $v \in E_{p,q,\beta}$ define the function

$$M(v) := \frac{1}{c_0}\sigma'(v_x). \tag{20}$$

It follows from (8) and continuity of σ' that $M(v) \in C(I \times J)$ for $v \in E_{p,q,\beta}$. Moreover, (9) and (19) imply

$$\|1 - M(v)\|_{C(I \times J)} \le \frac{\sigma_1 - \sigma_0}{\sigma_1 + \sigma_0}. \tag{21}$$

Denote also by the same symbol $M(v)$ the multiplication operator acting on $L^q(J)$:

$$M(v)u := \frac{1}{c_0}\sigma'(v_x)u, \ u \in L^q(J).$$

Let $\mathcal{M}(v)$ be its extension to $L^p(I; L^q(J))$, i.e. $\mathcal{M}(v) \in \mathcal{B}(L^p(I; L^q(J)))$.

Our ultimate goal is to find a solution $u \in E_{p,q,\beta}$ to the problem

$$\mathcal{L}_{\beta+1} u + \mathcal{M}(u)\mathcal{A}u = f. \tag{22}$$

We do this by applying the Schauder fixed point theorem to the equation

$$\mathcal{L}_{\beta+1} u + \mathcal{M}(v)\mathcal{A}u = f, \tag{23}$$

where $v \in E_{p,q,\beta}$. Set

$$\delta_{p,q,\beta} := \frac{\sigma_1 - \sigma_0}{\sigma_1 + \sigma_0} C_{p,q,\beta}, \tag{24}$$

where $C_{p,q,\beta}$ is the constant in (6). Thanks to assumption (10), $\delta_{p,q,\beta} < 1$. We prove successively

(a) for any $v \in E_{p,q,\beta}$ there exists a unique $u \in E_{p,q,\beta}$ satisfying (23). Denote the corresponding mapping $v \to u$ by $u =: Gv$. Let K be the set consisting of all $u \in E_{p,q,\beta}$ for which

$$\|\mathcal{L}_{\beta+1} u\|_{L^p(I;L^q(J))} + (1 - \delta_{p,q,\beta})\|\mathcal{A}u\|_{L^p(I;L^q(J))} \leq C_{p,q,\beta}\|f\|_{L^p(I;L^q(J))}.$$

Then K is a bounded closed convex set in $E_{p,q,\beta}$ and G maps it into itself;

(b) $G : E_{p,q,\beta} \to E_{p,q,\beta}$ is a continuous mapping;

(c) G maps bounded sets in $E_{p,q,\beta}$ in relatively compact sets in $E_{p,q,\beta}$.

(a) Define the operators $L(0)$ and $L(1)$ on $E_{p,q,\beta}$ by

$$L(0)u := \mathcal{L}_{\beta+1} u + \mathcal{A}u; \quad L(1)u := \mathcal{L}_{\beta+1} u + \mathcal{M}(v)\mathcal{A}u, \quad u, v \in E_{p,q,\beta}.$$

They are bounded linear operators from $E_{p,q,\beta}$ to $L^p(I; L^q(J))$. For $\rho \in [0, 1]$ set

$$L(\rho) := (1 - \rho)L(0) + \rho L(1).$$

We show that there exists a constant $C' > 1$, independent of ρ, such that

$$\|u\|_{E_{p,q,\beta}} := \|\mathcal{L}_{\beta+1} u\|_{L^p(I;L^q(J))} + \|\mathcal{A}u\|_{L^p(I;L^q(J))} \leq C'\|L(\rho)u\|_{L^p(I;L^q(J))}. \tag{25}$$

Indeed, let $L(\rho)u = f$, or, equivalently

$$\mathcal{L}_{\beta+1} u + \mathcal{A}u = f + \rho(1 - \mathcal{M}(v))\mathcal{A}u.$$

Applying (16), (21) and (24), we obtain

$$\|u\|_{E_{p,q,\beta}} \leq C_{p,q,\beta}\|f\|_{L^p(I;L^q(J))} + \rho\frac{\sigma_1 - \sigma_0}{\sigma_1 + \sigma_0}C_{p,q,\beta}\|\mathcal{A}u\|_{L^p(I;L^q(J))}$$

$$\leq C_{p,q,\beta}\|f\|_{L^p(I;L^q(J))} + \delta_{p,q,\beta}\|\mathcal{A}u\|_{L^p(I;L^q(J))}.$$

Therefore

$$\|\mathcal{L}_{\beta+1}u\|_{L^p(I;L^q(J))} + (1 - \delta_{p,q,\beta})\|\mathcal{A}u\|_{L^p(L^q)} \leq C_{p,q,\beta}\|f\|_{L^p(I;L^q(J))}, \tag{26}$$

and so, (25) is satisfied with $C' := \frac{C_{p,q,\beta}}{1-\delta_{p,q,\beta}}$. Since $L(0) : E_{p,q,\beta} \to L^p(I; L^q(J))$ is an isomorphism, applying the method of continuity ([4], Theorem 5.2, p. 70) it follows that $L(1) : E_{p,q,\beta} \to L^p(I; L^q(J))$ is also an isomorphism. Thus, for any $v \in E_{p,q,\beta}$ and $f \in L^p(I; L^q(J))$ (23) has a unique solution $u \in E_{p,q,\beta}$. Moreover, (26) shows that $u \in K$. The properties of the set K can be proved straightforwardly.

(b) Let $v_n \to v$ in $E_{p,q,\beta}$ and take $u_n := G(v_n)$, $u := G(v)$. Hence, Equation (23) is satisfied also with u replaced by u_n and v replaced by v_n. Therefore,

$$\mathcal{L}_{\beta+1}(u_n - u) + \mathcal{M}(v)\mathcal{A}(u_n - u) = (\mathcal{M}(v) - \mathcal{M}(v_n))\mathcal{A}u_n$$

and according to (26)

$$\|u_n - u\|_{E_{p,q,\beta}} \leq \frac{C_{p,q,\beta}}{1 - \delta_{p,q,\beta}}\|M(v) - M(v_n)\|_{C(I \times J)}\|\mathcal{A}u_n\|_{L^p(I;L^q(J))}. \tag{27}$$

Again by (26), $\|\mathcal{A}u_n\|_{L^p(I;L^q(J))}$ is bounded by a constant, which does not depend on n. Since $v_n \to v$ in $E_{p,q,\beta}$ and because of the embedding (8) and the continuity of σ', it follows that

$$\|M(v_n) - M(v)\|_{C(I \times J)} = \|\sigma'((v_n)_x) - \sigma'(v_x)\|_{C(I \times J)} \to 0, \quad n \to \infty. \tag{28}$$

So, the right-hand side of (27) tends to 0 as $n \to \infty$. Therefore $u_n \to u$ in $E_{p,q,\beta}$ and (b) is proved.

(c) Let v_n be a bounded sequence in $E_{p,q,\beta}$. Since (8) is a compact embedding, there exists a subsequence $v_{n,k}$, converging to some v in $C(I; C^1(J))$. Therefore, (28) is satisfied with v_n replaced by $v_{n,k}$. Applying now the same argument as in the proof of (b) we obtain from (27) (with v_n replaced by $v_{n,k}$, v by $v_{n,\ell}$, u_n by $u_{n,k}$, and u by $u_{n,\ell}$) that $u_{n,k}$ is a Cauchy sequence in $E_{p,q,\beta}$.

Properties (a), (b) and (c) imply that G is a completely continuous operator which maps the bounded closed convex set K into itself. According to the Schauder fixed point theorem there exists at least one function $u \in K$ such that $Gu = u$ and this is a solution of (22). In this way we proved the global smooth solvability of (1), (2). The uniqueness of the solution can be proved as in [3].

REMARK. Condition (7) shows that this approach is not applicable in the purely Hilbert space case $p = q = 2$.

Acknowledgment

The authors would like to thank the referee for his valuable comments.

REFERENCES

[1] BAJLEKOVA, E., *Fractional Evolution Equations in Banach Spaces*. Ph.D. Thesis, Eindhoven University of Technology, Eindhoven 2001.

[2] DA PRATO, G. and GRISVARD, P., *Sommes d'opérateurs linéaires et équations différentielles opérationelles*. J. Math. Pures Appl. *54* (1975), 305–387.

[3] ENGLER, H., *Global smooth solutions for a class of parabolic integro-differential equations*. Trans. Amer. Math. Soc. *348* (1996) no.1, 267–290.

[4] GILBARG, D. and TRUDINGER, N. S., *Elliptic Partial Differential Equations of Second Order*. Springer, Berlin 1983.

[5] GRIPENBERG, G., *Global existence of solutions of Volterra integrodifferential equations of parabolic type*. J. Diff. Eq. *102* (1993), 382–390.

[6] KOMATSU, H., *Fractional powers of operators*. Pacif. J. Math. *19* (1966), 285–346.

[7] PRÜSS, J., *Evolutionary Integral Equations and Applications*. Birkhäuser, Basel, Boston, Berlin 1993.

[8] PRÜSS, J. and SOHR, H., *Imaginary powers of elliptic second order differential operators in L^p-spaces*. Hiroshima Math. J. *23* (1993), 161–192.

[9] SOBOLEVSKII, P. E., *Fractional powers of coercively positive sums of operators*. Soviet Math. Dokl. *16* (1975).

[10] TRIEBEL, H., *Interpolation Theory, Function Spaces, Differential Operators*. North-Holland, Amsterdam 1978.

Emilia Bazhlekova
Institute of Mathematics
Bulgarian Academy of Sciences
Acad. G. Bontchev str., bl. 8
1113 Sofia
Bulgaria
e-mail: bajlekova@hotmail.com

Philippe Clément
Dept. of Applied Mathematical Analysis
Technische Universiteit Delft
Mekelweg 4
2628 CD Delft
The Netherlands
e-mail: ph.p.j.e.clement@its.tudelft.nl

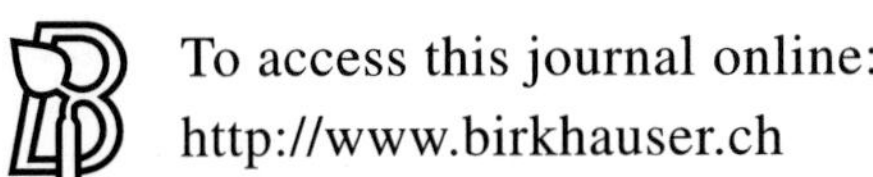

J.evol.equ. 3 (2003) 247 – 281
1424–3199/03/020247 – 35
DOI 10.1007/s00028-003-0094-y
© Birkhäuser Verlag, Basel, 2003

Journal of Evolution
Equations

On the uniqueness of solutions for nonlinear elliptic-parabolic equations

H. Gajewski and I. V. Skrypnik

Abstract. We prove a priori estimates in $L^2(0, T; W^{1,2}(\Omega))$ and $L^\infty(Q_T)$, existence and uniqueness of solutions to Cauchy-Dirichlet problems for elliptic-parabolic systems

$$\frac{\partial \sigma(u)}{\partial t} - \sum_{i=1}^{n} \frac{\partial}{\partial x_i} \left\{ \rho(u) b_i \left(t, x, \frac{\partial(u-v)}{\partial x} \right) \right\} + a(t, x, v, u) = 0,$$

$$-\sum_{i=1}^{n} \frac{\partial}{\partial x_i} \left[\kappa(x) \frac{\partial v}{\partial x_i} \right] + \sigma(u) = f(t, x), \quad (t, x) \in Q_T = (0, T) \times \Omega,$$

where $\rho(u) = \frac{\partial \sigma(u)}{\partial u}$. Systems of such form arise as mathematical models of various applied problems, for instance, electron transport processes in semiconductors. Our basic assumption is that $\log \rho(u)$ is concave. Such assumption is natural in view of drift-diffusion models, where σ has to be specified as a probability distribution function like a Fermi integral and u resp. v have to be interpreted as chemical resp. electrostatic potential.

1. Introduction

We prove a priori estimates, existence and uniqueness of weak solutions to initial boundary value problems of the form

$$\frac{\partial \sigma(u)}{\partial t} - \sum_{i=1}^{n} \frac{\partial}{\partial x_i} \left\{ \rho(u) b_i \left(t, x, \frac{\partial(u-v)}{\partial x} \right) \right\}$$

$$+ a(t, x, v, u) = 0, \quad (t, x) \in Q_T, \tag{1.1}$$

$$-\sum_{i=1}^{n} \frac{\partial}{\partial x_i} \left[\kappa(x) \frac{\partial v}{\partial x_i} \right] + \sigma(u) = f(t, x), \quad (t, x) \in Q_T, \tag{1.2}$$

$$u(t, x) = g_1(t, x), \quad (t, x) \in \Gamma = (0, T) \times \partial\Omega, \tag{1.3}$$

Mathematics Subject Classification (2000): 35B45, 35K15, 35K20, 35K65.
Key words and phrases: Nonlinear parabolic equations, bounded solutions, uniqueness, nonstandard assumptions, degenerate type.

$$v(t, x) = g_2(t, x), \quad (t, x) \in \Gamma = (0, T) \times \partial\Omega, \tag{1.4}$$

$$u(0, x) = h(x), \quad x \in \Omega, \tag{1.5}$$

where $\sigma(u) = \int_0^u \rho(s)ds$, Ω is a bounded open set in $\mathbb{R}^n$ and $Q_T = (0, T) \times \Omega$, $T > 0$.

Systems of the form (1.1), (1.2) arise as mathematical models of various applied problems, for instance reaction-drift-diffusion processes of electrically charged species, phase transition processes and transport processes in porous media. The investigation of nonlinear reaction–drift–diffusion systems has received much attention in recent years [1].

The Equation (1.1) is degenerate because the function $\rho(u)$ can tend to zero. Cauchy-Dirichlet problems for degenerate parabolic equations have been studied by many authors (see for example [2], [3], [11]). But the structure of the Equation (1.1) is different from that one considered in these papers. Boundary value problems for the equation of the structure (1.1) were studied by the authors in the stationary case in [8] and in the nonstationary case in [9].

The initial–boundary value problem for systems of the form (1.1)–(1.2) was studied in [5] under essentially stronger assumptions as in the presented paper. In [5] the solvability was proved for the special case $b_i(t, x, \xi) = \xi_i$, uniqueness was shown under the regularity assumption $v(x, t) \in L^\infty(0, T, W^{1,p}(\Omega))$, $p > n$.

We consider problem (1.1)–(1.5) under standard conditions for the functions $b_i(t, x, \xi)$ and some conditions for the function $a(t, x, v, u)$ to be formulated in Section 2. Our main specific assumption reads:

$\rho)$ $\ \rho \in (\mathbb{R}^1 \to \mathbb{R}^1)$ *with* $\rho(u) > 0$, $u \in \mathbb{R}^1$, *is continuous and has a piecewise continuous derivative* ρ' *such that* $\frac{\rho'(u)}{\rho(u)}$ *is nonincreasing on* $\mathbb{R}^1$.

For the semiconductor theory [5] relevant examples for functions ρ satisfying condition $\rho)$ are given by $\sigma = \mathcal{F}_{\gamma+1}, \rho = \sigma' = \mathcal{F}_\gamma$, where $\mathcal{F}_\gamma$ denotes the Fermi integral

$$\mathcal{F}_\gamma(u) = \frac{1}{\Gamma(\gamma + 1)} \int_0^\infty \frac{s^\gamma \, ds}{1 + \exp(s - u)}, \quad \gamma > -1. \tag{1.6}$$

Another example comes from phase separation problems [7], where the Fermi function

$$\sigma(u) = \frac{1}{1 + exp(-u)}, \quad \rho(u) = \sigma'(u) = \frac{1}{(1 + e^u)(1 + e^{-u})},$$

plays a role corresponding to $\mathcal{F}_{\gamma+1}$.

We formulate our assumptions and main results in Section 2. First a priori estimates for solutions u, v are given in Section 3. In that Section we prove also regularity properties of the function v, important for further considerations. An L^∞ estimate of u is given in Section 4. Section 5 is devoted to the existence proof for solutions of problem (1.1)–(1.5). Our main result, uniqueness of solutions, is proved in Section 6.

Note that our considerations can be carried over to the case of Neumann boundary conditions instead of the Dirichlet conditions (1.3), (1.4).

We are planning in forthcoming papers to apply our approach to more general reaction-drift-diffusion systems, including more than one species and the temperature.

2. Formulation of assumptions and main results

Let Ω be a bounded open set in $\mathbb{R}^n$ and $Q_T = (0, T) \times \Omega$, $T > 0$. We shall assume that $n > 2$. For $n \leq 2$ it is necessary to make simple changes in our conditions that are connected with Sobolev's embedding theorem.

We assume following regularity condition on the boundary $\partial\Omega$ of the set Ω:

∂) there exist positive numbers χ, R_0, such that for an arbitrary point $x \in \partial\Omega$ the inequality $\mathrm{meas}\{B(x, R) \setminus \Omega\} \geq \chi R^n$ holds, where $0 < R \leq R_0$ and $B(x, R)$ is a ball of radius R with center x.

Let the coefficients b_i, a, κ from (1.1), (1.2) satisfy following assumptions:

i) $a(t, x, v, u)$, $b_i(t, x, \xi)$, $i = 1, \ldots, n$, are measurable functions with respect to t, x for every $u, v \in \mathbb{R}^1$, $\xi \in \mathbb{R}^n$ and continuous with respect to $u, v \in \mathbb{R}^1, \xi \in \mathbb{R}^n$, for almost every $(t, x) \in Q_T$; $b_i(t, x, 0) = 0$; $\kappa(x)$ is measurable function of x;

ii) there exist positive constants v_1, v_2 such that for arbitrary $\xi', \xi'' \in \mathbb{R}^n, (t, x) \in Q_T$,

ii)$_1$ $\sum_{i=1}^n [b_i(t, x, \xi') - b_i(t, x, \xi'')](\xi_i' - \xi_i'') \geq v_1 |\xi' - \xi'|^2$,

ii)$_2$ $|b_i(t, x, \xi)| \leq v_2(|\xi| + 1), \quad i = 1, \ldots, n$,

ii)$_3$ $v_1 \leq \kappa(x) \leq v_2$;

iii) there exists a nonnegative function $\alpha \in L^{p_1}(Q_T)$, $p_1 > \frac{n+2}{2}$, such that for arbitrary $(t, x) \in Q_T$, $v, u, u', u'' \in \mathbb{R}^1$,

$[a(t, x, v, u') - a(t, x, v, u'')](u' - u'') \geq v_1 |u' - u''|^2$,

$|a(t, x, v, u)| \leq v_2(|v| + |u|) + \alpha(t, x)$.

We note some simple consequences from condition ρ). Let

$$\alpha_\pm = \lim_{u \to \pm\infty} \rho(u). \tag{2.1}$$

Then for nonconstant functions ρ at least one of the numbers α_-, α_+ is zero [8]. Studying the behavior of the solution to (1.1)–(1.5), we have to distinguish the cases of zero or non–zero value of $\alpha_\pm$. In order to include both cases, we assume

$$\alpha_- = 0, \quad \alpha_+ \neq 0. \tag{2.2}$$

The considerations for the case $\alpha_- = \alpha_+ = 0$ are analogous. We remark only that the assumptions for the function $a(t, x, v, u)$ are connected with the behavior of the function ρ and that the condition iii) corresponds to the case (2.2).

We consider problem (1.1)–(1.5) with data such that

$$f \in C([0, T]; L^{p_2}(\Omega)), \quad \frac{\partial f}{\partial t} \in L^2(0, T; [\mathring{W}^{1,2}(\Omega)]^*), \quad p_2 > \frac{n}{2}, \tag{2.3}$$

$$g_i \in L^\infty(Q_T) \cap L^\infty(0, T; W^{1,2}(\Omega)) \cap L^1(0, T; W^{1,\infty}(\Omega)),$$

$$\frac{\partial g_1}{\partial t} \in L^1(0, T; L^\infty(\Omega)), \quad \frac{\partial g_2}{\partial t} \in L^2(0, T; L^2(\Omega)), \tag{2.4}$$

$$h \in L^\infty(\Omega). \tag{2.5}$$

DEFINITION 2.1. A functions pair (u, v), $u, v \in L^2(0, T; W^{1,2}(\Omega))$, is called solution of problem (1.1)–(1.5) if:

i) $\sigma(u) \in C([0, T]; L^{\frac{2n}{n+2}}(\Omega))$,

$$\int_{Q_T} \int \rho(u) \left(\left| \frac{\partial u}{\partial x} \right|^2 + \left| \frac{\partial v}{\partial x} \right|^2 \right) dx\, dt < \infty, \tag{2.6}$$

the time derivative of $\sigma(u)$ in the sense of distributions satisfies

$$\frac{\partial \sigma(u)}{\partial t} \in L^2(0, T; [\mathring{W}^{1,2}(\Omega)]^*) \tag{2.7}$$

and the integral identities

$$\int_0^\tau \left\{ \left\langle \frac{\partial \sigma(u)}{\partial t}, \varphi \right\rangle \right.$$
$$\left. + \int_\Omega \left[\sum_{i=1}^n \rho(u) b_i \left(t, x, \frac{\partial(u - v)}{\partial x} \right) \frac{\partial \varphi}{\partial x_i} + a(t, x, v, u)\varphi \right] dx \right\} dt = 0, \tag{2.8}$$

$$\int_\Omega \left\{ \kappa(x) \sum_{i=1}^n \frac{\partial v}{\partial x_i} \frac{\partial \psi}{\partial x_i} + \sigma(u)\psi - f(t, x)\psi \right\} dx = 0 \tag{2.9}$$

$\forall \varphi \in C^\infty(\overline{Q}_T)$ vanishing near Γ, $\forall \psi \in C_0^\infty(\Omega)$ and almost every $\tau \in (0, T)$;

ii)　　$u - g_1 \in L^2(0, T; \mathring{W}^{1,2}(\Omega)), \quad v - g_2 \in L^2(0, T; \mathring{W}^{1,2}(\Omega)); \tag{2.10}$

iii) for functions φ, as in (2.8) and satisfying additionally $\varphi(\tau, x) = 0$ for $x \in \Omega$, the equality

$$\int_0^\tau \left\langle \frac{\partial \sigma(u)}{\partial t}, \varphi \right\rangle dt + \int_0^\tau \int_\Omega [\sigma(u) - \sigma(h)] \frac{\partial \varphi}{\partial t} \, dx\, dt = 0 \tag{2.11}$$

holds for $\tau \in (0, T)$.

In order to justify this definition it is sufficient to show that

$$\sigma(u) \in L^1(Q_T), \quad \rho(u) \in L^1(Q_T). \tag{2.12}$$

The first inclusion follows immediately from the assumption $\sigma(u) \in C(0, T; L^{\frac{2n}{n+2}}(\Omega))$. The second one follows from the inequality

$$\rho(u) \le \frac{\rho(1)}{\sigma(1)} \, \sigma(u) \quad \text{for} \quad u \ge 1, \tag{2.13}$$

which is a consequence of condition ρ) and

$$\frac{d}{du}\left(\frac{\sigma(u)}{\rho(u)}\right) = 1 - \frac{\rho'(u)}{\rho^2(u)} \int_0^u \rho(s)\, ds \ge 1 - \frac{1}{\rho(u)} \int_0^u \frac{\rho'(s)}{\rho(s)} \rho(s)\, ds = \frac{\rho(0)}{\rho(u)} > 0.$$

REMARK 2.2. Let (u, v) be a solution of problem (1.1)-(1.5). Since the set of functions from $C^\infty(\overline{Q}_T)$ vanishing near $\partial\Omega$ is dense in $L^2(0, T; \mathring{W}^{1,2}(\Omega, \rho(u)))$, the integral identity (2.8) holds for all $\varphi \in L^2(0, T; \mathring{W}^{1,2}(\Omega))$ such that

$$\int_{Q_T} \int \rho(u) \left|\frac{\partial\varphi}{\partial x}\right|^2 dx dt < \infty.$$

Analogously the identity (2.9) holds for arbitrary functions $\psi \in \mathring{W}^{1,2}(\Omega)$.

Besides of (1.1), (1.2) we consider the regularized system

$$\frac{\partial\sigma(u)}{\partial t} - \sum_{i=1}^n \frac{\partial}{\partial x_i}\left\{\rho_\delta(u)b_i\left(t, x, \frac{\partial(u - v)}{\partial x}\right)\right\} + a(t, x, v, u) = 0, \tag{2.14}$$

$$-\sum_{i=1}^n \frac{\partial}{\partial x_i}\left[\kappa(x)\frac{\partial v}{\partial x_i}\right] + \sigma(u) = f(t, x), \tag{2.15}$$

$$\rho_\delta(u) = \max\left\{\rho(u), \rho\left(-\frac{1}{\delta}\right)\right\} \quad \text{for} \quad \delta \in (0, 1], \quad \rho_0(u) = \rho(u). \tag{2.16}$$

We understand solutions of the auxiliary problem (2.14), (2.15), (1.3)–(1.5) in the sense of Definition 2.1. after replacing $\rho(u)$ by $\rho_\delta(u)$ in (2.6) and (2.8).

In what follows we understand as known parameters all numbers from the conditions ii), iii), norms of functions f, g_1, g_2, h, α in respective spaces and numbers that depend only on $n, \chi, R_0, \Omega, \rho$.

THEOREM 2.3. *Let the conditions* i)–iii), ρ), (2.3)–(2.5) *be satisfied. Then there exists a constant M_1 depending only on known parameters and independent of $\delta \in [0, 1]$ such that each solution (u, v) of problem* (2.14), (2.15), (1.3)–(1.5) *satisfies*

$$\mathop{ess\ sup}_{t \in (0,T)} \int_\Omega \left\{ \Lambda(u(t, x)) + \left| \frac{\partial v(t, x)}{\partial x} \right|^2 \right\} dx$$

$$+ \int_{Q_T} \int \rho_\delta(u) \left| \frac{\partial(u - v)}{\partial x} \right|^2 dt\, dx \leq M_1, \tag{2.17}$$

$$\Lambda(u) = \int_0^u s\, \rho(s)\, ds. \tag{2.18}$$

For proving regularity properties of the function v we need following growth condition

$$\rho_1^{-1}(u^\gamma + 1) \leq \rho(u) \leq \rho_1(u^\gamma + 1), \quad u > 0, \quad 0 \leq \gamma < \frac{2}{n - 2} \tag{2.19}$$

with some positive constant ρ_1. (2.19) implies $\sigma(u) \leq \rho_1(\frac{u^{\gamma+1}}{\gamma+1} + u)$ for $u > 0$ with $\gamma + 1 < \frac{n}{n-2}$. Remark that such type condition arised in [5] for $n > 2$ together with the stronger restriction $\gamma + 1 < \frac{2}{n-2}$.

THEOREM 2.4. *Let the assumptions of Theorem* 2.3 *and condition* (2.19) *be satisfied. Then there exists a constant M_2, depending only on known parameters and independent of $\delta \in [0, 1]$, such that each solution of problem* (2.14), (2.15), (1.3)–(1.5) *satisfies*

$$\int_{Q_T} \int \rho_\delta(u) \left\{ \left| \frac{\partial u}{\partial x} \right|^2 + \left| \frac{\partial v}{\partial x} \right|^2 \right\} dx dt \leq M_2. \tag{2.20}$$

THEOREM 2.5. *Let the assumptions of Theorem* 2.4. *be satisfied. Then the estimates*

$$\|v\|_{L^\infty(Q_T)} \leq M_3, \quad |v(t, x') - v(t, x'')| \leq H|x' - x''|^\eta \tag{2.21}$$

hold for arbitrary $t \in [0, T]$, $x', x'' \in \Omega$ and constants $M_3, H, \eta \in (0, 1)$, depending only on known parameters and independent of δ.

In order to prove a priori estimates for u we need additional conditions with repect to ρ and a. In view of our uniqueness result we assume stronger conditions for a than needed if proving a priori estimates only:

a) $\frac{a(t,x,v,u)}{\rho(u)}$ is nondecreasing with respect to $u \in \mathbb{R}^1$, for arbitrary $(t, x) \in Q_T, v \in \mathbb{R}^1$;

ρ') there exists a positive constant ρ_2 such that $\rho'(u) \leq \rho_2\, \rho(u)$ for $u < 0$.

THEOREM 2.6. *Let the conditions* i)–iii), ρ), ρ'), *a*), (2.3)–(2.5), (2.19) *be satisfied. Then there exists a constant M_4, depending only on known parameters and independent of $\delta \in [0, \frac{1}{M_4}]$, such that each solution u, v of problem* (2.14), (2.15), (1.3)–(1.5) *satisfies*

$$\text{ess } \sup \{|u(t, x)| : (t, x) \in Q_T\} \leq M_4. \tag{2.22}$$

THEOREM 2.7. *Let the conditions* i)–iii), ρ), ρ'), *a*), (2.3)–(2.5), (2.19) *be satisfied. Then the initial-boundary value problem* (1.1)–(1.5) *has at least one solution in the sense of Definition 2.1.*

THEOREM 2.8. *Let the conditions* i)–iii), ρ), ρ'), *a*), (2.3)–(2.5), (2.19) *be satisfied and assume additionally that the functions $b_i(t, x, \xi)$, $\rho'(u)$, $a(t, x, v, u)$ are locally Lipschitzian with respect to ξ, u, v respectively. Then the initial-boundary value problem* (1.1)–(1.5) *has a unique solution u, v in the sense of the Definition 2.1.*

COROLLARY 2.9. *Let the conditions of Theorem 2.8 be satisfied and assume additionally that the functions $f(t, x)$, $g_1(t, x)$, $g_2(t, x)$, $b_i(t, x, \xi)$, $a(t, x, v, u)$ are Lipschitzian with respect to t. Then the solution u of problem* (1.1)–(1.4) *is regular in the sense that*

$$t \to t \, \frac{\partial u}{\partial t} \in L^\infty(0, T; L^2(\Omega)) \cap L^2(0, T; W^{1,2}(\Omega)).$$

REMARK 2.10. Corollary 2.9 and Theorem 2.6 imply that $t \to t \, \frac{\partial \sigma(u)}{\partial t} \in L^\infty(0, T; L^2(\Omega))$. Consequently, (1.1) can be understand not only in the sense of distributions, but even as an equation in $L^2(0, T; L^2(\Omega))$.

Proofs of Theorems 2.3–2.5 are given in Section 3, proofs of Theorems 2.6–2.8 are given in Sections 4, 5, 6 respectively.

3. Regularity of the function v

We start this section proving firstly the priori estimate (2.17). Next we shall prove boundedness and Hölder continuity of the function v.

Proof of Theorem 2.3 Denote by $v_0(x)$ the solution of problem (2.15), (1.4) for $t = 0$ with $u(0, x)$ defined by (1.5) and let $(u(t, x), v(t, x))$ be the solution of problem (2.14), (2.15), (1.3)–(1.5). We extend functions $u(t, x)$, $v(t, x)$ by setting $u(t, x) = h(x), v(t, x) = v_0(x)$ for $t < 0$, $x \in \Omega$. In an analogous way we extend the functions $f(t, x)$, $g_2(t, x)$. Denote

$$\widetilde{u}(t, x) = u(t, x) - g_1(t, x), \quad \widetilde{v}(t, x) = v(t, x) - g_2(t, x).$$

Testing (2.9) with $\psi(x) = \tilde{v}(t+s, x) - \tilde{v}(t, x)$, we obtain for $\tau \in (0, T)$, $s \in (0, T-\tau)$

$$\int_{-s}^{\tau} \int_{\Omega} \left\{ \kappa(x) \sum_{i=1}^{n} \frac{\partial}{\partial x_i} [v(t+s), x) + v(t, x)] \frac{\partial}{\partial x_i} [\tilde{v}(t+s), x) - \tilde{v}(t, x)] \right.$$
$$+ [\sigma(u(t+s, x)) + \sigma(u(t, x)) - f(t+s, x)$$
$$\left. - f(t, x)][\tilde{v}(t+s), x) - \tilde{v}(t, x)] \right\} \, dx \, dt = 0.$$

Hence we get by simple calculations

$$\int_{\tau}^{\tau+s} \int_{\Omega} \kappa(x) \left| \frac{\partial v}{\partial x} \right|^2 dx \, dt - s \int_{\Omega} \kappa(x) \left| \frac{\partial v_0}{\partial x} \right|^2 dx \, dt$$
$$- \int_{-s}^{\tau} \int_{\Omega} \kappa(x) \sum_{i=1}^{n} \frac{\partial}{\partial x_i} [v(t+s), x + v(t, x)] \frac{\partial}{\partial x_i} [g_2(t+s), x$$
$$- g_2(t, x)] \, dx \, dt + \int_{0}^{\tau} \int_{\Omega} [\sigma(u(t-s, x)) + \sigma(u(t+s, x))$$
$$- f(t-s, x) + f(t+s, x)] \tilde{v}(t, x) \, dx \, dt$$
$$+ \int_{\tau}^{\tau+s} \int_{\Omega} [\sigma(u(t-s, x)) + \sigma(u(t, x)) - f(t-s, x)$$
$$- f(t, x)] \tilde{v}(t, x) \, dx \, dt - \int_{-s}^{0} \int_{\Omega} [\sigma(u(t, x)) - f(t, x) + \sigma(u(t+s, x))$$
$$- f(t+s, x)] \tilde{v}(t, x) \, dx dt = 0. \tag{3.1}$$

Dividing (3.1) by s and passing to the limit $s \to 0$, we obtain for almost every $\tau \in (0, T)$

$$\int_{\Omega} \kappa(x) \left| \frac{\partial v(\tau, x)}{\partial x} \right|^2 dx - \int_{\Omega} \kappa(x) \left| \frac{\partial v_0(x)}{\partial x} \right|^2 dx$$
$$- 2 \int_{0}^{\tau} \left\{ \int_{\Omega} \kappa(x) \sum_{i=1}^{n} \frac{\partial v}{\partial x_i} \frac{\partial}{\partial x_i} \frac{\partial}{\partial t} g_2 \, dx + \left\langle \frac{\partial \sigma(u)}{\partial t} - \frac{\partial f}{\partial t}, \tilde{v} \right\rangle \right\} dt$$
$$+ 2 \int_{\Omega} \{ [\sigma(u(\tau, x)) - f(\tau, x)] \tilde{v}(\tau, x) \, dx$$
$$- [\sigma(h) - f(0, x)][v_0(x) - g_2(0, x)] \} \, dx = 0. \tag{3.2}$$

Using (2.9) with $\psi(x) = \tilde{v}(\tau, x)$, we can rewrite the fifth term in (3.2) as

$$\int_{\Omega} [\sigma(u(\tau, x)) - f(\tau, x)] \tilde{v}(\tau, x) \, dx = - \int_{\Omega} \kappa(x) \sum_{i=1}^{n} \frac{\partial v}{\partial x_i} \frac{\partial \tilde{v}(\tau, x) \, dx}{\partial x_i} \, dx. \tag{3.3}$$

Remarking that it is simple to estimate the norm of $v_0(x)$ in $W^{1,2}(\Omega)$ and using the conditions (2.3), (2.4) and Cauchy's inequality, we infer from (3.2), (3.3)

$$\int_\Omega \kappa(x) \left|\frac{\partial v(\tau, x)}{\partial x}\right|^2 dx + \int_0^\tau \left\langle \frac{\partial \sigma(u)}{\partial t}, \tilde{v}\right\rangle dt$$

$$\leq c_1 \left\{1 + \int_0^\tau \int_\Omega \left|\frac{\partial v(t, x)}{\partial x}\right|^2 dx\, dt\right\}. \tag{3.4}$$

Here and in what follows c_i denote constants depending only on known parameters. The conditions (2.4), (2.6) and Remark 2.2 allow us to substitute $\varphi = \tilde{u} - \tilde{v}$ in the identity

$$\int_0^\tau \left\{\left\langle \frac{\partial \sigma(u)}{\partial t}, \varphi\right\rangle + \int_\Omega \left[\sum_{i=1}^n \rho_\delta(u)\, b_i\left(t, x, \frac{\partial(u-v)}{\partial x}\right)\frac{\partial \varphi}{\partial x_i}\right.\right.$$

$$\left.\left. + a(t, x, v, u)\varphi\right] dx\right\} dt = 0. \tag{3.5}$$

By (3.4) this gives

$$\int_0^\tau \left\langle \frac{\partial \sigma(u)}{\partial t}, u - g_1\right\rangle dt + \frac{1}{2}\int_\Omega \kappa(x)\left|\frac{\partial v(\tau, x)}{\partial x}\right|^2 dx$$

$$+ \int_0^\tau \int_\Omega \left\{\sum_{i=1}^n \rho_\delta(u)\, b_i\left(t, x, \frac{\partial(u-v)}{\partial x}\right)\frac{\partial(u-v)}{\partial x_i}\right.$$

$$\left. + [a(t, x, v, u) - a(t, x, v, v)](u - v)\right\} dx\, dt$$

$$\leq \int_0^\tau \int_\Omega \left\{\sum_{i=1}^n \rho_\delta(u)\, b_i\left(t, x, \frac{\partial(u-v)}{\partial x}\right)\frac{\partial(g_1 - g_2)}{\partial x_i} + a(t, x, v, u)[g_1 - g_2]\right.$$

$$\left. - a(t, x, v, v)(u - v)\right\} dx\, dt + c_1\left\{1 + \int_0^\tau \int_\Omega \left|\frac{\partial v(t, x)}{\partial x}\right|^2 dx\, dt\right\}. \tag{3.6}$$

We write the first integral from (3.6) in the form

$$\int_0^\tau \left\langle \frac{\partial \sigma(u)}{\partial t}, u - g_1\right\rangle dt$$

$$= \int_0^\tau \left[\left\langle \frac{\partial \sigma(u)}{\partial t}, [u]_{-m}^m - g_1\right\rangle + \left\langle \frac{\partial \sigma(u)}{\partial t}, u - [u]_{-m}^m\right\rangle\right] dt \tag{3.7}$$

with $m \geq \|g_1\|_{L^\infty(Q_T)}$, $[u]^m_{-m} = \max\{\min[u, m], -m\}$. Then we can evaluate the first and the second term of the right hand side of (3.7) by using Lemmas 3.2, 3.1 respectively [9]. So we obtain

$$\int_0^\tau \left\langle \frac{\partial \sigma(u)}{\partial t}, u - g_1 \right\rangle dt = \int_0^\tau \left\{ \int_0^{u(x,\tau)} s\rho(s)\, ds - \int_0^{h(x)} s\rho(s)\, ds \right\} dx$$

$$+ \int_0^\tau \int_\Omega [\sigma(u) - \sigma(h)] \frac{\partial g_1}{\partial t}\, dx\, dt - \int_\Omega [\sigma(u(\tau, x)) - \sigma(h(x))]\, g_1(\tau, x)\, dx. \qquad (3.8)$$

Immediately from the definition of $\Lambda(u)$ we deduce

$$\sigma(u) < \varepsilon \Lambda(u) + c_\varepsilon \quad \text{for} \quad u \geq 0 \qquad (3.9)$$

with arbitrary positive number ε and a constant c_ε depending only on ε and the function ρ. Using the conditions ii), (2.4), (2.5) and the inequalities (2.13), (3.9), we obtain with arbitrary positive number ε and some function $\mu(t) \in L^1(0, T)$:

$$\left| \int_0^\tau \int_\Omega \rho_\delta(u)\, b_i \left(t, x, \frac{\partial(u - v)}{\partial x} \right) \frac{\partial(g_1 - g_2)}{\partial x_i}\, dx\, dt \right|$$

$$\leq \varepsilon \int_0^\tau \int_\Omega \rho_\delta(u) \left| \frac{\partial(u - v)}{\partial x} \right|^2 dx\, dt + \frac{c_2}{\varepsilon} \int_0^\tau \int_\Omega \Lambda(u)\mu(t)\, dx\, dt,$$

$$\int_0^\tau \int_\Omega \sigma(u) \frac{\partial g_1}{\partial t}\, dx\, dt \leq c_2 \left\{ 1 + \int_0^\tau \int_\Omega \Lambda(u)\mu(t)\, dx\, dt \right\},$$

$$\int_\Omega \sigma(u(\tau, x))g_1(\tau, x)\, dx \leq c_2 \left\{ \varepsilon \int_\Omega \Lambda(u(\tau, x))\, dx + c_\varepsilon \right\}. \qquad (3.10)$$

We estimate the terms in (3.6) involving the function a in standard way by using (2.4) and the condition iii). Now from (3.6), (3.8), (3.10) and evident estimates for another terms in (3.8), we obtain

$$\int_\Omega \Lambda(u(\tau, x))\, dx + \int_\Omega \left| \frac{\partial v(\tau, x)}{\partial x} \right|^2 dx + \int_0^\tau \int_\Omega \rho(u) \left| \frac{\partial(u - v)}{\partial x} \right|^2 dx\, dt$$

$$\leq c_3 \left\{ 1 + \int_0^\tau \int_\Omega [1 + \mu(t)] \left[\Lambda(u) + \left| \frac{\partial v}{\partial x} \right|^2 \right] dx\, dt \right\}. \qquad (3.11)$$

Now the last inequality and Gronwall's lemma complete the proof of Theorem 2.3.

In order to prove Theorem 2.4 we need auxiliary estimates.

LEMMA 3.2. *Assume that the conditions of Theorem 2.3 are satisfied and*

$$\operatorname*{ess\ sup}_{t \in (0,T)} \int_\Omega \sigma^q(u(t, x))\, dx \leq K_1 \qquad (3.12)$$

with some numbers $q \in (\frac{2n}{n+2}, \frac{n}{2})$, K_1, depending only on known parameters. Then

$$ess \sup_{t \in (0,T)} \left\{ \int_\Omega |v(t,x)|^{\frac{pn}{n-2}} \, dx + \int_\Omega |v(t,x)|^{p-2} \left| \frac{\partial v(t,x)}{\partial x} \right|^2 dx \right\} \leq K_2, \tag{3.13}$$

where $p > 2$ is defined by the equality

$$p \frac{n}{n-2} = (p-1)\frac{q}{q-1} \tag{3.14}$$

and K_2 is a constant depending only on known parameters.

Proof. Denote

$$m_0 = \|g_1\|_{L^\infty(Q_T)} + \|g_2\|_{L^\infty(Q_T)} + \|h\|_{L^\infty(\Omega)} + 1 \tag{3.15}$$

and, for $k \in \mathbb{R}^1$ and arbitrary function w defined on Q_T,

$$w_k(t,x) = \min\{w(t,x), k\}, \quad w_+(t,x) = \max\{w(t,x), 0\}.$$

We test the integral identity

$$\sum_{i-1}^n \int_\Omega \kappa(x) \frac{\partial v}{\partial x_i} \frac{\partial \psi}{\partial x_i} \, dx + \int_\Omega [\sigma(u) - f]\psi \, dx = 0 \tag{3.16}$$

with $\psi = \operatorname{sign} v \, [|v|_k - m_0]^{p-1}$, $k > m_0$. Using the conditions ii), (2.3), (3.12) and Hölder's inequality, we obtain

$$\int_\Omega [|v|_k - m_0]_+^{p-2} \left| \frac{\partial v_k}{\partial x} \right|^2 dx \leq c_4 \left\{ \int_\Omega [|v|_k - m_0]_+^{(p-1)\frac{q}{q-1}} \, dx \right\}^{\frac{q-1}{q}}. \tag{3.17}$$

From this inequality and the embedding theorem we have

$$\left\{ \int_\Omega [|v|_k - m_0]_+^{\frac{pn}{n-2}} \, dx \right\}^{\frac{n-2}{n}} \leq c_5 \left\{ \int_\Omega [|v|_k - m_0]_+^{(p-1)\frac{q}{q-1}} \, dx \right\}^{\frac{q-1}{q}}. \tag{3.18}$$

Taking into account the restriction on q and the choice of p, we deduce (3.13) from (3.17), (3.18), (2.17) and the proof is completed. $\qquad\square$

Proof of Theorem 2.4. We assume firstly that $\frac{2+\gamma}{1+\gamma} < \frac{n}{2}$. It is simple to check ([8], inequality (8)) that the conditions ρ) and (2.2) imply

$$|\sigma(u)| \leq c_6 \quad \text{for} \quad u < 0. \tag{3.19}$$

From this and (2.19) we find

$$|\sigma(u)|^{q_0} \leq c_7[\Lambda(u) + 1] \quad \text{with} \quad q_0 = \frac{2+\gamma}{1+\gamma}. \tag{3.20}$$

By (3.20), (2.17) and Lemma 3.2, we obtain (3.13) with p_0 given by $p_0 \frac{n}{n-2} = (p_0 - 1)(2 + \gamma)$. Since p_0 satisfies $p_0 - 2 > \frac{n}{n-2} > \gamma$, (3.13), (2.19) imply

$$\int_{\{|u| \leq 2|v|\}} \int \rho_\delta(u) \left| \frac{\partial v}{\partial x} \right|^2 dx \, dt \leq c_8. \tag{3.21}$$

Here $\{|u| \leq 2|v|\} = \{(t, x) \in Q_T : |u(t, x)| \leq 2|v(t, x)|\}$ and analogous notations we shall use further. To establish an estimate analogous to (3.21) with respect to set $\{|u| > 2|v|\}$, we take into account that $\rho_\delta(u) \leq 1 + \rho(0)$ for $u < 0$. So we can restrict us to the set $\{u > 2|v|\}$ and put the test function

$$\psi = (|v|_k - g_2)\{[[u - |v|_k]_+]_k + |v|_k + m_0\}^{\widetilde{\gamma}} \operatorname{sign} v, \quad k > m_0, \ \widetilde{\gamma} > 0,$$

into (3.16). After standard calculations we obtain

$$I_1 := \int_{\{|v| < k\}} \int \{[[u - |v|]_+]_k + |v| + m_0\}^{\widetilde{\gamma}} \left| \frac{\partial v}{\partial x} \right|^2 dx \, dt \leq c_9(I_2 + I_3), \tag{3.22}$$

where

$$I_2 = \int_{\{|v|_k < u\}} \int , (|v|_k + m_0)\{[u - |v|_k]_k + |v|_k + m_0\}^{\widetilde{\gamma}-1} \left| \frac{\partial u}{\partial x} \right| \left| \frac{\partial v}{\partial x} \right| dx \, dt,$$

$$I_3 = \int_{Q_T} \int \{(u_+ + 1)^{\gamma-1} + |f(t, x)|\}(|v|_k + 1)\{[u_+]_{2k} + |v|_k + 1\}^{\widetilde{\gamma}} dx \, dt.$$

The integral I_2 will be estimated in different ways for $\widetilde{\gamma} \leq 1$ and for $\widetilde{\gamma} > 1$. For $\widetilde{\gamma} \leq 1$ we have

$$I_2 \leq \int_{\{|v|_k < u\}} \int (|v|_k + m_0)^{\widetilde{\gamma}} \left[\left| \frac{\partial u}{\partial x} \right|^2 + \left| \frac{\partial v}{\partial x} \right|^2 \right] dx \, dt$$

$$\leq 3 \int_{\{u > 0\}} \int \left\{ (u + m_0)^{\widetilde{\gamma}} \left| \frac{\partial (u - v)}{\partial x} \right|^2 \right.$$

$$\left. + (|v| + m_0)^{\widetilde{\gamma}} \left| \frac{\partial v}{\partial x} \right|^2 \right\} dx \, dt \leq c_{10}. \tag{3.23}$$

Here we used (3.13) and the inequality

$$\operatorname{ess} \sup_{t \in (0,T)} \int_\Omega u_+^{2+\gamma}(t, x) \, dx + \int_{\{u > 0\}} \int (1 + u)^\gamma \left| \frac{\partial (u - v)}{\partial x} \right|^2 dx \, dt \leq c_{11}, \tag{3.24}$$

that follows from (2.17), (2.19).

For $\widetilde{\gamma} > 1$ we estimate I_2 by using the evident inequality

$$[[u - |v|_k]_+]_k + |v|_k + m_0 \leq 2|v|_k + m_0$$

on the set $\{|v| \geq k\}$. Then we have

$$I_2 \leq \varepsilon I_1 + c_{12} \int_{\{u>0\}} \int \left\{ (u + m_0)^{\widetilde{\gamma}} \left| \frac{\partial(u-v)}{\partial x} \right|^2 \right.$$

$$\left. + \frac{1}{\varepsilon^{\widetilde{\gamma}-1}} (|v| + m_0)^{\widetilde{\gamma}} \left| \frac{\partial v}{\partial x} \right|^2 \right\} \, dx \, dt, \tag{3.25}$$

where the last integral can be estimated analogously to (3.23).

Using Hölder's inequality and the embedding theorem we obtain for $\delta \geq 0$

$$\int_{\Omega_T} \int |[u_+]_k - g_{1,+}|^{(2+\gamma)\frac{2}{n}+2+\delta} \, dx \, dt$$

$$\leq \int_0^T \left\{ \int_\Omega |[u_+]_k - g_{1,+}|^{2+\gamma} \, dx \right\}^{\frac{2}{n}} \left\{ \int_\Omega (|[u_+]_k - g_{1,+}|^{1+\frac{\delta}{2}})^{\frac{2n}{n-2}} \, dx \right\}^{\frac{n-2}{n}} dt$$

$$\leq c_{13} \left\{ ess \sup_{t \in (0,T)} \int_\Omega |[u_+]_k - g_{1,+}|^{2+\gamma} \, dx \right\}^{\frac{2}{n}}$$

$$\times \int_{\Omega_T} \int |[u_+]_k - g_{1,+}|^{\delta} \left| \frac{\partial}{\partial x}([u_+]_k - g_{1,+}) \right|^2 \, dx \, dt. \tag{3.26}$$

Choosing $\delta = 0$, the inequalities (2.17), (3.24) and condition (2.4) imply

$$\int_{Q_T} \int u_+^{(2+\gamma)\frac{2}{n}+2} \, dx \, dt \leq c_{14}. \tag{3.27}$$

We estimate I_3 by Young's inequality and condition (2.3) and obtain

$$I_3 \leq c_{15} \left\{ 1 + \int_{Q_T} \int u_+^{\gamma+\widetilde{\gamma}+2} \, dx \, dt + \int_{Q_T} \int |v|^{\gamma+\widetilde{\gamma}+2} \, dx \, dt \right\}. \tag{3.28}$$

The integral with v can be estimated by a constant in virtue of (3.13) in the case that $\widetilde{\gamma} \in [0, \gamma]$. If γ satisfies $2\gamma + 2 \leq (2+\gamma)\frac{2}{n} + 2$, the integral with u_+ and $\widetilde{\gamma} = \gamma$ in (3.28) can be also estimated by a constant because of (3.27). In the opposite case we choose $\widetilde{\gamma}$ such that $\gamma + \widetilde{\gamma} + 2 \leq (2+\gamma)\frac{2}{n} + 2$. For example we can take $\widetilde{\gamma} = \widetilde{\gamma}_1 = \frac{2}{n}$. Then we get from (3.22), (3.24), (3.25), (3.28) $I_1 \leq c_{16}$, which implies

$$\int \int_{\{u>|v|\}} (u - |v|)^{\widetilde{\gamma}} \left| \frac{\partial v}{\partial x} \right|^2 \, dx \, dt \leq c_{17}$$

and consequently

$$\int \int_{\{u>2|v|\}} [u(t,x)]^{\widetilde{\gamma}} \left| \frac{\partial v}{\partial x} \right|^2 \, dx \, dt \leq c_{18}. \tag{3.29}$$

From (2.17), (3.21), (3.29) we obtain

$$\int_{Q_T} \int \left\{ |u|^{\tilde{\gamma}} \left| \frac{\partial v}{\partial x} \right|^2 + |u|^{\tilde{\gamma}} \left| \frac{\partial u}{\partial x} \right|^2 \right\} dx \, dt \ \leq c_{19} \tag{3.30}$$

and this ends the proof of Theorem 2.4 in the case that $\frac{2+\gamma}{1+\gamma} < \frac{n}{2}$, $\tilde{\gamma} = \gamma$.

If $\tilde{\gamma} = \tilde{\gamma}_1 < \gamma$, we can iterate our arguments concerning $\tilde{\gamma}$. Using (3.30), we obtain from (3.26)

$$\int_{Q_T} \int u_+^{(2+\gamma)\frac{n}{2}+2+\tilde{\gamma}_1} \, dx \, dt \ \leq c_{20},$$

that allows us to choose $\tilde{\gamma}_2 = \min \{\gamma, \frac{4}{n}\}$. Repeating this argument, if necessary, we can choose $\tilde{\gamma}_3 = \gamma$ and we proved the Theorem if $\frac{2+\gamma}{1+\gamma} < \frac{n}{2}$.

If $\frac{2+\gamma}{1+\gamma} = \frac{n}{2}$ we can use Lemma 3.2 with $q' < q$ instead of q. We take q' such that the corresponding p' satisfies $p'-2 > \gamma$. Then we keep all discussions of the previous proof. If $\frac{2+\gamma}{1+\gamma} > \frac{n}{2}$, then the boundedness of solutions of Equation (2.15) under the conditions (2.3), (2.4), (3.12) and the assumption formulated above is well known [10]. In this case we can keep the previous discussions with corresponding simplification. The proof of Theorem 2.4 is completed.

LEMMA 3.4. *Assume that the conditions of Theorem 2.4 are satisfied and*

$$\operatorname{ess\ sup}_{t \in (0,T)} \int_{\Omega} \sigma^q(u_+(t, x)) \, dx \ + \int_{\{u>1\}} \int \rho_\delta^2(u) \sigma^{q-2}(u) \left| \frac{\partial u}{\partial x} \right|^2 dx \, dt \leq K_3 \tag{3.31}$$

holds with numbers $q \in [\frac{2+\gamma}{1+\gamma}, \frac{n}{2})$, K_3, *depending only on known parameters. Then there exist positive constants* β, K_4, *depending only on known parameters, such that*

$$\int_{\{u>1\}} \int \rho_\delta^2(u) \sigma^{q-2+\beta}(u) \left| \frac{\partial v}{\partial x} \right|^2 dx \, dt \leq K_4. \tag{3.32}$$

Proof. Theorem 2.4 implies (3.31) for $q = q_0 = \frac{2+\gamma}{1+\gamma}$. We want to prove (3.32) for this q. The proof of the lemma for $\frac{2+\gamma}{1+\gamma} < q < \frac{n}{2}$ would be the same. From Lemma 3.2 with $q = \frac{2+\gamma}{1+\gamma}$ we obtain analogously to (3.21)

$$\int_{\{|u|<2|v|\}} \int \rho_\delta^2(u) \sigma^{q_0-2+\beta_1}(u) \left| \frac{\partial v}{\partial x} \right|^2 dx \, dt \leq c_{21}, \quad \beta_1 = \frac{2-(n-2)\gamma}{(1+\gamma)(n-2)}. \tag{3.33}$$

For the proof of (3.32) it is sufficient to find a positive β_2 depending only on γ, n such that for $\tilde{\gamma} = \gamma + (1+\gamma)\beta_2$ the integral I_1 in (3.22) can be estimated by a constant. Since this estimation runs analogously to the estimation of (3.22), we make only some remarks.

We change the inequality (3.23) for $\tilde{\gamma} \le 1$, $\tilde{\gamma} \le \gamma + \frac{1}{2}(p_0 - 2 - \gamma)$, $p_0 = \frac{q_0(n-2)}{n-2q_0} > 2 + \frac{2}{n-2}$, in the following way

$$I_2 \le 3 \int_{\{u>0\}} \int \left\{ (u + m_0)^{\gamma} \left| \frac{\partial(u - v)}{\partial x} \right|^2 \right.$$

$$\left. + (|v| + m_0)^{p_0 - 2} \left| \frac{\partial v}{\partial x} \right|^2 \right\} \, dx \, dt \le c_{22} \tag{3.34}$$

after using Theorem 2.4 and Lemma 3.2. Analogously we change (3.25) for $\tilde{\gamma} > 1$.

In order to estimate I_3 we remark that (3.26) and Theorem 2.4 imply

$$\int_{Q_T} \int u_+^{(2+\gamma)(1+\frac{2}{n})} \, dx \, dt \le c_{23}. \tag{3.35}$$

From (3.28), (3.35), (3.13), we see that the integral I_3 can be estimated by a constant, provided

$$\gamma + \tilde{\gamma} + 2 \le (2 + \gamma)\left(1 + \frac{2}{n}\right), \quad \gamma + \tilde{\gamma} + 2 \le \frac{p_0 n}{n - 2}.$$

But both of these restrictions can be satisfied with $\tilde{\gamma} = \gamma + (1 + \gamma)\beta_3$ and some positive β_3 depending only on n, γ. Therefore we can choose positive β_2 such that the integral I_1 with $\tilde{\gamma} = \gamma + (1 + \gamma)\beta_2$ is estimated by a constant depending only on known parameters. That estimate and (3.33) verify (3.32). $\square$

LEMMA 3.5. *Assume that the conditions of Theorem 2.4 are satisfied. Then there exist numbers $\overline{q}$, K_5, depending only on known parameters, such that $\overline{q} > \frac{n}{2}$ and*

$$ess \sup_{t \in (0,T)} \int_{\Omega} \sigma^{\overline{q}}(u_+(t, x)) \, dx + \int_{\{u>1\}} \int \rho_{\delta}^2(u) \sigma^{\overline{q}-2}(u) \left| \frac{\partial u}{\partial x} \right|^2 \, dx \, dt \le K_5. \tag{3.36}$$

Proof. We put the function

$$\varphi = [\sigma(u_k) - \sigma(m_0)]_+^2 \{1 + [\sigma(u_k) - \sigma(m_0)]^3\}^r, \quad r \in \left(-\frac{2}{3}, \infty\right), \tag{3.37}$$

into the integral identity

$$\int_0^{\tau} \left\{ \left\langle \frac{\partial \sigma(u)}{\partial t}, \varphi \right\rangle + \int_{\Omega} \left[\sum_{i=1}^{n} \rho_{\delta}(u) b_i \left(t, x, \frac{\partial(u - v)}{\partial x} \right) \frac{\partial \varphi}{\partial x_i} \right. \right.$$

$$\left. \left. + a(t, x, v, u)\varphi \right] dx \right\} dt = 0. \tag{3.38}$$

Then, using Lemma 2 from [9], we can evaluate the first summand of (3.38) to obtain

$$\int_0^\tau \left\langle \frac{\partial \sigma(u)}{\partial t}, \varphi \right\rangle dt = \int_\Omega \Lambda^{(r)}(u(\tau, x))\, dx, \tag{3.39}$$

$$\Lambda^{(r)}(u) = \int_0^u \rho(s)[\sigma(s_k) - \sigma(m_0)]_+^2 \{1 + [\sigma(s_k) - \sigma(m_0)]^3\}^r\, ds$$

$$\geq \frac{1}{3(r+1)} \{1 + [\sigma(u_k) - \sigma(m_0)]^3\}^{r+1} - 1 \quad \text{for} \quad u > m_0. \tag{3.40}$$

Here $s_k = \min[s, k]$ and the value of u_k is analogous.

We write the derivative of φ in the form

$$\frac{\partial \varphi}{\partial x_i} = \left[\widetilde{\Phi}^{(r)}(u_k) \frac{\partial(u - v)}{\partial x_i} + \widetilde{\Phi}^{(r)}(u_k) \frac{\partial v}{\partial x_i} \right] \chi(m_0 < u < k), \tag{3.41}$$

where $\chi(m_0 < u < k)$ is the characteristic function of the set $\{m_0 < u < k\}$ and the function $\widetilde{\Phi}^r(u)$ satisfies for $r > -\frac{2}{3}$ the estimate

$$c_{24}k(r)\Phi^{(r)}(u)\rho(u) \leq \widetilde{\Phi}^{(r)}(u) \leq c_{25}(r+1)\Phi^{(r)}(u)\rho(u), \tag{3.42}$$

$$k(r) = \min(1, 2 + 3r),$$
$$\Phi^{(r)}(u) = [\sigma(u) - \sigma(m_0)]_+ \{1 + [\sigma(u) - \sigma(m_0)]^3\}^r. \tag{3.43}$$

Using (3.39)–(3.42) and conditions ii), iii), we obtain from (3.37), (3.38)

$$\int_\Omega \left\{ 1 + [\sigma(u_k(\tau, x)) - \sigma(m_0)]_+^3 \right\}^{r+1} dx$$

$$+ \int_0^\tau \int_\Omega \rho_\delta^2(u)\Phi^{(r)}(u_k)\chi(m_0 < u < k) \left| \frac{\partial u}{\partial x} \right|^2 dx\, dt$$

$$\leq c_{26} \left\{ \left[\frac{r+1}{\kappa(r)} \right]^2 \int_0^\tau \int_\Omega \rho_\delta^2(u)\Phi^{(r)}(u_k)\chi(m_0 < u < k) \left| \frac{\partial v}{\partial x} \right|^2 dx\, dt \right.$$

$$+ \frac{r+1}{\kappa(r)} \left[\int_0^\tau \int_\Omega (1 + |u| + |v| + \alpha(t, x)) \right.$$

$$\left. \left. [\sigma(u_k) - \sigma(m_0)]\Phi^{(r)}(u_k)\, dx\, dt + 1 \right] \right\}. \tag{3.44}$$

Let us assume now that for some $q \in [\frac{2+\gamma}{1+\gamma}, \frac{n}{2})$ the inequality (3.31) is fulfilled. Then Lemma 3.4 ensures that the first integral of the right hand side of (3.44) can be estimated by a constant independent on k for $r = \frac{1}{3}[q - 3 + \beta]$.

We shall check now that the second integral of the right hand site of (3.44) for $r = \frac{1}{3}[q - 3 + \beta']$ and some positive β' depending only on γ can be also estimated by a constant independent on k. Analogously to inequalities (3.26), (3.27) we obtain from (3.31)

$$\int_{Q_T}\int u_+^{q(1+\gamma)(1+\frac{2}{n})}\, dx\, dt \;\leq c_{27}. \tag{3.45}$$

From (3.31) and Lemma 3.2 we have

$$ess\ \sup_{t\in(0,T)} \int_\Omega |v(t,x)|^{\frac{qn}{n-2q}}\, dx \leq c_{28}. \tag{3.46}$$

(3.45), (3.46) imply the needed estimate for the last integral in (3.44) provided

$$\beta' \leq \frac{1}{1+\gamma}\left\{q(1+\gamma)\left(1+\frac{2}{n}\right)+\gamma\right\}-q, \quad \beta' \leq \frac{1}{1+\gamma}\left\{\frac{qn}{n-2q}+\gamma\right\}-q,$$

But, that holds for $\beta' = \frac{\gamma}{1+\gamma}$.

We proved that for $\overline{\beta} = \min(\beta, \beta')$ the left hand side of (3.44) is dominated by a constant depending only on known parameters if $r = \frac{1}{3}(q - 3 + \overline{\beta})$. This estimate implies that the inequality (3.31) is fulfilled with $q + \overline{\beta}$ instead of q. We can guarantee also by small change of $\overline{\beta}$ that the number $\frac{1}{\overline{\beta}}[\frac{n}{2} - \frac{2+\gamma}{1+\gamma}]$ is not integer, and denote by N its integer part. Recalling that the estimate (3.31) is fulfilled with $q = q_0 = \frac{2+\gamma}{1+\gamma}$ and choosing the sequence $q_i = q_0 + i\overline{\beta}$. We obtain after $N + 1$ iterations our previous discussions that the inequality (3.31) is fulfilled with $q = q_{N+1} > \frac{n}{2}$. Consequently the inequality (3.36) is satisfied with $\overline{q} = q_{N+1}$ and this ends the proof of Lemma 3.5. $\qquad\square$

Proof of Theorem 2.5. Theorem 2.5 follows immediately from (3.19), (3.36), the conditions ii), (2.3), (2.4) and the regularity assumption ∂) on the set Ω. It is necessary to apply only well known results on regularity of solutions of elliptic equations to equation (2.15) (see, for example, [10]).

4. Boundedness of the function u

We assume in this section that the conditions of Theorem 2.6 are satisfied. We shall prove estimates for u separately for the sets $\{u > 0\}$ and $\{u < 0\}$. These estimates will be given in Lemmas 4.1, 4.3.

LEMMA 4.1. *Let the conditions of Theorem 2.6 be satisfied. Then there exists a constant* M_5 *depending only on known parameters such that*

$$ess\ \sup \{u(t,x) : (t,x) \in Q_T\} \leq M_5. \tag{4.1}$$

Proof. We shall use the inequality (3.44). We start estimating the first integral of the right hand side of (3.44).

Let $\{\varphi_j^2(x)\}$, $j = 1, \ldots, J$, be a partition of unity such that

$$\sum_{j=1}^{J} \varphi_j^2(x) = 1, \quad \left| \frac{\partial \varphi_j}{\partial x} \right| \leq \frac{K_0}{R} \quad \text{for} \quad x \in \Omega,$$

$$\varphi_j(x) \in C^{\infty}(\mathbb{R}^n), \ \operatorname{supp} \varphi_j \subset B(x_j, R), \ J \leq \frac{K_0}{R^n}, \ R < 1, \tag{4.2}$$

where $B(x_j, R)$ is a ball of radius R with centre $x_j \in \Omega$, K_0 is a number depending only on n. The number R will be choosen later on.

We test the integral identity (3.16) with the function

$$\psi = \sum_{j=1}^{J} \rho_\delta^2(u_k)\Phi^{(r)}(u_k)[v - v_j]\varphi_j^2(x), \ v_j(t) = v(x_j, t). \tag{4.3}$$

Integration with respect to t yields

$$\int_{Q_\tau} \int \kappa(x)\rho_\delta^2(u_k)\Phi^{(r)}(u_k)\left| \frac{\partial v}{\partial x} \right|^2 dx \, dt = J_1 + J_2 + J_3, \tag{4.4}$$

where

$$J_1 = -\sum_{j=1}^{J}\sum_{i=1}^{n} \int_{Q_\tau} \int \kappa(x)\Phi_1^{(r)}(u_k)[v - v_j]\varphi_j^2 \frac{\partial u_k}{\partial x_i} \frac{\partial v}{\partial x_i} dx \, dt,$$

$$J_2 = -2\sum_{j=1}^{J}\sum_{i=1}^{n} \int_{Q_\tau} \int \kappa(x)\rho_\delta^2(u_k)\Phi^{(r)}(u_k)[v - v_j]\varphi_j \frac{\partial \varphi_j}{\partial x_i} \frac{\partial v}{\partial x_i} dx \, dt,$$

$$J_3 = -\sum_{j=1}^{J} \int_{Q_\tau} \int [\sigma(u) - f]\rho_\delta^2(u_k)\Phi^{(r)}(u_k)[v - v_j]\varphi_j^2(x) \, dx \, dt, \tag{4.5}$$

$$\Phi_1^{(r)}(u_k) = \rho_\delta^2(u_k)\rho(u_k)\left\{ \frac{1}{2} + [\sigma(u_k) - \sigma(m_0)]^3 \right\}^{r-1}$$

$$\times \left\{ \frac{1}{2} + (3r + 1)[\sigma(u_k) - \sigma(m_0)]^3 \right\}\chi(m_0 < u < k)$$

$$+ 2\,\rho_\delta(u_k)\rho'(u_k)\Phi^{(r)}(u_k)\chi(u < k). \tag{4.6}$$

Denote $u_0 = \sigma^{-1}[\sigma(m_0) + \frac{1}{2}]$. Analogously to (2.13) we obtain

$$\rho'(u)\sigma(u) \leq \rho^2(u) \quad \text{if } u > 0, \quad \rho(u) \leq 2\,\rho(u_0)[\sigma(u) - \sigma(m_0)] \quad \text{if } > u_0. \tag{4.7}$$

Hence we get for $r \geq -\frac{1}{2}$, $k > u_0$,

$$\Phi_1^{(r)}(u_k) \leq c_{29}(r+1)\{\rho_\delta^2(u_k)\Phi^{(r)}(u_k)\chi(u_0 < u < k) + \chi(m_0 < u \leq u_0)\}. \tag{4.8}$$

We assume further that $r \geq -\frac{1}{2}$ and choose the number R from (4.2) according to

$$R^\varepsilon = \frac{\varepsilon}{(r+1)^2}, \quad \varepsilon < \frac{1}{4}, \tag{4.9}$$

where ε will be specified later on. Using (2.21), (4.2), (4.8), (4.9), we obtain

$$|J_1| \leq \varepsilon\bigg\{\int_{Q_\tau}\int \kappa(x)\rho_\delta^2(u_k)\Phi^{(r)}(u_k)\left|\frac{\partial v}{\partial x}\right|^2 dx\,dt$$

$$+ c_{30}\frac{1}{(r+1)^2}\int_{Q_\tau}\int \kappa(x)\rho_\delta^2(u_k)\Phi^{(r)}(u_k)\chi(u_0 < u < k)\left|\frac{\partial u}{\partial x}\right|^2 dx\,dt$$

$$+ c_{30}\frac{1}{r+1}\int_{Q_\tau}\int \left[\left|\frac{\partial u}{\partial x}\right|^2 + \left|\frac{\partial v}{\partial x}\right|^2\right]\chi(u > m_0)\,dx\,dt\bigg\}. \tag{4.10}$$

By (4.2) and Cauchy's inequality we have

$$|J_2| \leq \int_{Q_\tau}\int \kappa(x)\rho_\delta^2(u_k)\Phi^{(r)}(u_k)\bigg\{\varepsilon\left|\frac{\partial v}{\partial x}\right|^2 + \frac{1}{\varepsilon}\frac{c_{31}}{R^{n+2}}\bigg\}dx\,dt. \tag{4.11}$$

From (4.4), (4.5), (4.9)–(4.11) we infer

$$\int_{Q_\tau}\int \kappa(x)\rho_\delta^2(u_k)\Phi^{(r)}(u_k)\left|\frac{\partial v}{\partial x_i}\right|^2 dx\,dt$$

$$\leq c_{32}\bigg\{\frac{\varepsilon}{(r+1)^2}\int_{Q_\tau}\int \kappa(x)\rho_\delta^2(u_k)\Phi^{(r)}(u_k)\chi(u_0 < u < k)\left|\frac{\partial u}{\partial x}\right|^2 dx\,dt$$

$$+ \frac{1}{r+1}\int_{Q_\tau}\int \left(\left|\frac{\partial u}{\partial x}\right|^2 + \left|\frac{\partial v}{\partial x}\right|^2\right)\chi(u > m_0)\,dx\,dt$$

$$+ \int_{Q_\tau}\int \rho_\delta^2(u_k)\Phi^{(r)}(u_k)\left[\sigma(u) + |f| + \frac{1}{\varepsilon}\left[\frac{(r+1)^2}{\varepsilon}\right]^{n+2}\right]dx\,dt. \tag{4.12}$$

Applying the last estimate to the first integral of the right hand side of (3.44) and choosing ε small enough, we get from (3.44), (2.20), (2.21), (2.13), (4.7)

$$\text{ess}\sup_{\tau \in (0,T)}\int_\Omega \left\{\frac{1}{2} + [\sigma(u_k(\tau, x)) - \sigma(m_0)]_+^3\right\}^{r+1}dx$$

$$+ \int_{Q_T}\int \rho_\delta^2(u)\Phi^{(r)}(u_k)\chi(m_0 < u < k)\left|\frac{\partial u}{\partial x}\right|^2 dx\,dt$$

$$\leq c_{33}(r+1)^{\lambda_1}\left\{\int_{Q_T}\int \Phi^{(r)}(u_k)[\sigma(u_k)-\sigma(m_0)]^2\right.$$

$$\left. [\sigma(u)+|\alpha|+|f|]\,dx\,dt\ +1\right\} \qquad (4.13)$$

with $\lambda_1 = 2(n+2)+2$.

We want to apply Moser iteration with respect to the integral

$$I_k(r) = \int_{Q_T}\int \Phi^{(r)}(u_k)[\sigma(u_k)-\sigma(m_0)]^2[\sigma(u)+|f|]\,dx\,dt. \qquad (4.14)$$

To this end we use the embedding inequality

$$\int_0^T\left\{\int_\Omega |v(t,x)|^{2(1+\frac{2p}{n})}\,dx\right\}^{\frac{1}{p}}dt$$

$$\leq C(n,p)\left\{ess\sup_{t\in(0,T)}\int_\Omega v^2(t,x)\,dx\right\}^{\frac{1}{p}+\frac{2}{n}-1}\int_{Q_T}\int\left|\frac{\partial v}{\partial x}\right|^2\,dx\,dt, \qquad (4.15)$$

which is fulfilled for $1 \leq p < \frac{n}{n-2}$ with a constant $C(n,p)$ depending only on n, p and with an arbitrary function $v \in L^\infty(0,T;L^2(\Omega)) \cap L^2(0,T;\overset{\circ}{W}{}^{1,2}(\Omega))$. From condition (2.3) and inequality (3.36) we have $\sigma(u)+|f| \in L^\infty(0,T,L^{p'}(\Omega))$ for some $p' > \frac{n}{2}$.

Applying Hölder's inequality to (4.14) we obtain

$$I_k(r) \leq c_{34}\int_0^T\left\{\int_\Omega [\Phi^{(r)}(u_k)[\sigma(u_k)-\sigma(m_0)]^2]^p\,dx\right\}^{\frac{1}{p}}dt$$

$$\leq c_{35}\left\{ess\sup_{t\in(0,T)}\int_\Omega [\Phi_2^{(r)}(u_k)]^2\,dx\right\}^{\frac{1}{p}+\frac{2}{n}-1}$$

$$\int_{Q_T}\int\left|\frac{\partial \Phi_2^{(r)}(u_k)}{\partial x}\right|^2\,dx\,dt, \qquad (4.16)$$

where

$$\Phi_2^{(r)}(u_k) = [\Phi^{(r)}(u_k)[\sigma(u_k)-\sigma(m_0)]^2]^{\frac{p}{2(1+\frac{2p}{n})}}. \qquad (4.17)$$

Simple calculations give

$$[\Phi_2^{(r)}(u_k)]^2 \leq \left\{\frac{1}{2}+[\sigma(u_k)-\sigma(u_0)]_+^3\right\}^{(r+1)\theta}, \quad \theta = \frac{p}{1+\frac{2p}{n}} < 1,$$

$$\left|\frac{\partial \Phi_2^{(r)}(u_k)}{\partial x}\right|^2 \leq c_{36}(r+1)^2\{\rho_\delta^2(u_k)\Phi^{(\theta r+\theta-1)}(u_k)+1\}\chi(m_0<u<k). \qquad (4.18)$$

For $r \geq -\frac{1}{2}$ we get from (4.16), (4.18) and (4.13)

$$\widetilde{I}_k(r) \leq c_{37}(r+1)^{\lambda_2}\{\widetilde{I}_k(\theta r + \theta - 1)\}^{\frac{1}{\theta}}, \quad \lambda_2 = 2 + \frac{\lambda_1}{\theta}, \quad \widetilde{I}_k(r) = I_k(r) + 1. \qquad (4.19)$$

We choose $r_j = \frac{1}{2}\theta^{-j} - 1$, $j = 0, 1, \ldots$, and obtain from (4.19)

$$\{\widetilde{I}_k(r_j)\}^{\theta^{-j}} \leq c_{37}^{\theta^{-j}}\theta^{-j\lambda_2\theta^{-j}}\{\widetilde{I}_k(r_{j-1})\}^{\theta^{j-1}}.$$

Iterating this estimate yields for arbitrary j

$$\{\widetilde{I}_k(r_j)\}^{\theta^{-j}} \leq c_{38}\,\widetilde{I}_k\left(-\frac{1}{2}\right). \qquad (4.20)$$

Now (4.16), (4.18) and (3.36) imply

$$\widetilde{I}_k\left(-\frac{1}{2}\right) \leq c_{39}\left\{1 + \left[ess \sup_{t\in(0,T)}\int_\Omega [\sigma(u_+(t,x))]^{\frac{3\theta}{2}}\,dx\right]^{\frac{1}{p}+\frac{2}{n}-1}\right.$$

$$\left. \times \int\int_{\{u>1\}} \rho_\delta^2(u)[\sigma(u)]^{\frac{3\theta}{2}-2}\left|\frac{\partial u}{\partial x}\right|^2 dx\,dt\right\} \leq c_{40}, \qquad (4.21)$$

where the constant c_{40} is independent of k. Recall that we consider the case $n > 2$, i. e., $\frac{3\theta}{2} < \frac{3}{2} \leq \overline{q}$, where $\overline{q}$ is the number from Lemma 3.4. Hence the desired estimate (4.1) follows from (4.20), (4.21). $\qquad\square$

For $k \in \mathbb{R}^1$ and arbitrary functions w defined on Q_T we use the notations

$$w^{(k)}(t,x) = \max\{w(t,x), k\}, \quad w_-(t,x) = \min\{w(t,x), 0\}. \qquad (4.22)$$

LEMMA 4.2. *Let the conditions of Theorem 2.6 be satisfied. Then there exists a constant M_6 depending only on known parameters such that*

$$ess \sup_{t\in(0,T)}\int_\Omega |u^{(k)}(t,x)|\,dx + \int_{Q_T}\int\left|\frac{\partial u^{(k)}}{\partial x}\right|^2 dx\,dt \leq M_6, \ for\ k > -\frac{1}{\delta}. \qquad (4.23)$$

Proof. We test the integral identity (3.5) with

$$\varphi = \frac{1}{\rho(u^{(k)})}[\sigma(u^{(k)}) - \sigma(-m_0)]_-|u^{(k)} + m_0|^r, \quad k < -m_0 - 1, \ r \geq 0,$$

to obtain

$$\int_0^\tau \left\langle \frac{\partial\sigma(u)}{\partial t}, \frac{1}{\rho(u^{(k)})}[\sigma(u^{(k)}) - \sigma(-m_0)]_-|u^{(k)} + m_0|^r\right\rangle dt$$

$$+ \sum_{i=1}^n \int_{Q_\tau}\int \rho_\delta(u)b_i\left(t, x, \frac{\partial(u-v)}{\partial x}\right)\frac{\partial(u-v)}{\partial x_i}\psi^{(r)}(u)\chi(k < u < -m_0)\,dx\,dt$$

$$+ \int_{Q_\tau} \int a(t, x, v, u) \frac{1}{\rho(u^{(k)})} [\sigma(u^{(k)}) - \sigma(-m_0)]_- |u^{(k)} + m_0|^r \, dx \, dt$$

$$+ \sum_{i=1}^{n} \int_{Q_\tau} \int \rho_\delta(u) b_i \left(t, x, \frac{\partial(u - v)}{\partial x} \right) \frac{\partial v}{\partial x_i} \psi^{(r)}$$

$$(u) \chi(k < u < -m_0) \, dx \, dt = 0, \tag{4.24}$$

where

$$\psi^{(r)}(u) = \left[1 + \frac{\rho'(u)}{\rho^2(u)} \int_u^{-m_0} \rho(s) \, ds \right] |u + m_0|^r$$

$$- r \frac{1}{\rho(u)} [\sigma(u) - \sigma(-m_0)] |u + m_0|^{r-1}. \tag{4.25}$$

We evaluate the first integral in (4.24) by Lemma 1 in [9] and find

$$\int_0^\tau \left\langle \frac{\partial \sigma(u)}{\partial t}, \frac{1}{\rho(u^{(k)})} [\sigma(u^{(k)}) - \sigma(-m_0)]_- |u^{(k)} + m_0|^r \right\rangle dt$$

$$= \int_\Omega \Lambda_-^{(r)}(u(\tau, x)) \, dx, \tag{4.26}$$

where

$$\Lambda_-^{(r)}(u) = \int_0^u \rho(s) \frac{1}{\rho(s^{(k)})} [\sigma(s^{(k)}) - \sigma(-m_0)]_- |s^{(k)} + m_0|^r \, ds, \quad s^{(k)} = \max\{s, k\}.$$

Simple calculations give for $u > -m_0$

$$\Lambda_-^{(r)}(u) \geq \frac{c_{41}}{r+1} |u^{(k)} + m_0|^{r+1} - c_{42}. \tag{4.27}$$

For $u < -m_0 T$ condition $\rho)$ implies

$$\frac{\rho'(u)}{\rho^2(u)} \int_u^{-m_0} \rho(s) \, ds \geq \frac{1}{\rho(u)} \int_u^{-m_0} \frac{\rho'(s)}{\rho(s)} \rho(s) \, ds = \frac{\rho(-m_0)}{\rho(u)} - 1 \tag{4.28}$$

and hence

$$\psi^{(r)}(u) \geq \frac{\rho(-m_0)}{\rho(u)} |u + m_0|^r. \tag{4.29}$$

Condition $\rho')$ and inequality (3.19) yield a corresponding estimate from above

$$\psi^{(r)}(u) \leq c_{43} \frac{r+1}{\rho(u)} [|u + m_0|^r + r]. \tag{4.30}$$

Further, condition $a)$, Theorem 2.5 and (3.19) imply the following estimate for the term involving a in (4.24):

$$\frac{1}{\rho(u)} a(t, x, v, u) \frac{\rho(u)}{\rho(u^{(k)})} [\sigma(u^{(k)}) - \sigma(-m_0)]_-$$

$$\geq \frac{a(t, x, v, 0)}{\rho(0)} \frac{\rho(u)}{\rho(u^{(k)})} [\sigma(u^{(k)}) - \sigma(-m_0)]_- \geq -c_{44}. \tag{4.31}$$

Using the inequalities (4.27), (4.29)–(4.31), we get from (4.24)

$$
\frac{1}{r+1} \int_\Omega |[u^{(k)} + m_0]_-|^{r+1}\, dx \;+\; \int_{Q_\tau} \int |u + m_0|^r \left|\frac{\partial u}{\partial x}\right|^2 \chi(k < u < -m_0)\, dx\, dt
$$

$$
\leq c_{45} \int_{Q_\tau} \int \left\{ (r+1)^2 |u + m_0|^r \left|\frac{\partial v}{\partial x}\right|^2 \chi(k < u < -m_0) \right.
$$

$$
+ r(r+1)\left(\left|\frac{\partial u}{\partial x}\right|^2 + \left|\frac{\partial v}{\partial x}\right|^2 \right) \chi(k < u < -m_0)
$$

$$
\left. + |[u^{(k)} + m_0]_-|^r \right\} dx\, dt. \tag{4.32}
$$

Finally, inequality (4.23) follows immediately from (4.32) with $r = 0$ and Theorem 2.3. $\square$

LEMMA 4.3. *Let the conditions of Theorem 2.6 be satisfied. Then a positive constant M_7, depending only on known parameters, exists such that*

$$
\mathrm{ess}\ \inf\{u(t, x) : (t, x) \in Q_T\} \geq -M_7 \quad \text{for } \delta \in \left[0, \frac{1}{M_7}\right]. \tag{4.33}
$$

Proof. We shall use inequality (4.32). To this end we start estimating the first integral of the right hand side of (4.32). We assume further that $k > -\frac{1}{\delta}$.

Let $\{\varphi_j^2(x)\}$, $j = 1, \ldots J$, be a partition of unity satisfying (4.2) with a number R to be fixed later on. We test the integral identity (3.16) with

$$
\psi = \sum_{j=1}^J [v - v_j] |[u^{(k)} + m_0]_-|^r \varphi_j^2(x), \quad r \geq 2,\ v_j(t) = v(t, x_j). \tag{4.34}
$$

After integration with respect to t we get

$$
\int_{Q_\tau} \int \kappa(x) |[u^{(k)} + m_0]_-|^r \left|\frac{\partial v}{\partial x}\right|^2 dx\, dt \;=\; J_1^- + J_2^- + J_3^-, \tag{4.35}
$$

where

$$
J_1^- = r \sum_{j=1}^J \sum_{i=1}^n \int_{Q_\tau} \int \kappa(x) |[u + m_0]_-|^{r-1} [v - v_j] \varphi_j^2 \frac{\partial u^{(k)}}{\partial x_i} \frac{\partial v}{\partial x_i}\, dx\, dt,
$$

$$
J_2^- = -2 \sum_{j=1}^J \sum_{i=1}^n \int_{Q_\tau} \int \kappa(x) [v - v_j] |[u + m_0]_-|^r \varphi_j \frac{\partial \varphi_j}{\partial x_i} \frac{\partial v}{\partial x_i}\, dx\, dt,
$$

$$
J_3^- = - \sum_{j=1}^n \int_{Q_\tau} \int [\sigma(u) - f][v - v_j] |[u^{(k)} + m_0]_-|^r \varphi_j^2\, dx\, dt. \tag{4.36}
$$

Repeating arguments used for estimating J_1 in Lemma 4.1 and choosing the R from (4.9), we get

$$|J_1^-| \leq \varepsilon \left\{ \int_{Q_\tau} \int \kappa(x) |[u^{(k)} + m_0]_-|^r \left| \frac{\partial v}{\partial x} \right|^2 dx\, dt \right.$$

$$\left. + c_{46} \frac{1}{(r+1)^2} \int_{Q_\tau} \int \kappa(x) |[u^{(k)} + m_0]_-|^r \left| \frac{\partial u^{(k)}}{\partial x} \right|^2 dx\, dt + c_{46} \right\}. \tag{4.37}$$

We apply Cauchy's inequality to J_2^-, J_3^-, use (4.37) and obtain from (4.35)

$$\int_{Q_\tau} \int \kappa(x) |[u^{(k)} + m_0]_-|^r \left| \frac{\partial v}{\partial x} \right|^2 dx\, dt$$

$$\leq c_{47} \left\{ \frac{\varepsilon}{(r+1)^2} \int_{Q_\tau} \int \kappa(x) |[u^{(k)} + m_0]_-|^r \left| \frac{\partial u^{(k)}}{\partial x} \right|^2 dx\, dt \right.$$

$$\left. + \left[1 + \frac{1}{\varepsilon} \left(\frac{(r+1)^2}{\varepsilon} \right)^{n+2} \right] \int_{Q_\tau} \int |[u^{(k)} + m_0]_-|^r dx\, dt \right\}. \tag{4.38}$$

Now (2.17), (4.23), (4.32) and the last estimate, taken with sufficiently small ε, imply

$$\underset{\tau \in (0,T)}{ess\ \sup} \int_\Omega |[u^{(k)}(\tau, x) + m_0]_-|^{r+1} dx$$

$$+ \int_{Q_T} \int |[u^{(k)} + m_0]_-|^r \left| \frac{\partial u^{(k)}}{\partial x} \right|^2 dx\, dt$$

$$\leq c_{48}(r+1)^{\lambda_3} \left\{ \int_{Q_T} \int |[u^{(k)} + m_0]_-|^r dx\, dt + 1 \right\}, \quad \lambda_3 = 2(n+3). \tag{4.39}$$

From this and Gronwall's Lemma we infer

$$\int_{Q_T} \int |[u^{(k)} + m_0]_-|^r dx\, dt \leq c(r) \tag{4.40}$$

for an arbitrary $r \geq 2$ and a constant $c(r)$ depending only on r and known parameters and independent of k. Using Moser's iteration process and inequality (4.40), we obtain

$$ess\ \sup\{|[u^{(k)}(t, x) + m_0]_-| : (t, x) \in Q_T\} \leq c_{49} \tag{4.41}$$

for $k > -\frac{1}{\delta}$ with a constant c_{49} depending only on known parameters. Inequality (4.41) means that the desired inequality (4.33) holds with $M_7 = m_0 + c_{49} + 1$, $0 \leq \delta \leq \frac{1}{M_7}$. $\square$

Proof of Theorem 2.6. The assertion of Theorem 2.6 follows immediately from Lemmas 4.1 and 4.3.

5. Existence proof

With the constant M_4 from Theorem 2.6 we modify the functions ρ and a such that

$$\rho^*(u) = \rho(\min[u, M_4]), \quad a^*(t, x, v, u) = a(t, x, v, \min[u, M_4]). \tag{5.1}$$

Note that ρ^*, a^* satisfy the conditions $\rho)$, $\rho')$, i), iii), (2.19), a) with the same parameters as ρ, a. Now we consider the system

$$\frac{\partial \sigma^*(u)}{\partial t} - \sum_{i=1}^{n} \frac{\partial}{\partial x_i} \left\{ \rho_\delta^*(u) b_i \left(t, x, \frac{\partial(u-v)}{\partial x} \right) \right\}$$

$$+ a^*(t, x, v, u) = 0, \quad \delta = \frac{1}{M_4}, \tag{5.2}$$

$$-\sum_{i=1}^{n} \frac{\partial}{\partial x_i} \left[\kappa(x) \frac{\partial v}{\partial x_i} \right] + \sigma^*(u) = f(t, x), \tag{5.3}$$

completed by the conditions (1.3)–(1.5). By Theorem 2.6 arbitrary solutions (u, v) of this initial boundary value problem satisfy the a priori estimate (2.22). From (5.1) and $\rho_\delta^*(u) = \max\{\rho^*(u), \rho^*(-M_4)\} = \max\{\rho(u), \rho(-M_4)\}$ we see that a solution of problem (5.2), (5.3), (1.3)–(1.5) is automatically a solution of problem (1.1)–(1.5).

We don't want to go into details of proving solvability of the problem (5.2), (5.3), (1.3)–(1.5). That could be done via Euler's backward time discretization. Such approach was used in [2], [5]. We remark only that solvability of the arising elliptic problem can be proved by using degree theory for operators of class (S_+) [13].

6. Proof of uniqueness

In order to prove uniqueness of solutions to problem (1.1)–(1.5) we assume that there exist two solutions (u_j, v_j), $j = 1, 2$, in the sense of Definition 2.1 and show that necessarily $u_1 = u_2$, $v_1 = v_2$.

By Theorems 2.4, 2.5, we have

$$\|u_j\|_{L^\infty(Q_T)} + \|v_j\|_{L^\infty(Q_T)} + \left\| \frac{\partial u_j}{\partial x} \right\|_{L^2(Q_T)} + \left\| \frac{\partial v_j}{\partial x} \right\|_{L^2(Q_T)} \leq M \tag{6.1}$$

with some constant M depending only on known parameters.

The proof of Theorem 2.8 will consists in four steps corresponding to four different choices of test functions in the integral identities (2.8), (2.9).

FIRST STEP.   We test (2.8) for $u = u_i$, $v = v_i$, $i = 1, 2$, with functions φ_i given by

$$\varphi_1 = \frac{1}{\rho(u_1)} [\sigma(u_1) - \sigma(u_2)], \quad \varphi_2 = u_1 - u_2.$$

Taking the difference of the obtained equalities we find

$$
\int_0^\tau \left\{ \left\langle \frac{\partial \sigma(u_1)}{\partial t}, \frac{1}{\rho(u_1)}[\sigma(u_1) - \sigma(u_2)] \right\rangle - \left\langle \frac{\partial \sigma(u_2)}{\partial t}, (u_1 - u_2) \right\rangle \right\} dt
$$

$$
+ \sum_{i=1}^n \int_{Q_\tau} \int \left\{ b_i \left(t, x, \frac{\partial(u_1 - v_1)}{\partial x} \right) \left[\left(\rho(u_1) - \frac{\rho'(u_1)}{\rho(u_1)} \int_{u_2}^{u_1} \rho(s)\, ds \right) \frac{\partial u_1}{\partial x_i} \right. \right.
$$

$$
\left. - \rho(u_2)\frac{\partial u_2}{\partial x_i} \right] - \rho(u_2) b_i \left(t, x, \frac{\partial(u_2 - v_2)}{\partial x} \right) \times \frac{\partial(u_1 - u_2)}{\partial x_i} \Bigg\} dx\, dt
$$

$$
+ \int_{Q_\tau} \int \left\{ a(t, x, v_1, u_1) \frac{1}{\rho(u_1)}[\sigma(u_1) - \sigma(u_2)] \right.
$$

$$
\left. - a(t, x, v_2, u_2)(u_1 - u_2) \right\} dx\, dt = 0. \tag{6.2}
$$

We shall evaluate the left hand side of (6.2) term by term. We start with the first integral applying Lemma 2 from [9] with respect to the function

$$
F(z_1, z_2) = F_1(z_1, z_2) = \int_{\sigma^{-1}(z_2)}^{\sigma^{-1}(z_1)} [\sigma^{-1}(z_1) - s]\rho(s)\, ds.
$$

We obtain by (6.1)

$$
\int_0^\tau \left\{ \left\langle \frac{\partial \sigma(u_1)}{\partial t}, \frac{1}{\rho(u_1)}[\sigma(u_1) - \sigma(u_2)] \right\rangle - \left\langle \frac{\partial \sigma(u_2)}{\partial t}, u_1 - u_2 \right\rangle \right\} dt
$$

$$
= \int_\Omega F_1(\sigma(u_1(\tau, x)), \sigma(u_2(\tau, x)))\, dx
$$

$$
= \int_\Omega \left\{ \int_{u_2(\tau,x)}^{u_1(\tau,x)} [u_1(\tau, x) - s]\rho(s)\, ds \right\} dx
$$

$$
\geq c_{50} \int_\Omega |u_1(\tau, x) - u_2(\tau, x)|^2\, dx. \tag{6.3}
$$

We shall estimate the second integral in (6.2) by using the inequalities

$$
\rho(u_1) - \frac{\rho'(u_1)}{\rho(u_1)} \int_{u_2}^{u_1} \rho(s)\, ds \geq \rho(u_1) - \int_{u_2}^{u_1} \frac{\rho'(s)}{\rho(s)} \rho(s)\, ds = \rho(u_2), \tag{6.4}
$$

$$
\left| \rho(u_1) - \rho(u_2) - \frac{\rho'(u_1)}{\rho(u_1)} \int_{u_2}^{u_1} \rho(s)\, ds \right| \leq c_{51} |u_1 - u_2|^2, \tag{6.5}
$$

that follow from condition $\rho)$ and the local Lipschitz condition for ρ', respectively.

From ii), (6.4), (6.5) and the local Lipschitz condition for the function b_i we get

$$\sum_{i=1}^{n} \int_{Q_\tau} \int \left\{ b_i\left(t, x, \frac{\partial(u_1 - v_1)}{\partial x}\right) \left[\left(\rho(u_1) - \frac{\rho'(u_1)}{\rho(u_1)} \int_{u_2}^{u_1} \rho(s)\, ds \right) \right. \right.$$

$$\times \left(\frac{\partial(u_1 - v_1)}{\partial x_i} + \frac{\partial v_1}{\partial x_i} \right) - \rho(u_2)\frac{\partial u_2}{\partial x_i} \right]$$

$$\left. - \rho(u_2) b_i\left(t, x, \frac{\partial(u_2 - v_2)}{\partial x}\right) \frac{\partial(u_1 - u_2)}{\partial x_i} \right\} dx\, dt$$

$$\geq \sum_{i=1}^{n} \int_{Q_\tau} \int \rho(u_2) \left[b_i\left(t, x, \frac{\partial(u_1 - v_1)}{\partial x}\right) - b_i\left(t, x, \frac{\partial(u_2 - v_2)}{\partial x}\right) \right]$$

$$\times \frac{\partial(u_1 - u_2 - v_1 + v_2)}{\partial x_i}\, dx\, dt$$

$$- c_{52} \int_{Q_\tau} \int \left\{ |u_1 - u_2|^2 \left[\left| \frac{\partial(u_1 - v_1)}{\partial x} \right| + 1 \right] \left| \frac{\partial v_1}{\partial x} \right| \right.$$

$$\left. + \left| \frac{\partial(u_1 - v_1 - u_2 + v_2)}{\partial x} \right| \left| \frac{\partial(v_1 - v_2)}{\partial x} \right| \right\} dx\, dt$$

$$\geq c_{53} \int_{Q_\tau} \int \left| \frac{\partial(u_1 - u_2)}{\partial x} \right|^2 dx\, dt$$

$$- c_{54} \int_{Q_\tau} \int \left\{ |u_1 - u_2|^2 \left[1 + \left| \frac{\partial(u_1 - v_1)}{\partial x} \right| \right] \left| \frac{\partial v_1}{\partial x} \right| \right.$$

$$\left. + \left| \frac{\partial(v_1 - v_2)}{\partial x} \right|^2 \right\} dx\, dt. \tag{6.6}$$

The last integral in (6.2) we estimate by using condition a), iii), the local Lipschitz condition for a and the inequality

$$\left| \frac{1}{\rho(u_2)}[\sigma(u_1) - \sigma(u_2)] - (u_1 - u_2) \right| \leq c_{55}|u - u_2|^2,$$

that follows from local boundedness of ρ'. We obtain

$$\int_{Q_\tau} \int \left\{ a(t, x, v_1, u_1)\frac{1}{\rho(u_1)}[\sigma(u_1) - \sigma(u_2)] - a(t, x, v_2, u_2)(u_1 - u_2) \right\} dx\, dt$$

$$\geq \int_{Q_\tau} \int \left\{ a(t, x, v_1, u_2) \left[\frac{1}{\rho(u_2)}(\sigma(u_1) - \sigma(u_2)) - (u_1 - u_2) \right] \right.$$

$$\left. + [a(t, x, v_1, u_2) - a(t, x, v_2, u_2)](u_1 - u_2) \right\} dx\, dt$$

$$\geq -c_{55} \int_{Q_\tau} \int \{ [1 + \alpha(t, x)](u_1 - u_2)^2 + (v_1 - v_2)^2 \}\, dx\, dt. \tag{6.7}$$

Now (6.2) and (6.3), (6.6), (6.7) and Poincaré's inequality imply

$$\int_\Omega |u_1(\tau, x) - u_2(\tau, x)|^2 \, dx \; + \int_{Q_\tau} \int \left| \frac{\partial(u_1 - u_2)}{\partial x} \right|^2 \, dx \, dt$$

$$\leq c_{56} \int_{Q_\tau} \int \left\{ \left| \frac{\partial(v_1 - v_2)}{\partial x} \right|^2 \right.$$

$$\left. + \left[\left(1 + \left| \frac{\partial(u_1 - v_1)}{\partial x} \right| \right) \left| \frac{\partial v_1}{\partial x} \right| + 1 + \alpha(t, x) \right] |u_1 - u_2|^2 \right\} \, dx \, dt. \tag{6.8}$$

SECOND STEP. We test the integral identity (2.9) for $u = u_j$, $v = v_j$, $j = 1, 2$, with $\psi_1 = v_1 - v_2$. Taking the difference of the obtained equalities, applying condition ii)$_3$ and the inequalities of Cauchy and Poincaré, we get

$$\int_\Omega \left| \frac{\partial(v_1 - v_2)}{\partial x} \right|^2 \, dx \leq c_{57} \int_\Omega |u_1 - u_2|^2 \, dx. \tag{6.9}$$

THIRD STEP. We test the integral identity (2.9) for $u = u_j$, $v = v_j$ with

$$\varphi_3 = \frac{1}{\rho(u_1)} [\exp(N\sigma(u_1)) - \exp(N\sigma(u_2))]_+, \quad \varphi_4 = N[u_1 - u_2]_+ \exp(N\sigma(u_2)),$$

where N is a positive number depending only on known parameters and satisfying

$$N\rho^2(s) + 2\rho'(s) \geq 1 \quad \text{for} \quad |s| \leq M \tag{6.10}$$

with the constant M from (6.1). Taking the difference of the obtained equalities we get

$$\int_0^\tau \left\{ \left\langle \frac{\partial \sigma_1}{\partial t}, \frac{1}{\rho(u_1)} [\exp(N\sigma(u_1)) - \exp(N\sigma(u_2))]_+ \right\rangle \right.$$

$$\left. - \left\langle \frac{\partial \sigma_2}{\partial t}, N[u_1 - u_2]_+ \exp(N\sigma(u_2)) \right\rangle \right\} \, dt \; + I^{(1)} + I^{(2)} + I^{(3)} = 0, \tag{6.11}$$

where with $Q_\tau^+ = \{(t, x) \in Q_\tau : u_1(t, x) > u_2(t, x)\}$

$$I^{(1)} = N \sum_{j=1}^n \int_{Q_\tau^+} \int b_i \left(t, x, \frac{\partial(u_1 - v_1)}{\partial x} \right) \left[\rho(u_1) \exp(N\sigma(u_1)) \frac{\partial u_1}{\partial x_i} \right.$$

$$- \rho(u_2) \exp(N\sigma(u_2)) \frac{\partial u_2}{\partial x_i} - \frac{\rho'(u_1)}{\rho(u_1)} \int_{u_2}^{u_1} \rho(s) \exp(N\sigma(s)) \, ds$$

$$\times \left. \left(\frac{\partial(u_1 - v_1)}{\partial x_i} + \frac{\partial v_1}{\partial x_i} \right) \right] \, dx \, dt, \tag{6.12}$$

$$I^{(2)} = -N \sum_{j=1}^{n} \int_{Q_\tau^+} \int \rho(u_2) b_i \left(t, x, \frac{\partial(u_2 - v_2)}{\partial x} \right)$$

$$\times \left[\frac{\partial(u_1 - u_2)}{\partial x_i} + N\rho(u_2)(u_1 - u_2) \frac{\partial u_2}{\partial x_i} \right] \exp(N\sigma(u_2)) \, dx \, dt, \tag{6.13}$$

$$I^{(3)} = \int_{Q_\tau^+} \int \left\{ \frac{a(t, x, v_1, u_1)}{\rho(u_1)} [\exp(N\sigma(u_1)) - \exp(N\sigma(u_2))] \right.$$

$$\left. - Na(t, x, v_2, u_2)(u_1 - u_2) \exp(N\sigma(u_2)) \right\} \, dx \, dt. \tag{6.14}$$

We shall evaluate the terms of the left hand side of (6.11). To the first one we apply Lemma 2 from [9] with respect to the function

$$F(z_1, z_2) = F_2(z_1, z_2) = N \left[\int_{\sigma^{-1}(z_2)}^{\sigma^{-1}(z_1)} (\sigma^{-1}(z_1) - s) \rho(s) \exp(N\sigma(s)) \, ds \right]_+ .$$

Using (6.1) we obtain

$$\int_0^\tau \left\{ \left\langle \frac{\partial \sigma_1}{\partial t}, \frac{1}{\rho(u_1)} [\exp(N\sigma(u_1)) - \exp(N\sigma(u_2))]_+ \right\rangle \right.$$

$$\left. - \frac{\partial \sigma_2}{\partial t}, N[u_1 - u_2]_+ \exp(N\sigma(u_2)) \right\} \, dt$$

$$= \int_\Omega F_2(\sigma(u_1(\tau, x)), \sigma(u_2(\tau, x))) \, dx$$

$$\geq c_{58} \int_\Omega [u_1(\tau, x) - u_2(\tau, x)]_+^2 \, dx. \tag{6.15}$$

As to the second summand in (6.11), we use the inequality

$$-\frac{\rho'(u_1)}{\rho(u_1)} \int_{u_2}^{u_1} \rho(s) \exp(N\sigma(s)) \, ds \geq - \int_{u_2}^{u_1} \rho'(s) \exp(N\sigma(s)) \, ds$$

$$= \rho(u_2) \exp(N\sigma(u_2)) - \rho(u_1) \exp(N\sigma(u_1))$$

$$+ N \int_{u_2}^{u_1} \rho^2(s) \exp(N\sigma(s)) \, ds, \tag{6.16}$$

that follows from condition ρ). We obtain

$$I^{(1)} \geq N^2 \sum_{j=1}^{n} \int_{Q_\tau^+} \int b_i \left(t, x, \frac{\partial(u_1 - v_1)}{\partial x} \right) \frac{\partial(u_1 - v_1)}{\partial x_i}$$

$$\times \int_{u_2}^{u_1} \rho^2(s) \exp(N\sigma(s)) \, ds \, dx \, dt + I_1^{(1)} + I_2^{(1)}, \tag{6.17}$$

where

$$I_1^{(1)} = N \sum_{j=1}^{n} \int_{Q_\tau^+} \int \rho(u_2) \exp(N\sigma(u_2)) b_i \left(t, x, \frac{\partial(u_1 - v_1)}{\partial x} \right) \frac{\partial(u_1 - u_2)}{\partial x_i} \, dx \, dt,$$

$$I_2^{(1)} = N \sum_{j=1}^{n} \int_{Q_\tau^+} \int b_i \left(t, x, \frac{\partial(u_1 - v_1)}{\partial x} \right) \frac{\partial v_1}{\partial x_i} \left\{ \rho(u_1) \exp(N\sigma(u_1)) \right.$$

$$\left. - \rho(u_2) \exp(N\sigma(u_2)) - \frac{\rho'(u_1)}{\rho(u_1)} \int_{u_2}^{u_1} \rho(s) \exp(N\sigma(s)) \, ds \right\} \, dx \, dt. \tag{6.18}$$

We transform the integral from (6.13) in the following way

$$I^{(2)} = -N^2 \sum_{i=1}^{n} \int_{Q_\tau^+} \int \rho^2(u_2) b_i \left(t, x, \frac{\partial(u_1 - v_1)}{\partial x} \right) \frac{\partial(u_1 - v_1)}{\partial x_i}$$

$$\times (u_1 - u_2) \exp(N\sigma(u_2)) \, dx \, dt + I_1^{(2)} + I_2^{(2)} + I_3^{(2)}, \tag{6.19}$$

where

$$I_1^{(2)} = -N \sum_{i=1}^{n} \int_{Q_\tau^+} \int \rho(u_2) \exp(N\sigma(u_2)) b_i \left(t, x, \frac{\partial(u_2 - v_2)}{\partial x} \right) \frac{\partial(u_1 - u_2)}{\partial x_i} \, dx \, dt,$$

$$I_2^{(2)} = -N^2 \sum_{i=1}^{n} \int_{Q_\tau^+} \int \rho^2(u_2) b_i \left(t, x, \frac{\partial(u_1 - v_1)}{\partial x} \right)$$

$$\frac{\partial v_1}{\partial x_i} (u_1 - u_2) \exp(N\sigma(u_2)) \, dx \, dt,$$

$$I_3^{(2)} = -N^2 \sum_{i=1}^{n} \int_{Q_\tau^+} \int \rho^2(u_2)(u_1 - u_2) \exp(N\sigma(u_2)) \left[b_i \left(t, x, \frac{\partial(u_2 - v_2)}{\partial x} \right) \frac{\partial u_2}{\partial x_i} \right.$$

$$\left. - b_i \left(t, x, \frac{\partial(u_1 - v_1)}{\partial x} \right) \frac{\partial u_1}{\partial x_i} \right] \, dx \, dt.$$

Now we shall estimate summands from $I^{(1)}$ and $I^{(2)}$ that arise from (6.17) and (6.19). In view of the choice of N ((6.10)) we have for $u_1 > u_2$

$$\int_{u_2}^{u_1} \rho^2(s) \exp(N\sigma(s)) \, ds - \rho^2(u_2) \exp(N\sigma(u_2))(u_1 - u_2)$$

$$= \int_{u_2}^{u_1} \int_{u_2}^{s} [2\rho(z)\rho'(z) + N\rho^3(z)] \exp(N\sigma(z)) \, dz \, ds \geq c_{58} |u_1 - u_2|^2. \tag{6.20}$$

Hence we get

$$N^2 \sum_{i=1}^{n} \int_{Q_\tau^+} \int b_i \left(t, x, \frac{\partial(u_1 - v_1)}{\partial x} \right) \frac{\partial(u_1 - v_1)}{\partial x_i} \left\{ \int_{u_2}^{u_1} \rho^2(s) \exp(N\sigma(s)) \, ds \right.$$

$$\left. - \rho^2(u_2) \exp(N\sigma(u_2))(u_1 - u_2) \right\} dx \, dt$$

$$\geq c_{59} \int_{Q_\tau^+} \int |u_1 - u_2|^2 \left| \frac{\partial(u_1 - v_1)}{\partial x} \right|^2 dx \, dt. \tag{6.21}$$

Using condition ii) we get

$$I_1^{(1)} + I_2^{(1)} \geq c_{60} \int_{Q_\tau^+} \int \left| \frac{\partial(u_1 - u_2)}{\partial x} \right|^2 dx \, dt$$

$$- c_{61} \int_{Q_\tau^+} \int \left| \frac{\partial(v_1 - v_2)}{\partial x} \right|^2 dx \, dt. \tag{6.22}$$

Since, as a consequence of the local Lipschitz continuity of ρ',

$$\left| \rho(u_1) \exp(N\sigma(u_1)) - \rho(u_2) \exp(N\sigma(u_2)) - \frac{\rho'(u_1)}{\rho(u_1)} \int_{u_2}^{u_1} \rho(s) \exp N\sigma(s) \, ds \right.$$

$$\left. - N\rho^2(u_2)(u_1 - u_2) \exp(N\sigma(u_2)) \right| \leq c_{62} |u_1 - u_2|^2, \tag{6.23}$$

condition ii) yields

$$|I_2^{(1)} + I_2^{(2)}| \geq c_{63} \int_{Q_\tau^+} \int \left(1 + \left| \frac{\partial(u_1 - v_1)}{\partial x} \right| \right) \left| \frac{\partial v_1}{\partial x} \right| |u_1 - u_2|^2 \, dx \, dt. \tag{6.24}$$

The next estimate follows from the local Lipschitz condition for b_i:

$$|I_3^{(2)}| \leq c_{64} \int_{Q_\tau^+} \int \left\{ \varepsilon |u_1 - u_2|^2 \left| \frac{\partial(u_1 - v_1)}{\partial x} \right|^2 + \frac{1}{\varepsilon} \left| \frac{\partial(u_1 - u_2)}{\partial x} \right|^2 \right.$$

$$\left. + \frac{1}{\varepsilon} \left| \frac{\partial(v_1 - v_2)}{\partial x} \right|^2 + \left| \frac{\partial v_1}{\partial x} \right|^2 |u_1 - u_2|^2 \right\} dx \, dt. \tag{6.25}$$

Here $\varepsilon \in (0, 1)$ is an arbitrary number. The term $I^{(3)}$ defined by (6.14) can be estimated analogously to (6.7) such that we get

$$I^{(3)} \geq -c_{65} \int_{Q_\tau^+} \int (1 + \alpha(t, x)) |u_1 - u_2|^2 \, dx \, dt. \tag{6.26}$$

Finally, we obtain from (6.11), (6.15), (6.17), (6.19), (6.21), (6.22), (6.24)–(6.26) for sufficiently small ε

$$\int_\Omega [u_1(\tau, x) - u_2(\tau, x)]_+^2 \, dx + \int_{Q_\tau^+} \int |u_1 - u_2|^2 \left| \frac{\partial(u_1 - v_1)}{\partial x} \right|^2 dx \, dt$$

$$\leq c_{66} \int_{Q_\tau^+} \int \left\{ \left| \frac{\partial(u_1 - u_2)}{\partial x} \right|^2 + \left| \frac{\partial(v_1 - v_2)}{\partial x} \right|^2 + \left| \frac{\partial v_1}{\partial x} \right|^2 |u_1 - u_2|^2 \right.$$

$$\left. + (1 + \alpha(t, x)) |u_1 - u_2|^2 \right\} dx \, dt. \tag{6.27}$$

Changing the places of u_1 and u_2 in the last inequality we get immediately

$$\int_\Omega |u_1(\tau, x) - u_2(\tau, x)|^2 \, dx + \int_{Q_\tau} \int |u_1 - u_2|^2 \left(\left| \frac{\partial u_1}{\partial x} \right|^2 + \left| \frac{\partial u_2}{\partial x} \right|^2 \right) dx \, dt$$

$$\leq c_{67} \int_{Q_\tau} \int \left\{ \left| \frac{\partial(u_1 - u_2)}{\partial x} \right|^2 + \left| \frac{\partial(v_1 - v_2)}{\partial x} \right|^2 + \left(\left| \frac{\partial v_1}{\partial x} \right|^2 + \left| \frac{\partial v_2}{\partial x} \right|^2 \right) |u_1 - u_2|^2 \right.$$

$$\left. + (1 + \alpha(t, x)) |u_1 - u_2|^2 \right\} dx \, dt. \tag{6.28}$$

FOURTH STEP. Let $\{\varphi_j(x)\}$, $j = 1, \ldots, J$, be a partition of unity satisfying the conditions (4.2) with a number R to be fixed chosen later on. We test the integral identity (2.9) for $u = u_1, v = v_1$ with

$$\psi = \sum_{j=1}^J [v_1 - v_{1,j}] \varphi_j^2 |u_1 - u_2|^2, \quad v_{1,j} = v_{1,j}(t) = v(t, x_j). \tag{6.29}$$

We obtain after integration with respect to t

$$\int_{Q_\tau} \int \kappa(x) |u_1 - u_2|^2 \left| \frac{\partial v_1}{\partial x} \right|^2 dx \, dt = J^{(1)} + J^{(2)} + J^{(3)}, \tag{6.30}$$

where

$$J^{(1)} = -2 \sum_{j=1}^J \sum_{i=1}^n \int_{Q_\tau} \int \kappa(x) \frac{\partial v_1}{\partial x_i} \frac{\partial(u_1 - u_2)}{\partial x_i} [v_1 - v_{1,j}](u_1 - u_2)\varphi_j^2 \, dx \, dt,$$

$$J^{(2)} = -2 \sum_{j=1}^J \sum_{i=1}^n \int_{Q_\tau} \int \kappa(x) \frac{\partial v_1}{\partial x_i} \frac{\partial \varphi_j}{\partial x_i} [v_1 - v_{1,j}]\varphi_j |u_1 - u_2|^2 \, dx \, dt,$$

$$J^{(3)} = - \sum_{j=1}^J \int_{Q_\tau} \int [\sigma(u_j) - f][v_1 - v_{1,j}]\varphi_j^2 |u_1 - u_2|^2 \, dx \, dt.$$

We estimate $J^{(1)}$, $J^{(2)}$, $J^{(3)}$ by Cauchy's inequality, Theorem 2.5 and (4.2) and obtain

$$|J^{(1)}| \leq \frac{1}{4} \int_{Q_\tau} \int \kappa(x) \left|\frac{\partial v_1}{\partial x}\right|^2 |u_1 - u_2|^2 \, dx \, dt$$

$$+ c_{68} R^\alpha \int_{Q_\tau} \int \left|\frac{\partial(u_1 - u_2)}{\partial x}\right|^2 \, dx \, dt,$$

$$|J^{(2)}| \leq \frac{1}{4} \int_{Q_\tau} \int \kappa(x) \left|\frac{\partial v_1}{\partial x}\right|^2 |u_1 - u_2|^2 \, dx \, dt + \frac{c_{69}}{R^2} \int_{Q_\tau} \int |u_1 - u_2|^2 \, dx \, dt,$$

$$|J^{(3)}| \leq c_{70} \int_{Q_\tau} \int (1 + |f|) |u_1 - u_2|^2 \, dx \, dt. \tag{6.31}$$

The equality (6.30) and inequalities (6.31) imply immediately

$$\int_{Q_\tau} \int |u_1 - u_2|^2 \left|\frac{\partial v_1}{\partial x}\right|^2 \, dx \, dt \leq c_{71} \int_{Q_\tau} \int \left\{ R^\alpha \left|\frac{\partial(u_1 - u_2)}{\partial x}\right|^2 \right.$$

$$\left. + \left(\frac{1}{R^2} + |f|\right) |u_1 - u_2|^2 \right\} \, dx \, dt. \tag{6.32}$$

End of the proof of Theorem 2.8. Applying Cauchy's inequality to the term in (6.8) involving the derivative of $u_1 - v_1$ and choosing a suitable value of R, we obtain from (6.8), (6.9), (6.28), (6.32)

$$\int_\Omega |u_1(\tau, x) - u_2(\tau, x)|^2 \, dx + \int_{Q_\tau} \int \left|\frac{\partial(u_1 - u_2)}{\partial x}\right|^2 \, dx \, dt$$

$$\leq c_{72} \int_{Q_\iota} \int (1 + |\alpha| + |f|) |u_1 - u_2|^2 \, dx \, dt. \tag{6.33}$$

We estimate the integral on the right hand site of (6.33) by Hölder's inequality and use the conditions on α, f to get

$$\underset{\tau \in (0,\theta)}{ess \sup} \int_\Omega |u_1(\tau, x) - u_2(\tau, x)|^2 \, dx + \int_{Q_\theta} \int \left|\frac{\partial(u_1 - u_2)}{\partial x}\right|^2 \, dx \, dt$$

$$\leq c_{73} \left\{ \int_{Q_\theta} \int |u_1 - u_2|^{2p_1'} \, dx \, dt \right\}^{\frac{1}{p_1'}}$$

$$+ c_{73} \int_0^\theta \left\{ \int_\Omega |u_1 - u_2|^{2p_2'} \, dx \right\}^{\frac{1}{p_2'}} \, dt \tag{6.34}$$

for an arbitrary $\theta \in (0, T)$. Estimating the first integral on the right hand site of (6.34) by Hölder's inequality, using the embedding $V^2(Q_T) \subset L^{2\frac{(n+2)}{n}}(Q_T)$ (comp. [11]) and setting $q_1 = n + 2 - p_1' n$, we find for arbitrary $\varepsilon \in (0, 1)$ and a constant c_{74} depending only on n

$$\left\{ \int_{Q_\theta} \int |u_1 - u_2|^{2p_1'} \, dx \, dt \right\}^{\frac{1}{p_1'}} \leq \left(\int_{Q_\theta} \int |u_1 - u_2|^2 \, dx \, dt \right)^{\frac{q_1}{2p_1'}}$$

$$\times \left\{ \int_{Q_\theta} \int |u_1 - u_2|^{\frac{2(n+2)}{n}} \, dx \, dt \right\}^{\frac{1}{p_1'} - \frac{q_1}{2p_1'}} \leq \varepsilon^{-\frac{2p_1'}{q_1}} \int_{Q_\theta} \int |u_1 - u_2|^2 \, dx \, dt$$

$$+ c_{74} \varepsilon^{\frac{2p_1'}{2p_1' - q_1}} \left\{ \mathrm{ess} \sup_{\tau \in (0,\theta)} \int_\Omega |u_1(\tau, x) - u_2(\tau, x)|^2 \, dx \right.$$

$$\left. + \int_{Q_\theta} \int \left| \frac{\partial (u_1 - u_2)}{\partial x} \right|^2 \, dx \, dt \right\}. \tag{6.35}$$

In analogous way we estimate the last integral in (6.34). We define γ to be solution of the equation $\frac{2-\gamma}{2p_2'-\gamma} = \frac{1}{2}(\frac{1}{p_2'} + \frac{n-2}{n})$. Then $q_2 = \frac{2p_2'-\gamma}{2-\gamma} < \frac{n}{n-2}$ and $2q_2 < 2(1 + \frac{2q_2}{n})$.

Estimating the last integral in (6.34) by Hölder's inequality and (6.35), we find

$$\int_0^\theta \left\{ \int_\Omega |u_1 - u_2|^{2p_2'} \, dx \right\}^{\frac{1}{p_2'}} dt \leq c_{75} \left(\int_0^\theta \int |u_1 - u_2|^2 \, dx \, dt \right)^{\frac{\gamma}{2p_2'}}$$

$$\times \left\{ \int_0^\theta \left[\int_\Omega |u_1 - u_2|^{2(1+\frac{2q_2}{n})} \, dx \right]^{\frac{1}{q_2}} dt \right\}^{(1+\frac{\gamma}{2p_2'}) \frac{nq_2}{n+2q_2}}$$

$$\leq c_{76} \left\{ \varepsilon^{-\frac{2p_2'}{\gamma}} \int_{Q_\theta} \int |u_1 - u_2|^2 \, dx \, dt \right.$$

$$+ \varepsilon^{\frac{2p_2'}{2p_2' - \gamma}} \left[\mathrm{ess} \sup_{\tau \in (0,\theta)} \int_\Omega |u_1(\tau, x) - u_2(\tau, x)|^2 \, dx \right.$$

$$\left. \left. + \int_{Q_\theta} \int \left| \frac{\partial (u_1 - u_2)}{\partial x} \right|^2 \, dx \, dt \right] \right\}. \tag{6.36}$$

The inequalities (6.34)–(6.36) imply with suitable ε

$$\int_\Omega |u_1(\theta, x) - u_2(\theta, x)|^2 \, dx \leq c_{77} \int_{Q_\theta} \int |u_1 - u_2|^2 \, dx \, dt \tag{6.37}$$

for arbitrary $\theta \in (0, T)$. Finally, Gronwall's lemma yields $u_1 = u_2$ and the equality $v_1 = v_2$ follows now from (6.9).

Proof of Corollary 2.9. Let u be the solution of (1.1)–(1.5), $u_1(t, x) = u(t, x)$, $u_2(t, x) = u(t + \delta, x)$ with $\delta \in (0, T - t)$. We test integral identity (2.8) with the functions

$$\varphi_1(t, x) = \frac{t^2}{\rho^*(u_1(t, x))} [\sigma^*(u_1(t, x)) - \sigma^*(u_2(t, x))],$$

$$\varphi_2(t, x) = t^2 (u_1(t, x) - u_2(t, x)).$$

Then, arguing essentially as in the proof of (6.34) and (6.37), we obtain

$$\tau^2 \int_\Omega |u_1(\tau, x) - u_2(\tau, x)|^2 \, dx + \int_{Q_\tau} t^2 \int \left| \frac{\partial (u_1 - u_2)}{\partial x} \right|^2 \, dx \, dt$$

$$\leq c_{64} \int_{Q_\tau} \int |u_1 - u_2|^2 \, dx \, dt.$$

Now dividing by δ^2, applying Gronwall's lemma and taking the limit $\delta \to 0$, the corollary follows.

REFERENCES

[1]   AMANN, H., *Nonhomogeneous linear and quasilinear elliptic and parabolic boundary value problems*, in: Function spaces, differential operators and nonlinear analysis (H. -J. Schmeisser, H. Triebel, eds.), B. G. Teubner Verlagsgesellschaft, Teubner-Texte Math. *133* Stuttgart 1993, 9–126.

[2]   ALT, H. W. and LUCKHAUS, S., *Quasilinear elliptic-parabolic differential equations*, Math. Z. *183* (1983), 311–341.

[3]   BENILAN, PH. and WITTBOLD, P., *On mild and weak solutions of elliptic-parabolic systems*, Adv. Differ. Equ. vol. *1* (1996), 1053–1073.

[4]   GAJEWSKI, H., *On a variant of monotonicity and its application to differential equations*, Nonlinear Anal. TMA, vol. *22* (1994), 73–80.

[5]   GAJEWSKI, H. and GRÖGER, K., *Reaction-diffusion processes of electrically charged species*, Math. Nachr. *177* (1996), 109–130.

[6]   GAJEWSKI, H. and ZACHARIAS, K., *Global behavior of a reaction-diffusion system modelling chemotaxis*, Math. Nachr. *195* (1998), 77–114.

[7]   GAJEWSKI, H. and ZACHARIAS, K., *On a nononlocal phase separation model*, Preprint 656 WIAS Berlin, 2001, Jnl. Math. Anal. Appl. (to appear).

[8]   GAJEWSKI, H. and SKRYPNIK, I. V., *To the uniqueness problem for nonlinear elliptic equations*, Nonlinear Analysis *52* (2003), 291–304.

[9]   GAJEWSKI, H. and SKRYPNIK, I.V., *On the uniqueness of solutions for nonlinear parabolic equations*, Preprint No. 658, WIAS (2001), Discrete and Continuous Dynamical Systems *9* (2003).

[10] LADYZHENSKAYA, O. A. and URALTSEVA, N. N., *Linear and quasilinear elliptic equations*, Nauka, Moscow 1973 (Russian).

[11] LADYZHENSKAJA, O. A., SOLONNIKOV, V. A. and URALTSEVA, N. N., *Linear and quasilinear equations of parabolic typ*, Transl. Math. Monographs, A.M.S. Providence, vol. *23* (1994).

[12] OTTO, F. L^1-*contraction and uniqueness for quasilinear elliptic-parabolic equations*, C.R. Acad. Sci. Paris, *318*, Serie 1 (1995), 1005–1010.

[13] SKRYPNIK, I. V., *Methods for analysis of nonlinear elliptic boundary value problems*, Transl. Math. Monographs, A.M.S. Providence, vol. *139* (1994).

H. Gajewski
Weierstraß–Institute for Applied Analysis and Stochastics
Mohrenstr. 39
D–10117 Berlin
Germany

I. V. Skrypnik
Institute for Applied Mathematics and Mechanics
Rosa Luxemburg Str. 74
340114 Donetsk
Ukraine

J.evol.equ. 3 (2003) 283 – 301
1424–3199/03/020283 – 19
DOI 10.1007/s00028-003-0095-x
© Birkhäuser Verlag, Basel, 2003

**Journal of Evolution
Equations**

Conservation laws with discontinuous flux functions and boundary condition

JOSÉ CARRILLO

Abstract. In a bounded domain Ω we study the existence and uniqueness of entropy solutions of $\partial u/\partial t +$ div $\Phi(u) = f$ with $u(0) = u_0$, where Φ is allowed to have some discontinuities of first type.

1. Introduction

Let Ω be a bounded domain in $\mathbb{R}^N$. For $N \geq 2$ we assume that Ω has a Lipschitz boundary. Let $T > 0$ and let $Q = (0, T) \times \Omega$, we consider the following problem

$$\begin{cases} \frac{\partial u}{\partial t} + \operatorname{div} \Phi(u) \ni f & \text{in } Q \\ u(0) = u_0 & \text{in } \Omega \end{cases} \tag{P(u_0, f)}$$

where the flux function Φ is allowed to have discontinuities of first type on a finite subset S of $\mathbb{R}$. More precisely, let $s \in S$, for $i = 1, 2 \ldots, N$, let $\Phi_i(s^+) = \lim_{r \to s, \, r > s} \Phi_i(r)$ and let $\Phi_i(s^-) = \lim_{r \to s, \, r < s} \Phi_i(r)$, we assume that, at least for some i, $\Phi_i(s^+) \neq \Phi_i(s^-)$ and we note $\gamma_{s_i} = \Phi_i(s+) - \Phi_i(s^-)$. We assume that Φ is normalized at 0: $\Phi_i(0^+) = -\Phi_i(0^-)$. In fact Φ_i can be considered as a jump function added to a continuous one. In order to get a rigorous formulation of (P(u_0, f)), we assume that Φ has the following structure:

$$\begin{cases} \Phi = \phi + J \\ \phi \in C(\mathbb{R}; \mathbb{R}^N), \quad \phi_i(0) = 0 \text{ for } i = 1, 2, \ldots, N \\ J(r) = \sum_{s \in S, \, s < 0} \gamma_s \operatorname{sign}^-(r - s) + \sum_{s \in S, \, s > 0} \gamma_s \operatorname{sign}^+(r - s) + \gamma_0 \operatorname{sign}(r) \end{cases} \tag{H1}$$

where $\gamma_s \in \mathbb{R}^N$ for any $s \in S$, $\gamma_0 = 0$ if 0 does not belong to S and

$$\operatorname{sign}(r) = \begin{cases} 1 & \text{if } r > 0 \\ [-1,1] & \text{if } r = 0 \\ -1 & \text{if } r < 0, \end{cases}$$

$$\operatorname{sign}^+(r) = \begin{cases} 1 & \text{if } r > 0 \\ [0,1] & \text{if } r = 0 \\ 0 & \text{if } r < 0, \end{cases}$$

and

$$\text{sign}^-(r) = \begin{cases} 0 & \text{if } r > 0 \\ [-1,0] & \text{if } r = 0 \\ -1 & \text{if } r < 0. \end{cases}$$

This work deals with existence and uniqueness of entropy solutions of $(P(u_0, f))$. Such equations have been studied for continuous, or more regular, flux functions Φ by many authors. Probably the modern approach of such equations is due to S. N. Kruzhkov [K1, K2]. Ph. Benilan has worked for many years in problems related to conservation laws in $(0, T) \times \mathbb{R}^N$([Be, BK, ABK]), in particular, in [BCW] the notion of renormalized entropy solution of conservation laws is introduced. In 1979 Bardos, Leroux and Nedelec proved existence and uniqueness of entropy solutions in bounded domain, for a smooth flux function Φ ([BLN]). Recently many authors have worked in problems related to conservation laws in bounded domain ([Ot, Va, PV]). In [Ca] the author proved the existence and uniqueness of entropy solutions for bounded data, with a continuous function Φ and, more recently, P. Wittbold and the author [CW] proved existence and uniqueness of renormalized entropy solutions for L^1 data. According to the previous works ([K1, K2, BLN, Ca]) dealing with similar problems for continuous flux functions Φ, we can define, "in a natural way", entropy solutions of $(P(u_0, f))$ as follows

DEFINITION 1.1. Let $(u_0, f) \in L^1(\Omega) \times L^1(Q)$, then $(u, (\chi_s)_{s \in S}) \in L^1(Q) \times (L^\infty(Q))^{Card\, S}$ is an entropy solution of $(P(u_0, f))$ if and only if

i) $\chi_s \in \text{sign}^-(u - s)$ for any $s < 0$, $\chi_s \in \text{sign}^+(u - s)$ for any $s > 0$, $\chi_0 \in \text{sign}(u)$ if $0 \in S$,

ii) $\phi(u) \in (L^1(Q))^N$,

iii) $\text{Ess}\lim_{t \to 0} \| u(t) - u_0 \|_{L^1(\Omega)} = 0$

iv) $(u, (\chi_s)_{s \in S})$ satisfy

$$\frac{\partial u}{\partial t} + \text{div}\left(\phi(u) + \sum_{s \in S} \gamma_s \chi_s\right) = f \quad \text{in } \mathcal{D}'(Q) \tag{1}$$

v) $(u, (\chi_s)_{s \in S})$ satisfy

$$\int_Q \left\{ (u - \lambda)^+ \xi_t \right.$$

$$+ \left(\text{sign}_0^+(u - \lambda)(\phi(u) - \phi(\lambda)) + \sum_{s \in S,\, s \geq \lambda} \gamma_s (\chi_s - \lambda_s)^+ \right) \cdot \nabla \xi$$

$$\left. + \text{sign}_0^+ \left(u - \lambda + \sum_{s \in S,\, s \geq \lambda} (\chi_s - \lambda_s) \right) f \xi \right\} dx\, dt \geq 0 \tag{2}$$

for any nonnegative $\xi \in \mathcal{D}((0, T) \times \mathbb{R}^N)$, for any nonnegative $\lambda \in \mathbb{R}$ and for any nonnegative $\lambda_s \in \mathrm{sign}^+(\lambda - s)$, where

$$\mathrm{sign}_0^+(s) = \begin{cases} 1 & \text{if } s > 0 \\ 0 & \text{if } s \leq 0. \end{cases}$$

and

$$\int_Q \left\{ (\lambda - u)^+ \xi_t \right.$$
$$+ \left(\mathrm{sign}_0^+(\lambda - u)(\phi(\lambda) - \phi(u)) + \sum_{s \in S, \, s \leq \lambda} \gamma_s (\lambda_s - \chi_s)^+ \right) \cdot \nabla \xi$$
$$\left. - \mathrm{sign}_0^+ \left(\lambda - u + \sum_{s \in S, \, s \geq \lambda} (\lambda_s - \chi_s) \right) f\xi \right\} \, dx \, dt \geq 0 \tag{3}$$

for any nonnegative $\xi \in \mathcal{D}((0, T) \times \mathbb{R}^N)$, for any nonpositive $\lambda \in \mathbb{R}$ and for any nonpositive $\lambda_s \in \mathrm{sign}^-(s - \lambda)$

The goal of this paper is to prove existence, comparison and uniqueness results for entropy solutions.

2. A different approach to $(\mathbf{P(u_0, f)})$

Let $(u, (\chi_s)_{s \in S})$ be an entropy solution of $(\mathrm{P}(u_0, \mathrm{f}))$; we set

$$v = u + \sum_{s \in S} \chi_s \tag{4}$$

then $v \in (I + \mathcal{G})(u)$ where $\mathcal{G}$ is a maximal monotone operator in $\mathbb{R}$:

$$\mathcal{G}(r) = \sum_{s \in S, \, s < 0} \mathrm{sign}^-(r - s) + \sum_{s \in S, \, s > 0} \mathrm{sign}^+(r - s) + \alpha_S \, \mathrm{sign}(r)$$

where α_S is 1 if $0 \in S$ and 0 otherwise. Then, $u = g(v)$ where $g = (I + \mathcal{G})^{-1}$ satisfies

$I - g$ is a nondecreasing, bounded and continuous function, $\hspace{2em}$ (A1)

$g \in C^{0,1}(\mathbb{R}), \quad g(0) = 0, \quad g$ is nondecreasing $\quad D(g^{-1}) = D(I + \mathcal{G}) = \mathbb{R} \hspace{1em}$ (A2)

(see [Br]).

One checks easily that for any $s \in S$ there exist 2 functions φ_s^1 and φ_s^2 such that

$$\begin{cases} \varphi_s^i \in \mathcal{C}(\mathbb{R}) & \text{for } i = 1, 2 \text{ and for any } s \in S \\ \varphi_s^1(0) = 0 & \text{for any } s \in S \\ \chi_s = \varphi_s^1(g(v)) + \varphi_s^2(g(v))v. \end{cases}$$

For instance, let us consider that the elements of S are ordered: $s_1 < s_2 < \ldots < 0 \le s_{i_0} < \ldots < s_k$ where s_{i_0} is the smallest element of S which is greater or equal than 0. Let φ_{s_k} be a nonnegative smooth function such that

$$\begin{cases} \varphi_{s_k}(r) = 1 & \text{for } r \ge s_k \\ \varphi_{s_k}(r) = 0 & \text{for } r \le \max(s_{k-1}, 0), \end{cases}$$

in particular $\varphi_{s_k}(0) = 0$. Then

$$\chi_{s_k} = \varphi_{s_k}(g(v))(v - g(v) - (k - i_0))$$

hence we set

$$\varphi_{s_k}^1(r) = -\varphi_{s_k}(r)(r + (k - i_0)), \quad \varphi_{s_k}^2(r) = \varphi_{s_k}(r).$$

Similarly, for any $s_i > 0$ let φ_{s_i} be a smooth and nonnegative function satisfying

$$\begin{cases} \varphi_{s_i}(r) = 1 & \text{for } r \ge s_i \\ \varphi_{s_i}(r) = 0 & \text{for } r \le \max(s_{i-1}, 0), \end{cases}$$

whence

$$\chi_{s_i} + \sum_{j=i+1}^{k} \chi_{s_j} = \varphi_{s_i}(g(v))(v - g(v) - (i - i_0))$$

and

$$\varphi_{s_i}^1(r) = -\varphi_{s_i}(r)(r + (i - i_0)) - \sum_{j=i+1}^{k} \varphi_{s_j}^1(r), \quad \varphi_{s_i}^2(r) = \varphi_{s_i}(r) - \sum_{j=i+1}^{k} \varphi_{s_j}^2(r).$$

Similarly, let φ_{s_1} be a nonnegative smooth function such that

$$\begin{cases} \varphi_{s_1}(r) = 1 & \text{for } r \le s_1 \\ \varphi_{s_1}(r) = 0 & \text{for } r \ge \min(s_2, 0), \end{cases}$$

then

$$\chi_{s_1} = \varphi_{s_1}(g(v))(v - g(v) + (\sigma - 1))$$

where $\sigma = i_0$ if $s_{i_0} = 0$, $\sigma = i_0 - 1$ if $s_{i_0} > 0$. Then we define

$$\varphi_{s_1}^1(r) = -\varphi_{s_1}(r)(r - (\sigma - 1)), \quad \varphi_{s_1}^2(r) = \varphi_{s_1}(r).$$

For any $s_i < 0$, let φ_{s_i} be a nonnegative smooth function satisfying

$$\begin{cases} \varphi_{s_i}(r) = 1 & \text{for } r \leq s_i \\ \varphi_{s_i}(r) = 0 & \text{for } r \geq \min(s_{i+1}, 0), \end{cases}$$

then

$$\chi_{s_i} + \sum_{j=1}^{i-1} \chi_{s_j} = \varphi_{s_i}(g(v))(v - g(v) + (\sigma - i)),$$

whence

$$\varphi_{s_i}^1(r) = -\varphi_{s_i}(r)(r - (\sigma - i)) + \sum_{j=1}^{i-1} \varphi_{s_i}^1(r), \quad \varphi_{s_i}^2(r) = \varphi_{s_i}(r) + \sum_{j=1}^{i-1} \varphi_{s_i}^2(r).$$

Finally, if $s_{i_0} = 0$ we have

$$\chi_0 = v - g(v) - \sum_{s \in S,\, s \neq 0} \chi_s = v - g(v) - \sum_{s \in S,\, s \neq 0} (\varphi_s^1(g(v)) + \varphi_s^2(g(v))v),$$

hence

$$\varphi_0^1(r) = -r - \sum_{s \in S,\, s \neq 0} \varphi_s^1(r), \quad \varphi_0^2(r) = 1 - \sum_{s \in S,\, s \neq 0} \varphi_s^2(r).$$

Therefore, we can define a map from $L^1(Q)$ into $L^1(Q) \times (L^\infty(Q))^{Card\,S}$ by

$$T(v) = (g(v), (\varphi_s^1(g(v)) + \varphi_s^2(g(v))v)_{s \in S}) \tag{5}$$

Then, let

$$\Psi(r) = \Psi^1(g(r)) + \Psi^2(g(r))r \tag{A3}$$

with

$$\begin{cases} \Psi^1(r) = \phi(r) + \sum_{s \in S} \gamma_s \varphi_s^1(r) \\ \Psi^2(r) = \sum_{s \in S} \gamma_s \varphi_s^2(r), \end{cases}$$

we check easily that

$$\Psi^i \in C(\mathbb{R}; \mathbb{R}^N) \text{ for } i = 1, 2 \text{ and } \Psi_j^1(0) = 0 \text{ for } j = 1, 2, \ldots, N \tag{A4}$$

and we define a "new" problem:

$$\begin{cases} \frac{\partial g(v)}{\partial t} + \operatorname{div} \Psi(v) = f & \text{in } Q, \\ g(v(0)) = u_0 & \text{in } \Omega. \end{cases} \qquad \tilde{P}(u_0, f)$$

According to [Ca] we have the following definition:

DEFINITION 2.1. Let (A2) and (A3) hold, then a function $v \in L^1(Q)$ is an entropy solution of (P(u_0, f)) if and only if

i) $\Psi(v) \in (L^1(Q))^N$,
ii) v satisfies

$$\frac{\partial g(v)}{\partial t} + \operatorname{div} \Psi(v) = f \quad \text{in } \mathcal{D}'(Q) \tag{6}$$

iii)
$$\int_Q \{(g(v) - g(\mu))^+ \xi_t$$
$$+ \operatorname{sign}_0^+(v - \mu)((\Psi(v) - \Psi(\mu)) \cdot \nabla \xi + f\xi)\} \, dx \, dt \geq 0 \tag{7}$$

for any nonnegative $\xi \in \mathcal{D}((0, T) \times \mathbb{R}^N)$ and for any nonnegative $\mu \in \mathbb{R}$, and

$$\int_Q \{(g(\mu) - g(v))^+ \xi_t$$
$$+ \operatorname{sign}_0^+(\mu - v)((\Psi(\mu) - \Psi(v)) \cdot \nabla \xi - f\xi)\} \, dx \, dt \geq 0 \tag{8}$$

for any nonnegative $\xi \in \mathcal{D}((0, T) \times \mathbb{R}^N)$ and for any nonpositive $\mu \in \mathbb{R}$.

REMARK 2.2. From (6), (7) and (8) we deduce easily that an entropy solution of ($\tilde{P}(u_0, f)$) still satisfies

$$\int_Q \{(g(v) - g(\mu))^+ \xi_t + \operatorname{sign}_0^+(v - \mu)((\Psi(v) - \Psi(\mu)) \cdot \nabla \xi + f\xi)\} \, dx \, dt \geq 0 \tag{9}$$

for any nonnegative $\xi \in \mathcal{D}(Q)$ and for any $\mu \in \mathbb{R}$, and

$$\int_Q \{(g(\mu) - g(v))^+ \xi_t + \operatorname{sign}_0^+(\mu - v)((\Psi(\mu) - \Psi(v)) \cdot \nabla \xi - f\xi)\} \, dx \, dt \geq 0 \tag{10}$$

for any nonnegative $\xi \in \mathcal{D}(Q)$ and for any $\mu \in \mathbb{R}$.

REMARK 2.3. We check easily that if v is an entropy solution of ($\tilde{P}(u_0, f)$) then $-v$ is an entropy solution of

$$\frac{\partial \tilde{g}(w)}{\partial t} + \operatorname{div} \tilde{\Psi}(w) = -f, \quad \tilde{g}(w(0)) = -u_0$$

where $\tilde{g}(r) = -g(-r)$ and $\tilde{\Psi}(r) = -\Psi(-r)$.

The problems (P(u_0, f)) and (P(u_0, f)) are equivalent. In fact

THEOREM 2.4. *Let v be an entropy solution of* ($\tilde{P}(u_0, f)$) *then $\mathcal{T}(v)$ is an entropy solution of* (P(u_0, f)) *Reciprocally, let $(u, (\chi_s)_{s \in S})$ be an entropy solution of* (P(u_0, f)) *then $v = u + \sum_{s \in S} \chi_s$ is an entropy solution of* ($\tilde{P}(u_0, f)$) .

3. Existence of entropy solutions of $(\tilde{P}(u_0, f))$

Let us consider the stationary equation

$$\operatorname{div} \Psi(v) = f \quad \text{in } \Omega. \tag{$\tilde{E}(f)$}$$

DEFINITION 3.1. Let $f \in L^1(\Omega)$, $v \in L^1(\Omega)$, then v is an entropy solution of $(\tilde{E}(f))$ if and only if:

i) $\Psi(v) \in (L^1(\Omega))^N$,

ii) v satisfies $\operatorname{div} \Psi(v) = f$ in $\mathcal{D}'(\Omega)$

iii) $\displaystyle\int_\Omega \{\operatorname{sign}_0^+(v - \lambda)((\Psi(v) - \Psi(\lambda)) \cdot \nabla\xi + f\xi)\}\, dx \geq 0$

    for any nonnegative $\xi \in \mathcal{D}(\mathbb{R}^N)$ and any nonnegative $\lambda \in \mathbb{R}$, and

$$\int_\Omega \{\operatorname{sign}_0^+(\lambda - v)((\Psi(\lambda) - \Psi(v)) \cdot \nabla\xi - f\xi)\}\, dx \geq 0$$

    for any nonnegative $\xi \in \mathcal{D}(\mathbb{R}^N)$ and any nonpositive $\lambda \in \mathbb{R}$

We define the operator $\mathcal{A}_{(g,\Psi)} \subset L^\infty(\Omega) \times L^\infty(\Omega)$ by $(u, f) \in \mathcal{A}_{(g,\Psi)}$ (or $f \in \mathcal{A}_{(g,\Psi)}(u)$) if and only if there exists $v \in L^\infty(\Omega)$ such that

$$u = g(v) \qquad \text{and } v \text{ is an entropy solution of } (\tilde{E}(f)).$$

Then we have

THEOREM 3.2. *Let g and Ψ be defined like above, more precisely, let* (A2), (A3) *and* (A4) *hold. Then $\mathcal{A}_{(g,\Psi)}$ satisfies:*

i) $\mathcal{A}_{(g,\Psi)}$ *is a T-accretive operator in* $L^1(\Omega)$, *i.e.*

$$\forall\, (u_i, f_i) \in \mathcal{A}_{(g,\Psi)}, \qquad \int_\Omega \kappa(f_1 - f_2)\, dx \quad \text{for some } \kappa \in \operatorname{sign}^+(u_1 - u_2).$$

ii) *For any $\lambda > 0$, $R(I + \lambda\mathcal{A}_{(g,\Psi)}) = L^\infty(\Omega)$.*

iii) $\overline{D(\mathcal{A}_{(g,\Psi)})} = L^1(\Omega)$.

iv) *For any $f \in L^\infty(\Omega)$ and any $\lambda > 0$, let u be such that $f \in (I + \lambda\mathcal{A}_{(g,\Psi)})(u)$, then*
    $\| u \|_{L^\infty(\Omega)} \leq \| f \|_{L^\infty(\Omega)}$.

Proof. (see [Ca], Theorem 10, p. 321) $\qquad\qquad\qquad\qquad\qquad\qquad\qquad\qquad$ $\square$

REMARK 3.3. From Theorem 3.2 we deduce that for any $\lambda > 0$ and for any $f \in L^\infty(\Omega)$ there exists a unique $u \in L^\infty(\Omega)$ such that $f \in (I + \lambda\mathcal{A}_{(g,\Psi)})(u)$ or, equivalently, $(f - u)/\lambda \in \mathcal{A}_{(g,\Psi)}(u)$

Moreover, from the theory of accretive operators in Banach spaces (see [Be], Propositions 1.28 and 2.2), we have:

THEOREM 3.4. *Let g and Ψ be defined like above, more precisely, let* (A2), (A3) *and* (A4) *hold. Then, for any $f \in L^1(Q)$ and for any $u_0 \in L^1(\Omega)$ there exists a unique integral solution u of the abstract problem*

$$\begin{cases} \frac{du}{dt} + \mathcal{A}_{(g,\Psi)}(u) \ni f, \\ u(0) = u_0, \end{cases} \qquad (\mathrm{P_A}(u_0, f))$$

and $u \in C([0, T]; L^1(\Omega))$. Moreover, let u_i be an integral solution of $(\mathrm{P_A}(u_{i_0}, f_i))$ for $i = 1, 2$. Then

$$\int_Q \kappa\{(u_1 - u_2)\zeta' + (f_1 - f_2)\zeta\}\, dx\, dt \geq 0$$

for any nonnegative function $\zeta \in \mathcal{D}(0, T)$ and for some $\kappa \in \mathrm{sign}^+(u_1 - u_2)$.

The way to prove existence of an entropy solution of $(\tilde{\mathrm{P}}(u_0, f))$ is to prove that for any integral solution u of $(\mathrm{P_A}(u_0, f))$ there exists a function $v \in L^1(Q)$, satisfying $g(v) = u$, and v is an entropy solution. In general, for arbitrary data $u_0 \in L^1(\Omega)$ and $f \in L^1(Q)$ we do not expect to have entropy solutions of $(\tilde{\mathrm{P}}(u_0, f))$ because the corresponding integral solution of $(\mathrm{P_A}(u_0, f))$, u, is unbounded, then any v such that $g(v) = u$ is unbounded whence the function $\Psi(v)$, in general, is not in $(L^1(Q))^N$.

THEOREM 3.5. *Let g and Ψ be like above, more precisely let* (A1), (A2), (A3) *and* (A4) *hold. Let $u_0 \in L^\infty(\Omega)$, let $f \in L^1((0, T); L^\infty(\Omega))$ and let u be the unique integral solution of $(\mathrm{P_A}(u_0, f))$. Then there exists an entropy solution, v, of $(\tilde{\mathrm{P}}(u_0, f))$ such that $g(v) = u$. Moreover $v \in L^\infty(Q)$.*

Proof. For any $m \in \mathbb{N}$ we define $\tau = T/m$. For $i = 0, 1, \dots, m$ we define $t_i = i \times \tau$. For $i = 1, 2, \dots, m$, let

$$f_i(x) = \tau^{-1} \int_{t_{i-1}}^{t_i} f(t, x)\, dt \quad \text{for a.e. } x \in \Omega,$$

then

$$\tau \sum_{i=1}^m \| f_i \|_{L^\infty(\Omega)} \leq \| f \|_{L^1((0,T); L^\infty(\Omega))},$$

and

$$\sum_{i=1}^m \int_{t_{i-1}}^{t_i} \| f(t) - f_i \|_{L^1(\Omega)} \overset{m \to \infty}{\to} 0.$$

For $i = 1, \ldots, m$, let u_i be the unique solution of $\tau f_i + u_{i-1} \in (I + \tau \mathcal{A}_{(g,\Psi)})(u_i)$ and let v_i be the corresponding entropy solution of $\tilde{E}(f_i + (u_{i-1} - u_i)/\tau)$ then

$$\|u_i\|_{L^\infty(\Omega)} \leq \|u_{i-1}\|_{L^\infty(\Omega)} + \tau \|f_i\|_{L^\infty(\Omega)}$$

$$\leq \|u_0\|_{L^\infty(\Omega)} + \tau \sum_{j=1}^{i} \|f_j\|_{L^\infty(\Omega)}$$

$$\leq \|u_0\|_{L^\infty(\Omega)} + \int_0^{t_i} \|f(t)\|_{L^\infty(\Omega)}.$$

Let us define $v^\tau = v_i$ and $f^\tau = f_i$ for $t_{i-1} \leq t \leq t_i$ and $1 \leq i \leq m$, and let $u^\tau = g(v^\tau)$ then

$$\int_Q \{(\tau^{-1}(g(v^\tau(t - \tau)) - g(v^\tau(t))) + f^\tau(t))\xi(t) + \Psi(v^\tau(t))) \cdot \nabla\xi(t)\} \, dx \, dt = 0 \quad (11)$$

for any nonnegative $\xi \in \mathcal{D}(Q)$ (where $g(v^\tau(s)) = u_0$ for $s \leq 0$). Moreover, we have

$$\int_Q \mathrm{sign}_0^+(v^\tau(t) - \mu)\{(\tau^{-1}(g(v^\tau(t - \tau)) - g(v^\tau(t))) + f^\tau(t))\xi(t)$$

$$+ (\Psi(v^\tau(t))) - \Psi(\mu)) \cdot \nabla\xi(t)\} \, dx \, dt \geq 0 \quad (12)$$

either for any nonnegative $\xi \in \mathcal{D}((0, T); \mathbb{R}^N)$ and for any $\mu \geq 0$, or for any nonnegative $\xi \in \mathcal{D}(Q)$ and any $\mu \in \mathbb{R}$; and

$$\int_Q \mathrm{sign}_0^+(\mu - v^\tau(t))\{-(\tau^{-1}(g(v^\tau(t - \tau)) - g(v^\tau(t))) - f^\tau(t))\xi(t)$$

$$+ (\Psi(\mu) - \Psi(v^\tau(t))) \cdot \nabla\xi(t)\} \, dx \, dt > 0 \quad (13)$$

either for any nonnegative $\xi \in \mathcal{D}((0, T); \mathbb{R}^N)$ and for any $\mu \leq 0$, or for any nonnegative $\xi \in \mathcal{D}(Q)$ and any $\mu \in \mathbb{R}$.

Let u be the integral solution of $(P_A(u_0, f))$, then (see [Be, CL])

$$\| u - u^\tau \|_{L^\infty((0,T);L^1(\Omega))} \overset{m \to \infty}{\to} 0,$$

and, since $u - u^\tau$ is uniformly bounded in $L^\infty(Q)$ the above convergence is still true in $L^2(Q)$. Moreover, from (A1) we deduce that v^τ is uniformly bounded in $L^\infty(Q)$, more exactly, $\mathcal{T}(v^\tau) = (u^\tau, (\chi_s^\tau)_{s \in S})$ where $\chi_s^\tau \in \mathrm{sign}^-(u^\tau - s)$ for $s \in S, s < 0$, $\chi_s^\tau \in \mathrm{sign}^+(u^\tau - s)$ for $s \in S, s > 0$ and $\chi_0^\tau \in \mathrm{sign}(u^\tau)$ if $0 \in S$, with $v^\tau = u^\tau + \sum_{s \in S} \chi_s^\tau$. Then, when τ tends to 0 we can consider that there exists a subsequence, still denoted by τ such that, for any $s \in S$, χ_s^τ converges weakly to some χ_s and, taking into account the strong convergence of u^τ in $L^2(Q)$, $\chi_s \in \mathrm{sign}^-(u - s)$ for $s < 0$, $\chi_s \in \mathrm{sign}^+(u - s)$ for $s > 0$ and $\chi_0 \in \mathrm{sign}(u)$ if $0 \in S$. Hence there exists a subsequence, still denoted by v^τ, converging in $L^\infty(Q)$ weak$\star$ to a function $v = u + \sum_{s \in S} \chi_s$ and $u = g(v)$ (see [Br]). By applying Lebesgue Convergence Theorem one checks easily that

$$\Psi^i(g(v^\tau)) = \Psi^i(u^\tau) \overset{m \to \infty}{\to} \Psi^i(u) = \Psi^i(g(v)) \quad \text{in } L^p(Q), \tag{14}$$

for any $1 \le p < \infty$, for $i = 1, 2$, whence $\Psi(v^\tau) = \Psi^1(g(v^\tau)) + \Psi^2(g(v^\tau))v^\tau$ converges to $\Psi(v)$ weakly in $L^q(Q)$ for any $1 \le q < \infty$ and

$$\int_Q \Psi(g(v^\tau)) \cdot \nabla \xi \, dx \, dt \overset{m \to \infty}{\to} \int_Q \Psi(g(v)) \cdot \nabla \xi \, dx \, dt$$

for any $\xi \in \mathcal{D}(Q)$. Moreover, for τ small enough

$$\int_Q (\tau^{-1}(g(v^\tau(t - \tau))) - g(v^\tau(t)) + f^\tau(t))\xi(t) \, dx \, dt$$

$$= \int_Q (g(v^\tau(t))\tau^{-1}(\xi(t + \tau) - \xi(t)) + f^\tau(t)\xi(t)) \, dx \, dt$$

$$\overset{m \to \infty}{\to} \int_Q (g(v)\xi_t + f\xi) \, dx \, dt$$

for any $\xi \in \mathcal{D}(Q)$, whence we deduce that v satisfies (6).

In order to prove that v satisfies the entropy inequalities (2) and (3), as a first step we consider the case where $\mu \in \mathbb{R} \backslash g^{-1}(S)$, $\mu \ge 0$. In this case $\operatorname{sign}_0^+(r - \mu) = \operatorname{sign}_0^+(g(r) - g(\mu))$ for any $r \in \mathbb{R}$. By considering that

$$\operatorname{sign}_0^+(v^\tau(t) - \mu)(g(v^\tau(t - \tau)) - g(v^\tau(t)))$$
$$= \operatorname{sign}_0^+(g(v^\tau(t)) - g(\mu))(g(v^\tau(t - \tau)) - g(\mu) + g(\mu) - g(v^\tau(t)))$$
$$\le (g(v^\tau(t - \tau)) - g(\mu))^+ - (g(v^\tau(t)) - g(\mu))^+$$

and by considering A3, from (12) we get

$$0 \le \frac{1}{\tau} \int_Q \{((g(v^\tau(t - \tau)) - g(\mu))^+ - (g(v^\tau(t)) - g(\mu))^+)\xi(t)\} \, dx \, dt$$

$$+ \int_Q \operatorname{sign}_0^+(g(v^\tau(t)) - g(\mu))(\Psi^1(g(v^\tau(t))) - \Psi^1(g(\mu))) \cdot \nabla \xi(t) \, dx \, dt$$

$$+ \int_Q \operatorname{sign}_0^+(g(v^\tau(t)) - g(\mu))\mu(\Psi^2(g(v^\tau(t))) - \Psi^2(g(\mu))) \cdot \nabla \xi(t) \, dx \, dt$$

$$+ \int_Q \Psi^2(g(v^\tau(t)))(v^\tau(t) - \mu)^+ \cdot \nabla \xi(t) \, dx \, dt$$

$$+ \int_Q \operatorname{sign}_0^+(g(v^\tau(t)) - g(\mu))f^\tau(t)\xi(t) \, dx \, dt \tag{15}$$

for any nonnegative $\xi \in \mathcal{D}((0, T) \times \mathbb{R}^N)$.

When τ tends to 0 we deduce easily that the first integral converges to

$$\int_Q (g(v) - g(\mu))^+ \xi_t \, dx \, dt.$$

Moreover, by taking into account (14) and the fact that $\mathrm{sign}_0^+(g(v^\tau) - g(\mu))$, or at least a subsequence, converges to some $\kappa \in \mathrm{sign}^+(g(v) - g(\mu)) = \mathrm{sign}^+(v - \mu)$ in $L^\infty(Q)$ weak$\star$, the second, the third and the fifth integrals converge, respectively, to

$$\int_Q \mathrm{sign}_0^+(v - \mu)(\Psi^1(g(v)) - \Psi^1(g(\mu))) \cdot \nabla\xi \, dx \, dt,$$

$$\int_Q \mu \, \mathrm{sign}_0^+(v - \mu)(\Psi^2(g(v)) - \Psi^2(g(\mu))) \cdot \nabla\xi \, dx \, dt$$

and to

$$\int_Q \kappa f \xi \, dx \, dt.$$

In order to get a limit to the fourth integral, let us observe that, for any $\mu \in \mathbb{R} \setminus g^{-1}(S)$, $\mu \geq 0$, we have $(r - \mu)^+ = (g(r) - g(\mu))^+ + \sum_{s \in S, s \geq 0}(r_s - \mu_s)^+$ where $r_s \in \mathrm{sign}^+(g(r) - s)$ and $\mu_s \in \mathrm{sign}^+(g(\mu) - s)$. Since $g(\mu) \neq s$ for any $s \in S$, we have $\mu_s = 1$ for $g(\mu) > s$ and $\mu_s = 0$ for $g(\mu) < s$, whence we have $(r - \mu)^+ = (g(r) - g(\mu))^+ + \sum_{s \in S, \, s > g(\mu)} r_s$. Then,

$$(v^\tau - \mu)^+ = (g(v^\tau) - g(\mu))^+ + \sum_{s \in S, \, s > g(\mu)} \chi_s^\tau \rightharpoonup (g(v) - g(\mu))^+$$

$$+ \sum_{s \in S, \, s > g(\mu)} \chi_s = (v - \mu)^+$$

in $L^\infty(Q)$ weak$\star$ and, therefore, the fourth integral converges to

$$\int_Q (v - \mu)^+ \Psi^2(g(v)) \cdot \nabla\xi \, dx \, dt.$$

Hence we have

$$\int_Q \{(g(v) - g(\mu))^+ \xi_t + \mathrm{sign}_0^+(v - \mu)(\Psi(v) - \Psi(\mu)) \cdot \nabla\xi + \kappa f \xi\} \, dx \, dt \geq 0 \qquad (16)$$

for any $\xi \in \mathcal{D}((0, T) \times \mathbb{R}^N)$ and any nonnegative $\mu \in \mathbb{R} \setminus g^{-1}(S)$ and for some $\kappa \in \mathrm{sign}^+(v - \mu)$. Arguing like above we prove that inequality (16) is still true for any $\mu \in \mathbb{R} \setminus g^{-1}(S)$ and for any nonnegative $\xi \in \mathcal{D}(Q)$ and for some $\kappa \in \mathrm{sign}^+(v - \mu)$.

In the case where $0 \in g^{-1}(S)$ (which is equivalent to $0 \in S$) the inequality (16) is still true with some modification. In fact we can observe that $\text{sign}_0^+(r)g(r) = (g(r))^+$, $\text{sign}_0^+(r)(\Psi^i(g(r)) - \Psi^i(0)) = \text{sign}_0^+(g(r))(\Psi^i(g(r)) - \Psi^i(0))$ and $\text{sign}_0^+(r) \in \text{sign}^+(g(r))$, whence, the limits of the first, second, third and fifth integrals of (15) are similar to the above limits. In the case of the fourth integral we have to consider that $(v^\tau)^+ = g(v^\tau)^+ + (\chi_0^\tau)^+ + \sum_{s \in S, s > 0} \chi_s^\tau$ where χ_0^τ, or at least a subsequence, still converges in $L^\infty(Q)$ weak$\star$ to a nonnegative function $\tilde{\chi}_0$. Since $(\chi_0^\tau)^+$ belongs to $\text{sign}^+(g(v^\tau))$ and $g(v^\tau)$ converges to $g(v)$ in $L^p(Q)$ for any $1 \le p < \infty$ we can deduce that $\tilde{\chi}_0$ belongs to $\text{sign}^+(g(v))$ and $\tilde{\chi}_0 \ge (\chi_0)^+$. Then the fourth integral converges to

$$\int_Q \tilde{v} \Psi^2(g(v)) \cdot \nabla \xi \, dx \, dt$$

where $\tilde{v} = g(v) + \tilde{\chi}_0 + \sum_{s \in S, s > 0} \chi_s \ge v^+$, hence we have

$$\int_Q \{g(v^+)\xi_t + (\Psi^1(g(v^+)) + \Psi^2(g(v^+))\tilde{v}) \cdot \nabla \xi + \kappa f \xi\} \, dx \, dt \ge 0 \tag{17}$$

for some $\kappa \in \text{sign}^+(g(v))$ and for any $\xi \in \mathcal{D}((0, T) \times \mathbb{R}^N)$.

Now, let us consider the case $g(\mu) = s$ for some positive $s \in S$, let $[\mu_{\min}, \mu_{\max}] = g^{-1}(s)$ and let $(s_m)_{m \in \mathbb{N}}$ and $(s^m)_{m \in \mathbb{N}}$ be 2 monotone sequences converging to s and such that $s_m < s < s^m$ for any $m \in \mathbb{N}$ and such that s_m and s^m are not in S. Let $\mu_m = g^{-1}(s_m)$ and $\mu^m = g^{-1}(s^m)$, then, obviously, $\mu_m < \mu_{\min} < \mu_{\max} < \mu^m$, and μ_m is an increasing sequence converging to $\mu_{\min}$ and μ^m is a decreasing sequence converging to $\mu_{\max}$.

We plug successively μ_m and μ^m in (16) and we let m tend to ∞. Then we get

$$\int_Q \{(g(v) - g(\mu))^+ \xi_t + \text{sign}_0^+(v - \mu)(\Psi^1(g(v)) - \Psi^1(g(\mu))) \cdot \nabla \xi$$
$$+ v \, \text{sign}_0^+(v - \mu)(\Psi^2(g(v)) - \Psi^2(g(\mu))) \cdot \nabla \xi$$
$$+ (v - \mu_{\min})^+ \Psi^2(g(\mu)) \cdot \nabla \xi + \text{sign}_{\max}^+(v - \mu_{\min})f\xi\} \, dx \, dt \ge 0$$

for any $\mu \in g^{-1}(s)$ and for any nonnegative $\xi \in \mathcal{D}((0, T) \times \mathbb{R}^N)$, where $\text{sign}_{\max}^+(r) = \max(\text{sign}^+(r))$. Moreover we get

$$\int_Q \{(g(v) - g(\mu))^+ \xi_t + \text{sign}_0^+(v - \mu)(\Psi^1(g(v)) - \Psi^1(g(\mu))) \cdot \nabla \xi$$
$$+ v \, \text{sign}_0^+(v - \mu)(\Psi^2(g(v)) - \Psi^2(g(\mu))) \cdot \nabla \xi$$
$$+ (v - \mu_{\max})^+ \Psi^2(g(\mu)) \cdot \nabla \xi + \text{sign}_0^+(v - \mu_{\max})f\xi\} \, dx \, dt \ge 0$$

for any $\mu \in g^{-1}(s)$ and for any nonnegative $\xi \in \mathcal{D}((0, T) \times \mathbb{R}^N)$. By applying Lemma 2 of [Ca], for any $\mu \in g^{-1}(s)$ we deduce the existence of a function $\theta \in \text{sign}^+(v - \mu)$ such that

$$\int_Q \{(g(v) - g(\mu))^+\xi_t + \text{sign}_0^+(v - \mu)(\Psi^1(g(v)) - \Psi^1(g(\mu))) \cdot \nabla\xi$$
$$+ v \, \text{sign}_0^+(v - \mu)(\Psi^2(g(v)) - \Psi^2(g(\mu))) \cdot \nabla\xi + (v - \mu)^+\Psi^2(g(\mu)) \cdot \nabla\xi$$
$$+ (\text{sign}_0^+(v - \mu_{\max})(1 - \theta) + \text{sign}_{\max}^+(v - \mu_{\min})\theta)f\xi\} \, dx \, dt \geq 0$$

for any nonnegative $\xi \in \mathcal{D}((0, T) \times \mathbb{R}^N)$. One checks easily that $(\text{sign}_0^+(v - \mu_{\max})(1 - \theta) + \text{sign}_{\max}^+(v - \mu_{\min})\theta) \in \text{sign}^+(v - \mu)$ whence we have

$$\int_Q \{(g(v) - g(\mu))^+\xi_t + \text{sign}_0^+(v - \mu)(\Psi(v) - \Psi(\mu)) \cdot \nabla\xi + \kappa f\xi\} \, dx \, dt \geq 0 \qquad (18)$$

for any positive $\mu \in \mathbb{R}$, for some $\kappa \in \text{sign}^+(v - \mu)$ and for any nonnegative $\xi \in \mathcal{D}((0, T) \times \mathbb{R}^N)$.

Now, let us consider $\mu \in g^{-1}(0)$. Let $[\mu_{\min}, \mu_{\max}] = g^{-1}(0)$, with $\mu_{\min} \leq 0 \leq \mu_{\max}$, ($\mu_{\min} = \mu_{\max} = 0$ corresponding to the case $0 \notin S$ and then, $\chi_0 = 0$). We consider 2 monotone sequences μ_m and μ^m such that $\mu_m < \mu_{\min} \leq \mu_{\max} < \mu^m$, converging, respectively, to μ_m and to μ^m. We plug μ^m in the inequality (16), with a nonnegative $\xi \in \mathcal{D}((0, T) \times \mathbb{R}^N)$ and we let m tend to ∞, then we get

$$\int_Q \{(g(v) - g(\mu))^+\xi_t + \text{sign}_0^+(v - \mu)(\Psi^1(g(v)) - \Psi^1(g(\mu))) \cdot \nabla\xi$$
$$+ v \, \text{sign}_0^+(v - \mu)(\Psi^2(g(v)) - \Psi^2(g(\mu))) \cdot \nabla\xi$$
$$+ (v - \mu_{\max})^+\Psi^2(g(\mu)) \cdot \nabla\xi + \text{sign}_0^+(v - \mu_{\max})f\xi\} \, dx \, dt \geq 0 \qquad (19)$$

for any $\mu \in g^{-1}(0)$ (we observe that $g(\mu) = 0$ in the above inequality). Similarly, we plug μ_m in the inequality (16) for any nonnegative $\xi \in \mathcal{D}(Q)$ and we let m tend to ∞, then

$$\int_Q \{(g(v) - g(\mu))^+\xi_t + \text{sign}_0^+(v - \mu)(\Psi^1(g(v)) - \Psi^1(g(\mu))) \cdot \nabla\xi$$
$$+ v \, \text{sign}_0^+(v - \mu)(\Psi^2(g(v)) - \Psi^2(g(\mu))) \cdot \nabla\xi$$
$$+ (v - \mu_{\min})^+\Psi^2(g(\mu)) \cdot \nabla\xi + \text{sign}_{\max}^+(v - \mu_{\min})f\xi\} \, dx \, dt \geq 0$$

for any $\mu \in g^{-1}(0)$ Then, by applying Lemma 2 of [Ca] with the 2 inequalities above we get

$$\int_Q \{(g(v) - g(\mu))^+\xi_t + \text{sign}_0^+(v - \mu)(\Psi^1(g(v)) - \Psi^1(g(\mu))) \cdot \nabla\xi$$
$$+ v \, \text{sign}_0^+(v - \mu)(\Psi^2(g(v)) - \Psi^2(g(\mu))) \cdot \nabla\xi$$
$$+ (v - \mu)^+\Psi^2(g(\mu)) \cdot \nabla\xi + \kappa_\mu f\xi\} \, dx \, dt \geq 0 \qquad (20)$$

for any $\mu \in g^{-1}(0)$ and for any nonnegative $\xi \in \mathcal{D}(Q)$ where $\kappa_\mu = \text{sign}_0^+(v - \mu_{\max})$ $(1 - \theta_\mu) + \text{sign}_{\max}^+(v - \mu_{\min})\theta_\mu$, where $\theta_\mu \in \text{sign}^+(v - \mu)$. One checks easily that $\kappa_\mu \in \text{sign}^+(v - \mu)$.

Now, by taking into account that, for any nonnegative $\mu \in g^{-1}(0)$ we have $(v - \mu_{\max})^+ \leq (v - \mu)^+ \leq v^+ \leq \tilde{v}$, from inequalities (17), (19) and (20) and from Lemma 3 of [Ca] we deduce that

$$\int_Q \{(g(v) - g(\mu))^+\xi_t + \text{sign}_0^+(v - \mu)(\Psi^1(g(v)) - \Psi^1(g(\mu))) \cdot \nabla\xi$$
$$+ v \, \text{sign}_0^+(v - \mu)(\Psi^2(g(v)) - \Psi^2(g(\mu))) \cdot \nabla\xi$$
$$+ (v - \mu)^+\Psi^2(g(\mu)) \cdot \nabla\xi + \kappa_\mu f\xi\} \, dx \, dt \geq 0$$

for any nonnegative $\mu \in g^{-1}(0)$, for some $\kappa_\mu \in \text{sign}^+(v - \mu)$ and for any nonnegative $\xi \in \mathcal{D}((0, T) \times \mathbb{R}^N)$.

Then, for any nonnegative $\mu \in \mathbb{R}$, for some $\kappa_\mu \in \text{sign}^+(v - \mu)$ and for any nonnegative $\xi \in \mathcal{D}((0, T) \times \mathbb{R}^N)$ we have

$$\int_Q \{(g(v) - g(\mu))^+\xi_t + \text{sign}_0^+(v - \mu)(\Psi(v) - \Psi(\mu)) \cdot \nabla\xi + \kappa_\mu f\xi\} \, dx \, dt \geq 0. \quad (21)$$

Then, for any $\mu \geq 0$ we take a decreasing sequence μ_m converging to μ as m tends to ∞. One checks easily that the corresponding sequence of κ_{μ_m} is an increasing sequence converging to $\text{sign}_0^+(v - \mu)$ whence, by changing μ by μ_m in the inequality (21) and letting m tend to ∞ we get (7). By arguing like above we get (8). $\qquad\square$

4. Comparison and uniqueness of entropy solutions

Entropy solutions satisfy the following global Kato type inequality:

THEOREM 4.1. *Let* (A2), (A3) *and* (A4) *hold. Let* $(u_{i_0}, f_i) \in L^1(\Omega) \times L^1((0, T); L^\infty(\Omega))$, *and let* v_i *be an entropy solution of* $(\tilde{P}(u_{i_0}, f_i))$ *for* $i = 1, 2$. *Then*

$$\int_Q \{(g(v_1) - g(v_2))^+\xi_t + \text{sign}_0^+(v_1 - v_2)(\Psi(v_1) - \Psi(v_2)) \cdot \nabla\xi$$
$$+ \kappa(f_1 - f_2)\xi\} \, dx \, dt \geq 0 \qquad (22)$$

for some $\kappa \in \text{sign}^+(v_1 - v_2)$ *and for any nonnegative* $\xi \in \mathcal{D}((0, T) \times \mathbb{R}^N)$.

Proof. This Theorem is, in fact, a particular case of Theorem 14 of [Ca]. However, since [Ca] is devoted to more general equations, here we perform a direct proof for first order equations.

The first step is to prove the inequality (22) for nonnegative test functions $\xi \in \mathcal{D}(Q)$. This step is similar to the proof of Kruzhkov [K1] for conservation laws in $(0, T) \times \mathbb{R}^N$. Let us consider $u_{1_0} = u_{1_0}(z)$, $v_1 = v_1(\theta, z)$ and $f_1 = f_1(\theta, z)$, and $u_{2_0} = u_{2_0}(y)$, $v_2 = v_2(\tau, y)$ and $f_2 = f_2(\tau, y)$. Let ξ be a nonnegative function of $\mathcal{D}(Q)$, let ρ_m be a sequence of mollifiers in $\mathbb{R}^{N+1}$ and let $\zeta(\theta, z, \tau, y) = \xi((\theta+\tau)/2, (z+y)/2)\rho_m((\theta-\tau)/2, (z-y)/2)$, with m large enough, such that $\zeta \in \mathcal{D}(Q \times Q)$. For almost every (τ, y) we write (9) for $v = v_1$, with $\mu = v_2(\tau, y)$ and we integrate over Q in (τ, y):

$$\int_{Q \times Q} \{(g(v_1) - g(v_2))^+ \zeta_\theta + \text{sign}_0^+(v_1 - v_2)(\Psi(v_1) - \Psi(v_2)) \cdot \nabla_z \zeta$$
$$+ \text{sign}_0^+(v_1 - v_2) f_1 \zeta\} \, dz d\theta \, dy \, d\tau \geq 0. \tag{23}$$

Similarly we write (10) for $v = v_2$, with $\mu = v_1(\theta, z)$ and we integrate over Q in (θ, z):

$$\int_{Q \times Q} \{(g(v_1) - g(v_2))^+ \zeta_\tau + \text{sign}_0^+(v_1 - v_2)(\Psi(v_1) - \Psi(v_2)) \cdot \nabla_y \zeta$$
$$- \text{sign}_0^+(v_1 - v_2) f_2)\zeta\} \, dy \, d\tau \, dz d\theta \geq 0. \tag{24}$$

We add (23) and (24) and we make the following change of variables: $x = (z + y)/2$, $t = (\theta + \tau)/2$, $\sigma = (z - y)/2$ and $\varsigma = (\theta - \tau)/2$ then we have:

$$\int_{\mathcal{B}(0,1/m)} \left(\int_Q \{(g(v_1) - g(v_2))^+ \xi_t + \text{sign}_0^+(v_1 - v_2)(\Psi(v_1) - \Psi(v_2)) \cdot \nabla_x \xi \right.$$
$$\left. + \text{sign}_0^+(v_1 - v_2)(f_1 - f_2)\zeta\} \, dx \, dt \right) \rho_m \, d\sigma d\varsigma \geq 0 \tag{25}$$

hence, by letting m tend to ∞ we get (22) for any nonnegative $\xi \in \mathcal{D}(Q)$.

Now, let us consider a nonnegative $\xi \in \mathcal{D}((0, T) \times \mathbb{R}^N)$, let $\Omega \subset \Omega_0 \cup (\cup_{i=1}^\ell B_i)$ where $\Omega_0 \subset\subset \Omega$ and $B_i \subset \mathbb{R}^N$ are balls satisfying that $B_i \cap \partial\Omega$ is a part of the graph of a Lipschitz continuous function. Let $(\varphi_i)_{i=0}^\ell$ be a partition of unity related to the above covering of Ω ($\varphi_0 \in \mathcal{D}(\Omega_0)$, $\varphi_i \in \mathcal{D}(B_i)$ for $i = 1, \ldots, \ell$). Let $\xi_i = \xi\varphi_i$, then, since $\xi_0 \in \mathcal{D}(Q)$, we just have to prove the inequality for each one of the functions ξ_i, for $i = 1, \ldots, \ell$. Then, without loss of generality we can consider ξ as a function of $\mathcal{D}((0, T) \times B_i)$ for some $1 \leq i \leq \ell$.

Let us consider $u_{1_0} = u_{1_0}(y)$, $v_1 = v_1(\tau, y)$ and $f_1 = f_1(\tau, y)$, and $u_{2_0} = u_{2_0}(x)$, $v_2 = v_2(t, x)$ and $f_2 = f_2(t, x)$. Let ρ_m be a sequence of mollifiers in $\mathbb{R}^{N+1}$ such that, for m large enough, $x \mapsto \xi(x, t)\rho_m((t - \tau)/2, (x - y)/2) \in \mathcal{D}(\Omega)$ and let $\zeta(t, x, \tau, y) = \xi(t, x)\rho_m((t - \tau)/2, (x - y)/2)$, then $\zeta \in \mathcal{D}(Q \times ((0, T) \times \mathbb{R}^N))$. Then, for almost every $(t, x) \in Q$ we write (7) with $v = v_1$ and $\mu = v_2^+(t, x)$ and we integrate over Q in (t, x):

$$\int_{Q \times Q} \{(g(v_1) - g(v_2^+))^+ \zeta_\tau + \text{sign}_0^+(v_1 - v_2^+)(\Psi(v_1) - \Psi(v_2^+)) \cdot \nabla_y \zeta$$
$$+ \text{sign}_0^+(v_1 - v_2^+) f_1 \zeta\} \, dy \, d\tau \, dx \, dt \geq 0. \tag{26}$$

Similarly, for almost every $(\tau, y) \in Q$ we write (10) with $v = v_2$ and $\mu = v_1^+(\tau, y)$, and we integrate over Q in (τ, y):

$$
\int_{Q \times Q} \{ (g(v_1^+) - g(v_2))^+ \zeta_t + \mathrm{sign}_0^+(v_1^+ - v_2)(\Psi(v_1^+) - \Psi(v_2)) \cdot \nabla_x \zeta
$$
$$
- \mathrm{sign}_0^+(v_1^+ - v_2) f_2 \zeta \} \, dx \, dt \, dy \, d\tau \geq 0. \tag{27}
$$

By taking into account that $(g(v_1^+) - g(v_2))^+ = (g(v_1^+) - g(v_2^+))^+ - g(-v_2^-)$, sign_0^+ $(v_1^+ - v_2)(\Psi(v_1^+) - \Psi(v_2)) = \mathrm{sign}_0^+(v_1^+ - v_2^+)(\Psi(v_1^+) - \Psi(v_2^+)) - \mathrm{sign}_0^+(v_2^-)\Psi(-v_2^-)$ and $\mathrm{sign}_0^+(v_1^+ - v_2) = \mathrm{sign}_0^+(v_2^-) + \mathrm{sign}_0^+(v_1^+ - v_2^+)(1 - \mathrm{sign}_0^+(v_2^-))$, from (27) we get

$$
\int_{Q \times Q} \{ (g(v_1^+) - g(v_2^+))^+ \zeta_t + \mathrm{sign}_0^+(v_1^+ - v_2^+)(\Psi(v_1^+) - \Psi(v_2^+)) \cdot \nabla_x \zeta
$$
$$
- \mathrm{sign}_0^+(v_1^+ - v_2^+)(1 - \mathrm{sign}_0^+(v_2^-)) f_2 \zeta \} \, dx \, dt \, dy \, d\tau
$$
$$
- \int_{Q \times Q} \mathrm{sign}_0^+(v_2^-) \{ g(v_2) \zeta_t + \Psi(v_2) \cdot \nabla_x \zeta + f_2 \zeta \} \, dx \, dt \, dy \, d\tau \geq 0. \tag{28}
$$

Then, by adding (26) and (28) and by taking into account the definition of ζ we get

$$
\int_{Q \times Q} \{ (g(v_1^+) - g(v_2^+))^+ \xi_t + \mathrm{sign}_0^+(v_1^+ - v_2^+)(\Psi(v_1^+) - \Psi(v_2^+)) \cdot \nabla_x \xi
$$
$$
+ \mathrm{sign}_0^+(v_1^+ - v_2^+)(f_1 - (1 - \mathrm{sign}_0^+(v_2^-)) f_2) \xi \} \rho_m \, dx \, dt \, dy \, d\tau
$$
$$
- \int_Q \mathrm{sign}_0^+(v_2^-) \{ g(v_2)(\xi \omega_m)_t + \Psi(v_2) \cdot \nabla_x(\xi \omega_m) + f_2(\xi \omega_m) \} \, dx \, dt \geq 0, \tag{29}
$$

where $\omega_m(t, x) = \int_Q \rho_m((t - \tau)/2, (x - y)/2) \, dy \, d\tau$. In order to let m tend to ∞ we observe that $0 \leq \xi \omega_m \leq \xi$, hence, by taking into account (8) we deduce that the last integral in the above inequality is bounded in the following way:

$$
0 \geq I_m = \int_Q \mathrm{sign}_0^+(v_2^-) \{ g(v_2)(\xi \omega_m)_t + \Psi(v_2) \cdot \nabla_x(\xi \omega_m) + f_2(\xi \omega_m) \} \, dx \, dt
$$
$$
\geq \int_Q \mathrm{sign}_0^+(v_2^-) \{ g(v_2) \xi_t + \Psi(v_2) \cdot \nabla_x \xi + f_2 \xi \} \, dx \, dt = I
$$

hence we deduce that I_m, or at least a subsequence of I_m, converges to some $\tilde{I}$ satisfying $0 \geq \tilde{I} \geq I$, when m tends to ∞. Then, by letting m tend to ∞, inequality (29) becomes

$$
\int_Q \{ (g(v_1^+) - g(v_2^+))^+ \xi_t + \mathrm{sign}_0^+(v_1^+ - v_2^+)(\Psi(v_1^+) - \Psi(v_2^+)) \cdot \nabla_x \xi
$$
$$
+ \kappa_+ (f_1 - (1 - \mathrm{sign}_0^+(v_2^-)) f_2) \, \mathrm{sign}_0^+(v_1^+) \xi \} \, dx \, dt - \tilde{I} \geq 0, \tag{30}
$$

where $\kappa_+ \in \mathrm{sign}^+(v_1^+ - v_2^+)$ and all the functions arising in the integral depend on (t, x).

By tacking into account Remark 2.3 we deduce that

$$\int_Q \{(g(-v_1^-) - g(-v_2^-))^+ \xi_t + \text{sign}_0^+(-v_1^- + v_2^-)(\Psi(-v_1^-) - \Psi(-v_2^-)) \cdot \nabla_x \xi$$
$$+ \kappa_-(f_1(1 - \text{sign}_0^+(v_1^+)) - f_2) \, \text{sign}_0^+(v_2^-)\xi\} \, dx \, dt + \tilde{J} \geq 0 \tag{31}$$

where $\kappa_- \in \text{sign}^+(-v_1^- + v_2^-)$ and $\tilde{J}$ is the limit of a subsequence of J_m where

$$0 \leq J_m = \int_Q \text{sign}_0^+(v_1^+)\{g(v_1)(\xi\omega_m)_t + \Psi(v_1) \cdot \nabla_x(\xi\omega_m) + f_1(\xi\omega_m)\} \, dx \, dt$$
$$\leq \int_Q \text{sign}_0^+(v_1^+)\{g(v_1)\xi_t + \Psi(v_1) \cdot \nabla_x\xi + f_1\xi\} \, dx \, dt = J.$$

Let $\tilde{\kappa} = \kappa_-(1 - \text{sign}_0^+(v_1^+)) \, \text{sign}_0^+(v_2^-) + \kappa_+ \, \text{sign}_0^+(v_1^+)$, we check easily that $\tilde{\kappa} = \kappa_+(1 - \text{sign}_0^+(v_2^-)) \, \text{sign}_0^+(v_1^+) + \kappa_- \, \text{sign}_0^+(v_2^-) \in \text{sign}^+(v_1 - v_2)$. Then, from (30) and (31) we have:

$$\int_Q \{(g(v_1) - g(v_2))^+ \xi_t + \text{sign}_0^+(v_1 - v_2)(\Psi(v_1) - \Psi(v_2)) \cdot \nabla_x \xi$$
$$+ \tilde{\kappa}(f_1 - f_2)\xi\} \, dx \, dt$$
$$\geq \lim_{m \to \infty} \int_Q \text{sign}_0^+(v_2^-)\{g(v_2)(\xi\omega_m)_t + \Psi(v_2) \cdot \nabla_x(\xi\omega_m) + f_2(\xi\omega_m)\} \, dx \, dt$$
$$- \lim_{m \to \infty} \int_Q \text{sign}_0^+(v_1^+)\{g(v_1)(\xi\omega_m)_t + \Psi(v_1) \cdot \nabla_x(\xi\omega_m) + f_1(\xi\omega_m)\} \, dx \, dt. \tag{32}$$

From the definition of ρ_n we deduce that $\xi\omega_n \in \mathcal{D}(Q)$. As $\xi = \xi\omega_n + \xi(1 - \omega_n)$, combining the preceding inequality and the local Kato inequality, we have

$$\int_Q \{(g(v_1) - g(v_2))^+ \xi_t + \text{sign}_0^+(v_1 - v_2)(\Psi(v_1) - \Psi(v_2)) \cdot \nabla_x \xi$$
$$+ \kappa(f_1 - f_2)\xi\} \, dx \, dt \geq \int_Q \{(g(v_1) - g(v_2))^+(\xi(1 - \omega_n))_t$$
$$+ \text{sign}_0^+(v_1 - v_2)(\Psi(v_1) - \Psi(v_2)) \cdot \nabla_x(\xi(1 - \omega_n))$$
$$+ \kappa(f_1 - f_2)(\xi(1 - \omega_n))\} \, dx \, dt \geq \int_Q (\kappa - \tilde{\kappa})(f_1 - f_2)(\xi(1 - \omega_n)) dx \, dt$$
$$+ \lim_{m \to \infty} \int_Q \text{sign}_0^+(v_2^-)\{g(v_2)(\xi(1 - \omega_n)\omega_m)_t$$
$$+ \Psi(v_2) \cdot \nabla_x(\xi(1 - \omega_n)\omega_m) + f_2(\xi(1 - \omega_n)\omega_m)\} \, dx \, dt$$
$$- \lim_{m \to \infty} \int_Q \text{sign}_0^+(v_1^+)\{g(v_1)(\xi(1 - \omega_n)\omega_m)_t$$
$$+ \Psi(v_1) \cdot \nabla_x(\xi(1 - \omega_n)\omega_m) + f_1(\xi(1 - \omega_n)\omega_m)\} \, dx \, dt.$$

Taking into account that $\omega_m(t, x)$ converges to 1, for any $(t, x) \in Q$, when m tends to ∞, we deduce that the first integral, at the right part of the above inequality, converges to 0 as m tends to ∞. Moreover, for m large enough we have $\omega_m \omega_n = \omega_n$ whence

$$\lim_{n\to\infty} \lim_{m\to\infty} \int_Q \mathrm{sign}_0^+(v_2^-)\{g(v_2)(\xi(1 - \omega_n)\omega_m)_t$$
$$+ \Psi(v_2) \cdot \nabla_x(\xi(1 - \omega_n)\omega_m) + f_2(\xi(1 - \omega_n)\omega_m\} \, dx \, dt$$
$$- \lim_{n\to\infty} \lim_{m\to\infty} \int_Q \mathrm{sign}_0^+(v_1^+)\{g(v_1)(\xi(1 - \omega_n)\omega_m)_t$$
$$+ \Psi(v_1) \cdot \nabla_x(\xi(1 - \omega_n)\omega_m) + f_1(\xi(1 - \omega_n)\omega_m\} \, dx \, dt$$
$$= \lim_{n\to\infty} \lim_{m\to\infty} \int_Q \mathrm{sign}_0^+(v_2^-)\{g(v_2)(\xi(\omega_m - \omega_n))_t$$
$$+ \Psi(v_2) \cdot \nabla_x(\xi(\omega_m - \omega_n)) + f_2(\xi(\omega_m - \omega_n))\} \, dx \, dt$$
$$- \lim_{n\to\infty} \lim_{m\to\infty} \int_Q \mathrm{sign}_0^+(v_1^+)\{g(v_1)(\xi(\omega_m - \omega_n))_t$$
$$+ \Psi(v_1) \cdot \nabla_x(\xi(\omega_m - \omega_n)) + f_1(\xi(\omega_m - \omega_n))\} \, dx \, dt = 0$$

hence

$$\int_Q \{(g(v_1) - g(v_2))^+ \xi_t + \mathrm{sign}_0^+(v_1 - v_2)(\Psi(v_1) - \Psi(v_2)) \cdot \nabla\xi$$
$$+ \kappa(f_1 - f_2)\xi\} \, dx \, dt \geq 0,$$

what achieves the proof. $\qquad\square$

COROLLARY 4.2. *Let* (A2), (A3) *and* (A4) *hold. Let* $(u_{i_0}, f_i) \in L^1(\Omega) \times L^1((0, T); L^\infty(\Omega))$, *and let* v_i *be an entropy solution of* $(\tilde{P}(u_{i_0}, f_i))$ *for* $i = 1, 2$. *Then*

$$\int_\Omega (g(v_1(t)) - g(v_2(t)))^+ \, dx \leq \int_\Omega (u_{1_0} - u_{2_0})^+ \, dx + \int_0^t \int_\Omega \kappa(f_1 - f_2)\xi \, dxds \quad (33)$$

for some $\kappa \in \mathrm{sign}^+(v_1 - v_2)$. *Hence, if* $u_{1_0} \leq u_{2_0}$ *almost everywhere in* Ω *and* $f_1 \leq f_2$ *almost everywhere in* Q, *we have* $g(v_1) \leq g(v_2)$ *almost everywhere in* Q. *In particular, we deduce that there exists no more than one* $g(v)$ *such that* v *is an entropy solution of* $(\tilde{P}(u_0, f))$.

Proof. We write inequality (22) for any function $\xi \in \mathcal{D}(0, T)$. Then we get

$$\frac{d}{dt} \int_\Omega (g(v_1) - g(v_2))^+ \, dx \leq \int_\Omega \kappa(f_1 - f_2) \, dx$$

whence we deduce the corollary. $\qquad\square$

REFERENCES

[ABK]   ANDREIANOV, B., BENILAN, PH. and KRUZHKOV, S. N., L^1 *theory of scalar conservation law with continuous flux function.* J. Funct. Anal. *171* (2000) n⁰.1, 15–33.

[BLN]   BARDOS, C., LEROUX, A. Y. and NEDELEC, J. C., *First order cuasilinear equations with boundary conditions.* Comm. in P.D.E. *4* (1979), 1017–1034.

[Be]   BENILAN, PH., *Equations d'évolution dans un espace de Banach quelconque et applications.* Thèse d'état, Orsay 1972.

[BCW]   BENILAN, PH., CARRILLO, J. and WITTBOLD, P., *Renormalized entropy solutions of scalar conservation laws.* Ann. Scuola Norm. Sup. Pisa Cl. Sci. (4) *29* (2000) n⁰. 2, 313–327.

[BK]   BENILAN, PH. and KRUZHKOV, S. N., *Conservation laws with continuous flux functions.* NoDEA Nonlinear Differential Equations Appl. *3* (1996) n⁰.4, 395–419.

[Br]   BREZIS, H., *Opérateurs maximaux monotones et semi-groupes de contractions dans les espaces de Hilbert.* North-Holland 1973.

[Ca]   CARRILLO, J., *Entropy solutions for nonlinear degenerate problems.* Arch. Rational Mech. Anal. *147* (1999), 269–361.

[CW]   CARRILLO, J. and WITTBOLD, P., *Renormalized entropy solutions of scalar conservation laws with boundary condition.* J. Diff. Eqs. *185* (2002), no. 1, 137–160.

[CL]   CRANDALL, M. G. and LIGGET, T., *Generation of semi-groups of nonlinear transformations in general Banach spaces.* Amer. J. Math. *93* (1971), 265–298.

[K1]   KRUZHKOV, S. N., *Generalized solutions of the Cauchy problem in the large for nonlinear equations of first order.* Dokl. Akad. Nauk SSSR *187* (1969), 29–32, (English transl. in Soviet Math. Dokl. *10*, (1969)).

[K2]   KRUZHKOV, S. N., *First order quasilinear equations in several independent variables.* Math. Sbornik, *81* (1970), 228–255, (English transl. in Math. USSR Sb. *10*, (1970)).

[PV]   PORRETTA, A. and VOVELLE, J., L^1 *solutions to first order hyperbolic equations in bounded domains.* To appear

[Ot]   OTTO, F., *Initial boundary value problem for a scalar conservation law.* C. R. Acad. Sci. Paris *322* Série I (1996) n⁰.8, 729–734.

[Va]   VALLET, G., *Dirichlet problem for a nonlinear conservation law.* Rev. Mat. Complutense. *13* (2000) n⁰.1, 231–250.

José Carrillo
Departamento de Matemática Aplicada
Universidad Complutense de Madrid
28040 Madrid
Spain
e-mail: jose_carrillo@mat.ucm.es

To access this journal online:
http://www.birkhauser.ch

J.evol.equ. 3 (2003) 303 – 319
1424–3199/03/020303 – 17
DOI 10.1007/s00028-003-0096-9
© Birkhäuser Verlag, Basel, 2003

Journal of Evolution
Equations

Regularity of solutions of nonlinear Volterra equations

VOLKER G. JAKUBOWSKI AND PETRA WITTBOLD

Dedicated to Philippe Bénilan

Abstract. We consider the regularity properties of solutions of the nonlinear Volterra equation

$$\frac{d}{dt}\left(u(t) - u_0 + \int_0^t k(t-s)\,(u(s) - u_0)\ ds\right) + Au(t) \ni f(t)$$

in Banach spaces X without the Radon-Nikodym property. Existence of strong solutions for an m-completely accretive operator A in a normal Banach space $X \subset L^1(\Omega; \mu)$ is shown for sufficiently smooth data.

1. Introduction

We study the regularity properties of generalized solutions u in the sense of Gripenberg, see [Gri85] and [CGL96], of the nonlinear Volterra equation

$$\frac{d}{dt}(u(t) - u_0 + (k * (u - u_0))(t)) + Au(t) \ni f(t), \qquad t \in [0, T). \tag{1}$$

Here, A is an m-accretive operator in a real Banach space X,

$$k : (0, \infty) \to \mathbb{R} \text{ is nonnegative, nonincreasing such that } k \in L^1_{\text{loc}}([0, \infty)), \tag{2}$$

and, as usual, $(k * v)(t) = \int_0^t k(t-s)v(s)ds$ denotes the convolution, where the integral is defined in the sense of Bochner for $v \in L^1(0, T; X)$. Moreover, we assume $u_0 \in \overline{D(A)}$ and $f \in L^1(0, T; X)$ for $0 < T < \infty$. For the inhomogeneous Cauchy problem, i.e. $k \equiv 0$, generalized solutions are in fact mild solutions.

Note that equation (1) is obtained when modelling heat flow in materials with memory, see e.g. [Nun71] and [CN81].

According to the existence results [Gri85, Theorem 1] and [CGL96, Theorem 1], the generalized solution u of (1) exists for the above choice of data u_0, f and is an element of $L^1(0, T; X)$. Moreover, for an arbitrary m-accretive operator A in a general real Banach

---

Mathematics Subject Classification 2000: 35D10, 45D05, 45N05, 47H06.

Key words and phrases: Nonlinear Volterra equation, inhomogeneous Cauchy problem, regularity of solutions, completely accretive operators.

space X, Lipschitz continuity of the generalized solution is known by [Gri85, Theorem 2] (see also [CGL96, Theorem 2]) if $u_0 \in \hat{D}(A)$ and $f \in L^1(0, T; X)$ is of bounded variation, i.e. if $f \in BV(0, T; X)$. Here, $\hat{D}(A) = \{x \in X \mid \sup_{\lambda>0} \|A_\lambda x\| < \infty\}$ denotes the generalized domain of the operator A, where $A_\lambda := \frac{1}{\lambda}(I - J_\lambda)$ for $\lambda > 0$ is the Yosida approximation of A and $J_\lambda := (I + \lambda A)^{-1}$ its resolvent.

Since the Radon-Nikodym property of the Banach space X is equivalent to the differentiability of absolutely continuous functions, generalized solutions of (1) are strong solutions in the sense of the following definition if X has the Radon-Nikodym property and $u_0 \in \hat{D}(A)$, $f \in BV(0, T; X)$.

DEFINITION 1.1. Let $\kappa \geq 0$, $k \in L^1_{\mathrm{loc}}([0, \infty))$ satisfying (2) and

$$\kappa + \int_0^t k(\tau)\, d\tau > 0, \qquad t > 0. \tag{3}$$

Then a function $u \in L^1(0, T; X)$ is called a strong solution of

$$\frac{d}{dt}\left(\kappa(u(t) - u_0) + (k * (u - u_0))(t)\right) + Au(t) \ni f(t), \quad t \in [0, T) \tag{4}$$

if the function $v : [0, T) \to X$ defined by $v := \kappa(u - u_0) + k * (u - u_0)$ is absolutely continuous and differentiable a.e. on $[0, T)$ with $v(0) = 0$, $u(t) \in D(A)$ a.e. $t \in [0, T)$, and, moreover, there exists a measurable function $w : [0, T) \to X$ with $w(t) \in Au(t)$ a.e. $t \in [0, T)$, such that $d/dt\, v(t) + w(t) = f(t)$ a.e. $t \in [0, T)$.

However, many important spaces for applications do not admit the Radon-Nikodym property-such as for instance L^1. Even in the special case of the homogeneous Cauchy problem

$$\frac{d}{dt}u(t) + Au(t) \ni 0, \qquad t \geq 0,$$
$$u(0) = u_0, \tag{5}$$

no general regularity result is known. However, for m-completely accretive operators A in a normal Banach space $X \subset L^1(\Omega, \mu)$, where $(\Omega, \mathcal{A}, \mu)$ shall, throughout the paper, always denote a σ-finite measure space, regularity results for mild solutions of (5) were achieved in [BC91].

In this respect, our main task will be to extend these results for m-completely accretive operators to generalized solutions of the nonhomogeneous Volterra problem (1). It will turn out that the results obtained are similar to those known in the case of maximal accretive operators in Hilbert spaces.

In Section 2 we recall some basic facts on ϕ-accretivity and complete accretivity. Most importantly, we derive an integral inequality, Proposition 2.2, for generalized solutions of (1) with a ϕ-accretive operator A. This integral inequality, which generalizes [Bén72, Proposition 1.27], will be the basis for the study of regularity.

Section 3 is devoted to the main result, Theorem 3.2, on the regularity of generalized solutions. In particular, we show that generalized solutions of the Volterra equation (1) for $f \in W^{1,1}(0, T; X)$ and $u_0 \in D(A)$ are strong solutions if the operator A is m-completely accretive in a normal Banach space $X \subset L^1(\Omega)$ (see Definition 3.1) satisfying the strong convergence condition (10).

In Section 4 we give a short overview on heat flow in materials with memory and derive a concrete example for (1).

2. Integral inequalities

In this section, we develop an integral inequality for generalized solutions of (4), where A is a ϕ-accretive operator in a real Banach space X, $\kappa \geq 0$, and $k \in L^1_{\mathrm{loc}}([0, \infty))$ satisfies (2) and (3). Recall that for a continuous convex function $\phi : X \to \mathbb{R}$ a possibly multivalued operator $A \subset X \times X$ is called ϕ-accretive if for all $(x, y), (\tilde{x}, \tilde{y}) \in A$

$$\phi'_+[x - \tilde{x}, y - \tilde{y}] \geq 0.$$

Here, $\phi'_+[x, y] := \lim_{\lambda \to 0+} \frac{1}{\lambda}(\phi(x+\lambda y) - \phi(x))$ denotes the right hand Gateaux derivative of ϕ at x in direction y.

In the special case $\phi := \|\cdot\|$ the right hand Gateaux derivative of the norm of the Banach space X is denoted by $[x, y]_+$ and a $\|\cdot\|$-accretive operator is just called accretive. Moreover, if in addition $R(I + \lambda A) = X$ for all $\lambda > 0$ then A is called m-accretive. As a further example consider the space

$$L_0(\Omega) := \left\{ u \in L^1(\Omega) + L^\infty(\Omega) \ \middle| \ \int (|u| - m)^+ \, d\mu < \infty \text{ for all } m > 0 \right\}$$

endowed with the norm $\|\cdot\|_{1+\infty}$ of $L^1(\Omega) + L^\infty(\Omega)$, defined by

$$\|u\|_{1+\infty} := \inf\{\|u_1\|_1 + \|u_2\|_\infty \mid u = u_1 + u_2, \ u_1 \in L^1(\Omega), \ u_2 \in L^\infty(\Omega)\}.$$

On the set $M(\Omega)$ of measurable functions we define $\phi_m^+, \phi_m^- : M(\Omega) \to \mathbb{R} \cup \{\infty\}$ for $m > 0$ by

$$\phi_m^+(u) := \int (u - m)^+, \quad \phi_m^-(u) := \int (u + m)^-, \qquad \text{for all } u \in M(\Omega). \tag{6}$$

For any real Banach space $X \subset L_0(\Omega)$ the restrictions of ϕ_m^+, ϕ_m^- to X again denoted by ϕ_m^+ and ϕ_m^-, respectively, are continuous convex $\mathbb{R}$-valued functions. Moreover, the right hand Gateaux derivatives of $\phi_m^+, \phi_m^- : X \to \mathbb{R}$ are given by

$$(\phi_m^+)'_+[u, v] = \int_{\{u>m\}} v \ + \int_{\{u=m\}} v^+, \quad (\phi_m^-)'_+[u, v] = -\int_{\{u<-m\}} v \ + \int_{\{u=-m\}} v^-.$$

We recall that an operator A in X is called completely accretive if A is ϕ-accretive for all $\phi = \phi_m^+$ and all $\phi = \phi_m^-$ with $m > 0$ (cf. [BC91]). Moreover, if in addition $R(I + \lambda A) = X$ for all $\lambda > 0$ then the operator A is called m-completely accretive.

In what follows we shall always assume that $\phi(u)$, $\phi_+'[u, v] \in L^1(0, T)$ for all $u, v \in L^1(0, T; X)$. Note that this is always satisfied for the functions ϕ in the examples above.

Next, we extend the concept of generalized solutions of (4) given for m-accretive A in [Gri85, CGL96] to equations with ϕ-accretive operators.

DEFINITION 2.1. Let $u_0 \in \overline{D(A)}$, $f \in L^1(0, T; X)$ and A be a ϕ-accretive operator in X. A function $u \in L^1(0, T; X)$ is called a generalized solution of (4) if for all sequences $\{k_n\}_{n \in \mathbb{N}}$ of functions k_n satisfying (2), (3) for $(0, k_n)$, $k_n(0+) := \lim_{t \to 0+} k_n(t) < \infty$, and

$$\int_0^t k_n(\tau) \, d\tau \to \kappa + \int_0^t k(\tau) \, d\tau \quad \text{as } n \to \infty \text{ for all } t > 0, \tag{7}$$

there exists a unique strong solution u_n of

$$\frac{d}{dt} \left(k_n * (u_n - u_0) \right)(t) + A u_n(t) \ni f(t), \qquad t \in [0, T), \tag{8}$$

for all $n \in \mathbb{N}$ such that $\|u_n - u\|_{L^1(0,T;X)} \to 0$ as $n \to \infty$.

Motivated by the integral inequality for solutions of inhomogeneous Cauchy problems, see [Bén72, Proposition 1.27], we develop an integral inequality for the comparison of two generalized solutions of (4) at two different points with different data for an arbitrary ϕ-accretive operator A. This generalizes earlier results of [Clé80, Theorem 1] and [KKM85, Lemma 3.1] for (4) with m-accretive operators.

PROPOSITION 2.2. *Let A be a ϕ-accretive operator in a Banach space X, $\kappa \geq 0$, $k \in L^1_{\mathrm{loc}}([0, \infty))$ satisfy (2) and (3), $u_0, v_0 \in \overline{D(A)}$, $f, g \in L^1(0, T; X)$. Let u be a generalized solution of (4), and v be a generalized solution of (4) with u_0, f replaced by v_0, g. Then a.e. $t \in [0, T)$*

$$\phi(u(t) - v(t)) \leq \phi(u_0 - v_0)$$
$$+ \int_{[0,t]} \phi_+'[u(t - s) - v(t - s), f(t - s) - g(t - s)] \, d\alpha(s).$$

Moreover, if there exists a sequence $\{u_n\}$ of solutions of (8) approximating u uniformly on $[0, T)$, such that (7) is satisfied then for all $0 < h < T$

$$\phi(u(t + h) - v(t)) \leq \sup_{\tau \in [0,h]} \phi(u(\tau) - v_0)$$
$$+ \int_{[0,t]} \phi_+'[u(t + h - s) - v(t - s), f(t + h - s) - g(t - s)] \, d\alpha(s)$$

a.e. $t \in [0, T - h)$. Here α is the resolvent of the first kind (cf. [Gri80]) of the pair (κ, k) defined by $\kappa \alpha([0, t]) + (k * \alpha([0, \cdot]))(t) = t$ for all $t > 0$.

Proof. In the first step, we consider the case of $\kappa = 0$, $k(0+) < \infty$. In this case the generalized solution $u \in L^1(0, T; X)$ is a strong solution of (4). In particular, for $0 \leq h < T$ fixed, we have $u(t + h) \in D(A)$ a.e. $t \in [0, T - h)$, and thus

$$f(t + h) - k(0+)(u(t + h) - u_0) - \int_{(0,t+h]} (u(t + h - s) - u_0)\, dk(s) \in Au(t + h).$$

Obviously, this holds as well for the generalized solution v, with u_0, f replaced by v_0, g and $t + h$ replaced by t. By the definition of ϕ-accretivity of the operator A, we have a.e. $t \in [0, T - h)$

$$0 \leq \phi'_+\Big[u(t + h) - v(t),\, f(t + h) - g(t) - k(0+)(u(t + h) - v(t) - (u_0 - v_0))$$

$$- \int_{(0,t+h]} (u(t + h - s) - u_0)\, dk(s) + \int_{(0,t]} (v(t - s) - v_0)\, dk(s)\Big]$$

$$\leq \phi'_+[u(t + h) - v(t),\, f(t + h) - g(t)]$$

$$- k(t + h)\{\phi(u(t + h) - v(t)) - \phi(u_0 - v_0)\}$$

$$+ \int_{(0,t]} \{\phi(u(t + h) - v(t)) - \phi(u(t + h - s) - v(t - s))\}\, dk(s)$$

$$+ \int_{(t,t+h]} \{\phi(u(t + h) - v(t)) - \phi(u(t + h - s) - v_0)\}\, dk(s)$$

$$= \phi'_+[u(t + h) - v(t),\, f(t + h) - g(t)] + k(t)\phi(u_0 - v_0)$$

$$- k(0+)\phi(u(t + h) - v(t)) - \int_{(0,t]} \phi(u(t + h - s) - v(t - s))\, dk(s)$$

$$- \int_{(t,t+h]} \{\phi(u(t + h - s) - v_0) - \phi(u_0 - v_0)\}\, dk(s).$$

Here, we used the properties of the Gateaux derivative ϕ'_+, in particular the continuity in the second variable and the fact that k is nonincreasing. The latter implies that the measure dk is a locally finite nonpositive measure on $(0, \infty)$. Since the resolvent of the first kind α

of the pair $(0, k)$ is a locally finite nonnegative measure on $[0, \infty)$ the convolution of the above inequality with the measure α yields

$$\phi(u(t + h) - v(t)) \leq \phi(u_0 - v_0)$$

$$- \int_{[0,t]} \int_{(t-s,t+h-s]} \{\phi(u(t + h - s - \sigma) - v_0) - \phi(u_0 - v_0)\} \, dk(\sigma) \, d\alpha(s)$$

$$+ \int_{[0,t]} \phi'_+[u(t + h - s) - v(t - s), f(t + h - s) - g(t - s)] \, d\alpha(s) \qquad (9)$$

a.e. $t \in [0, T - h)$. Note that the second term on the right hand side of (9) can be estimated by

$$\sup_{\tau \in [0,h]} \{\phi(u(\tau) - v_0) - \phi(u_0 - v_0)\} \cdot \int_{[0,t]} (k(t - s) - k(t + h - s)) \, d\alpha_n(s)$$

$$\leq \sup_{\tau \in [0,h]} \{\phi(u(\tau) - v_0) - \phi(u_0 - v_0)\}$$

This gives the assertion.

In the second step, we assume that $\kappa > 0$ or that $k(0+) = \infty$. According to the assumption we can approximate the generalized solutions u and v by sequences of strong solutions $\{u_n\}, \{v_n\}$ of (8) where $\{k_n\}_{n \in \mathbb{N}}$ is a sequence of functions in $L^1_{\mathrm{loc}}([0, \infty))$ satisfying (2), (3) for $(0, k_n)$, and (7). By the definition of generalized solutions we know that $u_n \to u$ and $v_n \to v$ in $L^1(0, T; X)$. Therefore, we can assume that $u_n(t) \to u(t)$ and $v_n(t) \to v(t)$ a.e. $t \in [0, T)$. As a result of the first step of the proof, for $0 \leq h < T$ fixed, the approximate solutions u_n, v_n satisfy the inequality

$$\phi(u_n(t + h) - v_n(t)) \leq \sup_{\tau \in [0,h]} \phi(u_n(\tau) - v_0)$$

$$+ \int_{[0,t]} \phi'_+[u_n(t + h - s) - v_n(t - s), f(t + h - s) - g(t - s)] \, d\alpha_n(s)$$

a.e. $t \in [0, T - h)$ and for all $n \in \mathbb{N}$, where α_n is the resolvent of the first kind of the pair $(0, k_n)$.

Due to the upper semicontinuity of the Gateaux derivative ϕ'_+ in the first variable, and the fact that the sequence of measures $\{\alpha_n\}$ converges in $\mathcal{D}'([0, \infty))$ to the resolvent of the first kind α of the pair (κ, k) as $n \to \infty$ (cf. [CGL96, Proposition 3]), we conclude that

$$\phi(u(t + h) - v(t)) \leq \left(\liminf_{n \to \infty} \sup_{\tau \in [0,h]} \phi(u_n(\tau) - v_0) \right)$$

$$+ \int_{[0,t]} \phi'_+[u(t + h - s) - v(t - s), f(t + h - s) - g(t - s)] \, d\alpha(s)$$

a.e. $t \in [0, T - h)$. This gives the assertion for $h = 0$ and in the case $0 < h < T$ with $u_n \to u$ uniformly on $[0, T)$. $\qquad \square$

Note that in special cases, such as $\phi = \|\cdot\|$, we can assume that $\phi(0) < \phi(x)$ for all $x \in X$ with $x \neq 0$. Thus Proposition 2.2 gives the uniqueness of generalized solutions.

3. Strong solutions

In this section, we shall always assume that A is an m-accretive operator in a Banach space X. Moreover, we assume that $\kappa \geq 0$, $k \in L^1_{\text{loc}}([0, \infty))$ satisfy (2), (3).

Recall that in case of $\kappa = 0$ and $k(0+) < \infty$ generalized solutions of (4) are strong solutions by definition. For regularity properties in case of $\kappa = 0$ and $k(0+) = \infty$, we refer to [CGL00, Theorem 1].

Therefore, we shall concentrate on the case $\kappa > 0$ and without loss of generality we always assume $\kappa = 1$. Even for the inhomogeneous Cauchy problem, which is contained as a special case, i.e. $k \equiv 0$, the results obtained are new.

It is well known (see [Gri85, Theorem 2]) that generalized solutions of (1) are Lipschitz continuous if $u_0 \in \hat{D}(A)$, $f \in BV([0, T]; X)$ and thus are strong solutions if the Banach space X has the Radon-Nikodym property.

In order to get regularity results in spaces without the Radon-Nikodym property one has to restrict the class of m-accretive operators. In [BC91] it has been shown that for the class of m-completely accretive operators in a normal Banach space regularity for mild solutions of the homogeneous Cauchy problem (5) can be obtained.

DEFINITION 3.1. A Banach space $\{0\} \neq X \subset L_0(\Omega)$ is called normal if

$$u \in X, \ v : \Omega \to \mathbb{R} \text{ measurable}, \ v \ll u \implies v \in X \text{ and } \|v\|_X \leq \|u\|_X.$$

Here, the relation $u \ll v$ for measurable functions $u, v \in M(\Omega)$ is defined by

$$u \ll v \iff \phi^+_m(u) \leq \phi^+_m(v) \text{ and } \phi^-_m(u) \leq \phi^-_m(v) \text{ for all } m > 0,$$

where ϕ^+_m and ϕ^-_m are given by (6).

Note that an operator A in $X \subset L_0(\Omega)$ is completely accretive if and only if

$$u - \tilde{u} \ll u - \tilde{u} + \lambda(v - \tilde{v}) \qquad \text{for all } (u, v), (\tilde{u}, \tilde{v}) \in A.$$

We now state the fundamental theorem on the regularity of solutions of (1) in a normal Banach space.

THEOREM 3.2. *Let A be an m-completely accretive operator in a normal real Banach space $X \subset L^1(\Omega)$ satisfying the strong convergence condition*

$$\left.\begin{array}{c} \{u_n\}_{n \in \mathbb{N}} \subset X, u \in L_0(\Omega), \\ \liminf_{n \to \infty} \|u_n\|_X < \infty, u_n \ll u, \\ u_n \to u \ a.e. \end{array}\right\} \implies u \in X \text{ and } \|u_n - u\|_X \to 0. \tag{10}$$

Then, for all $u_0 \in D(A)$, and $f \in W^{1,1}(0, T; X)$, the generalized solution u of (1) is locally Lipschitz continuous and differentiable a.e. on $[0, T)$ such that $u(t) \in D(A)$ and

$$-\frac{d}{dt}u(t) = \left(Au(t) - f(t) + \frac{d}{dt}(k * (u - u_0))(t)\right)^{\circ}$$

a.e. $t \in [0, T)$. Here, following [BC91], for a set $C \subset L_0(\Omega)$, C° denotes the minimal section of C defined by $C^{\circ} := \{u \in C \mid u \ll v \text{ for all } v \in C\}$, which has at most one element if C is convex. In particular, u is the unique strong solution of (1).

As an aside, we recall that the regularity result for mild solutions of homogeneous Cauchy problems (cf. [BC91, Theorem 4.2]) could be stated for normal Banach spaces only satisfying the convergence condition

$$u_n \ll u \in X \text{ for all } n \in \mathbb{N} \text{ and } u_n \to u \text{ a.e.} \implies \|u_n - u\|_X \to 0. \tag{11}$$

In this context, note that a normal Banach space $X \subset L_0(\Omega)$ satisfies the strong convergence condition (10) if and only if it satisfies the convergence condition (11) and has the Fatou property. In particular, $L^1(\Omega)$ satisfies the strong convergence condition (10).

In a first step we show the differentiability from the right at $t = 0$ using the straightforward idea of reducing the problem to the case of the homogeneous Cauchy problem in order to apply [BC91, Theorem 4.2].

LEMMA 3.3. *Let A be an m-completely accretive operator in a normal real Banach space $X \subset L_0(\Omega)$, satisfying the convergence condition (11), let $u_0 \in D(A)$, $f \in C([0, T); X)$. Then the unique generalized solution u of (1) is differentiable from the right at $t = 0$ and $(d/dt)^+ u(0) = -(Au_0 - f(0))^{\circ}$.*

Proof. Since A is m-completely accretive in X, one can easily verify that the operator $B = A - f(0)$ is m-completely accretive in X as well. Then, according to [BC91, Theorem 4.2], the mild solution v of the homogeneous Cauchy problem

$$\frac{d}{dt}v(t) + Bv(t) \ni 0, \qquad v(0) = u_0$$

is strongly differentiable from the right at $t = 0$. As v is a strong solution of the Cauchy problem, it is obvious that v is as well a strong solution of the Volterra equation

$$\frac{d}{dt}(v(t) - u_0 + (k * (v - u_0))(t)) + Av(t) \ni f(0) - g(t), \tag{12}$$

where $g \in L^1(0, T; X)$ is defined by $g(t) := d/dt (k * (v - u_0))(t)$ for $t \in [0, T)$.

By [Gri85, Theorem 1] and [CGL96, Theorem 1], v is the unique generalized solution of the Volterra equation (12). We apply Proposition 2.2 and obtain for all $0 < t < h < T$

$$\left\| \frac{u(t) - u_0}{t} - \frac{v(t) - u_0}{t} \right\| \le \frac{1}{t} \int_0^t \|f(t - s) - f(0)\| \, d\alpha(s)$$

$$+ \frac{1}{t} \int_0^t \left\| \frac{d}{dt}(k * (v - u_0))(t - s) \right\| d\alpha(s)$$

$$\le \frac{1}{t} \int_0^t \|f(s) - f(0)\| a(t - s) \, ds + \left\| \frac{d}{dt}^+ v \right\|_{L^\infty(0,h;X)} \|a\|_\infty \int_0^t k(s) \, ds. \tag{13}$$

Here, a denotes the Radon-Nikodym derivative of the resolvent of the first kind α of the pair $(1, k)$. Note that $a \le 1$ a.e. on $[0, T)$. Since v is locally Lipschitz continuous and differentiable from the right, we conclude $\|(d/dt)^+ v\|_{L^\infty(0,1;X)} < \infty$. By the continuity of f, and the fact that $k \in L^1(0, 1)$, one can pass to the limit for $t \to 0+$ in (13). Thus, we conclude that u is differentiable from the right at $t = 0$, and that

$$\frac{d^+}{dt} u(0) = \frac{d^+}{dt} v(0) = -(Au_0 - f(0))^\circ.$$

$\square$

While the preceding result used the method of reduction to the homogeneous case, it turns out that this is not applicable at $t_0 > 0$. Indeed, by defining $g := d/dt \, (k * (u - u_0))$, we easily derive a homogeneous Cauchy-problem for $B := A - f(t_0) + g(t_0)$. But since it is unclear whether $u(t_0) \in D(A)$ holds, we still can not apply the results of [BC91]. However, as in the proof of [BC91, Theorem 4.2], we will be able to show the weak sequential compactness of the differential quotient $\frac{1}{h}(u(t_0 + h) - u(t_0))$. But as we are in the inhomogeneous case the proof will require much more subtle methods, since we can only apply the integral inequality of Proposition 2.2. Therefore, it will become necessary to study Equation (1) in an appropriate Orlicz space (see e.g. [KR61] for the theory of Orlicz spaces). We will use the following Lemma to construct the appropriate Orlicz space.

LEMMA 3.4. *Let* $f_j \in L^1(\Omega)$ *for* $j = 1, \ldots, k$, *then there exists an N-function N satisfying the global Δ_2-condition such that* $f_j \in L_N(\Omega)$ *for all* $j = 1, \ldots, k$. *Here, we say N satisfies the global Δ_2-condition if there exists a constant $c > 0$ such that* $N(2r) \le cN(r)$ *for all* $r \ge 0$.

For the sake of completeness, we present the proof, which mainly follows the arguments given in [KR61, p. 60ff] for finite measure spaces.

Proof. For $n \in \mathbb{N}$ and $j = 1, \ldots, k$ we define $\Omega_n^{(j)} := \{n \le |f_j| < n + 1\}$. Since

$$\sum_{n=1}^{\infty} (n + 1)\mu(\Omega_n^{(j)}) \le \mu(\{1 \le |f_j|\}) + \int_{\{1 \le |f_j|\}} |f_j| \, d\mu < \infty \quad \text{for } j = 1, \ldots, k$$

there exist strictly increasing sequences $\{\alpha_n^{(j)}\}_{n\in\mathbb{N}}$ such that $\alpha_1^{(j)} \geq 3$, $\alpha_n^{(j)} \to \infty$ as $n \to \infty$ and $\sum_{n=1}^{\infty} \alpha_n^{(j)}(n+1)\mu(\Omega_n^{(j)}) < \infty$ for $j = 1, \ldots, k$. Defining $p : [0, \infty) \to \mathbb{R}$ by

$$p(r) := \begin{cases} r, & 0 \leq r < 1 \\ n+1, & 1 \leq n \leq r < n+1 \leq 3 \\ \min_{j=1,\ldots,k}(\alpha_n^{(j)}, 2p(n-1) - p(n-2)), & 3 \leq n \leq r < n+1, \end{cases}$$

it is clear that p is right-continuous, nondecreasing with $p(r) > 0$ for $r > 0$ and satisfies $\lim_{r\to 0+} p(r) = 0$ and $\lim_{r\to\infty} p(r) = \infty$. Thus, $N(r) := \int_0^{|r|} p(\rho)\, d\rho$ for $r \in \mathbb{R}$ defines an N-function. N satisfies the global Δ_2-condition, since $rp(r) \leq 4N(r)$ for all $r \geq 0$. Moreover, for all $j = 1, \ldots, k$

$$\int_{\Omega} N(|f_j|)\, d\mu \leq \int_{\{|f_j|<1\}} \frac{1}{2}|f_j|^2\, d\mu + \sum_{n=1}^{\infty}(n+1)p(n)\mu(\Omega_n^{(j)})$$

$$\leq \int_{\{|f_j|<1\}} |f_j|\, d\mu + \sum_{n=1}^{\infty}\alpha_n^{(j)}(n+1)\mu(\Omega_n^{(j)}) < \infty.$$

This gives the assertion. $\qquad\qquad\qquad\qquad\qquad\qquad\qquad\qquad\qquad\qquad\qquad\quad \square$

Proof of Theorem 3.2 We assume that $u_0 \in D(A)$, $f \in W^{1,1}(0, T; X)$, and that u is the unique generalized solution of (1) in X. Moreover, let $v_0 \in Au_0$. Since, by [Gri85, Theorem 2], u is Lipschitz continuous, [Gri85, Lemma 3.4] implies that the function $(k * (u - u_0))$ is absolutely continuous and differentiable a.e. on $[0, T)$, and that $g := d/dt\, (k * (u - u_0)) \in L^1(0, T; X)$. We remark that the set $L_r \subset [0, T)$ of right Lebesgue points of g, i.e. the set of all $t \in [0, T)$, such that

$$\lim_{h\to 0+} \frac{1}{h} \int_t^{t+h} \|g(\tau) - g(t)\|\, d\tau = 0, \tag{14}$$

is the complement of a nullset in $[0, T)$. We now prove that u is strongly differentiable at all $t_0 \in L_r$, and that at these points u satisfies Equation (1). This proof will consist of several steps.

(1) In the first step, we construct a particular Orlicz space $L_N(\Omega)$. Let us remark that for any N-function N satisfying the global Δ_2-condition, the Orlicz space $L_N(\Omega)$ equipped with the Luxemburg norm

$$\|u\|_N := \inf\left\{k > 0 \,\bigg|\, \int_{\Omega} N\left(\frac{|u|}{k}\right) \leq 1\right\}$$

is a normal Banach space satisfying the convergence condition (11).

By Lemma 3.4 there exists an N-function N satisfying the global Δ_2-condition, such that for $Q := (0, T) \times \Omega$ we have $|u_0|, |v_0|, |f|, |f'| \in L_N(Q)$. Here, we interpreted

u_0, v_0 as constant functions over $(0, T)$. By Fubini's theorem this implies u_0, $v_0 \in L_N(\Omega)$ and f, $f' \in L^1(0, T; L_N(\Omega))$.

(2) Now, our purpose is to show that for $t_0 \in [0, T)$ there exists a sequence $\{h_n\}_{n \in \mathbb{N}}$ with $h_n \to 0+$ as $n \to \infty$, such that the sequence $\{\frac{1}{h_n}(u(t_0 + h_n) - u(t_0))\}_{n \in \mathbb{N}}$ converges weakly in $L^1_{\mathrm{loc}}(\Omega)$.

Since $(u_0, v_0) \in A$, and u_0, $v_0 \in L_N(\Omega)$ the restriction of A to $Y = X \cap L_N(\Omega)$ given by

$$A_Y := \{(u, v) \in A \mid u, v \in L_N(\Omega)\}$$

is nonempty and obviously m-completely accretive in Y.

According to [Gri85, Theorem 1] and [CGL96, Theorem 1], the Volterra equation (1) in Y admits a unique generalized solution $v \in C([0, T); Y)$. Since the embeddings $Y \hookrightarrow X$ and $Y \hookrightarrow L_N(\Omega)$ are continuous, $v \equiv u$, and u is as well a generalized solution of (1) in the space $L_N(\Omega)$.

As u is continuous, we have using Proposition 2.2

$$\left\| \frac{u(t_0 + h) - u(t_0)}{h} \right\|_N$$
$$\leq \sup_{\tau \in [0,h]} \left\| \frac{u(\tau) - u_0}{\tau} \right\|_N + \int_0^{t_0} \frac{1}{h} \int_\tau^{\tau+h} \|f'(\sigma)\|_N \, d\sigma \, a(t_0 - \tau) \, d\tau \tag{15}$$

for $0 < h < T - t_0$. Here a denotes the Radon-Nikodym derivative of the resolvent of the first kind α of the pair $(1, k)$. By Lemma 3.3. applied to the space $Y \hookrightarrow L_N(\Omega)$, we already know that u is strongly differentiable from the right at $t = 0$. Therefore, applying Lebesgue's dominated convergence theorem, we can pass to the limit for $h \to 0+$ in (15) and obtain

$$\limsup_{h \to 0+} \left\| \frac{u(t_0 + h) - u(t_0)}{h} \right\|_N < \infty.$$

Since Ω is a σ-finite measure space, we may choose an increasing sequence $\omega_k \nearrow \Omega$ of measurable subsets of Ω, satisfying $\mu(\omega_k) < \infty$ for all $k \in \mathbb{N}$. Then, we can define the injection of $L_N(\Omega)$ into the Fréchet space $\prod_{k \in \mathbb{N}} L^1(\omega_k)$ by

$$\iota : L_N(\Omega) \hookrightarrow \prod_{k \in \mathbb{N}} L^1(\omega_k), \qquad f \mapsto (f \mathbf{1}_{\omega_k})_{k \in \mathbb{N}}.$$

By de la Vallée Poussin's theorem it is clear that $\iota(B)$ is weakly sequentially compact for all bounded subsets B of $L_N(\Omega)$. Thus, we can conclude that there exists a sequence $\{h_n\}$, with $h_n \to 0+$ as $n \to \infty$, and $(z_k)_k \in \prod_{k \in \mathbb{N}} L^1(\omega_k)$, such that

$$\frac{u(t_0 + h_n) - u(t_0)}{h_n} \mathbf{1}_{\omega_k} \rightharpoonup z_k \qquad \text{weakly in } L^1(\omega_k) \text{ for all } k \in \mathbb{N}. \tag{16}$$

Therefore, it is clear that there exists $z : \Omega \to \mathbb{R}$ measurable such that $z 1_{\omega_k} = z_k$ for all $k \in \mathbb{N}$.

(3) Our goal is to show that $u(t_0) \in D(\overline{A})$ and $f(t_0) - g(t_0) - z \in \overline{A}u(t_0)$ for all $t_0 \in L_r$, where $\overline{A}$ denotes the closure of A in the space $L_0(\Omega)$. For this purpose we first show that $z \in L_0(\Omega)$.

For all $m > 0$, it is clear that, for $0 < h < T - t_0$,

$$\left\| \left(\frac{u(t_0 + h_n) - u(t_0)}{h_n} - m \right)^+ \right\|_N \leq \left\| \frac{u(t_0 + h) - u(t_0)}{h} \right\|_N .$$

Thus, we may assume that

$$\left(\frac{u(t_0 + h) - u(t_0)}{h} - m \right)^+ 1_{\omega_k} \rightharpoonup z_m 1_{\omega_k} \quad \text{weakly in } L^1(\omega_n) \text{ for all } k \in \mathbb{N},$$

for a subsequence again denoted by $\{h_n\}$ and some $z_m : \Omega \to \mathbb{R}$ measurable. Since the weak limit in $L^1(\Omega)$ is order preserving, it follows from

$$\frac{u(t_0 + h_n) - u(t_0)}{h_n} - m \leq \left(\frac{u(t_0 + h_n) - u(t_0)}{h_n} - m \right)^+$$

that $z - m \leq z_m$. Since $z_m \geq 0$, we conclude $(z - m)^+ \leq z_m$ for all $m > 0$.

Choosing $\phi = \phi^+_{(h_n m)}$ in Proposition 2.2 we obtain for all $k \in \mathbb{N}$

$$\int_{\omega_k} (z - m)^+ \leq \int_{\omega_k} z_m = \lim_{n \to \infty} \int_{\omega_k} \left(\frac{u(t_0 + h_n) - u(t_0)}{h_n} - m \right)^+$$
$$\leq \int_{\Omega} \left(\frac{d}{dt}^+ u(0) - m \right)^+ + \int_0^{t_0} \int_{\Omega} |f'(\tau)| \, d\mu \, a(t_0 - \tau) \, d\tau.$$

Since, by definition, $\omega_k \nearrow \Omega$ as $k \to \infty$, this implies

$$\phi_m^+(z) = \int_{\Omega} (z - m)^+ < \infty \quad \text{for all } m > 0.$$

As one can apply exactly the same arguments used above for $r \mapsto (r + m)^-$ instead of $r \mapsto (r - m)^+$, we have shown that $z \in L_0(\Omega)$.

Before we proceed with the proof, we remark that the generalized solution u of the Volterra equation (1) is in fact a mild solution of the inhomogeneous Cauchy problem

$$\frac{d}{dt} v(t) + A v(t) \ni f(t) - g(t), \qquad t \in [0, T),$$
$$v(0) = u_0. \tag{17}$$

Now, let $(\xi, \eta) \in \overline{A}$, and for $m > 0$ let $w_m \in L_0(\Omega)' = L^1(\Omega) \cap L^\infty(\Omega)$ with $w_m \in \partial j_m^+(u(t_0) - \xi)$ a.e. in Ω, where j_m^+ is the convex function on $\mathbb{R}$ defined by $j_m^+(r) := (r - m)^+$

for all $r \in \mathbb{R}$, and ∂j_m^+ denotes the subdifferential of j_m^+. It is clear that $w_m = 1$ on $\{u(t_0) - \xi > m\}$, $w_m \in [0, 1]$ on $\{u(t_0) - \xi = m\}$, and $w_m = 0$ on $\{u(t_0) - \xi < m\}$. Since A is ϕ_m^+-accretive, and u is a mild solution of the inhomogeneous Cauchy problem (17), we can apply the integral inequality for ϕ_m^+-integral solutions (see [Bén72, Proposition 1.27])

$$\int_\Omega \frac{u(t_0 + h_n) - u(t_0)}{h_n} w_m$$

$$\leq \frac{1}{h_n} \int_\Omega [(u(t_0 + h_n) - \xi - m)^+ - (u(t_0) - \xi - m)^+]$$

$$\leq \frac{1}{h_n} \int_{t_0}^{t_0+h_n} (\phi_m^+)'_+[u(\tau) - \xi, f(\tau) - g(\tau) - \eta]\, d\tau$$

$$\leq \frac{1}{h_n} \int_{t_0}^{t_0+h_n} \int_\Omega |f(\tau) - f(t_0) - g(\tau) + g(t_0)|\, d\mu\, d\tau$$

$$+ \frac{1}{h_n} \int_{t_0}^{t_0+h_n} (\phi_m^+)'_+[u(\tau) - \xi, f(t_0) - g(t_0) - \eta]\, d\tau. \tag{18}$$

As f is continuous in t, and t_0 is a right Lebesgue point of g, and $(\phi_m^+)'_+$ is upper semicontinuous, we may pass to the limit at the right-hand side of (18). Moreover, since $w_m \in L^\infty(\Omega)$ and $\mu(\{w_m \neq 0\}) \leq \mu(\{u(t_0) - \xi > \frac{m}{2}\}) < \infty$, we can also pass to the limit at the left-hand side of (18), and we obtain

$$\int_\Omega z w_m \leq (\phi_m^+)'_+[u(t_0) - \xi, f(t_0) - g(t_0) - \eta].$$

Since $w_m \in L_0(\Omega)'$ with $w_m \in \partial j_m^+(u(t_0) - \xi)$ a.e. in Ω was chosen arbitrarily, it follows that

$$(\phi_m^+)'_+[u(t_0) - \xi, f(t_0) - g(t_0) - z - \eta] \geq 0.$$

The same arguments can be applied to $j_m^-(r) := (r + m)^-$ for all $m > 0$. Therefore, for all $\lambda > 0$ and all $(\xi, \eta) \in \overline{A}$ we have $u(t_0) - \xi \ll u(t_0) - \xi + \lambda(f(t_0) - g(t_0) - z - \eta)$. This implies $u(t_0) \in D(\overline{A})$ and $f(t_0) - g(t_0) - z \in \overline{A}u(t_0)$.

(4) We are now going to show that the right-hand side derivative of u exists in $L_0(\Omega)$ at all right Lebesgue points $t_0 \in L_r$ of g, and that

$$L_0(\Omega)\text{-}\frac{d^+}{dt}\, u(t_0) = (-\overline{A}u(t_0) + f(t_0) - g(t_0))^\circ.$$

To this end, we use a reduction to the homogeneous case, as we have already shown that $u(t_0) \in D(\overline{A})$. We define the operator $B \subset L_0(\Omega) \times L_0(\Omega)$ by $B := \overline{A} - f(t_0) + g(t_0)$.

It is obvious that B is m-completely accretive in $L_0(\Omega)$. Let v be the mild solution of the homogeneous Cauchy problem

$$\frac{d}{dt}v(t) + Bv(t) \ni 0, \qquad t \geq 0,$$
$$v(0) = u(t_0), \tag{19}$$

where $t_0 \in L_r$ is a right Lebesgue point of g. Then, as shown in step (3), $u(t_0) \in D(\overline{A})$, and by [BC91, Theorem 4.2] applied to the operator B in the space $L_0(\Omega)$, we know that v is differentiable from the right for all $t \geq 0$, and that $\lim_{t\to 0} \frac{1}{t}(v(t) - u(t_0)) = -(Bu(t_0))^\circ$ in $L_0(\Omega)$.

To be able to compare v and u, we first have to shift u by t_0 and then interpret this function as a solution of a Cauchy problem. We therefore define $w(t) := u(t + t_0)$ for $t \geq 0$. As we have already mentioned in Step (3), u is the mild solution of the inhomogeneous Cauchy problem (17) in X, and as the imbedding $X \hookrightarrow L_0(\Omega)$ is continuous, u is a mild solution of (17) in $L_0(\Omega)$ with A replaced by $\overline{A}$. Due to the translation invariance of Cauchy problems, w is the unique mild solution of the inhomogeneous Cauchy problem

$$\frac{d}{dt}w(t) + \overline{A}w(t) \ni f(t + t_0) - g(t + t_0), \quad t \in [0, T - t_0),$$
$$w(0) = u(t_0).$$

Since mild solutions satisfy the integral inequality, we have for all $0 < h < T - t_0$

$$\left\| \frac{w(h) - u(t_0)}{h} - \frac{v(h) - u(t_0)}{h} \right\|_{L_0(\Omega)}$$
$$\leq \frac{1}{h} \int_0^h \|f(t_0 + \tau) - f(t_0)\|_{L_0(\Omega)} \, d\tau + \frac{1}{h} \int_0^h \|g(t_0 + \tau) - g(t_0)\|_{L_0(\Omega)} \, d\tau.$$

As f is continuous, and t_0 is a right Lebesgue point of g, this implies that u is differentiable from the right in $L_0(\Omega)$ at t_0, and that

$$L_0(\Omega)\text{-}\frac{d^+}{dt}u(t_0) = L_0(\Omega)\text{-}\frac{d^+}{dt}v(0) = -(\overline{A}u(t_0) - f(t_0) + g(t_0))^\circ.$$

(5) The task is to show that $X\text{-}\lim_{h\to 0+} \frac{1}{h}(u(t_0 + h) - u(t_0))$ exists at all right Lebesgue points $t_0 \in L_r$. Therefore, we first note, that $u(t_0), f(t_0), g(t_0) \in X$. This implies, by the continuity of the embedding $X \hookrightarrow L_0(\Omega)$, that the solution v of the homogeneous Cauchy problem (19) in $L_0(\Omega)$ equals the mild solution of the inhomogeneous Cauchy problem in X, given by

$$\frac{d}{dt}v(t) + Av(t) \ni f(t_0) - g(t_0), \qquad t \geq 0,$$
$$v(0) = u(t_0). \tag{20}$$

In particular, $\frac{1}{h}(v(h) - u(t_0)) \in X$ for all $h > 0$. According to the result of Step (4), we already know that $z = L_0(\Omega)\text{-}\lim_{h \to 0+} \frac{1}{h}(v(h) - u(t_0))$ exists, and that $\frac{1}{h}(v(h) - u(t_0)) \ll z$ for all $h > 0$ small enough by [BC91, Theorem 4.2] applied to $B = A - f(t_0) + g(t_0)$. As v is the mild solution of (20), and $w = u(t_0 + \cdot)$ is the mild solution of (17), we can apply the integral inequality in X, and obtain

$$\left\| \frac{v(h) - u(t_0)}{h} \right\|_X$$

$$\leq \left\| \frac{v(h) - u(t_0)}{h} - \frac{w(h) - u(t_0)}{h} \right\|_X + \left\| \frac{u(t_0 + h) - u(t_0)}{h} \right\|_X$$

$$\leq \frac{1}{h} \int_0^h \| f(t_0 + \tau) - f(t_0) \|_X \, d\tau + \frac{1}{h} \int_0^h \| g(t_0 + \tau) - g(t_0) \|_X \, d\tau$$

$$+ \sup_{\tau \in [0,h]} \left\| \frac{u(\tau) - u_0}{h} \right\|_X + \int_0^{t_0} \left\| \frac{f(\tau + h) - f(\tau)}{h} \right\|_X a(t_0 - \tau) \, d\tau$$

$$< \infty.$$

Here, we used the facts that $f \in W^{1,1}(0, T; X)$, u is differentiable from the right at the point $t = 0$, and t_0 is a right Lebesgue point of g. Since convergence in $L_0(\Omega)$ implies a.e convergence of a subsequence, we can conclude by the strong convergence condition (10) that $z \in X$ and $z = X\text{-}\lim_{h \to 0+} \frac{1}{h}(v(h) - u(t_0))$, as all subsequences converge to the same limit z. Using the integral inequality, we conclude that

$$\left\| \frac{u(t_0 + h) - u(t_0)}{h} - \frac{v(h) - u(t_0)}{h} \right\|_X$$

$$\leq \frac{1}{h} \int_0^h \| f(t_0 + \tau) - f(t_0) \|_X \, d\tau + \frac{1}{h} \int_0^h \| g(t_0 + \tau) - g(t_0) \|_X \, d\tau$$

$$\to 0 \qquad \text{as } h \to 0+.$$

Thus we conclude that u is strongly differentiable a.e. on $[0, T)$.

4. Heat flow in materials with memory

We show that Equation (1) is obtained, when modelling nonlinear heat flow in materials with memory. Indeed, if we consider a rod of unit length over the interval $(0, 1)$ of a homogeneous material where $\vartheta = \vartheta(x, t)$ with $x \in (0, 1)$ and $t \in \mathbb{R}$ denotes the absolute temperature then by the law of conservation of energy under absence of deformation we have

$$\varepsilon_t + q_x = h.$$

Here, $\varepsilon = \varepsilon(x, t)$ denotes the internal energy of the material and $q = q(x, t)$ is the heat flux and $h = h(x, t)$ denotes a source term. This equation has to be supplemented by constitutive relations for the internal energy and the heat flux. Under a wide range of conditions Fourier's classical theory of heat conduction can be applied. It assumes that the internal energy and the heat flux depend on the temperature ϑ and the temperature gradient ϑ_x, respectively. In particular, considering the nonlinear heat flow, the constitutive relations are given by

$$\varepsilon = \vartheta, \qquad q = -\sigma(\vartheta_x).$$

Here, $\sigma \in C^1(\mathbb{R})$ with $0 < m \leq \sigma' \leq M$.

Note that Fourier's law predicts infinite speed of propagation for thermal disturbances. However, there are situations in which differences to the predictions of Fourier's law can be observed experimentally. In particular, "wavelike" pulses of heat that propagate with finite speed have been observed in certain dielectrics at very low temperatures (cf. [BH88] and the references therein). To overcome the problem of infinite speed of propagation several attempts have been made. In particular, [GP68] and [Nun71] introduce a model in which the constitutive relations for the internal energy and the heat flux, in difference to Fourier's law, depend on the history of the temperature and the temperature gradient, respectively. The constitutive relations introduced are of the form

$$\varepsilon(x, t) = \beta_0 \vartheta(x, t) + \int_0^t \beta(t - s)\vartheta(x, s)\, ds,$$

$$q(x, t) = -\gamma_0 \sigma(\vartheta_x(x, t)) + \int_0^t \gamma(t - s)\sigma(\vartheta_x(x, s))\, ds.$$

Here, β_0, γ_0 are positive constants and β, γ are assumed to be sufficiently smooth functions called the internal energy and heat relaxation functions.

For physically meaningful assumptions on β_0, γ_0, β, and γ (cf. [CN81]) this leads to an equation of the form (1) in the space $X = L^1(0, 1)$. Here, we define the operator A, which turns out to be m-completely accretive, as the $L^1(0, 1)$-closure of $A_0 := \{(u, w) \mid u \in W_0^{1,2}(0, 1) \cap L^\infty(0, 1), w = -\sigma(u_x)_x \in L^\infty(0, 1)\}$, which includes Dirichlet boundary conditions. Note that the operator $A = \overline{A_0}$ can be characterized in terms of entropy solutions for elliptic equations (cf. [BBG$^+$95]).

REFERENCES

[BBG$^+$95] BÉNILAN, PH., BOCCARDO, L., GALLOUËT, T., GARIEPY, R., PIERRE, M. and VAZQUEZ, J. L., *An L^1-theory of existence and uniqueness of solutions of nonlinear elliptic equations.* Ann. Sc. Norm. Super. Pisa, *22* (1995) (2), 241–273.

[BC91] BÉNILAN, PH. and CRANDALL, M. G., *Completely accretive operators.* In *Semigroup theory and evolution equations*, volume *135* of Lect. Notes Pure Appl. Math., Marcel-Dekker, 1991, 41–75.

[Bén72] BÉNILAN, PH., *Equation d'évolution dans un espace de Banach quelconque et applications.* Thèse d'Etat, Orsay, 1972.

[BH88]   BRANDON, D. and HRUSA, W. J., *Construction of a class of integral models for heat flow in materials with memory.* J. Integral Equations Appl. *1* (1988), 175–201.

[CGL96]   COCKBURN, B., GRIPENBERG, G. and LONDEN, S. -O., *On convergence to entropy solutions of a single conservation law.* J. Differ. Equations, *128* (1996), 206–251.

[CGL00]   CLÉMENT, PH., GRIPENBERG, G. and LONDEN, S. -O., *Smoothness in Fractional Evolution Equations and Conservation Laws.* Ann. Scuola Norm. Sup. Pisa Cl. Sci. *29* (2000), 231–251.

[Clé80]   CLÉMENT, PH., *On abstract Volterra equations with kernels having a positive resolvent.* Israel J. Math. *36* (1980), 193–200.

[CN81]   CLÉMENT, PH. and NOHEL, J. A., *Asymptotic behavior of solutions of nonlinear Volterra equations with completely positive kernels.* SIAM J. Math. Anal. *12* (1981), 514–535.

[GP68]   GURTIN, M. E. and PIPKIN, A. C., *A general theory of heat conduction with finite wave speeds.* Arch. Rational Mech. Anal. *31* (1968), 113–126.

[Gri80]   GRIPENBERG, G., *On Volterra equations of the first kind.* Integral Equations Oper. Theory, *3* (1980), 473–488.

[Gri85]   GRIPENBERG, G., *Volterra integro-differential equations with accretive nonlinearity.* J. Differ. Equations, *60* (1985), 57–79.

[KKM85]   KATO, N., KOBAYASI, K. and MIYADERA, I., *On the asymptotic behavior of solutions of evolution equations associated with nonlinear Volterra equations.* Nonlinear Anal. 9 (1985), 419–430.

[KR61]   KRASNOSEL'SKIJ, M. A. and RUTICKIJ, YA. B., *Convex functions and Orlicz spaces.* P. Noordhoff, Groningen, 1961.

[Nun71]   NUNZIATO, J. W., *On heat conduction in materials with memory.* Q. Appl. Math. *29* (1971), 187–204.

Volker G. Jakubowski
Fachbereich Mathematik
Universität Essen
D-45117 Essen
Germany
e-mail: Volker.Jakubowski@uni-essen.de

Petra Wittbold
UFR de Mathématiques
Université Louis Pasteur
F-67084 Strasbourg
France
e-mail: wittbold@math.u-strasbg.fr

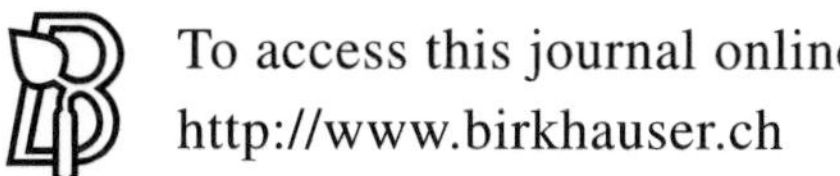

To access this journal online:
http://www.birkhauser.ch

J.evol.equ. 3 (2003) 321 – 331
1424–3199/03/020321 – 11
DOI 10.1007/s00028-003-0101-3
© Birkhäuser Verlag, Basel, 2003

Journal of Evolution
Equations

Nonautonomous heat equations with generalized Wentzell boundary conditions

JIN LIANG, RAINER NAGEL and TI-JUN XIAO

To the memory of Philippe Bénilan

Abstract. In this paper, we study the nonautonomous heat equation in $C[0, 1]$ with generalized Wentzell boundary conditions. It is shown, under appropriate assumptions, that there exists a unique evolution family for this problem and that the family satisfies various regularity properties. This enables us to obtain, for the corresponding inhomogeneous problem, classical and strict solutions having optimal regularity.

1. Introduction

For second order differential operators

$$\mathcal{A}(x, t) = a(x, t)\frac{d^2}{dx^2} + q(x, t)\frac{d}{dx} + r(x, t), \quad x \in [0, 1], \ t \in [0, T],$$

where $a(x, t) > 0$ ($x \in [0, 1]$, $t \in [0, T]$), we consider the following time dependent heat equation with a generalized Wentzell boundary condition

$$\begin{cases} \frac{\partial u}{\partial t} = \mathcal{A}(x, t)u, & 0 \le x \le 1, 0 \le s < t \le T, \\ u(x, s) = f(x), & 0 \le x \le 1, \\ \mathcal{A}(j, t)u(j, t) + \beta_j(t)\frac{\partial u}{\partial x}(j, t) + \gamma_j(t)u(j, t) = 0, j = 0, 1, 0 \le s < t \le T. \end{cases} \quad (1.1)$$

To the equation we associate the nonautonomous abstract Cauchy problem

$$\begin{cases} u'(t) = A(t)u(t), & 0 \le s < t \le T, \\ u(s) = f, \end{cases} \quad (NACP)$$

in the Banach space $C[0, 1]$, where the operators $A(t)$ are defined by

$$\begin{cases} (A(t)f)(x) = \mathcal{A}(x, t)f(x), & 0 \le x \le 1, \\ \mathcal{D}(A(t)) := \{f \in C^2[0, 1]; \quad \mathcal{A}(j, t)f(j) + \beta_j(t)f'(j) \\ \qquad\qquad\qquad + \gamma_j(t)f(j) = 0 \ \text{ at } \ j = 0, 1\}. \end{cases} \quad (1.2)$$

Mathematics Subject Classification (2000): 35K05, 47D03, 34G10.

Key words: Nonautonomous heat equations, generalized Wentzell boundary conditions, inhomogeneous Cauchy problems, classical and strict solutions.

The first author acknowledges support from the Max-Planck Society and from CAS and EMC. The third author acknowledges support from the Alexander-von-Humboldt Foundation and from CAS and NSFC.

Moreover, we assume for the coefficients that for some $\alpha \in (0, 1)$,

$$a, q, r \in C^{\alpha}([0, T]; C[0, 1]), \quad \beta_j, \gamma_j \in C^{\alpha}([0, T]; \mathbf{C}) \quad (j = 0, 1), \tag{1.3}$$

where $C^{\alpha}([0, T]; X)$ (for a Banach space X) is the space of Hölder continuous functions on $[0, T]$ defined by

$$\left\{ h \in C([0, T]; X); \quad \sup_{t,s \in [0,T],\ s<t} (t - s)^{-\alpha} \|h(t) - h(s)\|_X < +\infty \right\}.$$

With these assumptions we can show the wellposedness of (*NACP*) and (1.1).

The first result about the wellposedness of problem (1.1) with Robin boundary conditions (i.e. with $\mathcal{A}$ in the third line of (1.1) replaced by zero) was established in 1956 by T. Kato using C_0-semigroups ([12]). Later, T. Kato and H. Tanabe [13] sharpened this result using analytic semigroups. Recently, the wellposedness for the autonomous version of (1.1) and (*NACP*) has been studied by A. Favini, G. R. Goldstein, J. A. Goldstein and S. Romanelli ([9] and [10]), and most recently, the analyticity of the corresponding semigroup was also shown (see [7], [19], [20]). Using the known facts about the autonomous problem, one is faced in the nonautonomous case with difficulties caused by the variable domains $\mathcal{D}(A(t))$. So we will use suitable operator matrices on a product space to avoid this problem. Such matrices appeared before in abstract form as operator matrices with non-diagonal domain in [15] or as one-sided coupled operator matrices in [6]. Closer to our situation, such operator matrices were also used in [3], [4], [10] and [20].

We now define, for $t \in [0, T]$, linear operators $A_c(t) : C^2[0, 1] \subset C[0, 1] \to C[0, 1]$ by

$$(A_c(t)f)(x) := \mathcal{A}(x, t)f(x),$$

and linear operators $Q(t) : C^1[0, 1] \subset C[0, 1] \to \mathbf{C}^2$ by

$$Q(t)f := - \begin{pmatrix} \beta_0(t)f'(0) + \gamma_0(t)f(0) \\ \beta_1(t)f'(1) + \gamma_1(t)f(1) \end{pmatrix}.$$

The restriction of $A_c(t)$ to the subspace $\{f \in C^2[0, 1];\ f(0) = f(1) = 0\}$ is

$$A_{0c}(t) := A_c(t)\big|_{C_0[0,1] \cap C^2[0,1]}, \quad t \in [0, T].$$

2. Preliminary results

LEMMA 2.1. *Fix* $t \in [0, T]$. *For each* $\theta \in (\frac{\pi}{2}, \pi)$, *there exist constants* M_θ, $\omega_\theta > 0$ *such that*

$$\Sigma(\theta, \omega_\theta) := \{z \in \mathbf{C};\ z \neq \omega_\theta,\ |\arg(z - \omega_\theta)| < \theta\} \subset \rho(A_{0c}(t))$$

and

$$\|(\lambda - A_{0c}(t))^{-1}\|_{\mathcal{L}(C[0,1])} \leq M_\theta |\lambda|^{-1}, \tag{2.1}$$

$$\|(\lambda - A_{0c}(t))^{-1}\|_{\mathcal{L}(C[0,1],C^2[0,1])} \leq M_\theta \tag{2.2}$$

for $\lambda \in \Sigma(\theta, \omega_\theta)$.

Proof. The estimate (2.1) comes from [8, Chapter VI, Section 4b and the corresponding notes]. Pick $\mu \in \rho(A_{0c}(t))$. It is clear that $(\mu - A_{0c}(t))^{-1} \in \mathcal{L}(C[0, 1], C^2[0, 1])$. Hence, for $\lambda \in \Sigma(\theta, \omega_\theta)$, we obtain

$$\|(\lambda - A_{0c}(t))^{-1}\|_{\mathcal{L}(C[0,1],C^2[0,1])}$$

$$\leq \|(\mu - A_{0c}(t))^{-1}\|_{\mathcal{L}(C[0,1],C^2[0,1])} \|(\mu - A_{0c}(t))(\lambda - A_{0c}(t))^{-1}\|_{\mathcal{L}(C[0,1])}$$

$$\leq \text{const } \|\mu(\lambda - A_{0c}(t))^{-1} - \lambda(\lambda - A_{0c}(t))^{-1} + I\|_{\mathcal{L}(C[0,1])}$$

$$\leq \text{const,} \quad \text{by (2.1).}$$

$\square$

We now consider the product space $\mathcal{E} := C[0, 1] \times \mathbf{C}^2$ and operators thereon

$$\mathbb{A}(t) := \begin{pmatrix} A_c(t) & 0 \\ Q(t) & 0 \end{pmatrix}, \quad \mathcal{D}(\mathbb{A}(t)) := \left\{ \begin{pmatrix} f \\ y \end{pmatrix} \in C^2[0, 1] \times \mathbf{C}^2; \ Pf = y \right\},$$

where $Pf := \begin{pmatrix} f(0) \\ f(1) \end{pmatrix}$ for $f \in C[0, 1]$. For these operators we have an estimate analogous to (2.1).

LEMMA 2.2. *Let* $t \in [0, T]$. *For each* $\theta \in (\frac{\pi}{2}, \pi)$, *there exist constants* $M'_\theta, \omega'_\theta > 0$ *such that* $\Sigma(\theta, \omega'_\theta) \subset \rho(A_{0c}(t))$ *and*

$$\|(\lambda - \mathbb{A}(t))^{-1}\|_{\mathcal{L}(\mathcal{E})} \leq M'_\theta |\lambda|^{-1} \tag{2.3}$$

for $\lambda \in \Sigma(\theta, \omega'_\theta)$.

Proof. Fix $t \in [0, T]$, $\theta \in (\frac{\pi}{2}, \pi)$, and $\mu \in \rho(A_{0c}(t))$. In order to use perturbation arguments we write $\mathbb{A}(t) := \mathbb{A}_0(t) + \mathbb{Q}(t)$ with

$$\mathbb{A}_0(t) := \begin{pmatrix} A_c(t) & 0 \\ 0 & 0 \end{pmatrix} \quad \text{with } \mathcal{D}(\mathbb{A}_0(t)) := \mathcal{D}(\mathbb{A}(t)).$$

Since $P(\mathcal{D}(A_c(t))) = P(C^2[0, 1]) = \mathbf{C}^2$, $A_c(t) \in \mathcal{L}(C^2[0, 1], C[0, 1])$, and $P|_{C^2[0,1]} \in \mathcal{L}(C^2[0, 1], \mathbf{C}^2)$, we can define, similarly to [11, Lemmas 1.2 and 1.3] (see also [4, Lemma 2.2]), the Dirichlet operators with respect to $A_c(t)$ by

$$D_\lambda(t) := \left(P\big|_{\ker(\lambda - A_c(t))}\right)^{-1}, \quad \lambda \in \rho(A_{0c}(t)),$$

such that

$$D_\lambda(t) \in \mathcal{L}(\mathbf{C}^2, C^2[0, 1]) \text{ and } D_\lambda(t) = D_\mu(t) - (\lambda - \mu)\,(\lambda - A_{0c}(t))^{-1} D_\mu(t) \quad (2.4)$$

for $\lambda, \mu \in \rho(A_{0c}(t))$. Thus we have

$$\lambda - \mathbb{A}_0(t) = \begin{pmatrix} \lambda - A_{0c}(t) & 0 \\ 0 & \lambda \end{pmatrix} \begin{pmatrix} I & -D_\lambda(t) \\ 0 & I \end{pmatrix}$$

for $\lambda \in \rho(A_{0c}(t))$ and $t \in [0, T]$. Therefore, $\rho(A_{0c}(t)) \backslash \{0\} \subset \rho(\mathbb{A}_0(t))$ and

$$\left(\lambda - \mathbb{A}_0(t)\right)^{-1} = \begin{pmatrix} (\lambda - A_{0c}(t))^{-1} & \lambda^{-1} D_\lambda(t) \\ 0 & \lambda^{-1} \end{pmatrix}, \tag{2.5}$$

$$\mathbb{Q}(t)(\lambda - \mathbb{A}_0(t))^{-1} = \begin{pmatrix} 0 & 0 \\ Q(t)(\lambda - A_{0c}(t))^{-1} & \lambda^{-1} Q(t) D_\lambda(t) \end{pmatrix} \tag{2.6}$$

for $\lambda \in \rho(A_{0c}(t)) \backslash \{0\}$.

From (2.1), (2.4) and (2.5), we see that

$$\|(\lambda - \mathbb{A}_0(t))^{-1}\|_{\mathcal{L}(\mathcal{E})} \leq \text{const } |\lambda|^{-1}, \quad \lambda \in \Sigma(\theta, \omega_\theta). \tag{2.7}$$

We now estimate $\|\mathbb{Q}(t)(\lambda - \mathbb{A}_0(t))^{-1}\|_{\mathcal{L}(\mathcal{E})}$. To this purpose we use the fact (cf., e.g., [8, (2.2), p. 170]) that for each $\varepsilon > 0$ there exists b_ε such that

$$\|f'\|_{C[0,1]} \leq \varepsilon \|f''\|_{C[0,1]} + b_\varepsilon \|f\|_{C[0,1]}, \quad f \in C^2[0, 1].$$

Since $Q(t) \in \mathcal{L}(C^1[0, 1], \mathbf{C}^2)$, we then deduce by (2.1) and (2.2) that

$$\|Q(t)(\lambda - A_{0c}(t))^{-1} f)\|_{\mathbf{C}^2}$$

$$\leq \|Q(t)\|_{\mathcal{L}(C^1[0,1],\mathbf{C}^2)} \{\varepsilon \|(\lambda - A_{0c}(t))^{-1} f\|_{C^2[0, 1]} + b_\varepsilon \|(\lambda - A_{0c}(t))^{-1} f\|_{C[0,1]}\}$$

$$\leq \|Q(t)\|_{\mathcal{L}(C^1[0,1],\mathbf{C}^2)} (\varepsilon + b_\varepsilon |\lambda|^{-1})\, M_\theta \|f\|_{C[0,1]}$$

for all $\lambda \in \Sigma(\theta, \omega_\theta)$, $f \in C[0, 1]$.

Choose ε small enough such that $\varepsilon \| Q(t) \|_{\mathcal{L}(C^1[0,1],\mathbf{C}^2)} M_\theta \leq \frac{1}{4}$ and then choose $\omega_\theta' > \omega_\theta$ such that

$$\| Q(t) \|_{\mathcal{L}(C^1[0,1],\mathbf{C}^2)} b_\varepsilon M_\theta |\lambda|^{-1} \leq \frac{1}{4} \quad \text{for } \lambda \in \mathbf{C} \text{ with } |\lambda| > \omega_\theta'.$$

Hence

$$\| Q(t)(\lambda - A_{0c}(t))^{-1} \|_{\mathcal{L}(C[0,1],\mathbf{C}^2)} \leq \frac{1}{2}, \quad \lambda \in \Sigma(\theta, \omega_\theta'), \tag{2.8}$$

and, by (2.1), (2.2) and (2.4),

$$\| \lambda^{-1} Q(t) D_\lambda(t) \|_{\mathcal{L}(\mathbf{C}^2)} \leq \frac{1}{2}, \quad \lambda \in \Sigma(\theta, \omega_\theta').$$

This combined with (2.6) and (2.8) yields that

$$\| \mathbb{Q}(t)(\lambda - \mathbb{A}_0(t))^{-1} \|_{\mathcal{L}(\mathcal{E})} \leq \frac{1}{2}, \quad \lambda \in \Sigma(\theta, \omega_\theta').$$

So the operator $\lambda - \mathbb{A}(t) = [I - \mathbb{Q}(t)(\lambda - \mathbb{A}_0(t))^{-1}](\lambda - \mathbb{A}_0(t))$ is invertible with

$$(\lambda - \mathbb{A}(t))^{-1} = (\lambda - \mathbb{A}_0(t))^{-1}[I - \mathbb{Q}(t)(\lambda - \mathbb{A}_0(t))^{-1}]^{-1}, \quad \lambda \in \Sigma(\theta, \omega_\theta').$$

Thus we obtain (2.3) by recalling (2.7). $\qquad\square$

We now investigate the continuity of the map $t \mapsto \mathbb{A}(t)$. If we define

$$\mathcal{D} := \left\{ \begin{pmatrix} f \\ y \end{pmatrix} \in C^2[0, 1] \times \mathbf{C}^2; \ Pf = y \right\}$$

endowed with the norm

$$\left\| \begin{pmatrix} f \\ y \end{pmatrix} \right\|_{\mathcal{D}} := \| f \|_{C^2[0,1]},$$

we obtain the following.

LEMMA 2.3. *Under our assumptions, the map* $t \mapsto \mathbb{A}(t)$ *belongs to* $C^{\alpha}([0, T]; \mathcal{L}(\mathcal{D}, \mathcal{E}))$.

Proof. For $\begin{pmatrix} f \\ y \end{pmatrix} \in \mathcal{D}$ and $t, \ s \in [0, T]$, we estimate by (1.3) that

$$\left\| (\mathbb{A}(t) - \mathbb{A}(s)) \begin{pmatrix} f \\ y \end{pmatrix} \right\|_{\mathcal{E}}$$

$$= \left\| \begin{pmatrix} (A_c(t) - A_c(s))f \\ (Q(t) - Q(s))f \end{pmatrix} \right\|_{\mathcal{E}}$$

$$= \|(A_c(t) - A_c(s))f\|_{C[0,1]} + \|(Q(t) - Q(s))f\|_{\mathbb{C}^2}$$

$$\leq \|a(\cdot, t) - a(\cdot, s)\|_{C[0,1]} \|f''\|_{C[0,1]} + \|q(\cdot, t) - q(\cdot, s)\|_{C[0,1]} \|f'\|_{C[0,1]}$$

$$\quad + \|r(\cdot, t) - r(\cdot, s)\|_{C[0,1]} \|f\|_{C[0,1]}$$

$$\quad + \sum_{j=0,1} (|\beta_j(t) - \beta_j(s)||f'(j)| + |\gamma_j(t) - \gamma_j(s)||f(j)|)$$

$$\leq \text{const } |t - s|^{\alpha} \|f\|_{C^2[0,1]}$$

$$\leq \text{const } |t - s|^{\alpha} \left\| \begin{pmatrix} f \\ y \end{pmatrix} \right\|_{\mathcal{D}}.$$

$\square$

LEMMA 2.4. *For each* $t \in [0, T]$, *the Banach spaces* $\mathcal{D}$ *and* $[\mathcal{D}(\mathbb{A}(t)]$ *are isomorphic, and the constants* M_θ, $\omega_\theta > 0$ *in Lemma 2.2. can be chosen to be independent of* $t \in [0, T]$.

Proof. An isomorphism is easy to find. The independence of the constants is implied by Lemma 2.3. (cf. [5, Appendix]). $\square$

The following result covers the corresponding ones in [7, 19, 20] with a different approach.

PROPOSITION 2.5. *If* $A(t)$ *is as in* (1.2), *then it generates a strongly continuous analytic semigroup of angle* $\frac{\pi}{2}$ *satisfying*

$$\|e^{zA(t)}\| \leq M_\varphi e^{\omega_\varphi |z|}, \quad z \in \Sigma(\varphi, 0), \ t \in [0, T],$$

where M_φ, $\omega_\varphi > 0$ *are constants dependent on* $\varphi \in (0, \frac{\pi}{2})$ *but independent of* $t \in [0, T]$.

Proof. By Lemma 2.2, we infer (cf. [16]) that each $\mathbb{A}(t)$ generates an analytic semigroup $\{e^{s\mathbb{A}(t)}\}_{s \geq 0}$ on $\mathcal{E}$, and the restrictions of $e^{s\mathbb{A}(t)}$ to $\mathcal{E}_1 := \overline{\mathcal{D}(\mathbb{A}(t))}$ leave $\mathcal{E}_1$ invariant and become a strongly continuous analytic semigroup on $\mathcal{E}_1$, generated by the part $\mathbb{A}_1(t)$ of $\mathbb{A}(t)$ in $\mathcal{E}_1$. As a consequence, $\mathcal{D}(\mathbb{A}_1(t))$ is dense in $\mathcal{E}_1$. Clearly

$$\mathcal{E}_1 = \left\{ \begin{pmatrix} f \\ y \end{pmatrix} \in C[0, 1] \times \mathbf{C}^2; \; Pf = y \right\}.$$

It is not hard to see that

$$f \in \mathcal{D}(A(t)) \text{ if and only if } \begin{pmatrix} f \\ Pf \end{pmatrix} \in \mathcal{D}(\mathbb{A}_1(t)) \tag{2.9}$$

and that

$$(\lambda - A(t))^{-1} f = \pi_1 \left((\lambda - \mathbb{A}_1(t))^{-1} \begin{pmatrix} f \\ Pf \end{pmatrix} \right) \text{ for } \lambda \in \rho(\mathbb{A}_1(t)) \text{ and } f \in C[0, 1], \tag{2.10}$$

where π_1 is the canonical projection from $C[0, 1] \times \mathbf{C}^2$ onto $C[0, 1]$. From (2.9) we know that $\mathcal{D}(A(t))$ is dense in $C[0, 1]$ since $\mathcal{D}(\mathbb{A}_1(t))$ is dense in $\mathcal{E}_1$. Combining (2.9) and Lemma 2.4 yields that for each $\theta \in (\frac{\pi}{2}, \pi)$ there exist constants M_θ, $\omega_\theta > 0$ (independent of $t \in [0, T]$) such that

$$\|(\lambda - A(t))^{-1} f\|_{C[0,1]} \leq \left\| (\lambda - \mathbb{A}_1(t))^{-1} \begin{pmatrix} f \\ Pf \end{pmatrix} \right\|_{\mathcal{E}_1}$$

$$\leq M_\theta |\lambda|^{-1} (\|f\|_{C[0,1]} + \|Pf\|_{\mathbf{C}^2})$$

$$\leq 3 M_\theta |\lambda|^{-1} \|f\|_{C[0,1]}, \quad \lambda \in \Sigma(\theta, \omega_\theta), \; f \in C[0, 1].$$

This estimate implies the assertion. $\qquad\qquad\square$

3. Main results

We now return to (*NACP*) as well as to its inhomogeneous version

$$\begin{cases} u'(t) & = A(t)u(t) + F(t), \quad 0 \leq s < t \leq T, \\ u(s) & = f, \end{cases} \tag{\textit{INACP}}$$

where $F(\cdot)$ is a given function from $[0, T]$ to $C[0, 1]$.

Before stating the main result we briefly recall the basic concepts for nonautonomous abstract Cauchy problems (compare [14, Definition 6.0.1] or [8, Chapter VI, Definition 9.2]). We do so for arbitrary linear operators $A(t)$ ($t \in [0, T]$) in a Banach space X.

DEFINITION 3.1. A family of linear operators $\{U(t,s)\}_{0 \leq s \leq t \leq T} \subset \mathcal{L}(X)$ is called an evolution family for (*NACP*) if

    (I) $U(s,s) = I$ for $0 \leq s \leq T$,

    (II) $U(t,s)U(s,r) = U(t,r)$ for $0 \leq r \leq s \leq t \leq T$,

    (III) $(t,s) \to U(t,s)$ is strongly continuous for $0 \leq s \leq t \leq T$,

    (IV) $t \mapsto U(t,s)$ is strongly continuously differentiable in $(s,T]$ and

$$\frac{\partial}{\partial t} U(t,s) = A(t)U(t,s), \quad 0 \leq s < t \leq T.$$

DEFINITION 3.2. (i) Let $F(\cdot) \in C((s,T];X)$. A function $u(\cdot)$ is called a *classical solution* of (*INACP*) if $u(\cdot) \in C^1((s,T];X) \cap C([s,T];X)$ and (*INACP*) is satisfied.

    (ii) Let $F(\cdot) \in C([s,T];X)$. A function $u(\cdot)$ is called a *strict solution* of (*INACP*) if $u(\cdot) \in C^1([s,T];X)$ and (*INACP*) is satisfied.

We now prove our main results.

THEOREM 3.3. *Let $A(t)$ be as in (1.2). Then there exists a unique evolution family* $\{U(t,s)\}_{0 \leq s \leq t \leq T} \subset \mathcal{L}(C[0,1])$ *for (NACP) with the following properties.*

    (i) $\|A(t)U(t,s)\|_{\mathcal{L}(C[0,1])} \leq$ const $(t-s)^{-1}$ *and*

$$\|A(t)U(t,s)\|_{\mathcal{L}(C^2[0,1],C[0,1])} \leq \text{const}$$

    for $0 \leq s < t \leq T$.

    (ii) $U(\cdot,s)f \in C([s,T];C[0,1]) \cap C^{1+\alpha}([s+\varepsilon,T];C[0,1]) \cap C^{\alpha}([s+\varepsilon,T];C^2[0,1])$ *for* $f \in C[0,1]$, $s \in [0,T)$, $\varepsilon \in (0,T-s)$.

    (iii) $U(\cdot,s)f \in C^1([s,T];C[0,1])$ *for* $f \in \mathcal{D}(A(s))$, $s \in [0,T)$.

    (iv) $U(t,\cdot)f \in C([0,t];C[0,1])$ *for* $f \in C[0,1]$, $t \in (0,T]$.

    (v) $U(t,\cdot)f \in C^1([0,t];C[0,1]) \cap C([0,t];C^2[0,1])$ *for* $f \in \mathcal{D}(A(t))$, $t \in (0,T]$.

Proof. By Lemmas 2.2 and 2.3 there exists, in view of [1, 2] or [14, Sections 6.1 and 6.2], a family of linear operators $\{\mathbb{U}(t,s)\}_{0 \leq s \leq t \leq T} \subset \mathcal{L}(\mathcal{E})$ with the following properties.

    (a) For $\begin{pmatrix} f \\ Pf \end{pmatrix} \in \mathcal{E}_1$, the $\mathcal{E}$-valued function $\mathcal{U}(\cdot) := \mathbb{U}(\cdot,s) \begin{pmatrix} f \\ Pf \end{pmatrix}$ is the unique classical solution of the problem

$$\begin{cases} \mathcal{U}'(t) & = \mathbb{A}(t)\mathcal{U}(t), \quad s < t \leq T, \\ \mathcal{U}(s) & = \begin{pmatrix} f \\ Pf \end{pmatrix} \end{cases} \tag{3.1}$$

and belongs to $C([s,T];\mathcal{E}) \cap C^{1+\alpha}([s+\varepsilon,T];\mathcal{E}) \cap C^{\alpha}([s+\varepsilon,T];\mathcal{D})$ for each $\varepsilon \in (0,T-s)$.

(b) For $\begin{pmatrix} f \\ Pf \end{pmatrix} \in \mathcal{D}(\mathbb{A}_1(s))$, $\mathbb{U}(\cdot, s) \begin{pmatrix} f \\ Pf \end{pmatrix}$ belongs to $C^1([s, T]; \mathcal{E})$.

(c) For $0 \le s < t \le T$,

$$\|\mathbb{U}(t, s)\|_{\mathcal{L}(\mathcal{E})} \le \text{const},$$

$$\|\mathbb{A}(t)\mathbb{U}(t, s)\|_{\mathcal{L}(\mathcal{E})} \le \text{const} \, (t - s)^{-1}, \quad \|\mathbb{U}(t, s)\|_{\mathcal{L}(\mathcal{D}, \mathcal{E})} \le \text{const}. \tag{3.2}$$

(d) For $t \in (0, T]$,

$$\mathbb{U}(t, \cdot) \begin{pmatrix} f \\ Pf \end{pmatrix} \in C([0, t]; \mathcal{E}), \quad \begin{pmatrix} f \\ Pf \end{pmatrix} \in \mathcal{E}_1,$$

$$\mathbb{U}(t, \cdot) \begin{pmatrix} f \\ Pf \end{pmatrix} \in C^1([0, t]; \mathcal{E}) \cap C([0, t]; \mathcal{D}), \quad \begin{pmatrix} f \\ Pf \end{pmatrix} \in \mathcal{D}(\mathbb{A}_1(t)).$$

We now take the first coordinate of $\mathbb{U}(t, s) \begin{pmatrix} f \\ Pf \end{pmatrix}$ and define

$$U(t, s)f := \pi_1 \left(\mathbb{U}(t, s) \begin{pmatrix} f \\ Pf \end{pmatrix} \right) \quad \text{for } 0 \le s \le t \le T \text{ and } f \in C[0, 1].$$

Observe that $u(\cdot)$ is a classical solution of (*NACP*) if and only if $\mathcal{U}(\cdot) = \begin{pmatrix} u(\cdot) \\ Pu(\cdot) \end{pmatrix}$ is a classical solution of (3.1), and

$$f \in C[0, 1] \text{ if and only if } \begin{pmatrix} f \\ Pf \end{pmatrix} \in \mathcal{E}_1,$$

$$f \in C^2[0, 1] \text{ if and only if } \begin{pmatrix} f \\ Pf \end{pmatrix} \in \mathcal{D},$$

$$f \in \mathcal{D}(A(t)) \text{ if and only if } \begin{pmatrix} f \\ Pf \end{pmatrix} \in \mathcal{D}(\mathbb{A}_1(t)).$$

Accordingly, we obtain assertions (i)–(v), as well as (I) and (IV) in Definition 3.1, by the corresponding properties of $\mathbb{U}(t, s)$ listed above. Furthermore, we know that for each $f \in C[0, 1]$, (*NACP*) has a unique classical solution since the classical solution of (3.1) is unique.

Next, we show properties (II) and (I) in Definition 3.1. To this end, we let $f_1 \in C[0, 1]$ and $0 \le r \le s \le t \le T$. By (ii),

$$U(\cdot, r)f_1, \quad U(\cdot, s)U(s, r)f_1 \in C([s, T]; C[0, 1]).$$

This combined with (IV) in Definition 3.1 yields that $t \mapsto U(t, r) f_1$ and $t \mapsto U(t, s)$ $U(s, r) f_1$ are classical solutions of (*NACP*) with the same initial datum $U(s, r) f_1$ at $t = s$. Therefore (II) in Definition 3.1 is satisfied. The assertion (III) in Definition 3.1. follows from (II), (ii), (iv) and the uniform boundedness of $\|U(t, s)\|_{\mathcal{L}(C[0,1])}$ for $0 \le s < t \le T$ (derived from (3.2)).

Finally, the uniqueness of the classical solution of (*NACP*) implies the uniqueness of the evolution family for (*NACP*). $\qquad\square$

The inhomogeneous problem can be solved as follows.

THEOREM 3.4. *Let* $\beta \in (0, \alpha]$ *and* $F \in C^\beta([s, T]; C[0, 1])$.

1) *If* $f \in C[0, 1]$, *then (INACP) has a unique classical solution* $u(\cdot) \in C^{1+\beta}([s + \varepsilon, T]; C[0, 1]) \cap C^\beta([s + \varepsilon, T]; C^2[0, 1])$ *for every* $\varepsilon \in (0, T - s)$ *and is given by*

$$u(t) = U(t, s)f + \int_s^t U(t, \sigma)F(\sigma)\, d\sigma, \quad s \le t \le T.$$

2) *If* $f \in \mathcal{D}(A(s))$, *the above* $u(\cdot)$ *is a strict solution of (INACP)*.

Proof. Observe that $u(\cdot)$ is a classical (resp. strict) solution of (*INACP*) if and only if $\mathcal{U}(\cdot)\begin{pmatrix} u(\cdot) \\ Pu(\cdot) \end{pmatrix}$ is a classical (resp. strict) solution of the following inhomogeneous nonautonomous abstract Cauchy problem

$$\begin{cases} \mathcal{U}'(t) &= \mathbb{A}(t)\mathcal{U}(t) + \begin{pmatrix} F(t) \\ PF(t) \end{pmatrix}, \quad s < t \le T, \\ \mathcal{U}(s) &= \begin{pmatrix} f \\ Pf \end{pmatrix}. \end{cases} \tag{3.3}$$

Therefore, we obtain the desired conclusions from the corresponding results for (3.3) available because of Lemmas 2.2 and 2.3 (see the papers [1, 2] or the book [14, Corollary 6.1.6 (i) and (iii) and Corollary 6.2.4] stemming from the classical Sobolevskii-Tanabe work [17, 18] for abstract nonautonomous parabolic equations). $\qquad\square$

REMARK 3.5. In the same way, we can derive other properties of the evolution family $\{U(t, s)\}_{0 \le s \le t \le T}$ (of the solutions of (*INACP*), resp.) from the corresponding ones of $\{\mathbb{U}(t, s)\}_{0 \le s \le t \le T}$ (of (3.3), resp.).

REFERENCES

[1] ACQUISTAPACE, P. and TERRENI, B., *Some existence and regularity results for abstract nonautonomous parabolic equations*. J. Math. Anal. Appl. *99* (1984), 9–64.
[2] ACQUISTAPACE, P. and TERRENI, B., *Maximal space regularity for abstract linear nonautonomous parabolic equations*. J. Funct. Anal. *60* (1985), 168–210.

[3]   AMANN, H. and ESCHER, J., *Strongly continuous dual semigroups*. Ann. Mat. Pura Appl. *171* (1996), 41–62.

[4]   CASARINO, V., ENGEL, K. -J., NAGEL, R. and NICKEL, G., *A semigroup approach to boundary feedback systems*. Tübinger Berichte zur Funktionalanalysis, *10* (2000/2001), 30–43.

[5]   DA PRATO, G. and SINESTRARI, E., *Hölder regularity for nonautonomous abstract parabolic equations*. Israel J. Math. *42* (1982), 1–19.

[6]   ENGEL, K. -J., *Spectral theory and generator property for one-sided coupled operator matrices*, Semigroup Forum *58* (1999), 267–295.

[7]   ENGEL, K. -J., *Second order differential operators on C[0, 1] with Wentzell boundary conditions*. Preprint, 2002.

[8]   ENGEL, K. -J. and NAGEL, R., *One-Parameter Semigroups for Linear Evolution Equations*. GTM *194*, Springer-Verlag, Berlin, New York, 2000.

[9]   FAVINI, A., GOLDSTEIN, G. R., GOLDSTEIN, J. A. and ROMANELLI, S., *C_0-semigroups generated by second order differential operators with general Wentzell boundary conditions*. Proc. Amer. Math. Soc. *128* (2000), 1981–1989.

[10]   FAVINI, A., GOLDSTEIN, G. R., GOLDSTEIN, J. A. and ROMANELLI, S., *The heat equation with generalized Wentzell boundary condition*. J. Evolution Eqs. *2* (2002), 1–19.

[11]   GREINER, G., *Perturbing the boundary conditions of a generator*. Houston J. Math. *13* (1987), 213–229.

[12]   KATO, T., *On linear differential equations in Banach spaces*. Comm. Pure Appl. Math. *9* (1956), 479–486.

[13]   KATO, T. and TANABE, H., *On the analyticity of solutions of evolution equations*. Osaka J. Math. *4* (1967), 1–4.

[14]   LUNARDI, A., *Analytic Semigroups and Optimal Regularity in Parabolic Problems*. Birkhäuser Verlag, Basel, 1995.

[15]   NAGEL, R., *The spectrum of unbounded operator matrices with nondiagonal domain*. J. Funct. Anal. *89* (1990), 291–302.

[16]   SINESTRARI, E., *On the abstract Cauchy problem in spaces of continuous functions*. J. Math. Anal. Appl. *107* (1985), 16–66.

[17]   SOBOLEVSKII, P. E., *Equations of parabolic type in a Banach space*. Trudy Moskow Math. Obsc. *10* (1961), 297–350 (Russian). English transl.: Amer. Math. Soc. Transl. *49* (1964), 1–62.

[18]   TANABE, H., *Remarks on the equations of evolution in a Banach space*. Osaka Math. J. *12* (1960), 145–165.

[19]   WARMA, M., *Wentzell-Robin boundary conditions on C[0, 1]*. Semigroup Forum (to appear).

[20]   XIAO, T. J. and LIANG, J., Wave equations with generalized Wentzell boundary conditions. preprint, 2001.

Jin Liang, Ti-Jun Xiao
Department of Mathematics
University of Science and Technology of China
Hefei, Anhui 230026
Germany
and
Mathematisches Institut
Universität Tübingen
Auf der Morgenstelle 10
D-72076, Tübingen
Germany
e-mail: jili@fa.uni-tuebingen.de
           tixi@fa.uni-tuebingen.de

Rainer Nagel
Mathematisches Institut
Universität Tübingen
Auf der Morgenstelle 10
D-72076, Tübingen
Germany
e-mail: rana@fa.uni-tuebingen.de

J.evol.equ. 3 (2003) 332 – 359
1424–3199/03/020332 – 28
DOI 10.1007/s00028-003-0104-0
© Birkhäuser Verlag, Basel, 2003

**Journal of Evolution
Equations**

Maximal L^p-regularity for elliptic operators with VMO-coefficients

HORST HECK AND MATTHIAS HIEBER

In Memoriam Philippe Bénilan

Abstract. In this paper we show that the parabolic problem $u_t + Au = f$ on $\mathbb{R}^n$ associated with elliptic operators A having coefficients in VMO $\cap L^\infty$ has the property of maximal L^p-regularity.

1. Introduction

Maximal L^p-regularity results for linear parabolic problems are known to be very useful for nonlinear parabolic problems. This is certainly one reason why results in this direction gained considerable interest in recent years. This point of view, however, forces to look for minimal smoothness assumptions on the coefficients of the differential operators involved.

The case of nondivergence form elliptic operators with uniformly continuous and bounded coefficients subject to general boundary conditions was treated recently by Denk, Hieber and Prüss in [DHP01]. Indeed, maximal L^p-estimates for the solution of the parabolic boundary value problem were given by invoking the concept of $\mathcal{R}$-boundedness and an operator-valued version of Mikhlin's theorem due to Weis [Wei01], [Wei01a].

Another approach relies on the Dore-Venni-theorem [DV87] and the theory of bounded imaginary powers or, more generally, on the $\mathcal{H}^\infty$-calculus of the operators involved. Holomorphic functional calculus or bounded imaginary powers for nondivergence elliptic operators on L^p-spaces were investigated by many authors; see the work of Seeley [See67] and Duong [Duo90] for the case of C^∞-coefficients and [PS93], [AHS94], [DM96], [DR96], [DS97] and [DDHPV02] for the case of non smooth coefficients.

In this paper we consider nondivergence form elliptic operators on $\mathbb{R}^n$ of order m with coefficients belonging to VMO. It was shown by Angeletti, Mazet and Tchamitchian [AMT97] that the L^p-realization A of these operators generate holomorphic semigroups on $L^p(\mathbb{R}^n)$ and that $D(A) = W^{2,p}(\mathbb{R}^n)$ provided $m = 2$ and $1 < p < \infty$. Their approach is based on wavelet techniques. They rediscovered in particular the $W^{2,p}$-estimates for the

---

Mathematics Subject Classification 2000: 35J45, 35K55.
Key words and phrases: Maximal regularity, elliptic operators, VMO-coefficients.
Supported in part by Deutsche Forschungsgemeinschaft DFG.

solution of the elliptic equation $Au = f$ in $L^p(\mathbb{R}^n)$ due to Chiarenza, Frasca and Longo [CFL91], [CFL93]. Their proof is based on the theory of singular integrals and in particular on parameter depending Calderón-Zygmund theory. This method was lateron generalized to second order parabolic equation with VMO-coefficients subject to Dirichlet boundary conditions by Bramanti and Cerutti [BC93] and by Maugeri and Palagachev [MP98] to the case of the oblique derivative problem. Taira [Tai02] proved that in the case of Dirichlet boundary conditions these operators also generate Feller semigroups on $C_0(\overline{\Omega})$. A priori estimates for the elliptic problem subject to general boundary conditions were recently given by Guidetti [Gui02].

Using wavelet and singular integral techniques, Duong and Yan [DY01] proved that second order operators of this kind admit a bounded $\mathcal{H}^\infty$-calculus on $L^p(\mathbb{R}^n)$ provided the coefficients are small in the BMO-norm or, in particular, belong to the class VMO. As a consequence, they obtain maximal L^p-estimates for the parabolic problem under these smoothness assumptions.

The aim of this paper is to prove maximal L^p-estimates for the parabolic problem via the concept of $\mathcal{R}$-boundedness and not to use any wavelet techniques. Taking into account the fact that weighted estimates with respect to Muckenhoupt weights imply $\mathcal{R}$-boundedness of the family of operators in question (see [GR85], [HHH01]), we are aiming for weighted $W^{m,p}$-resolvent estimates for A. These weighted estimates are obtained by a technique which is much inspired by the one of Chiarenza, Frasca and Longo [CFL93]. Note, however, that no maximum principle technique is available in our situation since we are dealing in particuar with elliptic operators of arbitrary order m.

2. Preliminaries and main result

In this section we state our main result and introduce the notation being used throughout this paper. Furthermore we collect certain properties of functions belonging to the class $VMO(\mathbb{R}^n)$, discuss Muckenhoupt weights and their relationship to maximal L^p-regularity.

Let us start with the property of maximal L^p-regularity for parabolic evolution equations. To this end, let $-A$ be a sectorial operator of angle $\theta < \frac{\pi}{2}$ in a Banach space X and $1 < p < \infty$. For $f \in L^p(\mathbb{R}_+; X)$ consider the equation

$$u'(t) + Au(t) = f(t), \quad t > 0, \, u(0) = u_0. \tag{2.1}$$

We say that A belongs to the class $MR_p(X)$ if for given $f \in L^p(\mathbb{R}_+; X)$ the functions u' and Au belong also to $L^p(\mathbb{R}_+; X)$, where u denotes the solution of (2.1). It is well known that the class $MR_p(X)$ is independent of $p \in (1, \infty)$ and we therefore write $MR(X)$ instead of $MR_p(X)$. For this, definitions and for further information on maximal regularity, we refer to [DHP01].

It is known that the property of maximal L^p-regularity is closely related to weighted L^p-estimates with respect to Muckenhoupt weights. We therefore turn our attention to the

Muckenhoupt class A_p for $1 < p < \infty$: a function $0 \leq w \in L^1_{\text{loc}}(\mathbb{R}^n)$ is called an A_p-*weight* in the sense of Muckenhoupt, if there is a constant $C > 0$ such that

$$\left(\frac{1}{|I|} \int_I w \, dx \right) \left(\frac{1}{|I|} \int_I w^{-\frac{1}{p-1}} \, dx \right)^{p-1} \leq C,$$

for all cubes $I \in \mathbb{R}^n$. The smallest such C is called the A_p-*constant* $A_p(w)$ of w.

A locally integrable function f is said to belong to the class BMO (*bounded mean oscillation*) provided

$$\sup_{B \in \mathcal{B}} \frac{1}{|B|} \int_B |f(x) - f_B| \, dx = \sup_{B \in \mathcal{B}} |f - f_B|_B =: \|f\|_* < \infty.$$

Here $\mathcal{B}$ denotes the family of all balls in $\mathbb{R}^n$ and

$$f_B = \frac{1}{|B|} \int_B f(x) \, dx,$$

denotes the mean value of f over the ball B. It is well known that $(\text{BMO}/K, \| \cdot \|_*)$ is a Banach space, where K denotes the set of all constant functions. For further information on functions of bounded mean oscillation we refer to [Ste93, Chapter IV].

For $f \in \text{BMO}$ and $r \in \mathbb{R}$ we set

$$\sup_{B_\rho \in \mathcal{B}, \rho \leq r} \frac{1}{|B_\rho|} \int_{B_\rho} |f(x) - f_{B_\rho}| \, dx =: \eta(r),$$

where B_ρ denotes a ball in $\mathbb{R}^n$ of radius $\rho > 0$. A function $f \in \text{BMO}$ is said to belong to the class VMO (which means *vanishing mean oscillation*) if

$$\lim_{r \to 0} \eta(r) = 0.$$

In the following we consider elliptic operators with coefficients belonging to the class VMO. To this end, let $m \in \mathbb{N}$ and let $\mathcal{A}(D)$ be an operator of the form

$$\mathcal{A}(D) := \sum_{|\alpha|=m} a_\alpha D^\alpha,$$

where $D := -i(\partial_1, \ldots, \partial_n)$ and $a_\alpha \in \mathbb{C}$. Following [AHS94] we call a polynomial $\mathcal{A}(\xi) = \sum_{|\alpha|=m} a_\alpha \xi^\alpha$ (M, ϕ)-*parameter-elliptic* if there exists an angle $\phi \in [0, \pi)$ and a constant $M > 0$ such that

$$\sigma(\mathcal{A}(\xi)) \subset \Sigma_\phi \quad \text{and} \tag{2.2}$$

$$|\mathcal{A}(\xi)^{-1}| \leq M \quad \text{for all} \quad \xi \in \mathbb{R}^n, |\xi| = 1. \tag{2.3}$$

We then call

$$\phi_{\mathcal{A}} := \inf\{\phi : (2.2) \text{ holds}\} = \sup_{|\xi|=1} |\arg \sigma(\mathcal{A}(\xi))|$$

the *angle of ellipticity* of $\mathcal{A}$. We are now in the position to state the main result of this paper.

THEOREM 2.1. *Let $\mathcal{A}(x, D) = \sum_{|\alpha| \leq m} a_\alpha(x) D^\alpha$ be a differential operator of order m. Let $M > 0$ and $0 < \phi < \frac{\pi}{2}$ such that*

a) $a_\alpha \in \text{VMO}(\mathbb{R}^n) \cap L^\infty(\mathbb{R}^n)$, $|\alpha| = m$
b) *The symbol $\mathcal{A}(x, \xi) = \sum_{|\alpha|=m} a_\alpha(x)\xi^\alpha$ is (M, ϕ)-parameter-elliptic for almost all $x \in \mathbb{R}^n$.*
c) $a_\alpha \in L^\infty(\mathbb{R}^n)$, $|\alpha| < m$.

Let A denote the realization of $\mathcal{A}(x, D)$ in $L^p_w(\mathbb{R}^n)$ with domain $W^{m,p}_w(\mathbb{R}^n)$, where $w \in A_p$ and $1 < p < \infty$. Then there exists a constant $\mu \geq 0$ such that $\mu + A \in MR(L^p_w(\mathbb{R}^n))$. In particular, $\mu + A \in MR(L^p(\mathbb{R}^n))$.

3. Approximation of VMO-functions

The following lemma was established by Sarason [Sar75]

LEMMA 3.1. (Sarason). *Let $f \in \text{BMO}(\mathbb{R}^n)$. Then*

$$f \in \text{VMO}(\mathbb{R}^n) \iff f \in \overline{\text{UC} \cap \text{BMO}}^{\,\text{BMO}} \tag{3.1}$$

The following result will be important for the localization procedure described in Section 7. To this end we need the notion of the modulus of continuity $\omega_f(r)$ of a function f which is defined as

$$\omega_f(r) := \sup_{|x-y| \leq r} |f(x) - f(y)|.$$

LEMMA 3.2. *Let $f \in \text{VMO}$. Then for every $\varepsilon > 0$ there exists a radius $\rho(f, \varepsilon)$ such that for all $x_0 \in \mathbb{R}^n$ there is a function $g \in \text{VMO}$ with $\|g\|_* \leq \varepsilon$ and $f(x) = g(x)$ for all $x \in B_\rho(x_0)$.*

Proof. Let $\varepsilon > 0$. By (3.1) there exists $\varphi \in \text{UC}$ such that $\|f - \varphi\|_* < \frac{\varepsilon}{2}$. For this φ there exists $\rho > 0$ such that $\omega_\varphi(\rho) < \frac{\varepsilon}{2}$. For $x_0 \in \mathbb{R}^n$ we set

$$\psi(x) := \begin{cases} \varphi(x), & x \in B_\rho, \\ \varphi(x_0 + \rho \frac{x-x_0}{|x-x_0|}), & x \notin B_\rho \end{cases}$$

Hence $\sup_{r>0} \omega_\psi(r) = \omega_\varphi(\rho)$ and for a ball $B \subset \mathbb{R}^n$ we have

$$|\psi - \psi_B|_B = \frac{1}{|B|^2} \int_B \left| \int_B \psi(x) - \psi(y) \mathrm{d}y \right| \mathrm{d}x \leq \frac{1}{|B|^2} \int_B \int_B |\psi(x) - \psi(y)| \mathrm{d}y \mathrm{d}x$$

$$\leq \sup_{r>0} \omega_\psi(r) < \frac{\varepsilon}{2}.$$

Setting $g := f - \varphi + \psi$ we see that g fulfills the required properties. Indeed, $-\varphi(x) + \psi(x) = 0$ for $x \in B_\rho(x_0)$ and

$$\|g\|_* \leq \|f - \varphi\|_* + \|\psi\|_* < \varepsilon.$$

$\square$

4. Muckenhoupt weights

The main reason for considering Muckenhoupt weights in this context is that weighted resolvent estimates for sectorial operators $-A$ with respect to Muckenhoupt weights imply maximal regularity for the parabolic equation; see Proposition 4.1 below.

For operators acting in weighted L^p-spaces with w belonging to A_p the following *extrapolation theorem* (see [GR85]) is true. Let $1 < p, q < \infty$ and $\mathcal{T}$ be a family of operators such that for all $w \in A_p$ there exists a constant C depending only on the A_p-bound of w such that

$$\|Tf\|_{p,w} \leq C\|f\|_{p,w}, \qquad T \in \mathcal{T}.$$

Then it follows that the same inequality, with p replaced by q, holds for all $w \in A_q$. This extrapolation theorem allows to give the following sufficient criterion for maximal L^q-regularity on L^p-spaces, see [HHH01]. The proof combines a result due to Garcia-Cuerva and Rubio de Francia [GR85, Thm.V.6.4] with a recent Fourier multiplier result due to Weis [Wei01].

PROPOSITION 4.1. *Let* $1 < p, r < \infty$, $\Omega \subset \mathbb{R}^n$ *open and let* $-A$ *be a sectorial operator of angle* $\theta < \frac{\pi}{2}$ *in* $L_w^p(\Omega)$. *If*

$$\|\lambda(\lambda + A)^{-1}\|_{r,v} \leq C\|f\|_{r,v}, \qquad \lambda \in i\mathbb{R}, \ f \in L_v^r(\Omega) \cap L_w^p(\Omega)$$

for all $v \in A_r$ *with a constant* C *depending only on the* A_r-*bound of* v, *then* $A \in MR(L_w^p(\Omega))$.

For a proof we refer to [HHH01] or [Fro01]. Constants depending only on the A_p-constant of w are often called A_p-*consistent*. Precisely, the constant $C = C(w)$ is said to be A_p-consistent if

$$\sup\{C(w), C(w)^{-1} : w \in A_p, A_p(w) < c\} < \infty$$

for every $c > 0$. In the following we collect further properties of A_p-weights.

LEMMA 4.2. *Let w be an A_p-weight. Then there exists $s > 1$ and an A_p-consistent constant $C > 0$, such that*

$$\left(\frac{1}{|I|} \int_I w^s \, dx \right)^{\frac{1}{s}} \le C \frac{1}{|I|} \int_I w \, dx,$$

for all cubes $I \subset \mathbb{R}^n$.

For a proof see e.g. [GR85, Lem.IV.2.5].

LEMMA 4.3. *Let $1 < p < \infty$. For $w \in A_p$ there exists $p_1 < p$ such that $w \in A_{p_1}$ with A_p-consistent constant $A_{p_1}(w)$.*

Proof. We set $\frac{1}{p'} := 1 - \frac{1}{p}$ and $\sigma = w^{-p'/p} \in A_{p'}$. Choose s as in Lemma 4.2. and let $p_1 < p$ such that

$$s\frac{p'}{p} = \frac{s}{p-1} = \frac{1}{p_1 - 1} = \frac{p_1'}{p_1}.$$

Lemma 4.2 applied to σ implies

$$\left(\frac{1}{|I|} \int_I w^{-\frac{sp'}{p}} \right)^{\frac{p}{sp'}} \le C^{\frac{p}{p'}} \left(\frac{1}{|I|} \int_I w^{-\frac{p'}{p}} \right)^{\frac{p}{p'}}$$

and hence

$$\left(\frac{1}{|I|} \int_I w^{-\frac{p_1'}{p_1}} \right)^{\frac{p_1}{p_1}} \left(\frac{1}{|I|} \int_I w \right) \le C^{p/p'} \left(\frac{1}{|I|} \int_I w^{-\frac{p'}{p}} \right)^{\frac{p}{p'}} \left(\frac{1}{|I|} \int_I w \right)$$

$$\le C^{p/p'} A_p(w)$$

$\square$

The following lemma is also due to Garcia-Cuerva and Rubio de Francia [GR85, Lem.IV.2.2].

LEMMA 4.4. *Let $w \in A_p$. Then there exists a $\delta > 0$ and an A_p-consistent constant C such that*

$$\frac{w(S)}{w(I)} \le C \left(\frac{|S|}{|I|} \right)^{\delta}$$

for each cube I and each measurable set $S \subset I$, where $w(S) := \int_S w(x) \, dx$.

5. Calderón-Zygmund kernels and commutators in weighted L^p-spaces

In this section we extend the celebrated result on L^p-boundedness of commutators between Calderón-Zygmund operators and multiplication operators by BMO-functions due to Coifman, Rochberg and Weiss [CRW76] to weighted L^p-spaces, where the weight belongs to the Muckenhoupt class.

We start by making precise what we mean by a Calderón-Zygmund kernel. A function $k : \mathbb{R}^n \setminus \{0\} \to \mathbb{C}$ is called a *Calderón-Zygmund kernel* if there is a constant C such that the following conditions hold.

a) $\hat{k} \in L^\infty(\mathbb{R}^n)$,
b) $|k(x)| \le C|x|^{-n}$,
c) $|k(x - y) - k(x)| \le C|y||x|^{-(n+1)} \quad |x| > 2|y| > 0$.

It is well known that integral operators associated to kernels of the above type define bounded operators on weighted L^p spaces. Note that the inequalites b) and $|\nabla k(x)| \le C|x|^{-(n+1)}$ imply c). In particular, the following result is true; see [GR85, Thm.IV.3.1].

PROPOSITION 5.1. *Let $w \in A_p$ and k be a Calderón-Zygmund kernel. Then*

$$Kf := k * f \qquad \forall f \in \mathcal{S}(\mathbb{R}^n)$$

defines a bounded linear operator on $L_w^p(\mathbb{R}^n)$. Furthermore, there exists an A_p-consistent constant C such that $\|Kf\|_{p,w} \le C\|f\|_{p,w}$.

We now turn our attention to the commutator between the kernel operator associated to k and a multiplication operator with a BMO-function. To this end, let us introduce some notation. A *dyadic cube* is a cube in $\mathbb{R}^n$ with side length of the form 2^{-k} and corners of the form $(m_1/2^k, \ldots, m_n/2^k)$ where $m \in \mathbb{Z}^n$. We write D for the set of all dyadic cubes and W for the set of all cubes in $\mathbb{R}^n$. We will distinguish between the following four maximal functions.

DEFINITION 5.2. *Let $f \in L_{\mathrm{loc}}^1(\mathbb{R}^n)$. Then the maximal function Mf of f is defined as*

$$Mf(x) := \sup_{x \in I,\, I \in W} \frac{1}{|I|} \int_I |f(y)|\mathrm{d}y,$$

and the dyadic maximal function $M^\Delta f$ is given by

$$M^\Delta f(x) := \sup_{x \in I,\, I \in D} \frac{1}{|I|} \int_I |f(y)|\mathrm{d}y.$$

We also define the sharp function $f^\sharp$ of f as

$$f^\sharp(x) := \sup_{x \in I,\, I \in W} \frac{1}{|I|} \int_I |f(y) - f_I|\mathrm{d}y$$

and the dyadic sharp function $f^\sharp_\Delta$ as

$$f^\sharp_\Delta(x) := \sup_{x \in I,\, I \in D} \frac{1}{|I|} \int_I |f(y) - f_I|\mathrm{d}y.$$

LEMMA 5.3. *Let F and G be nonnegative functions on a σ-finite measure space (X, μ). If $F \in L^p(\mu)$ and*

$$\mu\{x : F(x) > \alpha;\; G(x) \le c\alpha\} \le a\mu\{x : F(x) > b\alpha,\}$$

for some constants a, b, c with $a < b^p$ and all $\alpha > 0$, then

$$\|F\|_p^p \le \frac{c^{-p}}{1 - ab^{-p}}\|G\|_p^p.$$

Proof. Note first that

$$\|F\|_p^p = p \int_0^\infty \alpha^{p-1}\mu\{F > \alpha\}\mathrm{d}\alpha.$$

By assumption and since

$$\{F > \alpha\} \subset \{F > \alpha,\; G \le c\alpha\} \cup \{G > c\alpha\},$$

the inequality

$$\mu\{F > \alpha\} \le a\mu\{F > b\alpha\} + \mu\{G > c\alpha\}$$

follows. Multiplying this inequality with $p\alpha^{p-1}$ and integrating with respect to α we get

$$\|F\|_p^p \le \int_0^\infty p\alpha^{p-1}(a\mu\{F > b\alpha\} + \mu\{G > c\alpha\})\mathrm{d}\alpha$$

$$= ab^{-p}\|F\|_p^p + c^{-p}\|G\|_p^p.$$

Since $ab^{-p} < 1$ the assertion follows. $\square$

PROPOSITION 5.4. *Let $a \in \mathrm{BMO}$, $w \in A_p$ and k be a Calderón-Zygmund kernel. Set $M_a f := af$ and $Kf(x) := \mathrm{p.v.}\int k(x - y)f(y)\mathrm{d}y$ for $x \in \mathbb{R}^n$. Then the operator*

$$C_a := [M_a, K] := M_a K - K M_a$$

is bounded on $L_w^p(\mathbb{R}^n)$. Furthermore, we have

$$\|C_a\| \le c\|a\|_*$$

with a constant c that is A_p-consistent.

Proof. We subdivide the proof in three steps.

STEP 1. We claim that $\|f\|_{p,w} \le C\|f^\sharp\|_{p,w}$ with an A_p-consistent constant C. It follows from [Ste93, IV.3.6] that

$$|\{x : M^\Delta f(x) > \alpha, \ f_\Delta^\sharp(x) \le c\alpha\}| \le a|\{x : M^\Delta f(x) > b\alpha\}|, \tag{5.1}$$

with $a = 2^n \frac{c}{1-b}$ for all $\alpha > 0$. By Lemma 4.4 we have

$$\frac{w(S)}{w(I)} \le C \left(\frac{|S|}{|I|}\right)^\delta \tag{5.2}$$

for any cube I and any measurable set $S \subset I$. Therefore

$$w(\{x : M^\Delta f(x) > \alpha, \ f_\Delta^\sharp(x) \le c\alpha\}) \le \tilde{a}w(\{x : M^\Delta f(x) > b\alpha\}). \tag{5.3}$$

To see this, we will have a closer look at the proof of (5.1). Following [Ste93] we first decompose the set $\{x : M^\Delta f(x) > b\alpha\}$ in "maximal" dyadic cubes and show then the inequality

$$|\{x \in I : M^\Delta f(x) > \alpha, \ f_\Delta^\sharp(x) \le c\alpha\}| \le a|I|,$$

where I denotes an arbitrary cube of the decomposition. The inequality (5.2) and summation over all cubes yield (5.3). Applying Lemma 5.3 we obtain

$$\|f\|_{p,w} \le \|M^\Delta f\|_{p,w} \le A\|f_\Delta^\sharp\|_{p,w} \le A\|f^\sharp\|_{p,w}.$$

since $|f(x)| \le M^\Delta f(x)$ a.e. and $f_\Delta^\sharp(x) \le f^\sharp(x)$.

STEP 2. Following an argument due to Strömberg we have

$$(C_a f)^\sharp(x) \le c\|a\|_*(M(|Kf|^r)(x)^{1/r} + M(|f|^s)(x)^{1/s}) \quad \text{for all} \quad r, s \in (1, \infty).$$

For a proof see e.g. [Tor86, XVI.4.1]

STEP 3.   We show that $\|(M(|f|^r(x))^{\frac{1}{r}}\|_{p,w} \le C\|f\|_{p,w}$ with an A_p-consistent constant C.

By Lemma 4.3 there exists p_1 depending on p, such that $w \in A_{p_1}$. Setting $r := \frac{p}{p_1} > 1$ it follows that

$$\|(M(|f|^r(x))^{\frac{1}{r}}\|_{p,w} = \|(M(|f|^r(x))\|_{\frac{p}{r},w}^{\frac{1}{r}} \le C\||f|^r(x)\|_{\frac{p}{r},w}^{\frac{1}{r}} = C\|f\|_{p,w}.$$

Combining the above three arguments with Proposition 5.1 the proof is complete. $\qquad\square$

In the following, we consider kernels which depend on a parameter. In [CFL91, Thm. 2.10] the following proposition was proved for the case $m = 2$ and $w \equiv 1$. As the proof carries over to our situation, we only point out the cornerstones of the proof and give details for the statements we need later in akin situations.

PROPOSITION 5.5. *Let w be an A_p-weight, $a \in$ BMO. Let $k : \mathbb{R}^n \times (\mathbb{R}^n \setminus \{0\}) \to \mathbb{C}$ be a function, such that*

a) *$k(x, \cdot)$ is homogeneous of degree 0 and $\int_{S^{n-1}} k(x, z)\mathrm{d}z = 0$.*
b) *$\|D_y^j k(x, y)\|_{L^\infty(\mathbb{R}^n \times S^{n-1})} \leq M \quad \forall |j| \leq 2n$.*

Then the operators K and C_a defined by

$$Kf(x) := \lim_{\varepsilon \to 0} K_\varepsilon f(x) := \lim_{\varepsilon \to 0} \int_{|x-y|>\varepsilon} k(x, x - y) f(y)\mathrm{d}y,$$

$$C_a f := [M_a, K]f$$

are continuous on $L_w^p(\mathbb{R}^n)$. Furthermore there exists an A_p-consistent constant C such that

$$\|Kf\|_{p,w} \leq C\|f\|_{p,w},$$
$$\|C_a f\|_{p,w} \leq C\|a\|_* \|f\|_{p,w}.$$

For the proof of the above proposition we need certain properties of spherical harmonics. We call a homogeneous polynomial p of degree m satisfying $\Delta p = 0$ *solid harmonic* of degree m. The restriction of p to the sphere is called *spherical harmonic*. We also set

$$\mathcal{H}_m := \{p : p \text{ is spherical harmonic of degree } m\} \text{ and}$$
$$\mathcal{H} := \bigcup_{m \in \mathbb{N}} \mathcal{H}_m.$$

If $p \in \mathcal{H}$, then $\tilde{p}(x) = p(\frac{x}{|x|})$, $x \in \mathbb{R}^n \setminus \{0\}$ defines a solid harmonic. Therefore we often do not distinguish between solid and spherical harmonics. The following properties of spherical harmonics will be used in the sequel. For further information concerning spherical harmonics see e.g. [SW71, IV.2]

LEMMA 5.6. *The space $L^2(S^{n-1})$ possesses a complete orthonormal system $Y_{km}\{Y_{k,m}\}$ consisting of functions in $\mathcal{H}$. Moreover, the following assertions are true:*

a) *$\dim \mathcal{H}_m =: g_m = \binom{m+n-1}{n-1} - \binom{m+n-3}{n-1} \leq c(n)m^{n-2}$;*
b) *$|Y_{k,m}| \leq c(n)m^{\frac{n}{2}-1}, \quad x \in S^{n-1}, \; Y_{k,m} \in \mathcal{H}_m, \; k = 1, \ldots, g_m$;*
c) *$Y_{k,m} = (-m)^{-r}(m + n - 2)^{-r}\Lambda^r Y_{k,m}$, where $\Lambda := M_{|x|^2}\Delta$ and $M_{|x|^2}$ denotes the multiplication operator with $|x|^2$.*

The next lemma is an easy consequence of Greens formula

LEMMA 5.7. *Let* $f, g \in C^{2r}(\mathbb{R}^n \setminus \{0\})$ *be homogeneous of degree zero. Then*

$$\int_{S^{n-1}} f \Lambda^r g \, \mathrm{d}\sigma = \int_{S^{n-1}} \Lambda^r f g \, \mathrm{d}\sigma.$$

We are now in the position to give an outline of the proof of Proposition 5.5.

Proof of Propostion 5.5. Let $f \in C_c^\infty(\mathbb{R}^n)$. In view of Lemma 5.6 there exists an orthonormal system $\mathcal{O} \subset \mathcal{H}$ of spherical harmonics in $L^2(S^{n-1})$. Denote the elements in $\mathcal{O}$ by $Y_{k,m} \in \mathcal{H}_m$, $k = 1, \ldots, g_m$. It follows that $\tilde{k}_x(z) := |z|^n k(x, z)$ admits a representation of the form

$$\tilde{k}_x(z) = \sum_{m=1}^\infty \sum_{k=1}^{g_m} b_{k,m}(x) Y_{k,m}(z),$$

where $b_{k,m}(x) = \langle \tilde{k}_x, Y_{k,m} \rangle_{L^2(S^{n-1})}$. We proceed in three steps.

STEP 1. Estimates for $\|b_{k,m}\|_\infty$
A short computation yields

$$\|\Lambda^n(|z|^n k(x, z))\|_{L^\infty(\mathbb{R}^n \times S^{n-1})} \leq c(n) \cdot M.$$

Combining this estimate with Lemmas 5.6.c) and 5.7 we obtain

$$|b_{k,m}(x)| = \left| (-m)^{-n}(m+n-2)^{-n} \int_{S^{n-1}} |z|^n k(x, z) \Lambda^n Y_{k,m}(z) \mathrm{d}z \right|$$

$$= \left| (-m)^{-n}(m+n-2)^{-n} \int_{S^{n-1}} (\Lambda^n(|z|^n k(x, z))) Y_{k,m}(z) \mathrm{d}z \right|$$

$$\leq m^{-n}(m+n-2)^{-n} \|\Lambda^n(|z|^n k(x, z))\|_{L^2(S^{n-1})}$$

$$\leq \tilde{c}(n) m^{-2n} M$$

STEP 2. Representation of K_ε and $C_{a,\varepsilon}$

For $|x - y| > \varepsilon$ we define

$$K_\varepsilon f(x) = \sum_{m=1}^\infty \sum_{k=1}^{g_m} b_{k,m}(x) \int_{|x-y|>\varepsilon} \frac{Y_{k,m}(x - y)}{|x - y|^n} f(y) \mathrm{d}y.$$

Similarly

$$C_{a,\varepsilon} f(x) = \sum_{m=1}^{\infty} \sum_{k=1}^{g_m} b_{k,m}(x) \int_{|x-y|>\varepsilon} \frac{Y_{k,m}(x-y)}{|x-y|^n} (a(x) - a(y)) f(y) \mathrm{d}y.$$

STEP 3. Norm bounds for K_ε and $C_{a,\varepsilon}$

For $x \in \mathbb{R}^n$ we set

$$R_{k,m,\varepsilon} f(x) := \int_{|x-y|>\varepsilon} \frac{Y_{k,m}(x-y)}{|x-y|^n} f(y) \mathrm{d}y,$$

and

$$S_{k,m,\varepsilon} f(x) := \int_{|x-y|>\varepsilon} \frac{Y_{k,m}(x-y)}{|x-y|^n} (a(x) - a(y)) f(y) \mathrm{d}y.$$

Since $\frac{Y_{k,m}(z)}{|z|^n}$ is a Calderón-Zygmund kernel, the operators

$$f \mapsto R_{k,m} f := \lim_{\varepsilon \to 0} R_{k,m,\varepsilon} f \text{ and } f \mapsto S_{a,k,m} f := \lim_{\varepsilon \to 0} S_{a,k,m,\varepsilon} f$$

are continuous on $L_w^p(\mathbb{R}^n)$, where their norms are depending only on n, p and $A_p(w)$ but not on k, m and ε. By the estimates for $\|b_{k,m}\|_\infty$ and since $g_m \le c(n) m^{n-2}$ we obtain

$$\sum_{m-1}^{\infty} \sum_{k-1}^{g_m} \|b_{k,m} R_{k,m,\varepsilon} f\|_{p,w} \le c(n, p, M, A_p(w)) \|f\|_{p,w} \sum_{m-0}^{\infty} m^{-2n} m^{n-2}$$

as well as

$$\sum_{m=1}^{\infty} \sum_{k=1}^{g_m} \|b_{k,m} S_{k,m,\varepsilon} f\|_{p,w} \le c(n, p, M, A_p(w)) \|f\|_{p,w} \|a\|_* \sum_{m=0}^{\infty} m^{-2n} m^{n-2}.$$

This shows that the above series are absolutely convergent with bounds independent of ε. Therefore we may define

$$Kf := \sum_{m=1}^{\infty} \sum_{k=1}^{g_m} b_{k,m} R_{k,m} f \text{ and } C_a f := \sum_{m=1}^{\infty} \sum_{k=1}^{g_m} b_{k,m} S_{k,m} f$$

and obviously we have

$$\|Kf\|_{p,w} \le c \|f\|_{p,w}, \quad \|C_a f\|_{p,w} \le c \|a\|_* \|f\|_{p,w}.$$

Since the series are absolutely convergent, we may interchange the limit and the sum and obtain thus $K_\varepsilon f \to Kf$, $C_\varepsilon f \to Cf$, for $\varepsilon \to 0$.

6. Elliptic operators with coefficients being small in BMO

In this section we prove a priori estimates for elliptic operators whose coefficients are bounded and small in the BMO-norm.

For $m \in \mathbb{N}$ consider operators $\mathcal{A}(x, D)$ of the form

$$\mathcal{A}(x, D) := \sum_{|\alpha|=m} a_\alpha(x) D^\alpha.$$

For the time being, we assume that for $|\alpha| = m$ the coefficients satisfy

$$a_\alpha \in L^\infty(\mathbb{R}^n).$$

We also assume that there exist $M > 0$ and $\phi_0 \in [0, \pi/2)$ such that the symbol $\mathcal{A}(x, \xi) = \sum_{|\alpha|=m} a_\alpha(x)\xi^\alpha$ is (M, ϕ_0)-parameter-elliptic for almost all $x \in \mathbb{R}^n$. We denote by E the set of all $x \in \mathbb{R}^n$ for which $\mathcal{A}(x, \xi)$ is (M, ϕ_0)-parameter-elliptic. We set

$$\mathcal{A}_{x_0} := \sum_{|\alpha|=m} a_\alpha(x_0) D^\alpha,$$

with coefficients being frozen at $x_0 \in E$.

Let $\phi > \phi_0$. By [DHP01], the fundamental solution $\gamma_\lambda^{x_0}$ of $\lambda + \mathcal{A}_{x_0}$ satisfies an estimate of the form

$$|D^\beta \gamma_\lambda^{x_0}(x)| \leq C_{\phi,k} |\lambda|^{\frac{n+k}{m}-1} p^n_{m,k}(c_\phi |\lambda|^{1/m}|x|), \quad x \in \mathbb{R}^n, \ |\beta| = k, \ \lambda \in \Sigma_{\pi-\phi}. \quad (6.1)$$

Here the functions $p^n_{m,k}$ are defined as in [DHP01] by

$$p^n_{m,k}(r) = \int_0^\infty \frac{s^{n-2}}{(1+s)^{m-k-1}} e^{-r(1+s)} ds;$$

they are completely monotone and $\int_0^\infty r^{n+\rho-1} p^n_{m,k}(r) dr < \infty$ holds if and only if $\rho > k - m$ and $\rho > -n$. We refer to [DHP01] for details. Note that

$$p^n_{m,k}(r) \leq c(n + \rho) r^{-(n+\rho)}, \quad (6.2)$$

since for $\tau \in (0, \infty)$ there is an $\eta \in (0, \tau]$ such that

$$p_{m,k}^n(\tau)(n+\rho)^{-1}\tau^{n+\rho} \le p_{m,k}^n(\eta)(n+\rho)^{-1}\tau^{n+\rho} = p_{m,k}^n(\eta)\int_0^\tau r^{n+\rho-1}dr$$

$$= \int_0^\tau r^{n+\rho-1}p_{m,k}^n(r)dr \le \int_0^\infty r^{n+\rho-1}p_{m,k}^n(r)dr < \infty.$$

LEMMA 6.1. *The functions* $\lambda^{1-\frac{k}{m}}D^\beta\gamma_\lambda^{x_0}$ *are Calderón-Zygmund kernels for* $x_0 \in E$, $\lambda \in \Sigma_{\pi-\phi}$ *and* β *with* $|\beta| = k \le m$.

Proof. We first show that $\mathcal{F}\lambda^{\frac{m-k}{m}}D^\beta\gamma_\lambda^{x_0}$ is bounded. Denote by $\mathcal{A}_{x_0}(\xi)$ the symbol of $\mathcal{A}_{x_0}$ and set $\tilde{\mathcal{A}}((\mu,\xi)) := (\mu^m + \mathcal{A}_{x_0}(\xi))$. Thus $\tilde{\mathcal{A}}$ is a homogeneous polynomial in $n+1$ variables. Since $\mathcal{A}_{x_0}$ is elliptic we obtain

$$|\lambda^{\frac{m-k}{m}}\mathcal{F}D^\beta\gamma_\lambda^{x_0}(\xi)| \le |\lambda|^{\frac{m-k}{m}}|\xi|^k\mathcal{F}\gamma_\lambda^{x_0}(\xi) = |\lambda|^{\frac{m-k}{m}}|\xi|^k|(\lambda + \mathcal{A}_{x_0}(\xi))^{-1}|$$

$$= |\xi|^k|\mu|^{m-k}|(\mu,\xi)|^{-m}|\tilde{\mathcal{A}}(\zeta)^{-1}|$$

$$\le C, \quad \lambda \in \Sigma_{\pi-\phi};$$

here $\lambda = \mu^m$ and $\zeta \in S^n$ is chosen such that $(\mu,\xi) = |(\mu,\xi)|\zeta$.

In order to complete the proof we show that for $|\beta| = k \le m$ there exists a constant $M > 0$ such that

$$|\lambda|^{\frac{m-k}{m}}|D^\beta\gamma_\lambda^{x_0}(x)| \le M|x|^{-n}, \qquad x \in \mathbb{R}^n$$

$$|\lambda|^{\frac{m-k}{m}}|D^{\beta+e_j}\gamma_\lambda^{x_0}(x)| \le M|x|^{-(n+1)}, \qquad x \in \mathbb{R}^n.$$

Consider first the case where $|\beta| < m$ and let $b \in \{0, 1\}$. Then, by (6.1) and (6.2)

$$|D^{\beta+be_j}\lambda^{\frac{m-k}{m}}\gamma_\lambda^{x_0}(x)| \le C_\phi|\lambda|^{\frac{m-k}{m}}|\lambda|^{-\frac{n+k+b}{m}-1}p_{m,k+b}^n(c_\phi|\lambda|^{\frac{1}{m}}|x|)$$

$$\le C_\phi|\lambda|^{\frac{n+b}{m}}(|\lambda|^{\frac{1}{m}}|x|)^{-(n+\rho)}$$

$$\le C_\phi|\lambda|^{\frac{n+b}{m}}|\lambda|^{\frac{-(n+\rho)}{m}}|x|^{-(n+\rho)}.$$

Setting $\rho = b$ the assertion follows. The remaining cases follow by the kernel estimate (6.1) and the fact that the functions $p_{m,k}^n$ are given by

$$p_{m,k}^n(r) = e^{-r}\int_0^\infty r^{m-k-1}(r+u)^{-m+k+1}r^{-n+1}u^{n-2}e^{-u}du, \quad r > 0$$

and satisfy an estimate of the form

$$p_{m,k}^n(r) \le Ce^{-r}r^{-n}, \qquad r > 0, k = |\beta| = m$$

$$p_{m,k}^n(r) \le Ce^{-r}r^{-(n+1)}, \qquad r > 0, k = |\beta| = m+1.$$

$\square$

LEMMA 6.2. *Let* $h \in C^\infty(\mathbb{R}^n \setminus \{0\})$ *be a homogeneous function of degree* 0. *Set* $Tf = \mathcal{F}^{-1} h \mathcal{F} f$ *for* $f \in L^2(\mathbb{R}^n)$. *Then there exists a homogeneous function* $k \in C^\infty(\mathbb{R}^n \setminus \{0\})$ *of degree* $-n$ *with* $\int_{S^{n-1}} k = 0$ *such that*

$$Tf = c \cdot f + \lim_{\varepsilon \to 0} \int_{|y| > \varepsilon} k(y) f(. - y) dy.$$

Moreover, c and the derivatives of k can be estimated in terms depending on h only.

For a proof, see e.g. [Ste70, III,§ 3.3.5 Thm. 6]. Note that k is a Calderón-Zygmund kernel.

To be more precise, we will have a closer look at $D^\alpha k$. The proof given by Stein involves a spherical harmonic expansion of the multiplier h, say $h = \sum_{k=0}^\infty \tilde{Y}_k$. He shows then, that $c = \tilde{Y}_0$ and $k(y) = \frac{\Omega(y)}{|y|^n}$ where $\Omega = \sum_{k=1}^\infty c_k^{-1} \tilde{Y}_k$ with an explicit formula for the constants c_k. He establishes also that $h \in C^\infty(S^{n-1})$ if and only if

$$\int_{S^{n-1}} |\tilde{Y}_k|^2 d\sigma = O(k^{-N}) \quad \forall N \in \mathbb{N}.$$

But this yields that (see Appendix C4 in [Ste70])

$$|D^\alpha h(x)| = \left| \sum_{k=0}^\infty D^\alpha \tilde{Y}_k(x) \right| \leq \sum_{k=0}^\infty |D^\alpha \tilde{Y}_k(x)| \leq C_r$$

for $|\alpha| < r$ and $x \in S^{n-1}$. Therefore we also have

$$|D^\alpha \Omega(x)| \leq \sum_{k=1}^\infty \left| \frac{D^\alpha \tilde{Y}_k(x)}{c_k} \right| \leq c \cdot C_r.$$

PROPOSITION 6.3. *Let* $\mathcal{A}(x, D) = \sum_{|\alpha|=m} a_\alpha(x) D^\alpha$ *be an* (M, ϕ_0)-*parameter-elliptic differential operator of order m with coefficients* $a_\alpha \in L^\infty(\mathbb{R}^n)$ *and* $\phi_0 < \frac{\pi}{2}$. *Let* $w \in A_p$ *and E be defined as above. Denote by* γ_λ^x *the kernel of* $\lambda + \mathcal{A}_x$, *where* $\lambda \in \Sigma_{\pi - \phi_0}$. *For* $|\alpha| = m$ *let* $k_\alpha(x_0, \cdot)$ *be the Calderón-Zygmund kernel corresponding to the multiplier* $h(\xi) := \xi^\alpha(\mathcal{A}_{x_0}(\xi))^{-1}$ *and denote by* $c_\alpha(x_0)$ *the constant, which is also associated to h. Then a function* $u \in C_c^\infty$ *may be represented for* $x \in E$ *as*
a)

$$D^\alpha u(x) = \text{p.v.} \int_{\mathbb{R}^n} k_\alpha(x, x - y)(\lambda + \mathcal{A}_x) u(y) dy + c_\alpha(x)(\lambda + \mathcal{A}_x) u(x)$$

$$- \int_{\mathbb{R}^n} \lambda \gamma_\lambda^x(x - y) \left(\text{p.v.} \int_{\mathbb{R}^n} k_\alpha(x, y - z)(\lambda + \mathcal{A}_x) u(z) dz \right) dy$$

$$- c_\alpha(x) \int_{\mathbb{R}^n} \lambda \gamma_\lambda^x(x - y)(\lambda + \mathcal{A}_x) u(y) dy,$$

for $|\alpha| = m$, and

b)

$$D^\alpha u(x) = \int_{\mathbb{R}^n} D^\alpha \gamma_\lambda^x (x - y)(\lambda + \mathcal{A}_x)u(y),$$

for $|\alpha| < m$.

Proof. First note that $\mathcal{F}u = (\lambda + \mathcal{A}_{x_0}(\xi))^{-1}\mathcal{F}(\lambda + \mathcal{A}_{x_0})u$ holds for arbitrary $x_0 \in E$. The case b) now follows by setting $x = x_0$, since $D^\alpha \gamma_\lambda^{x_0} \in L^1$. For a) we compute first

$$\begin{aligned}
\mathcal{F}D^\alpha u &= \xi^\alpha (\lambda + \mathcal{A}_{x_0}(\xi))^{-1}\mathcal{F}(\lambda + \mathcal{A}_{x_0})u \\[2mm]
&= \xi^\alpha (\mathcal{A}_{x_0}(\xi))^{-1}\mathcal{F}(\lambda + \mathcal{A}_{x_0})u \\[2mm]
&\quad + (\xi^\alpha (\lambda + \mathcal{A}_{x_0}(\xi))^{-1} - \xi^\alpha \mathcal{A}_{x_0}(\xi)^{-1})\mathcal{F}(\lambda + \mathcal{A}_{x_0})u \\[2mm]
&= \xi^\alpha (\mathcal{A}_{x_0}(\xi))^{-1}\mathcal{F}(\lambda + \mathcal{A}_{x_0})u \\[2mm]
&\quad - \lambda(\lambda + \mathcal{A}_{x_0}(\xi))^{-1}\xi^\alpha \mathcal{A}_{x_0}(\xi)^{-1}\mathcal{F}(\lambda + \mathcal{A}_{x_0})u.
\end{aligned}$$

Applying $\mathcal{F}^{-1}$ and using Lemma 6.2 we get for all $x \in E$

$$\begin{aligned}
D^\alpha u(x) = {}&\mathrm{p.v.}\int_{\mathbb{R}^n} k_\alpha(x_0, x - y)(\lambda + \mathcal{A}_{x_0})u(y)\mathrm{d}y + c_\alpha(x_0)(\lambda + \mathcal{A}_{x_0})u(x) \\[2mm]
&- \int_{\mathbb{R}^n} \lambda\gamma_\lambda^{x_0}(x - y)\left(\mathrm{p.v.}\int_{\mathbb{R}^n} k_\alpha(x_0, y - z)(\lambda + \mathcal{A}_{x_0})u(z)\mathrm{d}z\right)\mathrm{d}y \\[2mm]
&- \int_{\mathbb{R}^n} \lambda\gamma_\lambda^{x_0}(x - y)c_\alpha(x_0)(\lambda + \mathcal{A}_{x_0})u(y)\mathrm{d}y.
\end{aligned}$$

Setting $x_0 = x$ yields now the claim. $\qquad\square$

In the following, we use the above representation formula, to get a priori estimates for the operator $(\lambda + \mathcal{A})$, i.e. estimates of the form $\|\lambda^{\frac{m-|\alpha|}{m}} D^\alpha u\| \le C\|(\lambda + \mathcal{A})u\|$. Replacing $(\lambda + \mathcal{A}_x)$ by $(\lambda + \mathcal{A}) + (\mathcal{A}_x - \mathcal{A})$ in the above representation shows that we need to consider the following operators in detail. They are initially defined for $f \in \mathcal{S}$. We set

$$T_1 f(x) := \mathrm{p.v.}\int_{\mathbb{R}^n} k_\alpha(x, x - y)f(y)\mathrm{d}y$$

$$[T_1, a]f(x) := \mathrm{p.v.}\int_{\mathbb{R}^n} k_\alpha(x, x - y)(a(x) - a(y))f(y)\mathrm{d}y$$

$$T_2 f(x) := \int_{\mathbb{R}^n} \lambda \gamma_\lambda^x (x - y) \, \text{p.v.} \int_{\mathbb{R}^n} k_\alpha(x, y - z) f(z) dz dy$$

$$[T_2, a] f(x) := \int_{\mathbb{R}^n} \lambda \gamma_\lambda^x (x - y) \, \text{p.v.} \int_{\mathbb{R}^n} k_\alpha(x, y - z)(a(x) - a(z)) f(z) dz dy,$$

for $|\alpha| = m$ and

$$T_3^\alpha f(x) := \int_{\mathbb{R}^n} \lambda^{\frac{m - |\alpha|}{m}} D^\alpha \gamma_\lambda^x (x - y) f(y) dy$$

$$[T_3^\alpha, a] f(x) := \int_{\mathbb{R}^n} \lambda^{\frac{m - |\alpha|}{m}} D^\alpha \gamma_\lambda^x (x - y)(a(x) - a(y)) f(y) dy,$$

for $|\alpha| < m$, where a is assumed to be in $\text{BMO} \cap L^\infty$. Note that $[T_2, a]$ is not a commutator but on the account of the similar treatment to $[T_1, a]$ we have chosen a similar name. Estimates on the operators T_1 and $[T_1, a]$ were the matter of Proposition 5.5. We will see that we may deal with the other operators by similar techniques. More precisely, the operator T_2 and $[T_2, a]$ will be treated by combining T_1 with T_3^0 and by a spherical harmonic expansion. Hence, we start by considering T_3^α and $[T_3^\alpha, a]$, respectively.

PROPOSITION 6.4. *For an arbitrary cube $I \subset \mathbb{R}^n$ let $\mathcal{A}_I$ be the differential operator with constant coefficients defined by*

$$\mathcal{A}_I := \sum_{|\alpha| = m} \frac{1}{|I|} \int_I a_\alpha(x) dx \, D^\alpha.$$

Then $\mathcal{A}_I$ is $(\frac{M}{\cos \phi_0}, \phi_0)$-parameter elliptic.

Proof. First note that $\mathcal{A}$ is (M, ϕ_0)-parameter-elliptic if and only if $\mathcal{A}(x, \xi) \in \Sigma_{\phi_0} \backslash B_{\frac{1}{M}}(0)$ for all $x \in \mathbb{R}^n$, $|\xi| = 1$. Here $B_r(z) \subset \mathbb{C}$ denotes the ball with radius r and center z. Furthermore, for $M \subset \mathbb{R}^n$, $f, v \in L^1(M, \mathbb{C})$ with $v > 0$ we have

$$\int_M f(x) dx \in \int_M v(x) dx \cdot \overline{\text{conv}} \left\{ \frac{f(x)}{v(x)} : x \in M \right\}.$$

If not, there is a $z \in \mathbb{C}$ such that

$$\text{Re}\,(cz) < \text{Re} \left(\frac{\int f}{\int v} z \right) \qquad \forall c \in \overline{\text{conv}} \left\{ \frac{f(x)}{v(x)} : x \in M \right\}.$$

In particular, we have

$$\mathrm{Re}\left(\frac{f(x)}{v(x)}z\right) < \mathrm{Re}\left(\frac{\int f}{\int v}z\right), \quad x \in M.$$

Multiplication by $v(x)$ and integrating gives

$$\int\limits_M \mathrm{Re}\,(f(x)z)\mathrm{d}x < \int\limits_M \mathrm{Re}\left(\frac{\int f}{\int v}v(x)z\right)\mathrm{d}x = \mathrm{Re}\left(\int\limits_M f(x)\mathrm{d}x \cdot z\right),$$

which yields a contradiction.

Setting $M := I$, $v := \chi_I$ and $f(y) := \sum_{|\alpha|=m} a_\alpha(y)$ yields

$$\sum_{|\alpha|=m} \frac{1}{|I|}\int\limits_I a_\alpha(z)\mathrm{d}z\xi^\alpha = \int\limits_I \sum_{|\alpha|=m} a_\alpha(z)\xi^\alpha\mathrm{d}z \;\in\; \overline{\mathrm{conv}}\left\{\sum_{|\alpha|=m} a_\alpha(z)\xi^\alpha : z \in I\right\}$$

$$\subset \overline{\mathrm{conv}}\,\{\Sigma_{\phi_0}\setminus B_{\frac{1}{M}}(0)\}$$

$$\subset \Sigma_{\phi_0}\setminus B_{\frac{\cos\phi_0}{M}}(0)$$

Therefore $\mathcal{A}_I$ is $(\frac{M}{\cos\phi_0}, \phi_0)$-parameter-elliptic. $\qquad\qquad\square$

PROPOSITION 6.5. *Let* $|\alpha| < m$. *The operators* T_2, T_3^α *and* $[T_2, a]$, $[T_3^\alpha, a]$ *respectively, are bounded in* L_w^p. *Furthermore, the norms of these operators can be estimated by*

$$\|T_2\| + \|T_3^\alpha\| \le C \qquad and$$
$$\|[T_2, a]\| + \|[T_3^\alpha, a]\| \le C \max_{|\beta|=m}\{\|a\|_\infty, \|a_\beta\|_\infty\} \max_{|\beta|=m}\{\|a\|_*, \|a_\beta\|_*\},$$

where the constant C *can be chosen to be* A_p-*consistent.*

Proof. We consider first T_3^α. By (6.1) and using that $p_{m,k}^n$ is radially decreasing and integrable, we have

$$|T_3^\alpha f(x)| \le C\||\lambda|^{\frac{n}{m}} p_{m,|\alpha|}^n(c|\lambda|^{\frac{1}{m}}|.|) * f|(x) \le C\|p_{m,|\alpha|}^n\|_1 Mf(x). \tag{6.3}$$

For a proof of the last inequality see e.g. (16), p.57 of Stein [Ste93].

The boundedness of M in L_w^p yields immediately

$$\|T_3^\alpha f\|_{p,w} \le C\|f\|_{p,w},$$

where the constant C is A_p-consistent and independent of λ.

The commutator estimate for $[T_3^\alpha, a]$ will follow from a pointwise estimate for the sharp function of $[T_3^\alpha, a]f$. Take an arbitrary cube $I \subset \mathbb{R}^n$ with side length $d > 0$ and $x \in I$. We denote by a_I the mean value of a with respect to I. If $2^j I$ denotes the cube with the same center as I and side length $2^j d$ then

$$\left(\frac{1}{|2^j I|} \int_{2^j I} |a(z) - a_I|^r dz \right)^{\frac{1}{r}} \leq C(j+1)\|a\|_*, \ r \geq 1, \tag{6.4}$$

for $j \in \mathbb{N}$. This can be easily seen by induction using the John-Nirenberg inequality. In fact, for $j = 0$ (6.4) is exactly the assertion of the John-Nirenberg inequality. (See [Ste93, IV.1.3] for details.) Coming to $[T_3^\alpha, a]$, we firstly write

$$[T_3^\alpha, a]f(x) = (a(x) - a_I)T_3^\alpha f(x) - T_3^\alpha((a - a_I)\chi_{2I} f)(x)$$
$$- T_3^\alpha((a - a_I)\chi_{(2I)^c} f)(x)$$
$$= A(x) + B(x) + C(x)$$

and estimate the means of $|A|$, $|B|$ and $|C - C_I|$ over I separately.

By the inequalities of Hölder, Jensen and John-Nirenberg we firstly compute

$$\frac{1}{|I|} \int_I |A(z)| dz \leq \left(\frac{1}{|I|} \int_I |a(z) - a_I|^{r'} dz \right)^{\frac{1}{r'}} \left(\frac{1}{|I|} \int_I |T_3^\alpha f(z)|^r dz \right)^{\frac{1}{r}}$$
$$\leq C\|a\|_*(M|T_3^\alpha f|^r)^{\frac{1}{r}}(x)$$

and

$$\frac{1}{|I|} \int_I |B(z)| dz \leq \left(\frac{1}{|I|} \int_I |T_3^\alpha((a - a_I)f\chi_{2I})(z)|^q dz \right)^{\frac{1}{q}}$$

$$\leq \left(\frac{1}{|I|} \right)^{\frac{1}{q}} \|T_3^\alpha((a - a_I)f\chi_{2I})\|_q$$

$$\leq C \left(\frac{1}{|I|} \int_{2I} |a(z) - a_I|^{qu'} dz \right)^{\frac{1}{qu'}} \left(\frac{1}{|I|} \int_{2I} |f(z)|^{qu} dz \right)^{\frac{1}{qu}}$$

$$\leq C\|a\|_*(M|f|^s)^{\frac{1}{s}}(x). \qquad (s = qu)$$

In order to consider $|C - C_I|_I$, let x_I denote the center of I. Further we set $\lambda^{\frac{m-|\alpha|}{m}} D^\alpha \gamma_\lambda^x(z) = k(x, z)$ and denote the kernel of $\lambda^{\frac{m-|\alpha|}{m}} D^\alpha(\lambda + A_I)^{-1}$, where A_I is defined as in Proposition 6.4.

by k_{a_I}. Thanks to Proposition 6.4., we know that k_{a_I} is a Calderón-Zygmund kernel and that $k_{a_I} \in L^1$. Setting $c = \int k_{a_I}(x_I - y)(a(y) - a_I) f(y) \, dy$ we compute

$$|C(\cdot) - C_I|_I \le 2|C(\cdot) - c|_I$$

$$= \frac{2}{|I|} \int_I \left| \int_{\mathbb{R}^n} (k(z, z - y) - k_{a_I}(x_I - y)) \underbrace{(a(y) - a_I)\chi_{2I^c} f(y)}_{=:g(y)} \, dy \right| dz$$

$$= \frac{2}{|I|} \int_I \left| \int_{\mathbb{R}^n} (k(z, z - y) - k_{a_I}(z - y) + k_{a_I}(z - y) - k_{a_I}(x_I - y)) g(y) \, dy \right| dz$$

$$\le C \left(\frac{1}{|I|} \int_I C_1(z) \, dz + \frac{1}{|I|} \int_I C_2(z) \, dz \right).$$

Before estimating the mean of C_1, note that for $\varphi \in \mathcal{S}$

$$
\begin{aligned}
(k(x_0, \cdot) - k_{a_I}) * \varphi &= \mathcal{F}^{-1} |\lambda|^{\frac{m - |\alpha|}{m}} \xi^\alpha (\lambda + \mathcal{A}(x_0, \xi))^{-1} - (\lambda + \mathcal{A}_I(\xi))^{-1} \mathcal{F}\varphi \\
&= \sum_{|\beta|=m} ((a_\beta)_I - a_\beta(x_0)) \mathcal{F}^{-1} |\lambda|^{\frac{m - |\alpha|}{m}} \xi^\alpha (\lambda + \mathcal{A}(x_0, \xi))^{-1} \\
&\qquad \xi^\beta (\lambda + \mathcal{A}_I(\xi))^{-1} \mathcal{F}\varphi \\
&= \sum_{|\beta|=m} ((a_\beta)_I - a_\beta(x_0)) S_{\beta, x_0}\varphi.
\end{aligned}
$$

Furthermore, note that the operator S_β defined by $S_\beta\varphi(x) := S_{\beta, x}\varphi(x)$ is bounded in L_w^p with A_p-consistent norm. (In fact, the kernel of $\mathcal{F}^{-1} |\lambda|^{\frac{m - |\alpha|}{m}} \xi^\alpha (\lambda + \mathcal{A}(x_0, \xi))^{-1} \mathcal{F}$ has a Poisson bound, which is independent of x_0, with $p_{m, |\alpha|}^n$ as dominating function.) We then compute

$$
\begin{aligned}
\frac{1}{|I|} \int_I C_1(z) \, dz &= \frac{1}{|I|} \int_I \left| \sum_{|\beta|=m} ((a_\beta)_I - a_\beta(z)) S_\beta g(z) \right| dz \\
&\le \sum_{|\beta|=m} \frac{1}{|I|} \int_I |a_\beta(z) - (a_\beta)_I| |S_\beta(a\chi_{2I^c} f)(z)| \, dz \\
&\quad + \sum_{|\beta|=m} \frac{1}{|I|} \int_I |a_\beta(z) - (a_\beta)_I| |S_\beta(a_I \chi_{2I^c} f)(z)| \, dz.
\end{aligned}
$$

Setting $a_1 = a$, $a_2 = a_I$ and $j \in \{1, 2\}$ we may estimate all of the summands above simultaneously. We compute for $|\beta| = m$ as for $|A|_I$ and $|B|_I$

$$
\frac{1}{|I|} \int_I |a_\beta(z) - (a_\beta)_I| |S_\beta(a_j \chi_{2I^c} f)(z)| \mathrm{d}z
$$

$$
\leq \left(\frac{1}{|I|} \int_I |a_\beta(z) - (a_\beta)_I|^{r'} \mathrm{d}z \right)^{\frac{1}{r'}} \left(\frac{1}{|I|} \int_I |S_\beta(a_j f)(z)|^r \mathrm{d}z \right)^{\frac{1}{r}}
$$

$$
+ \left(\frac{1}{|I|} \int_I |a_\beta(z) - (a_\beta)_I|^{r'} \mathrm{d}z \right)^{\frac{1}{r'}} \left(\frac{1}{|I|} \int_I |S_\beta(a_j \chi_{2I} f)(z)|^r \mathrm{d}z \right)^{\frac{1}{r}}
$$

$$
\leq C \|a_\beta\|_* ((M|S_\beta a_j f|^r(x))^{1/r} + \tilde{C} \|a_j\|_\infty (M|f|^r(x))^{1/r}).
$$

Next, we estimate $|C_2(z)|$, for $z \in I$.

$$
|C_2(z)| \leq \int_{\mathbb{R}^n \setminus 2I} |k_{a_I}(z - y) - k_{a_I}(x_I - y)| |a(y) - a_I| |f(y)| \mathrm{d}y
$$

$$
\leq C \left(\int_{\mathbb{R}^n \setminus 2I} \frac{|x_I - z|}{|x_I - y|^{n+1}} |a(y) - a_I|^{s'} \mathrm{d}y \right)^{\frac{1}{s'}}
$$

$$
\left(\int_{\mathbb{R}^n \setminus 2I} \frac{|x_I - z|}{|x_I - y|^{n+1}} |f(y)| \mathrm{d}y \right)^{\frac{1}{s}},
$$

as $|z - y| > 2|x_I - z|$. These terms can be estimated analogously. So we first write

$$
\int_{\mathbb{R}^n \setminus 2I} \frac{|x_I - z|}{|x_I - y|^{n+1}} |a(y) - a_I|^{s'} \mathrm{d}y \leq \sum_{j=2}^{\infty} \frac{d}{(2^{j-1}d)^{n+1}} \int_{2^j I} |a(y) - a_I|^{s'} \mathrm{d}y
$$

$$
\leq C \sum_{j=2}^{\infty} 2^{-j} \frac{1}{|2^j I|} \int_{2^j I} |a(y) - a_I|^{s'} \mathrm{d}y \leq C \|a\|_*^{s'},
$$

and get in the same way that

$$\int_{\mathbb{R}^n \setminus 2I} \frac{|x_I - z|}{|x_I - y|^{n+1}} |f(y)| dy \leq C \sum_{j=2}^{\infty} 2^{-j} \frac{1}{|2^j I|} \int_{2^j I} |f(y)|^s dy \leq CM|f|^s(x).$$

Finally, notice that we have proved

$$|C - C_I|_I \leq C\left(\|a\|_* (M|f|^r(x))^{1/r} + \sum_{|\beta|=m} \|a_\beta\|_* ((M|S_\beta(af)|^r(x))^{1/r}) \right.$$

$$\left. + \|a_\beta\|_* \|a\|_\infty ((M|S_\beta f|^r(x))^{1/r} + (M|f|^r(x))^{1/r}) \right)$$

Collecting A, B, C and taking the supremum over all cubes with $x \in I$ we have

$$[T_3^\alpha, a]f^\#(x) \leq C\|a\|_* ((M|T_3^\alpha f|^r)^{1/r} + (M|f|^r)^{1/r})(x)$$

$$+ \sum_{|\beta|=m} \|a_\beta\|_* \|a\|_\infty ((M|f|^r)^{1/r}) + (M|S_\beta f|^r)^{1/r}))(x)$$

$$+ \|a_\beta\|_* ((M|S_\beta af|^r)^{1/r})(x)$$

Therefore, we are now in the position to estimate the L_w^p norm of $[T_3^\alpha, a]$. With r as in the third step of the proof of Proposition 5.4. we have

$$\|[T_3^\alpha, a]f\|_{p,w} \leq \|[T_3^\alpha, a]f^\#\|_{p,w}$$

$$\leq C \max_{|\beta|=m} \{\|a\|_*, \|a_\beta\|_*\}\left(\|T_3^\alpha f\|_{p,w} + \|f\|_{p,w} \right.$$

$$\left. + \sum_{|\beta|=m} \|S_\beta af\|_{p,w} + \|S_\beta f\|_{p,w} \right)$$

$$\leq C \max_{|\beta|=m} \{\|a\|_*, \|a_\beta\|_*\}\|f\|_{p,w},$$

where the constant C is A_p-consistent and depends only on p, r, w, n and the ellipticity constants of $\mathcal{A}$.

We now turn our attention to T_2. Using the same spherical harmonic expansion as in the proof of Proposition 5.5 we get (with analogous computations) the following representation of T_2:

$$T_2 f(x) = \sum_{m=1}^{\infty} \sum_{k=1}^{g_m} b_{k,m}(x) \underbrace{\int_{\mathbb{R}^n} \lambda \gamma_\lambda^x(x-y) \, \text{p.v.} \int_{\mathbb{R}^n} \frac{Y_{k,m}(y-z)}{|y-z|^n} f(z) dz dy}_{=T_3^0(K_Y f)}.$$

Therefore, we can finally estimate

$$\|T_2 f\|_{p,w} \le \sum_{m=1}^{\infty} \sum_{k=1}^{g_m} \|b_{k,m}(x)\|_{\infty} C \|T_3^0(K_Y f)\|_{p,w}$$

$$\le \sum_{m=1}^{\infty} \sum_{k=1}^{g_m} \|b_{k,m}(x)\|_{\infty} \tilde{C} \|f\|_{p,w} \le C \|f\|_{p,w}.$$

In almost the same manner, we can deal with the Operator $[T_2, a]$.

$$[T_2, a]f(x) = \int_{\mathbb{R}^n} \lambda \gamma_\lambda^x(x - y) \, \mathrm{p.v.} \int_{\mathbb{R}^n} k_\alpha(x, y - z)(a(x) - a(z)) f(z) \mathrm{d}z \mathrm{d}y$$

$$= \int_{\mathbb{R}^n} \lambda \gamma_\lambda^x(x - y)(a(x) - a(y)) \, \mathrm{p.v.} \int_{\mathbb{R}^n} k_\alpha(x, y - z) f(z) \mathrm{d}z \mathrm{d}y$$

$$+ \int_{\mathbb{R}^n} \lambda \gamma_\lambda^x(x - y) \, \mathrm{p.v.} \int_{\mathbb{R}^n} k_\alpha(x, y - z)(a(y) - a(z)) f(z) \mathrm{d}z \mathrm{d}y$$

$$= \sum_{m=1}^{\infty} \sum_{k=1}^{g_m} b_{k,m}(x)([T_3^0, a](K_Y f)(x) + T_3^0([K_Y, a]f)(x))$$

and therefore

$$\|[T_2, a]f\|_{p,w} \le \sum_{m=1}^{\infty} \sum_{k=1}^{g_m} \|b_{k,m}\|_{\infty} \|[T_3^0, a](K_Y f)(x) + T_3^0([K_Y, a]f)(x)\|_{p,w}$$

$$\le C \max_{|\beta|=m} \{\|a\|_{\infty}, \|a_\beta\|_{\infty}\} \max_{|\beta|=m} \{\|a\|_*, \|a_\beta\|_*\} \|f\|_{p,w}.$$

$$\square$$

After these preparations we now state the desired estimates for the operator $\mathcal{A}$.

THEOREM 6.6. *Let* $p \in (1, \infty)$, $w \in A_p$ *and* $m \in \mathbb{N}$. *Let* $\mathcal{A}(x, D) = \sum_{|\alpha|=m} a_\alpha(x) D^\alpha$ *be an* (M, ϕ_0)-*parameter-elliptic differential operator of angle* $\phi_0 < \frac{\pi}{2}$ *and order* m *with coefficients* $a_\alpha \in L^\infty(\mathbb{R}^n)$ *for* $|\alpha| = m$. *Let* $\phi > \phi_0$. *Then there exist constants* C *and* η *such that for* $u \in W_w^{m,p}$

$$\sum_{|\nu| \le m} \|\lambda^{\frac{m-|\nu|}{m}} D^\nu u\|_{L_w^p(\mathbb{R}^n)} \le C \|(\lambda + \mathcal{A})u\|_{L_w^p(\mathbb{R}^n)}, \quad \lambda \in \Sigma_{\pi - \phi}$$

provided $\|a_\alpha\|_* \le \eta$. *The constants* C *and* η *are only depending on the* A_p-*constant of* w *and on* ϕ.

Proof. We come back to our representation formula. First let $|\alpha| < m$. In view of Proposition 6.3 we write for $u \in C_c^\infty$

$$\lambda^{\frac{m-|\alpha|}{m}} D^\alpha u(x) = \int_{\mathbb{R}^n} \lambda^{\frac{m-|\alpha|}{m}} D^\alpha \gamma_\lambda^x(x-y)\left((\lambda + \mathcal{A}) + (\mathcal{A}_x - \mathcal{A})\right)u(y)\mathrm{d}y$$

$$= \int_{\mathbb{R}^n} \lambda^{\frac{m-|\alpha|}{m}} D^\alpha \gamma_\lambda^x(x-y)(\lambda + \mathcal{A})u(y)\mathrm{d}y$$

$$+ \int_{\mathbb{R}^n} \lambda^{\frac{m-|\alpha|}{m}} D^\alpha \gamma_\lambda^x(x-y) \sum_{|\beta|=m} (a_\beta(x) - a_\beta(y))D^\beta u(y)\mathrm{d}y$$

$$= (T_3^\alpha(\lambda + \mathcal{A}))u(x) + \sum_{|\beta|=m} ([T_3^\alpha, a_\beta]D^\beta)u(x).$$

With the estimates above

$$\|\lambda^{\frac{m-|\alpha|}{m}} D^\alpha u\|_{p,w} \le C_\alpha \|(\lambda + \mathcal{A})u\|_{p,w} + CM \max_{|\beta|=m}\{\|a_\beta\|_*\} \sum_{|\beta|=m} \|D^\beta u\|_{p,w}.$$

Noting that $c_\alpha \in L^\infty$ we get for $|\alpha| = m$

$$D^\alpha u(x) = (T_1(\lambda + \mathcal{A})u + c_\alpha(\lambda + \mathcal{A})u - T_2(\lambda + \mathcal{A})u - c_\alpha T_3^0(\lambda + \mathcal{A})u)(x)$$

$$+ \sum_{|\beta|=m} ([T_1, a_\beta]D^\beta u - [T_2, a_\beta]D^\beta u - c_\alpha[T_3^0, a_\beta]D^\beta u)(x).$$

Hence

$$\|D^\alpha u\|_{p,w} \le C_\alpha \|(\lambda + \mathcal{A})u\|_{p,w} + CM \max_{|\beta|=m}\{\|a_\beta\|_*\} \sum_{|\beta|=m} \|D^\beta u\|_{p,w}.$$

Summing over α with $|\alpha| \le m$ we get in the long run

$$\sum_{|\alpha|\le m} \|\lambda^{\frac{m-|\alpha|}{m}} D^\alpha u\|_{p,w} \le C\|(\lambda + \mathcal{A})u\|_{p,w},$$

if $CM \max_{|\beta|=m}\{\|a_\beta\|_*\} < 1$ $\hspace{2cm}\square$

THEOREM 6.7. *Let* $p \in (1, \infty)$, $w \in A_p$, $M > 0$, $\phi_0 \in [0, \frac{\pi}{2})$ *and* $m \in \mathbb{N}$. *Let* $\mathcal{A}(x, D)$ *be an* (M, ϕ_0)-*parameter-elliptic operator of order* m *with coefficients* $a_\alpha \in L^\infty(\mathbb{R}^n)$ *for* $|\alpha| = m$. *Let* $\phi > \phi_0$. *Then there exists* $\eta > 0$ *such that for* $f \in L_w^p(\mathbb{R}^n)$ *and* $\lambda \in \Sigma_{\pi-\phi}$ *there exists a unique* $u \in W_w^{m,p}(\mathbb{R}^n)$ *satisfying*

$$(\lambda + \mathcal{A}(x, D))u = f$$

provided $\max_{|\alpha|=m} \|a_\alpha\|_* \le \eta$.

Proof. The uniqueness result follows directly from Theorem 6.6. In order to prove the existence assertion, define for $t \in [0, 1]$ the operator

$$A_t := (1 - t)A_0 + tA,$$

where $A_0 = (-\Delta)^{\frac{m}{2}}$ denotes the $\frac{m}{2}$-th power of the Laplacian in $L_w^p(\mathbb{R}^n)$ with domain $W_w^{m,p}(\mathbb{R}^n)$ and $A := \mathcal{A}(x, D)$. Then A_t is parameter elliptic for each $t \in [0, 1]$ and there is a constant $C > 0$, not depending on t, such that

$$\|u\|_{W_w^{m,p}} \leq C\|(\lambda + A_t)u\|_{p,w}$$

Since $\lambda + A_0$ is onto, it follows from the continuity method that $\lambda + A$ is onto, also. $\square$

For $\mathcal{A}(x, D)$ as in Theorem 6.7. we are now in the position to define the realization of $\mathcal{A}(x, D)$ in $L_w^p(\mathbb{R}^n)$ where $1 < p < \infty$ and $w \in A_p$ as

$$Au := \mathcal{A}(x, D)u$$
$$D(A) := W_w^{m,p}(\mathbb{R}^n).$$

COROLLARY 6.8. *Under the assumptions of Theorem 6.6, the operator* $-A$ *generates a holomorphic* C_0 *semigroup on* $L_w^p(\mathbb{R}^n)$ *provided* $\phi_0 < \frac{\pi}{2}$. *In particular, there exists an* A_p-*consistent constant C such that for* $\phi > \phi_0$

$$\|(\lambda + A)^{-1}\|_{\mathcal{L}(L_w^p)} \leq \frac{C}{|\lambda|} \qquad \lambda \in \Sigma_{\pi - \phi}.$$

Proof. It follows from Theorem 6.7. that $(\lambda + A)$ is invertible for all $\lambda \in \Sigma_{\pi - \phi}$. The resolvent estimate is a consequence of Theorem 6.6. $\square$

7. Localization and lower order terms

In this section we treat parameter-elliptic operators with coefficients a_α belonging to the regularity class $\mathrm{VMO}(\mathbb{R}^n) \cap L^\infty(\mathbb{R}^n)$ which are no longer subject to the smallness condition described in the previous section. We do this via a countable localization procedure as described in [AHS94] and [HHH01]; also lower order terms are discussed. More precisely, we have the following result.

PROPOSITION 7.1. *Let* $\mathcal{A}(x, D) = \sum_{|\alpha| \leq m} a_\alpha D^\alpha$ *be an* (M, ϕ_0)-*parameter-elliptic operator of angle* $\phi_0 < \frac{\pi}{2}$ *and order m such that*

a) $a_\alpha \in \mathrm{VMO} \cap L^\infty(\mathbb{R}^n), \quad |\alpha| = m$
b) $a_\alpha \in L^\infty(\mathbb{R}^n), \quad |\alpha| < m.$

Let A denote the realization of $\mathcal{A}(x, D)$ in $L_w^p(\mathbb{R}^n)$ with domain $W_w^{m,p}(\mathbb{R}^n)$, where $w \in A_p$ and $1 < p < \infty$. Then the operator $-A$ generates a holomorphic C_0 semigroup on $L_w^p(\mathbb{R}^n)$ and there exist an A_p-consistent constant C and a constant $R > 0$ such that

$$\|(\lambda + A)^{-1}\|_{\mathcal{L}(L_w^p)} \leq \frac{C}{|\lambda|} \qquad \lambda \in \Sigma_{\pi - \phi}, |\lambda| > R.$$

Proof. Let $\rho > 0$ and let $\ell \in \mathbb{N}$ sufficiently large such that there exists a covering of $\mathbb{R}^n$ with balls of radius ρ such that every point $x \in \mathbb{R}^n$ is contained in at most ℓ balls. Denote such a covering by $\mathcal{B}_\rho := \{B_\rho(x_j) : j \in \mathbb{N}\}$. We now choose a decomposition of unity $(\psi_j)_{j \in \mathbb{N}} \subset C_c^\infty$ associated to the covering $\mathcal{B}_\rho$. Furthermore, we assume for the functions ψ_j that $\sup_{j \in \mathbb{N}} \sup_{|\alpha| \leq k} \|D^\alpha \psi_j\|_\infty \leq C_k$. We then define

$$\varphi_j := \psi_j \cdot \left(\sum_{k=1}^\infty \psi_k^2 \right)^{-1/2} . \tag{7.1}$$

Then $\varphi_j \in C_c^\infty(\mathbb{R}^n)$, $\sum_{j=1}^\infty \varphi_j^2 = 1$, $0 \leq \varphi_j \leq 1$, $\mathrm{supp}\,(\varphi_j) \subseteq B_\rho(x_j)$ and for any $k \in \mathbb{N}$ there exists a constant $C_k > 0$ such that $\sup_{j \in \mathbb{N}} \sup_{|\alpha| \leq k} \|D^\alpha \varphi_j\|_\infty \leq C_k$. To this family of functions we associate the following operators

$$P : \begin{cases} \ell^p(L_w^p(\mathbb{R}^n)) & \to \quad L_w^p(\mathbb{R}^n) \\ (f_j)_{j \in \mathbb{N}} & \mapsto \quad \sum_{j=1}^\infty \varphi_j f_j \end{cases} \qquad P^c : \begin{cases} L_w^p(\mathbb{R}^n) & \to \quad \ell^p(L_w^p(\mathbb{R}^n)) \\ f & \mapsto \quad (\varphi_j f)_{j \in \mathbb{N}} \end{cases}$$

These operators are bounded with norms indepedent of w. Moreover, the restriction of P to the space $\ell^p(W_w^{k,p}(\mathbb{R}^n))$ is for any $k \in \mathbb{N}$ continuous from $\ell^p(W_w^{k,p}(\mathbb{R}^n))$ to $W_w^{k,p}(\mathbb{R}^n)$ and the restriction of P^c to the space $W_w^{k,p}(\mathbb{R}^n)$ is continuous from $W_w^{k,p}(\mathbb{R}^n)$ to $\ell^p(W_w^{k,p}(\mathbb{R}^n))$, too.

Let η be the constant from Theorem 6.6. We then apply Lemma 3.2 to each function a_α with $|\alpha| = m$ and obtain $\rho_\alpha > 0$. Setting $\rho := \min_{|\alpha|=m}\{\rho_\alpha\}$ we choose a family of balls $\mathcal{B}_\rho$ and functions (φ_j) with the above described properties. For each ball $B_\rho(x_j)$ by Lemma 3.2 there are functions $a_{\alpha,j}$ such that $\|a_{\alpha,j}\|_* < \eta$ and $a_{\alpha,j} = a_\alpha$ on $B_\rho(x_j)$. Thus we define local operators as

$$A_j := \sum_{|\alpha|=m} a_{\alpha,j} D^\alpha.$$

By Corollary 6.6, every operator $-A_j$ generates a holomorphic C_0 semigroup on $L_w^p(\mathbb{R}^n)$ with A_p-consistent constants. We now set

$$S : \begin{cases} W_w^{m-1,p}(\mathbb{R}^n) & \to \ell^p(L_w^p) \\ u & \mapsto (\varphi_j Au - A_j \varphi_j u)_{j \in \mathbb{N}} =: (S_j u)_{j \in \mathbb{N}}, \end{cases}$$

$$T : \begin{cases} \ell^p(W_w^{m-1,p}(\mathbb{R}^n)) & \to L_w^p \\ (u_j)_{j \in \mathbb{N}} & \mapsto \sum_{j=1}^\infty (A\varphi_j - \varphi_j A_j) u_j \end{cases}$$

Theorem 6.6 implies the estimate

$$\|SP(\lambda + A)^{-1}\|_{\mathcal{L}(\ell^p(L_w^p))} \le \frac{C}{|\lambda|^{1/m}},$$

$$\|P^c T(\lambda + A)^{-1}\|_{\mathcal{L}(\ell^p(L_w^p))} \le \frac{C}{|\lambda|^{1/m}}.$$

Applying the perturbation theorem for holomorphic semigroups it follows that

$$-(\mu + A + SP) \quad \text{and} \quad -(\mu + A + P^c T)$$

are generators of bounded holomorphic semigroups for $\mu \ge 0$ big enough. By setting

$$L(\lambda) := P(\lambda + \mu + A + SP)^{-1} P^c \quad \text{and} \quad R(\lambda) := P(\lambda + \mu + A + P^c T)^{-1} P^c.$$

we obtain the left and right inverse of $(\lambda + \mu + A)$. $\qquad\qquad \square$

The assertion of our main result given in Theorem 2.1 follows finally by combining Proposition 7.1 with Proposition 4.1.

REFERENCES

[AHS94] AMANN, H., HIEBER, M., and SIMONETT, G., *Bounded H^∞-calculus for elliptic operators.* Differential Integral Equations *7* (1994), 613–653.

[AMT97] ANGELETTI, J. -M., MAZET, S., and TCHAMITCHIAN, PH., *Analysis of second order elliptic operators without boundary conditions and with* VMO *or Hölderian coefficients.* In: *Multiscale Wavelet Methods for PDEs*, W. Dahmen, A. Kurdilla, P. Oswald (eds.), Academic Press 1997, 495–539.

[BC93] BRAMANTI, M., and CERUTTI, M. C., $W_p^{1,2}$ *solvability for the Cauchy Dirichlet problem for parabolic equations with* VMO *coefficients.* Commun. Partial Differ. Equations *18* (1993), 1735–1763.

[CFL91] CHIARENZA, F., FRASCA, M., and LONGO, P., *Interior $W^{2,p}$ estimates for nondivergence elliptic equations with discontinuous coefficients.* Ricerche di Mat. *40* (1991), 149–168.

[CFL93] CHIARENZA, F., FRASCA, M., and LONGO, P., $W^{2,p}$-*solvability of the Dirichlet problem for nondivergence elliptic equations with VMO coefficients.* Trans. Amer. Math. Soc. *336* (1993), 841–853.

[CRW76] COIFMAN, R., ROCHBERG, R., and WEISS, G., *Factorization theorems for Hardy spaces in several variables.* Ann. of Math. *103* (1976), 611–635.

[DDHPV02] DENK, R., DORE, G., HIEBER, M., PRÜSS, J., and VENNI, A., *New Thoughts on old theorems of R. T. Seeley.* Preprint 2002.

[DHP01] DENK, R., HIEBER, M., and PRÜSS, J., $\mathcal{R}$-*boundedness, Fourier multipliers and problems of elliptic and parabolic type.* Memoirs Amer. Math. Soc., to appear.

[DV87] DORE, G., and VENNI, A., *On the closedness of the sum of two closed operators.* Math. Z. *196* (1987), 189–201.

[Duo90] DUONG, X. T., H_∞ *Functional calculus of elliptic operators with C^∞ coefficients on L^p spaces on smooth domains.* J. Austral. Math. Soc. Ser. A *48* (1990), 113–123.

[DM96] DUONG, X. T., and MCINTOSH, A., *Functional calculi for second order elliptic partial differential operators with bounded measurable coefficients.* J. Geom. Anal. *6* (1996), 181–205.

[DR96] DUONG, X. T., and ROBINSON, D. W., *Semigroup kernels, Poissons bounds, and holomorphic functional calculus.* J. Func. Anal. *142* (1996), 89–128.

[DS97]   DUONG, X. T., and SIMONETT, G., H^∞-calculus for elliptic operators with nonsmooth coefficients. Differential Integral Equations *10* (1997), 201–217.

[DY01]   DUONG, X. T., and YAN, L. X., *Bounded holomorphic functional calculus for non-divergence form differential operators*. Differential Integral Equations *15* (2002), 709–730.

[Fro01]   FRÖHLICH, A., *Stokes- und Navier-Stokes-Gleichungen in gewichteten Funktionenräumen*. Dissertation, TU Darmstadt, 2001.

[GR85]   GARCIA-CUERVA, J., and RUBIO DE FRANCIA, J. L., *Weighted Norm Inequalities and Related Topics*, North-Holland, 1985.

[Gui02]   GUIDETTI, D., *General linear boundary value problems for elliptic operators with VMO coefficients*. Math. Nachr. *237* (2002), 62–88.

[HHH01]   HALLER, R., HECK, H., and HIEBER, M., *Muckenhoupt weights and maximal regularity*. Archiv Math. to appear.

[MP98]   MAUGERI, A., and PALAGACHEV, D. K., *Boundary value problems with an oblique derivative for uniformly elliptic operators with discontinuous coefficients*. Forum Math. *10* (1998), 393–405.

[PS93]   PRÜSS, J., and SOHR, H., *Imaginary powers of elliptic second order differential operators in L^p-spaces*. Hiroshima Math. J. *23* (1993), 161–192.

[Sar75]   SARASON, D., *Functions of vanishing mean oscillation*. Trans. Amer. Math. Soc. *207* (1975), 391–405.

[See67]   SEELEY, R., *Complex powers of an elliptic operator*. In: Singular Integrals, Proc. Sympos. Pure Math. vol. *10*, American Mathematical Society 1967, 288–307.

[Ste70]   STEIN, E. M., *Singular Integrals and Differentiability Properties of Functions*. Princeton University Press, Princeton, 1970.

[Ste93]   STEIN, E. M., *Harmonic Analysis: Real-Variable Methods, Orthogonality and Oscillatory Integrals*. Princeton University Press, Princeton, 1993.

[SW71]   STEIN, E. M., and WEISS, G., *Introduction to Fourier Analysis on Euclidean Spaces*. Princeton University Press, Princeton, 1971.

[Tai02]   TAIRA, K., Singular integrals, Feller semigroups and Markov processes. Preprint 2002.

[Tor86]   TORCHINSKY, A., *Real-variable Methods in Harmonic Analysis*. Academic Press, New York, 1986.

[Wei01]   WEIS, L., *Operator-valued Fourier multiplier theorems and maximal L_p-regularity*, Math. Ann. *319* (2001), 735–758.

[Wei01a]   WEIS, L., *A new approach to maximal L_p-regularity*. In: Evolution Equ. and Appl. Physical Life Sciences, G. Lumer, L. Weis (eds.), Lect. Notes in Pure Appl. Math. vol. *215* Marcel Dekker, New York 2001, 195–214.

Horst Heck and Matthias Hieber
Technische Universität Darmstadt
Fachbereich Mathematik
Schlossgartenstr. 7
D-64289 Darmstadt
Germany
e-mail: heck@mathematik.tu-darmstadt.de
hieber@mathematik.tu-darmstadt.de

To access this journal online:
http://www.birkhauser.ch

J.evol.equ. 3 (2003) 361 – 373
1424–3199/03/020361 – 13
DOI 10.1007/s00028-003-0106-y
© Birkhäuser Verlag, Basel, 2003

**Journal of Evolution
Equations**

Linearized stability for nonlinear evolution equations

WOLFGANG M. RUESS

To the memory of Philippe Bénilan

Abstract. We present a general principle of linearized stability at an equilibrium point for the Cauchy problem $\dot{u}(t) + Au(t) \ni 0, t \geq 0, u(0) = u_0$, for an ω-accretive, possibly multivalued, operator $A \subset X \times X$ in a Banach space X, that has a linear 'resolvent-derivative' $\tilde{A} \subset X \times X$. The result is applied to derive linearized stability results for the case of $A = (B + G)$ under 'minimal' differentiability assumptions on the operators $B \subset X \times X$ and $G : cl\, D(B) \to X$ at the equilibrium point, as well as for partial differential delay equations.

1. Introduction

Our objective is to study principles of linearized stability for several classes of nonlinear evolution equations. The following is the prototype problem: Assume that $B \subset X \times X$ is a (generally, multivalued) operator in a Banach space X such that $(B + \omega I)$ is accretive for some $\omega \in \mathbb{R}$, and that $F : cl\, D(B) \to X$ is locally Lipschitz, and consider the associated initial value problem

$$\begin{cases} \dot{u}(t) + Bu(t) \ni F(u(t)), & t \geq 0 \\ u(0) = x \in cl\, D(B). \end{cases} \tag{EE}$$

Further, assume that equation (EE) has an equilibrium solution $u(t) \equiv x_e \in D(B)$, and that both B and F have 'Fréchet derivatives' $\tilde{B} \subset X \times X$ linear and $\tilde{F} \in B(X)$, respectively, at x_e. Under which conditions on B, $\tilde{B}$, F and $\tilde{F}$ is it true that (EE) is locally exponentially stable at x_e, provided the corresponding linearized equation

$$\begin{cases} \dot{v}(t) + \tilde{B}v(t) \ni \tilde{F}(v(t)), & t \geq 0 \\ v(0) = y \in cl\, D(\tilde{B}) \end{cases} \tag{EE}_{\text{lin}}$$

is exponentially stable?

Various authors, cf. [3, 6, 9, 10], have studied this problem for the semilinear case, where $B = \tilde{B}$ is a (single-valued) densely defined linear operator such that $-B$ generates

---

Mathematics Subject Classification (2000): 47J35, 47H06, 35R10, 34K20.

Key words: Accretive operators, nonlinear evolution equations, linearized stability, partial differential delay equations.

a C_0-semigroup of bounded linear operators on X, and the nonlinearity F is supposed to (be defined on all of X, and to) be Fréchet-differentiable on X, and such that the map $\{x \mapsto F'[x]\}$ from X to $B(X)$ ($=$ bounded linear operators from X to X) is Lipschitz continuous on bounded subsets of X ([3, Cor. 2.2]).

In this paper, we shall prove a principle of linearized stability for the nonlinear Cauchy problem

$$\begin{cases} \dot{u}(t) + Au(t) \ni 0, \quad t \geq 0 \\ u(0) = x \in cl\, D(A) \end{cases} \tag{CP}$$

for a general, possibly multivalued, ω-accretive operator $A \subset X \times X$: if A has a *resolvent-differential* $\tilde{A}$ at an equilibrium point $x_e \in D(A)$, i.e., there exists an operator $\tilde{A} \subset X \times X$ that is linear (or, more generally, simply with $(0, 0) \in \tilde{A}$) such that the Yosida-approximations $\tilde{A}_\lambda$ are $D(A)-$Fréchet-differentials at x_e of the A_λ's, uniformly over $\lambda > 0$ small enough, then equation (CP) is locally exponentially stable at x_e, provided the operator $(\tilde{A} - \tilde{\omega}I)$ is accretive for some $\tilde{\omega} > 0$ (Theorem 2.1).

This general linearization principle will allow for considerable extensions of the linearized stability results of [3, 6, 9, 10] for equations (CP) and (EE) in various directions, with, at the same time, simpler proofs. With regard to equation (CP), in contrast to [3, Thm. 2.1], we shall need neither single-valuedness of A or $\tilde{A}$, nor $\omega - m$-accretiveness, nor any differentiability of (the resolvents of) A away from the equilibrium – as implied by conditions (H1) and (H2) of [3, Thm. 2.1]. With regard to equation (EE), in contrast to the results of [3, 6, 9, 10], a.) the operator $B \subset X \times X$, once again, need only be a resolvent-differentiable ω-accretive operator, and thus neither linear, nor densely defined, nor single-valued, and b.) the nonlinear perturbation F will be allowed to be defined only on $cl\, D(B)$, Lipschitz on bounded sets, and need be $D(B)$-Fréchet-differentiable at the equilibrium x_e only (Theorem 3.1 and Corollary 3.2).

As a further illustration of Theorem 2.1, linearized stability results for partial differential delay equations will also be extended considerably (Theorem 4.1). Further applications to equations of age-dependent population dynamics, as well as to the differentiation of nonlinear semigroups, will be considered elsewhere.

Notation and terminology

Throughout the paper, X will denote a real Banach space, and $B(X)$ the space of bounded linear operators from X to X. Given a subset D of X, $cl\, D$ will denote its closure in X. Recall that a subset $C \subset X \times X$ is said to be *accretive* in X if for each $\lambda > 0$ and each pair $[x_i, y_i] \in C, i \in \{1, 2\}$, we have

$$\|(x_1 + \lambda y_1) - (x_2 + \lambda y_2)\| \geq \|x_1 - x_2\|,$$

and *m-accretive* in X if, in addition, $R(I + \lambda C) = X$ for all $\lambda > 0$. If ω is any real number, an operator $C \subset X \times X$ for which $(C + \omega I)$ is accretive will be called ω-accretive, and ω-m-accretive if, in addition, $R(I + \lambda C) = X$ for all $\lambda > 0$ with $\lambda\omega < 1$. If $C \subset X \times X$ is ω-accretive, then, for any $\lambda > 0$ with $\lambda\omega < 1$, $J_\lambda^C = (I + \lambda C)^{-1}$ denotes the resolvent of C. For all these notions and the general theory of accretive sets and evolution equations, the reader is referred to [1, 4].

2. Linearized stability for the Cauchy problem (CP)

In this section, we prove the following general linearization principle for the Cauchy problem (CP).

THEOREM 2.1. *Let $A \subset X \times X$ be ω-accretive for some $\omega \geq 0$, satisfying the range condition $R(I + \lambda A) \supset cl\, D(A)$ for all $\lambda > 0, \lambda\omega < 1$, and let $x_e \in D(A)$ such that $0 \in Ax_e$. Assume that there exists an operator $\tilde{A} \subset X \times X$, with $(0, 0) \in \tilde{A}$, such that $R(I + \lambda\tilde{A}) \supset (D(A) - x_e)$ for all $0 < \lambda \leq \lambda_0$, and such that $(\tilde{A} - \tilde{\omega}I)$ is accretive for some $\tilde{\omega} > 0$. Furthermore, assume that the following resolvent-differentiability condition holds:*

$$\begin{cases} \textit{For every } \epsilon > 0, \quad \textit{there exist } \delta > 0, \ \lambda_1 > 0, \ \textit{and a function} \\ \eta : (0, \lambda_1) \times D(A) \to \mathbb{R}^+, \ \textit{such that, if } x \in D(A), \ \textit{and } \|x - x_e\| < \delta, \\ \textit{then } \|J_\lambda^A x - x_e - J_\lambda^{\tilde{A}}(x - x_e)\| \leq \epsilon\lambda\|x - x_e\| + \lambda\eta(\lambda, x) \, \textit{for all} \\ 0 < \lambda < \lambda_1, \ \textit{where the function } \eta \textit{ is bounded on bounded sets, continuous} \\ \textit{in the second variable, and such that } \lim_{(\lambda, y) \to (0, y_0)} \eta(\lambda, y) = 0. \end{cases} \quad \text{(RD)}$$

Then the initial value problem (CP) is locally exponentially stable at the equilibrium x_e. More precisely: given any $0 < \omega_1 < \tilde{\omega}$, there exists $\delta > 0$ such that

$$\|S(t)x - x_e\| \leq e^{-\omega_1 t}\|x - x_e\| \quad \textit{for all } t \geq 0,$$
$$\textit{and all } x \in cl\, D(A) \ \textit{with } \|x - x_e\| < \delta,$$

where $S(\cdot)x : \mathbb{R}^+ \to X$ is the unique mild solution to (CP) with initial value $x \in cl\, D(A)$, with $(S(t))_{t \geq 0}$ denoting the strongly continuous semigroup of operators generated by $-A$ on $cl\, D(A)$.

REMARK 2.2. 1. So far, the conclusion of Theorem 2.1 has been shown in [3, Thm. 2.1] under the assumptions that the operator $A : D(A) \subset X \to X$ be single-valued, $\omega - m$-accretive and protodifferentiable with conditions (H1) and (H2) of [3, Section 2] (for which we refer the reader to [3]). These conditions particularly imply that the resolvents of A are actually continuously Gâteaux-differentiable in an open neighbourhood of x_e, and that the derivatives $J_\lambda^{\partial A(x_e)}$ at x_e even are uniform (over $\lambda > 0$ small enough) Fréchet-derivatives of $J_\lambda{}^A$ in the classical sense: there exist $r > 0$, $\lambda_1 > 0$, and $M \geq 1$, such that, if $x \in X$, and $\|x - x_e\| < r$ then $\|J_\lambda{}^A x - x_e - J_\lambda{}^{\partial A(x_e)}(x - x_e)\| \leq \lambda M \|x - x_e\|^2$ for all $0 < \lambda < \lambda_1$. Also, the protodifferentials $\partial A(x)$ are supposed to exist for $x \in D(A)$, $\|x - x_e\| < r$, and to be single-valued linear $\omega - m$-accretive operators. In contrast to these restrictions, aside from the added generality in the operator $A \subset X \times X$, condition (RD) asks for a differentiability condition only at the equilibrium point x_e, with the additional features that the analog of the protodifferential $\partial A(x_e)$ of [3, Thm. 2.1] is allowed to be any, possibly even multivalued, operator $\tilde{A} \subset X \times X$, not necessarily generating a semigroup-much less $\omega - m$-accretive-whose only resemblance to linearity stems from the requirement $(0, 0) \in \tilde{A}$.

In contrast to the added generality, the subsequent proof of Theorem 2.1 turns out short, simple, and direct, while the proof of its analog takes Section 3 and part of Section 4 in [3, pp. 2856–2860]. Altogether, this indicates that, in the presence of ω-accretivity, the notion of resolvent-differentiability is more natural than that of protodifferentiability for general operators, as it imposes conditions on the resolvents rather than on the operator itself-as is usual in operator-semigroup theory.

2. A further indication in this direction is the comparison with the discrete analog of the iterates of an operator. Recall that a fixed point $x_e \in X$ of an $X-$self-map T, defined on a neighbourhood of x_e, is attractive provided T has a Fréchet-derivative at x_e with spectral radius less than one, cf. [11, Thm. 4.C, p. 159]. In [2, Thm. 2.1], this result has been refined to show that the conclusion of Theorem 2.1 above holds in case the semigroup $(S(t))_{t \geq 0}$ generated by $-A$ has linear Fréchet-derivatives $S'(t)[x_e]$ at x_e such that this linearized semigroup is exponentially stable. On the level of the generator A, the analog-as well as extension-of the assumptions of both of these results is the requirement of Theorem 2.1 that there exists a "somewhat linear" resolvent-differential $\tilde{A} \subset X \times X$ for A at the equilibrium with its resolvents satisfying $\|J_\lambda{}^{\tilde{A}} x\| \leq (1 + \lambda \tilde{\omega})^{-1} \|x\|$ for some $\tilde{\omega} > 0$, $0 < \lambda$ small enough, and all $x \in R(I + \lambda \tilde{A})$. (Notice that, in applications, what is known is the generator A, but not the semigroup $(S(t))_{t \geq 0}$). Thus, in terms of the generator A, Theorem 2.1 appears to achieve that degree of generality that is just sufficient for the result to hold.

3. As a particular special case, we mention that, in Theorem 2.1, $\tilde{A}$ could be a Hille-Yosida operator of type $(-\tilde{\omega})$. Any Hille-Yosida operator of type $\alpha \in \mathbb{R}$, i.e., a not necessarily densely-defined single-valued linear operator satisfying the Hille-Yosida conditions on its resolvent set and its resolvent, is $\alpha - m$-accretive for an equivalent norm on X, and condition (RD) is invariant under equivalent renormings.

REMARK 2.3. For the sake of completeness, we further note that the above notion of resolvent-differentiability also subsumes the 'traditional' notion of Fréchet-differentiability. More generally, the following holds: Assume that $A : D(A) \subset X \to X$ is (single-valued and) ω-accretive, with $R(I + \lambda A) \supset D(A)$ for $\lambda > 0$, $\lambda \omega < 1$, $x_0 \in D(A)$, and that A is

$D(A)$–Fréchet-differentiable at x_0, i.e., there exists $\tilde{A} \in B(X)$ such that

$$\begin{cases} \text{For every } \epsilon > 0, \text{ there exists } \delta > 0, \text{ such that, if } x \in D(A), \\ \text{and } \|x - x_0\| < \delta, \text{ then } \quad \|Ax - Ax_0 - \tilde{A}(x - x_0)\| \leq \epsilon \|x - x_0\|. \end{cases} \quad \text{(FD)}$$

Then A is resolvent-differentiable at x_0 : For every $\epsilon > 0$, there exist $\delta > 0, \lambda_1 > 0$, and $M \geq 1$, such that, if $x \in D(A)$, and $\|x - x_0\| < \delta$, then $\|J_\lambda{}^A x - J_\lambda{}^A x_0 - J_\lambda{}^{\tilde{A}}(x - x_0)\| \leq \epsilon \lambda \|x - x_0\| + \lambda M \|J_\lambda{}^A x_0 - x_0\|$ for all $0 < \lambda < \lambda_1$. If, in addition, A is $\omega - m$-accretive, then this approximation holds for all $x \in X$ with $\|x - x_0\| < \delta$. (The proof is straightforward, and will be omitted.)

Proof of Theorem 2.1. Throughout this proof, we let, for $\lambda > 0$ small enough, $J_\lambda :=$ $J_\lambda{}^A$, and $\tilde{J}_\lambda := J_\lambda{}^{\tilde{A}}$. For $\epsilon := \tilde{\omega} - \omega_1$, choose $\delta > 0, \lambda_1 > 0, \eta : (0, \lambda_1) \times D(A) \longrightarrow \mathbb{R}^+$ according to (RD), with $\lambda_1 \leq \lambda_0$, and such that $\lambda_1 \omega \leq \frac{1}{2}$.

Let $x \in cl\, D(A), \|x - x_e\| < \delta$. For $\beta > 0$ such that $\|x - x_e\| < \delta - \beta$, choose $T_0 > 0$ such that $\exp(2\omega T_0)(\delta - \beta) < \delta$. Given $0 < \lambda < \lambda_1$, let N_λ be the largest integer such that $N_\lambda \lambda \leq T_0$, and define

$$x_k^\lambda = \begin{cases} x, & k = 0 \\ J_\lambda{}^k x, & 1 \leq k \leq N_\lambda, \end{cases} \quad (2.1)$$

and $u_\lambda : [0, N_\lambda \lambda] \longrightarrow X$ by

$$u_\lambda(t) = \begin{cases} x, & t = 0 \\ x_k^\lambda, & t \in ((k-1)\lambda, k\lambda], \ 1 \leq k \leq N_\lambda. \end{cases} \quad (2.2)$$

Note that

$$u_\lambda(\tau) \to S(\tau)x \text{ as } \lambda \to 0^+ \text{ uniformly over } [0, t] \text{ for any } 0 < t < T_0, \quad (2.3)$$

by the definition of a mild solution to (CP), c.f. [1]. By the choice of $0 < \lambda < \lambda_1$ and $T_0 > 0$,

$$\|J_\lambda{}^k x - x_e\| \leq (1 - \lambda\omega)^{-k}\|x - x_e\|$$

$$\leq \exp\left(\frac{k\lambda\omega}{1 - \lambda\omega}\right)\|x - x_e\| \leq \exp(2\omega T_0)(\delta - \beta) < \delta$$

for all $1 \leq k \leq N_\lambda$. (Recall the inequality $(1 - \rho)^{-k} \leq \exp(\frac{k\rho}{1-\rho})$ for all $k \in \mathbb{N} \cup \{0\}$, $0 \leq \rho < 1$). Thus, (RD) applies, so that, if we let

$$z_k^\lambda := J_\lambda x_{k-1}^\lambda - x_e - \tilde{J}_\lambda(x_{k-1}^\lambda - x_e),$$

then

$$\|z_k^\lambda\| \leq \epsilon\lambda\|x_{k-1}^\lambda - x_e\| + \lambda\eta(\lambda, x_{k-1}^\lambda) \quad \text{for all } 1 \leq k \leq N_\lambda. \quad (2.4)$$

Next, note that

$$x_k^\lambda - x_e = [J_\lambda x_{k-1}^\lambda - x_e - \tilde{J}_\lambda(x_{k-1}^\lambda - x_e)]$$
$$+ \tilde{J}_\lambda(x_{k-1}^\lambda - x_e) = z_k^\lambda + \tilde{J}_\lambda(x_{k-1}^\lambda - x_e).$$

Thus, as $(\tilde{A} - \tilde{\omega}I)$ is accretive, and $(0,0) \in \tilde{A}$,

$$\|x_k^\lambda - x_e\| \le \|z_k^\lambda\| + \frac{1}{1 + \lambda\tilde{\omega}} \|x_{k-1}^\lambda - x_e\|$$

$$= \|z_k^\lambda\| + \left(1 - \frac{\lambda\tilde{\omega}}{1 + \lambda\tilde{\omega}}\right) \|x_{k-1}^\lambda - x_e\|.$$

Using (2.4), this implies

$$\|x_k^\lambda - x_e\| - \|x_{k-1}^\lambda - x_e\|$$

$$\le \left(\epsilon - \frac{\tilde{\omega}}{1 + \lambda\tilde{\omega}}\right) \lambda \|x_{k-1}^\lambda - x_e\| + \lambda\eta(\lambda, x_{k-1}^\lambda). \tag{2.5}$$

Let $0 < s < t < T_0$ be arbitrary, and let $0 < \lambda < \min\{\lambda_1, s, \frac{1}{2}(t - s), T_0 - t\}$, and choose $2 \le k_s \le k_t \le N_\lambda$ such that $s \in ((k_s - 1)\lambda, k_s\lambda]$, and $t \in ((k_t - 1)\lambda, k_t\lambda]$. Summing (2.5) from $k = (k_s + 1)$ to $k = k_t$, and invoking (2.2), we arrive at

$$\|u_\lambda(t) - x_e\| - \|u_\lambda(s) - x_e\|$$

$$\le \left(\epsilon - \frac{\tilde{\omega}}{1 + \lambda\tilde{\omega}}\right) \sum_{k_s}^{k_t - 1} \lambda \|x_k^\lambda - x_e\| + \sum_{k_s}^{k_t - 1} \lambda\eta(\lambda, x_k^\lambda)$$

$$= \left(\epsilon - \frac{\tilde{\omega}}{1 + \lambda\tilde{\omega}}\right) \sum_{k_s}^{k_t - 1} \int_{(k-1)\lambda}^{k\lambda} \|u_\lambda(\tau) - x_e\| \, d\tau + \sum_{k_s}^{k_t - 1} \int_{(k-1)\lambda}^{k\lambda} \eta(\lambda, u_\lambda(\tau)) \, d\tau$$

$$\le \left(\epsilon - \frac{\tilde{\omega}}{1 + \lambda\tilde{\omega}}\right) \int_{(k_s-1)\lambda}^{(k_t-1)\lambda} \|u_\lambda(\tau) - x_e\| \, d\tau + \int_{(k_s-1)\lambda}^{t} \eta(\lambda, u_\lambda(\tau)) \, d\tau. \tag{2.6}$$

Letting $\lambda \to 0^+$, and invoking (2.3), the properties of the function η, and Lebesgue's Dominated Convergence Theorem, (2.6) implies that

$$\|S(t)x - x_e\| - \|S(s)x - x_e\| \le (\epsilon - \tilde{\omega}) \int_s^t \|S(\tau)x - x_e\| \, d\tau$$

$$= -\omega_1 \int_s^t \|S(\tau)x - x_e\| \, d\tau$$

for all $0 < s < t < T_0$. By continuity on either side of this inequality, this holds for all $0 \le s \le t \le T_0$. We conclude that

$$\|S(t)x - x_e\| \le e^{-\omega_1 t} \|x - x_e\| \qquad \text{for all} \quad 0 \le t \le T_0.$$

We let $T_1 := \sup\{t > 0 \mid \|S(s)x - x_e\| \le e^{-\omega_1 s}\|x - x_e\|$ for all $0 \le s \le t\}$, and claim that $T_1 = \infty$. Assuming, on the contrary, that $T_1 < \infty$, there exists $T_1 < t_1 \le T_1 + T_0$ such that $\|S(t_1)x - x_e\| > e^{-\omega_1 t_1}\|x - x_e\|$. However, from (2.7),

$$\|S(t_1)x - x_e\| = \|S(t_1 - T_1)S(T_1)x - x_e\| \le e^{-\omega_1(t_1 - T_1)}\|S(T_1)x - x_e\|$$
$$\le e^{-\omega_1 t_1}\|x - x_e\|,$$

since $\|S(T_1)x - x_e\| \le e^{-\omega_1 T_1}\|x - x_e\| < (\delta - \beta)$, and $t_1 - T_1 \le T_0$. This contradiction serves to complete the proof.

3. Linearized stability for the evolution equation (EE)

The general linearized stability principle for the Cauchy problem (CP) of Theorem 2.1 allows for considerable extensions of the linearized stability results of [3, 6, 9] for the evolution equation (EE). We start from the following assumptions:

(A1) $B \subset X \times X$ is m-accretive, $F : cl\, D(B) \to X$ is continuous such that $(B - F)$ is $\omega - m$-accretive for some $\omega \in \mathbb{R}$, and $x_e \in D(B)$ is such that $Fx_e \in Bx_e$.

(A2) There exists a linear operator $\tilde{B} \subset X \times X$ that is $\alpha - m-$accretive for some $\alpha \in \mathbb{R}$, such that the following resolvent-differentiability holds:

$$\left\{ \begin{array}{l} \text{For every } \epsilon > 0, \quad \text{there exist } \delta > 0, \lambda_0 > 0, \text{ and a function} \\ \eta_1 : (0, \lambda_0) \times X \to \mathbb{R}^+, \text{ such that, if } x \in X, \quad \text{and } \|x - x_e\| < \delta, \\ \text{then} \quad \|J_\lambda{}^B x - J_\lambda{}^B x_e - J_\lambda{}^{\tilde{B}}(x - x_e)\| \le \epsilon\lambda\|x - x_e\| + \lambda\eta_1(\lambda, x) \quad (\text{RD}_X) \\ \text{for all } 0 < \lambda < \lambda_0, \text{ with the function } \eta_1 \text{ of the same type as} \\ \text{the function } \eta \text{ in condition (RD) of Theorem 2.1.} \end{array} \right.$$

(A3) There exists $\tilde{F} \in B(X)$ that is a $D(B)-$Fréchet-derivative of F at x_e, i.e., given any $\epsilon > 0$, there exists $\delta > 0$ such that, if $x \in D(B)$, and $\|x - x_e\| < \delta$, then

$$\|Fx - Fx_e - \tilde{F}(x - x_e)\| \le \epsilon\|x - x_e\|.$$

THEOREM 3.1. *Under the assumptions* (A1)–(A3) *above, if there exists* $\tilde{\omega} > 0$ *such that the linearized operator* $(\tilde{B} - \tilde{F} - \tilde{\omega}I)$ *is accretive, then the initial value problem* (EE) *is locally exponentially stable at the equilibrium* x_e. *More precisely: Given any* $0 < \omega_1 < \tilde{\omega}$, *there exists* $\delta > 0$ *such that*

$$\|u_x(t) - x_e\| \le e^{-\omega_1 t}\|x - x_e\| \quad \text{for all } t \ge 0, \text{ and all}$$
$$x \in cl\, D(B) \text{ with } \|x - x_e\| < \delta,$$

where $u_x : \mathbb{R}^+ \to X$ *denotes the unique mild solution to* (EE) *with initial value* $x \in cl\, D(B)$.

Notice that, with regard to assumption (A1), if $B \subset X \times X$ is m-accretive, and $F : cl\, D(B) \to X$ is continuous, then $(B - F)$ is $\omega - m$-accretive for some $\omega \in \mathbb{R}$, provided it is ω-accretive, cf. [1, Ch. 16]. As a particular case, we can thus note the following linearized stability result for equation (EE) under the assumption that the operator $F : cl\, D(B) \to X$ is Lipschitz-continuous on bounded sets.

COROLLARY 3.2. *Let $B \subset X \times X$ be m-accretive, with $rcl\, D(B) \subset cl\, D(B)$ for all $0 \leq r \leq 1$, and let $F : cl\, D(B) \to X$ be Lipschitz-continuous on bounded sets, and assume that there exists $x_e \in D(B)$ such that $F x_e \in B x_e$, and such that (A3) is fulfilled. Moreover, assume that there exists $\tilde{B} \subset X \times X$ linear such that (A2) is fulfilled. Then we have: If there exists $\tilde{\omega} > 0$ such that the linearized operator $(\tilde{B} - \tilde{F} - \tilde{\omega}I)$ is accretive, then the evolution equation (EE) is locally exponentially stable at the equilibrium x_e. More precisely: Given any $0 < \omega_1 < \tilde{\omega}$, there exists $\delta > 0$ such that, if $x \in cl\, D(B)$, and $\|x - x_e\| < \delta$, then there exists a (unique) global mild solution $u_x : \mathbb{R}^+ \to X$ to (EE) such that $\|u_x(t) - x_e\| \leq e^{-\omega_1 t}\|x - x_e\|$ for all $t \geq 0$.*

REMARK 3.3. In the literature, equation (EE) has been considered in the following semilinear case: in [3, Cor. 2.2], [6, Thm. 11 (2)], [9, Thm. 11.22], and [10, Prop. 4.17], the operator $B : D(B) \subset X \to X$ is supposed to be (single-valued and) the infinitesimal generator of a C_0−semigroup of bounded linear operators on X, and $F : X \to X$ is supposed to be (globally defined and) Fréchet-differentiable on X, with the map $\{x \mapsto F'[x]\}$ Lipschitz on bounded sets. Obviously, Theorem 3.1 and Corollary 3.2 considerably widen the range of applicability:

1. Most importantly, the operator F need only be defined on $cl\, D(B)$, and need be $D(B)$-differentiable at the equilibrium x_e only. This is of interest in, for instance, equations of population dynamics, where F may only be defined on 'thin' subsets of X, such as nonnegative functions, or functions with values restricted to some interval.
2. According to Remark 2.3 above, the operator B in Theorem 3.1 and Corollary 3.2 can be any single-valued m-accretive operator in X that is $D(B)$−Fréchet-differentiable at the equilibrium in the sense of (FD) of Remark 2.3.
3. Obviously, Theorem 3.1 and Corollary 3.2 hold in particular if $B = \tilde{B} \subset X \times X$ is any linear m-accretive operator in X; it need neither be single-valued, nor densely-defined. In particular, it can be a Hille-Yosida operator, compare Remark 2.2.3. above.

Proof of Theorem 3.1. We show that property (RD) of Theorem 2.1 is fulfilled for the pair of operators $A = (B - F)$ and $\tilde{A} = (\tilde{B} - \tilde{F})$. To this end, we first observe that, for $x \in X$, and $\lambda > 0$ small enough,

$$J_\lambda{}^{B-F}x = J_\lambda{}^B(x + \lambda F(J_\lambda{}^{B-F}x)), \text{ and } x_e = J_\lambda{}^B(x_e + \lambda F x_e).$$

This yields:

$$
\begin{aligned}
J_\lambda{}^{B-F}x - x_e - &J_\lambda{}^{\tilde{B}-\tilde{F}}(x - x_e) \\
&= J_\lambda{}^{B}(x + \lambda F(J_\lambda{}^{B-F}x)) - J_\lambda{}^{B}x_e - J_\lambda{}^{\tilde{B}}(x - x_e + \lambda F(J_\lambda{}^{B-F}x)) \qquad (\mathrm{I}_1) \\
&\quad - [J_\lambda{}^{B}(x_e + \lambda Fx_e) - J_\lambda{}^{B}x_e - J_\lambda{}^{\tilde{B}}(\lambda Fx_e)] \qquad (\mathrm{I}_2) \\
&\quad + \lambda J_\lambda{}^{\tilde{B}}(F(J_\lambda{}^{B-F}x) - Fx_e - \tilde{F}(J_\lambda{}^{B-F}x - x_e)) \qquad (\mathrm{I}_3) \\
&\quad + \lambda J_\lambda{}^{\tilde{B}}\tilde{F}(J_\lambda{}^{B-F}x - x_e - J_\lambda{}^{\tilde{B}-\tilde{F}}(x - x_e)).
\end{aligned}
$$

Consequently,

$$
(I - \lambda J_\lambda{}^{\tilde{B}}\tilde{F})(J_\lambda{}^{B-F}x - x_e - J_\lambda{}^{\tilde{B}-\tilde{F}}(x - x_e)) = (\mathrm{I}_1) + (\mathrm{I}_2) + (\mathrm{I}_3).
$$

Thus, for $\lambda > 0$ small enough,

$$
\begin{aligned}
\| J_\lambda{}^{B-F}x &- x_e - J_\lambda{}^{\tilde{B}-\tilde{F}}(x - x_e)\| \\
&\leq \frac{1}{(1 - \lambda\|J_\lambda{}^{\tilde{B}}\tilde{F}\|)}\|(\mathrm{I}_1) + (\mathrm{I}_2) + (\mathrm{I}_3)\|
\end{aligned} \qquad (3.1)
$$

for all $x \in X$. Starting from (3.1), and invoking assumptions (A2) and (A3) of Theorem 3.1, it is easy to deduce that assumption (RD) of Theorem 2.1 is fulfilled for the operators $A = (B - F)$ and $\tilde{A} = (\tilde{B} - \tilde{F})$, with the function $\eta : (0, \lambda_0) \times D(A) \to \mathbb{R}^+$ given by $M[\eta_1(\lambda, x + \lambda F(J_\lambda{}^{B-F}x)) + \eta_1(\lambda, x_e + \lambda Fx_e) + \lambda\|Fx_e\|]$, for $0 < \lambda \leq \lambda_0$ small enough, $x \in D(A) = D(B)$, and some $M > 0$. This completes the proof.

Proof of Corollary 3.2. Under the assumptions of Corollary 3.2, we let $r := \|x_e\| + 1$, and define $F_r : cl\, D(B) \to X$ by

$$
F_r(x) = \begin{cases} Fx & \text{for } \|x\| \leq r \\ F(r\frac{x}{\|x\|}) & \text{for } \|x\| \geq r. \end{cases}
$$

Then, F_r is Lipschitz on $cl\, D(B)$, and, by [5, Thm. II], the operator $(B - F_r)$ is $\omega - m$-accretive for $\omega = $ Lipschitz-constant of F_r. We conclude from Theorem 3.1 that, given any $0 < \omega_1 < \tilde{\omega}$, there exists $\delta > 0$ such that, if $x \in cl\, D(B)$, and $\|x - x_e\| < \delta$, then there exists a (unique) global mild solution $u_x : \mathbb{R}^+ \to X$ to (EE), with F replaced by F_r, such that $\|u_x(t) - x_e\| \leq e^{-\omega_1 t}\|x - x_e\|$, for all $t \geq 0$. Thus, if, in addition, $\delta < 1$, then $\|u_x(t)\| < r$ and thus $F_r(u_x(t)) = F(u_x(t))$ for all $t \geq 0$, so that u_x is actually a mild solution of equation (EE). This completes the proof.

4. Linearized stability for partial differential delay equations

In this section, we apply Theorem 2.1 to partial functional differential equations with delay of the following general form:

$$\begin{cases} \dot{x}(t) + Bx(t) \ni F(x_t), & t \geq 0 \\ x_{|I} = \varphi \in \hat{E}. \end{cases} \tag{FDE}$$

$B \subset X \times X$ is an α-accretive operator in X, for some $\alpha \in \mathbb{R}$. As usual, for given $I = [-r, 0]$, $r > 0$ (finite delay), or $I = \mathbb{R}^-$ (infinite delay), and $t \geq 0$, $x_t : I \to X$ is the history of x up to $t : x_t(s) = x(t + s)$, $s \in I$, and $\varphi : I \to X$ is a given initial history out of a space E of functions from I to X. Moreover, F is a given history-responsive operator with domain $\hat{E} \subset E$ and range in X, Lipschitz on the (possibly 'thin') subset $\hat{E}$ of E.

We shall work with initial-history spaces of continuous functions. In the finite delay case $I = [-r, 0]$, $r > 0$, we take the initial-history space to be $E = C([-r, 0]; X)$, endowed with the sup-norm. In the infinite delay case $I = (-\infty, 0]$, E will be chosen to be a weighted sup-norm space of the type $E_v = \{\varphi \in C(\mathbb{R}^-, X) : v\varphi \in BUC(\mathbb{R}^-, X)\}$, with norm $\|\varphi\|_v := \sup\{v(s)\|\varphi(s)\| : s \in \mathbb{R}^-\}$, where the (weight-) function $v : \mathbb{R}^- \to (0, 1]$ has the following properties:

(v1) v is continuous, nondecreasing, and $v(0) = 1$;
(v2) There exists a constant $M_v \geq 0$ such that $|\frac{v(s+u)}{v(s)} - 1| \leq M_v|u|$ for all $s, u \leq 0$.

Typical such weight functions are $v(s) \equiv 1$ (with, in this case, $E_v = BUC(\mathbb{R}^-, X)$ with sup-norm), $v(s) = e^{\mu s}$, or $v(s) = (1 + |s|)^{-\mu}$, $\mu \geq 0$ (spaces of 'fading memory type'). (The Banach spaces E_v are sometimes called UC_g- spaces, $v = 1/g$, and have been considered by various authors.)

We first recall the solution theory for (FDE) from [7] under the following assumptions:

(B1) $B \subset X \times X$ is $\alpha-$accretive for some $\alpha \in \mathbb{R}$. $\hat{X} \subset X$ and $\hat{E} \subset E$ are closed subsets of X and E, respectively, such that

(1) $F : \hat{E} \to X$ is Lipschitz-continuous, with Lipschitz constant $M > 0$.
(2) For $x \in \hat{X}$, $\psi \in \hat{E}$, and $\lambda > 0$ with $\lambda\gamma < 1$, with $\gamma := \max\{0, M + \alpha\}$, if $\varphi_x \in E$ is the solution to

$$\varphi - \lambda\varphi' = \psi, \qquad \varphi(0) = x,$$

then $\varphi_x \in \hat{E}$.

(B2) If $\psi \in \hat{E}$ and $\lambda > 0$ with $\lambda\gamma < 1$, then

$$(\psi(0) + \lambda F(\varphi_x)) \in (I + \lambda B)(D(B) \cap \hat{X})$$

for each $x \in \hat{X}$.

We associate with (FDE) the operator A in E defined by

$$\begin{cases} D(A) = \{\varphi \in \hat{E} \mid \varphi' \in E, \varphi(0) \in D(B), \varphi'(0) \in F(\varphi) - B\varphi(0)\} \\ A\varphi := -\varphi' , \varphi \in D(A). \end{cases}$$

Then we have from [7, Thm. 2.1 and Remk. 2.3, 1.]:

(S) $-A$ generates a strongly continuous semigroup $(S(t))_{t\geq0}$ on $cl\, D(A) \subset \hat{E}$ such that, if $\varphi \in cl\, D(A)$, and $x_\varphi : I \cup \mathbb{R}^+ \to X$ is defined by

$$x_\varphi(t) = \begin{cases} \varphi(t) & t \in I \\ (S(t)\varphi)(0) & t \geq 0, \end{cases}$$

then x_φ is the unique (global) mild solution x to (FDE) with the property that $x_t \in \hat{E}$ for all $t \geq 0$.

In order to formulate the linearization principle for (FDE), we start from the following additional assumptions.

(B3) There exists an equilibrium solution $\varphi_e \in D(A)$ of the solution semigroup $(S(t))_{t\geq0}$ for (FDE) such that $x_e := \varphi_e(0) \in R(I + \lambda B)$ for all $\lambda > 0$ small enough. Moreoever, there exist a linear and $\beta - m$-accretive operator $\tilde{B} \subset X \times X$, some $\beta \in \mathbb{R}$, and a bounded linear operator $\tilde{F} : E \to X$ with the following properties:

$$(1) \begin{cases} \text{For every } \epsilon > 0, \quad \text{there exist} \quad \delta > 0, \ \lambda_0 > 0, \ \text{and a function} \\ \eta_1 : (0, \lambda_0) \times \bigcup \{R(I + \lambda B) \mid 0 < \lambda \leq \lambda_0\} \to \mathbb{R}^+, \ \text{such that} \\ \| J_\lambda{}^B x - J_\lambda{}^B x_e - J_\lambda{}^{\tilde{B}}(x - x_e)\| \leq \epsilon\lambda\|x - x_e\| + \lambda\eta_1(\lambda, x) \ \text{for all} \\ 0 < \lambda < \lambda_0, \ \text{and all } x \in R(I + \lambda B) \text{ with } \|x - x_e\| < \delta, \ \text{with the function} \\ \eta_1 \text{ of the same type as the function } \eta \text{ in condition (RD) of Theorem 2.1.} \end{cases}$$

(2) $\tilde{F}$ is a $D(A)$−Fréchet-derivative of F at φ_e, i.e., given any $\epsilon > 0$, there exists $\delta > 0$ such that, if $\varphi \in D(A)$, and $\|\varphi - \varphi_e\| < \delta$, then

$$\|F\varphi - F\varphi_e - \tilde{F}(\varphi - \varphi_e)\| \leq \epsilon\|\varphi - \varphi_e\|.$$

With assumption (B3) in place, we consider in E the solution operator $\tilde{A}$

$$\begin{cases} D(\tilde{A}) = \{\varphi \in E \mid \varphi' \in E, \varphi(0) \in D(\tilde{B}), \varphi'(0) \in \tilde{F}(\varphi) - \tilde{B}\varphi(0)\} \\ \tilde{A}\varphi := -\varphi', \varphi \in D(\tilde{A}), \end{cases}$$

associated with (FDE) with F and B being replaced by $\tilde{F}$ and $\tilde{B}$, respectively.

THEOREM 4.1. *Under the assumptions* (B1)–(B3) *above, if there exists $\tilde{\omega} > 0$ such that the linearized operator $(\tilde{A} - \tilde{\omega}I) \subset E \times E$ is accretive, then the initial history problem*

(FDE) *is locally exponentially stable at the equilibrium* φ_e. *More precisely: Given any* $0 < \omega_1 < \tilde{\omega}$, *there exists* $\delta > 0$ *such that*

$$\|(x_\varphi)_t - \varphi_e\| \le e^{-\omega_1 t}\|\varphi - \varphi_e\| \quad \text{for all } t \ge 0,$$

$$\text{and all } \varphi \in cl\, D(A) \text{ with } \|\varphi - \varphi_e\| < \delta,$$

where $x_\varphi : \mathbb{R}^+ \to X$ *denotes the unique mild solution to* (FDE) *with initial history* $\varphi \in cl\, D(A)$.

REMARK 4.2. 1. Obviously, Theorem 4.1 holds in particular for $B = \tilde{B} \subset X \times X$ any linear $\alpha - m$-accretive operator in X, in particular for a Hille-Yosida operator. For the case of $B : D(B) \subset X \to X$ the infinitesimal generator of a C_0−semigroup of bounded linear operators on X it has been shown by different methods in ([8, Thm. 2.4]).

2. Notice that, if $B \subset X \times X$ is $\alpha - m$-accretive, and F is globally Lipschitz, then (B1) (2) and (B2) are fulfilled automatically for $\hat{X} = X$, and $\hat{E} = E$. The above more general 'localized' assumptions are tailored for population models where, typically, the history-response F is only defined on thin subsets of the initial history space, such as truncated cones of nonnegative functions, compare [7, Section 5], and [8, Section 4].

3. Theorem 4.1 has been proved in [3, Thm. 6.1] in the finite-delay case for $B = \tilde{B}$ a single-valued linear $\alpha - m$-accretive operator, and F a globally defined and continuously Fréchet-differentiable operator with locally Lipschitz-continuous Fréchet-differentials. Notice that, in Theorem 4.1, these assumptions on F are strongly reduced in various directions, most noteworthy to differentiability at the equilibrium only.

Proof of Theorem 4.1. According to result (S) above, it suffices to show that the assumptions of Theorem 2.1 are fulfilled for the pair of operators A and $\tilde{A}$. Throughout this proof, we let, for $\lambda > 0$ small enough, $J_\lambda := J_\lambda{}^A$ and $\tilde{J}_\lambda := J_\lambda{}^{\tilde{A}}$. The necessary range conditions follow from Step 1 of the proof of Theorem 2.1 and Remark 2.3.3 of [7]. It remains to prove condition (RD). To this end, we let $\varphi \in cl\, D(A)$, and $\lambda > 0$ small enough, and first observe that, by the definition of the operators A and $\tilde{A}$,

$$\|J_\lambda\varphi - \varphi_e - \tilde{J}_\lambda(\varphi - \varphi_e)\| = \|(J_\lambda\varphi)(0) - x_e - (\tilde{J}_\lambda(\varphi - \varphi_e))(0)\|. \tag{4.1}$$

Then we recall from the proof of Theorem 2.1 of [7] that

$$J_\lambda\varphi(0) = J_\lambda{}^B(\varphi(0) + \lambda F(J_\lambda\varphi)), \text{ and } x_e = \varphi_e(0) = J_\lambda{}^B(x_e + \lambda F\varphi_e). \tag{4.2}$$

Next, observe that

$$J_\lambda{}^B(\varphi(0) + \lambda F(J_\lambda\varphi)) - J_\lambda{}^B(x_e + \lambda F\varphi_e) - J_\lambda{}^{\tilde{B}}((\varphi - \varphi_e)(0) + \lambda\tilde{F}(\tilde{J}_\lambda(\varphi - \varphi_e)))$$

$$= J_\lambda{}^B(\varphi(0) + \lambda F(J_\lambda\varphi)) - J_\lambda{}^B x_e - J_\lambda{}^{\tilde{B}}((\varphi - \varphi_e)(0) + \lambda F(J_\lambda\varphi))$$

$$- [J_\lambda{}^B(x_e + \lambda F\varphi_e) - J_\lambda{}^B x_e - J_\lambda{}^{\tilde{B}}(\lambda F\varphi_e)]$$

$$+ \lambda J_\lambda{}^{\tilde{B}}(F(J_\lambda\varphi) - F\varphi_e - \tilde{F}(J_\lambda\varphi - \varphi_e))$$
$$+ \lambda J_\lambda{}^{\tilde{B}}\tilde{F}(J_\lambda\varphi - \varphi_e - \tilde{J}_\lambda(\varphi - \varphi_e)).$$

Just as in the proof of Theorem 3.1 above, the desired conclusion follows from this equality, combined with (4.1), (4.2) and assumptions (B3) (1) and (2).

REFERENCES

[1] BÉNILAN, P., CRANDALL, M. G. and PAZY, A., *Evolution Equations Governed by Accretive Operators.* Monograph, in preparation.

[2] DESCH, W. and SCHAPPACHER, W., *Linearized stability for nonlinear semigroups.* In: Differential Equations in Banach Spaces (A. Favini and E. Obrecht, Eds.), 61–73, Lecture Notes in Math. vol. 1223, Springer, New York, 1986.

[3] KATO, N., *A principle of linearized stability for nonlinear evolution equations*, Trans. Amer. Math. Soc. *347* (1995), 2851–2868.

[4] MIYADERA, I., *Nonlinear Semigroups*, Transl. of Math. Monographs 109, Amer. Math. Soc. Providence, RI, 1992.

[5] PIERRE, M., *Perturbations localement Lipschitziennes et continues d'opérateurs m-accrétifs*, Proc. Amer. Math. Soc. *58* (1976), 124–128.

[6] RAUCH, J., *Stability of motion for semilinear equations.* In: Boundary Value Problems for Linear Evolution Partial Differential Equations (H.G. Garnir, Ed.), 319–349, D. Reidel Publ. Comp. Dordrecht, 1977.

[7] RUESS, W. M., *Existence and stability of solutions to partial functional differential equations with delay*, Adv. Diff. Eqns. *4* (1999), 843–876.

[8] RUESS, W. M. and SUMMERS, W. H., *Linearized stability for abstract differential equations with delay*, J. Math. Anal. Appl. *198* (1996), 310–336.

[9] SMOLLER, J., *Shock Waves and Reaction-Diffusion Equations*, Grundlehren Math. Wiss. 258, Springer, New York, 1983.

[10] WEBB, G.F., *Theory of nonlinear age dependent population dynamics*, Marcel-Dekker, New York, 1985.

[11] ZEIDLER, E., *Nonlinear Functional Analysis and Applications*, vol. I, Springer, New York, 1986.

Wolfgang M. Ruess
Fachbereich Mathematik
Universität Essen
D-45117 Essen
Germany
e-mail: w.ruess@uni-essen.de

To access this journal online:
http://www.birkhauser.ch

J.evol.equ. 3 (2003) 375 – 394
1424–3199/03/030375 – 20
DOI 10.1007/s00028-003-0099-5
© Birkhäuser Verlag, Basel, 2003

Journal of Evolution Equations

Nonlinear evolutions with Carathéodory forcing

DIETER BOTHE

Dedicated to the Remembrance of Philippe Bénilan

Abstract. Let A be an m-accretive operator in a real Banach space X and $f : J \times X \to X$ a function of Carathéodory type, where $J = [0, a] \subset \mathbb{R}$. This paper investigates the existence of mild solutions of the evolution system

$$u' + Au \ni f(t, u) \quad \text{on } J = [0, a],$$

satisfying additional time-dependent constraints $u(t) \in K(t)$ on J for a given tube $K(\cdot)$. Main emphasis is on existence results that are valid under minimal assumptions on f, K and X.

1. Introduction

Let X be a real Banach space and A be an m-accretive operator in X. Given $K : J = [0, a] \to 2^X \setminus \{\emptyset\}$ with closed values $K(t)$ such that $K_A(t) := K(t) \cap \overline{D(A)} \neq \emptyset$ on J and a forcing term $f : \mathrm{gr}(K_A) \to X$, we consider initial value problem

$$u' + Au \ni f(t, u) \quad \text{on } J, \quad u(0) = u_0. \tag{1}$$

In the sequel we always assume that f is at least *Carathéodory*, i.e. $f(\cdot, x)$ is strongly measurable for all x and $f(t, \cdot)$ is continuous for almost all t, such that

$$|f(t, x)| \leq c(t)(1 + |x|) \quad \text{on } \mathrm{gr}(K_A) \text{ with } c \in L^1(J). \tag{2}$$

Given any initial value $u_0 \in K_A(0)$, we look for a mild solution u of (1), by which we mean a continuous $u : J \to X$ such that u is the mild solution of the quasi-autonomous problem

$$u' + Au \ni w(t) \quad \text{on } J, \quad u(0) = u_0$$

with $w(\cdot) = f(\cdot, u(\cdot)) \in L^1(J; X)$. Notice that the time-dependent constraints $u(\cdot) \in K_A(\cdot)$ are incorporated into this problem by defining f on $\mathrm{gr}(K_A)$, only.

Mathematics Subject Classification 2000: Primary: 47J35, 35K90, Secondary: 34G25, 47H20.
Key words: Evolution equations, viability, flow invariance.

A tube $K(\cdot)$ is called *weakly positively invariant* (or *viable*) *for* $u' + Au \ni f(t, u)$ if

$$u' + Au \ni f(t, u) \quad \text{on } [t_0, a], \quad u(t_0) = u_0 \tag{3}$$

has a mild solution for every $t_0 \in [0, a)$ and every initial value $x_0 \in K_A(t_0)$. Consider this initial value problem (3) in the situation described above but with continuous right-hand side $f : \operatorname{gr}(K_A) \to X$. Then a necessary condition for weak positive invariance of $K(\cdot)$ can be obtained as follows. Suppose that (3) has mild solution u and let v be the mild solution of

$$v' + Av \ni f(t_0, u_0) \quad \text{on } [t_0, a], \quad v(t_0) = u_0.$$

By continuity of f and u it follows that

$$\frac{1}{h}|u(t_0 + h) - v(t_0 + h)| \le \frac{1}{h} \int_{t_0}^{t_0+h} |f(t, u(t)) - f(t_0, u_0)|dt \to 0 \quad \text{as } h \to 0+,$$

hence

$$\lim_{h \to 0+} h^{-1} \rho(S_{f(t_0, u_0)}(h)u_0, K_A(t_0 + h)) = 0,$$

where $S_z(\cdot)$ denotes the semigroup generated by $-A_z$ with $A_z x := Ax - z$ on $D(A_z) = D(A)$. Consequently,

$$f(t, x) \in T_K^A(t, x) \quad \text{for all } (t, x) \in \operatorname{gr}(K_A) \text{ with } t < a \tag{4}$$

is a necessary condition for weak positive invariance of $K(\cdot)$, where T_K^A is defined by

$$T_K^A(t, x) = \left\{ z \in X : \lim_{h \to 0+} h^{-1} \rho(S_z(h)x, K_A(t + h)) = 0 \right\}.$$

In the special case $A = 0$ this becomes

$$T_K(t, x) = \left\{ z \in X : \lim_{h \to 0+} h^{-1} \rho(x + hz, K(t + h)) = 0 \right\},$$

and if, in addition, $K(t) \equiv K$ holds then $T_K(t, x) = T_K(x)$ is the Bouligand contingent cone with respect to K at the point x; for the latter cone see e.g. §4 in [14].

Since all $K(t)$ are closed by assumption, it is also natural to assume that $\operatorname{gr}(K_A)$ is *closed from the left*, i.e.

$$(t_n) \subset J \text{ with } t_n \nearrow t \text{ and } x_n \in K_A(t_n) \text{ with } x_n \to x \text{ implies } x \in K_A(t);$$

notice that if there are mild solutions u_n with $u_n(t_n) = x_n$, then $K_A(t) \ni u_n(t) \to x$.

For continuous f it has been shown in [4] that the necessary "subtangential condition" (4) is sufficient for existence of (viable) solutions in several situations, in particular if $-A$

generates a compact semigroup. Since a fixed-point approach is not possible in presence of constraints, existence results are usually based on suitable approximate solutions. If f is only strongly measurable with respect to t, the construction of such approximate solutions is complicated by the fact that deviations in t have to be controlled carefully. In the present paper existence of mild solutions for Carathéodory forcing terms is established in general Banach spaces under the subtangential condition above in case of static constraints, i.e. if $K(t) \equiv K$. For time-dependent constraints the mild additional assumption of a separable Banach space is imposed.

Additional references to related work are given at the appropriate places later on. For possible application of such abstract results to reaction-diffusion systems with nonlinear diffusion of type $\Delta\varphi(u)$ see [4], [5] and the references given there.

Basic properties of m-accretive operators and evolution equations that are used here without further explanation can be found in [1] or [2].

2. Approximate solutions

In the sequel $u(\,\cdot\,; t_0, u_0, w)$ denotes the mild solution of

$$u' + Au \ni w(t) \quad \text{on } [t_0, a], \quad u(t_0) = u_0; \tag{5}$$

if $w \in L^1(J; X)$ (with $J = [0, a]$) then $u(\,\cdot\,; t_0, u_0, w)$ is short for $u(\,\cdot\,; t_0, u_0, w_{|[t_0,a]})$. With this notations, the semigroup property of solutions reads

$$u(t; \tau, u_0, w) = u(t; \overline{\tau}, u(\overline{\tau}; \tau, u_0, w), w) \quad \text{for all } 0 \le \tau \le \overline{\tau} \le t \le a.$$

If $t_0 = 0$ and u_0 is fixed, we simply write $u(\,\cdot\,; w)$ instead of $u(\,\cdot\,; t_0, u_0, w)$.

The subsequent result provides approximate solutions that are carefully adapted to situations when certain restrictions of f, such that t belongs to a closed subset J_0 of J, enjoy a better regularity. Observe also that, despite the fact that $K_A(t) \ne \emptyset$ on J is necessary for existence of mild solutions in $\mathrm{gr}\,(K)$, the latter assumption is not imposed explicitely. This will prove helpful for reduction to integrably bounded f later on.

Throughout this paper, let λ_1 denote the one-dimensional Lebesgue measure. For the proof of Lemma 2.2 below we need the following simple fact.

PROPOSITION 2.1. *Let* $J = [0, a] \subset \mathbb{R}$ *and* $c \in L^1(J)$ *with* $c \ge 0$. *Then there exists* $\hat{c} \in L^1(J)$ *having the following property: For every* $t_0 \in J$ *there is* $\delta = \delta(t_0) > 0$ *such that*

$$c(t_0) \le \hat{c}(t) \quad \text{for all } t \in (t_0 - \delta, t_0 + \delta) \cap J.$$

Proof. Let $B_m = \{t \in J : m - 1 \leq c(t) < m\}$ for $m \geq 1$ and $V_m \subset J$ open with $B_m \subset V_m$ and $\lambda_1(V_m) \leq \lambda_1(B_m) + 2^{-m}$. Define $\hat{c}$ by means of $\hat{c}(t) = \sum_{m \geq 1} m \chi_{V_m}(t)$, where χ_V denotes the characteristic function of V. Evidently

$$\int_J \hat{c}(t)dt \leq 1 + a + \sum_{m \geq 1}(m - 1)\lambda_1(B_m) \leq 1 + a + |c|_1 < \infty,$$

thus $\hat{c} \in L^1(J)$. Given $t_0 \in J$, there is $m \geq 1$ such that $t_0 \in B_m$. Hence $t_0 \in V_m$, and then $(t_0 - \delta, t_0 + \delta) \cap J \subset V_m$ for some $\delta > 0$. This yields $c(t_0) < m \leq \hat{c}(t)$ for all t from this set. $\square$

In the sequel, given $0 \leq c \in L^1(J)$, $\hat{c}$ always refers to the function provided by Proposition 1.

LEMMA 2.2. *Let A be m-accretive in a real Banach space X, $J = [0, a] \subset \mathbb{R}$ and $K : J \to 2^X$ with closed values be such that $K_A(0) \neq \emptyset$ and $G = gr(K_A)$ is closed from the left. Suppose that $f : G \to X$ satisfies (4) and $|f(t, x)| \leq c(t)$ on $gr(K_A)$ for some $c \in L^1(J)$. Let $J_0 \subset J$ be closed and $G_0 = [J_0 \times X] \cap G$. Then, given $u_0 \in K_A(0)$ and $\epsilon > 0$, there is a strongly measurable $w : J \to X$ such that*

$$|w(t)| \leq \hat{c}(t) \quad a.e.\ on\ J, \tag{6}$$

$$w(t) \in f([J_{t,\epsilon} \times \overline{B}_{\gamma\epsilon}(u(t; w))] \cap G_0) \quad on\ J_0, \tag{7}$$

$$w(t) \in f([J_{t,\epsilon} \times \overline{B}_{\gamma\epsilon}(u(t; w))] \cap G) \quad on\ J \tag{8}$$

with $\gamma = 1 + a$, where $J_{t,\epsilon} = [t - \epsilon, t]$.

Proof. Let $u_0 \in K_A(0)$ and $\epsilon > 0$, where it suffices to consider $\epsilon \leq 1$. Define a set M^ϵ of approximate solutions by means of

$M^\epsilon = \{(v, w, P, b) : b \in (0, a],$

$v : [0, b] \to X$ with $v(b) \in K_A(b)$, $v([0, b])$ relatively compact,

$w : [0, b] \to X$ strongly measurable such that (6), (7) and (8) hold on $[0, b]$,

$P \subset [0, b)$ with $0 \in P$ and $b \in \overline{P}$ such that $\tau \in P$ implies

$v(\tau) \in K_A(\tau)$ and $|v(t) - u(t; \tau, v(\tau), w)| \leq \epsilon(t - \tau)$ on $[\tau, b]\}$,

and notice that we are done if M^ϵ contains an element with $b = a$.

1. We claim that M^ϵ is nonempty. Since $z_0 := f(0, u_0) \in T_K^A(0, u_0)$, there is $h \in (0, \epsilon]$ such that $y_1 := S_{z_0}(h)u_0$ satisfies $\rho(y_1, K_A(h)) \leq \frac{1}{2}\epsilon h$, hence there is $u_1 \in K_A(h)$ such that $|e_0| \leq \epsilon$ for $e_0 := \dfrac{u_1 - y_1}{h}$. Let $t_0 = 0$, $t_1 = t_0 + h$ and

$$v(t) = S_{z_0}(t - t_0)u_0 + (t - t_0)e_0 \quad on\ [t_0, t_1].$$

We may assume $|v(t) - u_0| \leq \epsilon$ and $c(t_0) \leq \hat{c}(t)$ on $[t_0, t_1]$ if $h > 0$ is chosen small enough. By induction, we obtain sequences (t_k), (u_k), (z_k) and (e_k) such that

$$\left. \begin{array}{l} t_k \nearrow t_\infty \leq a, \quad u_k \in K_A(t_k), \quad z_k = f(t_k, u_k), \\[2mm] e_k = \dfrac{u_{k+1} - S_{z_k}(t_{k+1} - t_k)u_k}{t_{k+1} - t_k}, \quad |e_k| \leq \epsilon. \end{array} \right\} \tag{9}$$

For $k \geq 0$ we then let

$$v(t) = S_{z_k}(t - t_k)u_k + (t - t_k)e_k \quad \text{on } [t_k, t_{k+1}], \tag{10}$$

and may assume $t_{k+1} - t_k \leq \epsilon$ as well as $|v(t) - u_k| \leq \epsilon$ and $c(t_k) \leq \hat{c}(t)$ on $[t_k, t_{k+1}]$ by appropriate choice of the t_k. Let $P = \{t_k : k \geq 0\}$ and define $w : J \to X$ by means of $w(t) := z_k$ if $t \in [t_k, t_{k+1})$ and $w(t_\infty) = 0$. Then w is strongly measurable with $|w(t)| \leq \hat{c}(t)$ on $[0, t_\infty]$. Below, we will show that

$$|v(t) - u(t; t_k, u_k, w)| \leq \epsilon(t - t_k) \quad \text{on } [t_k, t_\infty) \quad \text{for all } k \geq 0. \tag{11}$$

Observe that (11) implies $|v(t) - u(t; w)| \leq \epsilon t$ on $[0, t_\infty)$, hence (8) holds on $[0, t_\infty)$ by definition of w. In order to obtain (7) on $[0, t_\infty)$, the choice of (t_k) above has to follow the subsequent additional rules. Notice first that by proper choice of the t_k we may assume $[t_0, t_\infty) \subset J \setminus J_0$ in case $t_0 \in J \setminus J_0$, since J_0 is closed. Consequently, with this choice (7) is no condition if $t_0 \notin J_0$. Consider now the case $t_0 \in J_0$. If $t_m \in J \setminus J_0$ for a first $m \geq 1$, we may choose the t_k for all $k > m$ in such a way that, in addition to the other properties above, $[t_m, t_\infty) \subset J \setminus J_0$. Then (7) is satisfied on $[t_0, t_\infty)$ if (11) holds. Finally, if the construction above yields $(t_k) \subset J_0$, then (7) obviously holds on $[t_0, t_\infty)$ if (11) is valid.

Evidently, (11) holds if

$$|v(t) - u(t; t_k, u_k, w)| \leq \epsilon(t - t_k) \quad \text{on } [t_j, t_{j+1}] \tag{12}$$

for all $j \geq k \geq 0$ and (12) is valid for $j = k$, by construction of v. Suppose that (12) holds for fixed $k \geq 0$ and $j = m - 1 \geq k$. Exploitation of

$$u(t; t_k, u_k, w) = u(t; t_m, u(t_m; t_k, u_k, w), z_m) \quad \text{on } [t_m, t_{m+1}]$$

and

$$v(t) = u(t; t_m, u_m, z_m) + (t - t_m)e_m \quad \text{on } [t_m, t_{m+1}]$$

yields

$$\begin{aligned} |v(t) - u(t; t_k, u_k, w)| &\leq |u_m - u(t_m; t_k, u_k, w)| + (t - t_m)|e_m| \\ &\leq (t_m - t_k)\epsilon + (t - t_m)\epsilon \end{aligned}$$

for all $t \in [t_m, t_{m+1}]$, hence (12) holds for $j = m$. By induction, (12) is valid for all $j \geq k \geq 0$.

Since (11) implies

$$v([0, t_\infty)) \subset C_k + (t_\infty - t_k)\overline{B}_\epsilon(0) \quad \text{for all } k \geq 0,$$

where $C_k := v([0, t_k]) \cup u([t_k, t_\infty]; t_k, u_k, w)$ is relatively compact, it follows that $v([0, t_\infty))$ is relatively compact. Let (u_{k_j}) be a convergent subsequence of $(u_k) = (v(t_k))$ and define $v(t_\infty) := \lim_{j \to \infty} u_{k_j}$. Then $v(t_\infty) \in K_A(t_\infty)$ since gr (K_A) is closed from the left, and therefore $v : [0, t_\infty] \to X$ has those properties required in the definition of M^ϵ. Moreover, it is easy to check that (11) is also valid on $[t_k, t_\infty]$. Therefore $(v, w, P, t_\infty) \in M^\epsilon$, if we redefine $w(t_\infty)$ by means of $w(t_\infty) = f(t_\infty, v(t_\infty))$; notice that w still satisfies (6) then.

2. $M^\epsilon \neq \emptyset$ by Step 1, and we shall use Zorn's lemma to obtain an element of M^ϵ with $b = a$. For this purpose define a partial ordering on M^ϵ by $(v, w, P, b) \leq (\overline{v}, \overline{w}, \overline{P}, \overline{b})$ if

$$b \leq \overline{b}, \quad v = \overline{v} \text{ on } [0, b], \quad w = \overline{w} \text{ a.e. on } [0, b], \quad P \subset \overline{P}.$$

To be able to apply Zorn's lemma we have to show that every ordered subset $M \subset M^\epsilon$ has an upper bound in M^ϵ. Let

$$b^* = \sup\{b \in (0, a] : (v, w, P, b) \in M \text{ for some } v, w, P\}.$$

In case the "sup" is actually a "max", i.e. if there is $(v, w, P, b^*) \in M$, we let

$$P^* = \{\tau \in [0, b^*) : \text{ there is } (v, w, P, b^*) \in M \text{ with } \tau \in P\}.$$

Evidently, (v, w, P^*, b^*) is an upper bound and $(v, w, P^*, b^*) \in M^\epsilon$ is easy to check.

In the remaining case there is a sequence $(v_n, w_n, P_n, b_n) \subset M$ with $b_n \nearrow b^*$, hence $P_n \subset P_{n+1}, v_{n+1} = v_n$ on $[0, b_n]$ and $w_{n+1} = w_n$ a.e. on $[0, b_n]$ for all $n \geq 1$. We then let

$$P^* = \bigcup_{n \geq 1} P_n, \quad v^*(t) = v_n(t) \text{ on } [0, b_n], \quad w^*(t) = w_n(t) \text{ on } [0, b_n].$$

Suppose, for the moment, that $v^*([0, b^*))$ is relatively compact. We let $v^*(b^*) = \lim_{j \to \infty} v^*(b_{n_j})$ where $(v^*(b_{n_j}))$ is a convergent subsequence of $(v^*(b_n))$, $w^*(b^*) := f(b^*, v^*(b^*))$ and claim that $(v^*, w^*, P^*, b^*) \in M^\epsilon$ is an upper bound for M. Evidently, (v^*, w^*, P^*, b^*) is an upper bound for M, since $(v, w, P, b) \in M$ implies $b < b_n$ for some $n \geq 1$, hence $(v, w, P, b) \leq (v_n, w_n, P_n, b_n)$. To check that $(v^*, w^*, P^*, b^*) \in M^\epsilon$ is also easy; notice in particular that $\tau \in P^*$ implies $\tau \in P_n$ and $v^*(\tau) = v_n(\tau)$ for all $n \geq n_\tau$. So, it remains to prove relative compactness of $v^*([0, b^*))$. But the latter follows by the corresponding arguments from Step 1, where this time we take any sequence $(t_k) \subset P^*$ with $t_k \nearrow b^*$ and $u_k := v^*(t_k)$; notice that (11) then holds with v^* instead of v. Consequently, there is a maximal element $(v^*, w^*, P^*, b^*) \in M^\epsilon$ and we are done if $b^* = a$. Suppose

$b^* < a$. We then let $t_0 = b^*$, $u_0 = v^*(b^*)$ and repeat the construction of Step 1 to obtain the sequences from (9) and function v from (10). Let

$$\overline{v}(t) = v^*(t) \text{ on } [0, b^*], \quad \overline{v}(t) = v(t) \text{ on } [b^*, t_\infty), \quad \overline{b} = t_\infty,$$

$$\overline{w}(t) = w^*(t) \text{ on } [0, b^*], \quad \overline{w}(t) = z_k \text{ on } [t_k, t_{k+1}], \quad \overline{P} = P^* \cup \{t_k : k \geq 0\}.$$

Then $\overline{v}([t_0, t_\infty))$ is relatively compact again, and we let $\overline{v}(t_\infty) := \lim_{j \to \infty} \overline{v}(t_{k_j})$ for an appropriate subsequence (t_{k_j}) and $\overline{w}(t_\infty) = f(t_\infty, \overline{v}(t_\infty))$ as before. To obtain $(\overline{v}, \overline{w}, \overline{P}, \overline{b}) \in M^\epsilon$ we show that $\tau \in P^*$ and $t \in (t_0, t_\infty)$ implies $|\overline{v}(t) - u(t; \tau, \overline{v}(\tau), w)| \leq \epsilon(t - \tau)$; the other cases as well as the remaining properties are rather obvious. Due to (11) and the properties of (v^*, w^*, P^*, b^*) we have

$$|\overline{v}(t) - u(t; \tau, \overline{v}(\tau), w)|$$
$$\leq |v(t) - u(t; t_0, u_0, w)| + |u(t; t_0, u_0, w) - u(t; t_0, u(t_0; \tau, v^*(\tau), w), w)|$$
$$\leq \epsilon(t - t_0) + |v^*(t_0) - u(t_0; \tau, v^*(\tau), w)| \leq \epsilon(t - t_0) + \epsilon(t_0 - \tau) = \epsilon(t - \tau),$$

hence $(\overline{v}, \overline{w}, \overline{P}, \overline{b}) \in M^\epsilon$ with $\overline{b} > b^*$, a contradiction. Consequently, $b^* = a$ for every maximal element of M^ϵ. $\qquad\square$

Let us note that in the situation described by Lemma 2.2, there are functions $w \in L^1(J; X)$, $\tau : J \to J$ and $v : J \to X$ such that

$$\tau(t) \in J_{t,\epsilon} \cap J \text{ on } J \quad \text{and} \quad \tau(t) \subset J_{t,\epsilon} \cap J_0 \text{ on } J_0, \tag{13}$$

$$v(t) \in K_A(\tau(t)) \text{ on } J \quad \text{and} \quad w(t) = f(\tau(t), v(t)) \text{ on } J, \tag{14}$$

$$|w(t)| \leq \hat{c}(t) \text{ a.e. on } J \quad \text{and} \quad |v(t) - u(t; w)| \leq \gamma\epsilon \text{ on } J. \tag{15}$$

Since no regularity of f is needed so far, Lemma 2.2 can be used to show that existence of subtangential vectors together with a mild integrability condition has remarkable implications concerning the regularity of $\mathrm{gr}(K_A)$. In particular, if X is separable then $\mathrm{gr}(K_A) \in \mathcal{L} \otimes \mathcal{B}(X)$, which will be important later on. Here $\mathcal{L}$ denotes the σ-algebra of Lebesgue measurable subsets of J and $\mathcal{B}(X)$ is the σ-algebra of all Borel subsets of X. Recall that $\phi : J \to \mathbb{R}$ is called *upper semicontinuous (usc) from the right* if $\overline{\lim}_{s \to t+} \phi(s) \leq \phi(t)$ for all $0 \leq t < a$.

PROPOSITION 2.3. *Let A be m-accretive in a real Banach space X, $J = [0, a] \subset \mathbb{R}$ and $K : J \to 2^X$ with closed values be such that $K_A(0) \neq \emptyset$ and $G = \mathrm{gr}(K_A)$ is closed from the left. Suppose there is $c \in L^1(J)$ such that*

$$T_K^A(t, x) \cap c(t)\overline{B}_1(0) \neq \emptyset \text{ for all } t \in [0, a), x \in K_A(t).$$

Then $K_A(t) \neq \emptyset$ on J. For every $x \in X$ the function $\rho(x, K_A(\cdot)) : J \to \mathbb{R}_+$ is usc from the right and especially measurable. Moreover, if X is separable then $\mathrm{gr}(K_A) \in \mathcal{L} \otimes \mathcal{B}(X)$.

Proof. 1. To prove the first assertion it suffices to consider $t_0 \in (0, a]$. Let $J_0 = [t_0, a]$, choose any $\epsilon > 0$ and let $f : \mathrm{gr}(K_A) \to X$ be any selection of $F(t, x) = T_K^A(t, x) \cap c(t)\overline{B}_1(0)$. Application of Lemma 2.2 yields functions $w \in L^1(J; X)$ and τ, v satisfying (13), (14), (15). Due to $\tau(t_0) \in [t_0 - \epsilon, t_0] \cap J_0 = \{t_0\}$ it follows that $v(t_0) \in K_A(t_0)$, hence $K_A(t_0) \neq \emptyset$.

2. Fix $x \in X$ and let $t_0 \in [0, a)$. Given $\epsilon > 0$, let $x_\epsilon \in K_A(t_0)$ be such that $|x - x_\epsilon| \leq \rho(x, K_A(t_0)) + \epsilon$. Redefine $K_A(t_0)$ as $\{x_\epsilon\}$, let f be given as in Step 1 and apply Lemma 2.2 with $J = [t_0, a]$ and $J_0 = [t_0 + h, a]$ for $h > 0$ such that $t_0 + h < a$. Then $\tau(t_0 + h) = t_0 + h$, hence $v(t_0 + h) \in K_A(t_0 + h)$. Therefore

$$
\begin{aligned}
\rho(x, K_A(t_0 + h)) &\leq |x - v(t_0 + h)| \leq |x - u(t_0 + h; t_0, x_\epsilon, w)| + (1 + a)\epsilon \\
&\leq |x - x_\epsilon| + |x_\epsilon - S(h)x_\epsilon| \\
&\quad + |S(h)x_\epsilon - u(t_0 + h; t_0, x_\epsilon, w)| + (1 + a)\epsilon \\
&\leq \rho(x, K_A(t_0)) + |x_\epsilon - S(h)x_\epsilon| + \int_{t_0}^{t_0+h} \hat{c}(t)dt + (2 + a)\epsilon \\
&\leq \rho(x, K_A(t_0)) + (3 + a)\epsilon \quad \text{for sufficiently small } h > 0,
\end{aligned}
$$

hence $\rho(x, K_A(\cdot))$ is usc from the right. Evidently $\phi = -\rho(x, K_A(\cdot))$ is lower semicontinuous from the right and such ϕ are measurable (see p. 70 in [14]). Hence $\rho(x, K_A(\cdot))$ is measurable for every $x \in X$.

3. Let X be separable and $\phi(t, x) = \rho(x, K_A(t))$ on $J \times X$. By the previous step, $\phi(\cdot, x)$ is measurable for every x, hence ϕ is Carathéodory. In this situation ϕ is $\mathcal{L} \otimes \mathcal{B}(X)$-measurable (see e.g. the Lemma in [17]) and therefore

$$
\mathrm{gr}(K_A) = \{(t, x) \in J \times X : \phi(t, x) \leq 0\} \in \mathcal{L} \otimes \mathcal{B}(X).
$$

$\square$

3. Almost continuous forcing terms

A basic step towards (1) with Carathéodory right-hand sides is to establish existence of mild solutions for functions $f : \mathrm{gr}(K_A) \to X$ having the following property: For every $\epsilon > 0$ there exists a closed $J_\epsilon \subset J$ with $\lambda_1(J \setminus J_\epsilon) \leq \epsilon$ such that

$$f_{|[J_\epsilon \times X] \cap \mathrm{gr}(K_A)} \text{ is continuous.}$$

Such an f is called *almost continuous* (or f is said to have the Scorza-Dragoni property).

Existence results for mild solutions of (1) in infinite dimensional Banach spaces require of course additional assumptions. For forcing terms that are not better than continuous, compactness of the semigroup $S(t)$ generated by $-A$ is especially fruitful for application to PDEs (on bounded domains). Actually, instead of compactness of $S(t)$, it suffices to assume that $S(t)_{|K_A(J)}$ is compact, i.e. $S(t)B$ is relatively compact for all $t > 0$ and bounded $B \subset K_A(J)$. Depending on the tube K, this may eventually be a much weaker assumption.

THEOREM 3.1. *Let A be m-accretive in a real Banach space X, $J = [0, a] \subset \mathbb{R}$ and $K : J \to 2^X$ with closed values be such that $K_A(0) \neq \emptyset$ and $G = gr\,(K_A)$ is closed from the left. Suppose that the semigroup $S(t)$ generated by $-A$ is such that $S(t)_{|K_A(J)}$ is compact. Let $f : G \to X$ be almost continuous satisfying (2) and (4). Then (1) has a mild solution for every $u_0 \in K_A(0)$.*

Proof. 1. In order to apply Lemma 2.2, we consider f on a certain smaller tube $\tilde{K}_A(\cdot)$ such that $|f(t, x)| \le \tilde{c}(t)$ on $gr\,(\tilde{K}_A)$ with some $\tilde{c} \in L^1(J)$. For this purpose fix $x_0 \in D(A)$ with $|u_0 - x_0| \le 1$, let $r(\cdot)$ be the solution of

$$r' = 1 + \hat{c}(t)(1 + r + |S(t)x_0|) \text{ on } J, \quad r(0) = 1$$

and define $\tilde{K}(\cdot)$ by means of $\tilde{K}(t) := K(t) \cap \overline{B}_{r(t)}(S(t)x_0)$ on J. Evidently $u_0 \in \tilde{K}(0)$, $gr\,(\tilde{K}_A)$ is closed from the left and $|f(t, x)| \le \tilde{c}(t)$ on $gr\,(\tilde{K}_A)$ for $\tilde{c} = \gamma c$ with sufficiently large $\gamma > 1$.

In order to show that (4) also holds for $\tilde{K}$ instead of K, notice first that r satisfies

$$D_+ r(t) \ge 1 + c(t)(1 + r(t) + |S(t)x_0|) \text{ for all } t \in [0, a),$$

where D_+ denotes the lower right Dini derivative. Indeed, given such t, there is $h_0 > 0$ such that $c(t) \le \hat{c}(s)$ for all $s \in [t, t + h_0]$. Then

$$\frac{r(t + h) - r(t)}{h} = \frac{1}{h} \int_t^{t+h} r'(s)ds = 1 + \frac{1}{h} \int_t^{t+h} \hat{c}(s)(1 + r(s) + |S(s)x_0|)ds$$

$$\ge 1 + \frac{1}{h} \int_t^{t+h} c(t)(1 + r(s) + |S(s)x_0|)ds \text{ for all } h \in (0, h_0]$$

implies

$$D_+ r(t) = \lim_{h \to 0+} \frac{r(t + h) - r(t)}{h} \ge 1 + c(t)(1 + r(t) + |S(t)x_0|).$$

Let $t \in [0, T)$, $x \in \tilde{K}_A(t)$ and $z := f(t, x)$. Due to $z \in T_K^A(t, x)$ there are sequences $h_n \to 0+$ and $e_n \to 0$ such that

$$S_z(h_n)x + h_n e_n \in K_A(t + h_n) \text{ for all } n \ge 1.$$

By means of the estimate

$$|S_z(h_n)x + h_n e_n - S(t + h_n)x_0|$$

$$\leq |S_z(h_n)x - S(h_n)x| + |x - S(t)x_0| + h_n|e_n| \leq h_n|z| + r(t) + h_n|e_n|$$

$$\leq r(t) + h_n c(t)(1 + r(t) + |S(t)x_0|) + h_n|e_n| \leq r(t + h_n),$$

which holds for all large $n \geq 1$ due to the above property of $D_+ r(t)$, this implies

$$S_z(h_n)x + h_n e_n \in \tilde{K}_A(t + h_n) \quad \text{for all large } n \geq 1,$$

hence (4) also holds for $\tilde{K}$. Consequently, all assumptions of Theorem 3.1 are also satisfied if K is replaced by $\tilde{K}$.

2. By Step 1 we may assume $|f(t, x)| \leq c(t)$ on $\mathrm{gr}\,(\tilde{K}_A)$ with $c \in L^1(J)$. Given a sequence $\epsilon_n \searrow 0$, there are closed $J_n \subset J$ with $\lambda_1(J \setminus J_n) \leq \epsilon_n$ such that $f_{|[J_n \times X] \cap \mathrm{gr}\,(K_A)}$ is continuous, where we may also assume that $J_n \subset J_{n+1}$ for all $n \geq 1$. Application of Lemma 2.2 with $J_0 := J_n$ yields functions $w_n \in L^1(J; X)$ and τ_n, v_n which satisfy (13), (14) and (15). In particular $\{v_n(t) : n \geq 1\} \subset K_A(t)$ is bounded for all $t \in J$. Consequently,

$$|u_n(t) - S(h)v_n(t - h)| \leq |u_n(t - h) - v_n(t - h)| + \int_{t-h}^{t} |w_n(s)| ds$$

$$\leq \gamma \epsilon_n + \int_{t-h}^{t} \hat{c}(s) ds \quad \text{for } 0 \leq t - h \leq t \leq a,$$

implies

$$\beta(\{u_n(t) : n \geq 1\}) = \beta(\{u_n(t) : n \geq p\})$$

$$\leq \gamma \epsilon_p + \int_{t-h}^{t} \hat{c}(s) ds \quad \text{for all } p \geq 1 \text{ and } 0 \leq t - h < t \leq a,$$

where $\beta(\cdot)$ denotes the Hausdorff measure of noncompactness. Therefore (u_n) has relatively compact sections, since $S(h)(\{v_n(t - h) : n \geq 1\})$ is relatively compact by the compactness assumption on the semigroup.

Given $0 \leq s \leq t, \bar{t} \leq T$, the inequality for integral solutions implies

$$|u_n(t) - u_n(\bar{t})| \leq |S(t - s)u_n(s) - S(\bar{t} - s)u_n(s)| + \int_s^t \hat{c}(\tau) d\tau + \int_s^{\bar{t}} \hat{c}(\tau) d\tau$$

$$\leq |S(|t - \bar{t}|)u_n(s) - u_n(s)| + \int_s^t \hat{c}(\tau) d\tau + \int_s^{\bar{t}} \hat{c}(\tau) d\tau.$$

Now if (u_n) is not equicontinuous, then $|u_n(t_n) - u_n(\bar{t}_n)| \geq \epsilon_0 > 0$ with $t_n \to t, \bar{t}_n \to t$ and $t = 0$ is not possible. Since $(u_n(s))$ is relatively compact for every $s \in J$, the estimate above yields the contradiction $\epsilon_0 \leq 2 \int_s^t \hat{c}(\tau) d\tau$ for all $s \in [0, t)$. Therefore (u_n) is equicontinuous, thus relatively compact in $C(J; X)$. Hence $u_{n_k} \to u$ in $C(J; X)$ for some subsequence of (u_n). We may also assume that $\tau_{n_k}(t) \nearrow t$. Due to $v_{n_k}(t) \in K_A(\tau_{n_k}(t))$ and $v_{n_k}(t) \to u(t)$ by (15), it follows that $u(t) \in K_A(t)$ since $\mathrm{gr}\,(K_A)$ is closed from the left.

Fix $m \geq 1$. Then $(\tau_{n_k}(t), v_{n_k}(t)) \in [J_m \times X] \cap \mathrm{gr}\,(K_A)$ for all large k and $(\tau_{n_k}(t), v_k(t)) \to (t, u(t))$ as $k \to \infty$ together with continuity of $f_{|[J_m \times X] \cap \mathrm{gr}\,(K_A)}$ yields

$$w_{n_k}(t) = f(\tau_{n_k}(t), v_{n_k}(t)) \to f(t, u(t)) \quad \text{for all } t \in J_m.$$

The latter holds for every $m \geq 1$, hence $w_{n_k}(t) \to f(t, u(t))$ a.e. on J. Since $|w_{n_k}(t)| \leq \hat{c}(t)$ a.e. on J, this implies $w_{n_k} \to w = f(\cdot, u(\cdot))$ in $L^1(J; X)$ by the dominated convergence theorem. Consequently $u(\cdot; w_{n_k}) \to u(\cdot; w)$ in $C(J; X)$, hence $u = u(\cdot; w)$ with $w = f(\cdot, u(\cdot))$, i.e. u is a mild solution of (1). $\qquad\square$

4. Carathéodory forcing terms

By means of reduction to the almost continuous case we are able to obtain mild solutions for Carathéodory right-hand sides. This approach relies on

LEMMA 4.1. *Let $J = [0, a] \subset \mathbb{R}$ and X be a separable Banach space. Let $G \in \mathcal{L} \otimes \mathcal{B}(X)$ and $f : G \to X$ be Carathéodory. Then f is almost continuous.*

Lemma 4.1 is a special case of Theorem 1 in [17]. Let us mention that the cited result applies since $(J, \mathcal{L}; X)$ has the so-called *projection property*, i.e. $\{t \in J : (t, x) \in G$ for some $x\} \in \mathcal{L}$ for every $G \in \mathcal{L} \otimes \mathcal{B}(X)$; see Chapter VIII in [13]. Now we have

THEOREM 4.2. *Let A be m-accretive in a real Banach space X, $J = [0, a] \subset \mathbb{R}$ and $K : J \to 2^X$ with closed values be such that $K_A(0) \neq \emptyset$ and $\mathrm{gr}\,(K_A)$ is closed from the left. Suppose that the semigroup $S(t)$ generated by $-A$ is such that $S(t)_{|K_A(J)}$ is compact. Let $f : \mathrm{gr}\,(K_A) \to X$ be Carathéodory satisfying (2) and (4). Then (1) has a mild solution for every $u_0 \in K_A(0)$ if also one of the following assumptions is fulfilled.*
(a) X is separable.　　(b) $K_A(t) \equiv K_A(0)$ on J.

Proof. Notice first that $G \in \mathcal{L} \otimes \mathcal{B}(X)$ is valid due to Proposition 2.3. Therefore Lemma 2.2 applies if X is separable, hence f is almost continuous then and Theorem 3.1 yields a mild solution of (1). Thus we are done if reduction to separable X is possible in case (b).

Given $u_0 \in K_A$, let $X_0 = \mathrm{span}\{u_0\}$. By induction, we define an increasing sequence of closed separable subspaces as follows. Given a separable subspace X_n with $u_0 \in X_n$, let $M_n = \{y_k : k \geq 1\}$ be a dense subset of X_n such that $K_A \cap M_n$ is dense in $K_A \cap X_n$. Since

$f(\cdot, x)$ is strongly measurable for all $x \in K_A \cap M_n$, there is a measurable subset J_n of J such that $f(J_n \times (K_A \cap M_n))$ is contained in a separable subspace. In addition, J_n can be chosen such that $f(t, \cdot)$ is continuous for all $t \in J_n$. For every $k \geq 1$ choose $z_k \in K_A$ such that $|y_k - z_k| \leq 2\rho(y_k, K_A)$, and let $K_n = \{z_k : k \geq 1\}$. Then X_{n+1} is defined as

$$X_{n+1} = \overline{\text{span}} \left(X_n \cup K_n \cup f(J_n \times (K_A \cap X_n)) \cup \bigcup_{\lambda > 0} (I + \lambda A)^{-1} X_n \right).$$

We then let

$$\hat{J} = \bigcap_{n \geq 0} J_n, \quad \hat{X} = \overline{\bigcup_{n \geq 0} X_n}, \quad \hat{K} = \overline{\bigcup_{n \geq 0} (K_A \cap X_n)};$$

notice that $\hat{K} \cap \overline{D(A)} = \hat{K}$. Evidently $\hat{J} \subset J$ is measurable with $\lambda_1(J \setminus \hat{J}) = 0$. To see that X_{n+1} (and consequently $\hat{X}$) is separable, recall that $\overline{\text{span}\, \overline{M}} = \overline{\text{span}\, M}$ for any $M \subset X$ and notice that $f(J_n \times (K_A \cap X_n)) \subset \overline{f(J_n \times (K_A \cap M_n))}$ by continuity of the $f(t, \cdot)$, and

$$\bigcup_{\lambda > 0} (I + \lambda A)^{-1} X_n \subset \overline{\{(I + \lambda A)^{-1} x : x \in M_n, 0 < \lambda \in \mathbb{Q}\}}.$$

Given $(t, x) \in \hat{J} \times \hat{X}$, there is a sequence (x_k) with $x_k \to x$ and $x_k \in X_{n_k}$ for some $n_k \geq 0$. Hence $y_k := f(t, x_k) \to f(t, x)$ and $y_k \in X_{n_k+1} \subset \bigcup_{n \geq 0} X_n$ implies $f(t, x) \in \hat{X}$. The same argument yields $(I + \lambda A)^{-1} x \in \hat{X}$ for any $\lambda > 0$, which also implies that the restriction of A to $\hat{X}$ is m-accretive in $\hat{X}$. To complete the reduction, fix $\tau \in \hat{J}$ and redefine f on $(J \setminus \hat{J}) \times K_A$ by means of $f(t, x) := f(\tau, x)$.

It remains to show that the restriction of f to $J \times \hat{K}$ satisfies the subtangential condition (4) with respect to $\hat{K}$. For this purpose, let us first show that

$$\rho(x, \hat{K}) \leq 2\rho(x, K_A) \quad \text{for every } x \in \hat{X}. \tag{16}$$

Fix $x \in \hat{X}$ and let $\epsilon > 0$. Then there is $y \in M_n$ for some $n \geq 0$ such that $|x - y| \leq \epsilon$. By definition of K_n there is $z \in K_n \subset K_A$ with $|y - z| \leq 2\rho(y, K_A)$. Hence $z \in K_A \cap X_{n+1} \subset \hat{K}$ and therefore

$$\rho(x, \hat{K}) \leq |x - z| \leq 2\rho(y, K_A) + |x - y|$$
$$\leq 2\rho(x, K_A) + 3|x - y| \leq 2\rho(x, K_A) + 3\epsilon.$$

This implies (16) since $\epsilon > 0$ was arbitrary.

Let $t_0 \in J$ with $t_0 < T$, $x_0 \in \hat{K}$ and $v = f(t_0, x_0)$. Due to $v \in T_K^A(x_0)$ there is a sequence $h_m \to 0+$ such that

$$\frac{1}{h_m} \rho(S_v(h_m) x_0, K_A) \to 0 \quad \text{as } m \to \infty.$$

Since $x_0 \in \hat{K} \subset \hat{X} \cap \overline{D(A)}$, $v \in \hat{X}$ and A (restricted to $\hat{X}$) is m-accretive in $\hat{X}$ it follows that $S_v(h_m)x_0 \in \hat{X}$ for all $m \geq 1$. Hence (16) implies

$$\frac{1}{h_m}\rho(S_v(h_m)x_0, \hat{K}) \leq \frac{2}{h_m}\rho(S_v(h_m)x_0, K_A) \to 0 \quad \text{as } m \to \infty,$$

i.e. $f(t_0, x_0) \in T_{\hat{K}}^A(x_0)$. $\qquad\qquad\qquad\qquad\qquad\qquad\qquad\qquad\qquad\qquad\square$

5. Measurable/locally lipschitz forcing terms

In applications one is of course also interested in uniqueness, which can be obtained under conditions of locally Lipschitz type. We call $f : J \times K_A \to X$ *locally Lipschitz with respect to x*, if for every $x_0 \in K_A$ there exist $\delta = \delta(x_0) > 0$ and $\omega = \omega(x_0) \in L^1(J)$ such that $f(t, \cdot)$ is Lipschitz of constant $\omega(t)$ on $B_\delta(x_0)$, i.e.

$$|f(t, x) - f(t, \overline{x})| \leq \omega(t)|x - \overline{x}| \quad \text{for a.a. } t \in J \text{ and all } x, \overline{x} \in B_\delta(x_0) \cap K_A.$$

It is rather obvious that the approximate solutions provided by Lemma 2.2 are not useful here due to deviations in t. For this reason we rest content with static constraints given by a fixed set K, since then an improved version of Lemma 2.2 holds. Observe that, despite this fact, the tube K below is allowed to depend on time. This is again needed for later reduction to integrably bounded f.

LEMMA 5.1. *Let A be m-accretive in a real Banach space X, $J = [0, T] \subset \mathbb{R}$ and $K : J \to 2^X$ with closed values be increasing with respect to inclusion and such that $K_A(0) \neq \emptyset$. Let $f : \mathrm{gr}\,(K_A) \to X$ be such that $|f(t, x)| \leq c(t)$ with $c \in L^1(J)$, $f(\cdot, x)$ is continuous from the right and (4) is satisfied. Then, given $u_0 \in K_A(0)$ and $\epsilon > 0$, there is $w \in L^1(J; X)$ satisfying (6) such that*

$$w(t) \in f(t, \overline{B}_{\gamma\epsilon}(u(t; w)) \cap K_A(t)) \quad \text{a.e. on } J \tag{17}$$

with $\gamma = 1 + a$.

Proof. Since the proof parallels the one given for Lemma 2.2, it suffices to explain the difference in the construction of the approximate solutions. Let $u_0 \in K_A(0)$ and $\epsilon \in (0, 1]$. We consider the set M^ϵ again, but with (17) instead of (7), (8). The basic idea in order to obtain (17) on a first interval $[0, h]$, is to replace $S_{f(0,u_0)}(\cdot)u_0$ by the mild solution $v_0(\cdot)$ of

$$v_0' + Av_0 \ni f(t, u_0) \quad \text{on } J, \quad v_0(0) = u_0.$$

This initial value problem obviously admits a mild solution if $g := f(\cdot, u_0) : J \to X$ is strongly measurable, and this holds if continuity from the right implies strong measurability. For the sake of completeness we include a short proof of this fact which belongs to "folklore".

Fix $\eta > 0$. Then, due to continuity of g from the right, for every $t \in J$ there is $\delta(t) > 0$ such that $|g(t) - g(s)| \leq \eta$ for $s \in [t, t + \delta(t)] \cap J$. Since

$$J = \bigcup_{t \in J} \ \bigcup_{0 < h \leq \delta(t)} [t, t + h]$$

is a Vitali cover of J, application of Vitali's covering theorem (see p.262ff in [15]) yields $t_k \in J$ and $h_k \in (0, \delta(t_k)]$ such that the $[t_k, t_k + h_k]$ are pairwise disjoint and $J_0 := \bigcup_{k \geq 1} [t_k, t_k + h_k]$ satisfies $\lambda_1(J \setminus J_0) = 0$. Then $g_\eta : J \to X$, defined by $g_\eta(t) = h(t_k)$ on $[t_k, t_k + h_k]$ and $g_\eta(t) = 0$ on $J \setminus J_0$, is a step function with $|g(t) - g_\eta(t)| \leq \eta$ a.e. on J. Therefore g is strongly measurable. Consequently, the initial value problem above has mild solution v_0, and

$$\frac{1}{h}|v_0(h) - S_{f(0, u_0)}(h)u_0| \leq \frac{1}{h} \int_0^h |f(t, u_0) - f(0, u_0)|dt \to 0 \text{ as } h \to 0+,$$

since $f(\cdot, u_0)$ is continuous from the right. Hence there is $h_0 \in (0, \epsilon]$ such that

$$\frac{1}{h}|v_0(h) - S_{f(0, u_0)}(h)u_0| \leq \frac{1}{3}\epsilon \text{ for all } h \in (0, h_0].$$

Since $z_0 := f(0, u_0) \in T_K^A(0, u_0)$, there is $h \in (0, h_0]$ such that $y_1 := S_{z_0}(h)u_0$ satisfies $\rho(y_1, K_A(h)) \leq \frac{1}{2}\epsilon h$, hence $\rho(v_0(h), K_A(h)) < \epsilon h$. Choose $u_1 \in K_A(h)$ such that $|e_0| \leq \epsilon$ for $e_0 := \dfrac{u_1 - y_1}{h}$. Let $t_0 = 0$, $t_1 = t_0 + h$ and

$$v(t) = v_0(t) + (t - t_0)e_0 \quad \text{on } [t_0, t_1],$$

where we may assume $|v(t) - u_0| \leq \epsilon$ and $c(t_0) \leq \hat{c}(t)$ on $[t_0, t_1]$. By induction, the same construction yields $t_k \nearrow t_\infty \leq a$, $u_k \in K_A(t_k)$ and mild solutions v_k of

$$v_k' + Av_k \ni f(t, u_k) \text{ on } [t_k, t_{k+1}], \quad v_k(t_k) = u_k$$

such that $e_k := \dfrac{u_{k+1} - v_k(t_{k+1})}{t_{k+1} - t_k}$ satisfy $|e_k| \leq \epsilon$. We then let

$$v(t) = v_k(t) + (t - t_k)e_k \text{ on } [t_k, t_{k+1}]$$

and may assume $t_{k+1} - t_k \leq \epsilon$ as well as $|v(t) - u_k| \leq \epsilon$ and $c(t_k) \leq \hat{c}(t)$ on $[t_k, t_{k+1}]$. Finally, we define $w \in L^1(J; X)$ by means of $w(t) = f(t, u_k)$ on $[t_k, t_{k+1}]$ and $w(t_\infty) = 0$. Now, the arguments given in the proof of Lemma 2.2 show that $(v, w, \{t_k : k \geq 1\}, t_\infty) \in M^\epsilon$, where (6) holds by construction and (17) follows from

$$w(t) = f(t, u_k) \in f(t, \overline{B}_\epsilon(v(t)) \cap K_A(t_k))$$

$$\subset f(t, \overline{B}_{\gamma\epsilon}(u(t; w)) \cap K_A(t)) \text{ on } [t_k, t_{k+1}].$$

Hence $M^\epsilon \neq \emptyset$, and a repetition of Step 2 of the proof of Lemma 2.2 implies the existence of a maximal element of the type (v, w, P, a) of M^ϵ. $\qquad\square$

Now we have

THEOREM 5.2. *Let A be m-accretive in a real Banach space X, $J = [0, a] \subset \mathbb{R}$ and $K \subset X$ closed with $K_A = K \cap \overline{D(A)} \neq \emptyset$. Let $f : J \times K_A \to X$ be strongly measurable with respect to t and locally Lipschitz with respect to x such that (2) and (4) hold. Then (1) has a unique mild solution $u(\cdot \,; u_0)$ for every $u_0 \in K_A$. In addition, this mild solution depends continuously on $u_0 \in K_A$.*

Proof. 1. We first reduce to integrably bounded f on a small increasing tube. For this purpose we repeat Step 1 of the proof of Theorem 3.1, with the following modification: Since $x_0 \in D(A)$ belongs to the generalized domain $\tilde{D}(A)$ of A, $S(\cdot)x_0$ is Lipschitz continuous of some constant L on J (see [2]). We then let $r(\cdot)$ be the solution of

$$r' = L + \hat{c}(t)(1 + r + |S(t)x_0|) \text{ on } J, \quad r(0) = 1$$

and $\tilde{K}(t) = K \cap \overline{B}_{r(t)}(S(t)x_0)$ on J. To see that $\tilde{K}(\cdot)$ is increasing, let $0 \leq s < t \leq a$ and $x \in \tilde{K}(s)$. Then

$$|x - S(t)x_0| \leq |x - S(s)x_0| + L(t - s) \leq r(s) + L(t - s) \leq r(t)$$

shows that $x \in \tilde{K}(t)$. The remaining arguments remain unchanged, hence (4) holds for $\tilde{K}$ instead of K. Consequently, given $\epsilon_n \searrow 0$, application of Lemma 5.1 yields $w_n \in L^1(J; X)$ with $|w_n(t)| \leq \hat{c}(t)$ a.e. on J such that (17) is satisfied.

Let $\delta > 0$ and $\omega \in L^1(J)$ be such that $f(t, \cdot)$ is Lipschitz of constant $\omega(t)$ on $B_\delta(u_0) \cap K_A$. Due to $|w_n(t)| \leq \hat{c}(t)$ a.e on J we find $b > 0$ such that $|u_n(t) - u_0| \leq \delta/2$ on $[0, b]$ for all $n \geq 1$. Exploitation of (17) yields $v_n : J \to K_A$ with $|u_n - v_n|_\infty \leq \gamma \epsilon_n$ such that $w_n(t) = f(t, v_n(t))$ a.e. on J, hence in particular $v_n(t) \in B_\delta(u_0) \cap K_A$ on $[0, b]$ for all $n \geq n_0$. Fix $p \geq n_0$ and consider $m, n \geq p$. Then the Lipschitz continuity of $f(s, \cdot)$ implies

$$\begin{aligned}
|u_n(t) - u_m(t)| &\leq \int_0^t |w_n(s) - w_m(s)| ds \\
&\leq \int_{[0,t] \cap J_p} |f(s, v_n(s)) - f(s, v_m(s))| ds + 2 \int_{J \setminus J_p} \hat{c}(s) ds \\
&\leq \int_0^t \omega(s) |u_n(s) - u_m(s)| ds + 2\gamma |\omega|_1 \epsilon_p \\
&\quad + 2 \int_{J \setminus J_p} \hat{c}(s) ds \text{ on } [0, b].
\end{aligned}$$

Application of Gronwall's lemma shows that $(u_{n|[0,b]})$ is Cauchy in $C([0, b]; X)$, and the limit is obviously a mild solution of (1) on $[0, b]$. Consequently, (1) has a unique noncontinuable mild solution u. In case u is defined on $[0, \tau)$ only, exploitation of (2) shows that $\lim_{t \to \tau-} u(t)$ exists and then u has an extension to $[0, \tau + b]$ with $b > 0$ by the arguments from above with J replaced by $[\tau, a]$. This contradiction shows that u is a mild solution of (1) on J. Uniqueness of u obviously follows by Gronwalls's lemma on a small interval, hence on all of J by continuation.

2. It remains to show that $u(\cdot \, ; u_0)$ depends continuously on $u_0 \in K_A$. Let $(x_n) \subset K_A$ with $x_n \to x_0$ and $u_n = u(\cdot \, ; x_n)$ for $n \geq 0$. By simple modifications of the arguments given in Step 1 it is clear that $u_n(t) \to u_0(t)$ uniformly on $[0, b]$ for some $b > 0$, and then either $u_n \to u_0$ in $C(J; X)$ or there is $\tau \in (0, a]$ such that $u_n(t) \to u_0(t)$ uniformly on $[0, \tau']$ for all $\tau' \in (0, \tau)$ but not on $[0, \tau]$. If the second case occurs we obtain $f(t, u_n(t)) \to f(t, u_0(t))$ a.e. on $[0, \tau)$, hence $f(\cdot, u_n(\cdot)) \to f(\cdot, u_0(\cdot))$ in $L^1([0, \tau]; X)$ by the dominated convergence theorem since $|u_n|_0 \leq R$ for all $n \geq 0$ with some $R > 0$. This implies $u_{n|[0,\tau]} \to u_{0|[0,\tau]}$ in $C([0, \tau]; X)$, a contradiction. $\qquad\square$

6. Additional remarks

In this final section we briefly discuss several related questions and try to give an overview concerning the more recent literature.

(a) *Sufficient Tangency Conditions.* In several applications it happens that for an appropriate choice of the $K(t)$ these sets are invariant under the resolvents of A. Then it is helpful to know that the subtangential condition can be separated, by which we mean that

$$\left. \begin{array}{l} J_\lambda K(t) \subset K(t) \text{ for } \lambda > 0, t \in [0, a) \text{ and} \\[2mm] f(t, x) \in T_K(t, x) \text{ for } t \in [0, a), x \in K_A(t) \end{array} \right\} \tag{18}$$

implies (4). Recall that conditions in terms of the resolvent of A correspond to properties of the solutions of stationary problems.

In fact, it is not difficult to show that (18) implies the "weak range condition"

$$\left. \begin{array}{c} \lim_{h \to 0+} h^{-1} \rho(x + hf(t, x), (I + hA)(K(t + h) \cap D(A))) = 0 \\[3mm] \text{for } t \in [0, a), x \in K_A(t), \end{array} \right\} \tag{19}$$

and the latter in turn implies (4) if f is continuous. This is the contents of Lemma 2.2 in [4]. For Carathéodory f, sufficiency of (18) or (19) is not clear in case of time-dependent constraints but holds if $K_A(t) \equiv K_A(0)$; the latter follows by exploitation of the above facts for $\tilde{K} := K(t)$ and $\tilde{f} = f(t, \cdot)$ and all $t \in [0, a)$. In case of static constraints it is also obvious that (4) can be slightly weakened: It suffices to have

$$f(t, x) \in T_K^A(t, x) \quad \text{for all } t \in [0, a) \backslash N, \, x \in K_A(t), \tag{20}$$

with some nullset $N \subset [0, a)$. For general tubes this is not possible, even if $A = 0$; a one-dimensional counter-example is $J = [0, 1]$, $K(t) = \{\phi(t)\}$ and $f = 0$, where $\phi : J \to J$

is the Cantor function, i.e. ϕ is continuous increasing with $\phi'(t) = 0$ a.e. and $\phi(J) = J$. Nevertheless, in case $A = 0$ this defect can be excluded if the assumption $T_K(t, x) \neq \emptyset$ for all $t \in N$ and $x \in K(t)$ is added; see [3].

If X and X^* are uniformly convex and $K(t) \equiv K$ is closed convex with nonempty interior, another possibility is to work with the explicit subtangential condition

$$f(t, x) - Ax \subset T_K(x) \quad \text{for all } t \in [0, a), \ x \in K \cap D(A). \tag{21}$$

Observe that (21) is a condition on $\partial K \cap D(A)$, only. In this setting, Lemma 2 in [7] shows that (21) implies (4) for continuous f. In the special case when X is a Hilbert space and $K = \overline{B}_R(0)$, this implication is implicitly contained in [12]. The latter paper gives an example in $L^2(\Omega)$ with $A = -\Delta_p$ that illustrates the advantage of condition (21) compared to the separated assumptions (18).

(b) *Related Existence Results.* In the recent paper [11] problem (1) has been considered for m-accretive A and locally integrably bounded Carathéodory forcing $f : [a, b) \times K \to X$, where K is a locally closed subset of $\overline{D(A)}$. In this situation of static constraints a local mild solution is obtained under the necessary subtangential condition (20), given that the semigroup generated by $-A$ is compact or K is relatively compact. This result is closely related to Theorem 4.2(b) above and extends a previous result for continuous f in [21], where a rather strong subtangential condition has been employed.

For dissipative (not necessarily continuous) $f : D(f) \subset X \to X$ and $K(t) \equiv K$ one may eventually apply invariance results for accretive operators to $A - f$. For example Theorem 2 in [20] implies that problem (1) with accretive A and s-dissipative f, considered as $u' + (A - f)u \ni 0$, has a mild solution if for every $x \in K_A := K \cap \overline{D(A)}$ and $\epsilon > 0$ there is $h \in (0, \epsilon]$, $x_h \in D(A) \cap D(f)$ and $y_h \in Ax_h$ such that

$$|x - x_h + h(f(x_h) - y_h)| \leq h\epsilon \quad \text{and} \quad \rho(x_h, K_A) \leq h\epsilon.$$

In case $D(f) = K$ this is just the weak range condition for $A - f$, and it becomes (19) if, in addition, f is continuous bounded and $\overline{K \cap D(A)} = K_A$.

(c) *Further Compactness Conditions.* The semigroup generated by an m-accretive operator is compact if and only if it is equicontinuous and the resolvents of A are compact (see [2]). In certain situations one can then replace compactness of the resolvents by a compactness condition on f, say

$$\beta(f(t, B)) \leq k(t)\beta(B) \quad \text{a.e. on } J \text{ for all bounded } B \subset K_A \tag{22}$$

with $k \in L^1(J)$, and still get existence of mild solutions. Notice that this is especially interesting for problems in Hilbert spaces with $A = \partial\varphi$, the subdifferential of a proper convex lsc function $\varphi : D_\varphi \subset X \to \mathbb{R}$, since the semigroup generated by $-A$ is equicontinuous then. The key to prove corresponding existence results is

LEMMA 6.1. *Let X be a real Banach space with uniformly convex dual and A be m-accretive in X such that $-A$ generates an equicontinuous semigroup. Let $J = [0, a] \subset \mathbb{R}$ and $(w_k) \subset L^1(J; X)$ such that $|w_k(t)| \leq \varphi(t)$ a.e. on J for all $k \geq 1$ with some $\varphi \in L^1(J)$. Then, given $t_0 \in J$ and $x_0 \in X$,*

$$\beta(\{u(t; t_0, x_0, w_k) : k \geq 1\}) \leq \int_0^t \beta(\{w_k(s) : k \geq 1\})\, ds \quad \text{on } J. \tag{23}$$

This result is from [6], where it is also shown that (23) is false without extra assumption on X^*. By means of this β-inequality, differential inequalities for $\phi(t) = \beta(\{u_n(t) : n \geq 1\})$ can be obtained, where $u_n = u(\cdot; w_n)$ are approximate solutions for (1) of the type provided by Lemma 5.1. If (22) holds and $K_A(t) \equiv K_A(0)$ this yields a mild solution of (1) in all of the corresponding situations considered in the present paper. Let us also note that Lemma 6.1 is valid in general Banach spaces if $A = A_0 + B$, where A_0 is linear, densely defined m-accretive and $B : X \to X$ is continuous and accretive. For more details see Chapter 2 in [8].

(d) *Semilinear Evolution Problems.* Let us briefly explain the relation between the m-accretive case and the semilinear case, i.e. when $-A$ generates a C_0-semigroup of bounded linear operators on X. First of all, notice that the existence results obtained in this paper remain valid if A is m-ω-accretive for some $\omega \in \mathbb{R}$, i.e. if $A_\omega := A + \omega I$ is m-accretive, since the corresponding result then applies to A_ω and $f_\omega := f + \omega I$ instead of A and f in each of these cases. Notice in particular that if $-A$ generates an equicontinuous or compact semigroup then this property is inherited to the semigroup generated by $-A_\omega$, which can be checked easily by means of the inequality for integral solutions and the fact that $(I + \lambda A_\omega)^{-1} = (I + \frac{\lambda}{1+\lambda\omega} A)^{-1} \circ \frac{1}{1+\lambda\omega} I$. Furthermore, $y \in T_K^A(t, x)$ implies $y + \omega x \in T_K^{A_\omega}(t, x)$ for $A_\omega = A + \omega I$. This leads to a mild solution of the original problem, since u is a mild solution of $u' + A_\omega u \ni w(t) + \omega u$ for $w \in L^1(J; X)$ iff u is a mild solution of $u' + Au \ni w(t)$.

Now suppose that $A : D(A) \subset X \to X$ is a closed, linear and densely defined operator such that $-A$ generates a C_0-semigroup $S(t)$. In this situation there is $\omega \in \mathbb{R}$ and $M \geq 1$ such that $|S(t)| \leq Me^{-\omega t}$ on $\mathbb{R}_+$, and X can be equipped with an equivalent norm $\|\cdot\|$ to achieve $M = 1$. Then A is m-ω-accretive in $(X, \|\cdot\|)$, hence we are within the framework of the present paper. Recall also that, in the semilinear case, $u \in C(J; X)$ is a mild solution of

$$u' + Au = f(t, u) \quad \text{on } J, \quad u(0) = u_0 \tag{24}$$

iff u satisfies the variation of constants formula. Therefore, it is easy to see that the necessary subtangential condition (4) is equivalent to

$$\lim_{h \to 0+} h^{-1} \rho(S(h)x + hf(t, x), K(t + h)) = 0 \quad \text{for } (t, x) \in \mathrm{gr}\,(K) \text{ with } t < a. \tag{25}$$

In [18] it has been shown that the necessary condition (25) is sufficient to get a local mild solution of (24) for continuous f in case of compact semigroup and static constraints. This result has recently been extended in [10] to Carathéodory f in reflexive Banach spaces; notice that this extension is a special case of Theorem 4.2(b), hence reflexivity is not needed.

Existence of mild solutions for perturbations of dissipative type has been obtained in [16] for continuous $f : \mathrm{gr}(K) \to X$ (with $\mathrm{gr}(K)$ closed from the left) such that (25) and

$$(f(t, x) - f(t, \overline{x}), x - \overline{x})_- \leq \omega(t, |x - \overline{x}|)|x - \overline{x}| \quad \text{for all } t \in J, x, \overline{x} \in K(t)$$

holds, where ω is a "uniqueness function" of Carathéodory type.

(e) *Multivalued Right-Hand Sides.* Notations for multivalued maps used below can be found in [14]. In [9] existence of solutions in closed sets is obtained for multivalued lsc perturbation under the following assumptions: A is m-accretive in a real Banach space X such that $-A$ generates a compact semigroup, $K \subset \overline{D(A)}$ is closed and $F : [0, a] \times K \to 2^X \setminus \{\emptyset\}$ is lsc and bounded with closed convex values such that $F(t, x) \subset T_K^A(x)$ on $[0, a) \times K$. The proof exploits the fact that a bounded lsc F admits a certain type of "directionally continuous" selections f, and existence of a mild solution of (1) for such a right-hand side f is obtained via approximate solutions. The construction of these approximate solutions is related to Lemma 2.2, but relies on compactness of the semigroup.

In Theorem 5.1.2 of [19] existence is proven for multivalued perturbations in the semi-linear case, where the assumptions are: X reflexive, compact semigroup, $K : J \to 2^X \setminus \{\emptyset\}$ a tube with locally closed graph and $F : \mathrm{gr}(K) \to 2^X \setminus \{\emptyset\}$ locally bounded and weakly usc with closed convex values such that (25) holds with $f(t, x)$ replaced by any $y \in F(t, x)$.

For m-accretive A, general tube $K(\cdot)$ and ϵ-δ-usc $F : \mathrm{gr}(K_A) \to 2^X \setminus \{\emptyset\}$, existence is obtained in Theorem 3.1 of [5], where the remaining assumptions are: F bounded with closed convex values such that (4) holds with $f(t, x)$ replaced by some $y \in F(t, x)$; either X^* uniformly convex and compact semigroup, or F maps bounded sets into weakly relatively compact sets and a stronger compactness assumption on A, which is motivated by application with $Au = -\Delta\varphi(u)$ where it holds if φ is continuous and strictly increasing.

Acknowledgment

This work is based on §4 of the authors Habilitation thesis. I am deeply grateful for the support that Philippe provided to me in particular concerning my Habilitation.

REFERENCES

[1] BARBU, V., *Nonlinear Semigroups and Differential Equations in Banach Spaces.* Leyden: Noordhoff 1976.

[2] BÉNILAN, PH., CRANDALL, M. G., and PAZY, A., *Nonlinear Evolution Equations in Banach Spaces.* (monograph in preparation).

[3] BOTHE, D., *Multivalued differential equations with time-dependent constraints.* pp. 1829–1839 in "Proc. of the First World Congress of Nonlinear Analysts, Tampa, Florida 1992" (V. Lakshmikantham, ed). W. de Gruyter 1996.

[4] BOTHE, D., *Flow invariance for perturbed nonlinear evolution equations.* Abstract and Applied Analysis *1* (1996), 379–395.

[5] BOTHE, D., *Reaction-diffusion systems with discontinuities. A viability approach.* Proc. 2nd World Congress of Nonlinear Analysts, Nonlinear Analysis *30* (1997), 677–686.

[6] BOTHE, D., *Multivalued perturbations of m-accretive differential inclusions.* Israel J. Math. *108* (1998), 109–138.

[7] BOTHE, D., *Periodic Solutions of a Nonlinear Evolution Problem from Heterogeneous Catalysis.* Differential and Integral Equations *14* (2001), 641–670.

[8] BOTHE, D., *Nonlinear Evolutions in Banach Spaces. Existence and Qualitative Theory with Applications to Reaction-Diffusion Systems.* Habilitation thesis, University Paderborn 1999.

[9] BRESSAN, A. and STAICU, V., *On Nonconvex perturbations of maximal monotone differential inclusions.* Setvalued Analysis *2* (1994), 415–437.

[10] CARJA, O. and MONTEIRO MARQUES, M. D. P., *Viability for nonautonomous semilinear differential equations.* J. Diff. Eqs. *166* (2000), 328–346.

[11] CARJA, O. and VRABIE, I. I., *Viable domains for differential equations governed by Carathéodory perturbations of nonlinear m-accretive operators.* pp. 109–130 in Lecture Notes in Pure and Appl. Math. *255*, Dekker 2002.

[12] CASCAVAL, R. and VRABIE, I. I., *Existence of periodic solutions for a class of nonlinear evolution equations.* Rev. Mat. Univ. Complutense Madr. *7* (1994), 325–338.

[13] COHN, D. L., *Measure Theory.* Birkhäuser 1980.

[14] DEIMLING, K., *Multivalued Differential Equations.* De Gruyter 1992.

[15] HEWITT, E. and STROMBERG, K., *Real and Abstract Analysis* (2nd ed.), Springer 1969.

[16] IWAMIYA, T., *Global existence of mild solutions to semilinear differential equations in Banach spaces.* Hiroshima Math. J. *16* (1986), 499–530.

[17] KUCIA, A., *Scorza Dragoni type theorems.* Fund. Math *138* (1991), 197–203.

[18] PAVEL, N. H., *Invariant sets for a class of semilinear equations of evolution,* Nonlinear Analysis *1* (1977), 187–196.

[19] PAVEL, N. H., *Differential Equations, Flow Invariance and Applications.* Res. Notes Math. 113, Pitman 1984.

[20] PIERRE, M., *Invariant closed subsets for nonlinear semigroups.* Nonlinear Analysis *2* (1978), 107–117.

[21] VRABIE, I. I., *Compactness methods and flow-invariance for perturbed nonlinear semigroups.* Anal. Stiin. Univ. Iasi *27* (1981), 117–124.

Dieter Bothe
Fachbereich Chemie und Chemietechnik
Universität Paderborn
Warburger Str. 100
D-33098 Paderborn
Germany
e-mail: mboth1@tc.uni-paderborn.de

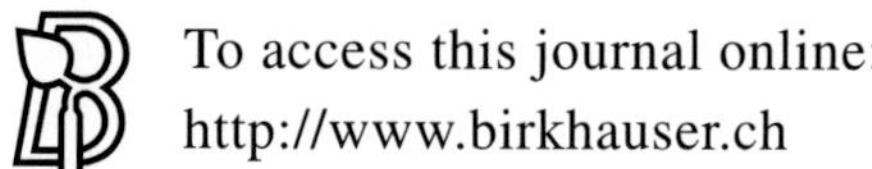

To access this journal online:
http://www.birkhauser.ch

J.evol.equ. 3 (2003) 395 – 406
1424–3199/03/030395 – 12
DOI 10.1007/s00028-003-0107-9
© Birkhäuser Verlag, Basel, 2003

**Journal of Evolution
Equations**

Linear parabolic equations with singular potentials

HERBERT AMANN

In memoriam Philippe Bénilan

Abstract. By means of maximal regularity techniques we study the solvability of a linear heat equation with a time-dependent potential under minimal regularity assumptions.

1. Introduction

Let Ω be a bounded smooth domain in $\mathbb{R}^n$ with $n \geq 3$. Set $\Gamma := \partial\Omega$, fix a positive number T, and put $J := (0, T)$. Suppose that a is a time-dependent function on Ω, a 'potential', and consider the initial boundary value problem

$$
\begin{aligned}
\partial_t u - \Delta u + au &= 0 \quad \text{in } \Omega \times J, \\
u &= 0 \quad \text{on } \Gamma \times J, \\
u(\cdot, 0) &= u^0 \quad \text{on } \Omega.
\end{aligned}
\tag{1.1}
$$

It is the purpose of this paper to discuss the well-posedness of (1.1) under minimal regularity requirements on a and u^0.

This problem has recently attracted some interest. More precisely, Brezis and Cazenave [7, Theorem A1] assume that

$$
\sigma > n/2, \quad a \in L_\infty(J, L_\sigma(\Omega)).
\tag{1.2}
$$

Then they show that, given $p \geq 1$ and $u^0 \in L_p(\Omega)$, problem (1.1) has a unique mild solution in the class

$$
C(\overline{J}, L_p(\Omega)) \cap L_{\infty,\mathrm{loc}}(J, L_\infty(\Omega)).
\tag{1.3}
$$

Hereby a mild solution in class (1.3) is an element u of (1.3) satisfying

$$
u(t) = e^{t\Delta_D} u^0 + \int_0^t e^{(t-\tau)\Delta_D} a(\tau) u(\tau)\, d\tau, \qquad t \in \overline{J},
$$

where Δ_D is the Dirichlet-Laplacian for Ω.

Mathematical Subject Classification 2000: 35K20, 35K05.
Key words: Second order linear parabolic boundary value problems, generalized solutions, optimal regularity.

They also show that uniqueness of mild solutions holds in the class

$$L_\infty(J, L_p(\Omega)), \qquad p > \sigma' := \sigma/(\sigma - 1). \tag{1.4}$$

Every mild solution of (1.1) is a distributional solution, that is, an integrable function u on $\Omega \times J$ satisfying

$$\int_J \langle -\partial_t \varphi - \Delta \varphi + a\varphi, u \rangle \, dt = \langle \varphi(0), u^0 \rangle \tag{1.5}$$

for all $\varphi \in \mathcal{D}(\Omega \times [0, T)) = \mathcal{D}([0, T), \mathcal{D}(\Omega))$, where $\mathcal{D}$ is the space of all smooth functions having compact support (in the indicated domain) and

$$\langle v, w \rangle := \int_\Omega vw \, dx.$$

In [7, Theorem A2] it is proved, in addition, that uniqueness of distributional solutions of (1.1) holds in the class (1.4), provided $p > n/(n - 2)$ and

$$a \in C(\overline{J}, L_{n/2}(\Omega)). \tag{1.6}$$

This result is optimal in the sense that, given (1.6), uniqueness fails if $p = n/(n - 2)$ (see [7, Remak A3]).

More recently, Hirata and Tsutsumi [10, Theorem 2.1] show that, given assumption (1.6), there exists a unique distributional solution of (1.1) in the class

$$C(\overline{J}, L_p(\Omega)) \cap L_p(J, L_{pn/(n-2)}(\Omega)) \cap L_{\infty,\mathrm{loc}}(J, L_q(\Omega))$$

for any $q < \infty$, provided $u^0 \in L_p(\Omega)$ and $p > 1$. These authors also show [10, Theorem 4.2] that there exist potentials

$$a \in L_\infty(J, L_{n/2}(\Omega))$$

such that uniqueness fails in the class (1.4) for $p > n/(n - 2)$. Furthermore, they show (see [10, Theorem 3.1]) that there are a satisfying (1.6) and $u^0 \in L_1(\Omega)$ such that (1.1) has no distributional solution u such that au is locally integrable on $\Omega \times J$ and $u \in C(\overline{J}, L_1(\Omega))$.

In both papers the proofs of the well-posedness assertions rely on approximation arguments, a priori estimates, and properties of the heat semigroup. In this paper we use a different approach, shedding new light on this problem. Namely, we employ maximal regularity techniques and, by this way, do not only get far reaching generalizations but also significant improvements of the above results.

2. Main results

For $s \in \mathbb{R}$ we denote by $H_q^s := H_q^s(\Omega)$, $1 < q < \infty$, the Bessel potential spaces, and by $B_{p,\rho}^s := B_{p,\rho}^s(\Omega)$, $1 \le p, \rho \le \infty$, the Besov spaces. (See [14] and [15] for precise definitions and the main properties of these spaces which we use without giving further references.)

For $q, r \in (1, \infty)$ we set

$$\mathbf{B}_{q,r}^s := \{ u \in B_{q,r}^s \; ; \; \gamma u = 0 \}, \qquad 1/q < s \leq 2,$$

where γ denotes the trace operator for Γ, and

$$\mathbf{B}_{q,r}^s := \begin{cases} B_{q,r}^s, & -2 + 1/q < s < 0, \\ (\mathbf{B}_{q',r'}^{-s})', & -2 \leq s \leq -2 + 1/q, \end{cases} \tag{2.1}$$

the dual spaces being determined by means of the duality pairing naturally induced by $\langle \cdot, \cdot \rangle$ (and usually again denoted by the same symbol).

We suppose that

- $n/(n-2) < q < \infty, \ n/2 < \sigma \leq \infty$;
- $a \in C(\bar{J}, L_{n/2}) + L_\infty(J, L_\sigma)$. $\tag{2.2}$

We also suppose that

- $0 \leq s < 2 - 1/q, \ 1 < r < \infty$;
- $((f, g), u^0) \in L_r(J, H_q^{-s} \times W_q^{-1/q}(\Gamma)) \times \mathbf{B}_{q,r}^{-2/r}$. $\tag{2.3}$

Then we consider the initial boundary value problem

$$\begin{aligned} \partial_t u - \Delta u + au &= f \quad \text{in } \Omega \times J, \\ u &= g \quad \text{on } \Gamma \times J, \\ u(\cdot, 0) &= u^0 \quad \text{on } \Omega. \end{aligned} \tag{2.4}$$

By an $L_r(L_q)$-**solution** of (2.4) we mean an element $u \in L_r(J, L_q)$ satisfying

$$\int_J \langle -\partial_t \varphi - \Delta \varphi + a\varphi, u \rangle \, dt = \int_J \{ \langle \varphi, f \rangle - \langle \partial_\nu \varphi, g \rangle_\Gamma \} \, dt + \langle \varphi(0), u^0 \rangle \tag{2.5}$$

for every $\varphi \in \mathcal{D}([0, T), \mathcal{D}_0(\overline{\Omega}))$, where

$$\mathcal{D}_0(\overline{\Omega}) := \{ \varphi \in \mathcal{D}(\overline{\Omega}) \; ; \; \varphi | \Gamma = 0 \},$$

and $\langle \cdot, \cdot \rangle_\Gamma$ is the duality pairing between $W_q^{-1/q}(\Gamma)$ and $W_q^{-1/q}(\Gamma)' = W_{q'}^{1/q}(\Gamma)$, naturally induced by

$$(v, w) \mapsto \int_\Gamma vw \, d\sigma, \qquad v, w \in C(\Gamma),$$

with $d\sigma$ denoting the volume measure of Γ. Of course, ∂_ν is the normal derivative with respect to the outer unit normal on Γ. Observe that (2.5) is formally obtained by multiplying (2.4) with a test function $\varphi \in \mathcal{D}([0, T), \mathcal{D}_0(\overline{\Omega}))$, integrating over $\Omega \times J$, integrating by parts, and using Green's formula and the initial and boundary conditions. By considering test functions φ in $\mathcal{D}(J, \mathcal{D}) = \mathcal{D}(\Omega \times J)$ only it follows from (2.5) that every $L_r(L_q)$-solution of (2.4) is a distributional solution of the first equation of (2.4).

Now we can formulate the first main result of this paper, the following existence, uniqueness, and continuity theorem.

THEOREM 2.1. *Let assumption* (2.2) *be valid. Then, given* $((f, g), u^0)$ *satisfying* (2.3), *there exists a unique* $L_r(L_q)$-*solution, u, of* (2.4). *Furthermore,*

$$u \in C(\overline{J}, \mathbf{B}_{q,r}^{-2/r}),$$

and the map $((f, g), u^0) \mapsto u$ *is linear and continuous from the space occurring in* (2.3) *to*

$$L_r(J, L_q) \cap C(\overline{J}, \mathbf{B}_{q,r}^{-2/r}).$$

If f, g, and u^0 are positive then u is positive as well.

COROLLARY 2.2. *Suppose that* $n/(n-1) < p < \infty$, *that* $n/2 < \sigma \leq \infty$, *and that*

$$a \in C(\overline{J}, L_{n/2}) + L_\infty(J, L_\sigma).$$

Put

$$q := \begin{cases} pn/(n-2) & \text{if } p \geq 2, \\ pn/(n-p) & \text{if } p < 2. \end{cases}$$

Then problem (1.1) *has for each* $u^0 \in L_p$ *a unique distributional solution in the class* $L_{p \vee 2}(J, L_q)$.

Proof. Set $r := p \vee 2$. Then, denoting by $\hookrightarrow$ continuous embedding,

$$L_p \hookrightarrow B_{p,r}^0 \hookrightarrow B_{q,r}^{-2/r}$$

thanks to $1/q = 1/p - 2/rn$. Since $-2/r > -2 + 1/q$ it follows that $B_{q,r}^{-2/r} = \mathbf{B}_{q,r}^{-2/r}$. Note that $p > n/(n-1)$ implies $q > n/(n-2)$. Thus Theorem 2.1 guarantees that (1.1) has a unique $L_{p \vee 2}(L_q)$-solution. Since the space $\mathcal{D} := \mathcal{D}(\Omega)$ is dense in $L_{p'}$ and in $L_{q'}$ it is obvious that u is an $L_{p \vee 2}(L_q)$-solution of (1.1) iff it is a distributional solution in the class $L_{p \vee 2}(J, L_q)$. $\qquad \square$

We denote by $\mathcal{M} := \mathcal{M}(\Omega)$, resp. $\mathcal{M}(\Gamma)$, the space of all bounded Radon measures on Ω, resp. Γ. Then we can formulate our second main result.

THEOREM 2.3. *Suppose that* $n/2 < \sigma \leq \infty$ *and* $a \in L_\infty(J, L_\sigma)$. *Then problem* (2.4) *has for each* $r \in (1, 2)$ *and*

$$((f, g), u^0) \in L_r(J, \mathcal{M} \times \mathcal{M}(\Gamma)) \times \mathcal{M}$$

a unique $L_r(L_q)$-*solution, for any* $q \in (1, n/(n-1))$.

COROLLARY 2.4. *Let the hypotheses of Theorem 2.3 be satisfied. Then (1.1) has for each $u^0 \in \mathcal{M}$ a unique distributional solution in $L_r(J, L_q)$, for any $q \in (1, n/(n-1))$.*

Proof. This follows from Theorem 2.3 and the density of $\mathcal{D}$ in $C_0 = \mathcal{M}'$. $\qquad\square$

Given the hypotheses of Theorems 2.1 and 2.3, respectively, it can be shown that the $L_r(L_q)$-solution is more regular on J if (f, g) has better regularity properties. For simplicity, we do not discuss these questions here.

The proofs of these theorems are given in Section 6. In fact, a more precise result than Theorem 2.3 is proven there (see Corollary 6.1.). We also refer to [3] for a detailed study of linear parabolic equations involving measures (with respect to time and space).

3. Maximal regularity

Given Banach spaces E and F, we denote by $\mathcal{L}(E, F)$ the Banach space of all bounded linear maps from E into F. Moreover, $\mathcal{L}\mathrm{is}(E, F)$ is the set of all isomorphisms in $\mathcal{L}(E, F)$. We write $[\cdot, \cdot]_\theta$ for the complex, and $(\cdot, \cdot)_{\theta,p}$, $1 \le p \le \infty$, for the real interpolation functors of exponent $\theta \in (0, 1)$ (cf. [2, Section I.2] for a summary of interpolation theory).

Let E_0 and E_1 be Banach spaces such that $E_1 \overset{d}{\hookrightarrow} E_0$, that is, E_1 is continuously and densely embedded in E_0. Then we put

$$\mathbb{W}_r^1(J, (E_0, E_1)) := L_r(J, E_1) \cap W_r^1(J, E_0).$$

It is a Banach space with the norm

$$\|u\|_{\mathbb{W}_r^1(J,(E_0,E_1))} := \|u\|_{L_r(J,E_1)} + \|\partial u\|_{L_r(J,E_0)},$$

where ∂ is the distributional derivative. Also, setting $E_{\theta,p} := (E_0, E_1)_{\theta,p}$ for $\theta \in (0, 1)$ and $1 \le p \le \infty$,

$$\mathbb{W}_r^1(J, (E_0, E_1)) \overset{d}{\hookrightarrow} C(\overline{J}, E_{1/r',r})$$

(cf. [2, Theorem III.4.10.2]). Hence $\gamma_0 u := u(0)$, the trace of $u \in \mathbb{W}_r^1(E_0, E_1)$ at $t = 0$, is well-defined. Moreover,

$$\mathbb{W}_{r,\tau}^1(J, (E_0, E_1)) := \{ u \in \mathbb{W}_r^1(J, (E_0, E_1)) \; ; \; u(\tau) = 0 \}$$

is for each $\tau \in \overline{J}$ a closed linear subspace of $\mathbb{W}_r^1(J, (E_0, E_1))$.

We suppose that $1 < r < \infty$ and denote by $\mathcal{MR}_r(E_1, E_0)$ the set of all operators A in $\mathcal{L}(E_1, E_0)$ such that, given any $f \in L_r(J, E_0)$, the Cauchy problem

$$\dot{u} + Au = f(t) \quad \text{in } J, \qquad u(0) = 0,$$

possesses a unique solution $u \in \mathbb{W}_r^1(E_0, E_1)$. It follows that $\mathcal{MR}_r(E_1, E_0)$ is open in $\mathcal{L}(E_1, E_0)$. Moreover, each $A \in \mathcal{MR}_r(E_1, E_0)$ is the negative infinitesimal generator of a strongly continuous analytic semigroup, $\{ e^{-tA} \; ; \; t \geq 0 \}$ on E_0, that is, in $\mathcal{L}(E_0)$.

THEOREM 3.5. *Suppose that* $\vartheta \in (0, 1)$, *that* E_ϑ *is an interpolation space of exponent* ϑ *between* E_0 *and* E_1, *and that*

$$A \in C(\overline{J}, \mathcal{MR}_r(E_1, E_0)), \quad B \in L_\infty(J, \mathcal{L}(E_1, E_\vartheta)).$$

Then

$$(\partial + A + B, \gamma_0) \in \mathcal{L}\mathrm{is}(\mathbb{W}_r^1(J, (E_0, E_1)), L_r(J, E_0) \times E_{1/r',r}).$$

For a proof of this theorem, as well as for the assertions preceding it, we refer to [5].

4. Stationary operators

Suppose that $1 < q < \infty$ and put

$$\mathbf{H}_q^s := \begin{cases} \{u \in H_q^s \; ; \; \gamma u = 0 \}, & 1/q < s \leq 2, \\ (\mathbf{H}_{q'}^{-s})', & -2 \leq s < -1 + 1/q, \end{cases}$$

where the dual spaces are determined by means of $\langle \cdot, \cdot \rangle$. It follows that

$$\mathbf{H}_q^s = H_q^s, \qquad -2 + 1/q < s < -1 + 1/q, \tag{4.1}$$

and that $L_q \xhookrightarrow{d} \mathbf{H}_q^{-2}$ (cf. [1, Section 7]). We define $\mathbf{A} \in \mathcal{L}(L_q, \mathbf{H}_q^{-2})$ by

$$\langle v, \mathbf{A}u \rangle_{\mathbf{H}_q^{-2}} := \langle -\Delta v, u \rangle, \qquad (v, u) \in \mathbf{H}_{q'}^2 \times L_q,$$

where, given a Banach space E, we write $\langle \cdot, \cdot \rangle_E$ for the duality pairing $E' \times E \to \mathbb{C}$.

PROPOSITION 4.1. $\mathbf{A} \in \mathcal{MR}_r(L_q, \mathbf{H}_q^{-2})$ *for* $1 < r < \infty$.

Proof. For Banach spaces E_0 and E_1 with $E_1 \xhookrightarrow{d} E_0$ we write $A \in \mathcal{BIP}(E_1, E_0)$, provided $A \in \mathcal{L}(E_1, E_0)$ and $-A$ generates a strongly continuous analytic semigroup on E_0, and if there exist $M, \omega \geq 0$ and $\vartheta \in (0, \pi/2)$ such that

$$\|(\omega + A)^{it}\|_{\mathcal{L}(E_0)} \leq M e^{\vartheta |t|}, \qquad t \in \mathbb{R}.$$

It is known that $-\Delta_D \in \mathcal{BIP}(\mathbf{H}_q^2, L_q)$ (cf. [12], [11], [13], [8], for example). Hence $\mathbf{A} \in \mathcal{BIP}(L_q, \mathbf{H}_q^{-2})$ by [2, Proposition V.1.5.5 and Theorem V.2.1.3]. From [2, Theorem III.4.5.2] and the fact that $\mathbf{H}_q^{-2}$ is isomorphic to L_q we infer that $\mathbf{H}_q^{-2}$ is a UMD space. Thus the assertion is a consequence of the Dore-Venni theorem [9] (see [2, Theorem III.4.10.8]). $\square$

Next we prove a simple technical lemma.

LEMMA 4.2. *Suppose that $n/2 \leq \sigma \leq \infty$ and $1 < q < \infty$. Also suppose that satisfies $0 \leq \tau < 1 + 1/q$ and either*

$$n/\sigma + \tau \leq 2 < n/q' + \tau \tag{4.2}$$

or

$$n - 2 + \tau < n/q \leq n/\sigma'. \tag{4.3}$$

Then

$$L_\sigma \times L_q \times \mathbf{H}_{q'}^{2-\tau} \to \mathbb{C}, \quad (b, u, v) \mapsto \langle v, bu \rangle$$

is a continuous trilinear map.

Proof. (i) Let (4.2) be satisfied. Then we deduce from Sobolev's embedding theorem that

$$\mathbf{H}_{q'}^{2-\tau} \hookrightarrow H_{q'}^{2-\tau} \hookrightarrow L_\xi, \qquad 1/\xi = 1/q' - (2 - \tau)/n.$$

Thus, since (4.2) implies

$$\frac{1}{\sigma} + \frac{1}{q} + \frac{1}{\xi} = \frac{1}{\sigma} + 1 - (2 - \tau)/n \leq 1,$$

the assertion follows from Hölder's inequality.

(ii) If (4.3) is satisfied then $\mathbf{H}_{q'}^{2-\tau} \hookrightarrow C_0$ thanks to $2 - \tau > n/q'$. Since (4.3) also implies $1/\sigma + 1/q \leq 1$, Hölder's inequality implies the assertion in this case also. $\square$

Given a function b, we write M_b for the multiplication operator $u \mapsto bu$. However, if no confusion seems likely, we simply denote M_b by b.

COROLLARY 4.3. *Let the hypotheses of Lemma 4.2 be satisfied. Then*

$$(b \mapsto M_b) \in \mathcal{L}(L_\sigma, \mathcal{L}(L_q, \mathbf{H}_q^{\tau-2})). \tag{4.4}$$

It is known that

$$\mathbf{B}_{q,r}^{2\theta-2} \doteq (\mathbf{H}_q^{-2}, L_q)_{\theta,r}, \qquad 0 < \theta < 1, \quad 1 < q, r < \infty, \tag{4.5}$$

where $\doteq$ means 'equal except for equivalent norms', and that

$$\mathbf{H}_q^{2\theta-2} \doteq [\mathbf{H}_q^{-2}, L_q]_\theta, \qquad 0 < 2\theta < 1 + 1/q, \quad 1 < q < \infty, \tag{4.6}$$

(cf. [1, Theorem 7.1]).

Now we can prove the following maximal regularity result.

THEOREM 4.4. *Suppose that $n/2 \le \sigma \le \infty$ and $b \in L_\sigma$. Also suppose that $1 < q < \infty$ and either*

$$2 < n/q' \tag{4.7}$$

or

$$n/\sigma \le n/q' < 2. \tag{4.8}$$

Then $\mathbf{A} + b \in \mathcal{MR}_r(L_q, \mathbf{H}_q^{-2})$.

Proof. (i) First suppose that $\sigma > n/2$. If (4.7) is satisfied then we can fix $\tau \in (0, 1)$ such that (4.2) is true. If (4.8) is valid then we fix $\tau \in (0, 1)$ such that (4.3) holds. Then Corollary 4.3 implies that $M_b \in \mathcal{L}(L_q, \mathbf{H}_q^{\tau-2})$. Thus, identifying $\mathbf{A}$ and M_b with the constant map

$$(t \mapsto \mathbf{A}) \in C(\overline{J}, \mathcal{L}(L_q, \mathbf{H}_q^{-2})) \quad \text{and} \quad (t \mapsto M_b) \in L_\infty(J, \mathcal{L}(L_q, \mathbf{H}_q^{\tau-2})),$$

respectively, the assertion follows from (4.6), Proposition 4.1, and Theorem 3.5.

(ii) Now suppose that $\sigma = n/2$. Proposition 4.1 and the fact that $\mathcal{MR}_r(L_q, \mathbf{H}_q^{-2})$ is open in $\mathcal{L}(L_q, \mathbf{H}_q^{-2})$ guarantee the existence of $\varepsilon > 0$ such that

$$\mathbf{A} + C \in \mathcal{MR}_r(L_q, \mathbf{H}_q^{-2}) \tag{4.9}$$

whenever

$$C \in \mathcal{L}(L_q, \mathbf{H}_q^{-2}), \qquad \|C\| \le \varepsilon. \tag{4.10}$$

Choose $d \in C(\overline{\Omega})$ such that $\|b - d\|_{L_{n/2}} \le \varepsilon/\kappa$, where κ is the norm of the map (4.4) for $\tau := 0$. Then it follows that M_{b-d} satisfies (4.10). Consequently, (4.9) is true for this choice of C. Since $d \in L_\infty$, we infer from (i) that $M_d \in \mathcal{L}(L_q, \mathbf{H}_q^{\tau-2})$ for some $\tau \in (0, 1)$. Thus, similarly as above,

$$\mathbf{A} + b = (\mathbf{A} + M_{b-d}) + M_d \in \mathcal{MR}_r(L_q, \mathbf{H}_q^{-2}).$$

This proves everything. $\qquad\qquad\square$

5. Nonautonomous problems

Throughout this section we suppose that

- $1 < r < \infty, \quad n/2 < \sigma \le \infty.$

Then we consider the following hypothesis:

- either $n/(n-2) < q < \infty$ and $a \in C(\overline{J}, L_{n/2}) + L_\infty(J, L_\sigma)$
- or $1 < q < n/(n-2)$ and $a \in L_\infty(J, L_\sigma)$. $\qquad\qquad (5.1)$

THEOREM 5.1. *Let assumption* (5.1) *be satisfied. Then*

$$(\partial + \mathbf{A} + a, \gamma_0) \in \mathcal{L}\mathrm{is}(\mathbb{W}_r^1(J, (\mathbf{H}_q^{-2}, L_q)), L_r(J, \mathbf{H}_q^{-2}) \times \mathbf{B}_{q,r}^{-2/r}).$$

Proof. (i) Suppose that $b \in C(\overline{J}, L_{n/2})$. Then we deduce from Corollary 4.3 and Theorem 4.4 that

$$\mathbf{A} + b \in C(\overline{J}, \mathcal{MR}_r(L_q, \mathbf{H}_q^{-2})).$$

(ii) Assume that $c \in L_\infty(J, L_\sigma)$. Since $\sigma > n/2$, part (i) of the proof of Theorem 4.4 shows that there exists $\tau \in (0, 1)$ such that

$$M_c \in L_\infty(J, \mathcal{L}(L_q, \mathbf{H}_q^{\tau-2})).$$

(iii) Finally, suppose that $a \in C(\overline{J}, L_{n/2}) + L_\infty(J, L_\sigma)$. Choose $b \in C(\overline{J}, L_{n/2})$ and $c \in L_\infty(J, L_\sigma)$ such that $a = b + c$. Then the theorem follows from (i), (ii), and Theorem 3.1.

$\qquad\qquad\qquad\qquad\qquad\qquad\qquad\qquad\qquad\qquad\qquad\qquad\qquad\qquad\quad \square$

PROPOSITION 5.2. *Let assumption* (5.1) *be satisfied. Then*

$$-\partial - \Delta_D + a \in \mathcal{L}\mathrm{is}(\mathbb{W}_{r',T}^1(J, (L_{q'}, \mathbf{H}_{q'}^2)), L_{r'}(J, L_{q'})).$$

Proof. First recall that $-\Delta_D \in \mathcal{MR}_{r'}(\mathbf{H}_{q'}^2, L_{q'})$. Second, if $b \in L_{n/2}$ then the preceding proofs show that $M_b \in \mathcal{L}(H_{q'}^2, L_{q'})$. Furthermore, if $b \in L_\sigma$ then we can find $\tau \in (0, 1)$ such that $M_b \in \mathcal{L}(H_{q'}^{2-\tau}, L_{q'})$. Thus an obvious modification of the proof of Theorem 5.1. implies that

$$\partial - \Delta_D + a \in \mathcal{L}\mathrm{is}(\mathbb{W}_{r',0}^1(J, (L_{q'}, \mathbf{H}_{q'}^2)), L_{r'}(J, L_{q'})).$$

Now the assertion follows by means of the transformation $t \mapsto T - t$ (cf. [2, Subsection V.2.5]). $\qquad\qquad\qquad\qquad\qquad\qquad\qquad\qquad\qquad\qquad\qquad\qquad\quad \square$

After these preparations we can prove the following equivalence theorem.

THEOREM 5.3. *Let assumption* (5.1) *be satisfied and suppose that* (F, u^0) *belongs to* $\mathbf{H}_q^{-2} \times \mathbf{B}_q^{-2/r}$. *Then the following are equivalent:*

(i) $u \in \mathbb{W}_r^1(J, (\mathbf{H}_q^{-2}, L_q))$ *and*

$$(\partial + \mathbf{A} + a)u = F, \quad u(0) = u^0;$$

(ii) $u \in L_r(J, L_q)$ *and*

$$\int_J \langle -\partial_t v - \Delta v + av, u \rangle \, dt = \int_J \langle v, F \rangle \, dt + \langle v(0), u^0 \rangle$$

for all $v \in \mathcal{D}([0, T), \mathcal{D}_0(\overline{\Omega}))$.

Proof. It follows from Proposition 5.2. and [2, Propositions V.2.6.2 and V.2.6.3] that (i) holds iff the integral relation in (ii) is true for all $v \in \mathbb{W}_{r', T}^1(J, (L_{q'}, \mathbf{H}_q^2))$. Thus the assertion is a consequence of the density of $\mathcal{D}([0, T), \mathcal{D}_0(\overline{\Omega}))$ in the latter space (cf. [3, Lemma 8(i)]). $\qquad\square$

6. Proof of the main theorems; generalizations

Now it is easy to prove Theorems 2.1 and 2.3. For this we first observe that the trace theorem implies $\partial_v \in \mathcal{L}(\mathbf{H}_{q'}^2, W_{q'}^{1/q}(\Gamma))$. Consequently,

$$(\partial_v)' \in \mathcal{L}(W_q^{-1/q}(\Gamma), \mathbf{H}_q^{-2}). \tag{6.1}$$

Proof. (i) Let assumptions (2.2) and (2.3) be satisfied. Put $F := f - (\partial_v)'g$. Then it follows from $H_q^{-s} = \mathbf{H}_q^{-s} \hookrightarrow \mathbf{H}_q^{-2}$ and (6.1) that $F \in L_r(J, \mathbf{H}_q^{-2})$. In fact,

$$((f, g) \mapsto f - (\partial_v)'g) \in \mathcal{L}(L_r(J, H_q^{-s} \times W_q^{-1/q}(\Gamma)), L_r(J, \mathbf{H}_q^{-2})).$$

Now the assertions of Theorem 2.1 — except for the positivity statement — follow from Theorems 5.1 and 5.3.

The positivity assertion is obtained by an obvious approximation argument (also see [6, Section 10]).

(ii) Let the assumptions of Theorem 2.3 be satisfied and suppose that

$$1 < q < n/(n-1).$$

Then $1/q > 1 - 1/n$ and, consequently, $W_{q'}^{1/q}(\Gamma) \overset{d}{\hookrightarrow} C(\Gamma)$. From this we deduce that

$$\mathcal{M}(\Gamma) \hookrightarrow W_q^{-1/q}(\Gamma).$$

Furthermore, since $1 > n/q'$,

$$\mathbf{H}^2_{q'} \overset{d}{\hookrightarrow} \mathbf{H}^1_{q'} \overset{d}{\hookrightarrow} C_0$$

from which we infer that

$$\mathcal{M} \hookrightarrow \mathbf{H}^{-1}_q \hookrightarrow \mathbf{H}^{-2}_q.$$

This and (6.1) imply that

$$((f, g) \mapsto f - (\partial_v)'g) \in \mathcal{L}(L_r(J, \mathcal{M} \times \mathcal{M}(\Gamma)), L_r(J, \mathbf{H}^{-2}_q))$$

for $1 < r < \infty$.

Recall that

$$\mathbf{B}^s_{q',r'} \overset{d}{\hookrightarrow} C_0, \qquad s > n/q', \quad 1 < r < \infty.$$

Thus $\mathcal{M} \hookrightarrow \mathbf{B}^{-s}_{q,r}$ for $s > n/q'$ and $1 < r < \infty$. Consequently,

$$\mathcal{M} \hookrightarrow \mathbf{B}^{-2/r}_{q,r}, \qquad 2/r + n/q > n.$$

Since $n/q > n - 1$ we see that

$$\mathcal{M} \hookrightarrow \mathbf{B}^{-2/r}_{q,r}, \qquad 1 < q < n/(n-1)$$

if $1 < r < 2$. Hence the assertion of Theorem 2.3 follows also from Theorems 5.1 and 5.3.
$\qquad\square$

COROLLARY 6.1. *Suppose that $1 < r < \infty$. Also suppose that $n/2 < \sigma \le \infty$ and $a \in L_\infty(J, L_\sigma)$. Then problem (2.4) has for each*

$$((f, g), u^0) \in L_r(J, \mathcal{M} \times \mathcal{M}(\Gamma)) \times \mathcal{M}$$

a unique $L_r(L_q)$-solution, u, provided $1 < q < n/(n-1)$ and $2/r + n/q > n$. The map

$$L_r(J, \mathcal{M} \times \mathcal{M}(\Gamma)) \times \mathcal{M} \to L_r(J, L_q), \quad ((f, g), u^0) \mapsto u$$

is linear and continuous. Furthermore, $u \ge 0$ if $((f, g), u^0) \ge 0$.

For simplicity, we have restricted ourselves to the model problem (2.4). However, the following generalizations are possible.

- The operator $-\Delta$ can be replaced by a general uniformly elliptic operator of the form

$$-\partial_j(a_{jk}\partial_k u) + a_j \partial_j u,$$

  with $a_{jk} = a_{kj} \in C^1(\overline{\Omega})$ and $a_j \in L_\infty(J, C^1(\overline{\Omega}))$ for $1 \le j, k \le n$ (summation convention).

- Neumann or, more generally, conormal boundary conditions can be handled also.
- Instead of a single equation one can treat strongly coupled normally parabolic systems in the sense of [1].
- Higher order systems can be studied as well.
- It suffices to assume that Ω has a compact boundary.

The main problem dealing with these generalizations is the proof of maximal regularity results for nonautonomous parabolic systems in a weak setting and with minimal regularity conditions for the coefficients. For this we refer to [5]. In [4] these results are then applied to nonlinear problems in order to identify and understand 'critical exponents' and derive optimal existence results.

REFERENCES

[1]   AMANN, H., *Nonhomogeneous linear and quasilinear elliptic and parabolic boundary value problems.* In H. J. Schmeisser, H. Triebel, editors, Function Spaces, Differential Operators and Nonlinear Analysis, pages 9–126. Teubner, Stuttgart, Leipzig, 1993.

[2]   AMANN, H., *Linear and Quasilinear Parabolic Problems, Volume I: Abstract Linear Theory.* Birkhäuser, Basel 1995.

[3]   AMANN, H., *Linear parabolic problems involving measures.* Rev. R. Acad. Cien. Serie A. Mat. (RACSAM), *95* (2001), 85–119.

[4]   AMANN, H., *Maximal regularity and critical growth in semilinear parabolic problems* 2002. To appear.

[5]   AMANN, H., *Maximal regularity and weak solutions of linear parabolic equations* 2002. To appear.

[6]   AMANN, H. and QUITTNER, P., *Semilinear parabolic equations involving measures and low-regularity data* 2002. Preprint.

[7]   BREZIS, H., and CAZENAVE, TH., *A nonlinear heat equation with singular initial data.* J. d'Analyse Math. *68* (1996), 277–304.

[8]   DENK, R., DORE, G., HIEBER, M., PRÜSS, J. and VENNI, A., *New thoughts on old results of R. T. Seeley* 2002. Preprint.

[9]   DORE, G. and VENNI, A., *On the closedness of the sum of two closed operators.* Math. Z. *196* (1987), 189–201.

[10]   HIRATA, D. and TSUTSUMI, M., *On the well-posedness of a linear heat equation with a critical singular potential.* Diff. Int. Equ. *14* (2001), 1–18.

[11]   PRÜSS, J. and SOHR, H., *On operators with bounded imaginary powers in Banach spaces.* Math. Z. *203* (1990), 429–452.

[12]   SEELEY, R. T., *Complex powers of an elliptic operator. In* singular integrals. Proc. Symp. Pure Math. AMS, Providence *10* (1968), 288–307.

[13]   SOHR, S., Beschränkter H_∞-Funktionalkalkül für elliptische Randwertsysteme. Doctoral dissertation, Universität GH Kassel, Germany 1999.

[14]   TRIEBEL, H., *Interpolation Theory, Function Spaces, Differential Operators.* North Holland, Amsterdam 1978.

[15]   TRIEBEL, H., *Theory of Function Spaces.* Birkhäuser, Basel 1983.

Herbert Amann
Institut für Mathematik
Universität Zürich
Winterthurerstr. 190
CH–8057 Zürich
Switzerland

J.evol.equ. 3 (2003) 407 – 418
1424–3199/03/030407 – 12
DOI 10.1007/s00028-003-0109-7
© Birkhäuser Verlag, Basel, 2003

**Journal of Evolution
Equations**

Some noncoercive parabolic equations with lower order terms in divergence form

LUCIO BOCCARDO[1], LUIGI ORSINA[2] AND ALESSIO PORRETTA[3]

Trois générations d'amis romains dédient cet article à Philippe

Abstract. This paper deals with existence and regularity results for the problem

$$\begin{cases} u_t - \operatorname{div}(a(x,t,u)\nabla u) = -\operatorname{div}(uE) & \text{in } \Omega \times (0,T), \\ u = 0 & \text{on } \partial\Omega \times (0,T), \\ u(0) = u_0 & \text{in } \Omega, \end{cases}$$

under various assumptions on E and u_0. The main difficulty in studying this problem is due to the presence of the term $\operatorname{div}(uE)$, which makes the differential operator non coercive on the "energy space" $L^2(0,T;H_0^1(\Omega))$.

1. Introduction

In this paper we are going to study the following parabolic problem

$$\begin{cases} u_t - \operatorname{div}(a(x,t,u)\nabla u) = -\operatorname{div}(uE) & \text{in } \Omega \times (0,T), \\ u = 0 & \text{on } \partial\Omega \times (0,T), \\ u(0) = u_0 & \text{in } \Omega, \end{cases} \tag{1}$$

where Ω is a bounded open subset of $\mathbf{R}^N$, $N \geq 2$ and $T > 0$. We assume that $a(x,t,s)$: $\Omega \times (0,T) \times \mathbf{R} \to \mathbf{R}$ is a function which is continuous with respect to s for almost every $(x,t) \in \Omega \times (0,T)$, measurable with respect to (x,t) for every $s \in \mathbf{R}$, and satisfies the following boundedness and coercivity assumptions:

$$0 < \alpha \leq a(x,t,s) \leq \beta, \tag{2}$$

for almost every $(x,t) \in \Omega \times (0,T)$, for every $s \in \mathbf{R}$, where α and β are two positive constants.

As far as the data are concerned, we assume that

$$u_0 \in L^1(\Omega), \qquad E \in (L^2(Q))^N.$$

Mathematics Subject Classification 2000: 35K10, 35K15, 35K65.
Key words: Nonlinear parabolic equations, noncoercive problems, infinite energy solutions.

The main difficulty in studying this problem is due to the presence of the term $\mathrm{div}(uE)$, which makes the differential operator non coercive on $L^2(0, T; H_0^1(\Omega))$. Since we do not assume that E is constant or that $\mathrm{div}(E) = 0$, problem (1) cannot be studied with the techniques used in [6], [4], [9], where this term can be managed thanks to the divergence theorem and the Dirichlet condition at the boundary.

Our assumption that $|E|$ may only belong to $L^2(Q)$ yields solutions with very little summability even if $u_0 \in L^\infty(\Omega)$, namely only $\log(1 + |u|)$ will be in $L^2(0, T; H_0^1(\Omega))$; thanks to the framework of entropy solutions, this is however enough to give a formulation to the problem. We also prove several existence and regularity results for problem (1) in the case $|E| \in L^r(0, T; L^q(\Omega))$ with r, q large enough, under various assumptions on u_0. Since the lack of regularity of the solutions u is due to the fact that E is singular, our results for (1) will be the same in presence of an additional term $f(x, t)$ in the right hand side, satisfying suitable summability assumptions (see Remark 2.2).

Our results extend those obtained in [3] for the stationary case; we also point out that the motivation for dealing with such equations comes from some applicative models studied in [7], where the term E is divergence free since it is the gradient of an harmonic function.

In the following, we denote by $Q_\sigma = \Omega \times (0, \sigma)$, $\sigma \in (0, T)$, and by $Q = \Omega \times (0, T)$, $\Sigma = \partial\Omega \times (0, T)$.

Moreover, for $k > 0$ and s in $\mathbf{R}$, we define the truncating function

$$T_k(s) = \min\{k, \max\{-k, s\}\},$$

and we set $G_k(s) = s - T_k(s)$.

In order to prove our existence and regularity results, we will consider the approximating problems

$$\begin{cases} (u_n)_t - \mathrm{div}(a(x, t, u_n)\nabla u_n) = -\mathrm{div}(T_n(u_n)\,E) & \text{in } Q, \\ u_n = 0 & \text{on } \Sigma, \\ u_n(0) = T_n(u_0) & \text{in } \Omega. \end{cases} \tag{3}$$

By known results (see, for instance, [8]), there exists at least a weak solution u_n of (3) which belongs to $L^2(0, T; H_0^1(\Omega)) \cap C^0([0, T]; L^2(\Omega))$.

We will then prove some results as follows; in Section 2 we deal with the case in which we have *a priori* estimates for u_n in $L^\infty(Q)$, hence the existence of bounded solutions of (1). In Section 3 we prove a priori estimates which yield unbounded solutions, which may, or may not, belong to $L^2(0, T; H_0^1(\Omega))$ depending on the summability of $|E|$ and u_0; using the unifying framework of entropy solutions we will then be able to pass to the limit in (3) and get existence results for (1) in the general case.

2. Bounded solutions

As stated in the Introduction, we are going to prove that if E has sufficient regularity, then (1) has a bounded solution. Note that the assumptions on E and u_0 in Theorem (2.1) below are such that the problem

$$u_t - \operatorname{div}(a(x, t, u)\nabla u) = -\operatorname{div}(E), \qquad u(0) = u_0,$$

has at least a bounded solution (see [1]); this is not surprising since E and uE have the same summability if u is bounded.

THEOREM 2.1. *Let $|E| \in L^r(0, T; L^q(\Omega))$ with $\frac{2}{r} + \frac{N}{q} < 1$ and $u_0 \in L^\infty(\Omega)$. Then there exists a function u in $L^\infty(Q) \cap L^2(0, T; H_0^1(\Omega))$ which is a weak solution of (1) in the sense that*

$$\langle u_t, \varphi \rangle + \int_Q a(x, t, u)\nabla u \cdot \nabla\varphi\, dx\, dt = \int_Q uE\, \nabla\varphi\, dx\, dt, \qquad (4)$$

for every φ in $L^2(0, T; H_0^1(\Omega))$, where $\langle \cdot, \cdot \rangle$ denotes the duality product between $L^2(0, T; H_0^1(\Omega))$ and $L^2(0, T; H^{-1}(\Omega))$.

Proof. Let u_n be a solution of (3), let τ in $(0, T)$, and choose $\varphi = \frac{G_k(u_n)}{1+|u_n|}\chi_{(0,\tau)}$ as test function in (3). Observe that $\nabla\varphi = (1+k)\frac{\nabla G_k(u_n)}{(1+|u_n|)^2}$; taking $k > \|u_0\|_{L^\infty(Q)}$ we get, after dividing by $1+k$,

$$\int_\Omega \left(\int_0^{u_n(\tau)} \frac{G_k(r)}{(1+k)(1+|r|)}\, dr \right) dx + \int_{Q_\tau} a(x, t, u_n)\frac{|\nabla G_k(u_n)|^2}{(1+|u_n|)^2}\, dx\, dt$$

$$\leq \int_{Q_\tau} |E|\, |u_n|\, \frac{\nabla G_k(u_n)}{(1+|u_n|)^2}\, dx\, dt,$$

which yields, using (2), Young inequality, and the fact that $\frac{|u_n|}{1+|u_n|} \leq 1$,

$$\int_\Omega \left(\int_0^{u_n(\tau)} \frac{G_k(r)}{(1+k)(1+|r|)}\, dr \right) dx + \int_{Q_\tau} \frac{|\nabla G_k(u_n)|^2}{(1+|u_n|)^2}\, dx\, dt$$

$$\leq c \int_{\{|u_n|>k\}} |E|^2\, dx\, dt. \qquad (5)$$

Here, and in the following, we will denote by c various constants (independent on n), whose value may vary from line to line. We have (since $e^s \geq 1 + s + \frac{s^2}{2}$ if $s \geq 0$)

$$\int_0^s \frac{G_k(r)}{(1+k)(1+|r|)}\, dr = \left[\frac{1+|s|}{1+k} - 1 - \log\left(\frac{1+|s|}{1+k}\right) \right]\chi_{\{|s|>k\}}$$

$$\geq \frac{1}{2}\left[\log^2\left(\frac{1+|s|}{1+k}\right) \right]\chi_{\{|s|>k\}},$$

so that (5) implies

$$\int_{\Omega} \left| \log\left(\frac{1+|u_n|}{1+k}\right) \right|^2 \chi_{\{|u_n|>k\}} \, dx \, dt + \int_{Q_\tau} \left| \nabla \log\left(\frac{1+|u_n|}{1+k}\right) \right|^2 \chi_{\{|u_n|>k\}} \, dx \, dt$$

$$\leq c \int_{\{|u_n|>k\}} |E|^2 \, dx \, dt, \quad \forall \tau \in (0,T).$$

Defining $v_n = \log(1+|u_n|)\,\mathrm{sign}(u_n)$ and $h = \log(1+k)$, the previous inequality gives

$$\sup_{[0,T]} \int_{\Omega} |G_h(v_n)|^2 \, dx + \int_{Q} |\nabla G_h(v_n)|^2 \, dx \, dt \leq c \int_{\{|v_n|>h\}} |E|^2 \, dx \, dt. \tag{6}$$

Since $|E| \in L^r(0,T;L^q(\Omega))$ with $\frac{2}{r} + \frac{N}{q} < 1$, from a result due to D.G. Aronson and J. Serrin (see [1]) it follows that there exists $M > 0$ such that $\|v_n\|_{L^\infty(Q)} \leq M$ for every n. The definition of v_n then implies that $\|u_n\|_{L^\infty(Q)} \leq e^M - 1$. Thus, since $T_n(u_n) = u_n$ as soon as $n > e^M - 1$, we have that u_n is a bounded solution of (1) which belongs to $L^2(0,T;H_0^1(\Omega))$. $\qquad\square$

REMARK 2.2. The same boundedness result for solutions can be proved if the right hand side of (1) is of the form $-\mathrm{div}(uE) + f(x,t)$, with f in $L^r(0,T;L^q(\Omega))$, and

$$\frac{N}{q} + \frac{2}{r} < 2, \qquad r \geq 1, \qquad q \geq 1.$$

3. Unbounded solutions

Our first result gives an *a priori* estimate on u_n under minimal regularity assumptions on E and u_0.

LEMMA 3.1. *Let $|E| \in L^2(Q)$, let $u_0 \in L^1(\Omega)$, and let u_n be a solution of (3) in the sense of (4). Then there exists a constant $c > 0$ such that, for every $n \in \mathbf{N}$,*

$$\|u_n\|_{L^\infty(0,T;L^1(\Omega))} + \|\log(1+|u_n|)\|_{L^2(0,T;H_0^1(\Omega))} \leq c,$$

and

$$\int_{Q} |\nabla T_k(u_n)|^2 \, dx \, dt \leq c k^2, \qquad \forall k \geq 1. \tag{7}$$

Proof. The use of $\frac{u_n}{1+|u_n|}$ as test function yields, reasoning as in the proof of Theorem 2.1,

$$\int_\Omega \left(\int_0^{u_n(T)} \frac{r}{1+|r|} \, dr \right) dx + \frac{\alpha}{2} \int_Q \frac{|\nabla u_n|^2}{(1+|u_n|)^2} \, dx \, dt$$

$$\leq c \int_Q |E|^2 \, dx \, dt + \int_\Omega \left(\int_0^{T_n(u_0)} \frac{r}{1+|r|} \, dr \right) dx \leq c \, \|E\|_{L^2(Q)}^2 + \|u_0\|_{L^1(\Omega)}.$$

Hence, the estimate on $\log(1+|u_n|)$ in $L^2(0, T; H_0^1(\Omega))$ follows. In particular

$$\frac{1}{(1+k)^2} \int_{\{|u_n| \leq k\}} |\nabla u_n|^2 \, dx \, dt \leq \int_Q \frac{|\nabla u_n|^2}{(1+|u_n|)^2} \, dx \, dt \leq c,$$

which yields (7).

Let $\varepsilon > 0$, let $\tau \in (0, T)$, and take $T_\varepsilon(u_n) \, \chi_{(0,\tau)}$ as test function in (3); using (2), and setting $\Theta_\varepsilon(s) = \int_0^s T_\varepsilon(r) \, dr$, we obtain

$$\int_\Omega \Theta_\varepsilon(u_n(\tau)) dx + \alpha \int_{Q_\tau} |\nabla T_\varepsilon(u_n)|^2 \, dx \, dt$$

$$\leq \int_{Q_\tau} |E| \, |u_n| \, |\nabla T_\varepsilon(u_n)| dx \, dt + \int_\Omega \Theta_\varepsilon(T_n(u_0)) dx.$$

Since we have $|u_n| \, |\nabla T_\varepsilon(u_n)| \leq \varepsilon |\nabla T_\varepsilon(u_n)|$, Young inequality implies

$$\int_\Omega \Theta_\varepsilon(u_n(\tau)) \, dx + \frac{\alpha}{2} \int_{Q_\tau} |\nabla T_\varepsilon(u_n)|^2 \, dx \, dt$$

$$\leq \frac{1}{2\alpha} \varepsilon^2 \int_{Q_\tau} |E|^2 \, dx \, dt + \int_\Omega \Theta_\varepsilon(T_n(u_0)) \, dx.$$

Since $\lim_{\varepsilon \to 0} \frac{\Theta_\varepsilon(r)}{\varepsilon} = |r|$, dividing by ε and then letting $\varepsilon \to 0$, we have

$$\|u_n(\tau)\|_{L^1(\Omega)} \leq \|u_0\|_{L^1(\Omega)},$$

for every $\tau \in [0, T]$, and this concludes the proof. $\square$

We are now going to deal with the *a priori* estimates in the case of u_0 unbounded, with assumptions on E similar to those of Theorem 2.1; however, we will no longer obtain bounded solutions.

LEMMA 3.2. *Let $|E|$ belong to $L^r(0, T; L^q(\Omega))$, with $r \geq 2$ and $q \geq N$ such that $\frac{2}{r} + \frac{N}{q} \leq 1$. Let u_0 be in $L^m(\Omega)$, with $1 < m < +\infty$. Then the sequence of the solutions u_n of (3) is bounded in $L^\infty(0, T; L^m(\Omega)) \cap L^\rho(0, T; W_0^{1,\rho}(\Omega))$, with $\rho = 2$ if $m \geq 2$, and $\rho = \frac{m(N+2)}{m+N}$ if $1 < m < 2$.*

Proof. We consider first the case in which $r < +\infty$ (or equivalently, $q > N$). We start by proving the following claim.

CLAIM. *Let v be a function such that*

$$\sup_{s\in[0,t]} \int_\Omega |v(s)|^2 \, dx + \int_{Q_t} |\nabla v|^2 \, dx \, ds \le c + c \int_{Q_t} |E|^2 \, |v|^2 \, dx \, ds, \tag{8}$$

for every $t \in [0, T]$.

Then there exist τ in $(0, T]$, depending only on E, and a constant c such that

$$\|v\|_{L^\infty(0,\tau;L^2(\Omega))} + \|v\|_{L^2(0,\tau;H_0^1(\Omega))} \le c. \tag{9}$$

Let τ belong to $(0, T]$, and let r and q be as in the statement. Assume for the moment that $q < +\infty$, which implies $r > 2$. Since $q > N \ge 2$,

$$2 \le \frac{2q}{q-2} < 2^*,$$

where $2^* = \frac{2N}{N-2}$ if $N > 2$ or any real number larger than $\frac{2q}{q-2}$ if $N = 2$, so that, for every t in $(0, \tau)$, by the interpolation inequality,

$$\left(\int_\Omega |v(x,t)|^{\frac{2q}{q-2}} \, dx \right)^{\frac{q-2}{2q}} \le \left(\int_\Omega |v(x,t)|^2 \, dx \right)^{\frac{q-N}{2q}} \left(\int_\Omega |v(x,t)|^{2^*} \, dx \right)^{\frac{N}{2^*q}}. \tag{10}$$

Thus, recalling the Sobolev inequality,

$$\left(\int_\Omega |v(x,t)|^{\frac{2q}{q-2}} \, dx \right)^{\frac{q-2}{q}\frac{r}{r-2}}$$

$$\le c \left(\int_\Omega |v(x,t)|^2 \, dx \right)^{\frac{q-N}{q}\frac{r}{r-2}} \left(\int_\Omega |\nabla v(x,t)|^2 \, dx \right)^{\frac{N}{q}\frac{r}{r-2}}. \tag{11}$$

Suppose now that $\frac{2}{r} + \frac{N}{q} = 1$, so that $\frac{N}{q}\frac{r}{r-2} = 1$. Integrating (11) from 0 to τ, by (8) we have

$$\int_0^\tau \left(\int_\Omega |v(x,t)|^{\frac{2q}{q-2}} \, dx \right)^{\frac{q-2}{q}\frac{r}{r-2}} dt \le c + c \left(\int_{Q_\tau} |E|^2 \, |v|^2 \, dx \, dt \right)^{\frac{r}{r-2}}.$$

Since, by Hölder inequality applied twice,

$$\int_{Q_\tau} |E|^2 \, |v|^2 dx \, dt \le \left(\int_0^\tau \left(\int_\Omega |E|^q \, dx \right)^{\frac{r}{q}} dt \right)^{\frac{2}{r}}$$

$$\left(\int_0^\tau \left(\int_\Omega |v(x,t)|^{\frac{2q}{q-2}} dx \right)^{\frac{q-2}{q}\frac{r}{r-2}} dt \right)^{\frac{r-2}{r}}, \tag{12}$$

setting

$$X(\tau) = \int_0^\tau \left(\int_\Omega |v(x,t)|^{\frac{2q}{q-2}}\, dx \right)^{\frac{q-2}{q}\frac{r}{r-2}} dt \tag{13}$$

we have

$$X(\tau) \le c + c \left(\int_0^\tau \left(\int_\Omega |E|^q\, dx \right)^{\frac{r}{q}} dt \right)^{\frac{2}{r-2}} X(\tau),$$

Choosing τ small enough in such a way that

$$c \left(\int_0^\tau \left(\int_\Omega |E|^q\, dx \right)^{\frac{r}{q}} dt \right)^{\frac{2}{r-2}} \le \frac{1}{2},$$

which is possible since $r < +\infty$ and since E belongs to $L^r(0,T; L^q(\Omega))$, we have $X(\tau) \le c$, so that (12) implies

$$\int_{Q_\tau} |E|^2\, |v|^2\, dx\, dt \le c,$$

and this yields (9) using (8). Suppose now that $\frac{2}{r} + \frac{N}{q} < 1$. Then $\frac{N}{q}\frac{r}{r-2} < 1$, so that from (11) and from the Hölder inequality applied to the term

$$\int_0^\tau \left(\int_\Omega |\nabla v(x,t)|^2\, dx \right)^{\frac{N}{q}\frac{r}{r-2}} dt,$$

one obtains

$$\int_0^\tau \left(\int_\Omega |v(x,t)|^{\frac{2q}{q-2}}\, dx \right)^{\frac{q-2}{q}\frac{r}{r-2}} dt \le c + c \left(\int_{Q_\tau} |E|^2\, |v|^2\, dx\, dt \right)^\sigma,$$

for some $\sigma < \frac{r}{r-2}$. Thus, defining $X(\tau)$ as in (13)

$$X(\tau) \le c + c \left(\int_0^\tau \left(\int_\Omega |E|^q\, dx \right)^{\frac{r}{q}} dt \right)^{\frac{2\sigma}{r}} X(\tau)^{\sigma\,\frac{r-2}{r}}$$

$$\le c + c\, \|E\|_{L^r(0,T;L^q(\Omega))}^{2\sigma}\, X(\tau)^{\sigma\,\frac{r-2}{r}}$$

and this yields $X(\tau)$ bounded. Thus one can choose $\tau = T$ to obtain (9). The case $q = +\infty$ is dealt with similarly, but we do not need to use the interpolation inequality. Indeed, since

$$\int_{Q_\tau} |E|^2\, |v|^2\, dx\, dt \le \left(\int_0^\tau \|E\|_{L^\infty(\Omega)}^2\, dt \right) \sup_{[0,\tau]} \int_\Omega |v|^2\, dx,$$

and since $\|E\|_{L^\infty(\Omega)}$ belongs to $L^r(0, T)$ with $r \geq 2$, choosing τ small enough allows to deduce (9) directly from (8) with $t = \tau$. Thus the claim is proved.

Let now $m \geq 2$, let $s \in (0, T)$, and choose $\varphi = |u_n|^{m-2} u_n \chi_{(0,s)}$ as test function in (3). We have

$$\int_\Omega |u_n(s)|^m \, dx + \int_{Q_s} |\nabla u_n|^2 \, |u_n|^{m-2} \, dx \, dt$$
$$\leq c \int_{Q_s} |E| \, |\nabla u_n| \, |u_n|^{m-1} \, dx \, dt + c \int_\Omega |T_n(u_0)|^m \, dx. \tag{14}$$

After applying the Young inequality to the first term of the right hand side, and using the fact that u_0 belongs to $L^m(\Omega)$, we have

$$\sup_{s \in (0,t)} \int_\Omega |u_n(s)|^m \, dx + \int_{Q_t} |\nabla u_n|^2 \, |u_n|^{m-2} \, dx \, ds \leq c + c \int_{Q_t} |E|^2 \, |u_n|^m \, dx \, ds.$$

Observing that $|\nabla u_n|^2 \, |u_n|^{m-2} = c \, |\nabla |u_n|^{\frac{m}{2}}|^2$, we have, setting $v_n = |u_n|^{\frac{m}{2}}$,

$$\sup_{s \in (0,t)} \int_\Omega |v_n(s)|^2 \, dx + \int_{Q_t} |\nabla v_n|^2 \, dx \, ds \leq c + c \int_{Q_t} |E|^2 \, |v_n|^2 \, dx \, ds,$$

which is exactly (8) for v_n. Thus, there exists τ in (0,T], depending only on the norm of E in $L^r(0, \tau; L^q(\Omega))$, such that ($v_n$ is bounded in $L^\infty(0, \tau; L^2(\Omega))$, hence) u_n is bounded in $L^\infty(0, \tau; L^m(\Omega))$.

To obtain the estimate in the whole Q_T, simply observe that, at $t = \tau$, u solves the same equation, with initial datum $u(\tau)$; therefore, split the interval $(0, T)$ in a finite number of subintervals (t_k, t_{k+1}) in such a way that the norm of E in $L^r(t_k, t_{k+1}; L^q(\Omega))$ is small enough to prove the claim. We then have that the sequence $\{u_n\}$ is bounded in $L^\infty(0, t_1; L^m(\Omega))$; this implies that it is bounded in $L^\infty(t_1, t_2; L^m(\Omega))$. Going on, we have thus proved the *a priori* estimate on $\{u_n\}$ in $L^\infty(0, T; L^m(\Omega))$ if $m \geq 2$.

If $1 < m < 2$, the choice of $\varphi = |u_n|^{m-2} u_n \chi_{(0,s)}$ is not admissible (since its gradient may not be in $L^2(\Omega)$). Anyway, it is possible to choose $\varphi = [(\varepsilon + |u_n|)^{m-1} - \varepsilon^{m-1}] \, \text{sign}(u_n) \chi_{(0,s)}$ as test function in (3), and then let ε tend to zero in order to obtain again (14).

As far as gradient estimates are concerned, observe that from (9) we have that $\{v_n\}$ is bounded in $L^2(0, T; H_0^1(\Omega))$; if $m \geq 2$, this implies

$$\int_{\{|u_n|>1\}} |\nabla u_n|^2 \leq \int_Q |u_n|^{m-2} |\nabla u_n|^2 \leq c,$$

which implies, together with (7) (with $k = 1$), that u_n is bounded in $L^2(0, T; H_0^1(\Omega))$.

If $1 < m < 2$, we have, again by (7) and (14),

$$\int_Q |\nabla u_n|^\rho \leq \int_{\{|u_n|\leq 1\}} |\nabla u_n|^\rho + \int_{\{|u_n|>1\}} |\nabla u_n|^\rho$$

$$\leq c + \int_{\{|u_n|>1\}} \frac{|\nabla u_n|^\rho}{|u_n|^{(2-m)\frac{\rho}{2}}} |u_n|^{(2-m)\frac{\rho}{2}}$$

$$\leq c + \left(\int_{\{|u_n|>1\}} \frac{|\nabla u_n|^2}{|u_n|^{(2-m)}} \right)^{\frac{\rho}{2}} \left(\int_{\{|u_n|>1\}} |u_n|^{\frac{\rho(2-m)}{2-\rho}} \right)^{1-\frac{\rho}{2}}$$

$$\leq c + c \left(\int_Q |u_n|^{\frac{\rho(2-m)}{2-\rho}} \right)^{1-\frac{\rho}{2}}. \tag{15}$$

Since $v_n = |u_n|^{\frac{m}{2}}$ is bounded in $L^\infty(0, T; L^2(\Omega)) \cap L^2(0, T; H_0^1(\Omega))$, by interpolation (see (10) with $q = N + 2$) we have that v_n is bounded in $L^{\frac{2(N+2)}{N}}(Q)$, hence u_n is bounded in $L^{\frac{m(N+2)}{N}}(Q)$. Choosing $\rho = \frac{m(N+2)}{m+N}$ we have $\frac{\rho(2-m)}{2-\rho} = \frac{m(N+2)}{N}$ and we deduce from (15) that the sequence $\{u_n\}$ is bounded in $L^\rho(0, T; W_0^{1,\rho}(\Omega))$. This concludes the proof for $r < +\infty$.

The case $r = +\infty$, which means $|E| \in L^\infty(0, T; L^N(\Omega))$, is slightly different and the previous argument does not apply. However, it is possible to use the same method of the elliptic case (see [3]). One has to choose $\varphi = |G_k(u_n)|^{m-2}G_k(u_n)$ (suitably modified near $u = 0$ if $m < 2$) as test function in (3) in order to obtain an estimate like (8) for $v_n = |G_k(u_n)|^{\frac{m}{2}}$. Instead of getting an estimate for small time intervals one obtains an estimate for k large enough using that meas $\{|u_n| > k\}$ tends to zero as k goes to infinity as a consequence of the estimate on u_n in $L^\infty(0, T; L^1(\Omega))$ (see Lemma 3.1); this estimate allows to conclude using the estimates on the truncates proved in Lemma 3.1. $\square$

In view of the previous estimates, we obtain the existence of a solution of (1) by passing to the limit in (3). However, since E and u_0 may be singular, solutions may have infinite energy and the distributional formulation of problem (1) may have no sense. A unifying framework for this class of singular problems is provided by the notion of entropy solution, first introduced in [2] for elliptic problems, then extended to parabolic equations in [10]. In order to introduce entropy solutions, we need first to suitably define the gradient of measurable functions whose truncates have finite energy.

DEFINITION 3.3. Let v be a measurable function on Q which is almost everywhere finite and such that $T_k(v)$ belongs to $L^2(0, T; H_0^1(\Omega))$ for every $k > 0$. Then (see [2], Lemma 2.1) there exists a unique measurable function $w : Q \to \mathbf{R}^N$ such that

$$\nabla T_k(v) = w \, \chi_{\{|v|<k\}}, \quad \text{almost everywhere in } Q, \text{ for every } k > 0.$$

We will define the gradient of v as the function w, and we will denote it by $w = \nabla v$.

We now give the definition of entropy solution. We denote by $\Theta_k(s) = \int_0^s T_k(r)\,dr$.

DEFINITION 3.4. A measurable function $u \in L^\infty(0, T; L^1(\Omega))$ is an entropy solution of (1) if $T_k(u) \in L^2(0, T; H_0^1(\Omega))$ for every $k > 0$ and u satisfies

$$\int_\Omega \Theta_k(u - \varphi)(t)\,dx - \langle \varphi_s, T_k(u - \varphi) \rangle + \int_0^t \int_\Omega a(x, s, u)\nabla u\, \nabla T_k(u - \varphi)\,dx\,ds$$

$$\leq \int_0^t \int_\Omega u E\, \nabla T_k(u - \varphi)\,dx\,ds + \int_\Omega \Theta_k(u_0 - \varphi(0))\,dx, \tag{16}$$

for almost every $t \in (0, T)$, for every $k > 0$ and every $\varphi \in L^2(0, T; H_0^1(\Omega)) \cap L^\infty(Q)$ such that $\varphi_t \in L^2(0, T; H^{-1}(\Omega)) + L^1(Q)$ (in (16) $\langle \cdot, \cdot \rangle$ denotes the duality between $L^2(0, t : H^{-1}(\Omega)) + L^1(Q_t)$ and $L^2(0, t; H_0^{-1}(\Omega)) \cap L^\infty(Q_t)$).

We now state and prove the main existence theorem of the paper.

THEOREM 3.5. *Let $|E| \in L^2(Q)$, and let $u_0 \in L^1(\Omega)$. Then there exists an entropy solution u of (1) which satisfies $\log(1 + |u|) \in L^2(0, T; H_0^1(\Omega))$.*
If $|E| \in L^r(0, T; L^q(\Omega))$, with $r \geq 2$ and $q \geq N$ such that $\frac{2}{r} + \frac{N}{q} \leq 1$, and u_0 is in $L^m(\Omega)$, with $m > 1$, then u is in $L^\infty(0, T; L^m(\Omega))$, and belongs to $L^\rho(0, T; W_0^{1,\rho}(\Omega))$ with $\rho = 2$ if $m \geq 2$, and $\rho = \frac{m(N+2)}{m+N}$ if $1 < m < 2$.

Proof. Thanks to Lemma 3.1, the estimates on the solutions u_n of (3), which only depend on $\|E\|_{L^2(Q)}$ and on $\|u_0\|_{L^1(\Omega)}$, imply that u_n is bounded in $L^\infty(0, T; L^1(\Omega))$, $\log(1 + |u_n|)$ is bounded in $L^2(0, T; H_0^1(\Omega))$ and $T_k(u_n)$ is bounded in $L^2(0, T; H_0^1(\Omega))$ for every $k > 0$. By using suitable test functions it is possible to prove (see [9]) that $\mathcal{T}_k(u_n)_t$ is bounded in $L^2(0, T; H^{-1}(\Omega)) + L^1(Q)$, where $\mathcal{T}$ is a regularized truncation; then, standard compactness results imply that $T_k(u_n)$ converges, up to subsequences, almost everywhere and in $L^1(Q)$. This fact, and the estimates on $\log(1 + |u_n|)$, imply that (a subsequence of) u_n almost everywhere converges to a function u. As a consequence, we have that $T_k(u_n)$ converges to $T_k(u)$ weakly in $L^2(0, T; H_0^1(\Omega))$ and almost everywhere in Q.

Let then $\varphi \in L^2(0, T; H_0^1(\Omega)) \cap L^\infty(Q)$ such that $\varphi_t \in L^2(0, T; H^{-1}(\Omega)) + L^1(Q)$, let $t \in (0, T)$, and take $T_k(u_n - \varphi)\chi_{(0,t)}$ as test function in (3). Since φ is bounded, we have $\nabla T_k(u_n - \varphi) = 0$ in the set $\{|u_n| > M\}$, where $M = k + \|\varphi\|_{L^\infty(Q)}$. Thus we get, integrating by parts,

$$\int_\Omega \Theta_k(u_n - \varphi)(t)\,dx - \langle \varphi_s, T_k(u_n - \varphi) \rangle$$

$$+ \int_{Q_t} a(x, s, u_n)\nabla T_M(u_n)\, \nabla T_k(T_M(u_n) - \varphi)\,dx\,ds$$

$$\leq \int_{Q_t} T_M(u_n)\, E\, \nabla T_k(T_M(u_n) - \varphi)\,dx\,ds + \int_\Omega \Theta_k(T_n(u_0) - \varphi(0))\,dx, \tag{17}$$

for $t \in (0, T)$. Since u_n is bounded in $L^\infty(0, T; L^1(\Omega))$ and Θ_k has linear growth, we have $\Theta_k(u_n - \varphi)$ bounded in $L^\infty(0, T; L^1(\Omega))$ as well; using that u_n almost everywhere converges to u and $\Theta_k \geq 0$ we get

$$\int_\Omega \Theta_k(u - \varphi)(t)dx \leq \liminf_{n \to \infty} \int_\Omega \Theta_k(u_n - \varphi)(t)dx, \tag{18}$$

for almost every $t \in (0, T)$. Since $T_k(u_n - \varphi)$ converges weakly in $L^2(0, T; H_0^1(\Omega))$, $*$–weakly in $L^\infty(Q)$ and almost everywhere in Q, there is no difficulty in passing to the limit in the second term of (17) as $\varphi_s \in L^2(0, T; H^{-1}(\Omega)) + L^1(Q)$. We also have

$$\int_{Q_t} a(x, s, u_n) \nabla T_M(u_n) \, \nabla T_k(T_M(u_n) - \varphi) \, dx \, ds$$
$$= \int_{Q_t} a(x, s, u_n) \, |\nabla T_k(T_M(u_n) - \varphi)|^2 \, dx \, ds$$
$$+ \int_{Q_t} a(x, s, u_n) \nabla \varphi \, \nabla T_k(T_M(u_n) - \varphi) \, dx \, ds,$$

so that, using the weak convergence of $T_k(T_M(u_n) - \varphi)$ to $T_k(T_M(u) - \varphi)$ in $L^2(0, T; H_0^1(\Omega))$, and the almost everywhere convergence of u_n, we get

$$\int_{Q_t} a(x, s, u) \nabla u \, \nabla T_k(u - \varphi) \, dx \, ds$$
$$= \int_{Q_t} a(x, s, u) \, |\nabla T_k(T_M(u) - \varphi)|^2 \, dx \, ds$$
$$+ \int_{Q_t} a(x, s, u) \nabla \varphi \, \nabla T_k(T_M(u) - \varphi) \, dx \, ds \tag{19}$$
$$\leq \liminf_{n \to \infty} \int_{Q_t} a(x, s, u_n) \nabla T_M(u_n) \, \nabla T_k(T_M(u_n) - \varphi) \, dx \, ds.$$

Similarly, thanks to the weak convergence of $T_k(T_M(u_n) - \varphi)$ we also have

$$\int_{Q_t} u E \, \nabla T_k(u - \varphi) \, dx \, ds = \int_{Q_t} T_M(u) \, E \, \nabla T_k(T_M(u) - \varphi) \, dx \, ds$$
$$= \lim_{n \to \infty} \int_{Q_t} T_M(u_n) \, E \, \nabla T_k(T_M(u_n) - \varphi) \, dx \, ds. \tag{20}$$

Therefore, by means of (18), (19), (20), and using the fact that $u_0 \in L^1(\Omega)$, we can pass to the limit in (17) and obtain that u is an entropy solution of (1).

The regularity of the solution u easily follows from the results of Lemma 3.1 and Lemma 3.2. $\qquad\square$

REFERENCES

[1] ARONSON, D. G. and SERRIN, J., *Local behavior of solutions of quasilinear parabolic equations.* Arch. Rational Mech. Anal. *25* (1967), 81–122.

[2] BENILAN, P., BOCCARDO, L., GALLOUËT, T., GARIEPY, R., PIERRE, M. and VAZQUEZ, J. L., *An L^1 theory of existence and uniqueness of solutions for nonlinear elliptic equations.* Ann. Scuola Norm. Sup. Pisa. *22* (1995), 240–273.

[3] BOCCARDO, L., *Some Dirichlet problems with lower order terms in divergence form,* preprint.

[4] BOCCARDO, L., *Some nonlinear Dirichlet problems in L^1 involving lower order terms in divergence form. Progress in elliptic and parabolic partial differential equations* (Capri, 1994), 43–57, Pitman Res. Notes Math. Ser. 350, Longman, Harlow 1996.

[5] BOCCARDO, L., DALL'AGLIO, A., GALLOUËT, T. and ORSINA, L., *Existence and regularity results for some nonlinear parabolic equations.* Adv. Mat. Sci. Appl. *9* (1999) no. 2, 1017–1031.

[6] BOCCARDO, L., DIAZ, J. I., GIACHETTI, D. and MURAT, F., *Existence and regularity of renormalized solutions for some elliptic problems involving derivatives of nonlinear terms.* J. Diff. Eq. *106* (1993), 215–237.

[7] FABRIE, P. and GALLOUËT, T., *Modelling wells in porous media flows.* Math. Models Methods Appl. Sci. *10* (2000), 673–709.

[8] LIONS, J. L., *Quelques méthodes de resolution des problèmes aux limites non linéaires.* Dunod, Gauthier-Villars, Paris 1969.

[9] PORRETTA, A., *Existence results for nonlinear parabolic equations via strong convergence of truncations.* Ann. Mat. Pura Appl. *177* (1999), 143–172.

[10] PRIGNET, A., *Existence and uniqueness of "entropy" solutions of parabolic problems with L^1 data.* Nonlinear Anal. *28* (1997), 1943–1954.

Lucio Boccardo
Dipartimento di Matematica
Università di Roma 1
Piazza A. Moro 2
00185 Roma
Italia

Luigi Orsina
Dipartimento di Matematica
Università di Roma "La Sapienza"
P. le A. Moro 2
00185 Roma
Italia

Alessio Porretta
Dipartimento di Matematica
Università di Roma "Tor Vergata"
Via della Ricerca Scientifica
00133 Roma
Italia

To access this journal online:
http://www.birkhauser.ch

J.evol.equ. 3 (2003) 419 – 441
1424–3199/03/030419 – 23
DOI 10.1007/s00028-003-0110-1
© Birkhäuser Verlag, Basel, 2003

**Journal of Evolution
Equations**

On the motion of rigid bodies in a viscous incompressible fluid

EDUARD FEIREISL*

Dedicated to the memory of Philippe Benilan

1. Introduction

1.1. *Classical formulation*

The motion of one or several rigid bodies in a viscous incompressible fluid has been a topic of numerous theoretical studies. The time evolution of the *fluid density* $\varrho^f = \varrho^f(t, \boldsymbol{x})$ and the *velocity* $\boldsymbol{u}^f = \boldsymbol{u}^f(t, \boldsymbol{x})$ is governed by the Navier-Stokes system of equations

$$\partial_t \varrho^f + \mathrm{div}(\varrho^f \boldsymbol{u}^f) = 0, \tag{1.1}$$

$$\partial_t(\varrho^f \boldsymbol{u}^f) + \mathrm{div}(\varrho^f \boldsymbol{u}^f \otimes \boldsymbol{u}^f) + \nabla p = \mathrm{div}\, \mathbb{T} + \varrho^f \boldsymbol{g}^f \tag{1.2}$$

satisfied in a region Q^f of the space-time occupied by the fluid. We focus on linearly viscous (Newtonian) incompressible fluids where the *stress tensor* $\mathbb{T}$ is determined through the constitutive relation

$$\mathbb{T} = \mathbb{T}(\boldsymbol{u}) \equiv 2\mu\, \mathbb{D}(\boldsymbol{u}), \ \mathbb{D}(\boldsymbol{u}) \equiv \frac{1}{2}(\nabla \boldsymbol{u} + \nabla \boldsymbol{u}^t), \ \mu > 0, \tag{1.3}$$

and the velocity satisfies the incompressibility condition

$$\mathrm{div}\, \boldsymbol{u}^f = 0. \tag{1.4}$$

The fluid contains a family of m rigid bodies, the position of which in the Euclidean space R^3 at any time $t \in [0, T]$ is represented by compact connected sets $\overline{S}^i(t)$, $i = 1, \ldots, m$. Denoting ϱ^{S^i} the *mass density* of each body we define the *total mass*

$$m^i \equiv \int_{\overline{S}^i(t)} \varrho^{S^i}(t, \boldsymbol{x})\, \mathrm{d}\boldsymbol{x},$$

Mathematics Subject Classification (2000): 35Q30, 35A05.
Key words: Navier-Stokes equations, rigid bodies in a viscous fluid.
*Work supported by the Grant A1019002 of GA AV ČR

the *center of mass*

$$X^i_g(t) \equiv \frac{1}{m^i} \int_{\overline{S}^i(t)} \varrho^{S^i}(t, x)\, x\, dx,$$

and the *inertial tensor*

$$\mathbb{J}^i a \cdot b \equiv \int_{\overline{S}^i(t)} \varrho^{S^i}(t, x)[a \times (x - X^i_g(t))] \cdot [b \times (x - X^i_g(t))]\, dx$$

for $i = 1, \ldots, m$.

The *rigid velocities* u^{S^i} at each point $x \in \overline{S}^i(t)$ can be written as

$$u^{S^i}(t, x) = V^i_g(t) + \mathbb{Q}^i(t)(x - X^i_g(t))\ \text{with}\ \frac{d}{dt} X^i_g(t) = V^i_g(t),$$

where V^i_g denotes the *translation velocity*, and $\mathbb{Q}^i$ the *angular velocity* of the body. The matrix $\mathbb{Q}^i$ is skew symmetric, therefore it can be represented by a vector w^i,

$$\mathbb{Q}^i(t)(x - X^i_g) = w^i \times (x - X^i_g).$$

Apart from the action of *volume forces* g^f, g^{S^i}, the motion results from the interaction of the solids with the fluid through the boundary $\partial \overline{S}^i$. It is a commonly accepted hypothesis that viscous fluids adhere to the boundary, more specifically, the velocity is continuous,

$$u^f(t, x) = u^{S^i}(t, x)\ \text{for}\ x \in \partial \overline{S}^i(t),\ i = 1, \ldots, m \tag{1.5}$$

provided there is no collision of two rigid objects.

In the case when the fluid is contained in a fixed open set $\Omega \subset R^3$, it is customary to append to (1.5) the no-slip boundary conditions for the velocity,

$$u^f|_{\partial \Omega} = 0. \tag{1.6}$$

Under the assumption of continuity of the stresses, the balance of *linear* and *angular momentum* for the body $\overline{S}^i$ read

$$m^i \frac{d}{dt} V^i_g(t) = \int_{\partial \overline{S}^i(t)} (\mathbb{T} - p\mathbb{I} n\, d\sigma) + \int_{\overline{S}^i(t)} \varrho^{S^i} g^{S^i}\, dx, \tag{1.7}$$

$$\mathbb{J}^i(t)\frac{d}{dt} w^i(t) = \mathbb{J}^i w^i(t) \times w^i(t) + \int_{\partial \overline{S}^i(t)} (x - X^i_g)$$

$$\times (\mathbb{T} - p\mathbb{I}) n\, d\sigma + \int_{\overline{S}^i(t)} \varrho^{S^i}(x - X^i_g) \times g^{S^i}\, dx,\ i = 1, \ldots, m. \tag{1.8}$$

The problem (1.1)–(1.8) is to be supplemented by the initial conditions for the density and the velocity

$$\left\{\begin{array}{l} \varrho^f(0, \boldsymbol{x}) = \varrho_0^f(\boldsymbol{x}), \ \varrho^{S^i}(0, \boldsymbol{x}) = \varrho_0^{S^i}(\boldsymbol{x}), \\ \boldsymbol{u}^f(0, \boldsymbol{x}) = \boldsymbol{u}_0^f(\boldsymbol{x}), \ \boldsymbol{V}_g^i(0) = \boldsymbol{V}_{g,0}^i, \ \boldsymbol{w}^i(0) = \boldsymbol{w}_0^i, \ i = 1, \ldots, m; \end{array}\right\} \tag{1.9}$$

as well as the initial position of the solids

$$\overline{S}^i(0) = \overline{S}^i, \ i = 1, \ldots, m \tag{1.10}$$

in order to obtain a (formally) well-posed problem.

The relations (1.1)–(1.10) represent a classical formulation of **Problem (P)** we shall deal with. Note that it has been derived under the fundamental hypothesis that there are no collisions of the rigid objects, i.e.,

$$\overline{S}^i(t) \cap \overline{S}^j(t) = \emptyset \text{ for } i \neq j, \ \overline{S}^i(t) \cap \partial\Omega = \emptyset \text{ for all } t \in [0, T].$$

On the other hand, it is not hard to believe that one has to abandon the concept of classical (smooth) solutions if these phenomena are to be taken into account. Indeed if the velocity $\boldsymbol{u}$ was smooth, in particular Lipschitz with respect to the spatial variable, then the position of the body $\overline{S}^i$ at a time $t \in [0, T]$ would be given as

$$\overline{S}^i(t) = \eta(t, \overline{S}^i(0)), \ t \in [0, T]$$

where $\eta(t, \boldsymbol{x}), t \in [0, T]$ is the characteristic curve determined uniquely through the initial-value problem

$$\frac{\partial}{\partial t}\eta(t, \boldsymbol{x}) = \boldsymbol{u}(t, \eta(t, \boldsymbol{x})), \ \eta(0, \boldsymbol{x}) = \boldsymbol{x}.$$

In particular, $\overline{S}^i(t) \cap \overline{S}^j(t) = \emptyset$ for all $t > 0$ whenever $\overline{S}^i(0) \cap \overline{S}^j(0) = \emptyset$.

The main goal of the present paper is to show that **Problem (P)** possesses a global-in-time variational (weak) solution (see Theorem 1.1 below). This and similar questions have been addressed by several authors. Desjardins and Esteban [2] discuss the existence of local-in-time solutions, more specifically, "local" means up to the first collision in two space dimensions, and up to the blow-up of the velocity gradient in a certain Sobolev norm in R^3. These ideas are further developed in [3] where both the incompressible and the compressible cases are studied.

The existence "up to the first collision" of one rigid body with the boundary of a bounded spatial domain $\Omega \subset R^3$ was proved by Gunzburger, Lee, and Seregin [7]. Similar problems were considered by Hoffmann and Starovoitov [8], Conca, San Martin, and Tucsnak [1].

To the best of our knowledge, the existence of global-in-time solutions regardless of possible contacts of two rigid bodies or a body with the boundary in *three space dimensions* is not known. There is a remarkable result of San Martin, Starovoitov, and Tucsnak [10], where the authors prove the existence of globally defined solutions for the problem in two space dimensions. They take advantage of the fact that the velocity u belongs to the space $L^2(0, T; W^{1,2}(\Omega))$, in particular, the velocity gradient lies in the Lebesgue space L^N with $N = 2$ - the dimension of the spatial domain Ω. The functions belonging to the Sobolev space $W^{1,N}$ are "almost continuous", and they enjoy many common properties with continuous functions. In particular, for the problem in question, one can deduce that only collisions with vanishing relative velocity and acceleration of two objects are allowed (see [10, Theorem 2.2]). In other words, the velocities of two solids *before* and *after* a possible contact must be the same, the velocity is (weakly) continuous with respect to time, and the set of all finite energy weak solutions is precompact in a suitable function space topology. Accordingly, the "real" collisions with non-zero relative velocities can occur only in three space dimensions.

1.2. *Variational (weak) formulation*

The variational structure of **Problem (P)** allows us to derive an alternative formulation where the required regularity of solutions is much lower than for the original system (1.1), (1.2). In fact, the density ϱ and the velocity u will belong to a phase space related to natural *a priori* estimates resulting from the energy inequality (cf. formula (1.16)).

To be more specific, the rigid objects will be represented by a family of sets $\overline{S}^i$ such that

$$\overline{S}^i = \mathrm{cl}(S^i) \text{ where } \left. \begin{array}{c} \overline{S}^i \text{ are compact, connected,} \\ S^i \equiv \mathrm{int}(\overline{S}^i) \neq \emptyset, \\ |\overline{S}^i \setminus S^i| = 0 \end{array} \right\} \tag{1.11}$$

for any $i = 1, \ldots, m$. The position $\overline{S}^i(t)$ of each solid at a time t will be given through a family of (affine) isometries η^i,

$$\overline{S}^i(t) = \eta^i(t, \overline{S}^i), \quad \eta^i = \eta^i(t, x), \quad \eta^i(t, \cdot) : R^3 \mapsto R^3, \quad i = 1, \ldots, m.$$

Now, given an open time interval I and an open set $\Omega \subset R^3$, we define the *solid region*

$$\overline{Q}^s = \left\{ (t, x) \,\middle|\, t \in \overline{I}, \; x \in \bigcup_{i=1}^{m} \overline{S}^i(t) \right\}$$

together with the *fluid region*

$$Q^f = (I \times \Omega) \setminus \overline{Q}^s.$$

Introducing

$$\varrho = \varrho^f + \sum_{i=1}^{m} \varrho^{S^i}, \quad u = \begin{cases} u^f & \text{in } Q^f, \\ u^{S^i} & \text{in } \overline{Q}^s \end{cases}$$

we can express the balance of mass in a concise form

$$\partial_t \varrho + \operatorname{div}(\varrho u) = 0 \text{ in } \mathcal{D}'(I \times R^3) \tag{1.12}$$

provided ϱ, u are extended to be zero outside Ω.

Similarly, the momentum equations (1.2), (1.7), (1.8) read

$$\int_I \int_{R^3} (\varrho u) \cdot \partial_t \varphi + [\varrho u \otimes u] : \mathbb{D}(\varphi) \, dx \, dt$$
$$= \int_I \int_{R^3} 2\mu \, \mathbb{D}(u) : \mathbb{D}(\varphi) - \varrho g \cdot \varphi \, dx \, dt \tag{1.13}$$

for any *test function* $\varphi \in \mathcal{T}(\overline{Q}^s)$, where

$$\begin{cases} \mathcal{T}(\overline{Q}^s) \equiv \{\varphi \in \mathcal{D}(I \times \Omega) \mid \operatorname{div} \varphi = 0 \text{ on } I \times \Omega, \\ \mathbb{D}(\varphi) = 0 \text{ on an open neighbourhood of the set } \overline{Q}^s \}. \end{cases}$$

The unpleasant feature of (1.13) is that the space $\mathcal{T}(\overline{Q}^s)$ depends on the solid region $\overline{Q}^s$, i.e., on the solution itself. There is an alternative way when one can fix a coordinate frame attached to one of the solids used for instance by GALDI [6] and Gunzburger et al. [7].

To close the system (1.12), (1.13), one has to specify the relation between the velocity u and the motion of solids characterized through the isometries η^i, $i = 1, \ldots, m$. Fixing $t \in I$ we can write

$$\eta^i(t, x) = X^i(t) + \mathbb{O}^i(t)x, \ \mathbb{O}^i \in SO(3), \ i = 1, \ldots, m.$$

In accordance with the continuity hypothesis (1.5), the velocity u must be *compatible* with the family $\{\overline{S}^i, \eta^i\}_{i=1}^m$, more specifically, we require the functions $t \mapsto X^i(t), t \mapsto \mathbb{O}^i(t)$ to be absolutely continuous on I, and

$$u(t, x) = u^{S^i}(t, x) \equiv V^i(t) + \mathbb{Q}^i(t)(x - X^i(t)) \text{ for a.a. } x \in \overline{S}^i(t), \tag{1.14}$$

where V^i, $\mathbb{Q}^i$ are given by

$$\frac{\mathrm{d}}{\mathrm{d}t} X^i(t) = V^i(t), \quad \left(\frac{\mathrm{d}}{\mathrm{d}t}\mathbb{O}^i(t)\right)(\mathbb{O}^i(t))^{-1} = \mathbb{Q}^i(t), \quad i = 1,\ldots,m. \tag{1.15}$$

Of course, both (1.14) and (1.15) are supposed to hold only for a.a. $t \in I$ which allows for possible discontinuities in the velocity field due to collisions.

Up to now we have not specified the classes of functions the variational solutions ϱ, u should belong to. Taking (formally) $\varphi = u$ in (1.13) we deduce the *energy inequality*

$$E(\tau) + \mu \int_0^\tau \int_\Omega |\nabla u|^2 \, \mathrm{d}x\, \mathrm{d}t \le E_0 + \int_0^\tau \int_\Omega \varrho g \cdot u \, \mathrm{d}x\, \mathrm{d}t \text{ for a.a. } \tau \in I \tag{1.16}$$

where

$$E(\tau) = E[\varrho, u](\tau) \equiv \frac{1}{2} \int_\Omega \varrho(\tau)|u(\tau)|^2 \, \mathrm{d}x.$$

Here, the quantity E_0 is obviously related to the "initial energy", i.e., the energy E at the time $t_0 = \inf I$.

In accordance with (1.16), any admissible velocity field should be looked for in the "Leray space" $u \in L^2(I; H^1(\Omega))$ provided I is bounded. Here, we have denoted

$$H^s(\Omega) = \{v \in W^{s,2}(\Omega) \mid \operatorname{div} v = 0 \text{ in } \mathcal{D}'(\Omega)\}, \quad s \in [0, 1],$$

where $W^{0,2}(\Omega) \equiv L^2(\Omega)$–the Lebesgue space of square integrable functions, $W^{1,2}(\Omega)$–the Sobolev space of square integrable functions with gradients in $L^2(\Omega)$, and $W^{s,2}(\Omega)$ defined by interpolation for $s \in (0, 1)$ (see e.g. Lions and Magenes [9]). For the sake of simplicity, we shall make no distinction between scalar and vector valued functions, i.e., $H^s \approx [H^s]^3$ etc. We shall write H instead of H^0.

Now, there are several ways how to express the no-stick boundary conditions (1.6). We adopt a formulation compatible with (1.12), (1.14), namely, we introduce the spaces

$$H_0^s(\Omega) = \operatorname{cl}_{H^s}\{v \in \mathcal{D}(\Omega) \mid \operatorname{div} v = 0 \text{ in } \Omega\},$$

and we require

$$u \in L^2(I; H_0^1(\Omega)). \tag{1.17}$$

Note that (1.17) implies that u vanishes on the boundary in the sense of traces provided $\partial\Omega$ is regular.

To conclude, we consider a time interval $I = (0, T)$ in order to formulate the initial conditions (1.9), (1.10). In accordance with the theory developed by Diperna and Lions [4], any bounded weak solution ϱ of the continuity equation (1.12) belongs to the class

$$\varrho \in C([0, T]; L^1(\Omega))$$

provided $\boldsymbol{u}$ satisfies (1.17) (cf. Lemma 2.1 below). Consequently, the instantaneous value of the density $\varrho(t)$ is well defined for any $t \in [0, T]$, and we can consider the initial condition

$$\varrho(0, \boldsymbol{x}) = \varrho_0(\boldsymbol{x}) \text{ for a.a. } \boldsymbol{x} \in \Omega. \tag{1.18}$$

As for the velocity $\boldsymbol{u}$, one can only deduce from the momentum equations (1.13) that the quantities

$$t \mapsto \int_\Omega \phi(\boldsymbol{x}) \cdot (\varrho \boldsymbol{u})(t, \boldsymbol{x}) \, \mathrm{d}\boldsymbol{x}$$

are continuous on a neighbourhood of any $\tau \in [0, T]$ for any test function $\phi \in T(\Omega_\tau)$, where

$$T(\Omega_\tau) \equiv \left\{ \phi \in \mathcal{D}(\Omega) \mid \operatorname{div} \phi = 0 \text{ in } \Omega, \right.$$

$$\left. \mathbb{D}(\phi) = 0 \text{ on a neighbourhood of the set } \bigcup_{i=1}^{m} \overline{S}^i(\tau) \right\}. \tag{1.19}$$

Accordingly, one can define the initial condition $(\varrho \boldsymbol{u})(0) = \varrho_0 \boldsymbol{u}_0$ through

$$\lim_{t \searrow 0} \int_\Omega \phi \cdot (\varrho \boldsymbol{u})(t) \, \mathrm{d}\boldsymbol{x} = \int_\Omega \phi \cdot (\varrho_0 \boldsymbol{u}_0) \, \mathrm{d}\boldsymbol{x} \text{ for any } \phi \in T(\Omega_0). \tag{1.20}$$

Finally, by virtue of the compatibility conditions (1.14) and the energy estimates (1.16), the isometries η^i are Lipschitz continuous with respect to $t \in [0, T]$. In particular, it seems convenient to "normalize" the system η^i taking

$$\eta^i(0, \boldsymbol{x}) = \boldsymbol{x} \text{ on } R^3.$$

Consequently, we can prescribe

$$\overline{S}^i(0) = \eta^i(0, \overline{S}^i) = \overline{S}^i \tag{1.21}$$

in agreement with (1.10).

1.3. *Main results*

In accordance with our previous considerations, we introduce the concept of variational (weak) solution to **Problem (P)**.

Let $I \subset R^1$ be a bounded open interval, $\Omega \subset R^3$ a bounded open set. We shall say that the quantities ϱ, $\boldsymbol{u}$, and $\{\overline{S}^i, \eta^i\}_{i=1}^m$ represent a *variational solution* of **Problem (P)** on the set $I \times \Omega$ if the following conditions hold.

- The density $\varrho = \varrho(t, \boldsymbol{x})$ is a bounded measurable function such that

$$0 < \underline{\varrho} \leq \varrho(t, \boldsymbol{x}) \leq \overline{\varrho} \text{ for a.a. } \boldsymbol{x} \in \Omega, \ t \in I$$

  for certain constants $\underline{\varrho}, \overline{\varrho}$
- The velocity field $\boldsymbol{u} = \boldsymbol{u}(t, \boldsymbol{x})$ belongs to the spaces

$$\boldsymbol{u} \in L^2(I; H_0^1(\Omega)) \cap L^\infty(I; H_0(\Omega)),$$

  in particular, $\boldsymbol{u}$ satisfies the no-stick boundary conditions in the sense of (1.17).
- The continuity equation (1.12) holds in $\mathcal{D}'(I \times R^3)$ provided ϱ, $\boldsymbol{u}$ were extended to be zero outside Ω.
- The variational form (1.3) of the momentum equation is satisfied for any test function $\varphi \in \mathcal{T}(\overline{Q}^s)$.
- The energy inequality (1.16) holds for a.a. $\tau \in I$ with a certain constant E_0.
- The mappings $\eta^i(t, \cdot) : R^3 \mapsto R^3$ are (affine) isometries, and the family $\{\overline{S}^i, \eta^i\}_{i=1}^m$ is compatible with $\boldsymbol{u}$ in the sense of (1.14), (1.15).

If there are no collisions, the family of the test functions ϕ introduced in (1.19) is likely to be rich enough to determine uniquely the value of the momentum $(\varrho\boldsymbol{u})$, and, consequently that of $\boldsymbol{u}$, at any time $\tau \in [0, T]$. On the other hand, it is easy to see that this is not the case when $\overline{S}^i(\tau) \cap \overline{S}^j(\tau) \neq \emptyset$, or if $\overline{S}^i(\tau) \cap \partial\Omega \neq \emptyset$ for certain $i \neq j$. Indeed while the test function ϕ must coincide with the same rigid velocity field on $\overline{S}^i(\tau) \cup \overline{S}^j(\tau)$ the velocity $\boldsymbol{u}$ need not. Thus one is given several possibilities how to continue the solution after the collision, all of them representing a variational solution of the same problem.

From the mathematical point of view, however, this "liberty of choice" can result in the loss of compactness in the velocity field which seems to be the main stumbling block to build up an existence theory of global-instime variational solutions in three space dimensions (cf. Gunzburger et al. [7], Desjardins and Esteban [3]). It is interesting to note that for viscous *compressible* fluids such a problem does not occur (see [5]).

The main result proved in this paper reads as follows.

THEOREM 1.4. *Let $\Omega \subset R^3$ be a bounded domain, $T > 0$. Assume that the initial position of the rigid bodies is given through a family $\overline{S}^i$, $i = 1, \ldots, m$ such that (1.11) holds. Moreover, let*

$$\overline{S}^i \cap \overline{S}^j = \emptyset \text{ for } i \neq j, \ \overline{S}^i \cap (R^3 \setminus \Omega) = \emptyset, \ i = 1, \ldots, m.$$

Let ϱ_0 be a measurable function satisfying

$$0 < \underline{\varrho} \leq \varrho_0(\boldsymbol{x}) \leq \overline{\varrho} \text{ for a.a } \boldsymbol{x} \in \Omega. \tag{1.22}$$

Finally, let $\boldsymbol{u}_0 \in H_0(\Omega)$, and $\boldsymbol{g} \in L^\infty((0, T) \times \Omega)$ be given.
Then there exists a variational solution ϱ, $\boldsymbol{u}$, $\{\overline{S}^i, \eta^i\}_{i=1}^m$ of **Problem (P)** *on the set $(0, T) \times \Omega$ satisfying the initial conditions (1.18), (1.20), and (1.21). Moreover, the energy inequality (1.16) holds with*

$$E_0 = \frac{1}{2} \int_\Omega \varrho_0 |\boldsymbol{u}_0|^2 \, d\boldsymbol{x}.$$

The rest of the paper will be devoted to the proof of Theorem 1.1. As already discussed above, the key point is the choice of appropriate "after collision" continuation condition for the velocity $\boldsymbol{u}$. Such a condition will play the same role as the entropy (shock) conditions for systems of conservation laws, namely, it will identify physically admissible solutions. Here, we adopt the simplest contact condition corresponding to "sticky" boundaries, more specifically, once two bodies touch one another, they will stay together forever (cf. Section 4). Such a condition may be viewed as a (very naive) form of the least energy principle.

The paper is organized as follows. In Section 2, we review some basic properties of the variational solutions which may be proved by available methods. In particular, we quote a local "up to collision" existence result in the spirit of Desjardins and Esteban [3], Gunzburger at al. [7] (see Proposition 2.1). In Section 3, we prove a general local existence result with no restriction on the smoothness of both the solids and the boundary of the spatial domain (see Proposition 2.1). Finally, we construct globally defined variational solutions by continuation (see Section 4).

2. Basic properties of variational solutions

In this section, we review some known results concerning the existence and some basic properties of the variational solutions introduced in Section 1. We begin with the continuity equation (1.12) investigated by Diperna and Lions [4].

LEMMA 2.1. *Assume that $\Omega \subset R^3$ is bounded domain. Let $\varrho \in L^\infty(I \times \Omega)$, $\boldsymbol{u} \in L^2(I; H_0^1(\Omega))$ satisfy the continuity equation (1.12) in $\mathcal{D}'((0, T) \times R^3)$ provided they were extended to be zero outside Ω.*

Then $\varrho \in C([0, T]; L^1(\Omega))$. Moreover, if

$$0 < \underline{\varrho} \le \varrho(0, \boldsymbol{x}) \le \overline{\varrho} \ \ for \ a.a. \ \boldsymbol{x} \in \Omega,$$

then

$$0 < \underline{\varrho} \le \varrho(t, \boldsymbol{x}) \le \overline{\varrho} \ \ for \ a.a. \ \boldsymbol{x} \in \Omega \ and \ for \ any \ t \in [0, T].$$

Finally, if

$$\varrho_n(0) \to \varrho_0 \ in \ L^1(\Omega),$$

$$\varrho_n \to \varrho \ weakly \ * \ in \ L^\infty((0, T) \times \Omega), \ and \ \boldsymbol{u}_n \to \boldsymbol{u} \ weakly \ in \ L^2(0, T; H_0^1(\Omega)),$$

where ϱ_n, $\boldsymbol{u}_n$ solve the continuity equation (1.12) in $\mathcal{D}'((0, T) \times R^3)$, then

$$\varrho_n \to \varrho \ in \ C([0, T]; L^1(\Omega)),$$

and the limits ϱ, $\boldsymbol{u}$ satisfy the same equation in $\mathcal{D}'((0, T) \times R^3)$.

Following [5, Section 5] we introduce the signed boundary distance $\mathbf{bd}_S$ for a set $S \subset R^3$,

$$\mathbf{bd}_S(\boldsymbol{x}) \equiv \mathrm{dist}(\boldsymbol{x}, \overline{R^3 \setminus S}) - \mathrm{dist}(\boldsymbol{x}, \overline{S}). \tag{2.1}$$

The next result concerns the possibility to extend a given variational solution beyond the existence time T.

LEMMA 2.2. *Let $\Omega \subset R^3$ be a bounded domain. Let ϱ, $\boldsymbol{u}$, and $\{\overline{S}^i, \eta^i\}_{i=1}^m$ be a variational solution of* **Problem (P)** *on $(0, T_1) \times \Omega$ such that*

$$r^s = \min_{i=1,\ldots,m} \left\{ \sup_{\mathbf{x} \in R^3} \mathbf{bd}_{\overline{S}^i}(\boldsymbol{x}) \right\} > 0. \tag{2.2}$$

Then the isometries η^i are Lipschitz continuous with respect to $t \in [0, T_1]$, more specifically,

$$\|\eta^i(t_1, \cdot) - \eta^i(t_2, \cdot)\|_{C(B; R^3)} \le c(B)|t_1 - t_2| \ for \ any \ bounded \ B \subset R^3, \ t_1, t_2 \in [0, T_1]$$

where the constant $c(B)$, besides the set B, depends solely on r^s, ϱ, the initial energy E_0, the amplitude of the function $\boldsymbol{g}$, and the length of the interval $[0, \overline{T_1}]$.

Moreover, there exists a function $\boldsymbol{u}(T_1) \in H_0(\Omega)$ such that

$$\lim_{t \nearrow T_1} \int_\Omega \phi \cdot (\varrho \boldsymbol{u})(t) \, \mathrm{d}\boldsymbol{x} = \int_\Omega \phi \cdot (\varrho(T_1) \boldsymbol{u}(T_1)) \, \mathrm{d}\boldsymbol{x} \ for \ any \ \phi \in \mathcal{T}(\Omega_{T_1}),$$

and

$$\frac{1}{2}\int_\Omega \varrho(T_1)|\boldsymbol{u}(T_1)|^2\,\mathrm{d}\boldsymbol{x} \le \operatorname*{ess\,lim\,inf}_{t\nearrow T_1} E(t).$$

*Finally, if ϱ, $\boldsymbol{u}$, $\{\overline{S}^i, \eta^i\}_{i=1}^m$ solve **Problem (P)** also on $(T_1, T_2)\times\Omega$ with the initial data*

$$\varrho(T_1),\ \ \boldsymbol{u}(T_1),\ \ \overline{S}^i(T_1) = \eta^i(T_1, \overline{S}^i(0)),$$

and the initial energy

$$E_{T_1} = \frac{1}{2}\int_\Omega \varrho(T_1)|\boldsymbol{u}(T_1)|^2\,\mathrm{d}\boldsymbol{x},$$

*then ϱ, $\boldsymbol{u}$, $\{\overline{S}^i, \eta^i\}_{i=1}^m$ is a variational solution of **Problem (P)** on $(0, T)\times\Omega$.*

Proof. To begin, observe that r^s is simply the radius of the largest ball which can be placed in the interior sets S^i, $i = 1, \ldots, m$. The Lipschitz continuity of the isometries η^i is then a straightforward consequence of the energy inequality (1.16) together with the compatibility conditions (1.14). Accordingly, we can set

$$\overline{S}^i(T_1) = \eta^i(T_1, \overline{S}^i)\ \text{for } i = 1, \ldots, m.$$

By virtue of the momentum equation (1.13), we have

$$\int_\Omega \boldsymbol{\phi}\cdot(\varrho\boldsymbol{u})(t)\,\mathrm{d}\boldsymbol{x} \to L\boldsymbol{\phi}\ \text{for } t \nearrow T_1$$

for any test function $\boldsymbol{\phi} \in \mathcal{T}(\Omega_{T_1})$, where L can be viewed as a linear form on the vector space $\mathcal{T}(\Omega_{T_1})$. Moreover,

$$|L\boldsymbol{\phi}|^2 \le 2\int_\Omega \varrho(T_1)|\boldsymbol{\phi}|^2\,\mathrm{d}\boldsymbol{x}\,(\operatorname*{ess\,lim\,inf}_{t\nearrow T_1} E(t)). \tag{2.3}$$

It follows from the Hahn-Banach theorem that L can be extended as a bounded linear form on the space $H_0(\Omega)$ endowed with a scalar product

$$\langle \boldsymbol{v}, \boldsymbol{w}\rangle = \int_\Omega \varrho(T_1)\boldsymbol{v}\cdot\boldsymbol{w}\,\mathrm{d}\boldsymbol{x}.$$

Thus making use of (2.3) and the Riesz representation theorem we can find a function $\boldsymbol{u}(T_1)$ such that

$$\boldsymbol{u}(T_1) \in H_0(\Omega),\ \frac{1}{2}\int_\Omega \varrho(T_1)|\boldsymbol{u}(T_1)|^2\,\mathrm{d}\boldsymbol{x} \le \operatorname*{ess\,lim\,inf}_{t\nearrow T_1} E(t),$$

and

$$\lim_{t\nearrow T_1}\int_\Omega \boldsymbol{\phi}\cdot(\varrho\boldsymbol{u})(t)\,\mathrm{d}\boldsymbol{x} = \int_\Omega \boldsymbol{\phi}\cdot(\varrho(T_1)\boldsymbol{u}(T_1))\,\mathrm{d}\boldsymbol{x}\ \text{for any } \boldsymbol{\phi} \in \mathcal{T}(\Omega_{T_1}).$$

The rest of the proof is straightfoward. $\qquad\square$

Now, given a system of compacts $\overline{S}^i(t)$, $i = 1, \ldots, m$, we define

$$d(\{\overline{S}^i(t)\}) \equiv \min \left\{ \min_{i,j=1,\ldots,m, i \neq j} \operatorname{dist}(\overline{S}^i(t), \overline{S}^j(t)); \quad \min_{i=1,\ldots,m} \operatorname{dist}(\overline{S}^i(t), R^3 \setminus \Omega) \right\}. \quad (2.4)$$

To conclude this introductory material, we report an existence result "up to the first collision" which can be proved by the methods introduced by Desjardins and Esteban [3], Gunzburger at al. [7], San Martin et al. [10].

PROPOSITION 2.3. *Let $\Omega \subset R^3$ be a bounded domain of class C^∞. Assume that the initial position of solids is given through a family of compact sets $\overline{S}^i$,*

$$\overline{S}^i = \operatorname{cl}(S^i), \ S^i \ \textit{bounded domains of class } C^\infty$$

such that $d(\{\overline{S}^i\}) > 0$. Let ϱ_0 be a measurable function satisfying (1.22). Finally, let $u_0 \in H_0(\Omega)$, $g \in L^\infty(R \times \Omega)$ be given.

Then **Problem (P)** *with the initial conditions (1.18), (1.20), and (1.21) admits a variational solution ϱ, u, $\{\overline{S}^i, \eta^i\}_{i=1}^m$ with*

$$E_0 = \frac{1}{2} \int_\Omega \varrho_0 |u_0|^2 \, dx$$

defined on some time interval $(0, T_0)$. The value of $T_0 > 0$ is bounded below away from zero by a constant which depends solely on r^s, $\underline{\varrho}$, the amplitude of g, and the initial distance of the rigid objects $d(\{\overline{S}^i\})$, where r^s is given by (2.2).

Combining the continuation Lemma 2.1 together with the local existence result stated in Proposition 2.3 we obtain the following corollary.

COROLLARY 2.4. *Under the hypotheses of Proposition 2.3,* **Problem (P)** *supplemented with the initial conditions (1.18), (1.20), (1.21) admits a variational solution ϱ, u, $\{\overline{S}^i, \eta^i\}_{i=1}^m$ defined on a maximal time interval $(0, T_{max})$, $T_{max} > 0$. Moreover, if $T_{max} < \infty$, we have $d(\{\overline{S}^i(T_{max})\}) = 0$.*

3. Local existence

The result presented in this section is the key tool for proving Theorem 1.1. We shall show that the conclusion of Proposition 2.3 as well as Corollary 2.4 hold without the hypothesis of smoothness of $\overline{S}^i$ and Ω. Even though this might seem a straightforward modification of the available techniques, it is not the case. The reason is that the proof uses some constructions based on solutions of several elliptic problems which are quite sensitive to rough boundary changes. A general existence result once proved makes it possible to continue the variational solutions "after collisions" when a union of two solids is treated as one (non-smooth) rigid object.

PROPOSITION 3.1. *Let $\Omega \subset R^3$ be a bounded domain. Let the initial positions of the solids be defined through a family of compacts $\overline{S}^i$, $i = 1, \ldots, m$, satisfying (1.11), and such that the initial distance $d(\{\overline{S}^i\})$ defined in (2.4) is strictly positive. Assume that ϱ_0 is a measurable function satisfying (1.22). Finally, let $u_0 \in H_0(\Omega)$, $g \in L^\infty(R \times \Omega)$ be given.*

Then **Problem (P)** *complemented by the initial conditions (1.18), (1.20), (1.21) admits a variational solution ϱ, u, $\{\overline{S}^i, \eta^i\}_{i=1}^m$ defined on some maximal time interval $(0, T_{max})$, $T_{max} > 0$. Moreover, if $T_{max} < \infty$, we have $d(\{\overline{S}^i(T_{max})\}) = 0$.*

REMARK 3.2. The conclusion of Proposition 3.1 remains valid provided $\Omega \subset R^3$ is an arbitrary (not necessarily connected) bounded open set. Indeed we can decompose Ω as a countable union of bounded domains on each of which Proposition 3.1 applies. Since there is only a finite number m of rigid bodies, only a finite number of these components contains one of the compacts $\overline{S}^i$ while on the rest the solution exists globally in time.

The rest of this section will be devoted to the proof of Proposition 3.1. This will be done in several steps.

STEP (i) Following [5, Section 5] we introduce convergence of a sequence of sets,

$$K_n \overset{b}{\to} K \quad \text{if } \mathbf{bd}_{K_n} \to \mathbf{bd}_K \text{ in } C_{loc}(R^3),$$

where the boundary distance $\mathbf{bd}$ is given by (2.1).

Now, there is a sequence $\{\overline{B}_n^i\}_{n=1}^\infty$ of compacts such that

$$\overline{B}_n^i = \mathrm{cl}(B_n^i), \quad \text{where } B_n^i \text{ are domains of class } C^\infty,$$

$$\overline{S}^i \subset B_{n+1}^i \subset B_n^i, \ \overline{B}_n^i \overset{b}{\to} \overline{S}^i \text{ as } n \to \infty \text{ for } i = 1, \ldots, m. \tag{3.1}$$

Similarly, we take a sequence of domains $\{\Omega_n\}_{n=1}^\infty$ such that

$$\Omega_n \text{ are of class } C^\infty, \ \Omega_n \subset \Omega_{n+1} \subset \Omega, \ \overline{\Omega}_n \overset{b}{\to} \overline{\Omega} \text{ as } n \to \infty. \tag{3.2}$$

Finally, we can find a sequence of initial velocities $\{u_{0,n}\}_{n=1}^\infty$ such that

$$u_{n,0} \in \mathcal{D}(\Omega_n), \ \mathrm{div}\, u_{n,0} = 0 \text{ in } \Omega_n, \ u_{n,0} \to u_0 \text{ in } H_0(\Omega), \tag{3.3}$$

and we set

$$\varrho_{n,0} = \varrho_0 \, 1_{\Omega_n}.$$

We introduce the quantities r_n^s, $d(\{\overline{B}_n^i\})$ replacing $\overline{S}^i$, Ω by $\overline{B}_n^i$, Ω_n in (2.2), (2.4) respectively. By virtue of (3.1), (1.11) we have

$$r_n^s \geq r^s > 0 \text{ for all } n = 1, 2, \ldots \tag{3.4}$$

while (3.1), (3.2) yield

$$d(\{\overline{B}_n^i\}) \geq \frac{d(\{\overline{S}^i\})}{2} > 0 \text{ for all } n \geq n_0. \tag{3.5}$$

In accordance with Proposition 2.1, **Problem (P)** considered on the spatial domain Ω_n and supplemented by the initial data $\varrho_{n,0}$, $\boldsymbol{u}_{n,0}$, $\overline{B}_n^i$, $i = 1, \ldots, m$ possesses a variational solution ϱ_n, $\boldsymbol{u}_n$, $\{\overline{B}_n^i, \eta_n^i\}_{i=1}^m$ defined on a time interval $(0, T_0)$. Moreover, it follows from (3.3)–(3.5), that $T_0 > 0$ can be found independent of n, and, moreover, we may suppose

$$d(\{\overline{B}_n^i(t)\}) \geq \frac{d(\{\overline{S}^i\})}{4} \text{ for all } t \in [0, T_0] \text{ independently of } n. \tag{3.6}$$

STEP (ii) As

$$H_0^s(\Omega_n) \subset H_0^s(\Omega_{n+1}) \subset H_0^s(\Omega) \text{ for all } n = 1, 2, \ldots,$$

it is clear that the sequence ϱ_n, $\boldsymbol{u}_n$ satisfies the hypotheses of Lemma 2.1. Consequently, we may suppose

$$\varrho_n \to \varrho \text{ in } C([0, T_0]; L^1(\Omega)), \tag{3.7}$$

$$\boldsymbol{u}_n \to \boldsymbol{u} \text{ weakly * in } L^\infty(0, T_0; H_0(\Omega)) \text{ and weakly in } L^2(0, T_0; H_0^1(\Omega)), \tag{3.8}$$

where the limit functions ϱ, $\boldsymbol{u}$ satisfy the continuity equation (1.12) together with the initial condition (1.18).

Next, thanks to the weak lower-semicontinuity of the integrals in (1.16), the limit functions ϱ, $\boldsymbol{u}$ satisfy the energy inequality for a.a. $\tau \in (0, T_0)$.

By virtue of Lemma 2.2 and (3.4), we have

$$\eta_n^i(t, \cdot) \to \eta^i(t, \cdot) \text{ in } C_{loc}(R^3, R^3) \text{ uniformly in } t \in [0, T_0],$$

$$\overline{B}_n^i(t) \overset{b}{\to} \overline{S}^i(t) \text{ as } n \to \infty \text{ uniformly in } t \in [0, T_0]$$

where $\overline{S}^i(t) = \eta^i(t, S^i)$, $i = 1, \ldots, m$. Moreover, the limit velocity $\boldsymbol{u}$ is compatible with the system $\{\overline{S}^i, \eta^i\}_{i=1}^m$ (see [5, Proposition 5.1]).

Finally, introducing the solid regions

$$\overline{Q}_n^s = \left\{(t, \boldsymbol{x}) \,\middle|\, t \in [0, T_0], \ \boldsymbol{x} \in \bigcup_{i=1}^m \overline{B}_n^i(t)\right\},$$

we have

$$\varphi \in \mathcal{T}(\overline{Q}_n^s) \text{ for all } n \geq n(\varphi) \text{ provided } \varphi \in \mathcal{T}(\overline{Q}^s), \tag{3.9}$$

and, similarly,

$$\phi \in \mathcal{T}(\Omega_{n,\tau}) \text{ for all } n \geq n(\phi) \text{ for any } \phi \in \mathcal{T}(\Omega_\tau), \ \tau \in [0, T_0].$$

In particular, we deduce that $\boldsymbol{u}$ satisfies the initial condition (1.20).

STEP (iii) In view of the previous discussion, our final task is to pass to the limit for $n \to \infty$ in the momentum equations (1.13). As a direct consequence of (3.7), (3.8), we obtain

$$\varrho_n \boldsymbol{u}_n \to \varrho\boldsymbol{u} \text{ weakly } * \text{ in } L^\infty(0, T_0; L^2(\Omega)),$$

and

$$\varrho_n \boldsymbol{u}_n \otimes \boldsymbol{u}_n \to \mathbb{M} \text{ weakly in } L^2(0, T_0; L^{\frac{3}{2}}(\Omega)).$$

Thus in view of (3.9), the limit functions ϱ, $\boldsymbol{u}$ will satisfy the momentum equation (1.13) provided we show

$$\boldsymbol{M} = \varrho\boldsymbol{u} \otimes \boldsymbol{u}. \tag{3.10}$$

Moreover, to show (3.10), it is enough to prove

$$\int_0^{T_0} \int_\Omega \varrho_n |\boldsymbol{u}_n|^2 \mathrm{d}\boldsymbol{x}\,\mathrm{d}t = \int_0^{T_0} \int_\Omega \varrho|\boldsymbol{u}|^2 \,\mathrm{d}\boldsymbol{x}\,\mathrm{d}t. \tag{3.11}$$

The rest of the section will be devoted to the proof of (3.11).

STEP (iv) Given the systems of compacts $\{\overline{S}^i\}_{i=1}^m$, $\{\overline{B}_n^i\}_{i=1}^m$ as above, we consider a family of regular domains

$$G^0 = R^3 \setminus \overline{\Omega}_1, \ G^i = B_1^i, \ i = 1, \ldots, m.$$

By virtue of (3.6) the sets G^i can be chosen in such a way that

$$\min_{i=0,1,\ldots,m,\ i \neq j} \mathrm{dist}(\overline{G}_n^i(t), \overline{G}_n^j(t)) > \frac{d(\{\overline{S}^i\})}{8} > 0 \text{ for all } t \in [0, T_0] \text{ and } n = 1, 2, \ldots \tag{3.12}$$

where, as usual,

$$G_n^i(t) = \eta_n^i(t, G^i), \ \overline{G}^i = \mathrm{cl}(G^i), \ i = 1, \ldots, m.$$

Next, we define

$$H_r^s[\delta](G \mid K) \equiv \left\{ \boldsymbol{v} \in H^s(G) \mid \mathbb{D}(\boldsymbol{v}) = 0 \text{ in } \mathcal{D}'(U_\delta(K)), \right.$$

$$\left. \int_\Gamma \boldsymbol{v} \cdot \boldsymbol{n}\,\mathrm{d}\sigma = 0 \text{ for any component } \Gamma \subset \partial G \right\}.$$

Here

$$U_\delta(K) \equiv \{x \in R^3 \mid \text{dist}(x, K) < \delta\}.$$

If the set K is connected, the space $H_r^s[\delta](G \mid K)$ consists of all solenoidal functions from $H^s(G)$ which coincide with a rigid velocity field on the δ neighbourhood of the set K, and such that

$$\int_G v \cdot \nabla\phi \, dx = 0 \text{ for any } \phi \in C^\infty(R^3), \ \nabla\phi = 0 \text{ on } R^3 \setminus G.$$

Similarly, we introduce

$$H_0^s[\delta](G^0 \mid R^3 \setminus \Omega) \equiv \left\{ v \in H^s(G^0) \mid v = 0 \text{ on } U_\delta(R^3 \setminus \Omega), \right.$$

$$\left. \int_\Gamma v \cdot n \, d\sigma = 0 \text{ for any component } \Gamma \subset \partial G^0 \right\},$$

and, finally,

$$H_r^s[0](G \mid K) \equiv \text{cl}_{H^s}\left(\bigcup_{\delta > 0} H_r^s[\delta](G \mid K) \right),$$

$$H_0^s[0](G^0 \mid R^3 \setminus \Omega) \equiv \text{cl}_{H^s}\left(\bigcup_{\delta > 0} H_0^s[\delta](G^0 \mid R^3 \setminus \Omega) \right).$$

LEMMA 3.1. *Let G, $K \subset G$, and $s \in [0, 1)$ be fixed. Denote*

$$P^{\delta,r}(G \mid K) : H^s(G) \mapsto H_r^s[\delta](G \mid K)$$

the H^s- orthogonal projection on the space $H_r^s[\delta](G \mid K)$, $\delta > 0$. Then there exists a function $h = h(\delta)$ such that

$$h(\delta) \searrow 0 \text{ for } \delta \searrow 0,$$

$$\| v - P^{\delta,r}(G \mid K)v \|_{H^s(G)} \le h(\delta)\|v\|_{H^1(G)}$$

for all $v \in H_r^1[0](G \mid K)$.

Proof. As $s < 1$, the embedding $H^1(G) \subset H^s(G)$ is compact. Consequently, given $\varepsilon > 0$ we can find a system of $k = k(\varepsilon)$ functions w_j such that

$$w_j \in H_r^1[0](G \mid K), \ \|w_j\|_{H^1(G)} \le 1, \ i = 1, \ldots, k,$$

and

$$\min_{j=1,\ldots,k} \| v - w_j \|_{H^s(G)} < \frac{\varepsilon}{3} \text{ for any } v \in H_r^1[0](G \mid K), \ \|v\|_{H^1(G)} \le 1.$$

Thus

$$\|\boldsymbol{v} - P^{\delta,r}(G \mid K)\boldsymbol{v}\|_{H^s(G)} \leq \|\boldsymbol{v} - \boldsymbol{w}_j\|_{H^s(G)}$$

$$+ \|P^{\delta,r}(G \mid K)(\boldsymbol{w}_j - \boldsymbol{v})\|_{H^s(G)} + \|\boldsymbol{w}_j - P^{\delta,r}(G \mid K)\boldsymbol{w}_j\|_{H^s(G)}$$

for a certain j, where

$$\|\boldsymbol{v} - \boldsymbol{w}_j\|_{H^s(G)} + \|P^{\delta,r}(G \mid K)(\boldsymbol{w}_j - \boldsymbol{v})\|_{H^s(G)} \leq \frac{2}{3}\varepsilon.$$

On the other hand, we have

$$\|\boldsymbol{w}_j - P^{\delta,r}(G \mid K)\boldsymbol{w}_j\|_{H^s(G)} = \inf_{\boldsymbol{v} \in H_r^s[\delta](G \mid K)} \|\boldsymbol{w}_j - \boldsymbol{v}\|_{H^s(G)}$$

$$\leq \inf_{\boldsymbol{v} \in H_r^s[\delta](G \mid K)} \|\boldsymbol{w}_j - \boldsymbol{v}\|_{H^1(G)} \leq \inf_{\boldsymbol{v} \in H_r^1[\delta](G \mid K)} \|\boldsymbol{w}_j - \boldsymbol{v}\|_{H^1(G)}.$$

As the functions $\boldsymbol{w}_j$ belong to $H_r^1[0](G \mid K)$, one can find $\delta = \delta(\varepsilon) > 0$ such that

$$\inf_{\boldsymbol{v} \in H_r^1[\delta](G \mid K)} \|\boldsymbol{w}_j - \boldsymbol{v}\|_{H^1(G)} < \frac{\varepsilon}{3} \text{ for any } j = 1, \ldots, k$$

which yields the desired conclusion. $\qquad\square$

In the same way, one can show the following assertion.

LEMMA 3.2. *Let G^0 and $s \in [0, 1)$ be fixed. Denote*

$$P^{\delta,0}(G^0 \mid R^3 \setminus \Omega) : H^s(G^0) \mapsto H_0^s[\delta](G^0 \mid R^3 \setminus \Omega)$$

the H^s-orthogonal projection on the space $H_0^s[\delta](G^0 \mid R^3 \setminus \Omega)$, $\delta > 0$.
Then there exists a function $h = h(\delta)$ such that

$$h(\delta) \searrow 0 \text{ for } \delta \searrow 0,$$
$$\| \boldsymbol{v} - P^{\delta,0}(G^0 \mid R^3 \setminus \Omega)\boldsymbol{v} \|_{H^s(G^0)} \leq h(\delta)\|\boldsymbol{v}\|_{H^1(G^0)}$$

for all $\boldsymbol{v} \in H_0^1[0](G^0 \mid R^3 \setminus \Omega)$.

Finally, we define

$$H^s[\delta](\Omega_{n,t}) \equiv \left\{ \boldsymbol{v} \in H^s(\Omega) \mid \mathbb{D}(\boldsymbol{v}) = 0 \text{ in } \mathcal{D}'\left(U_\delta\left(\bigcup_{i=1}^m \overline{S}_n^i(t)\right)\right), \right.$$

$$\left. \boldsymbol{v} = 0 \text{ a.a. on } U_\delta(R^3 \setminus \Omega) \right\}$$

together with the corresponding H^s orthogonal projection

$$P_n^\delta(t) : H^s(\Omega) \mapsto H^s[\delta](\Omega_{n,t}), \ t \in [0, T_0].$$

LEMMA 3.3. *Let $s \in (0, \frac{1}{2})$ be fixed. Then there exists a function $h = h(\delta)$ independent of n, $t \in [0, T_0]$ such that*

$$h(\delta) \searrow 0 \text{ for } \delta \searrow 0,$$

$$\|\boldsymbol{u}_n(t) - P_n^\delta(t)\boldsymbol{u}_n(t)\|_{H^s(\Omega)} \le h(\delta)\|\boldsymbol{u}_n(t)\|_{H^1(\Omega)} \tag{3.13}$$

for a.e. $t \in [0, T_0]$ and all $n = 1, 2, \ldots$.

Proof. The functions $\boldsymbol{u}_n(t)$ belong to the spaces

$$\boldsymbol{u}_n(t) \in H_0^1(\Omega_n), \ \mathbb{D}(\boldsymbol{u}_n(t)) = 0 \text{ on } B_n^i(t), \ i = 1, \ldots, m$$

for a.a. $t \in [0, T_0]$. In particular, we have

$$\boldsymbol{u}_n(t) \in H_0^1[0](G^0 \mid R^3 \setminus \Omega), \ \boldsymbol{u}_n(t) \in H_r^1[0] \, (G_n^i(t) \mid \overline{S}_n^i(t)), \ i = 1, \ldots, m,$$

where the sets G^i are defined at STEP (iv).

Thus we can use Lemmas 2.1, 2.2 to find functions

$$\boldsymbol{w}_0(t) \in H_0^s[\delta](G^0 \mid R^3 \setminus \Omega), \ \boldsymbol{w}_i(t) \in H_r^s[\delta](G_n^i(t) \mid \overline{S}_n^i(t)), \ i = 1, \ldots, m$$

such that

$$\max_{j=0,\ldots m} \|\boldsymbol{u}_n(t) - \boldsymbol{w}_j(t)\|_{H^s(G_n^j(t))} \le h(\delta)\|\boldsymbol{u}_n(t)\|_{H^1(\Omega)}. \tag{3.14}$$

Since η_n^i are isometries, the function $h = h(\delta)$ can be chosen as required in (3.13) independently of t, n.

Now, since the functions $\boldsymbol{u}_n(t)$ are solenoidal on Ω, and $\boldsymbol{w}_j$ satisfy

$$\int_\Gamma \boldsymbol{w}_j \cdot \boldsymbol{n} \, d\sigma = 0 \text{ for any component } \Gamma \subset \partial G_n^j(t), \ j = 0, \ldots, m,$$

we take a function $\boldsymbol{z}(t) = \nabla Y_n(t)$,

$$\boldsymbol{z}(t) \in H^s \left(\Omega \setminus \bigcup_{j=0}^m \overline{G}_n^j(t) \right),$$

where $Y_n(t)$ is the unique zero-mean solution of the Neumann problem

$$\Delta Y_n(t) = 0 \text{ in } \Omega \setminus \bigcup_{j=0}^m \overline{G}_n^i(t), \ \nabla Y_n(t) \cdot \boldsymbol{n}$$

$$= (\boldsymbol{u}_n(t) - \boldsymbol{w}_j(t)) \cdot \boldsymbol{n} \text{ on } \partial G_n^j(t), \ j = 0, 1, \ldots, m.$$

Note that $\Omega \setminus \cup_{j=0}^{m} \overline{G}_n^i(t)$ is an open set consisting of a finite number of components with boundaries of class C^{∞} (cf. (3.12)).

Since η_n^i are isometries and (3.12) holds, one can find a constant c independent of $t \in [0, T_0]$, n such that

$$\|z(t)\|_{H^s(\Omega \setminus \cup_{j=0}^{m} \overline{G}_n^i(t))} \leq c \max_{j=0,\ldots,m} \|u_n(t) - w_j(t)\|_{H^s(G_n^j(t))};$$

whence, by virtue of (3.14),

$$\|z(t)\|_{H^s(\Omega \setminus \cup_{j=0}^{m} \overline{G}_n^i(t))} \leq h(\delta)\|u_n(t)\|_{H^1(\Omega)} \text{ independedtly of } t \in [0, T_0], \ n.$$

Finally, we set

$$v_n(t) = \begin{cases} w_j(t) & \text{on } G_n^j(t), \ j = 0, \ldots, m, \\ u_n(t) - z(t) & \text{on } \Omega \setminus \cup_{j=1}^{m} \overline{G}_n^j(t). \end{cases}$$

Since $s < 1/2$, the function $v_n(t)$ belong to the space $H^s[\delta](\Omega_{n,t})$ and satisfies

$$\|u_n(t) - v_n(t)\|_{H^s(\Omega)} \leq h(\delta)\|u_n(t)\|_{H^1(\Omega)}$$

independently of $t \in [0, T_0]$. Observe that the normal trace of v_n is continuous on ∂G^j which ensures that div $v_n = 0$ in $\mathcal{D}'(\Omega)$. $\qquad \square$

Step (v) Similarly as above, we define spaces

$$H^s[\delta](\Omega_t) = \left\{ v \in H^s(\Omega) \mid \mathbb{D}(v) = 0 \text{ in } U_\delta \left(\bigcup_{i=1}^{m} \overline{S}^i(t) \right), \ v = 0 \text{ on } U_\delta(R^3 \setminus \Omega) \right\}$$

along with the corresponding projections

$$P^\delta(t) : H^s(\Omega) \mapsto H^s[\delta](\Omega_t).$$

We write $P^\delta v(t)$ for $P^\delta(t)v(t)$.

LEMMA 3.4. *We have*

$$\int_0^{T_0} \int_\Omega \psi \varrho_n u_n \cdot P^\delta u_n \, dx \, dt \to \int_0^{T_0} \int_\Omega \psi \varrho u \cdot P^\delta u \, dx \, dt \tag{3.15}$$

for any function $\psi \in \mathcal{D}(0, T_0)$.

Proof. As

$$\overline{B}_n^i(t) \overset{b}{\to} \overline{S}^i(t) \text{ uniformly in } t \in [0, T_0],$$

any $\tau \in [0, T]$ is contained in an open time interval $J = J(\delta)$ such that any function φ of the form

$$\varphi(t, \boldsymbol{x}) = \psi(t)\phi(\boldsymbol{x}), \quad \psi \in \mathcal{D}(J \cap (0, T_0)), \quad \phi \in \mathcal{D}(\Omega) \cap H\left[\frac{\delta}{4}\right](\Omega_\tau)$$

belongs to $\mathcal{T}(\overline{Q}_n^s)$, i.e., it is an admissible test function for (3.1), for all n large enough.

As $\mathcal{D}(\Omega) \cap H[\frac{\delta}{4}](\Omega_\tau)$ is certainly dense in $H[\frac{\delta}{2}](\Omega_\tau)$, we deduce from (3.1), (3.7), and (3.8) that

$$\varrho_n \boldsymbol{u}_n \to \varrho \boldsymbol{u} \text{ in } C\left(\overline{J} \cap [0, T_0]; \left[H\left[\frac{\delta}{2}\right](\Omega_\tau)\right]^*_{\text{weak}}\right), \tag{3.16}$$

specifically,

$$\int_\Omega \varrho_n \boldsymbol{u}_n \cdot \phi \, \mathrm{d}\boldsymbol{x} \to \int_\Omega \varrho \boldsymbol{u} \cdot \phi \, \mathrm{d}\boldsymbol{x} \text{ in } C(\overline{J} \cap [0, T_0]) \text{ for any } \phi \in H\left[\frac{\delta}{2}\right](\Omega_\tau).$$

On the other hand, one obtains from (3.8)

$$P^\delta \boldsymbol{u}_n \to P^\delta \boldsymbol{u} \text{ weakly in } L^2\left(J \cap (0, T_0); H^s\left[\frac{\delta}{2}\right](\Omega_\tau)\right). \tag{3.17}$$

Since the embedding $H^s(\Omega) \subset H(\Omega)$ is compact, the relations (3.16), (3.17) give rise to (3.15) for any $\psi \in \mathbb{D}(J \cap (0, T_0))$. As $[0, T_0]$ is compact, this yields the desired conclusion.

$$\square$$

STEP (vi) At this stage, we are able to show (3.11) to complete the proof of Proposition 3.1. To this end, we write

$$\int_0^{T_0} \int_\Omega \psi \varrho_n |\boldsymbol{u}_n|^2 \, dx \, dt = \int_0^{T_0} \int_\Omega \psi \varrho_n \boldsymbol{u}_n \cdot P^\delta \boldsymbol{u}_n \, \mathrm{d}\boldsymbol{x} \, dt$$
$$+ \int_0^{T_0} \int_\Omega \psi \varrho_n \boldsymbol{u}_n \cdot (\boldsymbol{u}_n - P^\delta \boldsymbol{u}_n) \, \mathrm{d}\boldsymbol{x} \, dt,$$

where, by virtue of Lemma 3.4,

$$\int_0^{T_0} \int_\Omega \psi \varrho_n \boldsymbol{u}_n \cdot P^\delta \boldsymbol{u}_n \, \mathrm{d}\boldsymbol{x} \, dt \to \int_0^{T_0} \int_\Omega \psi \varrho \boldsymbol{u} \cdot P^\delta \boldsymbol{u} \, \mathrm{d}\boldsymbol{x} \, dt \text{ as } n \to \infty. \tag{3.18}$$

On the other hand,

$$\left| \int_0^{T_0} \int_\Omega \psi \varrho_n \boldsymbol{u}_n \cdot (\boldsymbol{u}_n - P^\delta \boldsymbol{u}_n) \, \mathrm{d}\boldsymbol{x} \, dt \right|$$

$$\leq \overline{\varrho} \|\psi\|_{L^\infty(0, T_0)} \|\boldsymbol{u}_n\|_{L^2(0, T_0; H(\Omega))} \|\boldsymbol{u}_n - P^\delta \boldsymbol{u}_n\|_{L^2(0, T_0; H(\Omega))}. \tag{3.19}$$

Now, it is easy to see that

$$\| u_n - P^\delta u_n \|_{H(\Omega)} \leq \| u_n - P^\delta u_n \|_{H^s(\Omega)}$$
$$\leq \| u_n - P_n^{2\delta} u_n \|_{H^s(\Omega)} \text{ for all } n \text{ large enough,}$$

where right-hand side can be estimated by Lemma 3.1,

$$\| u_n - P_n^{2\delta} u_n \|_{H^s(\Omega)} \leq h(2\delta) \| u_n \|_{H^1(\Omega)}.$$

Consequently, the right-hand side of (3.19) is small for $\delta > 0$ small enough which, together with (3.18), yields (3.11). We have proved Proposition 3.1.

4. Proof of Theorem 1.1

Now, we are ready to prove Theorem 1.1. Let us fix $T > 0$. By virtue of Proposition 3.1, **Problem (P)** with the initial data as in Theorem 1.1 admits a variational solution ϱ_1, u_1, $\{\overline{S}_1^i, \eta_1^i\}_{i=1}^m$ defined on some maximal time interval $[0, T_1]$, $T_1 > 0$, and the spatial domain $\Omega_1 \equiv \Omega$.

Now either $T_1 \geq T$ and the proof is complete, or $T_1 < T$ in which case $d(\{\overline{S}_1^i(T_1)\}) = 0$. In the latter case, we consider a new problem with the initial data $\varrho(T_1)$, $u(T_1)$ is as in the continuation Lemma 2.2. Moreover, we can decompose the index set

$$\{1, \ldots, m\} = M_0 \cup \bigcup_{i=1}^k M_i, \ M_i \cap M_j = \emptyset \text{ for } i \neq j$$

so that the system of compacts

$$\overline{S}_2^i(T_1) = \bigcup_{j \in M_i} \overline{S}_1^j(T_1), \ i = 1, \ldots, k$$

together with the open set

$$\Omega_2 = \Omega_1 \setminus \bigcup_{j \in M_0} \overline{S}_1^j(T_1)$$

satisfy the hypotheses of Proposition 3.1 (cf. also Remark 3.1), in particular,

$$\overline{S}_2^i(T_1) \cap \overline{S}_2^j(T_1) = \emptyset \text{ for } i \neq j, \ \overline{S}_2^i(T_1) \cap \partial\Omega_2 = \emptyset.$$

Moreover, since $d(\{\overline{S}_1^i(T_1)\}) = 0$ we must have $k < m$.

Now, we apply once more Proposition 3.1 (see Remark 3.1) to deduce that **Problem (P)** endowed with the initial data $\varrho(T_1)$, $u(T_1)$, $\{\overline{S}_2^i\}_{i=1}^k$ possesses a variational solution ϱ_2, u_2, $\{\overline{S}_2^i, \eta_2^i\}_{i=1}^k$ defined on a maximal time interval (T_1, T_2), $T_2 > 0$ and the open set Ω_2.

Moreover, it is easy to check that ϱ_2, $\boldsymbol{u}_2$, and $\{\overline{S}_1^j, \eta_1^j\}_{j=1}^m$ is a variational solution of **Problem (P)** on (T_1, T_2) and the spatial domain Ω with the initial data $\varrho(T_1)$, $\boldsymbol{u}(T_1)$, $\{\overline{S}_1^j(T_1)\}_{j=1}^m$ where

$$\eta_1^j = \eta_2^i \text{ for } j \in M_i, \ i = 0, \ldots, k.$$

Indeed one has only to realize that any test function

$$\varphi \in \{\varphi \in \mathcal{D}(T_1, T_2) \times \Omega \mid \operatorname{div} \varphi = 0 \text{ on } (T_1, T_2) \times \Omega,$$

$$\mathbb{D}(\varphi) = 0 \text{ on an open neighbourhood of } \overline{Q}^s\}$$

belongs, in fact, to the set

$$\varphi \in \mathcal{D}((T_1, T_2) \times \Omega_2).$$

To see this it is enough to observe that the set

$$R^3 \setminus \Omega_2 = (R^3 \setminus \Omega) \cup \bigcup_{j \in M_0} \overline{S}_1^j(T_1) = (R^3 \setminus \Omega) \cup \bigcup_{j \in M_0} \overline{S}_1^j(t), \ \ t \in (T_1, T_2)$$

is connected, unbounded, and $\mathbb{D}(\varphi)$ vanishes on an open neighbourhood of this set, i.e., $\varphi \in \mathcal{D}((T_1, T_2) \times \Omega_2)$.

Now, using the continuation Lemma 2.2 we get that the quantities

$$\varrho(t) = \begin{cases} \varrho_1(t) \text{ for } t \in (0, T_1), \\ \varrho(t_2) \text{ for } t \in [T_1, T_2) \end{cases} ; \ \boldsymbol{u}(t) = \begin{cases} \boldsymbol{u}_1(t), \text{ for } t \in (0, T_1) \\ \boldsymbol{u}_2(t) \text{ for } t \in [T_1, T_2) \end{cases} ;$$

and $\overline{S}^i(t) = \overline{S}_1^i(t)$ represent a variational solution of **Problem (P)** on $(0, T_2) \times \Omega$. Moreover, if $T_2 < T$, we necessarily have $d(\{\overline{S}_2^i(T_2)\}) = 0$.

Since there is a finite number of solids contained in Ω at the initial time $t = 0$, this procedure yields a solution of **Problem (P)** on $(0, T) \times \Omega$ after a finite number of steps, i.e., for a finite number of T_j, $j = 1, 2, \ldots, n \le m$. Theorem 1.1 has been proved.

REFERENCES

[1] CONCA, C., SAN MARTIN, J. and TUCSNAK, M., *Existence of solutions for the equations modelling the motion of a rigid body in a viscous fluid.* Commun. Partial Differential Equations, *25* (2000), 1019–1042.

[2] DESJARDINS, B. and ESTEBAN, M. J., *Existence of weak solutions for the motion of rigid bodies in a viscous fluid.* Arch. Rational Mech. Anal., *146* (1999), 59–71.

[3] DESJARDINS, B. and ESTEBAN, M. J., *On weak solutions for fluid-rigid structure interaction: Compressible and incompressible models.* Commun. Partial Differential Equations, *25* (2000), 1399–1413.

[4] DIPERNA, R. J. and LIONS, P. -L., *Ordinary differential equations, transport theory and Sobolev spaces.* Invent. Math., *98* (1989), 511–547.

[5] FEIREISL, E., *On the motion of rigid bodies in a viscous compressible fluid.* Arch. Rational Mech. Anal., 2002. To appear.

[6]   GALDI, G. P., *On the steady self-propelled motion of a body in a viscous incompressible fluid*. Arch. Rat. Mech. Anal., *148* (1999), 53–88.

[7]   GUNZBURGER, M. D., LEE, H. C. and SEREGIN, A., *Global existence of weak solutions for viscous incompressible flow around a moving rigid body in three dimensions*. J. Math. Fluid Mech., *2* (2000), 219–266.

[8]   HOFFMANN, K. -H. and STAROVOITOV, V. N., *On a motion of a solid body in a viscous fluid. Two dimensional case*. Adv. Math. Sci. Appl., *9* (1999), 633–648.

[9]   LIONS, J. -L. and MAGENES, E., *Problèmes aux limites non homogènes et applications, I*. Dunod, Gautthier – Villars, Paris, 1968.

[10]   SAN MARTIN, J. A., STAROVOITOV, V. and TUCSNAK, M., *Global weak solutions for the two dimensional motion of several rigid bodies in an incopmressible viscous fluid*. Arch. Rational Mech. Anal., *161* (2002), 93–112.

Eduard Feireisl
Mathematical Institute AV ČR
Žitná 25
115 67 Praha 1
Czech Republic
e-mail: feireisl@math.cas.cz

To access this journal online:
http://www.birkhauser.ch

J.evol.equ. 3 (2003) 443 – 461
1424–3199/03/030443 – 19
DOI 10.1007/s00028-003-0111-0
© Birkhäuser Verlag, Basel, 2003

**Journal of Evolution
Equations**

Minimization problems for eigenvalues of the Laplacian

ANTOINE HENROT

Abstract. This paper is a survey on classical results and open questions about minimization problems concerning the lower eigenvalues of the Laplace operator. After recalling classical isoperimetric inequalities for the two first eigenvalues, we present recent advances on this topic. In particular, we study the minimization of the second eigenvalue among plane convex domains. We also discuss the minimization of the third eigenvalue. We prove existence of a minimizer. For others eigenvalues, we just give some conjectures. We also consider the case of Neumann, Robin and Stekloff boundary conditions together with various functions of the eigenvalues.

1. Introduction

Problems linking the shape of a domain to the sequence of its eigenvalues, or some of them, are among the most fascinating of mathematical analysis or differential geometry. In particular, problems of minimization of eigenvalues, or combination of eigenvalues, brought about many deep works since the early part of the twentieth century. Actually, this question appeared in the famous book of Lord Rayleigh "The theory of sound" (for example in the edition of 1894). Thanks to some explicit computations and "physical evidence", Lord Rayleigh conjectured that the disk should minimize the first Dirichlet eigenvalue λ_1 of the Laplacian among every open sets of given measure.

It was indeed in the 1920's that Faber [21] and Krahn [34] solved simultaneously the Rayleigh's conjecture using a rearrangement technique. This classical proof is given at Theorem 3.1 of Section 3 which is devoted to the first Dirichlet eigenvalue. We will also discuss the case of a multiconnected domain and present some open problems involving the first eigenvalue. For other problems and a complete bibliography, we refer to [2], [41], [42], [50], [59], [60]; this kind of question is often called "isoperimetric inequalities for eigenvalues" in standard works, see also [6], [38] [48], [49]. In Section 4, we investigate similar questions for the second eigenvalue. The open set, of given measure, which minimizes λ_2 is the union of two identical balls. This result is generally attributed to P. Szegö, as quoted by G. Pólya in [47]), but it was already contained (more or less explicitly) in one of the Krahn's papers, see [35]. In this section, we will also present very recent results about the

---

Mathematics Subject Classification 2000: 49Q10, 35P15, 49J20.
Key words: Eigenvalues, minimization, isoperimetric inequalities, optimal domain.

minimization of λ_2 among **convex plane domains**. In Section 5, we look at the remaining eigenvalues of the Dirichlet-Laplacian. Actually a very few things are known! We only know the existence of an optimal domain for λ_3 and the fact that this domain is connected in dimension 2 and 3. In Section 6, we will consider the so-called "Payne-Pólya-Weinberger" conjecture, solved by Ashbaugh and Benguria in the 90's, concerning the ratio of the two first eigenvalues λ_2/λ_1. We will also present some open problems on other ratios. Finally, in Section 7, we will present some results about other boundary conditions: Neumann, Robin and also the Stekloff problem. We have decided here to restrict ourselves to the Laplacian operator on open (Euclidean) sets. Now, there are also beautiful results and conjectures e.g. for the bi-Laplacian $\Delta(\Delta)$. A good overview is given by B. Kawohl in the book [32], see also [37] and [5]. There are also many similar results on manifolds, see e.g. [39] or [60].

2. Notations and prerequisites

For the basic facts we recall here, we refer to any textbook on partial differential equations. For example, [17] or [20] are good standard references. Let Ω be a bounded open set in $\mathbb{R}^N$. In the case of Dirichlet boundary conditions, the convenient functional space is the Sobolev space $H_0^1(\Omega)$ which is defined as the closure of C^∞ functions compactly supported in Ω for the norm $\|u\|_{H^1} := (\int_\Omega u(x)^2\,dx + \int_\Omega |\nabla u(x)|^2\,dx)^{1/2}$. The Laplacian on Ω with Dirichlet boundary conditions is a self-adjoint operator with compact inverse, so there exists a sequence of positive eigenvalues (going to $+\infty$) and a sequence of corresponding eigenfunctions that we will denote respectively $0 < \lambda_1(\Omega) \leq \lambda_2(\Omega) \leq \lambda_3(\Omega) \leq \ldots$ et $u_1, u_2, u_3, \ldots$. In other words, we have:

$$\begin{cases} -\Delta u_k = \lambda_k(\Omega)u_k & \text{in } \Omega \\ u_k = 0 & \text{on } \partial\Omega \end{cases} \tag{1}$$

We decide to normalize the eigenfunctions by the condition

$$\int_\Omega u_k(x)^2\,dx = 1. \tag{2}$$

The sequence of eigenfunctions defines an Hilbert basis of $L^2(\Omega)$. By hypo-analyticity of the Laplacian, each eigenfunction is analytic inside Ω, its behavior on the boundary is governed by classical regularity results for elliptic partial differential equations. From the maximum principle and the Krein-Rutman Theorem, it follows that the first eigenfunction u_1 is non negative in Ω and positive as soon as Ω is connected. In particular, since u_2 is orthogonal to u_1, it has to change the sign in Ω. The sets

$$\Omega_+ = \{x \in \Omega,\ u_2(x) > 0\} \quad \text{and} \quad \Omega_- = \{x \in \Omega,\ u_2(x) < 0\}$$

are called *the nodal domains* of u_2. According to the Courant-Hilbert Theorem, these two nodal domains are connected subsets of Ω. The set

$$\mathcal{N} = \{x \in \Omega,\ u_2(x) = 0\}$$

is called *the nodal line* of u_2. When Ω is a plane convex domain, this nodal line hits the boundary of Ω at exactly two points, see Melas [36], or Alessandrini [1]. For general simply connected plane domains Ω, it is still a conjecture, named after Larry Payne, the "Payne conjecture".

In the sequel, we are interested in minimization problems like

$$\min\{\lambda_k(\Omega),\ \Omega \text{ open subset of } \mathbb{R}^N,\ |\Omega| = A\}$$

(where $|\Omega|$ denotes the measure of Ω and A is a given constant). Sometimes, we will also consider other geometric or topologic constraints, this will be specified below.

3. The first eigenvalue of the Dirichlet-Laplacian

3.1. *The Rayleigh-Faber-Krahn inequality*

For the first eigenvalue, the basic result is (as conjectured by Lord Rayleigh):

THEOREM 3.1. (Rayleigh-Faber-Krahn) *Let Ω be any bounded open set in $\mathbb{R}^N$, let us denote by $\lambda_1(\Omega)$ its first eigenvalue for the Laplace operator with Dirichlet boundary conditions. Let B be the ball of the same volume as Ω, then*

$$\lambda_1(B) = \min\{\lambda_1(\Omega),\ \Omega \text{ open subset of } \mathbb{R}^N,\ |\Omega| = |B|\}.$$

The classical proof makes use of the *Schwarz spherical decreasing rearrangement*. Since such a rearrangement preserves any L^p norm and decreases the Dirichlet integral:

$$\int_B u^*(x)^2\, dx = \int_\Omega u(x)^2\, dx \qquad \int_B |\nabla u^*(x)|^2\, dx \leq \int_\Omega |\nabla u(x)|^2\, dx \tag{3}$$

the result follows using the variational characterization of the first eigenvalue (it minimizes the so-called Rayleigh quotient).

3.2. *The case of polygons*

We can ask the same question for the class of polygons with a given number n of sides. Actually, the result is known only for $n = 3$ and $n = 4$:

THEOREM 3.2. (Pólya) *The equilateral triangle has the least first eigenvalue among all triangles of given area. The square has the least first eigenvalue among all quadrilaterals of given area.*

The proof relies on the same technique as the Rayleigh-Faber-Krahn Theorem with the difference that is now used the so-called Steiner symmetrization (see e.g. [48] or [31]). This symmetrization is performed with respect to an hyperplane H: we transform a given set ω in a set ω^* symmetric w.r.t H by moving the center of each segment of ω orthogonal to H on H. Doing the same for the level set of a function allows to define the Steiner symmetrization of a given function. This symmetrization has the same properties (3) as the Schwarz rearrangement, therefore any Steiner symmetrization decreases the first eigenvalue. By a sequence of Steiner symmetrization with respect to the mediator of each side, a given triangle converges to an equilateral one. We can do the same for a quadrilateral by alterning symmetrization w.r.t. mediator of sides and diagonals. It will be transformed into a square with the means of an infinite sequence of Steiner symmetrization. This is the idea of Pólya's proof.

Unfortunately, for $n \geq 5$ (pentagons and others), the Steiner symmetrization increases, in general, the number of sides. This prevents us to use the same technique. So a beautiful (and hard) challenge is to solve the

Open problem 1 Prove that the regular n-gone has the least first eigenvalue among all the n-gone of given area for $n \geq 5$.

This conjecture is supported by the classical isoperimetric inequality linking area and length for regular n-gones, see e.g. Theorem 5.1 in Osserman, [38]. Another kind of result that can be proved on polygons has been stated by J. Hersch in [28]: *Among all parallelograms with given distances between their opposite sides, the rectangle maximizes λ_1.*

3.3. *Domains in a box*

Instead of looking at open sets just with a given volume, we could consider open sets constrained to lie into a given box D (and also with a given volume). In other words, we could look for the solution of

$$\min\{\lambda_1(\Omega), \ \Omega \subset D, \ |\Omega| = A \ (\text{given})\}. \tag{4}$$

According to the Theorem 5.2 of Buttazzo-DalMaso which will be stated below, the problem (4) has always a solution. Of course, if the constant A is small enough in such a way that the ball of volume A lies in the box D, it will provide the solution. Therefore, the interesting case is when the ball of volume A is "too big" to stay into D. In this case, we can prove, at least formally, that the optimal domain, say Ω^* has to touch the boundary of D. Indeed, if it was not the case, and assuming Ω^* to be regular, we can use classical Hadamard's formula for variations of eigenvalues, see e.g. [52], [53] to get an optimality condition. This formula is the following: if we deform the domain Ω^* thanks to a deformation field V such that if we set $\Omega_t = (Id + tV)(\Omega^*)$, then the differential quotients $(\lambda_1(\Omega_t) - \lambda_1(\Omega^*))/t$ have a limit when t goes to 0. Moreover, this limit is given by the formula:

$$d\lambda_1(\Omega^*, V) = -\int_{\partial\Omega^*} \left(\frac{\partial u_1}{\partial n}\right)^2 V.n\, d\sigma. \tag{5}$$

where $\frac{\partial u_1}{\partial n}$ denotes the normal derivative of the eigenfunction u_1 and $V.n$ is the normal displacement of the boundary induced by the deformation field V. We have a similar formula for the first variation of the volume Vol:

$$dVol(\Omega^*, V) = \int_{\partial\Omega^*} V.n\, d\sigma. \tag{6}$$

Therefore, the optimal domain must satisfy a Lagrange identity like

$$d\lambda_1(\Omega^*, V) = -c^2 dVol(\Omega^*, V)$$

for every deformation field V (with c^2 a Lagrange multiplier), which yields the following relation for the normal derivative of the first eigenfunction:

$$\frac{\partial u_1}{\partial n} = c. \tag{7}$$

Now, this relation (7) together with the p.d.e. (1) yields a well-known *overdetermined problem* whose only solution, according to J. Serrin cf [51], is a ball! Therefore, the optimal domain must touch ∂D. More precisely, the boundary of Ω^* has two kind of components:

- free components included in D,
- components lying on the boundary of D.

A natural question is to ask whether the free components are composed of pieces of spheres. We proved in a recent paper, see [24] that it is not the case:

PROPOSITION 3.3. *The free components of the domain Ω^* which solve problem (4) cannot be pieces of spheres unless the ball of volume A is the solution.*

Proof. Let us assume that $\partial\Omega^*$ contains a piece of sphere γ. On γ, Ω^* satisfies the optimality condition (7). We put the origin at the center of the corresponding ball and we introduce the functions

$$w_{i,j}(x, y) = x_i \frac{\partial u}{\partial x_j} - x_j \frac{\partial u}{\partial x_i}.$$

Then, we easily verify that

$$-\Delta w_{i,j} = \lambda_1 w_{i,j} \text{ in } \Omega^*$$

$$w_{i,j} = 0 \text{ in } \gamma$$

$$\frac{\partial w_{i,j}}{\partial n} = 0 \text{ in } \gamma.$$

Now we conclude, using Hölmgren uniqueness theorem, that $w_{i,j}$ must vanish in a neighborhood of γ, so in the whole domain by analyticity. Now, if all these functions $w_{i,j}$ are identically 0 in Ω^*, this would imply that u is radially symmetric in Ω^* and therefore that Ω^* is a ball. $\qquad\qquad\square$

Nevertheless, there are some interesting open questions for this very simple minimization problem. For example:

Open problem 2 Let Ω^* be a solution of the minimization problem (4). Prove that the free components of the boundary of Ω^* are C^∞ (or analytic). If D is convex, is it true that Ω^* is convex?

3.4. *Multi-connected domains*

This section could also be entitled "How to place an obstacle" (see [23]). Let us consider a multi-connected domain Ω with one or several holes whose boundaries are denoted by $\Gamma_0, \Gamma_1, \ldots$, the outer boundary of Ω being denoted by Γ. We can consider many problems, letting the boundary conditions varying on the outer boundary and/or the holes. Let me mention below the results known by the author on such minimization and maximization problems.

- One hole, Dirichlet boundary condition on Γ and Γ_0. J. Hersch in [27] proves:

  Of all such plane domains, with given area A, and given length L and L_0 of its outer and inner boundary satisfying $L^2 - L_0^2 = 4\pi A$, the annular domain (two concentric circles) maximizes λ_1.

  This result implies, in particular, that for a domain Ω of the kind $\Omega = B_1 \setminus B_0$ (difference of two disks of given radii), λ_1 is maximal when the disks are concentric. This particular result has been rediscovered later and extended to the N-dimensional case by several authors: M. Ashbaugh and T. Chatelain in 1997 (private communication), E. Harrel, P. Kröger and K. Kurata in [23], Kesavan, see [33]. They also proved that $\lambda_1(B_1 \setminus B_0)$ is a minimum when B_0 touches the boundary of B_1.

Open problem 3 Let Ω be a fixed domain and B_0 a ball of fixed radius. Prove that $\lambda_1(\Omega \setminus B_0)$ is minimal when B_0 touches the boundary of Ω (where?) and is maximum when

B_0 is centered at a particular point of Ω (at what point?). In [23], one can find some interesting partial answers assuming convexity and/or symmetry properties for Ω. They also give many illustrative examples. Actually, I think that the optimal center of B_0 depends on the radius and is not fixed (apart in the case of symmetries). When the radius of B_0 goes to zero, classical asymptotic formulae for eigenvalues of domains with small holes, see e.g. the review paper [22], lead one to think that the ball must be located at the maximal point of the first eigenvalue of the domain without hole. Of course, we can state the same question with a non circular hole of given measure: in such a case, we have to find not only the location but also the shape of the hole in order to minimize or maximize the first eigenvalue.

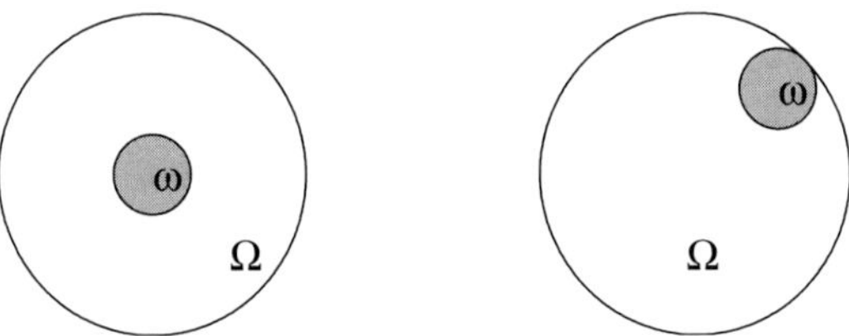

Figure 1   Position of the hole which maximizes $\lambda_1(\Omega\backslash\omega)$ (left); one position which minimizes $\lambda_1(\Omega\backslash\omega)$ (right).

- Several holes, Dirichlet boundary condition on the outer boundary Γ and Neumann boundary condition on the boundary of the holes. L. Payne and H. Weinberger proves in [46]:

  Among all multi-connected plane domains, with given area A, and given length L of its outer boundary, the annular domain (two concentric circles) maximizes the first eigenvalue λ_1 with Dirichlet boundary condition on the outer boundary Γ and Neumann boundary condition on the boundary of the holes.

- Several holes, Neumann boundary condition on the outer boundary Γ, Dirichlet boundary condition on one hole and Neumann boundary condition on the boundary of the other holes. J. Hersch proves in [27]:

  Among all multi-connected plane domains, with given area A, and given length L_0 of the first inner boundary, the annular domain (two concentric circles) maximizes the first eigenvalue λ_1 with Neumann boundary condition on the outer boundary Γ, Dirichlet boundary condition on one hole and Neumann boundary condition on the boundary of the other holes.

4. The second eigenvalue of the Dirichlet-Laplacian

For the second eigenvalue, the minimizer is not one ball, but two!

THEOREM 4.1. (Krahn-Szegö) *The minimum of $\lambda_2(\Omega)$ among bounded open sets of $\mathbb{R}^N$ with given volume is achieved by the union of two identical balls.*

Proof. Let Ω be any bounded open set, and let us denote by Ω_+ and Ω_- its nodal domains. Since u_2 satisfies

$$\begin{cases} -\Delta u_2 = \lambda_2 u_2 & \text{in } \Omega_+ \\ u_2 = 0 & \text{on } \partial\Omega_+ \end{cases}$$

$\lambda_2(\Omega)$ is an eigenvalue for Ω_+. But, since u_2 is positive in Ω_+, it is the first eigenvalue (and similarly for Ω_-):

$$\lambda_1(\Omega_+) = \lambda_1(\Omega_-) = \lambda_2(\Omega). \tag{8}$$

We now introduce Ω_+^* and Ω_-^* the balls of same volume as Ω_+ and Ω_- respectively. According to the Rayleigh-Faber-Krahn inequality

$$\lambda_1(\Omega_+^*) \leq \lambda_1(\Omega_+), \qquad \lambda_1(\Omega_-^*) \leq \lambda_1(\Omega_-). \tag{9}$$

Let us introduce a new open set $\tilde{\Omega}$ defined as

$$\tilde{\Omega} = \Omega_+^* \cup \Omega_-^*.$$

Since $\tilde{\Omega}$ is disconnected, we obtain its eigenvalues by gathering and reordering the eigenvalues of Ω_+^* and Ω_-^*. Therefore,

$$\lambda_2(\tilde{\Omega}) \leq \max(\lambda_1(\Omega_+^*), \lambda_1(\Omega_-^*)).$$

According to (8), (9) we have

$$\lambda_2(\tilde{\Omega}) \leq \max(\lambda_1(\Omega_+), \lambda_1(\Omega_-)) = \lambda_2(\Omega).$$

This shows that the minimum of λ_2 is to be obtained among the union of balls. But, if the two balls would have different radii, we would decrease the second eigenvalue by shrinking the largest one and dilating the smaller one (without changing the total volume). Therefore, the minimum is achieved by the union of two identical balls. $\qquad\square$

Being disappointed that the minimizer be not a connected set (it's hard to hit with one hand on a non-connected drum!), we could be interested in solving the minimization problem for λ_2 among **connected** sets. Unfortunately, a connectedness constraint does not really change the situation. Indeed, let us consider the following domain (see Figure 2) Ω_ε, obtained by joining the union of the two previous balls Ω by a thin pipe of width ε. We say that Ω_ε γ-converges to Ω if the resolvent operators T_ε associated with the Laplace-Dirichlet operator on Ω_ε simply converge to the corresponding operator T on Ω, see e.g. [18]. By a

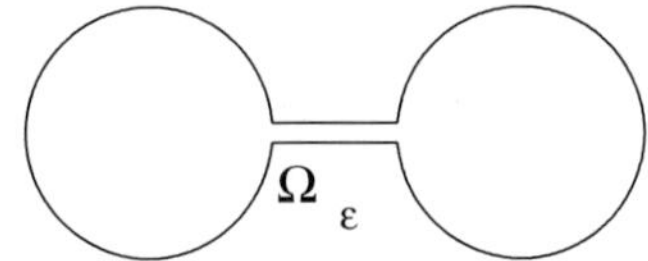
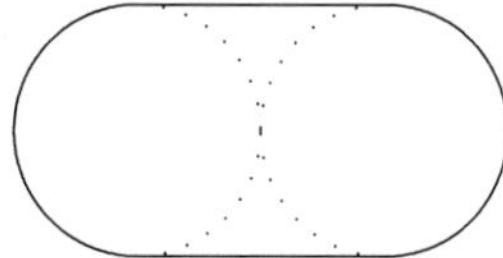

Figure 2   A minimizing sequence of connected domains (left), the stadium does not minimize λ_2 among convex sets of given volume (right)

compactness argument, see [12], [26] it can be proved that this simple convergence implies the convergence in the operator norm and therefore the convergence of the eigenvalues. Now, it is easy to verify, see [11], [26], that in the above situation Ω_ε γ-converges to Ω what yields $\lambda_2(\Omega_\varepsilon) \to \lambda_2(\Omega)$ and therefore:

$$\inf\{\lambda_2(\Omega), \Omega \subset \mathbb{R}^N, \Omega\text{connected}, |\Omega| = c\} = \min\{\lambda_2(\Omega), \Omega \subset \mathbb{R}^N, |\Omega| = c\}$$

what shows that this infimum is not achieved (actually, we can prove that the union of two balls is the unique minimizer of λ_2 up to displacements and zero-capacity subsets).

Now, the problem becomes again interesting if we ask the question to find the **convex** domain, of given area, which minimizes λ_2. For sake of simplicity, we restrict us here to the two-dimensional case. Existence of a minimizer Ω^* is easy to obtain (see [15] and [24], [25]). In a paper of 1973 [55], Troesch did some numerical experiments which led him to conjecture that the solution was *a stadium*: the convex hull of two identical tangent disks. It is actually the convex domain which is the closest to the solution without convexity constraint. In [24], we refute this conjecture:

THEOREM 4.2. (Henrot-Oudet) *The stadium, convex hull of two identical tangent disks, does not realize the minimum of λ_2 among plane convex domains of given area.*

Indeed, the proof is exactly the same as the proof of the above Proposition 3.3. Nevertheless, a more precise analysis and some numerical experiments show that the minimizer, say Ω^*, is very close to the stadium. Actually, we prove in [24], [25]:

THEOREM 4.3. (Henrot-Oudet)

Regularity *The minimizer Ω^* is at least C^1 and at most C^2.*

Simplicity *The second eigenvalue of Ω^* is simple.*

Geometry *The minimizer Ω^* has two (and only two) segments in its boundary and these segments are parallel.*

Open problem 4 Prove that a plane convex domain Ω^* which minimizes λ_2 (among convex domains of given area) has two perpendicular axes of symmetry.

5. Other eigenvalues of the Dirichlet-Laplacien

The minimization problem becomes much more complicated for the other eigenvalues! One of the only known result is the following, cf [10] and [58]:

THEOREM 5.1. (Bucur-Henrot and Wolff-Keller) *There exists a set Ω_3^* which minimizes λ_3 among the (quasi)-open sets of given volume. Moreover Ω_3^* is connected in dimension $N = 2$ or 3.*

The question of identifying the optimal domain Ω_3^* remains open. The conjecture is the following:

Open problem 5 Prove that the optimal domain for λ_3 is a ball in dimension 2 and 3 and the union of three identical balls in dimension $N \geq 4$.

Wolff and Keller have proved in [58] that the disk is a local minimizer for λ_3. There are two key-points in the existence proof of Theorem 5.1. The first one is a more general result of Buttazzo-Dal Maso (already cited earlier), see [12]:

THEOREM 5.2. (Buttazzo-Dal Maso) *Let D be a fixed ball in $\mathbb{R}^N$. For every fixed integer $k \geq 1$ and c fixed real number $0 < c < |D|$ the problem*

$$\min\{\lambda_k(\Omega); \quad \Omega \subset D, \quad |\Omega| = c\} \tag{10}$$

has a solution.
*More generally, the existence result remains valid for any function $\Phi(\lambda_1, \ldots, \lambda_k)$ of the eigenvalues **non decreasing** in each of its arguments.*

This theorem does not solve the general problem of existence of a minimizer for $\lambda_k(\Omega)$ since we assume to work with "confined" sets (that is to say, sets included in a box D). In order to remove this assumption in [10], we used a "concentration-compactness" argument together with the Wolff-Keller's result proving that the minimizer of λ_3 (if it exists) should be connected in dimension 2 and 3 (this is the second key-point). Here is the more general result they prove in [58]. Let us denote by Ω_n^* an open set which minimizes λ_n (among open sets of volume 1) and $\lambda_n^* = \lambda_n(\Omega_n^*)$ the minimal value of λ_n. We will also denote by $t\Omega$ the image of Ω by an homothety of ratio t. Then, we have:

THEOREM 5.3. (Wolf-Keller) *Let us assume that Ω_n^* is the union of (at least) two disjoints sets, each of them with positive measure. Then*

$$(\lambda_n^*)^{N/2} = (\lambda_i^*)^{N/2} + (\lambda_{n-i}^*)^{N/2} = \min_{1 \leq j \leq (n-1)/2} ((\lambda_j^*)^{N/2} + (\lambda_{n-j}^*)^{N/2}) \tag{11}$$

where, in the previous equality, i is a value of $j \leq (n-1)/2$ which minimizes the sum $(\lambda_j^*)^{N/2} + (\lambda_{n-j}^*)^{N/2}$. Moreover,

$$\Omega_n^* = \left[\left(\frac{\lambda_i^*}{\lambda_n^*} \right)^{1/2} \Omega_i^* \right] \bigcup \left[\left(\frac{\lambda_{n-i}^*}{\lambda_n^*} \right)^{1/2} \Omega_{n-i}^* \right]. \tag{12}$$

We have seen that the value of λ_n^* is not known unless for $n = 1$ or 2. Let us prove for example, that the optimal domain is connected in dimension 2. Indeed, if it was not connected, according to Theorem 5.3, we should have $\lambda_3^* = \lambda_1^* + \lambda_2^*$ ($i = 1$ is the only possible value here). Now $\lambda_1^* = \pi j_{0,1}^2 = 18.168$. ($j_{0,1}$ is the first zero of the Bessel function J_0) while, according to Theorem 4.1, $\lambda_2^* = 2\lambda_1^* = 36.336$. Therefore $\lambda_1^* + \lambda_2^* = 54.504$. But since λ_3^* is, by definition, lower or equal to the third eigenvalue of the unit disk $\lambda_3(D_1) = \pi j_{1,1}^2 = 46.125$, we see that it cannot be equal to $\lambda_1^* + \lambda_2^*$.

The same kind of computation works in dimension 3, but not in higher dimension. This is the reason why we think that the minimizer is the union of three identical balls in dimension greater than 4. For the fourth eigenvalue, it is conjectured that the minimum is attained by

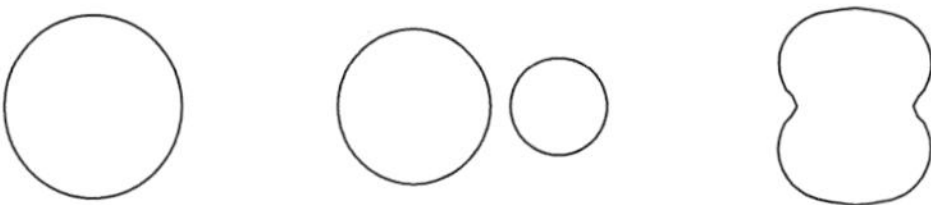

Figure 3   The disk probably minimizes λ_3 (left); two disks which probably minimize λ_4 (center); a domain candidate to minimize λ_5 (right).

the union of two balls whose radii are in the ratio $\sqrt{j_{0,1}/j_{1,1}}$ in dimension 2, where $j_{0,1}$ et $j_{1,1}$ are respectively the two first zeros of the Bessel functions J_0 et J_1, cf Figure 3, but it is not proved!

Open problem 6 Prove that the optimal domain for λ_4 is the union of two balls whose radii are in the ratio $\sqrt{j_{0,1}/j_{1,1}}$ in dimension 2.

Looking at the previous results and conjectures, P. Szegö asked the following question:

Is it true that the minimizer of any eigenvalue of the Laplace-Dirichlet operator is a ball or a union of balls?

The answer to this question is NO. For example, Wolff and Keller remarked that the thirteenth (!) eigenvalue of a square is lower than the thirteenth eigenvalue of any union of disks of same area. Actually, it is not necessary to go to the 13th eigenvalue. Numerical experiments, cf [40] and Figure 3, show that, for the n-th eigenvalue with n larger or equal to 5, the minimizer is no longer a ball or an union of balls.

Let us state now some new open problems:

Open problem 7 Prove that there exist a minimizer for λ_n among open sets of given volume. The technique we used in [10] allows us to prove such an existence result as soon as we are able to prove that the minimizers for λ_k, $k = 1 \ldots n - 1$ are indeed **bounded**.
Study the regularity and the geometric properties (e.g. symmetries) of such a minimizer.

6. Optimizing functions of eigenvalues

6.1. *Maximizing ratios of eigenvalues*

In 1955 L. Payne, G. Pólya and H. Weinberger in [43] considered the problem of bounding ratios of eigenvalues. In particular, they proved that the ratio λ_2/λ_1 is less than or equal to 3 (in dimension 2). They were led to conjecture that the optimal domain for this ratio is a disk. This conjecture has been proved 35 years later by M. Ashbaugh and R. Benguria, see [3] for the two-dimensional case and [4] for the N-dimensional case.

THEOREM 6.1. (Ashbaugh-Benguria) *The ball maximizes the ratio λ_2/λ_1.*

For ratios of eigenvalues, many problems remain open. A good overview and discussion on previous results is given in [2]. Below, some of them are listed

Open problem 8 (see [43]) Prove that the disk maximizes the quotient $(\lambda_2 + \lambda_3)/\lambda_1$ among plane domains of given area.
Prove that the ball maximizes the quotient $\sum_{i=2}^{N+1} \lambda_i/\lambda_1$ among domains of $\mathbb{R}^N$ with given area.

Open problem 9 Prove existence of a domain which maximizes the following ratios, study the geometric properties of such maximizers, if possible identify it

- $\frac{\lambda_3}{\lambda_1}$ (in the plane this is not the disk)
- $\frac{\lambda_{N+2}}{\lambda_1}$ in $\mathbb{R}^N$: it should be the ball
- $\frac{\lambda_{m+1}}{\lambda_m}$
- $\frac{\lambda_{2m}}{\lambda_m}$

6.2. *Other functions of eigenvalues*

We have already mentioned in Theorem 5.2. that any function of the kind $\Omega \mapsto \Phi(\lambda_1(\Omega), \lambda_2(\Omega))$ with Φ non decreasing with respect to each argument, admits a minimizer among

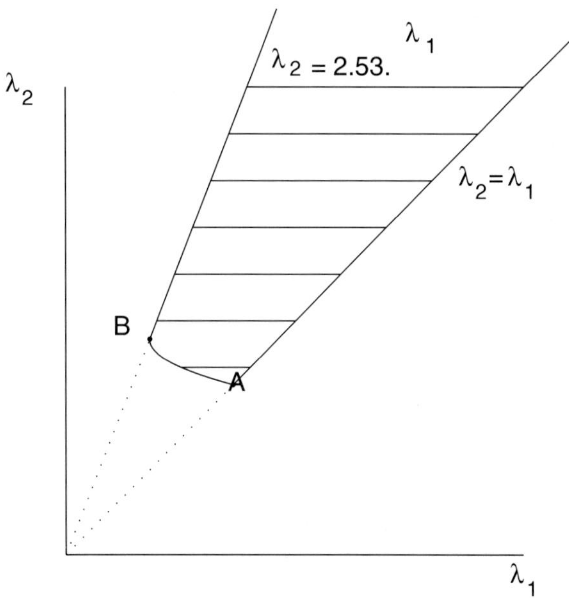

Figure 4   Range, in the plane (λ_1, λ_2), of the possible values for the two first eigenvalues of a domain of given area.

(quasi)-open sets of given volume. Note than none of the above ratios can be handled by this theorem. In an interesting paper [9], D. Bucur, G. Buttazzo and I. Figueiredo extended this result:

THEOREM 6.2. (Bucur, Buttazzo, Figueiredo) *Let* $\Phi : \overline{\mathbb{R}}^2 \to \mathbb{R}$ *be a lower semi-continuous function, D a given box and A a given constant. Then the problem*

$$\min\{\Phi(\lambda_1(\Omega), \lambda_2(\Omega)),\ \Omega \subset D, |\Omega| \leq A\}$$

has always a solution.

A more precise statement could be: either an optimal domain exists or for the minimizing sequence Ω_n we have $\Phi(\lambda_1(\Omega_n), \lambda_2(\Omega_n)) \to -\infty$. In this last case, we can choose as a minimizer the empty set. This is the case, for example for the gap function $\Phi(\lambda_1, \lambda_2) = \lambda_1 - \lambda_2$. The main ingredient of the proof of this Theorem is the closedness (in the plane) of the set $\mathcal{E} = \{(\lambda_1(\Omega), \lambda_2(\Omega)),\ \Omega \subset D, |\Omega| \leq A\}$. This set is represented in Figure 4. It is of course above the first bisectrix and, according to Ashbaugh-Benguria Theorem 6.1, it is below the line $y = 2.5387x$ (2.5387 is the value of the quotient for the disk). The point A in the Figure corresponds to the two identical balls (Krahn-Szegö Theorem 4.1.) while the point B corresponds to one ball (Rayleigh-Faber-Krahn). Indeed, they proved that this set $\mathcal{E}$ is convex in the x and the y direction.

Open problem 10 Prove that the set $\mathcal{E}$ defined above is convex.

Among various combinations of the two first eigenvalues, we can also consider for example $\lambda_1 + \lambda_2$: what is the set which minimizes the sum of the two first eigenvalues? It is not the disk, since it does not satisfy the generalized optimality conditions (see e.g. [14], [13]). More generally, we can ask this question for any convex combination of the two first eigenvalues of the kind $t\lambda_1 + (1-t)\lambda_2$. This question has a simple geometric interpretation: the wanted minimizer is indeed the first point of $\mathcal{E}$ we reach when making a line of equation $tx + (1-t)y = c$ approach the set $\mathcal{E}$ (by increasing c). In particular, for $t = 1$ the solution is a ball while for $t = 0$ it is given by two balls.

Open problem 11 For what value of t, the set which minimizes $t\lambda_1 + (1 - t)\lambda_2$ is no longer convex, no longer connected (perhaps $t = 1$ for that second question)?

7. Eigenvalues of the Laplacian with other boundary conditions

7.1. *Neumann boundary conditions*

The eigenvalues of the Laplacian with Neumann boundary conditions are also called the eigenvalues of the *free* membrane (in the case of Dirichlet boundary conditions, we speak about the *fixed* membrane). We will denote it by $0 = \mu_1(\Omega) \leq \mu_2(\Omega) \leq \mu_3(\Omega) \leq \ldots$ (the first eigenvalue is zero, corresponding to constant functions). They solve

$$\begin{cases} -\Delta u_k = \mu_k(\Omega)u_k & \text{in } \Omega \\ \frac{\partial u_k}{\partial n} = 0 & \text{on } \partial\Omega. \end{cases} \tag{13}$$

Minimizing the eigenvalues of the Laplacian with Neumann boundary conditions, with a volume constraint, is a trivial problem. Indeed, if we consider a long thin rectangle like $]0, L[\times]0, l[$, its n-th eigenvalue will be (for L large enough) $\mu_n = (n - 1)^2\pi^2/L^2$. Therefore, letting $L \to +\infty$, we see that $\inf\{\mu_n(\Omega), |\Omega| = A\} = 0$. Moreover, the infimum is attained for any open set which has at least n connected components. This shows that limiting the diameter of Ω does not improve the interest of the question! Now, if we assume that the domains must be convex and with a given diameter, then the infimum is not zero, but it is not achieved! Actually, L. Payne and H. Weinberger proved in [45] the following inequality for convex domains Ω in $\mathbb{R}^N$ with given diameter d:

$$\mu_2(\Omega) \geq \left(\frac{\pi}{d}\right)^2 .$$

This lower bound is optimal but not attained: any domain shrinking to a one-dimensional segment $[0, d]$ has its second eigenvalue which converges to the lower bound.

If we want to get a really interesting problem for eigenvalues of the Laplacian with Neumann boundary conditions, we must consider the problem of the **maximization** instead of the minimization:

THEOREM 7.1. (Szegö,Weinberger) *The ball maximizes the second Neumann eigenvalue among open sets of given volume.*

In the two-dimensional case, the proof (using conformal maps) was given by G. Szegö in [54]. It has been generalized to any dimension by H. Weinberger in [56]. We must also mention that Szegö and Weinberger in the above-mentioned papers have proved in two-dimensions that

$$\frac{1}{\mu_2} + \frac{1}{\mu_3} \quad \text{is minimal for the disk.}$$

Of course this result implies Theorem 7.1 since the second eigenvalue of the disk is double. Now, in higher dimensions, it is still an open problem:

Open problem 12 Prove that

$$\sum_{i=2}^{N+1} \frac{1}{\mu_i(\Omega)}$$

is minimal for the ball among all domains with a given volume.

More generally, the existence of a convex domain which maximizes the n-th Neumann eigenvalue μ_n (with given volume) has been proved in [16]. So, we are also led to the following open problem(s):

Open problem 13 Prove that there exists an open set (of given volume) which maximizes the n-th Neumann eigenvalue μ_n, for $n \geq 3$. If possible, identify this maximizer.

7.2. *Robin boundary condition*

The eigenvalues of the Laplacian with Robin boundary conditions are called the eigenvalues of the *elastically supported* membrane. We will denote them by $0 < \nu_1(\alpha, \Omega) \leq \nu_2(\alpha, \Omega) \leq \nu_3(\alpha, \Omega) \leq \ldots$ where α is a parameter, $0 < \alpha < 1$ (the cases $\alpha = 0$ or 1 obviously correspond to Neumann or Dirichlet conditions). The p.d.e. system is

$$\begin{cases} -\Delta u_k = \nu_k(\alpha, \Omega) u_k & \text{in } \Omega \\ \alpha u + (1 - \alpha) \frac{\partial u_k}{\partial n} = 0 & \text{on } \partial\Omega. \end{cases} \tag{14}$$

In two dimensions, we recover the Rayleigh-Faber-Krahn inequality; this result is not well known, it is due to M. H. Bossel in her thesis:

THEOREM 7.2. (Bossel) *The disk minimizes the first eigenvalue of the Robin problem among open sets with a given volume (for every value of $\alpha \in]0, 1]$).*

Her proof uses a new variational method, see [7]. This method is inspired by that of extremal length.

Open problem 14 Generalize Bossel's Theorem to dimension N.

Open problem 15 (see [44]) For what values of α, the ratio $\dfrac{\lambda_2}{\lambda_1}$ achieves its maximum for the disk?

Let us remark that Problem 14 is already stated in Daners, see [19] where a lower bound for ν_1 is given.

7.3. *Stekloff eigenvalue problem*

The Stekloff eigenvalue problem is the following:

$$\begin{cases} \Delta u = 0 & \text{in } \Omega \\ \frac{\partial u}{\partial n} = pu & \text{on } \partial\Omega. \end{cases} \tag{15}$$

We will denote its eigenvalues by $0 = p_1(\Omega) \le p_2(\Omega) \le p_3(\Omega) \le \ldots$ (the first eigenvalue is zero, corresponding to constant functions). Like in the Neumann case, it is the problem of **maximization** of the eigenvalues which is interesting here.

THEOREM 7.3. (Weinstock, Brock) *The ball maximizes the second Stekloff eigenvalue among open sets of given volume.*

R. Weinstock gave the proof of this theorem in the two-dimensional case in [56]. His proof was inspired by the one of Szegö for the free membrane problem. F. Brock in [8] proved actually a sharper inequality, namely:

Let Ω be a bounded domain in $\mathbb{R}^N$ and R the radius of the ball Ω^* of same volume than Ω, then

$$\sum_{i=2}^{N+1} \frac{1}{p_i(\Omega)} \ge NR \tag{16}$$

the equality sign in (16) is attained if Ω is a ball. It is clear that (16) implies the above Theorem since $p_2(\Omega^*) = 1/R$ has multiplicity N for the ball. I must also mention that J. Hersch and L. Payne have already proved (16) in two-dimensions in [29] and that they have also proved a sharper inequality, together with M.M. Schiffer in [30], namely:

the disk maximizes the product $p_2(\Omega)p_3(\Omega)$ among plane open sets of given volume.

Open problem 16 Study the maximization problem for other Stekloff eigenvalues.

Open problem 17 Prove that the N-ball maximizes the product $\Pi_{k=2}^{N+1} p_k(\Omega)$ among open sets in $\mathbb{R}^N$ with given volume.

Acknowledgements

The author wants to thank the referee who improves the preliminary version of this paper.

REFERENCES

[1]   ALESSANDRINI, G., *Nodal lines of eigenfunctions of the fixed membrane problem in general convex domains*, Comment. Math. Helv. *69* (1994) no. 1, 142–154.

[2]   ASHBAUGH, M. S., *Open problems on eigenvalues of the Laplacian*, Analytic and Geometric Inequalities and Their Applications, T. M. Rassias and H. M. Srivastava (editors), vol. 4787, Kluwer 1999.

[3]   ASHBAUGH, M. S. and BENGURIA, R., *Proof of the Payne-Pólya-Weinberger conjecture*, Bull. Amer. Math. Soc. *25* (1991) n°1, 19–29.

[4]   ASHBAUGH, M. S. and BENGURIA, R., *A sharp bound for the ratio of the first two eigenvalues of Dirichlet Laplacians and extensions*, Ann. of Math. *135* (1992) no. 3, 601–628.

[5]   ASHBAUGH, M. S. and BENGURIA, R., *On Rayleigh's conjecture for the clamped plate and its generalization to three dimensions*, Duke Math. J. *78* (1995), 1–17.

[6]   BANDLE, C., *Isoperimetric inequalities and applications.* Monographs and Studies in Mathematics, 7. Pitman, Boston, Mass.-London 1980.

[7]   BOSSEL, M. H., *Membranes élastiquement liées: extension du théorème de Rayleigh-Faber-Krahn et de l'inégalité de Cheeger*, C. R. Acad. Sci. Paris Sér. I Math. *302* (1986) no. 1, 47–50.

[8]   BROCK, F., *An isoperimetric inequality for eigenvalues of the Stekloff problem*, Z. Angew. Math. Mech. *81* (2001) no. 1, 69–71.

[9]   BUCUR, D., BUTTAZZO, G. and FIGUEIREDO, I., *On the attainable eigenvalues of the Laplace operator*, SIAM J. Math. Anal. *30* (1999) no. 3, 527–536.

[10] BUCUR, D. and HENROT, A., *Minimization of the third eigenvalue of the Dirichlet Laplacian*, Proc. Roy. Soc. London, *456* (2000), 985–996.

[11] BUCUR, D. and ZOLESIO, J. P., *N-dimensional shape optimization under capacitary constraints*, J. of Diff. Eq. *123* (1995), n°2, 504–522.

[12] BUTTAZZO, G. and DAL MASO, G., *An Existence Result for a Class of Shape Optimization Problems*, Arch. Rational Mech. Anal. *122* (1993), 183–195.

[13] CHATELAIN, T. and CHOULLI, M., *Clarke generalized gradient for eigenvalues*, Commun. Appl. Anal. *1* (1997) no. 4, 443–454.

[14] COX, S. J., *The generalized gradient at a multiple eigenvalue*, J. Funct. Anal. *133* (1995) no. 1, 30–40.

[15] COX, S. J. and ROSS, M., *Extremal eigenvalue problems for starlike planar domains*, J. Differential Equations, *120* (1995), 174–197.

[16] COX, S. J. and ROSS, M., *The maximization of Neumann eigenvalues on convex domains*, to appear.

[17] COURANT, R. and HILBERT, D., *Methods of Mathematical Physics,* vol. 1 et 2, Wiley, New York 1953 et 1962.

[18] DAL MASO, G., *An introduction to Γ-convergence*, Birkhäuser, Boston 1993.

[19] DANERS, D., *Robin boundary value problems on arbitrary domains*, Trans. AMS, *352* (2000), 4207–4236.

[20] DAUTRAY, R. and LIONS, J. L., (ed), *Analyse mathématique et calcul numérique,* Vol. I and II, Masson, Paris 1984.

[21] FABER, G., *Beweis, dass unter allen homogenen Membranen von gleicher Fläche und gleicher Spannung die kreisförmige den tiefsten Grundton gibt* , Sitz. Ber. Bayer. Akad. Wiss. 1923, 169–172.

[22]   FLUCHER, M., *Approximation of Dirichlet eigenvalues on domains with small holes*, J. Math. Anal. Appl. *193* (1995) no. 1, 169–199.

[23]   HARRELL, E. M., KRÖGER, P. and KURATA, K., *On the placement of an obstacle or a well so as to optimize the fundamental eigenvalue*, to appear in SIAM J. Math. Anal.

[24]   HENROT, A. and OUDET, E., *Le stade ne minimise pas λ_2 parmi les ouverts convexes du plan*, C. R. Acad. Sci. Paris Sr. I Math, *332* (2001) no. 4, 275–280.

[25]   HENROT, A. and OUDET, E., *Minimizing the second eigenvalue of the Laplace operator with Dirichlet boundary conditions*, to appear.

[26]   HENROT, A. and PIERRE, M., *Optimisation de forme,* book to appear.

[27]   HERSCH, J., *The method of interior parallels applied to polygonal or multiply connected membranes*, Pacific J. Math. *13* (1963), 1229–1238.

[28]   HERSCH, J., *Contraintes rectilignes parallels et valeurs propres de membranes vibrantes*, Z. Angew. Math. Phys. *17* (1966) 457–460.

[29]   HERSCH, J. and PAYNE, L. E., *Extremal principles and isoperimetric inequalities for some mixed problems of Stekloff's type*, Z. Angew. Math. Phys. *19* (1968), 802–817.

[30]   HERSCH, J., PAYNE, L. E. and SCHIFFER, M. M., *Some inequalities for Stekloff eigenvalues*, Arch. Rational Mech. Anal. *57* (1975), 99–114.

[31]   KAWOHL, B., Rearrangements and convexity of level sets in PDE, Lecture Notes in Mathematics, 1150. Springer-Verlag, Berlin, 1985.

[32]   KAWOHL, B., PIRONNEAU, O., TARTAR, L. and ZOLSIO, J. P., Optimal shape design, Lecture Notes in Mathematics, 1740. (Lectures given at the Joint C.I.M./C.I.M.E. Summer School held in Tróia, June 1–6, 1998, Edited by A. Cellina and A. Ornelas).

[33]   KESAVAN, S., *On two functionals connected to the Laplacian in a class of doubly connected domains*, to appear.

[34]   KRAHN, E., *Über eine von Rayleigh formulierte Minimaleigenschaft des Kreises*, Math. Ann. *94* (1924), 97–100.

[35]   KRAHN, E., *Über Minimaleigenschaften der Kugel in drei un mehr Dimensionen*, Acta Comm. Univ. Dorpat. *A9* (1926), 1–44.

[36]   MELAS, A., *On the nodal line of the second eigenfunction of the Laplacian in $\mathbb{R}^2$*, J. Diff. Geometry. *35* (1992), 255–263.

[37]   NADIRASHVILI, N. S., *Rayleigh's conjecture on the principal frequency of the clamped plate*, Arch. Rational Mech. Anal. *129* (1995), 1–10.

[38]   OSSERMAN, R., *The isoperimetric inequality*, Bull. AMS, *84*, (1978) no. 6, 1182–1238.

[39]   OSSERMAN, R., *Isoperimetric inequalities and eigenvalues of the Laplacian*, Proceedings of the International Congress of Mathematicians (Helsinki, 1978), pp. 435–442, Acad. Sci. Fennica, Helsinki, 1980.

[40]   OUDET, E., *Some numerical results about minimization problems involving eigenvalues*, to appear.

[41]   PAYNE, L. E., *Isoperimetric inequalities and their applications*, SIAM Rev. *9* (1967), 453–488.

[42]   PAYNE, L. E., *Some comments on the past fifty years of isoperimetric inequalities*, Inequalities (Birmingham, 1987), 143–161, Lecture Notes in Pure and Appl. Math. *129*, Dekker, New York 1991.

[43]   PAYNE, L. E., PÓLYA, G. and WEINBERGER, H. F., *On the ratio of consecutive eigenvalues*, J. Math. Phys. *35* (1956), 289–298.

[44]   PAYNE, L. E., PAYNE, L. E. and SCHAEFER, P. W., *Eigenvalue and eigenfunction inequalities for the elastically supported membrane*, Z. Angew. Math. Phys. 52 (2001) no. 5, 888–895.

[45]   PAYNE, L. E. and WEINBERGER, H. F., *An optimal Poincaré inequality for convex domains*, Arch. Rational Mech. Anal. *5* (1960), 286–292.

[46]   PAYNE, L. E. and WEINBERGER, H. F., *Some isoperimetric inequalities for membrane frequencies and torsional rigidity*, J. Math. Anal. Appl. 2 (1961), 210–216.

[47]   PÓLYA, G., *On the characteristic frequencies of a symmetric membrane*, Math. Z. *63* (1955), 331–337.

[48]   PÓLYA, G. and SZEGÖ, G., *Isoperimetric inequalities in mathematical physics*, Ann. Math. Studies, *27*, Princeton Univ. Press, 1951.

[49]   RASSIAS, T., *The isoperimetric inequality and eigenvalues of the Laplacian. Constantin Carathodory: an international tribute*, vol. I, II, 1146–1163, World Sci. Publishing, Teaneck, NJ, 1991.

[50]   SCHOEN, R. and YAU, S. -T., *Lectures on differential geometry*, Conference Proceedings and Lecture Notes in Geometry and Topology, I. International Press, Cambridge, MA, 1994.

[51]   SERRIN, J., *A symmetry problem in potential theory*, Arch. Rational Mech. Anal. *43* (1971), 304–318.

[52]   SIMON, J., *Differentiation with respect to the domain in boundary value problems*, Num. Funct. Anal. Optimz. *2* (1980), 649–687.

[53]   SOKOLOWSKI, J. and ZOLESIO, J. P., *Introduction to shape optimization: shape sensitity analysis*, Springer Series in Computational Mathematics, Vol. 10, Springer, Berlin 1992.

[54]   SZEGÖ, G., *Inequalities for certain eigenvalues of a membrane of given area*, J. Rational Mech. Anal. *3* (1954), 343–356.

[55]   TROESCH, B. A., *Elliptical Membranes with smallest second eigenvalue*, Math. of Computation, *27–124* (1973), 767–772.

[56]   WEINBERGER, H. F., *An isoperimetric inequality for the N-dimensional free membrane problem*, J. Rational Mech. Anal. *5* (1956), 633–636.

[57]   WEINSTOCK, R., *Inequalities for a classical eigenvalue problem*, J. Rational Mech. Anal. *3* (1954), 745–753.

[58]   WOLF, S. A. and KELLER, J. B., *Range of the first two eigenvalues of the Laplacian*, Proc. R. Soc. London A, *447* (1994), 397–412.

[59]   YAU, S. -T., *Problem section*, Seminar on Differential Geometry, pp. 669–706, Ann. of Math. Stud. 102, Princeton Univ. Press, Princeton, N.J. 1982.

[60]   YAU, S. -T., *Open problems in geometry. Differential geometry: partial differential equations on manifolds*, (Los Angeles, CA, 1990), 1–28, Proc. Sympos. Pure Math. 54, Part 1, Amer. Math. Soc. Providence, RI 1993.

Antoine Henrot
Ecole des Mines and Institut Elie Cartan Nancy
UMR 7502 CNRS and projet Corida
INRIA
B.P. 239
54506 Vandœuvre-lès-Nancy
France
e-mail: henrot@iecn.u-nancy.fr

J.evol.equ. 3 (2003) 463 – 484
1424–3199/03/030465 – 22
DOI 10.1007/s00028-003-1112-8
© Birkhäuser Verlag, Basel, 2003

Journal of Evolution
Equations

Rate of decay to equilibrium in some semilinear parabolic equations

ALAIN HARAUX, MOHAMED ALI JENDOUBI AND OTARED KAVIAN

Dedicated to Philippe Bénilan

Abstract. In this paper we prove, under various conditions, the so-called Lojasiewicz inequality $\|E'(u + \varphi)\| \geq \gamma |E(u + \varphi) - E(\varphi)|^{1-\theta}$, where $\theta \in (0, 1/2]$, and $\gamma > 0$, while $\|u\|$ is sufficiently small and φ is a critical point of the *energy functional* E supposed to be only C^2, instead of analytic in the classical settings. Here E can be for instance the energy associated to the semilinear heat equation $u_t = \Delta u - f(x, u)$ on a bounded domain $\Omega \subset \mathbb{R}^N$. As a corollary of this inequality we give the rate of convergence of the solution $u(t)$ to an equilibrium, and we exhibit examples showing that the given rate of convergence (which depends on the exponent θ and on the critical point φ through the nature of the kernel of the linear operator $E''(\varphi)$) is optimal.

1. Introduction and main results

Let Ω be a bounded, connected open subset of $\mathbb{R}^N$ with a Lipschitz continuous boundary and let us consider the semilinear parabolic equation

$$u_t - \Delta u + f(u) = 0 \quad \text{in} \quad \mathbb{R}^+ \times \Omega, \qquad u = 0 \quad \text{on} \quad \mathbb{R}^+ \times \partial\Omega \qquad (1.1)$$

where $f : \mathbb{R} \longrightarrow \mathbb{R}$ is a locally Lipschitz continuous function. According to the well known *La Salle's invariance principle*, a solution u of (1.1) which is uniformly bounded on $\mathbb{R}^+$, approaches the set of stationary solutions of (1.1) as $t \to \infty$, see for instance C. M. Dafermos [9], A. Haraux [12]. Convergence to an equilibrium has been established in many cases, for instance T. I. Zelenyak [26], H. Matano [22], L. Simon [25], P. L. Lions [19], J. Hale & G. Raugel [11], A. Haraux & P. Poláčik [16], P. Brunovský & P. Poláčik [5], but remains an open question in general (convergence may fail if the nonlinearity f depends on x, as shown in P. Poláčik & K. Rybakowski [23], see also P. Poláčik & F. Simondon [24]).

One of the means used in the proof of such convergence results to equilibrium is an inequality due to S. Lojasiewicz [20] asserting that for any analytic function $F : \mathbb{R}^n \longrightarrow \mathbb{R}$ with $F(0) = 0$ and $F'(0) = 0$, there exist constants $\gamma > 0$ and $\theta \in (0, 1/2]$ and a neighbourhood ω of the origin such that:

for any $x \in \omega$ one has $\|F'(x)\| \geq \gamma |F(x)|^{1-\theta}$.

The reason for which this inequality is crucial in the proof of convergence to equilibrium, and in estimating the rate of convergence, is quite simple and can be sketched briefly. If $x(t)$ is a global solution to $\dot{x}(t) + F'(x(t)) = 0$ with $x(0) = x_0$, one first proves that if the trajectory $x(t)$ enters at some time t_0 the neighbourhood ω, then it remains in that neighbourhood. Next, for $t \geq t_0$ we have (with $\xi \cdot \eta$ denoting the scalar product of $\xi, \eta \in \mathbb{R}^n$):

$$\frac{dF(x(t))}{dt} = F'(x(t)) \cdot \frac{dx(t)}{dt} = -\|F'(x(t))\|^2 \leq -\gamma^2 |F(x(t))|^{2(1-\theta)}.$$

Observing moreover that $F(x(t))$ is nonincreasing and that $F(x(t)) \downarrow F(0) = 0$, from this differential inequality one gets the estimate $F(x(t)) \leq C(\theta)(1+t)^{-1/(1-2\theta)}$ if $\theta < 1/2$, and $F(x(t)) \leq C \exp(-\alpha t)$ if $\theta = 1/2$, for some constants $C, \alpha > 0$. Now this decay estimate may be worked out to yield also a rate of convergence for $\|x(t)\|$ to zero as $t \to +\infty$.

The Lojasiewicz inequality, and the ideas behind the above analysis, have been extended to the case in which F maps an infinite dimensional space H into $\mathbb{R}$, and subsequently used in the proof of convergence results (see L. Simon [25]). In this paper we are going to present an elementary proof of the Lojasiewicz inequality in the framework of the energy associated to Equation (1.1), under simple circumstances but without assuming the analyticity of the nonlinearity f. It is also noteworthy to mention that our results are valid even in the case of a nonlinearity f depending on the variable $x \in \Omega$ (that is in Equation (1.1) one might have $f(x, u)$ instead of $f(u)$, see below). We also establish a simple criterion to ensure an exponential rate of convergence to the equilibrium, and give some examples in which the rate of convergence is exactly polynomial.

The exponential convergence to an equilibrium, corresponding to the case where the Lojasiewicz exponent $\theta = 1/2$, turns out to be a consequence of the following simple general result (see also Sections §2 and §3 below):

PROPOSITION 1.1. *Let V be a Banach space, and $E \in C^2(V, \mathbb{R})$ and set $\mathcal{M}(u) := E'(u)$ for $u \in V$. Let $\varphi \in V$ be a critical point of E, that is $\mathcal{M}(\varphi) = 0$ and assume that*

$$L := \mathcal{M}'(\varphi) \in \mathcal{L}(V, V') \text{ is an isomporphism.} \tag{1.2}$$

Then there exist two positive constants R and γ such that for all $u \in V$ with $\|u\|_V \leq R$ one has

$$\|\mathcal{M}(u + \varphi)\|_{V'} \geq \gamma |E(u + \varphi) - E(\varphi)|^{1/2}.$$

Proof. It is easy to see, using Taylor's expansion formula, that for $\|u\|_V$ small enough we have:

$$|E(u + \varphi) - E(\varphi)| \leq C\|u\|_V^2. \tag{1.3}$$

On the other hand, since $\mathcal{M}(u + \varphi) = \mathcal{M}(\varphi) + Lu + o(u) = Lu + o(u)$, we have

$$u = L^{-1}\mathcal{M}(u + \varphi) + o(u),$$

and therefore for any given $\varepsilon > 0$ we can find $\delta(\varepsilon) > 0$ such that if $\|u\|_V \leq \delta(\varepsilon)$ then

$$\|\mathcal{M}(u + \varphi)\|_{V'} \geq \|L^{-1}\|^{-1}\|u\|_V - \varepsilon\|u\|_V.$$

Choosing $\varepsilon := \varepsilon_0 := \|L^{-1}\|^{-1}/2$, we obtain for $\|u\|_V \leq \delta(\varepsilon_0)$

$$\|\mathcal{M}(u + \varphi)\|_{V'} \geq \varepsilon_0\|u\|_V. \tag{1.4}$$

The result follows by combining (1.3) and (1.4). $\square$

The remainder of this paper is organized as follows. In Section §2 we give the proof of the Lojasiewicz inequality under simple assumptions on the nonlinearity f. Section §3 is devoted to the proof of the fact that if a Lojasiewicz inequality holds in a neighbourhood of an equilibrium φ, then one can give an upper bound for the rate of convergence of a solution $u(t)$ of (1.1) to φ, and an example is given for solutions of a semilinear heat equation. In Section §4 we consider a one dimensional semilinear heat equation for which one can establish an exponential rate of convergence to an equilibrium (actually analogous heat equation in higher dimension can be treated in much the same way, but only for positive solutions). In Section §5 we give examples showing that the rates of convergence deduced from the Lojasiewicz inequality in our Section §3 are optimal.

2. A direct proof of the Lojasiewicz inequality for the energy functional

Let $\Omega \subset \mathbb{R}^N$ be a domain and denote by $(\cdot|\cdot)$ and $\|\cdot\|$ the scalar product and norm of $L^2(\Omega)$, $(A, D(A))$ a densely defined linear operator acting on $L^2(\Omega)$ satisfying

A is self-adjoint, having a compact resolvent,   (2.1)

its spectrum being denoted by $(\lambda_k)_{k\geq 1}$ and its eigenfunctions by $(\varphi_k)_{k\geq 1}$. We denote by $V := D(|A|^{1/2})$ a subspace of $L^2(\Omega)$, equipped with the natural graph norm, and for $v \in V$ we write $\langle Av, v\rangle$, or more often $(Av|v)$, to mean $\sum_k \lambda_k|(v|\varphi_k)|^2$. We consider a Caratheodory function $F : \Omega \times \mathbb{R} \longrightarrow \mathbb{R}$, assumed to be of class C^2 in the variable $s \in \mathbb{R}$ and we denote its derivative with respect to $s \in \mathbb{R}$ by $f(x, s) := \partial_s F(x, s)$. We assume that the growth of F and its derivatives in the variable s is so that the mapping

$$v \mapsto \int_\Omega F(x, v(x))dx \quad \text{is of class } C^2 \text{ on } V, \tag{2.2}$$

and actually we shall make more specific assumptions on F in the statement of our results. Indeed the growth conditions depend on the embedding of V in Lebesgue spaces $L^q(\Omega)$. To have a typical example in mind, consider the case in which V is the space $H_0^1(\Omega)$ (or $H^1(\Omega)$); this coresponds for example to the operator $Au := -\Delta u$ with $D(A) := \{u \in H_0^1(\Omega); -\Delta u \in L^2(\Omega)\}$. Then assuming that

$$|\partial_s f(x, s)| \leq a(x) + b|s|^{p_0}$$

for some $a \in L^\infty(\Omega) + L^{p_1}(\Omega)$ with $p_0 := 4/(N-2)$ and $p_1 := N/2$ if $N \geq 3$, and $p_0 < \infty$ and $p_1 > 1$ if $N = 2$, one has that $v \mapsto \int_\Omega F(x, v(x))dx$ is a C^2 mapping from $H_0^1(\Omega)$ into $\mathbb{R}$ (see for instance O. Kavian [18, Chapitre 1, Sections §16–17]; when $N = 1$ no growth assumption is needed).

Under these assumptions we may define the *energy functional* E

$$E(v) := \frac{1}{2}\langle Av, v \rangle + \int_\Omega F(x, v(x))dx = \frac{1}{2}(Av|v) + \int_\Omega F(x, v(x))dx, \tag{2.3}$$

which is a C^2 functional on V, namely for $\varphi, \psi \in V$ we have $E'(\varphi) = A\varphi + f(\cdot, \varphi)$, and $E''(\varphi)\psi = A\psi + \partial_s f(\cdot, \varphi)\psi$.

Our aim is the proof of the Lojasiewicz inequality asserting that, if $\varphi \in V \cap L^\infty(\Omega)$ is a critical point of E, that is if $E'(\varphi) = 0$, then for some $\theta \in (0, 1/2]$ and positive constants γ and R, for all $u \in V$ such that $\|u\|_V \leq R$ one has

$$\|E'(u + \varphi)\|_{V'} \geq \gamma |E(u + \varphi) - E(\varphi)|^{1-\theta}.$$

The dual of $L^2(\Omega)$ being identified to $L^2(\Omega)$, we have the continuous embeddings $V \subset L^2(\Omega) \subset V'$. When $u, v \in V$, by $(Au|v)$ we shall mean $\sum_{k\geq 1} \lambda_k (u|\varphi_k)(v|\varphi_k)$ where the eigenvalues and eigenfunctions (λ_k, φ_k) are defined above; equivalently Au can also be interpreted as an element of V' so that $(Au|v)$ would mean $\langle Au, v \rangle$ where $\langle \cdot, \cdot \rangle$ denotes the duality between V and V'. If $\varphi \in V$ is a critical point of E, we denote by $L := L_\varphi$ the linear operator

$$L\psi := L_\varphi\psi := E''(\varphi)\psi = A\psi + \partial_s f(x, \varphi)\psi. \tag{2.4}$$

We shall further assume that L has a compact resolvent and we denote by $(\mu_k)_{k\geq 1}$ and $(\psi_k)_{k\geq 1}$ its eigenvalues and eigenfunctions. Also we may write

$$\begin{cases} V = V_1 \oplus V_0 \oplus V_2, \text{ where} \\ V_1 := \bigoplus_{\mu_k < 0} \mathbb{R}\psi_k, \quad V_0 := \ker(L) := N(L), \quad V_2 := \bigoplus_{\mu_k > 0} \mathbb{R}\psi_k. \end{cases} \tag{2.5}$$

Our first result asserts that in the case of a *nondegenerate* critical point φ, that is when the operator L is a homeomorphism between V and V', in which case $V_0 = \{0\}$, the Lojasiewicz inequality holds with the optimal exponent $\theta = 1/2$ (and as we shall recall below, the decay rate of the energy $E(u(t)) - E(\varphi)$, and that of $\|u(t) - \varphi\|$, to zero is exponential if $u(t)$ is a global solution of the evolution equation $\partial_t u + Au + f(\cdot, u) = 0$).

THEOREM 2.1. *Assume that $(A, D(A))$ and F satisfy (2.1) and (2.2), that $\varphi \in V$ is a critical point of E such that $V_0 = \ker(L_\varphi) = \{0\}$. Then there exist two positive constants $R > 0$ and $\gamma > 0$ such that for all $u \in V$ with $\|u\|_V \leq R$ one has $\|u\|_V \leq C\|E'(u+\varphi)\|_{V'}$ and:*

$$\|E'(u + \varphi)\|_{V'} \geq \gamma |E(u + \varphi) - E(\varphi)|^{1-\theta}, \quad \text{with } \theta := \frac{1}{2}.$$

Proof. This result is an immediate consequence of Proposition 1.1. $\square$

Our next result concerns the case in which the critical point φ is the origin, regardless of the dimension of the kernel V_0: as one might expect, in this case the exponent θ is determined by the behaviour of $f(x, s)$ as $s \to 0$.

THEOREM 2.2. *Let $(A, D(A))$ and F satisfy (2.1) and (2.2). Assume that there exist $\alpha = \pm 1$, $a \in L^\infty(\Omega)$ with $a \not\equiv 0$ and $a \geq 0$ a.e., and $p > 1$ such that for some $q > p$ and all $\varepsilon > 0$ there exists $C_\varepsilon > 0$ so that:*

$$|f(x, s) - s\partial_s f(x, s) - \alpha a(x)|s|^{p-1}s| \leq \varepsilon|s|^p + C_\varepsilon|s|^q.$$

Assume also that the embedding $V \subset L^{q+1}(\Omega)$ is continuous and that

$$u_0 \in V_0 = \ker(L) \quad \text{and} \quad au_0 \equiv 0 \implies u_0 \equiv 0, \tag{2.6}$$

where $L\psi := A\psi + \partial_s f(\cdot, 0)\psi$. Then $E(0) = 0$ and $E'(0) = 0$, and there exist two positive constants $R > 0$ and $\gamma > 0$ such that for all $u \in V$ with $\|u\|_V \leq R$ one has:

$$\|E'(u)\|_{V'} \geq \gamma|E(u)|^{1-\theta}, \quad \text{where } \theta := \frac{1}{p+1}.$$

Proof. We are going to obtain a lower estimate on $\|E'(u)\|_{V'} = \sup_{\|w\|_V=1}\langle E'(u), w\rangle$. To this end set $w := v/\|u\|_V$ when $u \neq 0$ with $v := -u_1 + \alpha u_0 + u_2$ if $u = u_1 + u_0 + u_2$ and $u_j \in V_j$, the orthogonal projections of u on V_j, for $j = 0, 1, 2$ (note that $\|v\|_V = \|u\|_V$). We have

$$\langle E'(u), v\rangle = -(Lu_1|u_1) + (Lu_2|u_2) + r_1(u) \tag{2.7}$$

where the remainder term $r_1(u)$ is given by

$$r_1(u) := \int_\Omega [f(x, u(x)) - \partial_s f(x, 0)u(x)](-u_1(x) + \alpha u_0(x) + u_2(x))dx.$$

Denote $g(x, s) := f(x, s) - s\partial_s f(x, s) - \alpha a(x)|s|^{p-1}s$. Then it is clear that for $u \in V$ we have

$$\int_\Omega |g(x, u(x))v(x)|\, dx \leq \varepsilon\|u\|_{p+1}^{p+1} + C_\varepsilon\|u\|_{q+1}^{q+1},$$

and therefore, since the embeddings $V \subset L^2(\Omega) \cap L^{q+1}(\Omega) \subset L^{p+1}(\Omega)$ are continuous,

$$\int_\Omega g(x, u(x))v(x)dx = o(\|u\|_V^{p+1}). \tag{2.8}$$

Now using the above observation, it follows that, we have

$$r_1(u) = \alpha \int_\Omega a(x)|u(x)|^{p-1}u(x)(-u_1(x) + \alpha u_0(x) + u_2(x))dx + o(\|u\|_V^{p+1}). \tag{2.9}$$

By homogeneity, one easily sees that there exists a constant $C_1 > 0$ (depending only on p) such that for all $z_0, z_1, z_2 \in \mathbb{R}$

$$||z_0 + z_1 + z_2|^{p-1}(z_0 + z_1 + z_2) - |z_0|^{p-1}z_0|$$
$$\leq C_1(|z_1| + |z_2|)(|z_0|^{p-1} + |z_1|^{p-1} + |z_2|^{p-1}).$$

Hence by Young's inequality, for all $\delta > 0$ there exists $C_\delta > 0$ such that

$$||z_0 + z_1 + z_2|^{p-1}(z_0 + z_1 + z_2) - |z_0|^{p-1}z_0| \leq \delta|z_0|^p + C_\delta(|z_1|^p + |z_2|^p). \qquad (2.10)$$

On the other hand, observe that $V_0 = \ker(L)$ being finite dimensional, and $a \geq 0$ not identically zero, using the assumption (2.6), we know that

$$u_0 \mapsto \left(\int_\Omega a(x)|u_0(x)|^{p+1}dx \right)^{1/(p+1)}$$

is a norm, equivalent to the usual norm $u_0 \mapsto \|u_0\|_V$. It follows that, taking $\delta > 0$ small enough in (2.10), for some $\gamma_0 > 0$ and $C > 0$ we have:

$$\int_\Omega a(x)|u(x)|^{p-1}u(x)u_0(x)dx \geq \gamma_0\|u_0\|_V^{p+1} - C(\|u_1\|_V^{p+1} + \|u_2\|_V^{p+1}). \qquad (2.11)$$

On the other hand one has clearly:

$$\left| \alpha \int_\Omega a(x)|u(x)|^{p-1}u(x)(-u_1(x) + u_2(x))dx \right|$$
$$\leq K \int_\Omega (|u_0|^p + |u_1|^p + |u_2|^p)(|u_1| + |u_2|)\, dx$$
$$\leq \delta_0\|u_0\|_V^{p+1} + K(\delta_0)(\|u_1\|_V^{p+1} + \|u_2\|_V^{p+1}).$$

By combining this with inequality (2.11) and comparing with (2.9) and (2.7) we obtain for some constant $\gamma_1 > 0$ (after choosing $\delta_0 > 0$ small in the above inequality):

$$\langle E'(u), v \rangle \geq \gamma_1[-(Lu_1|u_1) + (Lu_2|u_2) + \|u_0\|_V^{p+1}]$$
$$-C(\|u_1\|_V^{p+1} + \|u_2\|_V^{p+1}), \qquad (2.12)$$

provided $\|u\|_V \leq R$ is small enough.

Finally, taking R still smaller if needed, from (2.12) we conclude that for some $\gamma_2 > 0$ we have

$$\langle E'(u), v \rangle \geq \gamma_2[\|u_1\|_V^2 + \|u_2\|_V^2 + \|u_0\|_V^{p+1}]. \qquad (2.13)$$

Since $\|u\|_V \leq R$, we have $\|u\|_V^{p+1} \leq C(\|u_1\|_V^2 + \|u_2\|_V^2 + \|u_0\|_V^{p+1})$, for some constant $C > 0$. Therefore, $\|u\|_V^{p+1} \leq C\langle E'(u), v \rangle \leq \|E'(u)\|_{V'}\|u\|_V$, and finally

$$\|u\|_V \leq C\|E'(u)\|_{V'}^{1/p}. \qquad (2.14)$$

Analogously we have

$$E(u) = \frac{1}{2}(Lu_1|u_1) + \frac{1}{2}(Lu_2|u_2) + r(u),$$

where the remainder term is given by (recall that $F(x,0) = f(x,0) = 0$)

$$r(u) = \int_\Omega \left[F(x, u(x)) - \frac{1}{2}\partial_s f(x,0)u(x)^2 \right] dx.$$

Thanks to the structure of the nonlinearity f, using the same arguments as the ones used in establishing (2.8), it follows that for $\|u\|_V \leq R_0$ small enough we have

$$|r(u)| \leq C(\|u_1\|_V^{p+1} + \|u_0\|_V^{p+1} + \|u_2\|_V^{p+1}).$$

Hence for some constant $C > 0$, for $\|u\|_V \leq R$ we have, provided R is small,

$$|E(u)| \leq C[-(Lu_1|u_1) + (Lu_2|u_2)] + C(\|u_1\|_V^{p+1} + \|u_0\|_V^{p+1} + \|u_2\|_V^{p+1})$$
$$\leq C[\|u_1\|_V^2 + \|u_2\|_V^2 + \|u_0\|_V^{p+1}]$$

Using inequalities (2.13) and (2.14) we infer that

$$|E(u)| \leq C\langle E'(u), v \rangle \leq C\|E'(u)\|_{V'}\|u\|_V \leq C\|E'(u)\|_{V'}^{(p+1)/p},$$

and the Lojasiewicz inequality claimed in Therorem 2.2 is proved with $1 - \theta = p/(p+1)$.
$$\square$$

REMARK 2.3. It is noteworthy to observe that the above proof shows also that when the nonlinearity satisfies the assumption of Theorem 2.2, then the trivial solution zero is isolated among all solutions of $A\varphi + f(\cdot, \varphi) = 0$ with $\varphi \in V$.

REMARK 2.4. Also we point out that the support of the function a (for instance when $f(x,s) := \lambda s \pm a(x)|s|^{p-1}s$) can be, in principle arbitrarily small (see Section §5 for an example). The condition (2.6) is satisfied in all practical examples and is in general a kind of unique continuation condition. For instance if $Au := -\Delta u$ and $D(A) := \{u \in H_0^1(\Omega) ; -\Delta u \in L^2(\Omega)\}$, then if $a > 0$ on an open subset $\Omega_0 \subset \Omega$, the condition (2.6) is satisfied and is precisely a unique continuation result for solutions of $-\Delta u + Vu = 0$: if such a u vanishes on Ω_0 then $u \equiv 0$ (see for instance N. Garofalo & F. Lin [10]).

In the proof of the above theorems we observe that the kernel V_0 of the operator L_φ plays a crucial role. In our next result we consider the case in which V_0 has dimension one.

THEOREM 2.5. *Assume that $(A, D(A))$ satisfies $(HypA1)$, that $F : \overline{\Omega} \times \mathbb{R} \longrightarrow \mathbb{R}$ is a Caratheodory function such that $s \mapsto F(x,s)$ is C^3 on $\mathbb{R}$. We assume moreover that for all*

$T > 0$ *there exists* $q > 2$ *such that for any* $\varepsilon > 0$ *there exists* $C_\varepsilon > 0$ *such that for* $|t| \le T$ *and* $s \in \mathbb{R}$

$$\left| f(x, s+t) - f(x, t) - s \partial_s f(x, t) - \frac{1}{2} s^2 \partial_s^2 f(x, t) \right| \le \varepsilon |s|^2 + C_\varepsilon |s|^q.$$

Assume also that the embedding $V \subset L^{q+1}(\Omega)$ *is continuous and that* $\varphi \in V \cap L^\infty(\Omega)$ *is a critical point of* E *such that the kernel of the operator* L *defined in (2.4) has dimension one, that is* $V_0 = \mathbb{R}\psi$ *for some* $\psi \ne 0$. *Then if*

$$\int_\Omega \partial_s^2 f(x, \varphi(x)) \psi(x)^3 dx \ne 0,$$

there exist two positive constants $R > 0$ *and* $\gamma > 0$ *such that for all* $u \in V$ *with* $\|u\|_V \le R$ *one has:*

$$\|E'(u + \varphi)\| \ge \gamma |E(u + \varphi) - E(\varphi)|^{1-\theta}, \quad \text{with } \theta := \frac{1}{3}.$$

Proof. Set $J(u) := E(u + \varphi) - E(\varphi)$, so that we have $J(0) = 0$, $J'(0) = 0$ and the inequality to be proved becomes:

$$\|J'(u)\|_{V'} \ge \gamma |J(u)|^{2/3} \tag{2.15}$$

for $\|u\|_V \le R$ small enough. Notice also that $J'(u) = Lu + f(\cdot, u + \varphi) - f(\cdot, \varphi) - \partial_s f(\cdot, \varphi)u$, while we have $J(u) = \frac{1}{2}(Lu|u) + r(u)$ where

$$r(u) := \int_\Omega \left[F(x, u(x) + \varphi(x)) - F(x, \varphi(x)) \right.$$

$$\left. - f(x, \varphi(x))u(x) - \frac{1}{2} \partial_s f(x, \varphi(x))u(x)^2 \right] dx. \tag{2.16}$$

As we did in the previous proof, we are going to obtain a lower estimate on $\|J'(u)\|_{V'}$ by estimating $\langle J'(u), w \rangle$ for $w := v/\|u\|_V$ when $u \ne 0$ with $v := -u_1 + \beta u_0 + u_2$, if $u = u_1 + u_0 + u_2$ and $u_j \in V_j$ being the orthogonal projections of u on V_j, for $j = 0, 1, 2$ (here $\beta = \pm 1$ will be chosen appropriately later on). We have

$$\langle J'(u), v \rangle = -(Lu_1|u_1) + (Lu_2|u_2) + r_1(u) \tag{2.17}$$

where the remainder term $r_1(u)$ is given by

$$r_1(u) := \int_\Omega [f(x, u(x) + \varphi(x)) - f(x, \varphi(x))$$

$$- \partial_s f(x, \varphi(x))u(x)](-u_1(x) + \beta u_0(x) + u_2(x))dx.$$

Now, φ being in $L^\infty(\Omega)$, for $s \in \mathbb{R}$ if we set

$$g(x, s) := f(x, s + \varphi(x)) - f(x, \varphi(x)) - \partial_s f(x, \varphi(x))s - \frac{1}{2} \partial_s^2 f(x, \varphi(x))s^2,$$

by the arguments used in the proof of (2.8) one sees that:

$$r_1(u) = \frac{1}{2} \int_\Omega \partial_s^2 f(x, \varphi(x)) u(x)^2 (-u_1 + \beta u_0 + u_2)(x) dx + o(\|u\|_V^3). \tag{2.18}$$

Using Hölder and Young inequalities, for all $\delta > 0$ we can find $C(\delta) > 0$ such that for $j = 1, 2$:

$$\int_\Omega \partial_s^2 f(x, \varphi(x)) u_j(x) u_0(x)^2 dx \leq C \|u_0\|_3^2 \|u_j\|_3 \leq \delta \|u_0\|_3^3 + C(\delta) \|u_j\|_3^3,$$

and

$$\int_\Omega \partial_s^2 f(x, \varphi(x)) u_j(x)^2 u_0(x) dx \leq C \|u_j\|_3^2 \|u_0\|_3 \leq \delta \|u_0\|_3^3 + C(\delta) \|u_j\|_3^3.$$

Next, we may expand the term $u^2 = (u_1 + u_2)^2 + 2(u_1 + u_2)u_0 + u_0^2$ in $(EgR1)$, then use the above two inequalities to obtain finally for some constants $C_1 > 0$ and $C_2 > 0$ depending only on φ and f (recall that $v := -u_1 + \beta u_0 + u_2$)

$$\int_\Omega \partial_s^2 f(x, \varphi(x)) u(x)^2 v(x) dx \geq \beta \int_\Omega \partial_s^2 f(x, \varphi(x)) u_0(x)^3 dx$$
$$- \delta C_1 \|u_0\|_3^3 - C_2(\|u_1\|_3^3 + \|u_2\|_3^3).$$

Now, as $V_0 = \mathbb{R}\psi$, we may set $u_0 = t\psi$ for some $t \in \mathbb{R}$, and choose

$$\beta = \text{sign}(t) \, \text{sign}\left(\int_\Omega \partial_s^2 f(x, \varphi(x))) \psi(x)^3 dx\right),$$

to obtain, provided $\delta > 0$ is chosen small enough, that for some positive constants γ_0, γ_1 and C

$$r_1(u) \geq \gamma_0 |t|^3 - C(\|u_1\|_V^3 + \|u_2\|_V^3) \geq \gamma_1 \|u_0\|_V^3 - C(\|u_1\|_V^3 + \|u_2\|_V^3).$$

Finally we get, for some $\gamma_2 > 0$ and $\|u\|_V \leq R$ small enough:

$$\langle J'(u), v \rangle \geq \gamma_2 [-(Lu_1 | u_1) + (Lu_2 | u_2) + \|u_0\|_V^3]. \tag{2.19}$$

Note also that the above inequality implies in particular that, for some constant $C > 0$, when $\|u\|_V \leq R$ we have $\|u\|_V^3 \leq C\langle J'(u), v \rangle \leq C \|J'(u)\|_{V'} \|u\|_V$, and finally

$$\|u\|_V \leq C \|J'(u)\|_{V'}^{1/2}. \tag{2.20}$$

Analogously $r(u)$ being defined in (2.16), we have $J(u) = \frac{1}{2}(Lu_1 | u_1) + \frac{1}{2}(Lu_2 | u_2) + r(u)$, and using the assumptions on the structure of the nonlinearity f it follows that for all $\varepsilon > 0$ there exists $C_\varepsilon > 0$ such that

$$\left| F(x, s + \varphi(x)) - F(x, \varphi(x)) - f(x, \varphi(x))s \right.$$
$$\left. - \frac{1}{2}\partial_s f(x, \varphi(x))s^2 - \frac{1}{6}\partial_s^2 f(x, \varphi(x))s^3 \right| \leq \varepsilon |s|^3 + C_\varepsilon |s|^{q+1},$$

and therefore, following the lines of (2.8), $|r(u)| \leq C\|u\|_V^3 \leq C(\|u_1\|_V^3 + \|u_2\|_V^3 + \|u_0\|_V^3)$, provided $\|u\|_V$ is small (we use here the assumption $V \subset L^2(\Omega) \cap L^{q+1}(\Omega) \subset L^3(\Omega)$).

Finally it follows that for some constants $R > 0$ small enough and $C > 0$, when $\|u\|_V \leq R$, we have

$$|J(u)| \leq C[-(Lu_1|u_1) + (Lu_2|u_2)] + C\|u_0\|_V^3 \leq C\langle J'(u), v\rangle.$$

Using inequality (2.20) we infer that

$$|J(u)| \leq C\langle J'(u), v\rangle \leq C\|J'(u)\|_{V'}\|u\|_V \leq C\|J'(u)\|_{V'}^{3/2},$$

and therefore the Lojasiewicz inequality claimed in the theorem is proved. $\qquad\square$

3. Some results on the rate of convergence to equilibrium

Using the Lojasiewicz inequality, in this section we give the rate of convergence to an equilibrium for a global solution of the evolution equation

$$\partial_t u + Au + f(\cdot, u) = 0, \qquad u \in C^1([0, +\infty), V),$$

or more generally for the solution of an evolution equation of the form

$$\frac{du}{dt} + E'(u(t)) = 0, \quad u(0) = u_0 \in V, \qquad u \in C^1([0, +\infty), V), \tag{3.1}$$

by the method used in A. Haraux & M. A. Jendoubi [15], but in a slightly more general setting. In the following we assume that H is a Hilbert space identified with its dual H', and that $V \subset H$ is another Hilbert space continuously and densely embedded in H, so that we have the continuous embeddings $V \subset H = H' \subset V'$.

We recall here that when E satisfies a Lojasiewicz inequality in a neighbourhood of each of its critical points, and the trajectory $\{u(t)\}_{t \geq t_0}$ is precompact in V, then one can show that the ω-limit set $\omega(u_0)$ is reduced to a single point φ, with $E'(\varphi) = 0$ and $\varphi \in V$ (cf. S. Lojasiewicz [21], L. Simon [25], M. A. Jendoubi [17]): if for a sequence one has $u(t_n) \to \varphi$ then $u(t) \to \varphi$ in H as $t \to +\infty$.

PROPOSITION 3.1. *Let* $E : V \longrightarrow \mathbb{R}$ *be of class* C^1 *and assume that* $\varphi \in V$ *is a critical point of* E *satisfying the Lojasiewicz inequality: there exists* $R > 0$ *and* $\gamma > 0$ *such that for all* $u \in V$ *with* $\|u - \varphi\|_V \leq R$ *one has*

$$\|E'(u)\|_{V'} \geq \gamma |E(u) - E(\varphi)|^{1/2}. \tag{3.2}$$

Then if $u \in C^1([0, +\infty), V)$ *is a solution of* (3.1) *such that* $u(t) \to \varphi$ *in* V *as* $t \to +\infty$, *there exists* $\delta > 0$ *depending only on* φ, *and a constant* $K > 0$ *depending on* u_0 *such that for all* $t > 0$ *one has:*

$$\|u(t) - \varphi\|_H \leq K e^{-\delta t}.$$

Proof. We have for almost all $t \geq 0$

$$\frac{d}{dt}[E(u) - E(\varphi)] = \left\langle E'(u), \frac{du}{dt} \right\rangle = -\|u'(t)\|_H^2 = -\|E'(u(t))\|_H^2.$$

We conclude first that $E(u(t)) - E(\varphi)$ is nonincreasing on $\mathbb{R}^+$ and, since $u(t) \to \varphi$ as $t \to \infty$, then for all $t \geq 0$ we have $E(u(t)) - E(\varphi) \geq 0$, and $E(u(t)) \to E(\varphi)$ when $t \to \infty$. Also, since we can fix $T > 0$ large enough so that $\|u(t) - \varphi\|_V \leq R$ for all $t \geq T$, by using (3.2), it follows that for $t \geq T$ we have $\|E'(u(t))\|_{V'}^2 \geq \gamma^2 |E(u(t)) - E(\varphi)| = \gamma^2 (E(u(t)) - E(\varphi))$. The continuous embedding of H in V' yields that for some constant $\alpha > 0$ depending on γ, for almost all $t \geq T$ we have

$$\frac{d}{dt}[E(u(t)) - E(\varphi)] \leq -\alpha[E(u(t)) - E(\varphi)],$$

and hence for all $t \geq T$, we get

$$E(u(t)) - E(\varphi) \leq [E(u(T)) - E(\varphi)] \exp(-\alpha(t - T)).$$

On the other hand, for all $t \geq 0$ we have

$$\int_t^\infty \|u'(s)\|_H^2 \, ds = E(u(t)) - E(\varphi).$$

In particular for $t \geq T$

$$\int_t^\infty \|u'(s)\|_H^2 \, ds \leq [E(u(T)) - E(\varphi)] \exp(-\alpha(t - T))$$

and this obviously implies that for some $M > 0$:

$$\forall t \geq 0, \quad \int_t^\infty \|u'\|_H^2 \, ds \leq M \exp(-\alpha t).$$

$$\square$$

Now we recall the following lemma (cf T. I. Zelenyak [26], A. Haraux & M. A. Jendoubi [15]):

LEMMA 3.2. *Assume that there exist two constants $\alpha > 0$ and $a > 0$ such that for all $t \geq 0$*

$$\int_t^{+\infty} \|u'(s)\|_H^2 \, ds \leq ae^{-\alpha t}.$$

Then setting $b := e^{\alpha/2}/(e^{\alpha/2} - 1)$, for all $\tau \geq t \geq 0$ we have:

$$\|u(t) - u(\tau)\|_H \leq \sqrt{a}\, b\, e^{-\alpha t/2}.$$

The result of Proposition (3.1) then follows from the above lemma.

Before considering the case of a critical point φ for which the exponent θ in the Lojasiewicz inequality is less than $1/2$, we establish first the following lemma (which is a variant of Lemma 3.2 for an *energy* $E(u(t))$ having a polynomial decay).

LEMMA 3.3. *Assume that for some $\alpha > 0$ and a constant $K > 0$, for all $t \geq 1$ we have:*

$$\int_t^\infty \|u'(s)\|_H^2 \, ds \leq K t^{-2\alpha - 1}$$

Then for all $\tau \geq t \geq 1$ we have:

$$\|u(t) - u(\tau)\|_H \leq \frac{\sqrt{K}}{1 - 2^{-\alpha}} \, t^{-\alpha}.$$

Proof. By Cauchy-Schwarz inequality, for all $t \geq 1$ we may write:

$$\int_t^{2t} \|u'(s)\|_H \, ds \leq \sqrt{t} \, (K t^{-2\alpha - 1})^{1/2} = \sqrt{K} \, t^{-\alpha},$$

hence

$$\int_t^\infty \|u'(s)\|_H \, ds = \sum_{k=0}^\infty \int_{2^k t}^{2^{k+1} t} \|u'(s)\|_H \, ds \leq \sqrt{K} \sum_{k=0}^\infty (2^k t)^{-\alpha} = \frac{\sqrt{K}}{1 - 2^{-\alpha}} \, t^{-\alpha}$$

and the result follows since $\|u(t) - u(\tau)\|_H \leq \int_t^\tau \|u'(s)\|_H \, ds \leq \int_t^\infty \|u'(s)\|_H \, ds$. $\qquad\square$

Now we consider a solution of (3.1) converging to an equilibrium φ for which the exponent θ is smaller than $1/2$.

PROPOSITION 3.4. *Let $E : V \longrightarrow \mathbb{R}$ be of class C^1 and assume that $\varphi \in V$ is a critical point of E satisfying the Lojasiewicz inequality: there exists $R > 0$ and $\gamma > 0$, and $\theta \in (0, 1/2)$ such that for all $u \in V$ with $\|u - \varphi\|_V \leq R$ one has*

$$\|E'(u)\|_{V'} \geq \gamma |E(u) - E(\varphi)|^{1-\theta}. \tag{3.3}$$

Then if $u \in C^1([0, +\infty), V)$ is a solution of (3.1) such that $u(t) \to \varphi$ in V as $t \to +\infty$, there exists a constant $K > 0$ depending on u_0 such that for all $t > 0$ one has:

$$\|u(t) - \varphi\|_H \leq K(1 + t)^b - \theta/(1 - 2\theta).$$

Proof. We have for almost all $t \geq 0$

$$\frac{d}{dt}[E(u) - E(\varphi)] = \left\langle E'(u), \frac{du}{dt} \right\rangle = -\|u'(t)\|_H^2 = -\|E'(u(t))\|_H^2.$$

We conclude first that $E(u(t)) - E(\varphi)$ is nonincreasing on $\mathbb{R}^+$ and, since $u(t) \to \varphi$ as $t \to \infty$, then for all $t \geq 0$ we have $E(u(t)) - E(\varphi) \geq 0$, and $E(u(t)) \to E(\varphi)$ when $t \to \infty$. In particular for $T > 0$ large enough, we have that $\|u(t) - \varphi\|_V \leq R$ for all $t \geq T$, and using (3.2), it follows that for $t \geq T$ we have $\|E'(u(t))\|_{V'}^2 \geq \gamma^2 |E(u(t)) - E(\varphi)|^{2(1-\theta)} = \gamma^2 (E(u(t)) - E(\varphi))^{2(1-\theta)}$. The continuous embedding of H in V' yields that for some constant $C > 0$ depending on γ, for $t \geq T$ we have

$$\frac{d}{dt}[E(u(t)) - E(\varphi)] \leq -C[E(u(t)) - E(\varphi)]^{2(1-\theta)},$$

and hence for $t \geq T$ we have

$$E(u(t)) - E(\varphi) \leq [(E(u(T)) - E(\varphi))^{2\theta-1} + C(t - T)]^{-1/(1-2\theta)}$$
$$\leq K(1 + t)^{-1/(1-2\theta)}$$

for some constant $K > 0$ depending on $u(T)$ and T. Now it is clear that Lemma 3.3 with $2\alpha + 1 = 1/(1 - 2\theta)$, that is $\alpha = \theta/(1 - 2\theta)$, gives the decay rate claimed here for $\|u(t) - \varphi\|_H$ and Proposition 3.4 is proved. $\qquad\square$

Regarding the convergence rates of the solutions to the semilinear heat equation

$$\begin{cases} \partial_t u - \Delta u + f(x, u) = 0 & \text{in } (0, \infty) \times \Omega \\ u(t, \sigma) = 0 & \text{on } \partial\Omega, \end{cases} \tag{3.4}$$

where $\Omega \subset \mathbb{R}^N$ is a bounded domain, we apply the above results to $H := L^2(\Omega)$, $V := H_0^1(\Omega)$ and $Av := -\Delta v$ for $v \in D(A) := \{u \in H_0^1(\Omega) \; ; \; -\Delta u \in L^2(\Omega)\}$. In order to be in a position to state and prove our result we begin with the following lemma.

LEMMA 3.5. *Let $\Omega \subset \mathbb{R}^N$ a domain, $c \in L^\infty(\mathbb{R}^+ \times \Omega)$ and $z \in C(\mathbb{R}^+, H_0^1(\Omega)) \cap C^1(\mathbb{R}^+, H^{-1}(\Omega))$ be a solution of*

$$\partial_t z = \Delta z + c(t, x)z.$$

Then we have for all $p \geq 1$

$$\|z(t)\|_{L^\infty(\Omega)} \leq C(t, p, c^+)\|z(0)\|_{L^p(\Omega)} \tag{3.5}$$

with $C(t, p, c^+) = (4\pi t)^{-N/2p} \exp(t\|c^+\|_{L^\infty(\mathbb{R}^+ \times \Omega)})$.

Proof. Consider $w := |z|$; by Kato's inequality we have $\Delta w \geq \text{sign}(z)\Delta z$ in the sense of distributions, while $\partial_t w = \text{sign}(z)\partial_t z$, and hence

$$0 = \text{sign}(z)(\partial_t z - \Delta z - cz) \geq \partial_t w - \Delta w - cw \geq \partial_t w - \Delta w - c^+ w.$$

Denoting $M := \|c^+\|_{L^\infty(\mathbb{R}^+ \times \Omega)}$, and letting v be the solution of the linear heat equation $\partial_t v - \Delta v - Mv = 0$ with the initial data $v(0) = |z(0)|$ and boundary conditions $v(t, \sigma) = 0$ on $\partial \Omega$, we have that

$$\partial_t w - \Delta w - Mw \leq \partial_t w - \Delta w - c^+ w \leq \partial_t v - \Delta v - Mv,$$

and thus by the maximum principle for parabolic equations we have $w = |z| \leq v$ on $\mathbb{R}^+ \times \Omega$. Now $v(t) = e^{Mt} \exp(t\Delta)v(0)$, and by the well known classical regularizing properties of the semi-group $\exp(t\Delta)$ (the so-called *ultra contractivity* of the heat semi-group) we have

$$\|v(t)\|_{L^\infty(\Omega)} \leq (4\pi t)^{-N/2p} e^{Mt} \|v(0)\|_{L^p(\Omega)}$$

for any $p \geq 1$. The proof of the lemma is over. $\qquad\square$

PROPOSITION 3.6. *Assume that Ω is a bounded domain, and that $u \in C^1((0, \infty), H_0^1(\Omega))$ is a global solution of equation (3.4) such that $u(t) \to \varphi$ in $H_0^1(\Omega)$ as $t \to +\infty$. Then under the hypotheses of Theorem 2.1 we have for all $t \geq 1$*

$$\|u(t) - \varphi\|_{L^\infty(\Omega)} \leq K e^{-\delta t}$$

with δ depending only on φ and K depending on the solution u.
Under the hypotheses of Theorem 2.2 we have for all $t \geq 1$

$$\|u(t)\|_{L^\infty(\Omega)} \leq K \, t^{-1/(p-1)}$$

with K depending only on the solution u.

Proof. We apply Lemma 3.5 with $p = 2$ and $z = u - \varphi$ in the first case. Note that $\partial_t z = \Delta z + c(t, x)z$, with $c(t, x) := -(f(u(t, x)) - \varphi(x))/(u(t, x) - \varphi(x))$ when $u(t, x) \neq \varphi(x)$, and $c(t, x) := -\partial_s f(x, u(t, x))$ when $u(t, x) = \varphi(x)$.

We find, using the fact that the estimates of Lemma 3.5 can be translated in time, that for all $t \geq 1$ we have

$$\|(u - \varphi)(t)\|_{L^\infty(\Omega)} \leq K_0 \|(u - \varphi)(t - 1)\|_{L^2(\Omega)},$$

with $K_0 := (4\pi)^{-N/4} \exp(\|c^+\|_{L^\infty((t-1,t)\times\Omega)})$. Then Proposition 3.1 gives the result.

The proof of the second part of the Proposition is analogous, using first Lemma 3.5 with $z := u$, and then applying Proposition 3.4. $\qquad\square$

REMARK 3.7. If $u(t) \to \varphi$, under the assumptions of Theorem 2.5, the rate of convergence is given by:

$$\|u(t) - \varphi\|_{L^\infty(\Omega)} \leq K(1 + t)^{-1},$$

which is a consequence of Proposition 3.4 with $\theta = 1/3$.

4. An alternative in some one dimensional cases

Consider the one dimensional semilinear heat equation

$$\begin{cases} u_t - u_{xx} + f(u) = 0 & \text{in } [0, +\infty) \times (0, \pi) \\ u(t, 0) = u(t, \pi) = 0 & \text{on } [0, +\infty) \\ u(0, x) = u_0(x) & \text{on } (0, \pi), \end{cases} \tag{4.1}$$

where $f : \mathbb{R} \longrightarrow \mathbb{R}$ is a C^1 function. It is well known that (see H. Matano [22], T. I. Zelenyak [26]) any solution u of this problem which is global and uniformly bounded on $[0, +\infty) \times (0, \pi)$ converges as $t \to +\infty$ to a solution φ of the elliptic problem

$$\begin{cases} -\varphi_{xx} + f(\varphi) = 0 \\ \varphi(0) = \varphi(\pi) = 0. \end{cases} \tag{4.2}$$

A natural question is to determine the rate at which $u(t, \cdot)$ converges to φ. Depending on whether φ is zero or not, we shall prove the following result:

THEOREM 4.1. *Assume that $f : \mathbb{R} \longrightarrow \mathbb{R}$ is a C^1 function such that $s \mapsto |s|^{-1} f(s)$ is increasing on $\mathbb{R}$, $f(0) = 0$ and $f'(s) > f(s)/s$ for s near to zero. Then any solution u of Equation (4.1) which is global and uniformly bounded on $\mathbb{R}^+ \times (0, \pi)$ either converges to 0, or converges exponentially as $t \to +\infty$ to a solution $\varphi \neq 0$ of the elliptic problem (4.2).*

Before proving this theorem we recall here a few results concerning elliptic equations such as (4.2). It is a well known fact that under the assumption made about f, if (4.2) has a positive solution φ then it is unique (see for instance A. Ambrosetti & G. Mancini [1], H. Brezis & S. Kamin [3], H. Brezis & L. Oswald [4]). In addition it is easily proved that φ is exponentially stable [13].

We recall also two other well known results concerning eigenvalues of operators such as L and nodal points of its eigenfunctions. First consider an open bounded domain $\Omega \subset \mathbb{R}^N$, and a potential $V \in L^\infty(\Omega)$. Then consider the self-adjoint operator $(L, D(L))$ defined on $L^2(\Omega)$ by

$$Lu := -\Delta u + Vu, \qquad \text{for } u \in D(L) := \{w \in H_0^1(\Omega) \ ; \ -\Delta w + Vw \in L^2(\Omega)\}$$

This operator has a sequence of eigenvalues which will be denoted by $\lambda_j(-\Delta + V, \Omega)$, for $j \geq 1$. Then we have the following result:

LEMMA 4.2. *Consider an open subdomain $\Omega_0 \subset\subset \Omega$. Then for any $j \geq 1$ one has*

$$\lambda_j(-\Delta + V, \Omega) < \lambda_j(-\Delta + V, \Omega_0).$$

The proof follows after a few simple observations. First the variational characterization of the eigenvalues states that if $B_j(\Omega)$ is the set of subspaces of $H_0^1(\Omega)$ having dimension j, then

$$\lambda_j(-\Delta + V, \Omega) = \min_{A \in B_j(\Omega)} \max_{u \in A \setminus \{0\}} \frac{J(u)}{(u|u)},$$

where $(\cdot|\cdot)$ denotes the usual scalar product in $L^2(\Omega)$ and $J(u) := (-\Delta u + Vu|u) = \int_\Omega (|\nabla u|^2 + V|u|^2)dx$. Recall also that $\lim_{j \to \infty} \lambda_j = +\infty$. Next, the fact that the extension by zero on $\Omega \setminus \Omega_0$ of a function $u \in H_0^1(\Omega_0)$ defines a natural embedding of $H_0^1(\Omega_0)$ into $H_0^1(\Omega)$, shows that we can consider $B_j(\Omega_0)$ as a subset of $B_j(\Omega)$. This implies that $\lambda_j(-\Delta + V, \Omega) \le \lambda_j(-\Delta + V, \Omega_0)$. Now, if we had $\lambda_j(-\Delta + V, \Omega) = \lambda_j(-\Delta + V, \Omega_0)$, calling this common value λ, then for any subdomain Ω_n such that $\Omega_0 \subset\subset \Omega_n \subset\subset \Omega$, we would have $\lambda_k(\Omega_n) = \lambda$. Let $(\Omega_n)_{n \ge 0}$ be an increasing sequence of such subdomains, and consider an eigenfunction $u \in H_0^1(\Omega_n)$ (with $u_n \ne 0$) such that

$$-\Delta u_n + V u_n = \lambda u_n, \qquad u_n \in H_0^1(\Omega_n). \tag{4.3}$$

(In order to have such a sequence of $(\Omega_n)_n$ in mind, one may set $\delta := \mathrm{dist}(\Omega_0, \Omega^c)$ and then consider $\Omega_n := \Omega_0 + B(0, (1 - 2^{-n})\delta))$. Denote again by $\widetilde{u}_n$ the extension by zero of u_n on Ω, so that $\widetilde{u}_n \in H_0^1(\Omega_p)$ for any two integers $p > n$. Multiplying by $\widetilde{u}_n$ the equation satisfied by u_p, with $p > n$, we have:

$$\int_{\Omega_p} \nabla u_p \cdot \nabla \widetilde{u}_n dx + \int_{\Omega_p} V u_p \widetilde{u}_n dx = \lambda \int_{\Omega_p} u_p \widetilde{u}_n dx,$$

and this amounts essentially to:

$$\forall n, p \in \mathbb{N}, \qquad \int_\Omega \nabla \widetilde{u}_p \cdot \nabla \widetilde{u}_n dx + \int_\Omega V \widetilde{u}_p \widetilde{u}_n dx = \lambda \int_\Omega \widetilde{u}_p \widetilde{u}_n dx. \tag{4.4}$$

It is easy to see that due to the unique continuation principle (see for instance N. Garofalo & F. Lin [10]), the support of each $\widetilde{u}_n$ is precisely $\overline{\Omega}_n$, since if $u_n = 0$ on a subdomain $\omega \subset \Omega_n$, then u_n satisfying the elliptic Equation (4.3) we would conclude $u_n \equiv 0$ on Ω_n, which is contrary to our assumption. This observation implies in particular that the functions $(\widetilde{u}_n)_n$ are linearly independent, and therefore for any $k \ge j$ the space $F_k := \mathrm{span}\{\widetilde{u}_i \; ; \; 1 \le i \le k\}$ has dimension k, that is $F_k \in B_k(\Omega)$. However for $k \ge j$ we have

$$\lambda = \lambda_j(-\Delta + V, \Omega) \le \lambda_k(-\Delta + V, \Omega) \le \max_{u \in F_k \setminus \{0\}} \frac{J(u)}{(u|u)} = \lambda, \tag{4.5}$$

since for $u = \sum_{i=1}^{k} \alpha_i \widetilde{u}_i$ using (4.4) we may write

$$J(u) = \sum_{i,\ell=1}^{k} \alpha_i \alpha_\ell \int_\Omega (\nabla \widetilde{u}_i \cdot \nabla \widetilde{u}_\ell + V \widetilde{u}_i \widetilde{u}_\ell) \, dx = \lambda \sum_{i,\ell=1}^{k} \alpha_i \alpha_\ell \int_\Omega \widetilde{u}_i \widetilde{u}_\ell dx = \lambda(u|u).$$

Finally, since (4.5) implies that for all $k \geq j$ we have $\lambda_k(-\Delta + V, \Omega) = \lambda$, we have a contradiction with the fact that the eigenvalues are unbounded, and the proof of the lemma is done.

Finally let us recall the following property and characterization of eigenfunctions of a Sturm-Liouville operator on $H_0^1(\alpha, \beta)$:

LEMMA 4.3. *Let $V \in L^\infty(0, \pi)$ and $(\alpha, \beta) \subset (0, \pi)$. Then, for any integer $j \geq 1$, a (non-zero) eigenfunction associated to the eigenvalue $\lambda_j(-\Delta + V, (\alpha, \beta))$ has exactly $(j - 1)$ nodal points.*

(This is a consequence of the Sturm theorem concerning the number of nodes of the solution u to the equation $-u'' + Vu = \lambda u$ with $u(0) = 0$ and $u'(0) = 1$: as λ increases, the number of nodes increases; see for instance R. Courant & D. Hilbert [8]).

Proof of Theorem 4.1. According to the abstract result proved in the previous section (cf. Theorem 2.1 and Proposition 3.6), we just need to show that for any solution $\varphi \neq 0$ of the elliptic problem (4.2), the self-adjoint unbounded operator $(L, D(L))$ defined on $H = L^2(0, \pi)$ by

$$D(L) := H^2(0, \pi) \cap H_0^1(0, \pi), \quad \text{and} \quad Lw := L_\varphi w := -w_{xx} + f'(\varphi)w,$$

has a kernel reduced to $\{0\}$. If φ is the positive (respectively negative) solution of Equation (4.2), by a result of A. Haraux [13] we know that $\lambda_1(L) > 0$ and in this case we are done. In the general case, it follows from H. Berestycki [2] that

$$\lambda_k(L) < 0 < \lambda_{k+1}(L)$$

where $k \geq 1$ is the number of nodal points of φ in $(0, \pi)$. For the convenience of the reader we recall the proof of this result. First $w := \varphi'$ has $k + 1$ nodal points in $(0, \pi)$ and, setting $V := f'(\varphi)$, we see that w satisfies

$$w \in C^2(0, \pi), \quad -w_{xx} + Vw = 0 \quad \text{in } (0, \pi).$$

In particular, if α, β are the extreme nodes of w in $(0, \pi)$, we have $0 < \alpha < \beta < \pi$, and by Lemma 4.3 we conclude that w corresponds to the eigenvalue $\lambda_k(-\Delta + V, (\alpha, \beta))$, that is $\lambda_k(-\Delta + V, (\alpha, \beta)) = 0$. Now using Lemma 4.2 we infer that $\lambda_k(-\Delta + V, (0, \pi)) < 0$.

Knowing that the eigenvalues of a Sturm-Liouville operator such as L on any interval are simple, if we can show that $\lambda_{k+1}(-\Delta + V, (0, \pi)) > 0$, then we are done: the spectrum of L does not contain zero, that is the kernel of L is reduced to zero.

Now, since φ has by definition $k \geq 1$ nodal points in $(0, \pi)$ and $(-\Delta + f(\varphi)/\varphi)\varphi = 0$, we have

$$\lambda_{k+1}(-\Delta + f(\varphi)/\varphi, (0, \pi)) = 0$$

Let $w \in H_0^1(0, \pi)$ be any solution of $-\Delta w + f'(\varphi(x))w = 0$ and let J be any interval of the form (α, β) where α and β are two consecutive zeroes of w. The classical Sturm oscillation theorem, since $f'(\varphi(x)) > f(\varphi(x))/\varphi(x)$ for x near the boundary of J, implies that φ must have at least one zero in each interval J. Denoting by r the number of nodes of w in $(0, \pi)$, it follows that φ has at least $r + 1$ zeroes in $(0, \pi)$, hence $r \leq k - 1$, according to Lemma 4.3 if $\lambda_p(-\Delta + V, (0, \pi)) = 0$ we have $p \leq k$. This means precisely that $\lambda_{k+1}(-\Delta + V, (0, \pi)) > 0$ as claimed.

5. Some cases of non exponential decay

In this section we show that the abstract results given in Section §2 and §3 are in some sense optimal: we are going to give examples in which the kernel of the linearized operator L is not zero and in which we have only a polynomial decay to equilibrium. Consider again the semilinear heat equation (analogous to (4.1)), in a bounded open domain $\Omega \subset \mathbb{R}^N$ with $N \geq 1$:

$$\begin{cases} u_t - \Delta u + f(x, u) = 0 & \text{in } [0, +\infty) \times \Omega \\ u(t, \sigma) = 0 & \text{on } [0, +\infty) \times \partial\Omega \\ u(0, x) = u_0(x) & \text{on } \Omega. \end{cases} \tag{5.1}$$

where the nonlinearity f is given by

$$f(s) := a(x)|s|^{p-1}s - \lambda_1 s \tag{5.2}$$

with λ_1 being the first eigenvalue of $-\Delta$ on $H_0^1(\Omega)$. Our first result of this section shows that if $u_0 \geq 0$ and $u_0 \not\equiv 0$, the rate of decay of u is of order $t^{-1/(p-1)}$.

THEOREM 5.1. *Let $a \in L^\infty(\Omega)$, with $a \geq 0$ a.e. and $a > 0$ a.e. on an open subset $\Omega_0 \subset \Omega$. Assume that f is given by (5.2) and that $u_0 \in H_0^1(\Omega)$ is not identically zero and $u_0 \geq 0$. Then the corresponding solution u of (4.1) converges to zero, more precisely for a positive constant C and all $t \geq 1$ we have*

$$\|u(t)\|_\infty \leq C \, t^{-1/(p-1)}.$$

On the other hand denoting by φ_1 a positive eigenfunction of the Laplacian on $H_0^1(\Omega)$, there exists a constant $c > 0$ such that for all $t \geq 1$ and all $x \in \Omega$ one has:

$$u(t, x) \geq ct^{-1/(p-1)}\varphi_1(x). \tag{5.3}$$

Proof. The convergence of all solutions of (5.1) to zero is easily shown by considering first the L^2 norm denoted by $\|\cdot\|$: indeed multiplying (5.1) by $u(t)$ and integrating over Ω we find

$$\frac{1}{2}\frac{d}{dt}\|u(t)\|^2 = \lambda_1\|u(t)\|^2 - \|\nabla u(t)\|^2 - \int_\Omega |u(t,x)|^{p+1}a(x)dx$$

$$\leq -\int_\Omega |u(t,x)|^{p+1}a(x)dx. \tag{5.4}$$

If $a \geq \varepsilon_0$ a.e. on Ω for some $\varepsilon_0 > 0$, then $\int_\Omega |u|^{p+1}a(x)dx \geq \varepsilon_0\|u(t)\|_{p+1}^{p+1} \geq c_1\|u(t)\|^{p+1}$, from which we deduce that $\frac{d}{dt}\|u(t)\|^2 \leq -c_1\|u(t)\|^{p+1}$ showing that the L^2 norm is bounded by $c_2 t^{-1/(p-1)}$.

In case a is not bounded away from zero, observe first that if $\varphi \in H_0^1(\Omega)$ satisfies $-\Delta\varphi + a|\varphi|^{p-1}\varphi = \lambda_1\varphi$, then $\varphi \equiv 0$. Indeed multiplying this equation by φ, we infer that $\int_\Omega a(x)|\varphi(x)|^{p+1}dx = 0$ and in particular $\varphi = 0$ on the open set Ω_0. Now the unique continuation principle implies (see for instance N. Garofalo & F. Lin [10])that $\varphi \equiv 0$. Going back to inequality (5.4), we see that $\|u(t)\|$ is decreasing, and a bootstrap argument implies that $\{u(t)\}_{t\geq 1}$ is precompact in $H_0^1(\Omega) \cap L^\infty(\Omega)$. Then using La Salle's invariance principle and the uniqueness of the sationary solutions of (5.1) just pointed out, we conclude that $u(t)$ converges to 0 in $H_0^1(\Omega) \cap L^\infty(\Omega)$ as $t \to +\infty$.

Finally, once one knows the convergence of $u(t)$ to zero, Theorem 2.2, together with Proposition 3.6 show that the rate of convergence of $\|u(t)\|_\infty$ to zero is bounded above by $c_2 t^{-1/(p-1)}$.

The second part of the theorem is a consequence of the parabolic maximum principle and the subsolution method: considering $v(t,x) = bt^{-\frac{1}{p-1}}\varphi_1(x)$, we are going to show that it is possible to choose $b > 0$ such that for all $t \geq 1$ we have $u(t,x) \geq v(t,x)$. Indeed by the maximum principle for $t > 0$ we have $u(t,x) > 0$; next, using the strong maximum principle for parabolic equations, we may fix some constant $c_0 > 0$ such that for all $x \in \Omega$ we have $u(1,x) \geq c_0\varphi_1(x)$.

Now a straightforward calculation shows that $v_t = -bt^{-p/(p-1)}\varphi_1/(p-1)$, and that $-\Delta v + a|v|^{p-1}v - \lambda_1 v = a(x)v^p = b^p a(x)t^{-p/(p-1)}\varphi_1^p$. Therefore it is clear that as soon as $b > 0$ is small enough so that $b^{p-1}\|a\|_\infty\|\varphi_1\|_\infty^{p-1} \leq (p-1)^{-1}$, we have:

$$v_t - \Delta v + a(x)|v|^{p-1}v - \lambda_1 v$$

$$= bt^{-p/(p-1)}\varphi_1(x)\left(b^{p-1}a(x)\varphi_1(x)^{p-1} - \frac{1}{p-1}\right) \leq 0.$$

This means that v is a subsolution of Equation (5.1) for any such b. In particular one may choose b still smaller so that $u(1,x) \geq b\varphi_1(x) = v(1,x)$, in such a way that the inequality $u(t,x) \geq v(t,x)$ follows for all times $t \geq 1$. From this it is clear that we can now find a constant $c > 0$ depending on b, p such that inequality (5.3) holds. $\qquad\square$

REMARK 5.2. It is noteworthy to observe that if the initial data u_0 does not have a constant sign, it may happen that $u(t)$ converges exponentially to zero in $L^\infty(\Omega)$ as $t \to +\infty$. For instance assume $\Omega := (0, \pi)$ and so $\lambda_1 = 1$. If we have $u_0(\pi - x) = -u_0(x)$ and $u_0 \geq 0$ on $(0, \frac{\pi}{2})$ (with $u_0 \not\equiv 0$ on this interval) the solution u of the Equation (5.1) will satisfy $u(t, \pi - x) = -u(t, x)$ for all $t \geq 0$. Therefore, due to the fact that the first eigenvalue of $-\Delta$ in $H_0^1(0, \pi/2)$ is $\lambda_1(-\Delta, (0, \pi/2)) = 4$, one can easily see that u is a positive solution of Equation (5.1) on $(0, \pi/2)$ and that one has the exponential decay (where $3 = \lambda_1(-\Delta, (0, \pi/2)) - \lambda_1$)

$$\|u(t)\|_{L^\infty(0,\pi/2)} \leq Ce^{-3t}.$$

Also the following extension of Theorem 5.1 shows that, in the one dimensional case, the optimality result extends to the nonlinearity $f(s) := |s|^{p-1}s - \lambda_k s$, for any eigenvalue $\lambda_k := \lambda_k(-\Delta, (0, \pi)) = k^2$ of $-\Delta$ in $H_0^1(0, \pi)$.

COROLLARY 5.3. *Let* $\Omega := (0, \pi)$ *and* $f(s) := |s|^{p-1}s - k^2 s$ *for some integer* $k \geq 1$. *Then there exist solutions u of* (5.1), *converging to zero in $L^\infty(0, \pi)$ as $t \to +\infty$ for which for all $t \geq 0$ one has*

$$u\left(t, \frac{\pi}{2k}\right) \geq \gamma(1 + t)^{-1/(p-1)}$$

for some positive constant $\gamma \leq (p - 1)^{-1/(p-1)}$.

Proof. We consider

$$v(t, x) = b(1 + t)^{-\frac{1}{p-1}} \sin(kx)$$

by the same calculation as above we find

$$v_t - v_{xx} + |v|^{p-1}v - k^2 v$$
$$= b(1 + t)^{-p/(p-1)} \sin(kx) \left(b^{p-1}|\sin(kx)|^{p-1} - \frac{1}{p-1}\right).$$

Now fix any $b > 0$, such that $b \leq (p - 1)^{-1/(p-1)}$ and consider the solution u of (5.1) with $u(0, x) = b \sin(kx)$. It is clear that $u(t, \cdot)$ is invariant by odd reflection with respect to the nodes $j\pi/k$ for $0 \leq j \leq k$ and satisfies

$$\forall t \geq 0, \quad \forall x \in \left(0, \frac{\pi}{k}\right), \quad u(t, x) \geq v(t, x).$$

The conclusion follows immediately, since the first part (that is convergence of $u(t)$ to zero as $t \to +\infty$) is a direct consequence of the fact that the solution u is characterized by its restriction to $\mathbb{R}^+ \times (0, \pi/k)$, and that $\lambda_1(-\Delta, (0, \pi/k)) = k^2$ (see Theorem 5.1). $\square$

REMARK 5.4. In a recent paper R. Chill [7] has proposed a general local decomposition method to reduce the proof of Lojasiewicz inequality to a finite dimensional (called "critical") manifold. His method gives a new proof of our Theorems 2.2 and 2.5. However up to now the case of a non-trivial equilibrium remains open for dimensions $N > 1$ by both methods.

Acknowledgement

The authors take the opportunity to thank Ralph Chill for pointing out to us a gap in a previous proof of Lemma 4.2.

REFERENCES

[1]   AMBROSETTI, A. and MANCINI, G., *Sharp nonuniqueness results for some nonlinear problems*, Nonlinear Analysis, Theory, Methods & Applications, *3* ♯ 5 (1979), 635–645.

[2]   BERESTYCKI, H., *Le nombre de solutions de certains problèmes semi-linéaires elliptiques*, J. Funct. Analysis, *40* (1981), 1–29.

[3]   BREZIS, H., and KAMIN, S., *Sublinear elliptic equations in* $\mathbb{R}^N$, Manuscripta Math. *74* (1992), 87–106.

[4]   BREZIS, H., and OSWALD, L., *Remarks on sublinear elliptic equations*, Nonlinear Anal. TMA *10* (1986), 55–64.

[5]   BRUNOVSKÝ, P. and POLÁČIK P., *On the local structure of ω-limits sets of maps*, Z. Angew. Math. Phys., *48* (1997), 976–986.

[6]   CAZENAVE, TH. and HARAUX, A., *An Introduction to Semilinear Evolution Equations*, Transl. by Yvan Martel. Revised ed. Oxford Lecture Series in Mathematics and its Applications. 13. Oxford: Clarendon Press. xiv (1998), 186.

[7]   CHILL, R., *On the Lojasiewicz-Simon gradient inequality*, J. of Functional Analysis, to appear.

[8]   COURANT, R., and HILBERT, D., *Methods of Mathematical Physics*, Wiley, New York 1953 (volume 1), 1962 (volume 2).

[9]   DAFERMOS, C. M., *Asymptotic behavior of solutions of evolution equations*, in Nonlinear Evolution Equations, M. G. Crandall Ed, Academic Press, New-York 1978, 103–123.

[10] GAROFALO, N. and LIN, F., *Unique continuation for elliptic operators: a geometric-variational approach*, Comm. Pure Appl. Math. *40* (1987), 346–366.

[11] HALE, J. and RAUGEL, G., *Convergence in gradient-like systems with applications to PDE*, Z. Angew Math. Phys. *43* (1992), 63–124.

[12] HARAUX, A., *Systèmes dynamiques dissipatifs et applications*, Collection R. M. A. *17*, Collection dirigée par P. G. Ciarlet et J. L. Lions, Masson, Paris 1991.

[13] HARAUX, A., *Exponentially stable positive solutions to a forced semilinear parabolic equation*, Asymptotic Analysis 7 (1993), 3–13.

[14] HARAUX, A., *Stability questions in PDE*, Lecture Notes, Georgia Tech. seminar-course, CDNS 92–90 (1992), 84 p.

[15] HARAUX, A., and JENDOUBI, M. A., *On the convergence of global and bounded solutions of some evolution equations*, (Preprint).

[16] HARAUX, A., and POLÁČIK, P., *Convergence to a positive equilibrium for some semilinear evolution equations in a ball*, Acta Math. Univ. Comenianae LXI 2 (1993), 129–141.

[17] JENDOUBI, M. A., *A simple unified approach to some convergence theorems of L. Simon*, J. Funct. Analysis, *153* (1998), 187–202.

[18] KAVIAN, O., *Introduction à la Théorie des Points Critiques et Applications aux Problèmes Elliptiques*, Série Mathématiques & Applications # *13*, Springer-Verlag, Paris-Berlin, 1993.

[19] LIONS, P. L., *Structure of the set of steady-state solutions and asymptotic behavior of semilinear heat equations*, J. Diff. Eq. *53* (1984), 362–386.

[20] LOJASIEWICZ, S., *Ensembles semi-analytiques*, I.H.E.S. notes 1965.

[21] LOJASIEWICZ, S., *Sur les trajectoires du gradient d'une fonction analytique*, Seminari di Geometria, (Bologna 1982–1983), 115–117. Universitá degli Studi di Bologna, (Bologna, Italia), 1984.

[22] MATANO, H., *Convergence of solutions of one-dimensional semilinear heat equations*, J. Math. Kyoto Univ *18* (1978), 221–227.

[23] POLÁČIK, P., and RYBAKOWSKI, K., *Nonconvergent bounded trajectories of semilinear heat equations*, J. Diff. Eq. *124* (1996), 472–494.

[24] POLÁČIK, P., and SIMONDON, F., *Nonconvergent bounded solutions of semilinear heat equations on arbitrary domains*, Preprint.

[25] SIMON, L., *Asymptotics for a class of non linear evolution equations, with applications to geometric problems*, Ann. of Math. *118* (1983), 525–557.

[26] ZELENYAK, T. I., *Stabilization of solutions of boundary value problems for a second-order parabolic equation with one space variable*, Differentsial'nye Uravneniya *4* (1968), 17–22.

Alain Haraux
Laboratoire J. L. Lions
Université P. & M. Curie
175, rue du Chevaleret
75013 Paris
France
e-mail: haraux@ann.jnssieu.pr

Mohamed Ali Jendoubi and Otared Kavian
Laboratoire de Mathématiques Appliquées
UMR 7641
Université de Versailles
45, avenue des Etats Unis
F-78035 Versailles cedex
France

J.evol.equ. 3 (2003) 485 – 498
1424–3199/03/030485 – 14
DOI 10.1007/s00028-003-0114-x
© Birkhäuser Verlag, Basel, 2003

**Journal of Evolution
Equations**

A new regularity result for Ornstein-Uhlenbeck generators and applications

G. Da Prato

Dedicated to Philippe Bénilan

1. Introduction and setting of the problem

Let H be a separable real Hilbert space (norm $|\cdot|$, inner product $\langle\cdot,\cdot\rangle$). We are given a linear operator $A\colon D(A) \subset H \to H$ such that

HYPOTHESIS 1.1.

 (i) *A is self–adjoint and there exists $\omega > 0$ such that*

$$\langle Ax, x\rangle \le -\omega|x|^2, \quad x \in D(A). \tag{1.1}$$

 (ii) *A^{-1} is of trace class.*

As well known, Hypothesis 1.1 implies that there exists a complete orthonormal basis $\{e_k\}$ in H and a sequence of real numbers $\{\alpha_k\}$ such that

$$Ae_k = -\alpha_k e_k, \quad \alpha_k \ge \omega,\ k \in \mathbb{N}$$

and

$$\mathrm{Tr}\,[-A]^{-1} := \sum_{k=1}^{\infty} \frac{1}{\alpha_k} < +\infty.$$

Under Hypothesis 1.1 we can consider the following Ornstein-Uhlenbeck semigroup, see [3],

$$R_t\varphi(x) = \int_H \varphi(e^{tA}x + y)N_{Q_t}(dy), \quad t > 0,\ x \in H,\ \varphi \in C_b(H). \tag{1.2}$$

Here N_{Q_t} is the Gaussian measure in H of mean 0 and covariance operator Q_t given by

$$Q_t = -\frac{1}{2}\,A^{-1}(1 - e^{2tA}), \quad t \ge 0.$$

Mathematics Subject Classification 2000: 35K90, 35R15, 46B70.
Key words: Ornstein-Uhlenbeck generators, maximal regularity.

Moreover $C_b(H)$ is the Banach space of all uniformly continuous and bounded functions from H into $\mathbb{R}$ endowed with the norm $\|\varphi\|_0 = \sup_{x \in H} |\varphi(x)|$.

The semigroup R_t is not strongly continuous when A is not identically equal to 0; however we can define its *infinitesimal generator L* as follows, see [7].

For any $h > 0$ we set

$$\Delta_h \varphi = \frac{1}{h} (R_h \varphi - \varphi), \quad \varphi \in C_b(H).$$

Then we define the generator L of R_t by setting

$$D(L) = \left\{ \varphi \in C_b(H) : \ \exists f \in C_b(H), \ \lim_{h \to 0^+} \Delta_h \varphi(x) = f(x), \right.$$

$$\left. \forall x \in H \text{ and } \sup_{h \in (0,1]} \|\Delta_h \varphi\|_0 < +\infty \right\},$$

and

$$L\varphi(x) = \lim_{h \to 0^+} \Delta_h \varphi(x) = f(x), \quad x \in H, \ \varphi \in D(L).$$

The following result is proved in [7].

PROPOSITION 1.2. *Assume that Hypothesis* 1.1 *holds. Then* $(0, +\infty)$ *belongs to the resolvent set of L and we have*

$$(\lambda - L)^{-1} \varphi(x) = \int_0^{+\infty} e^{-\lambda t} R_t \varphi(x) dt, \quad \varphi \in C_b(H), \ x \in H. \tag{1.3}$$

Moreover

$$\|(\lambda - L)^{-1} \varphi\|_0 \leq \frac{1}{\lambda} \|\varphi\|_0, \ \varphi \in C_b(H). \tag{1.4}$$

We can describe the generator L on the subspace $\mathcal{I}_A(H)$ of $C_b(H)$ defined a follows:

$$\mathcal{I}_A(H) = \text{linear span} \left\{ \int_0^a e^{i \langle e^{sA} x, h \rangle} ds : \ a > 0, \ h \in D(A) \right\}.$$

$\mathcal{I}_A(H)$ is not dense in $C_b(H)$; however it is not difficult to see that for any $\varphi \in C_b(H)$ there exists a sequence $\{\varphi_n\} \in \mathcal{I}_A(H)$ such that

  (i) $\lim_{n \to \infty} \varphi_n(x) = \varphi(x), \quad \forall \, x \in H,$
  (ii) $\|\varphi_n\|_0 \leq \|\varphi\|_0, \quad \forall \, n \in \mathbb{N}.$

Now, let us set

$$\varphi(x) = \int_0^a e^{i \langle e^{sA} x, h \rangle} ds,$$

where $a > 0$ and $h \in D(A)$.

Then, recalling the expression of the Fourier transform of a Gaussian measure, we see that

$$R_t\varphi(x) = \int_0^a e^{-\frac{1}{2}\langle Q_t e^{sA}h, e^{sA}h\rangle} e^{i\langle e^{(t+s)A}x,h\rangle}\,ds, \quad x \in H,\ t > 0. \tag{1.5}$$

Therefore we can check easily, using the very definition of L, that φ belongs to $D(L)$ and we have

$$L\varphi(x) = \frac{1}{2}\,\mathrm{Tr}\,[D^2\varphi(x)] + \langle x, AD\varphi(x)\rangle, \quad x \in H.$$

REMARK 1.3. The semigroup R_t is naturally related to the following differential stochastic equation

$$\begin{cases} dX(t) = AX(t)dt + dW(t), & t \geq 0, \\ X(0) = x \in H, \end{cases} \tag{1.6}$$

where $W(t)$ is a cylindrical Wiener process in some probability space $(\Omega, \mathcal{F}, \mathbb{P})$ taking values in H. Formally $W(t)$ is given by

$$W(t) = \sum_{k=0}^{\infty} \beta_k e_k,$$

where $\{\beta_k\}$ is a family of mutually independent standard Brownian motions.

We have

$$R_t\varphi(x) = \mathbb{E}[\varphi(X(t, x))], \quad t \geq 0,\ x \in H,\ \varphi \in C_b(H),$$

where $\mathbb{E}$ denotes the expectation.

In this paper we are interested in studying the restriction of the semigroup R_t to $C_b^\theta(H)$, $\theta \in (0, 1)$, the space of all θ-Hölder continuous and bounded real functions on H. It is easy to see that $C_b^\theta(H)$ is invariant for R_t.

We shall denote by R_t^θ the restriction of R_t to $C_b^\theta(H)$, and by L^θ the part of L in $C_b^\theta(H)$:

$$L^\theta\varphi = L\varphi, \quad \forall\,\varphi \in D(L^\theta) = \{\varphi \in D(L) \cap C_b^\theta(H): \ L\varphi \in C_b^\theta(H)\}.$$

The characterization of the domain of L^θ is still an open problem. However the following maximal regularity result, generalizing the well known Schauder estimates, is known, see [2].

PROPOSITION 1.4. *Assume that Hypothesis* 1.1 *holds. Let* $f \in C_b^\theta(H)$, *with* $\theta \in (0, 1)$ *and* $\lambda > 0$. *Set* $\varphi = (\lambda - L^\theta)^{-1}f$. *Then we have* $\varphi \in C_b^{2+\theta}(H)$ *and there exists* $M > 0$ *(independent on* λ *and on* f*) such that*

$$\|\varphi\|_{C_b^{2+\theta}(H)} \leq M\|f\|_{C_b^\theta(H)}. \tag{1.7}$$

The main result of this paper is another maximal regularity result. Let $f \in C_b^\theta(H)$, $\lambda > 0$ and set $\varphi = (\lambda - L^\theta)^{-1} f$. We are going to show that $D\varphi(x)$ belongs to $D((-A)^{1/2})$ for any $x \in H$ and that $(-A)^{1/2} D\varphi \in C_b^\theta(H)$.

This result can be applied to the following Kolmogorov equation

$$\lambda \varphi(x) - L^\theta \varphi(x) - \langle F(x), (-A)^{1/2} D\varphi(x) \rangle = f(x) \tag{1.8}$$

where $\lambda > 0$, $f \in C_b^\theta(H)$, and $F : H \to H$ is Hölder continuous and bounded. A typical example is presented in Section 3.

Let us finish this section by giving some notation and by recalling the definition of interpolation spaces needed in what follows.

Let E be a Banach space. We shall denote by $C_b(H; E)$ the Banach space of all uniformly continuous and bounded functions from H into E endowed with the norm $\|\varphi\|_0 = \sup_{x \in H} |\varphi(x)|_E$.

If $\theta \in (0, 1)$, we shall denote by $C_b^\theta(H; E)$ the subspace of $C_b(H; E)$ consisting of all functions $\varphi : H \to E$ such that

$$[\varphi]_\theta := \sup_{\substack{x, y \in H \\ x \neq y}} \frac{|\varphi(x) - \varphi(y)|}{|x - y|^\theta} < +\infty.$$

$C_b^\theta(H; E)$ is a Banach space with the norm

$$\|\varphi\|_\theta := \|\varphi\|_0 + [\varphi]_\theta, \quad \varphi \in C_b^\theta(H; E).$$

When $E = \mathbb{R}$ we shall set $C_b(H; \mathbb{R}) = C_b(H)$ and $C_b^\theta(H; \mathbb{R}) = C_b^\theta(H)$.

Let us finally recall the K definition of interpolation spaces, see e. g. [6]. Let X and Y be Banach spaces such that $Y \subset X$ with continuous embedding. Let us define

$$K(t, x) = \inf \{ \|a\|_X + t\|b\|_Y : x = a + b, \ a \in X, \ b \in Y \}.$$

Then, for arbitrary $\theta \in [0, 1]$, we set

$$[x]_{(X,Y)_{\theta,\infty}} = \sup_{t \in (0,1]} t^{-\theta} K(t, x),$$
$$(X, Y)_{\theta,\infty} = \{ x \in X : [x]_{(X,Y)_{\theta,\infty}} < +\infty \}.$$

As is easily seen $(X, Y)_{\theta,\infty}$, endowed with the norm

$$\|x\|_{(X,Y)_{\theta,\infty}} = \|x\|_X + [x]_{(X,Y)_{\theta,\infty}}, \quad x \in (X, Y)_{\theta,\infty},$$

is a Banach space.

It is not difficult to check that the following statements are equivalent.

(i) $x \in (X, Y)_{\theta,\infty}$ and $[x]_{(X,Y)_{\theta,\infty}} \leq L$.

(ii) For all $t \in (0, 1]$ there exist $a_t \in X$ and $b_t \in Y$ such that $x = a_t + b_t$ and

$$\|a_t\|_X + t\|b_t\|_Y \leq Lt^\theta.$$

Let us recall the basic interpolation theorem, see e. g. [8].

THEOREM 1.5. *Let X, X_1, Y, Y_1 be Banach spaces such that $Y \subset X, Y_1 \subset X_1$ with continuous embeddings. Let moreover T be a linear mapping $T : X \to X_1, \quad T : Y \to Y_1,$ such that for some $M, N > 0$*

$$\|Tx\|_{X_1} \leq M\|x\|_X, \quad \|Ty\|_{Y_1} \leq N\|y\|_Y.$$

Then T maps $(X, Y)_{\theta,\infty}$ into $(X_1, Y_1)_{\theta,\infty}$, and

$$[Tx]_{(X_1,Y_1)_{\theta,\infty}} \leq M^{1-\theta} N^\theta [x]_{(X,Y)_{\theta,\infty}}, \quad x \in (X, Y)_{\theta,\infty}.$$

We shall need also the following result, see [2].

THEOREM 1.6. *Let H be a separable Hilbert space. Then we have*

$$(C_b(H), C_b^1(H))_{\theta,\infty} = C_b^\theta(H), \quad \theta \in (0, 1). \tag{1.9}$$

Moreover there exists a positive constant κ_θ such that

$$\frac{1}{\kappa_\theta} \|\varphi\|_{C_b^\theta(H)} \leq \|\varphi\|_{(C_b(H),C_b^1(H))_{\theta,\infty}} \leq \kappa_\theta \|\varphi\|_{C_b^\theta(H)}. \tag{1.10}$$

2. The main result

We assume here that Hypothesis 1.1 holds. Let us recall that, under these assumptions, the semigroup R_t is regularizing. That is, for any $t > 0$ and any $\varphi = C_b(H)$ we have that $R_t\varphi \in C_b^\infty(H)$, see [3].

More precisely, set

$$\Lambda_t = Q_t^{-1/2} e^{tA} = \sqrt{2}\, (-A)^{1/2} e^{tA} (1 - e^{2tA})^{-1/2}$$

and notice that

$$\|\Lambda_t\| \leq c_1 t^{-1/2}, \qquad t > 0, \tag{2.1}$$

where

$$c_1 = \sup_{\xi > 0} \sqrt{2\xi}\, \frac{e^{-\xi}}{1 - e^{-2\xi}},$$

and

$$\|(-A)^{1/2}\Lambda_t\| \leq c_2 t^{-1}, \qquad t > 0, \tag{2.2}$$

where

$$c_2 = \sup_{\xi > 0} \sqrt{2\xi^2} \, \frac{e^{-\xi}}{1 - e^{-2\xi}}.$$

The following result is proved in [3].

PROPOSITION 2.1. *Let* $\varphi \in C_b(H)$ *and* $t > 0$. *Then* $R_t\varphi \in C_b^2(H)$ *and for any* $h, k \in H$ *we have*

$$\langle DR_t\varphi(x), h \rangle = \int_H \langle \Lambda_t h, Q_t^{-1/2} y \rangle \varphi(e^{tA}x + y) N_{Q_t}(dy), \ x \in H, \tag{2.3}$$

and

$$\langle D^2 R_t\varphi(x) \cdot h, k \rangle = \int_H \langle \Lambda_t h, Q_t^{-1/2} y \rangle \langle \Lambda_t k, Q_t^{-1/2} y \rangle \varphi(e^{tA}x + y) N_{Q_t}(dy)$$

$$-\langle \Lambda_t h, \Lambda_t k \rangle R_t\varphi(x), \ x \in H. \tag{2.4}$$

Now we show that for any $\varphi \in C_b(H)$ and any $t > 0$ we have that $DR_t\varphi(x) \in D((-A)^{1/2})$. Moreover we prove several lemmas and corollaries giving estimates for $(-A)^{1/2}DR_t\varphi$ in different norms as $C_b(H)$, $C_b^1(H)$ and $C_b^\theta(H)$. Finally, we obtain the desired result by interpolation.

LEMMA 2.2. *Let* $\varphi \in C_b(H)$ *and* $t > 0$. *Then* $DR_t\varphi(x) \in D((-A)^{1/2})$ *for all* $x \in H$. *Moreover* $(-A)^{1/2}DR_t\varphi \in C_b(H; H)$ *and we have*

$$\|(-A)^{1/2}DR_t\varphi\|_0 \leq c_1 t^{-1} \|\varphi\|_0, \quad t > 0. \tag{2.5}$$

Proof. Let $\varphi \in C_b(H), t > 0$ and $h \in H$. Then by (2.2) and (2.3) we see that $DR_t\varphi(x) \in D((-A)^{1/2})$ for all $x \in H$ and

$$\langle (-A)^{1/2}DR_t\varphi(x), h \rangle = \int_H \langle (-A)^{1/2}\Lambda_t h, Q_t^{-1/2} y \rangle \varphi(e^{tA}x + y) N_{Q_t}(dy), \ x \in H.$$

By the Hölder inequality it follows that

$$|\langle (-A)^{1/2}DR_t\varphi(x), h \rangle|^2 \leq \|\varphi\|_0^2 \int_H |\langle (-A)^{1/2}\Lambda_t h, Q_t^{-1/2} y \rangle|^2 N_{Q_t}(dy)$$

$$= \|\varphi\|_0^2 |(-A)^{1/2}\Lambda_t h|^2 \leq \|\varphi\|_0^2 \, c_1^2 t^{-2} |h|^2, \ x \in H.$$

Now the conclusion follows from the arbitrariness of h. $\qquad\qquad\square$

LEMMA 2.3. *Let $\varphi \in C_b^1(H)$ and $t > 0$. Then $DR_t\varphi(x) \in D((-A)^{1/2})$, for all $x \in H$. Moreover $(-A)^{1/2}DR_t\varphi \in C_b(H; H)$ and we have*

$$\|(-A)^{1/2}DR_t\varphi\|_0 \le c_3 t^{-1/2}\|\varphi\|_1, \quad t > 0, \tag{2.6}$$

where $c_3 = (2e)^{-1/2}$.

Proof. Let $\varphi \in C_b(H)$, $t > 0$. Then, differentiating (1.2) with respect to x we find that

$$DR_t\varphi(x) = \int_H e^{tA} D\varphi(e^{tA}x + y)N_{Q_t}(dy), \quad t > 0, \ x \in H.$$

Consequently $DR_t\varphi(x) \in D((-A)^{1/2})$, and we have

$$(-A)^{1/2}DR_t\varphi(x) = \int_H (-A)^{1/2} e^{tA} D\varphi(e^{tA}x + y)N_{Q_t}(dy), \quad t > 0, \ x \in H.$$

It follows

$$|(-A)^{1/2}DR_t\varphi(x)| \le \|\varphi\|_1 \|(-A)^{1/2}e^{tA}\| \le \|\varphi\|_1 (2te)^{-1/2},$$

that yields (2.6) $\square$

COROLLARY 2.4. *Let $\varphi \in C_b^\theta(H)$, $\theta \in (0, 1)$ and $t > 0$. Then $(-A)^{1/2}DR_t\varphi \in C_b(H; H)$ and we have*

$$\|(-A)^{1/2}DR_t\varphi\|_0 \le c_\theta t^{\theta/2-1} \|\varphi\|_\theta, \quad t > 0, \tag{2.7}$$

where $c_\theta = c_1^{1-\theta} c_3^\theta \kappa_\theta$ and κ_θ is defined in (1.10).

Proof. Let $t > 0$ be fixed and denote by γ the mapping

$$\gamma : \varphi \to (-A)^{1/2}DR_t\varphi.$$

From Lemmas 2.2 and 2.3 it follows that

    (i) γ maps $C_b(H)$ into $C_b(H; H)$ with norm $\le c_1 t^{-1}$,
    (ii) γ maps $C_b^1(H)$ into $C_b(H; H)$ with norm $\le c_3 t^{-1/2}$.

Consequently, by the basic interpolation theorem, Theorem 1.5, we have that γ maps $(C_b(H), C_b^1(H))_{\theta,\infty}$ into $C_b(H; H)$ with norm $\le (c_1 t^{-1})^{1-\theta}(c_3 t^{-1/2})^\theta$. Therefore

$$\|\gamma(\varphi)\|_{C_b(H;H)} \le (c_1 t^{-1})^{1-\theta}(c_3 t^{-1/2})^\theta \|\varphi\|_{(C_b(H),C_b^1(H))_{\theta,\infty}}.$$

On the other hand by Theorem 1.5 we have that

$$(C_b(H), C_b^1(H))_{\theta,\infty} = C_b^\theta(H),$$

and so the conclusion follows from (1.10). $\square$

LEMMA 2.5. *Let $\varphi \in C_b(H)$ and $t > 0$. Then $D^2 R_t\varphi(x)h \in D((-A)^{1/2})$, for all $x, h \in H$. Moreover $(-A)^{1/2}D^2 R_t\varphi \in C_b(H; L(H))$ and we have*

$$\|(-A)^{1/2}D^2 R_t\varphi\|_0 \le c_4 t^{-3/2}\|\varphi\|_0, \quad t > 0, \tag{2.8}$$

where $c_4 = (\sqrt{3}+1)c_1 c_2$.

Proof. Let $\varphi \in C_b(H)$, $t > 0$ and $h, k \in H$. Then by (2.5) we see that $D^2 R_t\varphi(x) \in D((-A)^{1/2})$, for all $x \in H$ and we have

$$\langle (-A)^{1/2}D^2 R_t\varphi(x) \cdot h, k \rangle = \langle D^2 R_t\varphi(x) \cdot h, (-A)^{1/2}k \rangle$$

$$= \int_H \langle \Lambda_t h, Q_t^{-1/2}y \rangle \langle (-A)^{1/2}\Lambda_t k, Q_t^{-1/2}y \rangle \varphi(e^{tA}x + y) N_{Q_t}(dy)$$

$$-\langle \Lambda_t h, (-A)^{1/2}\Lambda_t k \rangle R_t\varphi(x) := I_1 + I_2.$$

By the Hölder inequality it follows, recalling (2.2), that

$$|I_1|^2 \le \|\varphi\|_0^2 \int_H |\langle \Lambda_t h, Q_t^{-1/2}y \rangle|^2 \, |\langle (-A)^{1/2}\Lambda_t k, Q_t^{-1/2}y \rangle|^2 N_{Q_t}(dy)$$

$$\le 3\|\varphi\|_0^2|\Lambda_t h|^2|(-A)^{1/2}\Lambda_t k|^2 \le 3\|\varphi\|_0^2 c_1^2 c_2^2 t^{-3} \, |h|^2 \, |k|^2. \tag{2.9}$$

Concerning I_2 we have

$$|I_2| \le \|\varphi\|_0|\Lambda_t h| \, |(-A)^{1/2}\Lambda_t k| \le c_1 c_2\|\varphi\|_0 t^{-3/2} \, |h| \, |k|. \tag{2.10}$$

Now the conclusion follows from (2.9), (2.10) and the arbitrariness of h and k.   □

LEMMA 2.6. *Let $\varphi \in C_b^1(H)$ and $t > 0$. Then $D^2 R_t\varphi(x) \in D((-A)^{1/2})$, for all $x \in H$. Moreover $(-A)^{1/2}D^2 R_t\varphi \in C_b(H; L(H))$ and we have*

$$\|(-A)^{1/2}D^2 R_t\varphi\|_0 \le c_5 t^{-1}\|\varphi\|_1, \quad t > 0, \tag{2.11}$$

where $c_5 = c_1 c_3$.

Proof. Let $\varphi \in C_b^1(H)$, $t > 0$ and $h, k \in H$. Then differentiating (2.3) we find

$$\langle (-A)^{1/2}D^2 R_t\varphi(x)h, k \rangle$$

$$= \int_H \langle \Lambda_t h, Q_t^{-1/2}y \rangle \langle D\varphi(e^{tA}x + y), (-A)^{1/2}e^{tA}k \rangle N_{Q_t}(dy).$$

By the Hölder inequality it follows that

$$|\langle (-A)^{1/2}D^2 R_t\varphi(x)h, k \rangle|^2 \le |(-A)^{1/2}e^{tA}k|^2\|\varphi\|_1^2 \int_H |\langle \Lambda_t h, Q_t^{-1/2}y \rangle|^2 N_{Q_t}(dy)$$

$$\le c_3^2 t^{-1}|k|^2\|\varphi\|_1^2|\Lambda_t h|^2 \le c_1^2 c_3^2 t^{-2}\|\varphi\|_0^2|h|^2|k|^2.$$

Now the conclusion follows from the arbitrariness of h and k.   □

By interpolation we find the following result

COROLLARY 2.7. *Let* $\varphi \in C_b^\theta(H)$, $\theta \in (0, 1)$ *and* $t > 0$. *Then* $(-A)^{1/2} D^2 R_t \varphi \in C_b(H; L(H))$ *and we have*

$$\|(-A)^{1/2} D^2 R_t \varphi\|_0 \leq c_{1,\theta} t^{(\theta-3)/2} \|\varphi\|_\theta, \quad t > 0, \tag{2.12}$$

where $c_{\theta,1} = c_4^{1-\theta} c_5^\theta \kappa_\theta$.

Proof. Let $t > 0$ be fixed and denote by δ the mapping

$$\delta : \varphi \to (-A)^{1/2} D^2 R_t \varphi.$$

From Lemmas 2.5 and 2.6 it follows that

(i) δ maps $C_b(H)$ into $C_b(H; L(H))$ with norm $\leq c_4 t^{-3/2}$,
(ii) δ maps $C_b^1(H)$ into $C_b(H; L(H))$ with norm $\leq c_5 t^{-1}$.

Consequently, by Theorem 1.5, we have that δ maps $\left(C_b(H), C_b^1(H)\right)_{\theta,\infty}$ into $C_b(H; L(H))$ with norm $\leq (c_4 t^{-3/2})^{1-\theta} (c_5 t^{-1})^\theta$. Therefore

$$\|\delta(\varphi)\|_{C_b(H;L(H))} \leq (c_4 t^{-3/2})^{1-\theta} (c_5 t^{-1})^\theta \|\varphi\|_{\left(C_b(H),C_b^1(H)\right)_{\theta,\infty}}.$$

Now the conclusion follows from Theorem 1.5. $\qquad\qquad\square$

We are now ready to prove the main result of the paper.

THEOREM 2.8. *Assume that Hypothesis* 1.1 *holds. Let* $f \in C_b^\theta(H)$, $\theta \in (0, 1)$, $\lambda > 0$ *and set* $\varphi = (\lambda - L^\theta)^{-1} f$. *Then we have* $(-A)^{1/2} D\varphi \in C_b^\theta(H; H)$ *and there exists* $M_1 > 0$ *(independent on* λ *and on* f*) such that*

$$\|(-A)^{1/2} D\varphi\|_{C_b^\theta(H;H)} \leq M_1 \|f\|_{C_b^\theta(H)}. \tag{2.13}$$

Proof. Let $f \in C_b^\theta(H)$, $\lambda > 0$ and $\varphi = (\lambda - L^\theta)^{-1} f$. For any $h \in H$ set

$$\psi_h(x) = \int_0^{+\infty} e^{-\lambda s} \langle (-A)^{1/2} DR_s f(x), h\rangle ds, \quad x \in H.$$

Set moreover

$$a_t(x) = \int_0^{t^2} e^{-\lambda s} \langle (-A)^{1/2} DR_s f(x), h\rangle ds, \quad x \in H,$$

and

$$b_t(x) = \int_{t^2}^{+\infty} e^{-\lambda s} \langle (-A)^{1/2} DR_s f(x), h\rangle ds, \quad x \in H.$$

Then, by (2.7) we have

$$\|a_t\|_0 \leq c_\theta \|f\|_\theta \int_0^{t^2} s^{\theta/2-1} ds |h| = \frac{2}{\theta} c_\theta \|f\|_\theta |h| t^\theta. \tag{2.14}$$

Moreover, since

$$Db_t(x) = \int_{t^2}^{+\infty} e^{-\lambda s}(-A)^{1/2} D^2 R_s f(x) h ds, \quad x \in H,$$

by (2.12) we have

$$\|b_t\|_1 \leq c_{1,\theta} \|f\|_\theta \int_{t^2}^{+\infty} s^{(\theta-3)/2} ds |h| = \frac{2c_{1,\theta}}{1-\theta} \|f\|_\theta |h| t^{\theta-1}. \tag{2.15}$$

Therefore ψ_h belongs to $\left(C_b(H), C_b^1(H)\right)_{\theta,\infty}$ and so to $C_b^\theta(H)$ by Theorem 1.6. Moreover there exists $M_\theta > 0$ such that

$$\|\psi_h\|_\theta \leq M_\theta \|f\|_\theta \, |h|.$$

The conclusion follows from the arbitrariness of h. $\qquad\qquad\square$

3. A generalized Ornstein-Uhlenbeck semigroup

For any $z \in H$ we define the following semigroup on $C_b(H)$.

$$R_t^z \varphi(x) = \int_H \varphi(y) N_{e^{tA}x+(-A)^{-1}(z-e^{tA}z), Q_t}(dy), \quad \varphi \in C_b(H), \ x \in H. \tag{3.1}$$

Clearly we have

$$\begin{aligned} R_t^z \varphi(x) &= \int_H \varphi(e^{tA}x + (-A)^{-1}(z - e^{tA}z) + y) N_{Q_t}(dy), \\ & \qquad \varphi \in C_b(H), \ x \in H. \end{aligned} \tag{3.2}$$

REMARK 3.1. The semigroup R_t^z is naturally related to the following differential stochastic equation

$$\begin{cases} dY(t) = (AY(t) + z)dt + dW(t), & t \geq 0, \\ Y(0) = x \in H. \end{cases} \tag{3.3}$$

In fact we have

$$R_t^z \varphi(x) = \mathbb{E}[\varphi(Y(t, x))], \quad t \geq 0, \ x \in H, \ \varphi \in C_b(H).$$

We set

$$Q_t^{-1/2} e^{tA} x + (-A)^{-1}(z - e^{tA} z) = \Lambda_t x + G_t z,$$

where

$$G_t z = \sqrt{2}\,(1 - e^{tA})(1 - e^{2tA})^{-1/2}(-A)^{-1/2} z.$$

we shall denote by L^z the infinitesimal generator of R_t^z.

PROPOSITION 3.2. *Let* $\varphi = C_b(H)$, $z \in H$ *and* $t > 0$. *Then* $R_t^z \varphi \in C_b^2(H)$ *and for any* $h, k \in H$ *we have*

$$\langle DR_t^z \varphi(x), h \rangle = \int_H \langle \Lambda_t h, Q_t^{-1/2} y \rangle \varphi(e^{tA} x + (-A)^{-1}(z - e^{tA} z) + y) N_{Q_t}(dy),$$
$$x \in H, \tag{3.4}$$

and

$$\langle D^2 R_t^z \varphi(x) \cdot h, k \rangle = \int_H \langle \Lambda_t h, Q_t^{-1/2} y \rangle \langle \Lambda_t k, Q_t^{-1/2} y \rangle$$
$$\times \varphi(e^{tA} x + (-A)^{-1}(z - e^{tA} z) + y) N_{Q_t}(dy)$$
$$- \langle \Lambda_t h, \Lambda_t k \rangle R_t \varphi(x), \quad x \in H. \tag{3.5}$$

Proof. We first notice that, by the Cameron-Martin formula, we have

$$\rho(t, x, z, y) := \frac{dN_{e^{tA} x + (-A)^{-1}(z - e^{tA} z), Q_t}}{dN_{Q_t}}(y)$$
$$= e^{-\frac{1}{2}|\Lambda_t x + G_t z|^2 + \langle \Lambda_t x + G_t z, Q_t^{-1/2} y \rangle} \tag{3.6}$$

Taking into account (3.6) we find

$$\langle DR_t^z \varphi(x), h \rangle = \int_H \varphi(y) \rho(t, x, z, y) - \langle \Lambda_t x + G_t z, \Lambda_t h \rangle + \langle \Lambda_t h, Q_t^{-1/2} y \rangle N_{Q_t}(dy)$$
$$= \int_H \varphi(y) \rho(t, x, z, y) \langle \Lambda_t h, Q_t^{-1/2}(y - e^{tA} x + (-A)^{-1}(z - e^{tA} z))) \rangle N_{Q_t}(dy)$$
$$= \int_H \langle \Lambda_t h, Q_t^{-1/2} y \rangle \varphi(e^{tA} x + (-A)^{-1}(z - e^{tA} z)) N_{Q_t}(dy),$$

and (3.4) follows. (3.5) can be proved similarly. $\square$

By proceeding as in the proof of Theorem 2.8 we find the following result.

THEOREM 3.3. *Assume that Hypothesis* 1.1 *holds. Let* $f \in C_b^\theta(H)$, $\theta \in (0, 1)$, $z \in H$, $\lambda > 0$ *and set* $\varphi = (\lambda - L^z)^{-1} f$. *Then we have* $(-A)^{1/2} D\varphi \in C_b^\theta(H; H)$ *and there exists* $M_1 > 0$ (*independent on* λ *and on* f) *such that*

$$\|(-A)^{1/2} D\varphi\|_{C_b^\theta(H;H)} \leq M_1 \|f\|_{C_b^\theta(H)}. \tag{3.7}$$

4. Applications

We are here concerned with the Kolmogorov operator

$$N\varphi(x) = L^\theta \varphi(x) + \langle F(x), (-A)^{1/2} D\varphi(x)\rangle, \quad x \in H,$$

where $F \in C_b^\theta(H; H)$ for some fixed $\theta \in (0, 1)$ and with the following equation

$$\lambda\varphi(x) - L^\theta \varphi(x) - \langle F(x), (-A)^{1/2} D\varphi(x)\rangle = f(x) \tag{4.1}$$

where $\lambda > 0$ and $f \in C_b^\theta(H)$.

THEOREM 4.1. *Assume that Hypothesis 1.1 holds and that $F \in C_b^\theta(H; H)$ for some fixed $\theta \in (0, 1)$. Then there exists $c > 0$ such that if $\|F\|_\theta < c$ then Equation (4.1) has unique solution $\varphi \in C_b^{2+\theta}(H)$ such that $(-A)^{1/2} D\varphi \in C_b^\theta(H; H)$.*

Proof. Setting

$$\lambda\varphi(x) - L^\theta \varphi(x) = \psi(x), \quad x \in H,$$

Equation (4.1) reduces to

$$\psi - T\psi = f, \tag{4.2}$$

where

$$T\psi(x) = \langle F(x), (-A)^{1/2} D(\lambda - L^\theta)^{-1}\psi(x)\rangle. \tag{4.3}$$

By Theorem 2.8 it follows that

$$\|T\psi\|_\theta \leq M\|F\|_\theta\|\psi\|_\theta.$$

Setting $c = M^{-1}$, the conclusion follows from the contraction principle. $\qquad \square$

REMARK 4.2. It would be interesting to drop the condition that $\|F\|_\theta$ is small. For this we could consider the equation

$$\lambda\varphi(x) - L^\theta \varphi(x) - \alpha\langle F(x), (-A)^{1/2} D\varphi(x)\rangle = f(x) \tag{4.4}$$

where $\lambda > 0$, $f \in C_b^\theta(H)$, and $\alpha \in [0, 1]$. Setting

$$\Gamma = \{\alpha \in [0, 1] : (4.4) \text{ has a solution}\},$$

the set Γ is open in view of Theorems 4.1 and 3.3.

It remains to show that the set Γ is closed. For this it would be needed to prove an a priori estimate. The usual way is to proceed by using the maximum principle and localization of coefficients.

This procedure is classical in finite dimensions, see [5], and it works also for the heat equation in infinite dimensions, see [1], but we are not able at the moment to apply this idea to cover the present case.

REMARK 4.3. By a result due to L. Zambotti [9], it follows that, under the assumptions of Theorem 4.1, there exists a unique martingale solution to the differential stochastic equation

$$\begin{cases} dX(t) = [AX(t) + (-A)^{1/2}F(X(t))]dt + dW(t), & t \geq 0, \\ X(0) = x \in H, \end{cases} \tag{4.5}$$

where W is a cylindrical Wiener process taking values on H.

EXAMPLE 4.4. Let us consider the following stochastic partial differential equation

$$\begin{cases} dX(t, \xi) = [D_\xi^2 X(t, \xi) - X(t, \xi) + D_\xi[g(X(t, \xi))]]dt + dW(t), & t \geq 0, \\ X(t, 0) = X(t, 2\pi), & t \geq 0, \\ X(0) = x \in L^2(0, 2\pi), \end{cases} \tag{4.6}$$

where $g \in C_b^\theta(\mathbb{R})$ for some $\theta \in (0, 1)$ and $W(t)$ is a cylindrical Wiener process in some probability space $(\Omega, \mathcal{F}, \mathbb{P})$ taking values in $L^2(0, 2\pi)$.

Set $H = L^2(0, 2\pi)$ and let A be defined by

$$Ax = D_\xi^2 - \xi, \quad x \in D(A) = H_\#^2(0, 2\pi),$$

where $H_\#^2(0, 1)$ is the space of all functions of $H^2(0, 2\pi)$ which are periodic together with their first derivatives. A is self-adjoint and, setting

$$e_k(\xi) = \sqrt{\frac{1}{2\pi}} \, \sin k\xi, \quad k \in \mathbb{N} \cup \{0\},$$

we have

$$Ae_k = -(1 + k^2)e_k, \quad k \in \mathbb{N} \cup \{0\},$$

so that A^{-1} is of trace class and Hypothesis 1.1 holds with $\omega = 1$.

Set moreover

$$F(x) = D_\xi(-A)^{1/2}g(x), \quad x \in L^2(0, 2\pi).$$

Notice that the linear operator $D_\xi(-A)^{1/2}$ is bounded since

$$D_\xi(-A)^{1/2}e_k = \frac{ik}{\sqrt{1 + k^2}}, \quad k \in \mathbb{N} \cup \{0\}.$$

Consequently F belongs to $C_b^\theta(H; H)$ and Equation (4.6) is equivalent to Equation (4.5). So we can apply Theorem 2.8.

REFERENCES

[1]   CANNARSA, P. and DA PRATO, G., *Infinite dimensional elliptic equations with Hölder continuous coefficients,* Advances Diff. Equations, *1* (1996), 425–452.

[2]   CANNARSA, P. and DA PRATO, G., *Schauder estimates for Kolmogorov equations in Hilbert spaces,* Progress in elliptic and parabolic partial differential equations, A. Alvino, P. Buonocore, V. Ferone, E. Giarrusso, S. Matarasso, R. Toscano and G. Trombetti (editors), Research Notes in Mathematics, Pitman, *350* (1996), 100–111.

[3]   DA PRATO, G. and ZABCZYK, J., *Stochastic equations in infinite dimensions*, Cambridge University Press, 1992.

[4]   DA PRATO, G. and ZABCZYK, J., *Ergodicity for infinite dimensional systems*, London Mathematical Society Lecture Notes, *229*, Cambridge University Press, 1996.

[5]   LADYZHENSKAJA, O. A., SOLONNIKOV, V. A. and URAL'CEVA, N. N., *Linear and quasilinear equations of parabolic type*, Transl. Math. Monographs, Amer. Math. Soc. 1968.

[6]   LUNARDI, A., *Analytic semigroups and optimal regularity in parabolic problems*, Birkhäuser, 1995.

[7]   PRIOLA, E., *On a class of Markov type semigroups in spaces of uniformly continuous and bounded functions*, Studia Math. *136* (1999), 271–295.

[8]   TRIEBEL, H., *Interpolation theory, function spaces, differential operators*, North-Holland, 1978.

[9]   ZAMBOTTI, L., *A new approach to existence and uniqueness for martingale problems in infinite dimensions*, Probab. Th. Relat. Fields, *118* (2000), 147–168.

G. Da Prato
Scuola Normale Superiore di Pisa
Piazza dei Cavalieri 7
56126 Pisa
Italy
e-mail: daprato@sns.it

To access this journal online:
http://www.birkhauser.ch

J.evol.equ. 3 (2002) 499 – 521
1424–3199/03/030499 – 23
DOI 10.1007/s00028-003-0503-1
© Birkhäuser Verlag, Basel, 2003

Journal of Evolution
Equations

Global solution and smoothing effect for a non-local regularization of a hyperbolic equation

J. Droniou[1], T. Gallouët[2], J. Vovelle[2]

1. Introduction

We study the problem

$$\begin{cases} \partial_t u(t, x) + \partial_x (f(u))(t, x) + g[u(t, \cdot)](x) = 0 & t \in]0, \infty[, \ x \in \mathbb{R} \\ u(0, x) = u_0(x) & x \in \mathbb{R}, \end{cases} \tag{1.1}$$

where $f \in C^\infty(\mathbb{R})$ is such that $f(0) = 0$ (there is not loss of generality in assuming this), $u_0 \in L^\infty(\mathbb{R})$ and g is the non-local (in general) operator defined through the Fourier transform by

$$\mathcal{F}(g[u(t, \cdot)])(\xi) = |\xi|^\lambda \mathcal{F}(u(t, \cdot))(\xi) , \quad \text{with } \lambda \in]1, 2].$$

REMARK 1.1. We could also very well study a multi-dimensional scalar equation, that is to say on $\mathbb{R}^N$ instead of $\mathbb{R}$. All the methods and results presented below would apply; but this would lead to more technical manipulations so, for the sake of clarity, we have chosen to fully describe only the mono-dimensional case.

The interest of such an equation (namely Equation (1.1)) was pointed out to us by Paul Clavin in the context of pattern formation in detonation waves. The study of detonations leads, in a first approximation, to nonlinear hyperbolic equations. As it is well known, the solutions of such equations may develop discontinuities in finite time. A theory of existence and uniqueness of (entropy weak) solutions to Equation (1.1) with $g = 0$, in the L^∞ framework, is known since the work of Krushkov ([Kru70], see also [Vol67]). The case of a parabolic regularization (of a nonlinear hyperbolic equation) is often considered and used to prove the Krushkov result; it corresponds to (1.1) with $\lambda = 2$. In this case, existence and uniqueness of a solution is also well known along with a regularizing effect. However, it appears that the choice of $\lambda = 2$ is not quite natural, at least for the problem of detonation (see [CD02], [CH01], [CD01]) where it seems more natural to consider a nonlocal term as $g[u]$ with λ close to 1 but greater than 1 (although the case $\lambda = 1$ is also of interest but more complicated). This term corresponds to some spatial fractional derivative of u of order λ. The main motivation of this paper is therefore to prove existence and uniqueness of the solution to (1.1) in the L^∞ framework as well as a regularizing effect (a regularizing effect which is well-known in the case $\lambda = 2$, as it is said above). In particular, the solution will be C^∞ in space and time for $t > 0$. We also prove the so called "maximum principle", namely the fact that the solution takes values between the maximum and the minimum values of the initial data, and a property of "L^1 contraction" on the solutions, which is the fact that, for any time, the L^1 norm of the difference of two solutions with different initial data is bounded by the L^1 norm (if it exists) of the difference of the initial data.

A major difficulty is due to the nonlocal character of $g[u]$ if $\lambda \in]1, 2[$; this prevents the classical way to prove the maximum principle, which leads to an L^∞ *a priori* bound on the solution (a crucial estimate to obtain global solutions). It is interesting to notice that the hypothesis $\lambda \leq 2$ is necessary for the maximum principle. Indeed, the maximum principle is no longer true in general for $\lambda > 2$. However, the regularizing effect is still true for $\lambda > 2$, a property which is probably not verified if $\lambda < 1$. The case $\lambda = 1$ is not so clear and needs an additional work. Indeed, for the study of detonation waves, our result has to be viewed as a preliminary result or, at least, as a study of a very simplified case. Realistic models are much more complicated. In particular, it seems that λ is actually depending on the unknown and, even if $\lambda > 1$, λ is probably not bounded from below by some $\lambda_0 > 1$. The possibility to generalize our result to such a case is not manifest.

We first prove (Section 4) the uniqueness of a "weak" solution (solution in the sense of Definition 3.1. below). Then, assuming the existence of a "weak" solution, we prove (Section 5) the regularizing effect (the equation is then satisfied in a classical sense). The results of these two sections are in fact true for any $\lambda > 1$. In Section 6, the existence result is given, using a splitting method. The use of splitting methods is classical, in particular to define numerical schemes, but is not usual to prove an existence result as it is done here. In this section, the central argument is the proof of the maximum principle (which is limited to $\lambda \leq 2$).

Here is our main result.

THEOREM 1.1. *If $u_0 \in L^\infty(\mathbb{R})$, then there exists a unique solution u to (1.1) on $]0, \infty[$ (in the sense of Definition 3.1, see below). Moreover, this solution satisfies:*

i) *$u \in C^\infty(]0, \infty[\times\mathbb{R})$ and all its derivatives are bounded on $]t_0, \infty[\times\mathbb{R}$ for all $t_0 > 0$,*

ii) *for all $t > 0$, $\|u(t)\|_{L^\infty(\mathbb{R})} \leq \|u_0\|_{L^\infty(\mathbb{R})}$ and, in fact, u takes its values between the essential lower and upper bounds of u_0,*

iii) *u satisfies $\partial_t u + \partial_x(f(u)) + g[u] = 0$ in the classical sense ($g[u]$ being properly defined by Proposition 5.3).*

iv) *$u(t) \to u_0$, as $t \to 0$, in $L^\infty(\mathbb{R})$ weak-$*$ and in $L^p_{\mathrm{loc}}(\mathbb{R})$ for all $p \in [1, \infty[$.*

REMARK 1.2. In the course of our study of (1.1), we will also see that, if $u_0 \in L^\infty(\mathbb{R}) \cap L^1(\mathbb{R})$, then the solution u to (1.1) satisfies, for all $t > 0$: $\|u(t)\|_{L^1(\mathbb{R})} \leq \|u_0\|_{L^1(\mathbb{R})}$.

We will also see that (1.1) has a L^1 contraction property: if $(u_0, v_0) \in L^\infty(\mathbb{R})$ are such that $u_0 - v_0 \in L^1(\mathbb{R})$, then, denoting by u and v the solutions to (1.1) corresponding to initial conditions u_0 and v_0, we have, for all $t > 0$: $\|u(t) - v(t)\|_{L^1(\mathbb{R})} \leq \|u_0 - v_0\|_{L^1(\mathbb{R})}$.

2. Properties of the kernel of g

Using the Fourier transform, we see that the semi-group generated by g is formally given by the convolution with the kernel (defined for $t > 0$ and $x \in \mathbb{R}$)

$$K(t, x) = \mathcal{F}^{-1}(e^{-t|\cdot|^\lambda})(x) = \int_{\mathbb{R}} e^{2i\pi x\xi} e^{-t|\xi|^\lambda} \, d\xi = \mathcal{F}(e^{-t|\cdot|^\lambda})(x).$$

The function $\xi \in \mathbb{R} \to e^{-t|\xi|^\lambda}$ being real-valued and even, K is real-valued (in the sequel, we consider only real-valued solutions to (1.1)).

The most important property of K is its nonnegativity. For the sake of completeness, we give here a sketch of the proof of this result, but notice that it is a well-known result since a rather long time now. We refer to the work of Lévy for example [Lév25]. Also notice that we study the question of the non-negativity of the kernel K because it is the issue at stake in the analysis of a maximum principle for the equation $u_t + g[u] = 0$. From this point of view, we shall make reference to the work of Courrège and coworkers (see [BCP68] and references therein) who give a characterization of a large class of pseudo-differential operators satisfying the positive maximum principle and also, more recently, to the work of Farkas, Jacob, Schilling [FJS01] (see also Hoh [Hoh95]).

LEMMA 2.1. *If $\lambda \in]0, 2]$ then, for all $(t, x) \subset]0, \infty[\times \mathbb{R}$, we have $K(t, x) \geq 0$.*

Proof of Lemma 2.1. If $\lambda = 2$, it is well-known that $K(t, x) = (\pi/t)^{1/2} e^{-\frac{\pi^2}{t}\xi^2}$, which implies the result. Assume now that $\lambda \in]0, 2[$ and let $f(x) = A|x|^{-1-\lambda} \mathbf{1}_{\mathbb{R}\setminus]-1,1[}(x)$, with $A > 0$ such that $\int_{\mathbb{R}} f(x) \, dx = 1$. Since f is even with integral equal to 1, we have

$$\mathcal{F}(f)(\xi) = 1 + \int_{\mathbb{R}} (\cos(2\pi x\xi) - 1) f(x) \, dx = 1 + A|\xi|^\lambda \int_{|y|\geq|\xi|} \frac{\cos(2\pi y) - 1}{|y|^{1+\lambda}} \, dy.$$

Since $\cos(2\pi y) - 1 = \mathcal{O}(|y|^2)$ on the neighborhood of 0 and $\lambda < 2$, the dominated convergence theorem gives

$$\int_{|y|\geq|\xi|} \frac{\cos(2\pi y) - 1}{|y|^{1+\lambda}} \, dy \to I := \int_{\mathbb{R}} \frac{\cos(2\pi y) - 1}{|y|^{1+\lambda}} \, dy < 0 \quad \text{as } \xi \to 0.$$

Hence, $\mathcal{F}(f)(\xi) = 1 - c|\xi|^\lambda(1 + \omega(\xi))$ with $c = -AI > 0$ and $\lim_{\xi \to 0} \omega(\xi) = 0$.

Define $f_n(x) = n^{1/\lambda} f * f * \cdots * f(n^{1/\lambda} x)$, the convolution product being taken n times. By the properties of the Fourier transform with respect to the convolution product, we have, for all $\xi \in \mathbb{R}$,

$$\mathcal{F}(f_n)(\xi) = (\mathcal{F}(f)(n^{-1/\lambda}\xi))^n = (1 - cn^{-1}|\xi|^\lambda(1 + \omega(n^{-1/\lambda}\xi)))^n \to e^{-c|\xi|^\lambda}$$

as $n \to \infty$. Since $(\mathcal{F}(f_n))_{n\geq 1}$ is bounded by 1 (the L^1-norm of f_n for all $n \geq 1$), this convergence is also true in $\mathcal{S}'(\mathbb{R})$ and, taking the inverse Fourier transform, we see that $f_n \to \mathcal{F}^{-1}(e^{-c|\cdot|^\lambda}) = K(c, \cdot)$ in $\mathcal{S}'(\mathbb{R})$ as $n \to \infty$. f_n being nonnegative for all n, we deduce that K is nonnegative on $\{c\} \times \mathbb{R}$; the homogeneity property (2.1) below concludes then the proof of the lemma.

Here are some other important properties of K:

$$\forall (t, x) \in]0, \infty[\times\mathbb{R}, \quad K(t, x) = \frac{1}{t^{1/\lambda}} K\left(1, \frac{x}{t^{1/\lambda}}\right). \tag{2.1}$$

K is C^∞ on $]0, \infty[\times\mathbb{R}$ and, for all $m \geq 0$, there exists B_m such that

$$\forall (t, x) \in]0, \infty[\times\mathbb{R}, \quad |\partial_x^m K(t, x)| \leq \frac{1}{t^{(1+m)/\lambda}} \frac{B_m}{(1 + t^{-2/\lambda}|x|^2)}. \tag{2.2}$$

$(K(t, \cdot))_{t>0}$ is, as $t \to 0$, an approximate unit

$$\text{(in particular, } \|K(t, \cdot)\|_{L^1(\mathbb{R})} = 1 \text{ for all } t > 0). \tag{2.3}$$

$$\exists \mathcal{K}_1 \text{ such that, for all } t > 0, \ \|\partial_x K(t, \cdot)\|_{L^1(\mathbb{R})} = \mathcal{K}_1 t^{-1/\lambda}. \tag{2.4}$$

$$\forall (a, b) \in]0, \infty[, \ K(a, \cdot) * K(b, \cdot) = K(a + b, \cdot)$$

$$\text{and } K(a, \cdot) * \partial_x K(b, \cdot) = \partial_x K(a + b, \cdot). \tag{2.5}$$

Proof of these properties

Equation (2.1) is obtained thanks to the change of variable $\xi = t^{-1/\lambda}\eta$ in the integral defining K.

The regularity of K is an immediate application of the theorem of derivation under the integral sign. To prove the second part of (2.2), we write $\partial_x^m K(1, x) = \int_{\mathbb{R}} (2i\pi\xi)^m e^{-|\xi|^\lambda} e^{2i\pi x\xi} \, d\xi$; since $\lambda > 1$, the first two derivatives of $\xi \to \xi^m e^{-|\xi|^\lambda}$ are integrable on $\mathbb{R}$ and we can make two integrations by parts to obtain $\partial_x^m K(1, x) = \mathcal{O}(1/x^2)$ on $\mathbb{R}$; $\partial_x^m K(1, \cdot)$ being bounded on $\mathbb{R}$, we deduce the estimate of (2.2) for $t = 1$; the general case $t > 0$ comes from the case $t = 1$ and (2.1).

Since $K(1, \cdot) \geq 0$, we have $\|K(1, \cdot)\|_{L^1(\mathbb{R})} = \int_{\mathbb{R}} K(1, x) \, dx = \mathcal{F}(K(1, \cdot))(0) = e^{-|0|^\lambda} = 1$ and (2.3) is thus a consequence of (2.1).

The estimate (2.4) comes from the derivation of (2.1) and from the change of variable $y = t^{-1/\lambda}x$ in the computation of $\|\partial_x K(1, \cdot/t^{1/\lambda})\|_{L^1(\mathbb{R})}$.

The identity (2.5), which translates the fact that the convolution with $K(t)$ is the semigroup generated by g, can be directly checked via Fourier transform.

Let us also give some continuity results related to K.

LEMMA 2.2. i) *If $u_0 \in L^1(\mathbb{R})$, then $t \in [0, \infty[\to K(t, \cdot) * u_0$ is continuous $[0, \infty[\to L^1(\mathbb{R})$ (with value u_0 at $t = 0$).*

ii) *Let $T > 0$ and $(t_0, x_0) \in]0, T[\times \mathbb{R}$. If $v \in C_b(]0, T[\times \mathbb{R})$, then*

a) *for all $s_0 > 0$, $K(s, \cdot) * v(t, \cdot)(x) \to K(s_0, \cdot) * v(t_0, \cdot)(x_0)$ as $s \to s_0$, $t \to t_0$ and $x \to x_0$,*

b) *$K(s, \cdot) * v(t, \cdot)(x) \to v(t_0, x_0)$ as $s \to 0$, $t \to t_0$ and $x \to x_0$.*

All these properties are either classical results of approximate units or consequences of the estimate in (2.2) (with $m = 0$) and of the dominated convergence theorem. We do not give a precise proof of these results.

3. Definition and first properties of the solutions

The idea, to study (1.1), is to search for a solution to $\partial_t u + g[u] = -\partial_x(f(u))$ using Duhamel's formula: a solution to this equation is formally given by $u(t, x) = K(t) * u_0(x) - \int_0^t K(t - s) * \partial_x(f(u(s, \cdot)))(x)\, ds$. By putting the derivative of $f(u)$ on K, we are led to the following definition.

DEFINITION 3.1. Let $u_0 \in L^\infty(\mathbb{R})$ and $T > 0$ or $T = \infty$. A solution to (1.1) on $]0, T[$ is a function $u \in L^\infty(]0, T[\times \mathbb{R})$ which satisfies, for a.e. $(t, x) \in]0, T[\times \mathbb{R}$,

$$u(t, x) = K(t, \cdot) * u_0(x) - \int_0^t \partial_x K(t - s, \cdot) * f(u(s, \cdot))(x)\, ds. \tag{3.1}$$

The following proposition shows that all the terms in (3.1) are well-defined.

PROPOSITION 3.2. *Let $u_0 \in L^\infty(\mathbb{R})$ and $T > 0$. If $v \in L^\infty(]0, T[\times \mathbb{R})$, then*

$$u : (t, x) \in]0, T[\times \mathbb{R} \to K(t, \cdot) * u_0(x) + \int_0^t \partial_x K(t - s, \cdot) * v(s, \cdot)\, ds$$

defines a function in $C_b(]0, T[\times \mathbb{R})$ and we have, for all $t_0 \in]0, T[$, all $x \in \mathbb{R}$ and all $t \in]0, T - t_0[$,

$$u(t_0 + t, x) = K(t, \cdot) * u(t_0, \cdot)(x) + \int_0^t \partial_x K(t - s, \cdot) * v(t_0 + s, \cdot)(x)\, ds. \tag{3.2}$$

Proof of Proposition 3.2

STEP 1. First term of u.

Since $u_0 \in L^\infty(\mathbb{R})$ and, for $t > 0$, $K(t, \cdot) \in L^1(\mathbb{R})$, $K(t, \cdot) * u_0$ is well-defined and, by Young's inequalities for the convolution and (2.3), we have

$$\forall(t, x) \in]0, \infty[\times \mathbb{R}, \ |K(t, \cdot) * u_0(x)| \le \|u_0\|_{L^\infty(\mathbb{R})}. \tag{3.3}$$

Let $t_0 \in]0, T[$ and $x_0 \in \mathbb{R}$; for all $0 < t_0 < T < \infty$, by (2.2), we can write

$$\forall (t, x, y) \in]t_0, T[\times \mathbb{R} \times \mathbb{R}, \quad |K(t, x - y)| \leq \frac{C_1}{C_2 + |x - y|^2} \tag{3.4}$$

where $C_1 > 0$ and $C_2 > 0$ only depend on (t_0, T). We have $|y - x_0|^2 \leq 2|y - x|^2 + 2|x_0 - x|^2$, so that $|y - x|^2 \geq \frac{1}{2}|y - x_0|^2 - |x_0 - x|^2$. For all $x \in \mathbb{R}$ such that $|x - x_0|^2 \leq C_2/2$, for all $t \in]t_0, T[$ and all $y \in \mathbb{R}$, (3.4) gives

$$|K(t, x - y)| \leq \frac{C_1}{C_2 + \frac{1}{2}|x_0 - y|^2 - |x_0 - x|^2} \leq \frac{C_1}{(C_2/2) + \frac{1}{2}|x_0 - y|^2} = F(y)$$

with $F \in L^1(\mathbb{R})$. Since u_0 is bounded and K is continuous, the theorem of continuity under the integral sign gives the continuity of $(t, x) \to K(t, \cdot) * u_0(x)$.

STEP 2. The second term of u.

Define $G : \mathbb{R} \times \mathbb{R} \to \mathbb{R}$ and $H : \mathbb{R} \times \mathbb{R} \to \mathbb{R}$ by: for all $x \in \mathbb{R}$,

$$G(t, x) = \partial_x K(t, x) \mathbf{1}_{]0, T[}(t) \text{ if } t > 0, \quad G(t, x) = 0 \text{ if } t \leq 0,$$
$$H(t, x) = v(t, x) \text{ if } t \in]0, T[, \quad H(t, x) = 0 \text{ if } t \in \mathbb{R} \backslash]0, T[.$$

We notice that $G \in L^1(\mathbb{R} \times \mathbb{R})$; indeed, by Fubini-Tonelli's theorem and (2.4),

$$\int_{\mathbb{R} \times \mathbb{R}} |G(t, x)| \, dx dt \leq \mathcal{K}_1 \int_0^T t^{-1/\lambda} \, dt = \frac{\lambda \mathcal{K}_1}{\lambda - 1} T^{1 - \frac{1}{\lambda}} < \infty. \tag{3.5}$$

The function H is clearly in $L^\infty(\mathbb{R} \times \mathbb{R})$, being bounded by $\|v\|_{L^\infty(]0, T[\times \mathbb{R})}$. Thus, denoting by $\star$ the convolution in $\mathbb{R} \times \mathbb{R}$, $G \star H$ is well-defined, bounded and uniformly continuous on $\mathbb{R} \times \mathbb{R}$; moreover, by (3.5),

$$\|G \star H\|_{C_b(\mathbb{R} \times \mathbb{R})} \leq \|G\|_{L^1(\mathbb{R} \times \mathbb{R})} \|H\|_{L^\infty(\mathbb{R} \times \mathbb{R})} \leq \frac{\lambda \mathcal{K}_1}{\lambda - 1} T^{1 - \frac{1}{\lambda}} \|v\|_{L^\infty(]0, T[\times \mathbb{R})}. \tag{3.6}$$

By Fubini's theorem, one checks that, for all $t \in]0, T[$ and all $x \in \mathbb{R}$, $G \star H(t, x) = \int_0^t \partial_x K(t - s, \cdot) * v(s, \cdot)(x) \, ds$, and the second term of u is thus continuous and bounded on $]0, T[\times \mathbb{R}$.

We notice that, thanks to (3.3) and (3.6),

$$\|u\|_{C_b(]0, T[\times \mathbb{R})} \leq \|u_0\|_{L^\infty(\mathbb{R})} + \frac{\lambda \mathcal{K}_1}{\lambda - 1} T^{1 - \frac{1}{\lambda}} \|v\|_{L^\infty(]0, T[\times \mathbb{R})}. \tag{3.7}$$

STEP 3. To prove (3.2), we make the change of variable $\tau = t_0 + s$ in the last term of this equation, we use Fubini's theorem (thanks to (2.4)) to permute the convolution by $K(t, \cdot)$ and the integral sign in $u(t_0, \cdot)$ and we apply (2.5).

As an immediate consequence of this proposition, we have:

COROLLARY 3.3. *Let $u_0 \in L^\infty(\mathbb{R})$ and $T > 0$ or $T = \infty$. If u is a solution to (1.1) on $]0, T[$, then $u \in C_b(]0, T[\times\mathbb{R})$ and u satisfies (3.1) for all $(t, x) \in]0, T[\times\mathbb{R}$. Moreover, for all $t_0 \in]0, T[$ and all $(t, x) \in]0, T - t_0[\times\mathbb{R}$,*

$$u(t_0 + t, x) = K(t, \cdot) * u(t_0, \cdot)(x) - \int_0^t \partial_x K(t - s, \cdot) * f(u(t_0 + s, \cdot))(x) \, ds, \qquad (3.8)$$

i.e. $u(t_0 + \cdot, \cdot)$ is a solution to (1.1) on $]0, T - t_0[$ with $u(t_0, \cdot)$ instead of u_0.

To conclude this study of the first properties of the solutions, we prove item iv) of Theorem 1.1.

Proof of item iv) in Theorem 1.1

Suppose that u is a solution to (1.1) on $]0, T[$. Since $f(u)$ is bounded, we have, for all $(t, x) \in]0, T[\times\mathbb{R}$, by (2.4),

$$\left| \int_0^t \partial_x K(t - s, \cdot) * f(u(s, \cdot))(x) \, ds \right| \leq \mathcal{K}_1 \|f(u)\|_\infty \int_0^t \frac{1}{(t - s)^{1/\lambda}} \, ds = Ct^{1 - \frac{1}{\lambda}}$$

where C does not depend on t; hence, the last term of (3.1) tends to 0 in $L^\infty(\mathbb{R})$ as $t \to 0$. By classical properties of the approximate units, the first term in the right-hand side of (3.1) converges as wanted to u_0 and the proof is complete.

4. Uniqueness of the solution

THEOREM 4.1. *Let $u_0 \in L^\infty(\mathbb{R})$ and $T > 0$ or $T = \infty$. There exists at most one solution to (1.1) on $]0, T[$ in the sense of Definition 3.1.*

Proof of Theorem 4.1

STEP 1. We first prove a local uniqueness result. Denote by $\mathrm{Lip}_R(f)$ a lipschitz constant of f on $[-R, R]$. Let $T_1 > 0$. For all u and v solutions to (1.1) on $]0, T_1[$ bounded by R, by (3.1) and (2.4), we have

$$|u(t, x) - v(t, x)| \leq \frac{\lambda \mathcal{K}_1}{\lambda - 1} T_1^{1 - \frac{1}{\lambda}} \mathrm{Lip}_R(f) \|u - v\|_\infty = k(T_1, R) \|u - v\|_\infty.$$

There exists $T_0 > 0$ only depending on R such that, if $T_1 \leq T_0$, we have $k(T_1, R) < 1$; for $T_1 \leq T_0$, there exists therefore at most one solution to (1.1) on $]0, T_1[$ bounded by R.

STEP 2. Proof of the uniqueness result.

Let u and v be two solutions to (1.1) on $]0, T[$. Take $R = \max(\|u\|_\infty, \|v\|_\infty)$; let T_0 be given by Step 1 for R. By Step 1, since u and v are bounded by R, $u = v$ on $]0, \inf(T, T_0)[\times\mathbb{R}$.

Let $T' = \sup\{t \in]0, T[\mid u = v \text{ on }]0, t[\times\mathbb{R}\} \geq \inf(T, T_0)$, and suppose that $T' < T$. By definition of T', and since u and v are continuous on $]0, T[\times\mathbb{R}$, we have $u(T', \cdot) = v(T', \cdot)$ on $\mathbb{R}$. By Corollary 3.1, $u(T' + \cdot, \cdot)$ and $v(T' + \cdot, \cdot)$ are two solutions to (1.1) on $]0, T - T'[$ with the same initial condition $u(T', \cdot) = v(T', \cdot)$. These solutions being bounded by R, Step 1 shows that $u(T' + \cdot, \cdot) = v(T' + \cdot, \cdot)$ on $]0, \inf(T_0, T - T')[\times\mathbb{R}$, which is a contradiction with the definition of T'.

5. Regularizing effect

5.1. *Spatial regularity*

If we formally differentiate (3.1) with respect to x, we see that the spatial derivatives of u satisfy integral equations; the following theorem gives some properties on these integral equations.

PROPOSITION 5.1. *Let $M > 0$ and $F : (t, x, \zeta) \in]0, M[\times\mathbb{R} \times \mathbb{R} \to F(t, x, \zeta) \in \mathbb{R}$ be continuous; we suppose that $\partial_x F$, $\partial_\zeta F$, $\partial_\zeta \partial_x F$ and $\partial_\zeta \partial_\zeta F$ exist and are continuous on $]0, M[\times\mathbb{R} \times \mathbb{R}$; we also suppose that there exists $\omega :]0, \infty[\to \mathbb{R}^+$ such that, for all $L > 0$, F and these derivatives are bounded on $]0, M[\times\mathbb{R} \times [-L, L]$ by $\omega(L)$.*

Let $R_0 > 0$ and $R = (2 + \mathcal{K}_1)R_0$. Then there exists $T_0 > 0$ only depending on (R_0, ω) such that, if $T = \inf(M, T_0)$ and $v_0 \in L^\infty(\mathbb{R})$ satisfies $\|v_0\|_{L^\infty(\mathbb{R})} \leq R_0$, there exists a unique $v \in C_b(]0, T[\times\mathbb{R})$ bounded by R and such that

$$v(t, x) = K(t, \cdot) * v_0(x) + \int_0^t \partial_x K(t - s, \cdot) * F(s, \cdot, v(s, \cdot))(x)\, ds. \qquad (5.1)$$

Moreover, $\partial_x v \in C(]0, T[\times\mathbb{R})$ and, for all $a \in]0, T[$, $\|\partial_x v\|_{C_b(]a, T[\times\mathbb{R})} \leq Ra^{-1/\lambda}$.

Proof of Proposition 5.1.

The idea is to use a fixed point theorem. Let, for $T \in]0, M[$, $E_T = \{v \in C_b(]0, T[\times\mathbb{R}) \mid \partial_x v \in C(]0, T[\times\mathbb{R}) \text{ and } t^{1/\lambda}\partial_x v \in C_b(]0, T[\times\mathbb{R})\}$, endowed with its natural norm $\|v\|_{E_T} = \|v\|_\infty + \|t^{1/\lambda}\partial_x v\|_\infty$. We define, thanks to Proposition 3.1, $\Psi_T : C_b(]0, T[\times\mathbb{R}) \to C_b(]0, T[\times\mathbb{R})$ by

$$\Psi_T(v)(t, x) = K(t, \cdot) * v_0(x) + \int_0^t \partial_x K(t - s, \cdot) * F(s, \cdot, v(s, \cdot))(x)\, ds.$$

STEP 1. The first term of $\Psi_T(v)$ belongs to E_T.

The estimate (2.2) allows to see, as in Step 1 of the proof of Proposition 3.2, that, by derivation and continuity under the integral sign, $K(t, \cdot) * v_0$ is derivable on $\mathbb{R}$ and that

$(t, x) \in]0, T[\times\mathbb{R} \to \partial_x(K(t, \cdot) * v_0)(x) = \partial_x K(t, \cdot) * v_0(x)$ is continuous. By Young's inequalities and (2.4),

$$\|\partial_x(K(t, \cdot) * v_0)\|_{C_b(\mathbb{R})} \leq \mathcal{K}_1 t^{-1/\lambda}\|v_0\|_{L^\infty(\mathbb{R})},\tag{5.2}$$

which proves that $(t, x) \in]0, T[\times\mathbb{R} \to K(t, \cdot) * v_0(x)$ belongs to E_T.

STEP 2. We prove that, if $v \in E_T$, the second term of $\Psi_T(v)$ belongs to E_T.

Define $H(t, x) = \int_0^t \partial_x K(t-s, \cdot) * F(s, \cdot, v(s, \cdot))(x)\, ds$. Let $t \in]0, T[$ and $s \in]0, t[$. The function $x \in \mathbb{R} \to F(s, x, v(s, x))$ is in $C_b^1(\mathbb{R})$. We can thus differentiate under the integral sign to see that $\partial_x K(t - s, \cdot) * F(s, \cdot, v(s, \cdot))$ is C^1 with derivative $\partial_x K(t - s, \cdot) * (\partial_x F(s, \cdot, v(s, \cdot)) + \partial_\zeta F(s, \cdot, v(s, \cdot))\partial_x v(s, \cdot))$. Moreover, for all $x \in \mathbb{R}$,

$$|\partial_x K(t - s, \cdot) * (\partial_x F(s, \cdot, v(s, \cdot)) + \partial_\zeta F(s, \cdot, v(s, \cdot))\partial_x v(s, \cdot))(x)|$$
$$\leq \frac{\mathcal{K}_1\|\partial_x F(\cdot, \cdot, v(\cdot, \cdot))\|_\infty}{(t - s)^{1/\lambda}} + \frac{\mathcal{K}_1\|\partial_\zeta F(\cdot, \cdot, v(\cdot, \cdot))\|_\infty\|v\|_{E_T}}{s^{1/\lambda}(t - s)^{1/\lambda}}.\tag{5.3}$$

This last function is integrable with respect to $s \in]0, t[$, and we can thus apply the theorem of derivation under the integral sign to see that

$$\partial_x H(t, x) = \int_0^t \partial_x K(t - s, \cdot) * (\partial_x F(s, \cdot, v(s, \cdot))$$
$$+ \partial_\zeta F(s, \cdot, v(s, \cdot))\partial_x v(s, \cdot))(x)\, ds.\tag{5.4}$$

If $\partial_x v$ was bounded, the continuity of $\partial_x H$ would be a consequence of Proposition 3.2. We thus approximate $\partial_x v$ by bounded functions to conclude. Take $0 < \delta < T$ and define $w_\delta \in L^\infty(]0, T[\times\mathbb{R})$ by

$$w_\delta(t, x) = \partial_x F(t, x, v(t, x)) + \partial_\zeta F(t, x, v(t, x))\partial_x v(t, x)\mathbf{1}_{[\delta, T[}(t).$$

Denoting $A_\delta(t, x) = \int_0^t \partial_x K(t - s, \cdot) * w_\delta(s, \cdot)(x)\, ds$, (5.4) allows to see that, for all $t_0 \in]0, T[$, $A_\delta \to \partial_x H$ uniformly on $[t_0, T[\times\mathbb{R}$ as $\delta \to 0$; since, by Proposition 3.2, A_δ is continuous on $]0, T[\times\mathbb{R}$, we deduce that $\partial_x H$ is continuous on $]0, T[\times\mathbb{R}$. Moreover, by (5.4) and (5.3) and the change of variable $s = t\tau$ in the integrals on $]0, t[$, we have, for all $(t, x) \in]0, T[\times\mathbb{R}$,

$$|\partial_x H(t, x)| \leq C_0\mathcal{K}_1(\|\partial_x F(\cdot, \cdot, v(\cdot, \cdot))\|_\infty t^{1-\frac{1}{\lambda}}$$
$$+ \|\partial_\zeta F(\cdot, \cdot, v(\cdot, \cdot))\|_\infty\|v\|_{E_T} t^{1-\frac{2}{\lambda}})\tag{5.5}$$

where $C_0 = \max(\int_0^1 (1 - \tau)^{-1/\lambda}\, d\tau, \int_0^1 \tau^{-1/\lambda}(1 - \tau)^{-1/\lambda}\, d\tau)$, which proves that $H \in E_T$.

If v is bounded by R, the properties of F along with (5.2), (5.5) and (3.7), give

$$\|\Psi_T(v)\|_{E_T} \leq \|v_0\|_{L^\infty(\mathbb{R})} + \frac{\lambda\mathcal{K}_1}{\lambda - 1}T^{1-\frac{1}{\lambda}}\omega(R)$$
$$+ \mathcal{K}_1\|v_0\|_{L^\infty(\mathbb{R})} + C_0\mathcal{K}_1\omega(R)(T + T^{1-\frac{1}{\lambda}}\|v\|_{E_T}).\tag{5.6}$$

STEP 3. Fixed point.

We take, as in the proposition, $R = (2 + \mathcal{K}_1)R_0$ and we denote, for $T > 0$, $B_T(R)$ the closed ball in E_T of center 0 and radius R. Let $T_0 > 0$ be such that

$$R_0 + \frac{\lambda \mathcal{K}_1}{\lambda - 1} T_0^{1-\frac{1}{\lambda}} \omega(R) + \mathcal{K}_1 R_0 + C_0 \mathcal{K}_1 \omega(R)(T_0 + T_0^{1-\frac{1}{\lambda}} R) \leq R \tag{5.7}$$

$$\mathcal{K}_1 \omega(R) \left(\frac{\lambda}{\lambda - 1} T_0^{1-\frac{1}{\lambda}} + \frac{\lambda}{\lambda - 1} T_0 + C_0 R T_0^{1-\frac{1}{\lambda}} + C_0 T_0^{1-\frac{1}{\lambda}} \right) < 1 \tag{5.8}$$

(by definition of R, such a T_0 exists and only depends on (R_0, ω)).

Let $T = \inf(M, T_0)$. Take $v_0 \in L^\infty(\mathbb{R})$ bounded by R_0. Thanks to (5.6) and (5.7), Ψ_T sends $B_T(R)$ into $B_T(R)$. Let $(u, v) \in B_T(R)$. u and v are bounded by R and we have thus, for all $(t, x) \in]0, T[\times \mathbb{R}$, by (2.4) and the properties of F,

$$|\Psi_T(u)(t, s) - \Psi_T(v)(t, x)| \leq \mathcal{K}_1 \frac{\lambda}{\lambda - 1} T^{1-\frac{1}{\lambda}} \omega(R) \|u - v\|_\infty. \tag{5.9}$$

By (5.4) and the properties of F, we also have, for all $(t, x) \in]0, T[\times \mathbb{R}$,

$$t^{1/\lambda} |\partial_x \Psi_T(u)(t, x) - \partial_x \Psi_T(v)(t, x)|$$
$$\leq \mathcal{K}_1 \omega(R) \left(\frac{\lambda}{\lambda - 1} T + C_0 T^{1-\frac{1}{\lambda}} \|u\|_{E_T} + C_0 T^{1-\frac{1}{\lambda}} \right) \|u - v\|_{E_T}. \tag{5.10}$$

The properties (5.9), (5.10) and (5.8) ensure that Ψ_T is contracting on $B_T(R)$. Therefore, Ψ_T has a unique fixed point v in $B_T(R)$; v is a continuous and bounded solution to (5.1) such that $\partial_x v$ exists and is continuous on $]0, T[\times \mathbb{R}$. Moreover, since $v \in B_T(R)$, we have, for all $a \in]0, T[$ and all $(t, x) \in]a, T[\times \mathbb{R}$, $|\partial_x v(t, x)| \leq t^{-1/\lambda} \|v\|_{E_T} \leq a^{-1/\lambda} R$, which is the estimate on $\partial_x v$ stated in the proposition.

The inequalities (5.9) and (5.8) ensure that Ψ_T is contracting on the ball in $C_b(]0, T[\times \mathbb{R})$ of center 0 and radius R. Thus, Ψ_T can have only one fixed point in this ball, which is the uniqueness result of the proposition.

THEOREM 5.2. *Let $u_0 \in L^\infty(\mathbb{R})$ and $T > 0$ or $T = \infty$. If u is a solution to (1.1) on $]0, T[$ in the sense of Definition 3.1, then u is indefinitely derivable with respect to x. Moreover, for all $n \geq 0$ and all $t_0 \in]0, T[$, we have*

   i) $\partial_x^n u \in C_b(]t_0, T[\times \mathbb{R})$,
   ii) *for all $t \in]0, T - t_0[$,*

$$\partial_x^n u(t_0 + t, \cdot) = K(t, \cdot) * \partial_x^n u(t_0, \cdot) - \int_0^t \partial_x K(t - s, \cdot) * \partial_x^n(f(u(t_0 + s, \cdot))) \, ds$$

   iii) *if $R \geq \|u\|_{C_b(]0, T[\times \mathbb{R})}$, there exists C only depending on (R, t_0, n) such that $\|\partial_x^n u\|_{C_b(]t_0, T[\times \mathbb{R})} \leq C$.*

Proof of Theorem 5.2

We prove, by induction on n, that : u has spatial derivatives of order up to n which are continuous and bounded by $C(R, t_0, n)$ on $]t_0, T[\times\mathbb{R}$ for all $t_0 \in]0, T[$, item ii) is satisfied on $]0, T - t_0[\times\mathbb{R}$ for all $t_0 \in]0, T[$ and

$$\partial_x^n(f(u)) = U_n + (1 - \delta_{n,0})f'(u)\partial_x^n u + \delta_{n,0}f(u), \tag{5.11}$$

where $\delta_{n,0}$ is Krönecker's symbol, $U_0 = 0$ and, if $n \geq 1$, $U_n = G_n((\partial_x^k u)_{k\leq n-1})$ with G_n regular.

The validity of the property at the rank $n = 0$ is a consequence of Corollary 3.3. We suppose the induction hypothesis true up to a rank $n \geq 0$, and we prove it for the rank $n + 1$.

Let $b_0 \in]0, T[$. Take $b \in]b_0, T[$ and define $F : (t, x, \zeta) :]0, T - b[\times\mathbb{R}\times\mathbb{R} \to -U_n(b + t, x) - (1 - \delta_{n,0})f'(u)(b + t, x)\zeta - \delta_{n,0}f(\zeta)$. The function F satisfies the hypotheses of Proposition 5.1, with (by induction hypothesis) ω only depending on (R, b_0, n); we also have $\|\partial_x^n u\|_{C_b(]b_0, T[\times\mathbb{R})} \leq R_0$ where R_0 only depends on (R, b_0, n). Let T_0 only depending on (R_0, ω) (i.e. on (R, b_0, n)) be given by Proposition 5.1.

By induction hypothesis, $\partial_x^n u(b + \cdot, \cdot)$ is continuous and bounded by $R_0 \leq (2 + \mathcal{K}_1)R_0$ and satisfies (5.1) on $]0, T - b[\times\mathbb{R}$ for the preceding F and with $v_0 = \partial_x^n u(b, \cdot)$ bounded by R_0. Proposition 5.1 shows thus that $\partial_x^{n+1} u$ exists and is continuous and bounded by $(2 + \mathcal{K}_1)R_0 a^{-1/\lambda}$ on $]b + a, \inf(T, b + T_0)[\times\mathbb{R}$; this is true for all $b \in]b_0, T[$ and all $a \in]0, \inf(T - b, T_0)[$. Since T_0 does not depend on a or b, taking $t_0 \in]0, T[$, $b_0 = t_0/2$ and $a = \inf(t_0/2, T_0/2) < T - b_0$, we notice that the intervals $\{]b + a, \inf(T, b + T_0)[, \; b \in]b_0, T - a[\}$ cover $]b_0 + a, T[\supset]t_0, T[$ and we deduce that $\partial_x^{n+1} u$ has the regularity and satisfies the estimates we wanted to obtain.

Let us prove the formula for $\partial_x^{n+1} u$. By induction hypothesis,

$$\partial_x^n u(t_0 + t, \cdot) = K(t, \cdot) * \partial_x^n u(t_0, \cdot) - \int_0^t \partial_x K(t - s, \cdot) * \partial_x^n(f(u(t_0 + \cdot))) \, ds. \tag{5.12}$$

But we have just proved that $\partial_x^n u(t_0, \cdot) \in C_b^1(\mathbb{R})$; thus, we can write

$$\partial_x(K(t, \cdot) * \partial_x^n u(t_0, \cdot)) = K(t, \cdot) * \partial_x^{n+1} u(t_0, \cdot). \tag{5.13}$$

The function $(t, x) \in]0, T - t_0[\times\mathbb{R} \to \partial_x^n(f(u))(t_0 + t, x)$ and its first spatial derivative are continuous and bounded on $]0, T - t_0[\times\mathbb{R}$. The reasoning of Step 2 in the proof of Proposition 5.1 (with $\partial_x^n(f(u))(t_0 + \cdot, \cdot)$ instead of $F(\cdot, \cdot, v(\cdot, \cdot))$) allows to compute the spatial derivative of the last term in (5.12) by derivation under the integral sign, and, thanks to (5.13), proves item ii) for $\partial_x^{n+1} u$.

Property (5.11) for the derivative of order $n + 1$ simply comes from the derivation of this formula at rank n, and the induction is complete.

5.2. *Temporal regularity*

5.2.1. Preliminary: about the definition of g

The operator g has been formally defined by $\mathcal{F}(g[v])(\xi) = |\xi|^\lambda \mathcal{F}(v)(\xi)$; it can be shown that this definition makes sense for bounded functions, but we will not need it and we prefer to give here a simple formula for $g[v]$ which defines this operator on $C_b^\infty(\mathbb{R})$.

PROPOSITION 5.3. *There exists* $(g_1, g_2) \in (L^1(\mathbb{R}))^2$ *such that, for all* $v \in \mathcal{S}(\mathbb{R})$, $g[v] = g_1 * v + g_2 * v^{(4)}$. *This formula allows thus to define* $g[v]$ *for* $v \in C_b^\infty(\mathbb{R})$ (*and this definition does not depend on the choice of* g_1 *and* g_2 *as above*).

Proof of Proposition 5.3

Let $\chi \in C_c^\infty(\mathbb{R})$ be even and equal to 1 on a neighborhood of 0. By linearity of $\mathcal{F}^{-1}$, if $v \in \mathcal{S}(\mathbb{R})$,

$$g[v] = \mathcal{F}^{-1}(|\cdot|^\lambda \chi \mathcal{F}(v)) + \mathcal{F}^{-1}(|\cdot|^\lambda (1 - \chi) \mathcal{F}(v)) \tag{5.14}$$

(since $\mathcal{F}(v) \in \mathcal{S}(\mathbb{R})$, all these terms are well-defined as inverse Fourier transforms of integrable functions).

Let $h_1 : \xi \in \mathbb{R} \to |\xi|^\lambda \chi(\xi)$. The function h_1 is C^1 on $\mathbb{R}$, C^2 outside 0 and its first two derivatives are integrable on $\mathbb{R}$. We deduce, as in the proof of (2.2), that $\mathcal{F}^{-1}(h_1)(x) = \mathcal{O}(1/(1 + |x|^2))$ on $\mathbb{R}$. Hence, $\mathcal{F}^{-1}(h_1) \in L^1(\mathbb{R})$ and we can write $\mathcal{F}(\mathcal{F}^{-1}(h_1) * v) = h_1 \mathcal{F}(v)$, that is to say $\mathcal{F}^{-1}(h_1 \mathcal{F}(v)) = \mathcal{F}^{-1}(h_1) * v$.

Let $h_2 : \xi \in \mathbb{R} \to |\xi|^\lambda (1 - \chi(\xi))$; the function $h_2^\star : \xi \in \mathbb{R} \to (2i\pi\xi)^{-4} h_2(\xi)$ is C^∞ and all its derivatives are integrable on $\mathbb{R}$ (the p-th derivative of $h_2^\star$ behaves, on a neighborhood of the infinity, as $|\xi|^{-4-p+\lambda}$ and $-4 + \lambda < -1$ since $\lambda \leq 2$); thus, $\mathcal{F}^{-1}(h_2^\star) \in L^1(\mathbb{R})$ and

$$\mathcal{F}(\mathcal{F}^{-1}(h_2^\star) * v^{(4)})(\xi) = h_2^\star(\xi) \mathcal{F}(v^{(4)})(\xi)$$
$$= (2i\pi\xi)^4 h_2^\star(\xi) \mathcal{F}(v)(\xi) = h_2(\xi) \mathcal{F}(v)(\xi),$$

that is to say $\mathcal{F}^{-1}(h_2 \mathcal{F}(v)) = \mathcal{F}^{-1}(h_2^\star) * v^{(4)}$.

Identity (5.14) gives therefore $g[v] = g_1 * v + g_2 * v^{(4)}$, where $g_1 = \mathcal{F}^{-1}(h_1)$ and $g_2 = \mathcal{F}^{-1}(h_2^\star)$ are integrable on $\mathbb{R}$ (notice also that, since χ is even, h_1 and $h_2^\star$ are also even and real-valued, so that g_1 and g_2 are real-valued).

To prove that, if $v \in C_b^\infty(\mathbb{R})$, the definition of $g[v]$ by this formula does not depend on the choice of g_1 and g_2, we approximate v and its derivatives by functions in $C_c^\infty(\mathbb{R})$; this will be of no use to us in the sequel, so we do not detail this step.

The following proposition is quite natural, since K is the kernel associated to g. But the reasoning followed to obtain K was formal, so we must prove this result.

PROPOSITION 5.4. *If* $v \in C_b^\infty(\mathbb{R})$ *then, for all* $x \in \mathbb{R}$, $t \in]0, \infty[\to K(t, \cdot) * v(x)$ *is* C^1. *Moreover, for all* $t > 0$ *and all* $x \in \mathbb{R}$, *we have* $\frac{d}{dt}(K(\cdot, \cdot) * v(x))(t) = -g[K(t, \cdot) * v](x)$.

Proof of Proposition 5.4

We notice that $g[K(t, \cdot) * v]$ makes sense since $K(t, \cdot) * v \in C_b^\infty(\mathbb{R})$.

Suppose first that $v \in \mathcal{S}(\mathbb{R})$. The functions $K(t, \cdot)$ and v are integrable on $\mathbb{R}$, so, by definition of K, $K(t, \cdot) * v = \mathcal{F}^{-1}(e^{-t|\cdot|^\lambda} \mathcal{F}(v))$. A derivation under the integral sign shows that $t \in]0, \infty[\to K(t, \cdot) * v(x)$ is C^1 and that, for all $t > 0$ and all $x \in \mathbb{R}$,

$$\frac{d}{dt}(K(\cdot, \cdot) * v(x))(t) = -\mathcal{F}^{-1}(|\cdot|^\lambda e^{-t|\cdot|^\lambda} \mathcal{F}(v))(x). \tag{5.15}$$

Taking g_1 and g_2 as in the proof of Proposition 5.3, we can check, since $K(t, \cdot)$ and all the derivatives of v are integrable, that $\mathcal{F}(g[K(t, \cdot) * v]) = |\cdot|^\lambda e^{-t|\cdot|^\lambda} \mathcal{F}(v)$, that is to say $g[K(t, \cdot) * v] = \mathcal{F}^{-1}(|\cdot|^\lambda e^{-t|\cdot|^\lambda} \mathcal{F}(v))$ and (5.15) concludes the proof if $v \in \mathcal{S}(\mathbb{R})$

Take now $v \in C_b^\infty(\mathbb{R})$. We can find a sequence $(v_n)_{n \geq 1} \in \mathcal{S}(\mathbb{R})$ whose derivatives are bounded in $L^\infty(\mathbb{R})$ and converge to the corresponding derivatives of v.

Let $x \in \mathbb{R}$; define $F_n : t \in]0, \infty[\to K(t, \cdot) * v_n(x)$ and $F : t \in]0, \infty[\to K(t, \cdot) * v(x)$. By the convergence of $(v_n)_{n \geq 1}$ and the dominated convergence theorem, we see that $(F_n)_{n \geq 1}$ converges to F on $]0, \infty[$ and is bounded in $L^\infty(]0, \infty[)$. Therefore, the convergence is also true in the sense of the distributions on $]0, \infty[$ and we have $F_n' \to F'$ in $\mathcal{D}'(]0, \infty[)$.

But, since $v_n \in \mathcal{S}(\mathbb{R})$, we have seen that F_n is C^1 and that $F_n'(t) = -g[K(t, \cdot) * v_n](x) = -g_1 * K(t, \cdot) * v_n(x) - g_2 * K(t, \cdot) * v_n^{(4)}(x)$. Hence, $(F_n')_{n \geq 1}$ converges to $-g_1 * K(t, \cdot) * v(x) - g_2 * K(t, \cdot) * v^{(4)}(x) = -g[K(t, \cdot) * v](x)$ and is bounded in $L^\infty(]0, \infty[)$, which proves that $F_n' \to -g[K(t, \cdot) * v](x)$ in $\mathcal{D}'(]0, \infty[)$.

Identifying the limits of the derivatives of F_n, we find $F'(t) = -g[K(t, \cdot) * v](x)$ in $\mathcal{D}'(]0, \infty[)$; since $t \in]0, \infty[\to g[K(t, \cdot) * v](x) = g_1 * K(t, \cdot) * v(x) + g_2 * K(t, \cdot) * v^{(4)}(x)$ is continuous (Proposition 3.2 with $u_0 = g_1 * v$ or $u_0 = g_2 * v^{(4)}$), we deduce that $F : t \in]0, \infty[\to K(t, \cdot) * v(x)$ is in fact C^1 on $]0, \infty[$, which concludes the proof.

5.2.2. Proof of the temporal regularity

LEMMA 5.5. *Let $u_0 \in L^\infty(\mathbb{R})$ and $T > 0$ or $T = \infty$. If u is a solution to (1.1) on $]0, T[$ in the sense of Definition 3.1, then u is derivable with respect to t and $\partial_t u + \partial_x(f(u)) + g[u] = 0$ on $]0, T[\times \mathbb{R}$.*

Proof of Lemma 5.5

We can suppose that T is finite. Let $t_0 > 0$, $t \in]t_0, T[$ and $s \in]0, t[$. Using (3.4) (and an equivalent estimate for $\partial_x K$, obtained thanks to (2.2)), since $f(u(t_0 + s, \cdot))$ is in $C_b^1(\mathbb{R})$, we see that $\partial_x K(t - s, \cdot) * f(u(t_0 + s, \cdot)) = K(t - s, \cdot) * \partial_x(f(u(t_0 + s, \cdot)))$.

Defining $v : (t, x) \in]0, T - t_0[\times \mathbb{R} \to -\partial_x(f(u))(t_0 + t, x) \in \mathbb{R}$ (which is continuous and bounded, and has all its spatial derivatives continuous and bounded—see Theorem 5.2), we write, by (3.8),

$$u(t_0 + t, x) = K(t, \cdot) * u(t_0, \cdot) + \int_0^t K(t - s, \cdot) * v(s, \cdot)(x)\, ds. \tag{5.16}$$

Since $u(t_0, \cdot) \in C_b^\infty(\mathbb{R})$, Proposition 5.4 shows that $(t, x) \in]0, T - t_0[\times \mathbb{R} \to K(t, \cdot) * u(t_0, \cdot)(x)$ is derivable with respect to t, and has $-g[K(t, \cdot) * u(t_0, \cdot)](x)$ as derivative.

Proving the derivability of the second term of the right-hand side of (5.16) is more troublesome (because $K(t - s, \cdot)$ explodes as $s \to t$). Fix $x \in \mathbb{R}$ and $\delta_0 \in]0, T - t_0[$. Let $\delta \in]0, \delta_0[$ and, for $t \in]\delta_0, T - t_0[$, $H_\delta(t) = \int_0^{t-\delta} K(t - s, \cdot) * v(s, \cdot)(x)\, ds$. Also denote $H(t) = \int_0^t K(t - s, \cdot) * v(s, \cdot)(x)\, ds$. We have $|H_\delta(t) - H(t)| \le \delta \|v\|_{C_b(]0, T - t_0[\times\mathbb{R})}\, ds$, so that $H_\delta \to H$ uniformly on $]\delta_0, T - t_0[$ as $\delta \to 0$.

The function $\phi : (t, s) \in \{(t', s') \in]\delta, T - t_0[\times]0, T - t_0[\mid s' < t' - \delta/2\} \to K(t - s, \cdot) * v(s, \cdot)(x)$ is continuous (Lemma 2.2 ii)-a)) and bounded. By Proposition 5.4, ϕ is derivable with respect to t and $\partial_t \phi(t, s) = -g[K(t - s, \cdot) * v(s, \cdot)](x) = -g_1 * K(t - s, \cdot) * v(s, \cdot)(x) - g_2 * K(t - s, \cdot) * \partial_x^4 v(s, \cdot)(x)$; this formula and Lemma 2.2 ii)-a) show that $\partial_t \phi$ is continuous and bounded (because, by continuity under the integral sign, $(s, x) \to g_1 * v(s, \cdot)(x)$ and $(s, x) \to g_2 * \partial_x^4 v(s, \cdot)(x)$ are continuous and bounded on $]0, T - t_0[\times\mathbb{R})$. These properties allow to prove that H_δ is C^1 on $]\delta_0, T - t_0[$ and that

$$H_\delta'(t) = K(\delta, \cdot) * v(t - \delta, \cdot)(x) - \int_0^{t-\delta} g[K(t - s, \cdot) * v(s, \cdot)](x)\, ds.$$

The function $(s, x) \in]0, t[\times\mathbb{R} \to g[K(t - s, \cdot) * v(s, \cdot)](x) = g_1 * K(t - s, \cdot) * v(s, \cdot)(x) + g_2 * K(t - s, \cdot) * \partial_x^4 v(s, \cdot)(x)$ is continuous and bounded (Lemma 2.2 ii)-a)); thus, by Lemma 2.2 ii)-b), we see that H_δ' converges on $]\delta_0, T - t_0[$ to

$$F : t \in]0, T - t_0[\to v(t, x) - \int_0^t g[K(t - s, \cdot) * v(s, \cdot)](x)\, ds$$

while remaining bounded in $L^\infty(]\delta_0, T - t_0[)$. Since H_δ uniformly converges on $]\delta_0, T - t_0[$ to H, we deduce that $H' = F$ in $\mathcal{D}'(]\delta_0, T - t_0[)$. Since $g[K(t - s, \cdot) * v(s, \cdot)] = g_1 * K(t - s, \cdot) * v(s, \cdot) + g_2 * K(t - s, \cdot) * \partial_x^4 v(s, \cdot)$, the same reasoning as in Step 2 of the proof of Proposition 3.2 (with K instead of $\partial_x K$ and $(t, x) \to g_1 * v(t, \cdot)(x)$ or $(t, x) \to g_2 * \partial_x^4 v(t, \cdot)(x)$ instead of v) shows that F is in fact continuous. Hence, δ_0 being arbitrary, H is C^1 on $]0, T - t_0[$ and $H' = F$.

Coming back to (5.16), we see that $u(t_0 + \cdot, \cdot)$ is derivable with respect to t on $]0, T - t_0[\times\mathbb{R}$ and that

$$\partial_t u(t_0 + t, x) = -g[K(t, \cdot) * u(t_0, \cdot)](x) - \partial_x(f(u))(t_0 + t, x)$$
$$- \int_0^t g_1 * K(t - s, \cdot) * v(s, \cdot)(x) + g_2 * K(t - s, \cdot) * \partial_x^4 v(s, \cdot)(x)\, ds. \tag{5.17}$$

The time t_0 being arbitrary, this gives the temporal derivability of u on $]0, T[\times\mathbb{R}$. We now prove that the right-hand side of (5.17) is $-\partial_x(f(u)) - g[u]$.

Let $t \in]0, T - t_0[$. By Fubini, we have

$$\int_0^t g_1 * K(t - s, \cdot) * v(s, \cdot) + g_2 * K(t - s, \cdot) * \partial_x^4 v(s, \cdot)\, ds$$

$$= g_1 * \left[\int_0^t K(t - s, \cdot) * v(s, \cdot)\, ds \right] + g_2 * \left[\int_0^t K(t - s, \cdot) * \partial_x^4 v(s, \cdot)\, ds \right]. \quad (5.18)$$

With the same reasoning as in Step 2 of the proof of Proposition 5.1 (with K instead of $\partial_x K$ and v instead of $F(\cdot, \cdot, v(\cdot, \cdot))$), we prove by induction that $(t, x) \in]0, T - t_0[\times\mathbb{R} \to \int_0^t K(t - s, \cdot) * v(s, \cdot)(x)\, ds$ is indefinitely derivable with respect to x, has all its spatial derivatives continuous and bounded on $]0, T - t_0[\times\mathbb{R}$ and satisfies, for all $m \geq 0$,

$$\partial_x^m \left(\int_0^t K(t - s, \cdot) * v(s, \cdot)\, ds \right) = \int_0^t K(t - s, \cdot) * \partial_x^m v(s, \cdot)\, ds.$$

Thus, by (5.18),

$$\int_0^t g_1 * K(t - s, \cdot) * v(s, \cdot) + g_2 * K(t - s, \cdot) * \partial_x^4 v(s, \cdot)\, ds$$

$$= g_1 * \left[\int_0^t K(t - s, \cdot) * v(s, \cdot)\, ds \right] + g_2 * \partial_x^4 \left[\int_0^t K(t - s, \cdot) * v(s, \cdot)\, ds \right]$$

$$= g \left[\int_0^t K(t - s, \cdot) * v(s, \cdot)\, ds \right].$$

This equation, combined with (5.17) and (5.16), shows that u satisfies $\partial_t u + \partial_x(f(u)) + g[u] = 0$ on $]t_0, T[\times\mathbb{R}$ for all $t_0 > 0$, which concludes the proof.

Item i) of Theorem 1.1 is a direct consequence of Theorem 5.2, Lemma 5.5 and Proposition 5.3 (as well as the theorem of continuity under the integral sign and Young's inequalities which show that, if $v \in C_b(]t_0, T[\times\mathbb{R})$ and $w \in L^1(\mathbb{R})$, then $(t, x) \in]t_0, T[\times\mathbb{R} \to w * v(t, \cdot)(x)$ is continuous and bounded).

6. L^∞ estimate and global existence

We construct here a solution to (1.1) on $]0, \infty[$ which is bounded by $\|u_0\|_{L^\infty(\mathbb{R})}$ and satisfies the maximum principle, thus concluding the proof of Theorem 1.1.

We assume, in the three following subsections, that $u_0 \in C_c^\infty(\mathbb{R})$ (in fact, we just need $u_0 \in L^1(\mathbb{R}) \cap BV(\mathbb{R})$).

6.1. *Construction of an approximate solution by a splitting method*

Let $\delta > 0$. We construct, by induction, a function $u^\delta : [0, \infty[\times\mathbb{R} \to \mathbb{R}$ the following way: we let $u^\delta(0, \cdot) = u_0$ and, for all $n \geq 0$, we define

- u^δ on $]2n\delta, (2n+1)\delta] \times \mathbb{R}$ as the solution to $\partial_t u^\delta + 2g[u^\delta] = 0$ ([1]) with initial condition $u^\delta(2n\delta, \cdot)$, that is to say $u^\delta(t, x) = K(2(t - 2n\delta), \cdot) * u^\delta(2n\delta, \cdot)(x)$ for $(t, x) \in]2n\delta, (2n+1)\delta] \times \mathbb{R}$.
- u^δ on $](2n+1)\delta, 2(n+1)\delta] \times \mathbb{R}$ as the (entropy) solution to $\partial_t u^\delta + 2\partial_x(f(u^\delta)) = 0$ with initial condition $u^\delta((2n+1)\delta, \cdot)$.

Since $\|K(t, \cdot)\|_{L^1(\mathbb{R})} = 1$ for all $t > 0$, the regularizing operator does not increase the L^∞ norm (in fact, K being nonnegative, the maximum principle is satisfied), the L^1 norm and the BV semi-norm; it is a well-known result that the hyperbolic operator has the same properties. Moreover, the solutions to both equations are continuous with values in $L^1(\mathbb{R})$ (this is what states Lemma 2.2-i) for the regularizing equation). We have therefore defined $u^\delta \in C([0, \infty[; L^1(\mathbb{R}))$ such that $u^\delta(0, \cdot) = u_0$,

$$\forall t \geq 0, \ \|u^\delta(t, \cdot)\|_{L^\infty(\mathbb{R})} \leq \|u_0\|_{L^\infty(\mathbb{R})}, \ \|u^\delta(t, \cdot)\|_{L^1(\mathbb{R})} \leq \|u_0\|_{L^1(\mathbb{R})}$$

$$\text{and } |u^\delta(t, \cdot)|_{BV(\mathbb{R})} \leq \|u_0'\|_{L^1(\mathbb{R})}, \tag{6.1}$$

(in fact, u^δ takes its values between the minimum and maximum values of u_0) and, for all $n \geq 0$,

$$u^\delta(t, \cdot) = K(2(t - 2n\delta), \cdot) * u^\delta(2n\delta, \cdot) \text{ for all } t \in]2n\delta, (2n+1)\delta],$$

$$u^\delta \text{ satisfies } \partial_t u^\delta + 2\partial_x(f(u^\delta)) = 0 \text{ on}](2n+1)\delta, 2(n+1)\delta] \times \mathbb{R}. \tag{6.2}$$

By (2.2) (which also gives, through (2.1), estimates on the time derivatives of K) and the fact that $u^\delta(2n\delta, \cdot) \in L^\infty(\mathbb{R})$, we see that, for all $n \geq 0$, u^δ is C^∞ on $]2n\delta, (2n+1)\delta] \times \mathbb{R}$. Moreover,

$$\|\partial_x u^\delta((2n+1)\delta, \cdot)\|_\infty = \|\partial_x K(2\delta, \cdot) * u^\delta(2n\delta, \cdot)\|_\infty \leq \mathcal{K}_1\|u_0\|_\infty(2\delta)^{-1/\lambda}.$$

Hence, the time of regularity of u^δ on $](2n+1)\delta, 2(n+1)\delta] \times \mathbb{R}$ is at least

$$T^* \geq \frac{1}{2\|f''(u^\delta((2n+1)\delta))\partial_x u^\delta((2n+1)\delta)\|_{L^\infty(\mathbb{R})}} \geq C_0\delta^{1/\lambda}$$

where C_0 does not depend on δ or n (we have used (6.1) to bound $f''(u^\delta((2n+1)\delta))$). For δ small enough, this time of regularity is thus greater than δ.

The parameter δ being destined to tend to 0, we can always suppose that it is small enough (let us say $\delta \leq \delta_0$) in order that u^δ is regular on $]2n\delta, (2n+1)\delta] \times \mathbb{R}$ and on $[(2n+1)\delta, 2(n+1)\delta] \times \mathbb{R}$ for all $n \geq 0$ (the BV estimate of (6.1) turns then into a L^1 estimate on the first spatial derivative).

---

[1] The factor 2 comes from the fact that we solve the regularizing equation (and the hyperbolic equation) on half of the total time, so we must give it twice more weight.

REMARK 6.1. It is also possible to construct u^δ via a classical splitting method, i.e. to solve the regularizing equation (without the factor 2) on $[k\delta, (k+1)\delta]$ and then use the value thus obtained at $t = (k+1)\delta$ to solve the hyperbolic equation (still without the factor 2) on $[k\delta, (k+1)\delta]$ once again (and not on $[(k+1)\delta, (k+2)\delta])$. All the following reasoning can be done with such a construction; however, since the function thus defined is not continuous on $[0, \infty[$, more work is to be done.

6.2. *Compactness result on the sequence $(u^\delta)_{\delta>0}$*

PROPOSITION 6.2. *For all compact subset Q of $\mathbb{R}$ and all $T > 0$, $\{u^\delta,\ \delta \in]0, \delta_0]\}$ is relatively compact in $C([0, T]; L^1(Q))$.*

Proof of Proposition 6.2

Let Q be a compact subset of $\mathbb{R}$ and $T > 0$. For all $t \in [0, T]$, we have, by (6.1), $\|u^\delta(t)\|_{L^1(\mathbb{R})\cap BV(\mathbb{R})} \leq \|u_0\|_{W^{1,1}(\mathbb{R})}$ (we omit the space variable in u^δ); thus, by Helly's Theorem, $\{u^\delta(t, \cdot),\ \delta \in]0, \delta_0]\}$ is relatively compact in $L^1(Q)$

We will prove the equicontinuity of $\{u^\delta,\ \delta \in]0, \delta_0]\}$ in $C([0, \infty[; L^1(\mathbb{R}))$; this implies the equicontinuity in $C([0, T]; L^1(Q))$ and, thanks to Ascoli-Arzela's theorem, concludes the proof of the proposition.

It is classical that the solution to an hyperbolic equation is lipschitz-continuous $[0, \infty[\to L^1(\mathbb{R})$. Thanks to (6.1), we see that the lipschitz constant of u^δ on $[(2n+1)\delta, 2(n+1)\delta]$ does not depend on δ or $n \geq 0$: there exists C_0 such that, for all $\delta \in]0, \delta_0]$, for all $n \geq 0$ and all $(t, s) \in [(2n+1)\delta, 2(n+1)\delta]$,

$$\|u^\delta(t) - u^\delta(s)\|_{L^1(\mathbb{R})} \leq C_0|t - s|. \tag{6.3}$$

Taking into account that $u^\delta(s, \cdot) \in W^{1,1}(\mathbb{R})$ and the estimates of (6.1), some classical cuttings of integrals involving approximate units give, for all $\delta \in]0, \delta_0]$, all $t > 0$, all $s \geq 0$ and all $\eta > 0$,

$$\|K(t) * u^\delta(s) - u^\delta(s)\|_{L^1(\mathbb{R})} \leq 2\|u_0\|_{L^1(\mathbb{R})} \int_{|y|\geq\eta} K(t, y)\, dy + \eta\|u_0'\|_{L^1(\mathbb{R})}. \tag{6.4}$$

Let us now prove the equicontinuity of $\{u^\delta,\ \delta \in]0, \delta_0]\}$ in $C([0, \infty[; L^1(\mathbb{R}))$. Let $\delta \in]0, \delta_0]$ and $0 \leq t < s$. Let $p \leq q$ be integers such that $p\delta \leq t < (p+1)\delta$ and $q\delta \leq s < (q+1)\delta$; because of the different behaviours of u^δ (see (6.2)), we must separate the cases depending on the parity of p and q; since all these cases are similar, we study only one, for example p even and q odd.

The idea, to estimate $\|u^\delta(s, \cdot) - u^\delta(t, \cdot)\|_{L^1(\mathbb{R})}$ is to go from $u^\delta(q\delta)$ to $u^\delta((p+1)\delta)$ by the following technique: on the intervals where u^δ satisfies the regularizing equation, we use the formula $u^\delta(k\delta) = K(2\delta) * u^\delta((k-1)\delta)$ (hence for k odd) and, on the intervals where u^δ

satisfies the hyperbolic problem, we write $u^\delta(k\delta) = u^\delta((k-1)\delta) + (u^\delta(k\delta) - u^\delta((k-1)\delta))$ (k even), the second term being estimated by (6.3).

Applying this idea, using the semi-group property of the convolution by $K(t)$ and recalling that q is odd in our example, an induction allows to see that, for all $l \in [0, (q-1)/2]$,

$$u^\delta(q\delta) = K(2l\delta) * u^\delta((q-2l)\delta)$$

$$+ \sum_{j=1}^{l} K(2j\delta) * (u^\delta((q-2j+1)\delta) - u^\delta((q-2j)\delta))$$

(if $l = 0$, $\sum_{j=1}^{l}(...)$ is null and $K(2l\delta) * u^\delta((q-2l)\delta)$ is replaced by $u^\delta(q\delta)$). Taking $l = (q-p-1)/2 \in [0, (q-1)/2]$ (recall that q is odd and p is even and inferior to q, thus $p+1 \le q$) in this formula, we obtain

$$u^\delta(s) = u^\delta(s) - u^\delta(q\delta) + K((q-p-1)\delta) * u^\delta((p+1)\delta)$$

$$+ \sum_{j=1}^{(q-p-1)/2} K(2j\delta) * (u^\delta((q-2j+1)\delta) - u^\delta((q-2j)\delta)).$$

Since p is even, by definition of u^δ on $]p\delta, (p+1)\delta]$ and (2.5), we have $u^\delta((p+1)\delta) = K(2((p+1)\delta - t)) * (K(2(t-p\delta)) * u^\delta(p\delta)) = K(2((p+1)\delta - t)) * u^\delta(t)$. We can therefore write

$$u^\delta(s) - u^\delta(t) = u^\delta(s) - u^\delta(q\delta)$$

$$+ K((q-p-1)\delta + 2((p+1)\delta - t)) * u^\delta(t) - u^\delta(t)$$

$$+ \sum_{j=1}^{(q-p-1)/2} K(2j\delta) * (u^\delta((q-2j+1)\delta) - u^\delta((q-2j)\delta)).$$

On $[q\delta, s] \subset [q\delta, (q+1)\delta]$ and each $[(q-2j)\delta, (q-2j+1)\delta]$ for $j \in [1, (q-p-1)/2]$, u^δ satisfies the hyperbolic problem; thus, by (6.3) and (6.4), we have, for all $\eta > 0$,

$$\|u^\delta(s) - u^\delta(t)\|_{L^1(\mathbb{R})} \le C_0|s - q\delta| + \frac{q-p-1}{2} C_0\delta$$

$$+ 2\|u_0\|_{L^1(\mathbb{R})} \int_{|y|\ge\eta} K((q-p-1)\delta + 2((p+1)\delta - t), y)\, dy + \eta\|u_0'\|_{L^1(\mathbb{R})}.$$

But, since $t < (p+1)\delta \le q\delta \le s$, we have $(q-p-1)\delta = q\delta - (p+1)\delta \le s - t$, $2((p+1)\delta - t) \le 2(s-t)$ and $s - q\delta \le s - t$. Using these bounds in the preceding inequality, we obtain, for all $\delta \in]0, \delta_0]$, for all $0 \le t < s$ and for all $\eta > 0$,

$$\|u^\delta(s) - u^\delta(t)\|_{L^1(\mathbb{R})} \le \frac{3C_0}{2}|s - t|$$

$$+ 2\|u_0\|_{L^1(\mathbb{R})} \sup_{\tau \in]0, 3|s-t|]} \int_{|y|\ge\eta} K(\tau, y)\, dy + \eta\|u_0'\|_{L^1(\mathbb{R})} \tag{6.5}$$

(the same kind of formula can be obtained in the cases where p and q have other parities than the ones considered here).

Since, for all $\eta > 0$, $\sup_{\tau \in]0, 3|s-t|]} \int_{|y| \geq \eta} K(\tau, y)\, dy \to 0$ as $|s - t| \to 0$ (property of an approximate unit), (6.5) gives the desired equicontinuity and concludes the proof of Proposition 6.2.

6.3. *Passing to the limit $\delta \to 0$*

By Proposition 6.2, we can suppose, up to a subsequence, that, for all $T > 0$ and all Q compact subset of $\mathbb{R}$, $u^\delta \to u$ in $C([0, T]; L^1(Q))$ as $\delta \to 0$. For all $t \geq 0$, $u^\delta(t) \to u(t)$ in $L^1_{\mathrm{loc}}(\mathbb{R})$, hence almost everywhere on $\mathbb{R}$ up to a subsequence. We deduce thus from (6.1) and Fatou's lemma that, for all $t > 0$, $\|u(t)\|_{L^1(\mathbb{R})} \leq \|u_0\|_{L^1(\mathbb{R})}$, that

$$\forall t \geq 0, \quad \|u(t)\|_{L^\infty(\mathbb{R})} \leq \|u_0\|_{L^\infty(\mathbb{R})} \tag{6.6}$$

and that, as u^δ, the function u takes its values between the minimum and maximum values of u_0. Still using Fatou's lemma on subsequences (depending on s and t), we see that (6.5) is satisfied for all $(s, t) \in [0, \infty[$ with u instead of u^δ; hence, $u \in C([0, \infty[; L^1(\mathbb{R}))$ and, since $u^\delta(0) = u_0$ for all $\delta > 0$, we have $u(0) = u_0$.

We now show that u satisfies (1.1) if we use a formulation involving test functions.

PROPOSITION 6.3. *For all $\gamma \in C_c^\infty(]0, \infty[)$ and all $\varphi \in \mathcal{S}(\mathbb{R})$, we have*

$$\int_{\mathbb{R}^+ \times \mathbb{R}} u(t, x)\gamma'(t)\varphi(x) + f(u(t, x))\gamma(t)\varphi'(x) - u(t, x)\gamma(t)g[\varphi](x)\, dxdt = 0. \tag{6.7}$$

Proof of Proposition 6.3

Let $\delta \in]0, \delta_0]$. If p is an odd integer, u^δ is a regular solution to $\partial_t u^\delta + 2\partial_x(f(u^\delta)) = 0$ on $[p\delta, (p+1)\delta] \times \mathbb{R}$. Multiplying this equation by $\gamma(t)\varphi(x)$ and integrating by parts (recall that u^δ is bounded and that $\varphi \in \mathcal{S}(\mathbb{R})$), we find

$$0 = -\int_{p\delta}^{(p+1)\delta} \int_{\mathbb{R}} u^\delta(t, x)\gamma'(t)\varphi(x) + 2f(u^\delta(t, x))\gamma(t)\varphi'(x)\, dtdx$$

$$+ \int_{\mathbb{R}} u^\delta((p+1)\delta, x)\gamma((p+1)\delta)\varphi(x)\, dx - \int_{\mathbb{R}} u^\delta(p\delta, x)\gamma(p\delta)\varphi(x)\, dx. \tag{6.8}$$

If p is an even integer, then $u^\delta(t) = K(2(t - p\delta)) * u^\delta(p\delta)$ on $]p\delta, (p+1)\delta] \times \mathbb{R}$. Since $u^\delta(t) \in L^1(\mathbb{R})$ for all $t \geq 0$, Fubini's theorem allows to write

$$\mathcal{F}^{-1}(u^\delta(t)) = \mathcal{F}^{-1}(K(2(t - p\delta)))\mathcal{F}^{-1}(u^\delta(p\delta)) = e^{-2(t-p\delta)|\cdot|^\lambda}\mathcal{F}^{-1}(u^\delta(p\delta)).$$

Writing $\varphi = \mathcal{F}^{-1}(\mathcal{F}(\varphi))$ and $g[\varphi] = \mathcal{F}^{-1}(|\cdot|^\lambda \mathcal{F}(\varphi))$, since $u^\delta(t) \in L^1(\mathbb{R})$ for all $t \geq 0$, by Fubini's theorem we can put the inverse Fourier transform on u^δ and we thus check that

$$\int_{p\delta}^{(p+1)\delta} \int_{\mathbb{R}} u^\delta(t, x)\gamma'(t)\varphi(x) - 2u^\delta(t, x)\gamma(t)g[\varphi](x)\, dxdt$$

$$= \int_{\mathbb{R}} u^\delta((p+1)\delta, x)\gamma((p+1)\delta)\varphi(x) - u^\delta(p\delta, x)\gamma(p\delta)\varphi(x)\, dx. \tag{6.9}$$

Summing (6.8) on all odd integers p and (6.9) on all even integers p (notice that, since the support of γ is compact, these sums are finite), the boundary terms disappear (even for $p = 0$ since $\gamma(0) = 0$) and we find

$$\int_{\mathbb{R}^+} \int_{\mathbb{R}} u^\delta(t, x)\varphi(x)\gamma'(t)\, dxdt$$

$$+ \int_{\mathbb{R}^+} \int_{\mathbb{R}} 2f(u^\delta(t, x))\varphi'(x)\gamma(t)(1 - \chi_\delta(t))\, dxdt$$

$$- \int_{\mathbb{R}^+} \int_{\mathbb{R}} 2u^\delta(t, x)g[\varphi](x)\gamma(t)\chi_\delta(t)\, dxdt = 0 \tag{6.10}$$

where χ_δ is the characteristic function of $\cup_{\text{even } p} \,]p\delta, (p+1)\delta]$.

Taking $T \geq \max(\text{supp}(\gamma))$, we have, for all $A \geq 0$, thanks to (6.1) and (6.6)

$$\left| \int_{\mathbb{R}^+} \int_{\mathbb{R}} u^\delta(t, x)g[\varphi](x)\gamma(t)\, 2\chi_\delta(t)\, dxdt - \int_{\mathbb{R}^+} \int_{\mathbb{R}} u(t, x)g[\varphi](x)\gamma(t)\, dxdt \right|$$

$$\leq \left| \int_0^T \int_{-A}^A (2\chi_\delta(t)u^\delta(t, x) - u(t, x))g[\varphi](x)\gamma(t)\, dx\, dt \right| \tag{6.11}$$

$$+ 3\|u_0\|_{L^\infty(\mathbb{R})} T \|\gamma\|_{L^\infty(\mathbb{R}^+)} \int_{\mathbb{R}\backslash[-A,A]} |g[\varphi](x)|\, dx \tag{6.12}$$

Since $g[\varphi]$ is bounded on $\mathbb{R}$, $u^\delta \to u$ in $C([0, T]; L^1([-A, A]))$ and $\chi_\delta \to 1/2$ in $L^\infty(]0, \infty[)$ weak-$*$ as $\delta \to 0$, we see that (6.11) tends to 0 as $\delta \to 0$. By Proposition 5.3, we have $g[\varphi] = g_1 * \varphi + g_2 * \varphi^{(4)} \in L^1(\mathbb{R})$; hence, (6.12) tends to 0 as $A \to \infty$. We deduce thus that, as $\delta \to 0$,

$$\int_{\mathbb{R}^+} \int_{\mathbb{R}} u^\delta(t, x)g[\varphi](x)\gamma(t)\, 2\chi_\delta(t)\, dxdt \to \int_{\mathbb{R}^+} \int_{\mathbb{R}} u(t, x)g[\varphi](x)\gamma(t)\, dxdt.$$

The flux function f is lipschitz-continuous on $[-\|u_0\|_{L^\infty(\mathbb{R})}, \|u_0\|_{L^\infty(\mathbb{R})}]$ so that, by (6.1) and (6.6), $f(u^\delta) \to f(u)$ in $C([0, T]; L^1(Q))$ for all $T > 0$ and all compact subset Q of $\mathbb{R}$. Therefore, with the same kind of reasoning as before, we can pass to the limit $\delta \to 0$ in (6.10) to conclude that u satisfies (6.7).

We now prove that u is in fact a solution to (1.1) in the sense of Definition 3.1.

Recall that $u \in C([0, \infty[; L^1(\mathbb{R}))$. By (6.6) and the local lipschitz-continuity of f, we have $f(u) \in C([0, \infty[; L^1(\mathbb{R}))$ (recall that $f(0) = 0$). We deduce, since $\mathcal{F}^{-1} :$ $L^1(\mathbb{R}) \to C_b(\mathbb{R})$ is continuous, that $t \to \mathcal{F}^{-1}(u(t))$ and $t \to \mathcal{F}^{-1}(f(u(t)))$ are in $C([0, \infty[; C_b(\mathbb{R})) \subset C([0, \infty[\times\mathbb{R})$. Hence, for all $\gamma \in C_c^\infty(]0, \infty[)$, the function

$$w(\xi) = \int_{\mathbb{R}^+} \mathcal{F}^{-1}(u(t))(\xi)\gamma'(t) + 2i\pi\xi\mathcal{F}^{-1}(f(u(t)))(\xi)\gamma(t)$$
$$-\mathcal{F}^{-1}(u(t))(\xi)|\xi|^\lambda\gamma(t)\,dt,$$

is continuous on $\mathbb{R}$. Let $\psi \in C_c^\infty(\mathbb{R})$; applying (6.7) with $\varphi = \mathcal{F}^{-1}(\psi) \in \mathcal{S}(\mathbb{R})$ and using Fubini's theorem, we have $\int_\mathbb{R} w\psi = 0$; the function ψ being arbitrary, this implies $w \equiv 0$. Since this is true for all $\gamma \in C_c^\infty(]0, \infty[)$, we deduce that, for all $\xi \in \mathbb{R}$, $\frac{d}{dt}(\mathcal{F}^{-1}(u(\cdot))(\xi)) = -|\xi|^\lambda\mathcal{F}^{-1}(u(\cdot))(\xi)+2i\pi\xi\mathcal{F}^{-1}(f(u(\cdot)))(\xi)$ in $\mathcal{D}'(]0, \infty[)$. The right-hand side of this equation is a continuous function, and the equation is therefore a classical ODE; thus, for all $\xi \in \mathbb{R}$ and all $t \geq 0$,

$$\begin{aligned}
\mathcal{F}^{-1}(u(t))(\xi) &= e^{-t|\xi|^\lambda}\mathcal{F}^{-1}(u_0)(\xi) + \int_0^t 2i\pi\xi e^{-(t-s)|\xi|^\lambda}\mathcal{F}^{-1}(f(u(s)))(\xi)\,ds \\
&= \mathcal{F}^{-1}(K(t) * u_0)(\xi) - \int_0^t \mathcal{F}^{-1}(\partial_x K(t - s))\mathcal{F}^{-1}(f(u(s)))(\xi)\,ds \\
&= \mathcal{F}^{-1}(K(t) * u_0)(\xi) - \int_0^t \mathcal{F}^{-1}(\partial_x K(t - s) * f(u(s)))(\xi)\,ds
\end{aligned}$$

By (2.4) and since $f(u) \in C([0, \infty[; L^1(\mathbb{R}))$, Fubini's theorem gives then

$$\mathcal{F}^{-1}(u(t))(\xi) = \mathcal{F}^{-1}(K(t) * u_0)(\xi) - \mathcal{F}^{-1}\left(\int_0^t \partial_x K(t - s) * f(u(s))\,ds\right)(\xi).$$

$\mathcal{F}^{-1}$ being injective on $L^1(\mathbb{R})$, we deduce that u satisfies (3.1) on $]0, \infty[\times\mathbb{R}$.

Here is a summary of what we have proved so far in this section.

PROPOSITION 6.4. *If $u_0 \in C_c^\infty(\mathbb{R})$, then there exists a solution to (1.1) on $]0, \infty[$ which is bounded by $\|u_0\|_{L^\infty(\mathbb{R})}$ and takes its values between the minimum and maximum values of u_0.*

6.4. *Conclusion*

We now prove that if $u_0 \in L^\infty(\mathbb{R})$, then there exists a solution to (1.1) on $]0, \infty[$ which satisfies item ii) in Theorem 1.1, which concludes the proof of this theorem.

Let $u_0 \in L^\infty(\mathbb{R})$ and take $(u_0^n)_{n \geq 0} \in C_c^\infty(\mathbb{R})$ which converges a.e. on $\mathbb{R}$ to u_0 and such that, for all $n \geq 1$, u_0^n takes its values between the essential lower and upper bounds of u_0; in particular, $\|u_0^n\|_{L^\infty(\mathbb{R})} \leq \|u_0\|_{L^\infty(\mathbb{R})}$ for all $n \geq 0$. Denote by u^n a solution, given by Proposition 6.4, to (1.1) on $]0, \infty[$ with initial condition u_0^n instead of u_0; u^n is bounded by $\|u_0^n\|_{L^\infty(\mathbb{R})} \leq \|u_0\|_{L^\infty(\mathbb{R})}$. This bound and Theorem 5.2 show that, for all $t_0 > 0$ and all $m \geq 0$, $(\partial_x^m u^n)_{n \geq 1}$ is bounded on $]t_0, \infty[\times\mathbb{R}$; by Lemma 5.5 and Proposition 5.3, these bounds on the spatial derivatives imply that $(\partial_t u^n)_{n \geq 1}$ is also bounded on $]t_0, \infty[\times\mathbb{R}$.

Hence, by Ascoli-Arzela's theorem, up to a subsequence, we can suppose that there exists u such that $u^n \to u$ on $]0, \infty[\times\mathbb{R}$. Since $\|u^n\|_{L^\infty(]0,\infty[\times\mathbb{R})} \leq \|u_0\|_{L^\infty(\mathbb{R})}$, we also have $\|u\|_{L^\infty(]0,\infty[\times\mathbb{R})} \leq \|u_0\|_{L^\infty(\mathbb{R})}$ and, in fact, u takes (as each u^n) its values between the essential lower and upper bounds of u_0. The function u^n satisfies (3.1) with u_0^n instead of u_0; passing to the limit $n \to \infty$ in this equation, thanks to the dominated convergence theorem, we see that u is a solution to (1.1) on $]0, \infty[$.

REMARK 6.5. Since both the hyperbolic and regularizing equations satisfy the properties given in Remark 1.2, it is quite obvious, on our construction of a solution to (1.1), that (1.1) also satisfies the properties stated in Remark 1.2 (because we can always choose approximations of the initial conditions by regular data which satisfy the hypotheses of this remark).

REFERENCES

[BCP68]   BONY, J-M., COURRÈGE, P. and PRIOURET, P., *Semi-groupes de Feller sur une variété à bord compacte et problèmes aux limites intégro-différentiels du second ordre donnant lieu au principe du maximum*, Ann. Inst. Fourier (Grenoble) *18* (1968), no. fasc. 2, (1969), 369–521.

[CD01]   CLAVIN, P. and DENET, B., *Theory of cellular detonations in gases. part 2. mach-stem formation at strong overdrive*, C. R. Acad. Sci. Paris *t. 329* (2001), no. Série II b, 489–496.

[CD02]   CLAVIN, P. and DENET, B., *Diamond patterns in the cellular front of an overdriven detonation*, Phys. Rev. Letters *88* (2002).

[CH01]   CLAVIN, P. and HE, L., *Theory of cellular detonations in gases. part 1. stability limits at strong overdrive*, C. R. Acad. Sci. Paris *t. 329* (2001), no. Série II b, 463–471.

[FJS01]   FARKAS, W., JACOB, N. and SCHILLING, R. L., *Feller semigroups, L^p-sub-Markovian semigroups, and applications to pseudo-differential operators with negative definite symbols*, Forum Math. *13* (2001) no. 1, 51–90.

[Hoh95]   HOH, W., *Pseudodifferential operators with negative definite symbols and the martingale problem*, Stochastics Stochastics Rep. *55* (1995) no. 3–4, 225–252.

[Kru70]   KRUZHKOV, S. N., *First order quasilinear equations with several independent variables.*, Mat. Sb. (N.S.) *81 (123)* (1970), 228–255.

[Lév25]   LÉVY, P., *Calcul des Probabilités*, 1925.

[Vol67]   VOL'PERT, A. I., *Spaces* bv *and quasilinear equations*, Mat. Sb. (N.S.) *73* (*115*) (1967), 255–302.

J. Droniou
Département de Mathématiques
CC 051, Université Montpellier II
Place Eugène Bataillon
34095 Montpellier cedex 5
France
e-mail: droniou@math.univ-montp2.fr

T. Gallouët and J. Vovelle
CMI, Université de Provence
39 rue Joliot-Curie
13453 Marseille cedex 13
France
e-mail: gallouet@cmi.univ-mrs.fr
       vovelle@cmi.univ-mrs.fr

To access this journal online:
http://www.birkhauser.ch

J.evol.equ. 3 (2004) 523 – 548
1424–3199/03/040523 – 26
DOI 10.1007/s00030-003-0085-z
© Birkhäuser Verlag, Basel, 2004

Journal of Evolution Equations

Convergence to equilibrium for a parabolic problem with mixed boundary conditions in one space dimension

MARIA GOKIELI and FRÉDÉRIQUE SIMONDON

Dedicated to the memory of Philippe Bénilan

Abstract. We prove that any bounded non-negative solution of a degenerate parabolic problem with Neumann or mixed boundary conditions converges to a stationary solution.

1. Introduction.

We consider the following parabolic problem

$$
\text{(P)} \quad
\begin{cases}
u_t - \varphi(u)_{xx} = f(u) & \text{in } (0, \infty) \times (-L, L) \\
u(0, \cdot) = u_0 & \text{in } (-L, L) \\
\varphi(u)_x(t, L) + \beta(L)\,\varphi(u)(t, L) = h(L) & \text{for } t \in (0, \infty) \\
-\varphi(u)_x(t, -L) + \beta(-L)\,\varphi(u)(t, -L) = h(-L) & \text{for } t \in (0, \infty)
\end{cases}
$$

We study the asymptotic behavior, as $t \to \infty$, of its bounded non-negative solutions.

Main assumptions. The following set of conditions will be referred to as (H).

(H.1) $\varphi \in C[0, \infty) \cap C^3(0, \infty)$,
(H.2) $\varphi(0) = 0, \quad \varphi'(z) > 0 \quad \forall z > 0$,
(H.3) $f \in C^1[0, \infty)$,
(H.4) $f(0) \geq 0, \ h(\pm L) \geq 0, \ \beta(\pm L) \geq 0, \ u_0 \geq 0, \ u_0 \in L^\infty(-L, L)$.

Formulated in these terms, the problem includes the regular second–order parabolic equation, with the main part of strongly elliptic type, as well as degenerate equations, when $\varphi'(0) = 0$ or $\varphi'(0) = \infty$. We will define later on, in Section 2, the notion of weak solution for (P) and discuss the hypotheses and sufficient conditions for the solution to be global and bounded.

The main theorem that we are aiming at is:

2000 *Mathematics Subject Classification*: 35B40, 35K65, 35Q35.
Key words: Degenerate parabolic equation, convergence, large time behaviour, mixed boundary conditions.

THEOREM 1.1. *Assume* (H). *Any bounded global* (*weak*) *solution to problem* (P) *converges uniformly on* $[-L, L]$, *as* $t \to \infty$, *to one and only one stationary solution.*

In other words, the ω – limit set of any bounded solution is a singleton.

For the equation with the main part of form $(au_x)_x$ with $a = a(x)$ of strictly elliptic type ($a \geq \delta > 0$), the question of convergence for the one–dimensional problem has been answered independently by Zelenyak [23] and Matano [19]. Both methods are specific for the dimension one. In higher dimensions some partial answers have been given, see e.g. [22], [12], [11], but the question of convergence for space dimensions higher than one remains still open for general $f(u)$, even for regular equations.

As for degenerate parabolic equations, asymptotic stability in one space dimension with Dirichlet boundary conditions has been investigated by Aronson et al. in [3] for $\varphi(u) = u^m$, $m \geq 1$ and f a polynomial of degree 3; they noted the existence of a continuum of equilibrium solutions (now, if the stationary solutions are isolated, the result follows immediately by connectedness of the ω – limit set). A result analogous to ours for the homogeneous Dirichlet problem has been proved by Feireisl and Simondon in [8]; they have also obtained a partial result in n–space dimension in [9].

In this paper, we treat the second and third boundary problems for two cases of degeneracy: one corresponding to $\varphi'(0) = 0$ (e.g u^m with $m > 1$) and modelling the filtration of gas through a porous medium – the equation is also called then nonlinear diffusion equation with absorption; the other one corresponding to φ not differentiable in 0 (e.g u^m with $0 < m < 1$), modelling fast diffusion and arising in some problems of plasma physics or spread of biological populations.

The plan is as follows. The main results concerning the existence – uniqueness theory for the case of Neumann and mixed boundary conditions are gathered in Section 2. We refer to the appendix for proofs of this part. Next, in Section 3, we show that the ω – limit set is contained in the set of stationary solutions and we investigate its essential properties: compactness and connectedness in $C(\overline{\Omega})$ and cases when we can expect stronger convergence. Up to this point, the paper has no space dimension restriction, we consider (P) in a bounded domain $\Omega \subset R^N$. Section 4 is devoted to the analysis of stationary solutions. It comes out there that the degeneracy of φ and independence of x in f make appear a continuum of stationary states which are not strictly positive – this being, as noted above, the "bad property". The proof of convergence is finally done in Section 5. Its main concern is the Neumann boundary problem, which contains all the essential difficulties. It divides into two parts. We show at first that if the ω – limit set contains a strictly positive function, we can argue in a way analogous to the regular case by means of passing through an appropriate approximation. The second case is $\sup_{w \in \omega(u_0)} \min w = 0$. Here, we combine an argumentation inspired by [8] with the usual order argument. We relay here strongly on the results of Matano [20], Henry [14] and Angenent [2] dealing with the zero-set of the solution to a regular parabolic equation. This is why we also need a property called weak positivity of the solution, meaning $u(t_0, x_0) > 0 \implies u(t, x_0) > 0$ for $t \geq t_0$. It is well known that for regular problems this occurs in a much stronger form (namely $u_0 \not\equiv 0 \implies u(t) > 0$ for $t > 0$), but it is also known that even weak positivity is in general not true for degenerate problems (see [15], [16], [5] and references therein for the porous media equation, and e.g. [4], [21] for fast diffusion). Here, the question of weak positivity is answered in relation with the boundary conditions and the hypotheses on f.

2. Existence, uniqueness, regularity of solutions of (P).

In this section and in the next one, consider (P) in Ω, bounded domain of R^N; we also take here f depending additionally on t and x. We will use the following notion of solution:

DEFINITION 2.1. Let us take $u_0 \in L^\infty(\Omega)$ and write

$$(\mathrm{P}(\varphi, \mathrm{f}, \mathrm{h})) \quad \begin{cases} u_t(t, x) - \Delta\varphi(u)(t, x) = f(t, x, u) & \text{in } (0, \infty) \times \Omega \\ u(0, x) = u_0(x) & \text{in } \Omega \\ \frac{\partial\varphi(u)}{\partial n}(t, x) + \beta(x)\,\varphi(u)(t, x) = h(x) & \text{on } (0, \infty) \times \partial\Omega. \end{cases}$$

We call u a weak solution of $(\mathrm{P}(\varphi, \mathrm{f}, \mathrm{h}))$ on $(0, T)$ if

$$\varphi(u) \in L^2(0, T;\ H^1(\Omega))$$

$$u, f(\cdot, u) \in L^2(0, T;\ L^2(\Omega))$$

and for all $v \in L^2(0, T;\ H^1(\Omega)) \cap W_2^1(0, T;\ L^2(\Omega))$ such that $v(T) = 0$ we have

$$-\int_0^T \int_\Omega uv_t + \int_0^T \int_\Omega \nabla\varphi(u)\nabla v - \int_0^T \int_\Omega f(t, x, u)v$$
$$-\int_0^T \int_{\partial\Omega} (h - \beta(\varphi(u))v - \int_\Omega u_0 v(0) = 0.$$

We call u a subsolution (supersolution) of $(\mathrm{P}(\varphi, \mathrm{f}, \mathrm{h}))$ if we replace above the sign "=" by "$\leq$" ("$\geq$") and add $v \geq 0$ to conditions for the test function.

If u is a solution on $(0, T)$ for any T, it is said to be a global weak solution.

As far as general qualitative behaviour is concerned, namely in this and in the next section, we can relax the regularity conditions, but we need to make some assumptions on the global behaviour of φ and f. As our interest is limited to globally bounded solutions – these are the only for which the question of convergence can be asked – we can without loss of generality take into account in the sequel the following set of hypotheses that we denote by (H'):

(H.Ω) Ω is a bounded domain of R^N, $\partial\Omega$ is of class C^1,

(H.1') $\varphi \in C[0, \infty) \cap C^1(0, \infty)$;

(H.2') $\varphi'(r) > 0 \; \forall r > 0$ and $\varphi(r) \leq \frac{1}{2} C_\varphi (1 + r)$;

(H.3') $f \in L^\infty((0, T) \times \Omega \times (0, \infty))$ and $u \mapsto f(t, x, u)$ is uniformly Lipschitz continuous;

(H.4') $f(t, x, 0) \geq 0$, h, $\beta \in C^1(\partial\Omega)$, $u_0 \in L^\infty(\Omega)$ are non-negative.

An additional condition will be needed to get continuity of solutions in higher space dimensions:

(H.5')   either $\liminf\limits_{r \to 0} \varphi'(r) > 0$,

$$\text{or } \varphi'(0) = 0 \text{ and } \exists \delta_0 : \; \varphi' \text{ increases on } (0, \delta_0),$$

$$\varphi'(\delta_0) \leq \varphi'(z) \text{ for } z \geq \delta_0.$$

REMARK 2.2. Let us underline that (H.4') ensures automatically positivity of solutions to (P), so that our interest in non–negative solutions does not imply any further restriction. As for the additional condition (H.5'), its most general formulation can be found in [6]; roughly, it excludes the case $\varphi'(r_n) \to 0$, $\varphi'(z_n) > c > 0$ for two subsequences r_n, $z_n \to 0$, and for the case $\limsup_{r \to 0} \varphi'(r) = \varphi'(0) = 0$ it generalises the condition $\varphi'' > 0$, assumed e.g. in [5], [16]. As noted above, (H.5') is not required in one space dimension.

Finally, let us stress on the fact that our study includes the case $\varphi(r) = r^m$ for any real $m > 0$.

THEOREM 2.3. *Assume* (H'). $(P(\varphi, f, h))$ *admits a unique global weak solution* u, *which is non–negative. In addition, if* u *is globally bounded, i.e.* $u \in L^\infty((0, \infty) \times \Omega)$, *and* f *is independent of* t, *we have also, for any* $t_0 > 0$,

$$\varphi(u) \in L^\infty(t_0, \infty; \; H^1(\Omega)) \tag{2.1}$$

$$\gamma(u) \in W^{1,2}(t_0, \infty; \; L^2(\Omega)) \tag{2.2}$$

where $\gamma(z) = \int_0^z \varphi'(s)^{\frac{1}{2}} \, ds$. *If* $N = 1$ *or* $\partial\Omega$ *is of class* C^1 *and* (H.5') *is satisfied, then*

$$u \in C((0, \infty) \times \overline{\Omega})$$

and it satisfies

$$|u(t_1, x_1) - u(t_2, x_2)| \leq \omega_{t_0}(|x_1 - x_2| + |t_1 - t_2|^{\frac{1}{2}})$$
$$\forall (t_i, x_i) \in [t_0, \infty) \times \overline{\Omega}, \; i = 1, 2, \tag{2.3}$$

for some continuous function $\omega_{t_0} : \mathbb{R}^+ \longrightarrow \mathbb{R}^+$, *such that* $\omega_{t_0}(0) = 0$.

As usual, this solution is obtained as limit of solutions to regular problems, which is shown in the Appendix below. It will be of importance that the approximating sequence can be chosen to be decreasing (Lemma 6.4).

In the region where the solution stays strictly positive, we can provide more information about its regularity. The following proposition is a direct application of [17, Theorem V.6.1].

PROPOSITION 2.4. *Assume* (H). *Suppose also that* $u > 0$ *on* $K \times [t_1, t_2] \subset\subset \Omega \times (0, T]$, *where* K *is closed and* ∂K *is of class* $C^{2+\delta}$. *Then* u_{x_i}, $u_{x_i x_i}$, u_t *are Hölder continuous on* $K \times [t_1, t_2]$.

We will need in addition a precise form of the constant in the following estimate, which is proven in the Appendix.

PROPOSITION 2.5. *Assume* $\varphi \in C^1(0, \infty)$, *take* u *a globally bounded weak solution of* $(P(\varphi, f, h))$ *on* $(0, \infty)$, $u \geq \delta > 0$ *on some* $K \times [t_1, t_2]$, $K \subset \Omega$ *measurable and* $0 < t_0 \leq t_1 < t_2$. *Then there exists* $C = C(|t_2 - t_1|, |K|, \delta, t_0)$ *such that*

$$\int_{t_1}^{t_2} \int_K |\Delta\varphi(u)|^2 \, dx \, dt \leq C.$$

The dependence of C *on* $|t_2 - t_1|$ *is linear increasing.*

Finally, the following comparison principle will be crucial in our argumentation.

PROPOSITION 2.6 – **Comparison Principle**. *Let* $\underline{u}, \overline{u}$ *be solutions of, respectively,* $(P(\varphi, \underline{h}, \underline{f}))$ *and* $(P(\varphi, \overline{h}, \overline{f}))$ *with initial data* $\underline{u}_0, \overline{u}_0$. *Assume that* (H') *is satisfied for both and that* $\underline{h}, \overline{h} \in L^2(\Omega)$. *If*

$$\underline{u}_0 \leq \overline{u}_0 \ a.e. \ on \ \Omega, \quad \underline{h} \leq \overline{h}, \ a.e. \ on \ \partial\Omega; \quad \underline{f} \leq \overline{f} \ a.e. \ on \ [0, T] \times \Omega \times (0, \infty),$$

then

$$\underline{u} \leq \overline{u} \ a.e. \ on \ \Omega \times (0, T).$$

REMARK 2.7. The question what are sufficient or necessary conditions for the solutions to be globally bounded may naturally arise. The usual approach is to assume "f has good sign at ∞", which here would take the form

$$\exists b_0 \geq 0 : \sup_{x : \beta(x) \neq 0} (h(x)/\beta(x)) \leq b_0,$$

$$\text{and } \exists z \geq \max(\|u_0\|_\infty, \varphi^{-1}(b_0)) : f(z) \leq 0. \tag{2.4}$$

Other sufficient growth conditions are given e.g. in [1], they affect the respective behaviour of f and φ on (r_0, ∞). On the other hand, note that our study shows that, for the homogeneous Neumann boundary problem, we will have *no bounded solutions* if $f > 0$ everywhere, even for f bounded. Indeed, it is easy to see that in that case the stationary problem has no solutions, and on the other hand, we prove that any bounded solution approaches the set of stationary states.

3. Structure of the ω-limit set

We consider here and in what follows f depending only on u. Let us denote by $\mathcal{S}$ the semigroup in $L^2(\Omega)$, corresponding to the solution of (P) and $\omega(u_0)$ the omega–limit set corresponding to the initial data u_0, i.e.

$$\omega(u_0) = \omega(u) = \{w \in L^2(\Omega) : \mathcal{S}(t_n)u_0 \to w \text{ in } L^2(\Omega) \text{ for some } t_n \nearrow \infty\}$$

By Theorem 2.3, for any u bounded, $\omega(u) \neq \emptyset$, and we have

PROPOSITION 3.1. *Assume (H'), and either $N = 1$ or (H.5') is satisfied. For any u bounded,*

$$\omega(u) = \{w \in C(\overline{\Omega}) : \mathcal{S}(t_n)u_0 \to w \text{ in } C(\overline{\Omega}) \text{ for some } t_n \nearrow \infty\},$$

and $\omega(u)$ is compact and connected in $C(\overline{\Omega})$.

Let us note now

$$g = f \circ \varphi^{-1}.$$

Then g is continuous on $[0, \varphi_{\max})$, Lipschitz continuous on $[\delta, \varphi_{\max})$ for any $\delta > 0$, and $f(0) = g(0) \geq 0$. We consider the stationary problem corresponding to (P) in the following form:

$$\text{(S)} \quad \begin{cases} -\Delta v = g(v) & \text{on } \Omega \\ \frac{\partial v}{\partial n} + \beta v = h & \text{on } \partial\Omega. \end{cases}$$

In order to state that the ω–limit set is contained in the set of equilibrium solutions, we will follow the method of Langlais and Phillips, [18], because in higher space dimensions we are not sure of having continuity of the Liapunov functional, which makes the classical dynamical system approach difficult to apply.

THEOREM 3.2. *For any bounded global solution u, if $w \in \omega(u)$, then $\varphi(w)$ is a weak solution of (S). If $\partial\Omega$ is of class C^2 and (H.5') is satisfied, then $\varphi(w)$ is a classical solution of (S).*

Proof. In the same way as in [18], the proof relays on the uniform estimate (2.2).

Take $w \in \omega(u)$, $v = \varphi(w)$ and $t_n \to \infty$ the time sequence for which convergence occurs:

$$\lim_{n \to \infty} \|v - \varphi(u(t_n, \cdot))\|_{L^2(\Omega)} = 0.$$

Note that φ continuous and strictly monotone implies φ^{-1} continuous, and as u is bounded, by dominated convergence we get that $u(t_n, \cdot)$ converges in $L^2(\Omega)$ to $w = \varphi^{-1}(v)$ and $\gamma(u)(t_n, \cdot)$ converges in $L^2(\Omega)$ to $\gamma(w)$.

Let us consider, for $s \in (-1, 1)$

$$u^n(s, x) = u(t_n + s, x).$$

Then

$$\|\gamma(u^n) - \gamma(u)(t_n)\|^2_{L^2((-1,1)\times\Omega)} \leq 2 \int_\Omega \int_{t_n-1}^\infty |\gamma(u)_t|^2 dt\, dx$$

which tends to 0 as $t_n \to \infty$ by (2.2). As of course $\gamma(u(t_n, \cdot)) \to \gamma(\varphi^{-1}(v))$ in $L^2((-1, 1) \times \Omega)$ as well, it follows that

$$\gamma(u^n) \to \gamma(\varphi^{-1}(v)) \text{ in } L^2((-1, 1) \times \Omega).$$

As before, by continuity of γ^{-1}, boundedness of u and dominated convergence

$$u^n \to \varphi^{-1}(v), \ f(u^n) \to f(\varphi^{-1}(v)) \text{ in } L^2((-1, 1) \times \Omega) \tag{3.1}$$

Next, let us take the test functions

$$\rho \in C_0^2[-1, 1] \ : \ \rho \geq 0, \qquad \int_{-1}^1 \rho = 1; \tag{3.2}$$

$$\xi \in H^2(\Omega) \ : \ \frac{\partial \xi}{\partial n} = 0 \text{ on } \partial\Omega. \tag{3.3}$$

Taking $v = \rho(t - t_n)\xi(x)$ in Definition 2.1. and integrating by parts yields

$$-\int_{t_n-1}^{t_n+1} \rho'(t - t_n) \int_\Omega u(t, x)\xi(x)\, dx\, dt - \int_{t_n-1}^{t_n+1} \rho(t - t_n) \int_\Omega \varphi(u(t, x))\Delta\xi(x)\, dx\, dt$$

$$-\int_{t_n-1}^{t_n+1} \rho(t - t_n) \int_\Omega f(u(t, x))\xi(x)\, dx\, dt$$

$$-\int_{t_n-1}^{t_n+1} \rho(t - t_n) \int_{\partial\Omega} [h(x) - \beta(x)\varphi(u(t, x))]\xi(x)\, dx\, dt = 0;$$

if we change variables and pass to the limit in $L^2((-1, 1) \times \Omega)$, (3.1) yields

$$-\int_{-1}^{1} \rho'(s)\, ds \int_{\Omega} \varphi^{-1}(v)\,\xi = \int_{-1}^{1} \rho(s)\, ds \left(\int_{\Omega} v\Delta\xi + g(v)\,\xi + \int_{\partial\Omega} [h - \beta\varphi(v)]\,\xi \right).$$

Since $\int_{-1}^{1} \rho = 1$ and $\int_{-1}^{1} \rho' = 0$, we have, for any ξ verifying (3.2)

$$\int_{\Omega} v\, \Delta\xi + g(v)\,\xi + \int_{\partial\Omega} [h - \beta\varphi(v)]\,\xi = 0. \tag{3.4}$$

Thus, v is a solution of (S) in the sense of distributions. Let us investigate its regularity. As $v \in L^\infty(\overline{\Omega})$ and g is continuous on $[0, \infty)$, $g(v(x)) \leq C$. The equation (3.4) yields then $\int_{\Omega} v\Delta\xi \leq C\,(|\Omega|)\,\|\xi\|_{L^2(\Omega)}$ for all $\xi \in \mathcal{D}(\Omega)$. But this means that $\int_{\Omega} v\Delta\xi = \int_{\Omega} v^*\xi$ for some $v^* \in L^2(\Omega)$; by definition $v^* = \Delta v$. Thus

$$\Delta v \in L^2(\Omega).$$

If $\partial\Omega$ is of class C^2, this yields (cf. [10, Theorem 17.2]) $v \in H^2(\Omega)$ and we get

$$\int_{\Omega} \Delta v\,\xi - \int_{\partial\Omega} \left[\frac{\partial v}{\partial n} + h - \beta\varphi(v) \right] \xi + \int_{\Omega} g(v)\,\xi = 0 \tag{3.5}$$

for all ξ verifying (3.2). Take in particular $\xi \in \mathcal{D}(\Omega)$. We get that

$$\Delta v + g(v) = 0 \qquad \text{everywhere on } \overline{\Omega},$$

where 'everywhere' instead of 'a.e.' follows again by continuity of v (Proposition 3.1. and continuity of φ) and of g. Going back to (3.5), we obtain

$$\frac{\partial v}{\partial n} + \beta v = h$$

by density of traces ξ satisfying (3.2) in $L^2(\partial\Omega)$. $\square$

PROPOSITION 3.3. *Assume* (H') *and let* $N = 1$. *Take* u *globally bounded and* $w \in \omega(u)$ *such that* $w > 0$ *on some* $[a, b] \subset [-L, L]$. *Then*

$$\exists (s_n) \nearrow \infty : \quad \varphi(u)(s_n) \to \varphi(w) \text{ in } C^1([a, b]).$$

Proof. Let $\varphi(u)(t_n) \to w$ in $C[-L, L]$. Thanks to uniform continuity (2.3) of $\varphi(u)$ in $\Omega \times [1, \infty)$ there exist constants $\alpha, d > 0$, independent of n, such that

$$\varphi(u) \geq \alpha > 0 \text{ on } [a, b] \times (t_n, t_n + d)$$

for n large enough. Take now $u^n(s) = u(t_n + s)$ and note that u^n satisfies

$$(\mathrm{P}^n_\infty) \begin{cases} u^n_t - \varphi(u^n)_{xx} = f(u^n) & \text{in } (-L, L) \times (0, d) \\ u^n(0) = u(t_n) & \text{in } (-L, L), \\ \pm\varphi(u^n)_x(\pm L) + \beta(\pm L)\varphi(u^n)(\pm L) = h(\pm L) & \text{on } (0, d). \end{cases}$$

Also, u^n are uniformly bounded: $0 < \alpha \leq \varphi(u^n) \leq M$ on $[a, b] \times [0, d]$.
Let us take $s \in (0, d)$ and $x, y \in [a, b]$.

$$\begin{aligned} |\varphi(u^n)_x(x, s) - \varphi(u^n)_x(y, s)| &= \left| \int_x^y \varphi(u^n)_{xx}(\xi, s) \, d\xi \right| \\ &\leq \left(\int_a^b |\varphi(u^n(s))_{xx}|^2 \right)^{\frac{1}{2}} |y - x|^{\frac{1}{2}}. \end{aligned} \tag{3.6}$$

By Proposition 2.5

$$\int_0^d \int_a^b |\varphi(u^n)_{xx}|^2 = \int_{t_n}^{t_n+d} \int_a^b |\varphi(u)_{xx}|^2 \leq C(d, M, f, \varphi) = C(M, f, \varphi);$$

the dependence on d being increasing, we can omit it supposing that $d \leq 1$. We infer that

$$\forall\, \delta \in (0, d) \quad \exists \eta : \quad \forall n \quad \exists \tau_n \in (t_n, t_n + \delta) \text{ such that } \int_a^b |\varphi(u(\tau_n))_{xx}|^2 \leq \eta.$$

Going back to (3.6), we obtain that for any small δ, there exists a sequence (τ_n), $\tau_n \in (t_n, t_n + \delta)$, such that the family $\{\varphi(u^n)(\cdot, \tau_n)\}_n$ is relatively compact in $C[a, b]$; i.e., for some $\psi \in C[a, b]$

$$\forall\, \delta \in (0, d) \quad \exists(\tau_n) : \tau_n \in (t_n, t_n + \delta) \text{ such that } \forall\, \varepsilon > 0 \quad \exists K : \forall\, k \geq K$$

$$\sup_{x \in [a,b]} |\varphi(u^{n_k})_x(x, \tau_{n_k}) - \psi(x)| < \frac{\varepsilon}{2}. \tag{3.7}$$

Fix now $\varepsilon > 0$. By dominated convergence theorem $\varphi(u)(t_n) \to \varphi(w)$, let $N(\varepsilon)$ be fixed by:

$$n \geq N(\varepsilon) \implies \sup_{x \in [-L,L]} |\varphi(u)(x, t_n) - \varphi(w)(x)| < \frac{\varepsilon}{4}$$

and let $\delta = \delta(\varepsilon)$ be fixed by continuity of $\varphi(u)$:

$$|t - s| < \delta \implies \sup_{x \in [a,b]} |\varphi(u)(x, t) - \varphi(u)(x, s)| < \frac{\varepsilon}{4}.$$

For any (s_n) such that $|s_n - t_n| < \delta(\varepsilon)$ and for $n \geq N(\varepsilon)$ we have

$$\|\varphi(u)(s_n) - \varphi(w)\|_{C^1[a,b]} \leq \frac{\varepsilon}{2} + \sup_{x \in [a,b]} |\varphi(u)_x(x, s_n) - \varphi(w)_x(x)|.$$

Now, put $\delta = \delta(\varepsilon)$ in (3.7) and $s_n = t_{n_k} + \tau_{n_k}$, $\tau_n \in (t_n, t_n + \delta(\varepsilon))$ defined by (3.7). Of course we have now $\psi = \varphi(w)_x$. We have found, for any ε, a time sequence for which $\|\varphi(u)(s_n) - \varphi(w)\|_{C^1[a,b]} < \varepsilon$. It remains to take a sequence of $\varepsilon_n \to 0$ and to construct s_n for which convergence occurs. $\qquad\square$

4. Stationary solutions

We are investigating here the form of v, classical solution of (S) in one space dimension. We will need to refer to a homogeneous counterpart of (S) posed on whole $\mathbb{R}$:

$$(S_\mathbb{R}) \qquad \begin{cases} -v_{xx} = g(v) & \text{on } \mathbb{R} \\ \text{supp } v \text{ is compact in } \mathbb{R}. \end{cases}$$

LEMMA 4.1. (i) *Let v be a solution of $(S_\mathbb{R})$. If there exists $y \in (-L, L)$ such that $v(y) > 0$ and $v_x(y) = 0$, then v is symmetric around y, more precisely*

$$v(y - x) = v(y + x) \;\text{ for } x \in (0, a)$$

where

$$a = \sup\{x > 0 : \; v(y + x') \neq 0 \text{ for } x' \leq x\}.$$

(ii) *Also, if $\hat{v}$ is a solution of $(S_\mathbb{R})$ and if there exists y such that $v(y) = \hat{v}(y) > 0$ and $v_x(y) = \hat{v}_x(y)$ then $v \equiv \hat{v}$ on $(y - a, y + a)$.*

The same holds for v – solution to (S) with

$$a = \sup\{x > 0 : \; y \pm x \in [-L, L], \; v(y + x') \neq 0 \text{ for } x' < x\}.$$

Proof. Indeed, it is enough to consider $w(x) = v(y - x)$, $\tilde{w}(x) = v(y + x)$ and note that if $v'(y) = 0$ then (w, w_x) and $(\tilde{w}, \tilde{w}_x)$ satisfy the same Cauchy problem for the system of ordinary differential equations

$$(SO) \qquad \begin{cases} v' = u \\ u' = -g(v) \\ v(y), u(y) \quad \text{given} \end{cases}$$

As long as $v > 0$, the right-hand-side is locally Lipschitz continuous and the system has a unique solution, forward as well as backward i.e. on $(y, +\infty)$ but also on $(-\infty, y)$. $\qquad\square$

LEMMA 4.2. *For every solution v of (S) on $(-L, L)$*

$$(E) : \qquad \frac{1}{2}v_x^2(x) + G(v(x)) \equiv \text{const} \quad \text{for } x \in [-L, L],$$

where $G(z) = \int_0^z g(s)ds$. The constant is equal to 0 if $\min v = 0$.

Proof. Multiply (S) by v_x. Note that (E) remains true even if $v = 0$. $\qquad\square$

PROPOSITION 4.3. *Assume $h \equiv \beta \equiv 0$. Let v be a solution of (S) with $\min v = 0$. Then $v \equiv 0$ or*

$$v(x) = \sum_{i=1}^{n} v_g(x - x_i)\big|_{[-L,L]}$$

where v_g is the "basic ground state" solution to $(S_{\mathbb{R}})$, having the following properties:

$$\operatorname{supp} v_g = [-a, a]; \quad v_g(x) = v_g(-x); \quad v_g \text{ decreasing on } (0, L);$$

and $a \leq L$. The "basic ground state" solution, if it exists, is unique.

The set of admissible points $\{x_i\}_{i=1}^{N} \subset [-L, L]$ is determined by the following conditions:

$$|x_i - x_j| \geq 2a \text{ for } i \neq j,$$

$$|x_i + L| \geq a \text{ if } x_i \neq -L, |L - x_i| \geq a \text{ if } x_i \neq L;$$

and n is any natural number smaller than N.

Proof. We are in the case where every solution to (S) extended by 0 outside $[-L, L]$ is also a solution to $(S_{\mathbb{R}})$. Let $v \not\equiv 0$ be a solution of (S) such that $\min v = 0$. It is easy to see that $v(\cdot - y)$ is one also, if only the boundary condition $v'(\pm L - y) = 0$ is satisfied.

By continuity, v admits a maximum, at, say, $x_{\max}$. By Lemma 4.1, it is symmetric "around" $x_{\max}$. Assume that it attains also in some $x_{\min}$ a local minimum $\lambda > 0$. It is easy to see that then the function v_λ, which is periodic and equal to v on $[x_{\min}, x_{\max}]$ is also solution of (S). But this is a contradiction with Lemma 4.1 (ii).

So, the only local minimum of v is 0.

Suppose now v has two maxima $\mu^0 = v(x_{\max}^0)$, $\mu^1 = v(x_{\max}^1)$. Let us define v_g^i, for $i = 0, 1$, by

$$v_g^i(x) = \begin{cases} v(x + x_{\max}^i) & \text{for } x \in (-a^i, a^i), \\ 0 & \text{for } x \in [-L, L]\backslash(-a^i, a^i) \end{cases}$$

where $a^i = \sup\{x : x_{\max}^i - x, x_{\max}^i + x \in [-L, L] \text{ and } v(x_{\max}^i + x') \neq 0 \text{ for } x' < x\}$, as in Lemma 4.1. It is easy to see that v_g^0, v_g^1 are solutions of (S). By what precedes, their supports cannot intersect: $(x_0 - a_0, x_0 + a_0) \cap (x_1 - a_1, x_1 + a_1) = \emptyset$.

By (E), $G(\mu^0) = 0$. Let y^1 be the point where $v_g^1(y^1) = \mu^0$. By (E) again, $(v_g^1)_x(y^1) = 0$ and by Lemma 4.1 (i) μ^0 is the maximum of v_g^1, i.e. $\mu_0 = \mu_1$. But this yields $v_g^0 = v_g^1$ by Lemma 4.1 (ii). $\qquad\square$

PROPOSITION 4.4. *Take any β, $h \geq 0$. Let v be a solution to (S) with min $v = 0$. Define*

$$a_- = \inf\{x > 0: \quad v(-L + x) = 0\},$$
$$a_+ = \inf\{x > 0: \quad v(L - x) = 0\}.$$

Then v can be written as $v = v_0 + v_- + v_+$, with

$$\operatorname{supp} v_- = [-L, -L + a_-], \quad \operatorname{supp} v_+ = [L - a_+, L],$$
$$\operatorname{supp} v_0 \subset [-L - a_-, L - a_+],$$

(if $a_\pm = 0$ then $v_\pm = 0$), where v_0 is a solution of the problem with $h \equiv \beta \equiv 0$.

If the basic ground state v_g of $(S_\mathbb{R})$ exists, then v_+ and v_- are equal to v_g on their supports, up to a translation.

If not, v_- is decreasing, v_+ is increasing on their supports, and $v_0 = 0$.

Proof. We define $v_-(x) = v(x)$ if $x \in [-L, -L+a_-]$, $v(x) = 0$ otherwise; analogously for v_+. We note that $v - v_- - v_+$ is a solution of the homogenous Neumann boundary value problem.

Suppose at first that the ground state solution exists. There exist then points $s \in (-L, L)$, $x_0 \in \mathbb{R}$ such that $v_g(s - x_0) = v_-(s) > 0$, by symmetry of v_g we can also chose x_0 so that their derivatives in these points have the same sign. By (E), we have then $v_g'(s - x_0) = v_-'(s)$. We conclude by Lemma 4.1 (ii) and by uniqueness of the ground state that v_- and v_g are equal up to a translation. The same holds for v_+.

Now, if the ground state solution does not exist, suppose, arguing by contradiction, that $v(y) > 0$ and $v_x(y) = 0$ (recall that $v \in C^2[-L, L]$). Then z, the symmetrization of v around y extendeed by 0, satisfies (E). By derivation of (E) we have $z_x = 0$ or $z_{xx} + g(z) = 0$, which gives that z is a ground state solution to $(S_\mathbb{R})$. So, v can change its monotonicty character only when equal to 0.

Note that of course $v_0 \equiv 0$ and $v_\pm$ are monotone also if $\operatorname{supp} v_g \not\subset [-2L, 2L]$. $\square$

5. Convergence to a unique stationary solution.

This section is devoted to the proof of Theorem 1.1, which is divided into Proposition 5.1 and Proposition 5.4. We continue to work here in one space dimension.

PROPOSITION 5.1. *Assume (H') and take u globally bounded. Suppose there exists $w \in \omega(u)$ such that min $w > 0$. Then $\omega(u)$ is a singleton.*

Proof. We will proceed here in a way analogous to [19].

LEMMA 5.2. *Assume that $\omega(u)$ is not a singleton. Then there exists a time t_N such that $u(t_N, \cdot) - w(\cdot)$ has a finite number of zeros on $[-L, L]$ for any $w \in \omega(u)$.*

Proof. Let $\varphi(w_1)$, $\varphi(w_2)$ be two distinct solutions of (S) with $\min w_2 > 0$. Then $\min \varphi(w_2) > 0$ and by Lemma 4.1 $\varphi(w_1)$ and $\varphi(w_2)$ "cannot intersect" in the phase space of (SO); i.e.

$$\inf_{x \in [-L,L]} |\varphi(w_1)(x) - \varphi(w_2)(x)| + |\varphi(w_1)'(x) - \varphi(w_2)'(x)| \geq \delta > 0.$$

By Proposition 3.3 there exists $t_n \nearrow \infty$ such that

$$\sup_{x \in [-L,L]} |\varphi(u)_x(t_n, x) - \varphi(w_2)'(x)| + |\varphi(u)(t_n, x) - \varphi(w_2)(x)| < \frac{\delta}{2}.$$

Combining the two inequalities above we get

$$|\varphi(u)_x(t_n, x) - \varphi(w_1)'(x)| + |\varphi(u)(t_n, x) - \varphi(w_1)(x)| \geq \frac{\delta}{2} \quad \forall x \in [-L, L]. \qquad (5.1)$$

Therefore, $\varphi(u)(t_n, x) - \varphi(w_1)(x)$ cannot have infinitely many zeros in $[-L, L]$, otherwise the left-hand-side would vanish in each accumulation point of these zeros. Finally, by connectedness of $\omega(u)$ and compactness of the domain, there is a third element $w_3 \in \omega(u)$ with $\min w_3 > 0$, which allows to conclude that (5.1) holds for any $w \in \omega(u)$. $\qquad \square$

Let us note now, analogously to [19], [20]:

$$A^+(w) = \{(t, x) \in (t_0, \infty) \times [-L, L] \; : \; u(t, x) > w(x)\}$$
$$A^-(w) = \{(t, x) \in (t_0, \infty) \times [-L, L] \; : \; u(t, x) < w(x)\}$$
$$l_t = [-L, L] \times \{t\}.$$

LEMMA 5.3. *Let $w \in \omega(u)$ with $\min w > 0$; let C^- be any connected component of $A^-(w)$. Then $C^- \cap l_{t_0} \neq \emptyset$. The same assertion holds for C^+, connected component of $A^+(w)$.*

Proof. For $A^+(w)$, the conclusion follows immediately from [19], as the equation is regular in this region.

Consider C^- a connected component of $A^-(w)$. Take $(t, y) \in \mathrm{Int}\, C^-$. We want to show that (t, y) is connected by a curve $\gamma : [0, 1] \to C^-$ to some $(t_0, y_0) \in l_{t_0}$.

Let $A_n^- = \{(t, x) \; : \; u^n(t, x) < w(x)\}$, where u^n is the approximating sequence chosen as in Lemma 6.4. As u^n converges to u in $C(\overline{\Omega} \times [1, , \infty))$ we have

$$(t, y) \in \mathrm{Int}\, C^- \implies (t, y) \in \mathrm{Int}\, C_n^- \quad \text{for } n \geq N.$$

We state that (t, y) is connected to some (t_0, y_0) by a curve $\gamma \subset C_N^-$. Indeed, if C_N^- is not connected to l_{t_0}. its boundary divides into two parts, say Γ_1 and Γ_2, and on the first we have a Neumann/Robin condition, on the second a Dirichlet condition (by continuity of u and w, see Proposition 3.1):

$$\Gamma_1 = C_N^- \cap (\{-L, L\} \times [t_0, \infty)), \quad \Gamma_2 = \partial C_N^- \backslash \Gamma_1,$$

$$\frac{\partial(\varphi(u^N) - \varphi(w))}{\partial n} + \beta(\varphi(u^N) - \varphi(w)) = 0 \quad \text{on } \Gamma_1, \qquad u^N - w \equiv 0 \quad \text{on } \Gamma_2.$$

By the maximum principle for regular parabolic equations (cf. [10]), we get that $u^N - w = 0$ in Int C_N^-, which is a contradiction.

Now as (u^n) has been chosen to be decreasing, $A_N^- \subset A^-$, and $C_N^- \subset C^-$. So, $\gamma \subset C^-$, which ends the proof. □

End of the proof of Proposition 5.1. Lemmas 5.2 and 5.3 imply that for any $w \in \omega(u)$ such that min $w > 0$ the zeros of $u(-L, \cdot) - w(-L)$ are finite in number on some $[T, \infty)$. Now, if $\omega(u)$ is not a singleton, there exist, by connectedness of $\omega(u)$, three distinct elements w_1, w_2, $w_3 \in \omega(u)$, such that $0 \leq \min w_1 < \min w_3 < \min w_2$. Note that $w_1(L)$, $w_2(L)$, $w_3(L)$ are distinct by Lemma 4.1 (ii). So, $u(t, -L) - w_3(-L)$ stays either strictly positive or strictly negative for $t > T$, which contradicts w_1, $w_2 \in \omega(u)$.

PROPOSITION 5.4. *Assume* (H). *For any solution u of* (P) *which is globally bounded, suppose that* min $w = 0$ *for all* $w \in \omega(u)$. *Then* $\omega(u)$ *is a singleton.*

Proof. The proof will rely on the existing theory about the zero-set of regular parabolic equations, developed especially by Matano [20], Henry [14] and Angenent, [2]. We first need to determine the regions where the equation is regular, i.e, where u stays positive. This is the object of the following lemma.

LEMMA 5.5. *Suppose that* $(S_{\mathbb{R}})$ *admits the basic ground state solution* v_g. *Take* x_0 *and* t_0 *such that* $u(t_0, x_0) > 0$. *Then* $u(t, x) \geq c(t, x) > 0$ *for all* $t \geq t_0$ *and* $x \in I(x_0)$ *(some neighborhood of* x_0*), where c is a strictly positive, continuous function of x and t.*

Proof. Let $[-a, a]$ be the support of the basic ground state solution v_g and let us note $w_g = \varphi^{-1}(v_g)$; we have supp $w_g = [-a, a]$.

We consider first the case $a \leq L$. There exists then α such that

$$\text{supp } w_g(\cdot + \alpha) \subset [-L, L] \quad \text{and} \quad x_0 \in \text{Int supp } w_g(\cdot + \alpha).$$

We can also find a smooth function $w_0 \in C^\infty(\mathbb{R})$ such that

$$0 < \max_x w_0(x) = w_0(x_0),$$

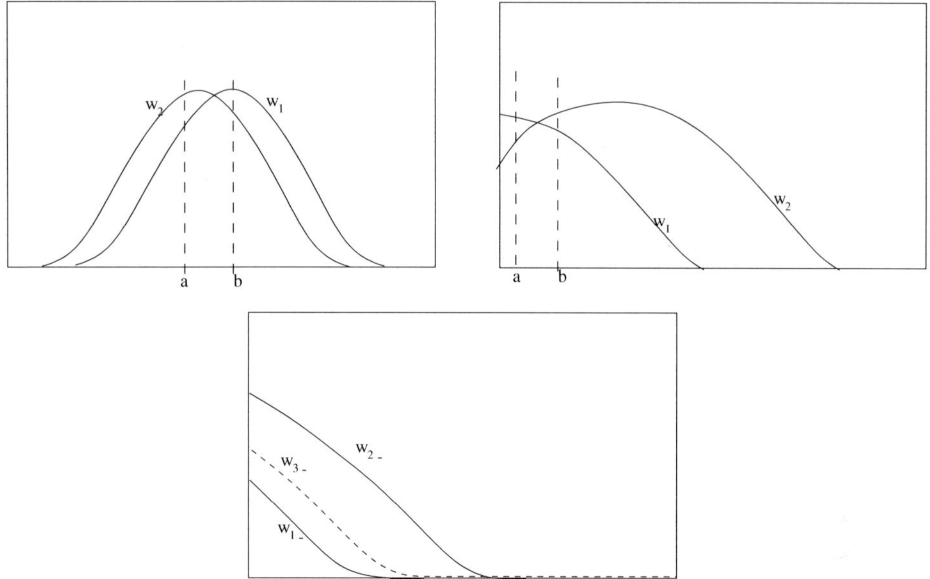

Figure 1   Functions introduced in the proof of Lemma 5.5; cases $a \leq L$ and $a > L$.

w_0 is symmetric with respect to $\{x = x_0\}$ and $w_0'(x) > 0$ for $x < x_0$;

$$w_0(x) \leq u(t_0, x),$$

and $\quad w_0(x) \leq w_g(x + \alpha) \quad$ for all $x \in \mathbb{R}$.

We take this w_0 as initial datum in the following equation

$$(\mathrm{P_M}) \quad \begin{cases} \bar{u}_t - \varphi(\bar{u})_{xx} = -M\bar{u} & \text{in } (t_0, \infty) \times (-L, L), \\ \bar{u}(t_0) = w_0 & \text{in } (-L, L), \\ \pm\varphi(\bar{u})_x(\pm L) + \beta(\pm L)\varphi(\bar{u})(\pm L) = 0. \end{cases}$$

Note that $(\mathrm{P_M})$ has a unique solution in the sense of Definition 2.1, see Section 2 and the Appendix; this solution is continuous. The comparison principle, Proposition 2.6, yields that $\bar{u}$ is non-negative and, moreover, globally bounded:

$$\bar{u}(t, x) \leq w_0(x_0) \qquad \text{for all } t \geq t_0, \ x \in [-L, L]. \tag{5.2}$$

Indeed, let $j(u) = \min(u, w_0(x_0))$ and $\widetilde{u}$ be solution of $\widetilde{u}_t - \varphi(\widetilde{u})_{xx} = -Mj(\widetilde{u})$, with the same boundary and initial conditions as in $(\mathrm{P_M})$. Then Proposition 2.6 yields $\widetilde{u} \leq w_0(x_0)$. In addition, by uniqueness of solution to $(\mathrm{P_M})$, $\widetilde{u} \equiv \bar{u}$.

Chose now $M > 0$ so that

$$f(z) \geq -Mz \quad \forall z \in [0, w_0(x_0))$$

which is possible as $f \in C^1[0, \infty)$. By the comparison principle we get then that

$$u(t, x) \geq \bar{u}(t, x), \quad \text{for all} \quad t \geq t_0, \ x \in [-L, L].$$

It is therefore sufficient to prove that $\bar{u}(x_0, t) > 0$.

In this aim, take $\bar{u}^\varepsilon$, solutions of approximating, regular problems (P_M^ε), where φ is replaced by $\varphi^\varepsilon \in C^\infty[0, \infty)$ such that $(\varphi^\varepsilon)' > c(\varepsilon) > 0$ and φ^ε converge to φ uniformly on compact sets; cf. also the Appendix. (P_M^ε) has a solution in the sense of Definition 6.1, let us take $v = 1$ there and obtain with the Gronwall lemma

$$\int_{-L}^{L} \bar{u}^\varepsilon(t, x) = \int_{-L}^{L} w_0(x) \, dx \, e^{-M(t-t_0)}$$
$$- \int_{t_0}^{t} [\beta \varphi^\varepsilon(\bar{u}^\varepsilon)(s, L) + \beta \varphi^\varepsilon(\bar{u}^\varepsilon)(s, -L)] e^{Ms} \, ds \, e^{-Mt} \tag{5.3}$$

Before passing to the limit, we note that the comparison principle applied to (P_M) yields further

$$\bar{u}(t, x) \leq w_g(x + \alpha) \quad \text{for all} \quad t \geq t_0, \ x \in [-L, L]; \tag{5.4}$$

it follows that $\bar{u}(t)$ has compact support included in $[\alpha - a, \alpha + a] \subset [-L, L]$ (recall that we are in the case $a \leq L$). As $\bar{u}^\varepsilon$ converges pointwise to $\bar{u}$, we have $\varphi^\varepsilon(\bar{u}^\varepsilon)(s, \pm L) \to 0$ and we obtain from (5.3)

$$\int_{-L}^{L} \bar{u}(t, x) \geq \int_{-L}^{L} w_0(x) \, dx \, e^{-M(t-t_0)} > 0. \tag{5.5}$$

In addition, x_0 is the maximum point for $\bar{u}(t, \cdot)$. Indeed, note at first that $\bar{u}(t, \cdot)$ is symmetric with respect to x_0: by (5.3), it can be prolongated by 0 outside $[-L, L]$ as solution of (P_M) on $\mathbb{R}$ and $\bar{u}(t, x_0 - x)$, $\bar{u}(t, x_0 + x)$ satisfy the same equation having a unique solution. Note then that $\varphi^\varepsilon(\bar{u}^\varepsilon)_x = \xi^\varepsilon$ satisfies on $(-L, x_0)$ the equation

$$\begin{cases} \xi_t^\varepsilon = (\varphi^\varepsilon)'(\bar{u}^\varepsilon) \xi_{xx}^\varepsilon + \frac{(\varphi^\varepsilon)''(\bar{u}^\varepsilon)}{(\varphi^\varepsilon)'(\bar{u}^\varepsilon)} \xi_x^\varepsilon \xi^\varepsilon - M(1 + \frac{\bar{u}^\varepsilon (\varphi^\varepsilon)''(\bar{u}^\varepsilon)}{(\varphi^\varepsilon)'(\bar{u}^\varepsilon)}) \xi^\varepsilon \\ \xi^\varepsilon(-L) = \xi^\varepsilon(x_0) = 0, \\ \xi^\varepsilon(t_0) \geq 0. \end{cases}$$

The classical comparison principle yields here $\xi^\varepsilon(t) \geq 0$ for all $t \geq t_0$. Analogously, we have $\xi^\varepsilon(t, x) \leq 0$ on (x_0, L). Thus, $\bar{u}$ is increasing on $(-L, x_0)$ and decreasing on (x_0, L) as limit of monotone functions.

With (5.5) this gives $\bar{u}(t, x_0) > 0$, which ends the proof for this case.

Consider now the case $a > L$, i.e. $[-L, L] \subset$ supp v_g. We chose then α_1, $\alpha_2 \in (-L, L)$ such that

$$\varphi(w_g)'(-L + \alpha_1) = \varphi(w_g)(-L + \alpha_1) = 0$$

and $\varphi(w_g)'(L + \alpha_2) = \varphi(w_g)(L + \alpha_2) = 0.$

Of course,

$$\varphi(w_g)'(L + \alpha_1) + \beta\varphi(w_g)(L + \alpha_1) \geq 0$$

and $-\varphi(w_g)'(-L + \alpha_2) + \beta\varphi(w_g)(-L + \alpha_2) \geq 0.$

We can also find w_0 as above, with $w_0(x) \leq w_g(x + \alpha_i)$, $i = 1, 2$, $x \in \mathbb{R}$. We consider the same equation (P_M) with this new w_0 as initial datum. By the comparison principle we obtain again $\bar{u}(t, x) \leq w_0(x_0)$ and $\bar{u}(t, x) \leq u(t, x)$ for all $t \geq t_0$, $x \in [-L, L]$. In addition, with the same choice of M and thanks to the boundary conditions stated above we have, as in (5.4)

$$\bar{u}(t, x) \leq w_g(x + \alpha_1) \quad \text{and} \quad \bar{u}(t, x) \leq w_g(x + \alpha_2) \quad \text{for all } t \geq t_0, \ x \in [-L, L];$$

so, $\bar{u}(t, \cdot)$ has compact support included in $[-L, L]$. We obtain now identically (5.3) and (5.5), which allow to conclude that $u(t, x_0) \geq \bar{u}(t, x_0) > 0$. $\qquad\square$

End of the proof of Proposition 5.4. We will prove that we may suppose that

1) — either

$$\exists\, w_1,\, w_2 \in \omega(u) \text{ s.t. } w_1 \not\equiv w_2,$$
$$\exists\, [a, b] \subset\subset \text{supp } w^1 \cap \text{supp } w^2 \cap (-L, L) : \quad w_x^1 > 0,\ w_x^2 < 0 \quad \text{on } [a, b],$$
$$\text{and the ground state exists;}$$

2) — or

$$\exists\, w_1,\, w_2 \in \omega(u) \text{ s.t. } w_1 \not\equiv w_2,$$
$$\exists\, i \in \{+, -\} \text{ s.t. } w_i^1 \not\equiv 0, \quad w_i^1, w_i^2 \text{ are monotone,} \quad w_i^1 < w_i^2 \text{ on supp } w_i^1.$$

Indeed, note at first that by conectedness of $\omega(u)$ we can assume that it contains two elements $w^1 \not\equiv w^2$ such that $w^1 \not\equiv 0$, $w^2 \not\equiv 0$. We will refer in what follows to Section 4, note that φ being monotone increasing and admitting its only zero in 0, the properties (in particular of decomposition) are analogous for w and for $v = \varphi(w)$.

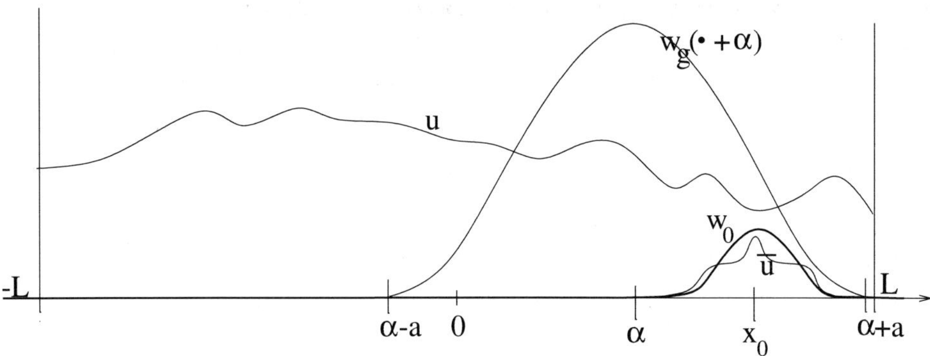

Figure 2   Cases considered in the proof of Proposition 5.4. The first two pictures correspond to Case 1) with homogeneous or non–homogeneous boundary conditions, the last one illustrates the Case 2); it can occur only if $w > 0$ on one of the sides.

If $w_0^1 \not\equiv w_0^2$, then at least one of them, say w^1, is non zero, and by conectedness of $\omega(u)$, there is another element $w^3 \in \omega(u)$ such that Int supp $w_0^1 \cap$ Int supp $w_0^3 \neq \emptyset$; then w^1, w^3 satisfy 1.

If $w_0^1 \equiv w_0^2$, we can suppose that $w_-^1 \not\equiv w_-^2$. If they are both monotone, they can intersect only when they are equal to 0, by (E) and Lemma 4.1 (ii). Suppose $w_-^1 \leq w_-^2$.

We note then that w_-^1, w_-^2 are of form $H(x + L - a_-^1)$, $H(x + L - a_-^2)$, where

$$H^{-1}(z) = \int_0^z \frac{du}{\sqrt{-2F(u)}}, \qquad F(z) = \int_0^z f(s)ds,$$

on their supports, H and H^{-1} are monotone increasing. So, $a_-^1 < a_-^2$. Therefore, $w_-^1 < w_-^2$ on supp w_-^1.

Finally if w_-^1 is not monotone, we are necessarily in the case when the ground state solution exists and again by conectedness we can chose w^2 so that w^1, w^2 satisfy conditions of the Case 1.

We start considering now the Case 2). Let $w_-^1(-L) < w_-^2(-L)$. By conectedness of $\omega(u)$, there exist $w^3 \in \omega(u)$ such that $w_-^1(-L) < w_-^3(-L) < w_-^2(-L)$. Again by (E) and Lemma 4.1 (ii) w_-^3 cannot intersect w_-^1 but in zero. If w_-^3 intersects w_-^2 then this means that w_-^3 is not monotone, that the ground state exists and we can chose by conectedness w^4 such that w^3, w^4 satisfy conditions of the Case 1. We can thus suppose that w_-^3 is monotone and that

$$w_-^1 < w_-^3 < w_-^2 \quad \text{on supp } w_-^3.$$

Take a sequence of time moments $\{t_n\}$ such that $u(t_n) \to w^2$ in $C[-L, L]$. There exists $t_N : u(t_N) \geq w_-^3$. But now, u satisfies (P) and w_-^3 satisfies on $[-L, L]$ $(P(\varphi, \underline{h}, f))$ with

$\underline{h}(-L) = h(-L)$ and $\underline{h}(L) = 0$. By the comparison principle, $u(t) \geq w_-^3$ for all $t \geq t_N$, which, as $w^1 < w_-^3$ on a compact intervall, leads to a contradiction with $w_1 \in \omega(u)$.

Let us consider the Case 1). By Proposition 3.3, there exist time sequences $(s_n^1) \nearrow \infty$, $(s_n^2) \nearrow \infty$ such that

$$u_x(s_n^1) \to w_x^1, \quad u_x(s_n^2) \to w_x^2 \qquad \text{in } C([a,b]).$$

This means in particular that u_x changes sign infinitely many times on (a,b). We will show that this is not possible.

In this aim, take $\alpha \in (a,b)$ and consider the function

$$v^\alpha(t,x) = u(t, x+\alpha) - u(t, -x+\alpha)$$

defined on $[-L + |\alpha|, L - |\alpha|]$. We note that by definition

$$v^\alpha(t,0) = 0 \quad \forall t \geq 0, \tag{5.6}$$
$$v_x^\alpha(t,0) = 2u_x(t,\alpha) \quad \forall t \geq 0. \tag{5.7}$$

It is also easy to see that v^α obeys to the equation

$$v_t^\alpha - [\varphi'(\xi(t,x,\alpha))v^\alpha]_{xx} = f'(\zeta(t,x,\alpha))\,v^\alpha$$

where $\xi(t,x,\alpha)$ and $\zeta(t,x,\alpha)$ lie between $u(t, \alpha - x)$ and $u(t, \alpha + x)$.

Take now any $T > 0$ finite. By Lemma 5.5 we have $u(t,x) \geq c_0 > 0$ for all $(t,x) \in [0,T] \times [a,b]$. It follows that v^α satisfies a regular equation on $[0,T] \times [-a_\alpha, a_\alpha]$ with $a_\alpha = \min(\alpha - a, -\alpha + b)$. (The condition $\alpha \in (a,b)$ is necessary and sufficient for the interval $[-a_\alpha, a_\alpha]$ to be well defined, included in the domain of definition of v^α and non $-$ trivial).

By [20, Theorem 5] and [2, Theorem D] we infer that on $[0,T] \times (-a_\alpha, a_\alpha)$

$v^\alpha(t, \cdot)$ has a finite number of zeroes N_t,

N_t does not increase in t,

N_t decreases strictly when one of the zeroes is multiple,

　　i.e. when $v^\alpha(t,z) = v_x^\alpha(t,z) = 0$. $\tag{5.8}$

Assume now that for some $\alpha \in (a,b)$, $u_x(\cdot, \alpha)$ has an infinite number of zeroes on $(0,\infty)$ and take T such that $u_x(\cdot, \alpha)$ has $N_0 + 1$ zeroes on $[0,T]$. Then by (5.6), (5.7) we see that we have $N_0 + 1$ multiple zeroes of $v^\alpha(\cdot, 0)$ on $[0,T]$. But each time that v^α has a multiple zero, the number of all zeroes has to diminish by (5.8); and this is a contradiction with (5.6). This ends the proof of Case 1).

We have shown that $w^1, w^2 \in \omega(u) \implies w^1 \equiv w^2$ which ends the proof of Proposition 5.4. $\qquad\qquad\square$

Theorem 1.1 is thus proved.

6. Appendix.

6.1. *Construction and main properties of approximating solutions.*

We consider a family of approximating problems, $(P^\varepsilon) = (P(\varphi^\varepsilon, h^\varepsilon, f^\varepsilon))$ with the following notion of solution.

DEFINITION 6.1. Let

$$L^\varepsilon(u, v) = \int_\Omega u_t v + \int_\Omega \nabla\varphi^\varepsilon(u)\nabla v - \int_\Omega f^\varepsilon(\cdot, u)v - \int_{\partial\Omega} \left[h - \beta\varphi^\varepsilon(u) \right] v.$$

We call a solution of (P^ε) a function $u^\varepsilon \in L^2(0, T; H^1(\Omega)) \cap W_2^1(0, T; L^2(\Omega))$ such that $u^\varepsilon(0) = u_0^\varepsilon$ a.e. in Ω, and

$$L^\varepsilon(u^\varepsilon, v) = 0 \quad \text{a.e. on } (0, T), \quad \text{for all } v \in L^2(0, T; H^1(\Omega)).$$

A subsolution (supersolution) of (P^ε) is a function u^ε of the same regularity such that $u^\varepsilon(0) \le u_0^\varepsilon$ $(u^\varepsilon(0) \ge u_0^\varepsilon)$, and

$$L^\varepsilon(u^\varepsilon, v) \le 0 \quad (L^\varepsilon(u^\varepsilon, v) \ge 0) \quad \text{for all } v \text{ as above, } v \ge 0.$$

Conformly to (H'), we chose the approximating functions so as to satisfy the following set of assumptions:

(A.1) $u_0^\varepsilon \in \mathcal{D}(\Omega)$, uniformly bounded in $C(\overline{\Omega})$, u_0^ε converge to u_0 in $L^2(\Omega)$;

(A.2) $\varphi^\varepsilon \in C^\infty(R) : \ 0 < c(\varepsilon) < (\varphi^\varepsilon)', \ \ \varphi^\varepsilon(r) \le C_\varphi(1 + r)$,
      φ^ε converge to φ uniformly on compact sets of $[0, \infty)$;

(A.3) $f^\varepsilon \in C^1([0, T] \times \overline{\Omega} \times [0, \infty))$, uniformly bounded : $\|f^\varepsilon\|_{L^\infty((0,T)\times\Omega\times(0,\infty))} \le C_f$,
      globally uniformly Lipschitz continuous with respect to u:
      $|f^\varepsilon(t, x, u) - f^\varepsilon(t, x, v)| \le L_f|u - v|$,
      and f^ε converge to f uniformly on compact sets;

(A.4) $h^\varepsilon \in C^1(\partial\Omega)$, $h^\varepsilon > 0$ and h^ε converges to h uniformly on $\partial\Omega$.

An approximation verifying (A.1)—(A.4) exists and is in general easy to construct.

It is also easy to verify that with these assumptions, (P^ε) admits a unique solution of regularity

$$u^\varepsilon \in C^{1+\delta}(0, T; H^1(\Omega)) \cap L^\infty((0, T) \times \Omega))$$

$$\Delta u^\varepsilon \in L^2(0, T; L^2(\Omega)),$$

and if $\partial\Omega$ is of class $C^{2+\delta}$,

$$u^\varepsilon \in C^\delta([0, T]; C^{2+\delta}(\Omega));$$

see [13, Theorems 3.3.3, 3.5.2] for results in Sobolev spaces, and [17, Theorem V.7.4] for continuity in space.

LEMMA 6.2. *Let $\underline{u}^{\varepsilon}$ be a subsolution of (P^{ε}) and $\overline{u}^{\varepsilon}$ a supersolution of (P^{ε}) with φ^{ε}, f^{ε} satisfying (A.2)—(A.3). Then for all $t \leq T$*

$$\int_{\Omega} [(\underline{u}^{\varepsilon}(t) - \overline{u}^{\varepsilon}(t))^{+}]^2 \leq e^{2L_f T} \int_{\Omega} [(\underline{u}_0 - \overline{u}_0)^{+}]^2,$$

This is checked by substracting the inequalities satisfied by $\underline{u}^{\varepsilon}, \overline{u}^{\varepsilon}$ and multiplying the difference by $\left(\underline{u}^{\varepsilon}(t) - \overline{u}^{\varepsilon}(t)\right)^{+}$.

LEMMA 6.3. *Let u^{ε} be a subsolution of (P^{ε}) f^{ε}, φ^{ε} satisfy (A.2)—(A.3). Then u^{ε} is bounded uniformly in ε in the space $L^{\infty}((0, T) \times \Omega)$:*

$$\sup_{(0,T)\times\Omega} |u^{\varepsilon}| \leq M(T).$$

We multiply this time (P^{ε}) by $(u^{\varepsilon} - M_0)^{+}$ and $(u^{\varepsilon} + M_0)^{-}$, $M_0 = \sup \|u_0^{\varepsilon}\|_{\infty}$.

LEMMA 6.4. *The approximation can be chosen to be a decreasing sequence of solutions*

$$u^{n+1} \leq u^n \qquad n = 1, 2, \ldots,$$

where we have denoted this time $u^n = u^{\frac{1}{n}}$.

Proof. It is sufficient to chose f^n, φ^n satisfying (A.2)—(A.3), and such that, in addition

$$\varphi^n \equiv \varphi \quad \text{on} \quad \left[\frac{1}{2n}, \infty\right), \tag{6.1}$$

and

$$f^n\left(t, x, \frac{1}{n}\right) \geq 0, \quad h^n \geq \beta\varphi^n\left(\frac{1}{n}\right), \quad f^{n+1}(t, x, \cdot) \leq f^n(t, x, \cdot), \quad h^{n+1} \leq h^n. \tag{6.2}$$

Such an approximation can be constructed: e.g.,

$$f^n(t, x, \cdot) = f(t, x, \cdot) + \sup_{(0, 1/n)} f^-(t, x, \cdot).$$

It is easy to see that $\frac{1}{n}$ is then a subsolution of (P^n) and thus, by the comparison principle (Lemma 6.2.)

$$u^n \geq \frac{1}{n}.$$

It follows by the condition (6.1) that $\varphi^{n+1}(u^n) = \varphi^n(u^n) = \varphi(u^n)$. Thus,

$$L_{n+1}(u^n, v) = L_n(u^n, v) + \int_\Omega f^n(t, x, u^n)v$$

$$- \int_\Omega f^{n+1}(t, x, u^n)v + \int_{\partial\Omega} (h^n - h^{n+1})v = \int_\Omega (f^n(t, x, u^n)$$

$$- f^{n+1}(t, x, u^n))v + \int_{\partial\Omega} (h^n - h^{n+1})v \geq 0$$

for all $v \geq 0$ by (6.2). Lemma 6.2. applied to L^{n+1} yields $u^{n+1} \leq u^n$. $\square$

Finally, DiBenedetto in [7, Theorems 6.2, 7.1] and [6, Theorems 3, 4] has proven equicontinuity of the family of approximating solutions under the assumption that they are uniformly bounded. We recall here the result.

PROPOSITION 6.5. *Assume that* (H') *holds and either* $N = 1$ *or* $\partial\Omega$ *is of class* C^1 *and* (H.5') *is satisfied. Then the family* $\{u^\varepsilon\}$ *is equicontinuous. Precisely, there exists a continuous function* $\omega_{t_0} : \mathbb{R}^+ \longrightarrow \mathbb{R}^+$, *such that* $\omega_{t_0}(0) = 0$ *and*

$$|u^\varepsilon(t_1, x_1) - u^\varepsilon(t_2, x_2)| \leq \omega_{t_0}(|x_1 - x_2| + |t_1 - t_2|^{\frac{1}{2}})$$

$$\forall(t_i, x_i) \in [t_0, T] \times \overline{\Omega}, \ i = 1, 2. \tag{6.3}$$

If u_0 *is continuous on* $\overline{\Omega}$, *one can take* $t_0 = 0$. *In addition, if* u *is uniformly bounded on* $\Omega \times (t_0, \infty)$, *then* ω_{t_0} *does not depend on* T.

Note that (6.3) holds also for $\varphi^\varepsilon(u^\varepsilon)$. The last statement of this proposition is implicit in [7], where the main concern is the case of T fixed; it follows however clearly from the remark at the end of the preliminary section, and of course from the proofs in [7].

6.2. *Proof of Theorem 2.3: existence and regularity part.*

We prove here uniform estimates on solutions of (P^ε). Note that we have already obtained a uniform $L^\infty((0, T) \times \Omega)$ bound on u^ε in Lemma 6.3. Note also that we do not use here the assumption on non–negativity of u_0, h and $f(0)$ (we keep however $\beta \geq 0$, so the reader should not be surprised when we omit this term in most estimates). This lack of restrictions will be of importance for the uniqueness proof.

As usual, we have by the first two energy estimates (where we use continuity of the embedding $H^1(\Omega) \hookrightarrow L^2(\partial\Omega)$ for the boundary term and the monotonicity of φ^ε):

LEMMA 6.6. *For all* $0 < T < \infty$ *there exist* ε_0, $C = C(T) > 0$ *such that for all* $\varepsilon \leq \varepsilon_0$

$$\int_0^T \int_\Omega |\nabla\varphi^\varepsilon(u^\varepsilon)|^2 \leq C, \quad \int_0^T \int_{\partial\Omega} \beta \, \varphi^\varepsilon(u^\varepsilon)^2 \leq C.$$

Assume in addition that f is independent of t and let

$$\gamma^\varepsilon(z) = \int_0^z (\varphi^\varepsilon)'(s)^{\frac{1}{2}}\, ds, \quad G^\varepsilon(x, z) = \int_0^z f^\varepsilon(x, s)\varphi^{\varepsilon\prime}(s)\, ds.$$

The functional

$$E^\varepsilon : v \mapsto \frac{1}{2}\int_\Omega |\nabla \varphi^\varepsilon(v)|^2 - \int_\Omega G^\varepsilon(x, v) - \int_{\partial\Omega}\left[h^\varepsilon - \frac{1}{2}\beta\,\varphi^\varepsilon(v)\right]\varphi^\varepsilon(v)$$

is a Liapunov function for (P^ε). For all $t_0 > 0$ and $T^* < \infty$ there exists C such that:

$$\sup_{t_0 \le t \le T^*}\int_\Omega |\nabla\varphi^\varepsilon(u^\varepsilon)|^2 \le C, \quad \int_{t_0}^{T^*}\int_\Omega \gamma^\varepsilon(u^\varepsilon)_t^2 \le C.$$

In addition, if u^ε are uniformly bounded in $L^\infty((0, \infty) \times \Omega)$, then the estimates above are also true for $T^* = \infty$.

Passing to the limit with $\varepsilon \to 0$ is done by dominated convergence, after having integrated by parts the time–derivative term

$$\int_0^T \int_\Omega u_t^\varepsilon\, v = -\int_0^T \int_\Omega u^\varepsilon\, v_t + \int_\Omega u^\varepsilon(0)\, v(0);$$

we use the convergence properties (A.1)–(A.4). Finally, from each subsequence of u^ε, we can chose a subsequence converging to u – weak solution of $(P(\varphi, f, h))$. We note that the first two of stated estimates are sufficient to get existence. It remains to show uniqueness of weak solution in the sense of Definition 2.1.

6.3.　Proof of Theorem 2.3: uniqueness. The Comparison Principle.

LEMMA 6.7. *Let u_1, u_2 be weak solutions of $(P(\varphi^1, f^1, h))$, $(P(\varphi^2, f^2, h))$ respectively. Then*

$$\int_0^T \int_\Omega (u^1(t) - u^2(t))[\varphi^1(u^1(t)) - \varphi^2(u^2(t))]\, dx\, dt$$

$$\le \int_0^T \int_\Omega (f^1 - f^2)\int_t^T [\varphi^1(u^1(\tau)) - \varphi^2(u^2(\tau))]\, d\tau\, dx\, dt$$

$$+ \int_\Omega (u_0^1 - u_0^2)\int_0^T [\varphi^1(u^1(t)) - \varphi^2(u^2(t))]\, dt\, dx$$

Proof. Indeed, we can take as test function in (P)

$$v(t) = \int_t^T \varphi^1(u^1(\tau)) - \varphi^2(u^2(\tau))d\tau, \quad \text{for } t \in [0, T];$$

note that $v \in W_2^1(0, T; H^1(\Omega))$ and $v_t = \varphi^2(u^2) - \varphi^1(u^1)$. We obtain then

$$-\int_0^T \int_\Omega (u^1 - u^2)\, v_t - \int_0^T \int_\Omega \nabla v_t\, \nabla v - \int_0^T \int_{\partial\Omega} \beta\, v_t\, v$$

$$= \int_0^T \int_\Omega (f^1 - f^2)\, v + \int_\Omega (u_0^1 - u_0^2)\, v(0).$$

As $-\int_0^T \int_\Omega \nabla v_t \nabla v = \frac{1}{2}\int_\Omega |\nabla v(0)|^2 \geq 0$ and $-\int_0^T \int_{\partial\Omega} \beta v_t v = \int_{\partial\Omega} \frac{\beta}{2} v(0)^2 \geq 0$, we have obtained the desired inequality. $\square$

LEMMA 6.8. *Assume* (H'). *Each u defined as in Definition 2.1 can be approximated by a sequence of solutions to* (P^ε).

Proof. Take u, weak solution of (P). Take also u_0^n, φ^n as in (A.1), (A.2) and $f^n \in L^\infty([0, T] \times \Omega)$ such that

$$f^n \to f(\cdot, u(\cdot)) \quad \text{in } L^2((0, T) \times \Omega).$$

Recall that $f(\cdot, u(\cdot)) \in L^2((0, T) \times \Omega)$ by definition and note that we work without the assumption $f(t, x, 0) \geq 0$, so we have here no restrictions on f as function of t, x. We have seen that u^n, solutions of $u_t^n - \Delta\varphi^n(u^n) = f^n$, exist, are unique, and by Lemmas 6.6 and 6.3, are uniformly bounded in $L^2((0, T) \times \Omega)$ and in $L^\infty((0, T) \times \Omega)$ by, say, M.

When applying Lemma 6.7. with

$$u^1 = u, \; f^1 = f(\cdot, u), \; \varphi^1 = \varphi, \qquad u^2 = u^n, \; f^2 = f^n, \; \varphi^2 = \varphi^n,$$

we obtain

$$\int_0^T \int_\Omega (u - u^n)[\varphi(u) - \varphi^n(u^n)] \to 0 \quad \text{as } n \to \infty. \tag{6.4}$$

As in addition ($\| \cdot \|$ below stands for $\| \cdot \|_{L^2((0,T)\times\Omega)}$)

$$\int_0^T \int_\Omega (u - u^n)[\varphi(u^n) - \varphi^n(u^n)] \leq (\|u\| + \|u^\varepsilon\|)\, (T|\Omega|)^{\frac{1}{2}} \sup_{z \in [0,M]} |\varphi(z) - \varphi^\varepsilon(z)| \to 0$$

by uniform convergence of φ^n to φ on compact sets, we obtain by adding this inequality to (6.4)

$$\int_0^T \int_\Omega (u - u^n)[\varphi(u) - \varphi(u^n)] \to 0 \quad \text{as } n \to \infty. \tag{6.5}$$

Now, monotonicity of φ allows to deduce that, for a subsequence, $u^n \to u$ a.e. on $(0, T) \times \Omega$. By uniform boundedness of (u^n) and the dominated convergence theorem we infer

$$u^n \to u, \quad \varphi(u^n) \to \varphi(u) \quad \text{in } L^2((0, T) \times \Omega).$$

We obtain therefore that u can be approximated by a sequence of solutions to $(P^{\frac{1}{n}})$. $\square$

From Lemmas 6.2 and 6.8 we deduce immediately the Comparison Principle, Proposition 2.6. This in its turn yields uniqueness of solution to (P).

6.4. *Proof of Proposition 2.5.*

Let $\sup u = M$. By (H.2'), $\min_{[\delta,M]} \varphi' = \delta' = \delta'(\delta) > 0$. By the last estimate of Lemma 6.6. we have

$$C \geq \int_{t_1}^{t_2} \int_K u_t^{\,2} (\varphi)'(u) \geq \delta' \int_{t_1}^{t_2} \int_K u_t^{\,2}. \tag{6.6}$$

where $C = C(\Omega, t_0)$, i.e. is independent of K, t_1, t_2. It follows that u_t and Δu belong to $L^2((t_1, t_2) \times K)$. We can multiply (P) by $u_t \, I_K$:

$$\int_{t_1}^{t_2} \int_K u_t^2 - \int_{t_1}^{t_2} \int_K \Delta\varphi(u)\, u_t = \int_{t_1}^{t_2} \int_K f(t, x, u)\, u_t.$$

With the Schwarz and Young inequalities applied to the last and the first terms, we obtain

$$\int_{t_1}^{t_2} \int_K \Delta\varphi(u)^2 \leq \frac{1}{4} \int_{t_1}^{t_2} \int_K f(t, x, u)^2 + 2 \int_{t_1}^{t_2} \int_K u_t^{\,2}. \tag{6.7}$$

Finally, (6.7) and (6.6) give together with (H.3')

$$\int_{t_1}^{t_2} \int_K \Delta\varphi(u)^2 \leq 2\,|K|\,|t_2 - t_1|\,C_f + \frac{2\,C(\Omega, t_0)}{\delta'} \tag{6.8}$$

which is the desired result.

REFERENCES

[1] ANDREU, F., MAZÓN, J. M., SIMONDON, F. and TOLEDO, J., *Global existence for a degenerate nonlinear diffusion problem with nonlinear gradient term and source*, Math. Ann., *314* (1999), pp. 703–728.

[2] ANGENENT, S., *The zeros set of a solution of a parabolic equation*, J. Reine Angew. Math., *390* (1988), pp. 79–96.

[3] ARONSON, D., CRANDALL, M. G. and PELETIER, L. A., *Stabilization of solutions of a degenerate nonlinear diffusion problem*, Nonlinear Anal., *6* (1982), pp. 1001–1022.

[4] BÉNILAN, PH. and CRANDALL, M. G., *The continuous dependence on φ of solutions of $u_t - \Delta\varphi(u) = 0$*, Indiana Univ. Math. J., *30* (1981), pp. 161–177.

[5] BERTSCH, M., KERSNER, R. AND PELETIER, L. A., *Positivity versus localization in degenerate diffusion equations*, Nonlinear Anal., *9* (1985), pp. 987–1008.

[6] DI BENEDETTO, E., *Interior and boundary regularity for a class of free boundary problems*, in Free boundary problems: theory and applications, vol. I, II (Montecatini, 1981), Pitman, Boston, Mass., 1983, pp. 383–396.

[7] DIBENEDETTO, E., *Continuity of weak solutions to a general porous medium equation*, Indiana Univ. Math. J., *32* (1983), pp. 83–118.

[8] FEIREISL, E. and SIMONDON, F., *Convergence for degenerate parabolic equations*, J. Differ. Equations, *152* (1999), pp. 439–466.

[9]   FEIREISL, E. and SIMONDON, F., *Convergence for semilinear degenerate parabolic equations in several space dimensions*, J. Dynam. Differential Equations, *12* (2000), pp. 647–673.

[10] FRIEDMAN, A., *Partial differential equations of parabolic type*, Prentice-Hall Inc., Englewood Cliffs, N.J., 1964.

[11] HALE, J. K. and RAUGEL, G., *Convergence in gradient-like systems with applications to PDE*, Z. Angew. Math. Phys., *43* (1992), pp. 63–124.

[12] HARAUX, A. and POLÁČIK, P., *Convergence to a positive equilibrium for some nonlinear evolution equations in a ball*, Acta Math. Univ. Comenian. (N.S.), *61* (1992), pp. 129–141.

[13] HENRY, D., *Geometric Theory of Semilinear Parabolic Equations*, vol. 840 of Lecture Notes in Mathematics, Springer–Verlag, 1980.

[14] HENRY, D., *Some infinite-dimensional Morse-Smale systems defined by parabolic partial differential equations*, J. Differential Equations, *59* (1985), pp. 165–205.

[15] KALASHNIKOV, S. A., *The propagation of disturbancies in problems of nonlinear heat conduction with absorption*, Zesh. Vychisl. Mat. i Mat. Phys., 1974.

[16] KERSNER, R., *Nonlinear heat conduction with absorption: space localization and extinction in finite time*, SIAM J. Appl. Math., *43* (1983), pp. 1274–1285.

[17] LADYZHENSKAYA, O., SOLONNIKOV, V. and URAL'CEVA, N., *Linear and quasi-linear equations of parabolic type*, vol. 23 of Translations of Mathematical Monographs, American Mathematical Society, 1968.

[18] LANGLAIS, M. and PHILLIPS, D., *Stabilization of solutions of nonlinear and degenerate evolution equations*, Nonlinear Analysis TMA, *9* (1985), pp. 321–333.

[19] MATANO, H., *Convergence of solutions of one-dimensional semilinear parabolic equations*, J. Math. Kyoto Univ., *18* (1978), pp. 221–227.

[20] MATANO, H., *Nonincrease of the lap-number of a solution for a one-dimensional semilinear parabolic equation*, J. Fac. Sci. Univ. Tokyo Sect. IA Math., *29* (1982), pp. 401–441.

[21] PELETIER, L. and ZHAO, J., *Source-type solutions of the porous media equation with absorption: The fast diffusion case*, Nonlinear Anal., Theory Methods Appl., *14* (1990), pp. 107–121.

[22] SIMON, L., *On the stabilization of solutions of nonlinear parabolic functional differential equations*, in Function spaces, differential operators and nonlinear analysis (Pudasjärvi, 1999), Acad. Sci. Czech Repub., Prague, 2000, pp. 239–250.

[23] ZELENYAK, T. I., *Stabilization of solutions of boundary value problems for a second-order parabolic equation with one space variable*, Differencial'nye Uravnenija, *4* (1968), pp. 34–45.

Maria Gokieli
ICM Warsaw University
al. Żwirlci wigury 93
02-089 Warsaw
Poland

Frédérique Simondon
Département de Mathématiques
Université de Henri Poincaré-Nancy 1
BP 239
54506 vandoeuvre les Nancy Cedex
France

J.evol.equ. 3 (2004) 549 – 576
1424–3199/03/040549 – 28
DOI 10.1007/s00030-003-0093-z
© Birkhäuser Verlag, Basel, 2004

**Journal of Evolution
Equations**

Analyticity of solutions to fully nonlinear parabolic evolution equations on symmetric spaces

JOACHIM ESCHER and GIERI SIMONETT

In memoriam Philippe Bénilan

Abstract. It is shown that solutions to fully nonlinear parabolic evolution equations on symmetric Riemannian manifolds are real analytic in space and time, provided the propagator is compatible with the underlying Lie structure. Applications to Bellman equations and to a class of mean curvature flows are also discussed.

1. Introduction

The smoothing property of solutions may be viewed as a characteristic feature of parabolic evolution equations. Roughly speaking, the smoothing property means that – under suitable assumptions – solutions to parabolic initial value problems enjoy more "spatial" regularity than the corresponding initial datum. This property is well-known for solutions to semilinear problems, see [18], and for solutions to classical quasilinear parabolic equations, see [10, 16, 21]. More recently, Amann [1] developed a theory for abstract quasi-linear problems in which the smoothing property of solutions appears as a cornerstone. Roughly speaking again, in these investigations smooth solutions are obtained by the property that parabolicity is preserved under (spatial) differentiation. This is obvious for linear problems. In the nonlinear situation considered in [1, 10, 16, 18, 21] this is due to the assumption that the nonlinearities are dominated by the parabolic linear part.

In this paper we are interested in fully nonlinear problems which we shall treat in the framework of maximal regularity. Here a completely different situation occurs. In fact, it can be viewed as a characteristic feature of this approach that it provides an existence and uniqueness theory without relying on a regularizing effect of solutions. This means that any form of smoothing of solutions to general fully nonlinear problems cannot be expected.

In order to be able to guarantee a smoothing action of solutions to fully nonlinear problems, one has to rely on additional structures of the problems under consideration. In this paper we consider the particular situation that the abstract equations come from nonlinear, possibly nonlocal operators acting on function spaces over a symmetric Riemannian manifold. Under the crucial assumption that these nonlinear operators are compatible with the underlying Lie structure we prove a strong regularizing property of solutions.

We illustrate the flexibility of our approach by discussing two different types of fully nonlinear parabolic evolution equations: First we treat a class of Bellman equations on $\mathbb{R}^m$, which arises in stochastic control theory. Secondly, we study a class mean curvature flow on spheres. A further example, occuring in the modelling of flows of incompressible fluids in rigid porous media, has been considered earlier in [11].

In the following we describe an important special case of our main result, Theorem 3.11. To make this more precise, let E_0 and E_1 be Banach spaces such that E_1 is continuously injected and dense in E_0. Assume further that $B \subset E_1$ is open and $P \in C^\omega(B, E_0)$. Given $u \in B$, let $\partial P(u) \in \mathcal{L}(E_1, E_0)$ denote the Fréchet derivative of P and assume that $\partial P(u)$ possess the property of maximal regularity in the sense of Da Prato-Grisvard, see Section 2 for a precise definition. Then, given $u_0 \in B$, the abstract evolution equation

$$\frac{d}{dt}u + P(u) = 0, \qquad u(0) = u_0, \tag{1.1}$$

possesses a unique solution

$$u := u(\cdot, u_0) \in C([0, t^+), B) \cap C^1([0, t^+), E_0), \tag{1.2}$$

where $t^+ := t^+(u_0) > 0$ stands for the maximal existence time of u, see again Section 2.

Assume now that M is a closed Riemannian manifold such that

$$E_1 \hookrightarrow buc^{1+\alpha}(M), \quad buc^\alpha(M) \hookrightarrow E_0 \hookrightarrow BUC(M), \tag{1.3}$$

for some $\alpha \in (0, 1)$, where $buc^s(M)$ denotes the closure of the smooth functions in the usual Hölder spaces $BUC^s(M)$, cf. Section 2. Furthermore, we assume that M is a globally symmetric space. This means that there is a Lie group G which acts as a transformation group on M. Let $\cdot : G \times M \to M$ denote the action of G on M and set

$$g \cdot v : M \to \mathbb{R}, \quad p \mapsto v(g \cdot p) \quad \text{for} \quad (g, v) \in G \times E_j,$$

where $j = 0, 1$. We call G a **strongly continuous transformation group** on E_j if $[v \mapsto g \cdot v] \in \mathcal{L}(E_j)$ for all $g \in G$ and if

$$a_v : G \to E_j, \qquad g \mapsto g \cdot v$$

is for any $v \in E_j$ continuous at e, the unit element in G. We further need a structural condition which connects the underlying geometry of M with the operator P. To make this precise, we say that P is **equivariant** with respect to G if there is a neighborhood U of e in G such that $U \cdot B \subset B$ and

$$P(g \cdot v) = g \cdot P(v) \quad \text{for} \quad (g, v) \in U \times B.$$

Finally, letting $\hat{u}(t, p) := u(t)(p)$ for $(t, p) \in [0, t^+) \times M$, we have the following result:

THEOREM 1.1. *Assume that G is a strongly continuous transformation group on E_j for $j = 0, 1$, and that P is equivariant with respect to G. Then the solution u to (1.1) is real analytic in space and time, i.e. $\hat{u} \in C^\omega((0, t^+) \times M)$.*

The above theorem is a special case of a more general result which is proved in the main body of this paper. We mention particularly that assumption (1.3), the assumption on M to be a symmetric space, as well as the equivariance of P can be weakened, see Theorem 3.11.

As mentioned above, the existence of solutions to (1.1) is obtained in the framework of continuous maximal regularity. We shall see in Section 3 that maximal regularity will also be instrumental in the proof of our main result, Theorem 3.11. In fact this property allows the application of the implicit function theorem in appropriate function spaces to show that, given $t \in (0, t^+)$, the mapping $(\lambda, X) \mapsto \exp(tX) \cdot u(\lambda t)$ is analytic on the Lie algebra $\mathbb{R} \times L(G)$ of $\mathbb{R} \times G$. This in turn, together with the analyticity of the exponential mapping, implies the analyticity of $\hat{u}$.

2. Continuous Maximal Regularity

In this section we briefly introduce the notion of maximal regularity in the sense of Da Prato-Grisvard. For this let E_0 and E_1 be Banach spaces such that E_1 is continuously injected and dense in E_0. Let $\mathcal{H}(E_1, E_0)$ denote the subset of all $A \in \mathcal{L}(E_1, E_0)$ such that $-A$, considered as a, in general, unbounded operator in E_0, generates a strongly continuous analytic semigroup on E_0. Let $B \subset E_1$ be open and assume that

$$P \in C^\omega(B, E_0) \quad \text{with} \quad \partial P(v) \in \mathcal{H}(E_1, E_0), \quad v \in B. \tag{2.1}$$

Given $T > 0$, set

$$\mathbb{E}_0 := C([0, T], E_0), \qquad \mathbb{E}_1 := C([0, T], E_1) \cap C^1([0, T], E_0),$$

and let $\gamma : \mathbb{E}_0 \to E_0$, $u \mapsto u(0)$ denote the trace operator in $\mathbb{E}_0$. We assume that $(\mathbb{E}_0, \mathbb{E}_1)$ is a pair of **maximal regularity** for $\partial P(v)$, this means we assume that

$$\left(\frac{d}{dt} + \partial P(v), \gamma \right) \in \mathcal{L}_{is}(\mathbb{E}_1, \mathbb{E}_0 \times E_1), \quad v \in B, \tag{2.2}$$

where $\mathcal{L}_{is}(X, Y)$ stands for the set of all bounded isomorphisms from the Banach space X into the Banach space Y. We are now ready to formulate the following existence and uniqueness result:

THEOREM 2.1. *Assume that (2.1) and (2.2) hold true. Then, given any $u_0 \in B$ and $f \in C(\mathbb{R}_+, E_0)$, there exist $t^+ := t^+(u_0) > 0$ and a unique maximal solution*

$$u := u(\cdot, u_0) \in C([0, t^+), B) \cap C^1([0, t^+), E_0) \tag{2.3}$$

of the initial value problem

$$\frac{d}{dt}u + P(u) = f, \qquad u(0) = u_0. \tag{2.4}$$

REMARKS 2.2. (a) Theorem 2.1 essentially goes back to Da Prato and Grisvard [9]. For some refinements and generalizations see also [5].

(b) Observe that assumption (2.2) and Theorem 2.1 coincide in the linear case, i.e., if $B = E_1$ and $P \in \mathcal{L}(E_1, E_0)$. Nevertheless, it is not at all clear whether or not property (2.2) can be verified if $E_1 \neq E_0$. In fact, it follows from a result of Baillon [7] that, in case $E_1 \neq E_0$, property (2.3) can only be expected if E_0 contains an isomorphic copy of the sequence space c_0. In particular, (2.3) will never be true in reflexive Banach spaces. However, in [9] the **continuous interpolation** functor $(\cdot, \cdot)^0_{\theta,\infty}$ was introduced, an interpolation method producing non-reflexive Banach spaces for which condition (2.2) can be verified.

(c) Let us briefly introduce an important scale of Banach spaces which may be realized as continuous interpolation spaces. Given $s \in \mathbb{R}$, define the **little Hölder spaces** to be

$$buc^s(\mathbb{R}^m) := \text{closure of } BUC^\infty(\mathbb{R}^m) \text{ in } B^s_{\infty,\infty}(\mathbb{R}^m),$$

where $B^s_{\infty,\infty}(\mathbb{R}^m)$ stands for the Besov spaces as defined in [26]. Note that the spaces $B^s_{\infty,\infty}(\mathbb{R}^m)$ coincides with the ususal Hölder spaces $BUC^s(\mathbb{R}^m)$, provided $s > 0$ is not an integer, see Theorem 2.5.7 and Remark 2.2.2.3 in [26]. Then it is shown in [22] Theorem 1.2.17 that

$$(BUC(\mathbb{R}^m), BUC^n(\mathbb{R}^m))^0_{\theta,\infty} = buc^{\theta n}(\mathbb{R}^m)$$

for all $n \in \mathbb{N}$ and $\theta \in (0, 1)$ such that $\theta n \notin \mathbb{N}$.

(d) Assume that M is a smooth Riemannian manifold with bounded curvature and positive radius of injectivity. Then Lemma 2.26 in [6] ensures the existence of a uniformly locally finite covering of geodesic balls $M(p_j, \delta)$ with $p_j \in M$, $j \in \mathbb{N}$ and $\delta > 0$. As before, the spaces $buc^s(M)$ are defined to be the closure of $BUC^\infty(M)$ in $BUC^s(M)$. Again we have that

$$(BUC(M), BUC^n(M))^0_{\theta,\infty} = buc^{\theta n}(M)$$

for all $n \in \mathbb{N}$ and $\theta \in (0, 1)$ such that $\theta n \notin \mathbb{N}$, cf. the proof of Corollary 1.2.19 in [22]. For simplicity we write $h^s(M) = buc^s(M)$ for $s \in \mathbb{R}$ if M is compact.

(e) Let M as above and fix $s_0, s_1 \in (0, \infty)$, $\theta \in (0, 1)$. Setting $s_\theta := (1 - \theta)s_0 + \theta s_1$, we have

$$(buc^{s_0}(M), buc^{s_1}(M))^0_{\theta,\infty} = buc^{s_\theta}(M),$$

provided s_0, s_1, and s_θ are not integers. This follows from (d), Theorem 7.4.4 in [27], and a density argument.

(f) A further scale of Banach spaces for which maximal regularity can be verified are the little Nikol'skii spaces. They can be realized as continuous interpolation spaces of Bessel potential spaces, cf. [9], Section 6 and [25], Section 6.

(g) Consider again the "linear" case $B = E_1$ and $P \in \mathcal{L}(E_1, E_0)$ and suppose in addition that $f \equiv 0$. Then problem (2.4) has for each $u_0 \in E_1$ a unique solution in the class $\mathbb{E}_1$ (for any $T > 0$, of course), provided $-P$ generates a strongly continuous semigroup, which does not need to be analytic. However, it is shown in [9] that the semigroup is automatically analytic if condition (2.2) is supposed to hold, see also Proposition III.3.1.1 in [2].

(h) A well-known characterization of generators of analytic semigroups yields that $A \in \mathcal{L}(E_1, E_0)$ belongs to $\mathcal{H}(E_1, E_0)$ if there are positive constants κ and ω such that $[\operatorname{Re} \lambda \geq \omega] \subset \rho(-A)$ and

$$|\lambda| \, \|(\lambda + A)^{-1}\|_{\mathcal{L}(E_0)} \leq \kappa, \qquad \operatorname{Re} \lambda \geq \omega.$$

(i) We mention that Theorem 2.1 remains true under a much weaker regularity assumption for P. Indeed, it suffices to assume that P is continuously Fréchet differentiable. Under this regularity assumption it can also be shown that the mapping

$$\bigcup_{x \in B} [0, t^+(x)) \times \{x\} \to B, \qquad (t, x) \mapsto u(t, x)$$

is a semiflow on B, provided f does not depend on t. However, since we are looking for possible smoothing properties of solutions, we presuppose analyticity of P from the very beginning.

(j) Let M be as in (d) and assume that $A \in \mathcal{H}(buc^{k+l+\beta}(M), buc^{k+\beta}(M))$ for some $k \in \mathbb{N}$, $l \in \mathbb{R}_+$, $\beta \in (0, 1)$ with $\beta + l \notin \mathbb{N}$. Let further $\alpha \in (\beta, 1)$ with $\alpha + l \notin \mathbb{N}$ and suppose that $buc^{k+l+\alpha}(M)$ is the domain of the $buc^{k+\alpha}(M)$-realization of A. Setting $E_0 := buc^{k+\alpha}(M)$ and $E_1 := buc^{k+l+\alpha}(M)$, it follows from Théorème 3.1 in [9] and (e) that $(\mathbb{E}_0, \mathbb{E}_1)$ is a pair of maximal regularity for the operator A, see also [2, 5].

3. The Smoothing Property

Let Σ be an analytic closed Riemannian manifold of dimension m and assume that E_0 and E_1 are Banach spaces of functions over Σ. More precisely, assume that E_1 is dense in E_0 and that

$$E_1 \hookrightarrow BUC(\Sigma), \qquad E_1 \hookrightarrow E_0 \hookrightarrow L_{1,\mathrm{loc}}(\Sigma). \tag{A_1}$$

Throughout this section we presuppose (2.1) and (2.2) and we let u denote the solution of (2.4) on $[0, t^+)$, where $u_0 \in B$ is given and where we assume for simplicity that $f \equiv 0$. Moreover, we set $\hat{u}(t, q) := u(t)(q)$ for $(t, q) \in [0, t^+) \times \Sigma$. Our goal is to show that u

enjoys a smoothing property. Hence, subdividing the interval of existence and using the semiflow property of u, see Remark 2.2(i), we may assume without loss of generality that $t^+ \leq 1$. Further, we fix $T \in (0, t^+)$ and set $I := [0, T]$.

From Theorem 2.1 we know that u belongs to $C(I, B) \cap C^1(I, E_0)$. Since we are dealing with nonlinear equations, including fully nonlinear partial differential equations involving nonlocal terms, there is no reason to expect u to have any further regularity, like

$$u \in C^\alpha(I \setminus \{0\}, E_1) \quad \text{or} \quad u \in C(I \setminus \{0\}, (E_1, E_2)_\alpha), \tag{3.1}$$

where E_2 stands for the domain of definition of $[\partial P(u_0)]^2$, equipped with the corresponding graph norm, and where $(\cdot, \cdot)_\alpha$ denotes a suitable interpolation method. However, it turns out that there is actually a strong smoothing property for solutions of problem (2.4), provided we impose suitable symmetry properties for the manifold Σ as well as for the nonlinear operator P.

It should be remarked that if P carries a quasilinear structure in the sense that Theorem 12.1 in [1] is applicable, then it can be shown that the corresponding solutions do in fact possess a smoothing property in the sense of (3.1) without any geometrical condition on Σ or on P.

Concerning the manifold Σ we shall assume that it is analytically diffeomorphic to a globally Riemannian symmetric space. More precisely, we assume that

there exists a globally symmetric Riemannian space

M and a $\Phi \in \mathrm{Diff}^\omega(M, \Sigma)$. $\hspace{2cm} (A_2)$

Recall that a Riemannian manifold M is called a globally symmetric space if it is connected and if for each $p \in M$ there is an involutive isometry $\sigma_p : M \to M$ such that p is an isolated fixed point of σ_p. Observe that σ_p reverses geodesics passing through the point p. This implies that M is complete and, by the Hopf-Rinow theorem, that the group $I(M)$ of all isometries acts transitively on M. Let g denote the metric on M and write $\tilde{g}$ for the metric on Σ induced by Φ and g. Then (A_2) implies that $(\Sigma, \tilde{g})$ is a globally symmetric Riemannian space. However, in view of applications, we prefer to keep the original metric on Σ.

Let now Φ^* and Φ_* denote the pull back and push forward operator induced by Φ. This means that, given $v \in L_{1,\mathrm{loc}}(\Sigma)$ and $w \in L_{1,\mathrm{loc}}(M)$, we have

$$\Phi^* v := v \circ \Phi \quad \text{and} \quad \Phi_* w := w \circ \Phi^{-1}.$$

For later purposes we need the following technical result.

LEMMA 3.1. $\Phi^* \in \mathcal{L}_{is}(L_{1,\mathrm{loc}}(\Sigma), L_{1,\mathrm{loc}}(M))$, $\Phi_* \in \mathcal{L}_{is}(L_{1,\mathrm{loc}}(M), L_{1,\mathrm{loc}}(\Sigma))$ and $[\Phi^*]^{-1} = \Phi_*$.

Proof. It follows from Theorem 2.2.26 and Corollary 2.2.21 in [19] that M has bounded curvature and a positive radius of injectivity $\delta > 0$. Hence Lemma 2.26 in [6] ensures that there exists a uniformly locally finite covering of geodesic balls $M(p_j, \delta)$ on M and a smooth partition of unity $\{\pi_j \,;\, j \in \mathbb{N}\}$ subordinated to $\{M(p_j, \delta)\,;\, j \in \mathbb{N}\}$. Let

$$\|w\|_{j,M} := \|\pi_j w\|_{L_1(M)}, \qquad \|v\|_{j,\Sigma} := \|\Phi^* v\|_{j,M}, \qquad j \in \mathbb{N},$$

for $v \in L_{1,\mathrm{loc}}(\Sigma)$ and $w \in L_{1,\mathrm{loc}}(M)$. Then, using the transformation theorem for the Lebesgue integral, it not difficult to see that

$$\{\|\cdot\|_{j,M} \,;\, j \in \mathbb{N}\} \quad \text{and} \quad \{\|\cdot\|_{j,\Sigma} \,;\, j \in \mathbb{N}\}$$

are separating families of seminorms, which induce the original topology on the spaces $L_{1,\mathrm{loc}}(M)$ and $L_{1,\mathrm{loc}}(\Sigma)$, respectively. The assertion follows now from the definition of $\|\cdot\|_{j,\Sigma}$. $\qquad\qquad\square$

For $j = 0,\ 1$, let

$$F_j := \{\Phi^* v \,;\, v \in E_j\}, \qquad \|w\|_{F_j} := \|\Phi_* w\|_{E_j}, \qquad w \in F_j.$$

Then it is not difficult to verify that $F_j := (F_j \,;\, \|\cdot\|_{F_j})$ are well-defined Banach spaces such that F_1 is continuously injected and dense in F_0. Moreover, we have $F_0 \subset L_{1,\mathrm{loc}}(M)$ and $F_1 \subset BUC(M)$. We next introduce

$$Q(w, \Phi) := \Phi^* P(\Phi_* w), \qquad v \in D,$$

where $D := \{\Phi^* v \,;\, v \in B\}$. Of course,

$$D \text{ is open in } F_1 \quad \text{and} \quad Q(\cdot, \Phi) \in C^\omega(D, F_0), \tag{3.2}$$

because of the fact that $\Phi^* : E_j \to F_j$ is an isometric isomorphism. We write $\partial_1 Q(w, \Phi) \in \mathcal{L}(F_1, F_0)$ for the Fréchet derivative of $Q(\cdot, \Phi)$. Further, we need the spaces

$$\mathbb{F}_0 := C(I, F_0), \qquad \mathbb{F}_1 := C(I, F_1) \cap C^1(I, F_0). \tag{3.3}$$

The pull back and push forward operator induced by Φ on $\mathbb{E}_0$ and $\mathbb{F}_0$ are defined pointwise with respect to $t \in I$, i.e., given $v \in \mathbb{E}_0$ and $w \in \mathbb{F}_0$, let

$$\Phi^* v : I \to F_0, \quad t \mapsto \Phi^* v(t), \qquad \Phi_* w : I \to E_0, \quad t \mapsto \Phi_* w(t).$$

Of course we do also not distinguish notationally between Φ^* and Φ_* and restrictions of these operators to linear subspaces of $\mathbb{E}_0$ and $\mathbb{F}_0$, respectively.

LEMMA 3.2. *The following assertions hold true:*

(i) $\Phi^* \in \mathcal{L}_{is}(\mathbb{E}_j, \mathbb{F}_j)$, $\Phi_* \in \mathcal{L}_{is}(\mathbb{F}_j, \mathbb{E}_j)$ *and* $[\Phi^*]^{-1} = \Phi_*$ *for* $j = 0, 1$.

(ii) $(\mathbb{F}_1, \mathbb{F}_0)$ *is a pair of maximal regularity for* $\partial_1 Q(\cdot, \Phi)$, *i.e.,*

$$\left(\frac{d}{dt} + \partial_1 Q(w, \Phi), \gamma \right) \in \mathcal{L}_{is}(\mathbb{F}_1, \mathbb{F}_0 \times F_1), \quad w \in D. \tag{3.4}$$

Proof. The first assertion follows from the construction of the spaces $\mathbb{F}_j$, $j = 0, 1$. The second one is a consequence of (i) and the chain rule. $\square$

Let $G := I_0(M)$ be the identity component of the group $I(M)$ of C^1-isometries on M. We already noticed that G acts transitively on M, see the remark after (A$_2$). Furthermore, it follows from (A$_2$) and Theorem I.4.6 in [20] that G acts analytically as a Lie transformation group on M, so that M is a homogeneous Riemannian space with respect to G. Fix $p_0 \in M$ and let $H := \{g \in G; \ g \cdot p_0 = p_0\}$ denote the isotropy group of p_0, where $\cdot : G \times M \to M$ denotes the action of G on M. Then G/H admits a real analytic structure and

$$j : G/H \to M, \qquad g \cdot H \mapsto g \cdot p_0$$

is a real analytic diffeomorphism, cf. Proposition I.4.2 in [20]. By means of this diffeomorphism we always identify M with the coset manifold G/H. Finally, recall that

- $\mathbb{R}^m$,
- the unit sphere $\mathbb{S}^m = SO(m + 1)/SO(m)$,
- products of Riemannian globally symmetric spaces

are Riemannian globally symmetric spaces.

Let Y be a nonempty set. Given $f \in Y^M$ and $g \in G$, we define $g \cdot f \in Y^M$ by $g \cdot f(p) := f(g \cdot p)$. The next result contains the transformation rule for the operator Q with respect to G which will be needed in the following.

LEMMA 3.3. *Let* $g \in G$ *be given. Then*

(i) $g \cdot \Phi \in \mathit{Diff}^\omega(M, \Sigma)$;

(ii) $g \cdot Q(w, \Phi) = Q(g \cdot w, g \cdot \Phi)$, $\quad w \in D$.

Proof. (i) This follows from the analyticity of the group action of G on M.

(ii) Given $g \in G$ we have

$$(g \cdot \Phi)^* v = g \cdot (\Phi^* v), \quad v \in E_0, \quad (g \cdot \Phi)_*(g \cdot w) = \Phi_* w, \quad w \in F_0. \tag{3.5}$$

Thus we find

$$Q(g \cdot w, g \cdot \Phi) = (g \cdot \Phi)^* P((g \cdot \Phi)_*(g \cdot w)) = (g \cdot \Phi)^* P(\Phi_* w)$$
$$= g \cdot (\Phi^* P(\Phi_* w)) = g \cdot Q(w, \Phi),$$

for any $w \in D$. $\square$

We next assume that

$$G \text{ is a strongly continuous transformation group on } F_j \text{ for } j = 0, 1. \qquad (A_3)$$

Writing $L(G)$ for the Lie algebra of G, we define

$$T_X(t)w := \exp(tX) \cdot w \quad \text{for} \quad (X, t, w) \in L(G) \times \mathbb{R} \times F_j.$$

It follows from (A_3) that $\{T_X(t) \,;\, t \in \mathbb{R}\}$ is for any $X \in L(G)$ a strongly continuous group on F_0. We write A_X for the infinitesimal generator of $\{T_X(t) \,;\, t \in \mathbb{R}\}$ and we assume that

$$F_1 \subset \mathrm{dom}(A_X) \quad \text{for any} \quad X \in L(G), \qquad (A_4)$$

where $\mathrm{dom}(A_X)$ is given the graph norm of A_X. Observe that (A_4) and the closed graph theorem imply that

$$A_X \in \mathcal{L}(F_1, F_0) \quad \text{for any} \quad X \in L(G). \qquad (3.6)$$

Recall that, given $w \in F_j$, we have set

$$a_w : G \to F_j, \qquad g \mapsto g \cdot w.$$

We write $d_g a_w$ for the differential of a_w at $g \in G$, provided a_w is differentiable, of course.

LEMMA 3.4. (i) *If $w \in F_j$ then $a_w \in C(G, F_j)$ for $j = 0, 1$.*

(ii) *If $w \in F_1$ then $a_w \in C^1(G, F_0)$ and, given $(g, X) \in G \times L(G)$, we have*

$$d_g a_w X = A_X(g \cdot w). \qquad (3.7)$$

Proof. (i) Let $g \in G$ and choose a sequence (g_k) in G such that $g_k \to g$. Since G is a strongly continuous transformation group on F_j we find

$$a_w(g_k) = a_{g \cdot w}(g^{-1} g_k) \to g \cdot w = a_w(g) \quad \text{in} \quad F_j.$$

(ii) Due to (i) and (3.6) it suffices to prove (3.7). Identifying $T_e G \cong L(G)$, we have that $\{g \cdot \exp(tX) \,;\, t \in \mathbb{R}\}$ is an integral curve on G through g with tangent vector X. Thus (A_3) yields

$$\lim_{t \to 0} \frac{a_w(g \cdot \exp(tX)) - a_w(g)}{t} = \lim_{t \to 0} \frac{\exp(tX) \cdot (g \cdot w) - (g \cdot w)}{t} = A_X(g \cdot w)$$

in F_0, since w belongs to $F_1 \subset \mathrm{dom}(A_X)$.

$$\square$$

REMARK 3.5. Let $M = \Sigma = \mathbb{R}$ and $E_j = BUC^j(\mathbb{R})$ for $j = 0, 1$. Then $F_j = E_j$ for $j = 0, 1$ and, given $w \in F_j$ and $\lambda \in \mathbb{R}$, we have that $a_w(\lambda) = \tau_\lambda w$, where $\tau_\lambda w$ denotes the left translation by $\lambda \in \mathbb{R}$ of w. This shows that the regularity of a_w with respect to G as stated in Lemma 3.4 is optimal.

We note that $L(G)$ is finite dimensional, as Theorem VI 3.3 and Theorem VI 3.4 (4) in [20] imply. Let $\mathcal{B} = \{X_1, \ldots, X_N\}$ be a fixed basis of $L(G)$. Given $(\mu_1, \ldots, \mu_N) \in \mathbb{R}^N$, we write

$$\mu\mathcal{B} := \sum_{k=1}^{N} \mu_k X_k, \quad T_\mu(t) := T_{\mu\mathcal{B}}(t), \quad A_\mu := A_{\mu\mathcal{B}}.$$

Let E, F, and G be Banach spaces. Writing $\mathcal{L}^2(E \times F, G)$ for the Banach space of all bilinear continuous mappings from $E \times F$ to G, we have

COROLLARY 3.6. $[(\mu, w) \mapsto A_\mu w] \in \mathcal{L}^2(\mathbb{R}^N \times \mathbb{F}_1, \mathbb{F}_0)$.

Proof. It follows readily from Lemma 3.4 that $[(\mu, w) \mapsto A_\mu w]$ is bilinear. Let $C := \sum_{k=1}^{N} \|A_{X_k}\|_{\mathcal{L}(F_1, F_0)}$ and observe that (3.6) implies that $C < \infty$. Now, given $w \in \mathbb{F}_1$, $\mu \in \mathbb{R}^N$, and $t \in I$, we have

$$\|A_\mu w(t)\|_{F_0} = \left\| d_e a_{w(t)} \sum_{k=1}^{N} \mu_k X_k \right\|_{F_0} \leq |\mu|_{\mathbb{R}^N} \sum_{k=1}^{N} \|A_{X_k} w(t)\|_{F_0} \leq C|\mu|\|w\|_{\mathbb{F}_1}.$$

Taking the maximum over $t \in I$, we get the assertion. $\square$

We shall now formulate the compatibility condition for the operator P with respect to G. For this we first have to introduce the following assumption:

D is invariant under $T_\mu(t)$ for all $(\mu, t) \in (-\varepsilon_0, \varepsilon_0)^N \times I$. (A_5)

Further, let $\mathbb{D}_1 := C(I, D) \cap C^1(I, F_0)$, pick $(\mu, w) \in (-\varepsilon_0, \varepsilon_0)^N \times \mathbb{D}_1$, and define

$$\mathbb{Q}(\mu, w)(t) := Q(w(t), T_\mu(t)\Phi), \quad t \in I,$$

where $T_\mu(t)\Phi := \exp(t\mu\mathcal{B}) \cdot \Phi$ for $(t, \mu) \in \mathbb{R} \times \mathbb{R}^N$. Observe that (A_5) ensures that this definition is meaningful. Moreover, we have

$$\mathbb{Q}(\mu, w)(t) = (T_\mu(t)\Phi)^* P((T_\mu(t)\Phi)_* w(t)), \quad t \in I.$$

Hence $\mathbb{Q}(\mu, w) \in F_0^I$. We need that this function belongs to $\mathbb{F}_0$ and that the operator $\mathbb{Q}(\mu, w)$ depends analytically on (μ, w). For this we first observe that

$\mathbb{D}_1$ is an open subset of $\mathbb{F}_1$. (3.8)

Indeed, this follows from the compactness of I and the fact that D is open in F_1. We now assume that there is a $r \in (0, \varepsilon_0)$ such that

$$[(\mu, w) \mapsto \mathbb{Q}(\mu, w)] \in C^\omega(\mathbb{B}_{\mathbb{R}^N}(0, r) \times \mathbb{D}_1, \mathbb{F}_0). \tag{A_6}$$

REMARKS 3.7. (a) Assume that $D \to F_0$, $w \mapsto \Phi^* P(\Phi_* w)$ is equivariant with respect to G. Then assumption (A_6) is satisfied. Indeed, recalling (3.5) we get

$$\begin{aligned}
\mathbb{Q}(\mu, w)(t) &= (T_\mu(t)\Phi)^* P((T_\mu(t)\Phi)_* w(t)) \\
&= T_\mu(t)\Phi^* P(\Phi_* T_\mu(-t)w(t)) \\
&= \Phi^* P(\Phi_* w(t)) = Q(w(t), \Phi)
\end{aligned}$$

for $w \in \mathbb{D}_1$ and $t \in I$. Now the assertion follows from (3.2).

(b) Assume that $\Sigma = M = \mathbb{R}^m$ and that $\Phi = id$. Furthermore, set

$$E_0 := buc^\alpha(\mathbb{R}^m), \quad B := E_1 := buc^{2+\alpha}(\mathbb{R}^m)$$

for some $\alpha \in (0, 1)$. We fix $a \in buc^\alpha(\mathbb{R}^m)$ and define

$$P(w) := a\Delta w \quad \text{for} \quad w \in E_1. \tag{3.9}$$

Obviously, P is only equivariant under translations if a is constant. In order to satisfy (A_6) for nonconstant coefficient functions a, observe that

$$\mathbb{Q}(\mu, w)(t) = (\tau_{t\mu} a)\Delta w(t) \quad \text{for} \quad w \in \mathbb{E}_1,$$

where $\tau_{t\mu} a$ stands for the translation of a by the vector $t\mu \in \mathbb{R}^m$. Assume now

$a \in C^\infty(\mathbb{R}^m)$ and there is a $M_0 > 0$ such that
$$\|\partial^\beta a\|_{BUC^\alpha(\mathbb{R}^m)} \leq M_0 \beta! \text{ for all } \beta \in \mathbb{N}^m. \tag{3.10}$$

Clearly, (3.10) implies that $a \in C^\omega(\mathbb{R}^m)$, but observe that (3.10) is in general stronger than pointwise analyticity. If (3.10) holds then P satisfies (A_6). Indeed, we first observe that the mapping $[(b, w) \mapsto b\Delta w] : \mathbb{E}_0 \times \mathbb{E}_1 \to \mathbb{E}_0$ is bilinear and bounded. Thus it remains to show that $[\mu \mapsto \tau_{t\mu} a] \in C^\omega(\mathbb{B}_{\mathbb{R}^m}(0, r), \mathbb{E}_0)$ for an appropriate number $r > 0$. An easy computation shows that

$$\|\partial_\mu^\beta \tau_{t\mu} a\|_{BUC^\alpha(\mathbb{R}^m)} = t^{|\beta|}\|\tau_{t\mu}\partial^\beta a\|_{BUC^\alpha(\mathbb{R}^m)} = t^{|\beta|}\|\partial^\beta a\|_{BUC^\alpha(\mathbb{R}^m)} \leq M_0 T^{|\beta|}\beta!$$

for all $(\beta, \mu) \in \mathbb{R}^m \times \mathbb{R}^m$ and $t \in I$. Recall that we have $T < 1$ by assumption. Hence we conclude that (A_6) is satisfied if $r = 1$, for instance.

(c) It has been shown in [15] that the strategy of using maximal regularity in conjunction with the implicit function theorem does also guarantee that the solutions to $\partial_t u - a\Delta u = 0$ are analytic in space and time, provided a satisfies the weaker assumption $a \in C^\omega(\mathbb{R}^m) \cap buc^\alpha(\mathbb{R}^m)$ instead of (3.10).

We next fix $\varepsilon_0 > 0$ such that $\lambda t \in [0, t^+)$ for all $\lambda \in (1 - \varepsilon_0, 1 + \varepsilon_0)$ and all $t \in [0, T]$. Given now any $(\lambda, \mu) \in (1 - \varepsilon_0, 1 + \varepsilon_0) \times \mathbb{R}^N$, we define $w_{\lambda,\mu} \in F_0^I$ by $w_{\lambda,\mu}(t) := T_\mu(t)\Phi^* u(\lambda t)$ for $t \in I$, where we recall that u is the solution to (2.4).

Our next result shows that the function $w_{\lambda,\mu}$ solves a parameter dependent evolution equation involving the operators $\mathbb{Q}$ and A_μ. In order to economize our notation, we set $\Pi := \Pi(\varepsilon_0) := (1 - \varepsilon_0, 1 + \varepsilon_0) \times (-\varepsilon_0, \varepsilon_0)^N$.

LEMMA 3.8. *Given* $(\lambda, \mu) \in \Pi$, *we have*

(i) $w_{\lambda,\mu} \in \mathbb{D}_1$.

(ii) $w_{\lambda,\mu}$ *solves the evolution equation*

$$\frac{d}{dt} w + \lambda \mathbb{Q}(\mu, w) = A_\mu w, \qquad w(0) = w_0, \tag{3.11}$$

where $w_0 := \Phi^* u_0$.

Proof. (i) This follows from (2.3), the definition of D, see (3.2), assumption (A_5), Lemma 3.4, and the analyticity of the exponential mapping. Furthermore, Lemma 3.4 implies that

$$\frac{d}{dt} w_{\lambda,\mu}(t) = T_\mu(t) A_\mu(\Phi^* u(\lambda t)) + \lambda (T_\mu(t)\Phi)^* \frac{d}{dt} u(\lambda t), \quad t \in I.$$

Observing (3.5) and the fact that $T_\mu(t)$ and A_μ commute on F_1 we get

$$\frac{d}{dt} w_{\lambda,\mu}(t) = A_\mu w_{\lambda,\mu}(t) + \lambda (T_\mu(t)\Phi)^* \frac{d}{dt} u(\lambda t), \quad t \in I. \tag{3.12}$$

(ii) Using (3.12) and (2.4) we now find

$$\frac{d}{dt} w_{\lambda,\mu}(t) = A_\mu w_{\lambda,\mu}(t) - \lambda (T_\mu(t)\Phi)^* P(u(\lambda t)), \quad t \in I.$$

From Lemma 3.3 (ii) and the definitions of the operators Q and $\mathbb{Q}$ we further conclude

$$(T_\mu(t)\Phi)^* P(\Phi_* \Phi^* u(\lambda t)) = T_\mu(t) Q(\Phi^* u(\lambda t), \Phi)$$

$$= Q(T_\mu(t)\Phi^* u(\lambda t), T_\mu(t)\Phi) = \mathbb{Q}(\mu, w_{\lambda,\mu})(t)$$

for $t \in I$. This completes the proof. $\square$

Our next lemma contains the key result to show via the implicit function theorem that the mapping $(\lambda, \mu) \mapsto w_{\lambda,\mu}$ is analytic.

LEMMA 3.9. *Given* $((\lambda, \mu), w) \in \Pi \times \mathbb{D}_1$, *let*

$$F((\lambda, \mu), w) := \left(\frac{d}{dt} w + \lambda \mathbb{Q}(\mu, w) - A_\mu w, w(0) - w_0 \right).$$

Then

$$F \in C^\omega(\Pi \times \mathbb{D}_1, \mathbb{F}_0 \times F_1) \tag{3.13}$$

and

$$\partial_2 F((1, 0), w) \in \mathcal{L}_{is}(\mathbb{F}_1, \mathbb{F}_0 \times F_1), \quad w \in \mathbb{D}_1, \tag{3.14}$$

where $\partial_2 F$ is the derivative of F with respect to $w \in \mathbb{D}_1$.

Proof. (i) Clearly, we have that

$$\left(\frac{d}{dt}, \gamma \right) \in \mathcal{L}(\mathbb{F}_1, \mathbb{F}_0 \times F_1).$$

Hence it follows from Corollary 3.6 that

$$\left[((\lambda, \mu), w) \mapsto \left(\frac{d}{dt} w - A_\mu w, w(0) - w_0 \right) \right] \in C^\omega(\Pi \times \mathbb{F}_1, \mathbb{F}_0 \times F_1).$$

Thus we obtain (3.13) from assumption (A_6).

(ii) Let $w \in \mathbb{D}_1$ and $h \in \mathbb{F}_1$ be given. Then we have

$$\partial_2 F((1, 0), w)h = \frac{d}{d\varepsilon} F((1, 0), w + \varepsilon h)\big|_{\varepsilon=0} = \left(\frac{d}{dt} h + \partial_1 Q(w, \Phi)h, h(0) \right).$$

Combining Lemma 3.2(ii) with Remark III 3.4.2(c) in [2] it follows that, given $(f, \varphi) \in \mathbb{F}_0 \times F_1$, there is a unique solution $h \in \mathbb{F}_1$ to the inhomogeneous evolution equation

$$\frac{d}{dt} h + \partial_1 Q(w(t), \Phi)h = f(t), \quad h(0) = \varphi.$$

(3.14) is now a consequence of the open mapping theorem. $\qquad\square$

We are now prepared to show that $w_{\lambda,\mu}$ depends analytically on the parameter (λ, μ).

PROPOSITION 3.10. *There is an $\varepsilon_0 > 0$ such that $[(\lambda, \mu) \mapsto w_{\lambda,\mu}] \in C^\omega(\Pi(\varepsilon_0), \mathbb{D}_1)$.*

Proof. Let F be given as in Lemma 3.9 and observe that $F((\lambda, \mu), w) = 0$ if and only if $w \in \mathbb{D}_1$ is a solution to

$$\frac{d}{dt} w + \lambda \mathbb{Q}(\mu, w) = A_\mu w, \quad w(0) = w_0.$$

Now the assertion follows from Lemma 3.8, Lemma 3.9, and the implicit function theorem in Banach spaces. $\qquad\square$

It remains to translate the above Proposition into the desired analyticity of $\hat{u}$, see the beginning of this section.

THEOREM 3.11. *Assume that* (A_1)–(A_6) *hold true. Then* $\hat{u} \in C^{\omega}((0, t^+) \times \Sigma)$.

Proof. (i) Let $\hat{w} := \Phi^* \hat{u}$, i.e. $\hat{w}(t, p) := \hat{u}(t, \Phi(p))$ for $(t, p) \in (0, t^+) \times M$. It suffices to show that $\hat{w} \in C^{\omega}((0, t^+) \times M)$. For this we fix $(t_0, p_0) \in (0, t^+) \times M$. Moreover there exists a subset $\{j_1, \ldots, j_m\}$ of $\{1, \ldots, N\}$ such that $\{X_{j_1}, \ldots, X_{j_m}\} \subset L(G)$ induces via the integral curves $[t \mapsto \exp(tX_{j_k}) \cdot p_0]$ a basis of $T_{p_0} M$ (recall that we identify M with the coset manifold G/H, see also Theorem IV.3.3(iii) in [17]). Without loss of generality we may assume that $j_1 = 1, \ldots, j_m = m$. Moreover, in the following we write $\hat{\mu} = (\mu_1, \ldots, \mu_m, 0, \ldots, 0) \in \mathbb{R}^N$ for $(\mu_1, \ldots, \mu_m) \in \mathbb{R}^m$ and we identify

$$\Pi_m := \Pi_m(\varepsilon_0) := (1 - \varepsilon_0, 1 + \varepsilon_0) \times (-\varepsilon_0, \varepsilon_0)^m \quad \text{with} \quad \Pi \cap (\mathbb{R}^{m+1} \times \{0\}).$$

Shrinking $\varepsilon_0 > 0$ if necessary, we have that

$$\varphi : \Pi_m \to (0, t^+) \times M, \qquad (\lambda, \hat{\mu}) \mapsto (\lambda t_0, T_{\hat{\mu}}(t_0) \cdot p_0)$$

is an analytic parametrization of an open neighborhood O of (t_0, p_0) in $(0, t^+) \times M$.

(ii) Observe that by assumption (A_1) we know that $\mathbb{D}_1 \subset C(I, BUC(M))$. Thus the evaluation mapping

$$\mathbb{D}_1 \to \mathbb{R}, \qquad w \mapsto w(t_0)(p_0)$$

is well-defined and clearly analytic. Combining this with Proposition 3.10 we find

$$[(\lambda, \hat{\mu}) \mapsto w_{\lambda, \hat{\mu}}(t_0)(p_0)] \in C^{\omega}(\Pi_m, \mathbb{R}).$$

But $\varphi^* \hat{w}(\lambda, \hat{\mu}) = w_{\lambda, \hat{\mu}}(t_0)(p_0)$ for $(\lambda, \hat{\mu}) \in \Pi_m$. This shows that $\hat{w} \in C^{\omega}(O, \mathbb{R})$ and completes the proof. $\qquad\qquad\square$

Proof of Theorem 1.1. Setting $\Sigma = M$ and $\Phi = id$, it is clear that the hypotheses of Theorem 1.1 on M and G imply that (A_1)–(A_3) are fulfilled. Moreover, (A_5) and, by Remark 3.7(a) also (A_6) are satisfied by the assumed equivariance of P with respect to G. Finally, let $X \in L(G)$ be given. Then, using local coordinates, it is not difficult to verify that

$$\lim_{t \to 0} \frac{\exp(tX) \cdot w - w}{t} = A_X w \quad \text{in} \quad buc^{\alpha}(M)$$

for all $w \in buc^{1+\alpha}(M)$. Since $buc^{\alpha}(M) \hookrightarrow E_0$ it follows from Lemma 3.4 that $buc^{1+\alpha}(M)$ is contained in $\text{dom}(A_X)$. Hence (1.3) implies that (A_4) is true as well. Now the assertion follows from Theorem 3.11.

4. Applications

This section is devoted to two applications of Theorem 3.11. First we consider the Bellman equation on $\mathbb{R}^m$, an equation occuring in stochastic control theory. Secondly, we discuss a class of mean curvature flow on spheres. For a further example of a fully nonlinear parabolic evolution equation on $\mathbb{R}^m$ which is equivariant with respect to translations we refer to [11].

4.1. *Bellman Equations*

In certain cases the Bellman equation occuring in stochastic control theory leads to the following fully nonlinear partial differential equation, cf. Section 8.5.5 in [22],

$$\partial_t u = \frac{1}{2} \sum_{j,k=1}^{m} a_{jk} \partial_j \partial_k u - \frac{1}{2} \langle (I + \partial^2 u)^{-1} \partial u | \partial u \rangle + \sum_{j=1}^{m} b_j \partial_j u, \qquad (4.1)$$

to be satisfied on $(0, T) \times \mathbb{R}^m$, where $T > 0$ is given. In (4.1), the Euclidean inner product on $\mathbb{R}^m$ is denoted by $\langle \cdot | \cdot \rangle$ and $I := id_{\mathbb{R}^m}$. Moreover, ∂u and $\partial^2 u$ stand for the gradient and the Hessian of u, respectively. Concerning the coefficients we assume that

$$a_{jk}, \; b_j \in buc^{\alpha}(\mathbb{R}^m) \quad \text{with} \quad a_{jk} = a_{kj} \qquad (4.2)$$

for $j, k = 1, \ldots, m$. Here, $\alpha \in (0, 1)$ is fixed. Given $x \in \mathbb{R}^m$, we set $A(x) := [a_{jk}(x)] \in \mathbb{R}^{m \times m}$ and assume the following ellipticity condition to hold: There is a constant $c > 0$ such that

$$\langle A(x)\xi | \xi \rangle \geq c|\xi|^2, \qquad (x, \xi) \in \mathbb{R}^m \times \mathbb{R}^m. \qquad (4.3)$$

The equation (4.1) is complemented by the initial condition

$$u(0, x) = u_0(x), \qquad x \in \mathbb{R}^m. \qquad (4.4)$$

Here we suppose that $u_0 \in buc^{2+\alpha}(\mathbb{R}^m)$ and that there is a $\delta > 0$ such that

$$\det(I + \partial^2 u_0(x)) \geq \delta, \qquad x \in \mathbb{R}^m. \qquad (4.5)$$

By continuity, there is an $\varepsilon > 0$ such that, given $v \in B := \mathbb{B}_{buc^{2+\alpha}(\mathbb{R}^m)}(u_0, \varepsilon)$, we have

$$\det(I + \partial^2 v(x)) \geq \delta/2, \qquad x \in \mathbb{R}^m. \qquad (4.6)$$

In order to economize our notation, we set $j(v) := (I+\partial^2 v)^{-1}$ for $v \in B, b = (b_1, \ldots, b_m)$, and

$$P(v) := \frac{1}{2}\mathrm{trace}(A\partial^2 v) - \frac{1}{2}\langle j(v)\partial v | \partial v \rangle + \langle b | \partial v \rangle, \quad v \in B.$$

It follows from (4.6) and the fact that inversion maps $\mathcal{L}_{is}(\mathbb{R}^m)$ analytically into $\mathcal{L}(\mathbb{R}^m)$, that

$$P \in C^\omega(B, buc^\alpha(\mathbb{R}^m)). \tag{4.7}$$

Moreover, given $v \in B$ and $h \in buc^{2+\alpha}(\mathbb{R}^m)$, we have $\partial j(v)h = -j(v)\partial^2 hj(v)$, and therefore

$$\partial P(v)h = \frac{1}{2}\text{trace}(A\partial^2 h) + \frac{1}{2}\langle \partial^2 hj(v)\partial v | j(v)\partial v\rangle - \langle j(v)\partial h|\partial v\rangle + \langle b|\partial h\rangle.$$

Let now

$$p(v) := \frac{1}{2}\left(A + (j(v)\partial v \otimes j(v)\partial v)\right), \qquad v \in B,$$

denote the principal symbol of the second order operator $\partial P(v)$. Then we have

$$\langle p(v)(x)\xi|\xi\rangle = \frac{1}{2}(\langle A(x)\xi|\xi\rangle + \langle \xi|j(v)(x)\partial v(x)\rangle^2), \quad (x, \xi) \in \mathbb{R}^m \times \mathbb{R}^m.$$

Hence (4.3) implies that $\partial P(v)$ is uniformly elliptic and we conclude from Theorem 4.2 and Remark 4.6 in [3] that

$$\partial P(v) \in \mathcal{H}(buc^{2+\beta}(\mathbb{R}^m), buc^\beta(\mathbb{R}^m)), \qquad \beta \in (0, \alpha], \ v \in B. \tag{4.8}$$

Setting $E_0 := buc^\alpha(\mathbb{R}^m)$ and $E_1 := buc^{2+\alpha}(\mathbb{R}^m)$, it follows from (4.7), (4.8), and Remark 2.2(j) that (2.1) and (2.2) hold true. Hence we obtain the following existence and uniqueness result for (4.1), (4.4).

PROPOSITION 4.1. *Let* $u_0 \in buc^{2+\alpha}(\mathbb{R}^m)$ *satisfy* (4.5) *and assume that* (4.2), (4.3) *hold true. Then there is a* $t^+ := t^+(u_0) > 0$ *such that the Bellman equation*

$$\begin{cases} \partial_t u = \frac{1}{2}trace(A\partial^2 u) - \frac{1}{2}\langle j(u)\partial u|\partial u\rangle + \langle b|\partial u\rangle, \ (t, x) \in (0, T) \times \mathbb{R}^m, \\ u(0, x) = u_0(x), \quad x \in \mathbb{R}^m \end{cases} \tag{4.9}$$

has a unique solution $u \in C([0, t^+), buc^{2+\alpha}(\mathbb{R}^m)) \cap C^1([0, t^+), buc^\alpha(\mathbb{R}^m))$.

Using classical Hölder spaces of parabolic type, a similar result is derived in [22].

In order to apply Theorem 3.11, we have to increase the regularity assumptions upon the coefficients in the following way. We suppose that

$$a_{jk}, \ b_j \in C^\infty(\mathbb{R}^m), \quad j, k = 1, \ldots, m,$$

and that there is a $M_0 > 0$ such that for all $\beta \in \mathbb{N}^m$:

$$\|\partial^\beta a_{jk}\|_{BUC^\alpha(\mathbb{R}^m)}, \ \|\partial^\beta b_j\|_{BUC^\alpha(\mathbb{R}^m)} \le M_0\beta!. \tag{4.10}$$

We are now prepared to show that the solutions to the Bellman equation (4.9) constructed in Proposition 4.1 are in fact real analytic in space and time.

THEOREM 4.2. *Let $u_0 \in buc^{2+\alpha}(\mathbb{R}^m)$ satisfy (4.5). Moreover, assume that (4.3) and (4.10) hold true and let $u \in C([0, t^+), buc^{2+\alpha}(\mathbb{R}^m)) \cap C^1([0, t^+), buc^\alpha(\mathbb{R}^m))$ be the solution to (4.9) as constructed in Proposition 4.1. Then*

$$[(t, x) \mapsto u(t)(x)] \in C^\omega((0, t^+) \times \mathbb{R}^m, \mathbb{R}).$$

Proof. Let $\Sigma = \mathbb{R}^m$ and $\Phi = id_{\mathbb{R}^m}$. Then it is not difficult to verify that the assumptions (A_1)–(A_5) of Section 3 are satisfied. Moreover, given $\mu \in \mathbb{R}^m$, $w \in \mathbb{E}_1$, and $t \in I$, we have

$$\mathbb{Q}(\mu, w)(t) = \frac{1}{2} \sum_{j,k=1}^{m} (\tau_{t\mu} a_{jk}) \partial_j \partial_k w(t) - \frac{1}{2} \langle (I + \partial^2 w(t))^{-1} \partial w(t) | \partial w(t) \rangle$$

$$+ \sum_{j=1}^{m} (\tau_{t\mu} b_j) \partial_j w(t).$$

Arguing as in Remark 3.7(b), one shows that also assumption (A_6) is satisfied and the assertion follows from Theorem 3.11. $\qquad\square$

4.2. *A Class of Mean Curvature Flows*

Consider the following nonlocal geometric evolution equation:

$$V(t) = -(-\Delta_{M(t)})^\beta (1 - \Delta_{M(t)})^\gamma H(t) \qquad \text{on } M(t) \text{ for } t > 0. \tag{4.11}$$

Here $M(t)$ is an unknown, with respect to time $t > 0$ evolving closed compact oriented hypersurface in $\mathbb{R}^{m+1}$. We write $\Delta_{M(t)}$ for the Laplace-Beltrami operator on $M(t)$ with respect to the Euclidean metric. The mean curvature of $M(t)$ is denoted by $H(t)$ and V stands for the normal velocity of the family $\{M(t) \,;\, t > 0\}$. Finally, β and γ are given real numbers. The evolution equation (4.11) does not depend on the local choice of orientation. However, if $M(t)$ encloses a domain $\Omega(t)$, we always choose the orientation such that $V(t)$ is positive if $\Omega(t)$ grows and such that $H(t)$ is positive if $M(t)$ is convex with respect to $\Omega(t)$. We mention that in the plane sometimes the opposite orientation is used.

For the values $\beta = 1$ and $\gamma = -1$ the evolution equation (4.11) is a special case of the so-called intermediate surface diffusion flow, introduced by Cahn and Taylor, see [8]. Assume next that $\beta = 0$. Then equation (4.11) reduces to the (negative) gradient flow of the area functional $\int_{M(t)} d\sigma(t)$ in the Sobolev space $H^{-\gamma}$. Also in the case $\gamma = 0$ the equation (4.11) is the (negative) gradient flow of the area functional in $H^{-\beta}$, but with the constraint that the volume of the domain $\Omega(t)$ enclosed by $M(t)$ is preserved during the evolution. Indeed, consider first the situation when $\beta = 0$. We assume that we are given a smooth solution $\{M(t) \,;\, t \in (0, T]\}$ to

$$V(t) = -(1 - \Delta_{M(t)})^\gamma H(t) \quad \text{on } M(t) \text{ for } t \in (0, T]. \tag{4.12}$$

We fix $t \in (0, T]$ and set $M := M(t)$. The normal field on M is denoted by ν_M. Given $h \in C^\infty(M)$, let

$$X_h : M \times (-\varepsilon, \varepsilon) \to \mathbb{R}^{m+1}, \qquad X_h(s, \tau) := s + \tau h \nu_M(s)$$

be the normal variation of M. Then $X_h(\cdot, \tau)$ is a smooth diffeomorphism from M onto its range $M_{\tau,h} := X_h(M, \tau)$, provided $\varepsilon > 0$ is chosen sufficiently small. Let

$$A_h(\tau) := m \int_{M_{\tau,h}} d\sigma(\tau), \quad \tau \in (-\varepsilon, \varepsilon),$$

denote the m-fold area functional of $M_{\tau,h}$. Here, the factor m in front of the integral is introduced just to simplify some of the calculations below. Using this notation, we say that the geometric evolution law

$$V(t) = F(M(t)), \quad t \in (0, T] \tag{4.13}$$

is the (negative) gradient flow of the area functional in $H^{-\gamma}$ if

$$A'_h(0) = \frac{d}{d\tau} \int_{M_{\tau,h}} d\sigma(\tau)|_{\tau=0} = (-F|h)_{H^{-\gamma}(M)}, \qquad h \in C^\infty(M).$$

Recall that $A'_h(0) = \int_M H_M h \, d\sigma$, where H_M stands for the mean curvature of M. Besides, the inner product in $H^{-\gamma}(M)$ is given by

$$(f|g)_{H^{-\gamma}(M)} = ((1 - \Delta_M)^{-\gamma/2} f | (1 - \Delta_M)^{-\gamma/2} g)_{L_2(M)},$$

see Section 4.2 in [23]. Consequently, we get

$$(H_M|h)_{L_2(M)} = -((1 - \Delta_M)^{-\gamma/2} F | (1 - \Delta_M)^{-\gamma/2} h)_{L_2(M)}.$$

Using the fact that $(1 - \Delta_M)^{-\gamma/2}$ is self-adjoint in $L_2(M)$ and the density of $C^\infty(M)$ in $L_2(M)$, we find that

$$F = -(1 - \Delta_M)^\gamma H_M.$$

We consider next (4.11) in the case when $\gamma = 0$. As before, let $\{M(t) \, ; \, t \in (0, T]\}$ be a smooth solution to (4.13) and assume in addition that each $M(t)$ encloses a well-defined domain $\Omega(t)$. Writing $vol(t) := \int_{\Omega(t)}$ for the $m + 1$-dimensional volume of $\Omega(t)$, we have

$$vol'(t) = \int_{M(t)} V(t) \, d\sigma(t) = \int_{M(t)} F(M(t)) \, d\sigma(t), \quad t \in (0, T]. \tag{4.14}$$

Hence the flow (4.13) preserves the volume if $\int_{M(t)} F(M(t)) \, d\sigma(t) = 0$. We fix again $t \in (0, T]$ and set $M := M(t)$. Moreover, we let

$$\mathbf{1}^\perp := \left\{ f \in L_2(M) \, ; \, \int_M f \, d\sigma = 0 \right\}$$

and write

$$P : L_2(M) \to \mathbf{1}^{\perp}, \quad f \mapsto f - \overline{f}$$

for the projection onto $\mathbf{1}^{\perp}$. Here, we used the notation $\overline{f} := \int_M f \, d\sigma / \int_M d\sigma$. Note that $f \mapsto \|(-\Delta_M)^{\beta/2} f\|_{L_2(M)}$ is an equivalent norm on $H^{\beta}(M) \cap \mathbf{1}^{\perp}$ and that the normal variation X_h is volume preserving, provided h belongs to $C^{\infty}(M) \cap \mathbf{1}^{\perp}$. Indeed, the normal velocity of the variation X_h is given by

$$V = (\partial_\tau X_h | \nu_M)|_{\tau=0} = h,$$

and the assertion follows from (4.14). According to these observations we say that (4.13) is the (negative) volume preserving gradient flow of the area functional in $H^{-\beta}$ if $\overline{F} = 0$ and if

$$A_h'(0) = -((-\Delta_M)^{-\beta/2} F | (-\Delta_M)^{-\beta/2} h)_{L_2(M)}, \qquad h \in C^{\infty}(M) \cap \mathbf{1}^{\perp}.$$

This implies that

$$\int_M HPg \, d\sigma = (PH|g)_{L_2(M)} = -((-\Delta_M)^{-\beta} F|Pg)_{L_2(M)} = -((-\Delta_M)^{-\beta} F|g)_{L_2(M)}$$

for all $g \in C^{\infty}(M)$, since $P = P^*$, $P(-\Delta_M)^{-\beta} P = (-\Delta_M)^{-\beta} P$, and $PF = F$. Using the density of $C^{\infty}(M)$ in $L_2(M)$, we therefore obtain

$$F = -(-\Delta_M)^{\beta} PH = -(-\Delta_M)^{\beta}(H - \overline{H}).$$

Finally, observe that $(-\Delta_M)^{\beta} \mathbf{1} = 0$ if $\beta > 0$. Hence we find

$$F = \begin{cases} -(-\Delta_M)^{\beta} H & \text{if } \beta > 0, \\ \overline{H} - H & \text{if } \beta = 0. \end{cases}$$

We remark that in the case $\beta = 0$ the evolution law $V = \overline{H} - H$ is known as the volume preserving mean curvature. In case $\beta = 1$, one calls $V = \Delta_M H$ the surface diffusion flow.

For simplicity, we restrict ourselves in the following to the case $\beta = 0$ and $\gamma \in [0, 1/2]$, i.e. we consider

$$V(t) = -(1 - \Delta_{M(t)})^{-\gamma} H(t), \quad t > 0, \quad M(0) = M_0. \tag{4.15}$$

The change of sign in the exponent has been made for simplicity only. We shall see that, given $\alpha \in (1/2, 1)$, the flow (4.15) is well-posed in $h^{\alpha-1+2\gamma}$ for initial data M_0 belonging to $h^{1+\alpha}$. For values of β, γ with $\beta \geq 0$ and $\beta - \gamma > 0$, the general equation (4.11) has to be treated in spaces with more regularity. If $\beta < 0$ one has to work in suitable subspaces of $\mathbf{1}^{\perp}$.

We parametrize (4.15) in a neighborhood of an analytic compact closed immersed oriented hypersurface Σ in $\mathbb{R}^{m+1}$. To make this precise, let ν denote the unit outer normal field on Σ. Moreover, given $a > 0$, choose a localization system $\{(U_l, \varphi_l) \,;\, l = 1, \ldots, n\}$ for Σ such that $\Sigma = \cup_{l=1}^{n} U_l$ and

$$\varphi_l \,:\, (-a, a)^m \to U_l, \quad l \in \{1, \ldots, n\},$$

is an analytic parametrization of U_l. Shrinking $a > 0$ if necessary, we may assume that

$$X_l \,:\, U_l \times (-a, a) \to \mathbb{R}^{m+1}, \qquad X_l(s, r) := s + r\nu(s).$$

is a smooth diffeomorphism onto its image $\mathcal{R}_l := \mathrm{im}(X_l)$, i.e.

$$X_l \in \mathrm{Diff}^{\omega}(U_l \times (-a, a), \mathcal{R}_l).$$

The inverse of X_l can be decomposed in the following way. Writing $S_l \in C^{\omega}(\mathcal{R}_l, U_l)$ and $\Lambda_l \in C^{\omega}(\mathcal{R}_l, (-a, a))$ for the metric projection of $\mathcal{R}_l$ onto U_l and for the signed distance function with respect to U_l, respectively, we have $X_l^{-1} = (S_l, \Lambda_l)$. In particular, observe that $\mathcal{R} := \cup_{l=1}^{n} \mathcal{R}_l$ consists of those points in $\mathbb{R}^{m+1}$ with distance less than a to Σ.

We now fix $\alpha > 1/2$, let

$$W(\Sigma) := W_a(\Sigma) := \{\rho \in h^{1+\alpha}(\Sigma) \,;\, \|\rho\|_{C^1(\Sigma)} \le a/2\},$$

and define

$$M_\rho := \bigcup_{l=1}^{n} \{X_l(s, \rho(s)) \,;\, s \in U_l\}$$

for $\rho \in W(\Sigma)$. Then M_ρ is a compact closed oriented immersed hypersurface in $\mathbb{R}^{m+1}$ of class $C^{1+\alpha}$, which can be seen as a graph in normal direction over Σ. Of course, ρ measures the signed distance of Σ to M_ρ. For convenience let us also introduce the mapping

$$\theta_\rho \,:\, \Sigma \to M_\rho, \quad s \mapsto X_l(s, \rho(s)) \quad \text{for} \quad s \in U_l.$$

Then θ_ρ is a well-defined global diffeomorphism of class $C^{2+\alpha}$ from Σ onto M_ρ. By means of this diffeomorphism we can pull back the Euclidean metric on M_ρ to Σ, producing in that way a Riemannian manifold which we denote in the following by $\Sigma(\rho)$. We now consider a family of hypersurfaces in $\mathcal{R}$. More precisely, let $T > 0$ be given, and define $I := [0, T]$, as well as

$$W(\Sigma_T) := W_a(\Sigma_T) := \{\rho \in C(I, h^{1+\alpha}(\Sigma)) \,;\, \|\rho\|_{C(I, C^1(\Sigma))} \le a/2\}.$$

Then, given $\rho \in W(\Sigma_T)$, we transform the evolution equation (4.15) for the family $\{M_{\rho(t)} \,;\, t \in [0, T]\}$ into an evolution equation on Σ. For this we first calculate the normal velocity of $[t \mapsto M_{\rho(t)}]$. We have, cf. [12],

$$V(t, s) = \partial_t \rho(t, s) \big/ |\nabla_x \Phi_\rho(x, t)|_{x = \theta_{\rho(t)}(s)} \qquad \text{for } (t, s) \in I \times \Sigma,$$

where we used the function

$$\Phi_\rho : \mathcal{R} \times [0, T] \to \mathbb{R}, \quad (x, t) \mapsto \Lambda(x) - \rho(t, S(x))$$

to represent $M_{\rho(t)}$ as the $0-$level set of $\Phi_\rho(\cdot, t)$, i.e. $M_{\rho(t)} = \Phi^{-1}(\cdot, t)(0)$. To shorten our notation, let

$$L(\rho)(t, s) := |\nabla_x \Phi_\rho(x, t)|_{x=\theta_{\rho(t)}(s)}, \quad (t, s) \in I \times \Sigma.$$

Moreover, we write $K(\rho) := \theta_\rho^* H$ and Δ_ρ for the mean curvature and Laplace-Beltrami operator of $\Sigma(\rho)$, respectively. Here, θ_ρ^* denotes the pull-back operator induced by the diffeomorphism θ_ρ, i.e. $\theta_\rho^* f = f \circ \theta_\rho$ for $f \in C(M_\rho)$. This means in particular that we have

$$\theta_\rho^* \Delta_{M_\rho} = \Delta_\rho \theta_\rho^*.$$

Given $\rho_0 \in W_a(\Sigma)$, consider now the following nonlinear nonlocal partial differential equation

$$\frac{d\rho}{dt} = -L(\rho)(1 - \Delta_\rho)^{-\gamma} K(\rho) \quad \text{in} \quad I \times \Sigma, \qquad \rho(0) = \rho_0 \quad \text{on} \quad \Sigma. \tag{4.16}$$

In order to treat (4.16) in the framework of Theorem 3.11, we set

$$P(\rho) := L(\rho)(1 - \Delta_\rho)^{-\gamma} K(\rho) \quad \text{for} \quad \rho \in W(\Sigma) \cap h^{2+\alpha}(\Sigma).$$

Our first result in this section shows that P can be extended to an analytic mapping with values in $h^{\alpha-1+2\gamma}(\Sigma)$.

LEMMA 4.3. *There exists an extension of P, again denoted by P, such that $P \in C^\omega(W(\Sigma), h^{\alpha-1+2\gamma}(\Sigma))$.*

Proof. (i) We first express the terms $L(\rho)$, $K(\rho)$, and Δ_ρ in local coordinates. To make this precise, let

$$\hat{\rho}_l(s) := \rho(\varphi_l(s)), \quad \hat{X}_l(s, r) := X_l(\varphi_l(s), r), \quad (s, r) \in (-a, a)^{m+1},$$

be the local representations of ρ_l and X_l with respect to U_l. In the following we do not always distinguish between ρ_l, X_l and their local representations $\hat{\rho}_l$, $\hat{X}_l$, as well as between local coordinates $s \in (-a, a)^m$ and the corresponding points $\varphi_l(s)$ on U_l. Moreover, we suppress the index $l \in \{1, \ldots, n\}$ if no confusion seems likely. Given $\rho \in W(\Sigma)$, define

$$w_{jk}(\rho)(s) := (\partial_j X | \partial_k X)|_{(s, \rho(s))}, \quad s \in (-a, a)^m,$$

for $j, k \in \{1, \ldots, m\}$, where $(\cdot|\cdot)$ stands for the Euclidean metric in $\mathbb{R}^{m+1}$ and ∂_j denotes the partial derivative with respect to the j-th variable of s. Since ρ belongs to $W(\Sigma)$, the matrix $[w_{jk}(\rho)]$ is invertible and we write $w^{jk}(\rho)$ for the entries of its inverse. Then we have

$$L(\rho) = \sqrt{1 + w^{jk}(\rho)\partial_j\rho\,\partial_k\rho} \tag{4.17}$$

cf. (2.3) in [14]. In (4.17) and in what follows we use summation convention over repeated indices. Moreover, we write

$$\Gamma^i_{jk}(\rho) := \frac{1}{2}w^{il}(\rho)(\partial_k(\partial_l X|\partial_j X) - \partial_l(\partial_j X|\partial_k X) + \partial_j(\partial_k X|\partial_l X))|_{(\cdot,\rho)},$$

for the corresponding Christoffel symbols. Then Lemma 2.1 in [14] shows that $K(\rho)$ carries a quasi-linear structure, i.e. given $\rho \in W(\Sigma)$, there are

$$K_1(\rho) \in \mathcal{L}(h^{2+\alpha}(\Sigma), h^\alpha(\Sigma)) \quad \text{and} \quad K_2(\rho) \in h^\alpha(\Sigma) \tag{4.18}$$

such that

$$K(\rho) = K_1(\rho)\rho + K_2(\rho) \quad \text{for} \quad \rho \in W(\Sigma) \cap h^{2+\alpha}(\Sigma). \tag{4.19}$$

In the chosen local coordinates these mappings are represented as:

$$\begin{aligned}
K_1(\rho) = \frac{1}{mL(\rho)^3}[\{&-L(\rho)^2 w^{jk}(\rho) + w^{jl}(\rho)w^{kn}(\rho)\partial_l\rho\,\partial_n\rho\}\partial_j\partial_k \\
&+ \{L(\rho)^2 w^{jk}(\rho)\Gamma^i_{jk}(\rho) + w^{jl}(\rho)w^{ki}(\rho)\Gamma^{m+1}_{jk}(\rho)\partial_l\rho \\
&+ 2w^{kn}(\rho)\Gamma^i_{(m+1)k}(\rho)\partial_n\rho - w^{jl}(\rho)w^{kn}(\rho)\Gamma^i_{jk}(\rho)\partial_l\rho\,\partial_n\rho\}\partial_i]
\end{aligned} \tag{4.20}$$

and

$$K_2(\rho) = -\frac{1}{mL(\rho)}w^{jk}(\rho)\Gamma^{m+1}_{jk}(\rho). \tag{4.21}$$

In order to express Δ_ρ in local coordinates, let η be the Euclidean metric and write $\sigma(\rho) := \theta^*_\rho\eta$ for the Riemannian metric on Σ induced by the diffeomorphism θ_ρ. This means that, using the above introduced notation, we have $\Sigma(\rho) = (\Sigma, \sigma(\rho))$. Let further $\sigma_{jk}(\rho)$ denote the components of $\sigma(\rho)$ in local coordinates and write $\sigma^{jk}(\rho)$ for the components of the inverse of $[\sigma_{jk}(\rho)]$. Using again summation convention over repeated indices, the Christoffel symbols of $\sigma(\rho)$ in the chosen coordinates are given by

$$\gamma^l_{jk}(\rho) = \frac{\sigma^{ln}(\rho)}{2}\left[\frac{\partial\sigma_{kn}(\rho)}{\partial s^j} + \frac{\partial\sigma_{jn}(\rho)}{\partial s^k} - \frac{\partial\sigma_{jk}(\rho)}{\partial s^n}\right],$$

and we have

$$\Delta_\rho = \sigma^{jk}(\rho)\left[\frac{\partial^2}{\partial s^j\partial s^k} - \gamma^l_{jk}(\rho)\frac{\partial}{\partial s^l}\right], \quad \rho \in W(\Sigma) \cap h^{2+\alpha}(\Sigma), \tag{4.22}$$

cf. the proof of Lemma 2.1 in [14].

(ii) It follows from [12, p. 1037] that $w_{jk}(\rho)$ is a quadratic polynomial in ρ. Moreover, we have that

$$[\rho \mapsto \theta_\rho - id_\Sigma] \in \mathcal{L}(h^{1+\alpha}(\Sigma), h^{1+\alpha}(\Sigma, \mathbb{R}^{m+1})).$$

This obviously implies that

$$w_{jk} \in C^\omega(W(\Sigma), h^{1+\alpha}(\Sigma)), \qquad \sigma_{jk} \in C^\omega(W(\Sigma), h^\alpha(\Sigma)), \tag{4.23}$$

and consequently:

$$w^{jk}, \; \Gamma^i_{jk} \in C^\omega(W(\Sigma), h^{1+\alpha}(\Sigma)), \quad \sigma^{jk} \in C^\omega(W(\Sigma), h^\alpha(\Sigma)). \tag{4.24}$$

(iii) Combining (4.24) and (4.17), we see that

$$L \in C^\omega(W(\Sigma), h^\alpha(\Sigma)). \tag{4.25}$$

Recall that we have assumed that $\alpha > 1/2$. Hence $\alpha > \max\{1-\alpha, \alpha-1\}$, and we conlcude from [26, Theorem 2.8.2] and a localization argument that

$$h^\alpha(\Sigma) \times h^{\alpha-1}(\Sigma) \to h^{\alpha-1}(\Sigma), \quad (f, g) \mapsto fg \tag{4.26}$$

is continuous and bilinear. Using this and Theorem 2.3.8 in [26], it follows from (4.20), (4.21), and (4.24) that

$$[\rho \mapsto K(\rho)] \in C^\omega(W(\Sigma), h^{\alpha-1}(\Sigma)). \tag{4.27}$$

(iv) Let $\rho_0 \in W(\Sigma)$ be given and choose $\varepsilon > 0$ such that

$$B_0 := \mathbb{B}_{h^{1+\alpha}(\Sigma)}(\rho_0, \varepsilon_0) \subset W(\Sigma).$$

For simplicity we set $X_0 := h^{\alpha-1}(\Sigma)$ and $X_1 := h^{1+\alpha}(\Sigma)$, and denote the X_0-realization of $1 - \Delta_\rho$ by $A(\rho)$. Shrinking $\varepsilon_0 > 0$, it follows from Theorem 4.2 in [3] and a localization argument that

$$A(\rho) \in \mathcal{H}(X_1, X_0), \qquad \rho \in B_0. \tag{4.28}$$

Moreover, we conclude from Theorem 7.4.3 and Remark 7.2.5.1 in [27] that $\mathbb{R}_+$ belongs to the resolvent set $\mathrm{res}(A(\rho))$ of $A(\rho)$ for all $\rho \in B_0$. Thus, given $\rho \in B_0$ and $s \in \mathbb{R}$ we conclude that $A(\rho)^s$ is a well-defined linear operator in X_0. Observing (4.22), we infer from (4.28) that

$$[\rho \mapsto A(\rho)A(0)^{-1}] \in C^\omega(B_0, \mathcal{L}(X_0)).$$

Hence we find

$$[\rho \mapsto A(0)^\gamma A(\rho)^{-\gamma}] \in C^\omega(B_0, \mathcal{L}(X_0)), \tag{4.29}$$

since $[B \mapsto B^{-\gamma}] \in C^{\omega}(\mathcal{L}(X_0))$, see Theorem VIII.7 in [28]. Recall that $A(0) = 1 - \Delta_0 = 1 - \Delta_{\Sigma}$ is a uniformly elliptic operator with smooth coefficients on the compact smooth manifold Σ. Thus Theorem III.4.7.5 in [2] implies that $1 - \Delta_0$ possesses bounded imaginary powers, i.e., there exists a $K > 0$ such that

$$\|(1 - \Delta_0)^{it}\|_{\mathcal{L}(X_0)} \leq K, \quad |t| \leq 1.$$

Hence it follows from Theorem 3 in [24] and the fact that the spaces $h^s(\Sigma)$ are stable under complex interpolation, cf. Theorem 7.4.4 in [27], that

$$A(0)^{-\gamma} \in \mathcal{L}(h^{\alpha-1}(\Sigma), h^{\alpha-1+2\gamma}(\Sigma)). \tag{4.30}$$

Combining this with (4.29) we get

$$[\rho \mapsto A(\rho)^{-\gamma}] \in C^{\omega}(B_0, \mathcal{L}(h^{\alpha-1}(\Sigma), h^{\alpha-1+2\gamma}(\Sigma))). \tag{4.31}$$

Recall that $\gamma \leq 1/2$. Thus pointwise multiplication maps $h^{\alpha-1+2\gamma}(\Sigma) \times h^{\alpha}(\Sigma)$ bilinearly and continuously into $h^{\alpha-1+2\gamma}(\Sigma)$ and it remains to combine (4.25), (4.27), and (4.31) to complete the argumentation.

In the following example we choose $\gamma = 1/2$. Given $\rho \in W(\Sigma)$, set

$$A_0(\rho) := L(\rho)(1 - \Delta_\rho)^{-1/2} K_1(\rho) \quad \text{and} \quad f(\rho) := -L(\rho)(1 - \Delta_\rho)^{-1/2} K_2(\rho).$$

Obviously, we have $P(\rho) = A_0(\rho)\rho - f(\rho)$ for $\rho \in W(\Sigma)$. Since also

$$A_0(\rho) \in \mathcal{L}(h^{1+\alpha}(\Sigma), h^{\alpha}(\Sigma)), \quad f \in C^{\omega}(W(\Sigma), h^{\alpha}(\Sigma)),$$

it follows that P consists of the quasilinear operator A_0 and the perturbation f. Nevertheless, the abstract equation

$$\frac{d\rho}{dt} + A_0(\rho)\rho = f(\rho)$$

cannot be treated in the framework of [1], since neither $W(\Sigma)$ nor the range of f is contained in any interpolation space of $h^{\alpha}(\Sigma)$ and $h^{1+\alpha}(\Sigma)$ obtained by an admissible interpolation functor in the sense of [1]. $\qquad\square$

We are now going to verify that the operator P satisfies the assumptions (2.1) and (2.2). To avoid too many technicalities, we consider here the particular situation that $M = \Sigma = \mathbb{S}^m$, where $\mathbb{S}^m$ denotes the standard unit sphere in $\mathbb{R}^{m+1}$. It is possible to study (4.16) in a more general situation, which will be done elsewhere. We fix $\alpha > 1/2$ and set

$$E_0 := h^{\alpha-1+2\gamma}(\mathbb{S}^m), \qquad E_1 := h^{1+\alpha}(\mathbb{S}^m).$$

Moreover, P denotes the operator constructed in Lemma 4.1 and $\Delta := \Delta_0$ stands for the Laplace-Beltrami operator on $\mathbb{S}^m$. Then we have

LEMMA 4.4. $\partial P(0) = \frac{1}{m}(1-\Delta)^{1-\gamma} - \frac{m+1}{m}(1-\Delta)^{-\gamma}$

Proof. Let $h \in E_1$ be given. Observe that $L(0) = 1$ and $K(0) = 1$, and that

$$\partial L(0)h = \frac{d}{d\varepsilon}\sqrt{1 + \varepsilon^2 w^{jk}\partial_j h \partial_k h}\Big|_{\varepsilon=0} = 0.$$

Further, we infer from (4.22) that $(s + 1 - \Delta_{\varepsilon h})(1/(s+1)) = 1$ for all $s \geq 0$ and all $\varepsilon \in \mathbb{R}$ which are in modulus sufficientlty small. Using the representation formula

$$(1 - \Delta_{\varepsilon h})^{-\gamma} = \frac{\sin(\pi\gamma)}{\pi}\int_0^\infty s^{-\gamma}(s + 1 - \Delta_{\varepsilon h})^{-1}\,ds, \qquad \gamma \in (0, 1/2],$$

and $\int_0^\infty s^{-\gamma}(1+s)^{-1}ds = \pi/\sin(\pi\gamma)$, see (III.4.6.9) and (III.4.6.10) in [2], this implies that $(1 - \Delta_{\varepsilon h})^{-\gamma}1 = 1$, and consequently

$$\partial(1 - \Delta_\rho)^{-\gamma}K(0)|_{\rho=0}h = \frac{d}{d\varepsilon}(1 - \Delta_{\varepsilon h})^{-\gamma}|_{\varepsilon=0}1 = 0.$$

Therefore we obtain

$$\partial P(0) = (1 - \Delta)^{-\gamma}\partial K(0). \tag{4.32}$$

But $\partial K(0) = -\frac{1}{m}(m + \Delta)$, see Lemma 3.1 in [13]. Combining this with (4.32) we easily get the assertion. $\qquad\square$

COROLLARY 4.5. *There exists an open neighborhood B of 0 in E_1 such that P satisfies* (2.1) *and* (2.2).

Proof. (i) Since $-(1-\Delta)$ generates a strongly continuous analytic semigroup on E_0, we know from Remark III.4.6.12 in [2] that the very same is true for $-(1-\Delta)^{1-\gamma}$. Repeating the arguments which lead to (4.30), we conclude that the domain of $-(1-\Delta)^{1-\gamma}$ is given by $h^{1+\alpha}(\mathbb{S}^m)$. This shows that $-(1-\Delta)^{1-\gamma} \in \mathcal{H}(E_1, E_0)$.

Since $(1 - \Delta)^{-\gamma} \in \mathcal{L}(E_0, E_0)$, it follows from Lemma 4.2 and a well-known perturbation result for generators of analytic semigroups that $\partial P(0) \in \mathcal{H}(E_1, E_0)$. But $\mathcal{H}(E_1, E_0)$ is open in $\mathcal{L}(E_1, E_0)$, see Theorem I.1.3.1 in [2]. Thus there is a open neighborhood V of $\partial P(0)$ such that $V \subset \mathcal{H}(E_1, E_0)$. From Lemma 4.1 we know that $\partial P \in C^\omega(W(\mathbb{S}^m), \mathcal{L}(E_1, E_0))$. Hence there is a open neighborhood B of 0 in $W(\mathbb{S}^m)$ such that $\partial P(B) \subset V$. In particular, we see that (2.1) is satisfied.

(ii) Let $\beta \in (1/2, \alpha)$ be fixed. Then the very same arguments as in (i) ensure that, given $\rho \in B$, we have that $\partial P(\rho) \in \mathcal{H}(h^{1+\beta}(\mathbb{S}^m), h^{\beta-1+2\gamma}(\mathbb{S}^m))$. Observing Remark 2.2(j), we see that assumption (2.2) is satisfied as well. $\qquad\square$

THEOREM 4.6. *Let $\alpha \in (1/2, 1)$ and $\gamma \in [0, 1/2]$. Let further $\rho_0 \in h^{1+\alpha}(\mathbb{S}^m)$ and assume that M_{ρ_0} is the graph of ρ_0 over $\mathbb{S}^m$ in normal direction. Then there exists a $t^+ > 0$ such that the mean curvature flow*

$$V = -(1 - \Delta_{M_{\rho(t)}})^{-\gamma} H(t), \quad M_{\rho(0)} = M_{\rho_0}$$

possesses a unique solution $\{M_{\rho(t)} ; t \in [0, t^+)\}$ with

$$\rho \in C([0, t^+), h^{1+\alpha}(\mathbb{S}^m)) \cap C^1([0, t^+), h^{\alpha-1+2\gamma}(\mathbb{S}^m)),$$

provided $\|\rho_0\|_{C^{1+\alpha}(\mathbb{S}^m)}$ is small enough. In addition,

$$[(t, p) \mapsto \rho(t)(p)] \in C^\omega((0, t^+) \times \mathbb{S}^m), \mathbb{R}).$$

Proof. (i) It follows from Remark 2.2 d) that (1.3) holds true.

(ii) Clearly, $SO(m + 1) \cdot BUC^\infty(\mathbb{S}^m) \subset BUC^\infty(\mathbb{S}^m)$. Moreover, it is not difficult to verify that $SO(m + 1)$ is a transformation group of isometries on $BUC^j(\mathbb{S}^m)$ for $j = 0, 1, 2$. By Remark 2.2(c) and a density argument we therefore conclude that $SO(m + 1)$ is a strongly continuous transformation group on $h^{j+\alpha}(\mathbb{S}^m)$ for $j = 0, 1$.

(iii) Observe that the metric on $\mathbb{S}^m$ is invariant under $SO(m + 1)$. This implies that the transformation group on $h^{1+\alpha}(\mathbb{S}^m)$ induced by $SO(m + 1)$ consists of isometries, i.e. given $R \in SO(m + 1)$, the mapping

$$h^{1+\alpha}(\mathbb{S}^m) \to h^{1+\alpha}(\mathbb{S}^m), \quad v \mapsto R \cdot v$$

is an isometry. Thus, replacing B by a sufficiently small ball in $h^{1+\alpha}(\mathbb{S}^m)$ around 0, we have that $SO(m + 1) \cdot B \subset B$.

(iv) Fix $\rho \in B$ and $R \in SO(m + 1)$. Obviously, we have

$$R \circ \theta_{R \cdot \rho} = R \cdot \theta_\rho. \tag{4.33}$$

Thus, given $p \in \mathbb{S}^m$, we get

$$\begin{aligned} K(R \cdot \rho)(p) &= H(\theta_{R \cdot \rho}(p)) = H(R^{-1}\theta_\rho(Rp)) \\ &= H(\theta_\rho(Rp)) = [R \cdot K(\rho)](p) \end{aligned} \tag{4.34}$$

since R is an isometry of the Euclidean space $(\mathbb{R}^{m+1}, \eta)$. Moreover, (4.33) implies that

$$R : (\mathbb{S}^m, (R \circ \theta_{R \cdot \rho})^* \eta) \to (\mathbb{S}^m, \theta_\rho^* \eta)$$

is an isometry as well. Hence we find that

$$R \cdot (\Delta_\rho v) = \Delta_{R \cdot \rho}(R \cdot v), \quad v \in h^{1+\alpha}(\mathbb{S}^m), \tag{4.35}$$

see e.g. Remark XI.6.9(c) in [4]. Consequently, we obtain

$$R \cdot ((1 - \Delta_\rho)^{-\gamma} w) = (1 - \Delta_{R \cdot \rho})^{-\gamma} (R \cdot w) \quad w \in h^{\alpha - 1 + 2\gamma}(\mathbb{S}^m). \tag{4.36}$$

Finally, it follows from the chain rule and (4.33) that $L(R \cdot \rho) = R \cdot L(\rho)$. Combining (4.34), (4.35), (4.36), and (iii) we see that

$$P : B \to h^{\alpha - 1 + 2\gamma}(\mathbb{S}^m), \quad \rho \mapsto L(\rho)(1 - \Delta_\rho)^{-\gamma} K(\rho)$$

is equivariant with respect to $SO(m + 1)$. Now the assertion follows from Theorem 1.1. $\square$

REFERENCES

[1]   AMANN, H., *Nonhomogeneous linear and quasilinear elliptic and parabolic boundary value problems*, in H. J. Schmeisser, H. Triebel, editors, Function Spaces, Differential Operators and Nonlinear Analysis, Teubner, Stuttgart, Leipzig, 1993, 9–126.

[2]   AMANN, H., *Linear and Quasilinear Parabolic Problems*, Vol. I, Birkhäuser, Basel, 1995.

[3]   AMANN, H., *Elliptic operators with infinite-dimensional state spaces*. J. Evol. Equs. *1* (2001), 143–188.

[4]   AMANN, H. and ESCHER, J., *Analysis III*, Birkhäuser, Basel, 2001.

[5]   ANGENENT, S., *Nonlinear analytic semiflows*. Proc. Roy. Soc. Edinburgh *115A* (1990), 91–107 .

[6]   AUBIN, T., *Nonlinear Analysis on Manifolds. Monge-Ampére Equations*. Springer, New York, 1992.

[7]   BAILLON, J. B., *Caractère borné de certains générateurs de semigroup linéaire dans les espaces de Banach*. C.R. Acad. Sc. Paris *290A* (1980), 757–760.

[8]   CAHN, J. W. and TAYLOR J. E., *Surface motion by surface diffusion*. Acta Metallurgica *42* (1994), 1045–1063.

[9]   DA PRATO, G. and GRISVARD, P., *Equations d'évolution abstraites nonlinéaires de type parabolique*. Ann. Mat. Pura Appl. *120* (1979), 329–396.

[10]   EIDELMAN, S. D., *Parabolic Systems*. North-Holland, Amsterdam, 1969.

[11]   ESCHER, J. and SIMONETT, G., *Analyticity of the interface in a free boundary problem*. Math. Ann. *305* (1996), 439–459.

[12]   ESCHER, J. and SIMONETT, G., *Classical solutions of multidimensional Hele-Shaw models*. SIAM J. Math. Anal. *28*, 1028–1047.

[13]   ESCHER, J. and SIMONETT, G., *A center manifold analysis for the Mullins-Sekerka model*. J. Differential Equations *143* (1998), 267–292.

[14]   ESCHER, J., MAYER, U. F. and SIMONETT, G., *The surface diffusion flow for immersed hypersurfaces*. SIAM J. Math. Anal. *29* (1999), 1419–1433.

[15]   ESCHER, J., SIMONETT, G. and PRÜSS, J., *A new approach to the regularity of solutions for parabolic equations*. In Evolution Equations: Proceedings in Honour of J. A. Goldstein's 60th Birthday, M. Dekker, to appear.

[16]   FRIEDMAN, A., *Partial Differential Equations*. Krieger, New York, 1976.

[17]   HELGASON, S., *Differential Geometry and Symmetric spaces*. Academic Press, New York, 1962.

[18]   HENRY, D., *Geometric Theory of Semilinear Parabolic Equations*. Lecture Notes in Math. *840*, Springer, Berlin (1981).

[19]   KLINGENBERG, W., *Riemannian Geometry*, de Gruyter, Berlin 1982.

[20]   KOBAYASHI, S. and NOMIZU, K., *Foundations of Differential Geometry*. Wiley, New York 1963.

[21]   LADYZENSKAJA, O. A., SOLONNIKOV, V. A. and URALCEVA N. N., *Linear and Quasilinear Equations of Parabolic Type*, AMS, Providence, Rhode Island, 1968.

[22]   LUNARDI, A., *Analytic Semigroups and Optimal Regularity in Parabolic Equations*. Birkhäuser, Basel, 1991.

[23]   PETERSEN, B. E., *Introduction to the Fourier Transform and Pseudo-Differential Operators*. Pitman, Boston, 1983.

[24]   SEELEY, R. T., *Norms and domains of the complex powers A_B^z*. Amer. J. Math. *93* (1971), 299–309.
[25]   SIMONETT, G., *Center manifolds for quasilinear reaction-diffusion systems*. Differential Integral Equations *8* (1995), 753–796.
[26]   TRIEBEL, H., *Theory of Function Spaces*. Birkhäuser, Basel, 1983.
[27]   TRIEBEL, H., *Theory of Function Spaces II*. Birkhäuser, Basel, 1992.
[28]   YOSIDA, K., *Functional Analysis*. Springer, Berlin, 1980.

J. Escher
Institute for Applied Mathematics
University of Hannover
D-30167 Hannover
Germany
e-mail: escher@ifam.uni-hannover.de

G. Simonett
Department of Mathematics
Vanderbilt University
Nashville, TN 37240
USA
e-mail: simonett@math.vanderbilt.edu

J.evol.equ. 3 (2004) 577 – 602
1424–3199/03/040577 – 26
DOI 10.1007/s00028-003-0097-8
© Birkhäuser Verlag, Basel, 2004

Journal of Evolution
Equations

Pointwise gradient estimates of solutions to onedimensional nonlinear parabolic equations

Philippe Bénilan[+] and Jesús Ildefonso Díaz

Abstract. We present here an improved version of the method introduced by the first author to derive pointwise gradient estimates for the solutions of one-dimensional parabolic problems. After considering a general qualinear equation in divergence form we apply the method to the case of a nonlinear diffusion-convection equation. The conclusions are stated first for classical solutions and then for generalized and mild solutions. In the case of unbounded initial datum we obtain several regularizing effects for $t > 0$. Some unilateral pointwise gradient estimates are also obtained. The case of the Dirichlet problem is also considered. Finally, we collect, in the last section, several comments showing the connections among these estimates and the study of the free boundaries associated to the solutions of the diffusion-convection equation.

1. Introduction

This paper deals with a very old task which seems to have its origin in the work by S. N. Bernstein the beginning of last century ([14]): find a priori pointwise estimates on the (spatial) gradient of solutions of nonlinear second order equations. To be more precise we shall consider onedimensional parabolic equations.

Our main purpose is to present here an improved version of a general method introduced, originally, by the first author (see Section I.2 of Bénilan [7]). The method has important differences with respect to the way in which D. G. Aronson [3] derived, in 1969, this type of estimates for the porous media equation. His technique was later applied by different authors to many other nonlinear degenerate equations (see references in the surveys [35] and [46]). Such a already classical technique consists of, firstly, introducing the variable $u = f(w)$ with a concave function of the form $f(r) = C_1 r(C_2 - r)$, for some suitable constants $C_i > 0$ and, secondly, estimating the value of $|v_x|$ at a maximum of the auxiliary function $z = \zeta^2 w_x^2$, with ζ a cutoff function.

The method here presented seems to be better adapted to recognize the different nonlinear terms involved in the equation and so, it allows to obtain *sharp* pointwise gradient estimates. Moreover, our method provides quantitative estimates. In many cases these estimates have an important physical meaning and allow to deduce many other qualitative properties of the solutions.

2000 *Mathematics Subject Classification*: 35K65, 35K10, 35R35, 76D27.
Key words: Pointwise gradient estimates, one-dimensional parabolic equations, non linear diffusion-convection equation, regularizing effects, unilateral estimates, interfaces.

We shall limit the exposition to a general class of equations in divergence form

$$u_t = \sigma(u, u_x)_x, \tag{1}$$

but the method also applies to more general equations (see the exposition made in [7]). In order to explain some additional details, we consider the special case of the nonlinear diffusion-convection equation

$$u_t = \varphi(u)_{xx} + \psi(u)_x. \tag{2}$$

Equation (2) appears in many different contexts. For instance, unsaturated soil-moisture flows are modelled by (2) under the condition $\varphi'(0) = 0$. In that case the equation becomes degenerate and due to that a very rich structure arise concerning many different aspects: finite speed of propagation, stability results, etc. In the context of gas flow in a porous medium, the expression

$$V(t, x) = -\frac{\varphi(u(x, t))_x + \psi(u(x, t))}{u(x, t)}$$

represents the Eulerian description of the velocity at the position x and the time t of the "homogenized gas" (see, [22], [18]). It seems natural to expect boundedness of V which, in fact, leads to "optimal gradient bounds" for the solution u of (2) in some special cases as, for instance, $\varphi(u) = u^m$, $\psi(u) = u^n$, $1 < m$, $1 \le n$. In that case, it is possible to find out some special solutions for which the modulus of continuity found via this gradient estimate can not be improved (see, e.g., [30], [25] and [33]).

The nature of "optimal gradient bounds" is much less evident in other contexts in which equation (2) arises (as, for instance, in Plasma Physics where $\varphi(u) = u^m$ and $0 < m < 1$ [15]) or even in the context of equation (2) when convection dominates over diffusion: $\varphi(u) = u^m$, $\psi(u) = u^n$ and $0 < n \le 1 < m$: [20]).

One of the main features of the method, as presented here, is that it endows us with a "simple" analytical argument that allows to find sharp gradient bounds independently on the physical arguments involved in the concrete formulation.

The main idea of the method is to find an auxiliary nonlinear parabolic operator $\mathcal{L}$ in such a way that a pointwise gradient estimate of the type

$$\sigma(u, u_x) \le k(t, x, u) \tag{3}$$

holds as a consequence of the comparison principle applied to a function of the form $p = \sigma(u, u_x)/k(., ., u)$, in the kernel of operator $\mathcal{L}$, and a constant function, i.e.

$$\mathcal{L}(p) = 0 \le \mathcal{L}(1). \tag{4}$$

In this way, the possible choices of the function $k(t, x, u)$ are determined by the condition $\mathcal{L}(1) \ge 0$. Another characteristic of the method is that it can be used *locally in time and space* and, in particular, it illustrates some regularizing effects (for instance some gradient bounds take place for any $t > 0$, independently of whether it already holds at $t = 0$ or not). In particular, we get some pointwise gradient estimates when $\varphi(u)$ is merely a maximal monotone graph (see Corollary 6).

The content of the paper is the following: Section 2 is devoted to a general presentation of the method for equation (1) and its special case (2) in the context of classical solutions. The application to the case of weaker notions of solution is carried out in Section 3. Pointwise gradient estimates are first obtained for the generalized solution of the Cauchy problem associated to (2) in Subsection 3.1. An unidirectional pointwise gradient estimate, that cease to be true if the direction of the spatial variable x is changed, is obtained in Subsection 3.2. The case of unbounded initial data and the proof of regularizing effects in the class of mild solutions are developed in Subsection 3.3. The derivation of pointwise gradient estimates for the case of a finite spatial interval is considered in Subsection 3.4. Finally, some comments on the connection between pointwise gradient estimates and the study of the free boundaries defined as the boundaries of the support of the solution are collected in Subsection 3.5.

Perhaps, some remarks on the long process of elaboration of this paper can be useful to some readers. The discussions on this subject by the authors were started after the stay of the first author at the University of Kentucky (during the 1980/1981 academic year). There he gave a general course on *Evolution Equations and Accretive Operators* (see [7]) and wrote a manuscript (without title and unpublished which we shall refer in the following as Bénilan [8]). In those days the authors were preparing their joint paper Bénilan-Díaz [10] submitted on May 18, 1981. The common point of view was a consequence of the results by the second author (in collaboration with R. Kersner) on the problem $u_t = \varphi(u)_{xx} + \psi(u)_x$ for a case including $\varphi(u) = u^m$, $\psi(u) = u^n$ with $0 < n < 1 \leq m$. The first version of the Díaz-Kersner paper [20] appeared as internal report of the MRC at Wisconsin-Madison in 1982 although the definitive version waited until in 1987. In the period from 1980 to 1985 the authors of the present paper visited each other at least once a year thanks to the economical support received from the French Embassy at Madrid (see the Acknowledgments of [10]).

The first application of the present technique, which was not included in [8] was an alternative proof to Theorem 1 of the paper Díaz-Kersner [20] (correspondence maintened in March, 20, 1983). A second written document on the elaboration of the present paper was a letter of Bénilan to Díaz containing the statement in of Theorem 1 (i.e. for any φ including the case of a maximal monotone graph). Simultaneously to this period, both authors developed their normal activity teaching postgraduate courses in their universities (for instance, a part of the results of this paper was exposed by the second author during the 1985/1986 academic year). In this way, the connection of the (at this time unpublished) results with the ones by other authors was becoming a little bit barroque. For instance, a first version of the results was sent to B. H. Gilding (among many other colleagues) who mentions the notes in his paper (Gilding [27] submitted on September 17, 1986) as P. Bénilan -J. I. Díaz, *Unpublished notes* (see reference [3] of [27]). A first typed draft of the paper was discussed by the authors at Montreal, in June of 1990 at the occasion of the participation by both authors at the Free Boundary Meeting. Then, they decided to add to the manuscript a part on the $L^1 - L^\infty$ smoothing effect incorporating some variations (occupying several pages written at Montreal and now in possession of the second author) to another unpublished manuscript by Bénilan (this time intitled by him as "On $u_t = \varphi(u)_{xx} + \psi(u)_x$: unfinished, 1978: [5]). Some related ideas of those modifications were published later in the paper by Bénilan and Bouillet [9]. The process of writing the definitive version of this paper was accelerated in the last five years before the death of Bénilan as the correspondence hold between both authors prove. In fact, Bénilan mentions this paper in the list of his publications (*paper No.*104) which he sent to the Departamento de Matematica Aplicada at the occasion of his nomination as Doctor Honoris Causa by the Universidad Complutense de Madrid which, although officially and definitively approved, it was programed for a too late date: two months after his death. The last meeting of both authors was in the "Journées Analyse non linéaire", held at occasion of the 60th birthday of Philippe Bénilan, in October 27–29, 2000 at Les Moussières, Jura (France). We were talking on the last details to incorporate in to the paper and decided to publish it in some (not yet decided) journal.

2. The general method

In this section we introduce the main idea of the method. As usual, pointwise gradient estimates will be obtained as *a priori* estimates for any possible solution u. So, at the first step, they will be obtained in the more favorable setting of smooth solutions. Since the resulting estimates will be independent of the regularity used in the proof, they can be extended to the framework of weaker solutions. This second step will be illustrated in the next section.

2.1. *A general quasilinear equation*

Let Q be an open regular set of $\mathbb{R} \times \mathbb{R}$ and let u be a classical solution of the equation

$$u_t = \sigma(u, u_x)_x \text{ in } Q. \tag{5}$$

We assume $u \in C^{2,3}(Q)$ such that

$$m_- \leq u(t, x) \leq m_+ \text{ on } Q, \tag{6}$$

for some $m_- \leq m+$. We also assume that $\sigma \in C^2([m_-, m_+] \times \mathbb{R})$.

Our main goal is to obtain pointwise estimates as (3) for suitable auxiliary functions k. As mentioned before, we start by assuming the regularity conditions: $k \in C^2(Q \times [m_-, m_+])$

and $k \neq 0$. The principal ingradient of the method is to find a (possibly nonlinear) parabolic operator, $\mathcal{L}$, vanishing when applied to the function

$$p(t, x) := \frac{\sigma(u(t, x), u_x(t, x))}{k(t, x, u)} \quad (t, x) \in Q \tag{7}$$

(some other choices of function p are also of interest: see [7]).

Using (5) we find

$$u_t = (kp)_x = (k_x + k_u + k_u u_x)p + kp_x \tag{8}$$

and differentiating (8) with respect to x,

$$u_{tx} = (k_{xx} + 2k_{xu}u_x + k_{uu}u_x^2 + k_u u_{xx})p + 2(k_x + k_u u_x)p_x$$
$$+ kp_{xx} = \sigma_u u_x + \sigma_{u_x} u_{xx}. \tag{9}$$

Differentiating (7) with respect to t we have

$$p_t k + p(k_t + k_u u_t) = \sigma_u u_t + \sigma_{u_x} u_{tx}.$$

Multiplying (8) by σ_u and (9) by σ_{u_x} and adding the resulting expressions we find that

$$\widetilde{\mathcal{L}}(p, u_x) = 0$$

where

$$\widetilde{\mathcal{L}}(p, u_x) = p_t k + pk_t + (pk_u - \sigma_u)\{(k_x + k_u u_x)p + kp_x\}$$
$$- \sigma_{u_x}[(k_{xx} + 2k_{xu}u_x + k_{uu}u_x^2)p + 2(k_x + k_u u_x)p_x + kp_{xx}]$$
$$- k_u p[p_x k + p(k_x + k_u u_x) - \sigma_u u_x].$$

In order to obtain an expression independent of u_x we assume

$$\sigma_{u_x} > 0. \tag{10}$$

Thus we can define $S(u, \cdot) := \sigma(u, \cdot)^{-1}$ and so $u_x = S(u, kp)$. Finally, in order to find an operator $\mathcal{L}$ such that $\mathcal{L}(p) = 0$ we define $\mathcal{L}$ by

$$\mathcal{L}(p) := \widetilde{\mathcal{L}}(p, S(u, kp))$$

(the dependence on u of the right hand side is understood as a (t, x)-dependent coefficient). The exact form of $\mathcal{L}(p)$ is irrelevant for our purposes. Our main aim is to compare p with 1 and so the relevant expression is $\mathcal{L}(1)$, i.e.

$$\mathcal{L}(1) = k_t + (k_u - \sigma_u(\widetilde{S}(k)))k_x - \sigma_{u_x}(\widetilde{S}(k))\{k_{xx} + 2k_{xu}\widetilde{S}(k) + k_{uu}\widetilde{S}(k)^2\} \tag{11}$$

where we used the notation $\sigma_u(\widetilde{S}(k)) = \sigma_u(u, \widetilde{S}(k))$ and $\widetilde{S}(k) := S(u, k)$.

The following result is a trivial consequence of the maximum principle (see, e.g., the presentation made in [6]):

LEMMA 1. *Assume* (10),

$$k(t, x, u) > 0 \text{ on } Q \times (m_-, m_+), \tag{12}$$

$$\mathcal{L}(1) \geq 0 \text{ in } Q \tag{13}$$

$$\sigma(u, u_x)(t, x) \leq k(t, x, u(t, x)), \quad (t, x) \in \partial_p Q, \tag{14}$$

where $\partial_p Q$ *denotes the parabolic boundary of* Q. *Then*

$$\sigma(u, u_x)(t, x) \leq k(t, x, u(t, x)) \text{ for any } (t, x) \in \overline{Q}. \tag{15}$$

A concrete boundary value problem for which it is easy to check assumption (14) corresponds to the case of Neumann type boundary conditions.

LEMMA 2. *Let* $I = (a, b)$ *with* $-\infty \leq a < b \leq +\infty$ *and let* $Q = (0, T) \times I$ *where* $0 < T \leq +\infty$. *Let* u *be a classical solution of* (5) *on* Q. *Assume* (10), (12), (13),

$$\sigma(u, u_x)(0, x) \leq k(u)(0, x) \text{ for all } x \in I, \tag{16}$$

$$\begin{cases} (u_t - k_x - k_u u_x)(t, a) \geq 0 \text{ if } a > -\infty, \text{ for all } t \in (0, T), \\ (u_t - k_x - k_u u_x)(t, b) \leq 0 \text{ if } b < +\infty, \text{ for all } t \in (0, T), \end{cases} \tag{17}$$

and

$$\left\| \frac{k_x + k_u u_x}{k} \right\|_{L^\infty((0,T)\times\partial I)} < \infty. \tag{18}$$

Then estimate (15) *holds.*

Proof. From (8) we have that

$$-p_x + p\left[-\left\{ \frac{k_x + k_u u_x}{k} \right\} \right] = \frac{-u_t}{k}$$

So, if v denotes the exterior "unit outer normal" to ∂I (i.e. $v(a) = -1$, $v(b) = 1$), we deduce that

$$-p_x v + p\left[-v\left\{ \frac{k_x + k_u u_x}{k} \right\} \right] = -\frac{u_t v}{k} \text{ on } (0, T) \times \partial I.$$

Since we assume (18) we can apply the comparison principle. Taking $\overline{p}(t, x) = 1$, assumption (16) implies that $p \leq \overline{p}$ on $\{0\} \times (a, b)$ and (17) implies that

$$-\overline{p}_x v + \overline{p}\left[-v\left\{ \frac{k_x + k_u u_x}{k} \right\} \right] \leq -\frac{u_t v}{k} \text{ on } (0, T) \times \partial I.$$

In consequence $p \leq \overline{p}$ on $[0, T] \times \overline{I}$. $\square$

In the special case of

$$k(t, x, u) = \theta(u), \tag{19}$$

we deduce, from (11), that

$$\mathcal{L}(1) = -\sigma_{u_x}(\widetilde{S}(\theta))\widetilde{S}(\theta)^2\theta'' \tag{20}$$

and then the following result holds:

COROLLARY 1. *Let u be a smooth solution of* (5) *satisfying* (6). *Assume* (10) *and let* $\theta \in C^2([m_-, m_+])$ *such that*

$$\theta > 0 \ and \ \theta'' \leq 0 \ on \ [m_-, m_+]. \tag{21}$$

Then

$$\frac{\sigma(u, u_x)}{\theta(u)} \leq \sup_{\partial_p Q} \frac{\sigma(u, u_x)}{\theta(u)}. \tag{22}$$

REMARK 1. Conclusion (22) expresses a "conservation of the regularity": if the estimate is true on the parabolic boundary of Q it also holds in the interior of Q. We emphasize the large generality on θ assumed in (21) and that, by passing to the limit, it is possible to extend the conclusion for θ concave such that $\theta \geq 0$. As a general remark, we point out that the Bernstein technique, as described before, still works in several space dimensions for radially symmetric solutions (see, for instance, [28], for a related result). We also mention that the method can be applied to a general class of quasilinear equations (see the unpublished notes by the first author [8] and the application made in [45]).

In the rest of this section we illustrate the applicability of the method by considering equation (2).

2.2. *A nonlinear diffusion-convection equation*

As it was already mentioned, a physically relevant special case of equation (1) is

$$u_t = \varphi(u)_{xx} + \psi(u)_x \ in \ Q, \tag{23}$$

where φ is a nondecreasing function with $\varphi(0) = 0$. For the sake of simplicity, we assume in this subsection that $\varphi \in C^2([m_-, m_+])$, $\varphi' > 0$ and $\psi \in C^1([m_-, m_+])$. In the applications we shall present here, we assume the (t, x)-dependence of function k of the form

$$k(t, x, u) = \rho(t, x)\theta(u). \tag{24}$$

Then, Lemma 1 leads to the following result

LEMMA 2. *Let $Q \subset \mathbb{R}^+ \times \mathbb{R}$. Let u be a smooth solution of (23) satisfying (6). Let $\rho > 0$ and $\theta > 0$ be smooth functions such that*

$$0 \leq \frac{\rho_t}{\rho} + \left[\theta' - \varphi'' \left(\frac{\rho\theta - \psi}{\varphi'} \right) - \psi'' \right] \frac{\rho_x}{\rho}$$

$$- \varphi' \left[\frac{\rho_{xx}}{\rho} + 2 \frac{\rho_x}{\rho} \frac{\theta'}{\theta} \left(\frac{\rho\theta - \psi}{\varphi'} \right) + \frac{\theta''}{\theta} \left(\frac{\rho\theta - \psi}{\varphi'} \right)^2 \right] \quad in \ Q \tag{25}$$

$$\varphi(u)_x + \psi(u) \leq \rho\theta(u) \ on \ \partial_\rho Q. \tag{26}$$

Then

$$\varphi(u)_x + \psi(u) \leq \rho\theta(u) \ on \ \ Q. \tag{27}$$

In the next section we shall give several applications of Lemma 2 to the case of weak solutions of (23). Nevertheless, some previous consequences for the cases in which condition (26) is trivially satisfied (the Cauchy and the Neumann problems) can be easily derived for the special choice $\rho(t, x) = 1/\sqrt{t}$.

COROLLARY 2. **A.** *Let $Q = (0, T) \times \mathbb{R}$ with $0 < T \leq \infty$ and $\theta \in C^2([m_-, m_+])$ satisfying $\theta > 0$ and*

$$\frac{\varphi'}{2} + \theta\theta'' \left(1 - \psi \frac{\sqrt{t}}{\theta} \right)^2 \leq 0 \ \ for \ any \ t \in (0, T] \ \ on \ \ [m_-, m_+]. \tag{28}$$

Then, if u is a smooth solution of (23) on Q satisfying (6), we have the estimate

$$\varphi(u)_x + \psi(u) \leq \frac{\theta(u)}{\sqrt{t}} \ on \ (0, T] \times \mathbb{R}. \tag{29}$$

B. *Let $I = (a, b)$ with $-\infty < a < b < +\infty$ and $Q = (0, T) \times I$. Assume that θ satisfies the conditions of part A and let u be a smooth solution of equation (23) on Q verifying (6) and the homogeneous Neumann boundary conditions*

$$0 = (\varphi(u)_x + \psi(u))(t, a) = (\varphi(u)_x + \psi(u))(t, b) \ for \ t \in (0, T). \tag{30}$$

Then inequality (29) holds on $(0, T] \times I$. The same conclusion holds if we replace the boundary condition (30) by

$$0 = \varphi(u)_x(t, a) = \varphi(u)_x(t, b) \tag{31}$$

and assume that $\psi \leq 0$ on (m_-, m_+).

REMARK 2. In contrast to Corollary 1 the above result shows the first "regularizing effect": estimate (29) holds for any $t > 0$, independently of either it takes place at $t = 0$ or not. Other regularizing effects will be given in Subsection 3.3.

Another interesting choice of function $\rho(t, x)$ leading to local effects, independently of the boundary conditions, is $\rho(t, x) = \xi(x)/\sqrt{t}$ with $\xi(x)$ vanishing on the boundary of the spatial domain.

COROLLARY 3. *Let* $I = (a, b)$ *with* $-\infty < a < b \leq +\infty$ *and* $Q = (0, T) \times I$. *Assume that* $\theta \in C^2([m_-, m_+])$, $\theta > 0$, $\xi \in C^2(I)$, $\xi > 0$ *on* I *and* $\xi = +\infty$ *on* ∂I. *Assume that*

$$0 \leq -\frac{1}{2t} + \left[\theta' - \varphi'' \left(\frac{\theta\xi/\sqrt{t} - \psi}{\varphi'} \right) - \psi'' \right] \frac{\xi_x}{\xi}$$
$$-\varphi' \left[\frac{\xi_{xx}}{\xi} + 2\frac{\xi_x}{\xi}\frac{\theta'}{\theta} \left(\frac{\theta\xi/\sqrt{t} - \psi}{\varphi'} \right) + \frac{\theta''}{\theta} \left(\frac{\theta\xi/\sqrt{t} - \psi}{\varphi'} \right)^2 \right] \tag{32}$$

Let u *be a smooth solution of equation* (23) *on* Q *satisfying* (6). *Then we have the inequality*

$$\varphi(u)_x + \psi(u) \leq \frac{\theta(u)\xi(x)}{\sqrt{t}} \quad \text{on } (0, T] \times I. \tag{33}$$

Condition (32) leads to interesting results once we know more about the functions φ and ψ, for instance, if $\varphi(u) = u^m$, $\psi(u) = u^n$ for some positive m and n (see Subsection 3.4).

3. Applications to generalized and mild solutions

Corollaries 2 and 3 will be used in this section for two different objectives: (i) to extend estimates (29) and (33) to weaker notions of solution of the Cauchy and Dirichlet problems associated to equation (23) (and so under much more generality on the data φ, ψ and u_0), and, (ii) to present explicitly different choices of the function θ according φ and ψ.

3.1. Generalized solutions

We start by considering the class of nonnegative bounded generalized solutions of the Cauchy problem

$$0 \leq u_0(x) \leq M_0 \quad \text{for any } x \in \mathbb{R}. \tag{34}$$

$$\begin{cases} u_t = \varphi(u)_{xx} + \psi(u)_x & \text{in } (0, T) \times \mathbb{R}, \\ u(0, x) = u_0(x) & x \in \mathbb{R}, \end{cases} \tag{35}$$

with

$$0 \le u_0(x) \le M_0 \quad \text{for } x \in \mathbb{R}, \tag{36}$$

for some M_0. We assume that

$$\varphi, \psi \in C^0([0, M_0]), \varphi(0) = 0, \varphi \text{ strictly increasing} \tag{37}$$

as well as the following technical conditions (see Subsection 3.3)

$$\begin{cases} \exists \delta_0 > 0 \text{ such that } \varphi, \psi \in C^2((0, M_0 + \delta_0]) \text{ and} \\ \varphi'' \text{ and } \psi'' \text{ are locally Hölder continuous on } (0, M_0 + \delta_0]. \end{cases} \tag{38}$$

We recall that given $0 < T \le +\infty$, a function $u(t, x)$ is said to be a *nonnegative bounded generalized solution* of equation (23) if $u \in C([0, T] \times \mathbb{R})$,

$$0 \le u(t, x) \le M_0 \text{ for } x \in \mathbb{R} \text{ and } t \in [0, T], \tag{39}$$

for some $m > 0$ and satisfies the integral identity

$$\int_{t_1}^{t_2} \int_{x_1}^{x_2} \{u\phi_t + \varphi(u)\phi_{xx} - \psi(u)\phi_x\}dxdt$$
$$= \int_{x_1}^{x_2} \{u(t_2, x)\phi(t_2, x) - u(t_1, x)\phi(t_1, x)\}dx$$
$$+ \int_{t_1}^{x_{t2}} \{u(t, x_2)\phi(t, x_2) - u(t, x_1)\phi(t, x_1)\}dt$$

for all non-empty bounded rectangles $(t_1, t_2) \times (x_1, x_2)$ and nonnegative functions $\phi \in C^{1,2}([t_1, t_2] \times [x_1, x_2])$ such that $\phi(t, x_1) = \phi(t, x_2) = 0$ for all $t \in [t_1, t_2]$.

We recall (see, e.g., [20]) that any generalized solution is a weak solution in the sense that the above integral identity can be replaced by the condition $u_t = \varphi(u)_{xx} + \psi(u)_x$ in $D'(Q)$.

Different results on the existence and uniqueness of a generalized solution u of the Cauchy Problem (35) are today available in the literature (see e.g. [13] [24] and [17]). For our present purposes we shall made use of the results of Gilding [27] requiring, merely, the continuity of the initial datum. We have:

THEOREM 1. *Assume that φ and ψ satisfy (37) and (38). Let $u_0 \in C^0(\mathbb{R})$ satisfying (36). Let $u \in C([0, T] \times \mathbb{R})$ be the generalized solution to the Cauchy problem (35). Assume that there exists a function $\lambda \in C^0([0, M_0])$, concave, increasing and such that*

$$M < \infty \tag{40}$$

where

$$M := \int_0^{M_0} \frac{d\varphi(s)}{\lambda(s)\lambda'(s)}.$$

Finally, assume that there exists a constant $C \geq 0$ such that

$$\psi(s) \leq C\lambda(s) \text{ for any } s \in (0, M_0). \tag{41}$$

Then

$$\varphi(u)_x + \psi(u) \leq \left[\sqrt{4C^2T + 2M} + 2C\sqrt{T}\right]\frac{\lambda(u)}{\sqrt{t}} \text{ in } D'((0, T) \times \mathbb{R}). \tag{42}$$

Before giving the proof we shall state two immediate consequences.

COROLLARY 4. *Let φ and ψ satisfying (37) and (38). Assume ψ be such that*

$$\psi(s) \leq C\sqrt{s} \text{ for any } s \in (0, M_0), \tag{43}$$

for some constant $C \geq 0$. Then, if $u_0 \in C^0(\mathbb{R})$ satisfies (36) and u is the generalized solution to the Cauchy Problem (35) we have the estimate

$$\varphi(u)_x + \psi(u) \leq 2\sqrt{\frac{u\varphi(M_0)}{t}} \text{ in } D'((0, T) \times \mathbb{R}). \tag{44}$$

COROLLARY 5. *Let φ and ψ satisfying (37) and (38). Let $u_0 \in C^0(\mathbb{R})$ satisfying (36) and u be the generalized solution of the Cauchy Problem (35). Then $\varphi(u(t, \cdot))_x \in L^\infty(\mathbb{R})$ and*

$$\|\varphi(u(t, \cdot))_x\|_{L^\infty(\mathbb{R})} \leq 2\left[\frac{\varphi(M_0)M_0}{2}\right]^{1/2} + N \tag{45}$$

for any $t > 0$, where

$$N = \sup_{[0, M_0]} \psi - \inf_{[0, M_0]} \psi. \tag{46}$$

Proof of Corollaries 4 and 5. Corollary 4 follows from the fact that function $\lambda(s) = \sqrt{s}$ satisfies (40) for any continuous nondecreasing function φ. To prove Corollary 5 we point out that u is also a solution of the equation

$$u_t = \varphi(u)_{xx} + \psi_d(u)_x, \tag{47}$$

where $\psi_d(s) := \psi(s) + d$ with $d \in \mathbb{R}$ arbitrary. By choosing $d = -d_1$ with $d_1 = \sup\{\psi(s) : s \in [0, M_0]\}$ condition (43) is verified for ψ_d (since by the maximum principle u we know that u satisfies $0 \leq u \leq M_0$). Then (44) implies

$$\varphi(u)_x \leq \left(\frac{2M_0\varphi(M_0)}{t}\right)^{1/2} + d_1 - d_2$$

where $d_2 = \inf\{\psi(s) : s \in [0, M_0]\}$. To obtain the estimate in the reverse sense we define $\widetilde{u}(t, x) = u(t, -x)$. Then $\widetilde{u}$ satisfies

$$\widetilde{u}_t = \varphi(\widetilde{u})_{xx} + \widetilde{\psi}_d(\widetilde{u})_x \tag{48}$$

with $\widetilde{\psi}_d(s) = -\psi(s) - d$. Then, by choosing $d = -d_2$, we conclude that

$$-\varphi(u)_x = \varphi(\widetilde{u})_x \leq 2 \left(\frac{M_0 \varphi(M_0)}{t} \right)^{1/2} + d_1 - d_2,$$

and so estimate (45) holds.

Proof of Theorem 1. To get the result we shall perform an approximation process according to classical ideas. Our main purpose is to approximate u by a sequence of smooth functions. Let δ_0 be given in (38). Fix $\delta \in (0, \delta_0)$ arbitrary and let $k_0 \in \mathbb{N}$ be such that

$$2^{-k_0+1} \leq \varphi(M_0 + \delta) - \varphi(M_0). \tag{49}$$

For $k \geq k_0$ we define the cylinder $Q_k = (0, T) \times (-k, k)$ and consider the boundary value problem

$$u_t = \varphi(u)_{xx} + \psi(u)_x \quad \text{in } Q_k, \tag{50}$$

$$u(t, \pm k) = M_0 \quad \text{for all } t \in (0, T), \tag{51}$$

$$u(0, x) = u_{0,k}(x) \text{ on } (-k, k), \tag{52}$$

where the sequence $\{u_{0,k}\}$ satisfies the following properties: (i) $u_{0,k}(x) \leq M_0 + \delta$ for all $x \in \mathbb{R}$, (ii) there exists $\varepsilon_k > 0$ such that $\varepsilon_k \leq u_{0,k}(x)$ for all $x \in \mathbb{R}$, (iii) there exists $\alpha_k \in (0, 1]$ such that $u_{0,k} \in C^{2+\alpha_k}(\mathbb{R})$, (iv) $u_{0,k+l}(x) \leq u_{0,k}(x)$ for all $x \in \mathbb{R}$, (v) $u_{0,k} \downarrow u_0$ as $k \uparrow \infty$ uniformly on compact subsets of $\mathbb{R}$, (vi) $u_{0,k}(x) = M_0 + \delta$ for all $|x| > k - 1/2$.

This approximation process was made explicit in [27]. Since properties (ii), (iii) and (vi) hold we can apply the theory of nondegenerate parabolic problems (see e.g. [38]) and so (50), (51), (52) has a unique classical solution $u_k(t, x)$. Moreover from (i) and (ii) we deduce that $\epsilon k \leq u_k(t, x) \leq M_0 + \delta$ for all $(t, x) \in Q_k$. By the maximum principle, property (iv) implies, that $u_{k+l}(t, x) \leq u_k(t, x)$ for all $(t, x) \in \overline{Q}_k$ (assumed $k \geq k_0$). Hence we can define

$$u^*(x, t) = \lim_{k \nearrow \infty} u_k(x, t) \quad \text{for all } (t, x) \in [0, T] \times \mathbb{R}.$$

It is easy to see that u^* is a generalized solution. It is clear that it remains to show that $u^* \in C^0([0, T] \times \mathbb{R})$ since by the uniqueness of the solution, $u = u^*$. The continuity of u^*

will be obtained similarly to the proof of Theorem 1 of [27]: as in the proof of Corollary 4, we first conclude the gradient estimate for $t \in (0, T]$

$$\|\varphi(u_k(t, \cdot))_x\|_{L^\infty((-k,k))} \leq 2 \left(\frac{\varphi(M_0 + \delta)(M_0 + \delta)}{t} \right)^{1/2} + N$$

with N given by (46). This implies the existence of a modulus of continuity of u_k in x and (by the results of Kruzkov [37] and [24]) also in t. This modulus of continuity is independent of k and so, by passing to the limit in k, it also applies to $u(t, x)$ for $t \in (0, T]$ and $x \in \mathbb{R}$. The continuity at $t = 0$, i.e. the inequalities

$$u_0(x_0) \leq \lim_{(t,x) \longrightarrow (0,x_0)} \inf u(t, x) \leq \lim_{(t,x) \longrightarrow (0,x_0)} \sup u(t, x) \leq u_0(x_0)$$

for any $x_0 \in \mathbb{R}$, were proved in [24] by using a local barrier function argument which literally applies to our case.

The rest of the proof consists in obtaining the gradient estimate (42), on $(0, T) \times (-k, k)$, for the smooth solutions u_k but replacing M_0 by M_δ in (42), where

$$M_\delta := \int_0^{M_0+\delta} \frac{d\varphi}{\lambda(s)\lambda'(s)}.$$

Then, passing to the limit in k, we deduce the same estimate for u (but now in the sense of $D'((0, T) \times \mathbb{R})$). Finally, as δ is arbitrary we get (42).

In order to estimate $\varphi(u_k)_x + \psi(u_k)$ we apply Lemma 2 with $k(s) = \theta(s)/\sqrt{t}$, where $\theta(s)$ is a convex, increasing and positive function to be determined. Notice that condition (16) trivially holds. On the other hand, from the boundary condition (51) we deduce that $(u_k(t, \pm k))_t = 0$ for $t \in (0, T)$. Moreover as $u_k(t, x) \leq M_0 + \delta$ we deduce that $u_x(t, -k) \leq 0$ and $u_x(t, k) \geq 0$ for all $t \in (0, T)$. Then, conditions (17) holds. Finally, since u_k is a positive classical solution, condition (18) is also satisfied.

Thus, it only remains to check condition (13) or, equivalently, (28). We introduce the function

$$\theta(s) := e\lambda(s - cj(c)) \tag{53}$$

where e and c are positive constants to be chosen later and j is the convex function defined by

$$j(s) := \int_0^s \int_0^r \frac{d\varphi(z)}{\lambda(z)\lambda'(z)} dr. \tag{54}$$

Our first condition on c is $c < 1/M$. So from the convexity of j we deduce that $\theta''(s) < 0$ and $\theta'(s) > 0$ if $s \in (0, M_0 + \delta]$. Moreover

$$\theta(s) \geq e\lambda((1 - cM)s) \geq e(1 - cM)\lambda(s) > 0 \quad \text{if } s \in (0, M_0 + \delta]. \tag{55}$$

In order to check condition (28), we observe that

$$\theta''(s) = e\lambda''(s - cj(s))(1 - cj''(s))^2 - ec\lambda'(s - cj(s))\frac{\varphi'(s)}{\lambda(s)\lambda'(s)}.$$

Then, (28) is equivalent to

$$ec\frac{\lambda(s - cj(s))}{\lambda(s)}\frac{\lambda'(s - cj(s))}{\lambda'(s)}\left(1 - \frac{\psi(s)\sqrt{t}}{e\lambda(s - cj(s))}\right)^2 \geq \frac{1}{2}. \tag{56}$$

But the concavity of λ implies that $\lambda'(s - cj(s)) \geq \lambda'(s)$. Then, from assumption (41)

$$\left(1 - \frac{\psi(s)\sqrt{t}}{e\lambda(s - cj(s))}\right) \geq \left(1 - \frac{C\sqrt{t}}{e(1 - cM_\delta)}\right).$$

So, condition (56) is verified if

$$e^2c(1 - cM_\delta)\left(1 - \frac{2C\sqrt{T}}{\varrho(1 - cM_\delta)}\right) \geq \frac{1}{2}. \tag{57}$$

Taking $c = 1/(2M_\delta)$, (57) reduces to the inequality

$$e^2 - 4eC\sqrt{T} - 2M_\delta \geq 0,$$

and, finally, the choice

$$e = 2C\sqrt{T} + \sqrt{4C^2T + 2M_\delta}$$

leads to the desired estimate (42) completing the proof of Theorem 1.

3.2. *Unidirectional estimates*

It is important to point out that sometimes, it is possible to get pointwise gradient estimates of *unidirectional type* for the solutions of the diffusion-convection Cauchy problem (35), i.e. inequalities which become false if we reverse the direction of the spatial variable x by making the transformation $\tilde{x} = -x$. This is illustrated in the next result which exhibits an alternative proof to a result derived in [19].

THEOREM 2. *Let $u_0 \in C^0(\mathbb{R})$ satisfies (36), with M_0 small enough, and let u be the generalized solution to the Cauchy problem (35) corresponding to*

$$\varphi(u) = u^m, \psi(u) = -u^n \text{ and } 0 < n < 1 \leq m.$$

Then

$$(u^m)_x \geq -K\frac{u^m}{t} \text{ in } D\prime((0, T) \times \mathbb{R}), \tag{58}$$

for some $K > 0$ depending only on $M_0, m,$ and n.

Proof. As in the proof of the previous theorem we can assume u to be positive and smooth enough. Let $\lambda(r)$ be a regular function, to be chosen later, such that $\lambda(r)$ is a smooth function transforming $[0, w_0]$ onto $[0, m \ln M_0]$, for some $w_0 > 0$ and such that

$$K = \sup_{w \in [0, w_0]} \lambda'(w) > \inf_{w \in [0, w_0]} \lambda'(w) = C > 0. \tag{59}$$

Then, the equation

$$m \ln u(t, x) = \lambda(w(t, x))$$

defines a regular function $w(t, x)$ such that $0 < w(t, x) \le w_0$ for all $(t, x) \in [0, T] \times \mathbb{R}$. In order to prove the estimate we introduce the function

$$z(t, x) = w_x(t, x)t.$$

Then, it is clear that the conclusion of the theorem will follow once we prove that $z(t, x) \ge -1$ on $[0, T] \times \mathbb{R}$, since

$$\frac{(u^m)_x}{u^m} = \lambda(w)_x = \lambda'(w)w_x \ge -\frac{K}{t}.$$

It is not difficult to see that z satisfies $\mathcal{L}(z) = 0$ where

$$\mathcal{L}(z) = z_t - mu^{m-1}z_{xx} - \frac{u^{m-1}}{t}\left[(3m - 1)\lambda'(w) + m\frac{\lambda''(w)}{\lambda'(w)}\right]zz_x$$

$$- nu^{n-1}z_x - \frac{u^{m-1}}{t^2}\left[(m - 1)\lambda'(w)^2 + m\left(\frac{\lambda''(w)}{\lambda'(w)}\right)_w + (2m - 1)\lambda''(w)\right]z^3$$

$$+ \frac{n(n - 1)}{mt}u^{n-1}\lambda'(w)z^2 - \frac{z}{t}.$$

Hence,

$$\mathcal{L}(-1) = \frac{u^{m-1}}{t^2}\left[(m - 1)\lambda'(w)^2 + m\left(\frac{\lambda''(w)}{\lambda'(w)}\right)_w + (2m - 1)\lambda''(w)\right]$$

$$- \frac{n(n - 1)}{mt}u^{n-1}\lambda'(w)z^2 + \frac{1}{t}.$$

So, we have to choose $\lambda(r)$ such that $\mathcal{L}(-1) \le 0$ (notice that the comparison at $t = 0$ is obviously satisfied and that we can argue starting with u_k and then passing to the limit as in the previous theorem). The above condition holds once that

$$(m - 1)\lambda'(w)^2 + m\left(\frac{\lambda''(w)}{\lambda'(w)}\right)_w + (2m - 1)\lambda''(w) \le 0 \tag{60}$$

and

$$-\frac{n(1 - n)C}{mM_0^{1-n}} + 1 \le 0, \tag{61}$$

with C given by (59). The condition (61) is satisfied by taking

$$C \geq \frac{m M_0^{1-n}}{n(1-n)}.$$

To simplify the study of condition (60) we shall take $\lambda(r)$ such that $\lambda''(r) \equiv -\gamma < 0$ for any $r \in [0, w_0]$. Then

$$\left(\frac{\lambda''(w)}{\lambda'(w)}\right)_w = \frac{-\gamma^2}{\lambda'(w)^2} < \frac{-\gamma^2}{K^2},$$

and (60) is satisfied once

$$(m-1)K^2 - m\frac{\gamma^2}{K^2} - (2m-1)\gamma \leq 0. \tag{62}$$

Let us take $\lambda(r) = ar^2 + br$, with $a = (-\gamma)/2$ and $b = K$. By taking $\gamma = K$, (62) is transformed into

$$(m-1)K^2 - (2m-1)K - m \leq 0,$$

and so, it if fullfilled once that

$$0 < K \leq 2 < \frac{(2m-1) + ((2m-1)^2 + 4m)^{1/2}}{2(m-1)}.$$

Finally, it is a routine matter to check that, among the many possible choices, $\lambda(r) = r(2-r)$ satisfies all the requirements once we assume

$$M_0^{1-n} < \min\left\{\frac{1}{2m}, \frac{1}{\frac{m}{2} + \frac{m}{2n(1-n)}}\right\}.$$

$\square$

As we shall comment later, in Subsection 3.4, estimate (58) ceases to be true if we reverse the direction of x.

3.3. *Mild solutions: smoothing effects*

In the case of $u_0 \in L^1(\mathbb{R})$ it is possible to show some regularizing effect for the, so called, *mild solution* of the Cauchy problem (35). To introduce this notion we consider the implicit time-discretization

$$(ITD) \begin{cases} u\dfrac{w_i - w_{i-1}}{t_i - t_{i-1}} = \varphi(w_i)_{xx} + \psi(w_i)_x + f_i \ \text{ in } D'(\mathbb{R}), \end{cases}$$

for a given $\epsilon > 0$ and a time discretization $t_0 = 0 < t_1 < \cdots < t_n \leq T$, $t_i - t_{i-1} < \epsilon$, $T - t_n < \epsilon$, and for given data w_0, $f_i \in L^1(\mathbb{R}) \cap L^\infty(\mathbb{R})$. We say that u is a mild solution of (35) if $u \in C([0, +\infty) : L^1(\mathbb{R}))$, $u(0, .) = u_0(.)$, and, for any $\epsilon > 0$ there exists $(t_0, t_1, \ldots, t_n, w_0, w_1, \ldots, w_n)$ satisfying (ITD) with

$$\|w_0 - u_0\|_1 \leq \epsilon, \quad \sum_i \int_{t_{i-1}}^{t_i} \|f_i\|_1 \, dt \leq \epsilon$$

and such that $\|u(t) - w_i\|_1 \leq \epsilon$ for any $t \in (t_{i-1}, t_i]$, $i = 1, \ldots, n$. The existence of a mild solution is due to Bénilan and Touré [13]. Moreover, it was proven there that, under suitable additional regularity on the data φ, ψ and u_0, the mild solution is also a weak solution.

THEOREM 3. *Let φ and ψ satisfy (37) and let $u_0 \in L^1(\mathbb{R})$, $u_0 \geq 0$. Let $u \in C([0, T] : L^1(\mathbb{R}))$ be the mild solution to the Cauchy problem (35). Assume that*

$$R(I) = \mathbb{R}, \tag{63}$$

where

$$I(u) = sign(u) \int_0^u (\varphi(u) - \varphi(r))dr.$$

Then

$$\|u(t, .)\|_{L^\infty(\mathbb{R})} \leq I^{-1}\left(\frac{\|u_0\|_{L^1(\mathbb{R})}^{3/2}}{4t}\right). \tag{64}$$

Moreover, $\varphi(u(t, \cdot))_x \in L^\infty(\mathbb{R})$ and

$$\|\varphi(u(t, \cdot))_x\|_{L^\infty(\mathbb{R})} \leq 2\left[\frac{\varphi(m(t))m(t)}{2}\right]^{1/2} + N \tag{65}$$

for any $t > 0$, with

$$m(t) = I^{-1}\left(\frac{\|u_0\|_{L^1(\mathbb{R})}^{3/2}}{4t}\right)$$

and N given by (46).

Proof. Using the continuity of the solution with respect to the data and the functions φ and ψ (see [12]) we may assume that $u_0, \varphi, \psi \in C^\infty(\mathbb{R})$ with $\varphi' > 0$, so that u is a classical solution with $u_x, u_t \in C([0, T] : L^1(\mathbb{R}))$. From the definition of I we deduce that $I(u)_x = |u| \, \varphi(u)_x$. Moreover

$$\frac{d}{dt} \int_\mathbb{R} \int_0^u I(r)dr = \int_\mathbb{R} I(u)u_t = -\int_\mathbb{R} I(u)_x \varphi(u)_x.$$

But, using Hölder inequality and the L^1-contraction of the associated semigroup ([13], [12]) we get that

$$\|I(u(t,.))\|_{L^\infty(\mathbb{R})}^2 \le \frac{1}{4}\|I(u)_x\|_{L^1}^2 \le \frac{1}{4}\left[\int_{\mathbb{R}} |u|^{1/2}\,|u|^{1/2}\,\varphi(u)_x\right]^2$$

$$\le \frac{1}{4}\|u_0\|_{L^1(\mathbb{R})}\left[\int_{\mathbb{R}} |u|\,\varphi(u)_x^2\right] = -\frac{1}{4}\|u_0\|_{L^1(\mathbb{R})}\int_{\mathbb{R}} I(u)_x\varphi(u)_x$$

$$= -\frac{1}{4}\|u_0\|_{L^1(\mathbb{R})}\frac{d}{dt}\int_{\mathbb{R}}\int_0^u I(r)dr.$$

On the other hand, if we define the function

$$l(t) = \int_{\mathbb{R}}\left(\int_0^{u(t,x)} I(r)dr\right)dx,$$

then

$$l(t) \le \|I(u(t,.))\|_{L^\infty(\mathbb{R})}\|u(t)\|_{L^1(\mathbb{R})} \le \|I(u(t,.))\|_{L^\infty(\mathbb{R})}\|u_0\|_{L^1(\mathbb{R})}.$$

So, $l(t)$ satisfies the ordinary differential inequality

$$l'(t) + \frac{4l(t)^2}{\|u_0\|_{L^1(\mathbb{R})}} \le 0.$$

Dividing by $l(t)^2$, integrating and using that $l(t_0) > 0$ for any $t_0 \ge 0$ we get that

$$l(t) \le \frac{\|u_0\|_{L^1(\mathbb{R})}^2}{4t},$$

and finally

$$\|I(u(t))\|_{L^\infty(\mathbb{R})} \le \frac{\|u_0\|_{L^1(\mathbb{R})}^{3/2}}{4t},$$

which leads to estimate (64). The rest is an obvious application of the arguments of Corollary 4. $\qquad\square$

REMARK 3. Some related results can be found in [9] and [39]. It is possible to apply the general method to the study of one-side estimates on u_t. Among the many papers devoted to this task, we only mention here the presentation made in [7] and the articles [4] and [45] (many other references can be found in the surveys [35] and [46]). This allows to prove stronger differentiabilty on the solutions of (2) leading to the existence of the, so called, *strong solutions* satisfying that u_t, $\varphi(u)_{xx}$ and $\psi(w_i)_x$ are in $L^1((0,T)\times\mathbb{R})$ (see also the different approach made in [11]).

3.4. *The case of a bounded interval*

Concerning the case in which the solution is not defined on $\mathbb{R}$ but on a bounded interval, we can get interior gradient estimates without using any concrete information on the boundary conditions. We illustrate it by considering the special case of $\psi \equiv 0$.

PROPOSITION 1. *Let $I = (a, b)$ with $-\infty < a < b < +\infty$ and $Q = (0, T) \times I$. Let $u \in L^\infty(Q)$ be a positive solution of*

$$u_t = (u^m)_{xx} \text{ in } Q.$$

Let $\xi \in C^2(I), \xi > 0$ on I and $\xi = +\infty$ on ∂I such that

$$|\xi'(x)| \le \xi(x)^2, |\xi''(x)| \le \xi(x)^3 \quad \text{for any } x \in I.$$

Then, if $M_0 := \sup_Q u$,

a) *if $0 < m < 1$,*

$$|(u^m)_x| \le \frac{\xi(x)u^m}{\sqrt{t}} \left(\frac{\sqrt{T}}{1 - m} + \sqrt{\frac{M_0}{2(1 - m)}} \right),$$

b) *if $m = 1$, for any $\varepsilon > 0$,*

$$|(u^m)_x| \le \frac{\xi(x)u}{\sqrt{t}} \left(\varepsilon + \log \frac{M_0}{u} \right) \left(2\sqrt{T} \left(1 + \sup\left(1, \frac{1}{\varepsilon}\right) \right) \right) + \sqrt{\frac{2}{\varepsilon}},$$

c) *if $m > 1$,*

$$|(u^m)_x| \le \frac{\xi(x)\lambda_0(\alpha u - u^m)}{\sqrt{t}}$$

with $\alpha > \alpha_0 := \sup_Q u^{m-1}$ and λ_0 such that

$$\lambda_0 \le \frac{2\sqrt{T}}{(m - 1)(\alpha - \alpha_0)}(m(\alpha - \alpha_0) + 2(m - 1)\alpha_0) + \sqrt{\frac{2}{(m - 1)(\alpha - \alpha_0)}}.$$

Proof. In the case a) we take $\theta(u) = \lambda u^m$. Then condition (32) is satisfied if.

$$-\lambda^2(1 - m) + \frac{u^{1-m}}{2} + \lambda\sqrt{t}\frac{\xi'}{\xi^2}(2m - 1) + mt\frac{\xi''}{\xi^3} \le 0,$$

which is implied by

$$-\lambda^2(1 - m) + \frac{M_0^{1-m}}{2} + |\lambda|\sqrt{T}\frac{\xi'}{\xi^2}|2m - 1| + mT \le 0.$$

This last condition holds once that

$$|\lambda| \geq \lambda_0 = \frac{1}{1-m}\left(\left|m-\frac{1}{2}\right|\sqrt{T} + \left(T\left|m-\frac{1}{2}\right|^2\right.\right.$$
$$\left.\left. + (1-m)\left(2mT + \frac{M_0^{1-m}}{2}\right)\right)^{1/2}\right).$$

A simple computation shows that

$$\lambda_0 \leq \frac{\sqrt{T}}{1-m} + \sqrt{\frac{M_0^{1-m}}{2(1-m)}}.$$

Case b). We take $\theta(u) = \lambda u \log \frac{M}{u}$ with $M > M_0$. Then, (32) leads to

$$-\lambda^2 \log\frac{M}{u} + \frac{1}{2} + \frac{\xi'}{\xi^2}\sqrt{t}\lambda\left(\log\frac{M}{u} - 1\right) + \frac{\xi''}{\xi^3}t \leq 0.$$

Assuming $|\lambda| \geq \sqrt{T}$, then

$$\frac{\xi'}{\xi^2}\sqrt{t}\lambda \leq \lambda^2$$

and the above condition becomes

$$-\lambda^2 \log\frac{M}{u} + \frac{1}{2} + \sqrt{T}\,|\lambda|\left|\log\frac{M}{u} - 1\right| + 2T \leq 0.$$

This is satisfied once that

$$|\lambda| \geq \lambda_0 = \frac{1}{\log\frac{M}{u}}\left(\left|\log\frac{M}{u} - 1\right|\sqrt{T} + \left(T\left(\log\frac{M}{u} - 1\right)^2\right.\right.$$
$$\left.\left. + 4\log\frac{M}{u}\left(2T + \frac{1}{2}\right)\right)^{1/2}\right).$$

Again, a simple computation shows that

$$\lambda_0 \leq 2\sqrt{T}\left(1 + \frac{1}{\log\frac{M}{u}}\right) + \left(\frac{2}{\frac{1}{\log\frac{M}{u}}}\right)^{1/2}.$$

Case c). We take $\theta(u) = \lambda(\alpha u - u^m)$ with $\alpha > \alpha_0 := \sup_Q u^{m-1}$. Then, (32) leads to

$$-\lambda^2(m-1)(\alpha - u^{m-1}) + \frac{1}{2} + \lambda\sqrt{t}\frac{\xi'}{\xi^2}(m\alpha - (2m-1)u^{m-1}) + mt\frac{\xi''}{\xi^3}u^{m-1} \leq 0.$$

Assuming $|\lambda| \geq \frac{(2m-1)}{m-1}\sqrt{T}$, then

$$(2m-1)\frac{\xi'}{\xi^2}\sqrt{t}\lambda \leq (m-1)\lambda^2$$

and the above condition becomes

$$-\lambda^2(m-1)(\alpha - \alpha_0) + \frac{1}{2} + |\lambda|\sqrt{T}\,|m\alpha - (2m-1)\alpha_0| + 2mT\alpha_0 \leq 0.$$

This condition is satisfied if

$$|\lambda| \geq \lambda_0 = \frac{1}{(m-1)(\alpha - \alpha_0)}\left(\sqrt{T}\,|m\alpha - (2m-1)\alpha_0| + \left(T\,|m\alpha - (2m-1)\alpha_0|^2\right.\right.$$
$$\left.\left. +4(m-1)(\alpha - \alpha_0)\left(\frac{1}{2} + 2mT\alpha_0\right)\right)^{1/2}\right),$$

and we have

$$\lambda_0 \leq \frac{2\sqrt{T}}{(m-1)(\alpha - \alpha_0)}(m(\alpha - \alpha_0) + 2(m-1)\alpha_0) + \left(\frac{2}{(m-1)(\alpha - \alpha_0)}\right)^{1/2}.$$

$\square$

In the case of homogeneous Dirichlet boundary conditions and for a general maximal monotone graph φ and $\psi \equiv 0$, it is not difficult to show that the conclusions of Corollary 4 and Proposition still hold for mild solutions. This leads to a gradient estimate which implies a regularizing effect stronger than the fact that $\varphi(u(t, \cdot))_x \in L^2(a, b)$, for any $t > 0$, shown by H. Brézis in [16] for initial data $u_0 \in H^{-1}(a, b)$. For the sake of the exposition, let us assume that

$$D(\varphi) = R(\varphi) = \mathbb{R}. \tag{66}$$

Then we have:

COROLLARY 6. *Let φ be a maximal monotone graph of $\mathbb{R}^2$ with $0 \in \varphi(0)$ and satisfying (66). Let $u_0 \in L^1(a, b)$, $u_0 \geq 0$. Then the mild solution u of problem*

$$\begin{cases} u_t = w_{xx}, w \in \varphi(u) & \text{in } (0, T) \times (a, b), \\ w(t, a) = w(t, b) = 0 & t \in (0, T), \\ u(0, x) = u_0(x) & x \in (a, b), \end{cases}$$

satisfies

$$\|w(t, \cdot)_x\|_{L^\infty(a,b)} \leq 2\left[\frac{\varphi^+(m(t))m(t)}{2}\right]^{1/2} \tag{67}$$

for any $t > 0$, with $\varphi^+(r) = \max\{s \in \mathbb{R} : s \in \varphi(r)\}$ and

$$m(t) = I^{-1}\left(\frac{\|u_0\|_{L^1(a,b)}^{3/2}}{4t}\right).$$

3.5. *Applications to the study of the free boundaries*

We shall end the paper with several comments on the connections between pointwise gradient estimates and the study of the free boundaries defined as the boundaries of the support of the solution of the equation (2).

As it was already mentioned in the introduction, in the context of the gas flow in porous media the expression

$$V(t, x) = -\frac{\varphi(u(x, t))_x + \psi(u(x, t))}{u(x, t)}$$

represents the Eulerian description of the velocity of the "homogenized gas" at the position x and the time t once we identify the parabolic equation with the continuity equation for the mass conservation. In particular, if we denote by $x = \zeta(t)$ any of those interfaces, by standard fluid mechanics arguments we deduce, at least formally, that if $\zeta(t)$ is differentiable, then the particles of the gas at the boundary of the support at time $t = 0$ must remain at the boundary of the support of the gas for any $t > 0$ and so they must move with velocity $V(t, \zeta(t))$. Hence if, for instance, $\zeta(t) = \sup\{x \in \mathbb{R} : u(t, x) > 0\}$ then

$$\zeta'(t) = -\frac{\varphi(u(t, \zeta(t)))_x + \psi(u(t, \zeta(t)))}{u(t, \zeta(t))}. \tag{68}$$

The question of justifying the above formula, in a rigorous way, is not an easy task. To start with, the occurrence of the free boundaries must be correctly proved without invoking any physical argument (recall that the equation holds for an homogeneized medium). Among many papers dealing with the study of the interfaces we would start by mentioning the short paper by A. S. Kalashnikov [34] in which he shows that the solutions of the Cauchy problem (35) with $\psi \equiv 0$ and $u_0 \geq 0$ with compact support remain with compact support for any $t > 0$ (a property that, in what follows, we shall call as *finite speed of propagation*), provided that the pointwise gradient estimate

$$|\varphi(u)_x| \leq Mu \tag{69}$$

holds. We recall that, by using completely different methods, it was shown that the finite speed of propagation holds if and only if

$$\int_{0+} \frac{d\varphi(s)}{s} < \infty \tag{70}$$

(see [41] for the sufficient part and [31] and [42] for the necessity part). So, we conclude that estimate (69) only remains true by assuming the condition (70). Although a direct derivation of estimate (69) was already shown in [3] and [32], we mention that this can be deduced also from our Theorem 1 by taking $\lambda(u) \equiv u$. Notice that condition (40) coincides with (70) and that the estimate can be also proved for the class of mild solutions when φ is a maximal monotone graph satisfying (70) since this condition make sense even if φ is not C^1.

In the case of $\psi \neq 0$ the analysis is more sophisticated since the equation loses its invariance after changing x by $-x$. The convection term privileges one direction in contrast with the other one. The *finite speed of propagation* (i. e. the existence of a minimal and maximal interface) was shown in [30] and [33] when $\varphi(u) = u^m$ and $\psi(u) = u^n$ with $m > 1$ and $n \geq 1$. Nevertheless, it was shown (by means of suitable pointwise gradient estimate) in [19] that if $\varphi(u) = u^m$ and $\psi(u) = -u^n$ with $0 < n < 1 \leq m$ then one of the interfaces $(\zeta_+(t) = \sup\{x \in \mathbb{R} : u(t, x) > 0\})$ does not appear $(\zeta_+(t) = +\infty$ for any $t > 0)$ although the other one $(\zeta_-(t) = \inf\{x \in \mathbb{R} : u(t, x) > 0\})$ exists $(-\infty < \zeta_-(t) < +\infty$ for any $t > 0)$. Notice that this can be also derived from Theorem 2 since, by a comparison argument, without loss of generality, we can assume M_0 small enough and from estimate (58) we deduce that

$$u(t, x)^m \geq u(t, x_0)^m e^{-\frac{K}{t}(x - x_0)}$$

for any $t > 0$ and $x \geq x_0$. Taking x_0 such that $u(t, x_0) > 0$ we get the result. We point out that the existence of the interface $\zeta_-(t)$ shows the impossibility to have the reverse directional symmetric pointwise gradient estimate

$$(u^m)_x \leq K \frac{u^m}{t} \quad \text{on } (0, T) \times \mathbb{R}.$$

The occurrence of the interfaces was studied by B. H. Gilding [26], in the general case of φ and ψ as in Theorem 1 He proved that the interface $\zeta_+(t)$ exists if and only if there is a real number σ such that

$$\sigma s + \psi(s) > 0 \quad \text{for all } s \in (0, M_0]$$

and

$$\int_{0+} \frac{\varphi'(s)}{\sigma s + \psi(s)} < \infty.$$

Notice that this proves the non existence of $\zeta_+(t)$ if $\varphi(u) = u^m$ and $\psi(u) = -u^n$ with $0 < n < 1 \leq m$ and that, by changing x by $-x$, it proves the existence of $\zeta_-(t)$. It was proved also in [26] that if $V(t, x)$ is bounded for $t = 0$ then it remains bounded for any $t > 0$ assumed (70) and

$$\int_{0+} \left[\frac{-\psi(s)}{s} \right] ds < \infty.$$

Nevertheless, when $\varphi(u) = u^m$ and $\psi(u) = -u^n$ with $0 < n < 1 \leq m$, it was shown in [20] that the best pointwise gradient estimate has the form

$$|(u^{m-n})_x| \leq M \tag{71}$$

so that estimate (69) fails. Estimate (71) was later extended in [26] to more general φ and ψ but satisfying additional conditions which are satisfied when $\varphi(u) = u^m$, $\psi(u) = -u^n$ and $0 < n < 1 \leq m$ (see also the work [29] for the study of the dependence of the Bernstein estimates on epsilon if one puts such a parameter in front of the diffusion term). We mention that, as kindly indicated in [24] and [26], the boundedness of V and the generalization of the estimate (71) were obtained by applying the general idea of the method presented in this paper (a preliminar version of it circulate, as unpublished notes, among the specialists in the field since the early eighties).

The justification of the differential equation was carried out, when $\varphi(u) = u^m$ and $\psi(u) = -u^n$ in [43] and [21] for the case $0 < n < 1 \leq m$ (see also the formal asymptotic analysis made in [36]). It was shown ([21]) that in ausence of *waiting time phenomena* ([2], [1]) $\zeta_- \in C([0, T]) \cap Lip(0, T)$,

$$\lim_{x \to \zeta_-(t)-0} \frac{\varphi(u(t, x))_x}{\psi(u(t, \zeta(t)))} = -1$$

and that formula (68) holds in the sense that

$$\zeta'(t) = - \lim_{x \to \zeta_-(t)-0} \frac{\varphi(u(t, x))_x + \psi(u(t, x))}{u(t, x)}$$

$$= - \lim_{x \to \zeta_-(t)-0} \left[\frac{\psi(u(t, x))}{u(t, x)} \right] \left[\frac{\varphi(u(t, x))_x}{\psi(u(t, x))} + 1 \right]$$

for *a.e.* $t \in [0, T]$. It is easy to see that applying (formally) the l'Hopital rule to the right-hand side of this interface equation we get a second-order equation. The interface equations of this type can also be derived by means of a special technique based on local comparison of the space profile of a solution with a set of the space profiles of a complete family of explicit solutions to the same equation, say, the travelling wave solutions: see [23], Section 5.

$C^{1+\alpha}$ and further regularity of the interface was obtained in [44] (see also [40]) under additional assumptions on the initial datum.

REFERENCES

[1]   ALVAREZ, L. and DÍAZ, J. I., *Sufficient and necessary initial mass conditions for the existence of a waiting time in nonlinear diffusion-convection processes*, Journal of Mathematical Analysis and Applications, *155* (1991), 378–392.

[2]   ALVAREZ, L., DÍAZ, J. I. and KERSNER, R., *On the initial growing of the interfaces in filtration processes*, in *Nonlinear Diffusion Equations and Their Equilibrium States I.*, W. M. Ni, L. A. Peletier and J. Serrin eds.), Springer-Verlag, New York 1988, 1–20.

[3]　ARONSON, D. G., *Regularity properties of flows through porous media*, SIAM J. Appl. Math. *17* (1969), 461–467.

[4]　ARONSON, D. G. and BÉNILAN, PH., Régularité des solutions de l'equation des mileux poroeux dans $\mathbb{R}^N$, C. R. Acad. Sc. Paris, *288* (1979), 103–105.

[5]　BÉNILAN, PH., On $u_t = \varphi(u)_{xx} + \psi(u)_x$, Unpublished 1978.

[6]　BÉNILAN, PH., *Opérateurs diffé rentiales linéaires, in Analyse mathématique et calcul numé rique pour les sciences et les techniques*, R. Dautray and J.-L. Lions eds., Masson, Paris 1984, 947–1085.

[7]　BÉNILAN, PH., *Evolution Equations and Accretive Operators*, Lecture Notes (taken by S. Lenhart), Univ. of Kentucky 1981.

[8]　BÉNILAN, PH., Personal Notes, Unpublished 1981.

[9]　BÉNILAN, PH. and BOUILLET, J., *On a slow fast diffusion equation*, Nonlinear Analysis, *26* (1996), 813–822.

[10]　BÉNILAN, PH. and DÍAZ, J. I., *Comparison of solutions of nonlinear evolutions equations with different nonlinear terms*, Israel Journal of Mathematics, *42* (1982), 241–257.

[11]　BÉNILAN, PH. and GARIEPY, R., *Strong solutions in L^1 of degenerate parabolic equations*, Journal of Differential Equations, *119* (1995), 473–502.

[12]　BÉNILAN, PH. and TOURÉ, H., Sur l'équation génerale $u_t = \varphi(., u)_{xx} - \psi(u)_x + v$, C.R. Acad.Sc. Paris, *299* (1984), 91–922.

[13]　BÉNILAN, PH. and TOURÉ, H., *Sur l'é quation génerale $u_t = a(., u, \varphi(., u)_x)_x + v$ dans L^1: II- Le problème d'évolution*, Ann. Inst. Henri Poincaré: Analyse non linéaire, *12*, (1995) 727–761.

[14]　BERNSTEIN, S. N., *Collected works, vol III: Differential equations, Calculus of Variations and Geometry* (1903 – 1947), Izdat. Akad. Nauk. SSSR, Moscow, 1960, (in russian).

[15]　BERRYMAN, J. G. and HOLLAND, C. J., *Stability of the separable solution for fast diffusion*, Arch. Rational Mech. Anal. *74* (1980), 379–388.

[16]　BRÉZIS, H., *Monotonicity Methods in Hilbert Spaces and Some Applications to Nonlinear Partial Differential Equations, in Contributions to Nonlinear Functional Analysis*, E. Zarantonello, ed., Academic Press, New York 1971, 101–156.

[17]　CARRILLO, J., *Entropy solutions for nonlinear degenerate problems*, Arch. Ration. Mech. Anal. *147* (1999), 269–361.

[18]　DÍAZ, J. I., *Two Problems in Homogenization of Porous Media*, Extracta Mathematica, *14* (1999), 141–155.

[19]　DÍAZ, J. I. and KERSNER, R., *Non existence d'une des frontières libres dans une équation dégénérée en theorie de la filtration*, C.R. Acad. Sc. París, *296* (1983), 505–508.

[20]　DÍAZ, J. I. and KERSNER, R., *On a nonlinear degenerate parabolic equation in filtration or evaporation through a porous medium*, Journal of Differential Equations, *69* (1987), 368–403.

[21]　DÍAZ, J. I. and SHMAREV, S. I., *On the behaviour of the interface in nonlinear processes with convection dominating diffusion via Lagrangian coordinates*, Advances in Mathematical Sciences and Applications, *1* (1992), 19–45. Corrections 2 (1993), 503–506.

[22]　GAGNEUX, G. and MADAUNE-TORT, M., *Analyse mathématique de modéles non linéaires de l'ingenierie petroliére. (Mathematical analysis of nonlinear models of petrol engineering)*. Mathematiques & Applications, *22* , Springer-Verlag, Paris, 1995.

[23]　GALAKTIONOV, V. A., SHMAREV, S. I. and VAZQUEZ, J. L., *Second-order interface equations for nonlinear diffusion with very strong absorption*, Commun. Contemp. Math., *1* (1999), 51–64.

[24]　GILDING, B. H., *Hölder continuity of solutions of parabolic equations*, J. London Math. Soc. *13* (1976), 103–106.

[25]　GILDING, B. H., *Properties of solutions of an equation in the theory of infiltration*. Arch. Ration. Mech. Anal. *65* (1977), 203–225.

[26]　GILDING, B. H., *The Ocurrence of Interfaces in Nonlinear Diffusion-Advection Processes*. Arch. Ration. Mech. Anal. *100* (1988), 243–263.

[27]　GILDING, B. H., *Improved Theory for a Nonlinear Degenerate Parabolic Equation*. Annali Scu. Norm. Sup. Pisa, Serie IV. *14* (1989), 165–224.

[28]　GILDING, B. H. and GONCERZEWICZ, J., *Localization of solutions of exterior domain problems for the porous media equation with radial symmetry*. SIAM, J. Math. Anal. *31* (2000), 862–893.

[29]　GILDING, B. H., NATALINI, R. and TESEI, A., *How parabolic free boundaries approximate hyperbolic fronts*. Trans. Amer. Math. Soc, *352* (2000), 1797–1824.

[30] GILDING, B. H. and PELETIER, L. A., *The Cauchy problem for an equation in the theory of infiltration.* Arch. Ration. Mech. Anal. *61* (1976), 127–140.

[31] KALASNIKOV, A. S., *Equations of the type of a nonstationary filtration with infinite rate of propagation of perturbations*, Vestnik Moskov. Univ. Ser. I Mat. Meh. *27* (1972), 45–49.

[32] KALASHNIKOV, A. S., *On the differential properties of generalized solutions of equations of the nonsteady-state filtration type*, Vestnik Moskow Univ. Ser. I Mat. Fiz. *29* (1974), 62–68.

[33] KALASHNIKOV, A. S., *The nature of the propagation of perturbations in processes that can be decribed by quasilinear degenerate parabolic equations*, Trudy Sem. Petrovsk., *1* (1975), 135–144, (in russian).

[34] KALASHNIKOV, A. S., *On the concept of finite rate of propagation of perturbations*, Uspekhi Mat. Nauk, *34* (1979), 199–200.

[35] KALASHNIKOV, A. S., *Some problems of the qualitative theory of second-order nonlinear degenerate parabolic equations*, Uspekhi Mat. Nauk, *42* (1987), 135–176.

[36] KAMIN, S. and ROSENAU, PH., *Thermal waves in an absorbing and convective medium*, Physica 8D, (1983) 717–737.

[37] KRUZHKOV, S. N., *Nonlinear parabolic equations with two independent variables*, Trudi Mosk. Matem. ob. (Proceedings of Moscow Math. Society), *6* (1967), 329–346.

[38] LADYZHENSKAJA, O. A., SOLONNIKOV, V. A., and URAL'TSEVA, N. N., *Linear and quasilinear equations of parabolic type*, American Mathematical Society, Providence, R.I., 1967.

[39] LAURENCOT, PH. and SIMONDON, F., *Source-type solutions to porous medium equations with convection*, Comm. in Appl. Anal., *1*, 1997.

[40] MEIRMANOV, A. M., PUKHNACHOV, V. V. and SHMAREV, S.I., *Evolution Equations and Lagrangian Coordinates*, Walter de Gruyter, Berlin, 1997.

[41] OLEINIK, O. A., KALASHNIKOV, A.S. and CHZHOU Y.-L., *The Cauchy problem and boundary problems for equations of the type of non-stationary filtration*, Izv. Akad. Nauk SSR, Ser. Mat. *22* (1958), 667–704 (in russian).

[42] PELETIER, L A., *A necessary and sufficient condition for the existence of an interface in flows through porous media*, Arch. Ration. Mech. Anal., *56* (1974), 183–190.

[43] SHMAREV S. I., *Instantaneous appearance of singularities of a solution of a degenerate parabolic equation*, Sib. Math. J., *31* (1990), 671–682.

[44] SHMAREV, S. I., *Global $C^{1+\alpha}$ regularity of interface for $u_t = (u^m)_{xx} + (u^\lambda)_x$*, Advances in Mathematical Sciences and Applications, *5* (1995), 1–29.

[45] SIMONDON, F., *Strong solutions for $u_t = (\varphi(u))_{xx} - f(t)(\psi(u))_x$*, Comm. in PDE, *13* (1988), 1337–1354.

[46] VAZQUEZ, J. L., *An introduction to the mathematical theory of the porous medium equation, in Shape Optimization and Free Boundaries*, M. Delfour, ed., Kluwer Acad. Publ., Dordrecht, Netherlands, 1992, 347–389.

Jesús Ildefonso Díaz
Facultad de Matemáticas
Universidad Complutense de Madrid
20040 Madrid
Spain

J.evol.equ. 3 (2004) 603 – 622
1424–3199/03/040603 – 20
DOI 10.1007/s00030-003-0105-z
© Birkhäuser Verlag, Basel, 2004

Journal of Evolution
Equations

Uniqueness of entropy solutions for nonlinear degenerate parabolic problems

MOHAMED MALIKI and HAMIDOU TOURÉ

In honour of Professor Philippe Bénilan

Abstract. We consider the general degenerate parabolic equation:

$$u_t - \Delta b(u) + div\, F(u) = f \quad \text{in } Q \in]0, T[\times\mathbb{R}^N, \quad T > 0.$$

We prove existence of Kruzkhov entropy solutions of the associated Cauchy problem for bounded data where the flux function F is supposed to be continuous. Uniqueness is established under some additional assumptions on the modulus of continuity of F and b.

1. Introduction and notations

We let $Q =]0, T[\times\mathbb{R}^N$ with $T > 0$, and consider initial data satisfying the following hypotheses:

$$\text{(H1)} \quad \begin{cases} 1) & u_0 \in L^\infty(\mathbb{R}^N), \\ 2) & f \in L^1_{\text{Loc}}(Q) \text{ for } a.e\ t \in]0, T[\ f(t) \in L^\infty(\mathbb{R}^N), \\ 3) & \int_0^T \|f(t)\|_{L^\infty(\mathbb{R}^N)} dt < \infty. \end{cases}$$

We consider the Cauchy Problem $(CP) = (CP)(b, F, f, u_0)$:

$$\text{(CP)} \quad \begin{cases} u_t - \Delta b(u) + div\, F(u) = f & \text{in } Q \\ u(0, .) = u_0 & \text{on } \mathbb{R}^N, \end{cases}$$

where $F \in C(\mathbb{R}, \mathbb{R}^N)$, $b \in C(\mathbb{R})$, b is nondecreasing. For normalization, we set $F(0) = 0$ and $b(0) = 0$. We define the operators H, H_0 and H_ϵ by:

$$H(s) = \begin{cases} 1 & \text{if } s > 0 \\ [0, 1] & \text{if } s = 0 \\ 0 & \text{if } s < 0 \end{cases} \qquad H_0(s) = \begin{cases} 1 & \text{if } s > 0 \\ 0 & \text{if } s \leq 0 \end{cases}$$

2000 *Mathematics Subject Classification*: 35K65, 35L65.

Key words: Parabolic equation, hyperbolic equation, weak solution, entropy solution.

Partially supported by The Abdus Salam International Centre for Theoritical Physics, under grant Net-47.

and

$$H_\epsilon = \min\left(\frac{s^+}{\epsilon}, 1\right).$$

Let γ be a maximal monotone operator defined on $\mathbb{R}$. We recall the definition of the main section γ_0 of γ

$$\gamma_0(s) = \begin{cases} \text{the element of minimal absolute value of } \gamma(s) \text{ if } \gamma(s) \neq \emptyset \\ +\infty \quad \text{if } [s, +\infty) \cap D(\gamma) = \emptyset \\ -\infty \quad \text{if } (-\infty, s] \cap D(\gamma) = \emptyset. \end{cases} \tag{1}$$

Several authors have studied the degenerate parabolic equation of the type we consider here (see e.g. [AL], [BG], [BT1], [BT2], [BW], [C2], [DT], [GT], [YJ]). Some of these authors proved the existence and uniqueness of weak solutions of the associated Cauchy Problems or the Dirichlet boundary value problems under various additional conditions. Amoung these results, the pioneering work of Alt and Luckhaus [AL] established existence and uniqueness of weak solutions under some energy condition provided $u_t \in L^1(Q)$.

Previously in 1969, Vol'pert and Hudjaev [HV], under some regularity assumptions on the data, in 1-dimensional space obtained uniqueness of entropy solutions for a class of functions of bounded variation $(BV(Q))$. More recently, Yin Jingxue [YH] obtained uniqueness and stability results of $BV(Q)$ solutions of (CP) in the case where b is strictly increasing. In fact, Bénilan and Gariepy [BG] proved that the problem considered by Yin Jingxue has a strong solution, that is u_t, $\Delta b(u)$ and div $F(u)$ are in $L^1_{\text{loc}}(Q)$.

In this paper, we present a proof of existence and uniqueness of the entropy solution which is simpler than that of several known results in the literature, and we also consider a more general setting. The motivation for this work is based on the result of Carrillo [C2] on elliptic-parabolic degenerate problems in bounded domains with Dirichlet homogeneous boundary conditions. The paper is divided in two main parts, in Section 2 we study the elliptic problem associated with the equation. Using the theory of evolution equations in $L^1(\mathbb{R}^N)$ in Section 3, we prove existence of integral solutions (mild solutions). We deduce from this general theory existence of entropy solutions and prove uniqueness.

2. Stationary problem

We consider the following stationary problem $(SP) = SP(b, F, f)$ defined by:

$$u - \Delta b(u) + div F(u) = f \quad \text{on} \quad \mathbb{R}^N.$$

2.1. *Existence of entropy solution*

We begin by introducing the following notions of weak and entropy solution for this problem.

DEFINITION 2.1 (Weak solution of $SP(b, F, f)$). Let $f \in L^\infty(\mathbb{R}^N)$. A weak solution of (SP) is a function $u \in L^\infty(\mathbb{R}^N)$ such that:

$$\begin{cases} b(u) \in H^{-1}_{\mathrm{loc}}(\mathbb{R}^N), \\ u - \Delta b(u) + div F(u) = f \quad \text{in} \quad \mathcal{D}'(\mathbb{R}^N). \end{cases} \tag{2}$$

It is well known, that there is no uniqueness of weak solutions in general. In order to get uniqueness, we may introduce entropy solutions following the notion of entropy of S.N. Kruzkhov for conservation laws (see e.g. [KA]).

DEFINITION 2.2 (Entropy solution $SP(b, F, f)$). Let $f \in L^\infty(\mathbb{R}^N)$. An entropy solution u of (SP) is a weak solution of (SP) such that:

$$0 \geq \int_{\mathbb{R}^N} H_0(u - s)\{(u - f)\xi + (\nabla b(u) - (F(s) - F(u))\nabla \xi\} \, dx \tag{3}$$

and

$$0 \leq \int_{\mathbb{R}^N} H_0(s - u)\{(u - f)\xi + (\nabla b(u) - (F(s) - F(u))\nabla \xi\} \, dx \tag{4}$$

for all $s \in \mathbb{R}$ and $\xi \in \mathcal{D}(\mathbb{R}^N)$, $\xi \geq 0$.

REMARK 2.3. It is easy to see that if u is an entropy solution of SP (b, F, f) then $(-u)$ is an entropy solution of SP $(\tilde{b}, \tilde{F}, -f)$ with $\tilde{b}(r) = -b(-r)$, $\tilde{F}(r) = -F(-r)$ (see [C2]).

For a b, we define : $E = \{r \in \mathrm{Im}(b)/; (b^{-1})_0$ is discontinuous at r$\}$ Since $(b^{-1})_0$ is a monotone function, E is a countable subset of $\mathbb{R}$ then we have:

$$\nabla b(u) = 0 \text{ a.e. in } O = \{ x \in \mathbb{R}^N /; b(u(x)) \in E\}. \tag{5}$$

If $b(s) \notin E$ then we have that $H_0(u - s) = H_0(b(u) - b(s))$ a.e. in $\mathbb{R}^N$. $\tag{6}$

THEOREM 2.4 (Existence of entropy solution). *For all $f \in L^\infty(\mathbb{R}^N)$, the problem $SP(b, F, f)$ has at least one entropy solution.*

Proof.

STEP 1. We suppose first that $f \geq 0$. We denote by B_n the ball centered at 0 of radius n. We consider the following stationary problem in a bounded domain $SP(b, F, f)_n$:

$$(SP)_n \quad \begin{cases} u - \Delta b(u) + div \, F(u) = f & \text{in } B_n \\ b(u) = 0 & \text{on } \partial B_n. \end{cases}$$

Using the results of J. Carrillo (see [C2]), the stationary problem $(SP)_n$ has a unique entropy solution u_n so that $u_n \in L^\infty(B_n), b(u_n) \in H_0^1(B_n)$; and $\|u_n\|_{L^\infty(B_n)} \leq \|f\|_{L^\infty(B_n)}$, a.e in $]0, T[$. (see also [AL], [BW], [CW]). Now denote by $\tilde{u}_n$ the extension of u_n to $\mathbb{R}^N$ by 0 out of B_n. Then by the comparison principle and the uniform bound of $\tilde{u}_n$ in $L^\infty(\mathbb{R}^N)$ (see [C2]), the sequence $\tilde{u}_n$ is nondecreasing. There exists $u \in L_{\text{loc}}^1(\mathbb{R}^N)$ such that $\lim_{n \to +\infty} \tilde{u}_n = u$ in $L_{\text{loc}}^1(\mathbb{R}^N)$, and such that $b(\tilde{u}_n)$ converges weakly in $H_{\text{loc}}^1(\mathbb{R}^N)$. Passing to the limit as n goes to $+\infty$, in the entropy inequalities (3), (4), one gets that u is an entropy solution of (SP).

STEP 2. (General case). Let $M := \|f^-\|_{L^\infty(\mathbb{R}^N)}$, where $h^- = \max(-h, 0)$. We set $\tilde{f} = f + M$, then we have $\tilde{f} \geq 0$. Let $\tilde{b}(r) = b(r - M) - b(-M)$ and $\tilde{F}(r) = F(r - M) - F(-M)$. Consider the problem $SP(\tilde{b}, \tilde{F}, \tilde{f})$: $u - \Delta\tilde{b}(u) + \text{div } \tilde{F}(u) = \tilde{f}$ in $\mathbb{R}^N$. Since $\tilde{f}$ is nonnegative, we are in the situation of Step 1. Thus there exists an entropy solution $\tilde{u}$ of $SP(\tilde{b}, \tilde{F}, \tilde{f})$. Then $u = \tilde{u} - M$ is an entropy solution of (SP). $\square$

2.2. *Uniqueness of entropy solution*

For the proof of uniqueness we need the following results, which we present as lemmas .

LEMMA 2.5. *Let* $f \in L^\infty(\mathbb{R}^N)$ *and let* u *be a weak solution of* (SP). *Then*

$$\begin{cases} \int_{\mathbb{R}^N} H_0(u - s)\{(\nabla b(u) + F(s) - F(u))\nabla\xi + (u - f)\xi\} \, dx \\ \quad = -\lim_{\epsilon \to 0} \int_{\mathbb{R}^N} |\nabla b(u)|^2 H_\epsilon'(b(u) - b(s))\xi \, dx \leq 0; \end{cases} \tag{7}$$

and

$$\begin{cases} \int_{\mathbb{R}^N} H_0(s - u)\{(\nabla b(u) + F(s) - F(u))\nabla\xi + (u - f)\xi\} \, dx \, dt \\ \quad = \lim_{\epsilon \to 0} \int_{\mathbb{R}^N} |\nabla b(u)|^2 H_\epsilon'(b(s) - b(u))\xi \, dx \geq 0; \end{cases} \tag{8}$$

for all $s \in \mathbb{R}$ *such that* $b(s) \notin E$ *and for all* $\xi \in \mathcal{D}(\mathbb{R}^N)$, $\xi \geq 0$.

Proof. Let $s \in \mathbb{R}$ be such that $b(s) \notin E$ and let $\xi \in \mathcal{D}(\mathbb{R}^N), \xi \geq 0$. Consider R big enough such that $supp\xi \subset B(0, R) =: \Omega$; in particular we have $\xi \equiv 0$ on $\partial\Omega$. By Remark 2.3 we know that u is a weak solution of (SP) on Ω, so that inequality (7) is equivalent to:

$$\begin{cases} \int_\Omega H_0(u - s)\{(\nabla b(u) + F(s) - F(u))\nabla\xi + (u - f)\xi\} \, dx \\ \quad = -\lim_{\epsilon \to 0} \int_{\mathbb{R}^N} |\nabla b(u)|^2 H_\epsilon'(b(u) - b(s))\xi \, dx \leq 0 \end{cases} \tag{9}$$

for all $s \in \mathbb{R}$ with $b(s) \notin E$ and $\xi \in \mathcal{D}(\mathbb{R}^N), \xi \geq 0$. We now apply Lemma 1 of [C2] and Theorem 1 of [C2] to obtain the desired regularity. By the same method of proof we can prove inequality (8).

We can now establish the so called Kato's inequality (cf. [AB]), for two entropy solutions of the stationary problem. $\square$

THEOREM 2.6 (Kato's inequality). *Let $f_1 \in L^\infty(Q)$, $f_2 \in L^\infty(Q)$. Let u_1, u_2 be entropy solutions with respect to $SP(b, F, f_1)$, $SP(b, F, f_2)$ respectively. Then there exists a measurable function v such that $v \in H(u_1 - u_2)$ a.e. so that:*

$$\begin{cases} \int_{\mathbb{R}^N} v(f_1 - f_2)\xi \, dx \geq \int_{\mathbb{R}^N} \{(u_1 - u_2)^+\xi + \nabla(b(u_1) - b(u_2))^+\nabla\xi \\ \qquad - H_0(u_1 - u_2)(F(u_1) - F(u_2))\nabla\xi\} \, dx \end{cases} \tag{10}$$

for all $\xi \in \mathcal{D}(\mathbb{R}^N)$, $\xi \geq 0$.

Proof. Let $s \in \mathbb{R}$ such that $b(s) \notin E$, let $\xi \in \mathcal{D}(\mathbb{R}^N)$, $\xi \geq 0$. Consider R big enough so that $supp\xi \subset B(0, R) =: \Omega$. Then we have that $\xi \equiv 0$ on $\partial\Omega$. Inequality (10) is equivalent to:

$$\begin{cases} \int_\Omega v(f_1 - f_2)\xi \, dx \geq \int_\Omega \{(u_1 - u_2)^+\xi + \nabla(b(u_1) - b(u_2))^+\nabla\xi \\ \qquad - H_0(u_1 - u_2)(F(u_1) - F(u_2))\nabla\xi\} \, dx. \end{cases} \tag{11}$$

The result is then a direct consequence of Theorem 8 of [C2]. $\qquad\square$

In the following theorem, we give an abstract result, which generalizes a previous result of Bénilan and Kruzkhov in the case of the scalar conservation law (cf. [BK]). Using this general result we will deduce from the Kato's inequality the L^1 contraction principle and then uniqueness.

THEOREM 2.7. *Let $\omega_1, \omega_2, \ldots \omega_l, \eta_1, \eta_2, \ldots \eta_l$ be nonnegative functions such that for all $i = 1, \ldots, l$*

$$\lim_{\epsilon \to 0} \frac{\omega_i(\epsilon)}{\epsilon} = \lim_{\epsilon \to 0} \frac{\eta_i(\epsilon)}{\epsilon} = +\infty; \quad \lim_{\epsilon \to 0} \omega_i(\epsilon) = \lim_{\epsilon \to 0} \eta_i(\epsilon) = 0; \tag{12}$$

$$\liminf_{\epsilon \to 0} \epsilon^{1-l} \prod_{i=1}^{l} [\eta_i(\epsilon) + (\eta_i^2(\epsilon) + 2\epsilon\omega_i(\epsilon))^{1/2}] < +\infty \quad \text{if } N > 2; \tag{13}$$

$$\liminf_{\epsilon \to 0} \epsilon^{1-l} \prod_{i=1}^{l} [\eta_i(\epsilon) + (\eta_i^2(\epsilon) + 2\epsilon\omega_i(\epsilon))^{1/2}] = 0 \quad \text{if } N = 2. \tag{14}$$

Let $h \in L^1_{\text{loc}}(\mathbb{R}^N)$ such that $h^+ = \max(h, 0) \in L^1(\mathbb{R}^N)$, let $W \in L^\infty(\mathbb{R}^N)$, $W \geq 0$, and $\lambda > 0$ so that:

$$\begin{cases} \int_{\mathbb{R}^N} W \, \xi \, dx \leq \sum_{j=1}^{l} \int_{\mathbb{R}^N} (W + \epsilon) \frac{\omega_j(\epsilon)}{\epsilon} \left| \frac{\partial^2\xi}{\partial x_j \partial x_j} \right| \, dx \\ \quad + \sum_{j=1}^{l} \int_{\mathbb{R}^N} (W + \epsilon) \frac{\eta_j(\epsilon)}{\epsilon} \left| \frac{\partial\xi}{\partial x_j} \right| \, dx \\ \quad + \sum_{j=l+1}^{N} \int_{\mathbb{R}^N} \lambda W \left(\left| \frac{\partial^2\xi}{\partial x_j \partial x_j} \right| + \left| \frac{\partial\xi}{\partial x_j} \right| \right) \, dx + \int_{\mathbb{R}^N} h \, \xi \, dx, \end{cases} \tag{15}$$

for all $\epsilon > 0$ and $\xi \in \mathcal{D}(\mathbb{R}^N)$, $\xi \geq 0$.
Then $h \in L^1(\mathbb{R}^N)$, $W \in L^1(\mathbb{R}^N)$, and $\int_{\mathbb{R}^N} W \, dx \leq \int_{\mathbb{R}^N} h \, dx$.

Proof. The proof relies on a particular choice of test functions ξ. We recall that $h^+ = \max(h, 0)$ and $h = h^+ - h^-$. For $x = (x_1, x_2, \ldots, x_N) \in \mathbb{R}^N$

we set

$$\rho(x) = \prod_{j=l+1}^{N} \phi\left(-\frac{\alpha}{\lambda}|x_j|\right) \quad \text{and} \quad \xi(x) = \rho(x) \prod_{j=1}^{l} \psi\left(\frac{|x_j|}{R_j}\right);$$

where $R_j = R_j(\epsilon, \eta)$, α, λ, η, are contants with $\eta \geq 1$. The functions $\phi \in \mathcal{C}^2(\mathbb{R})$, $\psi \in L^1(\mathbb{R}) \cap \mathcal{C}^2(\mathbb{R})$ verify the following:

  a) $\psi \equiv 1$ in $[-1, 1]$, $|\psi'| \leq \psi$, $|\psi''| \leq \psi$, $|\phi'| \leq \phi$, $|\phi''| \leq \phi$;
  b) $0 \leq \xi \leq \rho \leq 1$, in $\mathbb{R}^N$;
  c) $\xi(x) = \rho(x)$ for $x \in K := \prod_{j=1}^{l}[-R_j, R_j] \times \mathbb{R}^{N-l}$.

For example we can choose $\psi(r) = \frac{1}{(1+|r|)^2} * \exp(-(|r|-1)^+)$ and $\phi(r) = \frac{1}{(1+|r|)^2} * \exp(r)$. Set $C_0 = \int_0^{+\infty} \psi(x)\, dx$. If necessary substitute λ by $\tilde{\lambda} = \lambda + \epsilon^{\frac{-1}{N+1}}$ in inequality (14). We may always suppose that $l > 2$, (the case $N = 2$, $l = 1, 2$ will be treated separetely).

  Set $\tilde{K} = \mathbb{R}^N \setminus K = \bigcup_{j=1}^{l} (\{|x_j| > R_j\} \cap \mathbb{R}^N)$, and replace $|\frac{\partial \xi}{\partial x_j}| |\frac{\partial^2 \xi}{\partial x_j \partial x_j}|$ in (15) by their value to obtain:

$$\int_{\mathbb{R}^N} (W + \epsilon)\frac{\omega_j(\epsilon)}{\epsilon} \left|\frac{\partial^2 \xi}{\partial x_j \partial x_j}\right| dx \leq \frac{\omega_j(\epsilon)}{\epsilon R_j^2} \int_{\tilde{K}} W \xi\, dx + \frac{\omega_j(\epsilon)}{R_j^2} \int_{\tilde{K}} \xi\, dx. \tag{16}$$

$$\int_{\mathbb{R}^N} (W + \epsilon)\frac{\eta_j(\epsilon)}{\epsilon} \left|\frac{\partial \xi}{\partial x_j}\right| dx \leq \frac{\eta_j(\epsilon)}{\epsilon R_j} \int_{\tilde{K}} W \xi\, dx + \frac{\eta_j(\epsilon)}{R_j} \int_{\tilde{K}} \xi\, dx. \tag{17}$$

Inequality (15) gives:

$$\begin{cases} \int_{\mathbb{R}^N} W \xi\, dx \leq \sum_{j=1}^{l} \frac{1}{\epsilon R_j}\left(\frac{\omega_j(\epsilon)}{R_j} + \eta_j(\epsilon)\right) \int_{\mathbb{R}^N} W\xi\, dx \\ + \sum_{j=1}^{l} \frac{1}{R_j}\left(\frac{\omega_j(\epsilon)}{R_j} + \eta_j(\epsilon)\right) \int_{\mathbb{R}^N} \xi\, dx \\ + \sum_{l+1}^{N} \alpha(1 + \frac{\alpha}{\lambda}) \int_{\mathbb{R}^N} W\xi\, dx + \int_{\mathbb{R}^N} h\, \xi\, dx. \end{cases} \tag{18}$$

Now choose $R_j(\epsilon, \eta)$ as follow. First let $\tilde{R}_j(\epsilon)$ be such that:

$$\frac{1}{\epsilon \tilde{R}_j}\left(\frac{\omega_j(\epsilon)}{\tilde{R}_j} + \eta_j(\epsilon)\right) = \frac{1}{2}. \tag{19}$$

More precisely,

$$\tilde{R}_j = \frac{1}{\epsilon}[\eta_j(\epsilon) + (\eta_j^2(\epsilon) + 2\epsilon\omega_j(\epsilon))^{1/2}], \tag{20}$$

$$R_j(\epsilon) = \frac{\tilde{R}_j}{\eta} = \frac{1}{\epsilon\eta}[\eta_j(\epsilon) + (\eta_j^2(\epsilon) + 2\epsilon\omega_j(\epsilon))^{1/2}]. \tag{21}$$

Then $R_j(\epsilon)$ has the following propreties:

$$\lim_{\epsilon\to 0} R_j(\epsilon) = +\infty \quad \text{and for any } \epsilon > 0 \text{ one has, } 0 \le \frac{1}{\epsilon R_j}\left(\frac{\omega_j(\epsilon)}{R_j} + \eta_j(\epsilon)\right) \le \eta^2.$$

Let

$$\sum_{j=1}^{l} \frac{1}{\epsilon R_j}\left(\frac{\omega_j(\epsilon)}{R_j} + \eta_j(\epsilon)\right) \le \mu(\eta) = l\eta^2, \tag{22}$$

$$\delta(\alpha) = \sum_{j=l+1}^{N} \alpha\left(1 + \frac{\alpha}{\lambda}\right) = (N - l)\alpha\left(1 + \frac{\alpha}{\lambda}\right). \tag{23}$$

Inequality (18) becomes:

$$\int_{\mathbb{R}^N} ((1 - \mu(\eta) - \delta(\alpha))W + h^-)\xi\, dx \le \sum_{j=1}^{l} \frac{1}{R_j}\left(\frac{\omega_j(\epsilon)}{R_j} + \eta_j(\epsilon)\right)$$

$$\int_{\mathbb{R}^N} \xi\, dx + \int_{\mathbb{R}^N} h^+\,\xi\, dx.$$

We may now choose η, α so that $1 - \mu(\eta) - \delta(\alpha) \ge \frac{1}{2}$. Then we get:

$$\frac{1}{2}\int_K W\,\rho\, dx \le \int_{\mathbb{R}^N} (1 - \mu(\eta) - \delta(\alpha))W\,\xi\, dx,$$

and so

$$\frac{1}{2}\int_K W\,\rho\, dx \le \sum_{j=1}^{l} \frac{1}{R_j}\left(\frac{\omega_j(\epsilon)}{R_j} + \eta_j(\epsilon)\right)\int_{\mathbb{R}^N} \xi\, dx + \int_{\mathbb{R}^N} h^+\xi\, dx. \tag{24}$$

We also have

$$\Lambda(\epsilon) := \sum_{j=1}^{l} \frac{1}{R_j}\left(\frac{\omega_j(\epsilon)}{R_j} + \eta_j(\epsilon)\right)\int_{\mathbb{R}^N} \xi\, dx$$

$$= \sum_{j=1}^{l} \frac{1}{R_j}\left(\frac{\omega_j(\epsilon)}{R_j} + \eta_j(\epsilon)\right)\frac{\lambda^{N-l}C_0^{\,l}}{\alpha^{N-l}}2^N\prod_{i=1}^{l} R_i, \tag{25}$$

and then

$$\Lambda(\epsilon) = 2^N \frac{\lambda^{N-l} C_0^{\ l}}{\alpha^{N-l}} \sum_{j=1}^{l} \frac{\eta}{\tilde{R}_j} \left(\frac{\omega_j(\epsilon)\eta}{\tilde{R}_j} + \eta_j(\epsilon) \right) \prod_{i=1}^{l} \frac{\tilde{R}_i}{\eta}, \tag{26}$$

so that

$$\Lambda(\epsilon) = 2^N \frac{\lambda^{N-l} C_0^{\ l}}{\alpha^{N-l}\eta^l} \sum_{j=1}^{l} \frac{\eta}{\epsilon \tilde{R}_j} \left(\frac{\omega_j(\epsilon)\eta}{\tilde{R}_j} + \eta_j(\epsilon) \right) \epsilon \prod_{i=1}^{l} \tilde{R}_i. \tag{27}$$

Since

$$\epsilon \prod_{i=1}^{l} \tilde{R}_i = \epsilon^{1-l} \prod_{i=1}^{l} [\eta_i(\epsilon) + (\eta_i^2(\epsilon) + 2\epsilon\omega_i(\epsilon))^{1/2}],$$

and

$$\sum_{j=1}^{l} \frac{\eta}{\epsilon \tilde{R}_j} \left(\frac{\omega_j(\epsilon)\eta}{\tilde{R}_j} + \eta_j(\epsilon) \right) \leq l\eta^2,$$

we get that

$$\Lambda(\epsilon) \leq 2^N \frac{\lambda^{N-l} C_0^{\ l}}{\alpha^{N-l}\eta^{l-2}} \epsilon^{1-l} \prod_{i=1}^{l} [\eta_i(\epsilon) + (\eta_i^2(\epsilon) + 2\epsilon\omega_i(\epsilon))^{1/2}].$$

Using assumption (13), we may extract a subsequence so that

$$\lim_{k \to +\infty} \Lambda(\epsilon_k) \leq \frac{C}{\eta^{l-2}\alpha^{N-l}}.$$

Taking inequality (24) into account we obtain:

$$\frac{1}{2} \int_K W \rho \, dx \leq \Lambda(\epsilon) + \int_{\mathbb{R}^N} h^+ \xi \, dx.$$

Again using the subsequence ϵ_k, and the limit as k tends to infinity, we have that $W \rho \in L^1(\mathbb{R}^N)$. But $\int_{\tilde{K}} W \xi \, dx \leq \int_{\tilde{K}} W \rho \, dx$. Therefore

$$\forall \eta > 0 \quad \forall \alpha > 0 \quad \lim_{k \to +\infty} \int_{\tilde{K}} W \xi \, dx = 0. \tag{28}$$

Returning to inequality (18) and using the previous result, we have that

$$\begin{cases} \int_{\mathbb{R}^N} ((1 - \delta(\alpha))W + h^-)\xi \, dx \leq \sum_{j=1}^{l} \frac{1}{\epsilon R_j} (\frac{\omega_j(\epsilon)}{R_j} + \eta_j(\epsilon)) \int_{\tilde{K}} W \xi \, dx \\ \quad + \sum_{j=1}^{l} \frac{1}{R_j} (\frac{\omega_j(\epsilon)}{R_j} + \eta_j(\epsilon)) \int_{\mathbb{R}^N} \xi \, dx + \int_{\mathbb{R}^N} h^+ \xi \, dx. \end{cases} \tag{29}$$

Choose now $\alpha > 0$ such that $\delta(\alpha) < 1$ and let k tend to infinity. We obtain

$$\int_{\mathbb{R}^N} ((1 - \delta(\alpha))W + h^-)\rho \, dx \leq \frac{C}{\eta^{l-2}\alpha^{N-l}} + \int_{\mathbb{R}^N} h^+ \, dx$$

which implies that $h^-\rho \in L^1(\mathbb{R}^N)$. To complete the proof let η tend to $+\infty$, and then α tend to 0. $\qquad\square$

> REMARK 2.8. 1) The case $N = 2$ is similar but the constant C (in the proof) is zero, we have $\lim_{k \to +\infty} \Lambda(\epsilon_k) = 0$. For some choice of $\alpha > 0$ such that $\delta(\alpha) < 1$ and as k tends to infinity we obtain: $\int_{\mathbb{R}^N}((1 - \delta(\alpha))W + h^-)\rho \, dx \leq \int_{\mathbb{R}^N} h^+ \, dx$; and hence $h^-\rho \in L^1(\mathbb{R}^N)$. We complete the proof by letting η tend to infinity and then let α tend to 0.
> 2) For the case $N = 1$, condition (7) is always fulfilled. The proof is similar to the case $N = 2$.

We are now able to begin the proof of the comparison principle. Let us first begin with some notations. Let g be a continuous function. Denote the modulus of continuity of g by ω. Then ω is a non negative continuous function which is nondecreasing so that

$$\omega(0) = 0 \quad \text{and} \quad |g(x) - g(y)| \leq \omega(x - y).$$

For $\epsilon > 0, r > 0$ we get by euclidian division of r by ϵ, $r = k\epsilon + s$ where $k \in \mathbb{N}$, and $0 \leq s < \epsilon$. Then using properties of the modulus of continuity ω, we have that

$$\omega(r) < (k + 1)\omega(\epsilon) \leq (r + \epsilon)\frac{\omega(\epsilon)}{\epsilon}. \tag{30}$$

In general we have $\lim_{\epsilon \to 0} \frac{\omega(\epsilon)}{\epsilon} = +\infty$. In the particular case where g is Lipschitz continuous, we have $\frac{\omega(\epsilon)}{\epsilon} \leq \lambda$ where λ is some constant.

Changing coordinates if necessary, we can suppose that:

1) for $i = 1, \ldots, l$ the functions F_i are continuous but not Lipschitz continuous;
2) for $i = l + 1, \ldots, N$ the functions F_i are Lipschitz continuous, and denote their Lipschitz constants by λ_i, $i = l + 1, \ldots, N$;
3) we may replace λ_i by $\tilde{\lambda}_i = \lambda_i + \epsilon^{\frac{-1}{N+1}}$ if necessary, and we may always assume that $l > 1$. In the case where the function b is not Lipschitz continuous, we may suppose $l = N$.
4) Let us denote by

$$\begin{cases} \omega_i \text{ the modulus of continuity of } b \text{ for } i = 1, \ldots, N; \\ \eta_i \text{ the modulus of continuity of } F_i \text{ for } i = 1, \ldots, N. \end{cases}$$

We also suppose that

(H2) ω_i, η_i satisfy (13), (14) and (15).

THEOREM 2.9. *Let f_1, $f_2 \in L^\infty(\mathbb{R}^N)$ and let u_1, u_2 be entropy solutions with respect to SP (b, F, f_1), SP (b, F, f_2) respectively. Suppose that (H2) holds. Then:*

$$\int_{\mathbb{R}^N} (u_1 - u_2)^+ \, dx \le \int_{\mathbb{R}^N} v(f_1 - f_2) \, dx, \tag{31}$$

for $v \in H(u_1 - u_2)$ a.e.

Proof. Set $W = (u_1 - u_2)^+$. Then by Theorem 2.6, u_1, u_2 verify Kato's inequality. Since hypothesis (H2) is fulfilled, the result follows from Theorem 2.7.   □

COROLLARY 2.10. *Let $f \in L^\infty(\mathbb{R}^N) \cap L^1(\mathbb{R}^N)$, $\lambda > 0$. Suppose that (H1) and (H2) hold, and let u be the entropy solution of SP $(\lambda b, \lambda F, f)$. Then:*

$$\begin{cases} \|u\|_{L^\infty(\mathbb{R}^N)} \le \|f\|_{L^\infty(\mathbb{R}^N)}; \\ \int_{\mathbb{R}^N} b(u)u \, dx + \lambda \int_{\mathbb{R}^N} |\nabla b(u)|^2 \, dx \le \int_{\mathbb{R}^N} b(u)f; \end{cases} \tag{32}$$

so that $\|\nabla b(u)\|^2_{L^2(\mathbb{R}^N)} \le C(b, f)$; where $C(b, f)$ is a constant which depends only of b and f.

Proof. Returning to the proof of Theorem (2.4), using the same notation, we do not assume now that $f \ge 0$. Since u_n is a weak solution of SP $(\lambda b, \lambda F, f)$, using $b(u_n)$ as a test function it can be easily seen that:

$$\begin{cases} \|\tilde{u}_n\|_{L^p(\mathbb{R}^N)} \le \|f\|_{L^p(\mathbb{R}^N)} \text{ for any } p \text{ such that } 1 \le p \le \infty; \\ \int_{\mathbb{R}^N} b(\tilde{u}_n)\tilde{u}_n \, dx + \lambda \int_{\mathbb{R}^N} |\nabla b(\tilde{u}_n)|^2 \, dx \le \int_{\mathbb{R}^N} b(\tilde{u}_n)f. \end{cases}$$

The result is obtain by passing to the limit as n tends to $+\infty$.   □

3. Entropy solutions of the evolution problem

Let $Q =]0, T[\times \mathbb{R}^N$ with $T > 0$, and let u_0 and f be given such that (H1) is satisfied. We consider the Cauchy problem (CP) = (CP)(b, F, f, u_0):

(CP) $$\begin{cases} u_t - \Delta b(u) + \text{div } F(u) = f & \text{in } Q \\ u(0, .) = u_0 & \text{on } \mathbb{R}^N. \end{cases}$$

We begin by using nonlinear semi-group theory for the Cauchy problem (CP) = (CP) (b, F, f, u_0) for $f \in L^1(Q) \cap L^\infty(Q)$ and $u_0 \in L^1(\mathbb{R}^N) \cap L^\infty(\mathbb{R}^N)$.

The results obtained by this method will be used to establish our results for (u_0, f), such that (H1) is verified.

We define an operator $A_{(b,F)}$ in $L^1(\mathbb{R}^N)$ by:

$$\begin{cases} (u, f) \in A_{(b,F)} \text{ if and only if } u \in L^1(\mathbb{R}^N) \cap L^\infty(\mathbb{R}^N) \\ \text{and } u \text{ is an entropy solution of the stationary problem } SP(b, f, f + u). \end{cases}$$

We have the following result.

THEOREM 3.1. *The operator $A_{(b,F)}$ satisfies:*

(1) $A_{(b,F)}$ *is T-accretive operator in $L^1(\mathbb{R}^N)$.*

(2) *For all $\lambda > 0$ the range $R(I + \lambda A_{(b,F)})$ of $I + \lambda A_{(b,F)}$ is dense in $L^1(\mathbb{R}^N)$.*

(3) *The domain $\mathcal{D}(A_{(b,F)})$ of $A_{(b,f)}$, $\mathcal{D}(A_{(b,F)})$ is dense in $L^1(\mathbb{R}^N)$.*

Part 1 follows from Theorem 2.9, while part 2 and 3 can be deduced in the same way as in the proof of Theorem 3.2 of [C2].

From the Theory of Nonlinear Semigroups in Banach Spaces, (see [Be], [BCP]), we have the following result.

THEOREM 3.2. *Let $f \in L^1(Q)$ and $u_0 \in L^1(\mathbb{R}^N)$. Then there exists a unique integral solution u of*

$$(CP)(b, F, f, u_0) \qquad \begin{cases} \frac{du}{dt} + A_{(b,F)}u = f \\ u(0) = u_0 \end{cases}$$

with $u \in C([0, T], L^1(\mathbb{R}^N))$. In addition, if u_i is an integral solution of $(CP)(b, F, f_i, u_{0i})$, for $i = 1, 2$, then there exist $v \in H(u_1 - u_2)$ a.e. in Q so that

$$\int_Q v\{(u_1 - u_2)\xi' + (f_1 - f_2)\xi\}\, dx\, dt \geq 0,$$

for all $\xi \in \mathcal{D}(0, T), \xi \geq 0$. In particular

$$u_1 \leq u_2 \ a.e. in \ Q \ if \ u_{01} \leq u_{02}, \ and \ f_1 \leq f_2 \ a.e.$$

3.1. *Existence of entropy solutions*

Analogously to the notion of weak and entropy solution introduced in Section 2 for the stationary problem, we introduce the following definitions.

DEFINITION 3.3 (Weak solution of (CP)(b, F, f, u_0)). Let u_0 and f be such that (H1) is fulfilled. A weak solution of (CP) is a function $u \in L^\infty(Q)$ such that:

$$u_t \in L^2((0, T), H_{\mathrm{loc}}^{-1}(\mathbb{R}^N)) + L^1((0, T), L^\infty(\mathbb{R}^N)), \tag{33}$$

$$b(u) \in L^2((0, T), H_{\mathrm{loc}}^1(\mathbb{R}^N)) \tag{34}$$

$$u_t - \Delta b(u) + div F(u) = f \quad \text{in} \quad \mathcal{D}'(Q)$$

and

$$u(0, x) = u_0 \text{ on } \quad \mathbb{R}^N.$$

The last condition must be understood in the sense that

$$\int_0^T \langle u_t, \xi \rangle \, dt = - \int_Q u \, \xi_t \, dx \, dt - \int_{\mathbb{R}^N} u_0 \, \xi(0) \, dx, \tag{35}$$

for any $\xi \in L^2((0, T); \mathcal{D}(\mathbb{R}^N)) \cap W^{1,1}((0, T); L^\infty(\mathbb{R}^N))$ so that $\xi(T) = 0$ and $\langle \, , \rangle$ represents the duality product between $H^{-1}(\mathbb{R}^N)$ and $H^1(\mathbb{R}^N)$.

DEFINITION 3.4 (Entropy solution of (CP)(b, F, f, u_0)). Let u_0 and f verify (H1). An entropy solution u of (CP)(b, F, f, u_0) is a weak solution of (CP) such that:

$$\begin{cases} \int_Q H_0(u - s)\{\nabla b(u)\nabla\xi - (F(u) - F(s))\nabla\xi - (u - s)\xi_t - f\xi\} \, dx \, dt \\ \quad - \int_{\mathbb{R}^N} (u_0 - s)^+\xi(0) \, dx \leq 0 \end{cases} \tag{36}$$

and

$$\begin{cases} \int_Q H_0(s - u)\{\nabla b(u)\nabla\xi - (F(u) - F(s))\nabla\xi - (u - s)\xi_t - f\xi\} \, dx \, dt \\ \quad + \int_{\mathbb{R}^N} (s - u_0)^+\xi(0) \, dx \geq 0 \end{cases} \tag{37}$$

for any $s \in \mathbb{R}$ and $\xi \in \mathcal{D}(Q), \xi \geq 0$.

For b given as above, using definition in Section 2 of the set E, we have the following properties

$$\nabla b(u) = 0 \text{ a.e. in } O = \{ (x, t) \in Q/; b(u(x, t)) \in E\}. \tag{38}$$

If $s \in \mathbb{R}$ is such that $b(s) \notin E$ then we have that

$$H_0(u - s) = H_0(b(u) - b(s)) \text{ a.e. in } Q. \tag{39}$$

REMARK 3.5. It is easy to see that if u is an entropy solution of (CP)(b, F, f, u_0), then $(-u)$ is an entropy solution of (CP) $(\tilde{b}, \tilde{F}, -f, -u_0))$ with $\tilde{b}(r) = -b(-r)$ $\tilde{F}(r) = -F(-r)$ (see [C2]).

The following result gives the relationship between integral solutions (mild solutions) and entropy solutions when the data can be integrated.

THEOREM 3.6. *Let $f \in L^1(Q) \cap L^\infty(Q)$ and $u_0 \in L^1(\mathbb{R}^N) \cap L^\infty(\mathbb{R}^N)$. Let u be the unique integral solution of* $(CP)(b, F, f, u_0)$ *given by Theorem 3.2. Then u is an entropy solution of* $(CP)(b, F, f, u_0)$.

Proof. Let $M \in \mathbb{N}$ and define $\tau = \frac{T}{M}$. For $i = 0, 1, \ldots, M$, let $t_i = i \times \tau$, and $f_1, f_2, \ldots f_M \in L^1(\mathbb{R}^N) \cap L^\infty(\mathbb{R}^N)$ so that

$$\|f_i\|_{L^\infty(\mathbb{R}^N)} \leq \|f\|_{L^\infty(Q)}; \quad \|f_i\|_{L^1(\mathbb{R}^N)} \leq \frac{1}{T}\|f\|_{L^1(Q)}$$

and

$$\sum_{i=1}^{M} \int_{t_{i-1}}^{t_i} \|f(t) - f_i\|_{L^1(\mathbb{R}^N)}\, dt \longrightarrow 0 \text{ when } M \to +\infty.$$

For $i = 0, 1, \ldots, M$, let u_i be the semi-group solution of

$$\tau f_i + u_{i-1} = (I + \tau A_{(b,F)})u_i$$

more precisely u_i is the unique entropy solution of $SP(\tau b, \tau F, \tau f_i + u_{i-1})$. Then u_i verifies the following estimates:

$$\|u_i\|_{L^1(\mathbb{R}^N)} \leq \|u_0\|_{L^1(\mathbb{R}^N)} + \int_0^{t_i} \|f^\tau\|_{L^1(\mathbb{R}^N)}\, dt$$

and

$$\|u_i\|_{L^\infty(\mathbb{R}^N)} \leq \|u_0\|_{L^\infty(\mathbb{R}^N)} + \int_0^{t_i} \|f^\tau\|_{L^\infty(\mathbb{R}^N)}\, dt$$

where $f^\tau(t) = f_i$ for $t \in]t_{i-1}, t_i]$, $1 \leq i \leq M$. We denote by $u^\tau(t) = u_i$ for $t \in]t_{i-1}, t_i]$, $1 \leq i \leq M$. We have $\|u^\tau - u(t)\|_{L^\infty((0,T),L^1(\mathbb{R}^N))} \longrightarrow 0$ when $M \to +\infty$, and

$$\|u^\tau\|_{L^\infty((0,T)\times\mathbb{R}^N)} \leq \|u_0\|_{L^\infty(\mathbb{R}^N)} + T\,\|f\|_{L^\infty(Q)}.$$

It follows that

$$\|u^\tau - u(t)\|_{L^p(Q)} \longrightarrow 0 \text{ when } M \to +\infty, \text{ for all } 1 \leq p < \infty \text{ and a.e. on } Q.$$

But u_i is an entropy solution of $P_s(\tau b, \tau F, \tau f_i + u_{i-1})$, we can apply the estimate of Corollary 2.10, which gives:

$$\int_{\mathbb{R}^N} b(u_i)(u_i - u_{i-1})\, dx + \tau \int_{\mathbb{R}^N} |\nabla b(u_i)|^2\, dx \leq \tau \int_{\mathbb{R}^N} b(u_i)f_i\, dx.$$

We may write this estimate in the form

$$\int_{\mathbb{R}^N} b(u_i)(u_i - u_{i-1})\, dx + \int_{t_{i-1}}^{t_i} \int_{\mathbb{R}^N} |\nabla b(u_i)|^2\, dt\, dx \leq \int_{t_{i-1}}^{t_i} \int_{\mathbb{R}^N} b(u_i)f_i\, dt\, dx.$$

Set $J(r) = \int_0^r b(s)\, ds$, since $u_i \in L^1(\mathbb{R}^N) \cap L^\infty(\mathbb{R}^N)$, $J(u_i) \in L^1(\mathbb{R}^N)$. By definition of J we have $J(u_i) - J(u_{i-1}) \le b(u_i)(u_i - u_{i-1})$ so that

$$\int_{\mathbb{R}^N} (J(u_i) - J(u_{i-1}))\, dx + \int_{t_{i-1}}^{t_i} \int_{\mathbb{R}^N} |\nabla b(u_i)|^2\, dt\, dx \;\le\; \int_{t_{i-1}}^{t_i} \int_{\mathbb{R}^N} b(u_i) f_i\, dt\, dx.$$

Adding this inequality for $k = 1, \ldots . i$, we obtain the following

$$\int_{\mathbb{R}^N} J(u_i)\, dx + \int_0^{t_i} \int_{\mathbb{R}^N} |\nabla b(u^\tau)|^2\, dt\, dx \le \int_{\mathbb{R}^N} J(u_0)$$
$$+ \int_0^{t_i} \int_{\mathbb{R}^N} b(u^\tau) f^\tau\, dt\, dx;$$

so that $b(u^\tau)$ is uniformly bounded in $L^2(Q)$ since $J(r)$ is nonnegative. More precisely $\|\nabla b(u^\tau)\|_{L^2(Q)} \le C(T, u_0, f)$, and then there exists a subsequence $b(u^{\tau_k})$ converging weakly to $w = b(u)$ in $L^2(Q)$. We have also $F(u^{\tau_k})$ converges to $F(u)$ in $L^1_{\mathrm{loc}}(\mathbb{R}^N)$.

We may now use the fact that u_i is a weak solution of $P_s(\tau b, \tau F, \tau f_i + u_{i-1})$,

$$\int_{\mathbb{R}^N} \{(u_i - u_{i-1} - \tau_k f_i)\xi + \tau_k(\nabla b(u_i) - F(u_i))\nabla\xi\}\, dx = 0$$

for all $\xi \in \mathcal{D}(\mathbb{R}^N)$ which gives by computation

$$\begin{cases} \int_Q f^{\tau_k}(t)\xi(t)\, dxdt = \int_Q \{\frac{u^{\tau_k}(t) - u^{\tau_k}(t - \tau_k)}{\tau_k}\xi(t) \\ + (\nabla b(u^{\tau_k}(t)) - F(u^{\tau_k}(t)))\nabla\xi\}\, dxdt = \int_Q \{u^{\tau_k}(t)\frac{\xi(t) - \xi(t + \tau_k)}{\tau_k} \\ + (\nabla b(u^{\tau_k}(t)) - F(u^{\tau_k}(t)))\nabla\xi\}\, dxdt - \frac{1}{\tau_k}\int_0^{\tau_k}\int_{\mathbb{R}^N} u_0\,\xi(t)\, dxdt. \end{cases}$$

As τ_k tends to 0 we obtain

$$\int_Q f(t)\xi(t)\, dxdt = \int_Q \{-u\xi_t + (\nabla b(u) - F(u))\nabla\xi\}\, dxdt - \int_{\mathbb{R}^N} u_0\,\xi(0)\, dx$$

Thus $u_t \in L^2((0, T), H^{-1}_{\mathrm{loc}}(\mathbb{R}^N))$; and u is a weak solution of $CP(b, F, u_o, f)$, $\frac{u^{\tau_k}(t) - u^{\tau_k}(t - \tau_k)}{\tau_k}$ converges weakly to u_t in $L^2((0, T), H^{-1}_{\mathrm{loc}}(\mathbb{R}^N))$. Passing to the limit as τ_k tends to 0 in the entropy inequality we get that u is an entropy solution of the Cauchy problem $CP(b, F, u_o, f)$. $\qquad\square$

THEOREM 3.7. *For all u_0 and f verifying* (H1), *there exists an entropy solution of the problem* $CP(b, F, u_o, f)$.

Proof. For $C > \|u_0\|_{L^\infty(\mathbb{R}^N)} + \int_0^T \|f(t)\|_{L^\infty(\mathbb{R}^N)} dt$ we denote

$$w^{n,m}(x) = w^+ \chi_{B(0,n)}(x) + w^- \chi_{B(0,m)}(x).$$

Since $u_0^{n,m}$ and $f^{n,m}$ are in $L^1(\mathbb{R}^N) \cap L^\infty(\mathbb{R}^N)$ and $L^1(Q) \cap L^\infty(Q)$ respectively, we have existence and uniqueness of the integral solution $u_{n,m}$ of the Cauchy problem $(CP)(b, F, u_0^{n,m}, f^{n,m})$ which is also an entropy solution. We have also the estimate $\|u_{n,m}\|_{L^\infty(\mathbb{R}^N)} < C$, and the L^1 contraction principle. We deduce that

$$u_{n,m} \uparrow u_m \quad \text{when } n \longrightarrow +\infty \text{ and } u_m \downarrow \underline{u} \quad \text{when } m \longrightarrow +\infty.$$

We also have

$$u_{n,m} \downarrow u_n \quad \text{when } m \longrightarrow +\infty \text{ and } u_n \uparrow \overline{u} \quad \text{when } n \longrightarrow +\infty.$$

The sequences $u_{n,m}$, u_n, u_m are bounded and monotone. Following a method used before (see [MT]), we get that $\underline{u} \in C([0, T], L^1_{\text{loc}}(\mathbb{R}^N))$. Passing to the limit in the entropy inequality, we then obtain by Theorem 3.6, that $\underline{u}$ is in fact an entropy solution of $(CP)(b, F, u_0, f)$. $\qquad\square$

3.2. *Comparison principle of entropy solution*

To establish the result below, we follow the lines of the proof of Lemma 5 and Theorem 6.

LEMMA 3.8. *Let u_0, f be such that (H1) is fulfilled. Let u be a weak solution of (CP) then:*

$$\begin{cases} \int_Q H_0(u - s)\{(\nabla b(u) + F(s) - F(u))\nabla \xi + (s - u)\xi_t - f\xi\} \, dx \, dt \\ \quad - \int_{\mathbb{R}^N} (u_0 - s)^+ \xi(0) \, dx = \lim_{\epsilon \to 0} \int_Q |\nabla b(u)|^2 H'_\epsilon(b(s) - b(u))\xi \, dx \, dt; \end{cases} \tag{40}$$

and

$$\begin{cases} \int_Q H_0(s - u)\{(\nabla b(u) + F(s) - F(u))\nabla \xi + (s - u)\xi_t - f\xi\} \, dx \, dt \\ \quad - \int_{\mathbb{R}^N} (s - u_0)^+ \xi(0) \, dx = \lim_{\epsilon \to 0} \int_Q |\nabla b(u)|^2 H'_\epsilon(b(s) - b(u))\xi \, dx \, dt; \end{cases} \tag{41}$$

for all $s \in \mathbb{R}$ such that $b(s) \notin E$ and $\xi \in \mathcal{D}(Q), \xi \geq 0$.

Proof. Let $s \in \mathbb{R}$ such that $b(s) \notin E$, and let $\xi \in \mathcal{D}(Q), \xi \geq 0$. Consider R big enough such that $supp\xi \subset B(0, R) =: \Omega$. Since by Remark 3.5, u is a weak solution of (CP) in $[0, T) \times \Omega$, and inequality (40), is equivalent to:

$$\begin{cases} \int_{[0,T)\times\Omega} H_0(u - s)\{(\nabla b(u) + F(s) - F(u))\nabla \xi + (s - u)\xi_t - f\xi\} \, dx \, dt \\ \quad - \int_{\mathbb{R}^N} (u_0 - s)^+ \xi(0) \, dx = \lim_{\epsilon \to 0} \int_{[0,T)\times\Omega} |\nabla b(u)|^2 H'_\epsilon(b(s) - b(u))\xi \, dx \, dt \end{cases} \tag{42}$$

for all $s \in \mathbb{R}$ such that $b(s) \notin E$ and $\xi \in \mathcal{D}(Q), \xi \geq 0$.

Now, we may conclude our proof by using Lemma 4 and Lemma 5 of [C2]. By similar arguments, we get inequality (41). We can now establish the so called Kato's inequality (see [AB]), for two entropy solutions of the Cauchy problem. $\qquad\square$

THEOREM 3.9 (Kato's inequality.). Let (u_{10}, f_1), (u_{02}, f_2) satisfy (H1), let u_1, u_2 be entropy solutions with respect to (CP) (b, F, f_1, u_{01}), (CP) (b, F, f_2, u_{02}) respectively. Then

$$\begin{cases} \int_Q \{(\nabla(b(u_1) - b(u_2))^+ + H_0(u_1 - u_2)(F(u_2) - F(u_1)))\nabla\xi \\ \quad -(u_1 - u_2)^+ \xi_t\}\, dx\, dt - \int_{\mathbb{R}^N} (u_{01} - u_{02})^+ \xi(0)\, dx \\ \leq \int_Q \nu(f_1 - f_2)\xi\, dx\, dt; \end{cases} \tag{43}$$

for any $\nu \in H(u_1 - u_2)$ a.e. and $\xi \in \mathcal{D}([0, T[\times\mathbb{R}^N), \xi \geq 0$.

Proof. We follow the lines of the proof of Lemma 2.5. Let $s \in \mathbb{R}$, so that $b(s) \notin E$, and let $\xi \in \mathcal{D}([0, T[\times\mathbb{R}^N)$, $\xi \geq 0$, consider R big enough such that $supp\,\xi \subset B(0, R) =: \Omega$, then $\xi \equiv 0$ on $\partial\Omega$. We can then write inequality (43) as follow:

$$\begin{cases} \int_{[0,T)\times\Omega} \{(\nabla(b(u_1) - b(u_2))^+ + H_0(u_1 - u_2)(F(u_2) - F(u_1)))\nabla\xi \\ \quad -(u_1 - u_2)^+ \xi_t\}\, dx\, dt - \int_\Omega (u_{01} - u_{02})^+ \xi(0)\, dx \\ \leq \int_{[0,T)\times\Omega} \nu(f_1 - f_2)\xi\, dx\, dt. \end{cases} \tag{44}$$

Since u_1, u_2 are entropy solutions with respect to (CP) (b, F, f_1, u_{01}), (CP) (b, F, f_2, u_{02}), respectively, and they verify the inequalities of Definition 3.4, for any $\xi \in \mathcal{D}(]0, T[\times\Omega)$ the result is then a direct consequence of Theorem 3.6 of [C2]. $\square$

The following result is a version of Theorem 2.7 suitable for second order problems.

THEOREM 3.9. *Let* $\omega_1, \omega_2,\ldots \omega_l, \eta_1, \eta_2, \ldots \eta_l$, *be non negative functions so that for* $i = 1,\ldots, l$ *with* $l \leq N$;

$$\lim_{\epsilon\to 0} \frac{\omega_i(\epsilon)}{\epsilon} = \lim_{\epsilon\to 0} \frac{\eta_i(\epsilon)}{\epsilon} = +\infty; \quad \lim_{\epsilon\to 0} \omega_i(\epsilon) = \lim_{\epsilon\to 0} \eta_i(\epsilon) = 0; \tag{45}$$

$$\liminf_{\epsilon\to 0} \epsilon^{1-l} \prod_{i=1}^{l} [\eta_i(\epsilon) + (\eta_i^2(\epsilon) + 2\epsilon\omega_i(\epsilon))^{1/2}] < +\infty \quad \text{if } N > 2; \tag{46}$$

$$\liminf_{\epsilon\to 0} \epsilon^{1-l} \prod_{i=1}^{l} [\eta_i(\epsilon) + (\eta_i^2(\epsilon) + 2\epsilon\omega_i(\epsilon))^{1/2}] = 0 \quad \text{if } N = 2. \tag{47}$$

Let $h \in L^1_{loc}(Q)$ *such that* $h^+ = \max(h, 0) \in L^1(Q)$. *Let* $W_0 \in L^1(\mathbb{R}^N)$, $W \in L^\infty(Q)$, $W \geq 0$, *and let* $\lambda > 0$ *so that:*

$$\begin{cases} \int\int_Q W\frac{\partial\xi}{\partial t} + \sum_{j=1}^{l}(W + \epsilon)\frac{\omega_j(\epsilon)}{\epsilon}|\frac{\partial^2\xi}{\partial x_j\partial x_j}| + \sum_{j=1}^{l}(W + \epsilon)\frac{\eta_j(\epsilon)}{\epsilon}|\frac{\partial\xi}{\partial x_j}| \\ \quad + \sum_{j=l+1}^{N} \lambda W(|\frac{\partial^2\xi}{\partial x_j\partial x_j}| + |\frac{\partial\xi}{\partial x_j}|) + h\,\xi\, dx\, dt \geq 0, \end{cases} \tag{48}$$

for any $\epsilon > 0$ *and* $\xi \in \mathcal{D}(Q), \xi \geq 0$.

Suppose $(W(t, .) - W_0)^+ \longrightarrow 0$ in $L^1_{\mathrm{loc}}(\mathbb{R}^N)$, when $t \longrightarrow 0$ essentially. Then $h \in L^1(Q)$, $W \in L^\infty(0, T, L^1(\mathbb{R}^N))$ and

$$\int_{\mathbb{R}^N} W(\tau, x)\, dx \leq \int_{\mathbb{R}^N} W_0(x)\, dx + \int_{Q_\tau} h\, dx\, dt, \quad \text{for } \tau \in (0, T) \text{ a.e.}$$

with $Q_\tau =]0, \tau[\times \mathbb{R}^N$.

Proof. The proof of this result is similar to the proof of Theorem 2.7. Inequality (48) is equivalent to the following one.

$$\begin{cases} \int_{\mathbb{R}^N} W(\tau, .)\, \xi\, dx \leq \int_{\mathbb{R}^N} W(\sigma, .)\, \xi\, dx \\ + \int_\sigma^\tau dt \int_{\mathbb{R}^N} \sum_{j=1}^l (W + \epsilon)\frac{\omega_j(\epsilon)}{\epsilon} |\frac{\partial^2 \xi}{\partial x_j \partial x_j}|\, dx \\ + \int_\sigma^\tau dt \int_{\mathbb{R}^N} \sum_{j=1}^l (W + \epsilon)\frac{\eta_j(\epsilon)}{\epsilon} |\frac{\partial \xi}{\partial x_j}|\, dx \\ + \int_\sigma^\tau dt \int_{\mathbb{R}^N} \sum_{j=l+1}^N \lambda W(|\frac{\partial^2 \xi}{\partial x_j \partial x_j}| + |\frac{\partial \xi}{\partial x_j}|) + h\, \xi\, dx, \end{cases} \tag{49}$$

for any $\epsilon > 0$ and $\xi \in \mathcal{D}(\mathbb{R}^N), \xi \geq 0$; and for a.e. $\sigma, \tau \in (0, T)\, \sigma \leq \tau$. Letting σ tend to 0^+, this inequality implies that for any $\epsilon > 0, \xi \in \mathcal{D}(\mathbb{R}^N), \xi \geq 0$, and for $\tau \in (0, T)$ a.e.

$$\begin{cases} \int_{\mathbb{R}^N} (W(\tau, .) + \int_0^\tau h^-(t, .)\, dt)\, \xi\, dx \leq \int_{\mathbb{R}^N} W_0\, \xi\, dx + \int \int_{Q_\tau} h^+\, \xi\, dx\, dt \\ + \int_{\mathbb{R}^N} \sum_{j=1}^l (\int_0^\tau W(t, .)\, dt + \epsilon\tau)(\frac{\omega_j(\epsilon)}{\epsilon}|\frac{\partial^2 \xi}{\partial x_j \partial x_j}| + \frac{\eta_j(\epsilon)}{\epsilon}|\frac{\partial \xi}{\partial x_j}|)\, dx \\ + \int \int_{Q_\tau} \lambda W \sum_{j=l+1}^N (|\frac{\partial^2 \xi}{\partial x_j \partial x_j}| + |\frac{\partial \xi}{\partial x_j}|)\, dx\, dt. \end{cases} \tag{50}$$

Using the same choice of test function ξ, we obtain for $\tau \in (0, T)$ a.e.

$$\begin{cases} \int_{\mathbb{R}^N} (W(\tau, .) + \int_0^\tau h^-(t, .)\, dt)\, \xi\, dx \leq \int_{\mathbb{R}^N} W_0\, \xi\, dx + \int \int_{Q_\tau} h^+\, \xi\, dx\, dt \\ + \sum_{j=1}^l \frac{1}{\epsilon R_j}(\frac{\omega_j(\epsilon)}{R_j} + \eta_j(\epsilon)) \int_{\mathbb{R}^N} (\int_0^\tau W(t, .)\, dt)\, \xi\, dx \\ + \sum_{j=1}^l \frac{1}{R_j}(\frac{\omega_j(\epsilon)}{R_j} + \eta_j(\epsilon)) \int_{\mathbb{R}^N} \xi\, dx + \sum_{l+1}^N \alpha(1 + \frac{\alpha}{\lambda}) \int \int_{Q_\tau} W\xi\, dx\, dt. \end{cases} \tag{51}$$

Using again the same choice for $R_j(\epsilon, \eta) = R_j$ and with similar arguments as previously, we have that

$$\begin{cases} \int_{\mathbb{R}^N} (W(\tau, .) + \int_0^\tau h^-(t, .)\, dt)\, \xi\, dx \leq \int_{\mathbb{R}^N} W_0\, dx \\ + \int \int_{Q_\tau} h^+\, dx\, dt + \eta^2 \sum_{j=1}^l \int_{\tilde{K}} (\int_0^\tau W(t, .)\, dt)\, \xi\, dx \\ + (N - l)\alpha(1 + \frac{\alpha}{\lambda}) \int \int_{Q_\tau} W\xi\, dx\, dt + \frac{T\Lambda(\epsilon)}{\eta^{l-2}\alpha^{N-l}} \end{cases} \tag{52}$$

We choose η, α so that $\mu(\eta) + \delta(\alpha) = \frac{1}{2T}$ then we have

$$\begin{cases} \frac{1}{2} ess \sup_{\tau \in (0,T)} \int_K W(\tau, .)\rho \, dx \le \{1 - (\mu(\eta) + \delta(\alpha))T\} ess \sup_{\tau \in (0,T)} \int_{\mathbb{R}^N} W(\tau, .)\xi \, dx \\ \le ess \sup_{\tau \in (0,T)} \int_{\mathbb{R}^N} W(\tau, .)\xi \, dx - (\mu(\eta) + \delta(\alpha)) \int \int_Q W(\tau, .)\xi \, dxdt \\ \le \int_{\mathbb{R}^N} W_0 + \int \int_{Q_\tau} h^+ \, dx \, dt + \frac{T\Lambda(\epsilon)}{\eta^{l-2}\alpha^{N-l}}. \end{cases} \qquad (53)$$

We choose now a subsequence ϵ_k (as previously) and let $k$tend to infinity. We obtain:

$$W\rho \in L^{+\infty}(0, T, L^1(\mathbb{R}^N)), \quad \text{and} \quad \left(\int_0^\tau W(t, .) \, dt \right) \rho \in L^1(\mathbb{R}^N).$$

Using (50), $h^-\rho \in L^1(Q)$ and letting ϵ_k tend to 0 we obtain

$$ess \sup_{t \in (0,T)} \int_{\mathbb{R}^N} (1 - T\delta(\alpha))W(t, .)\rho + \int \int_{Q_\tau} h^- \rho dx \, dt$$
$$\le \int_{\mathbb{R}^N} W_0 + \int \int_{Q_\tau} h^+ \, dx \, dt + \frac{TC}{\eta^{l-2}\alpha^{N-l}}$$

for any $\tau \in [0, T]$, $\eta > 0$ and $\alpha > 0$ with $T\delta(\alpha) < 1$. The final result is obtained by letting η tend to infinity and α to 0. $\qquad \square$

REMARK 3.10. For the case $N = 2$, $N = 1$ we use a similar analysis as in the stationary problem. Then as a consequence of Theorem 3.9, we obtain the L^1 contraction principle which gives uniqueness. More precisely, we have the following result.

THEOREM 3.11. *Let* (u_{01}, f_1), (u_{02}, f_2) *verify* (H1), *and let* u_1, u_2 *be entropy solution with respect to* (CP) (b, F, f_1, u_{01}), *and* (CP) (b, F, f_2, u_{02}) *respectively. If* (H2) *holds, then:*

$$\int_{\mathbb{R}^N} (u_1(t) - u_2(t))^+ \, dx \le \int_{\mathbb{R}^N} (u_{01} - u_{02})^+ \, dx + \int_{Q_t} v(f_1 - f_2) \, dx \, dt; \qquad (54)$$

for $v \in H(u_1 - u_2)$*a.e so that:*

$$\|u_1(t) - u_2(t)\|_{L^1(\mathbb{R}^N)} \le \|u_{01} - u_{02}\|_{L^1(\mathbb{R}^N)} + \int_0^t \|f_1 - f_2\|_{L^1(\mathbb{R}^N)} \, ds;$$

which gives uniqueness of entropy solution.

Notice that the condition (H2) is optimal. Bénilan and Kruzkhov (see [BK]) have shown that if (H2) does not hold, there is no uniqueness of entropy solution with $b \equiv 0$. They constructed infinitely many entropy solutions for $N = 2$, $F_i(r) = r^{\alpha_i}$ $i = 1, 2$ with $\alpha_i \ge 0$ and $\alpha_1 + \alpha_2 < 1$. A direct consequence of this previous theorem is the comparison principle:

COROLLARY 3.12. *Let* (u_{01}, f_1), (u_{02}, f_2) *verify* (H1) *and* (H2). *Let* u_1, u_2 *be entropy solutions with respect to* (CP) (b, F, f_1, u_{01}), *and* (CP) (b, F, f_2, u_{02}) *respectively.*

If $u_{01} \leq u_{02}$ *a.e. in* $\mathbb{R}^N$ *and* $f_1 \leq f_2$ *a.e. on* Q *then* $u_1 \leq u_2$ *a.e. on* Q.

Acknowledgement

This paper is devoted to the memory of **Professor Philippe BÉNILAN**. We have the previlege to be his students and friends. Even after our studentship we continued to benefit from his adivse, his help and encouragement before his untimely passing away. Our honour for Philippe means that we have to continue following his footsteps in Mathematics. We thanks the referee for his kind remarks and suggestions.

REFERENCES

[ABK] ANDREIANOV, B. P., BÉNILAN, PH., KRUSKHOV, S. N., L^1 *theory of scalar conservation law with continuous flux function*, J. Funct. Anal *171* (2000), 15–33.

[AB] ARENDT, W., BÉNILAN, PH., *Inégalités de Kato et semi-groupes sous-Markoviens*, Revista Mathematica Complutense Madrid *5* (1992), 281–308.

[AL] ALT, H. W., LUCKHAUS, S., *Quasi-linear elliptic-parabolic differential equations*, Math.Z., *183* (1983), 311–341.

[BCP] BÉNILAN, PH., CRANDALL, M. G., PAZY, A., Evolution Equation Governed by Accretive Operators (book to appear).

[BG] BÉNILAN, PH., GARIEPY, B., *Strong solution* L^1 *of degenerate parabolic equation*, J. of Diff. Equat., *119* (1995), 473–502.

[BK] BÉNILAN, PH., KRUSKHOV, S. N., *Quasilinear first order equations with continuous non linearities*, Russian Acad. Sci. Dokl. Math. vol *50* (1995) no 3, 391–396.

[BT] BÉNILAN, PH., TOURÉ, H., *Sur l'équation générale* $u_t = a(., u, \varphi(., u)_x)_x$, *dans* L^1 II, *Le problème d'évolution*, Ann., Inst. Henri Poincaré, vol. *12* 6 (1995), 727–761.

[BW] BÉNILAN, PH., WITTBOLD, P., *On mild and weak solution of elliptic-Parabolic Problems*, Adv. in Diff. Equat. vol. *1* (1996) (6), 1053–1072.

[C1] CARRILLO, J., *On the uniquness of the solution of the evolution DAM problem, Nonlinear Analysis*, vol 22 (1999) no 5, 573–607.

[C2] CARRILLO, J., *Entropy solutions for nonlinear degenerate problems*, Arch. Ratio. Mech. Anal. *147* (1999), 269–361.

[C3] CARRILLO, J., *Unicité des solutions du type Kruskhov pour des problèmes elliptiques avec des termes de transport non linéaires*, C. R. Acad. Sc. Paris , t 33, Serie I, 1986 no 5.

[CW] CARRILLO, J., WITTBOLD, P., *Uniqueness of renormalized solutions of degenerate elliptic-parabolic problems*, J. Diff.Equation *156* (1999), 93–121.

[Di] DIBENEDETTO, E., *Continuity of weak solutions to a general porous medium equation*, IND. Univ. Math. J. vol. *32*, 1983 no 1.

[DT] DIAZ, J. I., THELIN, F., *On a nonlinear parabolic problem arising in some models related to turbulent flows*, SIAM. J. Math. Anal.; *25* (1994), 1085–1111.

[GT] GAGNEUX, G., TORT, M. M., *Unicité des solutions faibles d'équations de diffusion convection*, C. R. Acad. SC. Paris, t 318, Série I (1994) 919–924.

[HV] HUDJAEV, S. N., VOL'PERT, A. I., *Cauchy's problem for degenerate second order quasilinear parabolic equation*, Math. USSR-Sbornik, vol *7*, (1969) no 3, 365–387.

[KP]   KRUSKHOV, S. N., PANOV, E. YU., *Conservative quasilinear first order laws with an infinite domain of dependence on the initial data*, Soviet. Math. Dokl. vol. *42*, 2 (1991), 316–321.

[YJ]   YIN, J., *On the uniqueness and stability of BV solutions for nonlinear diffusion equations*, Comm. Part. Diff. Equat. *15*, 12 (1990), 1671–1683.

Mohamed Maliki
Équipe: modélisation
E.D.P. et Analyse numérique
F.S.T. Mohammédia
B.P. 146
Mohammédia
email: maliki@uh2m.ac.ma

Hamidou Touré
UFR/SEA Université de Ouagadougou
03 B.P.
7021 Ouagadougou 03
Burkina Faso
email: toureh@univ-ouaga.bf

To access this journal online:
http://www.birkhauser.ch

J.evol.equ. 3 (2003) 623 – 635
1424–3199/03/040623 – 13
DOI 10.1007/s00028-003-0113-z
© Birkhäuser Verlag, Basel, 2003

Journal of Evolution Equations

Oscillatory boundary conditions for acoustic wave equations

CIPRIAN G. GAL, GISÈLE RUIZ GOLDSTEIN and JEROME A. GOLDSTEIN

Dedicated to the memory of Philippe Bénilan

1. Introduction

In the textbook literature on theoretical acoustics, it was traditional to use the Robin boundary condition with the wave equation. But it was recognized that this was not the physically correct boundary condition. "Acoustic Boundary Conditions" (or ABC) were introduced in the monograph by Morse and Ingard [13, p. 263]. The presentation in [13] is not the usual approach to the wave equations, since the authors treat waves having definite frequency. The time dependent version of ABC was first formulated by Tom Beale and Steve Rosencrans [1] in a very interesting and original paper. ABC will be explained in detail in Section 2.

In the theory of Markov diffusion processes, one studies heat equations of the form

$$\frac{\partial u}{\partial t} = Lu$$

where L is a second order linear elliptic operator (e.g $L = \Delta$). A. D. Wentzell [15] introduced boundary conditions involving second derivatives as well as lower order (Robin type) terms. These parabolic problems were usually studied in spaces of continuous functions. But Favini, Goldstein, Goldstein and Romanelli [8] introduced a new approach to this problem, involving weighted L^p spaces and using the boundary as well as the domain.

For a specific example, consider the heat equation

$$\frac{\partial u}{\partial t} = c^2 \Delta u \ \ \text{in} \ \ \Omega \subset \mathbf{R}^n$$

with boundary condition

$$c^2 \Delta u + \beta_1 \frac{\partial u}{\partial n} + \gamma_1 u = 0 \ \ \text{on} \ \ \partial\Omega \tag{1.1}$$

2000 *Mathematics Subject Classifications*: Primary 35L05; Secondary 35L15, 35L20, 34G10, 76Q05
Key words: Acoustic waves, wave equation acoustic boundary conditions, general Wentzell boundary conditions, noncompact resolvent

where $\beta_1, \gamma_1 \in C(\partial\Omega)$ with $\beta_1 > 0$, $\gamma_1 \geq 0$ on $\partial\Omega$. The natural L^p space for this problem turns out to be

$$X_p = L^p(\Omega, dx) \oplus L^p \left(\partial\Omega, \frac{c^2}{\beta_1} dS \right), \quad 1 \leq p < \infty.$$

On X_2 the semigroup generator G for this problem (i.e. a suitable realization of $c^2\Delta$) is selfadjoint and nonpositive. The corresponding wave equation

$$u_{tt} = Gu$$

is governed by a unitary group on a four component space, based on $H^1(\Omega) \times L^2(\Omega)$ and two copies of $L^2(\partial\Omega, \frac{c^2}{\beta_1} dS)$. The energy space bears a formal resemblence to the four component energy space that Beale and Rosencrans associated with the wave equation with ABC.

Our goal in this paper is to show that these two versions of the wave equation are closely connected. We shall use these connections to derive new results about both wave equations. Interestingly, the form of the wave equation with ABC most closely associated with the energy conserving wave equation with boundary conditions (1.1) is a nonenergy conserving version; this will be explained in detail in the sequel. Furthermore, a new extension of the boundary condition (1.1) makes the resulting wave equation equivalent to the one with ABC.

The main results are contained in Theorems 1 and 2 (of Section 4) and 4 (of Section 5).

2. Acoustic Boundary Conditions

In this section we explain some of the results of Beale [2]. (The papers [2], [3] expand and develop the work begun in [1].)

Let Ω be a smooth bounded domain in $\mathbf{R}^n$, $n \geq 1$. (The paper [2] restricts n to be 3. For general n see Gal [11].). Fluid filling Ω is at rest except for acoustic wave motion. Let $\varphi : \bar{\Omega} \times \mathbf{R} \to \mathbf{R}$ be the velocity potential, so that $-\nabla\varphi(x, t)$ is the particle velocity. Thus φ satisfies the wave equation

$$\frac{\partial^2\varphi}{\partial t^2} = c^2\Delta\varphi \text{ in } \Omega \times \mathbf{R} \tag{2.1}$$

where $c > 0$ is the (constant) speed of propagation.

Each point x of $\partial\Omega$ is assumed to react to the excess pressure of the acoustic wave like a resistive harmonic oscillator or spring. The normal displacement $\delta(x, t)$ of the boundary into the domain satisfies

$$m(x)\delta_{tt}(x, t) + d(x)\delta_t(x, t) + k(x)\delta(x, t) + \rho\varphi_t(x, t) = 0 \tag{2.2}$$

on $\partial\Omega \times \mathbf{R}$, where ρ is the fluid density and $m, d, k \in C(\partial\Omega)$ with $m > 0, k > 0, d \geq 0$. If the boundary is impenetrable, continuity of the velocity on $\partial\Omega$ implies the compatibility condition

$$\delta_t(x, t) = \frac{\partial\varphi}{\partial n}(x, t) \tag{2.3}$$

on $\partial\Omega \times \mathbf{R}$, where $n = n(x)$ is the unit outer normal to $\partial\Omega$ at x. The energy of the solution is

$$E(t) = \int_\Omega \left(\rho |\nabla\varphi|^2 + \frac{\rho}{c^2} |\varphi_t|^2\right) dx + \int_{\partial\Omega} (k |\delta|^2 + m |\delta_t|^2) dS.$$

Moreover,

$$\frac{dE}{dt} = -2 \int_{\partial\Omega} d |\delta_t|^2 \, dS \leq 0,$$

and energy is conserved when the springs are all frictionless, i.e $d \equiv 0$.

The energy (Hilbert) space for this problem is

$$\mathcal{H} = H^1(\Omega) \oplus L^2(\Omega) \oplus L^2(\partial\Omega) \oplus L^2(\partial\Omega). \tag{2.4}$$

Its norm is determined by

$$\|u\|^2 = \int_\Omega \left(\rho |\nabla\varphi|^2 + \frac{\rho}{c^2} |\varphi_t|^2\right) dx + \int_{\partial\Omega} \left(k |\delta|^2 + m |\delta_t|^2\right) dS, \tag{2.5}$$

where $u = (u_1, u_2, u_3, u_4) = (\varphi, \varphi_t, \delta, \delta_t)$. The wave equation with (ABC), (2.1) $-$ (2.3) is equivalent to $u(t) \in \mathcal{D}(A)$ and $u_t = Au$, where

$$A \begin{pmatrix} u_1 \\ u_2 \\ u_3 \\ u_4 \end{pmatrix} = \begin{pmatrix} u_2 \\ c^2 \Delta u_1 \\ u_4 \\ -\frac{1}{m}(\rho u_2 + k u_3 + d u_4) \end{pmatrix} \tag{2.6}$$

and $\mathcal{D}(A) = \{u \in \mathcal{H} : \Delta u_1 \in L^2(\Omega), u_2 \in H^1(\Omega), \frac{\partial u_1}{\partial n} = u_4 \text{ on } \partial\Omega\}$. Here in $(Au)_4$, $u_2|_{\partial\Omega}$ makes sense as a member of $H^{\frac{1}{2}}(\partial\Omega)$ (in the trace sense) and $\frac{\partial u_1}{\partial n} = u_4$ is interpreted to mean:

$$\int_\Omega ((\Delta u_1) \psi + \nabla u_1 \cdot \nabla \psi) \, dx = \int_{\partial\Omega} (u_4 \psi) \, dS$$

for all $\psi \in H^1(\Omega)$. The operator A is densely defined in $\mathcal{H}$ and dissipative:

$$\operatorname{Re} \langle Au, u \rangle = - \int_{\partial\Omega} d |u_4|^2 \, dS \leq 0,$$

and A generates a (C_0) contraction semigroup on $\mathcal{H}$, which is a unitary group when $d \equiv 0$. The generator A has interesting spectral properties. Now let $n \geq 3$. (In [2] only $n = 3$ was considered; see Gal [11] for the extension.). Let

$$\sum := \{\lambda \in \mathbf{C} : m(x)\lambda^2 + d(x)\lambda + k(x)\lambda = 0 \text{ for some } x \in \partial\Omega\}.$$

$\sum$ is compact and symmetric about the real axis. Let R be the unbounded component of $\mathbf{C} \setminus \sum$; note that $0 \in R$. Beale proved that $\lambda \longmapsto (\lambda - A)^{-1}$ is meromorphic on R.

Suppose that m, d and k are constants. Then $\sum$ consists of two points (unless $d^2 = 4mk$), namely

$$\lambda_{\pm} = \frac{1}{2m}\left(-d \pm \sqrt{d^2 - 4mk}\right).$$

Then the essential spectrum of A is $\sum$ and the point spectrum $\sigma_P(A)$ consists of eigenvalue sequences (1) $\{\lambda_n\}$ with $\operatorname{Im}\lambda_n \to \infty$, $\operatorname{Re}\lambda_n \to 0$, (2)$\{\bar{\lambda}_n\}$, (3) $\{\mu_n^{\pm}\}$ with $\mu_n^{\pm} \to \lambda_{\pm}$, and (4) finitely many additional eigenvalues. In particular, $(\lambda - A)^{-1}$ is not a compact operator when $n \geq 2$. This is remarkable; most linear problems involving the Laplacian on bounded domains have generators with compact resolvents.

The operator A described by (2.6) has the matrix representation

$$A_1 = \begin{pmatrix} 0 & I & 0 & 0 \\ c^2\Delta & 0 & 0 & 0 \\ 0 & 0 & 0 & I \\ 0 & -\frac{\rho}{m}J & -\frac{k}{m}I & -\frac{d}{m}I \end{pmatrix}$$

where I is the identity operator (i.e. $I(u) = u$ for any u) and J means restriction to the boundary: $Ju = u|_{\partial\Omega}$. Using the compatibility condition (2.3), we can equally well represent A as the operator matrix

$$A_2 = \begin{pmatrix} 0 & I & 0 & 0 \\ c^2\Delta & 0 & 0 & 0 \\ 0 & 0 & 0 & I \\ -\frac{d}{m}\frac{\partial}{\partial n} & -\frac{\rho}{m}J & -\frac{k}{m}I & 0 \end{pmatrix}. \tag{2.7}$$

Again, this is just a restatement of (2.6) together with $\frac{\partial u_1}{\partial n} = u_4$. (See the definition of $\mathcal{D}(A)$.)

3. General Wentzell Boundary Conditions

Consider the heat equation

$$\frac{\partial u}{\partial t} = c^2\Delta u \text{ in } \Omega \times [0, \infty)$$

with the general Wentzell boundary condition (or GWBC)

$$\Delta u + \beta \frac{\partial u}{\partial n} + \gamma u = 0 \text{ on } \partial\Omega \times [0, \infty),$$

where $\beta, \gamma \in C(\partial\Omega)$ with $\beta > 0$, $\gamma \geq 0$ on $\partial\Omega$. This problem is governed by an analytic contraction semigroup on X_p, $1 \leq p \leq \infty$, where

$$X_p = L^p(\Omega, dx) \oplus L^p\left(\partial\Omega, \frac{1}{\beta}dS\right), 1 \leq p < \infty \text{ and } X_\infty = C(\bar{\Omega}),$$

with norm, for $u \in C(\bar{\Omega}) \subset X_p$:

$$\|u\|_{X_p}^p = \int_\Omega |u(x)|^p \, dx + \int_{\partial\Omega} |u(x)|^p \, \frac{dS}{\beta(x)}, 1 \leq p < \infty,$$

$$\|u\|_{X_\infty} = \lim_{p\to\infty} \|u\|_{X_p} = \|u\|_{L^\infty(\Omega)}.$$

This is proved in [8], except for the analyticity when $p = 1, \infty$, which is discussed in [9].

Let G be the generator of this semigroup on X_2 and let G_0 be G restricted to $C^2(\bar{\Omega}) \subset X_2$. Then G_0 is essentially selfadjoint on X_2. This follows as a very special case of the adjoint calculation in [9]. Here we explain it briefly. For $u, v \in \mathcal{D}(G_0)$,

$$\frac{1}{c^2} \langle G_0 u, v \rangle_{X_2} = \int_\Omega (\Delta u)\, \bar{v}\, dx + \int_{\partial\Omega} (\Delta u)\, \bar{v}\, \frac{dS}{\beta(x)}$$

$$= -\int_\Omega (\nabla u) \cdot (\nabla \bar{v}) dx + \int_{\partial\Omega} \left(\frac{\partial u}{\partial n}\right) \bar{v}\, dS + \int_{\partial\Omega} (\Delta u)\, \bar{v}\, \frac{dS}{\beta}$$

$$= \int_\Omega u \bar{\Delta v}\, dx - \int_{\partial\Omega} u \frac{\partial \bar{v}}{\partial n} dS - \int_{\partial\Omega} \gamma u \bar{v} \frac{dS}{\beta}$$

since $\Delta u + \beta \frac{\partial u}{\partial n} + \gamma u = 0$ on $\partial\Omega$, so

$$\frac{1}{c^2} \langle G_0 u, v \rangle_{X_2} = \int_\Omega u \bar{\Delta v}\, dx + \int_{\partial\Omega} u \bar{\Delta v}\, \frac{dS}{\beta} = \frac{1}{c^2} \langle u, G_0 v \rangle_{X_2}$$

since the GWBC holds for v as well.

So let us consider the wave equation with GWBC:

$$\frac{\partial^2 u}{\partial t^2} = c^2 \Delta u \text{ in } \Omega \times \mathbf{R}, \tag{3.1}$$

$$\Delta u + \beta \frac{\partial u}{\partial n} + \gamma u = 0 \text{ on } \partial\Omega. \tag{3.2}$$

Note that this boundary condition is identical to (1.1) when $\beta_1 = c^2\beta$ and $\gamma_1 = c^2\gamma$.

As usual, the wave equation (3.1), (3.2) can be written as

$$U_t = \begin{pmatrix} 0 & I \\ G & 0 \end{pmatrix} U = BU$$

(this defines B), where

$$G = c^2 \begin{pmatrix} \Delta & 0 \\ -\beta \frac{\partial}{\partial n} & -\gamma \end{pmatrix};$$

G acts on $X_2 = L^2(\Omega) \oplus L^2\left(\partial\Omega, \frac{dS}{\beta}\right)$ and $\begin{pmatrix} 0 & I \\ G & 0 \end{pmatrix}$ acts on

$$\mathcal{D}((-G)^{\frac{1}{2}}) \oplus X_2 = \widehat{\mathcal{H}}_{en}.$$

The norm in the energy Hilbert space $\widehat{\mathcal{H}}_{en}$ is given on $\mathcal{D}(G_0) \times \mathcal{D}(G_0)$ by

$$\left\| \begin{pmatrix} w_1 \\ w_2 \\ w_3 \\ w_4 \end{pmatrix} \right\|^2_{\widehat{\mathcal{H}}_{en}} = \left\| (-G)^{\frac{1}{2}} \begin{pmatrix} w_1 \\ w_2 \end{pmatrix} \right\|^2_{X_2} + \left\| \begin{pmatrix} w_3 \\ w_4 \end{pmatrix} \right\|^2_{X_2} \tag{3.3}$$

$$= \left\langle (-G) \begin{pmatrix} w_1 \\ w_2 \end{pmatrix}, \begin{pmatrix} w_1 \\ w_2 \end{pmatrix} \right\rangle_{X_2} + \left\langle \begin{pmatrix} w_3 \\ w_4 \end{pmatrix}, \begin{pmatrix} w_3 \\ w_4 \end{pmatrix} \right\rangle_{X_2}$$

$$= \left\{ c^2 \left\langle -\Delta w_1, w_1 \right\rangle_{L^2(\Omega)} + \left\langle \beta \frac{\partial w_1}{\partial n}, w_2 \right\rangle_{L^2\left(\partial\Omega, \frac{dS}{\beta}\right)} + \right.$$

$$\left. + \left\langle \gamma w_2, w_2 \right\rangle_{L^2\left(\partial\Omega, \frac{dS}{\beta}\right)} + \|w_3\|^2_{L^2(\Omega)} + \|w_4\|^2_{L^2\left(\partial\Omega, \frac{dS}{\beta}\right)} \right\}.$$

Here $w_1 = u|_\Omega$, $w_2 = u|_{\partial\Omega}$, $w_3 = \frac{\partial u}{\partial t}|_\Omega$ and $w_4 = \frac{\partial u}{\partial t}|_{\partial\Omega}$.

Thus

$$B \begin{pmatrix} w_1 \\ w_2 \\ w_3 \\ w_4 \end{pmatrix} = \begin{pmatrix} w_3 \\ w_4 \\ c^2 \Delta w_1 \\ -\beta \frac{\partial w_1}{\partial n} - \gamma w_2 \end{pmatrix}. \tag{3.4}$$

Define $\widehat{w} = \begin{pmatrix} \widehat{w_1} \\ \widehat{w_2} \\ \widehat{w_3} \\ \widehat{w_4} \end{pmatrix} = \begin{pmatrix} w_1 \\ w_3 \\ w_2 \\ w_4 \end{pmatrix}$ where $w = \begin{pmatrix} w_1 \\ w_2 \\ w_3 \\ w_4 \end{pmatrix}$ is in $\widehat{\mathcal{H}}_{en}$.

This enables us to identify $\widehat{\mathcal{H}}_{en}$ with $\mathcal{H}$. Write $Bw = v$ and define B_1 by $B_1\widehat{w} = \widehat{v}$, i.e. $(Bw)_j = v_j$, for $1 \le j \le 4$ and

$$\widehat{v} = \begin{pmatrix} v_1 \\ v_3 \\ v_2 \\ v_4 \end{pmatrix}.$$

Then

$$B_1\widehat{w} = \begin{pmatrix} w_3 \\ c^2 \Delta w_1 \\ w_4 \\ -\beta \frac{\partial w_1}{\partial n} - \gamma w_2 \end{pmatrix} = \begin{pmatrix} \widehat{w_2} \\ c^2 \Delta \widehat{w_1} \\ \widehat{w_4} \\ -\beta \frac{\partial \widehat{w_1}}{\partial n} - \gamma \widehat{w_3} \end{pmatrix}.$$

An operator representation of B_1 is

$$B_2 = \begin{pmatrix} 0 & I & 0 & 0 \\ c^2\Delta & 0 & 0 & 0 \\ 0 & 0 & 0 & I \\ -\beta \frac{\partial}{\partial n} & QJ & -\gamma I & -Q \end{pmatrix} \tag{3.5}$$

where Q is any multiplication operator acting on the boundary and $Ju = u|_{\partial\Omega}$, as before. Here B_2 acts on a Hilbert space $\mathcal{H}_{en}$ which is a closed subspace of $\mathcal{H}$, and the components of $\widehat{w} \in \mathcal{D}(B_2)$ are given by $\widehat{w} = (u_1, u_2, u_1|_{\partial\Omega}, u_2|_{\partial\Omega})$. More precisely,

$$\mathcal{H}_{en} = \{u = (u_1, u_2, u_3, u_4) \in \mathcal{H} : u_3 = u_1\,|_{\partial\Omega}\}. \tag{3.6}$$

Since u_1 is in $H^1(\Omega)$, it has a trace $u_1|_{\partial\Omega}$ in $H^{\frac{1}{2}}(\partial\Omega)$, and this determines u_3. Note that $\mathcal{H}_{en}$ coincides with a subspace of the space we previously called $\mathcal{H}$, except for the rearrangement of the components of its vectors.

If u is a solution of the wave equation with GWBC, then the fourth component of $B_2\widehat{w}$ is

$$\begin{aligned} (B_2\widehat{w})_4 &= -\beta \frac{\partial u}{\partial n}|_{\partial\Omega} + (Qu)|_{\partial\Omega} - \gamma u|_{\partial\Omega} - Qu|_{\partial\Omega} \\ &= \left(-\beta \frac{\partial u}{\partial n} - \gamma u\right)|_{\partial\Omega}, \end{aligned}$$

since Q is a multiplication operator. This a compatibility condition for the problem (analogous to (2.3)).

Note that $\mathcal{H}_{en}$ is a proper closed subspace of $\mathcal{H}$. Let

$$B_3 = B_2|_{\mathcal{H}_{en}} \oplus 0|_{\mathcal{H}_{en}^{\perp}}.$$

Then B_3 is densely defined on $\mathcal{H}$ and generates a (C_0) semigroup on $\mathcal{H}$.

Moreover, B_3 has the same matrix representation (3.5) as does B_2, except that it acts on a bigger domain.

Rather than compare A and B, which are defined on different spaces, we compare a unitarily equivalent version of A with a unitarily equivalent version of (an extension of) B. This enables us to compare A with B, even though $D(A)$ and $D(B)$ are quite different.

4. Comparing The Boundary Condition

While B_2 is defined on $\mathcal{H}_{en} \subset \mathcal{H}$, (3.5) enables us to extend B_2 and view its extension B_3 as a densely defined operator on $\mathcal{H}$.

We want to make the norm in (3.3) look as much as possible like the norm in (2.4), and make the operator B in (3.4) look as much like the operator A in (2.6) as possible.

The easiest way to do this is to compare the matrix representation A_2 for A in (2.7) and B_3 for B in (3.5) (also on $\mathcal{H}$). So we replace B_1 by B_3 so that B_3 and A_2 are both densely defined operators on $\mathcal{H}$. By making the identifications

$$\beta = \frac{d}{m}, \gamma = \frac{k}{m}, Q = -\frac{\rho}{m}, \tag{4.1}$$

we see that

$$B_3 - A_2 = K = \begin{pmatrix} 0 & 0 & 0 & 0 \\ 0 & 0 & 0 & 0 \\ 0 & 0 & 0 & 0 \\ 0 & 0 & 0 & Q \end{pmatrix}$$

where $Q = -\frac{\rho}{m}$ is a continuous negative function on $\partial\Omega$. Clearly K has operator norm equal to

$$\|K\| = \left\|\frac{\rho}{m}\right\|_{\infty} = \left\|\frac{\rho}{m}\right\|_{C(\partial\Omega)}$$

and K is compact when $n = 1$ since $L^2(\partial\Omega)$ is 2 dimensional. Finally K is a nonpositive selfadjoint operator on the energy Hilbert space $\mathcal{H}$; the norms are given by

$$\|U\|_{\mathcal{H},ABC}^2 = \int_{\Omega} \left(\rho |\nabla\varphi|^2 + \frac{\rho}{c^2} |\varphi_t|^2\right) dx + \int_{\partial\Omega} (k |\delta|^2 + m |\delta_t|^2) \, dS, \tag{4.2}$$

$$\|U\|_{\mathcal{H}_{en},GWBC}^2 = \int_{\Omega} \left(|\nabla u_1|^2 + \frac{1}{c^2} |u_2|^2\right) dx + \int_{\partial\Omega} (|u_3|^2 + |u_4|^2) \frac{dS}{\beta}, \tag{4.3}$$

where we identify $U = (u_1, u_2, u_3, u_4) = (u|_\Omega, u_t|_\Omega, u|_{\partial\Omega}, u_t|_{\partial\Omega})$ for the wave equation with GWBC with

$$U = \left(\sqrt{\rho}\varphi, \sqrt{\rho}\varphi_t, \sqrt{k\frac{d}{m}}\delta, \sqrt{d}\delta_t \right) \tag{4.4}$$

for the wave equation with ABC.

The norm of (4.3) is well defined for $u \in \mathcal{H}$. Our two problems are governed by (C_0) contraction semigroups $S = \{S(t) : t \geq 0\}$ on $\mathcal{H}$ and $T = \{T(t) : t \geq 0\}$ on $\mathcal{H}_{en}$, respectively. Let $\widehat{T}$ on $\mathcal{H}$ be the extension of T described above. The respective infinitesimal generators G_A of S and $G_W (= B_3)$ of $\widehat{T}$ differ by an operator K given in the discussion following (4.1). When we make these identifications, we assume that $d > 0$. We summarize now what the reductions have achieved.

THEOREM 1. *Consider the wave equation*

$$u_{tt} = c^2 \Delta u$$

associated with the acoustic boundary condition

$$m\delta_{tt} + d\delta_t + k\delta + \rho\varphi_t = 0, \ \delta_t = \frac{\partial\varphi}{\partial n},$$

and the wave equation with general Wentzell boundary condition

$$\Delta u + \beta\frac{\partial u}{\partial n} + \gamma u = 0.$$

Let (4.1) hold. These problems are governed by (C_0) contraction semigroups S on $\mathcal{H}$ and T on $\mathcal{H}_{en}$, respectively; let $\widehat{T}$, with generator $B_3(= G_W)$, be the extension of T to $\mathcal{H}_{en}$ described above. Then the generators $G_A(= A_2)$ of S and G_W of $\widehat{T}$ differ by an operator which is selfadjoint and bounded on $\mathcal{H}$ and is compact when the dimension of the underlying bounded set $\Omega \subset \mathbf{R}^n$ is one.

By our construction, B_3 is an extension of $G_A + K$, since in extending B_2 to B_3, we got rid of the restriction that $u_3 = u_1|_{\partial\Omega}$ in the domain of B_2. But for λ real and large, $\lambda \in \rho(B_3) \cap \rho(G_A + K)$, and so $B_3 = G_A + K$.

Our construction of B_3 was rather complicated. Associated with A is the compatibility condition (2.3), and this leads to many possible (operator) matrix representations of A. Similarly for B, since there are many choices for the Q in (3.5). Thus our conclusion above is that (a suitable matrix representation of) A is unitarily equivalent to a bounded perturbation of a proper extension of a matrix representation of B.

By semigroup perturbation theory, B_3 generates a (C_0) semigroup $\widehat{T}$ on $\mathcal{H}$ (with norm given by (4.2)) satisfying $\|\widehat{T}(t)\| \leq e^{\omega t}$, where $\omega \leq \|K\|$, for $t > 0$. Note that, by construction, B_2 generates a (C_0) contraction semigroup T on $\mathcal{H}_{en}$ and that $T(t)$ is the restriction of $\widehat{T}(t)$ to $\mathcal{H}_{en}$; equivalently $\widehat{T}(t)$ is an extension of $T(t)$ to $\mathcal{H}$.

We can make $G_W = G_A$ if we make our wave equation $u_{tt} = c^2 \Delta u$ have the modified GWBC given by

$$\Delta u + \beta \frac{\partial u}{\partial n} + \gamma u + \frac{\rho}{m} \frac{\partial u}{\partial t} = 0 \text{ on } \partial\Omega. \tag{4.5}$$

Let us interpret the GWBC

$$\Delta u + \beta \frac{\partial u}{\partial n} + \gamma u = 0 \text{ on } \partial\Omega$$

as

$$\frac{1}{c^2} u_{tt} + \beta \frac{\partial u}{\partial n} + \gamma u = 0 \text{ on } \partial\Omega$$

where $u_{tt} = c^2 \Delta u$ is assumed to hold on $\bar{\Omega}$. It seems natural to add to this a term involving $\frac{\partial u}{\partial t}|_{\partial\Omega}$; this is precisely what we did in (4.5). In this case we can identify the δ (in the ABC problem) with a multiple of $u\,|_{\partial\Omega}$ in GWBC problem incorporating (4.5).

When $m = k$ and $d = \rho$, then δ can be identified exactly with the restriction of u to the boundary.

THEOREM 2. *Suppose that $m = k$ and $d = \rho$. Then $\mathcal{H}_{en}$ is an invariant subspace of $\mathcal{H}$ for the wave equation with acoustic boundary conditions. Thus the solution at time t satisfies*

$$\varphi(t, \cdot)|_{\partial\Omega} = \delta(t, \cdot) \text{ for all } t \geq 0 \text{ if it holds at } t = 0.$$

Thus δ really is φ on the boundary in many cases in the Beale-Rosencrans theory.

5. Compactness Issues

Let A be a closed linear operator on a Banach space with nonempty resolvent set $\rho(A)$; and we let $\mathcal{R}(\lambda, A) = (\lambda I - A)^{-1}$ denote the resolvent operator of A for $\lambda \in \rho(A)$. Then A is called **resolvent compact** if $\mathcal{R}(\lambda, A)$ is compact for some $\lambda \in \rho(A)$ iff $\mathcal{R}(\lambda, A)$ is compact for all $\lambda \in \rho(A)$. The equivalence follows from the resolvent identity

$$\mathcal{R}(\lambda, A) - \mathcal{R}(\mu, A) = (\mu - \lambda)\,\mathcal{R}(\lambda, A)\,\mathcal{R}(\mu, A)$$

for all $\lambda, \mu \in \rho(A)$.

LEMMA 3. *Let $A_2 = A_1 - P$ where P is a bounded operator. Suppose $\rho(A_1) \cap \rho(A_2) \neq \emptyset$. Then A_2 is resolvent compact iff A_1 is.*

Proof. We can derive the following identity

$$\mathcal{R}(\lambda, A_2) = \mathcal{R}(\lambda, A_1) - \mathcal{R}(\lambda, A_1) \, P \mathcal{R}(\lambda, A_2) \tag{5.1}$$

for $\lambda \in \rho(A_1) \cap \rho(A_2)$. To prove this simply multiply

$$(\lambda I - A_1) = (\lambda I - A_2) - P$$

on the left by $\mathcal{R}(\lambda, A_1)$ and on the right by $\mathcal{R}(\lambda, A_2)$.

Now suppose $\mathcal{R}(\lambda, A_1)$ is compact. For $\lambda \in \rho(A_2)$ and P bounded, the right hand side of (5.1) (which equals $\mathcal{R}(\lambda, A_2)$) is compact. The equivalence in the lemma now follows by interchanging the indices 1 and 2. $\qquad\square$

Let G_{0W} [resp. G_A] be the generator of the (C_0) contraction semigroup governing the wave equation with general Wentzell [resp. acoustic] boundary conditions. Recall that $G_W = B_3$ is the generator of the semigroup $\widehat{T}$ on $\mathcal{H}$ extending $T = \{e^{tG_{0W}} : t \geq 0\}$. Then by the results of Section 4,

$$G_W - G_A = K$$

where K is a bounded operator. Moreover, K is compact when the dimension n is one and $\rho(G_W) \cap \rho(G_A) \neq \emptyset$ since each resolvent set contains a right half plane.

THEOREM 4. *G_W and G_A are both resolvent compact when $n = 1$. Neither G_W nor G_A is resolvent compact when $n \geq 2$.*

Proof. Binding, Brown and Watson [4]-[7] proved that $\Delta = \frac{d^2}{dx^2}$ with GWBC is resolvent compact when $n = 1$. The corresponding eigenvalue problem is

$$u'' = \lambda u \text{ in } \bar{\Omega} = [0, 1],$$

$$\lambda u + (-1)^{j+1} \beta_j u' + \gamma_j u = 0 \text{ at } x = j \in \{0, 1\}.$$

(Recall that $\frac{\partial}{\partial n} = (-1)^{j+1} \frac{d}{dx}$ at $x = j$ for $j = 0, 1$.) Binding, Brown and Watson made a systematic study of such Sturm-Liouville problems with eigenparameter λ in both the equation and the boundary condition. They established an orthonomal basis of eigenvectors in the space

$$\mathcal{H} = L^2(0, 1) \oplus \mathbf{C}^2$$

with norm

$$\|u\|_{\mathcal{H}}^2 = \int_0^1 |u(x)|^2 \, dx + \sum_{j=0}^{1} \frac{|u(j)|^2}{\beta_j}$$

and the real eigenvalues tend to $-\infty$. Thus A_W, the $1-$ dimensional Laplacian with GWBC, is selfadjoint and resolvent compact. The corresponding wave equation with the same GWBC is governed by a skewadjoint operator having an orthonormal basis of eigenvectors with eigenvalues $i\mu_n^{\pm}$ with $\mu_n^{\pm}$ real and $\mu_n^{\pm} \to \pm\infty$ as $n \to \infty$.

Thus G_{0W} is resolvent compact. The same conclusion (namely that the operator called G in Section 3 is resolvent compact) was reached independently by Kramar, Mugnolo and Nagel [12] who proved the compactness by a different method on X_p (and not just X_2) when the dimension n is one. In one dimension, the resolvent of G_W is a finite rank extension of the resolvent of G_{0W}; hence it is compact. By Lemma 3, G_A is compact in one dimension.

For $n = 3$, Beale and Rosencrans [1] $-$ [3] showed that G_A is not resolvent compact. In fact, let

$$\sum := \{\lambda \in \mathbf{C} : m(x)\lambda^2 + d(x)\lambda + k(x)\lambda = 0 \text{ for some } x \in \partial\Omega\}.$$

Thus G_W has eigenvalue sequences converging to $\pm i\infty$ and to $\sum$. Explicit calculations were given when m, d, k are constants and Ω is a ball, so that $\sum$ consists of two points ω_1, ω_2. Then both ω_1, ω_2 are limit points of eigenvalues of G_A and $\mathcal{R}(\lambda, G_A)$, which is meromorphic on $\mathbf{C} \setminus \sum$, has essential singularities at both ω_1 and ω_2. C. Gal [11] extended the Beale-Rosencrans results to dimension $n \geq 2$.

Independently of Gal's work [11], using a different method, Delio Mugnolo [14] recently showed that G_A is not resolvent compact in two or more dimensions. Mugnolo dealt with variable coefficients and worked in a very general context involving operator matrices.

So our conclusion is that G_A is not resolvent compact when $n \geq 2$. Neither is G_W by Lemma 3. $\qquad\qquad\Box$

We conjecture that G_{0W} is not resolvent compact in dimension two or more.

We thank Liang Jin, Xiao Ti-Jun and an anonymous referee for their helpful comments on our first draft of this paper.

REFERENCES

[1] BEALE, J. T. and ROSENCRANS, S. I., *Acoustic boundary conditions*, Bull. Amer. Math. Soc. *80* (1974), 1276–1278.

[2] BEALE, J. T., *Spectral properties of an acoustic boundary condition*, Indiana Univ. Math. J. *25* (1976), 895–917.

[3] BEALE, J. T., *Acoustic scattering from locally reacting surfaces*, Indiana Univ. Math. J. *26* (1977), 199–222.

[4] BINDING, P. A. and BROWNE, P. J., *Left definite Sturm-Liouville problems with eigenparameter dependent boundary conditions*, Differential Integral Equations *12* (1999), 167–182.

[5] BINDING, P. A. and BROWNE, P. J., *Sturm-Liouville problems with non-separated eigenvalue dependent boundary conditions*, Proc. Roy. Soc. Edinburgh Sect. A *130* (2000), 239–247.

[6] BINDING, P. A., BROWNE, P. J. and WATSON, B. A., *Inverse spectral problems for Sturm-Liouville equations with eigenparameter dependent boundary conditions*, J. London Math. Soc. *62*(2) (2000), 161–182.

[7]   BINDING, P. A., BROWNE, P. J. and WATSON, B. A., *Spectral problems for non-linear Sturm-Liouville equations with eigenparameter dependent boundary conditions*, Canad. J. Math. *52* (2000), 248–264.

[8]   FAVINI, A., GOLDSTEIN, G. R., GOLDSTEIN, J. A. and ROMANELLI, S., *The heat equation with generalized Wentzell boundary condition*, J. Evol. Equations *2* (2002) 1–19.

[9]   FAVINI, A., GOLDSTEIN, G. R., GOLDSTEIN, J. A. and ROMANELLI, S., to appear.

[10] GOLDSTEIN, J. A., *Semigroups of Linear Operators and Applications*, Oxford University Press, Oxford, New York, 1985.

[11] GAL, C. G., PhD thesis, in preparation.

[12] KRAMAR, M., MUGNOLO, D. and NAGEL, R., *Theory and applications of one-sided couplede operator matrices*, Conf. Sem. Matem. Univ. Bari, to appear.

[13] MORSE, P. M. and INGARD, K. U., *Theoretical Acoustics*, McGraw-Hill. New York, 1968.

[14] MUGNOLO, D., *Abstract wave equations with acoustic boundary conditions*, preprint.

[15] WENTZELL, A. D., On boundary conditions for multidimensional diffusion processes, Theory Prob. and its Appl. *4* (1959), 164–177.

Ciprian G. Gal, Gisèle Ruiz Goldstein and Jerome A. Goldstein
Department of Mathematical Sciences
University of Memphis
Memphis, Tennessee 38152
USA
e-mail: cgal@memphis.edu
*   ggoldste@memphis.edu*
*   jgoldste@memphis.edu*

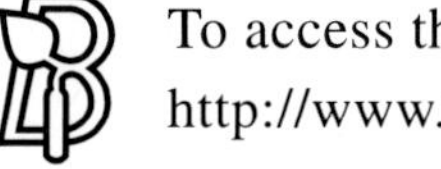

To access this journal online:
http://www.birkhauser.ch

J.evol.equ. 3 (2004) 637 – 652
1424–3199/03/040637 – 16
DOI 10.1007/s00030-003-0122-y
© Birkhäuser Verlag, Basel, 2004

Journal of Evolution
Equations

Existence and uniqueness results for large solutions of general nonlinear elliptic equations

MOSHE MARCUS AND LAURENT VÉRON

Dedicated to the memory of Philippe Bénilan

Abstract. We study under what condition there exists a solution of $-\Delta u + f(u) = 0$ in a domain Ω which blows-up on the boundary, independently of the regularity of the boundary, and we provide criteria for uniqueness. We apply our results to the case $f(u) = e^{au}$.

Introduction

Throughout this article Ω denotes a proper subdomain of $\mathbb{R}^N$, $N \geq 2$, and

$$\rho_{\partial\Omega}(x) = \mathrm{dist}\,(x, \partial\Omega), \quad \forall x \in \mathbb{R}^N.$$

If $f \in C(\mathbb{R})$, a function $u \in C^1(\Omega)$ is called a *large solution* of the equation

$$-\Delta u + f(u) = 0, \tag{0.1}$$

in Ω, if it satisfies

$$\lim_{\substack{\rho_{\partial\Omega}(x) \to 0 \\ x \in K}} u(x) = \infty, \tag{0.2}$$

for every bounded set $K \subset \Omega$. The existence of a large solution is related to the existence of a maximal solution $\bar{u}$ of (0.1) in Ω, which in turn depends on the so called Keller-Osserman condition. A function $f \in C(\mathbb{R}_+)$ satisfies the Keller-Osserman condition if there exists a positive non-decreasing function h such that

$$f(t) \geq h(t), \forall t \in \mathbb{R}_+ \quad \text{and} \quad \int_a^\infty \left(\int_0^t h(s)ds \right)^{-1/2} dt < \infty, \quad \forall a > 0. \tag{0.3}$$

It is known that if f is non-decreasing and satisfies the Keller-Osserman condition (in which case $h = f$) then a large solution exists in every bounded smooth domain. Uniqueness in smooth domains was established under some additional conditions on f (see e.g. [2]). The special case $f(u) = |u|^{q-1}u, q > 1$, is now well understood:

2000 *Mathematics Subject Classification*: 35J60.
Key words: Elliptic equations, Keller-Osserman a priori estimate, maximal solutions, super and sub solutions.

I. There always exists a maximal solution.

II. The maximal solution is a large solution if and only if $\partial\Omega$ satisfies some Wiener type criterion [15], which always holds if $1 < q < q_c = N/(N-2)$.

III. For every $q > 1$, uniqueness holds in every domain Ω with compact boundary such that $\partial\Omega$ is locally the graph of a continuous function [17]. If $1 < q < q_c$, uniqueness holds whenever $\partial\Omega = \partial\overline{\Omega}^c$ [28].

In this article we consider the problems of existence and uniqueness for Equation (0.1) under more general assumptions on f, in non-smooth domains. It will always be assumed that

$$f \in C(\mathbb{R}), \quad f(0) \geq 0 \quad \text{and} \quad f \text{ is non-decreasing.} \tag{0.4}$$

Following Brezis and Bénilan [4], if $N \geq 3$, we say that f satisfies the weak singularity condition if

$$\int_a^\infty f(s)s^{-2(N-1)/(N-2)}ds < \infty, \quad \forall a > 0. \tag{0.5}$$

When $N = 2$, this condition no longer applies, and in [24] Vazquez defines the exponential order of growth of f, $a_f^+ \in [0, \infty]$, by

$$a_f^+ = \inf\left\{a \geq 0 : \int_0^\infty f(s)e^{-as}ds < \infty\right\}. \tag{0.6}$$

We first prove

THEOREM 0.1. *Let Ω be a bounded subdomain of $\mathbb{R}^N$ and suppose that f satisfies (0.3) and (0.4). Then the maximal solution of (0.1) is a large solution if either*

(i) *$N \geq 3$ and f satisfies the weak singularity assumption,*

or

(ii) *$N = 2$ and the exponential order of growth of f is finite.*

Concerning uniqueness we have the following two theorems. The first one could appear as a technical result, but its range of applications is wide. It is settled upon the development of an idea introduced in [18] in a more regular case.

THEOREM 0.2. *Let Ω be a bounded domain in $\mathbb{R}^N$ and suppose that f satisfies (0.4) and f is convex. Let $\bar{u}$ be the maximal solution of (0.1) in Ω. If there exist two constants $K = K(\Omega, f) > 1, \delta = \delta(\Omega, f) > 0$ such that*

$$0 \leq \bar{u}(x) \leq Ku(x), \quad \forall x \in \Omega \text{ s.t. } \rho_{\partial\Omega}(x) \leq \delta, \tag{0.7}$$

for any large solution u, then Equation (0.1) admits at most one large solution in Ω.

The next result is a combination of [17, Theorem 2.2] and Theorem 0.2.

THEOREM 0.3. *Let Ω be a bounded domain in $\mathbb{R}^N$ such that $\partial\Omega$ is locally a continuous graph and suppose that f satisfies (0.4). Then there exists at most one large solution to (0.1) if one of the following conditions is satisfied:*

(i)　*f is convex and there exists a real number $L \geq 0$ such that*

$$f(a + b) \geq f(a) + f(b) - L, \quad \forall a,\, b \geq 0, \tag{0.8}$$

(ii)　*there exists a real number $L \geq 0$ such that*

$$f(ra + sb) \geq rf(a) + sf(b) - (r + s - 1)L, \quad \forall a,\, b \geq 0, \quad \forall r,\, s \geq 1. \tag{0.9}$$

Applying our results to the model case of equation

$$-\Delta u + e^{au} = 0, \tag{0.10}$$

in a N dimensional domain Ω, where $a > 0$, one obtains:

When $N = 2$, existence of a large solution holds if Ω is not everywhere dense, and the large solution is the unique one if Ω is bounded and $\partial\Omega = \partial\Omega^c$. This last condition which plays a crucial role for obtaining estimate (0.7) is almost necessary, since uniqueness does not holds if $\Omega = \Theta \setminus \{\omega\}$ for some $\omega \in \Theta$.

When $N \geq 3$, existence of a maximal solutions holds if Ω is bounded, and although this maximal solution may not be a large solution, there exists at most one large solution if $\partial\Omega$ is locally a continuous graph.

1. Existence of large solutions

The existence of large solutions to Equations (0.10) was known for a long time [3] [21]. The existence of large solutions for equations (or inequalities) with more general nonlinearities

$$-\Delta u + f(u) \leq 0, \tag{1.1}$$

was treated by Keller [14] and Osserman [20] in 1957. In these works they provided a condition on f which is necessary and sufficient in order that the set of solutions of (1.1) should be locally bounded from above.

DEFINITION 1.1. A continuous function f defined in $\mathbb{R}$ satisfies the Keller-Osserman condition if there exists a positive nondecreasing function h such that $f(r) \geq h(r)$ for all $r \in \mathbb{R}_+$ and

$$\int_a^\infty \left(\int_0^t h(s)\,ds \right)^{-1/2} dt < \infty, \quad \forall a > 0. \tag{1.2}$$

The result of Keller [14] and Osserman [20] states:

THEOREM 1.2. *Let f be a function satisfying the Keller-Osserman condition. There exists a nonincreasing nonnegative function g defined on $\mathbb{R}_+$ with the limits*

$$\lim_{\rho \to 0} g(\rho) = \infty, \tag{1.3}$$

$$\lim_{\rho \to \infty} g(\rho) = 0, \tag{1.4}$$

such that, for any domain $\Omega \subsetneq \mathbb{R}^N$ and any function $u \in C(\Omega)$ satisfying (1.1) in Ω,

$$u(x) \leq g(\rho_{\partial\Omega}(x)), \quad \forall x \in \Omega. \tag{1.5}$$

The following is a consequence of the above theorem:

THEOREM 1.3. *Let $\Omega \subset \mathbb{R}^N$ be a bounded domain, and $f \in C(\mathbb{R})$ a nondecreasing function satisfying the Keller-Osserman condition. Then there exists a maximal solution $\bar{u}$ to (0.1) in Ω. If the domain possesses a Lipschitz boundary, the maximal solution is a large solution.*

REMARK 1.4. Recently Labutin [15] showed that the Lipschitz condition on $\partial\Omega$ can be replaced by a Wiener type condition which is necessary and sufficient for the existence of a large solution.

REMARK 1.5. If f satisfies the conditions of Theorem 1.3. and if Ω is a domain in $\mathbb{R}^N$ (possibly unbounded) such that (0.1) possesses a subsolution v in Ω then a maximal solution exists. Indeed if Ω' is a smooth bounded subdomain of Ω then the large solution of (0.1) in Ω' is bounded below by v. Let $\{\Omega_n\}$ be a smooth exhaustion of Ω, i.e. Ω_n is a smooth bounded subdomain of Ω and

$$\Omega_n \subset \bar{\Omega}_n \subset \Omega_{n+1} \subset \Omega, \quad \bigcup_{n=0}^{\infty} \Omega_n = \Omega. \tag{1.6}$$

Then the corresponding sequence of large solutions $\{u_n\}$ is monotone decreasing and bounded below by v. Therefore $\bar{u} = \lim u_n$ is a solution of (0.1) in Ω and it dominates any other solution of the equation. Thus $\bar{u}$ is the maximal solution.

If f vanishes at some point r_1 then the function $v \equiv r_1$ is a solution of (0.1). Therefore in this case, a maximal solution exists in any domain Ω.

REMARK 1.6. When $N \geq 3$ and f is positive and satisfies the Keller-Osserman condition, there exists no radial solution of (0.1) in any exterior domain $\{x \in \mathbb{R}^N : |x| > R\}$. Since if ψ is such a solution, the function ϕ defined by

$$\phi(s) = s\psi(r), \quad s = r^{N-2}/(N-2);$$

satisfies

$$s^2\phi'' = C(N)s^{N/(N-2)} f(\phi/s) \quad \text{in } (S, \infty)$$

for some $C(N)$ and $S > 0$. By convexity, $\phi(s)/s$ is bounded from below by some δ, which implies

$$\phi'' \geq C(N)s^{N/(N-2)-2} f(\delta).$$

Thus

$$\liminf_{s \to \infty} s^{-N/(N-2)}\phi(s) > 0$$

by integrating this inequality twice. Therefore

$$\lim_{r \to \infty} \psi(r) = \infty.$$

However this contradicts Theorem 1.2.

REMARK 1.7. When $N = 2$, f is positive and satisfies the Keller-Osserman condition, there exists a radial solution of (0.1) in an exterior domain $\{x \in \mathbb{R}^N : |x| > R\}$ if and only if

$$\int_0^\infty f(\alpha t)e^{2t}\,dt < \infty, \tag{1.7}$$

for some $\alpha < 0$. For the necessary part, if ψ is a solution and $\phi(t) = \psi(r)$ with $t = \ln r$, then

$$-\phi'' + e^{2t} f(\phi) = 0 \quad \text{in } (\ln R, \infty) = (T, \infty).$$

Since ϕ is convex, $\phi(t) \geq \alpha t$ in $[T^*, \infty)$, for some $\alpha < 0$ and $T^* \geq T$. By the monotonicity of f,

$$\phi'(t) \geq \phi'(T^*) + \int_{T^*}^t e^{2s} f(\alpha s)\,ds \quad \text{in } [T^*, \infty).$$

Therefore (1.7) must hold, otherwise $\liminf_{t \to \infty} \phi'(t)$ would be infinite, contradicting Theorem 1.2. Conversely, we first notice that for any $m > 0$ there exists ϕ_m solution of

$$-\phi_m'' + e^{2t} f(\phi_m) = 0 \quad \text{in } (T, m), \quad \phi_m(T) = \phi_m(m) = 0. \tag{1.8}$$

Since (1.7) holds for some $\alpha < 0$, the quantity

$$t \mapsto \frac{1}{t - T} \int_T^t (t - s)e^{2s} f(\alpha(s - T))\,ds,$$

is uniformly bounded when $t \in [T, \infty)$. Let b be its supremum. If $a = \alpha - b$, there holds

$$a(t - T) + \int_T^t (t - s)e^{2s} f(\alpha(s - T))\,ds \leq \alpha(t - T), \quad \forall t \in [T, \infty). \tag{1.9}$$

If $\tilde{\phi}$ is defined by the formula

$$\tilde{\phi}(t) = a(t - T) + \int_T^t (t - s)e^{2s} f(\alpha(s - T))ds, \quad \forall t \in [T, \infty),$$

then $\tilde{\phi}(T) = 0$, $\tilde{\phi}''(t) = e^{2t} f(\alpha(t - T))$ and

$$-\tilde{\phi}''(t) + e^{2t} f(\tilde{\phi}(t)) = e^{2t}$$
$$f\left(a(t - T) + \int_T^t (t - s)e^{2s} f(\alpha(s - T))ds\right) - e^{2t} f(\alpha(t - T)). \tag{1.10}$$

Because f is nondecreasing, the right-hand side of (1.10) is nonpositive by (1.9). Therefore $r \mapsto \tilde{\psi}(r) = \tilde{\phi}(\ln r)$ is a nonpositive sub-solution of (0.1) defined in $[R, \infty)$ which lies below $r \mapsto \psi_m(r) = \phi_m(\ln r)$. By the maximum principle $m \mapsto \psi_m$ is decreasing. Thus

$$\lim_{m \to \infty} \psi_m = \psi$$

defines a nonnegative radial solution of (0.1) which vanishes for $|x| = R$.

The following result is a consequence of Remarks 1.5–1.7.

THEOREM 1.8. *Let $\Omega \subsetneq \mathbb{R}^N$ be an unbounded domain, and $f \in C(\mathbb{R})$ a nondecreasing function satisfying the Keller-Osserman condition. Assume moreover that $f(r_1) = 0$ for some real number r_1, or $N = 2$, Ω is not dense in $\mathbb{R}^2$, f is positive and (1.7) holds for some $\alpha < 0$. Then there exists a maximal solution $\bar{u}$ to (0.1) in Ω.*

The next definition was introduced by Bénilan and Brezis [4] for solving the semilinear elliptic equation

$$-\Delta u + f(u) = \mu \quad \text{in } \mathbb{R}^N. \tag{1.11}$$

with measure data $\mu \in \mathfrak{M}_+^b(\mathbb{R}^N)$ with $N > 2$.

DEFINITION 1.9. A continuous nondecreasing function f defined in $\mathbb{R}$ satisfies the *weak singularity assumption* in $\mathbb{R}^N$ $(N > 2)$ if $f > 0$ on $(0, \infty)$ and

$$\int_a^\infty f(s)s^{-2(N-1)/(N-2)}ds < \infty, \quad \forall a > 0. \tag{1.12}$$

If $f(r_1) = 0$ for some $r_1 \leq 0$, it is proved in [4] that for any bounded positive measure μ in $\mathbb{R}^N$ there exists a unique function $u \in M^{N/(N-2)}(\mathbb{R}^N)$ such that $f(u) \in L^1(\mathbb{R}^N)$ solution of (1.11).

In dimension two the class of nonlinearities for which Equation (1.11) admits solution is different from the one introduced by Brezis and Bénilan.

Following Vazquez [24] we introduce

DEFINITION 1.10. Let f be a continuous nondecreasing function defined on $\mathbb{R}$, positive on $(0, \infty)$. The *exponential order of growth* of f is defined by

$$a_f^+ = \inf \left\{ a \geq 0 : \int_0^\infty f(s) e^{-as} ds < \infty \right\}. \tag{1.13}$$

It is either a nonnegative real number or ∞.

If it is again assumed that f vanishes for some nonpositive value, it is proved in [24] that for any positive bounded measure μ in $\mathbb{R}^2$, with no atoms of mass larger than $2/a_f^+$, there exists a function $u \in L_{\mathrm{loc}}^q(\mathbb{R}^2)$ for any $q < \infty$ such that $f(u) \in L^1(\mathbb{R}^2)$ and u is a solution of (1.11).

The following related result was established by Véron [27].

PROPOSITION 1.11. *Suppose that f satisfies the conditions of Theorem I. Let $\beta \in \mathbb{R}$ and let R and k be positive numbers. Then there exists a unique radial solution $v = v_{k,\beta,R}$ of (0.1) in $B_R(0)$ such that $v(R) = \beta$ and*

$$\lim_{r \to 0} r^{N-2} v = k, \qquad \textit{if } N > 2,$$
$$\lim_{r \to 0} \left(\ln \tfrac{1}{r} \right)^{-1} v = k, \quad \textit{if } N = 2 \textit{ and } 0 \leq k < 2/a_f^+. \tag{1.14}$$

Notice that if $a_f^+ > 0$, k can be taken equal to $2/a_f^+$. As a consequence of the above we obtain,

COROLLARY 1.12. *Let $h \in L_+^\infty(\partial B_R(0))$. Then there exists a unique solution $v = v_{k,h,R}$ such that $v = h$ on $\partial B_R(0)$ and v satisfies (1.14).*

Proof. Let β_1 and β_2 denote the essential inf and the essential sup of h. Put $v_i = v_{k,\beta_i,R}$. Then, for every $s \in (0, R)$, thre exists a unique solution u_s of (0.1) in $s < |x| < R$ such that $u_s(R) = h$ and $u_s(s) = v_1(s)$. This solution satisfies $v_1 \leq u_s \leq v_2$. There exists a sequence $s_n \to 0$ such that $\{u_{s_n}\}$ converges to a solution u_0 of (0.1) in $0 < |x| < R$ such that $u_0(R) = h$ and $u_0(s)$ satisfies (1.14) at the origin. If u^* is another solution with these properties $u^*/u_1 \to 1$ at the origin. Hence, by the maximum principle, $u_0 = u^*$. $\square$

Proof of Theorem 0.1 Since Ω is bounded, there exists $R > 0$ such that for any $x_0 \in \partial\Omega$, $\partial\Omega \subset B_R(x_0)$. Let $\{\Omega_n\}$ be a smooth exhaustion of Ω. For each n, let u_n be the maximal solution of (0.1) in Ω_n. Then $\{u_n\}$ is an decreasing sequence. The following limit

$$\bar{u} = \lim_{n \to \infty} u_n,$$

exists by Theorem 1.2. and defines the maximal solution of (0.1) in Ω. For any $k \in (0, \infty)$, if $N \geq 3$, or $k \in (0, 2/a_f^+)$, if $N = 2$, let $v_{k,0,R}$ be the function defined in Proposition 1.11. Since f is nondecreasing, it follows by the comparison principle,

$$v_{k,0,R}(x - x_0) \leq u_n(x), \quad \forall x \in \Omega_n, \tag{1.15}$$

for any $x_0 \in \partial\Omega$ and $n \geq 0$. Letting $k \to \infty$, if $N \geq 3$ or $k \to 2/a_f^+$ if $N = 2$, we denote $v^* = v_{\infty,0,R}$ or $v^* = v_{2/a_f^+,0,R}$. Passing to the limit when $n \to \infty$ yields

$$v^*(x - x_0) \leq \bar{u}(x), \quad \forall x \in \Omega. \tag{1.16}$$

If $|x - x_0| = \rho_{\partial\Omega}(x)$, (1.16) becomes

$$v^*(\rho_{\partial\Omega}(x)) \leq \bar{u}(x), \quad \forall x \in \Omega. \tag{1.17}$$

Thus $\bar{u}$ is a large solution.

When Ω is not bounded the conclusion of Theorem 0.1 still holds under additional requirements on f.

THEOREM 1.13. *Let $\Omega \subsetneq \mathbb{R}^N$ be an unbounded domain and suppose that f satisfies (0.3) and (0.4). Then there exists a large solution if either*

(i) *$N \geq 3$, f satisfies the weak singularity assumption and $f(r_1) = 0$ for some $r_1 \leq 0$, or*

(ii) *$N = 2$, the exponential order of growth of f is finite and either $f(r_1)) = 0$ for some $r_1 \leq 0$, or f is positive and (1.7) holds for some $\alpha < 0$.*

Proof. By Theorem 1.8 there exists a maximal solution $\bar{u}$ of (0.1). In what follows k is a positive number which will be kept fixed, $0 < k < 2/a_f^+$ if $N = 2$. In the first case where there exists $r_1 \leq 0$ such that $f(r_1) = 0$, the constant function $x \mapsto r_1$ is a solution of (0.1) in $\mathbb{R}^N$. For every bounded smooth subdomain $G \subset \bar{G} \subset \Omega$, the maximal solution u_G of (0.1) in G satisfies

$$u_G(x) \geq \bar{u}(x) \geq r_1, \quad \forall x \in G. \tag{1.18}$$

In the second case where $N = 2$, f is positive, (1.7) holds and Ω is not dense, there exists $a \in \bar{\Omega}^c$ and $\rho_0 > 0$ such that

$$\bar{\Omega} \subset \{x \in \mathbb{R}^2 : |x - a| \geq \rho_0\}.$$

By Remark 1.7 and (1.7), for any $t \in (0, \rho_0)$, there exists a solution of $u_{t,a}^*$ of (0.1) in $\{x \in \mathbb{R}^2 : |x - a| > t\}$ which is radially symmetric with respect to a. Therefore (1.18) holds under the form

$$u_G(x) \geq \bar{u}(x) \geq u_{t,a}^*(x), \quad \forall t \in (0, \rho_0), \quad \forall x \in G. \tag{1.19}$$

We fix $R > 0$, $R = \rho_0/3$ in the second case. For a given $y \in \partial\Omega$ let $\{\Omega_n^y\}$ be an exhaustion of Ω such that $\Omega_n^y \cap B_R(y)$ is a finite union of smooth subdomains of $\Omega \cap B_R(y)$. We define the function $h = h_y \in C(\partial B_R(y))$ by $h_y \equiv r_1$ in the first case and $h_y = u_{R,a}^*|_{\partial B_R(y)}$ in the second one. Consequently $u_n \geq h_y$ on $\Omega_n \cap \partial B_R(y)$. Since u_n blows up on $\partial \Omega_n$ it follows that

$$u_n(x) \geq v_{k,h,R}(x - y), \quad \forall x \in \Omega_n \cap B_R(y).$$

This implies

$$\bar{u}(x) \geq v_{k,h,R}(x - y), \quad \forall x \in \Omega \cap B_R(y), \quad \forall y \in \partial \Omega, \tag{1.20}$$

which in turn implies that $\bar{u}$ is a large solution. $\square$

REMARK 1.14. In the case where f vanishes for some nonpositive value, the blow-up of the large solution is uniform on $\partial \Omega$. In the other case and if $\partial \Omega$ is not compact, this blow-up is not uniform since the functions h_y are not bounded from below.

2. Uniqueness

In order to prove Theorem 0.2, we recall a general existence result in presence of super and sub solutions (see e.g. [19], [22]).

PROPOSITION 2.1. *Let $\Omega \subset \mathbb{R}^N$ be any domain, $\mathcal{L}$ a second order elliptic operator with smooth coefficients and $h^*, h^+ h^*, h^\dagger \in C(\Omega \times \mathbb{R})$. We assume that $h = h^* + h^\dagger$, where $r \mapsto h^*(x, r)$ is nondecreasing for every $x \in \Omega$, and $(x, r) \mapsto h^\dagger(x, r)$ is locally Lipschitz continuous with respect to the r variable, uniformly when the x variable stays in a compact subset of Ω. Let u_* and u^* be two $C(\Omega) \cap W_{loc}^{1,2}(\Omega)$ functions satisfying*

(i) $\mathcal{L}u_* - h(x, u_*) \geq 0$ *in* Ω
(ii) $\mathcal{L}u^* - h(x, u^*) \leq 0$ *in* Ω $\qquad\qquad\qquad\qquad\qquad$ (2.1)
(iii) $u_* \leq u^*$ *in* Ω,

where the equations are understood in weak sense. Then there exists a $C^1(\Omega)$ function u which satisfies

(i) $\mathcal{L}u - h(x, u) = 0$ *in* Ω,
(ii) $\quad u_* \leq u \leq u^*$ *in* Ω. $\qquad\qquad\qquad\qquad\qquad\qquad$ (2.2)

Proof of Theorem 0.2. We denote $\Omega_\delta = \{x \in \Omega : \rho_{\partial \Omega}(x) \leq \delta\}$, and $\tilde{\Omega}$ be any smooth and bounded domain containing $\bar{\Omega}$. First we notice that any large solution u is bounded from below by the solution $v = \tilde{v}$ of the problem

$$\begin{aligned} -\Delta \tilde{v} + f(\tilde{v}) &= 0 \text{ in } \tilde{\Omega}, \\ \tilde{v} &= 0 \text{ on } \partial \tilde{\Omega}. \end{aligned} \tag{2.3}$$

We assume that $\bar{u} \neq u$ and put

$$w = u - \frac{1}{2K}(\bar{u} - u).$$

Then

$$-\Delta w + f(w) = f\left(\left(1 + \frac{1}{2K}\right)u - \frac{\bar{u}}{2K}\right) - \left(1 + \frac{1}{2K}\right)f(u) + \frac{1}{2K}f(\bar{u})$$

Since

$$\frac{2K}{1 + 2K}\left[\left(1 + \frac{1}{2K}\right)u - \frac{\bar{u}}{2K}\right] + \frac{1}{1 + 2K}\bar{u} = u,$$

by the convexity assumption,

$$\frac{2K}{1 + 2K}f\left(\left(1 + \frac{1}{2K}\right)u - \frac{\bar{u}}{2K}\right) + \frac{1}{1 + 2K}f(\bar{u}) \geq f(u).$$

Therefore w is a supersolution in Ω, larger than $(K + 1)u/2K$ in Ω_δ. For any $\theta \in [0, 1]$ the function $w_\theta = \theta u + (1 - \theta)\tilde{v}$, there holds

$$-\Delta w_\theta + f(w_\theta) = f(\theta u + (1 - \theta)\tilde{v}) - \theta f(u) - (1 - \theta)f(\tilde{v}) \leq 0,$$

again by convexity. If we fix $0 < \theta < (K + 1)/2K$, the function $x \mapsto (w_\theta - w)_+$ has compact support in Ω. Because f is nondecreasing, the inequality $w_\theta < w$ in Ω holds by the comparison principle. Then by Proposition 2.1 there exists a solution u_1 of (0.1) such that $w_\theta \leq u_1 \leq w$ in Ω. Therefore u_1 is a large solution; thus it satisfies (0.7). Moreover

$$u_1 \leq w = u - \frac{1}{2K}(\bar{u} - u) \Longrightarrow \bar{u} - u_1 \geq \left(1 + \frac{1}{2K}\right)(\bar{u} - u) \quad \text{in } \Omega. \tag{2.4}$$

Replacing u by u_1, which is a large solution, we introduce

$$w_1 = u_1 - \frac{1}{2K}(\bar{u} - u_1) \quad \text{and} \quad w_{1,\theta} = \theta u_1 + (1 - \theta)\tilde{v}.$$

The functions w_1 and $w_{1,\theta}$ are respectively supersolutions and subsolutions of (0.1), and $w_{1,\theta} \leq w_1$ provided $0 < \theta < (K + 1)/2K$. Therefore there exists a solution u_2 of (0.1) such that $w_{1,\theta} \leq u_2 \leq w_1$ in Ω. Furthermore

$$u_2 \leq u_1 - \frac{1}{2K}(\bar{u} - u_1) \Longrightarrow \bar{u} - u_2 \geq \left(1 + \frac{1}{2K}\right)(\bar{u} - u_1) \quad \text{in } \Omega. \tag{2.5}$$

Iterating this process, we construct a sequence of large solutions $\{u_n\}$ of (0.1) with $u_0 = u$ satisfying (0.7),

$$u_n \leq u_{n-1} - \frac{1}{2K}(\bar{u} - u_{n-1}) \Longrightarrow \bar{u} - u_n \geq \left(1 + \frac{1}{2K}\right)(\bar{u} - u_{n-1}) \quad \text{in } \Omega. \tag{2.6}$$

Therefore

$$\bar{u} - u_n \geq \left(1 + \frac{1}{2K}\right)^n (\bar{u} - u) \text{ in } \Omega. \tag{2.7}$$

Since $u_n \geq \tilde{v}$, we derive a contradiction by letting $n \to \infty$.

Proof of Theorem 0.3. We recall some of the notations used in [17, Theorem 2.2]. Since $\partial\Omega$ is locally the graph of a continuous function, for every boundary point P there exists an open neighborhood Q_P, a set of coordinates $\xi = (\xi_1, \ldots, \xi_N)$ obtained from x by rotation, and a function $F_P \in C(\mathbb{R}^{N-1})$ such that

$$Q_P \cap \Omega = Q_P \cap G(F_P) \text{ with } G(F_P) = \{\xi : \xi_N < F_P(\xi_1, \ldots, \xi_{N-1})\}. \tag{2.8}$$

We can assume that Q_P is a bounded cylindrical domain centered at P, with axis parallel to the ξ_N-axis:

$$Q_P = \{\eta = (\eta', \eta_N), \, |\eta'| < R_P, |\eta_N| < T_P\}, \tag{2.9}$$

where $\eta = \xi - P$ and $\eta' = (\eta_1, \ldots, \eta_{N-1})$. We can also suppose that $\partial\Omega$ is bounded away from the top and the bottom of the cylinder Q_P and that $\partial\Omega \cap \bar{Q}_P = \overline{\partial\Omega \cap Q_P}$. We denote

$$\Theta = \Omega \cap Q_P, \quad \Gamma_1 = Q_P \cap \partial\Omega, \quad \Gamma_2 = \partial Q_P \cap \bar{\Omega}. \tag{2.10}$$

We recall that any large solution of (0.1) is bounded from below by $\tilde{\ell}$ which is the minimum on $\bar{\Omega}$ of the solution $\tilde{v}$ of (2.3) in $\tilde{\Omega}$, and we can also assume that $\bar{Q}_P \subset \tilde{\Omega}$ for any P.

STEP 1. Assume there exist a large solution u. We claim that there exists a positive function $v \in C^2(\Theta)$ solution of (0.1) in Θ, with the property

(i) $v(x) \to \infty$ locally uniformly as $x \to \Gamma_1$,

(ii) $v(x) \to 0$ locally uniformly as $x \to \Gamma_2$, $\tag{2.11}$

where "$v(x) \to 0$ (or ∞) locally uniformly as $x \to \Gamma_i$" means that, for every compact subset K contained in the relative interior of Γ_i, there holds $v(x) \to 0$ (or ∞) as dist $(x, K) \to 0$. Set $\tilde{F}_P(\eta') = F_P(\eta' + P')$ and let $\{f_n\}$ be an increasing sequence of smooth positive functions defined in the closed $(N-1)$-ball $D_P = \{\eta' = (\eta_1, \ldots, \eta_{N-1}) : |\eta'| \leq R_P\}$, which converges uniformly in D_P to $\tilde{F}_P$. We set $\Theta_n = \{\eta : |\eta'| < R_P, -T_P < \eta_N < f_n(\eta')\}$, $\Gamma_{2,n} = \partial Q_P \cap \partial\Theta_n$ and $\Gamma_{1,n} = \partial\Theta_n \setminus \partial Q_P$. Let $v_{n,k} \in C(\Theta_n)$ be a positive solution of (0.1) in Θ_n such that

$$v_{n,k}(\eta) = k \text{ for } \eta \in \Gamma_{1,n},$$
$$v_{n,k}(\eta', \eta_N) = 0 \text{ for } \eta \in \Gamma_{2,n} \text{ s.t. } -T_P \leq \eta_N \leq f_n(\eta') - 1/n,$$
$$v_{n,k}(\eta', \eta_N) = n(\eta_N - f_n(\eta') + 1/n)k \text{ for } |\eta'|$$
$$= R_P, \, f_n(\eta') - 1/n \leq \eta_N \leq f_n(\eta'). \tag{2.12}$$

The sequence $\{v_{n,k}\}$ is increasing with respect to k, for n fixed. It converges, as $k \uparrow \infty$, to some function $v_n \in C(\Theta_n)$, solution of (0.1) in Θ_n. Moreover

   (i) $v_n(x) \to \infty$ locally uniformly as $x \to \Gamma_{1,n}$,

   (ii) $v_n(x) \to 0$ locally uniformly as $x \to \Gamma^*_{2,n}$,

where $\Gamma^*_{2,n} = \{\eta \in \Gamma_{2,n} : \eta_N \leq f_n(\eta') - 1/n\}$. The sequence $\{v_n\}$ is decreasing and converges locally uniformly in Θ to some continuous function v which satisfies (0.1) in Θ. Therefore v satisfies (2.11)–(ii). Moreover, v is bounded from below by ℓ^* which is the minimum of the value of the function v^* solution of

$$-\Delta v^* + f(v^*) = 0 \text{ in } Q_P,$$
$$\tilde{v}^* = 0 \text{ on } \partial Q_P. \tag{2.13}$$

In order to prove that v satisfies (2.11)–(i), let w be a large solution of (0.1) in Q_P (which exists since ∂Q_P is Lipschitz continuous, and is bounded from below by $\tilde{\ell}$ since $\bar{Q}_P \subset \tilde{\Omega}$). We put

$$\ell = \min\{\ell, \tilde{\ell}, 0\}, \quad v_\ell = v - \ell, \quad v_{n,\ell} = v_n - \ell, \quad w_\ell = w - \ell.$$

Then $v_{n,\ell}$ and w_ℓ are nonnegative and they satisfy

$$-\Delta(v_{n,\ell} + w_\ell) + f(v_{n,\ell} + w_\ell)$$
$$= f(v_{n,\ell} + w_\ell) - f(v_{n,\ell} + \ell) - f(w_\ell + \ell) \geq -L, \tag{2.14}$$

in Θ_n. Let $\tilde{\eta}$ be the solution of

$$-\Delta\tilde{\eta} = L \text{ in } \tilde{\Omega},$$
$$\tilde{\eta} = 0 \text{ on } \partial\tilde{\Omega}. \tag{2.15}$$

Since $\tilde{\eta}$ is nonnegative, the function $W_n = v_{n,\ell} + w_\ell + \tilde{\eta}$ satisfies

$$-\Delta W_n + f(W_n) \geq 0, \quad \text{in } \Theta_n. \tag{2.16}$$

and dominates u on $\partial\Theta_n$. Therefore $W_n \geq u$ in Θ_n. Letting $n \to \infty$ yields to $v_\ell + w_\ell \geq u$ in Θ. This implies (2.11)–(i).

STEP 2. Let $\bar{u}$ be the maximal solution and u any large solution of (0.1) in Ω, since $\partial\Omega$ is compact, there holds

$$\lim_{\rho_{\partial\Omega}(x) \to 0} \frac{\bar{u}}{u} = 1. \tag{2.17}$$

If assumption (i) holds, the function $2u$ satisfies

$$-\Delta(2u) + f(2u) \geq -L,$$

thus

$$-\Delta(2u + \tilde{\eta}) + f(2u + \eta) \geq 0.$$

Applying (2.17) and the comparison principle (since $(\bar{u} - 2u - \tilde{\eta})_+$ has compact support in Ω), infers

$$\bar{u} \leq 2u + \tilde{\eta} \quad \text{in } \Omega.$$

Therefore estimate (0.7) holds, and the proof follows from Theorem 0.2.

If assumption (ii) holds, for any $\varepsilon > 0$, the function $u_\varepsilon = (1 + \varepsilon)u$ satisfies

$$-\Delta u_\varepsilon + f(u_\varepsilon) \geq -\varepsilon L,$$

which implies

$$-\Delta(u_\varepsilon + \varepsilon\tilde{\eta}) + f(u_\varepsilon + \varepsilon\tilde{\eta}) \geq 0$$

Applying again (2.17) yields to

$$u_\varepsilon + \varepsilon\tilde{\eta} \geq \bar{u}.$$

Letting $\varepsilon \to 0$ infers

$$u \geq \bar{u},$$

and uniqueness follows.

3. Applications and open questions

In this section we apply our previous results to equation

$$-\Delta u + e^{au} = 0, \tag{3.1}$$

in a domain $\Omega \subset \mathbb{R}^N$, where $a > 0$ is arbitrary.

THEOREM 3.1. *Let Ω be a bounded domain in $\mathbb{R}^N$ ($N \geq 3$) such that $\partial\Omega$ is locally a continuous graph. Then there exists at most one large solution to (3.1).*

Proof. It is a direct application of Theorem 0.3 with $L = 1$ in (0.8). $\qquad\square$

Open question. The conditions under which the maximal solution is a large solution is not known. Is is likely that the condition should involve an extended notion of capacities associated to the exponential function. Exponential capacities have been introduced in [12] for studying boundary singularities of solutions of (3.1).

THEOREM 3.2. *Let Ω be a bounded domain in $\mathbb{R}^2$ such that $\partial\Omega = \partial\overline{\Omega}^c$. Then there exists one and only one large solution to* (3.1).

Proof. Let $R > 0$ such that for any $y \in \partial\Omega$, $\bar{\Omega} \subset B_R(y)$. From (0.6), the exponential order of growth of $r \mapsto e^{ar}$ is precisely a. Then by Theorem 0.2–(ii) the maximal solution $\bar{u}$ to (3.1) in Ω is a large solution. Let $v = v_{2/a,0,R}$ be the (radial) solution of

$$-\Delta v + e^{av} = 0 \quad \text{in } B_R(0),$$
$$\lim_{|x|\to 0} v(x)/\ln(1/|x|) = 2/a,$$
$$v(x) = 0 \ \text{ if } \ |x| = R. \tag{3.2}$$

Let $x \in \Omega$, $x_0 \in \partial\Omega$ such that $\rho_{\partial\Omega}(x) = |x - x_0|$ and $\{a_n\} \subset \overline{\Omega}^c$ such that $\lim_{n\to\infty} a_n = x_0$.

$$\liminf_{\rho_{\partial\Omega}(x)\to 0} u(x)/\ln(1/\rho_{\partial\Omega}(x)) = 2/a. \tag{3.3}$$

If u is any large solution, it is bounded from below by $y \mapsto v_{2/a,0,R}(y - a_n)$. Therefore

$$u(x) \geq v_{2/a,0,R}(x - x_0) = v_{2/a,0,R}(\rho_{\partial\Omega}(x)) \geq \frac{2}{a}\ln(1/\rho_{\partial\Omega}(x)) - D, \tag{3.4}$$

for some constant D. $\qquad\qquad\qquad\qquad\qquad\qquad\qquad\qquad\qquad\qquad\qquad\qquad\qquad\square$

The next estimate from above is a consequence of the proof of Keller's theorem, but for the sake of completeness we briefly recall an alternative direct proof. If $x \in \Omega$ let $0 < R < \rho_{\partial\Omega}(x)$. Set

$$\psi(y) = \lambda \ln(1/(R^2 - |y - x|^2) + \mu$$

where $\lambda > 0$ and μ are real numbers to be determined such that

$$-\Delta\psi + e^{a\psi} \geq 0 \quad \text{in } B_R(x). \tag{3.5}$$

Put $r = |x - y|$, then

$$-\Delta\psi + e^{a\psi} = -\frac{4\lambda R^2}{(R^2 - r^2)^2} + \frac{e^{a\mu}}{(R^2 - r^2)^{a\lambda}}.$$

If we choose $\lambda = 2/a$, and $\mu = a^{-1}\ln(8R^2/a)$, (3.5) holds. By the comparison principle, any solution u of (3.1) is bounded from above by ψ in $B_R(x)$, and in particular $u(x) \leq \psi(x)$. Letting $R \to \rho_{\partial\Omega}(x)$ the maximal solution $\bar{u}$ satisfies

$$\bar{u}(x) \leq \frac{2}{a}\ln(1/\rho_{\partial\Omega}(x)) + a^{-1}\ln(8/a). \tag{3.6}$$

Putting together (3.4) and (3.6) infers that (0.7) holds, and uniqueness follows by Theorem 0.2.

Open question. It has been recently noticed by Fabbri [8] that Theorem 0.2 applies for proving uniqueness of large solutions of

$$-\Delta u + u \left(\ln_+ u\right)^\alpha = 0, \tag{3.7}$$

for $\alpha > 2$, in a bounded domain $\Omega \subset \mathbb{R}^N$, $N \geq 2$, the boundary of which satisfies $\partial\Omega = \partial\bar{\Omega}^c$. The proof is based upon the asymptotic expansion of solutions with isolated singularities, see [23], and the precise computation of the Keller-Osserman a priori estimate. In the range $0 < \alpha \leq 2$ large solutions do not exist. A natural question is to find conditions on a real function f, besides the convexity, the monotonicity, the weak singularity condition and the Keller-Osserman condition for which inequality (0.7) in Theorem 0.2 holds. This problem reduces to a radial one, in proving that there exists some $K > 0$ such that the function g in the Keller-Osserman estimate (see Theorem 1.2) satisfies

$$g(r) \leq Ku_\infty(r), \quad \text{on } (0, \delta], \tag{3.8}$$

where u_∞ is the limit, when $k \to \infty$, of the u_k $(k > 0)$, where

$$u_k'' + \frac{N-1}{r}u_k' = f(u_k), \quad \text{on } (0, R_0),$$

$$\lim_{r \to 0} r^{N-2}u_k(r) = k, \quad u_k(R_0) = C,$$

for $R_0 > \text{diam}(\Omega)$, and some C. It is important to notice that $g(r) = \bar{u}_r(0)$, where $\bar{u}_r$ is a radial function, solution of (0.1) in the ball $B_r(0)$ and tending to infinity on $\partial B_r(0)$.

REFERENCES

[1] BANDLE, C. and MARCUS, M., *Large solutions of semilinear elliptic equations*: existence, uniqueness and asymptotic behavior, J. Anal. Math. *58* (1992), 9–24.

[2] BANDLE, C. and MARCUS, M., *Asymptotic behavior of solutions and their derivative for semilinear elliptic problems with blow-up on the boundary*, Ann. I.H.P., Analyse Non Linéaire *12* (1995), 155–171.

[3] BIEBERBACH, L., $\Delta u = e^u$ *und die automorphen funktionen*, Math. Annalen *77* (1916), 173–212.

[4] BREZIS, H., *Some variational problems of the Thomas-Fermi type, in variational Inequalities*, eds. R. W. Cottle, F. Giannessi and J. L. Lions, Wiley, Chichester 1980, 53–73.

[5] BREZIS, H. and VÉRON, L., *Removable singularities for some nonlinear elliptic equations*, Arch. Rat. Mech. Anal. *75* (1980), 1–6.

[6] CHASSEIGNE, E. and VAZQUEZ, J. L., *Theory of extended solutions for fast diffusion equations. Radiation from singularities*, Arch. Rat. Mech. Anal. *164* (2002), 133–187.

[7] DHERSIN, J. L. and LE GALL, J. F., *Wiener's test for super-Brownian motion and the Brownian snake*, Probab. Theory Rel Fields *108* (1997), 103–129.

[8] FABBRI, J. Personal communication 2002.

[9] FABBRI, J. and LICOIS, J. R., *Boundary behaviour of solutions of some weakly superlinear elliptic equations*, Advanced Nonlinear Studies *2* (2002), 147–176.

[10] FRANK, P. and VON MISES, R, *Die Differential und Integralgleichungen der Mechanik und Physik*, *I*, Second Edit., Rosenberg, New-York 1943.

[11] GILBARG, D. and TRUDINGER, N. S., *Partial Differential Equations of Second Order*, 2nd Ed. Springer-Verlag, Berlin/New-York 1983.

[12] GRILLOT, M. and VÉRON, L., *Boundary trace of solutions of the Prescribed Gaussian curvature equation*, Proc. Roy. Soc. Edinburgh *130 A* (2000), 1–34.

[13] ISCOE, I., *On the support of measure-valued critical branching Brownian motion*, Ann. Prob. *16* (1988), 200–221.

[14] KELLER, J. B., *On solutions of* $\Delta u = f(u)$, Comm. Pure Appl. Math. *10* (1957), 503–510.

[15] LABUTIN, D., *Wiener regularity for large solutions of Nonlinear equations*, preprint.

[16] LOEWNER, C. and NIRENBERG, L., *Partial differential equations invariant under conformal or projective transformations*, in *Contibutions to Analysis*, L. Ahlfors and al., eds. 1972, 245–272.

[17] MARCUS, M. and VÉRON, L., *Uniqueness and asymptotic behaviour of solutions with boundary blow-up for a class of nonlinear elliptic equations*, Ann. Inst. H. Poincaré *14* (1997), 237–274.

[18] MARCUS, M. and VÉRON, L., *The boundary trace of positive solutions of semilinear elliptlic equations: the subcritical case*, Arch. Rat. Mech. Anal. *144* (1998), 201–231.

[19] NI, W. M., *On the elliptic equation* $\Delta u + K(x)u^{(n+2)/(n-2)}$, Indiana Univ. Math. J *31* (1982), 493–539.

[20] OSSERMAN, R., *On the inequality* $\Delta u \geq f(u)$, Pacific J. Math. *7* (1957), 1641–1647.

[21] RADEMACHER, H., *Einige besondere Probleme partieller Differentialgleichungen*, see [10, pp. 838–845].

[22] RATTO, A., RIGOLI, M. and VÉRON, L., *Scalar curvature and conformal deformation of hyperbolic space*, J. Funct. Anal. *121* (1994), 543–572.

[23] RICHARD, Y. and VÉRON, L., *Isotropic singularities of solutions of some nonlinear elliptic inequalities*, Ann. Inst. H. Poincaré *6* (1989), 37–72.

[24] VAZQUEZ, J. L., *On a semilinear equation in* $\mathbb{R}^2$ *involving bounded measures*, Proc. Roy. Soc. Edinburgh *95A* (1983), 181–202.

[25] VÉRON, L., *Singularities of Solutions of Second Order Quasilinear Equations*, Pitman Research Notes in Math. *353* (1996), Addison Wesley Longman Inc.

[26] VÉRON, L., *Semilinear elliptic equations with uniform blow-up on the boundary*, J. Analyse Math. *59* (1992), 231–250.

[27] VÉRON, L., *Weak and strong singularities of nonlinear elliptic equations*, Proc. Symp. Pure Math. *45–2* (1986), 477–495.

[28] VÉRON, L., *Generalized boundary value problems for nonlinear elliptic equations*, Electr. J. Diff. Equ. Conf. *6* (2000), 313–342.

Moshe Marcus
Department of Mathematics
Israel Institute of Technology
Technion
Haifa 32000
Israel

Laurent Véron
Laboratoire de Mathématiques et Physique Théorique
CNRS UMR 6083
Université François Rabelais
Tours 37200
France

J.evol.equ. 3 (2004) 653 – 672
1424–3199/03/040653 – 20
DOI 10.1007/s00028-003-0146-3
© Birkhäuser Verlag, Basel, 2004

Journal of Evolution Equations

Another way to say caloric

MICHAEL G. CRANDALL and PEI-YONG WANG*

With warm memories of Philippe Bénilan

Abstract. This paper offers characterizations of subsolutions of the heat equation $u_t - \Delta u = 0$ (the subcaloric functions) and the infinity heat equation $u_t - \Delta_\infty u = 0$ (the infinity-subcaloric functions) by means of comparison properties with explicit families of solutions of the corresponding equations. The primary ingredients of functions in these families are translates of solutions which depend radially on the space variables. Results of independent interest include the presentation and study of the class of infinity-caloric functions employed in the characterization.

It was shown in Crandall and Zhang [5] that an upper-semicontinuous function u is subharmonic if and only if it enjoys a natural comparison property with certain explicit combinations of functions built from translates of the fundamental solution of the Laplacian. The combinations were sums of no more than n summands where n is the number of independent variables. This result of [5] was motivated by an earlier result of Crandall, Evans and Gariepy [3] characterizing the (viscosity) subsolutions of the "infinity-Laplace" equation

$$\Delta_\infty u = \sum_{i,j=1}^{n} u_{x_i} u_{x_j} u_{x_i, x_j} = 0$$

by means of comparison properties with "cone functions", that is functions of the form $c(x) = a|x - y|$ where $a \in \mathbb{R}$. The name "infinity-Laplace" equation arises in a natural way from noting that it is the result of letting $p \to \infty$ in the p-Laplace equation

$$\Delta_p u = \text{divergence}(|Du|^{p-2} Du) = 0,$$

as was first shown in Bhattacharya, DiBenedetto, and Manfredi [2]. Here "D" stands for the gradient. An attempt in [5] to generalize the results for $p = 2$ and $p = \infty$ to general p failed, but it did lead to the discovery of a surprising class of p-superharmonic functions.

Here we show that the positive results of [5] extend to the heat equation $u_t - \Delta u = 0$ and the infinity-heat equation $u_t - \Delta_\infty u = 0$. New "infinity- caloric" functions are introduced

2000 *Mathematics Subject Classification*: Primary 35K10, 35K65, 35K55, 35B50.
Key words: Maximum principle, heat equation, nonlinear parabolic.
* Partially supported by NSF DMS-0196526

to treat the latter case. As the characterization of infinity-subharmonic functions leads to most of the (currently known) core facts concerning infinity-harmonic functions and generalizes quite a lot, as shown in Aronsson, Crandall and Juutinen [1], one might hope that the results herein might prove similarly useful. However, this does not seem to be the case. In particular, our characterization of infinity-subcaloric functions does not reduce to comparion with cones when applied to time independent functions; rather it is then a less subtle characterization. This is interesting in itself, hinting that either there are other characterizations or that the infinity heat equation is not the "right" parabolic version of the infinity-Laplace equation. In any case, the infinity-heat equation stands as an interesting example of a degenerate nonlinear parabolic equation, and understanding it is useful. It is also employed in approximating infinity-harmonic functions, which are themselves used in image processing.

This paper is organized into three sections. Section 1 briefly treats preliminaries. Section 2 presents the generalizations of the results of [5] to the case of subcaloric functions. Section 3 contains the study of a class of solutions to the infinity heat equation and the attendant characterization of infinity subcaloric functions. Our bibliography is very short, and we rely on [1] and [5] for a more complete orientation and references to the stationary cases.

1. Preliminaries and Notation

We refer to [5] for orientation. We use without comment simple variants of things explained in detail in the preliminaries of [5]. The basic theory of viscosity solutions is also assumed, and all uses of the terms solutions, subsolutions and supersolutions herein are to be understood in the viscosity sense unless otherwise said. However, the reader should be able to follow most of the developments in any case, as the notions used are implicitly defined in the course of discussion.

We turn to notation. U is always an open subset of $\mathbb{R}^n$ and $\mathcal{U}$ is always an open subset of $\mathbb{R}^n \times \mathbb{R}$. $\mathcal{I}$ is always a subinterval, not necessarily open, of $\mathbb{R}$. We typically write things like $x \in U$ and $(x, t) \in \mathcal{U}$ with the obvious meaning. If $K \subset \mathbb{R}^n$, then a set of the form $K \times \mathcal{I}$ is called a "parabolic cylinder".

For any $A, B \subset \mathbb{R}^m$, ∂A is the boundary of A and $\overline{A}$ is the closure of A. Moreover, $A \subset\subset B$ means that $\overline{A}$ is a compact subset of B. If $K \subset \mathbb{R}^n$ and the boundary points of $\mathcal{I}$ are $a, b, a < b$, then parabolic boundary of the parabolic cylinder $K \times \mathcal{I}$ is

$$\partial_p(K \times \mathcal{I}) = \partial K \times [a, b] \cup \overline{K} \times \{a\}. \tag{1.1}$$

If $x, y \in \mathbb{R}^n$, then

$$\langle x, y \rangle = \sum_{j=1}^{n} x_j y_j$$

is the Euclidean inner-product of x and y and

$$|x| = \langle x, x \rangle^{\frac{1}{2}}$$

is the Euclidean length of x. We also employ:

$B_r(y)$ is the open ball of center y and radius r,
dom(u) is the domain of the function u,
USC$(\mathcal{U}) = \{$upper-semicontinuous $u : \mathcal{U} \to \mathbb{R}\}$,
$Dv = (v_{x_1}, \ldots, v_{x_n})$ and $D^2 v = (v_{x_i x_j})$.

In particular, "D", "D^2" denote the spatial gradient and spatial Hessian "operators", which are understood in various senses depending on the context in which they appear. E.g., "Du" is unambiguous and connotes a function if u is smooth, but in the relation $u_t + F(Du, D^2 u) \leq 0$ discussed below, the notation indicates that u is a viscosity subsolution of the corresponding equation and "Du" itself is not meaningful. Natural conventions are used with "dom"; e.g,

$$\text{dom}(u - v) = \text{dom}(u) \cap \text{dom}(v).$$

We will use the following definition.

DEFINITION 1. A function $u \in \text{USC}(U)$ is said to verify the parabolic maximum principle in $\mathcal{U}$ if for any parabolic cylinder

$$U \times [s, t] \subset\subset \mathcal{U}$$

one has

$$\max_{\overline{U} \times [s,t]} u = \max_{\partial_p (U \times [s,t])} u. \tag{1.2}$$

We use $u \in \text{MPar}(\mathcal{U})$ to indicate that u satisfies the parabolic maximum principle in $\mathcal{U}$.

REMARK 2. Suppose that u satisfies

$$u_t + F(Du, D^2 u) \leq 0 \text{ in } \mathcal{U}, \tag{1.3}$$

where F is degenerate elliptic in the sense of [4]. We recall that (1.3) and the statement that u is a (viscosity) subsolution of $v_t + F(Dv, D^2 v) = 0$ in $\mathcal{U}$ are equivalent and both require that $u \in \text{USC}(\mathcal{U})$. Assume moreover that $F(0, 0) = 0$. Examples include $F(p, X) = -\text{trace}(X)$, in which case (1.3) is $u_t - \Delta u \leq 0$, and $F(p, X) = -\langle Xp, p \rangle$, in which case (1.3) is $u_t - \Delta_\infty u \leq 0$. Then it follows from Theorem 8.2 of [4] that $u \in \text{MPar}(\mathcal{U})$. To see this, let $U \times [s, t] \subset\subset \mathcal{U}$ and $c = \max_{(\partial_p U \times [s,t])} u$ so that $u - c \leq 0$ on $\partial_p U \times [s, t]$. Then apply Theorem 8.2 to $u - c$ and $v \equiv 0$ (note that $v_t + F(Dv, D^2 v) \equiv 0$). Moreover, using the proof of Theorem 8.2, it also follows that if v is a solution of $v_t + F(Dv, D^2 v) \geq 0$ in $\mathcal{U}$, then $u - v \in \text{MPar}(\mathcal{U})$. Note that, according to the notions, u is to be upper-semicontinuous and v is to be lower-semicontinuous.

2. The Heat Equation

Theorem 3 below characterizes solutions of

$$Hu = u_t - \Delta u \leq 0,$$

that is, subsolutions of the heat equation or, equivalently, subcaloric functions, via comparison properties with linear combinations of translates of the heat kernel

$$\Phi(x, t) = t^{-\frac{n}{2}} e^{-\frac{|x|^2}{4t}}. \tag{2.1}$$

This result provides the parabolic analogues of characterizations of subharmonic functions given in [5].

We set $\mathrm{dom}(\Phi) = \mathbb{R}^n \times (0, \infty)$, and then

$$\mathrm{dom}(u - \Phi) = \mathrm{dom}(u) \cap \mathrm{dom}(\Phi) = \mathrm{dom}(u) \setminus \{\mathbb{R}^n \times (-\infty, 0]\},$$

and so on, for the functions used below.

The "test functions" used in the theorem are in two forms. The first form is

$$w(x, t) = \ell(x) - \sum_{j=1}^{n} \Phi(x - z^j, t - t_j) \tag{2.2}$$

where $z^j \in \mathbb{R}^n$, $t_j \in \mathbb{R}$ for $j = 1, 2, \ldots, n$ and ℓ is a linear function of x. The second form is

$$w(x, t) = -\sum_{j=1}^{n+1} a_j \Phi(x - z^j, t - t_j), \tag{2.3}$$

where $a_j \geq 0$ for $j = 1, 2, \ldots, n + 1$.

For brevity, when we are considering test functions given by (2.2) and (2.3) we will write

$$\Phi_j(x, t) = \Phi(x - z^j, t - t_j).$$

THEOREM 3. *Let $u \in \mathrm{USC}(\mathcal{U})$. Then the following are equivalent:*

  (i) *u is subcaloric in $\mathcal{U}$.*
  (ii) *$u - w \in \mathrm{MPar}(\mathrm{dom}(u - w))$ for every w of the form (2.2).*
  (iii) *$u - w \in \mathrm{MPar}(\mathrm{dom}(u - w))$ for every w of the form (2.3).*

REMARK 4. The sum in (2.2) ((2.3)) has *exactly n* (respectively, $n+1$) terms $-\Phi_j$. The result that (ii) implies (i) is false if the sums have fewer than n terms. For example, regarding (2.2), if $n = 1$ and we use $n - 1 = 0$ summands, test functions are linear functions in x.

However, $u(x, t) - cx \in \text{MPar}$ for every $c \in \mathbb{R}$ is true if $u(x, t)$ is convex in x for each t, and then u is not necessarily subcaloric. In the case $n = 2$, the function $u(x, t) = -3t - (x_1^2 + x_2^2)$ is strictly supercaloric, but one has $u(x, t) - \ell(x) + \Phi(x - z, t - s) \in \text{MPar}$ for every $z \in \mathbb{R}^2$, $s \in \mathbb{R}$. This is proved in Example 5 below, after we prove the theorem. However, we do not have examples indicating the minimality of the class of test functions of the form (2.3).

Proof. If u is subcaloric and w is a test function of either form, (2.2) or (2.3), then it follows from Remark 2 that

$$u - w \in \text{MPar}(\text{dom}\,(u - w))$$

because w is caloric (i.e., $Hw = 0$ on dom (w)).

To prove the other directions, which is the main point, we show that if u is not subcaloric, then there is a test function w such that $u - w \notin \text{MPar}(\text{dom}\,(u - w))$. The statement that u is not subcaloric at $(\hat{x}, \hat{t}) \in \mathcal{U}$ means that there is a smooth function ψ such that

$$u - \psi \text{ has a local maximum at } (\hat{x}, \hat{t}), \text{ but } \psi_t(\hat{x}, \hat{t}) - \Delta\psi(\hat{x}, \hat{t}) > 0. \tag{2.4}$$

To reduce notation, we translate coordinates so that $\hat{x} = 0$ and $\hat{t} = 1$. In all then, we assume that $(0, 1) \in \mathcal{U}$ and

(i) $\psi_t(0, 1) = a,\ D\psi(0, 1) = q,\ D^2\psi(0, 1) = -S$,

(ii) $u(x, t) - \psi(x, t) \leq u(0, 1) - \psi(0, 1) \quad$ for $\quad x, t \quad$ near $\quad (0, 1)$,

(iii) $H\psi \equiv a + \text{trace}\,(S) > 0$, $\tag{2.5}$

where $a \in \mathbb{R}$, $q \in \mathbb{R}^n$ and S is a symmetric $n \times n$ matrix.

We turn to the proof that Theorem 3 (ii) implies Theorem 3 (i). Assuming (2.5), in the following we are going to construct a test function w of the form (2.2) such that $(0, 1) \in \text{dom}(w)$ and

(i) $Dw(0, 1) = D\ell - \sum_{j=1}^{n} D\Phi_j(0, 1) = q = D\psi(0, 1)$,

(ii) $w_t(0, 1) = -\sum_{j=1}^{n} \Phi_{jt}(0, 1) < a = \psi_t(0, 1,)$,

(iii) $D^2 w(0, 1) = -\sum_{j=1}^{n} D^2\Phi_j(0, 1) > -S = D^2\psi(0, 1)$. $\tag{2.6}$

Recall that Dv, D^2v denote, respectively, the spatial gradient of v and the $n \times n$ symmetric (Hessian) matrix of the second partial derivatives of v with respect to the x_k. Also, of course, Φ_{jt} is the derivative of $\Phi_j(x, t)$ with respect to t.

The three conditions of (2.6) are sufficient for $\psi - w$ and hence for $u - w$ to violate the parabolic maximum principle near $(0, 1)$. Indeed, by (2.5), (2.6) (i) and Taylor expansion we find

$$u(x, t) - w(x, t) = u(x, t) - \psi(x, t) + \psi(x, t) - w(x, t)$$

$$\leq u(0, 1) - \psi(0, 1) + \psi(0, 1) - w(0, 1) + (a - w_t(0, 1))(t - 1)$$
$$- \frac{1}{2}\langle (S + D^2 w(0, 1))x, x\rangle + o(|t - 1| + |x|^2)$$
$$= u(0, 1) - w(0, 1) + (a - w_t(0, 1)(t - 1)$$
$$- \frac{1}{2}\langle (S + D^2 w(0, 1))x, x\rangle + o(|t - 1| + |x|^2). \tag{2.7}$$

It follows from this and (2.6) (ii), (iii) that $(0, 1)$ is the unique maximum point of $u - w$ over $B_\delta(0) \times (1 - \delta, 1]$ provided $\delta > 0$ is sufficiently small. Thus $u - w \notin \mathrm{MPar}(\mathcal{U})$.

Suppose that we construct a test function $w_0 = -\sum_{j=1}^{n} \Phi_j$ which satisfies (2.6)(ii) and (iii) in place of w. Then the test function

$$w(x) = \langle p, x\rangle + w_0(x, t) \quad \text{where} \quad p = q + \sum_{j=1}^{n} D\Phi_j(0, 1)$$

satisfies all of (2.6). Thus we now assume that

$$w(x, t) = -\sum_{j=1}^{n} \Phi_j(x, t)$$

and attempt to find $z^j \neq 0, t_j < 1$ such that (2.6) (ii), (iii) are satisfied.

A direct computation for the heat kernel gives

$$\Phi_t = \left(-\frac{n}{2} + \frac{|x|^2}{4t}\right) t^{-\frac{n+2}{2}} \exp\left(-\frac{|x|^2}{4t}\right),$$
$$D\Phi = -\frac{x}{2t} t^{-\frac{n}{2}} \exp\left(-\frac{|x|^2}{4t}\right),$$
$$D^2\Phi = \left(-\frac{1}{2}I + \frac{1}{4t} x \otimes x\right) t^{-\frac{n+2}{2}} \exp\left(-\frac{|x|^2}{4t}\right).$$

Thus, using the notation

$$\sigma_j = \frac{|z^j|^2}{2(1 - t_j)}, \quad \tau_j = \frac{1}{2}(1 - t_j)^{-\frac{n+2}{2}} \exp\left(-\frac{|z^j|^2}{4(1 - t_j)}\right), \tag{2.8}$$

we have

$$\Phi_{jt}(0, 1) = \left(-\frac{n}{2} + \frac{\sigma_j}{2}\right) 2\tau_j = -(n + \sigma_j)\tau_j \tag{2.9}$$

and

$$D^2\Phi_j(0, 1) = \left(-\frac{1}{2}I + \frac{\sigma_j}{2} \hat{z}^j \otimes \hat{z}^j\right) 2\tau_j = (-I + \sigma_j \hat{z}^j \otimes \hat{z}^j)\tau_j \tag{2.10}$$

where $\hat{y} = y/|y|$ for $y \neq 0$.

Note that the relations (2.8) may be inverted. Given $\sigma_j, \tau_j > 0$, (2.8) can be solved for $|z^j|^2, t_j < 1$:

$$(1 - t_j) = \left(\frac{\exp(-\frac{\sigma_j}{2})}{2\tau_j}\right)^{\frac{2}{n+2}}, \quad |z^j|^2 = 2\sigma_j \left(\frac{\exp(-\frac{\sigma_j}{2})}{2\tau_j}\right)^{\frac{2}{n+2}}. \tag{2.11}$$

Furthermore, (2.6) (ii) and (iii) are now equivalent to

(i) $a > \sum_{j=1}^n (n - \sigma_j)\tau_j,$

(ii) $S > -\sum_{j=1}^n (I - \sigma_j \hat{z}^j \otimes \hat{z}^j)\tau_j.$ \hfill (2.12)

Let

$$r = \left(\sum_{j=1}^n \tau_j\right)^{-1} \quad \text{and} \quad \delta_j = \sigma_j \tau_j r. \tag{2.13}$$

Multiplying the inequalities (2.12) by r and rearranging yields

(i) $ar > n - \sum_{j=1}^n \delta_j,$

(ii) $I + rS > \sum_{j=1}^n \delta_j \hat{z}^j \otimes \hat{z}^j.$ \hfill (2.14)

To proceed, let $\hat{z}^1, \hat{z}^2, \ldots, \hat{z}^n$ be an orthonormal basis of eigenvectors of S corresponding to eigenvalues λ_j:

$$S\hat{z}^j = \lambda_j \hat{z}^j.$$

Then

$$I + rS = \sum_{j=1}^n (1 + r\lambda_j)\hat{z}^j \otimes \hat{z}^j.$$

Thus, with this choice of the $\hat{z}^j$, (2.14) amounts to

(i) $ar > n - \sum_{j=1}^n \delta_j,$

(ii) $1 + r\lambda_j > \delta_j \quad \text{for} \quad j = 1, 2, \ldots, n.$ \hfill (2.15)

Suppose that we find $r > 0, \delta_j > 0$ so that (2.15) holds. Then we may choose any $\tau_j > 0$ so that the first equation of (2.13) holds and then solve the remaining equations for σ_j. Then (2.11) provides the $t_j, |z^j|$ which we need, along with $\hat{z}^j$, to determine Φ_j, and we are done.

It remains to solve (2.15). Choose $r > 0$ small enough so that

$$1 + r\lambda_j > 0 \quad \text{for} \quad j = 1, 2, \ldots, n. \tag{2.16}$$

In order that there then be $\delta_j > 0$, $j = 1, 2, \ldots, n$, such that (2.14) holds, it is clearly necessary that

$$n - ar < \sum_{j=1}^{n} (1 + r\lambda_j). \tag{2.17}$$

This condition holds for any $r > 0$ by (2.5) (iii), which may be restated as

$$a + \text{trace}(S) = a + \lambda_1 + \cdots + \lambda_n > 0$$

because the λ_j are the eigenvalues of S. Clearly this last equation and (2.17) are equivalent if $r > 0$.

Put

$$\delta_j = (1 + r\lambda_j) - \epsilon$$

where

$$0 < \epsilon < \min_{j=1,\ldots,n} (1 + r\lambda_j)$$

and

$$n - ar < \sum_{j=1}^{n} (1 + r\lambda_j) - n\epsilon.$$

This is possible by (2.16) and (2.17). This choice of the δ_j evidently solves (2.15).

We turn to the proof that Theorem 3 (iii) implies Theorem 3 (i). As above, we assume that (2.5) holds and seek a test function

$$w(x, t) = -\sum_{j=1}^{n+1} a_j \Phi(x - z^j, t - t_j), \tag{2.18}$$

with $a_j \geq 0$, $z^j \in \mathbb{R}^n$, $t_j \in \mathbb{R}$ for $j = 1, \ldots n + 1$ to be determined so that

$$Dw(0, 1) = q = D\psi(0, 1), \quad w_t(0, 1) < a = \psi_t(0, 1),$$
$$D^2w(0, 1) > -S = D^2\psi(0, 1).$$

In this case, these conditions rewrite to

(i) $\sum_{j=1}^{n} a_j \tau_j z^j + a_{n+1} \tau_{n+1} z^{n+1} = -q$

(ii) $a > \sum_{j=1}^{n} (n - \sigma_j)(a_j \tau_j) + (n - \sigma_{n+1})(a_{n+1} \tau_{n+1})$

(iii) $S > -\sum_{j=1}^{n} (a_j \tau_j)(I - \sigma_j \hat{z}_j \otimes \hat{z}_j) - a_{n+1} \tau_{n+1}(I - \sigma_{n+1} \hat{z}_{n+1} \otimes \hat{z}_{n+1})$,   (2.19) where σ_j and τ_j are still given by (2.8). The display above foreshadows that the term with index $n + 1$ in w will be treated as a perturbation of the first n terms.

First we ask when we might succeed with n terms, that is, with $a_{n+1} = 0$. Let $\hat{z}^j$ be an orthonormal basis of eigenvectors of S as in the preceding case. Then, with $a_{n+1} = 0$, (2.19) becomes

(i) $\sum_{j=1}^n a_j \tau_j z^j = -q$

(ii) $ar > n - \sum_{j=1}^n \delta_j$

(iii) $\sum_{j=1}^n (1 + r\lambda_j)\hat{z}^j \otimes \hat{z}^j > \sum_{j=1}^n \delta_j \hat{z}^j \otimes \hat{z}^j$ $\qquad(2.20)$

where $r = (\sum_{j=1}^n a_j \tau_j)^{-1}$ and $\delta_j = \sigma_j a_j \tau_j r$. The requirement (2.20) (ii) is that

$$a_j \tau_j |z^j| = -\langle q, \hat{z}^j \rangle \text{ for } j = 1, \ldots, n. \qquad(2.21)$$

If $-\langle q, \hat{z}^j \rangle \neq 0$, we can assume it is positive upon replacing $\hat{z}^j$ by $-\hat{z}^j$, and we do so if necessary. In the case that the $\hat{z}^j$ can be chosen so that

$$\langle q, \hat{z}^j \rangle < 0 \text{ for } j = 1, \ldots, n, \qquad(2.22)$$

we are done for the following reasons. Let $r, \delta_j > 0$ be given by the previous proof so that (2.20) (ii), (iii) hold. Let $\hat{z}^j$ be an orthonormal basis of eigenvectors of S with corresponding eigenvalues λ_j such that (2.22) holds. Then if $\tau_j, \sigma_j, |z_j|, a_j$ and t_j satisfy the following equation we are done:

(i) $(\sum_{j=1}^n a_j \tau_j)^{-1} = r,$

(ii) $\sigma_j a_j \tau_j = \frac{\delta_j}{r},$

(iii) $a_j \tau_j |z_j| = -\langle q, \hat{z}^j \rangle$

(iv) $\sigma_j = \frac{|z_j|^2}{2(1 - t_j)},$

(v) $\quad \tau_j = \frac{1}{2}(1 - t_j)^{-\frac{n+2}{2}} \exp\left(-\frac{|z_j|^2}{4(1 - t_j)}\right)$ $\qquad(2.23)$

Here is why. Choose $a_j, \tau_j > 0$ so that (i) holds. Then the $\sigma_j > 0$ and $|z_j| > 0$ are determined by (ii) and (iii). Note that τ_j and a_j may vary provided that the product $a_j \tau_j$ is fixed. The relation (iv) determines $1 - t_j$ (and hence t_j) as we already have $|z^j|$ and σ_j. Then (v) determines τ_j and therefore a_j via their product.

The analysis just done also suffices to handle the more general situation when $q \neq 0$ but some coordinate of q in any orthonormal basis of eigenvectors of S vanishes. This follows from a construction in [5] which verifies the existence of a symmetric matrix $\tilde{S}$ with the following properties

$$\tilde{S} \leq S, \ \tilde{S} \text{ has simple eigenvalues and } \langle q, z \rangle \neq 0 \text{ for every eigenvector } z \text{ of } \tilde{S}.$$

and

$$a + \operatorname{trace}(\tilde{S}) > 0.$$

Then we may replace ψ by

$$\tilde{\psi}(x, t) = \psi(x, t) + \frac{1}{2}\langle (S - \tilde{S})x, x \rangle$$

everywhere in the above proof. Note that $D^2\tilde{\psi}(0, 1) = -\tilde{S}$.

The extreme case is $q = 0$. In this case we use all $n + 1$ terms in the sum. Again let $r, \delta_j > 0$ for $j = 1, \ldots, n$ be chosen so that (2.20) (ii), (iii) hold. We are still letting $\hat{z}^j$, $j = 1, \ldots, n$, be an orthonormal basis of eigenvectors of S. Let $a_j = 1$ for $j = 1, \ldots, n$ and $\tau_j > 0$ be such that $\sum_{j=1}^n \tau_j = 1/r$. We seek $a_{n+1}, z^{n+1}, \tau_{n+1}$ such that (2.19) holds. The condition (2.19) (i) with $q = 0$ becomes

$$-a_{n+1}\tau_{n+1} z^{n+1} = \sum_{j=1}^n \tau_j z^j. \tag{2.24}$$

We choose z^{n+1} so that it is anti-parallel to the known vector on the right, and then (2.24) reduces to

$$a_{n+1}\tau_{n+1}|z^{n+1}| = k \tag{2.25}$$

where $k > 0$ is the norm of the vector on the right of (2.24). We put

$$r^* = \frac{1}{\sum_{j=1}^n \tau_j + a_{n+1}\tau_{n+1}} \quad \text{and} \quad \delta_j^* = \sigma_j \tau_j r^* \text{ for } j = 1, \ldots, n. \tag{2.26}$$

In the usual way, equations (2.19) (ii), (iii), now rewrite to

$$ar^* > n - \sum_{j=1}^n \delta_j^* - a_{n+1}\tau_{n+1}r^*,$$

$$\sum_{j=1}^n (1 + r^*\lambda_j)\hat{z}^j \otimes \hat{z}^j > \sum_{j=1}^n \delta_j^* \hat{z}^j \otimes \hat{z}^j + a_{n+1}\tau_{n+1}\sigma_{n+1} r^* \hat{z}^{n+1} \otimes \hat{z}^{n+1}. \tag{2.27}$$

Since $r^* \to r$, $\delta_j^* \to \delta_j$ when $a_{n+1}\tau_{n+1} \to 0$, we are done if we can arrange that

$$a_{n+1}\tau_{n+1} \to 0 \text{ and } a_{n+1}\tau_{n+1}\sigma_{n+1} \to 0 \tag{2.28}$$

while still enforcing the constraint (2.25). Since (2.27) holds (with strict inequality) in this limit, it will hold when the quantities above are sufficiently small. Since, by (2.25) and (2.8),

$$a_{n+1}\tau_{n+1}\sigma_{n+1} = a_{n+1}\tau_{n+1}|z^{n+1}|\frac{\sigma_{n+1}}{|z^{n+1}|} = k\frac{\sigma_{n+1}}{|z^{n+1}|} = k\frac{|z^{n+1}|}{2(1 - t_{n+1})} \tag{2.29}$$

all we need to do is to let $|z^{n+1}|, -t_{n+1} \to \infty$ in such a way that

$$\frac{|z^{n+1}|}{1 - t_{n+1}} \to 0.$$

Then (2.25), which defines a_{n+1} in terms of $|z^{n+1}|, t_{n+1}$, forces $a_{n+1} t_{n+1} \to 0$ while (2.29) shows that $a_{n+1} t_{n+1} \sigma_{n+1} \to 0$. $\qquad\square$

EXAMPLE 5. We verify the assertion of Remark 4 and show now that

$$w(x, t) = -3t - (x_1^2 + x_2^2),$$

which satisfies

$$Hw = -3 + 4 > 0,$$

also satisfies

$$v(x, t) := w(x, t) - \ell(x) + \Phi(x - z, t - s) \in \mathrm{MPar}(\{x \neq z\} \times (s, \infty))$$

for all $z \in \mathbb{R}^2, s \in \mathbb{R}$. Using calculations above,

$$v_t(x, t) = -3 + (-2 + \sigma)\tau,$$

$$D^2 v(x, t) = \begin{pmatrix} -2 & 0 \\ 0 & -2 \end{pmatrix} + \left(-I + \sigma \hat{y} \otimes \hat{y} \right) \tau, \tag{2.30}$$

where $y = x - z$ and

$$\sigma = \frac{|y|^2}{2(t - s)}, \qquad \tau = \frac{1}{2}(t - s)^{-\frac{n+2}{2}} \exp\left(-\frac{|y|^2}{4(t - s)} \right).$$

We claim that either $v_t < 0$ or an eigenvalue of $D^2 v$ is positive. This shows that $v \in \mathrm{MPar}$. Since $-2 + (-1 + \sigma)\tau$ is an eigenvalue of $D^2 v$, if neither condition holds, then

$$-3 + (-2 + \sigma)\tau \geq 0 \quad \text{and} \quad -2 + (-1 + \sigma)\tau \leq 0$$

which implies

$$-3 + (-2 + \sigma)\tau \geq -2 + (-1 + \sigma)\tau$$

or $-1 \geq \tau$. However, $t > s$ and $\tau > 0$, so we are done.

REMARK 6. We had to do considerable work to establish Theorem 3, in which we insisted that the test functions used be built primarily from translates of the fundamental solution. Another characterization, by means of simpler caloric functions, can be obtained quite easily. Test functions of the form

$$bt + \langle p, x \rangle - \frac{1}{2} \langle Rx, x \rangle,$$

where $b \in \mathbb{R}$, $p \in \mathbb{R}^n$ and R is a symmetric $n \times n$ matrix with

$$b + \text{trace}(R) = 0, \tag{2.31}$$

suffice. Note that this class is invariant (up to an additive constant) under translations. To see this, we assume (2.5) and then, via the above program, must solve

$$p = q, \; b < a \quad \text{and} \quad R < S \tag{2.32}$$

subject to (2.31). Putting $p = q$. $b = a - \epsilon$ and $R = S - \epsilon I$ with $\epsilon > 0$ guarantees the above, while

$$b + \text{trace}(R) = a + \text{trace}(S) - (n + 1)\epsilon$$

which vanishes if $\epsilon = (a + \text{trace}(S))/(n + 1)$, which is positive by assumption.

3. The Infinity Heat Equation

The infinity Laplacian Δ_∞ is defined on smooth functions via

$$\Delta_\infty u = \sum_{i,j=1}^{n} u_{x_i} u_{x_j} u_{x_i x_j} \tag{3.1}$$

and we call the equation

$$H_\infty u = u_t - \Delta_\infty u = 0 \tag{3.2}$$

the "infinity heat equation".

It is natural to attempt to find solutions of (3.2) which are suggested by scaling invariance. If u is a solution of $H_\infty u = 0$, then $v(x, t) = u(bx, ct)$ will also be a solution if

$$c = b^4. \tag{3.3}$$

This leads to the attempt to find solutions of the form

$$u(x, t) = H\left(\frac{|x|}{t^{1/4}}\right).$$

The solutions then found are, up to translations in x, t, of the type

$$u(x, t) = \pm \frac{|x|^2}{(-16t)^{1/2}}. \tag{3.4}$$

This solution is defined for $t < 0$, in contrast to the heat equation case, and blows up as $t \uparrow 0$. Of course, the scaling $x \to ix$, $t \to -t$ leaves the heat equation invariant, and takes the heat kernel into $\exp(|x|^2/4t)/((-t)^{n/2})$, which is also defined for $t < 0$ and blows up as $t \uparrow 0$. However, there is no scaling which takes the solutions (3.4) into a real-valued solution of the infinity heat equation defined for $t > 0$. Thus we seek other solutions.

Seeking separated solutions of the form $g(t) + f(|x|)$ leads to the expressions of the form

$$\Phi^{bd}(x,t) = b^3 \left(\frac{64}{81}\right) t + b(|x|+d)^{\frac{4}{3}}. \tag{3.5}$$

We will show that Φ^{bd} is a viscosity solution of $H_\infty \Phi^{bd} \geq 0$ on a set $\mathcal{W}_{b,d}$ which varies a bit with b, d. Note that the constants b, d are not necessarily positive and we regard $s^{4/3} = (s^4)^{1/3}$ as well defined for all reals s.

The formula

$$\Delta_\infty G(|x|) = G'(|x|)^2 G''(|x|), \tag{3.6}$$

which is valid in the classical sense if G is twice continuously differentiable on an interval containing $|x|$ and $x \neq 0$, makes the verification that $H_\infty \Phi^{bd} = 0$ easy on the set where it is C^2, that is, on

$$\mathcal{R}_d = \{(x,t) \in \mathbb{R}^n \times \mathbb{R} : x \neq 0, \ |x| \neq -d\}. \tag{3.7}$$

We have

$$\Delta_\infty \Phi^{bd} = b^3 \Delta_\infty ((|x|+d)^{\frac{4}{3}})$$

$$= b^3 \left(\frac{4}{3}(|x|+d)^{1/3}\right)^2 \left(\frac{4}{3}\frac{1}{3}(|x|+d)^{-2/3}\right) = b^3 \left(\frac{64}{81}\right),$$

while, obviously, $\Phi_t^{bd} = b^3(64/81)$.

We seek now the exact set on which Φ^{bd} is a viscosity solution of $H_\infty \Phi^{bd} \geq 0$. In other words, we seek the set of $(\hat{x}, \hat{t})$ such that

if ϕ is twice continuously differentiable near $(\hat{x}, \hat{t})$ and

$\Phi^{bd} - \phi$ has a local minimum at $(\hat{x}, \hat{t})$, then $H_\infty \phi(\hat{x}, \hat{t}) \geq 0.$ (3.8)

The answer depends on b, d. Note that one way for $(\hat{x}, \hat{t})$ to qualify for membership in $\mathcal{W}_{b,d}$ is for there to be no smooth ϕ such that $\Phi^{bd} - \phi$ has a minimum at $(\hat{x}, \hat{t})$. We will not need all of the information we find for the rest of the proceedings, but it is of independent interest.

PROPOSITION 7. *Let $\mathcal{W}_{b,d}$ be the set of $(\hat{x}, \hat{t})$ such that (3.8) holds. Then*

(i) $\mathcal{W}_{b,d} = \mathbb{R}^n \times \mathbb{R}$ *if* $bd \leq 0$.

(ii) $\mathcal{W}_{b,d} = \{(\hat{x}, \hat{t}) \in \mathbb{R}^n \times \mathbb{R} : \hat{x} \neq 0\}$ *if* $bd > 0$.   (3.9)

Proof. In both cases $\mathcal{W}_{bd} \supset \mathcal{R}_d$ where $\mathcal{R}_d$ is given by (3.7). Thus we need only be concerned with points $(\hat{x}, \hat{t})$ such that $\hat{x} = 0$ or $|\hat{x}| + d = 0$. As $\hat{t}$ will not play any role, for simplicity we set $\hat{t} = 1$ in what follows.

Assume that $(\hat{x}, 1)$ is a local minimum point of $\Phi^{bd} - \phi$, that is

$$b^3 \left(\frac{64}{81}\right) t + b(|x| + d)^{\frac{4}{3}} - \phi(x, t) \geq b^3 \left(\frac{64}{81}\right) + b(|\hat{x}| + d)^{\frac{4}{3}} - \phi(\hat{x}, 1)$$

near $(\hat{x}, 1)$. Equivalently,

$$b^3 \left(\frac{64}{81}\right)(t - 1) + b((|x| + d)^{\frac{4}{3}} - (|\hat{x}| + d)^{\frac{4}{3}}) \geq \phi(x, t) - \phi(\hat{x}, 1) \tag{3.10}$$

near $(\hat{x}, 1)$. We use (3.10) with $x = \hat{x}$ and then $t = 1$ to find

(i) $b^3 \left(\frac{64}{81}\right)(t - 1) \geq \phi(\hat{x}, t) - \phi(\hat{x}, 1)$,

(ii) $b((|x| + d)^{\frac{4}{3}} - (|\hat{x}| + d)^{\frac{4}{3}}) \geq \phi(x, 1) - \phi(\hat{x}, 1)$. $\tag{3.11}$

Clearly (3.11) (i), which holds for t near 1, implies

$$\phi_t(\hat{x}, 1) = b^3 \left(\frac{64}{81}\right). \tag{3.12}$$

We can assume that $b \neq 0$, since $\Phi^{0,d} \equiv 0$ is a classical solution of $H_\infty \Phi^{0,d} = 0$ on $\mathbb{R}^n \times \mathbb{R}$. Suppose now that

$$|\hat{x}| + d = 0. \tag{3.13}$$

Note that then $\hat{x} = 0$ if $d = 0$. Then (3.11) (ii) can be rewritten as

$$b(|x| - |\hat{x}|)^{\frac{4}{3}} \geq \phi(x, 1) - \phi(\hat{x}, 1). \tag{3.14}$$

This clearly implies

$$D\phi(\hat{x}, 1) = 0. \tag{3.15}$$

and therefore, using also (3.12),

$$H_\infty \phi(\hat{x}, 1) = \phi_t(\hat{x}, 1) = b^3 \left(\frac{64}{81}\right)$$

and $H_\infty \phi(\hat{x}, 1) > 0$ if $b > 0$. Thus if (3.13) holds, $(\hat{x}, 1) \in \mathcal{W}_{b,d}$ if $b \geq 0$. However, if $b < 0$, then (3.14) cannot hold for any smooth ϕ and we still have $(\hat{x}, 1) \in \mathcal{W}_{b,d}$. To verify this claim, note that if $D\phi(\hat{x}, 1) = 0$, then, by Taylor expansion,

$$\phi(x, 1) - \phi(\hat{x}, 1) = \frac{1}{2} \langle D^2 \phi(\hat{x})(x - \hat{x}), x - \hat{x} \rangle + o(|x - \hat{x}|^2) \geq -M|x - \hat{x}|^2$$

for a suitable M and x near $\hat{x}$. Thus, if $b < 0$ and (3.14) holds, we would have

$$M|x - \hat{x}|^2 \geq (-b)(|x| - |\hat{x}|)^{\frac{4}{3}} = \kappa(|x| - |\hat{x}|)^{\frac{4}{3}}, \tag{3.16}$$

where $\kappa = -b > 0$, for x near $\hat{x}$. Putting $x = (1 + \delta)\hat{x}$ in this relation yields

$$M(|\hat{x}|\delta)^2 \geq \kappa \left(|\hat{x}|\delta\right)^{\frac{4}{3}}$$

This cannot hold for small $\delta > 0$ unless $\hat{x} = 0$. If $\hat{x} = 0$ it is clear that (3.16) cannot hold for all small x. We have verified that:

$$\text{If } |\hat{x}| + d = 0, \text{ then } (\hat{x}, 1) \in \mathcal{W}_{bd} \text{ for all } b. \tag{3.17}$$

It remains to treat the situation $\hat{x} = 0$, $d \neq 0$, $b \neq 0$. We break this into two cases, $bd < 0$ and $bd > 0$. Here (3.11) (ii) becomes

$$b((|x| + d)^{\frac{4}{3}} - d^{\frac{4}{3}}) \geq \phi(x, 1) - \phi(0, 1) \tag{3.18}$$

for small x. But if $|x|$ is small, the left-hand side of (3.18) "looks like" $4bd^{1/3}|x|/3$. If $bd^{1/3} < 0$, (the appropriate perturbation of) the inequality $4bd^{1/3}|x|/3 \geq \phi(x, 1) - \phi(0, 1)$ cannot hold for small $|x|$ and any function ϕ which is differentiable at $(0, 1)$. Thus:

$$\text{If } bd < 0, \text{ then } (0, 1) \in \mathcal{W}_{b,d}. \tag{3.19}$$

However, if $bd > 0$, we have a problem. In this case, we take

$$\phi(x, t) = b^3 \left(\frac{64}{81}\right) t + \langle q, x \rangle + \frac{1}{2}\langle Sx, x \rangle + bd^{\frac{4}{3}},$$

where $q \in \mathbb{R}^n$ and S is a symmetric matrix, and note that

$$\Phi^{bd}(x, t) - \phi(x, t) = b\left((|x| + d)^{\frac{4}{3}} - d^{\frac{4}{3}}\right) - \langle q, x \rangle - \frac{1}{2}\langle Sx, x \rangle$$

has a minimum at $(0, 1)$ no matter what S is, provided that $|q| < 4bd^{1/3}/3$. This is because, as above, $b((|x| + d)^{\frac{4}{3}} - d^{\frac{4}{3}})$ looks like $4bd^{1/3}|x|/3$ for small $|x|$. Since

$$H_\infty\phi(0, 1) = b^3 \left(\frac{64}{81}\right) - \langle Sq, q \rangle$$

and we can take $q \neq 0$ while S is arbitrary, we can choose q, S so that $H_\infty\phi(0, 1) < 0$. We have verified that:

$$\text{If } db > 0, \text{ then } (0, 1) \notin \mathcal{W}_{bd}. \tag{3.20}$$

Taken together, (3.17), (3.19) and (3.20) imply (3.9). $\qquad\qquad\square$

We use the notation

$$\Phi^{bdz}(x,t) = \Phi^{bd}(x-z,t) \tag{3.21}$$

to denote spatial translates of Φ^{bd}. The domain dom (Φ^{bd}) of Φ^{bd} is hereafter defined to be the set $\mathcal{W}_{b,d}$ on which it is a viscosity solution of $H_\infty \Phi^{bd} \geq 0$ given in Proposition 7:

$$\mathrm{dom}(\Phi^{bd}) = \mathcal{W}_{b,d}.$$

Translates have the appropriate translated domain. Our characterization of infinity-subcaloric functions is:

THEOREM 8. *If $u \in \mathrm{USC}(\mathcal{U})$, then $H_\infty u \leq 0$ in the viscosity sense in $\mathcal{U}$ iff for all $b, d \in \mathbb{R}, z \in \mathbb{R}^n$*

$$u - \Phi^{bdz} \in \mathrm{MPar}(\mathcal{U} \cap \mathrm{dom}(\Phi^{bdz})). \tag{3.22}$$

Proof. The necessity of (3.22) follows from Proposition 7 and Remark 2. We turn to the proof that (3.22) implies $H_\infty u \leq 0$. Parallel to the preceding section, we assume that (3.22) holds, but that u is not a viscosity subsolution of the infinity heat equation at $(0,1) \in \mathcal{U}$ and then derive a contradiction. Assume that there exists $q \in \mathbb{R}^n$ and an $n \times n$ symmetric matrix S such that the following conditions on ψ, u hold:

(i) $\psi_t(0,1) = a, \ D\psi(0,1) = q, \ D^2\psi(0,1) = S$,

(ii) $u(x,t) - \psi(x,t) \leq u(0,1) - \psi(0,1)$ for (x,t) near $(0,1)$,

(iii) $H_\infty \psi(0,1) = a - \langle Sq, q \rangle > 0$. $\tag{3.23}$

We will show that then there is a choice of b, d, z such that

$$u - \Phi^{bdz} \notin \mathrm{MPar}(\mathrm{dom}(u \cap \Phi^{bdz})).$$

Note that the meaning of S has changed by a sign in contrast with (2.5).

To do this, as before, it suffices to produce b, d, z so that $(0,1) \in \mathrm{dom}(\Phi^{bdz})$ and

(i) $\Phi_t^{bdz}(0,1) < a = \psi_t(0,1)$,

(ii) $D\Phi^{bdz}(0,1) = q = D\psi(0,1)$,

(iii) $D^2\Phi^{bdz}(0,1) > S = D^2\psi(0,1)$. $\tag{3.24}$

We will not always succeed, and the case $q = 0$ will involve separate considerations.

First we record the relevant derivatives of Φ^{bd} at regular points $(x,t) \in \mathcal{R}_d$:

$$\Phi_t^{bd} = b^3 \left(\frac{64}{81} \right), \qquad D\Phi^{bd} = \frac{4b}{3}(|x|+d)^{\frac{1}{3}}\hat{x}$$

$$\tag{3.25}$$

$$D^2\Phi^{bd} = \frac{4b}{3}\frac{(|x|+d)^{\frac{1}{3}}}{|x|} \left(\left(\frac{1}{3}\frac{|x|}{|x|+d} - 1 \right) \hat{x} \otimes \hat{x} + I \right),$$

where $\hat{x} = x/|x|$. In view of (3.25) and (3.23) (i) and putting $r = |z|$, (3.24) amounts to

(i) $b^3 \left(\dfrac{64}{81}\right) < a,$

(ii) $-\dfrac{4b}{3}(r+d)^{\frac{1}{3}}\hat{z} = q,$

(iii) $\dfrac{4b}{3}\dfrac{(r+d)^{\frac{1}{3}}}{r}\left((\dfrac{1}{3}\dfrac{r}{r+d} - 1)\hat{z} \otimes \hat{z} + I\right) > S.$ (3.26)

Clearly, (3.26) (ii) cannot hold when $q = 0$ unless $b = 0$ (which is not useful) or $|z| + d = 0$, which invalidates the computation of $D^2 \Phi^{bdz}(0, 1)$. We deal with this case separately. By (3.23), if $q = 0$ then $a > 0$. Choose b so that

$$0 < b^3 \left(\frac{64}{81}\right) < a.$$

Then the function

$$\Phi^{b,0}(x, t) - \psi(x, t) = b^3 \left(\frac{64}{81}\right)t + b|x|^{4/3}$$
$$- \left(at + \frac{1}{2}\langle Sx, x\rangle + o(|t - 1| + |x|^2)\right)$$

attains its minimum value relative to $\{x : |x| < \epsilon\} \times (1 - \delta, 1]$ at the unique point $(0, 1)$ if $\epsilon, \delta > 0$ are sufficiently small. Since $d = 0$, $(0, 1) \in \mathrm{dom}(\Phi^{b,0})$ by Proposition 7. Thus, in view of (3.23) (ii),

$$\begin{aligned}
u(x, t) - \Phi^{b,0}(x, t) &\leq u(x, t) - \psi(x, t) + \psi(x, t) - \Phi^{b,0}(x, t) \\
&\leq u(0, 1) - \psi(0, 1) + \psi(0, 1) - \Phi^{b,0}(0, 1) \\
&= u(0, 1) - \Phi^{b,0}(0, 1)
\end{aligned}$$

in $\{x : |x| < \epsilon\} \times (1 - \delta, 1]$, with equality only at $x = 0$, $t = 1$. Hence $u - \Phi^{b,0} \notin$ MPar$(\mathcal{U} \cap \mathrm{dom}(\Phi^{b,0}))$, a contradiction to the assumptions.

Hereafter we can assume that $q \neq 0$. Our task is then to verify the existence of numbers $d, b \in \mathbb{R}$, $r \in (0, \infty)$ and a unit vector $\hat{z} \in \mathbb{R}^n$ such that (3.26) holds, provided that (3.23) (iii) holds.

The argument breaks into the two cases, $a > 0$ and $a \leq 0$. We assume now that $a > 0$ and seek to satisfy (3.26) with $b, d > 0$. Assuming this, (3.26) (ii) becomes

$$\hat{z} = -\hat{q}, \quad \text{and} \quad \frac{4b}{3}(r+d)^{\frac{1}{3}} = |q|.$$ (3.27)

Thus $\hat{z}$ and $b(r+d)^{1/3}$ are determined by (3.26) (ii).

Now we put

$$\mu = \left(\frac{4b}{3} \frac{(r+d)^{\frac{1}{3}}}{r} \right)^{-1} = \frac{3r}{4b(r+d)^{\frac{1}{3}}} \tag{3.28}$$

so that (3.26) (iii) may be rewritten

$$I > \mu S + \left(1 - \frac{1}{3} \frac{r}{r+d} \right) \hat{z} \otimes \hat{z}. \tag{3.29}$$

We ask, therefore, more generally, the restrictions on $\mu > 0$ which guarantee that

$$I > \mu S + \theta \hat{z} \otimes \hat{z} \tag{3.30}$$

given θ and information only on $\langle S\hat{z}, \hat{z} \rangle$ and the norm $\|S\|$ of S. Writing a general vector $y \in \mathbb{R}^n$ in the form $y = \alpha\hat{z} + z^{\perp}$ where $\langle z, z^{\perp} \rangle = 0$, we find

$$\begin{aligned}
\langle (\mu S &+ \theta\hat{z} \otimes \hat{z})(\alpha\hat{z} + z^{\perp}), \alpha\hat{z} + z^{\perp} \rangle \\
&= \alpha^2(\mu\langle S\hat{z}, \hat{z} \rangle + \theta) + 2\alpha\mu\langle S\hat{z}, z^{\perp} \rangle + \mu\langle Sz^{\perp}, z^{\perp} \rangle \\
&\leq \alpha^2(\mu\langle S\hat{z}, \hat{z} \rangle + \theta) + 2|\alpha|\mu\|S\||z^{\perp}| + \mu\|S\||z^{\perp}|^2 \\
&\leq \alpha^2(\mu\langle S\hat{z}, \hat{z} \rangle + \theta + \epsilon\mu) + \mu\|S\| \left(\frac{1}{\epsilon} + 1 \right) |z^{\perp}|^2
\end{aligned} \tag{3.31}$$

where $\epsilon > 0$. Thus (3.30) holds provided that

$$\mu\langle S\hat{z}, \hat{z} \rangle + \theta + \epsilon\mu\|S\| < 1 \text{ and } \mu\|S\| \left(\frac{1}{\epsilon} + 1 \right) < 1. \tag{3.32}$$

In our case, from (3.27) and (3.23) (iii)

$$\langle S\hat{z}, \hat{z} \rangle = \frac{9}{16} \frac{1}{b^2(r+d)^{\frac{2}{3}}} \langle Sq, q \rangle, \tag{3.33}$$

while

$$\theta = 1 - \frac{r}{3(r+d)} \text{ and } \mu = \frac{3r}{4b(r+d)^{\frac{1}{3}}}. \tag{3.34}$$

Hence (3.32), after some simplification, amounts to

$$\frac{27}{64} \frac{r}{b^3(r+d)} \langle Sq, q \rangle + \epsilon\|S\| \frac{r}{b(r+d)^{1/3}} < \frac{r}{3(r+d)} \text{ and}$$

$$\frac{r}{b(r+d)^{1/3}} \|S\| \left(\frac{1}{\epsilon} + 1 \right) < 1.$$

or

$$\frac{81}{64}\frac{1}{b^3}\langle Sq, q\rangle + 3\epsilon\|S\|\frac{(r+d)^{\frac{2}{3}}}{b} < 1 \text{ and}$$

$$\frac{r}{b(r+d)^{1/3}}\|S\|\left(\frac{1}{\epsilon}+1\right) < 1. \tag{3.35}$$

Now we choose $b > 0$ so that

$$\langle Sq, q\rangle < b^3\left(\frac{64}{81}\right) < a; \tag{3.36}$$

this is possible since we are assuming $a > 0$ and $\langle Sq, q\rangle < a$ by (3.23) (iii). In view of (3.36), the first term in (3.35) is less than 1. Since we have now fixed b and also have (3.27), $r + d$ is now determined. Next choose ϵ sufficiently small so that the first inequality of (3.35) holds. We are still free to choose r, and then determine d from the condition on $r+d$. Thus we may choose r sufficiently small so that the second inequality in (3.27) holds. We are done with the case $a > 0$.

To treat the case $a \leq 0$, we clearly require $b < 0$. We will also assume that $d < 0$ and work under the assumption that

$$r + d = |z| + d < 0.$$

Then μ given by (3.28) is still positive. The discussion leading to (3.35) still holds, with one sign change due to $(r + d) < 0$. The result is

$$\frac{81}{64}\frac{1}{b^3}\langle Sq, q\rangle + 3\epsilon\|S\|\frac{(r+d)^{\frac{2}{3}}}{b} > 1 \text{ and}$$

$$\frac{r}{b(r+d)^{1/3}}\|S\|\left(\frac{1}{\epsilon}+1\right) < 1. \tag{3.37}$$

This time when we choose b so that

$$\langle Sq, q\rangle < b^3\left(\frac{64}{81}\right) < a$$

we have, in view of the signs,

$$\frac{81}{64}\frac{1}{b^3}\langle Sq, q\rangle > 1;$$

the rest of the proof is the same. $\qquad\square$

REFERENCES

[1] ARONSSON, G., CRANDALL, M. G. and JUUTINEN, P., *A tour of the theory of absolutely minimizing functions*, preprint.

[2] BHATTACHARYA, T., DIBENEDETTO, E. and MANFREDI, J., *Limits as $p \to \infty$ of $\Delta_p u_p = f$ and related extremal problems*, Some topics in nonlinear PDEs (Turin, 1989). Rend. Sem. Mat. Univ. Politec. Torino 1989, Special Issue 15–68 (1991).

[3] CRANDALL, M. G., EVANS, L. C. and GARIEPY, R., *Optimal Lipschitz extensions and the infinity Laplacian*, Calc. Var. Partial Differential Equations. *13* (2001), no. 2, 123–139.

[4] CRANDALL, M. G., ISHII, H. and LIONS, P.-L., *User's guide to viscosity solutions of second order partial differential equations*, Bull. Am. Math. Soc. *27* (1992), 1–67.

[5] CRANDALL, M. G. and ZHANG, J., *Another way to say harmonic*, Trans. Amer. Math. Soc. *355* (2003), 241–263.

M.G. Crandall and P.-Y. Wang
Department of Mathematics
University of California
Santa Barbara, CA 93106
USA
e-mail: crandall@math.ucsb.edu
pywang@math.ucsb.edu

To access this journal online:
http://www.birkhauser.ch

J.evol.equ. 3 (2004) 673 – 770
1424–3199/03/040673 – 98
DOI 10.1007/s00028-003-0117-8
© Birkhäuser Verlag, Basel, 2004

Journal of Evolution
Equations

Nonlinear problems related to the Thomas-Fermi equation

PHILIPPE BÉNILAN and HAÏM BREZIS

Preface by Haïm Brezis

Most of the results in this work were obtained over the period 1975–77 and were announced at various meetings (see e.g. items [3], [4], [5] under Brezis [16]). This paper has a rather unusual history. Around 1972 I became interested in nonlinear elliptic equations of the form

$$-\Delta u + |u|^{p-1}u = f \quad \text{in a domain } \Omega \subset \mathbb{R}^N, \tag{P.1}$$

with zero Dirichlet condition, where $0 < p < \infty$ and $f \in L^1$. The motivation came from the study of the porous medium equation

$$\frac{\partial v}{\partial t} - \Delta(|v|^{m-1}v) = 0, \tag{P.2}$$

with $0 < m < \infty$.

The space L^1 is a natural functional space associated with (P.2) since (P.2) generates (at least formally) a contraction semi-group in L^1. When trying to apply the Crandall-Liggett theory in L^1 to (P.2) one is led to the question whether the nonlinear operator $Av = -\Delta(|v|^{m-1}v)$ is m-accretive in L^1, and in particular whether the equation

$$v - \lambda\Delta(|v|^{m-1}v) = f$$

admits a solution for every $f \in L^1$ and every $\lambda > 0$. Setting $u = |v|^{m-1}v$ and scaling out λ yields (P.1) with $p = 1/m$. In the sixties, equations of the type (P.1) had been extensively studied by F. Browder (see e.g. Browder [27]) and by J. L. Lions (see e.g. Leray-Lions [46]) using energy estimates and monotonicity methods which are suitable when $f \in H^{-1}$, but *not* when $f \in L^1$. No one in my circles was concerned with L^1 data for (P.1). The only result I had seen was stated in Stampacchia [56] and dealt with the *linear* elliptic equation in divergence form

$$Lu = -\sum \frac{\partial}{\partial x_j}\left(a_{ij}\frac{\partial u}{\partial x_i}\right) = \mu. \tag{P.3}$$

Stampacchia asserted that, given any $\mu \in L^1$ (or even measure), equation (P.3) admits a solution $u \in L^q$, $\forall q < N/(N-2)$; this was an easy consequence, via *duality*, of the DeGiorgi-Stampacchia estimate

$$\|v\|_{L^\infty} \leqslant C_p \|f\|_{L^p} \quad \forall p > N/2,$$

for the solution of $Lv = f$.

In 1972, Walter Strauss and I tackled (P.1) for $f \in L^1$. We proved, in Brezis-Strauss [25], that, for *every* $f \in L^1$ and *every* $0 < p < \infty$, equation (P.1) admits a unique solution $u \in L^p$. More generally, if $\beta : \mathbb{R} \to \mathbb{R}$ is any continuous nondecreasing function (such that $\beta(0) = 0$), we established that, given any $f \in L^1$, there exists a unique solution of

$$-\Delta u + \beta(u) = f \tag{P.4}$$

with $\beta(u) \in L^1$. We even dealt with maximal monotone graphs β in $\mathbb{R} \times \mathbb{R}$ and obtained the same conclusion for the multivalued equation.

$$-\Delta u + \beta(u) \ni f. \tag{P.5}$$

Later, we considered, in Bénilan-Brezis-Crandall [10], similar problems in all of $\mathbb{R}^N$ (instead of domains).

At the International Congress of 1974, I heard a lecture by E. Lieb reporting on the paper Lieb-Simon [48]. One of their results asserts that for some values of λ, $\lambda \geqslant 0$, the Thomas-Fermi equation

$$-\Delta u + [(u - \lambda)^+]^{3/2} = \sum_{i=1}^{\ell} m_i \delta_{a_i} \text{ in } \mathbb{R}^3, \tag{P.6}$$

with $m_i > 0$ and $\delta_{a_i} = $ Dirac mass at a_i, admits a solution. Of course, the function $\beta(t) = [(t - \lambda)^+]^{3/2}$ is nondecreasing, continuous and $\beta(0) = 0$ (since $\lambda \geqslant 0$). I became intrigued and decided that it would be interesting to study (P.1)(or(P.4)) for measures instead of L^1 functions. My initial intuition was that measures and L^1 functions are the same "creatures" from the point of view of estimates, and therefore the Brezis-Strauss theorem should extend easily to measures. On the other hand, the method of Lieb-Simon was totally different from ours. In their variational approach, equation (P.6) appears as the Euler equation of a "dual" convex minimization problem. Their technique could be adapted to solve (P.4) for a limited class of nonlinearities β and a limited class of measures f.

I mentioned the problem to Philippe Bénilan in the Spring of 1975 and he liked the idea of working together on this topic. Philippe had been my first Ph.D student, even though he was about four years older than me (he defended his Ph.D in 1972). He had been sent to me in 1970 by his mentor, Jacques Deny, who was one of the leaders of the French school in Potential Theory, jointly with M. Brelot and G. Choquet. He knew much better than me the fine properties of harmonic functions and of measure theory. He was the ideal partner on this project. We had been both invited the following summer to Madison, Wisconsin, by Mike Crandall. I have nostalgic memories from the long days we spent together working on the big tables outside the Memorial Union, facing the inspiring view of Lake Mendota. Philippe, who was an addicted smoker, felt free to finish pack after pack in this open-air environment. We managed rather quickly to prove that (P.1) has a solution for every measure f in the case where $p < N/(N-2)$ for $N > 2$ and no restriction on p for $N = 1, 2$ (see Theorem A.1 in Appendix A). Of course, this was sufficient to handle the Thomas-Fermi model since $N = 3$ and $3/2 < 3$. Still, we were puzzled and tried hard to remove the restriction $p < N/(N-2)$. For a few weeks we had no success, even on the simple equation

$$-\Delta u + u^3 = \delta_a \text{ in } \mathbb{R}^3. \tag{P.7}$$

I remember vividly the shiny day when we discovered, sitting at "our" table next to the lake, that (P.7) has *no* solution: this is the elementary computation in Remark A.4. We were stunned! There was indeed an unexpected difference between measures and L^1 and it was due to the nonlinear nature of the problem. Later, we decided to read carefully the paper of Lieb-Simon [48]. We thought about some of their open problems and succeeded in solving two of them (see Section 5 and 6 below). Then came the painful task of writing up our results. Philippe was a powerful and creative mathematician, able to analyze a concrete differential equation in its most minute details. However, when the time arrived to write a paper, he prefered to "hide" the simple illuminating examples and to present instead a grand abstract framework. He was still strongly influenced by the French school of Potential Theory whose program was to axiomatize Potential Theory "à la Bourbaki"–carefully hiding the Laplace operator! Philippe was a perfectionist, always eager to state a theorem in the most general setting, with minimal assumptions. He wrote a first, partial, draft of our paper (basically, Sections 1, 2, 3 below). I made drastic changes which he did not like, etc. After several divergent iterations we stopped and the paper was "buried" unfinished and unorganized. In the meantime we advertised some of the results through lectures, and some hand-written partial versions were circulated "under the coat" as "samizdats". In fact, our unpublished results gave an impetus to beautiful developments in numerous directions:

a) Solving nonlinear PDE's with L^1, or measures, as data became very fashionable. There is a vast and flourishing literature starting in the eighties. I have listed some of the references in Appendix A. Important connections with Probability Theory (E. Dynkin, J. M. LeGall and their students) have reinvigorated the whole subject in recent years.

b) The nonexistence aspect (e.g. for Dirac masses) has given rise to striking new results about *removable singularities* (e.g. point singularities). On the other hand, singular solutions have also been analyzed and classified; see some references in Appendix A.

c) Our approach turned out to be useful in other models arising in the density-functional theory of atoms and molecules; see e.g. Bénilan - G. Goldstein - J. Goldstein [11],

J. Goldstein - G. Rieder [38], G. Rieder [55], G. Goldstein - J. Goldstein - W. Jia [37], Breazna-G. Goldstein - J. Goldstein [15] and related references.

d) The need for new versions of the strong maximum principle in the case of "bad" coefficients stimulated new research in that direction; see Appendix C and the references therein.

e) The solution u of the Thomas-Fermi equation (P.6) tends to zero at infinity. The set where the density $\rho = [(u - \lambda)^+]^{3/2}$ is positive plays an important role. When $\lambda > 0$, this set is bounded. The regularity of its boundary has been studied by Caffarelli-Friedman [28].

After the tragic death of Philippe I decided that our work should not remain in a drawer. Out of respect for the memory of Philippe I have kept his style of presentation. Our notes were incomplete and the last time we touched them was in 1985. I have tried my best to put them in good order and fill in missing arguments. My apologies to the reader if there are still some inconsistencies. I have also added an extensive list of references published in recent years and which bear some relation to our work.

Haïm Brezis

0. Introduction

The principal motivation of this work comes from the important paper of Lieb-Simon [48]. One of their main results is the following. Given $I > 0$, let

$$K_I = \left\{ \rho \in L^1(\mathbb{R}^3); \rho \geqslant 0 \text{ a.e. and } \int \rho = I \right\}.$$

Consider the function

$$V(x) = \sum_{i=1}^{\ell} \frac{m_i}{|x - a_i|}, \quad m_i > 0, \ a_i \in \mathbb{R}^3, \tag{0.1}$$

and set for $\rho \in L^1 \cap L^{5/3}, \rho \geqslant 0$ a.e.,

$$\mathcal{E}(\rho) = \frac{3}{5} \int \rho^{5/3} - \int V\rho + \frac{1}{2} \iint \frac{\rho(x)\rho(y)}{|x - y|} dxdy. \tag{0.2}$$

It is not difficult to check that $\mathcal{E}(\rho)$ is well defined and bounded below. Consider the minimization problem

$$E(I) = \inf\{\mathcal{E}(\rho); \rho \in K_I \cap L^{5/3}\}. \tag{0.3}$$

THEOREM 0.1 (Lieb-Simon). *Set*

$$I_0 = \sum_{i=1}^{\ell} m_i \tag{0.4}$$

If $I \leqslant I_0$, the minimum in (0.3) is uniquely achieved by some ρ. Moreover there is a constant $\lambda \geqslant 0$ such that

$$\rho^{2/3} - V + B\rho = -\lambda \quad in \ [\rho > 0], \tag{0.5}$$

$$-V + B\rho \geqslant -\lambda \quad in \ [\rho = 0], \tag{0.6}$$

where

$$B\rho(x) = \int \frac{\rho(y)}{|x - y|} dy. \tag{0.7}$$

In the neutral case, $I = I_0$, one has $\rho > 0$ a.e. and $\lambda = 0$, so that ρ satisfies

$$\rho^{2/3} - V + B\rho = 0 \quad a.e. \ on \ \mathbb{R}^3. \tag{0.8}$$

The constant λ plays an important role; $-\lambda$ is called the chemical potential. It appears in the Euler "equation" (0.5)–(0.6), corresponding to the minimization problem (0.3), as a Lagrange multiplier associated with the constraint $\int \rho = I$. The dichotomy (0.5)–(0.6) is standard in variational inequalities involving a constraint of the type $\rho \geqslant 0$.

It is convenient to introduce

$$u = V - B\rho, \tag{0.9}$$

and then (0.5)–(0.6) may be rewritten as

$$-\frac{1}{4\pi} \Delta u = \sum m_i \delta_{a_i} - \rho, \tag{0.10}$$

$$\rho = [(u - \lambda)^+]^{3/2}, \tag{0.11}$$

where $r^+ = \max\{r, 0\}$ and $\delta_a = $ Dirac mass at a.

Hence we are led to the nonlinear PDE

$$-\Delta u + 4\pi[(u - \lambda)^+]^{3/2} = 4\pi \sum m_i \delta_{a_i} \tag{0.12}$$

coupled with a condition at infinity coming from (0.9),

$$u(x) \to 0 \quad as \ |x| \to \infty \tag{0.13}$$

(possibly to be understood in a weak sense). Note that here the constant $\lambda \geqslant 0$ is *not* given; it is part of the unknown. But we have instead the additional information

$$\int [(u - \lambda)^+]^{3/2} = I,$$

where $I \geqslant 0$ is *given*.

REMARK 0.1. When $I > I_0$, $E(I) = E(I_0)$ and the infimum in (0.3) is not achieved.

Our work goes in several directions. First, we replace the function $\frac{3}{5}\rho^{5/3}$ by a general convex function $j : \mathbb{R} \to [0, +\infty]$ such that $j(0) = 0$ and we incorporate the constraint $\rho \geqslant 0$ into j by assuming

$$j(r) = +\infty \quad \text{for } r < 0. \tag{0.14}$$

Next, we consider a general measurable function $V(x)$ instead of (0.1). We replace $\mathbb{R}^3$ by $\mathbb{R}^N$, $N \geqslant 3$, and we replace the Coulomb potential by the fundamental solution k of $(-\Delta)$, $k(x) = c_N/|x|^{N-2}$ with $c_N = 1/(N-2)\sigma_N$ and σ_N is the area of the unit space in $\mathbb{R}^N$.

The energy $\mathcal{E}$ takes the form

$$\mathcal{E}(\rho) = \int j(\rho) - \int V\rho + \frac{c_N}{2} \iint \frac{\rho(x)\rho(y)}{|x-y|^{N-2}} \, dxdy, \tag{0.15}$$

whenever it makes sense.

The minimization problem we tackle is

$$E(I) = \inf \left\{ \mathcal{E}(\rho); \int \rho = I \right\}. \tag{M_I}$$

The Euler equation (0.5)–(0.6) is replaced, at least formally, by a multivalued equation

$$\partial j(\rho) - V + B\rho \ni -\lambda \quad \text{a.e. on } R^N, \tag{0.16}$$

for some constant λ, where ∂j is the subdifferential of j.

Note that in the special case where j is C^1 on $(0, +\infty)$ we have

$$\partial j(r) = \begin{cases} j'(r) & \text{for } r > 0, \\ (-\infty, j'(0+)] & \text{for } r = 0, \\ \varnothing & \text{for } r < 0, \end{cases}$$

and thus (0.16) is equivalent to

$$j'(\rho) - V + B\rho = -\lambda \quad \text{in } [\rho > 0], \tag{0.17}$$

$$j'(0+) - V + B\rho \geqslant -\lambda \quad \text{in } [\rho = 0], \tag{0.18}$$

which is precisely (0.5)–(0.6) when

$$j(r) = \begin{cases} \frac{1}{p}r^p & \text{for } r \geqslant 0, \\ +\infty & \text{for } r < 0, \end{cases} \tag{0.19}$$

and $p = 5/3$.

Usually we will asume that $V(x) \to 0$ as $|x| \to \infty$ (at least in some weak sense-for example, meas $[|V| > \delta]$ is finite for every $\delta > 0$); we will also assume that $j'(0+) = 0$, and then (0.17)–(0.18) implies.

$$\lambda \geqslant 0. \tag{0.20}$$

As above, we introduce

$$u = V - B\rho, \tag{0.21}$$

so that we obtain

$$-\Delta u + \rho = -\Delta V \tag{0.22}$$

and

$$\partial j(\rho) \ni u - \lambda. \tag{0.23}$$

We now introduce the inverse maximal monotone graph, $\gamma = (\partial j)^{-1}$, which is also equal to ∂j^*, where j^* is the conjugate convex function of j (see e.g. item [2] under Brezis [16]). In the most important examples (see Section 4), γ is *singlevalued*.

Finally we arrive at the nonlinear multivalued PDE

$$-\Delta u + \gamma(u - \lambda) \ni -\Delta V, \tag{0.24}$$

$$u(x) \to 0 \text{ as } |x| \to \infty. \tag{0.25}$$

Again, λ is unknown, but we have the additional information

$$\int \Delta(u - V) = I, \tag{0.26}$$

with $I \geqslant 0$ given, or equivalently when γ is singlevalued

$$\int \gamma(u - \lambda) = I. \tag{0.27}$$

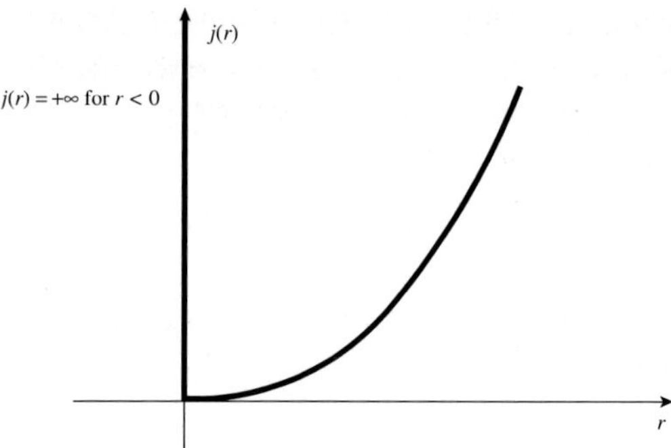

Figure 1   The function j.

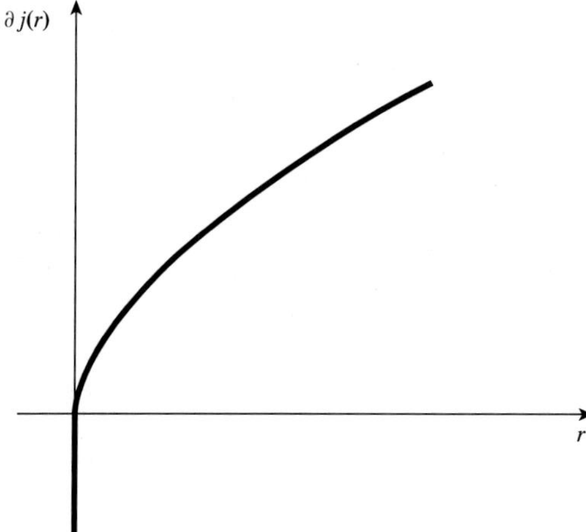

Figure 2   The graph of ∂j.

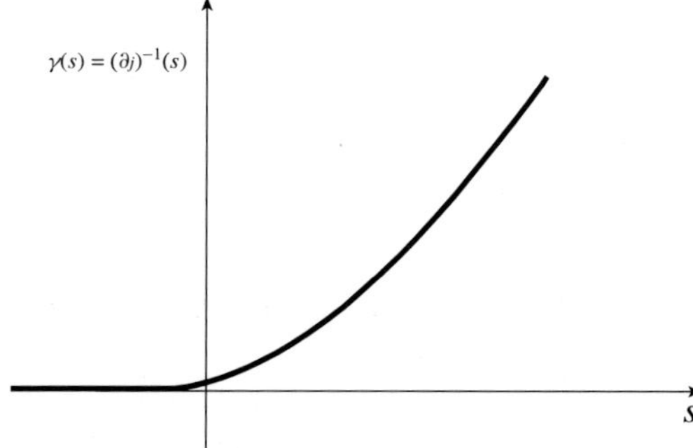

Figure 3   The graph of $\gamma = (\partial j)^{-1}$.

In Section 1 we study the relationship between the variational formulation (M_I) and the Euler equation (0.16). We prove in great generality (see Theorem 1) that if ρ is a minimizer for (M_I), then ρ satisfies the Euler equation (0.16). We establish the converse $(0.16) \Rightarrow (M_I)$ under the additional condition

$$j^*(V - M) \in L^1 \quad \text{for some constant } M, \tag{H}$$

which guarantees that $\mathcal{E}$ is bounded below. For example, when $N = 3$, $V(x) = \sum \frac{m_i}{|x - a_i|}$, and $j(r) = \frac{1}{p} r^p$ for $r \geqslant 0$, condition (H) corresponds to the restriction

$$p > 3/2. \tag{0.28}$$

In fact, when $1 < p \leqslant 3/2$, an easy computation (see Section 4) shows that $E(I) = -\infty$ for every $I > 0$. Despite this fact, we are going to see in Section 4 that the Euler equation (0.16) *does have* a solution when $p > 4/3$.

Therefore, we have a range of p's,

$$\frac{4}{3} < p \leqslant \frac{3}{2}, \tag{0.29}$$

where the *variational formation is meaningless while the PDE approach makes sense.* This is the reason why we have taken, in Sections 3, 4 and in Appendix A, a direct PDE route.

In Section 2 we make basically the following assumptions on j:

$$j \text{ is } C^1 \text{ on } (0, +\infty), \; j'(0+) = 0, \tag{0.30}$$

$$\lim_{r \to +\infty} \frac{j(r)}{r} = +\infty \tag{0.31}$$

On the function V we assume $V(x) \to 0$ as $|x| \to \infty$ in the weak sense that

$$\text{meas}[|V| > \delta] < \infty \quad \text{for every } \delta > 0. \tag{0.32}$$

and that
$$\omega = j^*((1 + \theta)(V - M)) \in L^1, \tag{H$^+$}$$

for some constants $\theta > 0$ and $M > 0$. It follows from (H$^+$) that

$$-V\rho \geqslant \; - j(\rho) - \omega - M\rho$$

and consequently $\mathcal{E}(\rho)$ is well defined in $(-\infty, +\infty]$ for every $\rho \in L^1, \rho \geqslant 0$ and $j(\rho) \in L^1$.

We then consider the auxilary problem, for every $\lambda \in \mathbb{R}$,

$$\inf \left\{ \mathcal{E}(\rho) + \lambda \int \rho; \rho \geqslant 0, \rho \in L^1 \text{ and } j(\rho) \in L^1 \right\}. \tag{P$_\lambda$}$$

We will also make the assumption

$$\operatorname*{ess\,sup}_{\mathbb{R}^N} V > 0, \tag{0.33}$$

which is quite natural. If it is not satisfied, then $V \leqslant 0$ a.e. on $\mathbb{R}^N$ and the unique minimizer in (P_λ) is $\rho = 0$.

In Section 2 we prove the following

THEOREM 0.2. *Assume* (H^+), *(0.30), (0.31), (0.32) and (0.33). Then, for every* $\lambda > 0$, (P_λ) *admits a unique minimizer* ρ_λ, *and* ρ_λ *satisfies (0.16). Set*

$$I(\lambda) = \int \rho_\lambda, \quad \lambda > 0.$$

Then the function $\lambda \longmapsto I(\lambda)$ *is nonincreasing, and continuous from* $(0, \infty)$ *into* $[0, \infty)$ *More precisely,*

$$I(\lambda) \text{ is decreasing on } (0, \operatorname*{ess\,sup}_{\mathbb{R}^N} V), \tag{0.34}$$

$$\begin{cases} I(\lambda) = 0 \quad \forall \lambda \geqslant \operatorname{ess\,sup}_{\mathbb{R}^N} V & \textit{if } \operatorname{ess\,sup}_{\mathbb{R}^N} V < \infty, \\ \lim_{\lambda \to \infty} I(\lambda) = 0 & \textit{if } \operatorname{ess\,sup}_{\mathbb{R}^N} V = \infty, \end{cases} \tag{0.35}$$

$$\begin{cases} I_0 = \lim_{\lambda \downarrow 0} I(\lambda) = \sup_{\lambda > 0} I(\lambda) < \infty & \textit{if and only if} \\ (P_0) \textit{ admits a minimizer} \rho_0, & \textit{and then } I_0 = \int \rho_0. \end{cases} \tag{0.36}$$

As a consequence, we easily derive,

COROLLARY 0.3. *Under the assumptions of Theorem 0.2, we have*

$$\begin{cases} \textit{for every } I \in (0, I_0) \textit{ problem } (M_I) \textit{ admits a unique minimizer} \\ \rho^I = \rho_\lambda, \textit{ where } \lambda > 0 \textit{ is the unique solution of } I(\lambda) = I; \end{cases} \tag{0.37}$$

if $I_0 < \infty$, *problem* (M_{I_0}) *admits* ρ_0 *as its unique minimizer;* $\tag{0.38}$

if $I_0 < \infty$ *and* $I > I_0$, *problem* (M_I) *admits no minimizer.* $\tag{0.39}$

In Section 3 we investigate situations where assumption (H) is *not* satisfied. For example $N = 3$, V of the form (0.1) and $j(r) = \frac{1}{p} r^p$, $r \geqslant 0$, with p in the range (0.29). We tackle *directly* the Euler equation (0.16), first with $\lambda \geqslant 0$ prescribed and then with λ free and $I = \int \rho$ prescribed.

The main result in Section 3 is

THEOREM 0.4. *Assume (0.30). Let V be any measurable function satisfying (0.32) and (0.33). Then, there exists $\lambda_1 \in [0, +\infty]$ such that*

for every $\lambda > \lambda_1$ (and $\lambda < +\infty$) there is a unique solution ρ_λ of (0.16) $\tag{0.40}$

for $\lambda < \lambda_1$ there is no solution of (0.16). $\tag{0.41}$

Moreover the function $I(\lambda) = \int \rho_\lambda$ *is nonincreasing continuous on* $(\lambda_1, +\infty)$, *and*

$$I_1 = \sup_{\lambda > \lambda_1} \int \rho_\lambda = \lim_{\lambda \downarrow \lambda_1} \int \rho_\lambda$$

is finite if and only if (0.16) *admits a solution for* $\lambda = \lambda_1$.

It may well happen that λ_1 in Theorem 0.4 is $+\infty$, meaning that there exists *no* λ for which (0.16) has a solution. Consider the case $N = 3$, $V(x)$ of the form (0.1) and $j(r) = \frac{1}{p} r^p, r > 0$, with $1 < p \leqslant \frac{4}{3}$. Then (0.16) is equivalent to

$$-\Delta u + [(u - \lambda)^+]^{1/(p-1)} = \sum m_i \delta_{a_i} \tag{0.42}$$

and we know from the nonexistence result (Remark A. 4) in Appendix A that for any $\lambda \geqslant 0$, (0.42) has no solution.

The numbers λ_1 and I_1 in Theorem 0.4 play a central role and it is important to determine their value in concrete situations. This is the content of Section 4. Here are some typical results

THEOREM 0.5. *Assume* (0.30). *Let V be any measurable function satisfying* (0.33) *and*

$$V = k * f \text{ for some } f \in L^1, \text{ i.e., } f = -\Delta V \in L^1. \tag{0.43}$$

Then, $\lambda_1 = 0$ *and*

$$0 < I_1 \leqslant \int f^+. \tag{0.44}$$

Under the additional assumption

$$j(r) \sim r^p \text{ near } r = 0 \text{ with } p \geqslant 2(N-1)/N, \tag{0.45}$$

then

$$\int f \leqslant I_1 \leqslant \int f^+ \tag{0.46}$$

and in particular

$$I_1 = \int f \quad \text{if } f \geqslant 0. \tag{0.47}$$

The proof of (0.46) relies heavily on the following ingredient established in Appendix B.

LEMMA 0.6. *Assume*

$$v = k * \mu, \text{ for some bounded measure in } \mathbb{R}^N \tag{0.48}$$

and

$$v^+ \in L^q(\mathbb{R}^N) \text{ for some } q \leqslant N/(N-2) \tag{0.49}$$

then

$$\int_{\mathbb{R}^N} \mu \leqslant 0. \tag{0.50}$$

Assumption (0.43) does not hold e.g. when $N = 3$ and V is given by (0.1) since ΔV is a measure and not an L^1 function. In this case we have

THEOREM 0.7. *Same assumptions as in Theorem* 0.5 *except that we replace* (0.43) *by*

$$V = k * f, \text{ for some bounded measure } f. \tag{0.43'}$$

Assume in addition that

$$j(r) \sim r^p \text{ as } r \to \infty, \text{ with } p > 2(N-1)/N. \tag{0.51}$$

Then all the conclusions of Theorem 0.5 *hold.*

Putting together all the above results, consider now the case where

$$j(r) = r^p, r \geqslant 0, \text{ with } p > 2(N-1)/N, \tag{0.52}$$

$$V = k * f \text{ for some bounded measure } f \geqslant 0, f \not\equiv 0. \tag{0.53}$$

Set

$$I_1 = \int f.$$

COROLLARY 0.8. *Assume* (0.52) *and* (0.53). *Given any* $I \in (0, I_1]$ *there exists a unique pair* $\rho \in L^1, \rho \geqslant 0,$ *and* $\lambda \geqslant 0,$ *denoted* $\rho_I, \lambda_I,$ *satisfying*

$$\rho^{p-1} - V + k * \rho = -\lambda \quad \text{in } [\rho > 0], \tag{0.54}$$

$$-V + k * \rho \geqslant -\lambda \quad \text{in } [\rho = 0], \tag{0.55}$$

$$\int \rho = I \tag{0.56}$$

When $I = I_1,$ *then* $\rho_I > 0$ *a.e. and* $\lambda_I = 0.$ *When* $I > I_1,$ *problem* (0.54)–(0.55)–(0.56) *has no solution. Under the stronger assumption* $p > N/2,$ ρ_I *is also the unique minimizer of* $\mathcal{E}(\rho)$ *subject to the constraint* $\{\rho \in L^1 \cap L^p, \rho \geqslant 0 \text{ and } \int \rho = I\}.$

In Sections 5 and 6 we solve two problems raised by Lieb-Simon [48]. The first one concerns the uniqueness of the extremal in some minmax principle.

THEOREM 0.9. *Consider for simplicity the setting of Corollary 0.8 with*

$$2(N-1)/N < p \leqslant 2. \tag{0.57}$$

Let $I \in (0, I_1]$, then λ_I given by Corollary 0.8 satisfies

$$\lambda_I = \max_{\substack{\rho \in L^1, \rho \geqslant 0 \\ \int \rho = I}} \operatorname*{ess\,inf}_{[\rho > 0]} \{V - \rho^{p-1} - k * \rho\} \tag{0.58}$$

and

$$\lambda_I = \min_{\substack{\rho \in L^1, \rho \geqslant 0 \\ \int \rho = I}} \operatorname*{ess\,sup}_{\mathbb{R}^N} \{V - \rho^{p-1} - k * \rho\}. \tag{0.59}$$

If $I < I_1$, the max (resp. min) in (0.58) (resp. (0.59)) is uniquely achieved by the solution ρ_I obtained in Corollary 0.8.

When $N = 3$ and $p = 4/3$, assertions (0.58) and (0.59) are due to Lieb-Simon [48]. They asked whether the max in (0.58) and the min in (0.59) are uniquely achieved. The answer is indeed positive when $I < I_1$. As we shall see in Section 5, the answer is *negative* when $I = I_1$. The proof of uniqueness in Theorem 0.9 involves a new form of the strong maximum principle with "bad" coefficients described in Appendix C.

Our last result concerns the asymptotic behavior of λ_I as $I \uparrow I_1$.

THEOREM 0.10. *Consider for simplicity the setting for Corollary 0.8, with*

$$2(N-1)/N < p < 2. \tag{0.60}$$

Then

$$\alpha = \lim_{I \uparrow I_1} \frac{\lambda_I}{(I_1 - I)^\tau} \quad exists, \tag{0.61}$$

where $\tau = 2(p-1)/(2-2N+pN)$ and the positive constant α can be computed explicitly via the solution of an elementary ODE.

The exact value of α is given in Theorem 9 (Section 6).

1. The variational problem and its Euler equation; conditions for equivalence

Let Ω be a σ-finite measure space with measure dx. Let $j : \Omega \times \mathbb{R} \to [0, +\infty]$ be a normal convex integrand, i.e., $j(x, r)$ is measurable and, for a.e. $x \in \Omega$, $j(x, \cdot)$ is convex l.s.c. (= lower semi-continuous). We assume that

$$j(x, 0) = 0 \text{ for a.e. } x \in \Omega \text{ and } j(x, r) = +\infty \text{ for a.e. } x \in \Omega \text{ and for all } r < 0. \tag{1.1}$$

Set

$$a(x) = \sup \{r \geqslant 0; \ j(x, r) < \infty\}. \tag{1.2}$$

Let $j^*(x, s)$ denote the conjugate convex function, that is,

$$j^*(x, s) = \sup_{r \in \mathbb{R}} \{sr - j(x, r)\} \text{ for a.e. } x \in \Omega, \text{ for all } s \in \mathbb{R}.$$

Note that

$$\begin{aligned} j^*(x, s) &= 0 &&\text{for a.e. } x \in \Omega, \quad \text{for all } s \leqslant 0, \\ j^*(x, s) &\geqslant 0 &&\text{for a.e. } x \in \Omega, \quad \text{for all } s \geqslant 0. \end{aligned}$$

Let $V : \Omega \to \mathbb{R}$ be a measurable function (so that $|V(x)| < \infty$ for a.e. $x \in \Omega$). The following assumption will sometimes play an important role:

$$\text{there exists a constant } M \text{ such that } j^*(\cdot, V(\cdot) - M) \in L^1(\Omega). \tag{H}$$

Note that assumption (H) holds for example if $V^+ \in L^\infty(\Omega)$.

Define the functional $J : L^1(\Omega) \to (-\infty, +\infty]$ to be

$$J(\rho) = \begin{cases} \int_\Omega \{j(x, \rho(x)) - V(x)\rho(x)\}\, dx & \text{if } j(\cdot, \rho) - V\rho \in L^1(\Omega), \\ +\infty & \text{otherwise.} \end{cases}$$

with

$$D(J) = \{\rho \in L^1(\Omega); \quad J(\rho) < +\infty\}.$$

In particular, if $\rho \in D(J)$, then $\rho(x) \geqslant 0$ for a.e. $x \in \Omega$.

REMARK 1. If (H) holds, then J is convex l.s.c. on $L^1(\Omega)$ and bounded below on bounded sets of $L^1(\Omega)$. This is a straightforward consequence of the fact that for every $\rho \in L^1(\Omega)$ we have $j(x, \rho(x)) - V(x)\rho(x) \geqslant -j^*(x, V(x) - M) - M\rho(x)$ for a.e. $x \in \Omega$ (so that we may use Fatou's lemma to check the lower semi-continuity of J). Note that if (H) does not hold, it may happen that $D(J)$ is not convex. Consider, for example $j(x, r) = r^2$ and a function V such that $V \geqslant 0$ a.e. and $V \notin L^2(\Omega)$; then $V \in D(J)$ while $\frac{1}{2}V \notin D(J)$.

Set

$$L_0^\infty(\Omega) = \{\rho \in L^\infty(\Omega); \rho = 0 \text{ outside a set of finite measure}\}.$$

Throughout the paper we shall assume that $k : \Omega \times \Omega \to \mathbb{R}$ is a measurable function satisfying

$$k(x, y) = k(y, x) \text{ and } k(x, y) \geqslant 0 \text{ for a.e. } x, y \in \Omega, \tag{1.3}$$

$$\begin{cases} \text{for every } \rho \in L_0^\infty(\Omega), \text{ then } k(x, y)\rho(x)\rho(y) \in L^1(\Omega \times \Omega) \\ \text{and } \iint_{\Omega \times \Omega} k(x, y)\rho(x)\rho(y)\,dx\,dy \geqslant 0 \end{cases} \tag{1.4}$$

(i.e., k is a nonnegative kernel).

Define the functional $K : L^1(\Omega) \to [0, +\infty]$ by

$$K(\rho) = \begin{cases} \frac{1}{2} \iint_{\Omega \times \Omega} k(x, y)\rho(x)\rho(y)\,dx\,dy & \text{if } \rho \in L^1(\Omega) \text{ and } \rho \geqslant 0 \text{ a.e. on } \Omega \\ +\infty & \text{otherwise,} \end{cases}$$

with

$$D(K) = \{\rho \in L^1(\Omega); \quad K(\rho) < +\infty\}.$$

Set

$$\mathcal{E}(\rho) = J(\rho) + K(\rho) \quad \text{for } \rho \in L^1(\Omega)$$

and

$$D(\mathcal{E}) = D(J) \cap D(K);$$

$\mathcal{E}$ is called the Thomas-Fermi energy functional.

Finally we introduce the mapping B defined for $\rho \in L^1(\Omega)$ with $\rho \geqslant 0$ a.e. on Ω, by

$$B\rho(x) = \int_\Omega k(x, y)\rho(y)\,dy.$$

The main result in this section is the following:

THEOREM 1. *Let $\rho_0 \in L^1(\Omega)$ with $\rho_0 \geqslant 0$ a.e. on Ω be such that*

$$0 < \int \rho_0(x)\,dx < \int a(x)\,dx, \tag{1.5}$$

and
$$\rho_0 \in D(\mathcal{E}) \text{ and } \mathcal{E}(\rho_0) \leqslant \mathcal{E}(\rho) \quad \forall \rho \in D(\mathcal{E}) \text{ with } \int \rho(x)\,dx = \int \rho_0(x)\,dx. \tag{M}$$

Then[1]
$$\begin{cases} \text{there exists a constant } \lambda \in \mathbb{R} \text{ such that} \\ \partial j(x, \rho_0(x)) + B\rho_0(x) \ni V(x) - \lambda \text{ for a.e. } x \in \Omega. \end{cases} \tag{E}$$

Conversely, when (H) *holds, then* (E) *implies* (M).

---

[1] $\partial j(x, r)$ denotes the subdifferential of $j(x, r)$ with respect to r.

REMARK 2. λ appears in (E) as a Lagrange multiplier corresponding to the constraint $\int \rho(x)\,dx = \int \rho_0(x)\,dx$ in the Euler equation (E) associated to the minimization problem (M).

In proving Theorem 1 we shall make use of the following:

LEMMA 1. *The functional K is convex l.s.c. on $L^1(\Omega)$. In addition*

$$\iint_{\Omega \times \Omega} k(x,y)\varphi(x)\psi(y)\,dxdy \leqslant K(\varphi) + K(\psi) \quad \forall \varphi, \psi \in D(K) \tag{1.6}$$

and equality in (1.6) holds if and only if $B\varphi = B\psi$. Moreover we have[2]

$$\text{if } \rho \in D(K) \text{ and } A \subset \Omega \text{ with } |A| < \infty, \text{ then } \chi_A B\rho \in L^1(\Omega). \tag{1.7}$$

Proof of Lemma 1. Let (Ω_n) be a nondecreasing sequence of measurable sets in Ω such that $|\Omega_n| < \infty \ \forall n$ and $\cup_n \Omega_n = \Omega$. Given $\varphi, \psi \in D(K) \cap L^\infty(\Omega)$ set

$$\varphi_n = \chi_{\Omega_n}\varphi \quad \text{and} \quad \psi_n = \chi_{\Omega_n}\psi.$$

By (1.4) we have

$$\iint_{\Omega \times \Omega} k(x,y)[\varphi_n(x) - \psi_n(x)][\varphi_n(y) - \psi_n(y)]\,dxdy \geqslant 0,$$

i.e.

$$\iint_{\Omega \times \Omega} k(x,y)\varphi_n(x)\psi_n(y)\,dxdy \leqslant K(\varphi_n) + K(\psi_n).$$

Using the monotone convergence theorem we obtain (1.6) for $\varphi, \psi \in D(K) \cap L^\infty(\Omega)$. The general case follows by truncation.

The function K is convex since for $\varphi, \psi \in D(K)$ and $t \in (0, 1)$ we have

$$\begin{aligned}
K((1-t)\varphi + t\psi) &= \frac{1}{2}\iint k(x,y)[(1-t)\varphi(x) + t\psi(x)][(1-t)\varphi(y) + t\psi(y)]\,dx\,dy \\
&= (1-t)^2 K(\varphi) + t^2 K(\psi) + t(1-t)\iint k(x,y)\varphi(x)\psi(x)\,dx\,dy \\
&\leqslant (1-t)^2 K(\varphi) + t^2 K(\psi) + t(1-t)[K(\varphi) + K(\psi)] \\
&= (1-t)K(\varphi) + tK(\psi).
\end{aligned}$$

The lower semi-continuity of K follows from Fatou's lemma.

---

[2] $|A| = \text{meas}(A)$ denotes the measure of A and χ_A denotes the characteristic function of A.

Next, let $\rho \in D(K)$ and $A \subset \Omega$ with $|A| < \infty$. We have

$$\int_A (B\rho)(x)\,dx = \iint_{\Omega \times \Omega} k(x, y)\rho(y)\chi_A(x)\,dxdy \leqslant K(\rho) + K(\chi_A).$$

Finally we show that equality in (1.6) holds if and only if $B\varphi = B\psi$.
First, suppose that $B\varphi = B\psi$; then we have

$$\int \varphi B\psi = \frac{1}{2}\int \varphi B\psi + \frac{1}{2}\int \psi B\varphi = K(\varphi) + K(\psi).$$

Conversely, assume that equality in (1.6) holds. Note that

$$\int (\psi + \zeta)B\varphi \leqslant K(\varphi) + K(\psi) + K(\zeta) + \int \zeta B\psi, \quad \forall \varphi, \psi, \zeta \in D(K),$$

and since (1.6) holds we obtain $\int \zeta B\varphi \leqslant K(\zeta) + \int \zeta B\psi$. Replacing ζ by $\lambda\zeta, \lambda > 0$,
we see that $\int \zeta B\varphi \leqslant \int \zeta B\psi \forall \zeta \in D(K)$. Reversing φ and ψ we find $\int \zeta B\varphi = \int \zeta B\psi$
$\forall \zeta \in D(K)$ and consequently $\int \zeta B\varphi = \int \zeta B\psi \forall \zeta \in L_0^\infty(\Omega)$. Therefore we have $B\varphi = B\psi$.

REMARK 3. The argument above shows that K is a strictly convex function on $D(K)$
if and only if B is injective.

Proof of Theorem 1. (E) $\Rightarrow$ (M) (*under assumption* (H)).
Indeed, by (E) and the definition of the subdifferential, we have for $\rho \in L^1(\Omega)$

$$j(\cdot, \rho) \geqslant j(\cdot, \rho_0) + (V - B\rho_0 - \lambda)(\rho - \rho_0) \quad \text{a.e. on } \Omega. \tag{1.8}$$

In particular, for $\rho \equiv 0$, we find

$$j(\cdot, \rho_0) - V\rho_0 + (B\rho_0)\rho_0 \leqslant -\lambda\rho_0 \text{ a.e. on } \Omega.$$

From (H) it follows that $j(\cdot, \rho_0) - V\rho_0$ is bounded below by some L^1 function; thus
$\rho_0 \in D(\mathcal{E})$. Now let $\rho \in D(\mathcal{E})$ with $\int \rho(x)\,dx = \int \rho_0(x)\,dx$; integrating (1.8) and using
(1.6) we obtain (M).
(M) $\Rightarrow$ (E) (*without assumption* (H), *but with* (1.5)).
　First let $\zeta \in D(\mathcal{E})$ with $\int \zeta(x)\,dx = \int \rho_0(x)\,dx$ and $V(\zeta - \rho_0) \in L^1(\Omega)$.
Let

$$\rho_t = (1 - t)\rho_0 + t\zeta \quad \text{with } 0 < t < 1.$$

We claim that

$$\rho_t \in D(J) \text{ and } J(\rho_t) \leqslant (1 - t)J(\rho_0) + tJ(\zeta). \tag{1.9}$$

Indeed we have a.e. on Ω,

$$j(\cdot, \rho_t) - V\rho_t \leqslant (1 - t)(j(\cdot, \rho_0) - V\rho_0) + t(j(\cdot, \zeta) - V\zeta). \tag{1.10}$$

On the other hand we have, a.e. on Ω,

$$j(\cdot, \rho_t) \geq \min\{j(\cdot, \rho_0), j(\cdot, \zeta)\}$$

– this follows from the monotonicity of $j(x, \cdot)$ on $[0, +\infty)$. Therefore we obtain a.e. on Ω,

$$j(\cdot, \rho_t) - V\rho_t \geq \min\{j(\cdot, \rho_0) - V\rho_0, j(\cdot, \zeta) - V\zeta\} - 2|V(\rho_0 - \zeta)|. \tag{1.11}$$

Combining (1.10) and (1.11) we see that $\rho_t \in D(J)$ and integrating (1.10) we find

$$J(\rho_t) \leq (1 - t)J(\rho_0) + tJ(\zeta).$$

It follows that

$$\mathcal{E}(\rho_t) \leq (1 - t)J(\rho_0) + tJ(\zeta) + (1 - t)^2 K(\rho_0) + t^2 K(\zeta) + t(1 - t)\int (B\rho_0)\zeta.$$

By assumption (M) we have

$$\mathcal{E}(\rho_0) = J(\rho_0) + K(\rho_0) \leq \mathcal{E}(\rho_t)$$

and thus

$$tJ(\rho_0) + (2t - t^2)K(\rho_0) \leq tJ(\zeta) + t^2 K(\zeta) + t(1 - t)\int (B\rho_0)\zeta.$$

Dividing by t and letting $t \to 0$ we find

$$\begin{cases} J(\rho_0) + \int (B\rho_0)\rho_0 \leq J(\zeta) + \int (B\rho_0)\zeta \\ \forall \zeta \in D(\mathcal{E}) \text{ with } \int \zeta(x)\,dx = \int \rho_0(x)\,dx \text{ and } V(\zeta - \rho_0) \in L^1(\Omega). \end{cases} \tag{1.12}$$

Set $\widetilde{V} = V - B\rho_0$ and define the functional $\widetilde{J} : L^1(\Omega) \to (-\infty, +\infty]$ by

$$\widetilde{J}(u) = \begin{cases} \int\{j(x, u(x)) - \widetilde{V}(x)u(x)\}\,dx & \text{if } j(\cdot, u) - \widetilde{V}u \in L^1(\Omega), \\ +\infty & \text{otherwise.} \end{cases}$$

It is clear that $\rho_0 \in D(\widetilde{J})$ (since $\rho_0 \in D(\mathcal{E})$).

We claim that

$$\begin{cases} \widetilde{J}(\rho_0) \leq \widetilde{J}(\zeta) \quad \forall \zeta \in D(\widetilde{J}) \\ \text{with } \int \zeta(x)\,dx = \int \rho_0(x)\,dx, \ (\zeta - \rho_0) \in L_0^\infty(\Omega) \text{ and } \widetilde{V}(\zeta - \rho_0) \in L^1(\Omega). \end{cases} \tag{1.13}$$

Indeed, suppose ζ satisfies the assumptions in (1.13), then ζ also satisfies the assumptions in (1.12). Note that:

a) $\zeta \in D(K)$, since $\zeta \leq \rho_0 + |\zeta - \rho_0| \in D(K) + D(K)$;
b) $\zeta \in D(\widetilde{J}) \cap D(K) \Rightarrow \zeta \in D(J)$, since $j(\cdot, \zeta) - V\zeta = j(\cdot, \zeta) - \widetilde{V}\zeta - (B\rho_0)\zeta$;
c) $V(\zeta - \rho_0) = \widetilde{V}(\zeta - \rho_0) + (B\rho_0)(\zeta - \rho_0)$.

We conclude the proof of Theorem 1 with the help of the next lemma applied to $\tilde{J}$ (instead of J).

LEMMA 2. *Assume $\rho_0 \in D(J)$ satisfies (1.5), as well as*

$$\begin{cases} J(\rho_0) \leqslant J(\rho) \quad \forall \rho \in D(J) \\ \text{with } \int \rho(x)\,dx = \int \rho_0(x)\,dx, \ (\rho - \rho_0) \in L_0^\infty(\Omega) \text{ and } V(\rho - \rho_0) \in L^1(\Omega). \end{cases} \quad (1.14)$$

Then there exists a constant $\lambda \in \mathbb{R}$ such that

$$V - \lambda \in \partial j(\cdot, \rho_0) \quad a.e. \ on \ \Omega. \tag{1.15}$$

Proof. Set

$$E_- = \{x \in \Omega \,;\, \rho_0(x) > 0\},$$
$$E_+ = \{x \in \Omega \,;\, \rho_0(x) < a(x)\}.$$

It follows from (1.5) that $|E_-| > 0$ and $|E_+| > 0$. Let (Ω_n) be as in the proof of Lemma 1 and set

$$\Omega_n' = \{x \in \Omega_n \,;\, |V(x)| + \rho_0(x) < n\},$$

so that $|E_- \cap \Omega_n'| \uparrow |E_-|$ and $|E_+ \cap \Omega_n'| \uparrow |E_+|$ as $n \to \infty$. Fix n_0 such that

$$|E_- \cap \Omega_{n_0}'| > 0 \quad \text{and} \quad |E_+ \cap \Omega_{n_0}'| > 0.$$

In what follows we choose $n \geqslant n_0$; for every $\lambda \in \mathbb{R}$ set

$$u_\lambda(x) = (I + \partial j(x, \cdot))^{-1}(V(x) + \rho_0(x) - \lambda) \quad \text{for } x \in \Omega, \tag{1.16}$$

$$I_\lambda = \int_{\Omega_n'} u_\lambda(x)\,dx. \tag{1.17}$$

Note that I_λ makes sense since $|u_\lambda(x)| \leqslant n + |\lambda|$ on Ω_n' and $|\Omega_n'| < \infty$. Clearly we have $u_\lambda(x) \uparrow a(x)$ as $\lambda \downarrow -\infty$ and $u_\lambda(x) \downarrow 0$ as $\lambda \uparrow +\infty$. Therefore

$$\lim_{\lambda \to -\infty} I_\lambda = \int_{\Omega_n'} a(x)\,dx \quad \text{and} \quad \lim_{\lambda \to +\infty} I_\lambda = 0.$$

On the other hand we have

$$0 < \int_{\Omega_n'} \rho_0(x)\,dx < \int_{\Omega_n'} a(x)\,dx$$

since $n \geqslant n_0$. Thus, there exists a constant $\lambda_n \in \mathbb{R}$ such that

$$I_{\lambda_n} = \int_{\Omega_n'} \rho_0(x)\,dx \tag{1.18}$$

(note that I_λ is a continuous function of λ and, in fact, $|I_\lambda - I_\mu| \leqslant |\lambda - \mu|\,|\Omega'_n|$). It follows from (1.16) that, a.e. on Ω,

$$u_{\lambda_n}(x) + \partial j(x, u_{\lambda_n}(x)) \ni V(x) + \rho_0(x) - \lambda_n \tag{1.19}$$

and so

$$j(x, \rho_0(x)) - j(x, u_{\lambda_n}(x)) \geqslant (V(x) + \rho_0(x) - \lambda_n - u_{\lambda_n}(x))(\rho_0(x) - u_{\lambda_n}(x)).$$

Hence a.e. on Ω we find

$$\begin{aligned}
j(x, u_{\lambda_n}(x)) &- V(x)u_{\lambda_n}(x) \\
&\leqslant j(x, \rho_0(x)) - V(x)\rho_0(x) - (\rho_0(x) - u_{\lambda_n}(x))^2 - \lambda_n(u_{\lambda_n}(x) - \rho_0(x)).
\end{aligned} \tag{1.20}$$

On the other hand we have, a.e. on Ω'_n,

$$j(x, u_{\lambda_n}(x)) - V(x)u_{\lambda_n}(x) \geqslant -V(x)u_{\lambda_n}(x) \geqslant -n(n + |\lambda_n|). \tag{1.21}$$

Combining (1.20) and (1.21) we see that

$$j(\cdot, u_{\lambda_n}) - Vu_{\lambda_n} \in L^1(\Omega'_n).$$

Set

$$\rho = \begin{cases} u_{\lambda_n} & \text{on } \Omega'_n, \\ \rho_0 & \text{on } \Omega \backslash \Omega'_n. \end{cases}$$

Therefore ρ satisfies all the assumptions in (1.14) and we deduce that

$$\int_{\Omega'_n} \{j(x, \rho_0(x)) - V(x)\rho_0(x)\}\, dx \leqslant \int_{\Omega'_n} \{j(x, u_{\lambda_n}(x)) - V(x)u_{\lambda_n}(x)\}\, dx. \tag{1.22}$$

Combining (1.20) and (1.22) we find

$$\rho_0 = u_{\lambda_n} \quad \text{a.e. on } \Omega'_n.$$

It follows from (1.19) that

$$V(x) - \lambda_n \in \partial j(x, \rho_0(x)) \quad \text{for a.e. } x \in \Omega'_n.$$

For every $n \geqslant n_0$, set

$$\Lambda_n = \{\lambda \in \mathbb{R};\ V(x) - \lambda \in \partial j(x, \rho_0(x)) \quad \text{for a.e. } x \in \Omega'_n\}.$$

We have just established that $\Lambda_n \neq \emptyset$. Clearly Λ_n is a closed interval. Moreover Λ_n is bounded; indeed if, for instance, Λ_n were unbounded below we would have $\rho_0(x) = a(x)$ for a.e. $x \in \Omega'_n$ – a contradiction with $|E_+ \cap \Omega'_n| > 0$. Since Λ_n decreases with n we obtain

$$\bigcap_{n \geqslant n_0} \Lambda_n \neq \emptyset$$

and the conclusion of Lemma 2 follows. $\qquad\square$

Our next lemma – which will be used later – is closely related to Theorem 1.

LEMMA 3. *Let $\rho_0 \in L^1(\Omega)$ with $\rho_0 \geqslant 0$ a.e. on Ω be such that*

$$\rho_0 \in D(\mathcal{E}) \text{ and } \mathcal{E}(\rho_0) \leqslant \mathcal{E}(\rho) \quad \forall \rho \in D(\mathcal{E}). \tag{1.23}$$

Then

$$\partial j(x, \rho_0(x)) + B\rho_0(x) \ni V(x) \quad \text{for a.e. } x \in \Omega. \tag{1.24}$$

Conversely, when (H) *holds, then* (1.24) *implies* (1.23).

REMARK 4. Note that assumption (1.5) is not required in Lemma 3.

Proof of Lemma 3. In order to prove that $(1.24) \Rightarrow (1.23)$ under assumption (H) one proceeds exactly as in the proof of (E) $\Rightarrow$ (M).

In order to prove that $(1.23) \Rightarrow (1.24)$ one uses the same $\widetilde{V}$ and $\widetilde{J}$ as in the proof of Theorem 1 and one shows that

$$\widetilde{J}(\rho_0) \leqslant \widetilde{J}(\zeta) \quad \forall \zeta \in D(\widetilde{J}) \text{ with } (\zeta - \rho_0) \in L_0^\infty(\Omega) \text{ and } \widetilde{V}(\zeta - \rho_0) \in L^1(\Omega). \tag{1.25}$$

Next one considers

$$\Omega_n' = \{x \in \Omega_n; \ |\widetilde{V}(x)| + \rho_0(x) < n\}$$

and one uses (1.25) with

$$\zeta = \begin{cases} u & \text{on } \Omega_n', \\ \rho_0 & \text{on } \Omega \backslash \Omega_n', \end{cases}$$

where $u(x) = (I + \partial j(x, \cdot))^{-1}(\widetilde{V}(x) + \rho_0(x))$. This leads to $\rho_0 = u$ a.e. on Ω_n' and $\widetilde{V}(x) \in \partial j(x, \rho_0(x))$ for a.e. $x \in \Omega_n'$.

REMARK 5. Suppose that (E) and (H) hold. Then we have, in fact, a stronger conclusion than (M), namely

$$\mathcal{E}(\rho_0) + \lambda \int \rho_0 \leqslant \mathcal{E}(\rho) + \lambda \int \rho \quad \forall \rho \in D(\mathcal{E}). \tag{1.26}$$

This follows from Lemma 3 applied with $(V - \lambda)$ instead of V. In particular, if we happen to know that $\lambda \geqslant 0$ (for example Lemma 8 implies that this holds when $V_\infty - j'(0+) \geqslant 0$, where V_∞ is defined at the beginning of Section 2), then we have

$$\mathcal{E}(\rho_0) \leqslant \mathcal{E}(\rho) \quad \forall \rho \in D(\mathcal{E}) \text{ with } \int \rho(x)dx \leqslant \int \rho_0(x)dx. \tag{1.27}$$

This explains why one can use a "relaxation" method (see Lieb-Simon [48] and also Proposition 3 in Section 2). In other words, the constraint $\int \rho(x)dx = I$ in the minimization problem is "relaxed" to $\int \rho(x)dx \leqslant I$.

REMARK 6. Assume (H) holds. Then we have

$$\overline{D(\mathcal{E})}^{L^1} = \{\rho \in L^1(\Omega); 0 \leqslant \rho(x) \leqslant a(x) \text{a.e. on } \Omega\} \tag{1.28}$$

and consequently, for every constant I with $0 \leqslant I < \int a(x)\,dx$, there is some $\rho \in D(\mathcal{E})$ such that $\int \rho(x)dx = I$. For this purpose, it suffices to show that every function $\rho \in L_0^\infty(\Omega)$ such that $0 \leqslant \rho(x) \leqslant a(x)$ a.e. on Ω, belongs to $\overline{D(\mathcal{E})}^{L^1}$. Indeed, set

$$\rho_\varepsilon(x) = \frac{(\rho(x) - \varepsilon)^+}{1 + \varepsilon j(x, (\rho(x) - \varepsilon)^+) + \varepsilon|V(x)|}, \quad \varepsilon > 0,$$

and note that $\rho_\varepsilon \in D(\mathcal{E})$ and $\rho_\varepsilon \to \rho$ in $L^1(\Omega)$ as $\varepsilon \to 0$.

2. Existence via the variational route

Given a constant I with $0 \leqslant I < \infty$ we set

$$K_I = \left\{\rho \in D(\mathcal{E}); \int \rho(x)dx = I\right\}.$$

In this section we are concerned with the following problem:

$$\text{find } \bar{\rho} \in K_I \text{ such that } \mathcal{E}(\bar{\rho}) \leqslant \mathcal{E}(\rho) \quad \forall \rho \in K_I. \tag{M_I}$$

For simplicity, we shall now assume that $j(x, r) = j(r)$ is independent of x and we set

$$a = \sup\{r \geqslant 0; \ j(r) < \infty\} \leqslant \infty.$$

Of course, we assume that $a > 0$.

We recall (see Remark 6) that $K_I \neq \emptyset$ for every $I < a|\Omega|$. When $I = a|\Omega|$ (assuming $a|\Omega| < \infty$), then either K_I is reduced to a single element $\{a\}$ or $K_I = \emptyset$ – so that problem (M_I) has no interest. Therefore we may always assume that $I < a|\Omega|$.

We shall encounter two different situations:

- in CASE I, a strong assumption (on V or Ω) implies that problem (M_I) has a solution for every $I < a|\Omega|$,
- in CASE II, problem (M_I) has a solution only for a *limited* range of I's, usually smaller than the interval $[0, a|\Omega|)$.

Throughout Section 2 we make an assumption slightly stronger than (H), namely

$$\text{there exist constants } \theta > 0 \text{ and } M \in \mathbb{R} \text{ such that } j^*((1+\theta)(V - M)) \in L^1(\Omega). \quad \text{(H}^+\text{)}$$

We also assume that j is coercive, i.e.,

$$\lim_{r \to +\infty} \frac{j(r)}{r} = +\infty. \tag{2.1}$$

Finally, we set[3]

$$V_\infty = \inf\{\alpha \in \mathbb{R}; \quad [V > \alpha] \text{ has finite measure}\}.$$

Note that there exist α's such that $[V > \alpha]$ has finite measure (this is so because (H) holds and $j^* \not\equiv 0$ since $a > 0$). Therefore we have either $V_\infty \in \mathbb{R}$ or $V_\infty = -\infty$. Of course if $|\Omega| < \infty$, then we have $V_\infty = -\infty$. In the special case where $\Omega = \mathbb{R}^N$ and $V(\infty) = \lim_{|x|\to\infty} V(x)$ exists, then $V_\infty = V(\infty)$.

CASE I. We assume here that

$$V_\infty = -\infty. \tag{2.2}$$

The main result is the following:

THEOREM 2. *Assume* (H$^+$), (2.1) *and* (2.2). *Then, for every* I *with* $0 \leqslant I < a|\Omega|$ *there exists a solution of* (M$_I$).

In the proof of Theorem 2 we shall use

LEMMA 4. *Assume* (H$^+$). *Let* (ρ_n) *be a sequence in* $D(J)$ *such that*

$$\int j(\rho_n) - V\rho_n \leqslant C_1 \text{ and } \int \rho_n \leqslant C_1 \quad \forall n, \text{ for some constant } C_1 > 0. \tag{2.3}$$

Then, there exists a constant C_2 *such that*

$$\int j(\rho_n) \leqslant C_2 \text{ and } \int |V\rho_n| \leqslant C_2 \quad \forall n. \tag{2.4}$$

Proof of Lemma 4. Set $\omega(x) = j^*((1 + \theta)(V(x) - M))$ so that $\omega \in L^1(\Omega)$ and $(1+\theta)(V - M)\rho_n \leqslant j(\rho_n) + \omega$. It follows that

$$V\rho_n \leqslant \frac{1}{1+\theta} j(\rho_n) + \omega + M\rho_n \tag{2.5}$$

and, using (2.3), we obtain

$$\int j(\rho_n) \leqslant C_1 + \frac{1}{1+\theta} \int j(\rho_n) + \int \omega + |M|\,C_1.$$

This leads to $\int j(\rho_n) \leqslant C_2$. Next, set

$$f_n = \frac{1}{1+\theta} j(\rho_n) + \omega + M\rho_n - V\rho_n$$

---

[3]We use the notations $[V > \alpha] = \{x \in \Omega; \ V(x) > \alpha\}$ and $[V \geqslant \alpha] = \{x \in \Omega; V(x) \geqslant \alpha\}$ etc...

so that $f_n \geqslant 0$ (by (2.5)). From (2.3) we have

$$\int j(\rho_n) + f_n - \frac{1}{1+\theta}\, j(\rho_n) - \omega - M\rho_n \leqslant C_1$$

and thus

$$\int |f_n| \leqslant C_1 + \int \omega + |M|\,C_1.$$

It follows that

$$\int |\nabla \rho_n| \leqslant C_1 + 2\int \omega + 2\,|M|\,C_1 + \int j(\rho_n)$$

and therefore we obtain a bound for $\int |\nabla \rho_n|$.

Proof of Theorem 2. From assumption (H) we have

$$j(\rho) - V\rho \geqslant -j^*(V - M) - M\rho \quad \forall \rho \in K_I,$$

so that

$$\mathcal{E}(\rho) \geqslant -\int j^*(V - M) - MI \quad \forall \rho \in K_I,$$

and consequently

$$E(I) = \inf_{\rho \in K_I} \mathcal{E}(\rho) > -\infty. \tag{2.6}$$

Let (ρ_n) be a minimizing sequence for (2.6). From Lemma 4 we deduce that

$$\int j(\rho_n) \leqslant C \tag{2.7}$$

and

$$\int |V|\rho_n \leqslant C, \tag{2.8}$$

for some constant C. We claim that the sequence (ρ_n) is equi-integrable in Ω, that is,

$$\forall \varepsilon > 0 \ \exists \delta > 0 \quad \text{such that} \int_A \rho_n < \varepsilon \ \forall n,$$
$$\forall A \subset \Omega \text{ measurable with } |A| < \delta, \tag{2.9}$$

and

$$\forall \varepsilon > 0 \ \exists \Omega' \subset \Omega \text{ measurable with } |\Omega'| < \infty \text{ such that } \int_{\Omega \setminus \Omega'} \rho_n < \varepsilon \ \forall n. \tag{2.10}$$

Verification of (2.9). Given any $k > 0$, there is a constant C_k such that

$$j(r) \geqslant kr - C_k \quad \forall r \geqslant 0$$

(this follows from (2.1)).

Consequently, we have for every measurable set $A \subset \Omega$

$$k \int_A \rho_n \leqslant \int_\Omega j(\rho_n) + C_k \, |A|$$

so that (by (2.7)) we obtain

$$\int_A \rho_n \leqslant \frac{C}{k} + \frac{C_k}{k} \, |A|.$$

Given $\varepsilon > 0$ we fix k large so that $\frac{C}{k} < \frac{\varepsilon}{2}$ and then we choose $\delta > 0$ so small that $\frac{C_k}{k} \delta < \frac{\varepsilon}{2}$.

Verification of (2.10). We recall that

$$\int |\nabla \rho_n| \leqslant C.$$

Choose $k > 0$ so large that $\frac{C}{k} < \varepsilon$ and set $\Omega' = [V > -k]$. It follows from assumption (2.2) that $|\Omega'| < \infty$. Clearly, we have

$$k \int_{\Omega \backslash \Omega'} \rho_n \leqslant \int_{\Omega \backslash \Omega'} |\nabla \rho_n| \leqslant C$$

and thus

$$\int_{\Omega \backslash \Omega'} \rho_n \leqslant \frac{C}{k} < \varepsilon \quad \forall n.$$

We may therefore apply the Dunford-Pettis theorem (see e.g. Dunford-Schwartz [30], Corollary IV.8.11) and conclude that there exists a subsequence (ρ_{n_k}) such that $\rho_{n_k} \rightharpoonup \bar\rho$ weakly in $L^1(\Omega)$. It follows that $\int \bar\rho = I$ and $\mathcal{E}(\bar\rho) \leqslant \inf_{\rho \in K_I} \mathcal{E}(\rho)$ (since $\mathcal{E}$ is convex and l.s.c. on $L^1(\Omega)$).

We now turn to Case II, which is the most important from the point of view of applications.

CASE II. We assume here that

$$V_\infty > -\infty.$$

This implies in particular that $|\Omega| = \infty$. For simplicity we will assume throughout the rest of this section that

$$V_\infty = 0. \tag{2.11}$$

This is just a normalization condition since in the problems of interest we may always add a constant to V. Note that (2.11) implies in particular that ess $\sup_\Omega V \geqslant 0$.

Concerning j we will assume that $j : \mathbb{R} \to [0, +\infty]$ is convex l.s.c.,

$$j(r) = +\infty \quad \text{for } r < 0 \text{ and } j(0) = 0, \tag{2.12}$$

$$j \text{ is finite and } C^1 \text{ on } (0, \infty), \tag{2.13}$$

$$j'(0+) = \lim_{r \downarrow 0} \frac{j(r)}{r} = 0. \tag{2.14}$$

In addition, we assume that

$$\mathcal{E} \text{ is strictly convex on } D(\mathcal{E}) \tag{2.15}$$

and

$$\text{for every } \rho \in D(\mathcal{E}) \text{ and every } \delta > 0, \text{ the set } [B\rho > \delta] \text{ has finite measure.} \tag{2.16}$$

Condition (2.16) says that, in some weak sense, $B\rho \to 0$ at "infinity".

In order to study problem (M_I), it will be extremely useful to introduce an auxiliary problem. For every $\lambda \in \mathbb{R}$, consider

$$\inf \left\{ \mathcal{E}(\rho) + \lambda \int \rho; \quad \rho \in D(\mathcal{E}) \right\}. \tag{P_λ}$$

The main result is the following:

THEOREM 3. *Assume* (H^+), *(2.1), (2.11), (2.12), (2.13), (2.14), (2.15), and (2.16). Then,*

$$\textit{for every } \lambda > 0, \textit{ problem } (P_\lambda) \textit{ admits a unique minimizer } \rho_\lambda, \tag{2.17}$$

$$\textit{for every } \lambda < 0, \textit{ the infimum in } (P_\lambda) \textit{ is } -\infty. \tag{2.18}$$

Set

$$I(\lambda) = \int \rho_\lambda, \quad \lambda > 0.$$

Then the function $\lambda \longmapsto I(\lambda)$ *is nonincreasing, and continuous from* $(0, \infty)$ *into* $[0, \infty)$ *More precisely,*

$$I(\lambda) \textit{ is decreasing on} (0, \text{ess} \sup_\Omega V), \tag{2.19}$$

$$\begin{cases} I(\lambda) = 0 \quad \forall \lambda \geqslant \text{ess} \sup_\Omega V & \textit{if ess} \sup_\Omega V < \infty, \\ \lim_{\lambda \to \infty} I(\lambda) = 0 & \textit{if ess} \sup_\Omega V = \infty, \end{cases} \tag{2.20}$$

$$\begin{cases} I_0 = \lim_{\lambda \downarrow 0} I(\lambda) = \sup_{\lambda > 0} I(\lambda) < \infty & \textit{if and only if} \\ (P_0) \textit{ admits a minimizer } \rho_0 \in D(\mathcal{E}), \textit{ and then } I_0 = \int \rho_0. \end{cases} \tag{2.21}$$

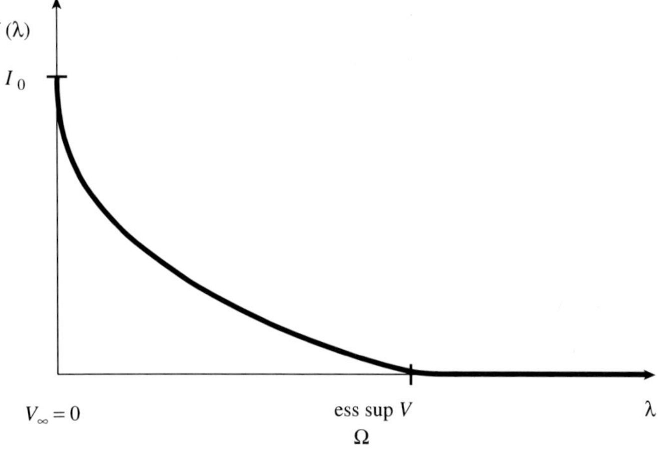

Figure 4   Typical shape of $I(\lambda)$.

The proof of Theorem 3 is based on several lemmas.

LEMMA 5. *Assume* (H$^+$), (2.1), (2.11), *and* (2.12). *Then, for every* $\varepsilon > 0$ *there exists a function* $\omega_\varepsilon \in L^1(\Omega)$ *such that*

$$j(r) - V(x)r + \varepsilon r \geqslant \omega_\varepsilon(x) \quad \text{for a.e. } x \in \Omega, \quad \forall r \geqslant 0. \tag{2.22}$$

Proof. Set $A = [V > \varepsilon]$ and so $|A| < \infty$ (since $V_\infty = 0$). For $x \in {}^cA = [V \leqslant \varepsilon]$ we have

$$j(r) - V(x)r + \varepsilon r \geqslant j(r) \geqslant 0$$

and thus we choose $\omega_\varepsilon(x) = 0$ on ${}^cA$.

Given any $k > 0$ (to be fixed later) there is a constant C_k such that

$$j(r) \geqslant kr - C_k \quad \forall r \geqslant 0.$$

(Here we have used (2.1)). For $x \in [\varepsilon < V \leqslant k]$ we have

$$j(r) - V(x)r + \varepsilon r \geqslant j(r) - kr \geqslant -C_k$$

and so we choose $\omega_\varepsilon(x) = -C_k$ on $[\varepsilon < V \leqslant k]$.

Finally, we consider the case where $x \in [V > k]$. We now use assumption (H$^+$) to write

$$(1 + \theta)(V(x) - M)r \leqslant j(r) + \omega(x)$$

where $\omega(x) = j^*((1+\theta)(V(x) - M))$. Therefore we have

$$j(r) - V(x)r + \varepsilon r \geqslant (1+\theta)(V(x) - M)r - \omega(x) - V(x)r + \varepsilon r$$
$$\geqslant r[-M + \theta V(x) - \theta M] - \omega(x) \geqslant -\omega(x)$$

provided we fix k so large that $-M + \theta k - \theta M \geqslant 0$. Hence we may choose $\omega_\varepsilon(x) = -\omega(x)$ on $[V > k]$. $\qquad\square$

LEMMA 6. *Same assumptions as in Lemma 5. Let (ρ_n) be a sequence in $D(\mathcal{E})$ such that $\int \rho_n \leqslant C$ and $\rho_n \rightharpoonup \bar\rho$ weakly in $L^1(\Omega_j)$ for each j.*[4] *Then*

$$\mathcal{E}(\bar\rho) \leqslant \liminf_{n\to\infty} \mathcal{E}(\rho_n).$$

Proof. For every $\varepsilon > 0$, let $\omega_\varepsilon(x)$ be as in Lemma 5. We have

$$\mathcal{E}(\rho_n) \geqslant \mathcal{E}(\rho_n \chi_{\Omega_j}) + \int_{\Omega\setminus\Omega_j} \omega_\varepsilon(x)\,dx - \varepsilon \int_\Omega \rho_n(x)\,dx.$$

For each j, $\rho_n \chi_{\Omega_j} \rightharpoonup \bar\rho\, \chi_{\Omega_j}$ weakly in $L^1(\Omega)$.
   Hence we obtain, for each j,

$$\liminf_{n\to\infty} \mathcal{E}(\rho_n) \geqslant \mathcal{E}(\bar\rho\, \chi_{\Omega_j}) + \int_{\Omega\setminus\Omega_j} \omega_\varepsilon(x)dx - \varepsilon C.$$

We conclude by letting $j \to \infty$ and then $\varepsilon \to 0$. $\qquad\square$

LEMMA 7. *Same assumptions as in Lemma 5. Then for every $\lambda > 0$ there is some $\bar\rho \in D(\mathcal{E})$ such that*

$$\mathcal{E}(\bar\rho) + \lambda \int \bar\rho \leqslant \mathcal{E}(\rho) + \lambda \int \rho \quad \forall \rho \in D(\mathcal{E}). \tag{2.23}$$

Proof. Applying Lemma 5 with $\varepsilon = \lambda/2$, we obtain some function $\omega \in L^1(\Omega)$ such that

$$j(r) - V(x)r + \frac{\lambda}{2}r \geqslant \omega(x) \text{ a.e. in } \Omega, \ \forall r \geqslant 0,$$

and so

$$j(r) - V(x)r + \lambda r \geqslant \frac{\lambda}{2}r + \omega(x) \text{ a.e. in } \Omega, \ \forall r \geqslant 0.$$

Therefore, for every $\rho \in D(\mathcal{E})$, we have

$$\mathcal{E}(\rho) + \lambda \int \rho \geqslant \frac{\lambda}{2} \int \rho - C.$$

---

[4] We recall that (Ω_j) is a nondecreasing sequence of measurable sets in Ω such that $|\Omega_j| < \infty \ \ \forall j$ and $\cup_j \Omega_j = \Omega$

Thus if (ρ_n) is a minimizing sequence for (2.23), then $\int \rho_n \leqslant C$ and also $\int j(\rho_n) - V\rho_n \leqslant C$. We deduce from Lemma 4 that $\int j(\rho_n) \leqslant C$. Therefore, the sequence (ρ_n) is equi-integrable on each Ω_j and we may extract a subsequence still denoted (ρ_n) such that $\rho_n \rightharpoonup \bar{\rho}$ weakly in $L^1(\Omega_j)$ for each j. We conclude with the help of Lemma 6 that (2.23) holds. $\qquad\square$

A final lemma,

LEMMA 8. *Assume* (2.12), (2.13), (2.14), (2.16), *and suppose* $\rho \in D(\mathcal{E})$. *Let W be any measurable function satisfying*

$$\partial j(\rho) + B\rho \ni W \quad a.e.\ on\ \Omega. \tag{2.24}$$

Then

$$W_\infty \leqslant 0. \tag{2.25}$$

Proof. Let $\alpha > 0$; we shall prove that $W_\infty \leqslant \alpha$. Indeed fix ε such that $0 < \varepsilon < \alpha$. By assumption (2.16) the set $\Omega_1 = [B\rho > \varepsilon]$ has finite measure. Since $\alpha - \varepsilon > 0$ there exists $\delta > 0$ such that $\partial j(r) \subset (-\infty, \alpha - \varepsilon]$ for $r \in [0, \delta]$. (Here we have used (2.14)). The set $\Omega_2 = [\rho > \delta]$ has also finite measure (since $\rho \in L^1(\Omega)$). Using (2.24) we see that

$$[W > \alpha] \subset \Omega_1 \cup \Omega_2$$

and thus the set $[W > \alpha]$ has finite measure. $\qquad\square$

Proof of Theorem 3. We split the proof into 5 steps.

STEP 1. The existence of a minimizer ρ_λ for (P_λ) when $\lambda > 0$ has been established in Lemma 7. We prove that $I(\lambda) = \int \rho_\lambda$ is nonincreasing and continuous on $(0, \infty)$.

Proof. Let $\lambda, \mu > 0$. We have

$$\begin{cases} \mathcal{E}(\rho_\lambda) + \lambda I(\lambda) \leqslant \mathcal{E}(\rho_\mu) + \lambda I(\mu) \\ \mathcal{E}(\rho_\mu) + \mu I(\mu) \leqslant \mathcal{E}(\rho_\lambda) + \mu I(\lambda) \end{cases}$$

and thus

$$(\lambda - \mu)(I(\lambda) - I(\mu)) \leqslant 0,$$

so that the function $\lambda \longmapsto I(\lambda)$ is nonincreasing. $\qquad\square$

We now prove that $I(\lambda)$ is continous on $(0, +\infty)$. Let $\lambda_n \to \bar{\lambda}$ with $\bar{\lambda} > 0$ and set $\rho_n = \rho_{\lambda_n}$. It is easy to see (as in the proof of Lemma 7) that $\int \rho_n \leqslant C$ and $\int j(\rho_n) \leqslant C$. Therefore we may extract a subsequence (ρ_{n_k}) such that $\rho_{n_k} \rightharpoonup \bar{\rho}$ weakly in $L^1(\Omega_j)$ for each j. We have

$$\mathcal{E}(\rho_{n_k}) + \lambda_{n_k} \int \rho_{n_k} \leqslant \mathcal{E}(\rho) + \lambda_{n_k} \int \rho \quad \forall \rho \in D(\mathcal{E}); \tag{2.26}$$

passing to the limit as $k \to \infty$ we find

$$\mathcal{E}(\bar{\rho}) + \bar{\lambda} \int \bar{\rho} \leqslant \mathcal{E}(\rho) + \bar{\lambda} \int \rho \quad \forall \rho \in D(\mathcal{E}),$$

so that $\bar{\rho}$ and $\rho_{\bar{\lambda}}$ are both minimizers for the problem $(P_{\bar{\lambda}})$. By (2.15) it follows that $\bar{\rho} = \rho_{\bar{\lambda}}$, $\int \bar{\rho} = I(\bar{\lambda})$. And also $\liminf_{k \to \infty} \int \rho_{n_k} \geqslant \int \bar{\rho} = I(\bar{\lambda})$. Next we have, from (2.26) (choosing $\rho = \bar{\rho}$)

$$\limsup_{k \to \infty} \lambda_{n_k} \int \rho_{n_k} \leqslant \mathcal{E}(\bar{\rho}) + \bar{\lambda} \int \bar{\rho} - \liminf_{k \to \infty} \mathcal{E}(\rho_{n_k}) \leqslant \bar{\lambda} \int \bar{\rho}.$$

We conclude that

$$\limsup_{k \to \infty} \int \rho_{n_k} \leqslant \int \bar{\rho}$$

and so $\lim_{k \to \infty} \int \rho_{n_k} = \int \bar{\rho} = I(\bar{\lambda})$. The uniqueness of the limit shows that, in fact, $\lim_{n \to \infty} I(\lambda_n) = I(\bar{\lambda})$.

STEP 2. Proof of (2.19).

Proof. Indeed let $\lambda, \mu \in (0, \operatorname*{ess\,sup}_{\Omega} V)$ be such that $I(\lambda) = I(\mu)$. We have

$$\mathcal{E}(\rho_\lambda) + \lambda \int \rho_\lambda \leqslant \mathcal{E}(\rho_\mu) + \lambda \int \rho_\mu,$$

$$\mathcal{E}(\rho_\mu) + \mu \int \rho_\mu \leqslant \mathcal{E}(\rho_\lambda) + \mu \int \rho_\lambda,$$

and therefore $\mathcal{E}(\rho_\lambda) = \mathcal{E}(\rho_\mu)$. We deduce from the strict convexity of $\mathcal{E}$ that $\rho_\lambda = \rho_\mu$. $\square$

On the other hand we have

$$\partial j(\rho_\lambda) + B\rho_\lambda \ni V - \lambda \quad \text{a.e.,}$$
$$\partial j(\rho_\mu) + B\rho_\mu \ni V - \mu \quad \text{a.e.,}$$

which means (since j is C^1 on $(0, \infty)$)

$$\begin{cases} j'(\rho_\lambda) + B\rho_\lambda = V - \lambda & \text{a.e. on } [\rho_\lambda > 0] \\ B\rho_\lambda \geqslant V - \lambda & \text{a.e. on } [\rho_\lambda = 0] \end{cases}$$

and similarly for ρ_μ.

If $\rho_\lambda = \rho_\mu = \rho$ is positive on a set of positive measure, then we have

$$V - \lambda - B\rho = V - \mu - B\rho,$$

and thus $\lambda = \mu$. Otherwise, $\rho_\lambda = \rho_\mu = \rho = 0$, and then $V - \lambda \leqslant 0$, $V - \mu \leqslant 0$, i.e., $\lambda \geqslant \operatorname{ess\,sup}_\Omega V$ and $\mu \geqslant \operatorname{ess\,sup}_\Omega V$, but this contradicts the assumption $\lambda, \mu \in (0, \operatorname{ess\,sup}_\Omega V)$.

STEP 3. Proof of (2.20).

Proof. By Lemma 3 we have

$$\partial j(\rho_\lambda) + B\rho_\lambda \ni V - \lambda \quad \text{a.e. on } \Omega$$

and thus

$$\rho_\lambda \in \partial j^*(V - \lambda - B\rho_\lambda).$$

It follows that

$$j^*(V - M) - j^*(V - \lambda - B\rho_\lambda) \geqslant \rho_\lambda(\lambda - M + B\rho_\lambda)$$

and therefore

$$\rho_\lambda \leqslant \frac{j^*(V - M)}{\lambda - M} \quad \text{for } \lambda > M.$$

Using assumption (H) we obtain $\lim_{\lambda \to +\infty} I(\lambda) = 0$. $\qquad\square$

From the relation $\partial j(\rho_\lambda) + B\rho_\lambda \ni V - \lambda$ a.e. onΩ we see that

$$[I(\lambda) = 0] \iff [\rho_\lambda = 0] \iff [V - \lambda \leqslant 0 \text{ a.e.}] \iff [\lambda \geqslant \operatorname*{ess\,sup}_\Omega V].$$

STEP 4. Proof of (2.21).

Proof. Suppose that (P$_0$) admits a minimizer $\rho_0 \in D(\mathcal{E})$. We have

$$\mathcal{E}(\rho_\lambda) + \lambda I(\lambda) \leqslant \mathcal{E}(\rho_0) + \lambda \int \rho_0 \quad \forall \lambda > 0,$$

and also

$$\mathcal{E}(\rho_0) \leqslant \mathcal{E}(\rho_\lambda)$$

so that $I(\lambda) \leqslant \int \rho_0$ and $I_0 \leqslant \int \rho_0 < \infty$. $\qquad\square$

Conversely, suppose $I_0 < \infty$, so that $\int \rho_\lambda \leqslant C \,\forall \lambda > 0$. It follows from Lemma 4 that $\int j(\rho_\lambda) \leqslant C \,\forall \lambda > 0$. Therefore, we may find, as in the proof of Theorem 2, a sequence $\lambda_n \to 0$ such that $\rho_{\lambda_n} \rightharpoonup \rho_0$ weakly in $L^1(\Omega_j)$ for each j. From Lemma 6 we easily see that ρ_0 is a minimizer for (P$_0$). Moreover, we have

$$\int \rho_0 \leqslant \liminf_{n \to +\infty} \int \rho_{\lambda_n} = \lim_{\lambda \downarrow 0} I(\lambda) = I_0.$$

Combining this with the above argument we find $\int \rho_0 = I_0$.

STEP 5. Proof of (2.18).

Proof. Suppose by contradiction that, for some $\lambda_0 < 0$, $\mathcal{E}(\rho) + \lambda_0 \int \rho$ is bounded below on $D(\mathcal{E})$. We deduce from Ekeland's principle (see Ekeland [32]) that for every $\varepsilon > 0$ there is some $\rho_\varepsilon \in D(\mathcal{E})$ such that

$$\mathcal{E}(\rho) + \lambda_0 \int \rho - \mathcal{E}(\rho_\varepsilon) - \lambda_0 \int \rho_\varepsilon + \varepsilon \int |\rho - \rho_\varepsilon| \geqslant 0 \quad \forall \rho \in D(\mathcal{E}).$$

Applying Lemma 3 and standard convex analysis we see that

$$\partial j(\rho_\varepsilon) + B\rho_\varepsilon \ni V - \lambda_0 - f_\varepsilon \quad \text{a.e. on } \Omega$$

for some function $f_\varepsilon \in L^\infty(\Omega)$ with $\|f_\varepsilon\|_{L^\infty} \leqslant \varepsilon$. We deduce from Lemma 8 that $(V - \lambda_0 - f_\varepsilon)_\infty \leqslant 0$ and consequently $V_\infty - \lambda_0 \leqslant \varepsilon$, so that $-\lambda_0 \leqslant \varepsilon$. Choosing $\varepsilon < -\lambda_0$ yields a contradiction. $\qquad\square$

We may now return to problem (M_I) described at the beginning of this section and state, using the notation introduced in Theorem 3, the following:

COROLLARY 1. *Under the assumptions of Theorem 3, we have*

$$\begin{cases} \textit{for every } I \in (0, I_0) \textit{ problem } (M_I) \textit{ admits a unique minimizer} \\ \rho^I = \rho_\lambda, \textit{ where } \lambda > 0 \textit{ is the unique solution of } I(\lambda) = I; \end{cases} \tag{2.27}$$

if $I_0 < \infty$, problem (M_{I_0}) admits ρ_0 as its unique minimizer; $\tag{2.28}$

if $I_0 < \infty$ and $I > I_0$, problem (M_I) admits no minimizer. $\tag{2.29}$

REMARK 7. If $I_0 < \infty$ and $I > I_0$, any minimizing sequence converges to ρ_0 weakly in $L^1(\Omega_j)$ for every j (this is proved at the end of the section). Note that the constraint $\int \rho = I$ is "lost" in the limit.

Proof of Corollary 1.

Proof of (2.27). We have, by construction,

$$\mathcal{E}(\rho_\lambda) + \lambda \int \rho_\lambda \leqslant \mathcal{E}(\rho) + \lambda \int \rho \quad \forall \rho \in D(\mathcal{E})$$

and thus

$$\mathcal{E}(\rho^I) + \lambda I \leqslant \mathcal{E}(\rho) + \lambda I \quad \forall \rho \in D(\mathcal{E}) \text{ with } \int \rho = I,$$

so that ρ_I is a minimizer for (M_I).

Proof of (2.28). If $I_0 < \infty$, we have

$$\mathcal{E}(\rho_0) \leqslant \mathcal{E}(\rho) \quad \forall \rho \in D(\mathcal{E})$$

and in particular

$$\mathcal{E}(\rho_0) \leqslant \mathcal{E}(\rho) \quad \forall \rho \in D(\mathcal{E}) \text{ with } \int \rho = I_0.$$

Therefore ρ_0 is a minimizer for (M_{I_0}).

Proof of (2.29). Indeed, suppose that problem (M_I) has a solution $\bar{\rho}$ for some $\bar{I} > I_0$. We deduce from Theorem 1 that there is a constant $\bar{\lambda} \in \mathbb{R}$ such that

$$\partial j(\bar{\rho}) + B\bar{\rho} \ni V - \bar{\lambda} \quad \text{a.e. on } \Omega.$$

Lemma 8 implies $V_\infty - \bar{\lambda} \leqslant 0$, i.e., $\bar{\lambda} \geqslant 0$ (since $V_\infty = 0$). From Lemma 3 we see that

$$\mathcal{E}(\bar{\rho}) + \bar{\lambda} \int \bar{\rho} \leqslant \mathcal{E}(\rho) + \bar{\lambda} \int \rho \quad \forall \rho \in D(\mathcal{E}).$$

Since $\mathcal{E}$ is strictly convex we must have $\bar{\rho} = \rho_{\bar{\lambda}}$ and thus $\int \bar{\rho} = \int \rho_{\bar{\lambda}} \leqslant I_0$. But, on the other hand, $\int \bar{\rho} = \bar{I} > I_0$ – a contradiction.

We gather some additional facts in the next propositions.

PROPOSITION 1. *Same assumptions as in Theorem 3. Then for every* $I \in (0, I_0)$ *we have*

$$[\rho^I > 0] \text{ has finite measure,} \tag{2.30}$$
$$\rho^I \leqslant \gamma^0(V) \quad \text{a.e.,} \tag{2.31}$$

where

$$\gamma^0(s) = (j^*)'(s - 0) = \lim_{t \uparrow s} \frac{j^*(t) - j^*(s)}{t - s}.$$

If $I_0 < \infty$, *we have*

$$\rho_0 \leqslant \gamma^0(V) \quad \text{a.e.} \tag{2.32}$$

and in particular

$$I_0 = \int \rho_0 \leqslant \int \gamma^0(V) \leqslant \infty. \tag{2.33}$$

Proof. Since $0 < I < I_0$ there is some $\bar{\lambda} > 0$ such that $\rho^I = \rho_{\bar{\lambda}}$ and thus we have

$$\partial j(\rho^I) + B\rho^I \ni V - \bar{\lambda} \quad \text{a.e. on } \Omega. \tag{2.34}$$

It follows from (2.34) and (2.14) that

$$[\rho^I > 0] \subset [V \geqslant \bar{\lambda}],$$

and so $[\rho^I > 0]$ has finite measure (since $V_\infty = 0$ and $\bar{\lambda} > 0$).
   We write (2.34) as

$$\rho^I \in \gamma(V - \bar{\lambda} - B\rho^I)$$

where $\gamma = \partial j^* = (\partial j)^{-1}$; (2.31) follows from the monotonicity of γ.
   When $I_0 < \infty$, the proof Theorem 3 (Step 4) shows that

$$\rho^I \underset{I \uparrow I_0}{\rightharpoonup} \rho_0 \quad \text{weakly in } L^1(\Omega_j) \forall j.$$

We deduce from (2.31) that

$$\rho_0 \leqslant \gamma^0(V) \quad \text{a.e. on } \Omega.$$

We now introduce two natural expressions

$$E(I) = \begin{cases} \inf\{\mathcal{E}(\rho); \rho \in D(\mathcal{E}) \text{ and } \int \rho = I\} & \text{if } I \geqslant 0 \\ +\infty & \text{if } I < 0, \end{cases} \tag{2.35}$$

and, for every $\lambda \in \mathbb{R}$,

$$\Phi(\lambda) = -\inf\left\{\mathcal{E}(\rho) + \lambda \int \rho; \rho \in D(\mathcal{E})\right\}. \tag{2.36}$$

$\square$

PROPOSITION 2. *Same assumptions as in Theorem 3. We have*

E is convex, l.s.c. on $\mathbb{R}$, *E*$(0) = 0$, (2.37)

E is strictly convex and decreasing on $(0, I_0)$, (2.38)

if $I_0 < \infty$, *then* $E(I) = E(I_0) = \mathcal{E}(\rho_0)$ *for* $I \geqslant I_0$, (2.39)

Φ *is convex, l.s.c. on* $\mathbb{R}$, (2.40)

$\Phi(\lambda) = \infty \quad \forall \lambda < 0,$ (2.41)

$\Phi(\lambda) \geqslant 0 \quad \forall \lambda \in \mathbb{R},$ (2.42)

Φ *is finite,* C^1, *nonincreasing on* $(0, \infty)$, (2.43)

$\Phi'(\lambda) = -I(\lambda) \quad \forall \lambda > 0,$ (2.44)

$$\begin{cases} \Phi(\lambda) = 0 \quad \text{for } \lambda \geqslant \text{ess sup}_\Omega V, & \text{if ess sup}_\Omega V < \infty \\ \lim_{\lambda \to \infty} \Phi(\lambda) = 0 & \text{if ess sup}_\Omega V = \infty, \end{cases} \tag{2.45}$$

$\Phi(\lambda) = E^*(-\lambda) \quad \forall \lambda \in \mathbb{R} \quad and \quad E(I) = \Phi^*(-I) \quad \forall I \in \mathbb{R}.$ (2.46)

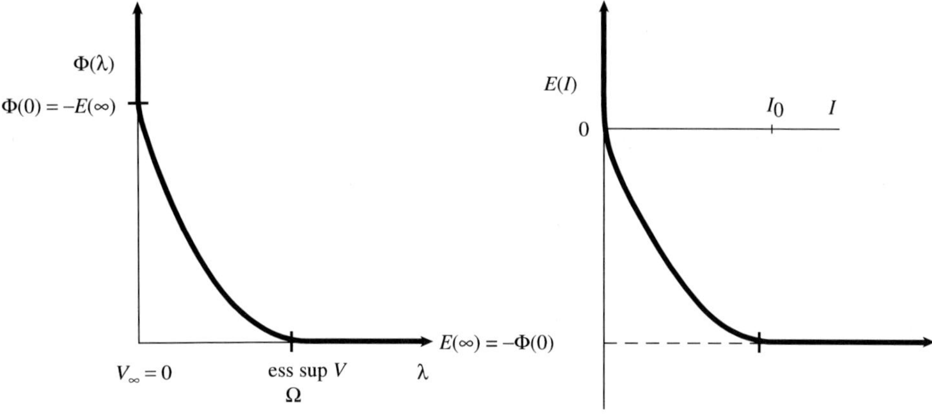

Figure 5   Typical shape of $\Phi(\lambda)$ and $E(I)$.

Proof.

Verification of (2.37). It follows from assumption (H) that $E(I) \geqslant -\int j^*(V-M) - MI$ and thus $E(I) \in \mathbb{R}$ for $I \geqslant 0$. Let $I_1, I_2 \geqslant 0$ and $t \in (0, 1)$. Given $\varepsilon > 0$ there is some $\rho_1 \in D(\mathcal{E})$ such that $\int \rho_1 = I_1$ and $\mathcal{E}(\rho_1) \leqslant E(I_1) + \varepsilon$ and there is some $\rho_2 \in D(\mathcal{E})$ such that $\int \rho_2 = I_2$ and $\mathcal{E}(\rho_2) \leqslant E(I_2) + \varepsilon$. Set $\bar{\rho} = t\rho_1 + (1-t)\rho_2$ so that $\int \bar{\rho} = tI_1 + (1-t)I_2$ and $\mathcal{E}(\bar{\rho}) \leqslant tE(I_1) + (1-t)E(I_2) + \varepsilon$. Therefore we obtain

$$E(tI_1 + (1-t)I_2) \leqslant tE(I_1) + (1-t)E(I_2) + \varepsilon,$$

and so E is convex.

In view of the convexity of E we already know that E is continuous on $(0, +\infty)$ and that $\limsup\limits_{I \downarrow 0} E(I) \leqslant E(0) = 0$. On the other hand if $I_n \to 0$ with $I_n \geqslant 0$, there is a sequence (ρ_n) in $D(\mathcal{E})$ such that $\int \rho_n = I_n$ and $\mathcal{E}(\rho_n) \leqslant E(I_n) + 1/n$. Since $\mathcal{E}$ is l.s.c. on L^1 we conclude that $\liminf\limits_{n \to \infty} E(I_n) \geqslant 0$. Therefore $\lim\limits_{I \downarrow 0} E(I) = E(0) = 0$.

Verification of (2.40), (2.41) **and** (2.42). It is clear that Φ is convex and l.s.c. since it is a sup of affine functions. (2.41) corresponds to assertion (2.18) in Theorem 3. (2.42) is obvious by choosing $\rho = 0$ as testing function in the definition of Φ.

Verification of (2.43) **and** (2.44). We have $\forall \lambda, \mu > 0$,

$$\mathcal{E}(\rho_\mu) + \mu \int \rho_\mu \leqslant \mathcal{E}(\rho_\lambda) + \mu \int \rho_\lambda$$

and thus,

$$-\Phi(\mu) \leqslant \mathcal{E}(\rho_\lambda) + \lambda \int \rho_\lambda + (\mu - \lambda) \int \rho_\lambda = -\Phi(\lambda) + (\mu - \lambda)I(\lambda).$$

Hence

$$\Phi(\mu) - \Phi(\lambda) + I(\lambda)(\mu - \lambda) \geq 0 \quad \forall \lambda, \mu > 0.$$

Changing λ and μ yields

$$|\Phi(\mu) - \Phi(\lambda) + I(\lambda)(\mu - \lambda)| \leq |I(\mu) - I(\lambda)| \, |\mu - \lambda| \quad \forall \lambda, \mu > 0.$$

Assertions (2.43) and (2.44) follow.

Verification of (2.45). We have $\forall \lambda \in \mathbb{R}$,

$$j(\rho) - (V - \lambda)\rho \geq -j^*(V - \lambda)$$

and thus, for $\rho \in D(\mathcal{E})$,

$$\mathcal{E}(\rho) + \lambda \int \rho \geq -\int j^*(V - \lambda),$$

so that

$$\Phi(\lambda) \leq \int j^*(V - \lambda).$$

If ess $\sup_\Omega V < \infty$, we see immediately that $\Phi(\lambda) \leq 0$ for $\lambda \geq$ ess $\sup_\Omega V$ (since $j^*(s) = 0$ for $s \leq 0$).

If ess $\sup_\Omega V = \infty$, we observe that $j^*(V - \lambda) \leq j^*(V - M)$ for $\lambda \geq M$ and $j^*(V - \lambda) \to 0$ a.e. as $\lambda \to +\infty$. It follows, by dominated convergence that $\int j^*(V - \lambda) \to 0$ as $\lambda \to +\infty$.

Verification of (2.46). It is clear that, for every $\lambda \in \mathbb{R}$,

$$\inf_{\rho \in D(\mathcal{E})} \left\{ \mathcal{E}(\rho) + \lambda \int \rho \right\} = \inf_{I \geq 0} \{E(I) + \lambda I\},$$

i.e., $\Phi(\lambda) = \sup_{I \geq 0} \{-\lambda I - E(I)\} = E^*(-\lambda)$. It follows that $E^{**}(I) = \Phi^*(-I) \, \forall I \in \mathbb{R}$. However $E^{**} = E$ since E is convex and l.s.c. on $\mathbb{R}$.

Verification of (2.38) **and** (2.39). Let $I_1, I_2 \in (0, I_0)$ with $I_1 \neq I_2$. We know from Theorem 3 that there exist $\rho_1, \rho_2 \in D(\mathcal{E})$ with $\int \rho_1 = I_1$ and $\int \rho_2 = I_2$, $E(I_1) = \mathcal{E}(\rho_1)$ and $E(I_2) = \mathcal{E}(\rho_2)$. Since $\mathcal{E}$ is strictly convex we have, for $t \in (0, 1)$,

$$E(tI_1 + (1 - t)I_2) \leq \mathcal{E}(t\rho_1 + (1 - t)\rho_2) < t\mathcal{E}(\rho_1) + (1 - t)\mathcal{E}(\rho_2)$$
$$= tE(I_1) + (1 - t)E(I_2).$$

On the other hand, we have from (2.46) and (2.41)

$$E(I) = \sup_{\lambda \in \mathbb{R}} \{-I\lambda - \Phi(\lambda)\} = \sup_{\lambda \geq 0} \{-I\lambda - \Phi(\lambda)\}.$$

For $\lambda \geqslant 0$ the function $I \mapsto (-I\lambda - \Phi(\lambda))$ is nonincreasing and thus the function $I \mapsto E(I)$ is also nonincreasing on $\mathbb{R}$. It follows that E is decreasing on $(0, I_0)$ since it is strictly convex on $(0, I_0)$. Finally, E is constant on $(I_0, +\infty)$. Indeed, if $I_0 < \infty$, there exists (by Theorem 3) some $\rho_0 \in D(\mathcal{E})$ with $\int \rho_0 = I_0$ and $\mathcal{E}(\rho_0) \leqslant \mathcal{E}(\rho)$ $\quad \forall \rho \in D(\mathcal{E})$, so that

$$E(I_0) = \mathcal{E}(\rho_0) \leqslant \mathcal{E}(\rho) \quad \forall \rho \in D(\mathcal{E}).$$

In particular, $E(I_0) \leqslant E(I) \forall I$. Since E is nonincreasing on $\mathbb{R}$ we conclude that $E(I) = E(I_0)$ for $I > I_0$. $\qquad \square$

REMARK 8. In all the examples related to Thomas-Fermi $I_0 < \infty$ (see Section 4). It would be illuminating to construct examples satisfying all the conditions of Theorem 3 such that $I_0 = \infty$. From the definitions of Φ and (2.46) we have

$$\Phi(0) = - \inf_{\rho \in D(\mathcal{E})} \mathcal{E}(\rho) = - \inf_{I \geqslant 0} E(I).$$

It would be useful to construct some examples where $I_0 = \infty$ and $\Phi(0) < \infty$, and other examples where $I_0 = \infty$ and $\Phi(0) = \infty$. From (2.44) we see that $\Phi(0) < \infty$ if and only if $\int_0^1 I(\lambda) \, d\lambda < \infty$.

The approach via relaxation

Another approach for proving Corollary 1 (without passing through Theorem 3) is the *relaxation* method used by Lieb-Simon [48].

Given a constant $0 \leqslant I < \infty$ set

$$\widehat{K}_I = \left\{ \rho \in D(\mathcal{E}); \int \rho(x) \, dx \leqslant I \right\}$$

and consider the relaxed minimization problem

$$\text{find } \widehat{\rho} \in \widehat{K}_I \text{ such that } \mathcal{E}(\widehat{\rho}) \leqslant \mathcal{E}(\rho) \quad \forall \rho \in \widehat{K}_I. \qquad (\widehat{\text{M}}_I)$$

Set, for every $I \geqslant 0$,

$$\widehat{E}(I) = \inf \left\{ \mathcal{E}(\rho); \rho \in D(\mathcal{E}) \text{ and } \int \rho \leqslant I \right\}. \qquad (2.47)$$

We keep the assumptions of Theorem 3. Clearly, the function $I \mapsto \widehat{E}(I)$ is convex, nonincreasing and continuous on $[0, \infty)$ (the argument is similar to the proof of (2.37) in Proposition 2). It is easy to see that for *every* $I \geqslant 0$ the infimum in (2.47) is *achieved* by some unique element, denoted $\widehat{\rho}_I$ (the argument is similar to the one used in the proof of Lemma 7). A simple consideration about convex functions shows that there exists some $\widehat{I}_0 \in [0, \infty]$ such that:

a) $\widehat{E}$ is *decreasing* on $[0, \widehat{I_0})$,

b) $\widehat{E}$ is constant on $[\widehat{I_0}, \infty]$ (assuming $\widehat{I_0} < \infty$).

PROPOSITION 3. *Under the assumptions of Theorem* 3, *this* $\widehat{I_0}$ *satisfies all the properties of* I_0 *described in Corollary* 1. *Moreover* $\widehat{E}(I) = E(I)$ $\forall I \geqslant 0$.

Proof. a) If $I \leqslant \widehat{I_0}$ we must have $\int \widehat{\rho}^I = I$, so that $\widehat{\rho}^I$ is a solution of (M_I). Otherwise, set $I' = \int \widehat{\rho}^I < I$. We have

$$\widehat{E}(I') = \mathcal{E}(\widehat{\rho}^{I'}) \leqslant \mathcal{E}(\rho) \quad \forall \rho \in \widehat{K}_{I'}.$$

Choosing $\rho = \widehat{\rho}^I$ we obtain $\widehat{E}(I') \leqslant \widehat{E}(I)$ – absurd.

b) If $I > \widehat{I_0}$, problem (M_I) has no solution. Indeed, suppose, by contradiction, that there is a solution $\bar{\rho}$ of (M_I) with $I > \widehat{I_0}$. We know, by Theorem 1, that there is a constant $\lambda \in \mathbb{R}$ such that

$$\partial j(\bar{\rho}) + B\bar{\rho} \ni V - \lambda \quad \text{a.e.}$$

and by Lemma 8 we find that $\lambda \geqslant 0$. Therefore we have

$$\mathcal{E}(\bar{\rho}) + \lambda \int \bar{\rho} \leqslant \mathcal{E}(\rho) + \lambda \int \rho \quad \forall \rho \in D(\mathcal{E}). \tag{2.48}$$

Choosing $\rho = \widehat{\rho}^{\hat{I}_0}$ in (2.48) we obtain

$$\mathcal{E}(\bar{\rho}) + \lambda I \leqslant \widehat{E}(\widehat{I_0}) + \lambda \widehat{I_0}.$$

However we have

$$\widehat{E}(\widehat{I_0}) = \inf_{\rho \in D(\mathcal{E})} \mathcal{E}(\rho)$$

and this infimum is achieved only when $\rho = \widehat{\rho}^{\hat{I}_0}$ so that $\mathcal{E}(\bar{\rho}) > \widehat{E}(\widehat{I_0})$. It follows that $\lambda < 0$ – absurd. $\qquad\qquad \square$

Proof of Remark 7. Let $I > I_0$ ($I_0 < \infty$); we have $E(I) = E(I_0)$. Thus if (ρ_n) is a minimizing sequence for (M_I) we have $\mathcal{E}(\rho_n) \to E(I_0)$. As in Lemma 7 we may extract a subsequence still denoted ρ_n such that $\rho_n \rightharpoonup \bar{\rho}$ weakly in $L^1(\Omega_j)$ for each j. By Lemma 6 we have $\mathcal{E}(\bar{\rho}) \leqslant E(I_0)$. Hence $\bar{\rho}$ is a minimizer for $\mathcal{E}$ on $D(\mathcal{E})$. By uniqueness we have $\bar{\rho} = \rho_0$.

REMARK 9. Throughout this section we have made assumption (2.1), i.e. j is coercive, and it played an essential role in applying the Dunford-Pettis theorem about weak convergence in L^1. When (2.1) does *not* hold it may be natural to extend the setting of problem (M_I) and to allow solutions ρ which are measures. This is an interesting direction of research.

3. A direct approach for solving the Euler equation

As in Section 2, let $j : \mathbb{R} \to [0, +\infty]$ be a convex l.s.c. function such that

$$j(0) = 0 \text{ and } j(r) = +\infty \quad \text{for all } r < 0,$$
$$j \text{ is finite and } C^1 \text{ on } (0, \infty),$$
$$j'(0+) = \lim_{r \downarrow 0} \frac{j(r)}{r} = 0.$$

Let $V : \Omega \to \mathbb{R}$ be a measurable function. Assume $k : \Omega \times \Omega \to \mathbb{R}$ is a measurable function satisfying (1.3) and (1.4). For every $\rho \in L^1(\Omega)$ with $\rho \geq 0$ a.e. we set

$$(B\rho)(x) = \int k(x, y)\, \rho(y) dy \leq +\infty.$$

We shall make further assumptions on B:

$$\int (B\rho - 1)^+ < \infty \quad \forall \rho \in L^1(\Omega) \text{ with } \rho \geq 0 \text{ a.e..} \tag{3.1}$$

It is equivalent to assume that for every $\rho \in L^1(\Omega)$ with $\rho \geq 0$ a.e. we have

$$\begin{cases} \int_A B\rho < \infty \quad \forall A \subset \Omega \text{ measurable with finite measure} \\ \text{and for every } \delta > 0 \text{ the set } [B\rho > \delta] \text{ has finite measure.} \end{cases} \tag{3.1'}$$

$$j(0) \text{ and } j(r) = -\infty \text{ for all } r < 0,$$
$$j \text{ is finite and } C^1 \text{ on } (0, \infty),$$
$$j'(0+) = \lim_{r \downarrow 0} \frac{j(r)}{r} = 0.$$

We may thus extend B as a linear operator from $L^1(\Omega)$ into $L^1(\Omega) + L^\infty(\Omega)$. Sometimes we shall use an assumption slightly stronger than (3.1):

$$\text{for every } M \geq 0, \ \sup\left\{ \int (B\rho - 1)^+; \rho \in L^1(\Omega), \rho \geq 0 \text{ a.e.}, \int \rho \leq M \right\} < \infty. \tag{3.2}$$

We shall also make an assumption related to the maximum principle:

$$\begin{cases} \int \rho\, p(B\rho) \geq 0 \quad \forall \rho \in L^1(\Omega), \quad \forall p \in \mathcal{P} \\ \text{and} \\ \int \rho\, p(B\rho) = 0 \quad \text{if and only if} \quad p(B\rho) = 0, \end{cases} \tag{3.3}$$

where

$$\mathcal{P} = \{ p \in C^\infty(\mathbb{R}; \mathbb{R}); 0 \leq p \leq 1, p' \geq 0 \text{ on } \mathbb{R}, p' \in L^\infty(\mathbb{R}), \text{ and } p(t) = 0 \text{ for } t \leq 1 \}.$$

Finally we suppose that

B is injective.   (3.4)

We are concerned with the following problem:

$$\begin{cases} \text{Given a constant } I, \text{ with } 0 < I < \infty, \text{ find a function } \rho \in L^1(\Omega) \text{ and a} \\ \text{constant } \lambda \in \mathbb{R} \text{ such that } \rho \geqslant 0 \text{ a.e., } \int \rho = I \text{ and } \partial j(\rho) + B\rho \ni V - \lambda \text{ a.e..} \end{cases} \quad (\mathrm{E}^I)$$

When assumption (H) holds, problem (E^I) is equivalent to problem (M_I) - which has been solved in Section 2. We emphasize that throughout Section 3 we do *not* assume (H) and we solve (E^I) by a direct method. Our main results are the following.

THEOREM 4. *Assume* (3.1), (3.3) *and* (3.4). *Then, there exists* I_1 *with* $0 \leqslant I_1 \leqslant \infty$ *such that:*

a) *for every* $0 < I \leqslant I_1 (\text{and } I < \infty)$ *there is a unique solution* ρ^I *of problem* (E^I),
b) *for* $I_1 < I < \infty$ *problem* (E^I) *has no solution.*

REMARK 10. It may well happen that there is *no* $I > 0$ whatsoever for which problem (E^I) admits a solution (see an elementary example in Section 4, Remark 15). In this case we say that $I_1 = 0$. In contrast with the situation of Theorem 3 (where the assumption (H^+) plays a central role), this may happen even if ess $\sup_\Omega (V - V_\infty) > 0$. (Again, in the example of Section 4, Remark 15 one has $V_\infty = 0$, ess $\sup_\Omega V = +\infty$ but assumption (H^+) fails).

In order to solve (E^I) we proceed as in Section 2 and introduce the auxiliary problem:

$$\begin{cases} \text{Given a constant } \lambda \in \mathbb{R}, \text{ find } \rho \in L^1(\Omega) \text{ with } \rho \geqslant 0 \text{ a.e. such that} \\ \partial j(\rho) + B\rho \ni V - \lambda \text{ a.e..} \end{cases} \quad (\mathrm{E}_\lambda)$$

Theorem 4 is a direct consequence of

THEOREM 5. *Assume* (3.1), (3.3) *and* (3.4). *Let V be any measurable function. Then, there exists* $\lambda_0 \in [V_\infty, +\infty]$ *such that:*

a) *for every* $\lambda > \lambda_0$ *(and* $\lambda < +\infty$) *there is a unique solution* ρ_λ *of* (E_λ),
b) *for* $\lambda < \lambda_0$ *there is no solution of* (E_λ).

The mapping $\lambda \mapsto \rho_\lambda$ *defined for* $\lambda \subset (\lambda_0, +\infty)$ *is nonincreasing and continuous with values into* $L^1(\Omega)$; *moreover* $\rho_\lambda \to 0$ *in* $L^1(\Omega)$ *as* $\lambda \to +\infty$. *Set*

$$I_1 = \sup_{\lambda > \lambda_0} \int \rho_\lambda = \lim_{\lambda \downarrow \lambda_0} \int \rho_\lambda.$$

If $\lambda_0 \in \mathbb{R}$ *the following are equivalent:*

(i) $I_1 < \infty$

(ii) (E_{λ_0}) *has a unique solution* ρ_{λ_0},

(iii) *there exist functions* $f \in L^1(\Omega)$, $f \geqslant 0$ *a.e., and* $U : \Omega \to \mathbb{R}$ *measurable with* $\gamma^0(U) \in L^1(\Omega)$ *such that* $V - \lambda_0 = U + Bf$.

where γ^0 *has been defined in Section 2, Proposition 1.*

In this case $\rho_\lambda \to \rho_{\lambda_0}$ *in* $L^1(\Omega)$ *as* $\lambda \downarrow \lambda_0$ *and*

$$I_1 \leqslant \int (\gamma^0(U) + f) \tag{3.5}$$

REMARK 11. Very often we will find that $\lambda_0 = V_\infty$ (see e.g. Theorem 6). However it may also happen sometimes that there is *no* $\lambda \in \mathbb{R}$ for which (E_λ) admits a solution (see Section 4, Remark 15).

We start with some lemmas:

LEMMA 9. *Assume* (3.1). *Let* (ρ_n) *be a sequence in* $L^1(\Omega)$ *such that* $\rho_n \rightharpoonup \rho$ *weakly in* $L^1(\Omega)$ *and* $|\rho_n| \leqslant f$ *for some* $f \in L^1(\Omega)$. *Then* $B\rho_n \to B\rho$ *a.e. and in* $L^1(\Omega) + L^\infty(\Omega)$.

Proof. We may always assume that $\rho = 0$. We recall that for a.e. $x \in \Omega$ the function $y \mapsto k(x, y)f(y)$ is integrable. We write, for $M > 0$

$$(B\rho_n)(x) = \int_{[k(x,\cdot) \leqslant M]} k(x, y)\rho_n(y)dy + \int_{[k(x,\cdot) > M]} k(x, y)\rho_n(y)dy \tag{3.6}$$

It follows that

$$\limsup_{n \to \infty} |B\rho_n(x)| \leqslant \int_{[k(x,\cdot) > M]} k(x, y)f(y)dy \quad \forall M > 0.$$

As $M \to \infty$ we see that $B\rho_n \to 0$ a.e. By dominated convergence we have

$$\int (|B\rho_n| - k)^+ \to 0 \quad \forall k > 0.$$

Finally we note that

$$\|B\rho_n\|_{L^1 + L^\infty} \leqslant k + \int (|B\rho_n| - k)^+ \quad \forall k > 0.$$

and thus

$$\limsup_{n \to \infty} \|B\rho_n\|_{L^1 + L^\infty} \leqslant k \quad \forall k > 0.$$

$\square$

LEMMA 10. *Assume* (3.1). *Then B is a bounded operator from $L^1(\Omega)$ into $L^1(\Omega) + L^\infty(\Omega)$ and from $L^1(\Omega) \cap L^\infty(\Omega)$ into $L^\infty(\Omega)$.*

Proof. Let (ρ_n) be a sequence in $L^1(\Omega)$ such that $\rho_n \to 0$ in $L^1(\Omega)$. We may extract a subsequence still denoted (ρ_n) such that $|\rho_n| \leqslant f$ a.e. with $f \in L^1(\Omega)$. We deduce from Lemma 9 that $B\rho_n \to 0$ in $L^1(\Omega) + L^\infty(\Omega)$. Thus B is a bounded operator from $L^1(\Omega)$ into $L^1(\Omega) + L^\infty(\Omega)$. It follows, by duality, that B is a bounded operator from $L^1(\Omega) \cap L^\infty(\Omega)$ into $L^\infty(\Omega)$. $\qquad\square$

LEMMA 11. *Assume* (3.1) *and* (3.3). *Let $\rho \in L^1(\Omega)$ and let $k \geqslant 0$ be a constant. Then we have*

$$\int_{[B\rho > k]} \rho \geqslant 0 \tag{3.7}$$

and

$$[B\rho \leqslant k \ a.e. \ on \ [\rho > 0]] \quad \Rightarrow \quad [B\rho \leqslant k \ a.e. \ on \ \Omega]. \tag{3.8}$$

Proof. It suffices to consider the case $k = 1$. We have

$$\int \rho \, p(B\rho) \geqslant 0 \quad \forall p \in \mathcal{P}$$

and we obtain (3.7) by choosing a sequence (p_n) in $\mathcal{P}$ such that $p_n(t) \to 1 \ \forall t > 1$. If $B\rho \leqslant 1$ on $[\rho > 0]$ we have for $p \in \mathcal{P}$

$$\int \rho \, p(B\rho) = \int_{[\rho \leqslant 0]} \rho \, p(B\rho) + \int_{[\rho > 0]} \rho \, p(B\rho) \leqslant 0$$

since $p(B\rho) = 0$ a.e. on $[\rho > 0]$. It follows that $p(B\rho) = 0$ a.e. on Ω, for every $p \in \mathcal{P}$ and thus $B\rho \leqslant 1$ a.e. on Ω. $\qquad\square$

LEMMA 12 (A comparison principle via L^∞). *Assume* (3.1) *and* (3.3). *Let V_1 and V_2 be two measurable functions. Let $\rho_1, \rho_2 \in L^1(\Omega)$ be such that $\rho_1 \geqslant 0$, $\rho_2 \geqslant 0$ and*

$$\begin{cases} \partial j(\rho_1) + B\rho_1 \ni V_1 \\ \partial j(\rho_2) + B\rho_2 \ni V_2. \end{cases} \tag{3.9}$$

Then

$$\|(B\rho_1 - B\rho_2)^+\|_{L^\infty} \leqslant \|(V_1 - V_2)^+\|_{L^\infty}. \tag{3.10}$$

In particular

$$[V_1 \leqslant V_2 \ a.e.] \quad \Rightarrow \quad [B\rho_1 \leqslant B\rho_2 \ a.e.]$$

and if B is injective

$$[V_1 = V_2 \ a.e.] \quad \Rightarrow \quad [\rho_1 = \rho_2 \ a.e.].$$

Proof. Set $k = \|(V_1 - V_2)^+\|_{L^\infty}$. On the set $[\rho_1 - \rho_2 > 0]$ we have, using (3.9), $V_1 - B\rho_1 \geqslant V_2 - B\rho_2$ and so $B(\rho_1 - \rho_2) \leqslant k$. It follows from (3.8) that $B(\rho_1 - \rho_2) \leqslant k$ a.e. on Ω. $\qquad\square$

LEMMA 13. *Assume (3.1) and (3.3). Suppose that there is some $\bar\rho \in L^1(\Omega)$ with $\bar\rho \geqslant 0$ such that*

$$\partial j(\bar\rho) + B\bar\rho \ni V \quad a.e.. \tag{3.11}$$

Then, for every $\lambda > 0$ there is some $\rho_\lambda \in L^1$ with $\rho_\lambda \geqslant 0$ such that

$$\partial j(\rho_\lambda) + B\rho_\lambda \ni V - \lambda \quad a.e.$$

and

$$\rho_\lambda \leqslant \bar\rho \quad a.e..$$

Proof. We divide the proof into 2 steps:

STEP 1. We claim that for every $\varepsilon > 0$ there is some $\rho^\varepsilon \in L^1(\Omega)$ with $\rho^\varepsilon \geqslant 0$ a.e. such that

$$\partial j(\rho^\varepsilon) + \varepsilon\rho^\varepsilon + B\rho^\varepsilon \ni V + \varepsilon\bar\rho - \lambda \quad a.e. \tag{3.12}$$

and

$$\rho^\varepsilon \leqslant \bar\rho \quad a.e. \tag{3.13}$$

Proof. In what follows $\varepsilon > 0$ is *fixed* and we set $V_n = \inf\{V + \varepsilon\bar\rho, n\}$. For every n there is a (unique) solution ρ_n of the problem

$$\partial j(\rho_n) + \varepsilon\rho_n + B\rho_n \ni V_n - \lambda \text{ a.e.}; \tag{3.14}$$

this is a consequence of Lemma 7. (Note that $V_\infty \leqslant 0$ – by (3.11) and Lemma 8 – and thus $(V_n)_\infty \leqslant 0$. An easy inspection of the proof shows that Lemma 7 still holds if one assumes $V_\infty \leqslant 0$ instead of $V_\infty = 0$). We have

$$\begin{aligned}
\rho_n &= (\partial j + \varepsilon I)^{-1}(V_n - \lambda - B\rho_n) \\
&\leqslant (\partial j + \varepsilon I)^{-1}(V + \varepsilon\bar\rho - \lambda) \\
&= (\partial j + \varepsilon I)^{-1}(V + \varepsilon\bar\rho - B\bar\rho + B\bar\rho - \lambda) \\
&\leqslant \bar\rho + \frac{1}{\varepsilon}(B\bar\rho - \lambda)^+ \in L^1(\Omega).
\end{aligned}$$

since $(\partial j + \varepsilon I)^{-1}$ is Lipschitz with constant $1/\varepsilon$. We may thus assume (for a subsequence) that

$$\rho_n \rightharpoonup \rho \quad \text{weakly in } L^1(\Omega)$$

and then, by Lemma 9,

$$B\rho_n \to B\rho \quad \text{a.e. and in } L^1(\Omega) + L^\infty(\Omega).$$

Using standard monotone analysis (see e.g. item [1], Lemma 3, under Brezis [16]) we can pass to the limit in (3.14) and conclude that ρ satisfies (3.12). Applying Lemma 12 to $(\partial j + \varepsilon I)$ we deduce from (3.11) and (3.12) that $0 \leqslant B\bar{\rho} - B\rho^\varepsilon \leqslant \lambda$. Therefore we obtain

$$\rho^\varepsilon = (\partial j + \varepsilon I)^{-1}(V + \varepsilon\bar{\rho} - B\rho^\varepsilon - \lambda) \leqslant (\partial j + \varepsilon I)^{-1}(V + \varepsilon\bar{\rho} - B\bar{\rho}) = \bar{\rho}.$$

$$\square$$

STEP 2. We let $\varepsilon \to 0$. It is easy to see that (for a subsequence $\varepsilon_n \to 0$)

$$\rho^\varepsilon \rightharpoonup \rho \quad \text{weakly in } L^1(\Omega)$$
$$B\rho^{\varepsilon_n} \to B\rho \quad \text{a.e. and in} L^1(\Omega) + L^\infty(\Omega)$$

and ρ satisfies

$$\partial j(\rho) + B\rho \ni V - \lambda \quad \text{a.e.}$$
$$\rho \leqslant \bar{\rho} \quad \text{a.e..}$$

$$\square$$

Proof of Theorem 5. Uniqueness follows from Lemma 12 since B is assumed to be injective. Let

$$\Lambda = \{\lambda \in \mathbb{R}; (E_\lambda) \text{ has a solution}\} \text{ and } \lambda_0 = \inf \Lambda$$

($\lambda_0 = +\infty$ if $\Lambda = \emptyset$). It follows from Lemma 13 that λ_0 has all the required properties; moreover the mapping $\lambda \mapsto \rho_\lambda$ is nonincreasing.

In order to check its continuity let $\lambda_n \to \lambda \in (\lambda_0, +\infty)$ be a monotone sequence so that $\rho_{\lambda_n} \to \rho$ in $L^1(\Omega)$ (by monotone convergence). It follows that $B\rho_{\lambda_n} \to B\rho$ in $L^1(\Omega) + L^\infty(\Omega)$ and thus ρ satisfies $\partial j(\rho) + B\rho \ni V - \lambda$ a.e., i.e., $\rho = \rho_\lambda$. As $\lambda \uparrow \infty$, $\rho_\lambda \downarrow \rho$ in $L^1(\Omega)$; since $\rho_\lambda = 0$ a.e. on the set $[V - j'(0+) < \lambda]$, we conclude that $\rho = 0$ a.e. on Ω.

For the last assertion in Theorem 5, we note that (i) $\Rightarrow$ (ii) and (ii) $\Rightarrow$ (iii) are straight forward (choose $f = \rho_{\lambda_0}$ and $U = V - \lambda_0 - B\rho_{\lambda_0}$). It remains to show that (iii) $\Rightarrow$ (i). For $\lambda > \lambda_0$ we have

$$\partial j(\rho_\lambda) + B\rho_\lambda \ni U + Bf + \lambda_0 - \lambda$$

so that $\rho_\lambda \leqslant \gamma^0(U + Bf - B\rho_\lambda)$ and therefore

$$\int_{[Bf - B\rho_\lambda \leqslant 0]} \rho_\lambda \leqslant \int \gamma^0(U). \tag{3.15}$$

On the other hand, we have, by Lemma 11,

$$\int_{[Bf-B\rho_\lambda>0]} (f - \rho_\lambda) \geqslant 0. \tag{3.16}$$

Combining (3.15) and (3.16) we see that

$$\int \rho_\lambda \leqslant \int (\gamma^0(U) + f).$$

We conclude with a rather general and useful result.

THEOREM 6. *Assume* (2.1), (2.12), (2.13), (2.14), (3.2), (3.3) *and* (3.4). *Assume in addition that there exist a function* $f \in L^1(\Omega)$, *and a measurable function* $U : \Omega \to \mathbb{R}$ *such that*

$$V = U + Bf \tag{3.17}$$

and

$$\int_\omega \gamma^0(U + t) < \infty \quad \forall t > 0, \quad \forall \omega \subset \Omega \text{ with } |\omega| < \infty. \tag{3.18}$$

Then, for every $\lambda > V_\infty$, *problem* (E$_\lambda$) *admits a solution, i.e., there exists a* $\rho_\lambda \in L^1$, $\rho_\lambda \geq 0$, *satisfying*

$$\partial j(\rho_\lambda) + B\rho_\lambda \ni V - \lambda.$$

In particular, if ess sup$_\Omega$ $V > V_\infty$, *problem* (E^I) *admits a solution for every* $I \in (0, I_1)$ *where*

$$0 < I_1 = \lim_{\lambda \downarrow V_\infty} \int \rho_\lambda \leqslant \infty.$$

Proof. Let $\lambda > V_\infty$ be fixed and let $V_n = \min\{V, n\}$. Let ρ_n be the solution of (3.19)

$$\partial j(\rho_n) + B\rho_n \ni V_n - \lambda. \tag{3.19}$$

The existence of ρ_n follows from Lemma 7. We claim that

$$\int \rho_n \leqslant C. \tag{3.20}$$

Indeed let μ be such that $\lambda > \mu > V_\infty$. We have

$$\rho_n \leqslant \gamma^0(V_n - \mu - B\rho_n) \leqslant \gamma^0(V - \mu - B\rho_n) = \gamma^0(U - \mu + Bf - B\rho_n)$$

and therefore

$$\int_{[Bf \leqslant B\rho_n]} \rho_n \leqslant \int \gamma^0(U - \mu) < \infty$$

since $[U > \mu]$ has finite measure (note that by (3.17), $U_\infty = V_\infty = 0$). On the other hand we have, by Lemma 12,

$$\int_{[Bf > B\rho_n]} (f - \rho_n) \geqslant 0.$$

It follows that

$$\int \rho_n \leqslant \int (\gamma^0(U - \mu) + f).$$

Clearly $\rho_n \leqslant \gamma^0(V - \mu)$, so that

$$\operatorname{Supp} \rho_n \subset \omega = [V > \mu] \tag{3.21}$$

and $|\omega| < \infty$.

Next, we claim that the sequence (ρ_n) is equi-integrable on ω. Indeed let $t > 0$ and let $A \subset \omega$ be measurable. We write

$$\int_A \rho_n \leqslant \int_{A \cap [Bf - B\rho_n \leqslant t]} \rho_n + \int_{[Bf - B\rho_n > t]} \rho_n.$$

As above we have

$$\int_{A \cap [Bf - B\rho_n \leqslant t]} \rho_n \leqslant \int_A \gamma^0(U - \mu + t)$$

and

$$\int_{[Bf - B\rho_n > t]} \rho_n \leqslant \int_{[Bf - B\rho_n > t]} f \leqslant \int_{[Bf > t]} f.$$

Consequently

$$\int_A \rho_n \leqslant \int_A \gamma^0(U - \mu + t) + \int_{[Bf > t]} f.$$

Given $\varepsilon > 0$ we first choose t large enough so that $\int_{[Bf > t]} f < \varepsilon$. Then we choose $\delta > 0$ small enough so that $|A| < \delta$ implies $\int_A \gamma^0(U - \mu + t) < \varepsilon$.

It follows from Lemma 12 that the sequence $(B\rho_n)$ is a nondecreasing. From (3.20) and assumption (3.2) we have

$$\int (B\rho_n - k)^+ \leqslant C(k) \quad \forall\, k > 0,\ \forall n.$$

Therefore $B\rho_n \uparrow u$ a.e. as $n \uparrow \infty$ and $\int (u - k)^+ < \infty \quad \forall k > 0$.

From (3.21) we deduce that (up to a subsequence)

$$\rho_n \rightharpoonup \rho \quad \text{weakly in } L^1(\Omega).$$

By Lemma 10 we have

$$B\rho_n \rightharpoonup B\rho \quad \text{weakly in } L^1(\Omega) + L^\infty(\Omega).$$

It follows that $B\rho_n \rightharpoonup B\rho$ weakly in $L^1(\Omega')$ for any $\Omega' \subset \Omega$ of finite measure. Since $B\rho_n \to u$ a.e. on Ω we deduce that $u = B\rho$ a.e. on Ω. Using Egorov's lemma and standard monotone analysis, we may now pass to the limit in (3.19) and conclude that

$$\partial j(\rho) + B\rho \ni V - \lambda \quad \text{a.e..}$$

$\square$

REMARK 12. Part of the argument used in the proof of Theorem 6 (e.g. the equi-integrability of ρ_n) is inspired by the papers quoted as items [2] [3] under Gallouët-Morel [35].

REMARK 13. If j is coercive, i.e., γ^0 is everywhere defined, then assumption (3.18) is weaker than (H^+). Indeed we write

$$j^*((1+\theta)(V - M)) - j^*(V + t) \geqslant \gamma^0(V + t)[\theta V - M - \theta M - t].$$

so that

$$\gamma^0(V + t) \leqslant j^*((1+\theta)(V - M)) \quad \text{on } [\theta V - M - \theta M - t \geqslant 1]$$

while

$$\gamma^0(V + t) \leqslant \gamma^0\left(\frac{1 + M + \theta M + t}{\theta} + t\right) \quad \text{on } [\theta V - M - \theta M - t < 1].$$

4. Some examples. Further properties of I_0 and I_1

In what follows and throughout the rest of the paper we assume that $\Omega = \mathbb{R}^N$ (with the Lebesgue measure dx) and $N \geqslant 3$.

We take $k(x, y) = k(x - y)$ where $k(x) = c_N/|x|^{N-2}$ with $c_N = 1/[(N-2)\sigma_N]$ and σ_N is the area of the unit sphere in $\mathbb{R}^N$, so that

$$k \in M^{N/(N-2)}(\mathbb{R}^N)$$

and

$$-\Delta k = \delta \quad \text{in the sense of } \mathcal{D}'(\mathbb{R}^N).$$

Here $M^p(\mathbb{R}^N)(1 < p < \infty)$ denotes the Marcinkiewicz (or weak L^p) space, i.e.,

$$M^p(\mathbb{R}^N) = \{u : \mathbb{R}^N \to \mathbb{R};\ u \text{ is measurable and } \|u\|_{M^p} < \infty\}$$

where the norm $\|u\|_{M^p}$ is defined by

$$\|u\|_{M^p} = \sup_{\substack{A \subset \mathbb{R}^N \\ |A| < \infty}} \frac{1}{|A|^{1/p'}} \int_A |u(x)|\, dx.$$

Some elementary properties of the spaces M^p are discussed in the Appendix of Bénilan-Brezis-Crandall [10]. In particular we recall that

$$a_p \|u\|_{M^p}^p \leqslant \sup_{\lambda > 0} \lambda^p \, \text{meas}\, [|u| > \lambda] \leqslant \|u\|_{M^p}^p \quad (a_p > 0).$$

We also recall that, for every $f \in L^1(\mathbb{R}^N)$,

$$Bf = k * f \in M^{N/(N-2)}$$

and

$$\|Bf\|_{M^{N/(N-2)}} \leqslant \|k\|_{M^{N/(N-2)}} \|f\|_{L^1}.$$

Moreover we have

$$-\Delta(Bf) = f \quad \text{in the sense of } \mathcal{D}'(\mathbb{R}^N)$$

and, in particular, B is *injective*. Therefore K defined in Section 1 is *strictly convex* (see Remark 3).

We claim that the kernel k satisfies properties (1.4), (3.2) and (3.3).

Verification of (3.2). Let $\rho \in L^1(\mathbb{R}^N)$ with $\rho \geqslant 0$ and $\|\rho\|_{L^1} \leqslant M$. We have

$$\int_{\mathbb{R}^N} (B\rho - 1)^+ \leqslant \int_{[B\rho > 1]} B\rho \leqslant \|B\rho\|_{M^p} |A|^{1/p'}$$

where $p = N/(N-2)$ and $A = [B\rho > 1]$. But $|A| \leqslant \|B\rho\|_{M^p}^p$ and therefore

$$\int_{\mathbb{R}^N} (B\rho - 1)^+ \leqslant \|B\rho\|_{M^p}^p \leqslant CM^p.$$

In order to check (1.4) and (3.3) it is convenient to use

LEMMA 14. *Let* $p \in C^1(\mathbb{R})$ *with* $p' \geqslant 0$ *and* $p(0) = 0$. *Let* $\rho \in L^1(\mathbb{R}^N)$ *be such that* $\rho\, p(B\rho) \in L^1(\mathbb{R}^N)$. *Then*

$$\int p'(B\rho)|\nabla(B\rho)|^2 \leqslant \int \rho\, p(B\rho).$$

Proof. We already know (by Lemma A.10 in Bénilan-Brezis-Crandall [10]) that the conclusion holds if, in addition, $p \in L^\infty(\mathbb{R})$. In the general case, let (p_n) be a sequence such that $p_n \in C^1(\mathbb{R}) \cap L^\infty(\mathbb{R})$, $p_n' \geq 0$, $p_n(0) = 0$, $|p_n(t)| \leq |p(t)| \, \forall t \in \mathbb{R}$, $p_n(t) \to p(t)$ $\forall t \in \mathbb{R}$ and $p_n'(t) \to p'(t) \, \forall t \in \mathbb{R}$.

We have

$$\int p_n'(B\rho)|\nabla(B\rho)|^2 \leq \int \rho \, p_n(B\rho)$$

and since $|\rho \, p_n(B\rho)| \leq |\rho \, p(B\rho)| \in L^1(\mathbb{R}^N)$ we conclude easily, using Fatou's Lemma and dominated convergence. $\qquad\qquad\square$

Verification of (1.4) **and** (3.3). Applying Lemma 14 with $p(t) = t$ we obtain (1.4) (note that $\int_A |B\rho| < \infty$ for every A with $|A| < \infty$). Suppose now $p \in C^1(\mathbb{R}) \cap L^\infty(\mathbb{R})$ with $p' \in L^\infty(\mathbb{R})$, $p' \geq 0$ and $p(0) = 0$. Let $\rho \in L^1(\mathbb{R}^N)$ be such that $\int \rho \, p(B\rho) = 0$. It follows from Lemma 14 that $p'(B\rho)|\nabla(B\rho)|^2 = 0$ and thus $\nabla p(B\rho) = p'(B\rho) \nabla(B\rho) = 0$. Therefore, $p(B\rho)$ is a constant. On the other hand, $B\rho \to 0$ as $|x| \to \infty$ in a weak sense (i.e., for every $\alpha > 0$ the set $[|B\rho| > \alpha]$ has finite measure) and so does $p(B\rho)$. It follows that $p(B\rho) = 0$.

We recall the main result of Section 3. Let $j : \mathbb{R} \to [0, +\infty]$ be any convex l.s.c. function such that

$$j(0) = 0 \text{ and } j(r) = +\infty \quad \text{for all } r < 0.$$

As above we set $\gamma = \partial j^* = (\partial j)^{-1}$.

Let $V : \mathbb{R}^N \to \mathbb{R}$ be any measurable function. We are concerned with the two problems

$$\begin{cases} \text{Given a constant } I \text{ with } 0 < I < \infty, \text{ find a function } \rho \in L^1(\mathbb{R}^N) \text{ and a} \\ \text{constant } \lambda \in \mathbb{R} \text{ such that } \rho \geq 0 \text{ a.e., } \int \rho = I \text{ and } \partial j(\rho) + B\rho \ni V - \lambda \text{ a.e.} \end{cases} \qquad (\mathrm{E}^I)$$

and

$$\begin{cases} \text{Given a constant } I \text{ with } 0 < I < \infty \text{ find a function} \\ \rho \in K_I = \{\rho \in D(\mathcal{E}); \int \rho = I\} \text{ which minimizes } \mathcal{E} \text{ on } K_I. \end{cases} \qquad (\mathrm{M}_I)$$

Corollary 1 says that, under some assumptions, there exists $0 \leq I_0 \leq \infty$ such that

a) for every $0 < I \leq I_0$ (and $I < \infty$) there is a unique solution ρ^I of problem (M_I),
b) if $I_0 < \infty$ and $I > I_0$ problem (M_I) admits no solution.

Theorem 4 asserts that there exists I_1 with $0 \leq I_1 \leq \infty$ such that:

a) for every $0 < I \leq I_1$ (and $I < \infty$), there is a unique solution ρ^I of problem (E^I),
b) if $I_1 < \infty$ and $I > I_1$, problem (E^I) has no solution.

In what follows we shall examine various examples of functions j and V, discuss the relation between problems (E^I) and (M_I) and describe some additional properties of I_0 and I_1.

Some specific examples of functions j are the following:

EXAMPLE 1. Let $1 < p < \infty$ and let

$$j(r) = \begin{cases} \frac{1}{p} r^p & \text{for } r \geq 0 \\ +\infty & \text{for } r < 0 \end{cases}$$

so that, with $\frac{1}{p} + \frac{1}{p'} = 1$,

$$j^*(s) = \begin{cases} \frac{1}{p'} s^{p'} & \text{for } s \geq 0 \\ 0 & \text{for } s < 0 \end{cases}$$

$$\partial j(r) = \begin{cases} r^{p-1} & \text{for } r > 0 \\ (-\infty, 0] & \text{for } r = 0 \\ \emptyset & \text{for } r < 0 \end{cases}$$

$$\gamma(s) = \partial j^*(s) = (\partial j)^{-1}(s) = \begin{cases} s^{p'-1} & \text{for } s \geq 0 \\ 0 & \text{for } s < 0. \end{cases}$$

The usual Thomas-Fermi problem (see e.g. Lieb-Simon [48], Lieb [47]) corresponds to the case $p = 5/3$.

EXAMPLE 2. Let $1 < p < \infty$ and let

$$j(r) = \begin{cases} \frac{1}{p}[(1+r)^p - 1 - pr] & \text{for } r \geq 0 \\ +\infty & \text{for } r < 0 \end{cases}$$

so that

$$j^*(s) = \begin{cases} \frac{1}{p'}(1+s)^{p'} - 1 - p's & \text{for } s \geq 0 \\ 0 & \text{for } s \leq 0 \end{cases}$$

$$\partial j(r) = \begin{cases} (1+r)^{p-1} - 1 & \text{for } r > 0 \\ (-\infty, 0] & \text{for } r = 0 \\ \emptyset & \text{for } r < 0 \end{cases}$$

$$\gamma(s) = \partial j^*(s) = (\partial j)^{-1}(s) = \begin{cases} (1+s)^{p'-1} - 1 & \text{for } s \geq 0 \\ 0 & \text{for } s < 0. \end{cases}$$

Such a j (with $p = 5/3$) occurs in the Thomas-Fermi theory of screening (see Lieb-Simon [48], Section VII). Note that $j(r) \sim r^p$ as $r \to +\infty$ while $j(r) \sim r^2$ as $r \to 0+$.

EXAMPLE 3. Let

$$j(r) = \begin{cases} 3\int_0^{r^{1/3}} t^2(\sqrt{1+t^2} - 1)\,dt & \text{for } r \geqslant 0 \\ +\infty & \text{for } r < 0 \end{cases}$$

so that

$$j^*(s) = \begin{cases} \int_0^s (2t + t^2)^{3/2}\,dt & \text{for } s \geqslant 0 \\ 0 & \text{for } s < 0 \end{cases}$$

$$\partial j(r) = \begin{cases} \sqrt{1 + r^{2/3}} - 1 & \text{for } r \geqslant 0 \\ (-\infty, 0] & \text{for } r = 0 \\ \varnothing & \text{for } r < 0 \end{cases}$$

$$\gamma(s) = \partial j^*(s) = (\partial j)^{-1}(s) = \begin{cases} (2s + s^2)^{3/2} & \text{for } s \geqslant 0 \\ 0 & \text{for } s < 0. \end{cases}$$

Such a j occurs in some relativistic Thomas-Fermi model (E. Lieb, personal communication). Note that $j(r) \sim r^{4/3}$ as $r \to +\infty$ while $j(r) \sim r^{5/3}$ as $r \to 0+$.

EXAMPLE 4. Let $1 < q < p < \infty$ and let

$$j(r) = \begin{cases} \frac{1}{p}r^p - \frac{a}{q}r^q + br & \text{for } r > 1 \\ 0 & \text{for } 0 \leqslant r \leqslant 1 \\ +\infty & \text{for } r < 0 \end{cases}$$

where $a = q(p-1)/p(q-1)$ and $b = (p-q)/p(q-1)$, so that

$$\partial j(r) = \begin{cases} r^{p-1} - ar^{q-1} + b & \text{for } r > 1 \\ 0 & \text{for } 0 < r \leqslant 1 \\ (-\infty, 0] & \text{for } r = 0 \\ \varnothing & \text{for } r < 0 \end{cases}$$

$$\gamma(s) = \begin{cases} 0 & \text{for } s < 0 \\ [0, 1] & \text{for } s = 0 \\ \text{singlevalued} & \text{for } s > 0. \end{cases}$$

Note that $j(r) \sim r^p$ as $r \to +\infty$ while $j(r) = 0$ for $0 < r < 1$ and $\gamma(s) \sim 1 + cs$, for $s > 0, s \sim 0$ with $c = \frac{p}{(p-1)(p-q)}$. Such a j occurs in Thomas-Fermi model with an "exchange correction" (see Benguria [7], Chapter 3).

In what follows we will assume that $N = 3$, but there are similar results for $N > 3$. Throughout the rest of this section we will assume (this is satisfied in all the examples above) that

$$j \text{ is } C^1 \text{ on } (0, \infty) \text{ with } j'(0+) = 0. \tag{4.1}$$

We will consider various types of functions V. In all cases we have $V_\infty = 0$.

TYPE I. $V = k * f$ for some $f \in L^1$.

Thus $V \in M^3$ and $-\Delta V = f$. In particular, we know that for every $\delta > 0$ the set $[|V| > \delta]$ has finite measure. This case is well adapted to the direct approach of Section 3. Indeed, the equation

$$-\Delta u_0 + \gamma(u_0) \ni f \quad \text{in } \mathbb{R}^3 \tag{4.2}$$

admits a unique solution $u_0 \in M^3$ (by Theorem 2.1 in Bénilan-Brezis-Crandall [10]), with $\gamma(u_0) \in L^1$ (more precisely $f + \Delta u_0 \in L^1$) and

$$\int \gamma(u_0) \leqslant \int f^+. \tag{4.3}$$

(Recall $\gamma(t) = 0$ for $t \leqslant 0$). If we set

$$\rho_0 = f + \Delta u_0 = \Delta(u_0 - V)$$

we see (from (4.2)) that

$$u_0 \in \gamma^{-1}(\rho_0) = \partial j(\rho_0)$$

and therefore

$$\partial j(\rho_0) + B\rho_0 \ni V \quad \text{a.e..} \tag{4.4}$$

More generally, for every $\lambda \geqslant 0$ there exists a unique solution $u_\lambda \in M^3$ of

$$-\Delta u_\lambda + \gamma(u_\lambda - \lambda) \ni f \quad \text{in } \mathbb{R}^3 \tag{4.5}$$

(since $\beta(t) = \gamma(t - \lambda)$ is a maximal monotone graph such that $0 \in \beta(0)$). Then

$$\rho_\lambda = f + \Delta u_\lambda \in L^1$$

satisfies

$$u_\lambda - \lambda \in \gamma^{-1}(f + \Delta u_\lambda) = \partial j(\rho_\lambda)$$

and therefore we have

$$\partial j(\rho_\lambda) + B\rho_\lambda \ni V - \lambda \quad \text{a.e..} \tag{E_λ}$$

Set

$$I_1 = \int f + \Delta u_0 = \int \gamma(u_0) \leqslant \int f^+.$$

Note that $I_1 > 0$ whenever $\text{ess sup}_\Omega V > 0$. (Indeed $[I_1 = 0] \Leftrightarrow [\gamma(u_0) = 0] \Leftrightarrow [u_0 \leqslant 0]$ because of assumption (4.1), and then by (4.2) we have $u_0 = V$).

COROLLARY 2. *For every $I \in (0, I_1]$ there exists a unique solution of problem (E^I). In addition, if we assume*

$$\int_0^1 \frac{\gamma^0(s)}{s^4} \, ds = \infty, \tag{4.6}$$

then

$$\int f \leqslant I_1 \leqslant \int f^+;$$

in particular, if $f \geqslant 0$ a.e. then

$$I_1 = \int f. \tag{4.7}$$

Proof. The conditions of Theorem 5 are satisfied with $\lambda_0 = V_\infty = 0$. Note that (E_λ) has no solution for $\lambda < 0$. (Indeed, if (E_λ) has a solution for some $\lambda \in \mathbb{R}$ we deduce from Lemma 8 and (4.1) that $(V - \lambda)_\infty = V_\infty - \lambda = -\lambda \leqslant 0$, i.e., $\lambda \geqslant 0$). Hence we have the first assertion of Corollary 2.

Next we assume (4.6). Applying Lemma B.1 and Theorem B.1 (from Appendix B) to the function u_0 we conclude that

$$\int \Delta u_0 \geqslant 0.$$

Therefore, $I_1 = \int f + \Delta u_0 \geqslant \int f$. $\qquad\qquad\square$

REMARK 14. We emphasize that the first assertion in Corollary 2 applies to Example 1 *without any restriction* on p. The second assertion holds only under the restriction

$$p \geqslant \frac{4}{3} \tag{4.8}$$

(this is an assumption about j *near zero*). It is clearly satisfied for the standard Thomas-Fermi exponent $p = 5/3$.

On the other hand if (4.8) fails, i.e., if $p < 4/3$, then for $f \geqslant 0$ with compact support, $f \not\equiv 0$, we have $I_1 < \int f$. Indeed in this case $\gamma(s) \sim s^q$ as $s \to 0$ with $q = p' - 1 > 3$. Applying a result of Véron (see item [3], Théorème 4.1, under Véron [59]) we see that $u_0(x) \sim c/|x|$ as $|x| \to \infty$ with $c > 0$. Therefore (by Theorem B.1) we have $\int \Delta u_0 < 0$ and $I_1 = \int f + \Delta u_0 < \int f$.

Alternatively, we could also try to apply the variational route of Section 2. This is indeed possible in Example 1 when

$$p > 3/2 \tag{4.9}$$

((4.9) is now an assumption about j *near infinity*). Note that (4.9) holds for the standard Thomas-Fermi exponent $p = 5/3$. However (4.9) *does not hold in Example* 2 (relativistic Thomas-Fermi).

Indeed, the basic condition (H) (or H$^+$) says that for some constant $C \in \mathbb{R}$

$$(V - C)^+ \in L^{p'}. \tag{4.10}$$

Recall that $V \in M^3$ and thus $V|_\omega \in L^q(\omega)$ for any $q < 3$ and any set ω with finite measure. If we take $C > 0$ and $\omega = [|V| > C]$ we see that (4.10) holds provided $p' < 3$, i.e., $p > 3/2$.

When condition (4.9) fails—for example $j(r) = r^p$ with $p \leqslant 3/2$—the functional

$$\mathcal{E}(\rho) + \lambda \int \rho = \int j(\rho) - V\rho + \lambda\rho + \frac{1}{2} \int \rho B\rho \tag{4.11}$$

is usually *unbounded from below for any* $\lambda > 0$. This means that the variational route used in Section 2 is *not* practicable for a general $V = Bf$, $f \in L^1$.

Here is a sketch of the argument. Suppose that we have a lower bound. Then

$$\int V\rho \leqslant \int \rho^p + \frac{1}{2} \int \rho B\rho + C \int \rho + C. \tag{4.12}$$

It is easy to see from Young's inequality on convolutions or the L^p regularity theory that $\|B\rho\|_{L^6} \leqslant C\|\rho\|_{L^{6/5}}$ and thus

$$\int \rho B\rho \leqslant \|\rho\|_{L^{6/5}} \|B\rho\|_{L^6} \leqslant C\|\rho\|_{L^{6/5}}^2.$$

Since $p < 2$ we deduce from (4.12) that

$$\int V\rho \leqslant C(\|\rho\|_{L^p}^2 + \|\rho\|_{L^{6/5}}^2 + C)$$

and by scaling we find

$$\int V\rho \leqslant C(\|\rho\|_{L^p} + \|\rho\|_{L^{6/5}}).$$

Hence

$$V \in L^{p'} + L^6$$

so that

$$V \in L^q_{\text{loc}} \quad \text{with } q = \min(p', 6).$$

Since $p \leqslant 3/2$ we have $p' \geqslant 3$ and then $q \geqslant 3$. On the other hand B does not map L^1 into L^3 (only into M^3) [otherwise B would also map $L^{3/2}$ into L^∞ and then $k \in L^3$ – impossible]. Hence there are some f's in L^1 such that $V = Bf \notin L^3$. For such V's the functional (4.11) is unbounded below.

TYPE II. $V = k * \mu$ for some bounded measure μ.

This case is especially important in the Thomas-Fermi setting because it includes functions $V(x)$ of the form

$$V(x) = \sum_{i=1}^{\ell} \frac{m_i}{|x - a_i|}, \quad m_i \in \mathbb{R}, \tag{4.13}$$

which play a central role in the analysis of Lieb-Simon [48]. Here we have

$$V = k * \mu \quad \text{and} \quad \mu = 4\pi \sum_{i=1}^{\ell} m_i \delta_{a_i}.$$

Again it is well suited to the direct approach of Section 3 *provided we make the additional assumption*

$$\int_{|x|<1} \gamma^0 \left(\frac{1}{|x|} \right) = C \int_1^{\infty} \frac{\gamma^0(s)}{s^4} \, ds < \infty \tag{4.14}$$

which is required in order to apply Theorem A.1 (in Appendix A). In the framework of Examples 1, 2, 4 this corresponds to the condition

$$p > \frac{4}{3}. \tag{4.15}$$

Assumption (4.14) is an assumption about j near *infinity*. It is satisfied for the standard Thomas-Fermi exponent $p = 5/3$. However (4.14) *fails* in Example 2 (relativistic Thomas-Fermi).

As above we solve the equation

$$-\Delta u_0 + \gamma(u_0) \ni \mu \quad \text{in } \mathbb{R}^3 \tag{4.16}$$

with the help of Theorem A.1 and we set

$$\rho_0 = \mu + \Delta u_0 \in L^1$$

and

$$I_1 = \int \mu + \Delta u_0 = \int \gamma(u_0) \leqslant \int \mu^+.$$

Again $I_1 > 0$ whenever ess $\sup_{\mathbb{R}^3} V > 0$.

Using the same strategy as in Corollary 2, we have

COROLLARY 3. *Assume* (4.14). *Then for every* $I \in (0, I_1]$ *there exists a unique solution of problem* (E^I).

In addition, if we assume (4.6), *then*

$$\int \mu \leqslant I_1 \leqslant \int \mu^+;$$

in particular if $\mu \geqslant 0$, then

$$I_1 = \int \mu.$$

REMARK 15. Condition (4.15) is *absolutely essential*. When it is not satisfied there is usually *no I* whatsoever such that problem (E^I) admits a solution. Take, for example, $j(r) = \frac{3}{4}r^{4/3}$ and then $\gamma(s) = (s^+)^3$. Let $V(x) = 1/|x|$ (so that $-\Delta V = 4\pi \delta_0$). If we had a solution of (E^I) for some I, it would satisfy

$$\partial j(\rho) + B\rho \ni V - \lambda.$$

Necessarily $\lambda \geqslant 0$ (by Lemma 8) and $u = \frac{c}{|x|} - B\rho$ satisfies

$$-\Delta u + \left[(u - \lambda)^+\right]^3 = \delta_0$$

with $(u - \lambda)^+ \in L^3$. But this is impossible, even locally near 0; see the discussion in Remark A.4. In particular, for the relativistic Thomas-Fermi model (Example 3 above) with the Coulomb potential $V(x) = 1/|x|$, there is *no I* such that problem (E^I) admits a solution; existence holds provided the potential is slightly more "diffuse".

REMARK 16. As above, we see that the variational route discussed in Section 2 holds in Example 1 when $p > 3/2$. If $p \leqslant 3/2$ and $V(x) = \sum_i \frac{m_i}{|x - a_i|}$, the functional $\mathcal{E}(\rho) + \lambda \int \rho$ is *unbounded below*.

TYPE III. $V \in M^3(\mathbb{R}^3)$.

Clearly this situation is more general than Type II (since $k * \mu \in M^3$). Here we cannot anymore rely on Appendix A to solve

$$-\Delta u_0 + \gamma(u_0) \ni -\Delta V$$

since ΔV need not be a measure. Instead we will rely on Theorem 6. The conclusion is less precise since we have little information about I_1 (we suspect that I_1 might sometimes be infinite).

COROLLARY 4. *Assume again* (4.14). *Let $V \in M^3(\mathbb{R}^3)$ be such that* ess $\sup_{\mathbb{R}^3} V > 0$. *Then there exists $0 < I_1 \leqslant \infty$ such that*

a) *for every $I \in (0, I_1)$ there is a unique solution of problem (E^I),*

b) *if $I_1 < \infty$, problem (E^{I_1}) admits a solution, and problem (E^I) has no solution when*
    $I > I_1$.

Proof. Apply Theorem 6 with the decomposition $V = U + Bf$ and $f = 0$. We have to verify (3.18), i.e.,

$$\int_\omega \gamma^0(V + t) < \infty \quad \forall t > 0, \; \forall \omega \subset \Omega \text{ with } |\omega| < \infty.$$

This follows immediately from assumption (4.14) and Lemma A.1 applied to the function $u_n \equiv V + t \in M^3$ on ω. $\square$

REMARK 17. There are *many* variants of Corollary 4. For instance, in the standard Thomas-Fermi theory (Example 1 with $p = 5/3$), it suffices to assume, for example, that for every $\delta > 0$, the set $[V > \delta]$ is bounded and $V \in L_{\text{loc}}^{3/2}$ (singularities such as $|x|^{-\alpha}$, $\alpha < 2$ are admissible).

    In the relativistic Thomas-Fermi (Example 2), it suffices to assume that for every $\delta > 0$, the set $[V > \delta]$ is bounded and that $V = V_1 + V_2$ with $V_1 \in L_{\text{loc}}^3$ and $V_2 \in L_{\text{loc}}^1$ with $\Delta V_2 \in L_{\text{loc}}^1$. Note that the singularity $V(x) = 1/|x|$ is excluded, but this is consistent with the discussion in Remark 15 (see also Remark A. 4).

5. A min-max principle for the Lagrange multiplier λ; uniqueness of the extremals

Throughout this section we take $\Omega = \mathbb{R}^N$, $N \geq 3$ and $B\rho = k * \rho$ as in Section 4.
Let $j : \mathbb{R} \to [0, +\infty]$ be a convex l.s.c. function such that

$$j(0) = 0 \text{ and } j(r) = +\infty \text{ for all } r < 0 \tag{5.1}$$

$$j \text{ is } C^1 \text{ on } (0, \infty), \quad \text{and} \quad j'(0+) = 0. \tag{5.2}$$

Let $V : \Omega \to \mathbb{R}$ be a measurable function such that $V_\infty = 0$. Recall (see Theorem 5) that exists $\lambda_0 \in [0, +\infty]$ such that for every $\lambda > \lambda_0$ problem

$$\partial j(\rho) + B\rho \ni V - \lambda \quad \text{a.e.} \tag{E_λ}$$

admits a unique solution $\rho_\lambda \in L^1$, $\rho_\lambda \geq 0$. As in the previous sections we set

$$I(\lambda) = \int \rho_\lambda \text{ and } I_1 = \sup_{\lambda > \lambda_0} I(\lambda) = \lim_{\lambda \downarrow \lambda_0} I(\lambda) \leq \infty.$$

Note that $I(\lambda) > 0$ if and only if $\lambda < \text{ess sup}_{\mathbb{R}^N} V$. Recall that (E_λ) has no solution for $\lambda < \lambda_0$ and (E_{λ_0}) admits a unique solution if and only if $I_1 < \infty$.

THEOREM 7. *For any $\lambda_0 < \lambda < \text{ess sup}_{\mathbb{R}^N} V$ we have*

$$\lambda = \max_{\substack{\rho \in L^1, \rho \geqslant 0 \\ \int \rho = I(\lambda)}} \text{ess inf}_{[\rho > 0]} \{V - B\rho - j'(\rho)\} \tag{5.3}$$

and

$$\lambda = \min_{\substack{\rho \in L^1, \rho \geqslant 0 \\ \int \rho = I(\lambda)}} \text{ess sup}_{\mathbb{R}^N} \{V - B\rho - j'(\rho)\}. \tag{5.4}$$

Conclusion (5.3) holds for $\lambda = \lambda_0 < \infty$ provided $I_1 < \infty$; conclusion (5.4) holds for $\lambda = \lambda_0$ provided $\lambda_0 = 0$ and $I_1 < \infty$.

In (5.4) we use the convention that $j'(0) = j'(0+)(= 0)$.

REMARK 18. The conclusion of Theorem 7 were obtained by Lieb-Simon [48] (Theorems II.28 and II.29) in the context of the standard Thomas-Fermi model (see Example 1 in Section 4 with $p = 5/3$, and $V(x)$ given by (4.13) with $m_i > 0 \ \forall i$).

Proof. If we take $\rho = \rho_\lambda$ we have on the set $A = [\rho_\lambda > 0]$ (which has positive measure because of the assumption $\lambda < \text{ess sup}_{\mathbb{R}^N} V$),

$$j'(\rho_\lambda) + B\rho_\lambda = V - \lambda, \tag{5.5}$$

so that

$$\text{ess inf}_{[\rho_\lambda > 0]} \{V - B\rho_\lambda - j'(\rho_\lambda)\} = \lambda.$$

Moreover, on the set $[\rho_\lambda = 0]$ we have

$$V - \lambda - B\rho_\lambda \leqslant 0.$$

and in particular

$$V - B\rho_\lambda - j'(\rho_\lambda) \leqslant \lambda.$$

Thus

$$\text{ess sup}_{\mathbb{R}^N} \{V - B\rho_\lambda - j'(\rho_\lambda)\} = \lambda.$$

To conclude the proof it remains to show that for every $\rho \in L^1, \rho \geqslant 0$, with $\int \rho = I(\lambda)$ we have

$$\text{ess inf}_{[\rho > 0]} \{V - B\rho - j'(\rho)\} \leqslant \lambda \tag{5.6}$$

and

$$\operatorname*{ess\,sup}_{\mathbb{R}^N} \{V - B\rho - j'(\rho)\} \geqslant \lambda. \tag{5.7}$$

$\square$

Proof of (5.6). Suppose, by contradiction, that there is some $\bar{\rho} \in L^1, \bar{\rho} \geqslant 0$, with $\int \bar{\rho} = I(\lambda)$, such that

$$\lambda^* = \operatorname*{ess\,inf}_{[\bar{\rho}>0]} \{V - B\bar{\rho} - j'(\bar{\rho})\} > \lambda. \tag{5.8}$$

Let $\rho^* = \rho_{\lambda^*}$ be the unique solution of (E_{λ^*}), i.e.,

$$\partial j(\rho^*) + B\rho^* \ni V - \lambda^*. \tag{5.9}$$

Set

$$W = \begin{cases} j'(\bar{\rho}) + B\bar{\rho} & \text{on } [\bar{\rho} > 0], \\ \min\{B\bar{\rho}, V - \lambda^*\} & \text{on } [\bar{\rho} = 0]. \end{cases}$$

Clearly we have

$$\partial j(\bar{\rho}) + B\bar{\rho} \ni W \quad \text{a.e. on} \mathbb{R}^N, \tag{5.10}$$

and

$$W \leqslant V - \lambda^* \quad \text{a.e. on } \mathbb{R}^N. \tag{5.11}$$

We deduce from (5.9), (5.10), (5.11) and Lemma 12 that

$$B\bar{\rho} \leqslant B\rho^*. \tag{5.12}$$

Applying Theorem B.1 with $u = B(\bar{\rho} - \rho^*) \leqslant 0$, we see that $\int (\bar{\rho} - \rho^*) \leqslant 0$, i.e.,

$$\int \bar{\rho} = I(\lambda) \leqslant I(\lambda^*) = \int \rho^*. \tag{5.13}$$

Let ρ_λ be the solution of (E_λ). From Theorem 5 we know that

$$\rho_{\lambda^*} \leqslant \rho_\lambda. \tag{5.14}$$

Combining (5.14) with (5.13) we deduce that

$$\rho^* = \rho_\lambda. \tag{5.15}$$

Recall that $A = [\rho_\lambda > 0] = [\rho^* > 0]$ has positive measure. Applying (E_λ) and (E_{λ^*}) on A we find

$$V - \lambda = V - \lambda^* \quad \text{a.e. on } A,$$

and thus $\lambda = \lambda^*$— a contradiction.

Proof of (5.7). Suppose, by contradiction, that there is some $\bar{\rho} \in L^1, \bar{\rho} \geqslant 0$, with $\int \bar{\rho} = I(\lambda)$ such that

$$\mu^* = \operatorname*{ess\,sup}_{\mathbb{R}^N}\{V - B\bar{\rho} - j'(\bar{\rho})\} < \lambda. \tag{5.16}$$

Fix μ such that $\max\{\mu^*, \lambda_1\} < \mu < \lambda$. Set

$$W = j'(\bar{\rho}) + B\bar{\rho},$$

so that

$$\partial j(\bar{\rho}) + B\bar{\rho} \ni W \tag{5.17}$$

and

$$W \geqslant V - \mu^* > V - \mu. \tag{5.18}$$

Let ρ_μ be the solution of

$$\partial j(\rho_\mu) + B\rho_\mu \ni V - \mu \tag{5.19}$$

(which exists since $\mu > \lambda_1$). Combining (5.17), (5.19) and (5.18), we deduce from the comparison principle in Lemma 12 that $B\rho_\mu \leqslant B\bar{\rho}$. Applying Theorem B.1 once more yields $\int (\rho_\mu - \bar{\rho}) \leqslant 0$, i.e.,

$$I(\mu) = \int \rho_\mu \leqslant \int \bar{\rho} = I(\lambda).$$

We conclude that $\rho_\lambda = \lambda_\mu$ and obtain a contradiction as above.

In the limiting case $\lambda = \lambda_0$, the proof of (5.6) is unchanged. But we cannot use the above proof for (5.7). In this case we simply observe that

$$\operatorname*{ess\,sup}_{\mathbb{R}^N}\{V - B\rho - j'(\rho)\} \geqslant V_\infty = 0 = \lambda_0.$$

Lieb and Simon [48] have conjectured the uniqueness of the maximizer in (5.3) and the minimizer in (5.4) (see Problem 4 in the Introduction and the discussion in Section II.7). We will prove that the conjecture is *true* when $\lambda_0 < \lambda < \operatorname{ess\,sup}_{\mathbb{R}^N} V$ for a large class of problems including the standard Thomas-Fermi model: Example 1 in Section 4 with $p = 5/3$. (With the notations of Lieb-Simon [48] this means that the conjecture holds when $N < Z$). A basic ingredient is a sharp form of strong maximum principle described in Appendix C.

However we will see that the conjecture *fails* (even for the standard Thomas-Fermi model) in the "neutral" case $\lambda = 0$ (i.e., $N = Z$ with the notations of Lieb-Simon [48]).

A counter example in the neutral case.

Consider for simplicity the case $N = 3$ and the Example of Section 4 with $p > 4/3$. In the neutral case, the Thomas-Fermi ρ is the unique solution of the equations

$$\rho^{p-1} + B\rho = V = \sum_i \frac{m_i}{|x - a_i|} \tag{5.20}$$

with $m_i > 0 \ \forall i$. In other words $u = \rho^{p-1}$ is the unique positive solution of

$$-\Delta u + u^{1/(p-1)} = 4\pi \sum_i m_i \delta_{a_i}. \tag{5.21}$$

Moreover we have, by Corollary 3,

$$\int \rho = 4\pi \sum m_i. \tag{5.22}$$

Clearly the function ρ satisfies $\rho > 0$, $\int \rho = I = 4\pi \sum m_i$, and

$$\operatorname*{ess\,inf}_{\mathbb{R}^N} (V - B\rho - \rho^{p-1}) = 0 \tag{5.23}$$

$$\operatorname*{ess\,sup}_{\mathbb{R}^N} (V - B\rho - \rho^{p-1}) = 0. \tag{5.24}$$

We will now construct two functions ρ_1, ρ_2, distinct from ρ, satisfying $\rho_1 > 0, \rho_2 > 0$, $\int \rho_1 = \int \rho_2 = I$,

$$\operatorname*{ess\,inf}_{\mathbb{R}^N} (V - B\rho_1 - \rho_1^{p-1}) = 0, \tag{5.25}$$

$$\operatorname*{ess\,sup}_{\mathbb{R}^N} (V - B\rho_2 - \rho_2^{p-1}) = 0. \tag{5.26}$$

Given $k > 0$, let $u_k > 0$ be the solution of

$$-\Delta u_k + k u_k^{1/(p-1)} = 4\pi \sum m_i \delta_{a_i}, \tag{5.27}$$

and set

$$\rho_k = k u_k^{1/(p-1)}. \tag{5.28}$$

From the results of Appendix B we deduce that

$$\int \rho_k = I = 4\pi \sum m_i \quad \forall k.$$

On the other hand, we see from (5.27) and (5.28) that

$$(k^{-1}\rho_k)^{p-1} + B\rho_k = V = \sum \frac{m_i}{|x - a_i|}$$

and therefore

$$V - B\rho_k - \rho_k^{p-1} = (k^{-(p-1)} - 1)\rho_k^{p-1}.$$

We obtain the desired ρ_1 and ρ_2 satisfying (5.25) and (5.26) by choosing $\rho_1 = \rho_{k_1}$ and $\rho_2 = \rho_{k_2}$ with $k_1 < 1$ and $k_2 > 1$.

Uniqueness of the extremals in the "ionic" case, $0 < I < I_0$

In addition to the standard assumptions (5.1) and (5.2) on j, we assume here that

$$j' \text{ is concave on } (0, \infty), \tag{5.29}$$

and

$$\lim_{r \to \infty} \frac{j(r)}{r} = +\infty. \tag{5.30}$$

As a result, it is easy to see that $\gamma = (\partial j)^{-1}$ is a continuous nondecreasing function on $\mathbb{R}$ such that

$$\gamma(s) = 0 \text{ for } s \leqslant 0, \tag{5.31}$$

and

$$\gamma \text{ is convex on } \mathbb{R}, \tag{5.32}$$

so that $\gamma'(s-)$ exists at every $s \in \mathbb{R}$, and will be denote simply $\gamma'(s)$.
A typical example is

$$j(r) = \begin{cases} \frac{1}{p}r^p & \text{for } r \geqslant 0, \\ +\infty & \text{for } r < 0, \end{cases} \tag{5.33}$$

with $1 < p < 2$ and then $\gamma(r) = (r^+)^{p'-1}$; recall that the standard Thomas-Fermi model corresponds to $p = 5/3$ and then $\gamma(r) = (r^+)^{3/2}$.
Let $\lambda > 0$ and let V be any measurable function such that, for some $R > 0$,

$$V(x) \leqslant \lambda \text{ for a.e. } x \text{ with } |x| > R. \tag{5.34}$$

We will assume that

$$\gamma'(V - \lambda) \in L^1(\mathbb{R}^N). \tag{5.35}$$

The standard Thomas-Fermi model corresponds to $V(x) = \sum \frac{m_i}{|x-a_i|}$ in $\mathbb{R}^3$ and satisfies all the required assumptions (any $p > 5/4$ would be acceptable).

Let $\rho \in L^1$, $\rho \geqslant 0$, be a solution of the problem

$$\partial j(\rho) + B\rho \ni V - \lambda \quad \text{a.e. on } \mathbb{R}^N. \tag{5.36}$$

Suppose now that ρ_1 is a maximizer for (5.3), i.e., $\rho_1 \in L^1$, $\rho_1 \geqslant 0$, satisfies

$$\int \rho_1 = \int \rho \tag{5.37}$$

and

$$\operatorname*{ess\,inf}_{[\rho_1 > 0]} \{V - B\rho_1 - j'(\rho_1)\} = \lambda. \tag{5.38}$$

Similarly, suppose that ρ_2 is a minimizer for (5.4), i.e., $\rho_2 \in L^1$, $\rho_2 \geqslant 0$, satisfies

$$\int \rho_2 = \int \rho \tag{5.39}$$

and

$$\operatorname*{ess\,sup}_{\mathbb{R}^N} \{V - B\rho_2 - j'(\rho_2)\} = \lambda \tag{5.40}$$

(with the convention that $j'(0) = 0$).

THEOREM 8. *Assume* (5.1), (5.2), (5.29), (5.30), (5.34)–(5.40). *Then*

$$\rho_1 = \rho_2 = \rho.$$

The key ingredient in the proof is the following:

LEMMA 15. *Assume* (5.1), (5.2), (5.29) *and* (5.30). *Let* $\psi_1, \psi_2 \in L^1$ *with* $\psi_1 \geqslant 0$ *a.e. on* $\mathbb{R}^N$, $\psi_2 \geqslant 0$ *a.e. on* $\mathbb{R}^N$ *be such that*

$$\int \psi_1 = \int \psi_2. \tag{5.41}$$

Let f_1, f_2 *be measurable functions on* $\mathbb{R}^N$ *such that*

$$f_1 \leqslant f_2 \quad \text{a.e. on } \mathbb{R}^N, \tag{5.42}$$

$$f_1(x) \leqslant 0 \quad \text{for a.e. } x, |x| > R, \tag{5.43}$$

$$\gamma'(f_1) \in L^1. \tag{5.44}$$

Assume

$$\partial j(\psi_1) + B\psi_1 \ni f_1 \quad \textit{a.e. on } \mathbb{R}^N, \tag{5.45}$$

and

$$\partial j(\psi_2) + B\psi_2 \ni f_2 \quad \textit{a.e. on } \mathbb{R}^N. \tag{5.46}$$

Then

$$\psi_1 = \psi_2. \tag{5.47}$$

Proof of Lemma 15. From Lemma 12 and (5.42) we already know that

$$B\psi_1 \leqslant B\psi_2. \tag{5.48}$$

Set $u = B(\psi_2 - \psi_1) \geqslant 0$ and

$$a = \begin{cases} \dfrac{\gamma(f_1 - B\psi_1) - \gamma(f_1 - B\psi_2)}{u} & \text{on } [u > 0], \\ 0 & \text{on } [u = 0], \end{cases}$$

so that $a \geqslant 0$ a.e.

Clearly we have

$$\begin{aligned} -\Delta u + au &= (\psi_2 - \psi_1) + au \\ &= \gamma(f_2 - B\psi_2) - \gamma(f_1 - B\psi_1) + au \\ &\geqslant \gamma(f_1 - B\psi_2) - \gamma(f_1 - B\psi_1) + au \equiv 0. \end{aligned} \tag{5.49}$$

From the convexity of γ we see that

$$\gamma(f_1 - B\psi_2) - \gamma(f_1 - B\psi_1) \geqslant \gamma'(f_1 - B\psi_1)(B\psi_1 - B\psi_2),$$

and thus, by (5.44),

$$a(x) \leqslant \gamma'(f_1) \in L^1. \tag{5.50}$$

On the other hand $u \in M^{N/(N-2)}$, $\Delta u \in L^1$ and $\int \Delta u = 0$ (by (5.41)); moreover

$$-\Delta u = \psi_2 - \psi_1 \geqslant -\gamma(f_1),$$

since $\psi_2 \geqslant 0$ and $\psi_1 = \gamma(f_1 - B\psi_1) \leqslant \gamma(f_1)$. From (5.43) we infer that

$$-\Delta u \geqslant 0 \quad \text{for a.e. } x, |x| > R. \tag{5.51}$$

Applying Corollary B.3 we see that $u \equiv 0$ in $[|x| > R]$. We may then invoke Theorem C. 1 to conclude that $u \equiv 0$, i.e., $\psi_1 = \psi_2$.

We may now go to the

Proof of Theorem 8. Set

$$W = j'(\rho_2) + B\rho_2 \text{ a.e. on } \mathbb{R}^N,$$

so that

$$\partial j(\rho_2) + B\rho_2 \ni W \text{ a.e.}$$

and by (5.40)

$$W \geqslant V - \lambda \text{ a.e.}$$

Applying Lemma 15 to $\psi_1 = \rho$, $f_1 = V - \lambda$, $\psi_2 = \rho_2$ and $f_2 = W$, we find that $\rho = \rho_2$. Next, letting

$$Z = \begin{cases} j'(\rho_1) + B\rho_1 & \text{on } [\rho_1 > 0], \\ \min\{B\rho_1, V - \lambda\} & \text{on } [\rho_1 = 0], \end{cases}$$

we see that

$$\partial j(\rho_1) + B\rho_1 \ni Z \quad \text{a.e. on } \mathbb{R}^N$$

and

$$W \leqslant V - \lambda \quad \text{a.e. on } \mathbb{R}^N.$$

Applying Lemma 15 to $\psi_1 = \rho_1$, $f_1 = W$, $\psi_2 = \rho$ and $f_2 = V - \lambda$ we find that $\rho_1 = \rho$.

6. Asymptotic estimates for $I(\lambda)$ as $\lambda \downarrow 0$; behavior of the chemical potential in the weakly ionized limit

In this section we assume that (where the symbol $\sim$ means, as usual, that the ratio tends to 1),

$$\gamma(s) \sim s^q \quad \text{as } s \downarrow 0, \text{ for some } 1 < q < \frac{N}{N-2}, \tag{6.1}$$

and

$$f = -\Delta V \quad \text{is a nonnegative, nonzero, measure in } \mathbb{R}^N \text{ with compact support}, \tag{6.2}$$

where $V \in M^{N/(N-2)}(\mathbb{R}^n)$. If $f \notin L^1(\mathbb{R}^N)$, we suppose, in addition, that

$$\gamma^0 \left(\frac{1}{|x|^{N-2}} \right) \in L^1_{\text{loc}}(\mathbb{R}^N). \tag{6.3}$$

Using Theorem 2.1 in Bénilan-Brezis-Crandall [10] if $f \in L^1(\mathbb{R}^N)$, or Theorem A.1 in Appendix A if $f \notin L^1(\mathbb{R}^N)$, we know that for every $\lambda \geqslant 0$, there exists $(u_\lambda, \rho_\lambda) \in M^{N/(N-2)} \times L^1$ such that

$$\rho_\lambda \in \gamma(u_\lambda - \lambda) \quad \text{a.e.} \quad \text{and} \quad -\Delta u_\lambda + \rho_\lambda = f \quad \text{in } \mathcal{D}'(\mathbb{R}^N). \tag{6.4}$$

We start with a result which is basically known (see e.g. Hille [39], Lieb-Simon [48], item [3] under Véron [59]):

PROPOSITION 4. *We have*

$$u_0(x) \sim \left(\frac{B}{|x|}\right)^k \quad as \ |x| \to \infty, \tag{6.5}$$

where

$$k = \frac{2}{q-1} \quad and \quad B = B(k, N) = (k(k - N + 2))^{1/2}. \tag{6.6}$$

We now set

$$I(\lambda) = \int \rho_\lambda(x)\, dx,$$

$$\overline{R}_\lambda = \inf \{r > 0; \ u_\lambda(x) < \lambda \text{ a.e. on } [|x| > r]\},$$

$$\underline{R}_\lambda = \sup \{r > 0; \ u_\lambda(x) > \lambda \text{ a.e. on } [|x| < r]\}.$$

Clearly, we have $\underline{R}_\lambda \leqslant \overline{R}_\lambda$, $\text{supp}\,\rho_\lambda \subset [|x| \leqslant \overline{R}_\lambda]$, and $\rho_\lambda(x) > 0$ a.e. on $[|x| < \underline{R}_\lambda]$.
The main result of this section is the following

THEOREM 9. *We have, as $\lambda \downarrow 0$,*

$$\overline{R}_\lambda \sim \underline{R}_\lambda \sim B\left(\frac{A_0}{\lambda}\right)^{1/k}, \quad I_0 - I(\lambda) \sim a A_0^\theta \lambda^{1-\theta}, \tag{6.7}$$

with

$$\theta = \frac{N-2}{k}, \quad a = (N-2)B^{N-2}\sigma_N, \tag{6.8}$$

where $\sigma_N = |S^{N-1}|$, k and B are given by (6.6), $A_0 = (2k - N + 2)A^{1/2}(N-2)^{-1}$, and $A = h(0)$ is a constant, depending only on q and N, defined via the solution of an ODE described in Lemmas 17 and 18.

In order to prove Proposition 4, we need the following lemma, essentially due to Hille [39, Theorem 4] (see also Lemme 2.2 in item [3] under Véron [59]):

LEMMA 16. *Let $N \geqslant 3$, $1 < q < \frac{N}{N-2}$, $R_0 > 0$, $\ell > 0$, $\phi_0 > 0$, and $v_0 \in C^2([R_0, \infty))$, $v_0 \geqslant 0$, be the solution of*

$$\begin{cases} v_0'' + \frac{N-1}{r} v_0' = \ell v_0^q & \text{in } [R_0, \infty), \\ v_0(R_0) = \phi_0. \end{cases} \tag{6.11}$$

Then

$$v_0(r) \sim \left(\frac{B}{\ell^{1/2} \, r} \right)^k \qquad \text{as } r \to \infty, \tag{6.12}$$

where k and B are given by (6.6).

It is well-known (see item [8] under Brezis [16]) that (6.11) has a unique solution, even without prescribing a condition at infinity. Moreover, there exists a constant $C > 0$ (depending on the given data) such that

$$v_0(r) \leqslant \frac{C}{r^k} \qquad \forall r \geqslant R_0. \tag{6.13}$$

Proof of Lemma 16. By a simple scaling argument, it suffices to prove the lemma for $\ell = 1$. Set $v_0(r) = (\frac{B}{r})^k w_0(r^n)$, with

$$n = 2k - (N - 2), \tag{6.14}$$

so that $w_0 \in C^2([\sigma_0, \infty))$, $w_0 \geqslant 0$, satisfies

$$\begin{cases} \sigma^2 w_0'' = L w_0(w_0^{q-1} - 1) & \text{in } [\sigma_0, \infty), \\ w_0(\sigma_0) = \psi_0, \end{cases} \tag{6.15}$$

where $\sigma_0 = R_0^n$, $\psi_0 = \phi_0(\frac{R_0}{B})^k$, and $L = (\frac{B}{n})^2$. Clearly, in order to prove (6.12), it suffices to show that

$$\lim_{\sigma \to \infty} w_0(\sigma) = 1. \tag{6.16}$$

Note that the function $(w_0 - 1)^2$ is convex; indeed,

$$\frac{1}{2} \frac{d^2}{d\sigma^2} (w_0 - 1)^2 \geqslant (w_0 - 1) w_0'' \geqslant 0$$

by (6.15).

Suppose, by contradiction, that (6.16) does not hold. Since $(w_0 - 1)^2$ is convex and bounded (for this last property we just apply (6.13)), there would exist a $\delta > 0$ small enough so that $(w_0 - 1)^2 \geqslant \delta^2$ on $[\sigma_0, \infty)$. We now split the argument into two cases:

CASE 1. $w_0(\sigma_0) > 1$.

In this case, one has $w_0 \geqslant 1 + \delta$ on $[\sigma_0, \infty)$ and

$$\sigma^2 w_0'' \geqslant \bar{\delta} = L(1 + \delta)((1 + \delta)^{q-1} - 1) > 0 \quad \text{on } [\sigma_0, \infty). \tag{6.17}$$

In particular, w_0 itself is convex and bounded. Thus it is also decreasing. We then conclude that

$$\lim_{\sigma \to \infty} \sigma w_0'(\sigma) = 0. \tag{6.18}$$

In fact, by the convexity of w_0, we can write

$$0 \leqslant -\sigma w_0'(\sigma) \leqslant 2(w_0(\sigma/2) - w_0(\sigma)) \quad \text{for } \sigma \geqslant 2\sigma_0.$$

Since $w_0(\sigma)$ converges as $\sigma \to \infty$, (6.18) follows.

On the other hand, it follows from (6.17) that

$$-w_0'(\sigma) = \int_\sigma^\infty w_0''(\tau)\, d\tau \geqslant \frac{\bar{\delta}}{\sigma} \quad \forall \sigma \geqslant \sigma_0,$$

which contradicts (6.18). This proves (6.16) in Case 1.

CASE 2. $w_0(\sigma_0) < 1$.

We have $0 < w_0 \leqslant 1 - \delta$ on $[\sigma_0, \infty)$, so that w_0 is concave. We deduce that w_0 is increasing,

$$\lim_{\sigma \to \infty} \sigma w_0'(\sigma) = 0 \quad \text{and} \quad \sigma^2 w_0''(\sigma) \leqslant -\bar{\delta}$$

for some $\bar{\delta} > 0$. As before, this gives a contradiction.

Proof of Proposition 4. By the maximum principle, we have $0 \leqslant u_0 \leqslant V$ on $\mathbb{R}^N$. Since V is harmonic outside some large ball, $\lim_{|x|\to\infty} V(x) = 0$. Then for any pair of positive numbers $\bar{\ell}, \underline{\ell}$ with $0 < \bar{\ell} < 1 < \underline{\ell}$, there exists $R_0 > 0$ such that

$$\bar{\ell} u_0^q \leqslant \rho_0 \leqslant \underline{\ell} u_0^q \quad \text{a.e. on } [|x| > R_0].$$

We may also assume that the support of f is contained in $[|x| < R_0/2]$; in particular, u_0 is C^2 on $[|x| \geqslant R_0]$ (see e.g. Theorem 3 in item [8] under Brezis [16]).

Set $\bar{\phi}_0 = \max_{|x|=R_0} u_0(x)$, and consider the solution $\bar{v}_0 \in C^2([R_0, \infty))$, $\bar{v}_0 \geqslant 0$, of (6.11) with ℓ and ϕ_0 replaced by $\bar{\ell}$ and $\bar{\phi}_0$, respectively. By the maximum principle, we have $u_0(x) \leqslant \bar{v}_0(|x|)$ on $[|x| \geqslant R_0]$, so that, by Lemma 16,

$$\limsup_{|x|\to\infty} \left[\left(\frac{|x|}{B}\right)^k u_0(x) \right] \leqslant \left(\frac{1}{\bar{\ell}}\right)^{k/2}. \tag{6.19}$$

We now claim that $u_0 > 0$ on $[|x| \geqslant R_0]$. For a.e. $x \in \mathbb{R}^N$, let

$$a(x) = \begin{cases} \dfrac{\rho_0(x)}{u_0(x)} & \text{if } u_0(x) \neq 0, \\ 0 & \text{if } u_0(x) = 0, \end{cases}$$

so that u_0 satisfies

$$-\Delta u_0 + a u_0 = f \geqslant 0 \quad \text{in } \mathcal{D}'(\mathbb{R}^N). \tag{6.20}$$

Using (6.1), we deduce that $a \in L^1(\mathbb{R}^N)$; moreover, a is bounded on $[|x| \geqslant R_0]$. By the strong maximum principle, then either $u_0 > 0$ on $[|x| \geqslant R_0]$, or $u_0 \equiv 0$ on $[|x| \geqslant R_0]$. Suppose, by contradiction, that $u_0 \equiv 0$ on $[|x| \geqslant R_0]$; in this case, Theorem C.1 in Appendix C would imply that $u_0 \equiv 0$ in $\mathbb{R}^N$, which is not possible because, by assumption (6.2), f is a nonzero measure. We deduce that $u_0 > 0$ on $[|x| \geqslant R_0]$, as claimed.

Set $\underline{\phi}_0 = \min_{|x|=R_0} u_0(x) > 0$, and consider the solution $\underline{v}_0 \in C^2([R_0, \infty))$, $\underline{v}_0 \geqslant 0$, of (6.11) corresponding to $\underline{\ell}$ and $\underline{\phi}_0$. We have $u_0(x) \geqslant \underline{v}_0(|x|)$ on $[|x| \geqslant R_0]$, and then

$$\liminf_{|x| \to \infty} \left(\frac{|x|}{B} \right)^k u_0(x) \geqslant \left(\frac{1}{\underline{\ell}} \right)^{k/2}. \tag{6.21}$$

Since (6.19) and (6.21) hold for every $0 < \overline{\ell} < 1 < \underline{\ell}$, the proposition follows.

In order to prove Theorem 9, we need the following

LEMMA 17.   *Let $K \in C^1([0, 1])$ with $K > 0$ on $(0, 1)$, and $K'(1) < 0$. Then there exists a unique solution $h \in C^1([0, 1])$ of*

$$\begin{cases} \frac{1}{2} h'(\xi) + h(\xi)^{1/2} + K(\xi) = 0 & \text{in } [0, 1], \\ h(1) = 0, \quad h(\xi) \geqslant 0 \quad \forall \xi \in [0, 1]. \end{cases} \tag{6.22}$$

Proof. (We present a modification due to M. Crandall of our original proof). Given $\varepsilon > 0$, set

$$F_\varepsilon(s) = \begin{cases} s^{1/2} & \text{if } s \geqslant \varepsilon, \\ \dfrac{s}{\varepsilon^{1/2}} & \text{if } 0 < s < \varepsilon, \\ 0 & \text{if } s \leqslant 0. \end{cases}$$

Then F_ε is Lipschitz continuous, and there exists a (unique) solution $h_\varepsilon \in C^1([0, 1])$ of

$$\begin{cases} \frac{1}{2} h_\varepsilon'(\xi) + F_\varepsilon(h_\varepsilon(\xi)) + K(\xi) = 0 & \text{in } [0, 1], \\ h_\varepsilon(1) = \varepsilon. \end{cases}$$

Since $h_\varepsilon' \leqslant 0$, we have $h_\varepsilon \geqslant \varepsilon$, and

$$\frac{1}{2} h_\varepsilon'(\xi) + h_\varepsilon(\xi)^{1/2} + K(\xi) = 0 \quad \forall \xi \in [0, 1].$$

Moreover, $\varepsilon \longmapsto h_\varepsilon(\xi)$ is increasing, and the limit h_0 of h_ε as $\varepsilon \downarrow 0$ is a solution of (6.22).

We now turn to uniqueness. Let $\tilde{h}$ be any solution of (6.22). Since $\tilde{h}' < 0$ on $(0, 1)$, we have $\tilde{h} > 0$ on $[0, 1)$; also, $\tilde{h} < h_\varepsilon$ on $[0, 1)$ for every $\varepsilon > 0$, and so $\tilde{h} \leqslant h_0$ on $[0, 1]$.

Take $\xi_0 \in [0, 1)$ so that $K' < 0$ on $[\xi_0, 1]$. For $0 < \delta < 1 - \xi_0$, let h^δ be a function defined on $[\xi_0 + \delta, 1]$ by $h^\delta(\xi) = \tilde{h}(\xi - \delta)$. We have

$$\frac{dh^\delta}{d\xi}(\xi) + h^\delta(\xi)^{1/2} + K(\xi) = K(\xi) - K(\xi - \delta) \leqslant 0,$$

$$h^\delta(1) = \tilde{h}(1 - \delta).$$

Thus if we take $\varepsilon = \tilde{h}(1 - \delta) > 0$, then $h^\delta \geqslant h_\varepsilon$ on $[\xi_0 + \delta, 1]$. At the limit as $\delta \downarrow 0$, $\tilde{h} \geqslant h_0$ on $[\xi_0, 1]$, and so $\tilde{h} = h_0$ on $[\xi_0, 1]$. In particular, if we now choose $\varepsilon = \tilde{h}(\xi_0) = h_0(\xi_0)$, then both $\tilde{h}$ and h_0 satisfy the initial value problem:

$$\begin{cases} \frac{1}{2}h'(\xi) + F_\varepsilon(h(\xi)) + K(\xi) = 0 & \text{in } [0, \xi_0], \\ h(\xi_0) = \varepsilon, \end{cases}$$

since $\tilde{h}, h_0 \geqslant \varepsilon$ on $[0, \xi_0]$, and $F_\varepsilon(s) = s^{1/2}$ if $s \geqslant \varepsilon$. By uniqueness, we conclude that $\tilde{h} = h_0$ on $[0, \xi_0]$, and hence on the entire interval $[0, 1]$. $\qquad\square$

We now prove the following

LEMMA 18. *Let $N \geqslant 3$, $1 < q < \frac{N}{N-2}$, $R_0 > 0$, $\ell > 0$, $\phi_0 > 0$, $\lambda > 0$, and $v_\lambda \in C^2([R_0, \infty))$ be the solution of*

$$\begin{cases} v_\lambda'' + \frac{N-1}{r}v_\lambda' = \ell[(v_\lambda - \lambda)^+]^q & \text{in } [R_0, \infty), \\ v_\lambda(R_0) = \phi_0, \quad \lim_{r \to \infty} v_\lambda(r) = 0. \end{cases} \tag{6.23}$$

Then $v_\lambda(r)$ is decreasing with respect to r on $[R_0, \infty)$.
For every $0 < \lambda \leqslant \phi_0$, let $R_\lambda \in [R_0, \infty)$ be such that $v_\lambda(R_\lambda) = \lambda$. We have

$$-v_\lambda'(R_\lambda) = \frac{(N-2)\lambda}{R_\lambda} \sim \frac{nA^{1/2}}{R_\lambda}\left(\frac{B}{\ell^{1/2}R_\lambda}\right)^k \quad \text{as } \lambda \downarrow 0, \tag{6.24}$$

with k and B given by (6.6), n given by (6.14), and $A = h(0)$, where h is the solution of (6.22) corresponding to

$$K(\xi) = \left(\frac{B}{n}\right)^2 \xi(1 - \xi^{q-1}).$$

Proof. By a simple scaling argument, it suffices to prove the lemma for $\ell = 1$. Firstly, we have

$$\frac{d}{dr}(r^{N-1}v_\lambda'(r)) = r^{N-1}[(v_\lambda(r) - \lambda)^+]^q \geqslant 0 \quad \forall r \geqslant R_0.$$

In particular, since $v_\lambda(R_0) > 0$ and $\lim_{r \to \infty} v_\lambda(r) = 0$, it follows from the maximum principle that $v_\lambda > 0$ in $[R_0, \infty)$. We claim that $v_\lambda' < 0$ in $[R_0, \infty)$. In fact, if $v_\lambda'(r_0) \geqslant 0$ for some $r_0 \geqslant R_0$, then we would have

$$r^{N-1}v_\lambda'(r) \geqslant r_0^{N-1}v_\lambda'(r_0) \geqslant 0 \quad \text{for every } r \geqslant r_0.$$

In other words, $v_\lambda'(r) \geq 0$ for $r \geq r_0$, and so

$$\liminf_{r \to \infty} v_\lambda(r) \geq v_\lambda(r_0) > 0.$$

But this contradicts $\lim_{r \to \infty} v_\lambda(r) = 0$. We then deduce that $v_\lambda' < 0$ in $[R_0, \infty)$.

For each $0 < \lambda \leq \phi_0$, it follows that there exists a unique $R_\lambda \in [R_0, \infty)$ such that $v_\lambda(R_\lambda) = \lambda$. Moreover, if $r \geq R_\lambda$, then $\frac{d}{dr}\left(r^{N-1}v_\lambda'(r)\right) = 0$. Thus

$$v_\lambda(r) = \lambda \left(\frac{R_\lambda}{r}\right)^{N-2} \quad \text{and} \quad v_\lambda'(r) = -\frac{(N-2)\lambda}{R_\lambda}\left(\frac{R_\lambda}{r}\right)^{N-1}.$$

In particular,

$$v_\lambda'(R_\lambda) = -\frac{(N-2)\lambda}{R_\lambda}. \tag{6.25}$$

Now set, as in Lemma 16, $v_\lambda(r) = (\frac{B}{r})^k w_\lambda(r^n) + \lambda$, so that w_λ satisfies

$$\begin{cases} w_\lambda \in C^2([\sigma_0, \sigma_\lambda]), \quad w_0 \geq 0, \\ \sigma^2 w_\lambda'' = L w_\lambda(w_\lambda^{q-1} - 1) \quad \text{in } [\sigma_0, \sigma_\lambda], \\ w_\lambda(\sigma_0) = \psi_\lambda, \quad w_\lambda(\sigma_\lambda) = 0, \end{cases} \tag{6.26}$$

where $\sigma_0 = R_0^n$, $\sigma_\lambda = R_\lambda^n$, $\psi_\lambda = (\phi_0 - \lambda)(\frac{R_0}{B})^k$, and $L = (\frac{B}{n})^2$. Using this notation, we can rewrite (6.25) as

$$v_\lambda'(R_\lambda) = \frac{n}{R_\lambda}\left(\frac{B}{R_\lambda}\right)^k \sigma_\lambda w_\lambda'(\sigma_\lambda).$$

In order to establish (6.24), it suffices to show that

$$\lim_{\lambda \downarrow 0}(\sigma_\lambda w_\lambda'(\sigma_\lambda))^2 = A. \tag{6.27}$$

Before proving (6.27), we first remark that if v_0 and w_0 are the functions introduced in Lemma 16, it follows from the standard maximum principle that

$$v_\lambda \downarrow v_0 \quad \text{and} \quad v_\lambda - \lambda \uparrow v_0 \quad \text{as } \lambda \downarrow 0, \tag{6.28}$$

so that

$$\sigma_\lambda \uparrow \infty \quad \text{and} \quad w_\lambda \uparrow w_0 \quad \text{as } \lambda \downarrow 0. \tag{6.29}$$

As in Lemma 16, we split the proof of (6.27) into two cases:

CASE 1. $w_0(\sigma_0) \leq 1$.

Since $(w_0 - 1)^2$ is convex and $\lim_{\sigma \to \infty} w_0(\sigma) = 1$, we have $w_0 \leq 1$, and then $w_\lambda < 1$ for $\lambda > 0$. It follows from (6.26) that w_λ is strictly concave.

Let $m_\lambda = \max w_\lambda$, and $\overline{\sigma}_\lambda \in [\sigma_0, \sigma_\lambda]$ be such that $w_\lambda(\overline{\sigma}_\lambda) = m_\lambda$. We have $w'_\lambda < 0$ on $(\overline{\sigma}_\lambda, \sigma_\lambda]$, and, by (6.29), $m_\lambda \uparrow 1$.

Define $\varphi_\lambda : [0, m_\lambda] \to [\overline{\sigma}_\lambda, \sigma_\lambda]$ to be the inverse function of $w_\lambda|_{[\overline{\sigma}_\lambda, \sigma_\lambda]}$. Set

$$h_\lambda(\xi) = [w'_\lambda(\varphi_\lambda(\xi)) \, \varphi_\lambda(\xi)]^2.$$

We have

$$\varphi'_\lambda(\xi) \, w'_\lambda(\varphi_\lambda(\xi)) = 1,$$
$$\varphi_\lambda(\xi) \, w'_\lambda(\varphi_\lambda(\xi)) = -h_\lambda(\xi)^{1/2},$$
$$\varphi_\lambda(\xi)^2 \, w''_\lambda(\varphi_\lambda(\xi)) = L\xi(\xi^{q-1} - 1),$$

so that h_λ satisfies

$$\begin{cases} \frac{1}{2}h'_\lambda(\xi) + h_\lambda(\xi)^{1/2} + L\xi(1 - \xi^{q-1}) = 0 & \text{in } [0, m_\lambda], \\ h_\lambda(m_\lambda) = 0. \end{cases}$$

Since $h_\lambda(0) = (\sigma_\lambda w'_\lambda(\sigma_\lambda))^2$ and $m_\lambda \uparrow 1$, the lemma follows in this case.

CASE 2. $w_0(\sigma_0) > 1$.

For $\lambda > 0$ small enough, $w_\lambda(\sigma_0) > 1$ by (6.29). It then follows from the convexity of $(w_\lambda - 1)^2$ that there exists a unique $\overline{\sigma}_\lambda \in (\sigma_0, \sigma_\lambda)$ such that $w_\lambda(\overline{\sigma}_\lambda) = 1$, and $w'_\lambda < 0$ on $[\overline{\sigma}_\lambda, \sigma_\lambda]$.

Define $\varphi_\lambda : [0, 1] \to [\overline{\sigma}_\lambda, \sigma_\lambda]$ to be the inverse function of $w_\lambda|_{[\overline{\sigma}_\lambda, \sigma_\lambda]}$. As before, set

$$h_\lambda(\xi) = [w'_\lambda(\varphi_\lambda(\xi)) \, \varphi_\lambda(\xi)]^2,$$

so that h_λ satisfies

$$\frac{1}{2}h'_\lambda(\xi) + h_\lambda(\xi)^{1/2} + L\xi(1 - \xi^{q-1}) = 0 \quad \text{in } [0, 1].$$

Note that

$$h_\lambda(0) = (\sigma_\lambda w'_\lambda(\sigma_\lambda))^2.$$

By Lemma 17, to conclude this second case it suffices to show that $\lim_{\lambda\downarrow 0} h_\lambda(1) = 0$; in other words, we only need to prove that

$$\lim_{\lambda\downarrow 0} \overline{\sigma}_\lambda w'_\lambda(\overline{\sigma}_\lambda) = 0. \tag{6.30}$$

The convexity of w_λ on $[\sigma_0, \overline{\sigma}_\lambda]$ implies that

$$0 \leqslant w'_\lambda(\overline{\sigma}_\lambda)(r - \overline{\sigma}_\lambda) \leqslant w_\lambda(r) - w_\lambda(\overline{\sigma}_\lambda) \quad \forall r \in [\sigma_0, \overline{\sigma}_\lambda];$$

consequently

$$0 \geqslant \overline{\sigma}_\lambda w'_\lambda(\overline{\sigma}_\lambda) \geqslant \frac{\overline{\sigma}_\lambda}{\overline{\sigma}_\lambda - r}[1 - w_\lambda(r)] \quad \forall r \in [\sigma_0, \overline{\sigma}_\lambda].$$

Taking $\lambda \downarrow 0$, and then $r \to \infty$, we get (6.30) as desired. $\qquad\square$

Proof of Theorem 9. Let $0 < \overline{\ell} < 1 < \underline{\ell}$. Since $0 \leqslant u_0 \leqslant V$, there exists $R_0 > 0$ such that

$$\overline{\ell}[(u_\lambda - \lambda)]^q \leqslant \rho_\lambda \leqslant \underline{\ell}[(u_\lambda - \lambda)]^q \quad \text{a.e. on } [|x| > R_0].$$

We may also assume that the support of f is contained in $[|x| < R_0/2]$.

Set

$$\underline{\varphi}_0 = \min_{|x|=R_0} u_0(x) \quad \text{and} \quad \overline{\varphi}_0 = \max_{|x|=R_0} u_0(x),$$

and consider $\underline{v}_\lambda, \overline{v}_\lambda \in C^2([R_0, \infty))$ to be the corresponding solutions given by Lemma 18. By the maximum principle, we have

$$\underline{v}_\lambda(|x|) \leqslant u_\lambda(x) \leqslant \overline{v}_\lambda(|x|) \quad \text{on } [|x| \geqslant R_0]. \tag{6.31}$$

It is clear that

$$\underline{R}'_\lambda = \underline{v}_\lambda^{-1}(\lambda) \leqslant \underline{R}_\lambda \leqslant \overline{R}_\lambda \leqslant \overline{v}_\lambda^{-1}(\lambda) = \overline{R}'_\lambda,$$

and then, by Lemma 18,

$$\frac{nA^{1/2}}{N-2}\left(\frac{B}{\underline{\ell}^{1/2}}\right)^k \leqslant \liminf_{\lambda \downarrow 0} \lambda \underline{R}_\lambda^k \leqslant \limsup_{\lambda \downarrow 0} \lambda \overline{R}_\lambda^k \leqslant \frac{nA^{1/2}}{N-2}\left(\frac{B}{\overline{\ell}^{1/2}}\right)^k.$$

Since the estimates above hold for any $0 < \overline{\ell} < 1 < \underline{\ell}$, we conclude that

$$\underline{R}_\lambda \sim \overline{R}_\lambda \sim B\left(\frac{nA^{1/2}}{(N-2)\lambda}\right)^{1/k} \quad \text{as } \lambda \downarrow 0.$$

Take $\underline{u}_\lambda, \overline{u}_\lambda \in M^{N/(N-2)}$, with $\Delta\underline{u}_\lambda, \Delta\overline{u}_\lambda \in \mathcal{M}$, to be any extensions inside $[|x| < R]$ of $\underline{v}_\lambda(|x|)$ and $\overline{v}_\lambda(|x|)$, respectively. By Corollary B1 in Appendix B, and by (6.31), we have

$$\int_{\mathbb{R}^N} \Delta\underline{u}_\lambda \geqslant \int_{\mathbb{R}^N} \Delta u_\lambda \geqslant \int_{\mathbb{R}^N} \Delta\overline{u}_\lambda. \tag{6.32}$$

Then, for $\lambda > 0$ sufficiently small (so that $\underline{R}'_\lambda > R_0$),

$$\int_{\mathbb{R}^N} \Delta\underline{u}_\lambda = \int_{|x|<\underline{R}'_\lambda} \Delta\underline{u}_\lambda = \sigma_N(\underline{R}'_\lambda)^{N-1}\underline{v}'_\lambda(\underline{R}'_\lambda), \tag{6.33}$$

since $\Delta \underline{u}_\lambda = 0$ on $|x| > \underline{R}'_\lambda$. Similarly,

$$\int_{\mathbb{R}^N} \Delta \overline{u}_\lambda = \int_{|x| < \overline{R}'_\lambda} \Delta \overline{u}_\lambda = \sigma_N (\overline{R}'_\lambda)^{N-1} \overline{v}'_\lambda (\overline{R}'_\lambda). \tag{6.34}$$

Thus, for $\lambda > 0$ small enough, it follows from (6.32), (6.33) and (6.34) that

$$\sigma_N (\underline{R}'_\lambda)^{N-1} \underline{v}'_\lambda (\underline{R}'_\lambda) \geqslant \int_{\mathbb{R}^N} \Delta u_\lambda \geqslant \sigma_N (\overline{R}'_\lambda)^{N-1} \overline{v}'_\lambda (\overline{R}'_\lambda).$$

Using Lemma 18, at the limit as $\underline{\ell} \downarrow 1$ and $\overline{\ell} \uparrow 1$, we obtain

$$-\int_{\mathbb{R}^N} \Delta u_\lambda \sim a A_0^\theta \lambda^{1-\theta} \quad \text{as } \lambda \downarrow 0, \tag{6.35}$$

where a is given by (6.8) and $A_0 = \frac{nA^{1/2}}{N-2}$. We now apply Corollary B.2 and Lemma B.1 (with $p = N/(N-2)$) in Appendix B. By (6.1), we have $\int_{\mathbb{R}^N} \Delta u_0 = 0$, so that

$$I_0 - I(\lambda) = \int_{\mathbb{R}^N} (\Delta u_0 - \Delta u_\lambda) = -\int_{\mathbb{R}^N} \Delta u_\lambda. \tag{6.36}$$

Combining (6.35) and (6.36), we conclude that

$$I_0 - I(\lambda) \sim a A_0^\theta \lambda^{1-\theta} \quad \text{as } \lambda \downarrow 0.$$

Behavior of the chemical potential in the weakly ionized limit

We consider the standard Thomas-Fermi model and we follow now the notations of Lieb-Simon [48] (except that we set $\mu = -\varepsilon_F$, instead of φ_0, where ε_F is the chemical potential). The functions φ_μ and ρ_μ satisfy, with $\mu > 0$,

$$-\Delta \varphi_\mu + 4\pi [(\varphi_\mu - \mu)^+]^{3/2} = 4\pi \sum z_i \delta_{a_i} \tag{6.37}$$

and

$$\rho_\mu = [(\varphi_\mu - \mu)^+]^{3/2}. \tag{6.38}$$

Set

$$J(\mu) = \int \rho_\mu$$

and

$$J(0) = \int \rho_0 = \sum z_i = Z$$

Lieb-Simon [48, Problem 5] raised the following problem: prove that

$$\lim_{\mu \downarrow 0} \frac{\mu}{[Z - J(\mu)]^{4/3}} \quad \text{exists.} \tag{6.39}$$

The answer is indeed positive and can be easily derived from Theorem 9.

COROLLARY 5. *We have*

$$\lim_{\mu \downarrow 0} \frac{\mu}{[Z - J(\mu)]^{4/3}} = \left(\frac{\pi^2}{63A^{1/2}}\right)^{1/3} \tag{6.40}$$

where $A = h(0)$, and h is the unique solution $h > 0$ of the differential equation

$$\begin{cases} \frac{1}{2}h'(\xi) + h(\xi)^{1/2} + \frac{12}{49}\xi(1 - \xi^{1/2}) = 0 & \text{in } (0, 1), \\ h(1) = 0. \end{cases} \tag{6.41}$$

REMARK 19. Solving numerically (6.41) yields $A = h(0) = 1.129359\ldots$ and then

$$\lim_{\mu \downarrow 0} \frac{\mu}{[Z - J(\mu)]^{4/3}} = 0.52826\ldots; \tag{6.42}$$

with the notation of Lieb-Simon [48], (6.42) reads

$$\lim_{N \uparrow Z} \frac{\varepsilon_F(N)}{[Z - N]^{4/3}} = -0.52826\ldots. \tag{6.43}$$

This exact value is consistent with a lower bound for $-\varepsilon_F(N)$ near $N = Z$ obtained by Benguria-Yáñez [8] with the help of a new variational characterization for ε_F; we refer the reader to the paper of Benguria-Yáñez [8] for other comments on this question.

Proof of Corollary 5. Let $M = 1/16\pi^2$ and set

$$u = M^{-1}\varphi_\mu. \tag{6.44}$$

From (6.37) we obtain

$$-\Delta u + \left[\left(u - \frac{\mu}{M}\right)^+\right]^{3/2} = (4\pi/M)\sum z_i \delta_{a_i}. \tag{6.45}$$

We may apply Theorem 9 with $\lambda = \mu/M$, $q = 3/2$, $k = 4$, $B = (12)^{1/2}$, $\theta = 1/4$, $a = 4\pi(12)^{1/2}$, $A_0 = 7A^{1/2}$, and we obtain

$$I_0 - I(\lambda) \sim 4\pi(12)^{1/2}(7A^{1/2})^{1/4}\lambda^{3/4}. \tag{6.46}$$

Here $I_0 = \frac{4\pi}{M}Z = (4\pi)^3 Z$ and

$$I(\lambda) = \int \left[\left(u - \frac{\mu}{M}\right)^+\right]^{3/2}. \tag{6.47}$$

Note that

$$J(\mu) = \int \rho_\mu = \int [(\varphi - \mu)^+]^{3/2} = \int [(Mu - \mu)^+]^{3/2} = M^{3/2}I(\lambda),$$

and

$$J(0) = Z = M^{3/2} I_0.$$

Thus, by (6.46),

$$\begin{aligned}
J(0) - J(\mu) &= M^{3/2}[I_0 - I(\lambda)] \\
&\sim \frac{1}{(4\pi)^2}(12)^{1/2}(7A^{1/2})^{1/4}[16\pi^2\mu]^{3/4} \\
&= (4\pi)^{-1/2}(12)^{1/2}(7A^{1/2})^{1/4}\mu^{3/4},
\end{aligned}$$

which is the desired result (6.39).

7. Another dual variational formulation

In this section we assume that (H) holds, and that meas$[V > \delta] < \infty$ for every $\delta > 0$. Set

$$L = \left\{ (u, \lambda) \left| \begin{array}{l} u \text{ is a measurable function, } \lambda \geqslant 0, \\ j^*(u - \lambda) \in L^1, \\ u - V \in M^{N/(N-2)} \text{ and } \nabla(u - V) \in L^2 \end{array} \right. \right\}.$$

Fix $I > 0$; consider the following convex functional defined on L:

$$\Phi(u, \lambda) = \frac{1}{2} \int |\nabla(u - V)|^2 + \int j^*(u - V) + \lambda I.$$

THEOREM 10. *Let (u_0, λ_0) be such that*

$$\begin{cases} u_0 \text{ is measurable, } \lambda_0 > 0, \\ u_0 - V \in M^{N/(N-2)}, \quad \Delta(u_0 - V) \in L^1 \text{ and } \int \Delta(u_0 - V) = I, \\ -\Delta(u_0 - V) + \gamma(u_0 - \lambda_0) \ni 0 \quad a.e. \end{cases}$$

Then

$$(u_0, \lambda_0) \in L \quad and \quad \Phi(u_0, \lambda_0) \leqslant \Phi(u, \lambda) \quad \forall(u, \lambda) \in L.$$

Proof. Set $\rho_0 = \Delta(u_0 - V)$, so that $\rho_0 \in L^1$, $\rho_0 \geqslant 0$, $\int \rho_0 = I$, and

$$\partial j(\rho_0) + B\rho_0 \ni V - \lambda \quad a.e..$$

By Theorem 1, $\rho_0 \in D(K)$; it follows from Lemma 14 that $\nabla B\rho_0 = \nabla(V - u_0) \in L^2$, and

$$\int \rho_0 B\rho_0 = \int |\nabla(u_0 - V)|^2 = -\int (u_0 - V)\Delta(u_0 - V).$$

Using the fact that $\gamma = \partial j^*$, we have, for any $(u, \lambda) \in L$,

$$
\begin{aligned}
j^*(u - \lambda) - j^*(u_0 - \lambda_0) &\geqslant \Delta(u_0 - V)\left[(u - \lambda) - (u_0 - \lambda_0)\right] \\
&= \Delta(u_0 - V)\left[(u - V) - (u_0 - V)\right] \\
&\quad + (\lambda_0 - \lambda)\,\Delta(u_0 - V).
\end{aligned}
\tag{7.1}
$$

Take first $u = V$ and $\lambda \geqslant 0$ such that $j^*(V - \lambda) \in L^1$ (here we use assumption (H)); we deduce from (7.1) that

$$
j^*(u_0 - \lambda_0) \leqslant j^*(V - \lambda) + \rho_0\,(B\rho_0 + (\lambda - \lambda_0)) \in L^1,
$$

and thus $(u_0, \lambda_0) \in L$.

Now suppose $(u, \lambda) \in L$ is such that $u - V \in L^\infty$. In this case, all the functions in (7.1) are integrable, and we find

$$
\begin{aligned}
\int j^*(u - \lambda) - \int j^*(u_0 - \lambda_0) \\
\geqslant \int \Delta(u_0 - V)\,(u - V) + \int |\nabla(u_0 - V)|^2 + (\lambda_0 - \lambda)I \\
= -\int \nabla(u_0 - V) \cdot \nabla(u - V) + \int |\nabla(u_0 - V)|^2 + (\lambda_0 - \lambda)I \\
\geqslant -\frac{1}{2} \int |\nabla(u - V)|^2 + \frac{1}{2} \int |\nabla(u_0 - V)|^2 + (\lambda_0 - \lambda)I,
\end{aligned}
$$

i.e., $\Phi(u_0, \lambda_0) \leqslant \Phi(u, \lambda)$.

[Here we have used the fact that

$$
-\int (\Delta\varphi)\psi = \int \nabla\varphi \cdot \nabla\psi
$$

$\forall \varphi, \psi$ with $\varphi \in M^{N/(N-2)}, \Delta\varphi \in L^1, \nabla\varphi \in L^2, \psi \in L^\infty \cap M^{N/(N-2)}, \nabla\psi \in L^2$; and this may be easily justified by a smoothing argument.]

For a general $(u, \lambda) \in L$, set $u_n = T_n(u - V) + V$, where

$$
T_n(r) = \begin{cases} n & \text{if } r > n \\ r & \text{if } |r| \leqslant n \\ -n & \text{if } r < -n \end{cases}
$$

so that $u_n - V \in M^{N/(N-2)} \cap L^\infty$, $\nabla(u_n - V) \in L^2$, and $\nabla(u_n - V) \to \nabla(u - V)$ in L^2 as $n \to \infty$. Moreover,

$$
\begin{aligned}
j^*(u_n - \lambda) &\leqslant j^*(u - \lambda) && \text{on } [u - V \geqslant -n], \\
j^*(u_n - \lambda) &= j^*(V - n - \lambda) && \text{on } [u - V < -n].
\end{aligned}
$$

Note that for $n \geqslant M - \lambda$, we have $j^*(V - n - \lambda) \leqslant j^*(V - M) \in L^1$ (by (H)); thus, for n sufficiently large,

$$j^*(u_n - \lambda) \in L^1 \quad \text{and} \quad j^*(u_n - \lambda) \to j^*(u - \lambda) \quad \text{in } L^1 \text{ as } n \to \infty.$$

Therefore, $(u_n, \lambda) \in L$ for n large, and

$$\Phi(u_0, \lambda_0) \leqslant \Phi(u_n, \lambda) \to \Phi(u, \lambda) \quad \text{as } n \to \infty.$$

This completes the proof of the theorem. $\qquad\square$

Appendix A

The equation $-\Delta u + \beta(u) \ni \mu$ with μ measure

Let β be a maximal monotone graph on $\mathbb{R}$ with $0 \in \beta(0)$. Let $\mathcal{M}(\mathbb{R}^N)$ denote the space of bounded measures on $\mathbb{R}^N$ with the usual norm:

$$\|\mu\|_{\mathcal{M}} = \sup\left\{\int \varphi \, d\mu; \ \varphi \in C_0(\mathbb{R}^N) \text{ and } \|\varphi\|_{L^\infty} \leqslant 1\right\},$$

where $C_0(\mathbb{R}^N)$ is the space of continuous functions on $\mathbb{R}^N$ tending to zero at infinity. In this Appendix we assume that $N \geqslant 3$.

THEOREM A.1. *Assume β satisfies*

$$D(\beta) = \mathbb{R} \quad and \quad \beta^0\left(\pm \frac{1}{|x|^{N-2}}\right) \in L^1_{loc}(\mathbb{R}^N). \tag{A.1}$$

Then, for every measure $\mu \in \mathcal{M}(\mathbb{R}^N)$ there exists a unique solution $u \in M^{N/(N-2)}(\mathbb{R}^N)$ of the problem

$$-\Delta u + \beta(u) \ni \mu \quad a.e. \text{ in } \mathbb{R}^N \tag{A.2}$$

such that

$$w \equiv \Delta u + \mu \in L^1(\mathbb{R}^N). \tag{A.3}$$

Moreover if $\widehat{u}$ is the solution corresponding to $\widehat{\mu}$ we have

$$\|u - \widehat{u}\|_{M^{N/(N-2)}} + \|\nabla(u - \widehat{u})\|_{M^{N/(N-1)}} \leqslant C\|\mu - \widehat{\mu}\|_{\mathcal{M}}, \tag{A.4}$$

$$\|(w - \widehat{w})^+\|_{L^1} \leqslant \|(\mu - \widehat{\mu})^+\|_{\mathcal{M}}, \tag{A.5}$$

and

$$[\mu \leqslant \widehat{\mu} \quad in \ \mathcal{M}] \quad \Rightarrow \quad [u \leqslant \widehat{u} \quad a.e.]. \tag{A.6}$$

The proof of Theorem A.1 relies on the following

LEMMA A.1. *Let Ω be a measurable space of finite measure. Let (u_n) be a bounded sequence in $M^p(\Omega)$ for some $1 < p < \infty$. Let $\beta : \mathbb{R} \to \mathbb{R}$ be a nondecreasing function such that*

$$\int_{|s|>1} \frac{|\beta(s)|}{|s|^{p+1}}\, ds < \infty. \tag{A.7}$$

Then $(\beta(u_n))$ is bounded in $L^1(\Omega)$ and equi-integrable.

Proof. We may always assume that $\beta(0) = 0$. Set $\gamma(s) = \beta(s+0) - \beta(-s-0)$ for $s \geqslant 0$, so that γ is also nondecreasing and satisfies

$$\int_1^\infty \frac{\gamma(s)}{s^{p+1}}\, ds < \infty.$$

Let

$$\alpha_n(\lambda) = \operatorname{meas}[|u_n| > \lambda].$$

Since (u_n) is bounded in $M^p(\Omega)$ there is a constant C such that $\alpha_n(\lambda) \leqslant C/\lambda^p$, $\forall \lambda > 0$. Let $A \subset \Omega$ be measurable and let $t > 0$. We have

$$
\begin{aligned}
\int_A |\beta(u_n)|\, dx &\leqslant \int_A \gamma(|u_n|)\, dx \\
&\leqslant \int_{A \cap [|u_n| \leqslant t]} \gamma(|u_n|) + \int_{[|u_n|>t]} \gamma(|u_n|) \\
&\leqslant |A|\,\gamma(t) + \alpha_n(t)\gamma(t) + \int_t^\infty \alpha_n(\lambda)\, d\gamma(\lambda) \\
&\leqslant |A|\,\gamma(t) + C\left(\frac{\gamma(t)}{t^p} + \int_t^\infty \frac{1}{\lambda^p}\, d\gamma(\lambda)\right) \\
&= |A|\,\gamma(t) + C_p \int_t^\infty \frac{\gamma(\lambda)}{\lambda^{p+1}}\, d\lambda.
\end{aligned}
$$

Given $\varepsilon > 0$ we fix t_0 large enough so that

$$C_p \int_{t_0}^\infty \frac{\gamma(\lambda)}{\lambda^{p+1}}\, d\lambda < \varepsilon/2.$$

Then we have $\int_A |\beta(u_n)| \leqslant \varepsilon$ for every A such that $|A| < \delta \equiv \varepsilon/2\,\gamma(t_0)$. $\qquad\square$

Proof of Theorem A.1.
Uniqueness. Let $\widehat{u}$ be another solution. We have

$$-\Delta(u - \widehat{u}) + w - \widehat{w} = 0 \quad \text{in } \mathbb{R}^N$$

and thus, by Kato's inequality (see Kato [43]), we obtain

$$-\Delta|u - \widehat{u}| + (w - \widehat{w})\,\text{sign}\,(u - \widehat{u}) \leqslant 0 \quad \text{in } \mathcal{D}'(\mathbb{R}^N).$$

Since $w \in \beta(u)$ and $\widehat{w} \in \beta(\widehat{u})$ it follows that $(w - \widehat{w})\,\text{sign}\,(u - \widehat{u}) \geqslant 0$. Therefore, the function $\varphi = |u - \widehat{u}|$ is subharmonic and, for a.e. x_0, we have

$$\varphi(x_0) \leqslant \frac{1}{|C_n(x_0)|} \int_{C_n(x_0)} \varphi(x)\,dx$$

where $C_n(x_0) = \{x \in \mathbb{R}^N; n < |x - x_0| < 2n\}$. From the fact that $\varphi \in M^{N/N-2}$ we deduce that

$$\int_{C_n(x_0)} \varphi(x)\,dx \leqslant C|C_n(x_0)|^{2/N}.$$

Letting $n \to \infty$ we conclude that $\varphi = 0$ a.e.

Existence. We already know (see Bénilan-Brezis-Crandall [10, Theorem 2.1]) that if $\mu \in L^1(\mathbb{R}^N)$ all the conclusions of Theorem A.1 hold, even without assumption (A.1). If $\mu \in \mathcal{M}(\mathbb{R}^N)$, we let $f_n = \rho_n * \mu$ where (ρ_n) is a sequence of mollifiers. We have

$$f_n \in L^1 \cap C^\infty, \quad \|f_n\|_{L^1} \leqslant \|\mu\|_{\mathcal{M}} \quad \text{and} \quad f_n \rightharpoonup \mu \text{ in the } w^* - \text{topology of } \mathcal{M}.$$

Let $u_n \in M^{N/(N-2)}(\mathbb{R}^N)$ be the (unique) solution of

$$-\Delta u_n + \beta(u_n) \ni f_n$$

with $w_n = \Delta u_n + f_n \in L^1(\mathbb{R}^N)$. We already know that

$$\|u_n\|_{M^{N/(N-2)}} + \|\nabla u_n\|_{M^{N/(N-1)}} \leqslant C\|f_n\|_{L^1}$$

and

$$\|w_n\|_{L^1} \leqslant \|f_n\|_{L^1}.$$

It follows that (u_n) is relatively compact in $L^1_{\text{loc}}(\mathbb{R}^N)$. On the other hand, assumption (A.1) implies (A.7) with $p = N/(N-2)$. We deduce from Lemma A.1 that (w_n) is equi-integrable on every bounded set of $\mathbb{R}^N$. Applying the Dunford–Pettis theorem (see e.g. Dunford-Schwartz [30, Corollary IV.8.11]) we may choose a subsequence such that

$$u_{n_k} \to u \quad \text{in } L^1_{\text{loc}}(\mathbb{R}^N),$$
$$w_{n_k} \rightharpoonup w \quad \text{weakly in } L^1_{\text{loc}}(\mathbb{R}^N).$$

We have $u \in M^{N/(N-2)}(\mathbb{R}^N)$, $w \in L^1(\mathbb{R}^N)$ and (by standard monotone analysis; see e.g. Lemma 3 in item [1] under Brezis [16]) $w \in \beta(u)$ a.e. Therefore u is a solution of (A.2)–(A.3). Properties (A.4), (A.5) and (A.6) follow easily[5] from the corresponding properties for $u_n, \widehat{u}_n$. Indeed $f_n - \widehat{f}_n = \rho_n * (\mu - \widehat{\mu})$ so that

$$\|f_n - \widehat{f}_n\|_{L^1} \leq \|\mu - \widehat{\mu}\|_{\mathcal{M}} \quad \text{and} \quad \|(f_n - \widehat{f}_n)^+\|_{L^1} \leq \|(\mu - \widehat{\mu})^+\|_{\mathcal{M}}.$$

REMARK A.1. The case $N = 2$ has been investigated by J.L. Vázquez item [1] under Vázquez [58].

REMARK A.2. Let $\Omega \subset \mathbb{R}^N$ be a bounded domain with smooth boundary. Under assumption (A.1), the same method as above shows that, for every bounded measure μ on Ω, there exists a unique solution $u \in W_0^{1,1}(\Omega)$ of the problem

$$\begin{cases} -\Delta u + \beta(u) \ni \mu & \text{in } \Omega \\ u = 0 & \text{on } \partial\Omega \end{cases}$$

with $w = \Delta u + \mu \in L^1(\Omega)$.

REMARK A.3. Local regularity. Assume ω is an open subset of $\mathbb{R}^N$. Suppose $\mu \in L_{\text{loc}}^q(\omega)$ for some $1 < q < \infty$, then the solution u of (A.2) satisfies $u \in W_{\text{loc}}^{2,q}(\omega)$ (see Brezis [8, Theorem 3]).

REMARK A.4. Non existence without (A.1). Assume $D(\beta) = \mathbb{R}$ but (A.1) does not hold—for example

$$\int_{|x|<1} \beta^0 \left(\frac{1}{|x|^{N-2}} \right) dx = \infty. \tag{A.8}$$

Then, for each $c > 0$, problem

$$-\Delta u + \beta(u) \ni c\delta \quad \text{in } \mathbb{R}^N \tag{A.9}$$

has *no* solution (with $w = c\delta + \Delta u \in L^1$).

Indeed, suppose u is a solution of (A.9), then u is radial (by uniqueness) and $u \in C^1(\mathbb{R}^N \setminus \{0\})$ (by Remark A.3). We have

$$\int_{|x|<r} w \, dx = c + \int_{|x|<r} \Delta u \, dx = c + \sigma_N r^{N-1} u'(r)$$

and therefore

$$u'(r) = -\frac{c}{\sigma_N r^{N-1}} + o\left(\frac{1}{r^{N-1}} \right) \quad \text{as } r \to 0.$$

---

[5] Note that if (v_n) is a bounded sequence in $M^p \, (1 < p < \infty)$ such that $v_n \rightharpoonup v$ weakly in L_{loc}^1, then $v \in M^p$ and $\|v\|_{M^p} \leq \liminf \|v_n\|_{M^p}$. This is clear since $\{v \in M^p; \ \|v\|_{M^p} \leq C\}$ is a closed convex set in L_{loc}^1 by Fatou's lemma.

It follows that

$$u(r) = \frac{c}{\sigma_N (N-2) r^{N-2}} + o\left(\frac{1}{r^{N-2}}\right) \quad \text{as } r \to 0.$$

This contradicts (A.8) since $\int |\beta^0(u)| < \infty$.

In the special case where $\beta(u) = |u|^{q-1} u$ assumption (A.1) holds if and only if $q < N/(N-2)$. When $q \geqslant N/(N-2)$ the nonexistence of solutions for $\mu = c\delta$ may also be viewed as a consequence of results about removable singularities (see Brezis-Véron [26] and also item [1] under Baras-Pierre [3]). When $q \geqslant N/(N-2)$, the measures μ for which the equation $-\Delta u + |u|^{q-1} u = \mu$ has a solution $u \in L^q$ have been completely characterized; see item [1] under Baras-Pierre [3] (and also item [2] under Gallouët-Morel [35]). The result of Baras-Pierre asserts that, for $1 < q < \infty$, the equation

$$-\Delta u + |u|^{q-1} u = \mu \quad \text{in } \mathbb{R}^N \tag{A.10}$$

has a solution $u \in L^q(\mathbb{R}^N)$ if and only if the bounded measure μ satisfies

$$\mu(E) = 0 \quad \forall E \subset \mathbb{R}^N \text{ such that } \mathrm{cap}_{2,q'}(E) = 0, \tag{A.11}$$

where $\mathrm{cap}_{2,q'}$ is the capacity associated to the Sobolev space $W^{2,q'}$, and $q' = q/(q-1)$.

An equivalent form asserts that (A.10) has a solution if and only if

$$\mu \in L^1 + W^{-2,q}. \tag{A.12}$$

Prior to our study very few authors had considered nonlinear PDE's involving measures as data (see however the pioneering nonexistence result in item [1] under Kamenomost-skaia [40] and the paper of Bamberger [2]). Theorem A.1 and the nonexistence result stated above has been the starting point and the motivation for many subsequent works in various directions:

A) Removable singularities. A typical result is the following (see e.g. Brezis-Véron [26], items [6], [7] under Brezis [16]).

Assume $0 \in \Omega \subset \mathbb{R}^N$ and $q \geqslant N/(N-2)$. Let $f \in L^1(\Omega)$, and suppose $u \in L^q_{\mathrm{loc}}(\Omega \backslash \{0\})$ satisfies

$$-\Delta u + |u|^{q-1} u = f \quad \text{in } \mathcal{D}'(\Omega \backslash \{0\}).$$

Then $u \in L^q_{\mathrm{loc}}(\Omega)$, and we have

$$-\Delta u + |u|^{q-1} u = f \quad \text{in } \mathcal{D}'(\Omega).$$

A similar result has been established by Baras-Pierre (see item [1] under Baras-Pierre [3]), when the point 0 is replaced by a closed set $E \subset \Omega$ with $\mathrm{cap}_{2,q'}(E) = 0$ (following earlier works by Loewner-Nirenberg [49] and Véron [1]).

B) Classification of singularities. When the singularities are not removable it is an important task to understand the nature of the singularities and possibly classify them.

A remarkable result of Véron asserts that if $u \in C^2(\Omega \backslash \{0\})$, $u \geqslant 0$, and u satisfies

$$-\Delta u + u^q = 0 \quad \text{in } \Omega \backslash \{0\} \tag{A.13}$$

with $1 < q < N/(N-2)$, then:

a) either u is smooth at 0;
b) or $\lim_{|x| \to 0} |x|^{N-2} u(x) = c$, where c is an arbitrary positive constant;
c) or $\lim_{|x| \to 0} |x|^{\frac{2}{q-1}} u(x) = C(q, N)$, where $C(q, N)$ is an explicit constant such that $C(q, N)|x|^{-\frac{2}{q-1}}$ is an exact solution of (A.13). For example, if $q = 3/2$ and $N = 3$, then $C(q, N) = 144$. Following the terminology introduced in Brezis-Peletier-Terman [22], this type of solution is called *very singular* (VSS).

For the proof we refer to item [2] under Véron [59]; see also Brezis-Oswald [21]. A variety of other results are presented in the book of Véron (item [4] under Véron [59]).

C) Measures as boundary data. Similar questions can be asked for nonlinear equations involving measures as boundary condition. A typical example is the problem

$$-\Delta u + |u|^{q-1} u = 0 \quad \text{in } \Omega, \tag{A.14}$$

$$u = \mu \quad \text{on } \partial\Omega, \tag{A.15}$$

where μ is a positive Borel measure on $\partial\Omega$. The detailed investigation of such questions was initiated by Gmira-Véron [36], and has vastly expanded in recent years; see the papers of Marcus-Véron [50]. Important motivations coming from the theory of probability—and the use of probabilistic methods—have reinvigorated the whole subject; see the pioneering papers of LeGall [45], the recent book of Dynkin [31], and the numerous references therein.

D) Singular solutions and removable singularities for other nonlinear problems. Questions concerning the existence (or nonexistence) of solutions with measure data, removable singularities, and classification of singularities have been investigated for a large variety of nonlinear problems (elliptic and parabolic), such as

$$\frac{\partial u}{\partial t} - \Delta u + |u|^{q-1} u = f,$$

$$\frac{\partial u}{\partial t} - \Delta(|u|^{m-1} u) = f,$$

$$- \operatorname{div}(a(x, \nabla u)) + |u|^{q-1} u = f,$$

$$-\Delta u + u|\nabla u|^2 = f,$$

$$\frac{\partial u}{\partial t} + \gamma \left| \frac{\partial u}{\partial t} \right| - \Delta u = 0, \quad \text{with } 0 < |\gamma| < 1,$$

see Brezis-Friedman [17], Baras-Pierre (item [2] under Baras-Pierre [3]), Brezis-Peletier-Terman [22], Brezis-Nirenberg [20], Boccardo-Gallouët [13], Boccardo-Gallouët-Orsina

[14], Boccardo-Dall' Aglio-Gallouët-Orsina [12], Oswald [52], Pierre [54], and the numerous references in these papers. The study of nonlinear parabolic equations with a Dirac mass as initial data is closely related to the analysis of self-similar solutions; see Barenblatt [4], Barenblatt-Sivashinski [5], Friedman-Kamin [33], Kamenomostskaia (item [2] under [40]), Kamin-Peletier [41], Kamin-Peletier-Vázquez [33], [42], Pattle [53], and Zel'dovich-Kompaneec [60].

E) "Forcing" solutions to exist. Assume $\beta : \mathbb{R} \to \mathbb{R}$ is continuous and nondecreasing, with $\beta(0) = 0$. We make *no* assumption about the behavior of β at infinity, so that (A.1) may fail. Our goal is to solve

$$-\Delta u + \beta(u) = \mu \quad \text{in } \Omega, \tag{A.16}$$

$$u = 0 \quad \text{on } \partial\Omega. \tag{A.17}$$

In general, (A.16)–(A.17) need not have a solution, but we may still consider approximations of (A.16)–(A.17), and try to understand how they fail to converge to a solution of (A.16)–(A.17). There are several natural approximations. For example, we may solve

$$-\Delta u_n + \beta_n(u_n) = \mu \quad \text{in } \Omega, \tag{A.18}$$

$$u_n = 0 \quad \text{on } \partial\Omega, \tag{A.19}$$

where (β_n) is a sequence of continuous nondecreasing functions with $\beta_n(0) = 0$, such that $\beta_n \to \beta$, e.g. uniformly on compact sets. Assume that each β_n has at most a linear growth at infinity, e.g. $\beta_n = \beta$ truncated at $\pm n$, or β_n is the Yosida approximation of β. Then (A.18)–(A.19) admits a unique solution. Another reasonable approximation is

$$-\Delta u_n + \beta(u_n) = \rho_n * \mu 0 \quad \text{in } \Omega, \tag{A.20}$$

$$u_n = 0 \rho_n * \mu \quad \text{on } \partial\Omega, \tag{A.21}$$

where (ρ_n) is a sequence of mollifiers.

Let us start with the case $\beta(u) = |u|^{q-1}u$. It is not difficult to see that if $q < N/(N-2)$, then the solutions (u_n) of (A.18)–(A.19) or (A.20)–(A.21) converge to the solution of (A.16)–(A.17), which exist for every measure μ. The difficulty arises when $q \geqslant N/(N-2)$ and (A.16)–(A.17) has no solution, e.g. when $\mu = \delta_a (a \in \Omega)$. In this case, it has been proved by H. Brezis (see item [7] under reference [16]) that $u_n \to 0$. More generally, if $\mu = f + \delta_a$ with $f \in L^1$, then $u_n \to u^*$, where u^* is the solution of

$$-\Delta u^* + |u^*|^{q-1}u^* = f \quad \text{in } \Omega,$$

$$u^* = 0 \quad \text{on } \partial\Omega.$$

Observe that u^* does *not* satisfy $-\Delta u^* + |u^*|^{q-1}u^* = f + \delta_a$. An interesting aspect to the same phenomenon is that when $\beta(u) = |u|^{q-1}u$ and $q \geqslant N/(N-2)$, the solution of (A.16)–(A.17)—assuming it exists—is "not sensitive" to large perturbation of the data μ, provided these perturbations are localized on sets of small capacity (in the sense of $\mathrm{cap}_{2,q'}$); this is quantified in a recent estimate of Labutin [44] (see also Marcus-Véron, item [4] under reference [50]). For a general measure $\mu \geqslant 0$, it has been proved in Brezis-Marcus-Ponce [19] that $u_n \to u^*$, where u^* is the unique solution of

$$-\Delta u^* + |u^*|^{q-1} u^* = \mu^* \quad \text{in } \Omega,$$
$$u^* = 0 \quad \text{on } \partial\Omega.$$

Here, μ^* denotes the "regular" part μ_1 of μ in the decomposition

$$\mu = \mu_1 + \mu_2,$$

where $\mu_1(E) = 0$, $\forall E$ with $\mathrm{cap}_{2,q'}(E) = 0$, and μ_2 is concentrated on a set Σ with $\mathrm{cap}_{2,q'}(\Sigma) = 0$; recall that this decomposition exists and is unique—see e.g. Fukushima-Sato-Taniguchi [34].

Returning to a general continuous nondecreasing function $\beta : \mathbb{R} \to \mathbb{R}$, the convergence of the sequences (u_n) has been thoroughly investigated for a general measure $\mu \geqslant 0$ in Brezis-Marcus-Ponce [19]. The sequences (u_n) always converge to a well-defined limit u^* independent of the approximation method. In addition, $\beta(u^*) \in L^1$ and Δu^* is a bounded measure, so that one may define the "reduced" measure

$$\mu^* = -\Delta u^* + \beta(u^*).$$

The measure μ^*, which is a kind of "projection" of μ on the class of "admissible" measures, has a number of remarkable properties. It is the largest measure ν such that $\nu \leqslant \mu$ and

$$-\Delta v + \beta(v) = \nu \quad \text{in } \Omega,$$
$$v = 0 \quad \text{on } \partial\Omega,$$

admits a solution, and therefore u^* is the largest subsolution of (A.16)–(A.17). Moreover, $(\mu - \mu^*)$ is concentrated on a set Σ with $\mathrm{cap}_{1,2}(\Sigma) = 0$.

Applying a result of Vázquez (item [1] under reference [58]), one may identify the measure μ^* when $N = 2$ and $\beta(t) = (e^t - 1)$. The identification of μ^* in more general situations is an interesting direction of research:

OPEN PROBLEM 1. What is μ^* when $\beta(t) = (e^t - 1)$ and $N \geqslant 3$? What is μ^* when $\beta(t) = (e^{t^2} - 1)$ and $N \geqslant 2$?

Similar questions arise when β admits vertical asymptotes. Suppose for example that $\beta : (-1, 1) \to \mathbb{R}$ is a continuous nondecreasing function such that $\beta(0) = 0$ and $\lim_{t \to \pm 1} \beta(t) = \pm\infty$.

OPEN PROBLEM 2. What are the properties of the mapping $\mu \mapsto \mu^*$ in this case ?

Other multivalued graphs β are of interest—for example the graphs

$$\beta(r) = \begin{cases} 0 & \text{if } r < a \\ [0, \infty) & \text{if } r = a \\ \emptyset & \text{if } t > a \end{cases}$$

(for some $a \geqslant 0$), and

$$\beta(r) = \begin{cases} \emptyset & \text{if } r < -1 \text{ and } r > 1 \\ (-\infty, 0] & \text{if } r = -1 \\ 0 & \text{if } -1 < r < 1 \\ [0, \infty) & \text{if } r = 1. \end{cases}$$

They correspond respectively to one-sided and two-sided variational inequalities. The objective is to solve in some natural "weak sense" the equation

$$-\Delta u + \beta(u) \ni \mu,$$

where μ is a given bounded measure. There are some partial results; see e.g. Baxter [6], Dall'Aglio-Dal Maso [29], Orsina-Prignet [51], Brezis-Serfaty [24], and the references therein.

Appendix B

Some properties of $\int \Delta u$

It is clear that if a (smooth) function u decays "very fast" at infinity on $\mathbb{R}^N$ – for example if u has compact support – then $\int \Delta u = 0$; on the other hand, if u decays at infinity like $1/|x|^{N-2}$ then $\int \Delta u \neq 0$. In this paragraph we investigate the relation between $\int \Delta u$ and the behavior of u at infinity. Throughout this Appendix we take $N \geqslant 3$.

THEOREM B.1. *Assume* $u \in M^{N/(N-2)}(\mathbb{R}^N)$ *with* $\Delta u \in \mathcal{M}(\mathbb{R}^N)$. *Then*

$$\lim_{\lambda \downarrow 0} (\lambda^{N/(N-2)} \operatorname{meas}[u > \lambda]) \quad \text{exists and equals} \quad d_N \left[\left(-\int_{\mathbb{R}^N} \Delta u \right)^+ \right]^{N/(N-2)}$$

where d_N *is a positive constant depending only on* N.

Before proving Theorem B.1 we deduce some corollaries

COROLLARY B.1. *Assume* $u \in M^{N/(N-2)}(\mathbb{R}^N)$ *with* $\Delta u \in \mathcal{M}(\mathbb{R}^N)$. *If*

$$\liminf_{\lambda \downarrow 0} (\lambda^{N/(N-2)} \operatorname{meas}[u > \lambda]) = 0,$$

then

$$\int \Delta u \geqslant 0.$$

In particular if $u(x) \leqslant 0$ for $|x| > R$, then $\int \Delta u \geqslant 0$.

COROLLARY B.2. *Assume $u \in M^{N/(N-2)}(\mathbb{R}^N)$ with $\Delta u \in \mathcal{M}(\mathbb{R}^N)$. Then*

$$\lim_{\lambda \downarrow 0} (\lambda^{N/(N-2)} \, \text{meas}[|u| > \lambda]) \quad \text{exists and equals} \quad d_N \left| \int \Delta u \right|^{N/(N-2)}.$$

Proof of Corollary B.2. Without loss of generality we may assume that $\int \Delta u \leqslant 0$. By Theorem B.1 we have

$$\lim_{\lambda \downarrow 0} (\lambda^{N/(N-2)} \, \text{meas}[u > \lambda]) = d_N \left| \int \Delta u \right|^{N/(N-2)}$$

and

$$\lim_{\lambda \downarrow 0} (\lambda^{N/(N-2)} \, \text{meas}[-u > \lambda]) = 0.$$

The conclusion follows since

$$\text{meas}[|u| > \lambda] = \text{meas}[u > \lambda] + \text{meas}[-u > \lambda].$$

It is convenient, in the proof of Theorem B.1, to use the following notations:

$$p = N/(N-2),$$
$$\overline{M}(u) = \limsup_{\lambda \downarrow 0} (\lambda^p \, \text{meas}[u > \lambda]),$$
$$\underline{M}(u) = \liminf_{\lambda \downarrow 0} (\lambda^p \, \text{meas}[u > \lambda]).$$

Notice that, for any functions u_1, u_2, we have

$$\overline{M}(u_1 + u_2) \leqslant \frac{1}{t^p}\overline{M}(u_1) + \frac{1}{(1-t)^p}\overline{M}(u_2) \quad \forall t \in (0, 1) \tag{B.1}$$

and

$$\underline{M}(u_1 + u_2) \leqslant \frac{1}{t^p}\underline{M}(u_1) + \frac{1}{(1-t)^p}\overline{M}(u_2) \quad \forall t \in (0, 1). \tag{B.2}$$

These relations follow from the fact that

$$[u_1 + u_2 > \lambda] \subset [u_1 > t\lambda] \cup [u_2 > (1-t)\lambda] \quad \forall t \in (0, 1).$$

Proof of Theorem B.1. Set $A = -\int \Delta u$. Given $\varepsilon > 0$, we fix R large enough so that

$$\int_{|x| \geqslant R} |\Delta u| < \varepsilon.$$

Let $k(x) = c_N / |x|^{N-2}$ where $c_N = 1/(N-2)\sigma_N$ and σ_N is the area of the unit sphere in $\mathbb{R}^N$ (so that $-\Delta k = \delta_0$). Set

$$f_1 = (-\Delta u)\chi_{B_R} \quad \text{and} \quad f_2 = (-\Delta u)(1 - \chi_{B_R}),$$
$$u_1 = k * f_1 \quad \text{and} \quad u_2 = k * f_2,$$

where χ_{B_R} is the characteristic function of $B_R = \{x \in \mathbb{R}^N \,;\, |x| < R\}$.

We have

$$u_1 + u_2 = k * (-\Delta u) = u$$

and

$$\|u_2\|_{M^p} \leqslant \|k\|_{M^p} \|f_2\|_{\mathcal{M}} \leqslant C\varepsilon. \tag{B.3}$$

We claim that there is some $\overline{R} > R$ such that

$$\left| u_1(x) - \frac{A c_N}{|x|^{N-2}} \right| \leqslant \frac{2\varepsilon c_N}{|x|^{N-2}} \quad \text{for } |x| > \overline{R}. \tag{B.4}$$

Indeed we have

$$u_1(x) = \int_{B_R} \frac{c_N}{|x-y|^{N-2}} f_1(y)\,dy$$

and thus

$$u_1(x) - \frac{A c_N}{|x|^{N-2}} = c_N \int_{B_R} \frac{1}{|x-y|^{N-2}} f_1(y)\,dy - \frac{c_N}{|x|^{N-2}} \int_{B_R} f_1(y)\,dy$$
$$+ \frac{c_N}{|x|^{N-2}} \left(\int_{B_R} f_1(y)\,dy - A \right).$$

It follows that

$$\left| u_1(x) - \frac{A c_N}{|x|^{N-2}} \right| \leqslant c_N \int_{B_R} \left| \frac{1}{|x-y|^{N-2}} - \frac{1}{|x|^{N-2}} \right| |f_1(y)|\,dy + \frac{\varepsilon c_N}{|x|^{N-2}}.$$

On the other hand, we have

$$\left| \frac{1}{|x-y|^{N-2}} - \frac{1}{|x|^{N-2}} \right| \leqslant \frac{(N-2)R}{(|x|-R)^{N-1}} \quad \text{provided } |y| < R < |x|$$

[it suffices to write that $|\varphi(1)-\varphi(0)| \leqslant \int_0^1 |\varphi'(s)|ds$ with $\varphi(t) = 1/|x-ty|^{N-2}$]. Therefore, we obtain

$$\left| u_1(x) - \frac{Ac_N}{|x|^{N-2}} \right| \leqslant \frac{C}{(|x| - R)^{N-1}} + \frac{\varepsilon c_N}{|x|^{N-2}} \quad \text{provided } |x| > R$$

and we deduce (B.4) easily.

We now distinguish two cases:

(i) $\mathbf{A} \leqslant 0$

(ii) $\mathbf{A} > 0$.

CASE (i). It follows easily from (B.4) that

$$\overline{M}(u_1) \leqslant (2\,\varepsilon\,c_N)^p b_N \tag{B.5}$$

where b_N denotes the measure of the unit ball in $\mathbb{R}^N$. Using (B.1), (B.3) and (B.5) we find

$$\overline{M}(u) \leqslant C\varepsilon^p$$

and since ε is arbitrary we conclude that $\overline{M}(u) = 0$.

CASE (ii). It follows easily from (B.4) that

$$[(A - 2\varepsilon)c_N]^p b_N \leqslant \underline{M}(u_1) \leqslant \overline{M}(u_1) \leqslant [(A + 2\varepsilon)c_N]^p b_N \tag{B.6}$$

provided $\varepsilon < A/2$. Using (B.1), (B.3) and (B.6) we find

$$\overline{M}(u) \leqslant \frac{1}{t^p}[(A + 2\varepsilon)c_N]^p b_N + \frac{1}{(1-t)^p} C\,\varepsilon^p \quad \forall t \in (0, 1)$$

Letting $\varepsilon \to 0$ and then $t \to 1$ we are led to

$$\overline{M}(u) \leqslant A^p c_N^p b_N.$$

On the other hand we have (by (B.2))

$$\underline{M}(u_1) \leqslant \frac{1}{t^p}\underline{M}(u) + \frac{1}{(1-t)^p}\overline{M}(-u_2) \quad \forall t \in (0, 1),$$

which implies

$$[(A - 2\varepsilon)\,c_N]^p b_N \leqslant \frac{1}{t^p}\underline{M}(u) + \frac{1}{(1-t)^p}C\varepsilon^p \quad \forall t \in (0, 1).$$

Letting $\varepsilon \to 0$ and then $t \to 1$ we are led to

$$\underline{M}(u) \geqslant A^p c_N^p b_N.$$

We conclude that

$$\underline{M}(u) = \overline{M}(u) = A^p c_N^p b_N.$$

This establishes Theorem B.1 with

$$d_N = c_N^p b_N = \frac{1}{(N-2)^p \sigma_N^p} \cdot \frac{\sigma_N}{N} = \frac{1}{N(N-2)^p \sigma_N^{p-1}}.$$

Here is another useful application.

COROLLARY B.3. *Assume $u \in M^{N/(N-2)}(\mathbb{R}^N)$, $\Delta u \in \mathcal{M}(\mathbb{R}^N)$ and*

$$\int_{\mathbb{R}^N} \Delta u = 0. \tag{B.7}$$

Suppose that, for some $R > 0$,

$$u \geqslant 0 \quad a.e. \ in \ [|x| > R] \tag{B.8}$$

and

$$-\Delta u \geqslant 0 \quad a.e. \ in \ [|x| > R]. \tag{B.9}$$

Then

$$u \equiv 0 \quad in \ [|x| > R].$$

Proof. From (B.8), (B.9) and the strong maximum principle we know that either

$$u \equiv 0 \quad in \ [|x| > R]$$

and the proof is finished, or

$$u > 0 \quad in \ [|x| > R]. \tag{B.10}$$

More precisely, for every open set ω with compact closure in $[|x| > R]$ there is a constant $\delta_\omega > 0$ such that

$$u \geqslant \delta_\omega \quad \text{a.e. in } \omega.$$

We will show that (B.10) is impossible. Suppose that (B.10) holds. Fix $R_1 > R$; then for some $\delta > 0$ we have

$$u \geqslant \delta \quad \text{a.e. in } [R_1 < |x| < 2R_1].$$

Fix $\varepsilon > 0$ so that

$$u(x) \geqslant \frac{\varepsilon}{|x|^{N-2}} \quad \text{a.e. in } [R_1 < |x| < 2R_1]. \tag{B.11}$$

Note that by (B.9) we have

$$-\Delta\left(u - \frac{\varepsilon}{|x|^{N-2}}\right) \geq 0 \quad \text{in } [|x| > R].$$

Applying the maximum principle in the region $[R_1 < |x| < \rho]$ with $\rho > 2R_1$ we see that

$$u(x) - \frac{\varepsilon}{|x|^{N-2}} \geq -\frac{\varepsilon}{\rho^{N-2}} \quad \text{in } [R_1 < |x| < \rho].$$

As $\rho \to \infty$ we conclude that

$$u(x) \geq \frac{\varepsilon}{|x|^{N-2}} \quad \text{in } [|x| > R_1].$$

From Corollary B.1 applied with $v(x) = \frac{\varepsilon}{|x|^{N-2}} - u(x)$ we obtain $\int \Delta v \geq 0$. But $\Delta v = -\varepsilon \delta_0 / c_N - \Delta u$, and thus by (B.7), $\int \Delta v = -\varepsilon / c_N < 0$. A contradiction. $\qquad\square$

It is sometimes convenient to combine Theorem B.1 with the following:

LEMMA B.1. *Let Ω be a measurable space (with $|\Omega| \leq \infty$). Let $\beta : \mathbb{R} \to \mathbb{R}$ be a nondecreasing function such that*

$$\beta(0) = 0 \quad and \quad \int_0^1 \frac{\beta(s)}{s^{p+1}} \, ds = \infty \quad for \ some \ 1 < p < \infty. \tag{B.12}$$

Let $u : \Omega \to \mathbb{R}$ be a measurable function such that

$$\int_\Omega \beta(u^+(x)) \, dx < \infty.$$

Then

$$\liminf_{\lambda \downarrow 0} (\lambda^p \, \text{meas}[u > \lambda]) = 0. \tag{B.13}$$

REMARK B.1. Condition (B.12) is also necessary. More precisely, if u satisfies (B.13) one can show that there exists a function $\beta : \mathbb{R} \to \mathbb{R}$, convex, nondecreasing, Lipschitz continuous, such that $\beta(s) = 0$ for $s \leq 0$, $\int_0^1 \frac{\beta(s)}{s^{p+1}} \, ds = \infty$ and $\int_\Omega \beta(u^+(x)) \, dx < \infty$.

Proof of Lemma B.1. Assume, by contradiction, that

$$\liminf_{\lambda \downarrow 0} (\lambda^p \, \text{meas}[u > \lambda]) > 0.$$

There exist $\lambda_0 > 0$ and $\varepsilon > 0$ such that

$$\alpha(\lambda) = \text{meas}[u > \lambda] \geq \frac{\varepsilon}{\lambda^p} \quad \text{for } 0 < \lambda < \lambda_0.$$

We have, for $0 < \delta < \lambda_0$,

$$\int_{[\delta < u < \lambda_0]} \beta(u(x))\, dx = -\int_\delta^{\lambda_0} \beta(\lambda)\, d\alpha(\lambda)$$

$$= -\beta(\lambda_0)\alpha(\lambda_0) + \beta(\delta)\alpha(\delta) + \int_\delta^{\lambda_0} \alpha(\lambda)\, d\beta(\lambda)$$

$$\geqslant -\beta(\lambda_0)\alpha(\lambda_0) + \frac{\varepsilon}{\delta^p}\beta(\delta) + \int_\delta^{\lambda_0} \frac{\varepsilon}{\lambda^p}\, d\beta(\lambda)$$

$$= -\beta(\lambda_0)\alpha(\lambda_0) + \frac{\varepsilon}{\lambda_0^p}\beta(\lambda_0) + \varepsilon p \int_\delta^{\lambda_0} \frac{1}{\lambda^{p+1}}\beta(\lambda)\, d\lambda.$$

It follows that $\int_{[\delta < u < \lambda_0]} \beta(u(x))\, dx \to +\infty$ as $\delta \to 0$. A contradiction.

Appendix C

A form of the strong maximum principle for $-\Delta + a(x)$ with $a(x) \in L^1$

The strong maximum principle asserts that if u is smooth, $u \geqslant 0$ and $-\Delta u \geqslant 0$ in a domain $\Omega \subset \mathbb{R}^N$, then either $u \equiv 0$ in Ω or $u > 0$ in Ω. The same conclusion holds when $-\Delta$ is replaced by $-\Delta + a(x)$ with $a \in L^p(\Omega)$, $p > N/2$ (this is a consequence of Harnack's inequality; see e.g. Stampacchia [56], and also Trudinger [57], Corollary 5.3). Another formulation of the same fact says that if $u(x_0) = 0$ for some point $x_0 \in \Omega$, then $u \equiv 0$ in Ω. A similar conclusion fails when $a \notin L^p(\Omega)$, $p > N/2$. For example $u(x) = |x|^2$ satisfies $-\Delta u + a(x)u = 0$ with $a = \frac{2N}{|x|^2} \notin L^{N/2}$.

However if u vanishes on a larger set, not just at one point, one may still hope to conclude that $u \equiv 0$ in Ω. Here is such a result.

THEOREM C.1. *Assume* $u \in L^1_{\mathrm{loc}}(\mathbb{R}^N)$ *with* $u \geqslant 0$ *a.e. and* $\Delta u \in L^1_{\mathrm{loc}}(\mathbb{R}^N)$. *Let* $a \in L^1_{\mathrm{loc}}(\mathbb{R}^N)$, $a \geqslant 0$ *a.e. Assume* u *has compact support and satisfies*

$$-\Delta u + au \geqslant 0 \quad \textit{a.e. in } \mathbb{R}^N. \tag{C.1}$$

Then $u \equiv 0$.

Proof. (We present a modification due to R. Jensen of our original proof). Set

$$a_n(x) = \min\{a(x), n\}$$

and

$$g_n = -\Delta u + a_n u, \tag{C.2}$$

so that g_n is a nondecreasing sequence of functions in $L^1(\mathbb{R}^N)$ and

$$g_n \uparrow g = -\Delta u + au \quad \text{a.e.}.$$

Note that g need not belong to L^1; g is just measurable and $g \geqslant 0$. Fix R sufficiently large, so that $u(x) = 0$ for $|x| > R - 1$. Solve

$$\begin{cases} \Delta b_n = a_n & \text{in } B_R = [|x| < R] \\ b_n = 0 & \text{on } \partial B_R = [|x| = R], \end{cases}$$

so that $b_n \in W^{2,p}(B_R) \,\forall p < \infty$, $b_n \leqslant 0$ in B_R, $0 \leqslant e^{b_n} \leqslant 1$ in B_R, with

$$\Delta e^{b_n} = e^{b_n}(|\nabla b_n|^2 + \Delta b_n).$$

As $n \to \infty$, $b_n \to b$ in $W^{1,p}(B_R) \,\forall p < \frac{N}{N-1}$, where b is the solution of

$$\begin{cases} \Delta b = a & \text{in } B_R \\ b = 0 & \text{on } \partial B_R. \end{cases}$$

From (C.2) we have

$$-\int_{B_R} u \Delta \zeta + \int_{B_R} a_n u \zeta = \int_{B_R} g_n \zeta \quad \forall \zeta \in W^{2,q}(B_R) \text{ for some } q > N/2. \tag{C.3}$$

Note that the first integral in (C.3) makes sense since $u \in L^r \,\forall r < \frac{N}{N-2}$ (recall that $\Delta u \in L^1$). [One may first prove (C.3) for $\zeta \in C^2(\overline{B}_R)$ and then argue by density.]

Choosing $\zeta = e^{b_n}$ in (C.3) yields

$$-\int_{B_R} u e^{b_n}(|\nabla b_n|^2 + \Delta b_n) + \int_{B_R} (\Delta b_n) u e^{b_n} = \int_{B_R} g_n e^{b_n}$$

and, in particular,

$$\int_{B_R} g_n e^{b_n} \leqslant 0.$$

Therefore

$$\int_{B_R} (g_n - g_1) e^{b_n} \leqslant -\int_{B_R} g_1 e^{b_n} \leqslant \int_{B_R} |g_1|. \tag{C.4}$$

Since $g_n - g_1 \geqslant 0$ for $n \geqslant 1$, we conclude by Fatou's lemma that $(g - g_1)e^b \in L^1$ and thus $g e^b \in L^1$. Returning to (C.4) we also have

$$\int_{B_R} (g - g_1) e^b \leqslant -\int_{B_R} g_1 e^b$$

and thus

$$\int_{B_R} g e^b \leqslant 0.$$

Since $g \leqslant 0$ a.e. (by hypothesis (C.1)) we deduce that $g \equiv 0$ and consequently $-\Delta u \leqslant 0$. Therefore $u \leqslant 0$ a.e. By assumption, $u \geqslant 0$ a.e. and thus $u \equiv 0$.   □

REMARK C.1. Theorem C.1 is a special case of a much more general result due to Ancona [1]:

THEOREM (Ancona [1]). *Assume $u \in L^1_{\mathrm{loc}}(\Omega)$, $\Omega \subset \mathbb{R}^N$ open connected, $u \geqslant 0$ a.e., $\Delta u \in \mathcal{M}(\Omega)$, $a \in L^1_{\mathrm{loc}}(\Omega)$, $a \geqslant 0$ a.e., satisfy*

$$\Delta u \leqslant au \quad \textit{in the sense of measures,}$$

i.e.,

$$\int_E \Delta u \leqslant \int_E a\,u \quad \textit{for every Borel set } E \subset \Omega. \tag{C.5}$$

(Note that the integral on the right-hand side is well-defined in $[0, \infty]$ since $au \geqslant 0$ a.e.). Assume that u vanishes on a set $E \subset \Omega$ of positive measure, then $u \equiv 0$.

The proof of Ancona relies on Potential Theory. The interested reader will find another proof based on PDE techniques in Brezis-Ponce [23].

There are several interesting questions related to Theorem C.1:

OPEN PROBLEM 3. Can one replace in Theorem C.1 the assumption $a \in L^1_{\mathrm{loc}}$ by a weaker condition, for example $a^{1/2} \in L^1_{\mathrm{loc}}$ (or $a^{1/2} \in L^p_{\mathrm{loc}}$ for some $p > 1$)?

Note that one cannot hope to go below $L^{1/2}$. For instance the C^2 function u given by

$$u(x) = \begin{cases} (1 - |x|^2)^4 & \text{for } |x| \leqslant 1 \\ 0 & \text{for } |x| > 1 \end{cases}$$

satisfies $-\Delta u + au \geqslant 0$ for some function $a(x)$ such that $a(x) \sim \frac{1}{(1-|x|)^2}$ for $|x| < 1$ and $|x|$ close to 1. Here $a^\alpha \in L^1$, $\forall \alpha < 1/2$, but $a^{1/2} \notin L^1$.

Still one more:

OPEN PROBLEM 4. Assume $u \in C^0$, $\Delta u \in L^1_{\mathrm{loc}}$, $u \geqslant 0$, $a \in L^q_{\mathrm{loc}}$ for some $q \geqslant 1$, $a \geqslant 0$ a.c., satisfy (C.1). Assume that $u = 0$ on a set E with $\mathrm{cap}_{1,2q}(E) > 0$, where $\mathrm{cap}_{1,2q}$ refers to the capacity associated with the Sobolev space $W^{1,2q}$. Can one conclude that $u \equiv 0$?

Ancona [1] (see also Brezis-Ponce [23]) has shown that the answer is positive when $q = 1$. The answer is again positive when $q > \frac{N}{2}$ by the strong maximum principle mentioned above.

Acknowledgments

We thank M. Crandall for simplifying our original proof of Lemma 17 and we thank R. Jensen for useful discussions concerning the version of the strong maximum principle presented in Appendix C. Juan Davila and Augusto Ponce helped with the figures and the typescript of part of the paper. Augusto Ponce also checked carefully the arguments in Section 6 and made the numerical computations leading to Corollary 5. The second author (H. B) is partially sponsored by an EC Grant through the RTN Program "Front-Singularities", HPRN-CT-2002-00274. He is also a member of the Institut Universitaire de France. Most of this work was done during visits of the authors at the Mathematics Research Center of the University of Wisconsin in Madison, over the years 1975–1977. They thank M. Crandall and P. Rabinowitz for their invitation and warm hospitality.

REFERENCES

[1] ANCONA, A., *Une propriété d'invariance des ensembles absorbants par perturbation d'un opérateur elliptique*, Comm. PDE *4* (1979), 321–337.

[2] BAMBERGER, A., *Étude de deux équations nonlinéaires avec une masse de Dirac au second membre*, Rapport 13 du Centre de Mathématiques Appliquées de l'École Polytechnique, Oct. 1976.

[3] BARAS, P. and PIERRE, M., [1], *Singularités éliminables pour des équations semi-linéaires*, Ann. Inst. Fourier (Grenoble) *34* (1984), 185–206; [2], *Problèmes paraboliques semi-linéaires avec données mesures*, Applicable Anal. *18* (1984), 111–149.

[4] BARENBLATT, G. I., *"Scaling, self-similarity, and intermediate asymptotics"*. Cambridge Texts in Applied Mathematics, 14. Cambridge University Press, Cambridge, 1996.

[5] BARENBLATT, G. I. and SIVASHINSKI, G. I., *Self-similar solutions of the second kind in nonlinear filtration*, Prikl. Mat. Meh. *33* (1969), 861–870 (Russian); translated as J. Appl. Math. Mech. *33* (1969), 836–845.

[6] BAXTER, J. R., *Inequalities for potentials of particle systems*, Illinois J. Math. *24* (1980), 645–652.

[7] BENGURIA, R., *The von Weizsäcker and exchange corrections in the Thomas-Fermi theory*, Ph.D. Dissertation, Princeton Univ. 1979.

[8] BENGURIA, R. and YÁÑEZ, J., *Variational principle for the chemical potential in the Thomas-Fermi model*, J. Phys. A *31* (1998), 585–593.

[9] BENGURIA, R., BREZIS, H. and LIEB, E., *The Thomas-Fermi-von Weizsäcker theory of atoms and molecules*, Comm. Math. Phys. *79* (1981), 167–180.

[10] BÉNILAN, PH., BREZIS, H. and CRANDALL, M., *A semilinear equation in $L^1(\mathbb{R}^N)$*, Ann. Sc. Norm. Sup. Pisa, Serie IV, *2* (1975), 523–555.

[11] BÉNILAN, PH., GOLSTEIN, G. and GOLDSTEIN, J., *A nonlinear elliptic system arising in electron density theory*, Comm. PDE *17* (1992), 2907–2917.

[12] BOCCARDO, L., DALL'AGLIO, A., GALLOUËT, T. and ORSINA, L., *Nonlinear parabolic equations with measure data*, J. Funct. Anal. *147* (1976), 237–258.

[13] BOCCARDO, L. and GALLOUËT, T., [1], *Nonlinear elliptic and parabolic equations involving measure data*, J. Funct. Anal. *87* (1989), 149–169; [2] *Nonlinear elliptic equations with right-hand side measures*, Comm. PDE *17* (1992), 641–655.

[14] BOCCARDO, L., GALLOUËT, T. and ORSINA, L., [1], *Existence and uniqueness of entropy solutions for nonlinear elliptic equations with measure data*, Ann. Inst. H. Poincaré, Anal. Non Linéaire *13* (1996), 539–551; [2], *Existence and nonexistence of solutions for some nonlinear elliptic equations*, J. Anal. Math. *73* (1997), 203–223.

[15] BREAZNA, A., GOLDSTEIN, G. and GOLDSTEIN, J., *Parameter dependence in Thomas-Fermi theory*, Comm. Appl. Anal. *5* (2001), 421–432.

[16]   BREZIS, H., [1], *Monotonicity methods in Hilbert spaces and some applications to nonlinear partial differential equations*, in "*Contributions to nonlinear functional analysis*" (E. Zarantonello ed.), Acad. Press 1971, 101–156; [2],"*Opérateurs maximaux monotones et semigroupes de contractions dans les espaces de Hilbert*", North-Holland, Amsterdam, 1973; [3], *Nonlinear problems related to the Thomas-Fermi equation* in "Contemporary developments in continuum mechanics and partial differential equations" (Proc. Internat. Sympos., Rio de Janeiro, 1977), (de la Penha and Medeiros ed.), North-Holland, 1978, pp. 81–89; [4], *A free boundary problem in quantum mechanics: Thomas-Fermi equation*, in "Free boundary problems vol II, (Proc. Sympos. Pavia, 1979), Ist. Naz. Alta Mat., Rome, 1980, pp. 85–91; [5], *Some variational problems of the Thomas-Fermi type*, in "Variational inequalities and complementary problems: theory and applications", (Proc. Internat. School, Erice, 1978) (R. W. Cottle, F. Giannessi and J.-L. Lions ed.), Wiley, 1980, 53–73; [6], *Problèmes elliptiques et paraboliques non linéaires avec données mesures*, Seminaire Goulaouic-Meyer-Schwartz, 1981–82, XX.1–XX.12; [7], *Nonlinear elliptic equations involving measures*, in "Contributions to Nonlinear Partial Differential Equations" (Madrid, 1981), (C. Bardos, A. Damlamian, J. I. Diaz and J. Hernandez ed.), Pitman, 1983, 82–89; [8], *Semilinear equations in* $\mathbb{R}^N$ *without conditions at infinity*, Appl. Math. Optim. *12* (1984), 271–282.

[17]   BREZIS, H. and FRIEDMAN, A., *Nonlinear parabolic equations involving measures as initial conditions*, J. Math. Pures Appl. *62* (1983), 73–97.

[18]   BREZIS, H. and LIEB, E., *Long range atomic potentials in Thomas-Fermi theory*, Comm. Math. Phys. *65* (1979), 231–246.

[19]   BREZIS, H., MARCUS, M. and PONCE, A. C., *Nonlinear elliptic equations with measures revisited* (to appear).

[20]   BREZIS, H. and NIRENBERG, L., *Removable singularities for nonlinear elliptic equations*, Topol. Methods Nonlinear Anal. *9* (1997), 201–219.

[21]   BREZIS, H. and OSWALD, L., *Singular solutions for some semilinear elliptic equations*, Arch. Rat. Mech. Anal. *99* (1987), 249–259.

[22]   BREZIS, H., PELETIER, L. and TERMAN, D., *A very singular solution of the heat equation with absorption*, Arch. Rat. Mech. Anal. *95* (1986), 185–209.

[23]   BREZIS, H. and PONCE, A. C., *Remarks on the strong maximum principle*, Diff. Int. Equations *16* (2003), 1–12.

[24]   BREZIS, H. and SERFATY, S., *A variational formulation for the two-sided obstacle problem with measure data*, Commun. Contemp. Math. *4* (2002), 357–374.

[25]   BREZIS, H. and STRAUSS, W., *Semilinear second-order elliptic equations in* L^1, J. Math. Soc. Japan *25* (1973), 565–590.

[26]   BREZIS, H. and VÉRON, L., *Removable singularities of some nonlinear elliptic equations*, Arch. Rat. Mech. Anal. *75* (1980), 1–6.

[27]   BROWDER, F., *Nonlinear elliptic boundary value problems*, Bull. Amer. Math. Soc. *69* (1963), 862–874.

[28]   CAFFARELLI, L. and FRIEDMAN, A., *The free boundary in the Thomas-Fermi model*, J. Diff. Eq. *32* (1979), 335–356.

[29]   DALL'AGLIO, P. and DAL MASO, G., *Some properties of the solutions of obstacle problems with measure data*, Ricerche Mat. suppl. *48* (1999), 99–116.

[30]   DUNFORD, N. and SCHWARTZ, J. T., "*Linear operators*", Part I, Wiley Interscience, 1958.

[31]   DYNKIN, E., "*Diffusions, superdiffusions and partial differential equations*", American Mathematical Society Colloquium Publications, *50*. American Mathematical Society, Providence, RI, 2002.

[32]   EKELAND, I., *On the variational principle*, J. Math. Anal. Appl. *47* (1974), 324–353.

[33]   FRIEDMAN, A. and KAMIN, S., *The asymptotic behavior of gas in an n-dimensional porous medium*, Trans. Amer. Math. Soc. *262* (1980), 551–563.

[34]   FUKUSHIMA, M., SATO, K. and TANIGUCHI, S., *On the closable part of pre-Dirichlet forms and the fine supports of underlying measures*, Osaka Math. J. *28* (1991), 517–535.

[35]   GALLOUËT, TH. and MOREL, J. M., [1], *On some properties of the solution of the Thomas-Fermi problem*, Nonlinear Anal. *7* (1983), 971–979; [2], *Resolution of a semilinear equation in* L^1, Proc. Roy. Soc. Edinburgh, *96A* (1984), 275–288; [3], *On some semilinear problems in* L^1, Boll. Un. Mat. Ital. *4* (1985), 123–131.

[36] GMIRA, A. and VÉRON, L., *Boundary singularities of solutions of nonlinear elliptic equations*, Duke J. Math. *64* (1991), 271–324.

[37] GOLDSTEIN, G., GOLDSTEIN, J. and JIA, W., *Thomas-Fermi theory with magnetic fields and the Fermi-Amaldi correction*, Diff. Int. Equations *8* (1995), 1305–1316.

[38] GOLDSTEIN, J. and RIEDER, G., [1], *A rigorous Thomas-Fermi theory for atomic systems*, J. Math. Phys *28* (1987), 1198–1202; [2], *Spin polarized Thomas-Fermi theory*, J. Math. Phys *29* (1988), 709–716; [3], *Thomas-Fermi theory with an external magnetic field*, J. Math. Phys. *32* (1991), 2901–2917.

[39] HILLE, E., *Some aspects of the Thomas-Fermi equation*, J. Anal. Math. *23* (1970), 147–170.

[40] KAMENOMOSTSKAIA (KAMIN), S. L., [1], *Equation of the elastoplastic mode of filtration*, Prikl. Mat. Meh. *33* (1969), 1076–1084 (Russian); translated as J. Appl. Math. Mech. *33* (1969) 1042–1049; [2], *The asymptotic behavior of the solution of the filtration equation*, Israel J. Math. *14* (1973), 76–87.

[41] KAMIN, S. and PELETIER, L. A., *Singular solutions of the heat equation with absorption*, Proc. Amer. Math. Soc. *95* (1985), 145–158.

[42] KAMIN, S., PELETIER, L. A. and VÁZQUEZ, J. L., [1], *On the Barenblatt equation of elastoplastic filtration*, Indiana Univ. Math. J. *40* (1991), 1333–1362; [2], *Classification of singular solutions of a nonlinear heat equation*, Duke Math. J. *58* (1989), 601–615.

[43] KATO, T., *Schrödinger operators with singular potentials*, Israel J. Math. *13* (1972), 135–148.

[44] LABUTIN, D., *Wiener regularity for large solutions of nonlinear equations*, Arkiv för Math. (to appear).

[45] LEGALL, J. F., [1], *The Brownian snake and solutions of $\Delta u = u^2$ in a domain*, Probab. Theory Related Fields, *102* (1995), 393–432; [2], *A probabilistic Poisson representation for positive solutions of $\Delta u = u^2$ in a planar domain*, Comm. Pure Appl. Math. *50* (1997), 69–103.

[46] LERAY, J. and LIONS, J. L., *Quelques résultats de Visik sur les problèmes elliptiques nonlinéaires par les méthodes de Minty-Browder*, Bull. Soc. Math. Fr. *93* (1965), 97–107.

[47] LIEB, E. H., [1], *The stability of matter*, Rev. Mod. Phys. *48* (1976), 553–569; [2], *Thomas-Fermi and related theories of atoms and molecules*, Rev. Mod. Phys. *53* (1981), 603–641; [3], *"The stability of matter: from atoms to stars"*, Selecta of E. Lieb (W. Thirring ed.) Springer, second edition, 1997.

[48] LIEB, E. H. and SIMON, B., *The Thomas-Fermi theory of atoms, molecules and solids*, Advances in Math. *23* (1977), 22–116.

[49] LOEWNER, C. and NIRENBERG, L., *Partial differential equations invariant under conformal or projective transformations*, in "Contributions to Analysis", Acad. Press, 1974, 245–272.

[50] MARCUS, M. and VÉRON, L., [1], *The boundary trace of positive solutions of semilinear elliptic equations: the subcritical case*, Arch. Rat. Mech. Anal. *144* (1998), 201–231;[2], *The boundary trace of positive solutions of semilinear elliptic equations: the supercritical case*, J. Math. Pures Appl. *77* (1998), 481–524;[3], *Removable singularities and boundary traces*, J. Math. Pures Appl. *80* (2001), 879–900;[4], *Capacitary estimates of solutions of a class of nonlinear elliptic equations*, C. R. Acad. Sc. Paris *336* (2003), 913–918.

[51] ORSINA, L. and PRIGNET, A., *Nonexistence of solutions for some nonlinear elliptic equations involving measures*, Proc. Royal Soc. Edinburgh *130* (2000), 561–592.

[52] OSWALD, L., *Isolated positive singularities for a nonlinear heat equation*, Houston J. Math. *14* (1988), 543–572.

[53] PATTLE, R. E., *Diffusion from an instantaneous point source with a concentration-dependent coefficient*, Quart. J. Mech. Appl. Math. *12* (1959), 407–409.

[54] PIERRE, M., *Uniqueness of the solutions of $u_t - \Delta\varphi(u) = 0$ with initial datum a measure*, Nonlinear Anal. *6* (1982), 175–187.

[55] RIEDER, G., *Mathematical contributions to Thomas-Fermi theory*, Houston J. Math. *16* (1990), 179–201.

[56] STAMPACCHIA, G., *"Équations elliptiques du second ordre à coefficients discontinus"*, Presses de l'Université de Montréal, 1966.

[57] TRUDINGER, N., *Linear elliptic operators with measurable coefficients*, Ann. Scuola Norm. Sup. Pisa *27* (1973), 265–308.

[58] VÁZQUEZ, J. L., [1], *On a semilinear equation in $\mathbb{R}^2$ involving bounded measures*, Proc. Roy. Soc. Edinburgh, *95A* (1983), 181–202; [2], *A strong maximum principle for some quasilinear elliptic equations*, Appl. Math. Optim. *12* (1984), 191–202.

[59]   VÉRON, L., [1], *Singularités éliminables d'équations nonlinéaires*, J. Differential Equations *41* (1981), 87–95; [2], *Singular solutions of some nonlinear elliptic equations*, Nonlinear Anal. *5* (1981), 225–242; [3], *Comportement asymptotique des solutions d'équations elliptiques semi-linéaires dans* $\mathbb{R}^N$, Ann. Mat. Pura Appl. *127* (1981), 25–50; [4],*"Singularities of solutions of second order quasilinear equations"*, Pitman Research Notes, vol. *353*, Longman, 1996.

[60]   ZEL'DOVICH, Y. B. and KOMPANEEC, A. S., *On the theory of propagation of heat with the heat conductivity depending upon the temperature*, Collection in honor of the seventieth birthday of academician A. F. Ioffe, pp. 61–71. Izdat. Akad. Nauk USSR, Moscow, 1950.

Philippe Bénilan
Departement de Mathématiques
Université de Franche-comté
25030 Besançon Cedex

Haïm Brezis
Analyse Numérique
Université P. et M. Curie, B. C. 187
4 Pl. Jussieu
75252 Paris Cedex 05

Rutgers University
Dept. of Math.
Hill Center, Busch Campus
110 Frelinghuysen RD
Piscataway, NJ 08854
USA
e-mail: brezis@ccr.jussieu.fr;
*   brezis@math.rutgers.edu*

To access this journal online:
http://www.birkhauser.ch

Existence of attractors in $L^\infty(\Omega)$ for a class of reaction-diffusion systems

PHILIPPE BENILAN and HALIMA LABANI

Let us consider as an example, the reaction-diffusion system named "Brusselator":

$$u_t - d_1 \Delta u = u^2 v - (B+1)u + A \quad \text{in} \quad (0,T) \times \Omega \tag{0.1}$$

$$v_t - d_2 \Delta v = -u^2 v + Bu \quad \text{in} \quad (0,T) \times \Omega \tag{0.2}$$

where Ω is smooth bounded open subset of IR^n and $T > 0$, with boundary conditions

$$\lambda_1 \frac{\partial u}{\partial n} + (1 - \lambda_1)u = \alpha_1 \quad \text{on} \quad \partial\Omega \tag{0.3}$$

$$\lambda_2 \frac{\partial v}{\partial n} + (1 - \lambda_2)v = \alpha_2 \quad \text{on} \quad \partial\Omega \tag{0.4}$$

where d_1, d_2, B, A are positive constants, $0 \le \lambda_1$, $\lambda_2 \le 1$ and α_1, $\alpha_2 \ge 0$. Here u, v are functions of (t,x) with $x \in \Omega$.

In various papers, R.H. Martin and M. Pierre (cf. [9], [11]) studied global existence in time of bounded solutions to (0.1)–(0.4) with initial data u_0, $v_0 \in L^\infty(\Omega)$, u_0, $v_0 \ge 0$.

In the case $\lambda_1 = \lambda_2$, we proved in [6] the existence of a maximal (or universal) attractor in $L^\infty(\Omega)$ for the semi-group associated to this evolution problem; we also announced an abstract version of this result developed in [6].

In this paper, we prove the following result.

Theorem 0.1. *Assume $d_1 > 0$ and*

(H1) $d_1(1 - \lambda_2) + a(\lambda_1 + \alpha_1) > 0$
(H2) $d_2\lambda_1 = 0$ *or* $\lambda_2 > 0$.

Then the evolution problem (0.1)–(0.4) has a maximal attractor in $L^\infty(\Omega)_+^2$.

We will make precise the conclusion of this statement in Section 3.

In this paper, we show the existence of a maximal attractor for an abstract version of reaction-diffusion systems of the type (0.1)–(0.4); namely, we consider a system

$$(S) \qquad \begin{cases} \frac{du}{dt} + A_1(u - \underline{u}) = f(., u, v) \\ \frac{dv}{dt} + A_2(v - \underline{v}) = g(., u, v) \\ \quad u(0) = u_0, \ v(0) = v_0 \end{cases}$$

where $-A_1, -A_2$ are infinitesimal generators of a continuous semi-group of positive linear operators on $L^2(\Omega)$ uniformly bounded in $L^\infty(\Omega)$, (Ω, β, μ) being an abstract measured set with finite measure μ.

The data $\underline{u}, \underline{v}$ belong to $L^\infty(\Omega)$ and satisfy

$$e^{-tA_1}\underline{u} \le \underline{u}, \quad e^{-tA_2}\underline{v} \le \underline{v} \qquad \text{a.e. on } \Omega \text{ for } t \ge 0 \tag{0.5}$$

and $f, g : \Omega \times \mathbb{R}^{+2} \to \mathbb{R}$ are measurable in $x \in \Omega$, locally Lipschitz continuous in $(u, v) \in \mathbb{R}^{+2}$ uniformly for $x \in \Omega$ with $f(., 0, 0), g(., 0, 0) \in L^\infty(\Omega)$ and satisfy the quasi-positivity assumption

$$f(x, 0, v) \ge 0, \ \ g(x, u, 0) \ge 0 \quad \text{for } (x, u, v) \in \Omega \times \mathbb{R}^{+2}. \tag{0.6}$$

Under these assumptions, there is local existence in time of a unique mild solution to (S) for any initial data $u_0, v_0 \in L^\infty(\Omega)$, $u_0, v_0 \ge 0$ (cf. for instance [3]). Recall the following definition:

Definition 0.2. *For $0 < t \le \infty$, a mild solution of (S) in $[0, T]$ is a pair $(u, v) \in L^\infty((0, T) \times \Omega)^2$ such that $u \ge 0$, $v \ge 0$ and*

$$u(t, .) = \underline{u} + e^{-tA_1}(u_0 - \underline{u}) + \int_0^t e^{(s-t)A_1} f(., u(s), v(s))ds \tag{0.7}$$

$$v(t, .) = \underline{v} + e^{-tA_2}(v_0 - \underline{v}) + \int_0^t e^{(s-t)A_2} g(., u(s), v(s))ds \tag{0.8}$$

a.e. on Ω for a.a. $t \in (0, T)$.
Notice that $u, v \in C([0, T[; L^2(\Omega))$.

In order to show global existence in time and existence of a maximal attractor in $L^\infty(\Omega)$, one needs compactness assumptions and estimates in $L^\infty(\Omega)$ of the mild solution (u, v) of the type

$$u(t) + v(t) \le \Phi(\|u_0 + v_0\|_\infty, t) \quad \text{a.e. on } \Omega, \tag{0.9}$$

for $t \in [0, T)$, where $\Phi : [0, \infty[\times]0, \infty[\to \mathbb{R}^+$ satisfies

$$\begin{cases} \Phi(r, t) \text{ is nondecreasing in } r \ge 0, \text{ nonincreasing in } t > 0 \text{ and} \\ \qquad \limsup_{\substack{t \to \infty \\ r \ge 0}} \Phi(r, t) < \infty. \end{cases} \tag{0.10}$$

For simplification, we will use the following definition.

Definition 0.3. *An estimate of attractor type is a function $\Phi : [0, \infty[\times]0, \infty[\to \mathbb{R}^+$ satisfying (0.10).*

In this paper, we will mainely show that, under assumptions on data A_1, A_2, f, g, if there is an estimate of attractor type Φ_0 such that

$$v(t) \leq \Phi_0(\|u_0 + v_0\|_\infty, t) \quad \text{a.e. on } \Omega \text{ for } 0 < t < T, \tag{0.11}$$

then there exists an estimate of attractor type Φ such that (0.9) holds (cf. Theorem 3 for the precise statement). Notice that this type of result has been developed in [9] for existence of bounded global solutions.

We will make precise in Section 1 the complete set of assumptions on the data; but we may mention in this introduction that we will mainly assume

$$\begin{cases} f(x, u, v) + g(x, u, v) \leq L(v) \quad \text{for } (x, u, v) \in \Omega \times \mathbb{R}^{+2} \\ \qquad \text{where } L : \mathbb{R}^+ \to \mathbb{R}^+ \text{ is non-decreasing} \end{cases} \tag{0.12}$$

$$\begin{cases} \qquad \text{The operator } A_1 \text{ is defined by a coercive} \\ \text{continuous bilinear form } a : V \times V \to \mathbb{R}, \text{ where} \\ \qquad V \text{ is a Hilbert space embedded} \\ \qquad \text{in } L^q(\Omega) \text{ with } 2 < q < \infty; \end{cases} \tag{0.13}$$

and

$$A_1^{-1} A_2 \in \pounds(L^\infty(\Omega), L^p(\Omega)) \text{ for } 2 \leq p < \infty. \tag{0.14}$$

The assumption (0.13) is standard and in particular satisfied in concrete situations by the operator

$$A_1 u = -d_1 \Delta u + w_1 u$$

with $D(A_1) = \{u \in H^2(\Omega), \lambda_1 \frac{\partial u}{\partial n} + (1 - \lambda_1)u = 0\}$ if $d_1 > 0$, $w_1 > 0$, $0 \leq \lambda_1 \leq 1$, $w_1 + (1 - \lambda_1) > 0$.

Notice that we do not make extra assumptions on A_2: in particular the result applies with $A_2 = 0$. The assumption (0.13) has to be compared with the assumption made in [11]; in particular, such an assumption allows to consider, in concrete situation, different boundary conditions.

The plan of this paper is the following: in Section 1, we make precise the notations and assumptions, and state the main abstract results; Section 2 is devoted to the proof of the abstract results; in Section 3, we show how these abstract results apply to the "Brusselator"; then we prove Theorem 0.1.

1. Assumptions and main results

Let Ω be a given measured set with finite measure; for $1 \leq p \leq \infty$, $L^p(\Omega)$ is the real Lebesgue space with norm $\|.\|_p$. Let $a : V \times V \to \mathbb{R}$ be a continuous coercive bilinear form on a given real Hilbert space V, with $\|u\|_V^2 = a(u, u)$.

We assume that V is densely embedded in $L^q(\Omega)$ with

$$2 < q < \infty \tag{1.1}$$

and that

$$u \in V \Rightarrow \begin{cases} u \wedge 1 = \inf(u, 1) \in V \text{ and} \\ \qquad a(u, u \wedge 1) \geq 0. \end{cases} \tag{1.2}$$

We consider A_1 the linear operator in $L^2(\Omega)$ induced by the bilinear form a, defined by

$$\begin{cases} u \in D(A_1), \ A_1 u = w \text{ if } u \in V, \ w \in L^2(\Omega) \\ \text{and } \int_\Omega w\xi = a(u, \xi) \text{ for any } \xi \in V. \end{cases} \tag{1.3}$$

It is well known (cf. for instance [1]) that $-A_1$ generates an analytic exponentially stable semigroup on $L^2(\Omega)$: there are constants $C \geq 0$, $w > 0$ such that

$$\|A_1 e^{-tA_1}\|_{\mathcal{L}(L^2(\Omega))} \leq \frac{C}{t}, \quad \|e^{-tA_1}\|_{\mathcal{L}(L^2(\Omega))} \leq e^{-wt}. \tag{1.4}$$

The assumption (1.2) exactly means (cf. [1]) that

$$e^{-tA_1} \geq 0 \text{ and } \|e^{-tA_1}u\|_p \leq \|u\|_p \quad \text{for } u \in L^\infty(\Omega), \ 1 \leq p < \infty. \tag{1.5}$$

The operator A_1 is one to one from $D(A_1)$ onto $L^2(\Omega)$ and

$$A_1^{-1} = \int_0^\infty e^{-tA_1} dt \geq 0, \quad C_0 = \|A_1^{-1}1\|_2 \leq \frac{|\Omega|^{1/2}}{w}. \tag{1.6}$$

Finally, $V \subset L^q(\Omega)$ with (1.1) implies (cf. for instance [1]) that the semigroup is hypercontractive, more precisely, there is a constant still denoted by C, such that

$$\|e^{-tA_1}u\|_\infty \leq \frac{C\|u\|_1}{t^\varrho} \quad \text{with } \varrho = \frac{q}{q-2}. \tag{1.7}$$

We assume $-A_2$ is a given infinitesimal generator of a continuous semigroup on $L^2(\Omega)$ satisfying

$$e^{-tA_2} \geq 0 \text{ and } \limsup_{t \to 0} \|e^{-tA_2}1\|_\infty < \infty. \tag{1.8}$$

We also assume (0.13), namely

$$\|A_1^{-1}A_2 w\|_p \leq C_p\|w\|_\infty \text{ for } w \in D(A_2) \cap L^\infty(\Omega), \ 2 \leq p < \infty. \tag{1.9}$$

Let $f, g : \Omega \times \mathbb{R}^{+2} \to \mathbb{R}$ be measurable in $x \in \Omega$, locally Lipschitz continuous in $(u, v) \in \mathbb{R}^{+2}$, uniformly for $x \in \Omega$, with $f(., 0, 0), g(., 0, 0) \in L^\infty(\Omega)$, satisfying the quasi-positivity assumption (0.6), the property (0.12) and

$$f(x, u, v) \leq L(v)(1 + u)^{m(v)} \text{ for } (x, u, v) \in \Omega \times \mathbb{R} \tag{1.10}$$

where $L, m : \mathbb{R}^+ \to \mathbb{R}^+$ are nondecreasing.

At last $\underline{u}, \underline{v} \in L^\infty(\Omega)$ are given satisfying (0.5).

Now we state the main result of this paper.

Theorem 1.1. *With the assumptions above on data A_1, A_2, f, g, $\underline{u}$, $\underline{v}$; assume that*

$$\begin{cases} \text{there exists an estimate of attractor type } \Phi_0 \text{ such that} \\ v(t) \leq \Phi_0(\|u_0 + v_0\|_\infty, t) \ a. \ e. \text{ on } \Omega \text{ for } t \in (0, T) \\ \text{for } (u, v) \text{ mild solution on } [0, T) \text{ of } (S) \\ \text{for any } u_0, v_0 \in L^\infty(\Omega), \ u_0 \geq 0, \ v_0 \geq 0 \text{ and } 0 < T < \infty. \end{cases} \tag{1.11}$$

Then for any u_0, $v_0 \in L^\infty(\Omega)$, u_0, $v_0 \geq 0$, there exists a unique mild solution (u,v) on $[0,\infty[$ of (S). Moreover, there exists a constant R_0 such that

$$\left\{ \begin{array}{l} \forall r > 0, \exists T(r), \forall u_0, v_0 \in L^\infty(\Omega), u_0, v_0 \geq 0, u_0 + v_0 \leq r \\ \text{the mild solution } (u,v) \text{ on } [0,\infty[\text{ of (S) satisfies} \\ u(t) + v(t) \leq R_0 \text{ a.e. on } \Omega \text{ for } t \geq T(r). \end{array} \right. \tag{1.12}$$

In other words, under the assumptions of Theorem 1.1, we may associate to the system (S), a semi-group $(S(t))_{t\geq 0}$ on $L^\infty(\Omega)^{+2}$ defined by

$$S(t)(u_0, v_0) = (u(t), v(t)) \tag{1.13}$$

where $(u,v) \in L^\infty((0,\infty) \times \Omega)^2 \times C([0,\infty[; L^2(\Omega))^2$ is the mild solution on $[0,\infty[$ of (S); moreover this semi-group possesses an absorbing set in $L^\infty(\Omega)^{+2}$,

$$E_0 = \{(u,v) \in L^\infty(\Omega)^{+2}; u + v \leq R_0, \text{ a.e. on } \Omega\};$$

which is defined by,

$$\left\{ \begin{array}{l} \text{for any bounded set } E \text{ in } L^\infty(\Omega)^{+2}, \text{ there} \\ \text{exists } T > 0 \text{ such that } S(t)E \subset E_0 \text{ for } t \geq T \end{array} \right. \tag{1.14}$$

A maximal attractor in $L^\infty(\Omega)^{+2}$ is a set $\mathcal{M}$ in $L^\infty(\Omega)^{+2}$ satisfying

$$\text{(i)} \quad \mathcal{M} \text{ is compact in } L^\infty(\Omega)^2$$

$$\text{(ii)} \quad S(t)\mathcal{M} = \mathcal{M} \text{ for } t \geq 0 \tag{1.15}$$

$$\text{(iii)} \quad \lim_{t \to \infty} \sup_{\substack{u_0, v_0 \geq 0, u_0 + v_0 \leq r \\ u_0, v_0 \in L^\infty(\Omega)^2.}} dist_{L^\infty(\Omega)^2}(S(t)(u_0, v_0), \mathcal{M}) = 0 \text{ for } r > 0. \tag{1.16}$$

Under the assumptions and conditions of Theorem 1.1, the existence of a maximal attractor is equivalent to asymptotic compactness of the semi-group $(S(t))_{t\geq 0}$. Let us only state here the following corollary.

Corollary 1.2. *With the assumptions of Theorem 1.1, assume moreover that*

$$e^{-tA_i} \text{ is compact in } L^\infty(\Omega) \text{ for } i = 1, 2, \quad t > 0. \tag{1.17}$$

Then there exists a maximal attractor in $L^\infty(\Omega)^{+2}$ for the semigroup $(S(t))_{t\geq 0}$ associated to (S).

2. Proof of Theorem 1.1 and Corollary 1.2

This section is mainly devoted to the proof of Theorem 1.1 and Corollary 1.2. We will give the proof of Theorem 1.1 in different steps stated as lemmas.

Lemma 2.1. *Under assumptions of Theorem 1.1, there exists an estimate of attractor type Φ_1 such that*

$$\|A_1^{-1}u(t)\|_{L^2} \leq \Phi_1(\|u_0 + v_0\|_\infty, t) \quad \text{for } 0 < t < T \tag{2.1}$$

if (u,v) is mild solution on $[0,T)$ of (S).

Proof. Let $w(t) = A_1^{-1}(u(t) + v(t))$. We have

$$w(t) = e^{-tA_1}w_0 + \int_0^t e^{(s-t)A_1}H(s)ds \tag{2.2}$$

with

$$w_0 = A_1^{-1}(u_0 + v_0),$$

$$H(s) = A_1^{-1}(f(., u(s), v(s)) + g(., u(s), v(s))) + (I - A_1^{-1}A_2)v(s) + \underline{u} + A_1^{-1}A_2\underline{v}.$$

According to (1.6),

$$||w_0||_2 \le C_0 r \qquad \text{with} \qquad r = ||u_0 + v_0||_\infty$$

and using (0.12), (1.11), (1.9),

$$||H(s)||_2 \le \gamma(r, s)$$

with

$$\gamma(r, s) = C_0 L(\Phi_0(r, s)) + (|\Omega|^{1/2} + C_2)\Phi_0(r, s) + ||\underline{u}||_2 + C_2||\underline{v}||_\infty. \tag{2.3}$$

Since $0 \le A_1^{-1}u(t) \le w(t)$, we deduce that (0.1) holds with

$$\Phi_1(r, t) = e^{-wt}r + \sup_{\tau \ge t} \int_0^\tau e^{w(s-\tau)}\gamma(r, s)ds. \tag{2.4}$$

Since γ is clearly an estimate of attractor type, it is the same for Φ_1 with

$$\Phi_1(r, \infty) = \frac{\gamma(r, \infty)}{w}. \tag{2.5}$$

$\square$

Lemma 2.2. *Under the assumptions of Theorem 1.1, for any $2 \le p < \infty$ and $\delta > 0$, there exists an estimate of attractor type $\Phi_{p,\delta}$ such that*

$$||u||_{L^p(t,t+\delta)\times\Omega} \le \Phi_{p,\delta}(||u_0 + v_0||_\infty, t) \quad \text{for } 0 < t < T - \delta \tag{2.6}$$

if (u, v) is a mild solution on $[0, T[$ of (S).

Proof. Let, for $\tau > 0$ and $0 < t_0 < T - \delta$,

$$w(\tau) = u(t_0 + \tau) + A_1^{-1}A_2 v(t_0 + \tau) \tag{2.7}$$

$$w_1(\tau) = e^{-\tau A_1}(u(t_0) + A_1^{-1}A_2 v(t_0)) + (I - e^{-\tau A_1})(\underline{u} + A_1^{-1}A_2\underline{v}) \tag{2.8}$$

$$w_2(\tau) = \int_0^\tau e^{(s-\tau)A_1}(f(., u, v) + g(., u, v))(t_0 + s)ds \tag{2.9}$$

$$W(\tau) = \int_0^\tau (w(s) - w_1(s) - w_2(s))ds. \tag{2.10}$$

We have

$$W \in C^1([0, t - t_0 + \delta], L^2(\Omega))$$

and

$$W'(\tau) + AW(\tau) = h(\tau), W(0) = 0$$

with

$$h(\tau) = (A_1^{-1}A_2 - I)(v(t_0 + \tau) - v(t_0)).$$

Using (1.9), (1.11),
$$\|h(\tau)\|_p \le 2(C_p + |\Omega|^{1/p})\Phi_0(r,t_0)$$
with
$$r = \|u_0 + v_0\|_\infty,$$
and then
$$\|h\|_{L^p((0,t-t_0+\delta)\times\Omega)} \le 2(t - t_0 + \delta)^{1/p}(C_p + |\Omega|^{1/p})\Phi_0(r,t_0).$$

Using the Theorem of Lamberton [7], there exists a constant C'_p such that
$$\|W'\|_{L^p((0,t-t_0+\delta)\times\Omega)} \le C'_p(t - t_0 + \delta)^{1/p}\Phi_0(r,t_0). \tag{2.11}$$

On the other hand, using (1.4), (1.5), (1.7) and Lemma 2.1 for $t_0 < t_1 < t$ and $\tau \in [t - t_0, t - t_0 + \delta]$, one has
$$\left\{ \begin{aligned} \|e^{-\tau A_1}u(t_0)\|_p &= e^{(t-t_0-\tau)A_1} e^{(t_0-t_1)A_1} A_1 e^{(t_1-t)A_1} A_1^{-1}u(t_0)\|_p \\ &\le \left[\frac{C}{(t_1-t_0)^\varrho}\right]^{1/2-1/p} \frac{C}{t-t_0}\Phi_1(r,t_0) \; ; \end{aligned} \right. \tag{2.12}$$

Then, since
$$\|e^{-\tau A_1} A_1^{-1} A_2 v(t_0) + (I - e^{-\tau A_1})(\underline{u} + A_1^{-1} A_2\underline{v})\|_p$$
$$\le C_p\Phi_0(r,t_0) + 2(\|\underline{u}\|_p + C_p\|\underline{v}\|_\infty),$$

there exists a constant C''_p such that
$$\|w_1\|_{L^p((t-t_0,t-t_0+\delta)\times\Omega)} \le C''_p\delta^{1/p}\left[\frac{\Phi_1(r,t_0)}{(t-t_0)^{1+\varrho(1/2-1/p)}} + \Phi_0(r,t_0) + 1\right]. \tag{2.13}$$

Finally,
$$\|w_2(\tau)\|_p \le \tau|\Omega|^{1/p}L(\Phi_0(r,t_0)), \tag{2.14}$$
$$\|A_1^{-1}A_2 v(t_0 + \tau)\|_p \le C_p\Phi_0(r,t_0). \tag{2.15}$$
Since $u(t_0 + \tau) = w_1(\tau) + w_2(\tau) + W(\tau) - A_1^{-1}A_2 v(t_0 + \tau)$, we deduce from (2.12), (2.13), (2.14) and (2.15) that
$$\|u\|_{L^p((t,t+\delta)\times\Omega)} \le \gamma(r,t_0,t)$$
with
$$\left\{ \begin{aligned} \gamma(r,t_0,t) &= C'_p(t - t_0 + \delta)^{1/p}\Phi_0(r,t_0) \\ &+ \delta^{1/p}\left[C''_p(\frac{\Phi_1(r,t_0)}{(t-t_0)^{1+\rho(1/2-1/p)}} + 1) + (C''_p + C_p)\Phi(r,t_0)\right] \\ &+ \delta^{2/p}(\frac{|\Omega|}{2})^{1/p}L(\Phi_0(r,t_0)). \end{aligned} \right. \tag{2.16}$$
We deduce that (2.6) holds with
$$\Phi_{p,\delta}(r,t) = \sup_{\tau \ge t} \inf_{t_0 \in (0,\tau)} \gamma(r,t_0,\tau). \tag{2.17}$$

It is clear that $\Phi_{p,\delta}$ is an estimate of attractor type since
$$\left\{ \begin{aligned} \Phi_{p,\delta}(r,\infty) &\le \inf C'_p(\sigma + \delta)^{1/p}\Phi_0(r,\infty) \\ &+ \delta^{1/p}\left[C''_p(\frac{\Phi_1(r,\infty)}{\sigma^{1+\rho(1/2-1/p)}} + 1) + (C''_p + C_p)\Phi_0(r,\infty)\right] \\ &+ \delta^{2/p}(\frac{|\Omega|}{2})^{1/p}L(\Phi_0(r,\infty)). \end{aligned} \right. \tag{2.18}$$
$$\square$$

Lemma 2.3. *Under the assumptions of Theorem 1.1, there exists an estimate of attractor type* Φ *such that*

$$u(t) \leq \Phi(\|u_0 + v_0\|_\infty, t) \quad \text{for } 0 < t < T, \tag{2.19}$$

if (u, v) *is a mild solution on* $[0, T]$ *of (S).*

Proof. Let $0 < t_0 < t < T$. One has

$$u(t) = e^{(t_0-t)A_1}u(t_0) + \int_{t_0}^t e^{(s-t)A_1} f(., u(s), v(s))ds. \tag{2.20}$$

Using (1.4), (1.5), (1.7) and Lemma 2.2 as in (2.12) one has

$$e^{(t_0-t)A_1}u(t_0) \leq C_0' \frac{\Phi_1(r, t_0)}{(t - t_0)^{1+\varrho/2}} \tag{2.21}$$

with $C_0' = \min\limits_{0<\lambda<1}(\frac{C}{\lambda\varrho})^{1/2}\frac{C}{(1-\lambda)}$. On the other hand, using (1.10),

$$f(., u(s), v(s)) \leq L(\Phi_0(r, t_0)(1 + u(s))^{m(\Phi_0(\Omega, t_0))} \quad \text{for } t_0 < s < t.$$

But using (1.5), (1.7), one has for $m_0 \geq 1$

$$\| \int_{t_0}^t e^{(s-t)A_1}(1 + u(s))^{m_0}ds\|_\infty$$

$$\leq \int_0^{t-t_0} (\frac{C}{(t - t_0 - \tau)^\varrho})^{\frac{1}{\varrho+2}}\|1 + u(t_0 + \tau)\|_{m_0(\varrho+2)}^{m_0}ds$$

$$\leq C_0''(t - t_0)^{\frac{1}{\varrho+2}}\|1 + u\|_{L^{m_0(\varrho+2)}((t_0,t)\times\Omega)}^{m_0}$$

with

$$C_0'' = C^{\frac{1}{\varrho+2}} \sup_{\frac{\varrho}{\varrho+2}<\theta<\frac{\varrho+1}{\varrho+2}} (1 - \frac{\varrho}{\theta(\varrho + 2)})^\theta.$$

Then using Lemma 3.2, one has

$$\int_{t_0}^t e^{(s-t)A_1} f(., u(s), v(s))ds \leq \gamma(r, t_0, t)$$

with

$$\begin{cases} \gamma(r, t_0, t) = C_0''(t - t_0)^{\frac{1}{\varrho+2}} L(\Phi_0(r, t_0)) \\ \{((t - t_0)|\Omega|)^{\frac{1}{\varrho+2}} + (\varrho + 2)m(\Phi_0(r, t_0) \\ \quad \Phi_{m(\Phi_0(r,t_0))(\varrho+2)),t-t_0}(r, t_0)\}. \end{cases} \tag{2.22}$$

We deduce that (2.19) holds with

$$\Phi(r, t) = \sup_{\tau \geq t} \inf_{t_0 \in [0,\tau], r_0 < r} \left[C_0' \frac{\Phi_1(r, t_0)}{(t - t_0)^{1+p/2}} + \gamma(r, t_0, t) \right]. \qquad \square$$

Proof of Theorem 1.1. This is a direct consequence of Lemma 2.3. $\qquad \square$

Now to prove Corollary 1.2, we need the following lemma.

Lemma 2.4. *Let (E, d) be a metric space, $S : \mathbb{R}_+ \times E \to E$ a semigroup, and τ a Hausdorff topology on E such that $S : \mathbb{R}_+ \times (E, d) \to (E, \tau)$ is continuous. Then S has a maximal attractor in E, if the two following properties are satisfied:*

(i) S has an absorbing set in E, that is a bounded set $\mathcal{A}$ in E such that

$$\begin{cases} \forall\, B \text{ bounded set in } E, \exists\, T > 0 \text{ such that} \\ \quad S(t, B) \subset \mathcal{A} \quad \text{ for any } t \geq T \end{cases}$$

(ii) S is asymptotically compact in E, i.e. for any sequences $\{w_m\} \in E$ bounded and $\{t_m\} \in \mathbb{R}_+$ with $t_m \to +\infty$, there is a subsequence $(m_k)_k$ such that the sequence $\{S(t_{mk}, w_{mk})\}$ converges in E.

Moreover a maximal attractor is unique and if $\mathcal{A}$ is an absorbing set,

$$m = \bigcap_{\tau \geq 0} \overline{\bigcup_{t \geq \tau} S(t, \mathcal{A})}.$$

Everything in the statement above is classical (cf. [13]) except that we do not assume S continuous in (E, d), however it is immediate to verify that the continuity of S in (E, τ) is enough (see [3],[12]).

The proof of Lemma 2.4 is given in [8].

3. Proof of Theorem 0.1

We consider the example of the reaction-diffusion system, the so called "Brusselator", introduced by Lefever-Prigogine [10], as the basic model of highly instable chemical reactions.

$$u_t - d_1 \Delta u = u^2 v - (b + 1)u + a, \qquad \text{in } (0, T) \times \Omega \tag{3.1}$$

$$v_t - d_2 \Delta v = -u^2 v + bu, \qquad \text{in } (0, T) \times \Omega \tag{3.2}$$

$$u(0) = u_0, \quad v(0) = v_0 \qquad \text{in } \Omega, \tag{3.3}$$

where Ω is an open bounded subset in $\mathbb{R}^n$, $T > 0$, d_1, d_2 the positive coefficients respectively of u and v, $a > 0$, $v > 0$, u_0 and v_0 are given positive functions in $L^\infty(\Omega)$ with the boundary conditions

$$\lambda_1 \frac{\partial u}{\partial n} + (1 - \lambda_1)u = \alpha_1, \qquad \text{on } \partial\Omega, \tag{3.4}$$

$$\lambda_2 \frac{\partial v}{\partial n} + (1 - \lambda_2)v = \alpha_2, \qquad \text{on } \partial\Omega, \tag{3.5}$$

$0 \leq \lambda_1 \leq 1$, $0 \leq \lambda_2 \leq 1$, α_1 and α_2 are positive functions on $\partial\Omega$.

We assume $d_1 > 0$ and $d_2 \geq 0$. To apply the abstract result of Sections 1 and 2, we define the operators A_1 and A_2 according to different conditions:

(i) If $d_2 = 0$, we set $A_2 = 0$.

(ii) If $d_2 > 0$,

    a) if $\lambda_2 = 0$, then $V = H_0^1(\Omega)$ and $A_2 v = -d_2 \Delta v$ in $D'(\Omega)$.

    b) if $0 < \lambda_2 < 1$, then $V = H^1(\Omega)$ and for $h \in L^2(\Omega)$, $v \in V$, $A_2 v = h$ is equivalent to $\int_\Omega hw\,dx = d_2 \int_\Omega \nabla h \nabla v\,dx$ for all $w \in V$.

780 $\qquad$ PHILIPPE BENILAN and HALIMA LABANI

(iii) a) If $\lambda_1 = 0$, then $V = H^1_0(\Omega)$ and $A_1 u = a(u, .)$, where

$$a(u_1, u_2) = d_1 \int_\Omega \nabla u_1 \nabla u_2 dx$$

for $u_1, u_2 \in V$.

b) If $\lambda_1 > 0$, then $V = H^1(\Omega)$ and $A_1 u = a(u, .)$, where

$$a(u_1, u_2) = \int_\Omega [d_2 \nabla u_1 \nabla u_2 + w u_1 u_2]\, dx + \frac{(1 - \lambda_1)}{\lambda_1} \int_\Omega u_1 u_2 d\sigma$$

for $u_1, u_2 \in V$, $0 < w < 1$ for $0 < \lambda_1 \leq 1$ and $w = 0$ for $\lambda_1 = 0$.

Now, the reaction terms f and g are defined by

$$\begin{cases} g(x, u, v) = bu - u^2 v \\ f(x, u, v) = -g(x, u, v) + a + (w - 1)u \end{cases}$$

where $w = 1$ if $\lambda_1 = 0$ and $w = 0$ if $0 \leq \lambda < 1$. These functions satisfy (0.12) and (0.10) as well as the "basic assumptions".

We denote by G the Green operator in $L^2(\Omega)$ defined by

$$\begin{cases} -\Delta(Gf) = f & \text{in} \quad \Omega \\ \\ \lambda_2 \frac{\partial Gf(x)}{\partial n} + (1 - \lambda_2)Gf(x) = 0 & \text{for } x \in \partial\Omega, \end{cases}$$

and $\|G1\|$ denotes the norm in $L^\infty(\Omega)$.

The proof of Theorem 0.1 is based on three lemmas.

Lemma 3.1. *Assume $d_1 > 0$ and the hypothesis*

(H1) $$d_2(1 - \lambda_2) + a(\lambda_1 + \alpha_1) > 0.$$

Then v has an (E.A.T.) (see (1.11)).

Proof. If $d_2 > 0$, let us set $z = v^2 - \alpha_2^2$; then the function $z \in V$ satisfies

$$\begin{cases} \frac{\partial z}{\partial t} \leq d_2 \Delta z + \frac{b^2}{2}, & \text{on } (0, T) \times \Omega \\ \lambda_2 \frac{\partial z}{\partial n} + (1 - \lambda_2)z \leq 2\lambda_2 v \alpha_2, & \text{on } (0, T) \times \partial\Omega \\ z(0) = v_0^2 - \alpha_2^2. \end{cases}$$

Now, by principle of comparison, v has a priori (E.A.T.) as there exists $\theta > 0$ such that

$$z(t) \leq e^{-\theta t}\|z_0\|_\infty + \frac{b^2}{\alpha w}\|G1\| + (2\alpha_2\lambda_2 + 1)\|v\|_\infty.$$

Then

$$v(t)$$
$$\leq \max\left(\alpha_2\lambda_2 + \frac{1}{2}, \frac{\alpha_2^2}{2}\|G1\|_\infty + 2e^{-\theta t}\max(\|\alpha_2^2, \|v_0\|_\infty^2) + (\alpha_2^2 + \alpha_2^2\lambda_2^2 + \frac{1}{4})\right).$$

Lemma 3.1 follows and Φ_0 is defined by

$$\Phi_0(||u_0 - v_0||_\infty, t)$$
$$= \max\left(\left(\alpha_2\lambda_2 + \frac{1}{2}\right)^{\frac{1}{2}}, \left(\frac{\alpha_2^2}{2}||G1||_\infty + (\alpha_2^2 + \alpha_2^2\lambda_2^2 + \frac{1}{4})\right)^{\frac{1}{2}}\right).$$

If $d_2 = 0$, then (H1) is satisfied when $\lambda_1 + \alpha_1 > 0$.
   The system (3.1)–(3.4) is written

$$\frac{\partial u}{\partial t} + d_1 A_1 u = f(u, v) \tag{3.6}$$

$$\frac{\partial v}{\partial t} = g(u, v). \tag{3.7}$$

We may write

$$v(x,t) = v_0(x)e^{-\int_0^t u^2(x,s)ds} + \int_0^t e^{-\int_s^t u^2(x,s)ds} u(x,s)ds$$
$$= K_1(x,t) + K_2(x,t).$$

We have

$$K_1(x,t) \le v_0(x)$$

and by Schwarz inequality

$$K_2(x,t) \le \left[\int_0^t e^{-\int_s^t u^2(x,s)ds}ds\right]^{1/2} \times \left[\int_0^t e^{-\int_s^t u^2(x,s)ds} u^2(x,s)ds\right]^{1/2}.$$

On the other hand,

$$\int_0^t e^{-\int_s^t u^2(x,s)ds} u^2(x,s)ds = -\int_0^t \frac{d}{ds}\left(e^{-\int_0^t u^2(x,s)ds}\right)$$
$$= 1 - e^{-\int_0^t u^2(x,\tau)d\tau}$$
$$\le 1,$$

and there exists γ a function of attractor type, $T > 0$ (see [6]), such that

$$u \ge \gamma, \quad \text{for } t \ge T.$$

Then

$$\int_0^T e^{-\int_s^t u^2(x,s)ds}ds \le \int_0^T e^{-\gamma_1(t)}dt,$$

where $\gamma_1(t) = \int_0^t \gamma^2(s)ds$, non-positive and non-increasing. The term $\int_0^T e^{-\gamma_1(t)}dt$ satisfies (E.A.T). $\qquad\square$

Remark. The hypothesis $(H1)$ is necessary and sufficient; indeed, for $\lambda_1 = 0$ ($\alpha_1 = 0$) and $\lambda_2 = 0$ the solution blows-up in finite time. We apply the same proof as in [2].

Lemma 3.2. *Assume $d_1 > 0$ and the hypothesis*

(H2) $$d_2\lambda_1 = 0 \ \text{or} \ \lambda_2 > 0.$$

Then the condition

(1.9) $$A_1^{-1}A_2 \in \pounds(L^\infty, L^p), \quad 2 \le p < \infty$$

is satisfied.

Proof. (i) If $d_2 = 0$ then $A_1^{-1}A_2 \equiv 0$. Then Lemma 3.2 is obvious.

(ii) If $d_2 > 0$, to prove Lemma 3.2 we use a duality method. First, we assume $\lambda_1 = 0$. For $2 \le p < \infty$, $v \in L^\infty \cap D(A_2)$ and $w \in L^q(\Omega)$, where q is the conjuguate of $p(\frac{1}{p} + \frac{1}{q} = 1)$, as A_1 is symmetric in $L^p(\Omega)$, we have

$$
\begin{aligned}
\int_\Omega wA_1^{-1}A_2 v\,dx &= \int_\Omega A_1^{-1}wA_2 v\,dx \\
&= \int_\Omega \zeta(-A_2 v)\,dx \\
&= d_2\left(\int_{\partial\Omega} (v\frac{\partial\zeta}{\partial n} - \zeta\frac{\partial v}{\partial n})d\sigma + \int_\Omega vw\,dx\right),
\end{aligned}
$$

where ζ is the solution of

$$
\begin{cases}
-\Delta\zeta = w, & \text{on } \Omega \\
\lambda_2\frac{\partial\zeta}{\partial n} + (1 - \lambda_2)\zeta = 0, & \text{on } \partial\Omega.
\end{cases}
$$

Then, we deduce that

$$
\int_\Omega wA_1^{-1}A_2 v\,dx \le ||v||_\infty\left[\left\|\frac{\partial\zeta}{\partial n}\right\|_{L^1(\partial\Omega)} + ||w||_{L^1(\Omega)}\right].
$$

The estimate (cf. Dautray Lions [5])

$$
\left\|\frac{\partial\zeta}{\partial n}\right\|_{L^1(\partial\Omega)} \le k||w||_{L^1(\Omega)}
$$

gives (1.9).

Now, assume $\lambda_1 > 0$, then the hypothesis (H2) implies $\lambda_2 > 0$. As the operators A_1, A_2 are symmetric in $L^p(\Omega)$, $2 \le p < \infty$, we can apply Proposition 3.3 from [11]; then condition (1.9) is satisfied. $\qquad\square$

To prove Theorem 0.1, we need a compactness result: the linear semigroup associated to (0.1)–(0.2) is compact and the nonlinear term is locally Lipschitz continuous. We deduce the compactness of the nonlinear semigroup $(S(t))_{t\ge 0}$ defined by (1.13) for any $t > 0$ in $(L^\infty(\Omega))^2$ (see [8]).

Remark. If $d_2 = 0$ and $\lambda_1 + \alpha_1 > 0$, the system (0.1)–(0.2) can be written as

$$
\begin{cases}
u_t - d_1\Delta u = u^2 v - (B + 1)u + A \ \text{in} \ (0, T) \times \Omega \\
v_t = -u^2 v + Bu \ \text{in} \ (0, T) \times \Omega.
\end{cases}
$$

Then v is a function of u,

$$(*) \qquad v(x,t) = v_0(x)e^{-\int_0^t u^2(x,s)ds} + \int_0^t e^{-\int_s^t u^2(x,s)ds} u(x,s)ds.$$

We estimate u from below (cf. [6]); we use Schwarz inequality, then v has an estimate of (E.A.T).

Now the result of compactness follows from $(*)$.

References

[1] V. Barbu. Nonlinear semigroups and differential equations in Banach spaces, Noordhoff International Publishing Leyden The Netherlands, 1976.

[2] J. Bebernes and A. Lacey. Finite time blow-up semilinear reactive-diffusive systems, *J. Diff. Eqns.* 95 (1992), 105–129.

[3] Ph. Bénilan, H. Labani, Systèmes de réaction-diffusion abstraits, P. M. B. de Besançon, 1994.

[4] Ph. Bénilan, H. Labani, Existence of attractors for Brusselator in Recent Advances in nonlinear elliptic and parabolic problems, Pitman Research Notes in Mathematics Series 209, 1989.

[5] R. Dautray-J. Lions, Analyse mathématiques et calcul numérique pour les sciences et les techniques, Vol. 2, 1987.

[6] H. Labani, Contribution à l'étude de systèmes de réaction-diffusion, Thèse de l'Université de Franche-Comté, Besançon, 1989.

[7] D. Lamberton, Equations d'évolution linéaires associées à des semi-groupes de contraction dans les espaces L^p, *Journal of Functional Analysis* 72 (1987), 252–262.

[8] H. Labani, Comportement Asymptotique de Certaines Equations de Réaction-Diffusion et d'une Classe D'équations des Ondes, Doctorat d'Etat, 2002.

[9] S. L. Hollis, R. H. Martin, M. Pierre, Global existence and boundedness in reaction-diffusion systems, *SIAM. J. Math. Anal.* 18 n°3 (1987), 744–761.

[10] R. Lefever, I. Prigogine, Symmetry breaking instabilities in dissipative systems II, *J. Chemical Phys.* 48, n°4(1968).

[11] R. H. Martin, M. Pierre, Influence of mixed boundary conditions in some reaction-diffusion system, *Proc. Roy. Soc. Edinburgh*, section A 127 (1997), 1053–1066.

[12] L. Veron, J. C. Hassan, P. Baras, Compacité de l'opérateur définissant la solution d'une équation dévolution non homogène, *C.R.A.S. Paris*, Série A 284 (1977), 799–802.

[13] M. I. Vischik, A. V. Babin, Attractors of partial differential evolution equations and estimates of their dimension, *Russian Math, Surveys* 38, 4 (1983), 151–213.

PHILIPPE BENILAN
Université de Franche-Comté
Faculté des Siences et Techniques
Laboratoire de Mathématiques
Route de Gray 25000
Besançon, France

HALIMA LABANI
Université Chouaib Doukkali
Faculté des Siences
Departement de Mathématiques et Informatique
B.P:20, El Jadida
24000 Marocco

J.evol.equ. 4 (2004) 273 – 295
1424–3199/04/020273 – 23
DOI 10.1007/s00028-004-0143-1
© Birkhäuser Verlag, Basel, 2004

**Journal of Evolution
Equations**

Uniqueness for an elliptic-parabolic problem with Neumann boundary condition

BORIS P. ANDREIANOV and FOUZIA BOUHSISS

Dedicated to the memory of Philippe Bénilan

Abstract. We consider the problem $b(u) - \Delta u + \operatorname{div} F(u) = f$ in a smooth bounded domain $\Omega \subset \mathbb{R}^N$, as well as the corresponding evolution equation $b(u)_t - \Delta u + \operatorname{div} F(u) = f, b(u(0, .)) = b^0$. For the stationary equation we show existence results, then we adapt the techniques of doubling of variables to the case of the homogeneous Neumann boundary conditions and obtain the appropriate L^1-contraction principle and uniqueness. Subsequently, we are able to apply the nonlinear semigroup theory and prove the L^1-contraction principle for the associated evolution equation.

1. Introduction

Let $\Omega \subset \mathbb{R}^N$ be a bounded domain with Lipschitz boundary. We consider the Neumann problems

$$\begin{cases} b(u) - \Delta u + \operatorname{div} F(u) = f & \text{in } \Omega \\ (\nabla u - F(u)) \cdot \nu = 0 & \text{on } \partial\Omega \end{cases} \qquad P(b, F)(f)$$

and

$$\begin{cases} b(u)_t - \Delta u + \operatorname{div} F(u) = f & \text{in } Q = (0, T) \times \Omega \\ (\nabla u - F(u)) \cdot \nu = 0 & \text{on } (0, T) \times \partial\Omega \\ b(u)(t = 0) = b^o & \text{in } \Omega, \end{cases} \qquad E(b, F)(f, b^o)$$

where ν is the exterior unit normal vector to $\partial\Omega$.

Here,

$b : \mathbb{R} \longrightarrow \mathbb{R}$ is increasing, normalised by $b(0) = 0$,

$F : \mathbb{R} \longrightarrow \mathbb{R}^N$ is continuous, normalised by $F(0) = 0$.

Received: 31 July 2003

Mathematics Subject Classification (2000): 35J65, 35K60, 35K65, 47H06, 47H20.

Key words: Degenerate parabolic equations, Neumann boundary condition, doubling of variables, nonlinear semigroup theory.

For the Dirichlet or mixed Dirichlet-Neumann boundary conditions, different uniqueness results for elliptic-parabolic equations of the form $b(u)_t + \operatorname{div} a(u, \nabla u) = f$ have been obtained by Alt-Luckhaus [2], Diaz-de Thélin [13], Otto [18], Bénilan-Wittbold [7], assuming in particular that $a(\cdot, \cdot)$ is Hölder continuous in the first argument of order $\alpha \geq \operatorname{const} > 0$. These results cover the case of the equations we consider if $\alpha \geq 1/2$.

Results for the stationary problem and the corresponding abstract evolution problem are given by Simondon [19], Bénilan-Touré [5], [6] and by Bénilan-Wittbold [7].

Developping the Kruzhkov's ideas of doubling of variables ([15]), Carrillo proves in [10] uniqueness results for hyperbolic-elliptic-parabolic equations without any Hölder continuity assumption, in the case of homogeneous Dirichlet boundary conditions; further results on renormalized solution are given in [11]. In this paper, we show how one can use the doubling of variables in case of Neumann boundary conditions. We use the techniques of [10] and [7], the existence of solutions to the stationary problem which are regular up to the boundary, and the general nonlinear semigroup theory. To have regular solutions, we assume the local Hölder continuity of F of unrestricted order $\alpha > 0$ (here we use a result of Lieberman [16]).

Let us give the main definitions and results of this paper.

First we consider the stationary problem $P(b, F)(f)$.

DEFINITION 1. Let $f \in L^1(\Omega)$. A function $u \in H^1(\Omega)$ such that $b(u) \in L^1(\Omega)$ and $F(u) \in L^2(\Omega)$ is a weak solution of $P(b, F)(f)$ if for all $\xi \in H^1(\Omega) \cap L^\infty(\Omega)$, one has

$$\int_\Omega b(u)\,\xi + \int_\Omega (\nabla u - F(u)) \cdot \nabla \xi = \int_\Omega f\,\xi. \tag{1}$$

An existence result for weak solutions of $P(b, F)(f)$ with $f \in L^2(\Omega)$ is shown in Section 2, under the assumptions

$$\begin{cases} \text{there exist } c > 0, \delta > 0 \text{ such that} \\ |F(z)|^2 \leq c(1 + |z|^2 + (zb(z))^{1-\delta}) \text{ for all } z \in \mathbb{R}, \end{cases} \tag{H1}$$

and

$$b(+\infty) = +\infty, \quad b(-\infty) = -\infty.$$

We rewrite this last hypothesis under the equivalent form:

$$\begin{cases} \text{there exists a function } \beta : \mathbb{R}^+ \longrightarrow \mathbb{R}^+ \text{ such that} \\ \beta(r) \to +\infty \text{ as } r \to +\infty \text{ and } |b(z)| \geq \beta(|z|) \text{ for all } z \in \mathbb{R}. \end{cases} \tag{H2}$$

In Section 3, we prove the uniqueness of $b(u)$ such that u is a weak solution of $P(b, F)(f)$ and the L^1contraction principle (2) below. In case F is Hölder continuous with exponent $\alpha \geq 1/2$, (2) is easy to prove. In this paper, we assume

$$\begin{cases} \text{for all compact set } K \subset \mathbb{R} \text{ there exists } \alpha > 0 \text{ such that} \\ F \text{ is Hölder continuous of order } \alpha \text{ on } K. \end{cases} \tag{H3}$$

We show that a weak solution satisfies entropy inequalities in the spirit of Kruzhkov [15] and Carrillo [10], and adapt their techniques to establish:

THEOREM 1. *Suppose $(H1), (H2), (H3)$ hold, and $\partial\Omega \in \mathcal{C}^2$. Let $f, g \in L^1(\Omega)$ and u, v be weak solutions of $P(b, F)(f)$ and $P(b, F)(g)$, respectively. Then*

$$\|b(u) - b(v)\|_{L^1(\Omega)} \leq \|f - g\|_{L^1(\Omega)}. \tag{2}$$

In particular, there exists at most one function $b(u)$ such that u is a weak solution of $P(b, F)(f)$.

In the proof, we first show (2) in the case one of the two solutions is regular up to the boundary. The fact that there exists a weak solution v of $P(b, F)(g)$ which belongs to $\mathcal{C}^1(\overline{\Omega})$ is ensured by the hypotheses $(H1)$, $(H2)$, $(H3)$ whenever $g \in L^\infty(\Omega)$ (cf. Proposition 2). The result then follows by a density argument.

Next we consider the elliptic-parabolic problem $E(b, F)(f, b^o)$.

DEFINITION 2. Let $f \in L^1(Q)$, $b^o \in L^1(\Omega)$. A function $u \in L^2(0, T; H^1(\Omega))$ such that $b(u) \in L^1(Q)$ and $F(u) \in L^2(Q)$ is a weak solution of $E(b, F)(f, b^o)$ if for all $\xi \in L^2(0, T; H^1(\Omega)) \cap L^\infty(Q)$ such that $\xi_t \in L^\infty(Q)$ and $\xi(T) = 0$, one has

$$\int_0^T \int_\Omega (b^o - b(u))\, \xi_t + \int_0^T \int_\Omega (\nabla u - F(u)) \cdot \nabla\xi = \int_0^T \int_\Omega f\,\xi. \tag{3}$$

Different existence results for weak solutions of $E(b, F)(f, b^o)$ under additional assumptions on f and b^o were obtained, using in particular the techniques of [2] (eg., cf. [14]).

For the uniqueness, we need one more hypothesis

$$\begin{cases} \text{there exist } c > 0, \delta > 0 \text{ such that} \\ |F(z)|^2 \leq c(1 + |z|^{2(1-\delta)} + zb(z)) \text{ for all } z \in \mathbb{R}. \end{cases} \tag{H4}$$

In this case, the results obtained for the stationary problem permit to apply the non-linear semigroup theory and obtain the existence and uniqueness for integral solutions of the associated abstract evolution problem (the problem $S(b, F)(f, b^o)$ in Section 4). This problem is closely related to $E(b, F)(f, b^o)$. More exactly, using again the techniques of doubling of variables in space, we show that if u is a weak solution of $E(b, F)(f, b^o)$, then $w = b(u)$ is an integral solution of $S(b, F)(f, b^o)$. These results, gathered in Section 4, yield

THEOREM 2. *Suppose $(H1), (H2), (H3), (H4)$ hold and $\partial\Omega \in \mathcal{C}^2$. Let $f, \hat{f} \in L^1(Q), b^o, \hat{b}^o \in L^1(\Omega)$ and $u, \hat{u}$ be weak solutions of $E(b, F)(f, b^o)$ and $E(b, F)(\hat{f}, \hat{b}^o)$ respectively. Then for a.a $t \in (0, T)$,*

$$\|b(u(t)) - b(\hat{u}(t))\|_{L^1(\Omega)} \leq \|b^o - \hat{b}^o\|_{L^1(\Omega)} + \int_0^t \|f(\tau) - \hat{f}(\tau)\|_{L^1(\Omega)}d\tau.$$

In particular, there exists at most one function $b(u)$ such that u is a weak solution of $E(b, F)(f, b^o)$.

At the present stage, it is not clear to the authors whether the uniqueness of $b(u)$ for $P(b, F)(f)$ and $E(b, F)(f, b^o)$ holds without the hypothesis $(H3)$, that is, for only continuous F. Note that this is true for $N = 1$, since all weak solution of $P(b, F)(f)$ is in $C^1(\overline{\Omega})$ in this case.

2. Existence of weak and regular solutions to the stationary problem

First we establish some a priori estimates.

LEMMA 1. *Suppose $(H1)$, $(H2)$ hold. Let $f \in L^2(\Omega)$ and u be a weak solution of $P(b, F)(f)$. Then $u\,b(u) \in L^1(\Omega)$ and there exists a constant C which depends only on Ω, β, $\|f\|_{L^2(\Omega)}$, c and δ such that $\|u\|_{H^1(\Omega)} \leq C$ and $\|u\,b(u)\|_{L^1(\Omega)} \leq C$.*

Throughout the paper, we abbreviate the notations for sets; for instance, the set $\{x \in \Omega, |u(x)| > k\}$ is denoted by $\{|u| > k\}$ and its characteristic function is denoted by $\chi_{\{|u|>k\}}$. For a set $E \subset \mathbb{R}^N$, we denote by $|E|$ its Lebesgue measure. The notation $H_\varepsilon(\cdot)$ is used for the approximation of $\text{sign}(\cdot)$ given by

$$H_\varepsilon(r) = \begin{cases} 1 & \text{if } r \geq \varepsilon \\ r/\varepsilon & \text{if } -\varepsilon \leq r \leq \varepsilon \\ -1 & \text{if } -\varepsilon \leq r. \end{cases}$$

We denote $H_\varepsilon'(r) = \frac{1}{\varepsilon}\chi_{\{|r|<\varepsilon\}}$; for all $w \in H^1(\Omega)$, one has $\nabla H_\varepsilon(w) = H_\varepsilon'(w)\nabla w$ in $L^2(\Omega)$.

Finally, we denote by 2^* the number $\frac{2N}{N-2}$ if $N > 2$ and an arbitrary number $2^* > 2$ if $N = 1$ or $N = 2$.

REMARK 1. Let $u, \xi \in H^1(\Omega)$ and $k \in \mathbb{R}$. Then

$$\int_\Omega (F(u) - F(k)) \cdot \nabla u\, H_\varepsilon'(u - k)\, \xi \to 0 \text{ as } \varepsilon \to 0.$$

Proof. Set $\psi_\varepsilon(r) = \int_{k-\varepsilon}^r (F(s) - F(k))\, H_\varepsilon'(s - k)\, ds$. Note that $|\psi_\varepsilon(r)| \leq 2\,\omega(\varepsilon)$, where $\omega : \varepsilon \in \mathbb{R} \mapsto \sup_{|s-k|<\varepsilon} |F(s) - F(k)|$ is the modulus of continuity of F at the point k.

It follows by the Green-Gauss Formula that

$$\left| \int_\Omega H_\varepsilon'(u - k)\, \nabla u \cdot (F(u) - F(k))\, \xi \right| = \left| \int_\Omega \text{div}(\psi_\varepsilon(u))\, \xi \right|$$

$$\leq 2\,\omega(\varepsilon) \left(\int_\Omega |\nabla \xi| + \int_{\partial\Omega} |\xi| \right) \to 0 \text{ as } \varepsilon \to 0. \qquad \square$$

Proof of Lemma 1. We denote by C all constant depending only on Ω, $\|f\|_{L^2(\Omega)}$, β in $(H2)$ and c, δ in $(H1)$.

CLAIM 1. $|\{|u| > \kappa\}| \leq \frac{C}{\beta(\kappa)}$ for all $\kappa > 0$.

Take $H_\varepsilon(u) \in H^1(\Omega) \cap L^\infty(\Omega)$ as a test function in (1). We have

$$\int_\Omega b(u) H_\varepsilon(u) + \int_\Omega |\nabla u|^2 H_\varepsilon'(u) - \int_\Omega F(u) \cdot \nabla u \, H_\varepsilon'(u) = \int_\Omega f \, H_\varepsilon(u).$$

The second term is non-negative; in the other terms we use Remark 1, the Lebesgue dominated convergence theorem, and pass to the limit as $\varepsilon \to 0$ to get

$$\int_\Omega |b(u)| \leq \int_\Omega |f| \leq C. \tag{4}$$

Therefore by $(H2)$, for all $\kappa \geq 0$ we have $\beta(\kappa)|\{|u| > \kappa\}| \leq \int_\Omega |f| \leq C$. Hence Claim 1 follows.

CLAIM 2. for all $\kappa \geq 0$, one has

$$\delta \int_{\{|u|>\kappa\}} u \, b(u) + \int_{\{|u|>\kappa\}} |\nabla u|^2 \leq C \left(\kappa + 1 + \int_{\{|u|>\kappa\}} |u|^2 \right).$$

First note that for all increasing globally Lipschitz continuous function ϕ, we have $b(u)\phi(u) \in L^1(\Omega)$ and

$$\int_\Omega b(u) \, \phi(u) + \int_\Omega (\nabla u - F(u)) \cdot \nabla \phi(u) = \int_\Omega f \, \phi(u). \tag{5}$$

Indeed, one can take $T_k(\phi(u))$ as a test function in (1), where

$$T_k(r) = \begin{cases} r & \text{if } |r| < k \\ k \, \text{sign}(r) & \text{if } |r| \geq k, \end{cases} \tag{6}$$

and pass to the limit as $k \to \infty$ using Levy's and Lebesgue's theorems to obtain (5).

In particular, with $\phi(u) = (|u| - \kappa)^+ \text{sign}(u)$, we get

$$\int_{\{|u|>\kappa\}} |b(u)|(|u|-\kappa)^+ + \int_{\{|u|>\kappa\}} |\nabla u|^2 \leq \int_{\{|u|>\kappa\}} F(u) \cdot \nabla u + \int_{\{|u|>\kappa\}} |f| \, |u|.$$

Using the Young inequality, $(H1)$ and the Hölder inequality, we obtain

$$\int_{\{|u|>\kappa\}} |b(u)|(|u|-\kappa)^+ + \frac{1}{2} \int_{\{|u|>\kappa\}} |\nabla u|^2$$

$$\leq C \left(1 + \int_{\{|u|>\kappa\}} |u|^2 \right) + \int_{\{|u|>\kappa\}} \frac{c}{2} (u b(u))^{1-\delta}.$$

Applying the Young inequality in the last term and using (4), we finally get Claim 2.

CLAIM 3. $\int_{\Omega} |u|^{2^*} \leq C.$

For all $v \in H^1(\Omega)$, the Sobolev inequality (e.g., cf. [1]) can write

$$\int_{\Omega} |v|^{2^*} \leq C \left(\left(\int_{\Omega} |\nabla v|^2 \right)^{\frac{2^*}{2}} + \left(\int_{\Omega} |v|^2 \right)^{\frac{2^*}{2}} \right).$$

Take $v = \phi(u)$. By Claim 2 and the Hölder inequality, we have

$$I_\kappa = \int_{\Omega} ((|u| - \kappa)^+)^{2^*} \leq C \left(\kappa^{2^*} + 1 + \left(\int_{\{|u|>\kappa\}} |u|^{2^*} \right) |\{|u| > \kappa\}|^{\frac{2^*}{2}-1} \right).$$

Since $|u| \leq (|u| - \kappa)^+ + \kappa$, we get

$$I_\kappa \leq C(\kappa^{2^*} + 1 + (I_\kappa + \kappa^{2^*})|\{|u| > \kappa\}|^{\frac{2^*}{2}-1}).$$

By Claim 1, $|\{|u| > \kappa\}| \to 0$ as $\kappa \to +\infty$ and we can choose κ_o (depending on β) such that

$$\frac{1}{2} I_{\kappa_o} \leq C \left(1 + \frac{3}{2}(\kappa_o)^{2^*} \right) \leq C.$$

Since $|u| \leq (|u| - \kappa_o)^+ + \kappa_o$, Claim 3 follows.

Finally, we deduce from Claim 3 that $\int_{\Omega} |u|^2 \leq C$, and take $\kappa = 0$ in Claim 2 to obtain the desired estimates. $\qquad\qquad\square$

PROPOSITION 1. *Suppose (H1) and (H2) hold. Let $f \in L^2(\Omega)$. Then there exists a weak solution to $P(b, F)(f)$.*

In the proof, we use the following

LEMMA 2. *Let V be a reflexive separable Banach space, and A be an operator from V to its dual V'. Suppose that A is coercive, i.e, $\dfrac{\langle Av, v \rangle_{V',V}}{\|v\|_V}$ tends to $+\infty$ as $\|v\|_V$ tends to $+\infty$, and that A is continuous for the weak topologies of V and V'. Then A is surjective.*

This lemma can be proved as in [17] Chap. 2, Theorem 2.1, using Galerkin approximations. Also recall the De La Vallée Poussin Lemma:

LEMMA 3. *Let $\Omega \subset \mathbb{R}^N$ be of finite measure, $f_n \to f$ a.e. on Ω and $|f_n| \leq \mathcal{L}(|g_n|)$ for some sequence $(g_n)_n$ bounded in $L^q(\Omega)$, $1 \leq q < \infty$ and some function $\mathcal{L} : \mathbb{R}^+ \mapsto \mathbb{R}^+$ such that $\mathcal{L}(r)/r \to 0$ as $r \to +\infty$. Then $f_n \to f$ in $L^q(\Omega)$.*

The proof of this lemma consists in showing that the sequence $(|f_n|^q)_n$ is equi-integrable on Ω, and then applying the Egorov theorem.

Proof of Proposition 1. For $n \in \mathbb{N}$, set $F_k(z) = F(T_k(z))$, where $T_k(\cdot)$ is the cut-off function in (3), and set

$$b_k(z) = \begin{cases} z + k + b(-k), & z \leq -k, \\ b(z), & |z| < k, \\ z - k + b(k), & z \geq k. \end{cases}$$

Note that $F_k \to F$ and $b_k \to b$ uniformly on all compact subset of $\mathbb{R}$. In addition,

$$\min_{k \in \mathbb{N}} |b_k(z)| = \min_{|\zeta| \leq |z|} (|z| - |\zeta| + |b(\zeta)|) \geq \min\{|z|/2, \min_{|z|/2 \leq |\zeta| \leq |z|} |b(\zeta)|\},$$

which tends to infinity as $z \to \infty$, because b satisfies $(H2)$. Therefore $(H2)$ is satisfied by $b_k(\cdot)$ with a function $\beta(\cdot)$ independent of k; also $(H1)$ is satisfied by $b_k(\cdot)$, $F_k(\cdot)$ with c, δ independent of k. Furthermore, there exist $M_k > 0$, $R_k > 0$ such that $|F_k(z)| \leq M_k$ and $z b_k(z) \geq |z|^2 - R_k$, for all $z \in \mathbb{R}$.

Define $A_k : H^1(\Omega) \longrightarrow (H^1(\Omega))'$ by

$$\langle A_k(u), \xi \rangle_{(H^1(\Omega))', H^1(\Omega)} = \int_\Omega b_k(u)\,\xi + (\nabla u - F_k(u)) \cdot \nabla \xi$$

for $u, \xi \in H^1(\Omega)$; note that the integral in the right-hand side always makes sense.

It can be easily checked that A_k satisfy the hypotheses of Lemma 2. Hence there exists $u_k \in H^1(\Omega)$ such that

$$\int_\Omega b_k(u_k)\xi + (\nabla u_k - F_k(u_k)) \nabla \xi = \int_\Omega f \xi \quad \text{for all } \xi \in H^1(\Omega). \tag{7}$$

Since $u_k b_k(u_k) \in L^1(\Omega)$ and $F_k(u_k) \in L^2(\Omega)$, u_k is a weak solution to $P(b_k, F_k)(f)$. By Lemma 1, $\|u_k\|_{H^1(\Omega)}$ and $\|u_k b_k(u_k)\|_{L^1(\Omega)}$ are bounded by a constant independent of k. There exists $u \in H^1(\Omega)$ and a subsequence, which we still denote $(u_k)_k$, such that $\nabla u_k \to \nabla u$ weakly in $L^2(\Omega)$ and $u_k \to u$ a.e on Ω; moreover, by the Sobolev injection theorem, $(u_k)_k$ is bounded in $L^{2^*}(\Omega)$. Due to the hypothesis $(H1)$ and the uniform on all compact set convergence of F_k to F, we can apply Lemma 3 and obtain that $F_k(u_k) \to F(u)$ in $L^2(\Omega)$. Finally, note that $|b_k(z)| \leq |z| + |b(z)|$ for all z, k; therefore there exists a function $\mathcal{L}(\cdot)$ such that $\mathcal{L}(r)/r \to 0$ as $r \to +\infty$ and $|b_k(z)| \leq \mathcal{L}(z b_k(z))$ for all k, z. Hence by Lemma 3 we also have $b_k(u_k) \to b(u)$ in $L^1(\Omega)$.

Now for all $\xi \in H^1(\Omega) \cap L^\infty(\Omega)$, we can pass to the limit as $k \to \infty$ in (7) to obtain that u is a weak solution of $P(b, F)(f)$. $\qquad\square$

Now let us consider the case of bounded function f.

LEMMA 4. *Suppose (H1) and (H2) hold. Let $f \in L^{\infty}(\Omega)$ and u be a weak solution of $P(b, F)(f)$. Then $u \in L^{\infty}(\Omega)$ and there exists a constant C which depends only on Ω, β, $\|f\|_{L^{\infty}(\Omega)}$, c and δ such that $\|u\|_{L^{\infty}(\Omega)} \leq C$.*

Proof. We use the Moser iteration techniques.

CLAIM 1. Let $p \geq 2$. If $u \in L^p(\Omega)$, then $u \in L^{\gamma p}(\Omega)$, where $\gamma = 2^*/2 > 1$, and

$$\|u\|_{L^{\gamma p}(\Omega)} \leq (C_o p^{2+1/\delta})^{1/p}\|u\|_{L^p(\Omega)} \text{ in case } \|u\|_{L^p(\Omega)} \geq 1.$$

Take $T_k(|u|^{p-2}u)$ as a test function in (1), where T_k is the cut-off function defined in (6). Since $b(u)T_k(|u|^{p-2}u)$is non-negative, we get

$$\int_{\{|u|^{p-1}<k\}} |u|^{p-2}ub(u) + (p-1)\int_{\{|u|^{p-1}<k\}} |\nabla u|^2|u|^{p-2}$$

$$\leq (p-1)\int_{\{|u|^{p-1}<k\}} F(u)\cdot\nabla u|u|^{p-2} + \|f\|_{L^{\infty}(\Omega)}\int_{\Omega} |u|^{p-1}.$$

Since $F(u)\cdot\nabla u|u|^{p-2} \leq \frac{1}{2}|F(u)|^2|u|^{p-2} + \frac{1}{2}|\nabla u|^2|u|^{p-2}$, by $(H1)$ we have

$$\int_{\{|u|^{p-1}<k\}} |u|^{p-2}ub(u) + \frac{(p-1)}{2}\int_{\{|u|^{p-1}<k\}} |\nabla u|^2|u|^{p-2}$$

$$\leq C(p-1)\int_{\Omega} (|u|^{p-2} + |u|^{p-1} + |u|^p)$$

$$+ \frac{c}{2}(p-1)\int_{\{|u|^{p-1}<k\}} |u|^{p-2}(ub(u))^{1-\delta}.$$

Using the Young inequality for the last term, by the Fatou Lemma we deduce that $|\nabla u|^2|u|^{p-2} \in L^1(\Omega)$; since $p \geq 2$ and δ in $(H1)$ can be taken less than or equal to 1, we get

$$\frac{(p-1)}{2}\int_{\Omega} |\nabla u|^2|u|^{p-2} \leq C(p-1)^{1/\delta}\int_{\Omega} |u|^{p-2} + |u|^{p-1} + |u|^p, \tag{8}$$

where C is independent of p. Moreover, if $\|u\|_{L^p(\Omega)} \geq 1$, (8) implies that

$$\int_{\Omega} |\nabla u|^2|u|^{p-2} \leq Cp^{1/\delta}\int_{\Omega} |u|^p, \tag{9}$$

where C is still independent of p.

Note that $|\nabla u||u|^{\frac{p}{2}-1} = \frac{2}{p}|\nabla v|$, where $v = |u|^{\frac{p}{2}-1}u$. From (9) and the Sobolev injection $\|v\|^2_{L^{2^*}(\Omega)} \leq C(\|\nabla v\|^2_{L^2(\Omega)} + \|v\|^2_{L^2(\Omega)})$, it follows that $u \in L^{\gamma p}(\Omega)$ and $\|u\|^p_{L^{\gamma p}(\Omega)} = \|v\|^2_{L^{2^*}(\Omega)} \leq C_o p^{2+1/\delta}\|u\|^p_{L^p(\Omega)}$ in case $\|u\|_{L^p(\Omega)} \geq 1$. This proves Claim 1.

Now suppose that ess $\sup|u| > 1$ (otherwise, there is nothing to prove). Then there exists $p_o \geq 2$ such that $\|u\|_{L^p(\Omega)} \geq 1$ for all $p \geq p_o$. Iterating the estimate in Claim 1, we get

$$\|u\|_{L^{\gamma^k p_o}(\Omega)} \leq \prod_{n=0}^{\infty} (C_o p_o^{2+1/\delta} \gamma^{(2+1/\delta)n})^{1/(p_o \gamma^n)} \|u\|_{L^{p_o}(\Omega)}$$

for all $k \in \mathbb{N}$ large enough. Since in the right-hand side the infinite product converges, we have

$$\text{ess} \sup|u| = \liminf_{p \to \infty} \|u\|_{L^p(\Omega)} \leq \liminf_{k \to \infty} \|u\|_{L^{\gamma^k p_o}(\Omega)} < \infty.$$

Thus $u \in L^{\infty}(\Omega)$.

Finally, iterating more carefully the estimate (8), one can obtain that $\|u\|_{L^{\infty}(\Omega)}$ is bounded by a constant depending only on Ω, $\|f\|_{L^{\infty}(\Omega)}$, c, δ and $\|u\|_{L^2(\Omega)}$. Therefore Lemma 4 follows from Lemma 1. $\qquad\square$

It follows from a result of Lieberman [16] that any bounded weak solution of $P(b, F)(f)$ is regular up to the boundary provided $(H3)$ holds. Combining Proposition 1 with Lemma 4, we have

PROPOSITION 2. *Suppose $(H1)$, $(H2)$, $(H3)$ hold, and $\partial\Omega \in C^{1,\gamma}$, $\gamma > 0$. Let $f \in L^{\infty}(\Omega)$. Then there exists a weak solution to $P(b, F)(f)$ which belongs to $C^1(\overline{\Omega})$.*

3. The contraction principle for the stationary problem

In this section we prove Theorem 1. Since we use the Kruzhkov's idea of doubling of variables ([15]), let us remark that weak solution of $P(b, F)(f)$ satisfies the appropriate entropy formulation.

REMARK 2. Let u be a weak solution of $P(b, F)(f)$. Then for all $\xi \in H^1(\Omega) \cap L^{\infty}(\Omega)$, $\xi \geq 0$, for all $k \in \mathbb{R}$ one has

$$\int_{\Omega} \text{sign}\,(u - k)\,b(u)\,\xi + \int_{\Omega} \text{sign}\,(u - k)\{\nabla u - F(u) + F(k)\} \cdot \nabla \xi$$

$$\leq \int_{\Omega} \text{sign}\,(u - k)\,f\xi + \int_{\partial\Omega} \text{sign}\,(u - k)\,F(k) \cdot \nu\,\xi \qquad (10)$$

Proof. Let $\xi \in H^1(\Omega) \cap L^{\infty}(\Omega)$, $\xi \geq 0$. Take $H_\varepsilon(u - k)\xi \in H^1(\Omega) \cap L^{\infty}(\Omega)$ for the test function in (1). Adding the term

$$0 = \int_{\Omega} H_\varepsilon(u - k)\,F(k) \cdot \nabla \xi - \int_{\partial\Omega} H_\varepsilon(u - k)\,F(k) \cdot \nu\,\xi$$

$$+ \int_{\Omega} H_\varepsilon'(u - k)\,\nabla u \cdot F(k)\,\xi,$$

we obtain

$$\int_\Omega H_\varepsilon(u-k)\{b(u)-f\}\xi + \int_\Omega H_\varepsilon(u-k)\{\nabla u - F(u) + F(k)\}\cdot\nabla\xi$$

$$-\int_{\partial\Omega} H_\varepsilon(u-k)\,F(k)\cdot\nu\,\xi + \int_\Omega H_\varepsilon'(u-k)\,|\nabla u|^2\,\xi$$

$$-\int_\Omega H_\varepsilon'(u-k)(F(u)-F(k))\cdot\nabla u\,\xi = 0. \tag{11}$$

By Remark 1, the last term in (11) tends to zero as $\varepsilon \to 0$. Further, the fourth term in (11) is nonnegative. Finally, we pass to the limit in the first three terms using the Lebesgue dominated convergence theorem and obtain (10). $\qquad\square$

In the sequel, let $f,\,g \in L^1(\Omega)$ and $u,\,v$ be weak solutions of $P(b,F)(f)$ and $P(b,F)(g)$, respectively. For a function ξ of two variables $(x,y) \in \mathbb{R}^N \times \mathbb{R}^N$ we denote by $\nabla_x\xi$, $\nabla_y\xi$ its gradients with respect to x and y, respectively. We have the following basic lemma.

LEMMA 5. *Suppose $v \in L^\infty(\Omega)$. Then for all $\xi \in \mathcal{D}(\overline{\Omega} \times \overline{\Omega})$, $\xi|_{\overline{\Omega}\times\partial\Omega}=0$, $\xi \geq 0$, one has*

$$\iint_{\Omega\times\Omega} |b(u(x)) - b(v(y))|\,\xi + \iint_{\Omega\times\Omega} \mathrm{sign}\,(u(x)-v(y))$$

$$\times\{\nabla u(x) - \nabla v(y) - F(u(x)) + F(v(y))\}\cdot(\nabla_x\xi + \nabla_y\xi)$$

$$\leq \iint_{\Omega\times\Omega} \mathrm{sign}\,(u(x)-v(y))(f(x)-g(y))\,\xi$$

$$-\iint_{\partial\Omega\times\Omega} \mathrm{sign}\,(u(x)-v(y))\{\nabla v(y) - F(v(y))\}\cdot\nu_x\,\xi, \tag{12}$$

where for $(x,y) \in \partial\Omega \times \Omega$, ν_x is the exterior unit normal vector to $\partial\Omega$ at the point x.

Proof. Take $\xi \in \mathcal{D}(\overline{\Omega} \times \overline{\Omega})$, $\xi \geq 0$ such that $\xi|_{\overline{\Omega}\times\partial\Omega} = 0$. The function u is a weak solution of $P(b,F)(f)$; in particular, for a.e. $y \in \Omega$, u satisfies (1) with the test function $H_\varepsilon(u(\cdot) - v(y))\,\xi\,(\cdot,y) \in H^1(\Omega)\cap L^\infty(\Omega)$. Integrating in $y \in \Omega$, we obtain

$$\iint_{\Omega\times\Omega} H_\varepsilon(u(x)-v(y))\{b(u(x)) - f(x)\}\,\xi$$

$$+ \iint_{\Omega\times\Omega} H_\varepsilon(u(x)-v(y))\{\nabla u(x) - F(u(x))\}\cdot\nabla_x\xi$$

$$+ \iint_{\Omega\times\Omega} H_\varepsilon'(u(x)-v(y))|\nabla u(x)|^2\,\xi$$

$$-\iint_{\Omega\times\Omega} H_\varepsilon'(u(x)-v(y))F(u(x))\cdot\nabla u(x)\,\xi = 0. \tag{13}$$

In addition, we have by the Green-Gauss formula

$$\iint_{\Omega\times\Omega} H_\varepsilon(u(x)-v(y))(\nabla u(x) - F(u(x))) \cdot \nabla_y \xi$$

$$- \iint_{\Omega\times\Omega} H_\varepsilon{}'(u(x)) - v(y))\nabla u(x) \cdot \nabla v(y)\,\xi$$

$$+ \iint_{\Omega\times\Omega} H_\varepsilon{}'(u(x)-v(y))\,F(u(x)) \cdot \nabla v(y)\,\xi$$

$$= \iint_{\Omega\times\partial\Omega} H_\varepsilon(u(x)-v(y))(\nabla u(x) - F(u(x))) \cdot v_y\,\xi. \tag{14}$$

Due to the special choice of the test function ξ, the right-hand side of (14) is zero. In (13) and (14), we can substitute f by g, exchange u with v and x with y. We obtain two equalities, which we add to (13) and (14) and rearrange the terms.

This yields

$$\iint_{\Omega\times\Omega} H_\varepsilon(u(x)-v(y))\{b(u(x))-b(v(y)) - f(x) + g(y)\}\xi$$

$$+ \iint_{\Omega\times\Omega} H_\varepsilon(u(x)-v(y))\{\nabla u(x) - \nabla v(y) - F(u(x)) + F(v(y))\} \cdot (\nabla_x\xi+\nabla_y\xi)$$

$$+ \iint_{\partial\Omega\times\Omega} H_\varepsilon(u(x)-v(y))(\nabla v(y) - F(v(y))) \cdot v_x\,\xi$$

$$= \iint_{\Omega\times\Omega} H_\varepsilon{}'(u(x)-v(y))\{-|\nabla u(x)|^2 - |\nabla v(y)|^2 + 2\nabla u(x) \cdot \nabla v(y)\}\,\xi$$

$$+ \iint_{\Omega\times\Omega} H_\varepsilon{}'(u(x)-v(y))(F(u(x)) \quad F(v(y))) \cdot \nabla u(x)\,\xi$$

$$- \iint_{\Omega\times\Omega} H_\varepsilon{}'(u(x)-v(y))(F(u(x))-F(v(y))) \cdot \nabla v(y)\,\xi. \tag{15}$$

The first term in the right-hand side of (15) is negative or zero. Further, we have required that $\|v\|_{L^\infty(\Omega)} \leq +\infty$. Proceeding for a.e. $y \in \Omega$ as in the proof of Remark 1, we get

$$\left|\iint_{\Omega\times\Omega} H_\varepsilon{}'(u(x)-v(y))(F(u(x))-F(v(y))) \cdot \nabla u(x)\,\xi\right|$$

$$\leq \int_{y\in\Omega}\left(\int_{x\in\Omega} |\nabla_x\xi| + \int_{x\in\partial\Omega} |\xi|\right) 2\,\omega(\varepsilon) \to 0$$

as $\varepsilon \to 0$, where $\omega(\varepsilon) = \sup_{r,s\in[-M,M],|r-s|\leq\varepsilon} |F(r) - F(s)|$ and $M = \|v\|_{L^\infty(\Omega)} + \varepsilon$. Similarly, the last term in (15) vanishes as $\varepsilon \to 0$. Now we pass to the limit as $\varepsilon \to 0$ in the left-hand side of (15), using the Lebesgue dominated convergence theorem. Hence (12) follows. $\qquad\square$

Now we will make y tend to x. Note that heuristically, the last term in (12) tends to zero, due to the Neumann boundary condition on the solution v. We can justify it requiring additional regularity of v.

PROPOSITION 3. *Let $\partial\Omega \in C^2$ and suppose $v \in C^1(\overline{\Omega})$. Then for all $\eta \in \mathcal{D}(\overline{\Omega})$, $\eta \geq 0$,*

$$\int_\Omega |b(u) - b(v)|\eta + \int_\Omega \text{sign}\,(u - v)\{\nabla u - \nabla v - F(u) + F(v)\} \cdot \nabla\eta$$

$$\leq \int_\Omega \text{sign}\,(u - v)(f - g)\,\eta + \int_{\{u=v\}} |f - g|\,\eta. \tag{16}$$

Note that the right-hand side of (16) is exactly the bracket in $L^1(\Omega)$ between $(u - v)$ and $(f - g)\eta$. For $w, h \in L^1(\Omega)$, the bracket is defined by

$$[w, h]_{L^1(\Omega)} = \int_\Omega \text{sign}\,(w)\,h + \int_{\{w=0\}} |h| \tag{17}$$

(cf. e.g. [4]). For the bracket in $\mathbb{R}^N$, we just use the notation $[\cdot, \cdot]$.

For $n \in \mathbb{N}$ and any $m \in \mathbb{N}$, we will abusively use the notation $\delta_n(s)$ for $n^m \prod_{i=1}^m \delta_1(ns_i)$ with $s = (s_1, \ldots, s_m) \in \mathbb{R}^m$, where $\delta_1 \in C_0^\infty([-1, 1], \mathbb{R})$ is a nonnegative even function such that $\int_\mathbb{R} \delta_1(r)\,dr = 1$.

LEMMA 6. *Let for all $s \in \mathbb{R}^N$, $w_n(\cdot, s) \to w(\cdot)$ and $h_n(\cdot, s) \to h(\cdot)$ in $L^1(\mathbb{R}^N)$ as $n \to \infty$. If in addition $\|h_n(\cdot, s)\|_{L^1(\mathbb{R}^N)}$ is bounded uniformly in n and s, then*

$$\limsup_{n\to\infty} \iint_{\mathbb{R}^N \times \mathbb{R}^N} \text{sign}\,w_n(x, s)\,h_n(x, s)\delta_1(s) \leq [w, h].$$

Moreover, if for all $n \in \mathbb{N}$ and a.e. $s \in \mathbb{R}^N$ $h_n(\cdot, s) = 0$ a.e. on $\{w_n(\cdot, s) = 0\}$ and if $h = 0$ a.e. on $\{w = 0\}$, then there exists

$$\lim_{n\to\infty} \iint_{\mathbb{R}^N \times \mathbb{R}^N} \text{sign}\,w_n(x, s)\,h_n(x, s)\delta_1(s) = \int_{\mathbb{R}^N} \text{sign}\,(w)\,h.$$

Proof. The first claim follows from the definition and the upper semicontinuity of the bracket (cf. [4]), the definition of δ_1 and the Fatou lemma. The second claim follows by applying the first one to w_n, h_n and to $-w_n$, h_n. $\square$

It is well known that for $g \in L^1(\mathbb{R}^N)$, $g(x + \sigma_n) \to g(x)$ in $L^1(\mathbb{R}^N)$, if $\sigma_n \to 0$ as $n \to +\infty$. We need the following generalisation of this property:

LEMMA 7. *Let $g \in L^1(\mathbb{R}^N)$ and $g_n(x', x_N) = g(x' + \sigma_n, x_N + \phi_n(x'))$, where $\sigma_n \in \mathbb{R}^{N-1}$, $\phi_n : \mathbb{R}^{N-1} \mapsto \mathbb{R}$ are such that $|\sigma_n| \to 0$ and $\phi_n(x' - \sigma_n) \to 0$ as $n \to \infty$, for a.e. $x' \in \mathbb{R}^{N-1}$. Then $\|g_n\|_{L^1(\mathbb{R}^N)} = \|g\|_{L^1(\mathbb{R}^N)}$ for all n, and $g_n \to g$ in $L^1(\mathbb{R}^N)$ as $n \to \infty$.*

Proof. Setting $z = x' + \sigma_n$, we have

$$\iint_{\mathbb{R}^{N-1}\times\mathbb{R}} |g - g_n| \leq \iint_{\mathbb{R}^{N-1}\times\mathbb{R}} |g(x', x_N) - g(x' + \sigma_n, x_N)|$$

$$+ \iint_{\mathbb{R}^{N-1}\times\mathbb{R}} |g(z, x_N) - g(z, x_N + \phi_n(z - \sigma_n))| = I_n^1 + I_n^2.$$

Clearly, $I_n^1 \to 0$ as $n \to \infty$. Also $I_n^2 \to 0$ as $n \to \infty$ by the Lebesgue dominated convergence theorem. Indeed, $I_n^2 = \int_{\mathbb{R}^{N-1}} \Delta_n(z)$, where

$$\Delta_n(z) = \int_{\mathbb{R}} |g(z, x_N) - g(z, x_N + \phi_n(z - \sigma_n))|.$$

One has $\Delta_n(z) \leq 2 \int_{\mathbb{R}} |g(z, x_N)| \in L^1(\mathbb{R}^{N-1})$ and $\Delta_n(z) \to 0$ as $n \to \infty$ for a.e. $z \in \mathbb{R}^{N-1}$, by the standard translation property. $\square$

Proof of Proposition 3. Take $\eta \in \mathcal{D}(\overline{\Omega})$, $\eta \geq 0$. We construct a sequence $\xi_n \in \mathcal{D}(\overline{\Omega} \times \Omega)$ of test functions in (12) as follows.

Since $\partial\Omega$ is of class C^2, we can cover $\partial\Omega$ by the union of open sets $\mathcal{O}_1, \ldots, \mathcal{O}_l \subset \mathbb{R}^N$ such that, upon rotating and relabeling the coordinate axes for each $i = 1, \ldots, l$, there exist $\Phi_i \in C_0^2(\mathbb{R}^{N-1}, \mathbb{R})$ such that $\partial\Omega \cap \mathcal{O}_i = \{x = (x', x_N) \in \mathcal{O}_i, x_N = \Phi_i(x')\}$ and $\Omega \cap \mathcal{O}_i = \{x = (x', x_N) \in \mathcal{O}_i, x_N > \Phi_i(x')\}$. In addition, there exists an open set $\mathcal{O}_0 \subset \Omega$ such that $\overline{\Omega} \subset \cup_{i=0}^l \mathcal{O}_i$. Take nonnegative functions $\alpha_0, \ldots, \alpha_l \in \mathcal{D}(\mathbb{R}^N, \mathbb{R})$ such that $\operatorname{supp} \alpha_i \subset \mathcal{O}_i$ for $i = 0, \ldots, l$ and $\sum_{i=0}^l \alpha_i(x) = 1$ for all $x \in \overline{\Omega}$.

Let $x, y \in \mathbb{R}^N$; set $\rho_n^0(x, y) = \delta_n(x - y)$, and for $i = 1, \ldots, l$, set

$$\rho_n^i(x, y) = \delta_n(x' - y')\delta_n\left(x_N - \Phi_i(x') - y_N + \Phi_i(y') + \frac{2}{n}\right).$$

Here and below we abusively keep the same notation for different coordinate systems corresponding to $i = 0, \ldots, l$. Take

$$\xi_n(x, y) = \sum_{i=0}^l \eta(x)\alpha_i(x)\rho_n^i(x, y). \tag{18}$$

CLAIM 1. $\xi_n \in \mathcal{D}(\overline{\Omega} \times \overline{\Omega})$, $\xi_n \geq 0$ and $\xi_n = 0$ on $\overline{\Omega} \times \partial\Omega$ for all n sufficiently large.

The first two properties are evident. Next, observe that there exists $d > 0$ such that for $i = 0, \ldots, l$ one has $\rho_n^i(x, y) = 0$ for $|x - y| > d/n$. Let $n > d \times \min_{i=0,\ldots,l}$ dist $(\operatorname{supp} \alpha_i, \mathbb{R}^N \backslash \mathcal{O}_i)$. We have $\rho_n^i(x, y) = 0$ whenever $x \in \operatorname{supp} \alpha_i$ and $y \notin \mathcal{O}_i$. In particular, for $(x, y) \in \overline{\Omega} \times \partial\Omega$ the sum in (18) reduces to the sum over $i \in \{1, \ldots, l\}$ such that both x and y belong to $\mathcal{O}_i$. But in this case, $x_N - \Phi_i(x') \geq 0$ and $y_N - \Phi_i(y') = 0$. Thus $\rho_n^i(x, y) = 0$, which ends the proof of Claim 1.

Denote $\eta(x)\alpha_i(x)$ by $\eta_i(x)$ for $i = 0, \dots, l$. Taking $\xi = \xi_n$ in (12), we get

$$\sum_{i=0}^{l} I_n^i + \sum_{i=1}^{l} J_n^i \leq \sum_{i=0}^{l} K_n^i + \sum_{i=0}^{l} L_n^i, \tag{19}$$

where

- for $i = 0, \dots, l$,

$$I_n^i = \iint_{\Omega \times \Omega} \text{sign}\,(u(x) - v(y))\{(b(u(x)) - b(v(y)))\eta_i(x) \\ + (\nabla u(x) - \nabla v(y) - F(u(x)) + F(v(y))) \cdot \nabla \eta_i(x)\}\, \rho_n^i(x, y);$$

- for $i = 1, \dots, l$,

$$J_n^i = \iint_{\Omega \times \Omega} \text{sign}\,(u(x) - v(y))\{(\nabla u(x) - \nabla v(y) - F(u(x)) + F(v(y)))\eta_i(x)\} \\ \cdot (\nabla_x + \nabla_y)\rho_n^i(x, y);$$

- for $i = 0, \dots, l$,

$$K_n^i = \iint_{\Omega \times \Omega} \text{sign}\,(u(x) - v(y))\{(f(x) - g(y))\eta_i(x)\}\, \rho_n^i(x, y);$$

- for $i = 0, \dots, l$,

$$L_n^i = \iint_{\Omega \times \partial\Omega} \text{sign}\,(u(x) - v(y))\{\eta_i(x)(\nabla v(y) - F(v(y))) \cdot \nu_x\}\, \rho_n^i(x, y).$$

In I_n^i, J_n^i, K_n^i we extend the domain of integration to $\mathbb{R}^N \times \mathbb{R}^N$ by formally setting u, v, ∇u, ∇v, f, g to be zero outside Ω.

In I_n^0, let us perform the change of variables $(x, y) \to (x, s)$, where $s = n(x - y)$. We obtain

$$I_n^0 = \iint_{\Omega \times \Omega} \text{sign}\, w_n(x, s)\, h_n(x, s)\delta_1(s)$$

with $w_n(x, s) = u(x) - v(x - \frac{s}{n})$ and

$$h_n(x, s) = \left(b(u(x)) - b\left(v\left(x - \frac{s}{n}\right)\right)\right)\eta_0(x) \\ + \left\{\nabla u(x) - \nabla v\left(x - \frac{s}{n}\right) - F(u(x)) + F\left(v\left(x - \frac{s}{n}\right)\right)\right\} \cdot \nabla \eta_0(x).$$

Note that for all $s \in \mathbb{R}^N$, $h_n(\cdot, s) = 0$ a.e. on $\{w_n(\cdot, s) = 0\}$. Hence by Lemma 6, there exists

$$\lim_{n \to \infty} I_n^0 = \int_{\Omega} \text{sign}\,(u - v)\{(b(u) - b(v))\eta_0 + (\nabla u - \nabla v - F(u) + F(v)) \cdot \nabla \eta_0\}. \tag{20}$$

Similarly, we get

$$\lim_{n\to\infty} K_n^0 \leq [u - v, (f - g)\eta_0]. \tag{21}$$

For $i = 1, \ldots, l$ we use the change of variables $(x, y) = (x', x_N, y', y_N) \to (x, s) = (x', x_N, s', s_N)$, where $s' = n(x' - y')$ and $s_N = n(x_N - \Phi_i(x') - y_N + \Phi_i(y') + \frac{2}{n})$. It yields

$$K_n^i = \iint_{\Omega\times\Omega} \text{sign } w_n(x, s)\, h_n(x, s)\delta_1(s)$$

with

$$w_n(x, s) = u(x', x_N) - v\left(x' - \frac{s'}{n}, x_N - \frac{s_N}{n} - \Phi_i(x') + \Phi_i\left(x' - \frac{s'}{n}\right) + \frac{2}{n}\right)$$

and

$$h_n(x, s) = \left\{f(x', x_N) - g\left(x' - \frac{s'}{n}, x_N - \frac{s_N}{n} - \Phi_i(x') + \Phi_i\left(x' - \frac{s'}{n}\right) + \frac{2}{n}\right)\right\}\eta_i(x', x_N).$$

Fix $s \in \mathbb{R}^N$; it follows by Lemma 7 that $w_n(\cdot, s) \to u - v$ and $h_n(\cdot, s) \to f - g$ in $L^1(\mathbb{R}^N)$ as $n \to \infty$ and $\|h_n(\cdot, s)\|_{L^1(\mathbb{R}^N)} = \|f - g\|_{L^1(\mathbb{R}^N)}$. Hence we can apply Lemma 6 to obtain

$$\limsup_{n\to\infty} K_n^i \leq [u - v, (f - g)\eta_i]. \tag{22}$$

Similarly, we show that there exists

$$\lim_{n\to\infty} I_n^i = \int_\Omega \text{sign}\, (u - v)\{(b(u) - b(v))\eta_i + (\nabla u - \nabla v - F(u) + F(v)) \cdot \nabla \eta_i\}. \tag{23}$$

CLAIM 2. There exists $\lim_{n\to\infty} J_n^i = 0$ for $i = 1, \ldots, l$.

For simplicity, let us treat only the first of $(N - 1)$ terms in the scalar product in J_n^i (remark that the Nth term is zero). This term equals

$$J_n^{i,1} = -\iint_{\Omega\times\Omega} \text{sign}\, (u(x) - v(y))\{(\partial_1 u(x) - \partial_1 v(y) - F_1(u(x)) + F_1(v(y)))\eta_i(x)\}$$

$$\times \delta_n(x' - y')\delta_n'\left(x_N - \Phi_i(x') - y_N + \Phi_i(y') + \frac{2}{n}\right)(\partial_1\Phi_i(x) - \partial_1\Phi_i(y)),$$

where ∂_1 means the derivation with respect to the first variable, and F_1 means the first component of F. Note that for all x', y' there exists ξ' such that $\partial_1\Phi_i(x') - \partial_1\Phi_i(y') = \partial_1(\nabla\Phi) \cdot (x' - y')$ and $|\xi' - x'| \leq |y' - x'|$. With the same change of variables as above, we calculate that

$$J_n^{i,1} = -\int \delta_1'(s_N)\delta_1(s')\, s' \cdot R_n(s', s_N),$$

where

$$R_n(s', s_N) = \int_{x \in \mathbb{R}^N}$$

$$\cdot \operatorname{sign}\left(u(x', x_N) - v\left(x' - \frac{s'}{n}, x_N - \frac{s_N}{n} - \Phi_i(x') + \Phi_i\left(x - \frac{s}{n}\right) + \frac{2}{n}\right)\right)$$

$$\times \left\{\left(\partial_1 u(x', x_N) - \partial_1 v\left(x' - \frac{s'}{n}, x_N - \frac{s_N}{n} - \Phi_i(x') + \Phi_i\left(x' - \frac{s'}{n}\right) + \frac{2}{n}\right)\right.\right.$$

$$-F_1(u(x', x_N)) + F_1\left(v\left(x' - \frac{s'}{n}, x_N - \frac{s_N}{n} - \Phi_i(x')\right.\right.$$

$$\left.\left.\left.+ \Phi_i\left(x' - \frac{s'}{n}\right) + \frac{2}{n}\right)\right)\right) \eta_i(x', x_N)\right\} \partial_1(\nabla\Phi_i)(\xi').$$

As in the proof of Lemma 6, it follows by Lemma 7 and the continuity of $\partial_1(\nabla\Phi_i)$ that $R_n(s', s_N)$ converges in $L^1(\mathbb{R}^N, \mathbb{R}^{N-1})$ to the bracket

$$[u - v, (\partial_1 u - \partial_1 v - F_1(u) + F_1(v))\eta_i \partial_1(\nabla\Phi_i)],$$

which does not depend on s. Hence $J_n^{i,1} \to 0$ because $\int \delta_1'(s_N)\delta_1(s')s' = 0$. This proves Claim 2.

Finally, due to the fact that $v \in C^1(\overline{\Omega})$, we have

CLAIM 3. There exists $\lim_{n \to \infty} L_n^i = 0$ for $i = 0, \ldots, l$.

Indeed, since the function $\nabla v - F(v)$ is continuous on $\overline{\Omega}$, it is easy to show that the Neumann condition $(\nabla v - F(v)) \cdot v = 0$ is satisfied pointwise on $\partial\Omega$. Set $\omega(\varepsilon) = \sup_{x,y \in \overline{\Omega}, |x-y| \le \varepsilon} |\nabla v(y) - F(v(y)) - \nabla v(x) + F(v(x))|$. We have

$$|L_n^i| = \left|\iint_{\partial\Omega \times \Omega} \{(\nabla v(y) - F(v(y))) \cdot v_x - (\nabla v(x) - F(v(x))) \cdot v_x\}\eta_i(x)\rho_n^i(x, y)\right|$$

$$\le \max_{x \in \partial\Omega} |\eta_i(x)| \int_{x \in \partial\Omega} \int_{y \in \Omega} \omega(|x - y|)\rho_n^i(x, y) \le \max_{x \in \partial\Omega} |\eta_i(x)| \omega\left(\frac{d}{n}\right) \operatorname{meas}(\partial\Omega),$$

where $\operatorname{meas}(\partial\Omega)$ is the $(N - 1)$–dimensional Hausdorff measure of $\partial\Omega$, and the constant d is introduced in the proof of Claim 1. Thus Claim 3 follows.

Now since $\eta = \sum_{i=0}^{l} \eta_i$, (16) follows from Claim 2, Claim 3 and (20)–(23). $\qquad\square$

Proof of Theorem 1. Let $f, g \in L^1(\Omega)$. There exists a sequence $(g_n)_n \subset L^\infty(\Omega)$ which converges to g in $L^1(\Omega)$. By virtue of Proposition 2, for all n there exists $v_n \in C^1(\overline{\Omega})$ a weak solution of $P(b, F)(g_n)$. Applying Proposition 3 with the test function $\eta \equiv 1$ on Ω to u, v_n and to v, v_n, we obtain

$$\int_\Omega |b(u) - b(v)| \leq \int_\Omega |b(u) - b(v_n)| + \int_\Omega |b(v) - b(v_n)|$$

$$\leq \int_\Omega |f - g_n| + \int_\Omega |g - g_n|.$$

At the limit as $n \to \infty$, we get (2).　　　　$\square$

Finally we can state the well-posedness result for $P(b, F)(f)$.

COROLLARY 1. *Assume* $\partial\Omega \in C^2$, $(H1)$, $(H2)$ *and* $(H3)$ *hold. Then*

(i) *for all* $f \in L^2(\Omega)$, *there exists a unique* $b(u)$ *such that* u *is a weak (entropy) solution of* $P(b, F)(f)$;

(ii) *the corresponding mapping* $f \mapsto b(u)$ *is a contraction for the* L^1 *norm on* Ω.

4. Uniqueness for the evolution problem

In this section we prove Theorem 2, applying the nonlinear semigroup theory.

Let us define the (possibly multivalued) operator $A_{b,F} : L^1(\Omega) \longrightarrow L^1(\Omega)$ by its graph:

$$(\beta, g) \in A_{b,F} \iff \begin{cases} \text{there exists } v \in C^1(\overline{\Omega}) \text{ such that } \beta = b(v) \text{ and} \\ v \text{ is a weak solution of } P(b, F)(g + \beta). \end{cases} \tag{24}$$

For an operator $A : L^1(\Omega) \longrightarrow L^1(\Omega)$, denote by $R(A)$ its range, by $D(A)$ its domain and by $\overline{R(A)}$, $\overline{D(A)}$ their closures in $L^1(\Omega)$ respectively. Recall (cf. [4]) that A is accretive if

$$[\beta - \hat{\beta}, g - \hat{g}]_{L^1(\Omega)} \geq 0, \quad \text{for all } (\beta, g), (\hat{\beta}, \hat{g}) \in A,$$

where the bracket $[\cdot, \cdot]_{L^1(\Omega)}$ is given by (17).

If A is accretive and $R(I + \lambda A) = L^1(\Omega)$ for some $\lambda > 0$, then A is m-accretive.

PROPOSITION 4. *Suppose* $(H1)$, $(H2)$, $(H3)$, $(H4)$ *hold and* $\partial\Omega \in C^2$. *Then*

(i) $A_{b,F}$ *is accretive in* $L^1(\Omega)$

(ii) $R(I + \lambda A_{b,F}) \supset L^\infty(\Omega)$ *for all* λ *sufficiently small;*

(iii) $\overline{D(A_{b,F})} = L^1(\Omega)$.

Proof. (i) Let (β, g), $(\hat{\beta}, \hat{g}) \in A_{b,F}$. There exist v, $\hat{v} \in C^1(\overline{\Omega})$ such that $\beta = b(v)$, $\hat{\beta} = b(\hat{v})$, and v, $\hat{v}$ are weak solutions to $P(b, F)(g + \beta)$, $P(b, F)(\hat{g} + \hat{\beta})$ respectively. Applying Proposition 3 with $\eta \equiv 1$ on Ω, we have

$$\|b(v) - b(\hat{v})\|_{L^1(\Omega)} \leq [v - \hat{v}, (g - \hat{g}) + b(v) - b(\hat{v})]_{L^1(\Omega)}.$$

Using (17), we deduce that $[b(v) - b(\hat{v}), g - \hat{g}]_{L^1(\Omega)} \geq [v - \hat{v}, g - \hat{g}]_{L^1(\Omega)} \geq 0$.

(ii) For $g \in L^\infty(\Omega)$ and $\lambda > 0$, consider the problem $P(\frac{b}{\lambda}, F)(\frac{g}{\lambda})$. By Proposition 2, there exists $u_\lambda \in C^1(\overline{\Omega})$ a weak solution to $P(\frac{b}{\lambda}, F)(\frac{g}{\lambda})$. By (24), $(b(u_\lambda), \frac{g - b(u_\lambda)}{\lambda}) \in A_{b,F}$. Hence $g \in R(I + \lambda A_{b,F})$.

(iii) First take $g \in L^\infty(\Omega)$. By $(H2)$, there exists $u \in L^\infty(\Omega)$ such that $g = b(u)$. One can approach u a.e on Ω by a bounded in $L^\infty(\Omega)$ sequence of $H^1(\Omega)$ functions. Therefore the set $\widetilde{D} = \{g \in L^\infty(\Omega), \ g = b(u), \ u \in H^1(\Omega) \cap L^\infty(\Omega)\}$ is dense in $L^\infty(\Omega)$ for the L^1norm; hence it is dense in $L^1(\Omega)$.

It remains to show that $D(A_{b,F})$ is dense in $\widetilde{D}$. We have already shown that for all $g \in \widetilde{D}$ there exists a weak solution u_λ to $P(\frac{b}{\lambda}, F)(\frac{g}{\lambda})$, and $b(u_\lambda) \in D(A_{b,F})$.

In the sequel, C denotes a constant independent of λ.

CLAIM 1. for all λ sufficiently small, $\int_\Omega u_\lambda \, b(u_\lambda) \leq C$.

Here we can relax $(H4)$ by taking $\delta = 0$. We argue as in the proof of Lemma 1, using in addition that $|g \, u_\lambda| \leq C + \frac{1}{2} u_\lambda b(u_\lambda)$ since $g \in L^\infty(\Omega)$ and $(H2)$ holds. For all λ sufficiently small, we get $\|u_\lambda\|_{L^{2^*}(\Omega)} \leq C/\sqrt{\lambda}$ and, finally, $\int_\Omega u_\lambda b(u_\lambda) \leq C$.

CLAIM 2. for all λ sufficiently small, $\|u_\lambda\|_{H^1(\Omega)} \leq C$.

Take $(u_\lambda - u) \in H^1(\Omega) \cap L^\infty(\Omega)$ as a test function for $P(\frac{b}{\lambda}, F)(\frac{g}{\lambda})$. Since $g = b(u)$, we get

$$\frac{1}{\lambda} \int_\Omega (b(u_\lambda) - b(u))(u_\lambda - u) + \int_\Omega (\nabla u_\lambda - F(u_\lambda)) \cdot \nabla(u_\lambda - u) = 0.$$

The first term is non-negative; by the Young inequality, we deduce that

$$\int_\Omega |\nabla(u_\lambda - u)|^2 \leq \int_\Omega |F(u_\lambda) - \nabla u|^2.$$

Using $(H4)$ and Claim 1, we obtain that

$$\int_\Omega |\nabla u_\lambda|^2 \leq C \left(1 + \int_\Omega |u_\lambda|^{2(1-\delta)} \right) \tag{25}$$

for all λ sufficiently small. By Claim 1 and $(H2)$, we also have $\int_\Omega |u_\lambda| \leq C$. Substitute (25) in the Sobolev inequality $\int_\Omega |u_\lambda|^2 \leq C(\int_\Omega |\nabla u_\lambda|^2 + (\int_\Omega |u_\lambda|)^2)$; by the Young inequality we deduce $\int_\Omega |u_\lambda|^2 \leq C$. Now Claim 2 follows from (25).

By Claim 2, there exist $v \in L^2(\Omega)$ and a sequence $\lambda_m \to 0$ such that $u_{\lambda_m} \to v$ a.e. on Ω. Hence $b(u_{\lambda_m}) \to b(v)$ a.e. on Ω. By Claim 1 and Lemma 3 we get $b(u_{\lambda_m}) \to b(v)$ in $L^1(\Omega)$. Moreover, by Claim 1, Claim 2 and $(H4)$, $\lambda(\nabla u_\lambda - F(u_\lambda)) \to 0$ in $L^2(\Omega)$ as $\lambda \to 0$. Taking $\xi \in \mathcal{D}(\Omega)$ as a test function for $P(\frac{b}{\lambda}, F)(\frac{g}{\lambda})$, and passing to the limit as $\lambda \to 0$, we obtain that $b(u_\lambda) \to g$ in $\mathcal{D}'(\Omega)$. Hence $b(u_{\lambda_m}) \to g$ in $L^1(\Omega)$, which concludes the proof. $\qquad\square$

DEFINITION 3. Let $f \in L^1(Q)$, $b^o \in L^1(\Omega)$. A function $w \in L^1(Q)$ is an integral solution of the problem

$$w_t + A_{b,F}(w) \ni f, \ w(t=0) = b^o, \quad S(b, F)(f, b^o)$$

if there exists ess $\lim_{t \to 0} w(t) = b^o$ in $L^1(\Omega)$ and for all $(\beta, g) \in A_{b,F}$

$$\frac{d}{dt}\|w(t) - \beta\|_{L^1(\Omega)} \le [w(t) - \beta, f(t) - g]_{L^1(\Omega)} \text{ in } \mathcal{D}'(0, T). \tag{26}$$

By Proposition 4, the closure of the operator $A_{b,F}$ is m-accretive densely defined in $L^1(\Omega)$. By the general theory of non-linear semigroups (cf. [12], [8], [4] and [3], Theorem 4), we have the following result:

COROLLARY 2. *Suppose* $(H1)$, $(H2)$, $(H3)$, $(H4)$ *hold and* $\partial\Omega \in C^2$. *Then for all* $f \in L^1(Q)$, $b^o \in L^1(\Omega)$, *there exists a unique integral solution of* $S(b, F)(f, b^o)$. *Moreover, if* $\hat{f} \in L^1(Q)$, $b^o \in L^1(\Omega)$ *and* w, $\hat{w}$ *are integral solutions of* $S(b, F)(f, b^o)$, $S(b, F)(\hat{f}, \hat{b}^o)$, *respectively, then for a.e.* $t \in (0, T)$

$$\|w(t) - \hat{w}(t)\|_{L^1(\Omega)} \le \|b^o - \hat{b}^o\|_{L^1(\Omega)} + \int_0^t \|f(\tau) - \hat{f}(\tau)\|_{L^1(\Omega)} d\tau.$$

Now Theorem 2 follows immediately from (iii) of Proposition 4 and

PROPOSITION 5. *Suppose* $\partial\Omega \in C^2$. *Let* $f \in L^1(Q)$, $b^o \in \overline{D(A_{b,F})}$. *If* u *is a weak solution of* $E(b, F)(f, b^o)$, *then* $w = b(u)$ *is an integral solution of* $S(b, F)(f, b^o)$.

Proof. We first argue as in [7], Theorem 4.4.

Let us define the functions

$$B_{\varepsilon,k} : r \in \mathbb{R} \mapsto \int_k^r H_\varepsilon(r-k) \, db(s)$$

for all $\varepsilon > 0, k \in \mathbb{R}$.

Note that for all $r, \hat{r} \in \mathbb{R}$,

$$(b(r) - b(\hat{r}))H_\varepsilon(r - k) \ge B_{\varepsilon,k}(r) - B_{\varepsilon,k}(\hat{r}), \tag{27}$$

and for all $r \in \mathbb{R}$,

$$\begin{cases} B_{\varepsilon,k}(r) \to |b(r) - b(k)| \text{ as } \varepsilon \to 0, \\ 0 \le B_{\varepsilon,k}(r) \le |b(r) - b(k)|. \end{cases} \tag{28}$$

Take u a weak solution of $E(b, F)(f, b^o)$; we extend u by u^o for $t \le 0$, where u^o is a measurable function such that $b^o = b(u^o)$.

Take positive functions $\eta \in \mathcal{D}(\Omega)$, $\mu \in \mathcal{D}(-\infty, T)$, and $h > 0$ such that $\operatorname{supp}\mu \subset (-\infty, T-h)$. Set $\phi = H_\varepsilon(u-k)\eta\mu$ and take $\phi^h = \frac{1}{h}\int_t^{t+h}\phi(s)\,ds$ as a test function in (3). For the first term, using the change of variable $t \to t+h$ and (27), we get

$$\int_0^T\int_\Omega (b^o - b(u))\phi_t^h = \int_0^T\int_\Omega (b^o - b(u))\frac{\phi(t+h)-\phi(t)}{h}$$

$$= \frac{1}{h}\int_0^T\int_\Omega (b(u(t-h))-b(u(t)) \times H_\varepsilon(u(t)-k)\mu(t)\eta$$

$$\geq \frac{1}{h}\int_0^T\int_\Omega (B_{\varepsilon,k}(u(t)) - B_{\varepsilon,k}(u(t-h)))\mu(t)\eta.$$

Using the inverse change of variable, we obtain

$$\int_0^T\int_\Omega (b^o - b(u))\phi_t^h \geq \int_0^T\int_\Omega (B_{\varepsilon,k}(u^o) - B_{\varepsilon,k}(u))\frac{\mu(t+h)-\mu(t)}{h}\eta. \tag{29}$$

By (3) and (29), since $\phi^h \to \phi$ in $L^2(0, T; H^1(\Omega))$ and $\frac{1}{h}(\mu(\cdot+h) - \mu(\cdot)) \to \mu_t$ in $L^\infty(0, T)$ as $h \to 0$, we get

$$\int_0^T\int_\Omega (B_{\varepsilon,k}(u^o) - B_{\varepsilon,k}(u))\mu_t\eta \leq -\int_0^T\int_\Omega (\nabla u - F(u)) \cdot \nabla(H_\varepsilon(u-k)\eta)\mu$$

$$+ \int_0^T\int_\Omega f H_\varepsilon(u-k)\eta\mu. \tag{30}$$

We are now in a position to perform the doubling of variables. As in Lemma 5, take $\xi \in \mathcal{D}(\overline{\Omega} \times \overline{\Omega})$ such that $\xi|_{\overline{\Omega}\times\partial\Omega} = 0, \xi \geq 0$.

Consider $g \in L^1(\Omega)$ and $v \in C^1(\overline{\Omega})$ such that v is a weak solution of $P(b, F)(g+b(v))$. For a.e. $(t, x) \in Q$, take $H_\varepsilon(v(\cdot) - u(t, x))\xi(x, \cdot)\mu(t)$ as a test function for $P(b, F)$ $(g + b(v))$. Integrating in $(t, x) \in Q$, we get

$$\int_0^T\iint_{\Omega\times\Omega} (\nabla v(y) - F(v(y))) \cdot \nabla_y(H_\varepsilon(v(y)-u(t, x))\xi(x, y))\mu(t)$$

$$= \int_0^T\iint_{\Omega\times\Omega} g(y)H_\varepsilon(v(y)-u(t, x))\xi(x, y)\mu(t). \tag{31}$$

Similarly, for a.e. $y \in \Omega$ let us take $k = v(y)$ and $\eta(\cdot) = \xi(\cdot, y)$ in (30) and integrate in $y \in \Omega$. Summing the obtained inequality with (31), arguing as in Lemma 5, we pass to the limit as $\varepsilon \to 0$. Note that by (28) and the Lebesque dominated convergence theorem we have

$$\lim_{\varepsilon\to 0}\int_0^T\iint_{\Omega\times\Omega} (B_{\varepsilon,v(y)}(u^o(x)) - B_{\varepsilon,v(y)}(u(t, x)))\mu_t\,\xi$$

$$= \int_0^T\iint_{\Omega\times\Omega} (|b^o(x)-b(v(y))|-|b(u(t, x))-b(v(y))|)\mu_t\,\xi.$$

We infer

$$\int_0^T \iint_{\Omega\times\Omega} (|b^o(x)-b(v(y))|-|b(u(t,x))-b(v(y))|)\mu_t\,\xi$$

$$\leq -\int_0^T \iint_{\Omega\times\Omega} \mathrm{sign}\,(u(t,x)-v(y))\{\nabla u(t,x)-\nabla v(y)-F(u(t,x))+F(v(y))\}\cdot$$

$$\cdot(\nabla_x\xi+\nabla_y\xi)\mu + \int_0^T \iint_{\Omega\times\Omega} \mathrm{sign}\,(u(t,x)-v(y))(f(t,x)-g(y))\xi\mu$$

$$-\int_0^T \iint_{\partial\Omega\times\Omega} \mathrm{sign}\,(u(t,x)-v(y))\{\nabla v(y)-F(v(y))\}\cdot v_x\,\xi\mu. \tag{32}$$

Now we make y tend to x.

CLAIM 1. For all u weak solution of $E(b,F)(f,b^o)$, $g\in L^1(\Omega)$ and $v\in C^1(\overline{\Omega})$ weak solution of $P(b,F)(g+b(v))$ we have

$$\int_0^T \int_\Omega (|b^o-b(v)|-|b(u)-b(v)|)\mu_t\eta$$

$$+\int_0^T \int_\Omega \mathrm{sign}\,(u-v)(\nabla u-\nabla v-F(u)+F(v))\cdot\nabla\eta\,\mu$$

$$\leq \int_0^T \mu\,[u-v,(f-g)\eta]_{L^1(\Omega)}$$

for all positive $\eta\in\mathcal{D}(\overline{\Omega})$, $\mu\in\mathcal{D}(-\infty,T)$.

We should slightly modify the Proof of Proposition 3, because of the extra integral in t. Take $\xi=\xi_n$ in (32), where ξ_n is defined in (18). Consider for instance the left-hand side of (32). As in the Proof of Proposition 3, for a.e. $t\in(0,T)$ we get

$$I_n(t) = \iint_{\Omega\times\Omega} (|b^o(x)-b(v(y))|-|b(u(t,x))-b(v(y))|)\xi_n(x,y)$$

$$\longrightarrow \int_\Omega (|b^o(x)-b(v(x))|-|b(u(t,x))-b(v(x))|)\eta(x)$$

as $n\to\infty$.

Moreover,

$$|I_n(t)| \leq (\|b^o\|_{L^1(\Omega)} + 2\|b(v)\|_{L^1(\Omega)} + \|b(u(t,\cdot)\|_{L^1(\Omega)})\|\eta\|_{L^\infty(\Omega)} \in L^1(0,T).$$

Multiplying by μ_t and applying the Lebesque dominated convergence theorem, we get

$$\lim_{n\to\infty} \int_0^T I_n(t)\mu_t = \int_0^T (|b^o-b(v)|-|b(u)-b(v)|)\eta\mu_t.$$

In the same way, we pass to the limit as $n\to\infty$ in the right-hand side of (32) and obtain Claim 1.

Now for all $(\beta, g) \in A_{b,F}$ there exists $v \in C^1(\overline{\Omega})$ such that $\beta = b(v)$ and v is a weak solution to $P(b, F)(g+b(v))$. Since $[u-v, f-g]_{L^1(\Omega)} \leq [b(u)-b(v), f-g]_{L^1(\Omega)}$, applying Claim 1 with $\eta \equiv 1$ on Ω, we get (26).

Finally, consider $\hat{t} \to 0^+$. Take $h \in (0, T - \hat{t})$ and $\mu^h \in \mathcal{D}(-\infty, T)$ such that $\mu^h \equiv 1$ on $(-\infty, \hat{t})$, $\mu^h \equiv 0$ on $(\hat{t} + h, T)$ and $0 \leq -\mu_t^h \leq 2/h$. Applying Claim 1 with $\mu = \mu^h$ and $\eta \equiv 1$ and passing to the limit as $h \to 0$, for a.e. $\hat{t} \in (0, T)$ we get

$$\int_\Omega |b(u(\hat{t})) - \beta| - |b^o - \beta| \leq \int_0^{\hat{t}} |f(\tau) - g|\, d\tau. \tag{33}$$

Since $b^o \in \overline{D(A_{b,F})}$, for all $\varepsilon > 0$ there exists $(\beta, g) \in A_{b,F}$ such that $\|b^o - \beta\|_{L^1(\Omega)} \leq \varepsilon/3$. For $\hat{t}$ sufficiently small, the right-hand side of (33) is less than $\varepsilon/3$. Hence $\|b(u(\hat{t})) - b^o\|_{L^1(\Omega)} \leq \varepsilon$ for a.a $\hat{t}$ sufficiently small, which concludes the proof. $\square$

Acknowledgement. Philippe Bénilan has suggested this problem and contributed to the solution through numerous discussions and constant encouragement. This work is to his memory.

REFERENCES

[1] ADAMS, R. A., *Sobolev spaces*. Academic Press, New York-London, 1975. Pure and Applied Mathematics, vol. 65.

[2] ALT, H. W. and LUCKHAUS, S., *Quasilinear elliptic-parabolic differential equations*. Math. Z., *183(3)*, (1983) 311–341.

[3] BARTHÉLÉMY, L. and BÉNILAN, PH., *Subsolutions for abstract evolution equations*. Potential Anal., *1(1)*, (1992) 93–113.

[4] BÉNILAN, PH., CRANDALL, M. G. and PAZY, A., *Nonlinear evolution equations in banach spaces*. book to appear.

[5] BÉNILAN and PH., TOURÉ, H., *Sur l'équation générale $u_t = a(\cdot, u, \phi(\cdot, u)_x)_x + v$ dans L^1. I. Étude du problème stationnaire*. In Evolution equations (Baton Rouge, LA, 1992), pp. 35–62. Dekker, New York, 1995.

[6] BÉNILAN and PH., TOURÉ, H., *Sur l'équation générale $u_t = a(\cdot, u, \phi(\cdot, u)_x)_x + v$ dans L^1. II. Le problème d'évolution*. Ann. Inst. H. Poincaré Anal. Non Linéaire, *12(6)*, (1995) 727–761.

[7] BÉNILAN, PH. and WITTBOLD, P., *On mild and weak solutions of elliptic-parabolic problems*. Adv. Differential Equations, *1(6)*, (1996) 1053–1073.

[8] PHILIPPE BÉNILAN, *Equations d'évolution dans un espace de Banach quelquonques et applications*. Thèse d'état, 1972.

[9] BOUHSISS, F., *Etude d'un problème parabolique par les semi-groupes non linéaires*. P.M.B. Analyse non linéaire, *15* (1995/97) 133–141.

[10] CARRILLO, J., *Entropy solutions for nonlinear degenerate problems*. Arch. Ration. Mech. Anal., *147(4)*, (1999) 269–361.

[11] CARRILLO, J. and WITTBOLD, P., *Uniqueness of renormalized solutions for degenerate elliptic-parabolic problems*. J. Diff. Eq., *156(1)*, (1999) 93–121.

[12] CRANDALL, M. G. and LIGGETT, T. M., *Generation of semi-groups of nonlinear transformations on general Banach spaces*. Amer. J. Math., *93* (1971) 265–298.

[13] DIAZ, J. I. and DE THÉLIN, F., *On a nonlinear parabolic problem arising in some models related to turbulent flows*. SIAM J. Math. Anal., *25(4)*, (1994) 1085–1111.

[14] FILO, J. and KAČUR, J., *Local existence of general nonlinear parabolic systems*. Nonlinear Anal., *24(11)*, (1995) 1597–1618.

[15] KRUŽKOV, S. N., *First order quasilinear equations with several independent variables*. Mat. Sb. (N.S.), *81(123)*, (1970) 228–255.

[16] LIEBERMAN, G. M., *Boundary regularity for solutions of degenerate elliptic equations*. Nonlinear Anal., *12(11)*, (1988) 1203–1219.

[17] LIONS, J.-L., *Quelques méthodes de résolution des problèmes aux limites non linéaires*. Dunod, 1969.

[18] OTTO, F., L^1*-contraction and uniqueness for quasilinear elliptic-parabolic equations*. J. Differential Equations, *131(1)*, (1996) 20–38.

[19] SIMONDON, F., *Étude de l'équation $\partial_t bu - \operatorname{div} a(bu, \nabla u) = 0$ par la méthode des semi-groupes dans L^1*. P.M.B. Analyse non linéaire, 1983.

Boris Andreianov
Centre de Mathématiques et Informatique
Université de Provence
39, rue F. Joliot-Curie
13453 Marseille
France
e-mail: borisa@math.univ-fcomte.fr

Fouzia Bouhsiss
Laboratoire de Mathématiques
Université de Franche-Comté
16, route de Gray
25030 Besançon
France
e-mail: bouhsiss@math.univ-fcomte.fr

To access this journal online:
http://www.birkhauser.ch

First published in 2001
1 volume per year
4 issues per volume
Format: 17 x 24 cm

Also available in electronic form.
For further information and electronic
sample copy please visit:
www.birkhauser.ch

For orders and additional information
originating from all over the world
exept USA and Canada:

Birkhäuser Verlag AG
c/o SAG GmbH & Co.
Customer Service Journals
Haberstrasse 7
D-69129 Heidelberg/ Germany
Tel. ++49 6221 345 0
Fax ++49 6221 345 4229
e-Mail: info@birkhauser.ch

Customers located in USA and Canada:

Springer-Verlag New York Inc.
Journal Fulfillment
333 Meadowlands Parkway
Secaucus, NJ 07094-2491
Phone: +1 201 348-4033
Fax: +1 201 348-4505
e-mail: journals@birkhauser.com
(Toll-free-number for customers in USA:
+1 800 777 4643)

Aims and Scope

Journal of Evolution Equations (JEE) publishes high-
quality, peer-reviewed papers on equations dealing
with time dependent systems and ranging from
abstract theory to concrete applications.
Research articles should contain new and important
results. Survey articles on recent developments are
also considered as important contributions to the
field.

Abstracted/Indexed in

CompuMath Citation Index, Current
Contents/Physical, Chemical and Earth Sciences, DB
Math, ISI Alerting Services, Mathematical Reviews,
MathScinet, SciSearch, Zentralblatt für Mathematik

Editors in Chief

Wolfgang Arendt
Universität Ulm
Abteilung Angewandte Analysis
D-89069 Ulm, Germany
e-mail: arendt@mathematik.uni-ulm.de

Rainer Nagel
Universität Tübingen
Mathematische Fakultät
Auf der Morgenstelle 10
D-72076 Tübingen, Germany
e-mail: rana@giotto.mathematik.uni-tuebingen.de

Subscription Information for 2004

Volume 4 (2004):
Institutional subscriber: € 188.–
Single issue: € 55.–
Postage: € 34.–
Prices are recommended retail prices.
Back volumes are available.
ISSN 1424-3199 (printed edition)
ISSN 1424-3202 (electronic edition)

www.birkhauser.ch

MATHEMATICS WITH BIRKHAUSER

Birkhäuser